W0232025

4th INTERNATIONAL SYMPOSIUM ON CERAMIC MATERIALS AND COMPONENTS FOR ENGINES

Proceedings of the 4th International Symposium on Ceramic Materials and Components for Engines, organized by the Swedish Ceramic Society and held at Göteborg, Sweden, 10–12 June 1991

Co-sponsored by

American Ceramic Society
Australian Ceramic Society
Ceramic Society of Japan
European Ceramic Society
International Ceramic Federation

4th INTERNATIONAL SYMPOSIUM ON CERAMIC MATERIALS AND COMPONENTS FOR ENGINES

Edited by

R. CARLSSON

T. JOHANSSON

and

L. KAHLMAN

Swedish Ceramic Society,
Goteborg, Sweden

SPRINGER-SCIENCE+BUSINESS MEDIA, B.V.

WITH 174 TABLES AND 932 ILLUSTRATIONS

© 1992 Springer Science+Business Media Dordrecht
Originally published by ELSEVIER SCIENCE PUBLISHERS LTD in 1992
Softcover reprint of the hardcover 1st edition 1992

British Library Cataloguing in Publication Data

Ceramic Materials and Components for
Engines: 4th
International Symposium on Ceramic
Materials and Components for Engines
I. Title. II. Carlsson, R. III. Johansson, T.
IV. Kahlman, L.
666

ISBN 978-94-010-5280-1 ISBN ISBN 978-94-011-2882-7 (eBook)
DOI 10.1007/978-94-011-2882-7

Library of Congress CIP data applied for

No responsibility is assumed by the Publisher for any injury and/or damage to persons or property as a matter of products liability, negligence or otherwise, or from any use or operation of any methods, products, instructions or ideas contained in the material herein.

Special regulations for readers in the USA

This publication has been registered with the Copyright Clearance Center Inc. (CCC), Salem, Massachusetts. Information can be obtained from the CCC about conditions under which photocopies of parts of this publication may be made in the USA. All other copyright questions, including photocopying outside the USA, should be referred to the publisher.

All rights reserved. No part of this publication may be reproduced, stored in a retrieval system, or transmitted in any form or by any means, electronic, mechanical, photocopying, recording, or otherwise, without the prior written permission of the publisher.

PREFACE

The 4th International Symposium on Ceramic Materials and Components for Engines, organized by the Swedish Ceramic Society in Göteborg on 10–12 June 1991, continues the series of conferences which started in 1983 at Hakone, Japan, followed by Travemünde, Germany, in 1986 and Las Vegas, USA, in 1988. The aim of these conferences is to bring together engineers, scientists and students working on ceramics for engine applications to discuss the state-of-the-art and recent developments. Three hundred and twenty-five participants from 23 countries attended the conference, which indicates the worldwide interest in the subject.

The success of a conference depends primarily on the quality of the papers presented and on the participants. At the 4th symposium 173 papers were presented, divided into 12 invited papers, 91 oral presentations in three concurrent sessions and 70 posters. This book contains the written version of 142 of these papers. Every paper has been refereed by two experts. About one-third of the papers come from the industry and two-thirds from universities, institutes and government laboratories.

The Local Organizing Committee would like to thank all the authors for the effort put into their presentations at the conference and during the preparation of their manuscripts. We also acknowledge the International Advisory Committee for their continuous support and the co-sponsoring organizations for their marketing efforts. Special thanks go to all the members of the Committee of Referees for their very important work in assessing and checking all the manuscripts. Finally, we wish to express our gratitude to Ulla-Britt Jigholm and Margareta Jansson of the Swedish Ceramic Institute for their outstanding job in dealing with all practical details before, during and after the conference.

The 5th symposium will be organized in China in 1994 and the 6th symposium is provisionally planned to take place in Japan. The new President of the International Advisory Committee is Professor T. Yen of Academia Sinica, China.

ROGER CARLSSON
THOMAS JOHANSSON
LARS KAHLMAN

CONTENTS

Applications

INTERNATIONAL ADVISORY COMMITTEE

T. Johansson	Höganäs Eldfast AB/Luleå University of Technology, Sweden (*President*)
G. N. Babini	NRC-Research Institute for Ceramics Technology, Italy
B. Bertrand	GIE.PSA Etudes et Recherches, DRAS, France
C. Bonnet	SEP, France
D. Broussaud	Rhône-Poulenc Centre de Recherches, France
K. Esaki	Japan Fine Ceramics Center, Japan
M. K. Ferber	Oak Ridge National Laboratory, USA
X. Fu	Shanghai Institute of Ceramics, China
H. Hausner	Technische Universität Berlin, Germany
W. P. Holbrook	American Ceramic Society, USA
M. H. Lewis	University of Warwick, UK
R. Metselaar	Eindhoven University of Technology, The Netherlands
M. Morita	Japan Fine Ceramics Center, Japan
R. Neumann	Preussag AG, Germany
D. E. Niesz	Rutgers University, USA
G. Petzow	Max-Planck-Institut für Metallforschung, Germany
H. B. Probst	NASA Lewis Research Center, USA
S. Saito	Kanagawa Academy of Science and Technology, Japan
V. Shevchenko	Academy of Sciences, Russia
S. Sōmiya	The Nishi Tokyo University, Japan
C. Sorrell	Sydney University of Technology, Australia
R. M. Spriggs	Alfred University, USA
V. J. Tennery	Oak Ridge National Laboratory, USA
E. Tiefenbacher	Daimler Benz, Germany
M. H. Van de Voorde	CEC Joint Research Centre, The Netherlands
P. Vincenzini	NRC-Research Institute for Ceramics Technology, Italy
T. Yen	Academia Sinica, China

LOCAL ORGANIZING COMMITTEE

T. Johansson	Höganäs Eldfast AB/Luleå University of Technology, Sweden (*President*)
M. Bergström	Scania AB, Sweden
R. Carlsson	Swedish Ceramic Institute, Sweden (*Secretary*)
L. Kahlman	Swedish Ceramic Institute, Sweden (*Assistant Secretary*)
H. T. Larker	ABB Cerama AB, Sweden
L. Pejryd	Volvo Flygmotor AB, Sweden
J. Rehn	United Turbine AB, Sweden
D. J. Rowcliffe	Royal Institute of Technology, Sweden
A. Wendel	AB Volvo, Sweden

COMMITTEE OF REFEREES

Babini, Gian N.	CNR-IRTEC, Faenza, Italy
Bellosi, Alida	CNR-IRTEC, Faenza, Italy
Bergström, Magnus	Saab-Scania AB, Södertälje, Sweden
Bönsch, Christof	Fraunhofer Institute, IPT, Aachen, Germany
Brandt, Gunnar	AB Sandvik Coromant, Stockholm, Sweden
Breder, Kristin	Royal Institute of Technology, Stockholm, Sweden

Brinkman, Charles	Martin Marietta Energy Systems Inc., Oak Ridge, TN, USA
Broussaud, Daniel	Rhône-Poulenc Recherches, Aubervilliers, France
Brown, Ian	DSIR, Chemistry, Lower Hutt, New Zealand
Burström, Martin	Institute for Production Engineering Research, Luleå, Sweden
Cales, Bernard	Céramiques Techniques Desmarquest, Evreux, France
Cannon, Roger	Rutgers University, Center for Ceramic Research, Piscataway, NJ, USA
Carlsson, Lennart	Swedish National Testing Laboratory, Borås, Sweden
Carlsson, Roger	Swedish Ceramic Institute, Göteborg, Sweden
Clegg, William	ICI Solid State Science Group, Runcorn, UK
Collin, Marianne	AB Sandvik Coromant, Stockholm, Sweden
Cutler, Raymond	Ceramatec Inc., Salt Lake City, UT, USA
Demaestri, Pier Paolo	Centro Ricerche Fiat, Orbassano, Italy
Dunlop, Gordon	University of Queensland, Brisbane, Australia
Dworak, Ulf	Elektroschmelzwerk Kempten GmbH, Kempten, Germany
Ekberg, Inga-Lill	Swedish Ceramic Institute, Göteborg, Sweden
Ekström, Thommy	AB Sandvik Hard Materials, Stockholm, Sweden
Engell, John	Technical University of Denmark, Copenhagen, Denmark
Engström, Håkan	AB Sandvik Hard Materials, Stockholm, Sweden
Ericsson, Torsten	Linköping University of Technology, Linköping, Sweden
Ernstsson, Marie	Institute for Surface Chemistry, Stockholm, Sweden
Fabbri, Luciano	Eniricerche SpA, Monterotondo, Italy
Falk, Lena	Chalmers University of Technology, Göteborg, Sweden
Fu, Xiren	Shanghai Institute of Ceramics, Shanghai, China
Funatani, Kiyoshi	Japan Fine Ceramics Center, Nagoya, Japan
Furey, Michael	Virginia Polytechnic Institute and State University, Blacksburg, VA, USA
Ge, Changchun	University of Science and Technology, Beijing, China
Gugel, Ernst	Cremer Forschungsinstitut GmbH & Co. KG, Rödental, Germany
Hampshire, Stuart	University of Limerick, Limerick, Ireland
Hausner, Hans	Technische Universität Berlin, Berlin, Germany
Heinrich, Jürgen G.	Hoechst CeramTec AG, Selb, Germany
Hermansson, Leif	Doxa Certex AB, Uppsala, Sweden
Hirschfeld, Deirdre	Virginia Polytechnic Institute and State University, Blacksburg, VA, USA
Jack, Kenneth	The Cookson Group plc, Wallsend, UK
Jiang, Dongliang	Shanghai Institute of Ceramics, Shanghai, China
Johansen, Knut	Elkem A/S, Keramer, Kristiansand, Norway
Johansson, Thomas	Höganäs Eldfast AB, Höganäs, Sweden
Kahlman, Lars	Swedish Ceramic Institute, Göteborg, Sweden
Kamo, Roy	Adiabatics Inc., Columbus, IN, USA
Karlsson, Sven	Swedish Ceramic Institute, Göteborg, Sweden
Knutson-Wedel, Maria	Chalmers University of Technology, Göteborg, Sweden
Larker, Hans T.	ABB Cerama AB, Robertsfors, Sweden
Leuchs, Martin	MAN Technologie AG, München, Germany
Lewis, M. H.	University of Warwick, Coventry, UK
Linde, Kerstin	AB Sandvik Hard Materials, Stockholm, Sweden
Lundberg, Robert	Volvo Flygmotor AB, Trollhättan, Sweden
McLaren, Malcolm	Rutgers University, Center for Ceramic Research, Piscataway, NJ, USA
Mäntylä, Tapio	Tampere University of Technology, Tampere, Finland
Natansohn, Samuel	GTE Laboratories Inc., Waltham, MA, USA
Niesz, Dale	Rutgers University, Center for Ceramic Research, Piscataway, NJ, USA

Nygren, Mats	University of Stockholm, Stockholm, Sweden
Okada, Akira	Nissan Motor Co., Central Eng. Lab., Yokosuka, Japan
Olsson, Per-Olof	Defence Research, Stockholm, Sweden
Pejryd, Lars	Volvo Flygmotor AB, Trollhättan, Sweden
Persson, Åke	Dynamec Research AB, Södertälje, Sweden
Persson, Michael	Eka Nobel AB, Bohus, Sweden
Petzow, Günther	Max-Planck-Institute for Metals Research, Stuttgart, Germany
Rowcliffe, David	Royal Institute of Technology, Stockholm, Sweden
Rundgren, Kent	Swedish Ceramic Institute, Göteborg, Sweden
Schulz, Robert	US Department of Energy, Washington, DC, USA
Sjöberg, Jörgen	Chalmers University of Technology, Göteborg, Sweden
Sōmiya, Shigeyuki	The Nishi Tokyo University, Tokyo, Japan
Spriggs, Richard M.	Alfred University, Center for Advanced Ceramic Technology, Alfred, NY, USA
Swab, Jeffrey J.	US Army Materials Technology Lab., Watertown, MA, USA
Tennary, Victor	Oak Ridge National Laboratory, Oak Ridge, TN, USA
Van de Voorde, Marcel	CEC Joint Research Centre, Petten, The Netherlands
Warren, Richard	Chalmers University of Technology, Göteborg, Sweden
Wendel, Agneta	AB Volvo, Technical Development, Göteborg, Sweden
Yen, T. S.	Shanghai Institute of Ceramics, Shanghai, China
Yust, Charles S.	Oak Ridge National Laboratory, Oak Ridge, TN, USA

EC-ADVENTURE IN MOBILITY

H.E. Ambassador Ivo Dubois
Head of Delegation
Commission of the European Communities, Stockholm, Sweden

INTRODUCTION

Mankind is always on the way. And so is man.

The need for movement is a necessity he cannot withstand very long. As such, he is part of the ever moving creation, part of elements, moving and interacting, coming together and falling apart, establishing patterns of new forms of life, or leaving traces like the wind creates ripples in the sand, the water imprints its movement on the sandy bottom of the seas. This archetypal movement is like respiration, heaving up and down, inside out, repeating itself and hardly evolving, it seems, like the amoeba contracting and expanding, over and over again. Thus movement, while tending to go places, is confined in preprogrammed fashions within the limitations nature itself imposes.

Is it then human hybris to explore the conditions for mobility ? And to pursue change ? Obviously not, as human beings it is our task to explore the universe and further its evolution by using the means at our disposal. But belonging to this earth and being part of the creation, we are bound by its logic, by its built-in wisdom. We should constantly renew our awareness of this basic truth. Accelerating movement in all fields, technical, even political, might make us forget this. More importantly, it might cause us to identify as the objective what essentially is the means, the intermediate stage, the subordinated realisation leading on to further advances. On many occasions, it appears indicated to make this remark. It may also be relevant today, starting out with your discussions on the passionate subject of the advances in science and technology in the field of ceramics.

Engineering of movement, technical development, itself confined within the parameters of nature, needs a sense of direction to create and maintain dynamism and make the effort worthwhile. The same is true for creating conditions for movement in the social sense, in the "societal sphere" to use a controversial neologism. It seems not difficult to admit that, in both cases, **Man** is the ultimate logic within our creation, that Man is the cause, the "causa efficiens" of our involvement with movement and mobility. Respect for the inherent values of nature and the need for a man centred sense of direction should be present in our minds, also today at the occasion of making a few remarks on the subject of creating conditions for movement, of mobility, on EC...

WHAT IS, ON THIS BASIS, OUR ADVENTURE IN MOBILITY ?

Movement in terms of progress, with a sense of direction, is being achieved in the field of Research and Development on ceramics. But I will not deal with that specific subject for which, I am sure you know, I have no qualification whatsoever.

I thought the idea was rather, against the general background I gave in my introduction, for me to bring together under this one heading and discuss briefly, a number of other subjects which, each in their own way, concern the "societal" dynamics of the European Community.

THE FIRST BASIC SUBJECT IS ABOUT THE EC AS A DYNAMIC ENTERPRISE

What is EC ?

Most of you know that it was founded quite a while ago, in 1951/1952 in the form of a first step towards the setting up of what we can call today a "European Union". It soon proved to be a major achievement because it broke with rhetoric and instead set up a realistic plan of assuring peace in Western Europe. Realistic, since the founding 6 nations agreed to go ahead in a progressive way, creating solidarity in concrete terms, through common legislative measures, in the field of economics and social affairs. These laws were made by independent bodies, and applied directly in the six participating member states.

This first step was soon followed by another, two more Treaties were set up, in 1957, the Economic Community - EEC - and Euratom - The Atomic Energy Community. So there was, and still is today, a lot of movement, as we shall see. But I would point out already at this stage that the construction was successful in that it conformed to the conditions for positive movement, or "development" which I said a short while ago.

- it respected the principle of gradual movement towards progress, mindful of what in a natural way, can be the subject of integration;
- development had a human face and was achieved in patient negotiations "uniting peoples, not states", as one of the Communities great wonders, Jean Monnet, put it.

Moreover, the construction had an additional quality: once a decision is taken, it is irrevocable, becomes part of common rules that are virtually impossible to do away with. The EC as a dynamic enterprise was successful, also in withstanding the forces of anti-dynamics, of regression. This became apparent in the 1970s, when contrary movements, with a different quality, assailed the community construction. Difficulties within some of our member states, oil crises, political "misunderstandings" and temporary weaknesses of an organisation which had to digest a number of new members. These were the reasons behind the anti-forces. But soon became clear that the absence of movement, an immobile Community, was detrimental to the interests of the states it comprises. Renewed movement occurred in the 1980s, triggered by interacting, natural dynamic forces. Soon the Community was again on the move. A new installment in the adventure of mobility in the socio-political sense went underway: setting up the EC internal market. As you know, the internal market project is today well advanced, envied may be, and respected.

And its success is considered certain to the point that a long and difficult negotiation process was started in 1989 with the EFTA countries in order to achieve the establishment of a European Economic Area.

TOWARDS MORE MOBILITY: EUROPE 1992

How far have we come today with the preparations for our internal market ?

Clearly, the project is in essence exactly what one could call "an adventure in mobility": after so many years of hesitations and fiscal fantasies by our member states in order to keep up with relative advantages, hiding behind their national borders in order to wage commercial war on their Community neighbours on **their** conditions, reason has prevailed, they now agree that the borders of our realms or republics no longer are the determinant factor for establishing the law which determines the conditions for dealings between persons or enterprises within the Community.

As of January 1, 1993, it is Community law which prevails. To achieve that, we first had to determine what we did **not** want,: all that limited the free flow of goods, people, capital and services. Thus, Europe 1992 is above all, an "adventure in deregulation", as VP Frans Adriessen has called it.

Once that was clear, we did not forget, that freedom is defined in relation to the limitations which give freedom its positive connotation. Like in nature movement is confined within the borders of archetypal imprint, this basic principle is being mirrored in the regulations which accompany this movement towards the four freedoms. These regulations are the rules of play which make the match worth playing for the human beings who play it and take home the points. The principles on which these rules are established can be found in our Treaties: non-discrimination, good conditions of competition, the rule of law.

Establishing those rules is not always easy. Now that we have come to the final stage of our preparations, hardcore problems still remain to be solved, like the ones on indirect taxation. Deciding the rules require unanimity, reflecting the basic importance for our member states of matters relating to their adopted political philosophy and the revenue of the state. But we have now in principle agreed on the way value added taxes and accises will be levied after the 1st of January 1993. And the rates we will apply will be close enough, within a few years, to neutralise a distorting impact on the implementation of the four freedoms, as was all but decided on June 3.

WHAT MORE IS REQUIRED TO MAKE THAT NEW MOBILITY WORK ?

Obviously, a number of factors must be right in order to create the conditions for the freedom of movement to have effect, like

* the material conditions to facilitate transportation of goods and persons;
* or the conditions of fair competition I have already mentioned;
* or the conditions of overall economic and monetary policy

Let us start with transportation which is part of the services sector in the establishment of the internal market, but which has also an important role to play in making real exchange happen, and creates mobility, which is one of the material conditions for movement. The picture here is one of serious preoccupation.

All reports concur in stating that traffic will increase; in 20 years overall traffic will double, but transportation of goods by road will already double within 15 years, and for air traffic we need only 10 years for doubling the traffic.

In the meantime, investment is below 1 % of GNP and decreasing, which is about half of what is needed. One can improve the use of the existing infrastructure, e.g. management of road traffic on motorways but this does not solve the problem.

There is a phenomenon in the working of the internal market concerning the free flow of goods, which is casting its shadows ahead: the system of "just-in-time", feasible in a commercial space without borders, has considerable impact on the intensity of transportation, and this is of course only an example.

It is thus not exaggerated to say that the functioning of our industry is at stake: can it really reap the benefits of the internal market ? And what bout the desire of our citizens to exploit the Communities' new openness ? Is not the free movement of people and ideas largely dependent on the possibilities of physical movement ?

The EC Commission has the intention, in addition to the efforts we are making to implement a common transport policy as the EEC Treaty has foreseen from the outset, to come with a "White Book" with concrete proposals by the end of the year. This is the follow-up of the report which was established by the "Group 2000" at the request of the Commission. This "White Book" will include reflections made internally within the Commission on the subject of transportation and the environment as well as the economic cost of the non-realisation of the necessary measures concerning infrastructure.

The Commission considers that it is not acceptable to hide behind so called principles of free trade, talking in negative terms of the "Brussels bureaucracy", and to continue to refuse the so called "cabotage" which makes 30 % of all trucks go empty in international traffic, that rail traffic is still a monopoly, and that airlines are a national(istic) business where a country's flag must fly. While here there is indeed mobility, it is of a bad quality because it is economically aberrant.

With regard to infrastructure, the situation is not much better since most systems are national. Yes, there is basic agreement now on a European network of fast trains but there is no coordination of efforts on the two sides of the Alps on tunnel building, to give only one example. There is also a lack of commitment and funds, although some signs of improvement can be noted. Much remains to be achieved in this field, real mobility is at a price.

And what about competition ?

Respect for the rule of law is the basic element to avoid discriminatory praxises and to establish and maintain a sound climate of competition in an area of great freedom of movement. It is not so much a regulatory function: the principles on which the Commissions' competences in the field are based, are simple. They are contained in a few Treaty provisions: cartels and concerted practices are prohibited, mergers can be stopped if the result is limitation of competition. Important is rather that the Commission has the power to implement these principles in an efficient way, making decisions and levying heavy fines in order to convince those responsible that it is real freedom of movement the Community is after.

A third element required to create and maintain real mobility is concerned with economic and monetary policy: the internal market is of course part of the Communities' and member states' economic system, today more or less integrated.

The need is to continue this movement towards further integration of our economies, to create progressively a macro-economic framework in which the internal market forces can securely operate. This also requires to do away with differences in fiscal and monetary policies and to create at least the effect one single currency has in such a context. On this last point alone it has been calculated that an amount of 20,000 million ECU can be economised every year by having one single currency instead of the present 11.

Most of you know that these matters are the subject today of intergovernmental discussions between the Communities' member states. We hope to conclude in the fall with the signing of a Treaty, creating the Economic and Monetary Union; the chances are good. This will mean, among other things, that we will establish by the mid-nineties, a European Central bank and achieve monetary union and one single currency before the year 2000.

Already now, the movement is under way, since on 1 July 1990, action towards greater economic convergence of the member states economies started officially as the first stage of what the Treaty will confirm as a 3 stage project.

I would invite your attention, in passing, to the political importance of this development. I will come back to this subject in a few moments.

TO WHAT DEGREE IS THE COMMUNITIES' MOBILITY AND ITS RESULTS CONFINED ?

The question is relevant for several reasons:

* to take the internal market, is it exclusively inside looking, reserving benefits to its primary actors, living and working inside the Community ?
* what is the Community doing with its emerging political identity, result of its internal dynamics ?

On the first question, a number of indications show the character of non-exclusivity of our internal, dynamic enterprise:

* the negotiations of an EEA - agreement with EFTA demonstrates the Communities' intent to favour a close relationship and sharing of benefits with the other industrialised countries in its region. The negotiations are not concluded yet, but it is possible today to state without too great a risk, that the conditions for extending the benefits of our increased internal mobility are good. On the other hand it is also true that only the functioning of the complicated agreement we are now hopefully reaching can ultimately confirm the veracity of this statement.
* the character of non-exclusivity of our internal market framework has come to light at various occasions. To give but one example: the liberal requirements within the Community of the establishment of Banks, is of great benefit also to non-Community financial institutions. Our rules compare quite favourably with those of our main trading partners, inside and outside Europe.

* the Communities' participation in the worldwide economic cooperation effort and its positive contribution to it, indicate our readiness to work along the lines agreed to by all the participants in the process. UN, OECD, GATT, only to name a few.

(Allow me to open here a parenthesis:

The Communities' position in the process of the Uruguay round has been wrongly termed as unhelpful (to paraphrase somewhat diplomatically other epithets). Indeed, while much progress was achieved in many of the 15 areas of negotiation, it so happened that too much emphasis was put on agriculture. In our context today of the Community dynamics we should all agree that the Communities' agricultural policies do not fit in.)

Our emerging political identity has been briefly mentioned already when talking about the establishment of EMU which is under way. Obviously, dynamics within the Community have made it of relevance for the world at large, and particularly within Europe.

How have we reacted to this development ?

For the time being, the Communities' external policy positions are identified and elaborated essentially by our member states in intergovernmental consultation, with the participation of the Commission. This "political cooperation" is not part of the institutional framework of the EC. But this does not concern the external aspects of the Communities' activities on the basis of the EC Treaties which are being handled essentially by the Commission as the executive arm of the Communities. As already mentioned, in this way the Community is present in the international economic cooperation (UN, OECD, GATT etc.). A (second) intergovernmental conference is now under way on the establishing of a "Union" and external policy is part of these discussions, as all of you, I suppose, know.

While the Commission is actively participating in this conference and has made important proposals, it is up to the Communities' member states to define, in the last analysis, the content of the consensus needed to establish the base for a new Treaty on "European Union". This also with regard to the establishment of an **overall** EC common foreign policy. i.e. including commercial, economic, development - and purely political, aspects.

For the Community as a dynamic enterprise, to live up to its obligations towards the world, the Commission is of the opinion that all the aspects of the "Union" to be established should be based on the existing Community Treaties. This means that also foreign policy matters, including foreign security, should become a **Community** matter, instead of remaining in the hands of the member states.

Whatever the outcome of this discussion, as well as of all the other points under consideration now, it must suffice to mention the importance of the link between EC's economic relevance and its overall political impact. A case in point is what we said earlier on the political signification of EMU: the economic and monetary strength of the Community is necessary in order to live up to the expectations the world has of the Community. This is particularly evident in the case of the countries of Central and Eastern Europe who count on the Communities' economic strength to safeguard their existence as independent democratic states.

CONCLUSION

It is time to conclude.

Our adventure in mobility is full of suspense, sometimes leading to unsuspected consequences. Still we must intensify our effort to well understand the rationale of our enterprise, in order to keep a firm grip on the project we embarked upon. We must analyze the natural limits of movement to create the right conditions leading to mobility and lasting change. The determinant factor is to continuously realise ourselves that our objective is a man-centred, a humane society we are building.

Our "Projet de société", our societal project - forgive me the term once more - as President Delors named it, is said to achieve Unity, Union. But this, as the President added, cannot be itself the ultimate goal, but is the condition for people and ideas to move around freely. In the late 18th century in Europe, cosmopolitism was something for the few, the wealthy and educated. Today, in our democratic Europe, we would want to achieve cosmopolitism for all.

That is the deep sense or our adventure in mobility.

Because, as the chorus Sings in **Sophocles' Antigone**

"Wonders are many on earth, and the greatest of these
"Is man, who rides the ocean and takes his way
"Through the deeps, through wind swept valleys or perilous seas,
"That surge and sway".

(Translation of E.F. Watling, the Penguin Classics".)

THE STUDY OF A HEAT INSULATED ENGINE CONSTRUCTED BY CERAMICS ENGINE PARTS

HIDEO KAWAMURA
Executive Director
Isuzu Ceramics Research Institute Co., Ltd.
Fujisawa, JAPAN

ABSTRACT

The development of ceramic engine that started from the 1970s has been gradually toned down despite continuous enthusiastic expectation held to date. The most reason for making us to be disappointed with development of the ceramic engine is that its characteristic required with respect to engine technology does not coincide with the basic property of ceramics. The author has advanced the development of ceramic engine and its parts fot the last 10 years or so. In recent years, a systematic evaluation of the ceramic material that satisfies the strength required of ceramic engine components and a survey of strength of the ceramic parts after grinding have been executed, while at the same time as a result of thorough calculation of the stress of engine members by using F. E. M., the author has successfully developed the parts satisfying mechanical and thermal requirements and successfully executed the operation of a heat insulated engine without cooling system by using the superior characteristic of ceramics in thermal resistance. Further, the frictional characteristic of the ceramics has been investigated for the development of a low frictional engine.

On the other hand, as a result of operating the heat insulated engine using heat resistive ceramics in the inner wall of the combustion chamber, it was found that the air temperature rose, the combustion and the exhaust gas deteriorated. In order to resolve these matters, the author clarified the transition of high temperature combustion, examined a means of improving the combustion efficiency, and searched for the feasibility of the improvement.

INTRODUCTION

A development of ceramic engine and engine parts started in the second half of the 1970s[1], and two or three parts were commercialized and put to practical use.[2][3][4] However, research institutes showed weak desire to develop them in the latter half of the 1980s, and few institutes have been actively involved in the development.

Using a ceramic material which is superior to metal in heat resistance, heat insulation and wear and abrasion resistance, the present writers have developed automobile engine and engine parts[5]. In developing them, since we aim at creating new functions by making use of characteristics of ceramic materials, their design and shape may be far different from those of conventional engines and engine parts. Purposes of the development of the ceramic engines are to remove a cooling system from an engine, to substantially improve fuel efficiency and to develop multi-fuel engines. Therefore, the development project has become very large, and various problems have been present to be solved.

However, there are not sufficient necessary research data about strength, fatigue, wear and abrasive resistance, heat resistance, etc. of ceramic materials which are indispensable to complete the ceramic engines, and we have to conduct basic research necessary for engine technology. This paper clarifies various problems now we have and the status of the development.

Evaluation of Ceramic Materials for Engines and Designing Method

Ceramic engine parts have been manufactured in the same shape as metal parts, and their performance have been evaluated. However, if strength and fracture toughness of materials are different, it is natural for the shape and structure to be changed, and new structures which meet performance of ceramic materials have to be considered.

Figure-1 shows a piston structure and calculated results mechanical stress and thermal stress which would be generated on a piston structure of a heat insulated engine. The mechanical stress is generated by inertia force and impact of piston slap due to reciprocating motion of a piston and compression load of combustion gas.

When the combustion gas acts on an upper face of a piston, the mechanical stress is mainly caused on the reverse side of the piston. Thrust load is applied on a side of a piston, and the piston slap force is applied as an impulse force on the side of a piston, and it generates large stress.

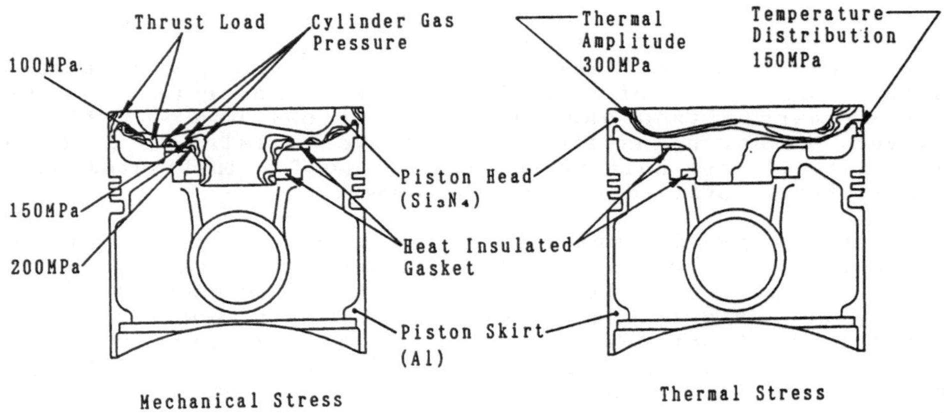

Fig.1 Mechanical & Thermal Stress in Ceramics Heat Insulated Piston Made of Si_3N_4

The thermal stress is caused on a surface of a cylinder wall by temperature swing of a cylinder wall which rises when the upper face of a piston is exposed to combustion gas in a cylinder and cooling effect of intake air.

In addition, the thermal stress is also induced by temperature gradient which is produced while heat is conducted from a surface of the piston exposed to the combustion gas to a skirt. The stress generated on the surface is not large because thermal conductivity and the Young's modulus of silicon nitride are respectively 2 times of those of steel. These calculation are analyzed by a finite element method.

A piston made of silicon nitride has a gasket which has low thermal conductivity in order to reduce the quantity of heat transferring from an upper face of a piston to a skirt between the two parts. A strength of the piston made of silicon nitride in this structure is calculated, and shape of parts on which stress level is found to be high are modified in order to drop the stress.

Nitride Silicon Materials

Since silicon nitride (Si_3N_4) has high strength and good thermal shock resistance, it is easy for us to use it as a material of an internal combustion engine. When the present writers succeeded in mass-producing glow plugs in 1981[2] and hot plugs made of silicon nitride in 1983, those parts were sintered at low pressure. Crystallized grains of the low-pressure sintered silicon nitride are found to be massive and relatively large when the micro structure of the silicon nitride is observed under an electron microscope(EM-002B), and Y_2O_3 and Al_2O_3 from a grain boundary layer in a glass layer between crystallized grains. Furthermore, pores are found to scatter between grains.

A high-pressure sintering method has been perfected recently, and it extremely improves toughness and strength. The micro structure is cross-linked by fine crystal grains and silicon nitride grains developed in a needle shape, and closely fills space, and crystal are thinly overlapped with each other in a glass layer of a grain boundary. No large pores are found, and no defect which can be a starting point of fracture is observed.

Silicon nitride of which toughness values are large has large crystal, and the crystal grains are closely cross-linked by fine crystal, and distributed. Figure-2 shows correlation between micro structure and its strength. Though bending strength of silicon nitride which has large grain size is low, a fracture toughness value (KIC) is large, and it means that this material with large KIC is easier to design in parts due to less stress condition induced by sharp and less strength inconsistency resulted from surface roughness. Therefore, the fracture toughness value may be allowed to be small if the bending strength becomes large, and even though the bending strength becomes small, the material can be used as machine parts only if KIC is large.(Figure-3) Since shape of engine parts are complicated, stress concentration is also complicated when various types of load are applied, and depending on the bending strength, the fracture toughness value has more influence than bending strength upon strength reliability.

The Griffith's fracture theory(6) which attributes causes of fracture to the size and quantity of scattered internal pores has had difficulties in explaining the relationship, and the cause of fracture should be clarified by making research about correlation between crystal structure or molecular dispersibility and strength(7).

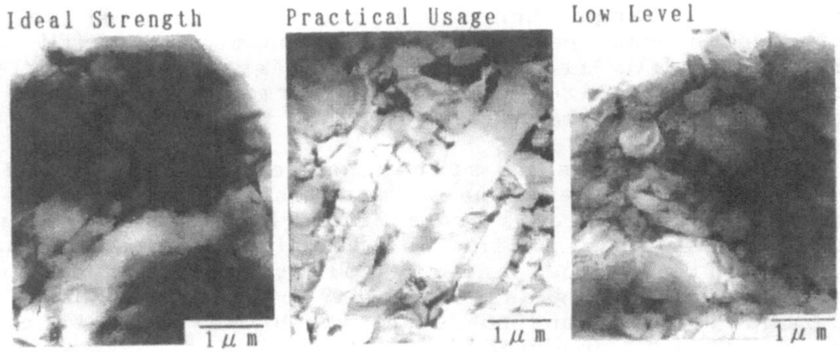

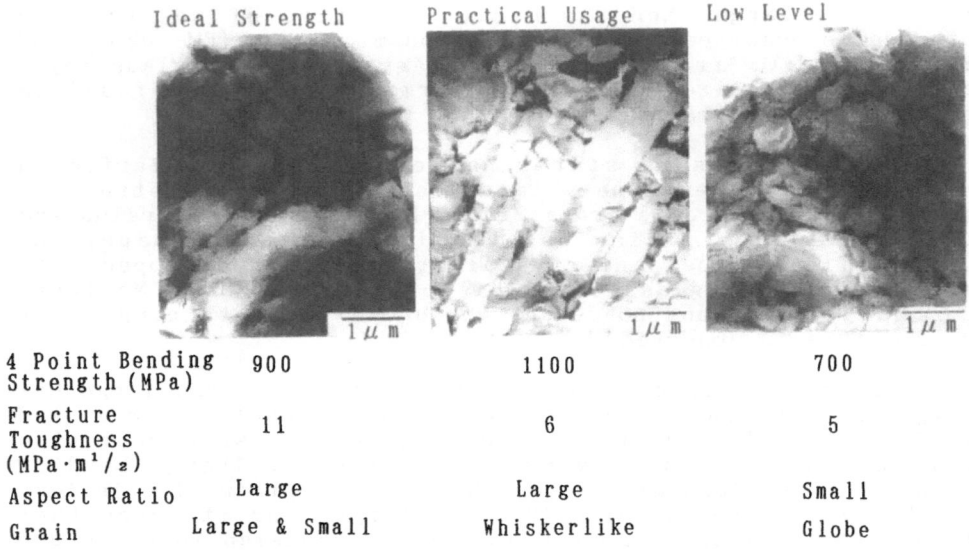

	Ideal Strength	Practical Usage	Low Level
4 Point Bending Strength (MPa)	900	1100	700
Fracture Toughness ($MPa\cdot m^{1/2}$)	11	6	5
Aspect Ratio	Large	Large	Small
Grain	Large & Small	Whiskerlike	Globe

Fig.2 Microstructure of Si_3N_4 for Ceramics Heat Insulated Engine Observed by Transmission Electron Microscope.

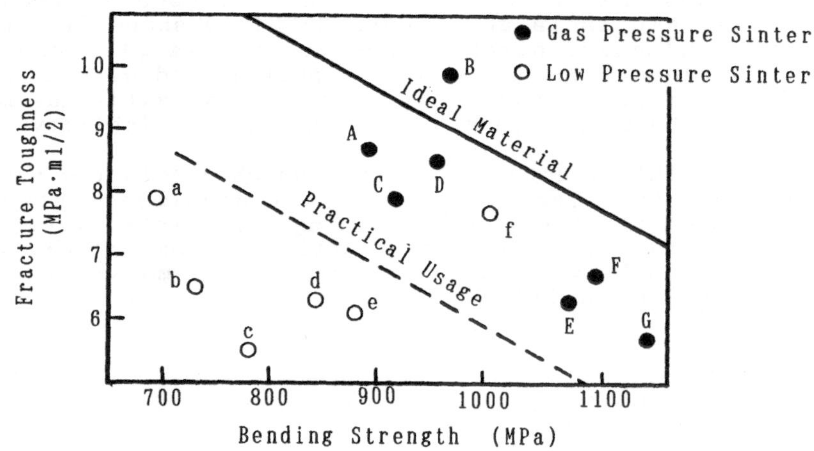

Fig.3 Ideal Ceramics Materials for Heat Insulated Engine of Si_3N_4 (Mark from a to f & from A to G are shown ceramics makers made Si_3N_4)

Structure of Heat Insulated Engine

In the past, a method to form a zirconia film on a surface of a combustion chamber was applied to many heat insulated engine in order to reduce the size of a cooling system by 20% - 30%[8].

A heat insulated engine which the present writers are developing is fundamentally different from conventional engines since we try to remove a cooling system using cooling water from an engine. Figure-4 shows a basic structure of the heat insulated engine. A combustion chamber is made up of an upper part of a cylinder liner, a head of a cylinder and an upper face of a piston, and since this part is exposed to hot combustion gas, a radiating structure is devised to reduce heat flow going to the outside to the minimum level.

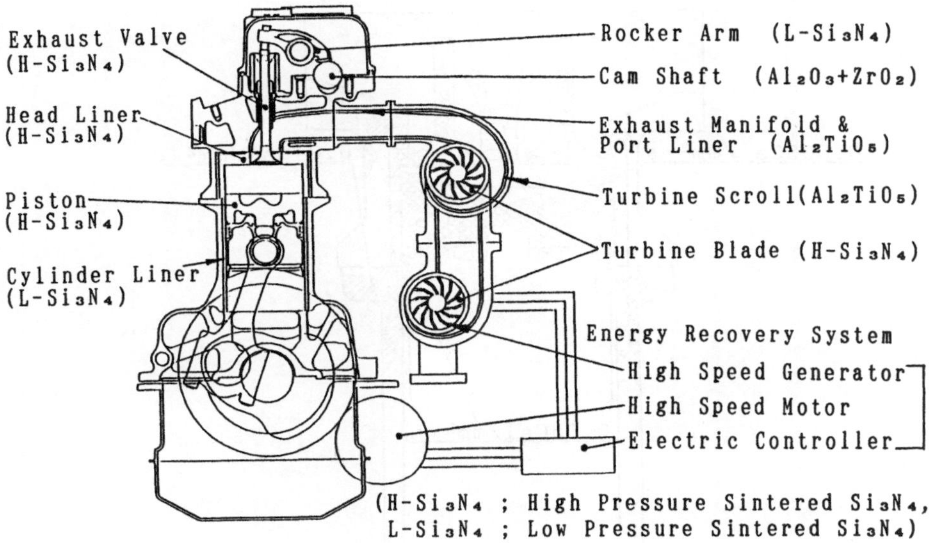

Fig.4 Construction of Turbo Compound Ceramics Heat Insulated Engine and Ceramics Engine Parts.

The cylinder liner and a piston skirt are blocked heat flow coming from the combustion chamber which is exposed to hot gas by using a gasket of low thermal conductivity in order to maintain temperature at an appropriate level so as to keep reciprocating motion of the piston smooth.

A heat flow calculation of the finite element method is applied to this heat insulated structure in designing its structure and selecting materials so that the temperature of an engine outer wall can be maintained at 130℃ or less, and that the temperature of the cylinder liner can be maintained at the optimum level for sliding.

Figure-5 shows calculation results of heat flux. The heat flux is obtained by calculating quantity of heat transferred from a combustion chamber of an engine by using the Pluem's heat transfer expression(9) for each crank angle, and the calculation on changes in heat flow and temperature distribution which change in response to engine operation continues by moderating coefficient of heat transfer until their values are converged.

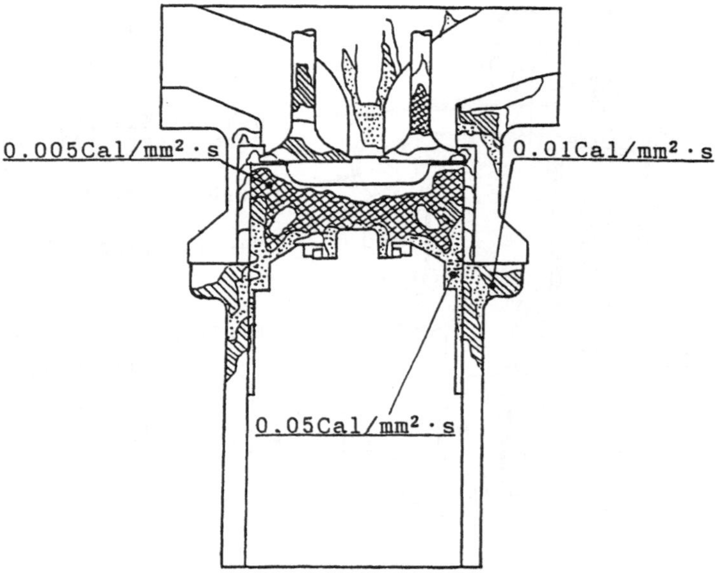

Fig.5 Heat Flux in Ceramics Heat Insulated Engine
Calculated those at Maximum Heat Release Condition.

The heat flux reaches to the maximum level around a connected section of a piston head and a piston skirt as well as a contacting section of a top and a bottom of a cylinder liner, and gaskets used for the sections should be made of materials of low thermal conductivity. The inner wall of a combustion chamber of the heat insulated engine is made up of silicon nitride, and a gasket made of aluminum titanate and an air layer are provided outside the wall. Between the piston head and the piston skirt, and between the top and the bottom of the cylinder, gaskets manufactured with partially-stabilized zirconium are mounted to heat insulation.

Calculation results on the above-mentioned structure are shown in Figure-6[10]. With design for a heat insulated structure and application of heat insulating materials, engine outer temperature can be maintained at 130℃ or less. However, since the thermal conductivity of silicon nitride is not so low, we cannot expected higher heat insulating effect if the material is used for the inside of a combustion chamber. Therefore, if a gasket made of a material with low thermal conductivity is provided to the outside in order to heat insulation, it raises the temperature in the combustion chamber, and aggravates air-intake and combustion efficiencies of the engine.

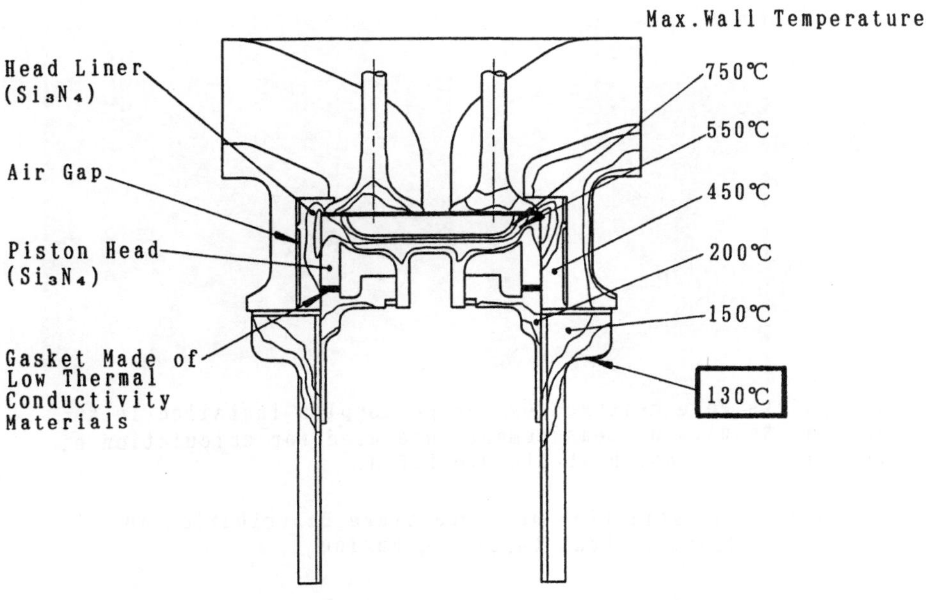

Fig.6 Temperature Distribution at Full Load Condition in Ceramics Heat Insulated Engine.

In comparison with values measured on an engine, the temperature distribution obtained from calculated results proves that assumptions of the calculation method are correct.

Figure-7 shows a temperature distribution under engine operation. With thermocouples attached at 130 spots for temperature measurement, changes in temperature at each spot against load and engine speed are measured. The coefficient of heat transfer of a combustion chamber surface and thermal conductivity coefficient of contacting sections of parts are modified to agree with these measures values. For this heat insulated engine, silicon nitride which has superior bending strength is used. However, since linear expansion coefficient of silicon nitride is large, and its thermal conductivity is about 1.5 times that of steel, heat stress acts on its surface. Since materials used for a combustion chamber are more influenced by thermal stress to the extent that their linear expansion coefficient becomes large and thermal conductivity small, thermal stress level influencing the surface depends on these physical constants. Figure-8 is about thermal stress of various types of ceramic materials calculated based on thermal conductivity and temperature swing obtained from the quantity of heat exposed in a combustion gas of an engine.

130 points were measured by thermo couples installed in an engine. Results of measurement were used for calculation of thermal stress and heat flux to F.E.M.

Fig.7 Investigation of Temperature Distribution in Ceramics Heat Insulated Engine

Aluminum titanate and cordierite are suitable as a material of the heat insulated engine. However, engine parts which are subjected to mechanical load should have high bending strength, and a ceramic material which satisfies conditions of low linear expansion coefficient and small thermal conductivity as heat insulating materials are sought for.

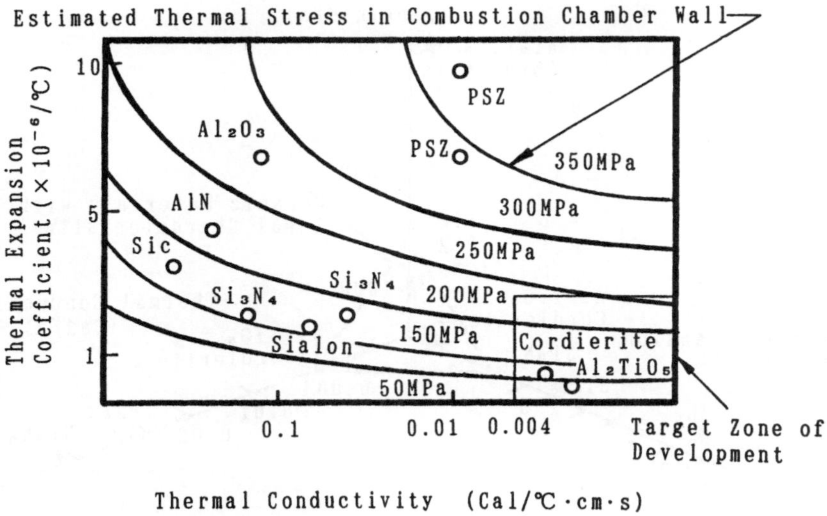

Fig.8 Ideal Ceramics Materials for Heat Insulated Engine

Aluminum titanate and cordierite has many pores in it, and its thermal conductivity is small. However, those pores have high possibility of becoming a starting point of fracture, and its strength tends to drop. Ideal materials do not exist on the earth. What is necessary is to make research to produce an ideal material by combining ceramic materials and other methods.(Figure-9)

Tribology of Ceramic

Among research and development of ceramic engines, a research of reducing friction force is important. Mechanical friction force of a piston motion is said to be the largest among that of reciprocated engines, and of the friction force is investigated[11]. Figure-10 shows the results that the load produced by the cylinder movement is measured, which the cylinder is floated so as to move freely. They are measured friction force of engines which is made of Si_3N_4 and a steel ring.

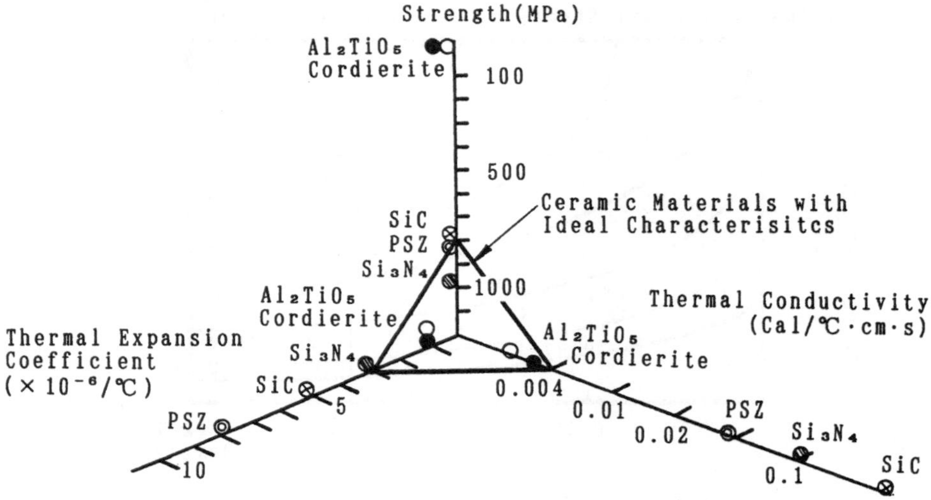

Fig.9 Ideal Ceramics with High Strength & Heat Insulation

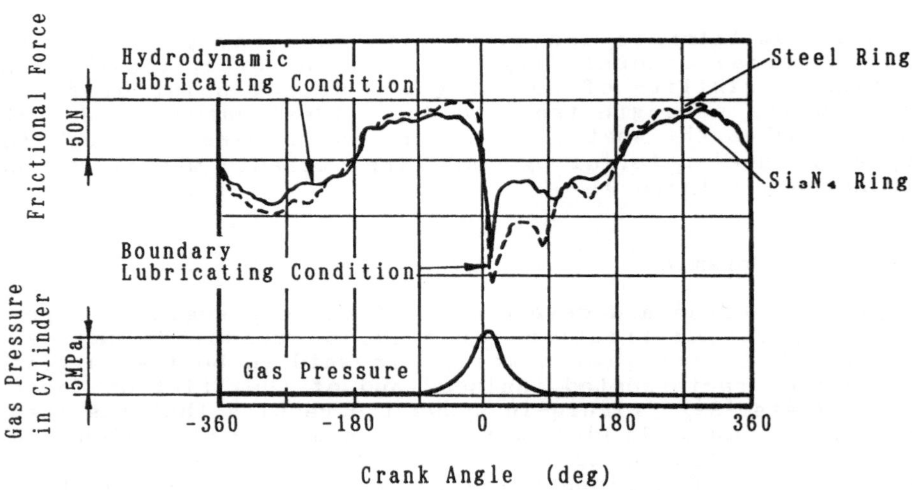

Fig.10 Frictional Force of Reciprocationg Engine Measured in Water Cooled Engine which Cylinder Liner is Made of Cast Iron

A cylinder of the engine is made of steel,and 10W-30 oil is used as a lubricant which is the same as one generally used. According to this measurement, it is found that large friction force is generated at a top dead center of an engine. This friction force is in the range of boundary lubrication since its speed is slow and load is large, and it is presumed that the friction force is in a fluid lubrication conditions in other range. In order to reduce this friction force, it is necessary to investigate the effect of friction coefficient in the boundary lubrication range by using different materials.

Figure-11 shows measured friction coefficient of various types of ceramic materials. SiC of which covalent bonding property is strong and which have less additives such as oxidized metal has large friction coefficient, and Si_3N_4 added with oxidized metal has small friction coefficient. As for boundary lubrication of metal, friction force is proved to be reduced due to absorption film of a lubricant, and similar phenomena can be expected as for ceramic materials as well.

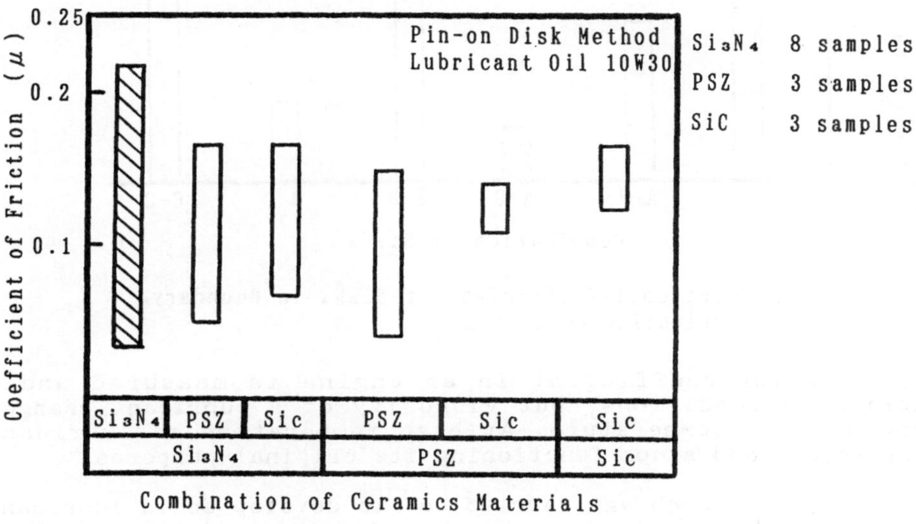

Fig.11 Frictional Coefficient of Ceramics Materials in Boundary Lubricating Condition

Detailed measurement about friction force of Si_3N_4 are shown in Figure-12. Friction coefficient becomes large when the same types of silicon nitride are combined, but the combination of one silicon nitride which is mixed with oxidized metals and has 20-50μ m of pores on its surface and another silicon nitride of which surface is very fine and crystal grains have less additives shows the lowest friction coefficient. The friction coefficient on boundary conditions is measured by contacting a round bar of 10ϕ on a silicon nitride plate, applying load on it, rotating it at a certain constant speed in a reciprocating motion.

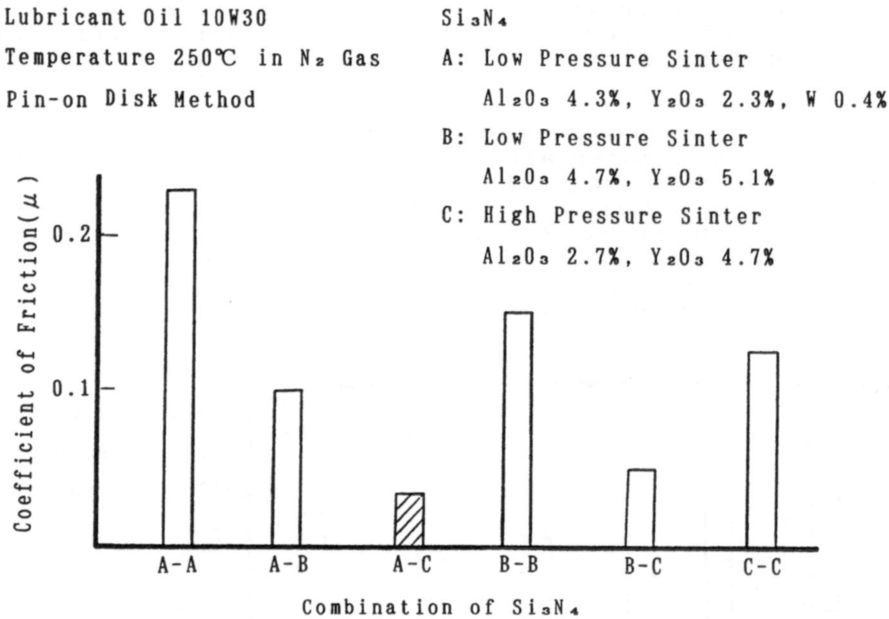

Fig.12 Frictional Coefficient of Si_3N_4 in Boundary Lubricating Condition

Friction coefficient in an engine is measured under lubricated conditions, but viscosity of a lubricant changes depending on temperature. At high temperature, a lubricant evaporates, and stops functioning its original purposes.

Then, research was carried out to develop solid lubricant which would contain friction coefficient and wear even if ceramic materials contacted with each other under hot conditions[12]. Powder of CaF_2 was added to silicon nitride, and put to the friction test. It was found that the friction coefficient declined as the atmospheric temperature rose, and that the friction coefficient was below that of steel under boundary lubrication conditions when it was 800℃ . (Figure-13) An observation with an electron proved micro analyzer found that the surface was coated with $Ca_8Si_5O_{18}$ and Ca_2SiO_4 , and that the coated layer acted with silicon nitride and diffused into the inside. In order to reduce friction coefficient of ceramic materials, it is important to form a layer produced by acting with lubricant, and to form the layer, it is necessary to add to ceramic components which have reactivity.

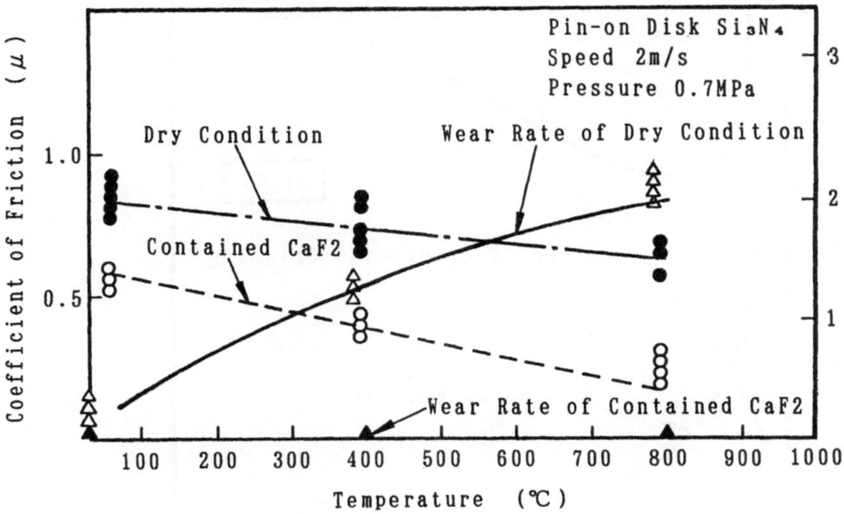

Fig.13 Possibility of Solid Lubrication in Si_3N_4 (Correlation between Temperature and Coefficient of Friction and Wear Rate)

In the above section, strength of ceramic materials used for heat insulated engines, a structure of the heat insulated engine and friction force are explained. Below, we will study about machining efficiency which may have large influence upon cost of ceramic engines and influence of strength on machining conditions.

For heat insulated engines, materials of high strength, high toughness materials and materials of small friction coefficient are used for purposes of various engine parts. Silicon nitride materials sintered with high-pressure gas and low-pressure gas are mixed with different additives, and their crystal compositions are largely different. When a ceramic material is machined, diamond grains of the wheel collide with structure of a ceramic materials and abrade the materials. An affected layer with micro cracks which is different in grinding conditions remains, and its depth influences bending strength. For the reason, in grinding silicon nitride materials, if a rough machining which raises machining efficiency and a precision machining which removes the affected layers caused by the rough machining are combined, parts can be efficiently machined with high strength[13]. (Figure-14)

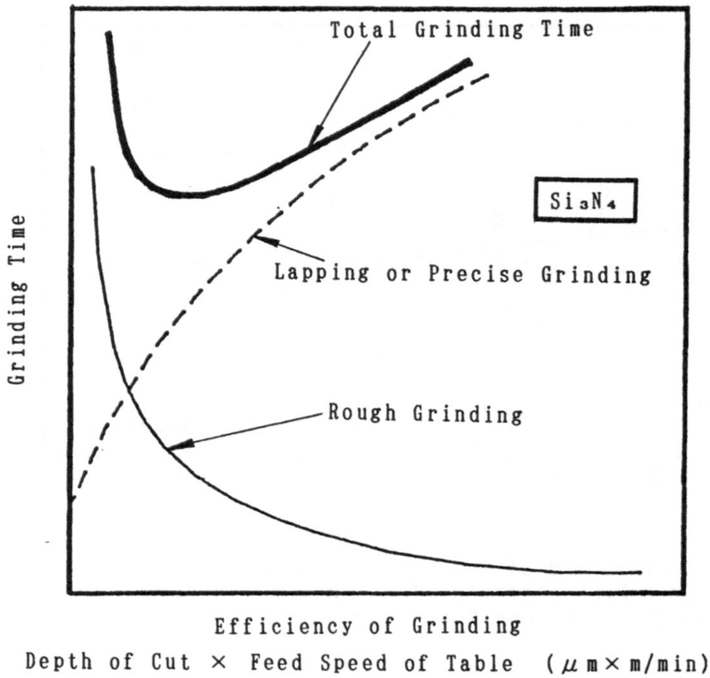

Fig.14 Effective Machining Condition for Achieving High Strength & Reducing Grinding Time.

If two types of silicon nitride of which structures are different with each other are machined under the same conditions, less difference of bending strength after rough and precision machining are found on a low-pressure sintered silicon nitride which contains lot of additives and pores of 50μm diameter in grain boundary, however strength after the rough machining and the precision machining are largely different in gas pressured sinter silicon nitride. An optimum machining method makes it possible to produce engine materials of which bending strength is large. (Figure-15)

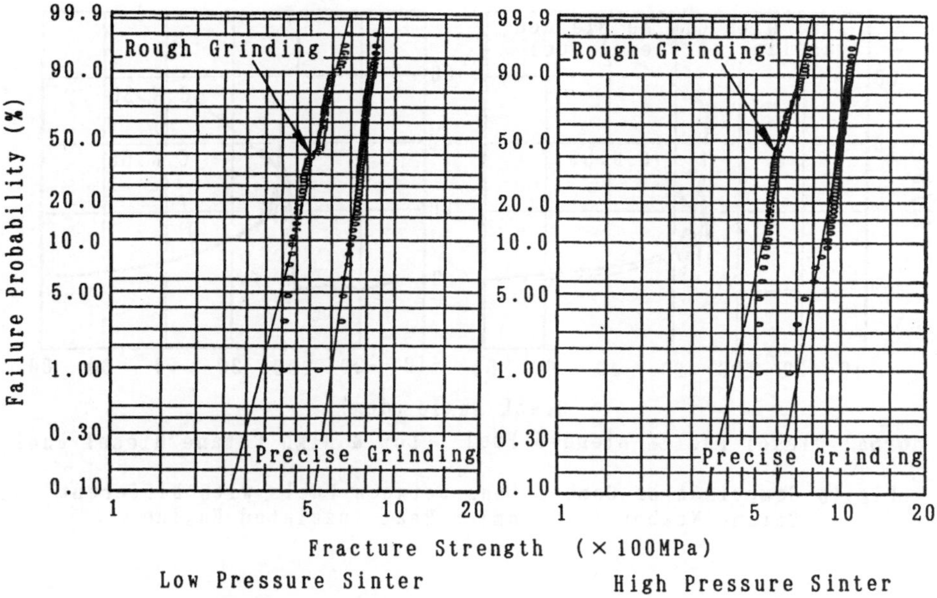

Fig.15 Strength of Si_3N_4 Plates Machined Different Grinding Conditions

Combustion of Heat Insulated Ceramic Engine

On a heat insulated engine removed of a cooling system, temperature in a combustion chamber rises, and combustion is changed because the temperature of air goes up by heat transfered from the combustion chamber. If fuel is injected into high temperature and high pressure air, combustion becomes deteriorated since the penetration of the fuel into a combustion chamber is restricted due to large air viscosity, the time of delayed ignition is shortened, and rate of pre-mixed combustion becomes smaller. Figure-16 is about heat-release rate which are calculated based on pressure measured with a piezoelectric transducer located in a cylinder. This picture shows the deterioration of combustion mentioned above. However, since detailed information about the development of flame in a combustion chamber can be obtained through observing combustion in an engine, we observed it by using a photographic system used high speed camera shown in Figure-17(14).

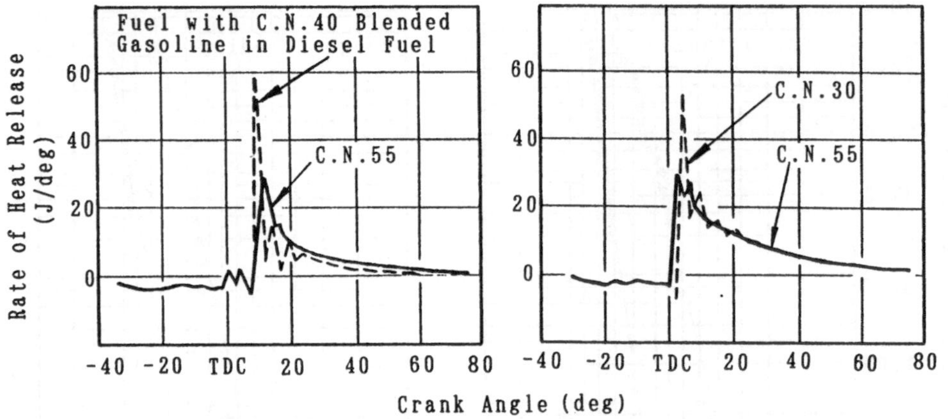

Normal Diesel Fuel & Blended Fuel   Low & High Cetane Diesel Fuel

Fig.16 Comprison of Combustions between Fuels with Different Cetane Number in Ceramics Heat Insulated Engine

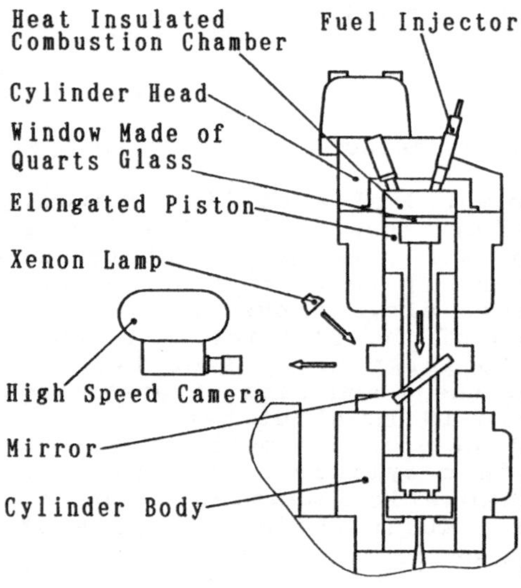

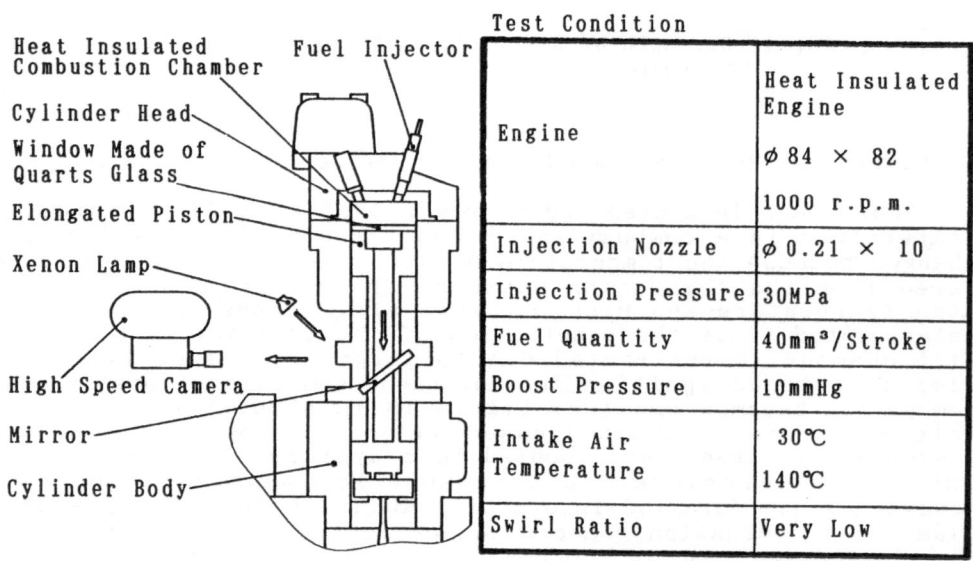

Test Condition

Engine	Heat Insulated Engine φ 84 × 82 1000 r.p.m.
Injection Nozzle	φ 0.21 × 10
Injection Pressure	30MPa
Fuel Quantity	40mm³/Stroke
Boost Pressure	10mmHg
Intake Air Temperature	30℃ 140℃
Swirl Ratio	Very Low

Fig.17 Photographic System for Observation of Combustion Progress

Figure-18 shows pictures taken by the high-speed camera. Under an atmosphere of high temperature and high pressure, combustion starts early, and lot of flames are observed while air-fuel mixture is burning. Soot is observed in the second half of the combustion. Therefore, it is presumed that air viscosity increases in high temperature and high pressure air, that mist of fuel dose not diffuse and that air entrain of fuel jet is deteriorated.

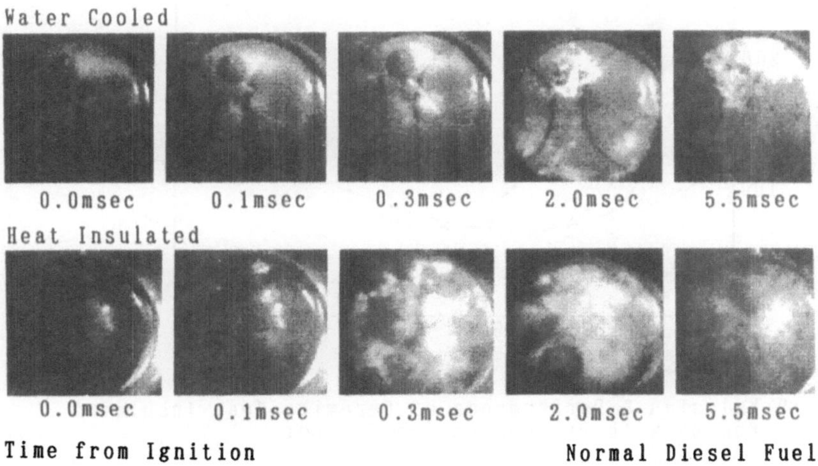

Fig.18 Observation of Combustion Progress in Ceramics Heat Insulated Engine

Fuel and air are mixed by strong air flow in a combustion chamber which has swirl type of pre-combustion chamber, and flame is blown off into a main combustion chamber through a throat. Since mixture in the pre-chamber is rich and development of nitrogen oxides is restrained, hydrocarbon around a high temperature cylinder wall burns and the volume of unburned hydrocarbon is reduced as well (Figure-19). By optimizing a combustion chamber, deterioration of high temperature combustion can be improved, and the development of a heat insulated ceramic engine has proceeded to the next step of how to extract exhaust enthalpy increased by heat insulated combustion chamber as motive power.

Figure-20 shows an outline of a ceramic heat insulated turbo compound engine. The heat insulated structure prevents heat transfer from a combustion chamber and an exhaust system to engine construction.

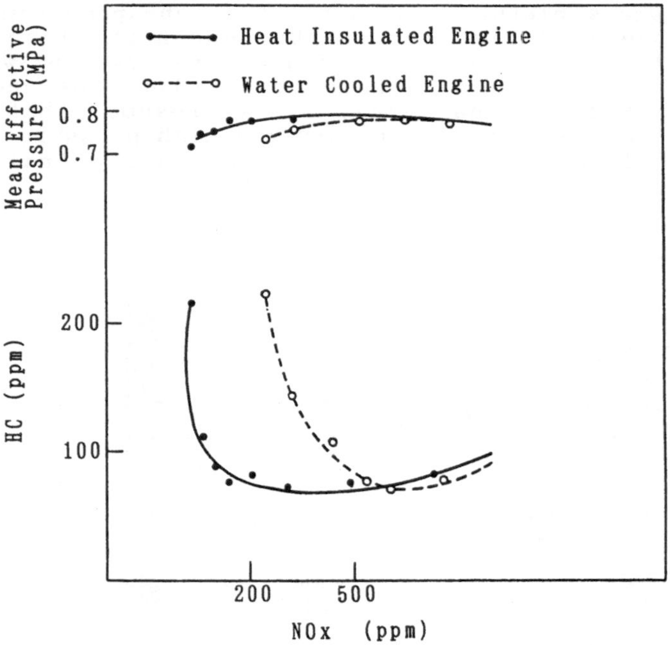

Fig.19 Emission & Performance of Ceramics Heat Insulated Engine with Pre-combustion Chamber

Exhaust gas coming out of an exhaust manifold works in a turbochargers to send air to engine. The gas function again at an energy-recovery turbine where its workload is converted into electric energy, and the power is sent to a motor which is directly connected with a turbine shaft. On this turbo compound system, an ultra-high speed generator is directly connected with a turbine shaft, and energy is efficiently recovered. The structure is shown in Figure-21. Its turbine blades and bearings are made of Si_3N_4, and the turbine scroll of Al_2TiO_5 for heat insulating material. Its revolving speed is 100,000rpm, generating 10Kw. Because the revolving speed is high, its rotor is made of neodymium-steel permanent magnet with high strength for resisting a centrifugal force.

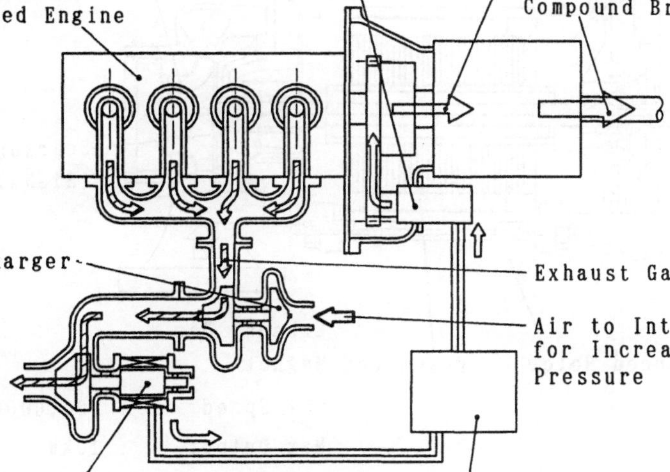

Fig.20 Turbo Compound System in Ceramics Heat Insulated Engine

The heat insulated turbo compound engine can be completed based on the above mentioned development of technology. However, since improvement of fuel efficiency can be accomplished through the accumulation of improvements, it is necessary to modify technology of each items. Figure-22 shows rates of improvement in fuel efficiency. When load is high, fuel efficiency is largely improved due to increase in workload of the energy recovery turbine. At low load, much effects of reduction in friction in an engine can be observed.

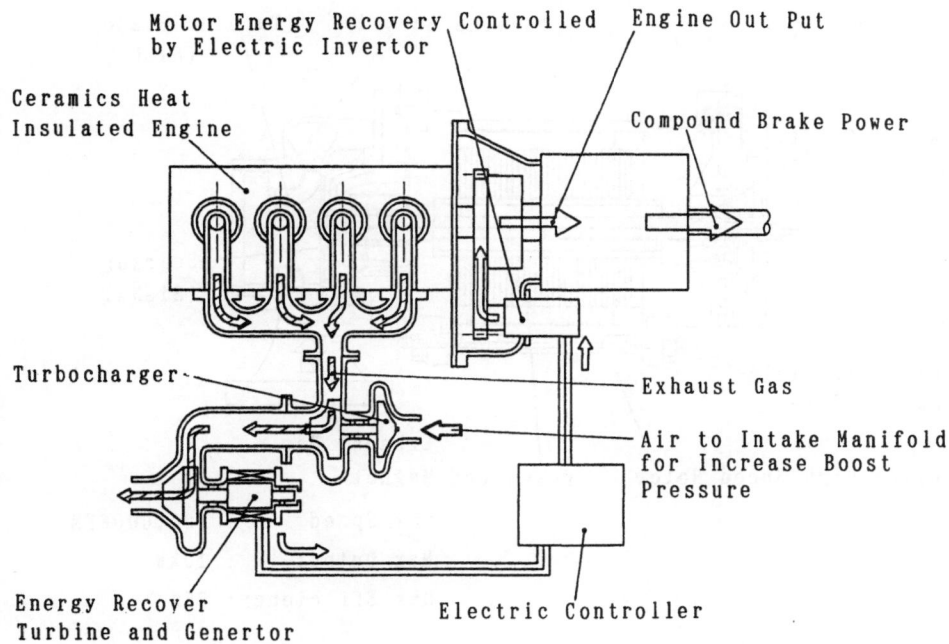

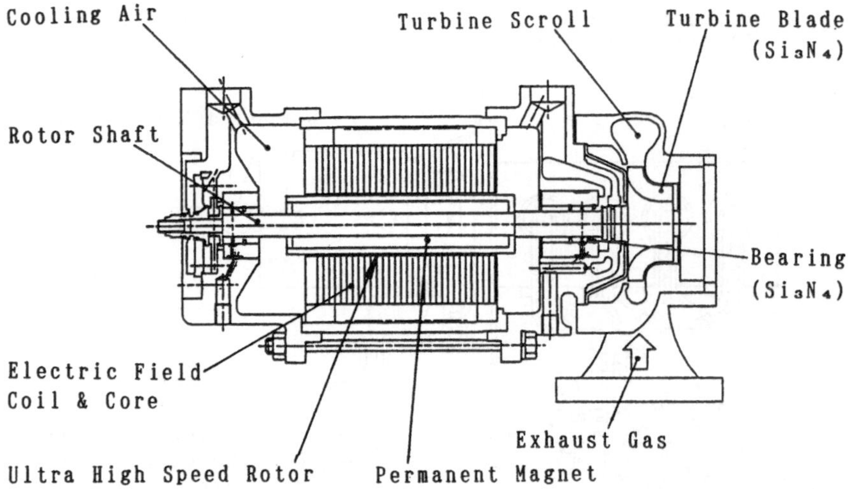

Max. Speed : 100,000RPM

Max. Output : 10kW

Max. Efficiency: 85%

Fig.21 High Speed Generator & Turbine for Recovering Energy from Exhaust Gas

Items	Improved Percentage	
	High Load	Low Load
Engine Friction	3%	15%
Cooling Device	5%	3%
Work of Turbocharger	10%	—
Energy Recovered from Exhaust Gas	12%	—
Combustion Improvement	2%	5%
Other	—	5%
Summation	32%	28%

Fig.22 Rate of Improvement of Thermal Efficiency

New technology should be developed in a wide range of fields such as ceramic materials, heat insulating structure, tribology, high temperature combustion, energy recovery of exhaust gas, etc. Demands for ceramic engines have changed from its original one, improvement in thermal efficiency, in order to correspond to the shortage of fuel in the oil shock. However, demands for energy saving has been further strengthened.

Simplification of engine structure through removing a cooling system will cause large changes in design and structure of a whole vehicle. For example, if a fan and other equipment are removed, the engine could be compact like a capsule, and mounted at any place on a vehicle. Figure-23 is a picture of a vehicle mounted with a ceramic engine, which is designed by a young car designer.

Fig.23 Future Vehicle Installed Ceramic Engine

CONCLUSION

1. We have finished the development of turbo compound heat insulated engine using existing ceramic materials for engine components.

2. Yet we should develop new ceramic materials with high strength and low thermal conductivity to complete the ideal ceramic engine.

3. Development of new ceramic material which provides an absorption film between lubricant oil and its surface is needed in order to reduce a frictional force in engine.

4. We anticipate that turbo compound ceramic heat insulated engine with high thermal efficiency and low emissions will be realized in near future.

5. Development of turbo compound ceramic engine with high thermal efficiency, compact in size and capable of using multi fuels will reform technological fields of automotive engines.

REFERENCES

(1) R. Kamo, W. Bryzik and P. Glance "TACOM/Cummins Adiavatic Engine Program" SAE Paper 830314

(2) H. Kawamura & S. Yamamoto "Improvement of Diesel Engine Startability by Ceramic Glow Plug Start System" SAE Paper 830580

(3) H. Matsuoka, H. Kawamura and S. Toeda "Development of Ceramic Pre-Combustion Chamber for Automotive Diesel Engine" SAE Paper 840426

(4) K. Katayama, M. Sasaki and T. Itoh "Development of Ceramic Turbine Rotors" ASME 88-GT-282

(5) H. Kawamura "Study of Construction and Tribology in Heat Insulated Ceramic Engine" SAE Paper 900624

(6) A. A. Griffith "Phili. Trans. Roy, Soc., 221 (1920)"

(7) G. Ziegler, J. Heinrich and G. Wolling "Review Relationships between Processing, Microstructure and Properties of Dense and Reaction-bonded Silicon Nitride" Journal of Materials Science 22 (1987) 3041-3086

(8) P. C. Glance, W. Bryzik, J. Mahishi and J. Sphar "Engine Compornent Design Methodology for Ceramic and Ceramic-Matrix Composite Materials" SAE Paper 880193

(9) F. Nagao "Internal Combustion Engine" Yohkendo Japan

(10) H. Matsuoka and H. Kawamura "Structure of Heat Insulated Ceramics Engine and Heat Insulating Performance" JSAE October 1991

(11) S. Furuhama and S. Sasaki "New Device for the Measurement of Piston Frictional Forces in Small Engines" SAE Paper 831284

(12) H. Kawamura and H. Kita "Study of Solid Lubrication for

Ceramics Engine" The Symposium of Solid Lubrication in Japan, Society of Japanease Tribology September 1990

(13) H. Kawamura "Study of Grinding Process and Strength for Ceramics Heat Insulated Engine" SUPER ABRASIVES '91 - The International Diamond/C&N Conference & Exposition, SME June 10-14

(14) H. Kawamura, S. Sekiyama and K. Hirai "Observation of the Combustion Process in a Heat Insulated Engine" SAE Paper 910462

PROSPECTS FOR CERAMICS IN AIRBORNE GAS TURBINE ENGINES

E.G. BUTLER+ AND M.H. LEWIS*
Rolls Royce plc, Advanced Ceramic Centre+
University of Warwick, UK.
Centre for Advanced Materials Technology*
University of Warwick, UK.

ABSTRACT

Ceramics offer major advantages in relation to efficiency and performance of gas turbines; their potential for higher operating temperature reduces the need for cooling and their lower density reduces both weight and stresses created in moving components.
A survey of mechanical properties for optimised monolithic ceramics or their dispersion toughened variants, based mainly on Si_3N_4 and SiC chemistries, indicates a need for design criteria based on the statistics of brittle failure. This is not generally acceptable for high-risk airborne turbine components, thus emphasising the need for fibrous composites with well defined material parameters such as matrix microcracking stress, ultimate failure stress and work of fracture, some of which may be used as design criteria. A study of available ceramic matrix composites based on SiC (Nicalon) fibres within glass-ceramic or CVI SiC matrices shows that whilst they may satisfy the stressing conditions for modest temperature components, they are susceptible to higher temperature time dependant degradation, especially in oxidising conditions. The microstructural origins of such problems are largely understood such that the experimental development of model fibre, interface and matrix combinations will be a priority in achieving a high degree of ceramic/metal substitution in a new generation of turbines.

INTRODUCTION

The aero engine market is highly competitive and is driven by the continuing desire for improvement in engine performance. Performance increases can be achieved through improved design methods and the application of new technology. For example over the last 50 years improvements in materials and aerothermal technology have led to increases in thrust to weight ratios from 3:1 to 10:1 and take off thrusts have increased from 1,000 lbs to over 50,000lbs. If present trends continue, then by the turn of the century gas temperatures will approach 2,000°C (stoichiometric combustion) thrust to weight ratios will be of the order of 20:1 and take-off thrust capability will have exceeded 80,000lbs. However the potential for aerodynamic efficiency improvements is becoming less so that further progress can only be achieved by operating at increased temperatures and pressures. The scope for achieving these targets would be severely limited if the current metallic materials were the only alternative. However the development of ceramic materials with high specific stiffness and strength at high temperature offer significant advantages which may be exploited in order to achieve the desired performance improvements.

In view of an apparent maturity in the new era of structural ceramics, with a number of turbine-oriented national programmes initiated more than 10 years ago, it is hence surprising that few ceramic components have reached other than demonstrator-engine status. However it has become clear that for such high-risk applications the use of statistical failure criteria, intrinsic to brittle monolithic ceramics, is not generally acceptable without significant increases in fracture toughness (K_c).

In this paper an initial survey of the advantages of ceramic/metallic substitution, in terms of engine performance, is followed by a review of available materials and their property limitations relative to design targets.

Design Goals

There are a number of factors which have to be considered when optimising the design of civil and military gas turbines. The relative importance of these factors may vary depending on the engine duty. However among the major factors driving the material requirements are thrust, weight, fuel efficiency and cost. Table I illustrates the development of the military gas turbine over the last 50 years and projects future requirements into the 21st century.

TABLE 1
Development of Aero Engine Requirements

Year	1940	1987	2000+
Thrust/weight	3:1	10.1	20:1
Compression ratio	4.1	30:1	40.1
Turbine entry temperature (TET)	800°C	1400°C	2000°C

These design goals require the attainment of turbine entry temperatures (TET's) in excess of 2,000°C and compression ratios of about 40:1 in a package which provides smaller frontal area and much lower weight.

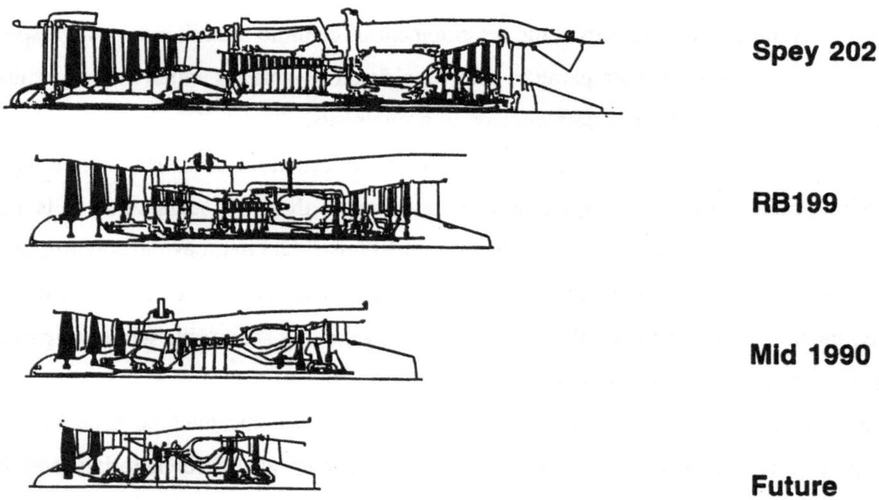

Fig.1 Military engine evolution.

Figure 1 shows the progressive downsizing of military engines in which fewer components must work harder and at higher temperatures. The engine of the future will have to operate at these high TET's with less available cooling air in order to reduce parasitic losses. Currently metallic components can only operate at today's TET's by employing as much as 20% of the compressor delivering air in cooling. Since a 1% reduction in cooling air at a fixed TET is worth 1% increase in power it is clear there are large gains to be made in this area. Provided TET's can be raised without incurring parasitic losses through the use of cooling air there are also large gains to be made in engine thermal efficiency. Figure 2 shows a family of thermal efficiency curves as a function of overall pressure ratio.

The full potential in thermal efficiency can only be realised at high pressure ratios if TET's in excess of 1800°K can be employed. (The design goal for the stoichiometric engine is a pressure ratio of about 40:1).

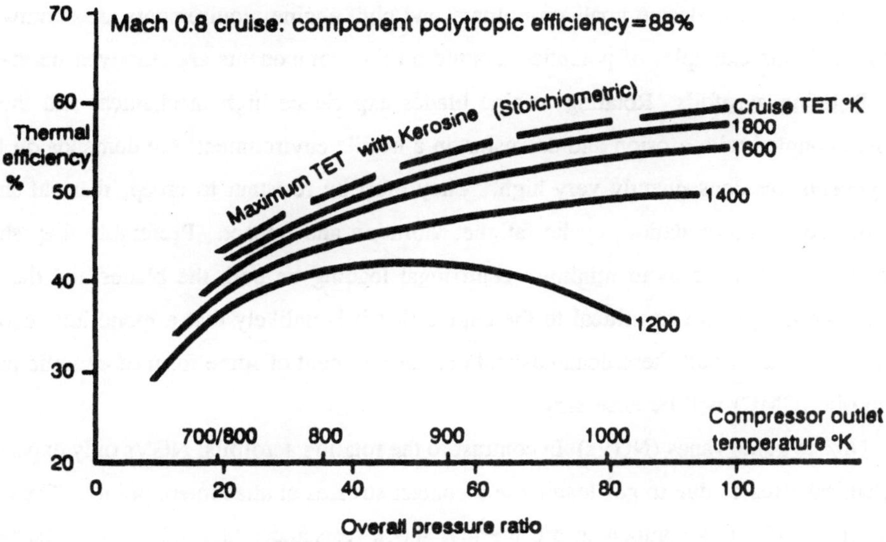

Fig.2 Effect of cycle pressure and temperature on thermal efficiency.

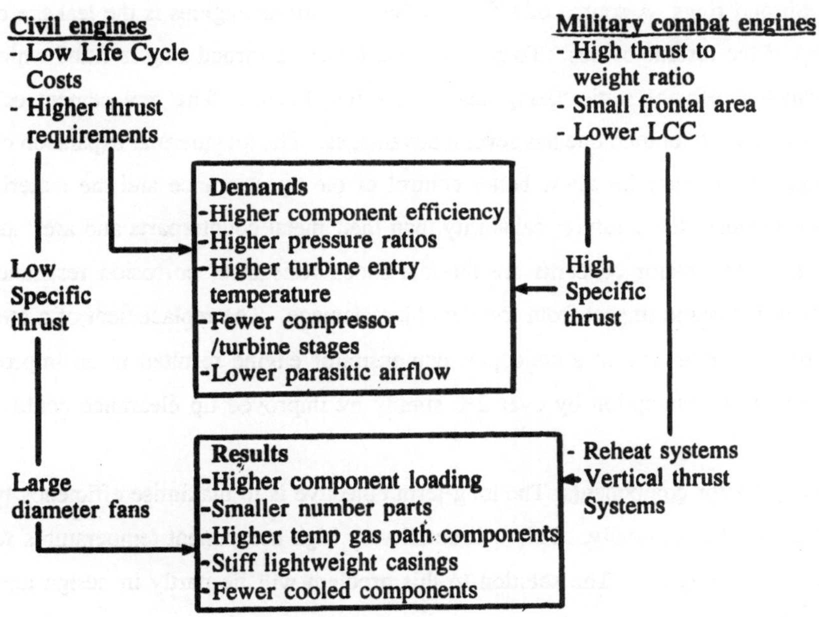

Fig.3 Influence of design goals on military and civil engine features.

The impact of these design goals on military and civil engine requirements are summarised in fig.3. Some examples of potential ceramic turbine components are surveyed here:-

(i) Rotating aerofoils Rotating turbine blades experience high mechanical and thermal stresses coupled with erosion and corrosion in a hostile environment: the demands on these components are consequently very high. They must be resistant to creep, thermal shock, erosion, corrosion/oxidation, cyclic fatigue, vibration and impact. Preferably they should be of low density, so as to minimise centrifugal loading on both the blades and the disc. These components are so critical to the engine that it is unlikely that a monolithic ceramic material will satisfy all these demands and the development of some form of ceramic matrix composite (CMC) will be necessary.

(ii) Nozzle guide vanes (NGVs) In contrast to the rotating aerofoils, NGVs only experience mechanical stresses due to gas loading and contact stresses at attachment points. The major requirements for this component are thermal shock resistance, thermal cycling resistance, oxidation, corrosion and erosion resistance and impact resistance. Again, this is a critical component in the engine and failure could have severe consequences for components downstream.

(iii) Shroud rings A source of efficiency-loss in turbine engines is the leakage of gas past the tip of the turbine blades. To reduce these losses, a shroud ring is fitted which controls the gap between the static casing and the rotating blades. The replacement of the metal component by a ceramic one has several advantages. The low thermal expansion coefficients of engineering ceramics allow better control of the tip clearance and the materials have a higher uncooled temperature capability than their metal counterparts and are more erosion resistant. The major concerns are thermal shock resistance, corrosion resistance and the ability to withstand impact from foreign object damage. The replacement of a metal shroud ring by a ceramic one in a helicopter demonstrator engine resulted in an improvement in specific fuel consumption by over 2% simply by improved tip clearance control (Benger, [1]).

(iv) Combustor components The long-term objective is to maximise efficiency by burning the fuel stoichiometrically. This will entail very high component temperatures without the benefit of cooling air. The solution to this problem will lie partly in design and partly in material development.

The material requirements are low thermal conductivity, thermal shock resistance, erosion and corrosion resistance. The mechanical stresses are likely to be low, consisting mainly of combustion gas pressure and attachment stresses. In the short term, the need is likely to be satisfied by non-structural fibrous insulating tiles supported on a structural backing. Ultimately, self supporting hybrid structures composed of dense load-bearing and fibrous insulating materials will be required.

(v) Re-heat components Re-heat is a method of augmenting the thrust of a turbine engine by burning fuel in the exhaust section and utilising the oxygen from unburned cooling air to support combustion. In order to stabilise the flame, an eddy-current generator is placed in the combustion region. This component is known as a flame holder and can be in the form of an annulus or a series of radial fingers. It has to withstand quite severe thermal shocks since the temperature rise once the re-heater is switched on is very rapid. The mechanical loads are quite low, with contact and bending stresses being the most severe.

Apart from thermal shock resistance, the major requirements are for oxidation/corrosion resistance, impact and erosion resistance and the ability to withstand high frequency vibrations induced by the combustion process.

(vi) Exhaust components These consist of the exhaust cone, the jet pipe and the nozzle. Currently, the gas temperatures encountered in this section are between 550 and 850°C, rising to 1700°C in the re-heat mode. The metallic components are air- cooled and there are significant gains in efficiency to be made if this cooling air can be eliminated.

The exhaust temperature in stoichiometric burning engines is likely to rise to 1850°C. A prime function of the jet pipe is to insulate external components from the exhaust so that in order to eliminate cooling air, the jet pipe material must be insulating. Since the structure is load bearing to some degree, it must also be relatively strong and stiff. This partially conflicting requirements could be met by employing hybrid structures consisting of a dense lightweight backing coupled with a low density fibrous insulating layer covered with hard dense skin to provide erosion resistance.

The variable exhaust nozzles also have to withstanding fretting and abrasion as the elements slide over each other during operation.

Material Requirements

It is clear that the improvements which led to the development of powder metallurgical technology and single crystal nickel based alloys will be incapable of meeting the future challenges in engine design requiring increased specific strength, stiffness and temperature capability. However, any replacement will have to demonstrate predictable behaviour and guaranteed component life. In addition both materials and manufacturing costs levels will have to be maintained and preferably reduced.

In summary the materials must exhibit:-

(1) Increased specific strength

(2) Increased specific stiffness

(3) Increased temperature capability

(4) Predictable behaviour

(5) Contained cost levels

AVAILABLE MATERIALS AND CURRENT LIMITATIONS

Monolithic Ceramics

The major contenders for high-temperature turbine application have been the 'Si-based' ceramics (SiC, Si_3N_4 and the Sialon derivatives) based on the intrinsic properties of passive oxidation, low thermal expansion and high specific stiffness. Microstructures are normally diphasic, resulting from the necessity for sintering additions. For example, B & C are added to SiC as surface-active solid state sintering catalysts and excess C frequently remains in dispersed form (Fig.4a). Silicon nitrides and sialons are sintered with a liquid silicate minor phase which normally remains as a glass. This liquid phase may influence growth morphology of the major (βSi_3N_4 or β'Sialon) phase with consequent improvement in fracture toughness over that for equiaxed microstructures typified by solid-state sintered ceramics or those with transient liquid sintering aids (Fig.4 b,c).

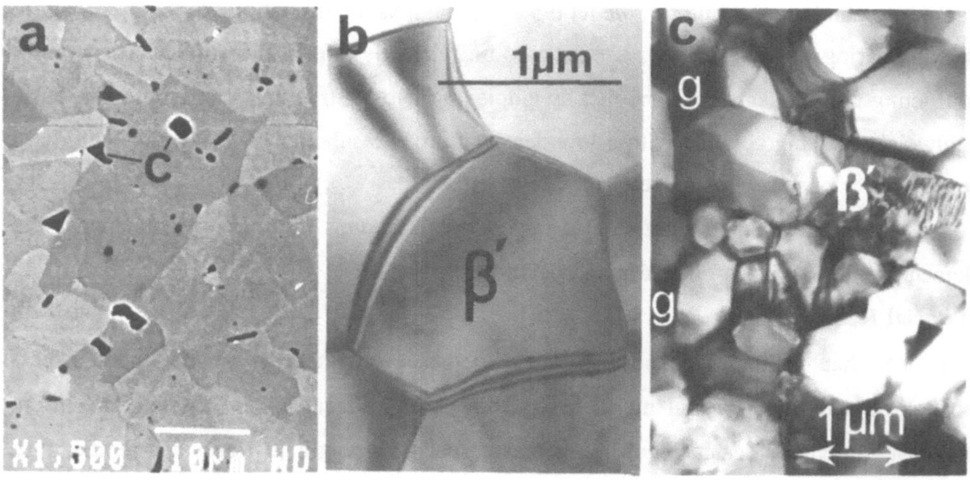

Fig.4 Examples of monolithic ceramic microstructures; (a) solid-state-sintered SiC (courtesy Dr. G. Leng-Ward), (b) transient-liquid-sintered monophase sialon, (c) liquid-phase-sintered diphasic sialon.

A major limitation for these ceramics is a fracture-toughness (K_c) within the range 3-8 MPam$^{1/2}$, the highest values resulting from mechanisms such as crack deflection, bridging or pull-out associated with elongated grain morphologies. In terms of brittle fracture mechanics these values mean that for service stresses of 100-300 MPa flaw sizes of 50-100μm are critical. The inability to easily detect or eliminate inhomogeneities or machining flaws of these dimensions, together with the susceptibility of environmental impact damage, necessitates the application of probabilistic fracture mechanics within design philosophies. Using the well-established Weibull modulus (m) as a measure of flaw size distribution a probability of component failure may be estimated, with varying design stress. Fig. 5 exemplifies the high level of m (>30) required for a 'better-than-0.001%' probability of fracture at a design stress of 200 MPa in a small (~2cm^3) component using a ceramic with mean fracture-stress of 300 MPa. Values of m are typically in the range 10-15 such that failure probabilities are unacceptably high for turbine applications, except for very small components operating at low stresses with little prospect of transient overloads during impact. Even thermally-induced stresses in non-rotating components (e.g. combustor tiles or nozzle guide vanes) may be beyond such limits.

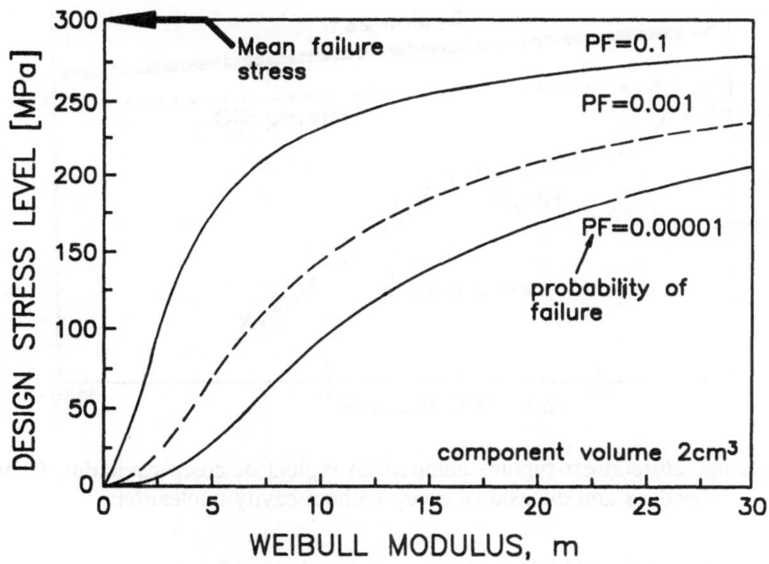

Fig.5 Brittle failure probabilities for varying Weibull moduli (m) and design-stresses.

Whereas most design philosophies have so far considered 'zero-time', athermal data within probabilistic analyses it is essential to include time and temperature dependent properties of creep and stress-rupture. An important step in the evolution of Si_3N_4-based ceramics has been the suppression of short-time stress-rupture due to creep cavitation nucleated within glassy grain boundary residues (Fig.6). The importance of transient liquids or post-sintering glass crystallisation in reducing cavity nucleation sites to sub-critical size was first demonstrated for Sialon microstructures [2,3]. Similar improvements may be obtained in compositionally tailored Si_3N_4 ceramics [4,5] resulting in the characteristic 'threshold' stress below which cavitation is absent and pre-existing flaws do not extend but are 'blunted' by diffusional creep (Fig 6). The solid-state sintered SiC ceramics exhibit similar characteristics. Creep rates in such microstructures do not present design problems and are generally in the 10^{-9} - 10^{-8} s^{-1} range for 100-200 MPa stresses in sialons, HIP Si_3N_4 and SiC at temperatures of 1300°C, 1400°C and 1500°C, respectively.

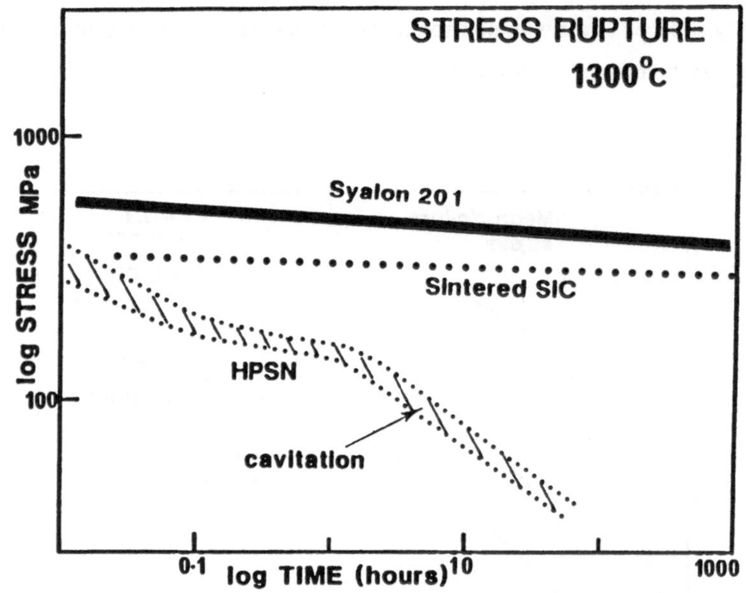

Fig.6 High-temperature stress-rupture comparison typical of creep-cavitation from glassy residues and diffusional creep without cavity nucleation.

Environmentally-induced high-temperature failure is significant, especially for diphasic Sialon ceramics which undergo a reversion of the intergranular (oxide) phase to the liquid state by reaction with the surface SiO_2 oxidation layer above ~ 1300°C (the silicate eutectic temperature). Oxidation kinetics increase rapidly in this regime due to metallic ion transport to the SiO_2 layer. Intergranular disilicate phases in HIP Si_3N_4 ceramics normally form stable couples with SiO_2 up to eutectic temperature above 1500° and hence have higher temperature limits (1400°C-1500°C) at which oxidation kinetics and creep rates become significant.

The operation of turbine ceramics in salt-laden atmospheres or with impure fuels may lead to accelerated oxidation due to viscosity-reduction in the protective SiO_2 layer. The development of localised oxidation pits may also have an influence on time-dependent strength (Fig.7)

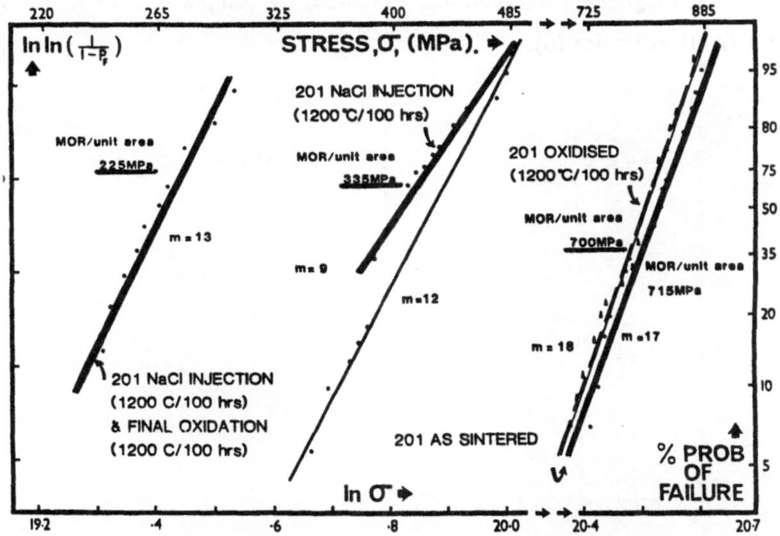

Fig.7 Oxidation/corrosion influence of salt-containing flame conditions on strength (MOR) of a β' sialon ceramic.

Dispersion-Toughened Ceramics

The simplest microstructures are formed by the conventional mixing/sintering of thermodynamically stable phase combinations. The matrix remains the main load-bearing component and typically occupies 60-80% of the microstructure; the restriction in dispersed phase content is dictated by impaired 'sinterability; and the non-linear relation with property change. Particle, whisker and platelet dispersions, normally SiC have been used to toughen matrices which have been prominent as monolithics, viz. silicon nitrides, sialons and oxides of alumina or zirconia. Transition metal compounds have been used mainly in particulate-dispersed, the best example being TiB_2 in SiC matrices.

The limitations described for monolithic ceramics, in turbine component probabilistic design methods, largely apply to the dispersion-toughened variants. Fracture-toughness values based on theoretical modelling of mechanisms such as crack deflection, crack bridging, 'pull-out' and thermally-induced microcracking indicate maxima of ~ 15 $MPa^{1/2}$ with optimistic assumptions concerning interface fracture energies relative to that for the dispersed phase, with the latter in high-aspect morphology [e.g. 6,7]. However, values of K_c in excess of 8 $MPa.m^{1/2}$ are rarely achieved and normally require some processing - induced anisotropy in whisker/platelet orientation (e.g. Fig.8). Low interface cohesion is important for most toughening mechanisms, evidenced by comparisons of K_c for whisker-pretreated SiC/Al_2O_3 ceramics [6].

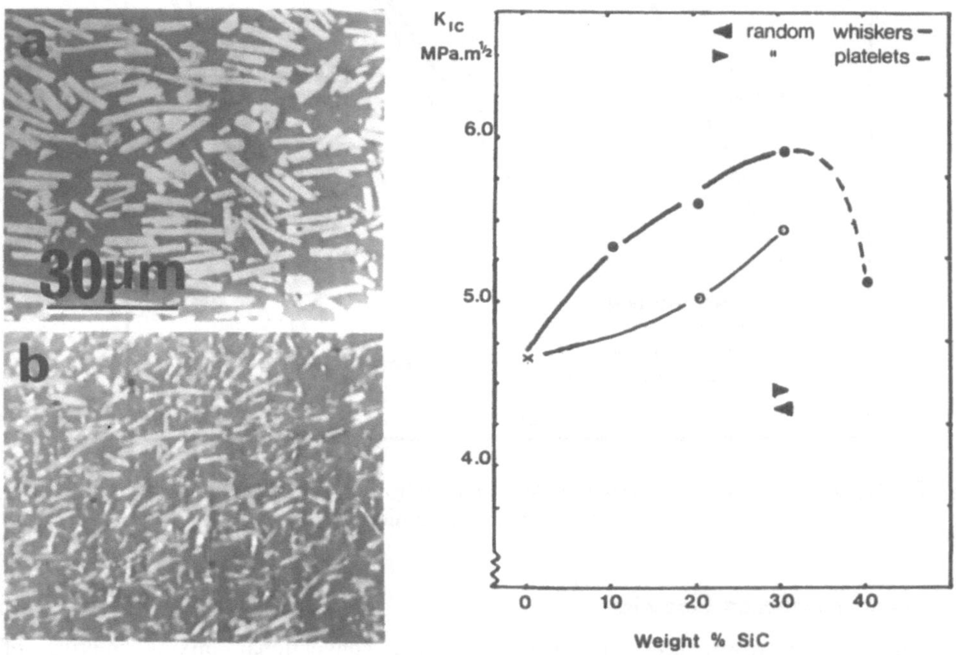

Fig.8 The influence of dispersoids (platelets and whiskers imaged in (a) and (b) respectively) on transverse fracture toughness (K_c) of a Si_3N_4 ceramic.

A secondary influence of dispersions, which may have a greater impact on design, is an occasionally reported improvement in Weibull modulus of up to m = 30. This is presumably related to a unification of flaw size to that limited by dispersion geometry (or to machining flaws restricted to interparticle spacings) and is often accompanied by a reduction in mean strength, even with enhanced K_c.

High-temperature limitations may be more severe than for the isolated matrices; oxidation resistance may be impaired due to reaction between oxidation products of the separate phases, e.g. SiC/Al_2O_3, TiB_2/SiC, but is retained in SiC/Si_3N_4 combinations which have a common passive film chemistry (SiO_2[8]. The limited creep and stress-rupture data shows little change from matrix characteristics in some systems (eg SiC/Si_3N_4 [9] with considerable reduction in creep rates for matrices which normally exhibit crystal plasticity in addition to diffusional creep (SiC/Al_2O_3 [10].

Novel processes of ceramic synthesis and fabrication have recently appeared, such as the 'Lanxide' process for directed oxidation or nitriding of liquid metals [11]. These microstructures represent a variant of dispersion-toughening via constrained plasticity of residual metallics in a ceramic matrix (e.g. Al in Al_2O_3). K_c values in excess of 20 MPa.m$^{1/2}$ may be obtained. These matrices have been used in composites with SiC particles or fibres (du Pont-Lanxide, USA) and appear to offer some attractive solutions to both fabrication/shaping and toughness problems for selected turbine components. However, high temperature limitations are likely to relate to creep cavitation from the residual metallic content or lack of synergism in oxidation for the diverse phase chemistry.

Long-Fibre CMCs

The concept of 'damage-tolerance' via load-transfer to high-modulus semi-continuous aligned fibres during matrix-microcracking in a service stress transient has been demonstrated experimentally and by theoretical modelling [12,13,14]. The classical CMC response is that of 'graceful', non-brittle, failure with high fracture energy due to fibre pull-out and crack-bridging. Design and materials development for airborne gas turbines has to be based largely on this class of CMC . However there are currently major problems concerning the difficulties and cost of fabrication together with availability of fibres which are stable during fabrication or high temperature service and have the necessary interfacial properties within a range of matrices. Some of these limitations are exemplified here for CMCs at an advanced development stage, with SiC fibres in silicate glass-ceramic, silicon nitride or silicon carbide matrices.

Fine (10-20μm dia.), polymer-precursor, SiC-based fibres have given a major impetus to CMC development following earlier experiments with, oxidation-susceptible, carbon fibres. Nicalon (Nippon Carbon) and Tyranno (Ube) fibres have a fortuitous combination of composition (Si-C-0) and non-crystalline structure which produces a controlled reaction with silicate matrices, during fabrication, resulting in carbon-rich interface layers Fig.9.

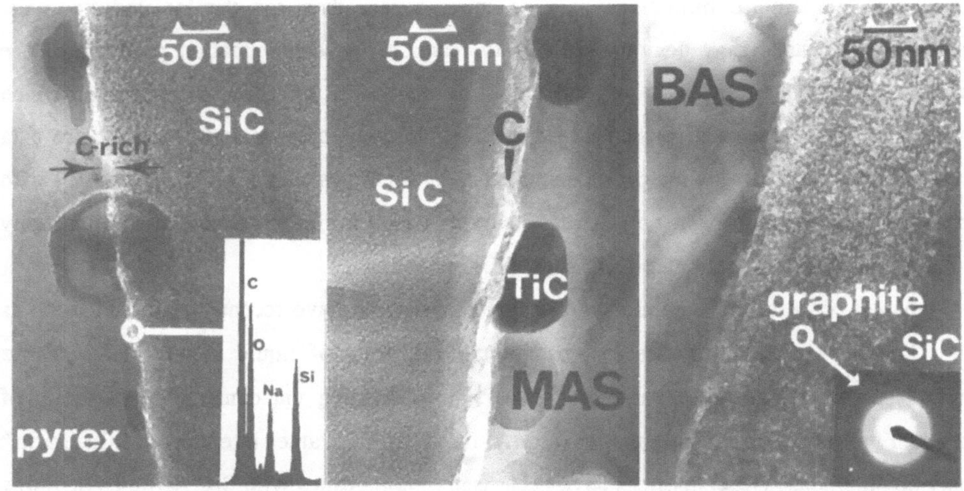

Fig.9 C-rich interfaces formed during processing reactions between Nicalon SiC fibres and silicate matrices; (a) Nicalon-Pyrex (glass) [15], (b) Nicalon-MAS [16], (c) Nicalon-BAS [17].

These 'in-situ' interfaces debond readily during matrix microcracking but are prone to oxidation above ~500°C. These fibres also undergo structural and mechanical degradation above ~1200°C which must represent an application limit and is near to a fabrication maximum. In non-silicate matrices, such as SiC produced by CVI (chemical vapour infiltration) the necessary low-cohesion interface may be formed by fibre precoating (e.g. Nicalon is available in a carbon coated state). CVD pseudocrystalline SiC monofilaments (Textron, USA), 140μm dia., are structurally stable to ~1300°C and will tolerate short excursions to 1700°C; hence they have been used as reinforcements for high temperature sintered matrices (Si_3N_4) using thick, partially sacrificial, carbon coatings to constitute the low cohesion interface (Fig.10), [18].

The selection of silicate matrices is based on fabrication route, utilising the viscous flow property of the parent glass, and the ability to tailor thermal expansion near to that of the fibres. An example is the low-expansion borosilicate glass (Pyrex) which may be fabricated at ~1000°C but has an application limit near to its softening point (~600°C). The glass-ceramic route has extended the matrix range and temperature capability, typically to aluminosilicate compositions such as Li_20-Al_20_3-$Si0_2$ (LAS), $Mg0$-Al_20_3-$Si0_2$ (MAS) and $Ca0$-Al_20_3-$Si0_2$ (CAS) in which low-expansion phases crystallise during fabrication and on subsequent heat treatment.

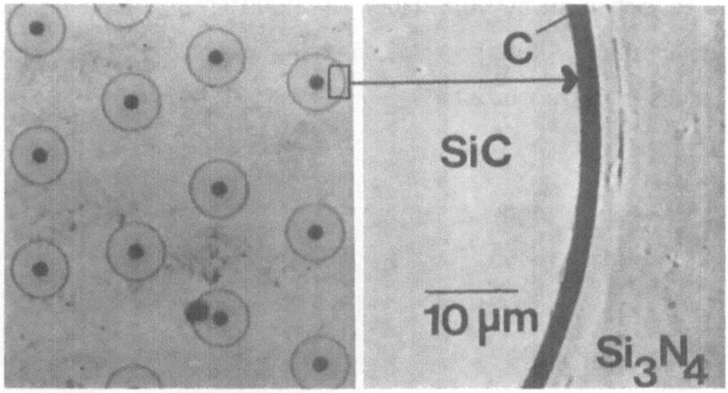

Fig.10 C-precoated SiC monofilaments within a sintered Si_3N_4 matrix (courtesy, A.G. Razzell, University of Warwick).

There is currently some limitation in achieving an ideal fabrication cycle for these glass-ceramic matric composites because of the conflicting requirements for densification, fibre stability, crystallisation and interface reaction kinetics. The latter is important for control of the interval between ultimate fracture stress σ_u and matrix-microcracking stress σ_m. The value of σ_m is determined, for unidirectional composites, by the interface shear resistance τ after debonding together with matrix and fibre parameters [12];

$$\sigma_m = \left[6\ G_m\ \frac{\tau}{r}\ \frac{E_m E_f}{E} \cdot \frac{V_f}{1+V_f}\right]^{\frac{1}{3}}$$

where G_m = matrix fracture energy, r = fibre radius, E_m, E_f, E are Young moduli of matrix, fibre and composite with fibre volume fraction V_f.

For critical turbine components σ_m must be a design limit, especially with oxidation-susceptible interfaces. Values of τ may vary from <5 MPa, for high-temperature glass ceramic fabrication with well-developed graphitic layers (Fig.9b,c), to >50 MPa for interfaces formed at lower temperatures and composed of non-crystalline carbon enriched layers (Fig.9a). Apart from fabrication temperature, time and matrix chemistry, the τ value may be adjusted with selection of matrix thermal expansion to induce radial compression on the fibre and hence modify the frictional coefficient after interface debonding.

For CVI matrices, such as SiC,. there is more control over interface structure and properties via fibre precoating or C-coating of woven preforms. However, although CVI is a relatively low-temperature method for fabricating high-temperature matrices such as SiC, the temperature restriction set by fibre stability (~1200°C) may limit the matrix to non-stoichiometric compositions. It is also time-consuming, expensive, limited to small sections and generally results in matrix porosity.

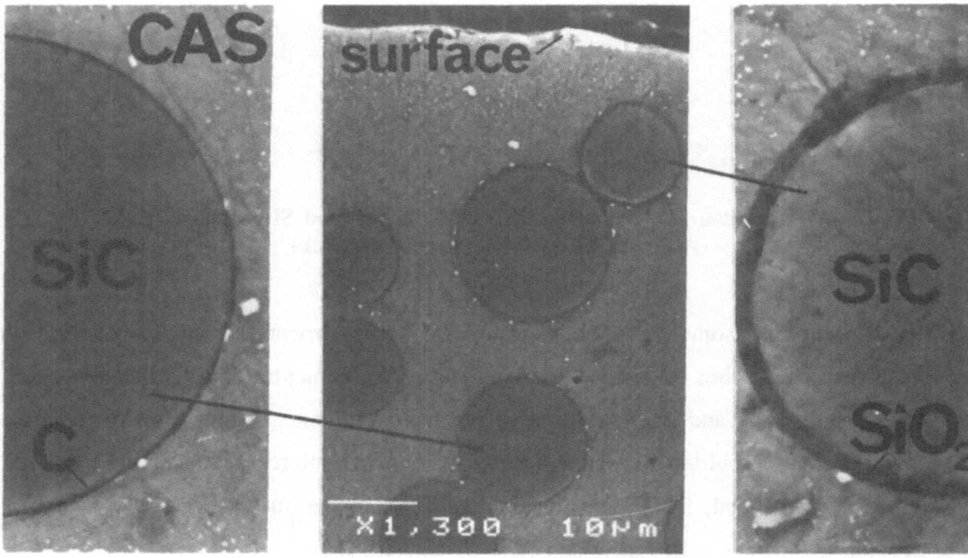

Fig.11 Oxidative diffusional degradation of C-rich interfaces, in Nicalon SiC/CAS composite, confined to outer fibre layers at 1200°C (courtesy, Dr. M. Pharoah, University of Warwick.

All C-rich interfaces suffer oxidative degradation, with oxygen transport via matrix diffusion, matrix microcracks or porosity. An additional mechanism is that of 'pipe'-reaction from exposed fibre ends. For dense matrices below the microcracking limit there is little interface degradation since oxygen diffusion rates are low, even for crystalline silicate matrices up to 1200°C.Fig.11 illustrates diffusive degradation via oxidation and $Si0_2$ bridging of C-rich interfaces, in a CAS matrix, which is restricted to the near-surface fibres. A more significant problem is the piped reaction which removes the C-interface for large distances from fibre ends and reduces σ_m and σ_u following intermediate temperature exposure (500-800°C; Fig.12). The retention of properties after higher-temperature oxidation is due to a blocking of the reaction by a passive $Si0_2$ layer at fibre ends. Hence it appears that a pre-oxidation surface treatment may be effective in extending application temperatures of glass-ceramic matrix composites up to the fibre-instability limit. Similar interface oxidation phenomena appear in CVI SiC/SiC composites; heat-treatments at 800°C reduce σ_m and σ_u there is little property change (Fig.12b - data from Ref.[20]). CVI SiC/SiC composites from SEP (Bordeaux) are supplied with glassy coatings, presumably to inhibit the intermediate temperature instability. These materials have been successfully run in turbine tests as exhaust components.

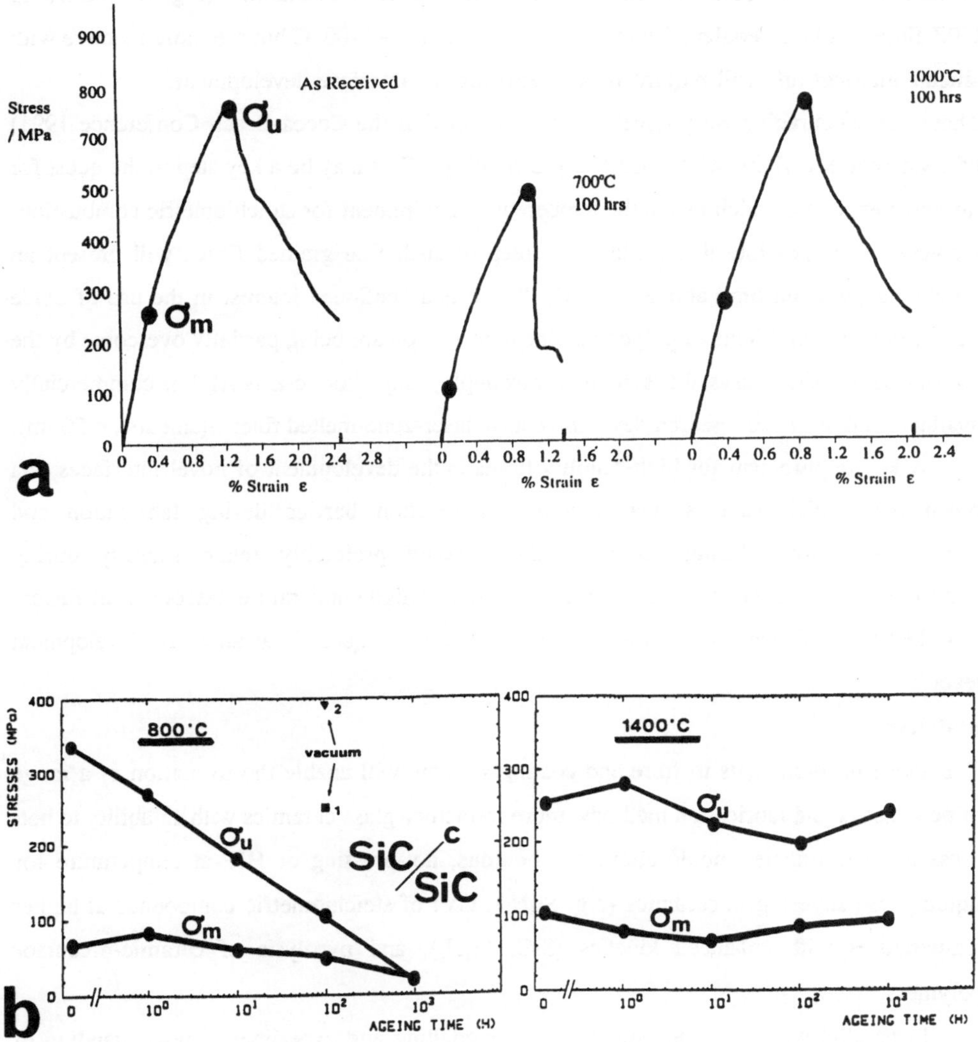

Fig.12 Intermediate temperature property-degradation due to piped oxidation of C-rich interfaces in (a)Nicalon-CAS [19] and (b)Nicalon-CVI SiC [20].

NEW DEVELOPMENT REQUIREMENTS FOR CMCs

Fibres and Interfaces

A recurrent constraint in relation to fabrication temperatures or application conditions is the lack of small diameter, weaveable, fibres which are structurally and mechanically stable to at least 1600°C, which may be manufactured at low cost from polymer precursors. Non - crystalline silicon carbo-nitride fibres, which have restricted availability (e.g. Dow-Corning HPZ fibre or Rhone-Poulenc 'Fiberamic') are stable to ~1400°C but are more reactive with silicate matrices and will require fibre precoating for interface development.

There are encouraging early developments (reported at the Cocoa-Beach Conference 1991) of crystalline SiC fibres with 1700-1800°C stability. This may be a key step in the quest for turbine composites which match the temperature requirement for stoichiometric combustion. However, it is possible that diffusional creep of such fine grained fibres will present an ultimate application limit above 1600°C. There is a continued interest in the use of oxide fibres, in which problems of polycrystalline deformation are being partially overcome by the use of oriented single crystal filaments; for example, 'Saphikon' c-axis Al_2O_3 is commercially available and there are research developments in laser-zone-melted finer filaments ($<50\mu m$).

A key requirement for high-stability fibres is the development of novel interfaces, via precoating, which have a dual function of reaction barrier during fabrication and debond/shear layer during service. They should preferably retain stability during atmospheric exposure due to matrix-microcracking or high temperature gaseous in-diffusion. This key area of technology, using PVD or CVD techniques, is at an early development level.

Matrices

The above developments in fibre and coating systems will enable the formation of a range of new matrix and fabrication methods; more refractory glass-ceramics with an ability to hot-press or glass-transfer mould above the liquidus, hot-pressing or HIP at temperature for liquid phase sintering in ceramics (e.g. Si_3N_4), CVI of stoichiometric compounds at higher temperatures with enhanced kinetics (SiC, Si_3N_4), and pyrolysis of ceramic-precursor polymers.

In parallel with these developments the modelling and experimental understanding of deformation and failure modes must be extended to a range of fibre architectures and time-temperature conditions typical of turbine cycles.

REFERENCES

1. Benger, B. (1986) Rolls-Royce Private Communication.
2. Karunaratne, B.S.B. and Lewis, M.H., J.Mater.Sci. 15 449 (1980).
3. Lewis, M.H., Mason, S. and Szweda, A., in 'Non-Oxide Engineering and Technical Ceramics' ed. S.Hampshire (Elsevier 1986) 175.
4. Tuersley, I.P., Leng-Ward, G and Lewis, M.H., in 'Ceramic Materials and Components for Engines' ed. V.J. Tennery, (Amer. Ceram. Soc. Inc. 1989) 856.
5. Tuersley I.P., Leng-Ward, G. and Lewis, M.H., in 'Engineering with Ceramics' (Inst. of Ceramics 1990) 231.
6. Becher P.F., Hsueh C.H., Angelini P. and Tiegs T.N., J. Amer. Ceram. Soc. 71 1050 (1988).
7. Evans, A.G., J. Amer. Ceram. Soc. 73 187 (1990).
8. Ketchion, S.M., Leng-Ward G. and Lewis M.H., in press, Proc. 2nd Int. Ceram. Science and Technology Congress, Orlando 1990, (Amer. Ceram. Soc.).
9. Ketchion, S.M., Leng-Ward G. and Lewis M.H., in press, 'Ceramic Materials and Components for Engines' ed. R. Carlsson (Elsevier 1991).
10. Porter, J.R., Lange, F.F. and Chokshi, A.H., Am. Ceram. Soc. Bull. 66 343 (1987).
11. Newkirk, M.S., Urqhart, A.W. and Zwicker, H.R., J. Mater. Res. 1, 81 (1986).
12. Evans, A.G. and Marshall, D.B., Acta. Metal. 37 2567 (1989).
13. Prewo, K.M., Brennan, J.J. and Layden G.K., Amer. Ceram. Soc. Bull. 65 305 (1986).
14. Briggs, A. and Davidge, R.W., Mat. Sci. and Eng. 109 363 (1989).
15. Murthy, V.S.R., Pharaoh, M.W. and Lewis, M.H., Mat.Lett. 10 161 (1990).
16. Murthy, V.S.R., Li Jie and Lewis M.H., Ceram. Eng. Sci. Proc. 10 938 (1989).
17. Murthy, V.S.R. and Lewis, M.H. Trans. Brit. Ceram. Soc. 89 173 (1990).
18. Razzell, A.G. and Lewis, M.H., in press, Ceram. Eng. Sci. Proc. 12 (1991).
19. Pharaoh, M.P. and Lewis, M.H., in press, Mat. Lett. (1991).
20. Frety, N and Boussuge, M., J. Comp. Sci. and Technology 37 177 (1990)

METAL CERAMIC JOINING FOR HIGH TEMPERATURE APPLICATIONS

Lars Pejryd
Volvo Flygmotor AB
S-461 81 TROLLHÄTTAN
SWEDEN

ABSTRACT

The joining of metals and ceramics is one of the key questions for making wide use of ceramic materials in engines possible. This paper is focusing on the users questions and demands for metal ceramic joining. The need for application oriented work is stressed. Questions such as service conditions and compatibility, both chemical, thermal and mechanical are dealt with. Joining techniques such as brazing, diffusion bonding, friction welding, riveting and shrink fitting are discussed. The need for development of design and NDE methods is also pointed out.

INTRODUCTION

Ceramic materials have a great potential for use in heat engine applications due to their high temperature capability. It can however not be expected that they will be the only materials used in the engine, but they will rather be parts of an integrated construction. The need for development of joining techniques for metal/ceramic joining is thus obvious. The brittle nature of ceramics will however make it necessary to use ceramics only where they are absolutely needed, eg at high temperatures. This is the driving force for trying to obtain as high service temperatures as possible for metal/ceramic joints.

Potential processes for obtaining high temperature joints are brazing, diffusion bonding and friction welding. Each of these processes have their advantages and drawbacks in the form ease of processing, high temperature capability, stresses etc. Joining methods that are more seldom discussed in the literature are mechanical methods such as shrink fitting and riveting.

Much work during recent years has been performed on chemical bonding techniques, especially on brazing and diffusion bonding. Several international conferences focusing on joining of ceramics to metals have been held and an increasing number of papers is being published in the field. However, this paper has not the ambition to cover all the joining literature but rather to focus on the questions and demands of users concerning joining of metals and ceramics for application at high temperatures.

SERVICE CONDITIONS

What are the service conditions that a metal ceramic joint must withstand? The answer to that question will obviously vary depending on the actual application and the variations in expected life, stress levels, temperature, and gas phase composition. It is however very important already at the R&D stage to have solid knowledge of the expected service environment of a joint for instance to avoid solving one problem by creating another. An example of this is the use of W–interlayers to solve problems of thermal mismatch in diffusion bonding of high temperature joints. Such joints cannot work at high temperature in oxidizing environments since tungsten oxidize rapidly at relatively low temperatures giving a gaseous oxide!

In most combustion engines, the atmosphere is predominantly oxidizing, but in some parts one can encounter both oxidizing and reducing environments. Figure 1. shows some aspects of the environment in a jet engine, based on equilibrium calculations. The overall variation in conditions shown also apply to other engines, the actual levels may vary. The major environmental differences between a jet engine and a car engine are that the temperatures in the latter are lower and that the expected life and time between overhaul is much longer. In rocket engines on the other hand, some parts might work under more reducing conditions but the main differences are that more parts will work at high temperatures and with a shorter expected life.

In a jet engine, or a gas turbine, ceramic materials can be expected to be used in combustors, turbine parts and afterburner components(1). Ceramic bearings are another application but this will not be discussed here. As can be seen from Figure 1, the above components encounter, very high temperatures and both oxidizing and reducing conditions.

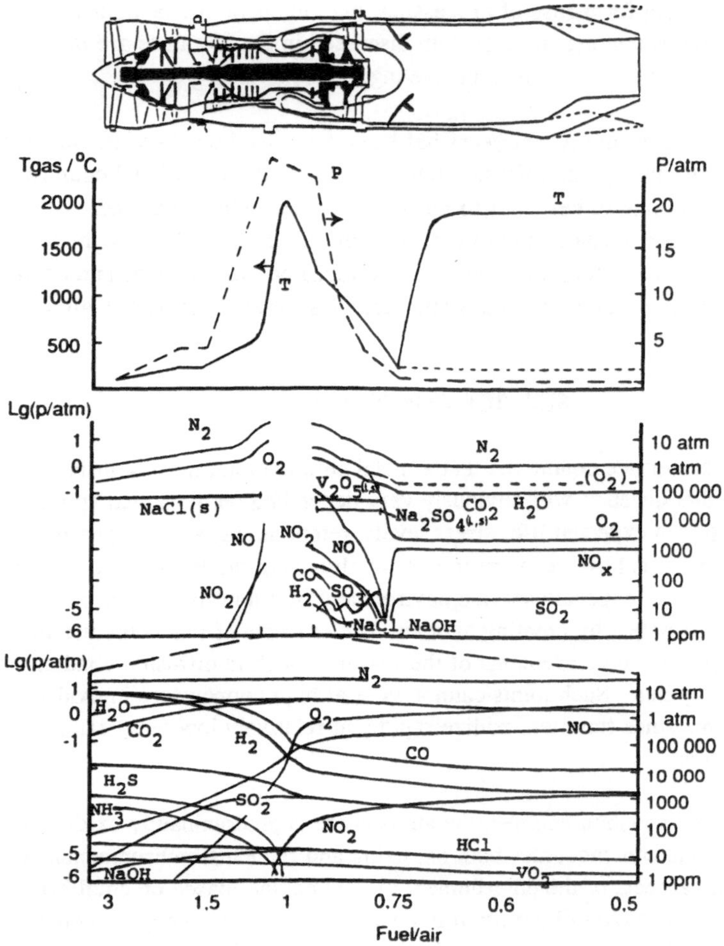

Figure 1. T, P and gas phase composition in a jet engine. The gas phase composition is obtained through equilibrium calculations.

The mechanical loads will vary even more between different applications and are therefore not easy to summarize. The joints will however in many cases be a weak spot in the construction and stresses in the joint region should therefore be minimized. Metal/ceramic joints will of course differ from metal/metal joints but some rules of thumb derived from the joining of metals are worth mentioning. Brazed joints are used mainly in shear loading conditions while welded joints also can be used in tensile load situations. The joints are if possible placed in low stress regions of the construction. Bolted and riveted joints on the other hand can be used in rather high stress situations for joined metals.

Summarizing the conditions under which a metal/ceramic joint will have to function, it is obvious that the demands on the joining techniques for possible application in engines are stringent.

COMPATIBILITY

Materials combinations that are to be joined must be at least moderately compatible with each other. Problems of compatibility become more severe the higher the temperature of both joining and service. The problem of chemical compatibility is twofold. On one hand some chemical reaction between the materials is desirable to obtain a strong bond, on the other hand too strong reactions and especially continued reaction during service is undesirable. The materials and the processes must also be thermally compatible. This means e.g. that the heat treatment cycle of a joining process must be compatible with the heat treatment cycle of the alloy to be joined to the ceramic. This is generally not a problem for the ceramic. Mechanical compatibility is mainly a problem of differences in thermal expansion coefficient between metals and ceramics, leading to high residual mismatch stresses.

Chemical Compatibility

Chemical reactions between the different materials in a joint are as already mentioned desirable but only to a limited extent. In order to evaluate the possible reactivity of a certain material combination, phase diagrams and chemical thermodynamics are invaluable tools. The easiest way to obtain information about a system is to study the relevant phase diagrams. Several books containing systematized phase diagram information are now available. Efforts have also been made to collect and/or to determine phase diagrams of relevant metal ceramic systems, of special interest to joining. Among these Warren & Andersson (2) concerning Me–Si–C systems, Schuster (3, 4) concerning Me–Si–N systems and a more comprehensive study of metal ceramic systems by Hultman et.al. (5) can be mentioned.

In the absence of phase diagrams, thermodynamic calculations using computer programs like SOLGASMIX (6) and thermodynamic data from e.g. JANAF tables (7) , Barin & Knacke (8, 9) or Barin (10) can give the same kind of information.

Metal/ceramic systems can be classified as either stable or reactive depending on the equilibrium conditions as proposed in ref. (2). This classification was used in (5) and the result is shown in Table 1. This demonstrates that in many systems consisting of metals combined with the most common engineering ceramics (Al2O3, SiC, Si3N4) the metal and ceramic will react with each other at high temperatures, and are thus "reactive" systems. In practice it is however not always that simple, as can e.g. be illustrated by the SiC–Ni and Si3N4–Ni systems. From the phase diagrams, Figure 2 (valid for 1 atm. total pressure), the carbide system can be identified as a reactive system, while the nitride system is a stable system. In a work by Brito (11)

Table 1. Stable and reactive metal/ceramic systems, from (5).

ceramic \ metal	Ag	Al	Co	Cr	Cu	Fe	Li	Mg	Mo	Ni	Ti
SiC	-	S[1]	-	R	S	R[2]	-	R	R	R	R
TiC	-	S	S	S	-	S	S	S	S	S	B
ZrC	-	R	S	S	-	S	-	S	S	S	R
Si_3N_4	-	R	S	R	S	S	-	-	R	S	R
TiN	-	-	-	S	-	-	-	-	S	-	B
ZrN	-	R	-	S	-	-	-	-	S	-	-
Al_2O_3	-	B	R	-	-	S	R	R	-	S[3]	S
SiO_2	-	R	-	-	-	S	-	R	-	-	-
TiO_2	-	-	-	-	-	S	-	-	S	S	B
Y_2O_3	-	-	-	-	-	S	-	-	-	-	-
ZrO_2	-	R	-	S	-	S	-	R	S	-	R

S = Stable system, R = Reactive system, - = No diagram obtained in this study,
B = Binary system.

1) If temperature exceeds aluminiums melting point, the system is stable if aluminium is stabilized with silicon.

2) Stable if iron stabilized with silicon.

3) Stable if oxygen activity is low.

reactions have been identified also in the nitride system at low pressures, where Si3N4 decomposes and forms a Si–Ni–N solution. The possibility of such reactions at low nitrogen activity is also shown by Klomp (12) using thermodynamic calculations. Thus it must be remembered that a particular phase diagram shows the equilibrium conditions only for a specific set of parameters, T,P composition etc.

Other information that is not obtainable through phase diagrams and thermodynamics is the rate of the reactions taking place. Kinetic data is not as easily obtained as thermodynamic data, and the information concerning reaction paths and kinetics must be searched in the published literature on joining of the system of interest. Some general aspects of kinetics in joining have been discussed by Klomp (12), and work to identify diffusion mechanisms using marker experiments have been performed by e.g. Schiepers et.al.(13). The area is however rather unexplored and much work is needed before "the best" joining processes can be developed.

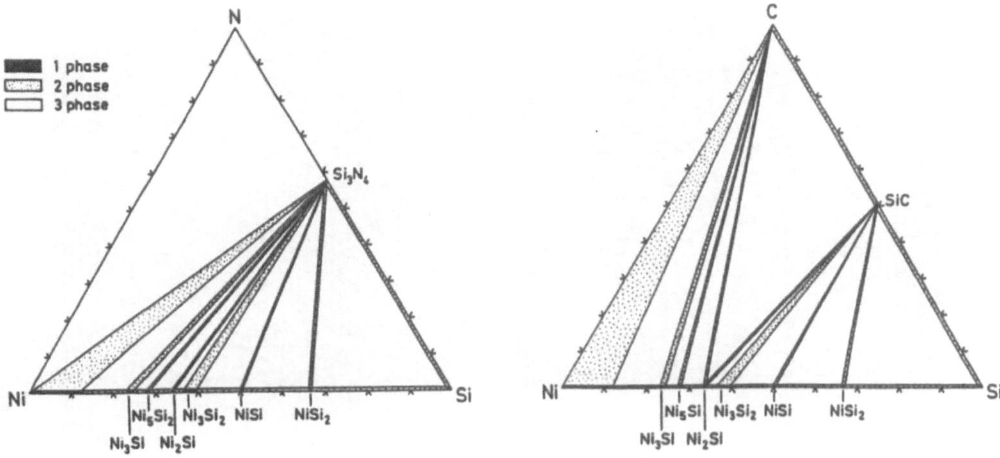

Figure 2. a) The phase diagram Ni–Si–N, b) The phase diagram Ni–Si–C, from (5).

Another aspect of chemical compatibility are possible reactions during service. This is a question which has not been addressed significantly in reported studies and much work therefore remains in order to be able to safely use ceramic metal joints at high temperatures.

To conclude this brief discussion of chemical compatibility it can be said that since reaction rates normally increase dramatically with temperature the "reactivity" of ceramic metal system is a potential hindrance for obtaining joints that can work at very high temperatures.

Thermal Compatibility

A problem not often discussed in joining of metals and ceramics is what could be described as the problem of thermal compatibility. This is the problem of designing the joining process in such a way that the heat treatment cycle is compatible with the heat treatment response of the materials to be joined. The problem is more severe for the metal. When for example brazing metals one of the key questions in selecting a braze is the heat treatment cycle of the metals to be brazed. For precipitation hardening materials brazing is normally performed at the solution treatment temperature for the alloy so as to minimise the internal stresses in the materials. After joining, the material is then aged at specified temperatures and times in order to obtain an optimal microstucture. The effect of different heat treatments on the microstructure of the high temperature Fe–Ni–alloy, Incoloy 909 can be seen in Figure 3. The effect of a brazing cycle on the tensile strength of a superalloy can be large as has eg been shown by Schwartz (14). In Figure 4 the effect of two different brazing cycles on the strength of Incoloy 909 is shown. As can be seen both the UTS and the yield behaviour is strongly affected.

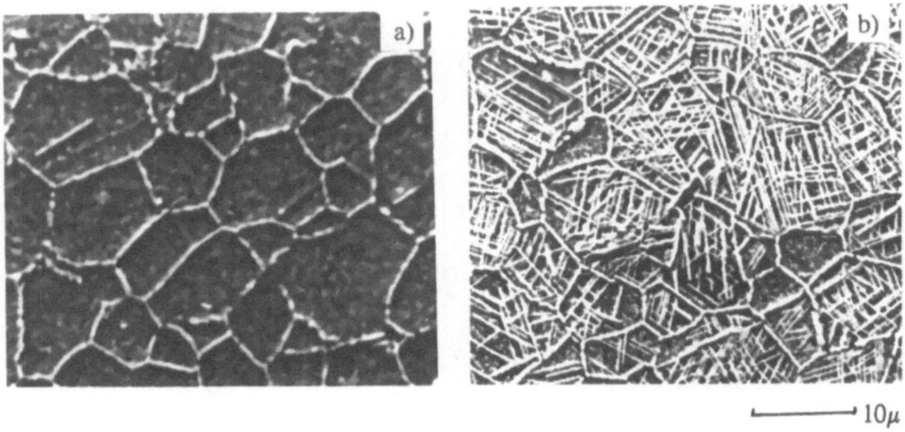

Figure 3. Microstructure of IN 909 a) as recived, b) heat treated at 800 C for 4h.

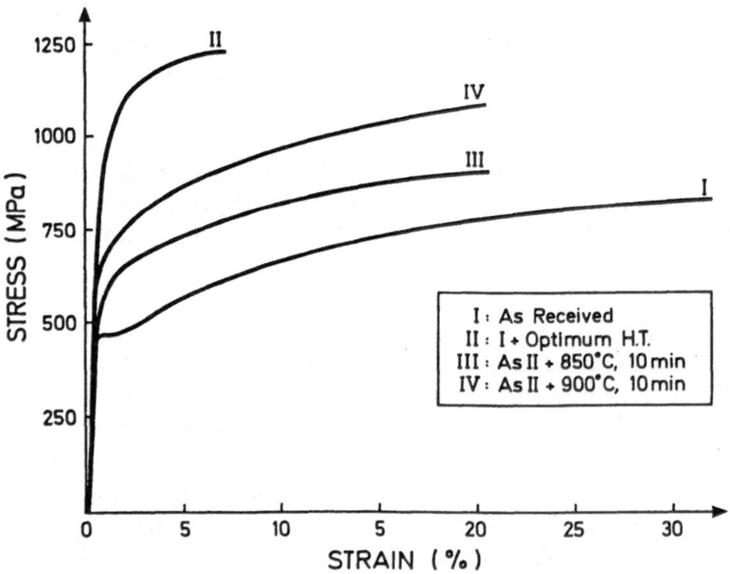

Figure 4. Tensile test curves for Incoloy 909 submitted to different brazing heat treatment cycles.

Most of the joining processes described in literature, both brazing and diffusion bonding, are in the temperature ranges of aging of e.g. superalloy materials. Without taking the heat treatment of the metals for which the process is developed into account the result of the work will be of less importance.

Mechanical Compatibility

Problems of mechanical compatibility comes from the fact that metals and ceramics generally have a large difference in thermal expansion. This leads to a greater shrinkage in the metal than in the ceramic parts of a joint when cooling down from the joining temperature, which leads to high stresses in the bond and even higher (tensile) stress in the ceramic. In many cases the stresses exceed the strength of the ceramic material and the ceramic part will fracture close to the joint. This can be seen in quite a few of the papers published on joining. The high stresses have been confirmed by modeling of the stress situation, in most cases using elastic (15, 16, 17, 18) but also using elastic–plastic (19, 20) analyses. Both methods give similar results namely that for joints of butt end type, the highest tensile stresses are obtained at the surface of the ceramic part close to the joining area. The elastic–plastic analysis givies somewhat lower (about 20%) and more realistic stress levels. High tensile stresses in the surface region of the ceramic have also been measured using X–ray techniques (21).

The problem of stresses due to differences in thermal expansion becomes greater the higher the joining temperature and thus also the desired application temperature. This factor works against the use of the joints at very high temperatures. An approach to solving these problems is the use of interlayers (22, 23, 24, 25), either soft metals, low expansion metals or graded joints (26). One must however always bear in mind the need to avoid solving one problem by creating another, as mentioned above. Soft interlayers, with low yield strength, can for example give problems with fatigue behaviour. Thus, the optimization of such interlayers is an intricate task. The major conclusions concerning mechanical compatibility are that the problem becomes more severe the higher the desired application temperature. Low expansion metals and soft metals are seldom suitable for use in high temperature joints in oxidizing environments. A more promising approach might be graded joints, but here much work remains to be done.

JOINING METHODS

Brazing

Brazing is a rather attractive joining method which is not restricted to simple geometries and which can be performed in relatively simple equipment such as a vacuum furnace. It is either accomplished by metalizing the ceramic surface (27, 28) or by using an active braze (29, 30, 31). This is to overcome the problems of low otherwise poor wetting on ceramic surfaces by the filler metals. Brazed joints are often used at temperatures close to the brazing temperature. Typical brazing temperature ranges for different filler metals are shown in Figure 5.

Most active brazes are based on copper, and this restricts the use of the joints to lower temperatures than would be possible with e.g. Ni–brazes. However, Ni–brazes

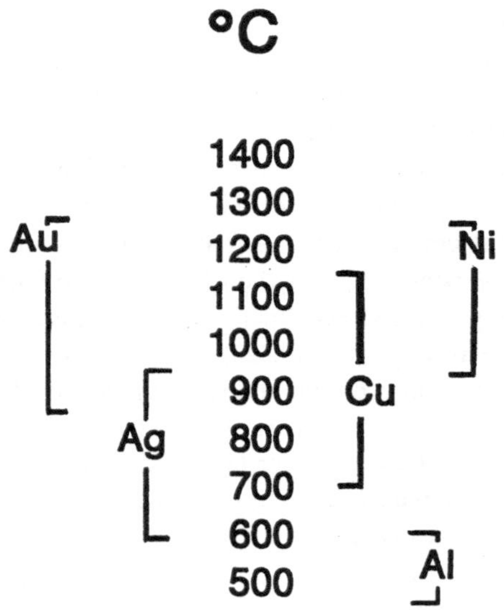

Figure 5. Temperature ranges for different filler metal systems.

often react severely with the ceramic (cf. chemical compatibility) producing phases which are very brittle. The microstructure of a brazed Si3N4–IN 909 joint is shown in Figure 6.

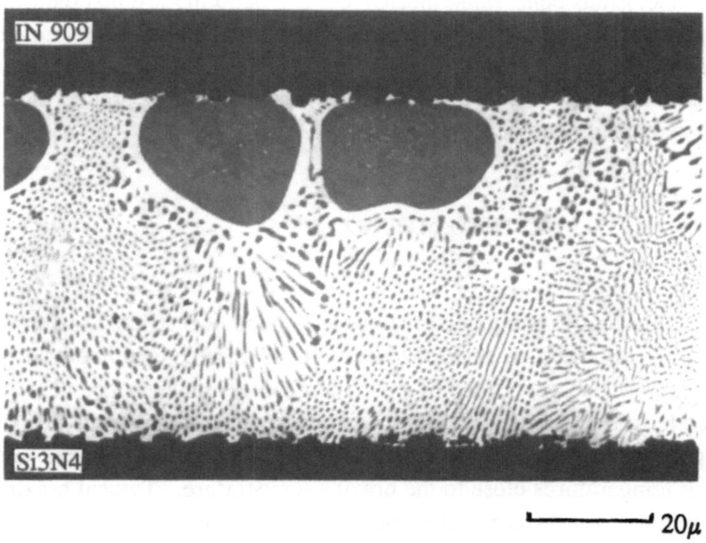

Figure 6. Microstructure of a brazed Si3N4–IN 909 joint, using a Cu–Ag–Ti filler metal.

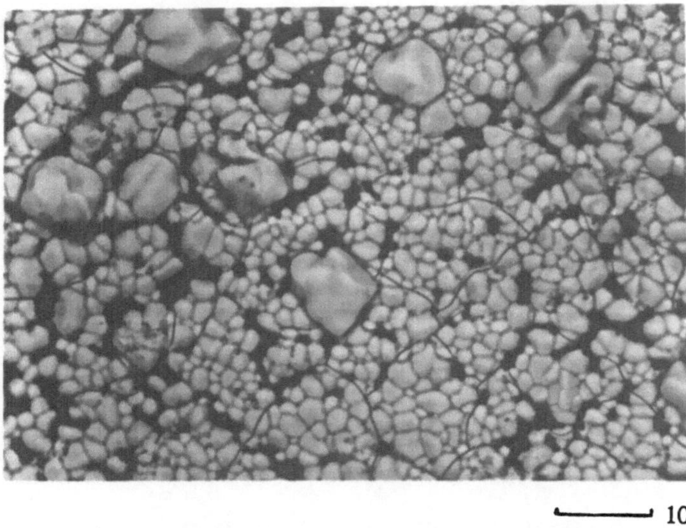

Figure 7. Reaction products on the Si3N4 surface of a brazed Si3N4–IN 909 joint revealed by using the method of Ljungberg et al(33).

For brazing, wetting is a central problem, much work has therefore been devoted to study wetting and reactions during wetting. It should however be noted that reactions during brazing can differ from what could be expected from results obtained in wetting experiments (32). In order to evaluate phases formed during brazing, Ljungberg et al (33) have developed a method that makes it easier to identify phases formed. Phases formed on Si3N4 during active brazing are shown in Figure 7. Brazing is one of the methods of high temperature joining for which one reports on applications are found in the literature. In most cases it is a combination of brazing and shrink fitting that is used (ref 33, 35). A typical component being a turbocharger rotor where the application temperature in the joint area is below 500C.

Diffusion Bonding

Diffusion bonding has a potential for giving higher temperature joints than brazing. Since diffusion bonding is normally performed at higher temperatures than brazing it suffer from more severe stress problems due to thermal mismatch. Work on diffusion bonding is therefore directed towards reducing the process temperature using HIP (36) or interdiffusion layers (37) and to minimizing stresses using interlayers either soft, low expansion or graded, as described above.

Another problem with diffusion bonding is reactivity. On the one hand, a reactive system is required to provide a driving force for the diffusion; on the other hand, if the reactivity is too high, excessive amounts of brittle phases may form the joint region. As discussed above SiC reacts strongly with Ni, this is exemplified in Figure 8 showing a SiC/SiC–Ni diffusion bonded joint. High reactivity also creates the risk of continued reaction during service. Some of the problems with too high reactivity might be solved by the use of diffusion barriers, although this adds one more complication to the problem.

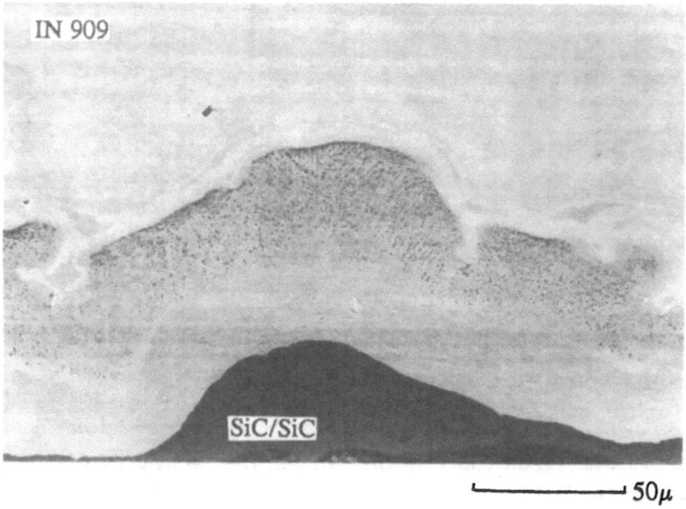

Figure 8. Microstructure of a diffusion bonded SiC/SiC– IN 909 joint.

Most diffusion bonding processes involve application of a high pressure in the joint region. This makes them more complicated and therefore less attractive to apply, especially to large components of complicated geometry. Nevertheless, if the stress problems can be solved with e.g. graded joints diffusion bonding might become the most attractive method for obtaining high temperature joints.

Friction Welding
Friction welding is a method which has received little attension. An advantage is that the process times are very short (38). The process is effective for soft metals such as aluminium and copper but successful joining with other metals of more interest to high temperature and high stress situations is yet to be proven.

Mechanical Joining
Mechanical joining methods are also seldom discussed although they are quite often used with good results. One method already mentioned above is shrink fitting. The method is sometime used alone but quite often in combination with brazing or diffusion bonding. The chemical bonding is used to increase the strength of the joint especially at high temperatures since the strength contribution of the shrink fit clearly decreases with temperature. In ref (34), Mitzuhara et al. have demonstrated the combination of brazing and mechanical joining in a compression joint for a shaft-like component.

Compression joints have also been used by United Turbine AB in their Mk I engine, to fit the turbine wheel to the shaft. Other components, like the combustor were not joined tightly, but rather fastened and centred in position through an outer metallic system.

Figure 9. Shrink fitted turbine wheel shaft (United Turbine AB).

In their development of jet engine components such as nozzle petals in ceramic composite materials SNECMA have used bolting and riveting techniques(39). Riveting of CMC to metals is currently being studied at Volvo Flygmotor (Figure 10).

Figure 10. Riveted SiC/SiC–Hastelloy X joint.

TESTING AND NDE

Vital for the successful use of joinig in construction work is the availability of test methods that can produce reliable engineering data. For the typical stress situations for metallic joints discussed above, there is a lack of standardized testing methods. For brazed joint two standards can be mentioned, ISO 55187–1985 for tensile strength, and AMS C3.2–82 for shear strength. The situation for metal ceramic joints is that no standardized test methods exist. Many of the test methods applied were developed for the study of phenomena rather than producing engineering data. Thus, this is an area in which much work is needed.

The reliability of joints and joining processes is also a very important issue. For the production of jet engine components much of the process control is based on the possibility of using NDE methods on the final products. For combinations of materials having large differences in properties, such as ceramics and metals, most NDE methods suffer from interference in the contact area between the materials. This might create large problems in the use of NDE methods for metal/ceramic joints.

PROSPECTS FOR THE FUTURE

The need for higher efficiency and cleaner engines will increase demands for high temperature capability put on the materials to be used. However due to the cost and to the brittle nature of ceramics they cannot be expected to be the only materials used in an engine. Thus the question of how to join ceramics to other materials with very different properties is obviously one of the key questions in making wider use of ceramics in engines.

In developing metal/ceramic joints capable of working at high temperatures it is vital to always keep the overall system in mind, taking chemical, thermal and mechanical compatibility as well as service conditions in to account. The problem is very complicated and each joint will probably require a special solution. The work must therefore be rather application oriented.

What are the prospects for the different joining methods? The method that hither to has proved to work rather well in high temperature applications is mechanical joining. This is not one method but rather a collection of many different methods such as shrink fitting, riveting etc. Riveting has the possibility of working at high temperatures. The limit is probably set by the working temperature of the metal. The technique is probably not suitable for monolithic ceramics but restricted to tough CMCs. Shrink fitting produces joints that are stronger at low temperatures introducing the need of chemical joining methods to strengthen the effect at higher temperatures. The most successful joints will probably be the ones using a combination of mechanical and chemical methods.

Of the chemical joining methods brazing is the easier and more flexible method. From an industrial point of view, active brazing is the most interesting chemical joining method. Much work is however needed to develop filler metal alloys optimized for different metal ceramic combinations, and the application temperatures will of course be restricted by the melting temperature of the braze, (see Figure 5). A potential for reaching higher temperatures is offered by diffusion bonding techniques. These methods are often rather complicated, restricted in geometry and require expensive equipment. This makes diffusion bonding less attractive than brazing except for the highest temperatures. It must however be remembered that at high temperatures, diffusion barriers might be needed, and also that residual stress levels increase with service temperature. Of the chemical joining processes, friction welding has yet to prove its high temperature capability.

The residual stress problem is one of the major problems of metal/ceramic joining. The stresses often exceed the strength of the ceramic and the stresses increase with joining temperature. Here much work is needed both on design methods and on stress reducing interlayers, always bearing in mind all aspects of the joint system. To reveal flaws and poor bonding etc. in these highly stressed regions adequate NDE techniques are also required. This calls for considerable development of such methods especially designed for studying interphase phenomena.

CONCLUSIONS

In conclusion it can be said that metal ceramic joints will always be needed. The joints should preferably rely on both mechanical and chemical joining methods, with active brazing as the more flexible and diffusion bonding as the more high temperature method. The joints must always be designed and the joining method must always be chosen on the basis of a sound knowledge of the application conditions.

ACKNOWLEDGMENTS

I thank Dr R Warren, Mr L Ljungberg and Ms L Pettersson for valuable discussions on the problems of joining. Dr Warren is also thanked for the linguistic corrections.

REFERENCES

1. Lewis M H, Butler E G. Proc this conference (1991).

2. Warren R, Andersson C–H. Composites, 15 no2 101–111 (1984).

3. Schuster J. J. Mat. Sci. 23 2792–2796 (1988).

4. Schuster J. Joining Ceramics, Glass and Metal, Ed. W Kraft DGM Informationsverlag, Oberursel, Germany, 131–316 (1989).

5. Hultman L, Pejryd L, Warren R, Andersson C–H. To be published.

6. Eriksson G. Chem. Scr., 8 100–103 (1975).

7. JANAF Thermochemical tables, Second Edition (Washington, 1971).

8. Barin I, Knacke O. Thermochemical Properties of Inorganic Substances (Sringer–Verlag, Berlin, 1973).

9. Barin I, Knacke O, Kubaschewski O. Ibid, Supplement (1977).

10. Barin I. Thermochemical Data of Pure Substances (VCH Verlag, Weinheim, 1989).

11. Brito M E. Thesis, Dept. of Mat. Sci. & Eng., Techn. Univ. of Nagaoka, Kamitomioka, Nagaoka, Japan (1989).

12. Klomp J T. Joining Ceramics, Glass and Metal, Ed. W Kraft DGM Informationsverlag, Oberursel, Germany, 55–64 (1989).

13. Schiepers R C J, Van Loo F J J, De With G. J. Am. Ceram. Soc. 71 no6 C284–C287 (1988).

14. Schwartz M M. Welding Research Council Bull. 340 ISSN 0043–2326 (1989).

15. Iancu O T, Munz D. Joining Ceramics, Glass and Metal, Ed. W Kraft DGM Informationsverlag, Oberursel, Germany, 257–264 (1989).

16. Xian A–P, Z–Y S. J Am. Ceram. Soc. 73 no11 3462–65 (1990).

17. Mullen R L, Padovan J, Braun M J, Chung B T F, McDonald G, Hendricks R C. High Tech Ceramics, Ed. P Vincenzini, Elsevier, Amsterdam, 2547–2569 (1987).

18. Evans A G, Lu M C, Schmauder S, Ruhle M. Acta Metall. 34 no8 1643–1655 (1986).

19. Stoop B T J, den Ouden G. In Joining Ceramics, Glass and Metal, Ed W Kraft DGM Informationsverlag, Oberursel, Germany, 235–242 (1989).

20. Häggblad H–Å. to be presented at 2nd European Colloquium: Designing Ceramic Interfaces, Petten The Netherlands, November (1991).

21. Eigenmann B, Scholtes B, Macherauch E. Joining Ceramics, Glass and Metal, Ed W Kraft DGM Informationsverlag, Oberursel, Germany, 249–256 (1989).

22. Suganuma K, Okamoto T, Miyamoto Y, Shimada M, Koizumi. Mat. Sci. and Tech. 1156–1161 (1986).

23. Lugscheider E, Boretius M. Joining Ceramics, Glass and Metal, Ed W Kraft DGM Informationsverlag, Oberursel, Germany, 25–32 (1989).

24. Jaquot P, Coll B, Gabriel M, Speri R. Joining Ceramics, Glass and Metal, Ed W Kraft DGM Informationsverlag, Oberursel, Germany, 45–52 (1989).

25. Courbiere M, Kinoshita M, Kondho I. Joining Ceramics, Glass and Metal, Ed W Kraft DGM Informationsverlag, Oberursel, Germany, 95–102 (1989).

26. Batfalsky P, Godziemba–Maliszewski J, Lison R. Joining Ceramics, Glass and Metal, Ed W Kraft DGM Informationsverlag, Oberursel, Germany, 325–332 (1989).

27. Rice R. Adv. in Joining Tech. Ed. J J Burke, Book Hill Publ. Co. Boston (1975).

28. Santella M L. Adv. Ceram Mat, 3, no5, 457–462 (1988).

29. Mizuhara H, Huebel E. Welding Journal October (1986).

30. Nicolas M G. Joining Ceramics, Glass and Metal, Ed W Kraft DGM Informationsverlag, Oberursel, Germany, 3–16 (1989).

31. Nicolas M G, Peteves S D. Proc this conference (1991).

32. Ljungberg L, Pejryd L, Warren R. Proc this conference (1991).

33. Ljungberg L, Warren R, Li C–H. J. Mat. Sci. Let. 9, 1316–1318 (1990).

34. Mizuhara H, Huebel E, Oyama. Am. Ceram. Soc. Bull. 68, no 9, 1591–1599 (1989).

35. Butler E, Lewis M H, Hey A, Sharples R V. Ceramic Joining in Japan, Repport DTI (1986).

36. Larker R, Loberg B, Johansson T. Ceram. Mat. & Comp. for Eng. Ed V J Tennery, Am. Ceram. Soc. 503–512 (1989).

37. Iino Y, Taguchi N. J. Mat. Sci. Let. 7. 891–982 (1988).

38. Grunauer H, Horn H, Weiss H. Joining Ceramics, Glass and Metal, Ed W Kraft DGM Informationsverlag, Oberursel, Germany, 185–190 (1989).

39. Gauthier G, Bessenay G, Honnorat Y. Proc this conference (1991).

MASS PRODUCTION OF CERAMIC PORTLINERS FOR PISTON ENGINES

JÜRGEN HEINRICH
Hoechst CeramTec AG
W-8672 Selb, Germany

ABSTRACT

Ceramic portliners in the cylinder head of an engine reduce the cooling efforts, improve the response behaviour of the catalyst and increase the degree of efficiency of the turbocharger. Because of the brittleness of the material the design of this component has to be adjusted to the ceramic materials properties in order to enable casting around with liquid metal. The manufacturing process has to be designed in a way that at a low production cost level the extremely tight dimensional tolerances demanded for ceramic parts in automotive constructions can be met reproducably. Using the slip casting process as an example the process steps design, mould preparation, powder processing, shaping, sintering and finishing will be explained and the quality control and quality assurance concepts necessary for mass production will be discussed.

INTRODUCTION

New materials to be introduced into traditional markets ought to be at least partially superior to classical materials, a requirement to be met in order to overcome reluctance and scepticism on behalf of potential users. Ceramics are a group of materials which are possessing physical properties that could lead to considerable substitution of metals in certain areas of application. Some ceramic materials like aluminum titanate, show a low thermal conductivity and good thermal shock resistance, making it suitable for heat insulation purposes.

A concrete example shall illustrate the need for a better heat insulation in automotive engines. In a heat balance of a Daimler-Benz M 110 E Otto engine operating at rated power it is seen that only 30% of the fuel energy can be used for direct car propulsion. The main energy losses are caused by heat transfer to the cooling media (15%) and to the exhaust gas (42%). If the engine is operating at a lower load, the effective power is even decreased. The losses by cooling and heat convection are increased, whereas the heat energy of the exhaust gas is almost halfened [1]. Insulation measures could help reducing energy losses caused by heat emission and -convection and energy flow to the cooling media. Heat insulating materials should have low thermal conductivity, a small thermal expansion coefficient, good thermal shock and corrosion resistance, requirements that are met especially by aluminum titanate.

A typical approach for the usage of low heat conduction ceramic materials is the insulation of certain engine compartments, where considerable heat losses are to be expected. Such a position is the cylinder head exhaust channel. Lining of this channel with ceramic heat insulating materials should lead to technical and economical improvements in engine performance.

At the cylinder head foundry those portliners are fixed in the mould and cast around with liquid alumium. The portliner is now fixed form- and force-closed and constitutes the thermal insulation of the exhaust port (fig. 1).

In this contribution the advantages of ceramic portliners, the design adjusted to ceramics and the manufacturing process with all possibilities of influencing the dimensional aberrations of those extremely tight dimensioned components will be discussed.

BENEFITS OF CERAMIC PORTLINERS

For manufacturing cylinder heads with ceramic portliners aluminum or cast iron has to be cast around the ceramic portliner. Therefore a ceramic materials has to be used which can widstand the compressive stresses that arise during solidification of the metal. Aluminum titanate is possessing a low specific weight (table 1) which is in the range of aluminum, thus favoring aluminium titanate in terms of weight reduction. The low coefficient of thermal expansion is responsible for the good thermal shock resistance of aluminum titanate. Microcrack phenomena are responsible for the very low Young's modulus, which is leading to a pseudo-elastic behaviour of aluminum titanate. Due to this fact, it is possible that metals can be casted around aluminum titanate parts. Aluminum titanate possesses very low heat conductivities, ranging between 1,5 and 2,5 $W/m^{-1}K$, thus 20-30 times lower compared to cast iron and about 100 times lowered with respect to aluminum. Fig. 2 shows the heat flux through a ceramic layer or its corresponding insulation as a function of the wall thickness. Its essential information is that low heat conductivity is decreasing the wall thickness needed for obtaining a certain degree of insulation.

What are the technical benefits of ceramic portliners? To answer this question it is necessary to know something about the influence of portliners on the engine energy flow. First of all, there will be no changes - in fuel consumption and effective power, as combustion processes are not altered by portlining. Convection- and emission losses will be influenced only in a marginal way. But there is a certain kind of redistribution of energy between cooling and exhaust gas. In a Porsche 944 engine ceramic portliners of about 3 mm thickness are reducing the heat flow from the exhaust port to the cooling by up to 7 kW, thus causing a 13% decrease of cooling efforts, thereby rendering some weight reduction of the total cooling system [2]. The exhaust gas energy flow is enhanced by 8 kW at full load, leading to an exhaust gas energy increase of about 3% [2] compared to an engine without portliners. This is especially favourable in an engine equipped with a turbocharger unit. In fig. 3 comparing the exhaust gas temperature of a Porsche 944 Turbo engine with and without portliner one can see that the exhaust gas is warmed up by more than 30°C at full load. Increased exhaust gas temperatures shorten the light off time of the exhaust gas catalysts. Due to the use of portliners the hydrocarbon level in the exhaust gas has been reduced by 16% and the CO and NO_x contents are lowered by about 10%.

As diesel engines also have low exhaust gas temperatures ceramic portliners could find application in diesel passenger cars, heavy-duty engines and tractors, especially when the engines are equipped with turbochargers.

The beneficial results of portliners have to be considered in connection with higher costs per cylinder head. The introduction into serial production can be enhanced by mass production feasibilities guaranteeing high piece numbers. But a thorough problem oriented use/cost analysis always is necessary.

DESIGN OF CERAMIC PORTLINERS

The aspects of design adjusted to ceramics shall be explained exemplary by discussing the development of the Porsche 944 Turbo engine cylinder head. The original design (Fig. 4) did not include a portliner unit. Therefore it was necessary to adapt the portliner design to the cylinder head geometry of the originate design, the valve size and the requirements imposed by fluid mechanics. Because of its low strenght aluminum titanate cannot be used as a construction material on its own. It has to be combined with a supporting material. This could be achieved by casting around with aluminum or cast iron. Casting around a ceramic portliner designed according to fig. 4 would introduce heavy flexural- and tensile stress caused by shrinking phenomena during the soldification of the metal, thus leading to destruction of the element. Therefore it was necessary to create a portliner geometry which is only exposed to forces of pressure. Such a design, not completely optimized for fluid-mechanical requirements and cylinder head geometry, can be seen in fig. 5. The key feature of this design in the concave shape of all curves. Portliners designed according to fig. 5 were suited for the casting procedure. This portliner design was only marginally alterated in some details for mass production. Before reaching serial production level the portliners had to survive several tests to prove their good reliability. A similar adjusted of the portliner shape to a design meeting ceramic properties is necessary for each newly designed portliner.

PROCESSING

For portliners, very narrow tolerances have to be met in production. As for many other multi-step processes alterations in a variety of parameters are resulting in dimensional deviations. Raw materials are playing a key role for accurate production. Strict and specified control of the starting material is therefore a very important measure. The casting process requires the production of plaster moulds with very narrow tolerances. During shaping leaching processes alter the behaviour of the slip necessary for the casting process, thus leading to shrinkage varaiations and thereby to further dimensional aberrations. This is the same case for the influence of the sintering atmosphere and the temperature-time-program during sintering. Thus some sophistication of the firing process is necessary to reach appropriate shrinkage. Furthermore it is necessary to develop a sophisticated quality control and -assurance system in order to be able to produce ceramic portliners reliable and reproducable meeting the tolerances demanded by the automotive industry. The optimizing of the manufacturing processes is in the focus of all efforts.

From the view of minimized passage periods, of manufacturing control and of quality control for optimizing the process results one has to be in command of the material- and communication flows (fig. 6). The manufacturing- and quality control level has to be joined to a roughly planning and disposing PPS-System. The PPS-System has to releave procurement orders within the material management. At this location all informations of the production segments of a plant are founded. The coordination of all production systems also has to seize the capacities of production engineering. This includes also the design department, the production planning and the shops for tools- and device construction. The production engineering business requires about 50% of the order passage time. Therefore future PPS-systems have to enable the disposal and the preparatory control of design and production planning.

From an article drawing a synoptic drawing can be produced by CAD-systems to relief conventional tools construction as well as NC programming taking place before ceramic production. The external programming of NC machines by an integrated CAD/CAP-system increases the output of CNC-machines (fig. 6).

Subsequently various production steps in the manufacturing of ceramic portliners and emerging problems will be discussed in detail.

Raw Materials

The ceramic raw materials alumina and titania are homogenized in ball mills together with organic binders, defloculants and plastisizers (fig. 7). Despite many powder characteristics like grain size, grain size distribution, specific surface area and degree of impurity are very tightly tolerated, various raw materials lots behave very different in ceramic production. Obviously this is caused by different surface conditions of those powders, which are manifuting in different surface potentials (fig. 8). Despite constant grain size distribution, specific surface area and degree of impurity the Point of Zero Charge is eg. at pH values between about 7 and 10 for various alumina lots. In titanium dioxide powders commonly smaller fluctuations of the Points of Zero Charge are detected (fig. 9). The addition of binders, deflocculants and plastisizers in different concentrations also lead to distinct changes of zeta-potentials (fig. 10). Those alterations of powder surface conditions are leading to fluctuating processing properties, especially to changes of viscosity and body forming rate in the slip casting process. In the Point of Zero Charge area slips are tending for flocculation and sedimentation. For all those reasons it is urgently necessary to specify the properties of the powders used together with the raw material supplier to reach a most extensive elimination of process fluctuations caused by raw materials.

Casting Process

The portliner casting takes place in fully automatized casting stations long term introduced in ceramics. As already mentioned the portliner has to be designed adjusted to ceramic properties with regard to absorption of shrinkage stresses in casting around of the portiner in the metal foundry. For producing plaster moulds commonly from a customer drawing a three dimensional CAD drawing is generated and resolved to several sections. After optimizing the cross section with regard to absorption of shrinkage stresses at casting around with metal a three dimensional model is produced. From the model drawing a NC-program is generated. Subsequently a model is milled on a CNC-machine from which plaster moulds can be reproduced (fig. 11). This procedure enables the manufacturing of very tightly tolerated plaster moulds for the casting process.

Within the life cycle of a plaster mould dimensional alterations occur, which are due to abrasive wear during casting and to chemical processes. Such temporal dimensional alterations of plaster moulds lead to a substitution within regular periods. For example calcium- and sulphate ions are released from the plaster moulds according to pH-value and time of exposition which naturally have an influence on the liquifying behaviour of the slip, thereby influencing the body forming rate and dimensional tolerances of the portliners. Fig. 12 shows the calcium ion concentration in an aluminum titanate slip as a function of the time a plaster mould has been used. As calcium ions change the rheological behaviour of slips, the total ion concentration has to be controlled and kept constant very carefully.

Drying

The drying of the portliners takes place in conventional swing dryers at temperatures between 40°C and 60°C. Different surface conditions of various raw materials lots lead to different green densities after drying because of changed viscosity and body forming rate (fig. 13). Those green density fluctuations naturally lead to a different shrinkage behaviour at sintering thereby influencing the dimensional tolerances required. As mentioned before, attention has to be directed for this reasons controlling and specifying of raw material properties.

Sintering

Either the gas atmcsphere or the temperature-time-program during sintering influence the dimensional alterations and the property profile of aluminum titanate portliners (fig. 14). The sintering temperature and the gas atmosphere during sintering have influence on the amount of reaction product between Al_2O_3 and TiO_2. Thereby the properties of aluminum titanate are varied in a broad range. The formation of nuclei of aluminum titanate precumably takes place at temperatures around 1300°C by a very fast occuring diffusion of aluminum ions in TiO_2 [3, 4]. At higher temperatures the reaction is returded by the distinctirely slower diffusion rate of aluminum and titanium in aluminum titanate. The formation of aluminum titanate is combined with a volume expansion of about 10%, which besides the shrinkage leads to an increase of the density in this sintering process. Furthermore the property profile of aluminum titanate can be influenced by the Al_2O_3/TiO_2 ratio [5]. Especially exess alumina within the aluminum titanate microstructure leads to an increase of density, strength and coefficient of thermal expansion.

Provided that such alterations in the property profile have no influence on the application of the final product the total shrinkage can be influenced and raw material quality fluctuations can be compensated in a certain range.

Machining

Finishing of the portliners usually takes place in the sintered state by sawing and drilling. Thereby the observance of position- and shape tolerances with regard to the valve bore-hole and the connection of the cylinder respectively the waste gas collecting unit. Finishing is done with diamand tools as usual in ceramics. Finishing of aluminum titanate in the sintered state is not essentially more expensive as in the green state, because the strength of this material is comporatively low and porosity is comporatively high.

The very narrow dimensional tolerances of a portliner regarding inlet-, exhaust diameter and position of valve bore-hole are checked by random tests.

QUALITY CONCEPT

The high quality requirements for portliners only can be met by a corporate concept on assuring and continous improving of quality which is supported by the management and all co-workers. Quality has to be defined as full satisfaction of all internal and external "suppliers" and "customers" cooperating in the quality line in respect to product performance, on-time delivery, costs and cooperation. Systematical quality techniques like design of experiments and target orientation are essential tools. Comprehensive education of the management and of all co-workers are as important as project group work.

The quality concept at Hoechst CeramTec is based upon the functional units quality planning, quality control, quality proof and quality improvement. The tools used for process improvement are listed in fig. 15. Especially important are the design of experiments with statistical methods eg. by Shainin and Taguchi [5, 6] and the statistical process control (SPC). The aim of both methods named first is not to understand the scientific relations between various parameters, but to extract by systematic techniques main parameters out of a large number of possible variables and to reveal the regularity of the "response behaviour" of the whole process or of a single process step to the variation of those main parameters. If this is achieved, only the main parameters of the process have to be observed. For other parameters tolerances often can be broadened considerably.

SUMMARY

The mass production of portliners with high quality requirements needs most modern production processes, a quality assurance system with quality planning, quality control and an independent quality proof, as well as continous methodically implemented improvements. Thereby the observance of highest levels on quality requirements at high numbers of pieces is assured.

REFERENCES

1. J. Huber, J. Heinrich: Ceramics in Internal Combustion Engines; Proceedings 2nd European Symposium on Engineering Ceramics, London 1987

2. M. Körkemeyer: Erfahrungen mit Portlinern im Otto-Motor; Materialien zur Tagung Nr. T-30-313-056-7, Haus der Technik e.V., Essen

3. B. Freudenberg, A. Moccellin: Aluminium Titanate Formation by Solid-State Reaction of Fine Al_2O_3 and TiO_2 Powders. J. Am. Ceram. Soc. 70 (1987), S. 33-38

4. B. Freudenberg, A. Moccellin: Aluminium Titanate Formation by Solid-State Reaction of Course Al_2O_3 and TiO_2 Powders. J. Am. Ceram. Soc. 71 (1988), S. 22-28

5. J. Heinrich, P. Stingl, R. Heinl, W. Benker: Entwicklung und Erprobung keramischer Komponenten aus SiSiC und Al_2TiO_5 für einen schadstoffarmen Öldampf-Keramik-Motor. BMFT-Abschlußbericht 03 28688 A, Bonn 1989

6. K. Bhote: World Class Quality. ASA Membership Publications Div., Amer. Managmt. Ass., 1988

7. G. Taguchi: System of Experimental Design. American Supplier Inst., Dearborn, Mich., 1987

TABLE 1
Physical properties of aluminum titanate, zirconia, cast iron and aluminum

	Al_2TiO_5	Zirconia	Cast Iron	Aluminum
Density [g/cm³]	3.15	5.8	7.25	2.7
Young's Modulus [GPa]	15	200	120	70
Heat Conductivity [W/mK]	2	2.5	58	220
Coefficient of Expansion [$1/K \cdot 10^{-6}$]	1	10	10.5	23.8
Bending Strength [MPa]	30	500	250-600	100-600

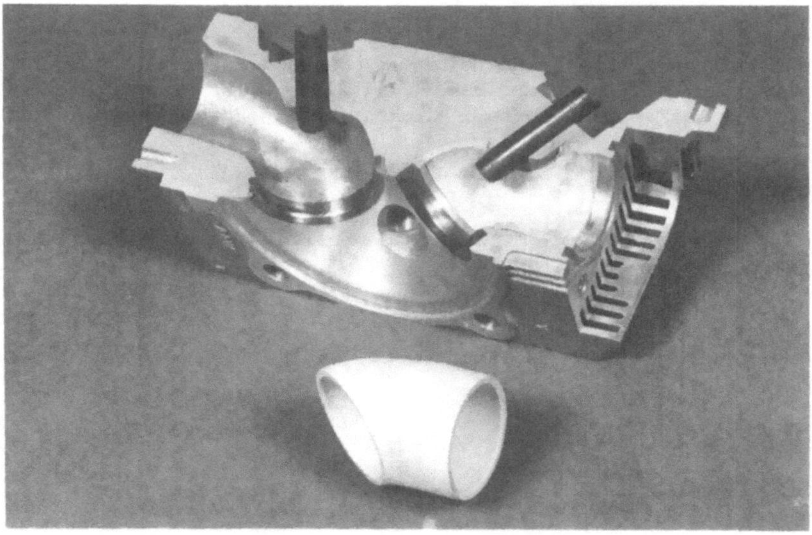

Figure 1: Section of a cylinder head showing the thermal insulation of the exhaust port

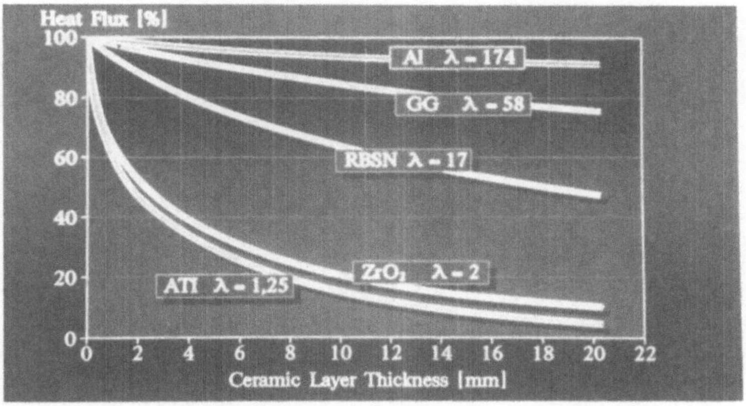

Figure 2: Heat flux vs ceramic layer thickness of the thermal insulation material

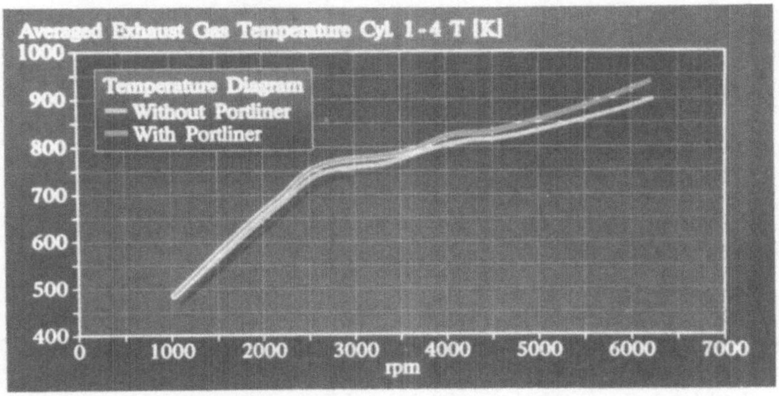

Figure 3: Exhaust gas temperature of Porsche 944 turbo engine at full load

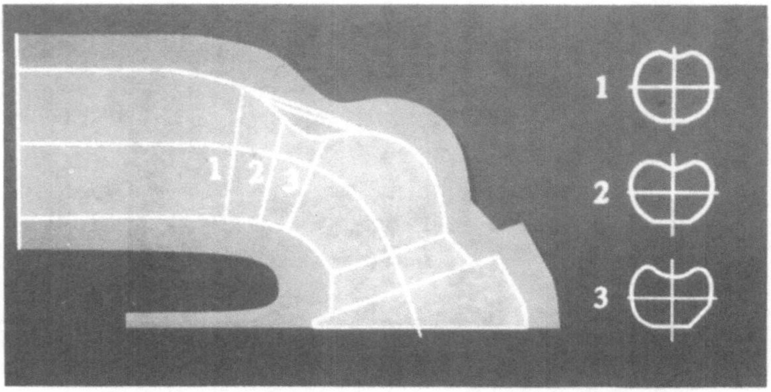

Figure 4: Section of the exhaust valve of the Porsche 944 turbo engine

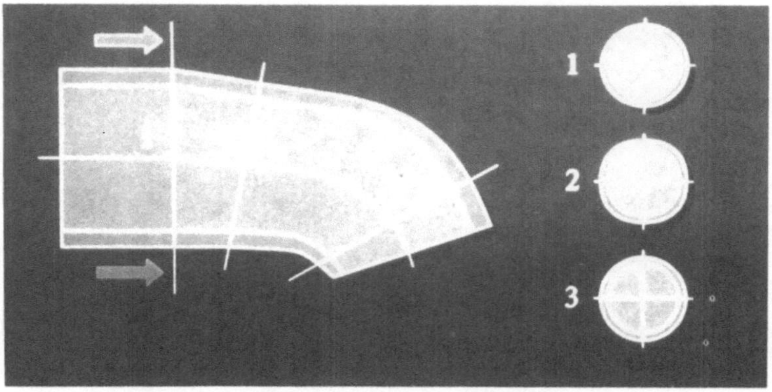

Figure 5: Section of a portliner design adjusted to ceramic material properties

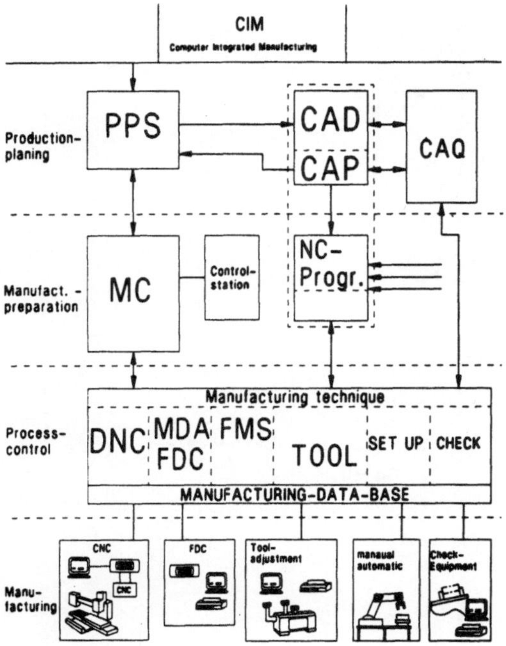

Figure 6: Material and communication flow in computer integrated manufacturing

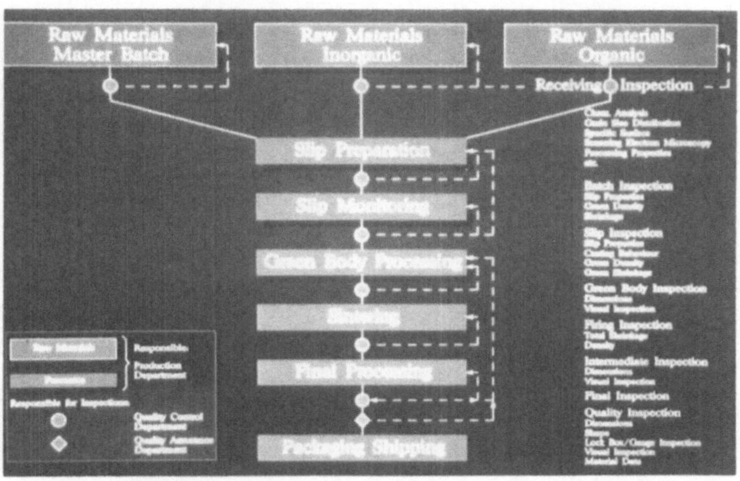

Figure 7: Flow diagram for the slip casting process

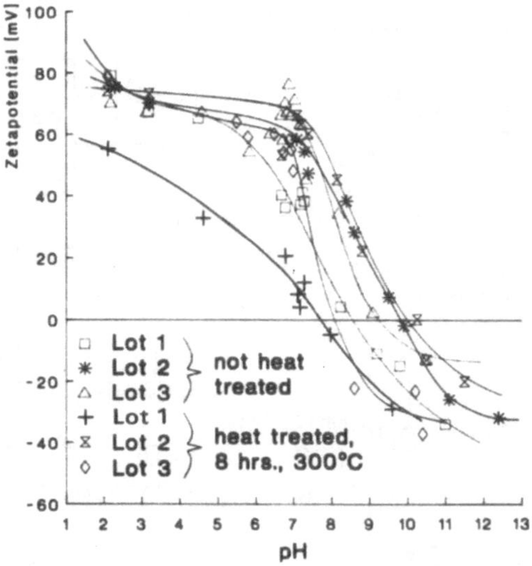

Figure 8: Zetapotential as a function of pH for different alumina powder lots

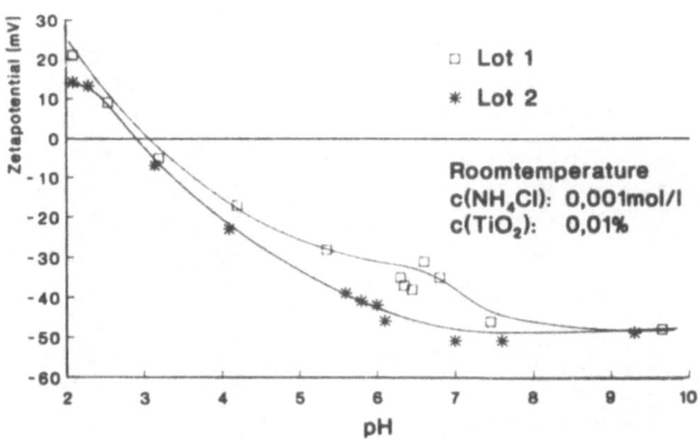

Figure 9: Zetapotential as a function of pH for different titania powder lots

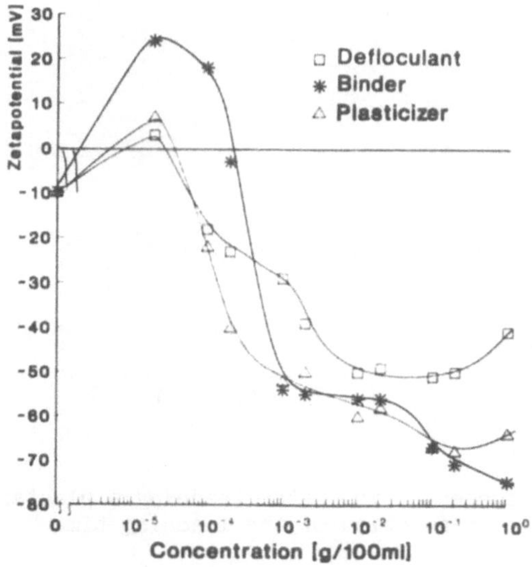

Figure 10: Zetapotential as a function of the organic additive concentration in alumina (lot 1) slurries

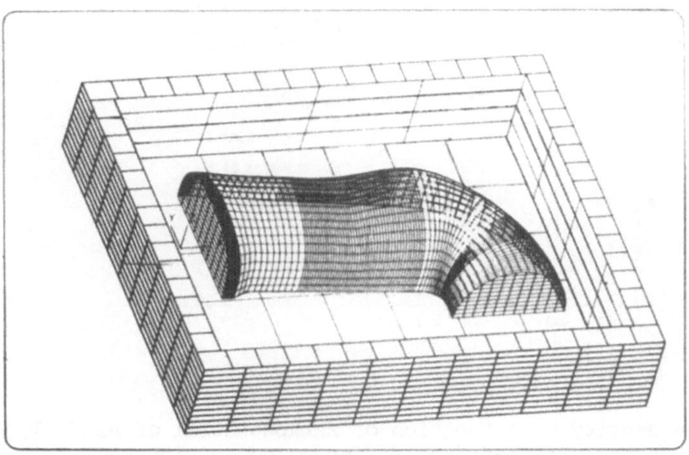

Figure 11: 3-D drawing for the NC-program for manufacturing the plaster mould model

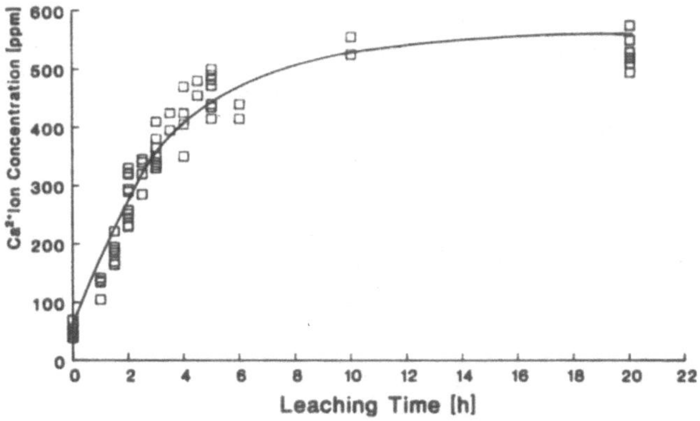

Figure 12: Ca^{2+}ion concentration after leaching a plaster mould in water as a function of the leaching time

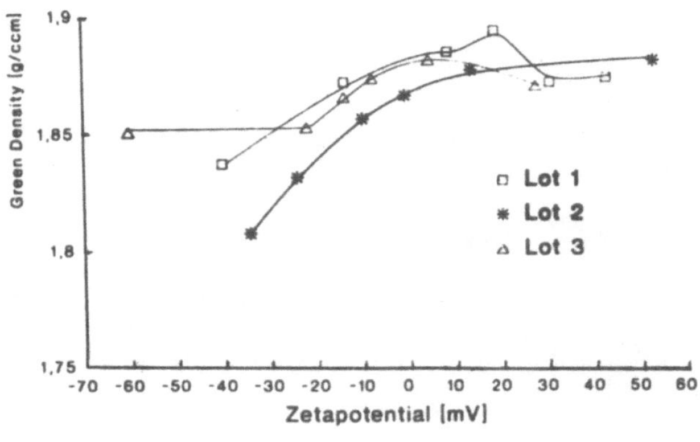

Figure 13: Green density as a function of zetapotential of Al_2O_3/TiO_2 mixtures prepared from different powder lots

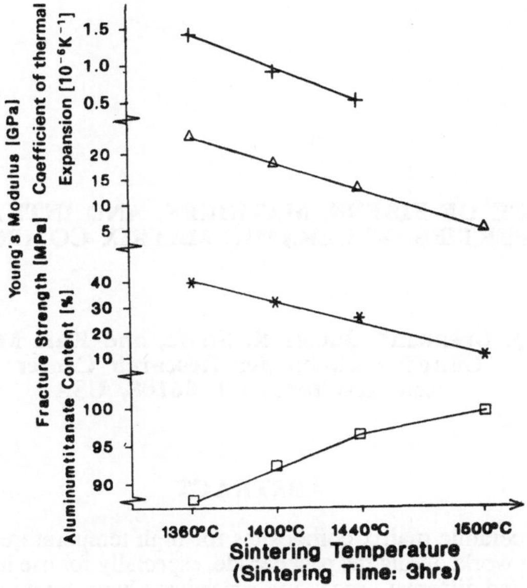

Figure 14: Properties of aluminum titanate as a function of the sintering temperature

Objective	Tool
□ System Analysis Problem Description Stratification	o Brainstorming o Metaplan o 7-M-Tools
□ Process Representation	o Flow Charts - functions - quantities - value o Process Flow Tables
□ Identification of Parameters	o Brainstorming o Cause & Effect - Diagramm (Ishikawa-, Fishbone-Diagramm) o FMEA o Method 635 o Morphological Boxes
□ Search for Causes	o Sampling of Information / Data Collection o Check Sheet o Frequency Distributions (Histograms) o Pareto Analysis o Control Charts o Multi-Vari Charts o Paired Comparision o Component Search
□ Design of Experiments	o Variables Search (Shainin) o Full Factorial Experiments o Orthogonal Arrays (Taguchi) o Deterministic Methods, o Genetic Methods
□ Robust Process	o Scatter Plots o Signal to Noise Analysis o Poka Yoke
□ Confirmation & Validation	o B versus C Test o Statistical Process Control (SPC) o Positrol o Audits

Figure 15: Tools for process improvement

THE INFLUENCE OF FIBERS, MATRICES, AND INTERFACES ON THE PROPERTIES OF CERAMIC MATRIX COMPOSITES

John J. Brennan*, James R. Strife, and Karl M. Prewo
United Technologies Research Center
East Hartford, CT 06108, USA

ABSTRACT

With the interest in ceramic matrix composites for high temperature structural applications increasing around the world during the past decade, especially for use in heat engines, a myriad of different systems and different processing procedures have emerged. Among the types of CMC's under investigation are: whisker reinforced crystalline and glass-ceramics, and continuous carbide, oxide, and nitride fiber reinforced ceramics. The range of fabrication methods is diverse and includes hot-press densification of glasses, glass-ceramics, and crystalline ceramics, sol-gel infiltration and pyrolysis of ceramics, polymer precursor infiltration and pyrolysis, reactive oxidation of metals, reactive sintering, and chemical vapor infiltration (CVI) of silicon based ceramics.

In all of these composite systems, it has been found that in order to achieve high strength and, in particular, high toughness, the characteristics of the fiber/matrix interface must be controlled such that the bond is strong enough to allow load transfer from the matrix to the fibers under stress, but weak enough so that an advancing matrix crack can be deflected by the fibers. In addition, the nature of the fiber/matrix interface must include high temperature stability and resistance to environmental attack. Examples of CMC's with both weak and strong interfacial bonding will be discussed, with emphasis on SiC and Si_3N_4 type continuous fibers in glass and glass-ceramic matrices, as well as SiC, C, and oxide continuous fibers in CVI SiC matrices.

Progress in the application of ceramic matrix composites to heat engines will also be discussed.

INTRODUCTION

The application of advanced ceramic materials as structural components for heat engines began in earnest in the late 1960's and early 1970's, primarily in the US, UK, and Germany, with a variety of programs established to apply ceramics such as silicon nitride to both internal combustion diesel engines and gas turbines. An example of one of these programs is the US Department of Defense (DoD) Advanced Projects Research Agency's (DARPA) effort to develop a design capability with brittle materials that focused on demonstrating ceramic hardware in hot flow path components that would operate at 1370°C uncooled for 200 hrs in a vehicular engine (Ford Motor Co.), and 100 peaking cycles in a electrical power generating turbine test rig (Westinghouse Electric Co.)[1]. While these programs and others that followed could be considered to be at least partially successful, by 1980 the US DoD had spent over $60 million dollars pursuing the use of monolithic ceramics for gas turbine engines, with only limited success[2]. In the years since, considerably more has been invested with promising, but not proven, results.

The primary problem of monolithic ceramic materials such as silicon nitride or silicon carbide is their lack of toughness, i.e., their susceptibility to catastrophic fracture when subjected to impact loading or thermal or mechanical overstress, especially in the presence of a flaw. For many years researchers have sought to develop "tough" ceramics whose performance characteristics retain the best properties of their parent ceramics and have the additional quality of not being susceptible to catastrophic fracture under stress. The addition of fibers to ceramics has been known for many years to be one approach for achieving this goal, as exemplified in fiber reinforced cement. Extension of this concept into higher performance ceramic matrices has proven to be much more difficult.

Early experiments performed during the late 1960's and early 1970's in England and the US[3-8], however, demonstrated that high performance fibers could be successfully incorporated into glass matrices to achieve high strength, tough, composite materials. By 1973, carbon fiber reinforced glasses and glass-ceramics had been demonstrated to achieve strengths of over 700 MPa. By 1975, toughening of silicon nitride by the addition of refractory metal wires had been demonstrated[9]. However, for various reasons, one of which was the inherent lack of oxidation resistance of carbon fibers and refractory metal wires, these early developments were not carried further.

In the late 1970's, two new developments revived the interest in ceramic matrix composites. First, carbon fiber reinforced epoxy matrix composites were accepted as reliable structural materials and thus examples for the use of other fibrous composites. Second, new fibers such as the organometallic polymer derived silicon carbide yarn and the CVD derived SiC monofilament, as well as various sol-gel derived oxide fibers, became available and thus permitted the potential creation of composites with superior oxidation resistance.

During the past decade, the interest in ceramic matrix composites for high temperature structural applications, especially for use in heat engines, has increased to the point that a large number of industrial organizations as well as universities and government laboratories throughout the world are actively performing research with a myriad of different systems and different processing procedures for these materials. Among the types of ceramic matrix composites under investigation are whisker reinforced glasses and glass-ceramics as well as whisker reinforced crystalline ceramics, and continuous fiber reinforced ceramics produced by methods that include hot-pressing of glasses and glass-ceramics, sol-gel infiltration and pyrolysis of ceramics, polymer precursor infiltration and pyrolysis, reactive oxidation of metals, reactive sintering and hot-pressing, and chemical vapor infiltration (CVI) of silicon based ceramics.

A comparison of methods utilized for the densification of ceramic matrix, metal matrix, and polymer matrix composites is presented in Table 1. The most numerous routes to matrix densification are associated with ceramic processing. These ceramic matrix densification methods include derivatives of the resin and metal matrix composite technologies as well as novel processing specific to the creation of ceramics.

TABLE 1
Comparison of fibrous composite densification methods

Polymer Matrix	Metal Matrix	Ceramic Matrix
Polymer infiltration	Hot pressing	Hot pressing
Thermoplastic molding	Liquid metal infiltration	Sintering
	Directional solidification	Melt infiltration
		Sol-gel
		Polymer pyrolysis
		Gas-metal reaction
		Viscous phase consolidation
		CVI

While a wide variety of processing technologies are available, the various processing routes are generally applied to specific families of ceramic matrices as shown in Table 2. Traditional hot-pressing has been applied primarily to the silicon-based ceramics, while sol-gel is applied exclusively to oxides. Polymer pyrolysis has also been most extensively applied to silicon-based ceramics, although oxides can also be processed. Gas-metal reaction has been utilized to create the specific oxide and nitride matrices listed. Viscous phase consolidation is restricted to the family of glasses and glass-ceramics. CVI is the most flexible processing route allowing creation of all major families of ceramic materials.

TABLE 2
Examples of processable matrices

Process	Residual Porosity	Matrices
Hot pressing	<2%	SiC, Si_3N_4
Melt infiltration	<2%	Si/SiC
Sol-gel	20-30%	Oxides
Polymer pyrolysis	15-30%	Si_xC_y, Si_x-C_y-N_z, Si_xN_y
Gas-metal reaction	<5%	Al_2O_3, AlN, ZrN, TiN
Viscous phase consolidation	<2%	Glasses, glass-ceramics
CVI	15-30%	Carbides, nitrides, oxides, borides

A number of potential reinforcing fibers and/or whiskers are available for the fabrication of ceramic matrix composites utilizing the above matrix densification methods. While not intended to be totally inclusive, examples of the primary types of available fibers and whiskers are given in Table 3. The fibers range from relatively large monofilaments of SiC and Al_2O_3 to small multifilament tow fibers, primarily based on oxides of Al_2O_3 or mullite, and mixed composition Si-C-N-O fibers that are either based on SiC or Si_3N_4. The large diameter fibers are processed either by CVD of SiC on a C or W core or by single crystal Al_2O_3 growth from the melt, while the small diameter fibers are usually processed by spinning of sol-gels (oxides) or polymers (Si-C-N-O) followed by pyrolyzation. Carbon fibers are also important, especially for the CVI processed composites.

A variety of whiskers are available for composite fabrication, only a few of which are listed in Table 3. The whisker reinforced ceramic matrix composites do not exhibit the toughness and graceful failure that continuous fiber reinforced ceramics can, but have found utilization as cutting tools for metals and in wear resistant applications. Alumina and silicon nitride reinforced with SiC whiskers are the most common systems available. While whisker reinforced ceramic matrix composites may have applications in certain components of heat engines, they will not be discussed in detail in this paper.

It has been found in all of the above-mentioned ceramic composites that in order to achieve high strength and, in particular, high toughness, the bonding at the fiber/matrix interface must be controlled such that the bonding is strong enough to allow load transfer from the matrix to the fibers under stress but weak enough so that an advancing matrix crack can be deflected by the fibers. In addition, the nature of the fiber/matrix interface must include resistance to oxidation at elevated temperature as well as resistance to other environmental effects.

In the following sections, the most widely utilized generic processing routes for the fabrication of ceramic matrix composites will be discussed and specific examples of a variety of fiber/matrix combinations that fall under each processing route will be given. In addition, examples of the influence of the bonding and chemistry of the fiber/matrix interface on composite properties will be discussed for the glass-ceramic and CVI SiC matrix composite systems. The comparative mechanical performance of these composites will also be discussed with reference to what the authors feel is the primary objective in creating ceramic composites - to achieve more graceful failure modes and thus enhanced reliability when compared with the unreinforced ceramic matrix. Finally, examples of the use of these two types (glass-ceramic and CVI SiC matrix) of composites in gas turbine engine applications will be presented.

TABLE 3
Candidate fibers for ceramic matrix composites

Type	Designation	Manufacturer	Composition	Diam µm	UTS (GPa)	E (GPa)	Density g/cc	CTE 10^{-6}/°C	Source
Monofil.	SCS-6	Textron	SiC/C	143	3.9	406	3.0	4.4	(1,2)
"	Sigma	BP	SiC/W	100	3.5	400	3.4	4.6	(3)
"	Saphikon	Saphikon	Al_2O_3	125	~2.6	414	4.0	~8	(4)
Multifil. Oxide	Nextel 440	3M	Mullite+2%B_2O_3	11	2.1	189	3.05	4.5	(1)
"	Altex	Sumitomo	Al_2O_3+15%SiO_2	15	~2.2	~230	3.2	8.8	(1)
"	FP	DuPont	Al_2O_3	20	1.4	370	3.9	5.7	(1)
"	PRD-166	DuPont	Al_2O_3+ZrO_2	20	2.1	380	4.2	9.0	(5)
Multifil. Carbide	NICALON	Nippon Carbon	SiC+C+O	15	~2.9	~200	2.55	3.1	(1)
"	Tyranno	Ube Industries	SiC+C+O+Ti	10	>2.9	>200	~2.4	3.1	(1)
Multifil. Nitride	HPZ	Dow Corning	Si-N-C-O	10	~2.3	~160	2.35	3.0	(1,6)
"	Fiberamic	Rhone-Poulenc	Si-N-C-O	15	1.8	220	2.4	3.1	(7)
Multifil. Carbon	HMU	Hercules	graphite	8	2.8	380	1.84	-0.7	(8)
"	FT-700	Tonen	graphite	10	3.3	700	2.16	-1.5	(9)
Whiskers	VLS	Los Alamos N. L.	β-SiC	4-6	8.4	580	3.2	-	(2)
"	TWS-100	Tokai Carbon	"	0.3-0.6	14	400-700	3.2	5	(10)
"	TWS-400	"	"	1.0-1.4	"	"	"	"	"
"	SNW	Tateho	α-Si_3N_4	0.2-0.5	-	-	3.18	2.5	(11)

(1) Ceramic Source Guide, Am. Cer. Soc., pp 380-381.
(2) Prewo, K.M., Brennan, J.J., and Layden, G.K.: Fiber-Reinforced Glasses and Glass-Ceramics for High Performance Applications, Am. Cer. Soc. Bull., 65 [2] 1986,305-314.
(3) Sigma Data Sheet, BP Chemicals, Feb. 1991.
(4) Saphikon Data Sheet, July, 1990.
(5) Romine, J.C.: New High-Temperature Ceramic Fiber, Cer. Engr. Sci. Proc. 8 [7-8], 1987, 755-765.
(6) Dow Corning HPZ Fiber Data Sheet, 1990.
(7) Rhone-Poulenc Announcement from Paris Air Show and Data Sheet, June, 1989.
(8) Hercules Product Data Sheet #851 (1981).
(9) Tonen Product Data (1990).
(10) Tokai Carbon Data (1989).
(11) Tateho Data Sheet (1984).

CERAMIC COMPOSITE SYSTEMS AND PROCESSING METHODS

Hot-Pressing

The ceramic matrix composites utilizing continuous fiber reinforcement that are consolidated by hot-pressing have been confined primarily to Si_3N_4 matrices reinforced with CVD derived SiC monofilaments, although some work has been done utilizing single crystal sapphire fibers in oxide matrices[10]. In addition, Japanese researchers have utilized a novel approach to create ceramic matrix composites by hot-pressing fibers themselves with no matrix present. Glass and glass-ceramic matrix composites are also commonly densified by hot-pressing, but they will be discussed separately in a later section of this paper under viscous phase densification.

A considerable amount of work has been done at Textron Specialty Materials, Lowell, MA, in the development of hot-pressed silicon nitride matrix composites utilizing their own SCS-6 SiC monofilament fibers as reinforcement[11]. The matrix consisted of 93.75 wt% Si_3N_4 powder (Starck LC-12), with 5.00% Y_2O_3 and 1.25% MgO as densification aids. The 30 vol% unidirectional SCS-6 SiC fiber composite was densified at 1700°C, 70 MPa, 1 hr, in vacuum. These composites have been evaluated by workers at the University of Michigan[12]. As shown in Fig. 1, these composites exhibit a proportional limit signifying a first matrix microcracking event, followed by a capability to bear increasing loads when tested in tension.

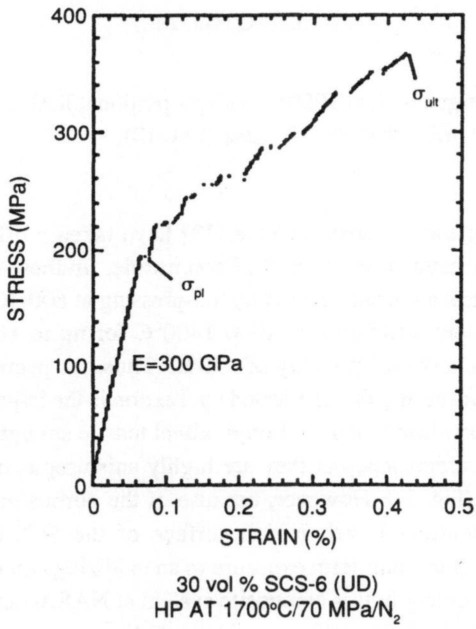

Figure 1. Tensile behavior at 1000°C of unidirectional SCS-6 SiC fiber/Si_3N_4 matrix composite (Ref. 12).

The significance of this microcracking is shown in Fig. 2, where results of elevated temperature tension-tension fatigue tests in air are summarized. It is clear that cycles to failure are limited by the proportional limit stress. Stresses above the proportional limit lead to increased microcracking and, since the SCS-6 SiC fibers have a ~3 μm thick carbon rich surface coating, oxidation of this carbon layer results in degradation of composite properties. The effects of the fiber/matrix interface on ceramic matrix composite properties will be discussed in greater detail in a later section of this paper.

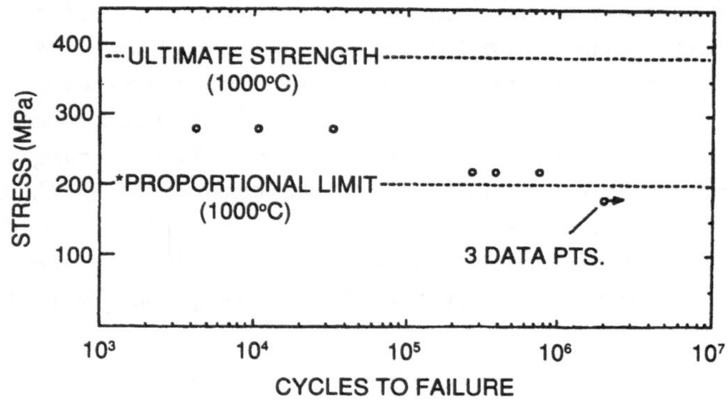

Figure 2. Relationship of fatigue limit at 1000°C with proportional limit stress for SCS-6 SiC fiber reinforced Si_3N_4 (Ref. 12).

Researchers at NASA-Lewis Research Center[13] have taken a different approach to fabricating SCS-6 SiC monofilament-reinforced silicon nitride. In these composites, silicon powder is utilized as the matrix and consolidated by hot-pressing at 600-1000°C, 27-138 MPa pressure, for 1 hr. Subsequent nitridation at 1000-1400°C for up to 100 hrs converts the matrix to silicon nitride with a typical porosity of ~30%. Thus, the primary microstructural difference between this composite and that fabricated by Textron is the large amount of residual porosity present in the silicon nitride matrix. Longitudinal tensile strengths in excess of 500 MPa are achieved in these composites, but they are highly anisotropic, exhibiting very low transverse tensile strengths (Fig. 3). However, because of the porous nature of the matrix, combined with the aforementioned carbon rich surface of the SCS-6 fibers, a loss in mechanical properties occurs after long-term exposure in an oxidizing environment in the 600-1000°C range. Work is continuing in this composite system at NASA-Lewis that is aimed at densifying the matrix by post-nitridation HIP consolidation[14].

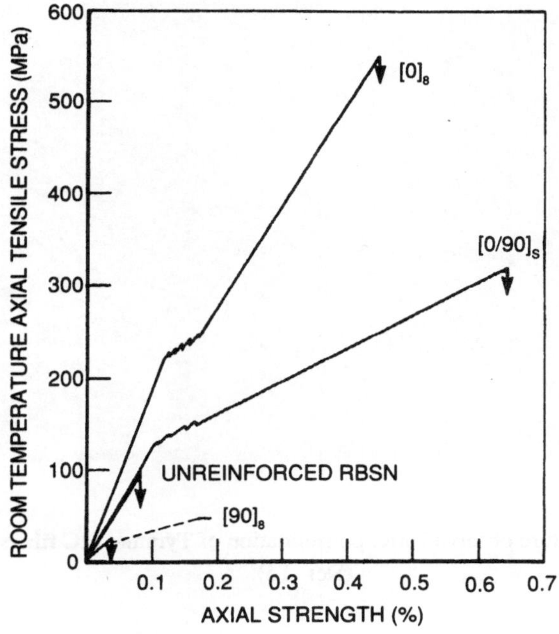

Figure 3. Tensile behavior of SCS-6 SiC fiber reinforced reaction bonded Si_3N_4 (Ref. 13).

Lundberg, et al[15], at the Swedish Institute for Silicate Research, Goteborg, have utilized both carbon fibers in a silicon nitride matrix that was densified by hot isostatic pressing (HIP), and NICALON SiC fibers in a silicon nitride matrix that consists of silicon nitride powder plus sintering aids, as well as silicon powder that is subsequently converted to silicon nitride by a nitridation step. The former composite resulted in a relatively strong material with good fracture toughness, while the latter resulted in a quite weak composite, probably due to fiber degradation during the 1350°C nitriding step.

A novel hot-pressed composite is being developed by Yamamura and coworkers at Ube Industries in Japan[16]. Unlike the previous examples that utilized a monofilament SiC fiber for reinforcement, these composites are processed using SiC yarn (Tyranno) derived from polytitanocarbosilane polymer. In this approach, woven fiber cloth arrays are hot-pressed at 1800-2100°C either by themselves or combined with powder derived from the same polymer. An example of the type of microstructure obtained is shown in Fig. 4. The fiber cross-section has assumed a hexagonal shape, but has not been destroyed by the high hot-pressing temperatures. The flexural behavior at 1400°C is shown in Fig. 5, and indicates that these hot-pressed fiber arrays have substantially improved the load-deflection characteristics relative to the matrix.

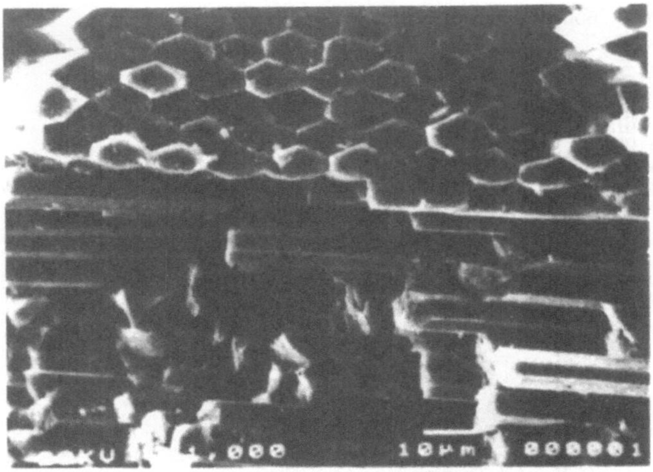

Figure 4. Microstructure observed after consolidation of Tyranno SiC fibers at 1800-2100°C (Ref. 16).

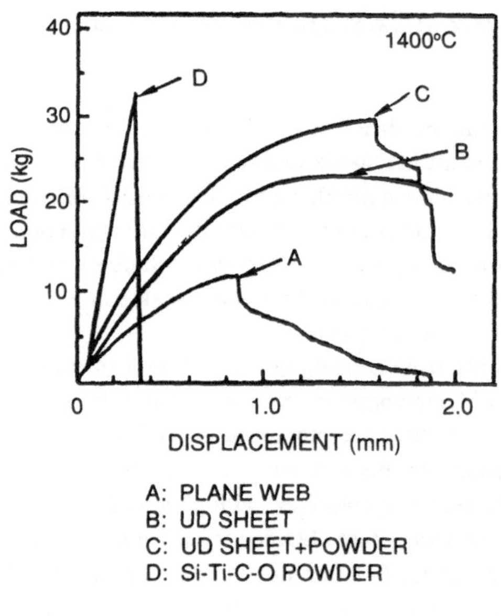

Figure 5. Effect of precursor format on the flexural behavior at 1400°C for hot pressed Tyranno SiC fibrous arrays (Ref. 16).

Sol-Gel Infiltration and Pyrolysis

The sol-gel technique was originally developed for the preparation of unreinforced oxides; however, the technique can also be applied to composite fabrication[17,18]. The method includes infiltrating a fibrous preform with a solution normally containing various alkoxide compounds with different catalysts, dissolved salts, and other additives. The gel is formed from the liquid sol by further hydrolysis with subsequent dehydration and polymerization. A number of useful chemistries to form single and mixed oxides exist, and are detailed by Fitzer, et al[17]. Following drying, the gels are converted to ceramics by calcination or pyrolysis normally at temperatures above 600°C. Primary disadvantages of this process are the very large shrinkage during processing and residual fine scale porosity. Thus, many impregnation cycles are required to achieve reasonably dense bodies. In fact, the most structurally performant composites have been fabricated by inclusion of a final hot-pressing step[18], even though this negates one of the most important reasons for utilizing the sol-gel approach in the first place; that of forming a ceramic matrix composite to net shape without resorting to high temperature and pressure densification routes.

An example of a composite fabricated by the sol-gel approach is shown in Fig. 6. In Fig. 6, the reinforcement with either FP alumina or NICALON SiC fibers of a GeO_2 modified silica glass results in dramatic improvements in strength, modulus, and failure strain of the composite relative to the matrix. Workers at GEC Research, Ltd., in the UK[19] have developed a sol-gel approach to fabricating carbon fiber reinforced silica tubes that reportedly are capable of operation to 1500°C in a non-oxidizing atmosphere, with strength values of over 700 MPa.

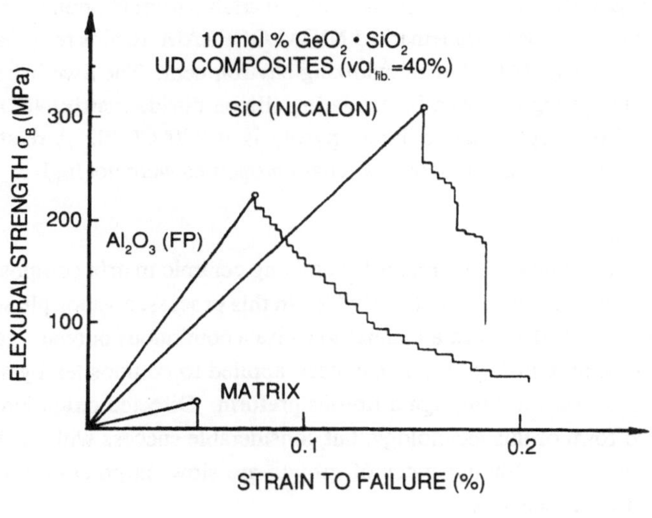

Figure 6. Sol-gel derived GeO_2 modified SiO_2 composite flexural properties (Ref. 17).

Polymer Precursor Infiltration and Pyrolysis

The use of polymer pyrolysis to densify ceramic composites is a rapidly growing research area which has been stimulated in part by the many research activities aimed at creating ceramic fibers, such as NICALON, Tyranno, and HPZ, from preceramic polymers. Polymers such as polycarbosilane and methylvinylsilanes for SiC, polysilazanes for silicon nitride, and vinyl modified polysilazanes for Si-C-N compounds have all been developed. The interesting aspect of this processing route is that it directly parallels certain aspects of polymer matrix composites and also carbon/carbon processing[20]. Thus, there is a large experience base for the layup and composite preforming technologies.

Because the matrix pyrolysis step is accompanied by significant shrinkage and gas evolution, five to ten reimpregnation cycles under pressure are commonly used for densification. Most polymers are fully converted to ceramic after heating to 800-900°C, but the reaction products are amorphous in nature, with higher temperature heat-treatment being required to crystallize the matrix. It has been found that many polymer derived matrices can be substantially off stoichiometry and are usually contaminated with excess carbon and oxygen which can affect their stability at high temperatures in oxidizing environments.

A NICALON SiC fiber reinforced silicon oxycarbide matrix composite (unidirectional reinforcement) synthesized from a 50/50 phenyl/methyl silsesquioxane copolymer and pyrolyzed at 1400°C[21], was found to have a room temperature tensile strength of 246 MPa, elastic modulus of 141 GPa, and a strain-to-failure of 0.17%. At United Technologies Research Center (UTRC), very promising results were obtained on a carbon fabric-reinforced silicon carbide matrix composite, where the matrix is derived from a vinylmethylsilane[22]. As shown in Fig. 7, a room temperature tensile strength of over 300 MPa was achieved, with a strain-to-failure of over 0.6%. A NICALON fiber reinforced SiC matrix composite derived from a polycarbosilane precursor is now commercially available from Nippon Carbon Co.[23]. As shown in Fig. 8, this composite, trademarked NICALOCERAM, retains reasonable flexural strength after exposure up to 1000°C in an oxidizing environment. The Swedish Institute for Silicate Research in Goteborg has been investigating silicon nitride matrix/NICALON fiber composites fabricated by polysilazane polymer pyrolysis at 900°C[15]. A relatively tough fracture mode was found, although composite strength properties were not high.

Gas-Metal Reaction

One of the most innovative new technologies for creating ceramic matrix composites was the development of the Lanxide™ process in the 1980's. In this process, a vapor phase reactant is contacted with a molten metal alloy in a manner to cause a continuous outward growth of the reaction product. As shown in Fig. 9, this has been applied to composites by allowing the reaction product to grow outward through a fibrous preform. Directed oxidation is the most frequently referenced form of this technology, but considerable success with nitride matrices has been achieved as well. Although rates of growth are slow, large complex shapes are readily achievable with this technology.

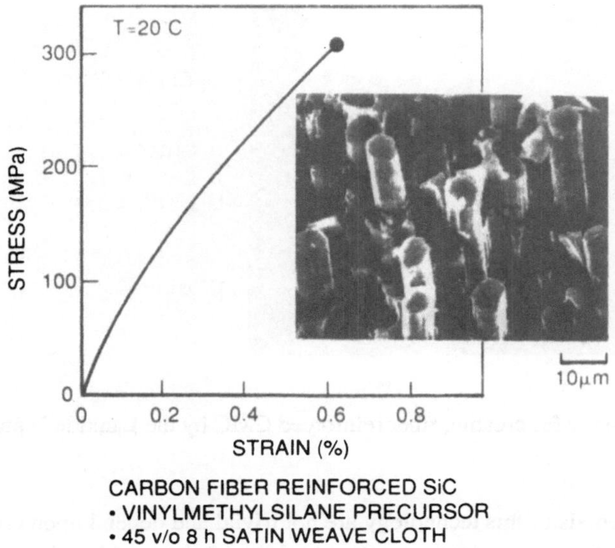

Figure 7. Tensile behavior of T-300 carbon fabric reinforced SiC matrix derived from vinylmethylsilane polymer (Ref. 22).

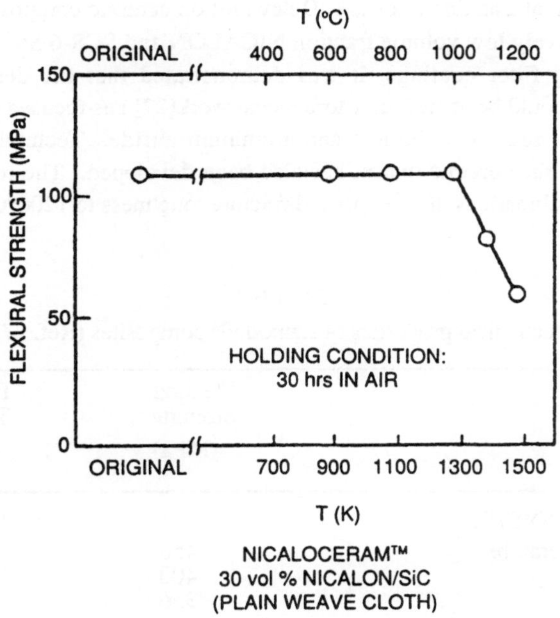

Figure 8. RT flexural strength of NICALOCERAM™ after elevated temperature exposure (Ref. 23).

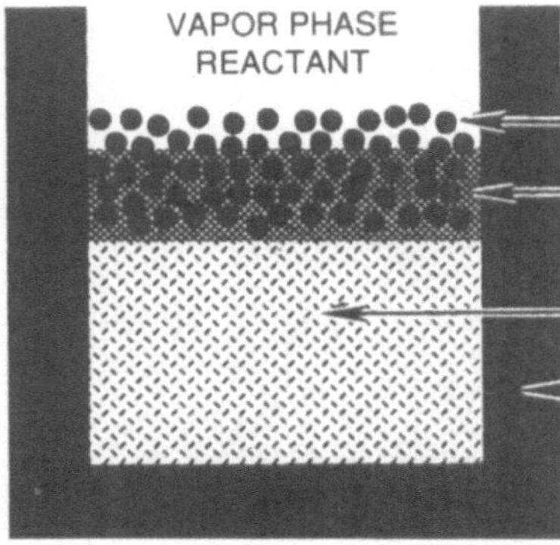

Figure 9. Schematic for creating fiber reinforced CMC by the Lanxide™ process (Ref. 24).

The fundamentals of this technology are not trivial and depend upon creating the proper surface energy balances to maintain a network of liquid phase within the reaction product to feed the outward growing interface. For example, in the aluminum oxide system there are critical temperature ranges and alloy compositions which optimize the kinetics of oxide growth[25].

The initial work at Lanxide (Newark, Delaware) on ceramic composites[26] focused on the creation of relatively low volume fraction NICALON and SCS-6 SiC reinforced alumina composites. Using fiber coatings, it was demonstrated that the desired resistance to catastrophic failure could be achieved. More recent work[27] has focused on the utilization of NICALON fabric to reinforce alumina and aluminum nitride. Technology to remove the residual metal from the ceramic matrix has also been developed. The resultant composites exhibit excellent combinations of strength and fracture toughness to 1200°C, as summarized in Table 4.

TABLE 4
Mechanical properties of Lanxide™ composites (Ref. 27)

	Flexural Strength (MPa)	Fracture Toughness ($MPa\sqrt{m}$)
NICALON™/Al_2O_3 SYSTEM		
Room Temperature	450	21
1000°C	400	23
1200°C	350	18
NICALON™/AlN SYSTEM		
Room Temperature	440	20
1000°C	340	14

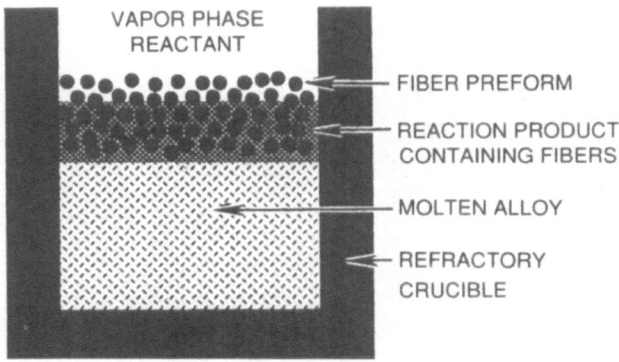

Viscous Phase Consolidation

As mentioned in the Introduction, the formation of glass and glass-ceramic matrix composites through viscous phase consolidation has been practiced in both the United States and the United Kingdom since the late 1960's. This technique is a direct extension of polymer matrix technology and is a direct parallel to thermoplastic molding. As a technology development, it has been extremely significant since it has provided the materials for study which have been utilized to define many of the guiding principles for the creation of useful fiber-reinforced ceramics over the last two decades. Currently, work in the area of glass and glass-ceramic matrix composites is being performed in the US, UK, Germany, and France[28-54]. For the purposes of this paper, research conducted at UTRC in the area of glass and glass-ceramic matrix composites will be utilized to illustrate the variety of fibers, matrices, and processing procedures used for these composites, and the typical mechanical performance that can be obtained.

In contrast to many of the other types of ceramic matrix composites discussed in this paper, glass and glass-ceramic matrix composites follow traditional resin and metal matrix composite practice in that higher elastic modulus fibers have been incorporated into a lower elastic modulus matrix to achieve structural reinforcement. In addition, the glass matrix can be readily deformed at elevated temperatures and flowed in its low viscosity state around the reinforcing fibers to achieve full density without damaging them. Glass-ceramics provide the unique capability to densify a composite in the glassy state and then subsequently crystallize the matrix to achieve high temperature capability.

During the years since glass matrix composite research first began at United Technologies Research Center in 1974, most of the fiber and whisker reinforcements listed in Table 3 have been utilized in a variety of glass and glass-ceramic matrices. The matrices have ranged from relatively low temperature borosilicate and aluminosilicate glasses, to much higher temperature capability glass-ceramics based on the lithium aluminosilicate (LAS), magnesium aluminosilicate (MAS), and barium-magnesium aluminosilicate (BMAS) systems. Early experiments demonstrated that a NICALON fiber reinforced borosilicate glass matrix composite system is capable of achieving excellent mechanical properties at temperatures up to 600°C[28] and a high silica glass matrix system achieved peak strength at 1000°C[29]. More recent work in the LAS glass-ceramic matrix/NICALON fiber system[33-36,54] has demonstrated tensile strengths in excess of 400 MPA and failure strains exceeding 1% for 0/90° ply layup composites tested at 1300°C, as shown in Fig. 10. The tensile failure surface was also fully fibrous, as shown in Fig. 11, indicating retention of NICALON fiber strength within the composite at this test temperature. Large diameter fibers such as the SCS-6 SiC monofilament have also been successfully utilized for glass-ceramic matrix composites with 0/90 tensile strengths exceeding 700 MPa and a very tough, fibrous fracture mode, as seen in Fig. 12.

However, the effect of environment plays a significant role in the fracture characteristics of these composites, and indeed in many other ceramic matrix composites as well. For example, the effect of an oxidizing vs an inert test environment is demonstrated in Fig. 13, where even in the short time frame of this test, the LAS matrix/NICALON fiber composite failure strain and strength were reduced significantly due to the oxidation of the fiber/matrix interface caused by air permeating through matrix microcracks that occur at stress levels above the proportional limit. This oxidation effect will be discussed in more detail in a later section of this paper dealing with the influence of the fiber/matrix interface in ceramic matrix composites.

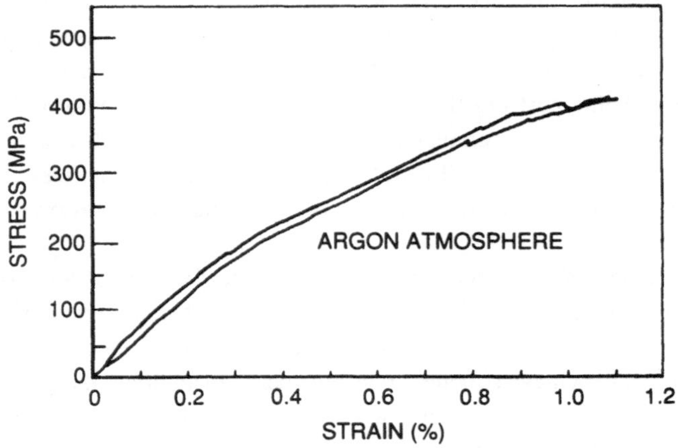

Figure 10. Tensile behavior of a 0/90 NICALON SiC fiber reinforced LAS III matrix composite at 1300°C (Ref. 54).

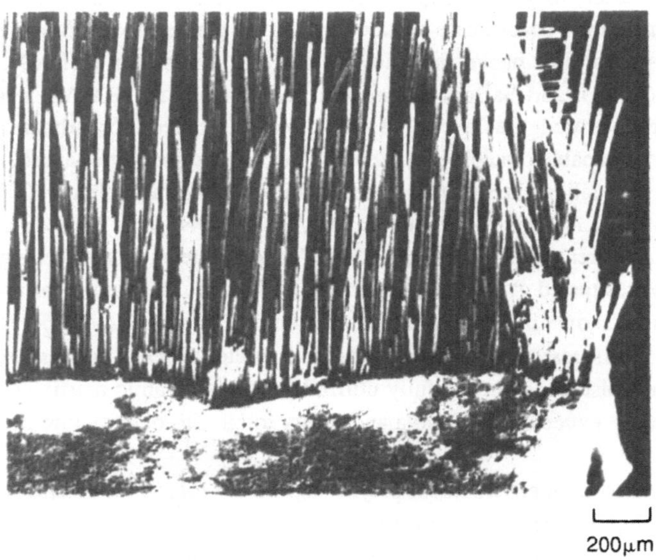

Figure 11. 1300°C tensile failure surface for NICALON SiC fiber reinforced LAS III (Ref. 54).

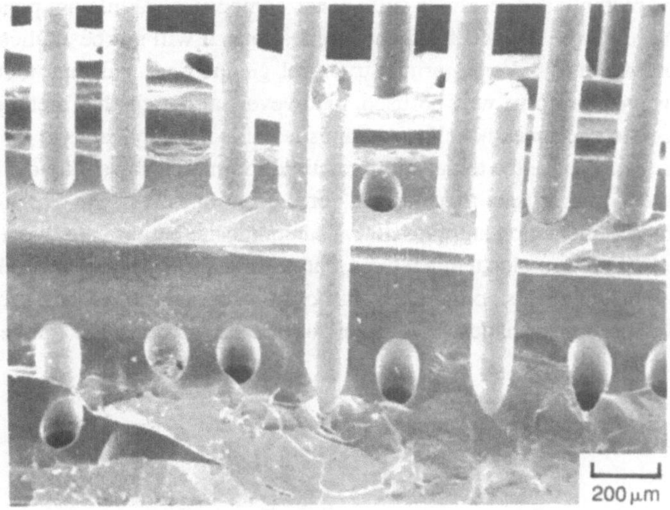

Figure 12. 0/90 SCS-6 fiber/LAS matrix composite tensile fracture surface.

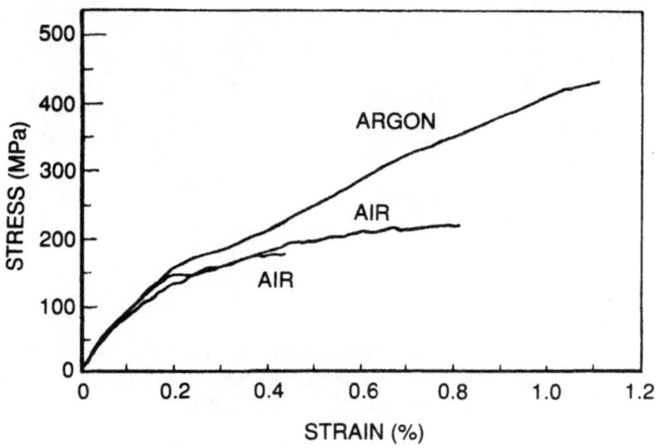

Figure 13. Tensile behavior at 1000°C of NICALON SiC fiber reinforced LAS III (Ref. 54).

One of the most significant attributes of glass and glass-ceramic matrix composites is the ease with which they can be fabricated. Processes utilized for the fabrication of these composites consist of hot-pressing of glass powder infiltrated unitape and fabrics layups, hot matrix transfer molding into woven fiber preforms, and hot injection molding of chopped fiber/matrix compounds into shaped molds. Hot-pressing of pre-infiltrated tapes is the most routine fabrication method, and results in fully dense composites. While hot pressing of tape layups may be the preferred method for many applications, it will not be applicable in some cases, because of external and/or internal structural geometries, and thus matrix transfer molding into a woven preform will be used. However, although this method offers the potential for processing complex net shape composites, it is not applicable to all fiber/matrix combinations. This is because the temperature required for matrix flowability during injection is often substantially higher than that utilized for normal hot-pressing, which can lead to excessive fiber/matrix interaction and thus degraded composite properties. Injection molding can be done at normal hot-pressing temperatures, but is limited to relatively short chopped fiber lengths (<1 cm) and thus yields composites with only moderate mechanical properties. As shown in Fig. 14, complex shapes have been fabricated utilizing the above processing procedures.

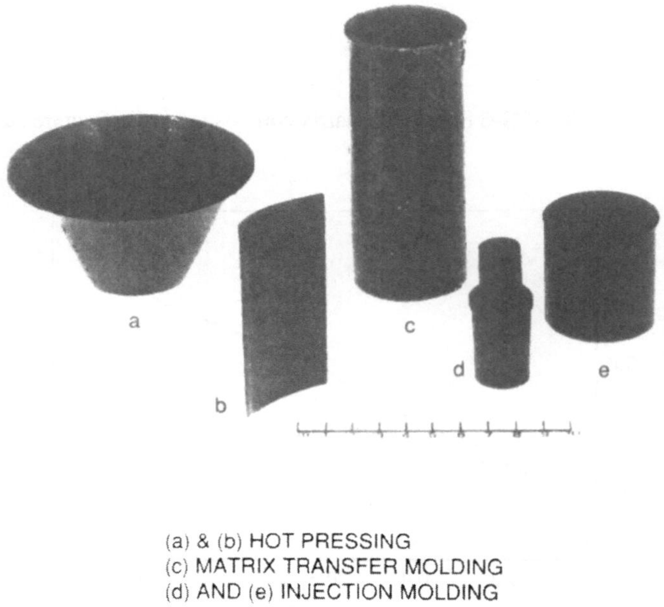

Figure 14. Consolidation techniques demonstrated with viscous phase processing (Ref. 33).

Chemical Vapor Infiltration (CVI)

The use of chemical vapor deposition processes to densify fibrous preforms has grown rapidly during the last decade. Much of this growth can be attributed to the industrial application of CVI to carbon-carbon composites and the successful demonstration of applying SiC CVD techniques to CVI composites. From a matrix formation viewpoint, chemical vapor processes provide the means to create the widest variety of ceramic materials. As shown in Table 5, precursors for carbides, borides, nitrides, and oxides are readily available.

TABLE 5
Useful CVD reaction chemistries (Ref. 55)

Refractory materials	Gaseous precursors
Carbon	CH_4; C_3H_8; C_2H_2; C_6H_6
Boron	BCl_3-H_2; BBr_3-H_2
SiC	CH_3SiCl_3-H_2; $(CH_3)_2SiCl_2$-H_2; $SiCl_4$-CH_4-H_2
Si_3N_4	$SiCl_4$-NH_3; SiF_4-NH_3
B_4C	BCl_3-CH_4-H_2; BBr_3-CH_4-H_2
TiC, ZrC	MCl_4-CH_4-H_2 (M=Ti,Zr)
TiB_2, ZrB_2	MCl_4-BCl_3-H_2 (M=Ti,Zr)
BN	BCl_3-NH_3-H_2; BF_3-NH_3
Al_2O_3, ZrO_2	$AlCl_3$-H_2-CO_2; $ZrCl_4$-H_2-CO_2

Two primary approaches to CVI composites are currently under development. Isothermal chemical vapor deposition was developed for carbon-carbon densification and has been extended by Societe Europeenne de Propulsion (SEP), Bordeaux, France, on a commercial basis to SiC CVI composite fabrication[56]. In addition, CVI under the conditions of thermal and pressure gradients has been developed at the Oak Ridge National Laboratory (ORNL) in Knoxville, Tennessee[57,61]. This process, referred to as "forced CVI" is schematically shown in Fig. 15 and offers the promise of thick-section densification and shorter densification times.

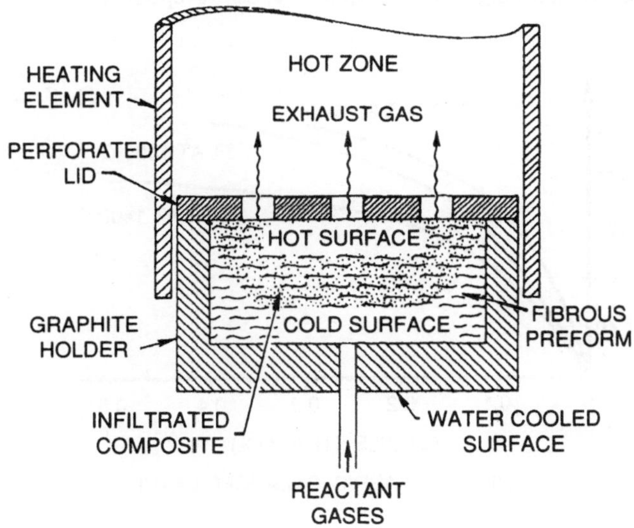

Figure 15. Thermal/pressure gradient CVD (forced CVI) (Ref. 60).

A number of ceramic composites have been fabricated using the forced CVI technique on a laboratory scale. Fabric preforms of NICALON and Tyranno SiC, as well as alumina and mullite type fibers were densified with SiC[60]. In general, the SiC fiber reinforcements resulted in much higher strength composites than the oxide fibers, although all of the composites exhibited fibrous failure to some degree.

The largest commercial endeavor in the ceramic matrix composites field to date has been the development of isothermal CVI SiC matrix composites reinforced with both carbon and NICALON SiC fibers by SEP. This technology has been recently licensed to DuPont, who is now a commercial supplier of these composites in the US. Typical data from DuPont on NICALON fiber reinforced SiC are shown in Figs. 16 and 17. Shown in Fig. 16 are the tensile stress-strain characteristics of this composite material as a function of temperature. It can be seen that good tensile strengths are maintained to 1400°C. Cyclic loading effects at room temperature (Fig. 17), show that significant decreases in elastic modulus can occur when the stress exceeds the composite proportional limit stress, likely due to matrix microcracking. Significant differences are also observed in the stress-strain characteristics of the composite after various oxidative exposures to 1200°C. These results indicate that combined effects of loading and environment can affect the microstructural features of the composite that control the desirable non-catastrophic failure modes. It should be noted that the ultimate tensile stress of the composite does not change as a result of these oxidative exposures. The oxidative stability of these composites is highly dependent on a protective coating applied to the composite. Without a protective coating, these types of composites, which rely on a weak carbon coating at the fiber/matrix interface for their tough fracture behavior, can be very prone to oxidative degradation. Examples of this will be given in the following section on the influence of the fiber/matrix interface in controlling ceramic matrix composite properties.

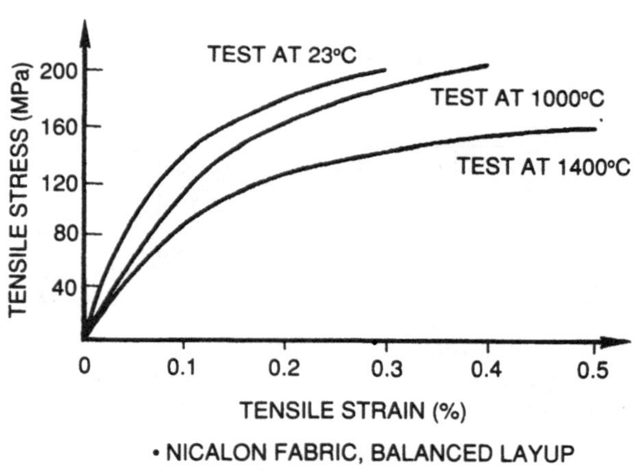

Figure 16. Stress-strain behavior of CVI NICALON SiC fiber/SiC matrix composites (Ref. 62).

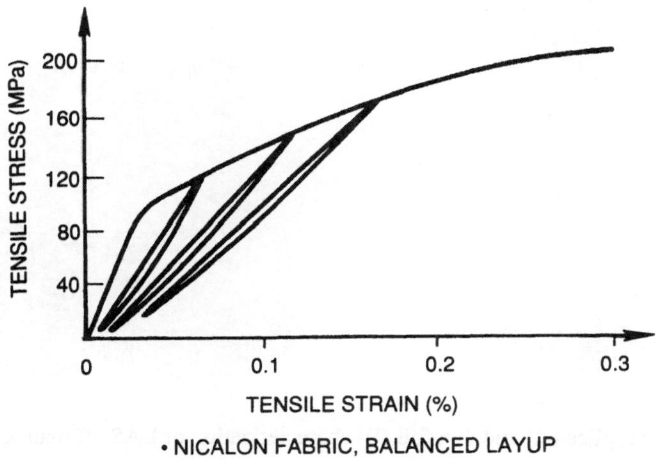

Figure 17. Cyclic loading effects on SiC/SiC at 20°C (Ref. 62).

INFLUENCE OF THE FIBER/MATRIX INTERFACE IN CERAMIC MATRIX COMPOSITES

As mentioned previously in the Introduction section of this paper, it has been found in all of the ceramic matrix composites discussed that in order to achieve high strength and, in particular, high toughness, the bonding at the fiber/matrix interface must be controlled such that the interface is strong enough to allow load transfer from the matrix to the fibers under stress, but weak enough so that an advancing matrix crack can be deflected by the fibers. If the bonding is too strong, the crack will proceed directly from the matrix into and through the fiber with the result being a very weak and brittle composite. In this section, examples of the effect of interfacial chemistry and bonding on composite properties will be discussed for the glass-ceramic matrix and CVI SiC matrix composite systems.

Glass-Ceramic Matrix Composite Interfaces

From studies at UTRC during the past few years[31,33,36,39], it has been found that polymer derived SiC type fibers such as NICALON and Tyranno that contain excess carbon and oxygen over stoichiometric SiC, form a carbon rich fiber/matrix interfacial layer when incorporated into glass-ceramic matrices at elevated temperature. This carbon rich interfacial layer can be seen in the TEM replica and thin foil micrographs in Fig. 18, for a NICALON fiber in a LAS matrix. This particular LAS matrix (LAS-III) also contained a small amount of niobium oxide, which reacted during composite fabrication to form a layer of NbC particles between the carbon rich layer and the LAS matrix. Energy dispersive X-ray spectroscopy and selected area electron diffraction of the interfacial region in Fig. 18B confirmed that the particles were NbC and the interfacial layer was carbon.

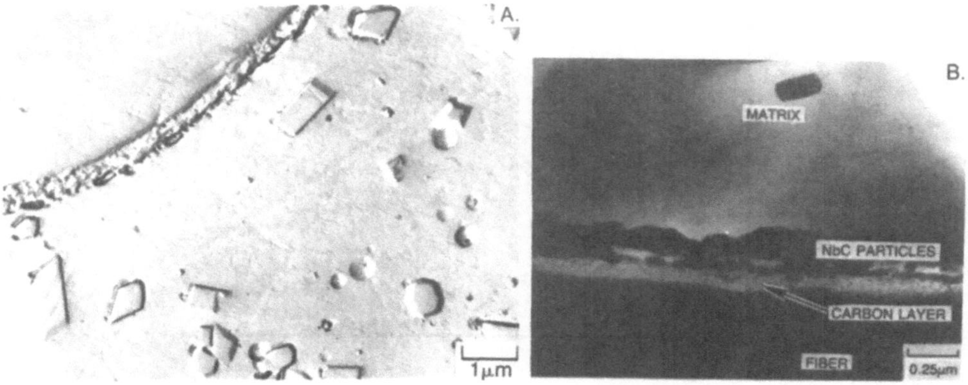

Figure 18. TEM replica (A) and thin foil (B) characterization of LAS-III matrix/NICALON SiC fiber composite - as-pressed.

Confirmation of the chemistry of the interfacial region of this composite was also done utilizing a scanning Auger electron microprobe (SAM). While an as-received desized NICALON fiber exhibits an oxygen rich surface (Fig. 19), depth profiling a fractured composite by Ar ion sputtering a fiber surface and a matrix trough from which a fiber has debonded and then combining the two profiles into a composite interfacial analysis (Fig. 20),

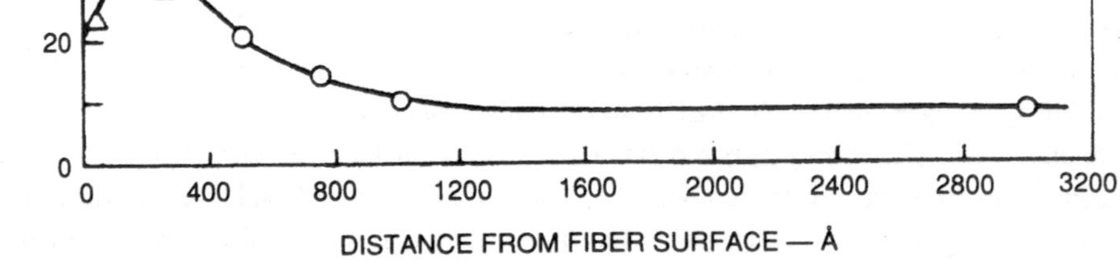

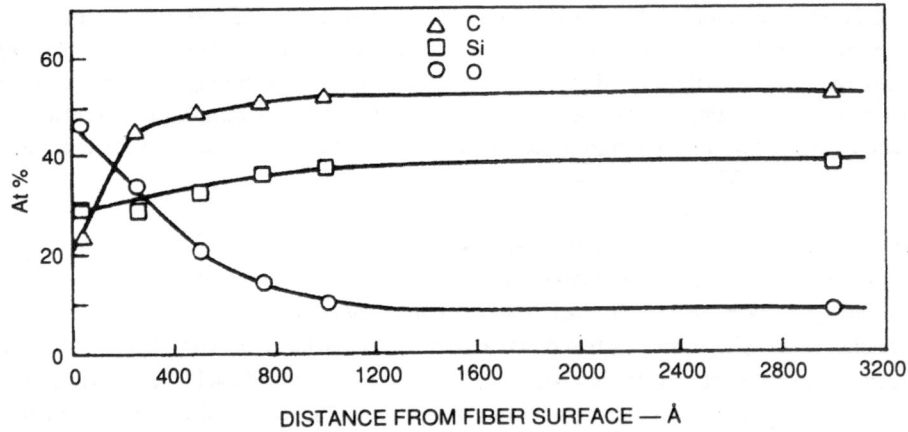

Figure 19. SAM depth profile of as-received NICALON fiber (flame de-sized).

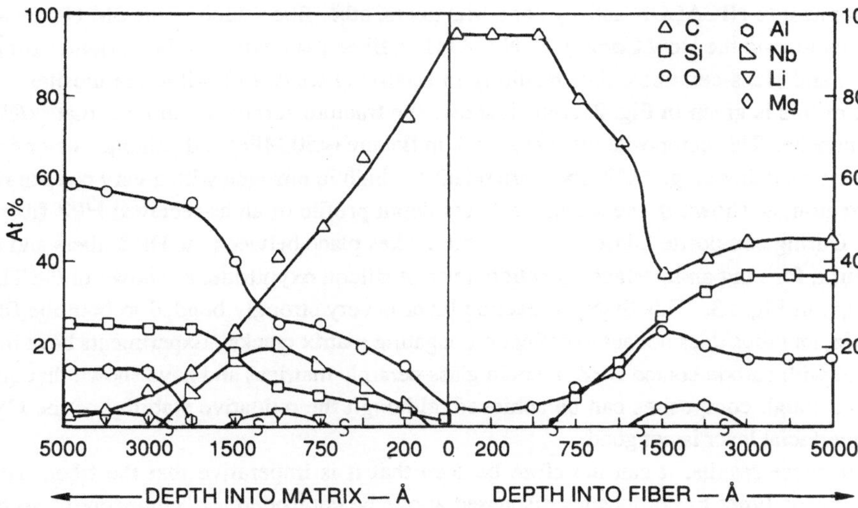

Figure 20. SAM interfacial depth profile for LAS-III matrix/NICALON fiber composite.

verifies that a carbon rich interfacial layer has indeed formed during composite fabrication. In addition to the 500-1000Å thick carbon rich layer, which usually adheres to the fiber surface during fracture of the composite, it can be seen from Fig. 20 that the NbC particle layer remains in the matrix trough and appears to be on the order of 3000Å in thickness, similar to that found from TEM analysis. In addition, aluminum diffusion into the fiber can be seen, but no Al was detected in the carbon layer. The presence of this carbon layer in acting as a matrix crack diverter gives these composite materials their unusually high toughness, as depicted in the fractured composite surface that was illustrated in Fig. 11. The reactions which lead to the formation of this carbon layer also lead to a decrease in NICALON fiber strength, with the degree of degradation found to depend on the matrix composition[34].

It has been postulated by Cooper, et al[38] and Bonney, et al[53] that the source of this layer in NICALON/glass-ceramic composites is due to the reaction:

$$SiC(s) + O_2(g) \rightarrow SiO_2(s) + C(s) \qquad (1)$$

According to Cooper, et al[38], of all the chemical equilibria which describe oxidation of SiC, this reaction has the most negative free energy and, most likely, the most rapid reaction kinetics since other oxidation reactions require the diffusion of a gaseous species away from the reaction interface. Thus, the carbon that constitutes the interface in NICALON/glass-ceramic composites consists of the carbon formed by reaction (1) plus the condensed carbon that is inherently part of the NICALON fiber. The silica formed as a result of reaction (1) can be dissolved into the glass-ceramic matrix, provided that this matrix is not already saturated with silica. Very similar interfaces form in Tyranno SiC fiber/LAS matrix composites, with the exception that small Ti rich precipitates also form in or near the carbon rich layer.

In contrast to NICALON and Tyranno SiC fibers, other fibers such as the 3M Nextel and DuPont oxides and the Dow Corning Si-N-C-O HPZ fiber, react with and bond quite strongly to glasses and glass-ceramics, thus resulting in relatively weak and brittle composites. An example of this is given in Fig. 21, which shows the fracture surface of an LAS matrix/HPZ fiber composite. This composite was very weak in flexure (<50 MPa) and exhibited essentially no fiber/matrix debonding. HPZ fibers are relatively high in nitrogen with a very oxygen rich surface region, as shown in the scanning Auger depth profile of an as-received HPZ fiber in Fig. 22. During composite fabrication, a reaction takes place between the HPZ fibers and the LAS matrix, forming an interfacial reaction layer of silicon oxynitride, as shown in the TEM micrograph in Fig. 23. This Si_2N_2O reaction layer is very strongly bonded to both the fiber and matrix, and thus does not act to deflect propagating matrix cracks. Experiments have been performed with carbon coated HPZ fibers in glass-ceramic matrices and have shown that quite strong and tough composites can be achieved, although the oxidative stability of the CVD carbon interfacial layer is not good.

From these results, it can therefore be seen that it is imperative that the fiber/matrix interface in the types of composites discussed above be controlled, or "engineered," so that relatively weak interfacial bonding exists for matrix crack deflection while maintaining oxidative stability. This might be accomplished in the carbon interfacial layer systems by doping, such that the carbon layer becomes more resistant to oxidation. Another approach could be through the use of fiber coatings that are applied to the fiber surfaces prior to composite fabrication. The purpose of these coatings would be to weaken the interfacial bonding in those composite systems that bond too strongly for tough composite behavior to be achieved, or to improve the oxidation characteristics of those systems that suffer from oxidative instability at the fiber/matrix interface. In addition, in those systems where a reaction occurs between fiber and matrix, such as the HPZ fiber/glass-ceramic matrix systems, the fiber coating could act as a diffusion and reaction barrier.

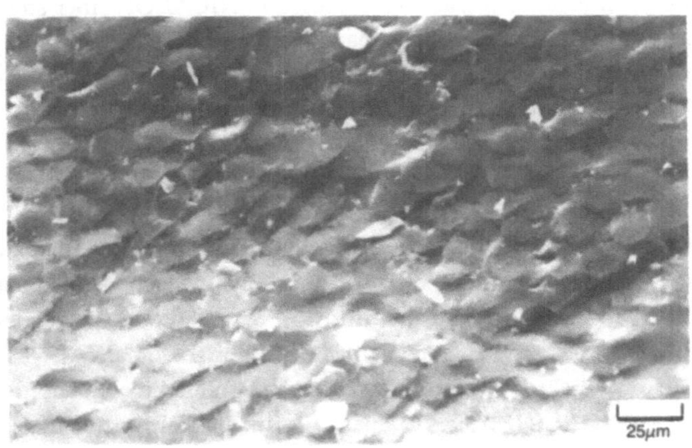

Figure 21. Fracture surface of an LAS-III matrix/HPZ fiber composite.

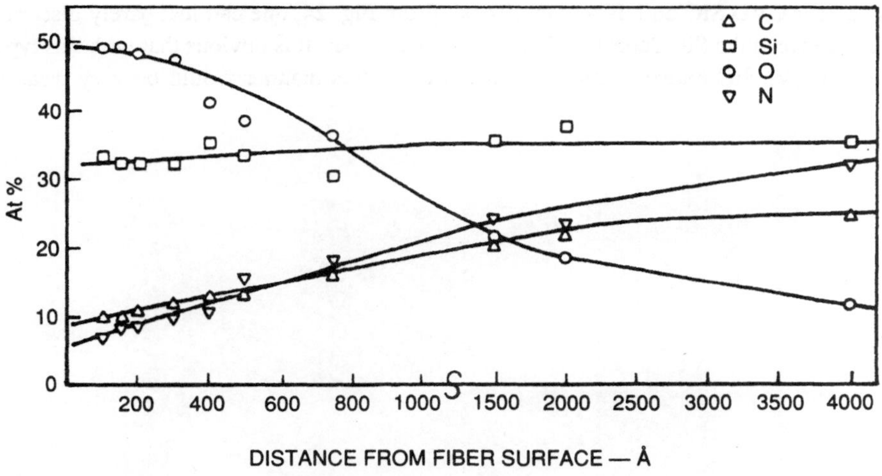

Figure 22. SAM depth profile of as-received HPZ fiber.

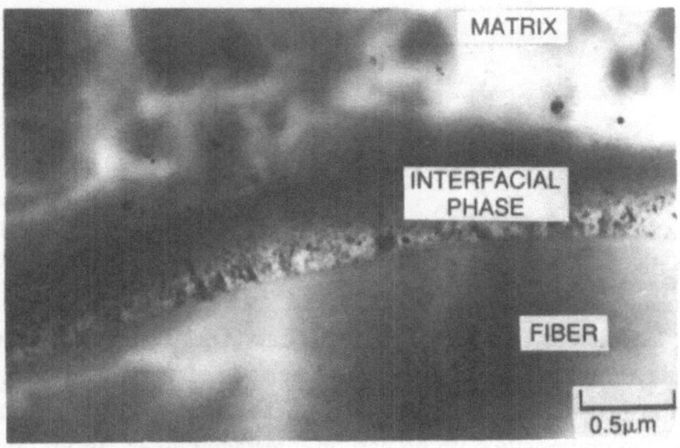

Figure 23. TEM thin foil analysis of LAS matrix/HPZ fiber composite (as-pressed).

CVI SiC Matrix Composite Interfaces

During the late 1980's, a study was conducted at UTRC aimed at the investigation of fiber/matrix interfacial bonding and reactivity in the ceramic matrix composite system of a CVI SiC matrix (isothermal deposition) reinforced with both NICALON SiC and Nextel 440 mullite fibers[63]. The CVI SiC precursor was either methyldichlorosilane (MDS) or methyltrichlorosilane (MTS) plus hydrogen. The results of this study showed that, in a very clean reactor chamber, CVD SiC bonded very strongly to either NICALON or Nextel 440 fibers, as shown in Fig. 24 for CVD SiC on NICALON fibers. From Fig. 24, one can just barely discern the interface between the SiC deposit and the NICALON fiber. It is obvious that with this type of bonding, a CVI SiC matrix composite processed in this manner would be very weak and brittle.

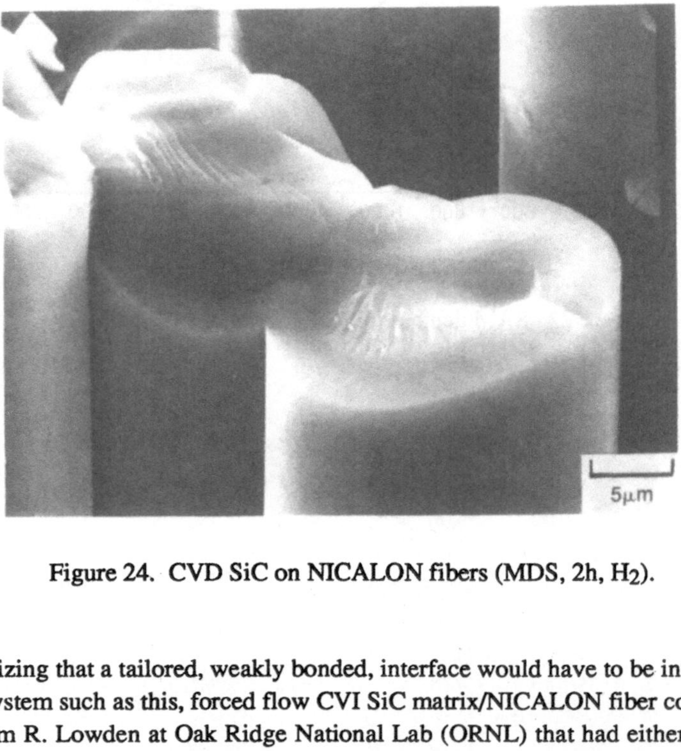

Figure 24. CVD SiC on NICALON fibers (MDS, 2h, H_2).

Recognizing that a tailored, weakly bonded, interface would have to be introduced into a composite system such as this, forced flow CVI SiC matrix/NICALON fiber composites were acquired from R. Lowden at Oak Ridge National Lab (ORNL) that had either a deliberately deposited carbon layer at the fiber/matrix interface, or an identical composite with no carbon layer. Both composites utilized a woven NICALON fiber preform. Figure 25 shows the fracture surface of the composite with no carbon interfacial layer. From this figure, it can be seen that the fracture mode was quite brittle, with little or no fiber pullout and a rather well bonded fiber/matrix interface. The room temperature flexural strength of this composite was approximately 90 MPa. In contrast, Fig. 26 shows the fracture surface of a composite with a carbon interfacial layer of approximately 1 µm in thickness. This composite exhibited a very tough fracture surface with a large amount of fiber pullout and a flexural strength of 380 MPa. It is apparent that the presence of a weakly bonded carbon interface drastically alters the fracture behavior and increases the strength and toughness in CVI SiC matrix/NICALON fiber composites.

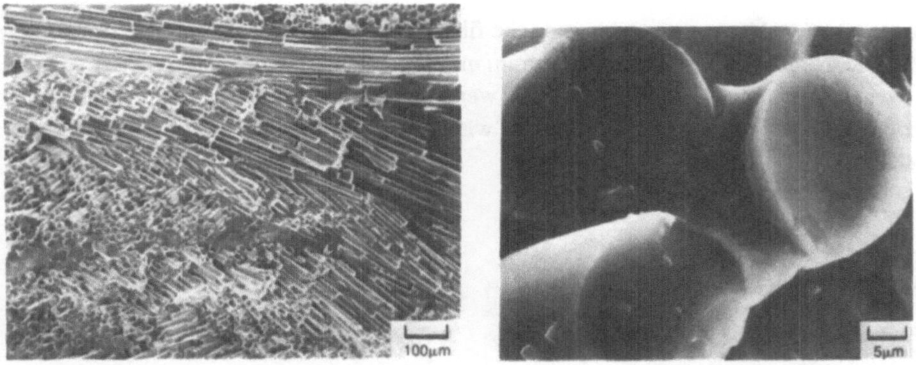

Figure 25. Fracture surface of ORNL CVI SiC matrix/NICALON fiber composite (RTσ=90 MPa) (Ref. 63).

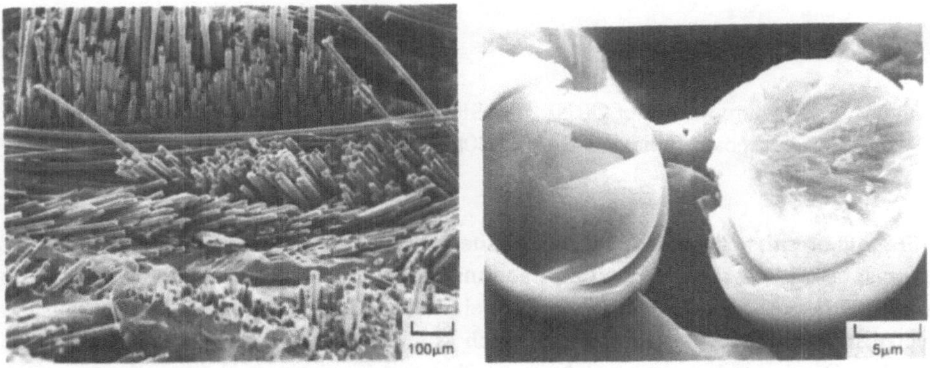

Figure 26. Fracture surface of ORNL CVI SiC matrix/carbon coated NICALON fiber composite (RTσ=380 MPa) (Ref. 63).

While carbon interfaces have been found to also result in strong and tough composites in glass-ceramic matrix composites reinforced with NICALON and other fibers, as discussed previously, the presence of this type of interface can result in severe degradation of composite properties when the composite is exposed to oxidizing environments at elevated temperatures. This is also the case for carbon interfaces in CVI SiC matrix composites. Figure 27 shows the fracture surface of an ORNL composite with a carbon interfacial layer that has been exposed to flowing oxygen at 1000°C for 70 hrs. It can be seen that while the fracture surface of this composite is still quite tough in appearance with a large amount of fiber pullout, the strength has dropped drastically to 77 MPa. At high magnification, it is apparent that the carbon layer has oxidized away leaving a gap between the fibers and matrix. While this gap results in a "tough" fracture appearance, load transfer from matrix to fiber cannot take place, resulting in a very weak composite. Similar results were found by Frety, et al, on CVI SiC matrix/NICALON fiber composites, with and without carbon fiber coatings, manufactured by SEP[64].

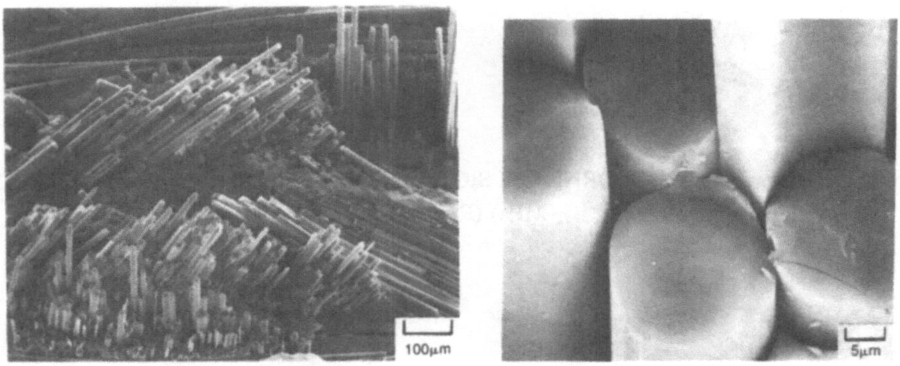

Figure 27. Fracture surface of ORNL CVI SiC matrix/carbon coated NICALON fiber composite, oxidized 1000°C, 20h (RTσ=77 MPa) (Ref. 63).

From the results of this CVI SiC composite interfacial study, it is apparent that a weak bond must be created between the fibers and matrix for tough and strong composites to be obtained. However, this weak interface must either be oxidation resistant itself, or must be protected from oxidation by other means, such as an impermeable coating over the entire composite. One has to be concerned, however, that utilizing the composite coating approach can lead to oxidative degradation of composite properties from such aspects of service as coating spallation due to impact or thermal shock, or cracking of the coating under stress, allowing oxygen penetration to the fiber/matrix interface. All of the CVI SiC matrix composite fabricators are aware of these concerns, and are implementing strategies to address and overcome them.

CERAMIC MATRIX COMPOSITE APPLICATIONS TO HEAT ENGINES

Of all the ceramic matrix composite types discussed in this paper, the ones processed by either viscous phase consolidation (glass and glass-ceramic matrix) or CVI have received the widest attention internationally and have provided a basis for both extensive laboratory experience and heat engine component demonstration.

A variety of components have been produced both in France at SEP and in the US at DuPont, utilizing the CVI SiC matrix composite approach with both carbon and NICALON SiC fibers[62,65,66]. At SEP, rocket engine nozzles and thrusters and a variety of gas turbine engine components including exhaust nozzle inner and outer flaps and seals, combustors, turbine rotors, center bodies, and flame holders have been produced and tested. Tests conducted have included rig and engine testing, with the C/SiC and SiC/SiC rocket nozzles withstanding temperatures of over 1500°C for >900 sec test durations, and the gas turbine engine components accumulating over 300 engine test hours at temperatures of 850° to 1200°C, depending on the component. Examples of SEP CVI SiC composite gas turbine engine components are shown in Fig. 28. DuPont has produced similar gas turbine engine CVI SiC composite components, including turbine engine exhaust seals and flaps, combustor elements, and rotor components.

SILICON CARBIDE COMPOSITE NOZZLE PETALS

Figure 28. SEP CVI SiC matrix composite gas turbine components.

The most noteworthy engine testing to date for ceramic matrix composites has been the SEP/SNECMA joint development and testing program that included a flight demonstration at the 1989 Paris Air Show of CVI SiC matrix composite exhaust nozzle flaps in a SNECMA M-53 engine on a Mirage 2000 fighter. Since then, both flaps and seals have been flight tested for over 50 hrs.

Gas turbine engine component testing of glass-ceramic matrix composites has been much less aggressive than for the SEP CVI SiC matrix composites. However, Pratt and Whitney has tested a segmented combustor fabricated at UTRC that was made from NICALON SiC fiber reinforced LAS glass-ceramic, as shown in Fig. 29. This segmented combustor was rig tested for a total of 18 hrs at typical combustor conditions, with no failures of any of the combustor segments. To further establish the effects of actual turbine operating conditions, UTRC and Pratt and Whitney are continuing to perform extensive gas turbine related testing of fiber reinforced glass-ceramic matrix composites.

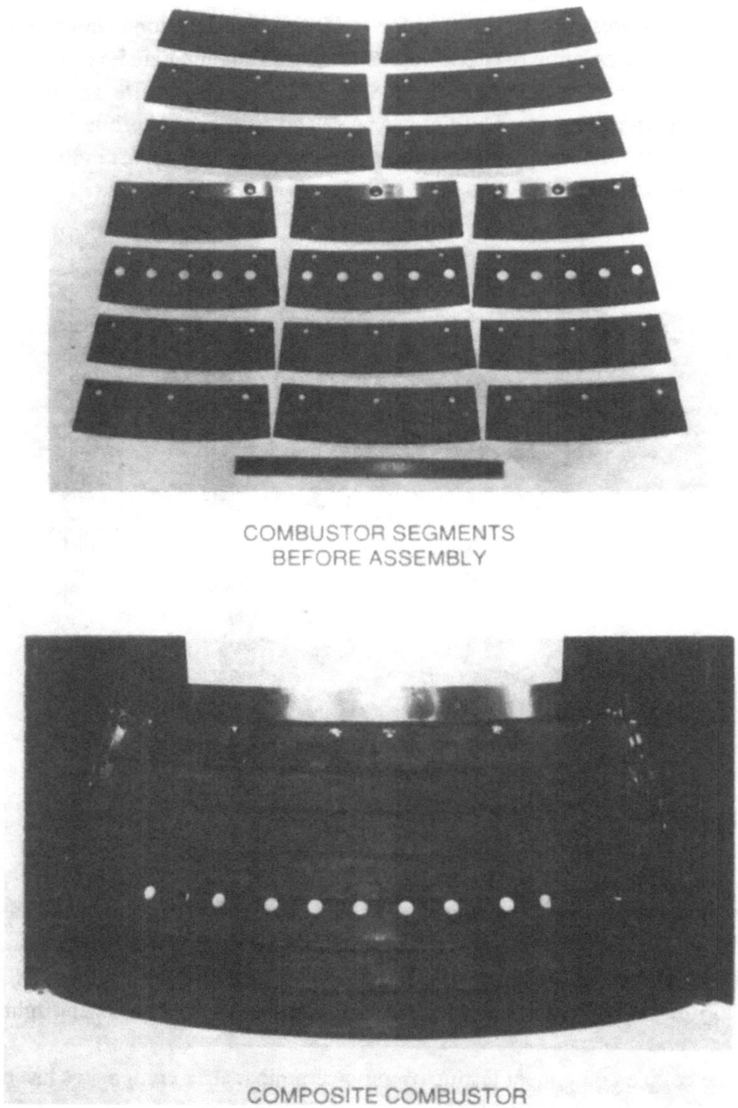

Figure 29. UTRC/Pratt & Whitney LAS matrix/NICALON fiber composite segmented combustor components.

SUMMARY

This overview of the available fiber compositions and properties, and the matrix processing techniques that are applicable to the fabrication of ceramic matrix composites, has shown that there exists a myriad of different systems and different processing procedures. Some generic systems are only applicable to specific fiber/matrix combinations, while others offer an abundance of possible composite compositions. It should be noted that the question of the cost of fabricating ceramic matrix composites has not been addressed in this paper. Some of the approaches outlined may have inherent cost advantages, while others may be prohibitively expensive.

In all of the systems discussed, the ability to create ceramic matrix composites exhibiting graceful failure modes, ie, "toughness," was demonstrated. This has been achieved through the use of either artificial or *in situ* fiber/matrix interface optimization, where the interface has been controlled such that the bonding is strong enough to allow load transfer from the matrix to the fibers under stress, but weak enough so that an advancing matrix crack can be deflected by the fibers. As has been demonstrated, the nature of the fiber/matrix interface must include high temperature stability and resistance to environmental attack. While impressive mechanical properties have been demonstrated, significant concerns must be raised by the damage accumulation mechanisms that occur for these composites and the subsequent interaction with oxidizing environments. The combined effects of thermo-mechanical fatigue and the actual service environment on the ultimate utilization of ceramic matrix composites are yet to be determined.

While some of the approaches to fabricating ceramic matrix composites have engine component test experience, many do not. Even the ones that do have some experience, such as the CVI and glass-ceramic matrix composite approaches, are really still in the early stages of component design, fabrication, and testing. From experience in transitioning new metal alloys into actual engine bill-of-material components, a twenty year time span is not unusual. The unique characteristics of ceramic matrix composites may add further to the time required for these materials to be accepted as reliable, cost effective, high temperature engineering structural components.

ACKNOWLEDGMENTS

The authors would like to Dr. Steven Fishman of the US Office of Naval Research for his sponsorship of the glass-ceramic and CVI SiC composite work at UTRC covered in this paper.

REFERENCES

1. Katz, N.R.: Applications of Nitrogen Ceramics--Gas Turbines: U.S. National Programs, Nitrogen Ceramics, edited by F. L. Riley, Noordhoff Publishing, Netherlands, 1977, 643-646.

2. Persh, J: An Assessment of the Department of Defense Ceramic Materials and Structures Research and Development, Cer. Engr. and Sci. Proc., V 1, 7-8, 1980, 491-494.

3. Crivelli-Visconti, and Cooper, G.A.: Mechanical Properties of a New Carbon Fiber Material, Nature, 221, 1969, 754-755.

4. Sambell, R.A., Bowen, D., and Phillips, D.C.: Carbon Fiber Composites with Ceramic and Glass Matrices, Part 1, Discontinuous Fibers, J. Mat. Sci., 7, 1972, 663-675.

5. Sambell, et al: Carbon Fiber Composites with Ceramic and Glass Matrices, Part 2, Continuous Fibers, *ibid.*, 676-681.

6. Phillips, D.C., Sambell, R.A., and Bowen, D.H.: The Mechanical Properties of Carbon Fiber Reinforced Pyrex, *ibid.*, 1454-1464.

7. Phillips, D.C.: Interfacial Bonding and Toughness of Carbon Fiber Reinforced Glass and Glass-Ceramics, *ibid.*, 9, 1974, 1847-1954.

8. Levitt, S.R.: High Strength Graphite Fiber-LAS, *ibid.*, 8, 1973, 793-806.

9. Brennan, J.J.: Increasing the Impact Strength of Si_3N_4 Through Fibre Reinforcement, Special Ceramics 6, edited by P. Popper, BCRA, McCorquodale Printers, Manchester, England, 1975, 123-134.

10. Matson, L.E., and Hay, R.S.: Stability of a Sapphire/Yttrium Aluminum Garnet Composite System, *ibid*, [10], 7-8, 1989, 764.

11. Foulds, W., LeCostaouec, J.F., Landry, C., and DiPietro, S.: Tough Silicon Nitride Matrix Composites Using Textron Silicon Carbide Monofilaments, *ibid.*, [10], 9-10, 1989, 1083-1099.

12. Holmes, J. W., Kotil, T. and Foulds, W.: High-Temperature Fatigue of SiC Fiber-Reinforced Si_3N_4 Ceramic Composites, Symp. of High-Temperature Composites, Am. Soc. for Comp., 1989, 176-182.

13. Bhatt, R.T.: Mechanical Properties of SiC Fiber-Reinforced Reaction Bonded Si_3N_4 Composites, Mat. Sci. Res., Vol. 20, Tailoring Multiphase and Composite Ceramics, Plenum Press, NY (1986) 675-686.

14. Bhatt, R.T., and Kiser, J.D.: Matrix Density Effects on the Mechanical Properties of SiC Fiber-Reinforced Silicon Nitride Matrix Properties, Ceram. Engr. Sci. Proc. 11 [7-8], 1990, 974-994.

15. Lundberg, R., Pompe, R., and Carlsson, R.: Fibre Reinforced Silicon Nitride Composites, Comp. Sci. and Tech. 37, 1990, 165-176.

16. Yamamura, T., Ishikawa, T., Sato, M., Shibuya, M., Ohtsubo, H., Nagasawa, T., and Okamura, K.: Characteristics of a Ceramic Matrix Composite using a Continuous Si-Ti-C-O Fiber, Ceram. Engr. Sci. Proc., [11]. 9-10, 1990, 1648-1660.

17. Fitzer, E. and Gadow, R.: Fiber Reinforced Composites Via the Sol/Gel Route, Materials Sci. Res., Vol. 20, Plenum Press, NY (1986) 571-608.

18. Fitzer, E.: Fiber-Reinforced Ceramics, Whisker and Fiber Toughened Ceramics Conf. Proc. ASM Intl., 1988, 165-192.

19. Hyde, A.R.: Fibre Reinforced Glass and Ceramic Composites, GEC Journal of Research, V. 6, #1, 1988, 44-49.

20. Mah. T., Yu, Y.F., Hermes, E.E., and Mazdiyazni, K.S.: Ceramic Fiber Reinforced Metal-Organic Precursor Matrix Composites, Chapter 10 in Fiber Reinforced Ceramic Composites, edited by K. S. Mazdiyasni, Noyes Publications, Park Ridge, NJ., 1990.

21. Hurwitz, F.I., Gyekenyesi, J.Z., Conroy, P.J., and Rivera, A.L.: NICALON/Siliconoxycarbide Ceramic Composites, Ceram. Engr. Sci. Proc., 11, [7-8, 1990, 931-946.

22. Strife, J.R., Brennan, J.J., and Prewo, K.M.: Status of Continuous Fiber-Reinforced Ceramic Matrix Composite Processing Technology, *ibid.*, 11,{7-8}, 1990, 871-919.

23. Ichikawa, H., Mitsuno, S., Imai, Y., and Ishikawa, T.: Mechanical and Electrical Properties of SiC Fiber (NICALON) and Their Composites, Proc. of the First Japan Int. SAMPE Symp., 1989, 923-928.

24. Newkirk, M.S., Lesher, H.D., White, D.R., Kennedy, C.R., Urquhart, A.W., and Claar, T.D.: Preparation of Lanxide Ceramic Matrix Composites: Matrix Formation by the Directed Oxidation of Molten Metals, Cer. Engr. Sci. Proc., [8],7-8, 1987, 879-885.

25. Newkirk, M.S., Urquhart, A.W., Zwicker, H.R., and Breval, E.: Formation of Lanxide Ceramic Composite Materials, J. Mat. Sci., [1], 1986, 81-89.

26. Antolin, P.B., Schiroky, G.H., and Andersson, C.A.: Properties of Fiber-Reinforced Alumina Matrix Composites, Cer. Engr. Sci. Proc., [9], 7-8, 1988, 759-766.

27. Fareed, A.S., Sonuparlak, B., Lee, C.T., Fortini, A.J., and Schirocky, G.: Mechanical Properties of 2D NICALON Fiber-Reinforced Lanxide Al_2O_3 and AlN Matrix Composites, *ibid.*, [11], 7-8, 1990, 782-794.

28. Prewo, K.M. and Brennan, J.J.: High Strength Silicon Carbide Fiber-Reinforced Glass Matrix Composites, J. Mat. Sci. 15 (1980) 463-468.

29. Prewo, K.M. and Brennan, J.J.: Silicon Carbide Yarn Reinforced Glass Matrix Composites, J. Mat. Sci. 17 (1982) 1201-1206.

30. Brennan, J.J. and Prewo, K.M.: Silicon Carbide Fiber Reinforced Glass-Ceramic Matrix Composites Exhibiting High Strength and Toughness, J. Mat. Sci. 17 (1982) 2371-2383.

31. Brennan, J.J.: Interfacial Characterization of Glass and Glass-Ceramic Matrix/Nicalon SiC Fiber Composites, Proc. of the Conf. on Tailoring Multiphase and Composite Ceramics, Penn St. Univ. (July 1985). Materials Science Research Vol 20, Plenum Press, New York (1986) 549-560.

32. Minford, E.J. and Prewo, K.M.: Fatigue Behavior of SiC Fiber Reinforced LAS Glass-Ceramic, *ibid.*, 561-570.

33. Prewo, K.M., Brennan, J.J. and Layden, G.K.: Fiber Reinforced Glasses and Glass-Ceramics for High Performance Applications, Am. Cer. Soc. Bull. Vol. 65, No. 2 (Feb. 1986).

34. Prewo, K.M.: Tension and Flexural Strength of Silicon Carbide Fiber-Reinforced Glass-Ceramics, J. Mat. Sci., 21, 1986, 3590-3600.

35. Prewo, K.M.: Fatigue and Stress Rupture of Silicon Carbide Fiber-Reinforced Glass-Ceramics, *ibid.*, 22, 1987, 2695-2701.

36. Brennan, J.J.: Interfacial Chemistry and Bonding in Fiber Reinforced Glass and Glass-Ceramic Matrix Composites, Proc. of the Conf. on Ceramic Microstructures '86: Role of Interfaces, Univ of Calif, Berkeley (July 1986) Materials Science Res Vol. 21, Plenum Press, NY (1987) 387-400.

37. Mah, T., Mendiratta, M.G., Katz, A.P., Ruh, R. and Mazdiyasni, K.S.: Room Temperature Mechanical Behavior of Fiber-Reinforced Ceramic-Matrix Composites, J. Am. Cer. Soc. 68 [1] (1985) C-27-30.

38. Cooper, R.F. and Chyung, K.: Structure and Chemistry of Fibre-Matrix Interfaces in SiC Fibre-Reinforced Glass-Ceramic Composites: An Electron Microscopy Study, J. Mat. Sci. 22 (1987) 3148-3160.

39. Brennan, J.J.: Interfacial Characteristics of Glass-Ceramic Matrix/SiC Fiber Composites, Journal de Physique, Colloque C5, supplement au n° 10, Tome 49, Oct. 1988, 791-809.

40. Michalske, T.A., and Hellman, J.R.: Strength and Toughness of Continuous Alumina-Fiber-Reinforced Glass-Matrix Composites, J. Am. Cer. Soc., 71 [9] 1988, 725-731.

41. Cho, K., Kerans, R.J., and Jepsen, K.A.: Selection, Fabrication, and Failure Behavior of SiC Monofilament-Reinforced Glass Composites, Cer. Engr. Sci. Proc., [9], 7-8, 1988, 815-824.

42. Lankford, J.: Strength of Monolithic and Fiber-Reinforced Glass-Ceramics at High Rates of Loading and Elevated Temperatures, *ibid.*, 843-852.

43. Kim, R.Y., and Katz, A.P.: Mechanical Behavior of Unidirectional SiC/BMAS Ceramic Composites, *ibid.*, 853-860.

44. Murthy, V.S.R., Jie, L., and Lewis, M.H.: Interfacial Microstructure and Crystallization in SiC-Glass Ceramic Composites, *ibid.*, [10], 7-8, 1989, 938-954.

45. Pannhorst, W., Spallek, M., Bruckner, R., Hegeler, H., Reich, C., Grathwohl, G., Meier, B., and Spelmann, D.: Fiber-Reinforced Glasses and Glass-Ceramics Fabricated by a Novel Process, *ibid.*, [11], 7-8, 1990, 947-963.

46. Parlier, M., Ritti, M.H., Stohr, J.F., and Vignesoult, S.: Silicon Carbide Fibre-Reinforced Glass-Ceramic Matrix Composites: A High Temperature Material for High Performance Application, Proc. of ICAS 90, Stockholm, Sweden, Sept. 9-14, 1990.

47. Hegeler, H., and Bruckner, R.: Fibre-Reinforced Glasses: Influence of Thermal Expansion of the Glass Matrix on Strength and Fracture Toughness of the Composites, J. Mat. Sci., [25], 1990, 4836-4846.

48. Phillips, D.C., Park, N., and Lee, R.J.: The Impact Behavior of High Performance, Ceramic Matrix Fibre Composites, Comp. Sci. & Tech., [37], 1990, 249-265.

49. Homeny, J., VanValzah, J.R., and Kelley, M.A.: Interfacial Characterization of Silicon Carbide Fiber/Lithia-Alumina-Silica Glass Matrix Composites, J. Am. Cer. Soc. 73 [7], 1990, 2054-59.

50. Ponthieu, C., Lancin, M.,Thibault-Desseaux, J., and Vignesoult, S.: Microstructure of Interfaces in SiC/Glass Composites of Different Tenacity, Colloque de Physique, C1, Supplement au n°1, Tome 51, Janvier 1990, 1021-1026.

51. Laube, B.L., and Brennan, J. J.: Scanning Auger Electron Spectroscopy of the Fiber/Matrix Interface of SiC Fiber/Silicate Glass Matrix Composites, J. Vac. Sci. Tech. A 81 (3) May/June 1990, 2096-2100.

52. Brennan, J. J.: Glass and Glass-Ceramic Matrix Composites, Chapter 8 in Fiber Reinforced Ceramic Composites, edited by K. S. Mazdiyasni, Noyes Publications, Park Ridge, NJ, 1990.

53. Bonney, L.A., and Cooper, R.F.: Reaction-Layer Interfaces in SiC-Fiber-Reinforced Glass-Ceramics: A High-Resolution Scanning Transmission Electron Microscopy Analysis, J. Am. Cer. Soc., 73 [10] (1990) 2916-21.

54. Prewo, K.M.., Johnson, B., and Starrett, S.: Silicon Carbide Fibre Reinforced Glass-Ceramic Composite Behavior at Elevated Temperature, J. Mat. Sci., 24, 1989, 1373-1379.

55. Naslain, R. and Langlais, F.: CVI-Processing of Ceramic-Ceramic Composite Materials, Conf on Tailoring Multiphase and Composite Ceramics, Penn St. Univ. (July 1985) Plenum Press, NY, 145-164.

56. Lamicq, P.J., Bernhart, G.A., Dauchier, M.M., and Mace, J.G.: SiC/SiC composite Ceramics, Am. Cer. Soc. Bull., 65 [2] 1986, 336-338.

57. Caputo, A.J. and Lackey, W.J.: Fabrication of Fiber-Reinforced Ceramic Composites by Chemical Vapor Infiltration, Cer. Eng. Sci. Proc. 5 [7-8] (1984) 654-67.

58. Caputo, A.J., Lackey, W.J. and Stinton, D.P.: Development of a New, Faster Process for the Fabrication of Ceramic Fiber-Reinforced Ceramic Composites by Chemical Vapor Infiltration, *ibid* 6 [7-8] (1985) 694-706.

59. Stinton, D.P., Caputo, A.J. and Lowden, R.A.: Synthesis of Fiber-Reinforced SiC Composites by Chemical Vapor Infiltration, *ibid*, 347-50.

60. Caputo, A.J., Stinton, D.P., Lowden, R.A. and Besmann, T.M.: Fiber-Reinforced SiC Composites with Improved Mechanical Properties, Am. Cer. Soc. Bull. 66 [2] (1987) 368-72.

61. Stinton, D.P., Besmann, T.M. and Lowden, R.A.: Advanced Ceramics by Chemical Vapor Deposition Techniques, Am. Cer. Soc. Bull. 67 [2] (1988) 350-356.

62. DuPont Company Preliminary Engineering Data, Wilmington, DE, 1989.

63. Brennan, J. J.: Interfacial Studies of Chemical-vapor-infiltrated Ceramic Matrix Composites, Matls. Sci. and Engr., A126, 1990, 203-223.

64. Frety, N., and Boussuge, M.: Relationship Between High-Temperature Development of Fibre-Matrix Interfaces and the Mechanical Behavior of SiC-SiC Composites, Comp. Sci. and Tech., 0266-3538, 1989, 177-189.

65. SEP Literature, Solid Propulsion and Composites Division, France, 1989.

66. DuPont Brochure on Ceramic Matrix Composites, DuPont Advanced Composites, Newark, DE, 1991.

CERAMICS FOR RECIPROCATING ENGINES: AN APPLICATION REVIEW

H. Takao, A. Okada, M. Ando, Y. Akimune and N. Hirosaki
Central Engineering Laboratories, Nissan Motor Co., Ltd.,
1, Natsushima-cho, Yokosuka 237, Japan

ABSTRACT

The application of ceramics to reciprocating engines is reviewed with the aim of predicting near-future applications. The exceptional properties of ceramics enabling them to meet the changing social environment are now clearly known to enhance their further application, and the material cost is shown to influence the maximum amount of materials designed for automobile use. Candidates for further application of ceramics to automobiles are thought to be classified into two categories. One is aimed at substituting such conventional metal parts as the tappet, piston pin and cam lobe, where cost reduction facilitates expanding application. The second category involves use as a key material in helping to ease ecological problems such as decreasing CO_2 emission by decreasing fuel consumption. Extensive work being undertaken to develop energy recovery system as well as to further clarify combustion behaviour in reciprocating engines is indispensable for advancing the application of ceramics to heat insulation components of the combustion chamber, while novel ceramic processing that remarkably reduces production cost is strongly required to generate applications that reflect its light weight and outstanding wear resistance properties. Because of the strong potential of ceramics for use in high-thermal efficiency engines, broad-ranging application of ceramics in the near future will be achieved through close cooperation between diverse fields of research.

INTRODUCTION

Development of reciprocating engines has continued at great strength ever since the invention of the first automobile to use a gasoline engine. Development to produce advanced engines featuring high reliability, optimum power, and minimal fuel consumption has subsequently led to the use of wide-ranging materials. In this regard, the usage of ceramics which had previously been limited to spark plug insulators expanded to include sensor materials such as the temperature sensor, oxygen sensor, and sensor utilizing the piezo-electric property [1]. In particular, the combination of a cordierite catalyst honeycomb and a zirconia oxygen sensor has played an important role in the Japanese emission regulations formulated in 1978. In addition, silicon nitride was used in such engine components as the turbocharger rotor in the 1980s.

In the present paper, we describe the current status of ceramics application, requirements for new materials for application to automobiles, and trends in automotive engine development for the purpose of predicting the near-future ceramics application.

CERAMICS IN AUTOMOBILES

The major ceramic parts presently used in automobiles are listed in Table 1. Alumina ceramics for spark plug insulators are used because of their excellent durability for ensuring electric insulation at high temperatures. Requirements for prompt detection of oxygen partial pressure at high temperatures led to the use of ceramics for oxygen sensors. PZT ceramics, having an excellent piezo-electric property, have been applied to knock sensors, road clearance sensors and rear obstacle sensors. Recent application of silicon nitride ceramics to high-performance engines can be recognized as the by-product for the subsequent extensive work aimed toward producing a ceramic gas turbine.

Oxide ceramics

Application of oxide ceramics has been based on the traditional technologies developed for clay-based ceramics, which have made it possible to apply them to automobile components characterized by high reliability and reasonable cost.

Spark plug insulator: Although the insulator was previously made of traditional porcelain since J. J. E. Lenoer invented the spark plug in 1860, a considerable change in materials with the use of high-content alumina ceramics occurred following the wide-spread use of tetraethyl lead addition to gasoline [2, 3]. Reflected by the recent social requirements, a spark plug involving

TABLE 1. Major ceramic parts used in automobiles

Material	Applications
Alumina	Spark plug, mechanical seal, IC substrate
Cordierite	Catalyst honeycomb
Zirconia	Oxygen sensor
Titania	Oxygen sensor
PZT	Knock sensor, road clearance sensor, rear obstacle sensor
NTC thermistor	Temperature sensor for cooling water and intake air, Fuel level sensor
PTC thermistor	Inlet gas heater, steering wheel warmer
Al_2TiO_5	Port liner
Silicon nitride	Glow plug, swirl chamber, rocker arm pad, turbocharger rotor

a ceramic resistor was developed to suppress electric noise by reducing electromagnetic wave generation [4, 5].

Oxygen sensor: Air-fuel ratio sensors have been used to control the air-fuel ratio close to 14.7 corresponding to the stoichiometric ratio for complete combustion [6-8]. A typical tube-shaped sensor, which is made of Y_2O_3-partially stabilized zirconia, yields a sudden change in electromotive force near the stoichiometric air-fuel ratio. This is because the electromotive force due to the oxygen-ion conductivity depends on the oxygen partial pressure ratio of the air on the inside to the exhaust gas on the outside. Titania sensors, utilizing an electric resistance dependence on the oxygen partial pressure in the atmosphere, are characterized by less degradation for the tetraethyl lead addition to gasoline [9-12]. In addition, a limiting current-type zirconia sensor has also been used to control a leaner burning combustion engine [13].

Honeycomb catalyst substrate: The three-way catalyst system assisted by feedback control using an oxygen sensor has been employed for cleaning the exhausted gas emissions since NO_x emission was regulated in the 1970s [1]. A cordierite honeycomb substrate [14, 15] has been employed because of its high-temperature durability at the maximum temperature of 900°C, good thermal shock resistance, and a sufficiently high strength to endure vibration common during driving.

Knock sensor: The knock phenomenon is caused by the sudden combustion of the air-fuel mixture before ignition so that a strong shock wave is generated in the cylinder. The frequency of the shock wave is generally in the 5 to 10 kHz range. Suppression of the knock phenomenon can be achieved, for example, through delaying ignition timing or by decreasing the compression ratio. However, these modifications run counter to the principle of fuel economy. Essential then is developing fine engine control to maintain a high compression ratio without generating the knock phenomenon. Knock sensors [16-19], which detect high-frequency vibration, are capable of automatically maintaining optimum engine combustion. A piezo-electric knock sensor of the resonance type exhibits good capability of easily detecting light knocking because the frequency of the piezo-electric element is adjusted to properly match the knock vibration frequency of the engine.

NTC thermistor: Negative temperature coefficient (NTC) thermistors, which have generally been made from mixed oxides of MnO, CoO and NiO, have commonly been used for temperature detection of cooling water and intake air. A fuel level sensor [20] using an NTC thermistor utilizes self-heating characteristics derived from the heat radiation coefficient dependence on the surrounding medium of air or fuel. High-temperature thermistors made of zirconia and spinel have been used for detecting catalyst honeycomb overheating [1].

PTC thermistor: Using the characteristics of a sudden increase in electric resistance above the critical temperature, positive temperature coefficient (PTC) thermistors have been applied to constant temperature heating such as the intake air heater, automatic choking heater, evaporation heater for diesel engines, and steering wheel warmer for drivers during the winter season [1, 21].

Non-oxide ceramics

Recent advances in ceramics application have been achieved through the development of silicon nitride. The status of these applications is summarized in Table 2. It should be noted that these ceramic parts have been supplied by the major traditional ceramic companies of Japan principally in the 1980s. In addition, these companies have been aggressive in working to develop new technologies, and are fully capable of utilizing redeemed equipment to produce non-oxide ceramics.

Glow plug: The ceramic glow plug, which had the advantage of quick starting diesel engines, was the first silicon nitride ceramic component applied to automobiles [22]. The element was produced by hot pressing silicon nitride powder around a tungsten coil. The preheating time for starting an engine using ceramic glow plugs could be considerably shortened over that of traditional metallic glow plugs which required about 20 to 30 seconds. Advantages of using

TABLE 2. Silicon nitride application to reciprocating engines

Ceramic part	Automobile company	Ceramics company	Year
Glow plug	Isuzu	Kyocera	1981
	Mitsubishi	Kyocera	1983
	Nissan	NGK Spark Plugs	1985
Swirl chamber	Isuzu	Kyocera	1983
	Toyota	Toyota	1984
	Matsuda	NGK Insulators	1986
Rocker arm pad	Mitsubishi	NGK Insulators	1984
	Nissan	NGK Insulators	1987
	Nissan	NGK Spark Plugs	1988
Turbocharger rotor	Nissan	NGK Spark Plugs	1985
	Nissan	NGK Insulators	1985
	Isuzu	Kyocera	1989
	Toyota	Toyota	1989
	Toyota	Kyocera	1990

ceramic glow plugs have been reported to be reduction in white smoke and mechanical noise because they could be continually heated after the engine was started.

Swirl Chamber: The ceramic swirl chamber has been reported to have several advantages. In 1983, Isuzu Motor developed a silicon nitride lower chamber to improve combustion behaviour during low-speed, low-load driving [23]. The advantages were indicated to include reducing the mechanical noise level during idling, bettering the starting characteristics, and reducing HC emission. A silicon nitride lower chamber developed by Toyota Motor has been demonstrated to have the advantage of augmenting the power performance of turbocharged diesel engines [24]. An all-ceramic swirl chamber developed by Matsuda Motor has been reported to have the advantage of reducing particulate emission from diesel engines [25].

Rocker arm pad: A ceramic rocker arm pad developed principally for taxi-cab use has advantages in being maintenance-free and in performing well during high-speed driving due to its light weight [26]. Evaluation of the wear resistance indicates that wear of ceramic pads is much smaller than that of traditional sintered ferrous alloy due to the mirror-like finish produced on the wear surface of ceramics without them incurring adhesive wear.

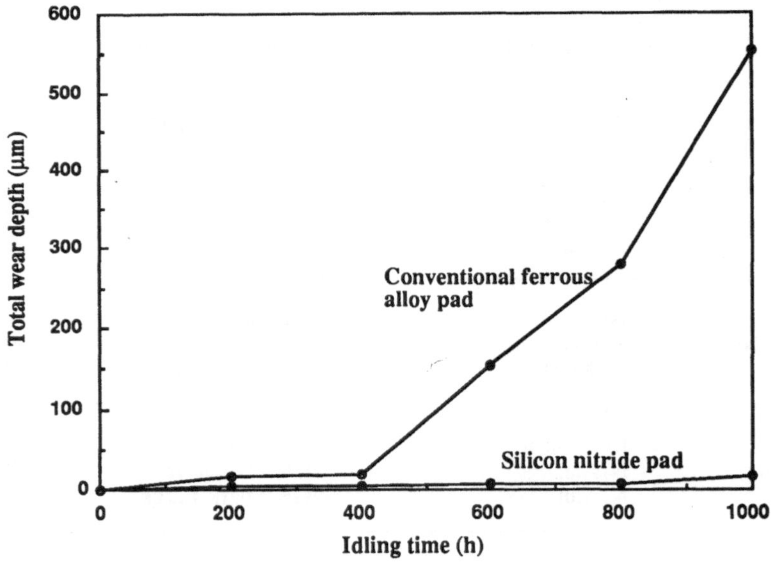

Figure 1. Total wear depth of a cam and a rocker arm pad during idling. Comparison is made for a conventional ferrous alloy pad and a silicon nitride pad operated with a mating cam of chilled cast iron [26].

The wear resistance of a ceramic rocker arm pad tested with a mating cam of chilled cast iron is presented in Fig. 1 [26]. The total wear depth of the mating cam and the rocker arm pad during idling indicates that the silicon nitride pad results in a much smaller total wear depth than does the conventional sintered ferrous alloy pad. Replacement of sintered alloy pads with ceramic ones has improved valve train performance at high speed because the inertial forces generated in the rocker arm were significantly reduced. The load reduction on the rocker arm in the high-speed range has been reported to result in a raising of the valve bounce speed by 200 rpm [27].

Turbocharger rotor: The major purpose behind using ceramics for the turbocharger was to reduce the moment of inertia, thereby improving acceleration response [28]. Reductions in the moment of inertia by employing ceramics relative to that of GMR 235 Nickel alloy as a rotor and as an assembly were 45% and 34% respectively.

The step response characteristics of metal and ceramic rotors are given in Fig. 2 [28]. From comparison of the time required to reach 10 x 10^4 rpm, lowering the inertia moment definitely improved the acceleration response. The first ceramic turbocharger was used in a 2-litre twin cam, 24-valve, in-line six-cylinder engine. In reporting transient performance of the engine, boost pressure rising time to 170 kPa was 14 percent shorter for a ceramic rotor than

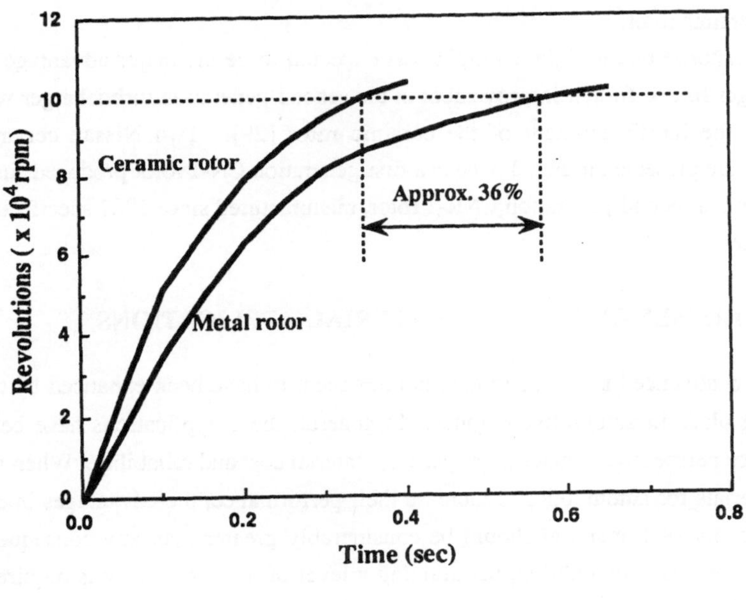

Figure 2. Step response characteristics of metal and ceramic rotors [28].

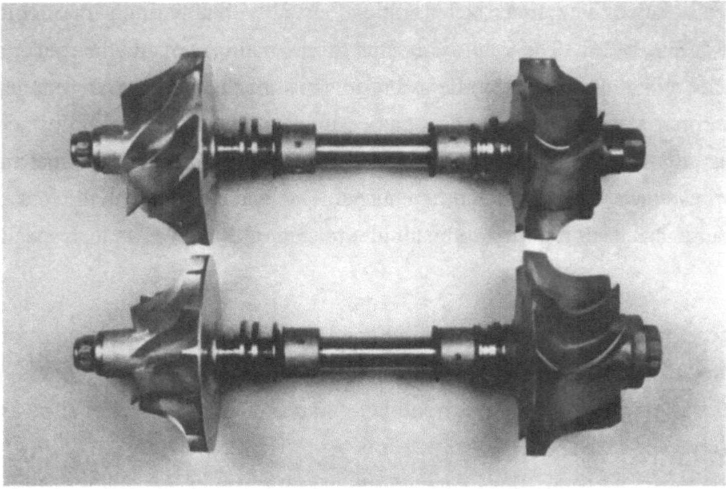

Figure 3. Two ceramic turbocharger rotors. The lower one is a first-generation CN-1 turbocharger rotor that has been produced since 1985. The upper one, which is a second-generation CNR-1 turbocharger rotor produced since 1987, has further improved response through a decrease in the rotor weight.

for a metallic one when both were operated under full acceleration under road load conditions in the 4th gear from 40 km/h, and the acceleration time from 40 km/h to 70 km/h was 10 percent shorter using the ceramic rotor.

Since quick response due to lighter weight was expected to be the major advantage of ceramic turbocharger rotors, further improvement in the second-generation turbocharger was made to decrease the inertia moment of the ceramic rotor [29]. Two Nissan ceramic turbocharger rotors are presented in Fig. 3: One is a first-generation CN-1 rotor produced since 1985, and the other is a second-generation CNR-1 rotor, manufactured since 1987 specifically to improve response.

REQUIREMENTS FOR NEW MATERIAL APPLICATIONS

Applications of new advanced materials to automobiles seem to have been enhanced by the innovations taking place in automotive engines. In general, these applications have been evaluated from three perspectives: material properties, material cost and reliability. When we use advanced materials for automobiles to increase their performance, the advantages in car performance using any such material should be considerably greater than any consequent increase in production cost. In addition, maintaining a level of high reliability is required. From this point of view, ceramics are materials generally difficult to apply widely to automobiles because of their high production cost and low reliability.

Automobiles are virtually very cheap products when viewed from their price against their weight. Therefore, the material cost is averaged as low as possible with the use of expensive materials in automobiles being quite limited. Figure 4 shows plots of material weight used in automobiles against their prices [30], where one dollar equals 135 yen. As is evident, the higher is the price of a material, the less it is used in automobiles. Except for platinum, the material weight can be directly related to the material price:

$$W \propto P^{-2} \tag{1}$$

where W is the averaged material weight in an automobile and P is the averaged material price. The upper limit of the material weight can be estimated from the material cost. It should be noted that the weight for platinum is much higher than that expected from the relationship because precious metals such as platinum, palladium and rhodium are used as catalysts to ensure that emissions are clean under present regulations. Thus, automobile companies are concentrating considerable effort on decreasing the amount of precious metals used for the catalyst systems to bring about a closer material-to-cost relationship.

Considering the application of silicon nitride, the maximum amount to be used in an automobile is supposed to be around 100 to 200 g provided that the price of high-purity silicon

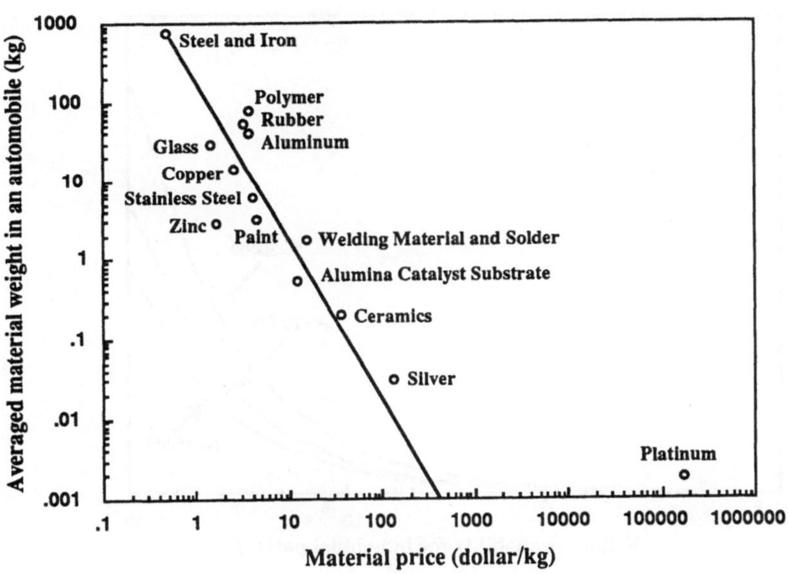

Figure 4. Plots of material weight used in automobiles against the price [30].

nitride powder is \$50/kg. The weight of ceramic components commercially used in an automobile is lighter than the limit weight. For example, a turbocharger rotor is in the range of 70 to 100 g, a rocker arm pad is about 4.5 g and a glow plug is about 1.5 g. Realizing significant application to large components such as the exhaust valve, piston, cam lobe and cylinder head plate, however, will require much greater effort in cost reduction because the estimated weight of the components is much greater than the limiting weight estimated from today's silicon nitride powder price.

Reliability exceeding 10^{-6} is generally required for individual automobile parts. This is because an automobile comprises some 15,000 to 20,000 parts [1]. The failure probability of the system, F, can be estimated as:

$$F = 1 - (1 - f)^N \tag{2}$$

where f is the failure probability of individual parts and N is the number of parts. The calculated results for the relationship between the failure probabilities of a system and an individual part for the cases of N = 15,000, 20,000, and 30,000 are given in Fig. 5. Maintaining a low-failure probability of the system requires a very-low-failure probability for a part smaller than 10^{-6}, and the use of the additional components in automobiles will result in a

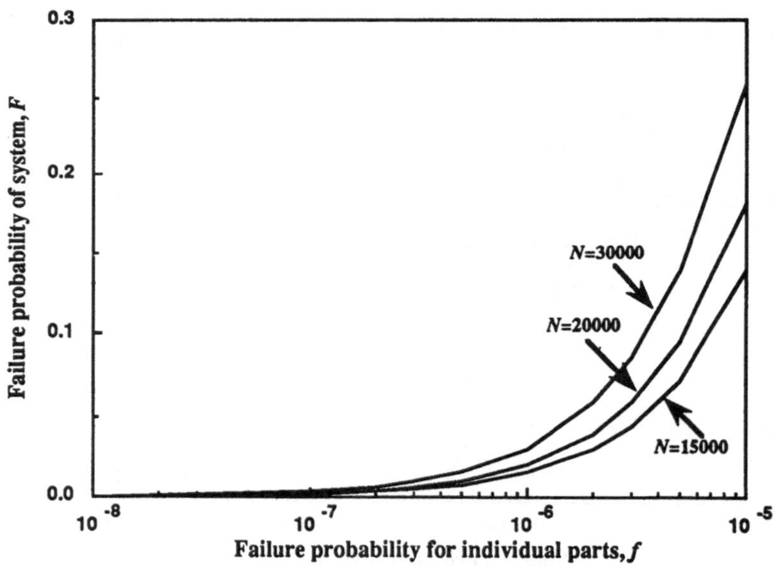

Figure 5. System reliability, F, against part reliability, f, as a function of the number of parts, N.

requirement for an ever much lower failure probability of the part because increasing the number of parts leads to degradation of system reliability.

ADVANCES IN RECIPROCATING ENGINES

Market needs reflecting the social condition virtually govern the direction of automotive engine development because the direction is chosen to meet the market needs to ensure that the huge investment in manufacturing equipment is easily recovered. Consequently, a change in the prevailing social conditions inevitably changes the development direction. For example, the approach to engine development, which had previously been focused toward higher performance, increased fuel economy and higher reliability, witnessed a change to decreasing emissions when air pollution was first realized as being a serious problem in the 1970s. Such major alternations have continued to the present. In order to solve the recent issue of global warming resulting in part from the emission of CO_2, a study for improving the thermal efficiency of engines has been started [31].

The market needs governing the direction of improving the reciprocating engine can roughly be classified into customer and social needs. The demand by people using cars represents the customer needs, and that by the society possessing cars represents the social needs. Customer needs comprise reduction in fuel consumption through improving thermal efficiency and reducing friction loss, and improving engine performance yielding greater power, quicker response and longer-term durability. Social needs include producing cleaner emissions of the exhaust gases, a quieter car, and more effective use of resources taking into consideration the recycling of used materials. It should be noted, however, that the social needs become more important as the number of automobiles on the road increases.

In this light, it is interesting to follow the chronological sequence of market needs. Specifically, the pursuit for higher performance and the better fuel economy in gasoline engines led to an increase in the compression ratio although the maximum ratio was limited by the occurrence of the knock phenomenon. The development in the reciprocating engines offering a high compression ratio was accelerated after gasoline containing tetraethyl lead as an anti-knocking agent was employed in 1923. Since the addition of tetraethyl lead was found to degrade spark plug insulators, development of insulator materials was accelerated. As a result, high-purity alumina ceramics were chosen for the insulator material although they had already been invented in 1932. The study for sintering at lower temperatures without degrading electric properties was conducted by adding SiO_2, CaO and MgO because the high-purity alumina ceramics required a very high temperature of 1800°C for densification [2, 3]. This might represent the first case in which customer needs enhanced the application of ceramics to the automobile. It is noteworthy in this regard that the issues of production cost and quality control were assumed to be less serious problems than those of today because the price of automobiles was relatively high.

In seeking to resolve air pollution problems in the 1970s, a range of methods for decreasing emissions from automobiles was investigated to develop a system that would meet the new exhaust gas emission regulations. This led to the subsequent development and implementation of the catalyst system. Since tetraethyl lead was revealed to decrease the activity of the catalyst [32, 33], the compound was eliminated from gasoline. A knock sensor, detecting shock waves generated during light knocking, was employed in its place to achieve efficient combustion to sufficiently adjust ignition timing [1, 16-19].

The emission regulations in the 1970s essentially facilitated the engine improvement initiated by the social needs. As a decrease in the three kinds of the emission species, namely, NO_x, HC and CO was required, a three-way catalyst system combined with the restrictive control of oxygen partial pressure in the exhaust gases was proposed [5-9]. An oxygen sensor using such ceramics as Y_2O_3 partially stabilized zirconia was adopted for detecting the oxygen partial pressure in the high-temperature exhaust gases to maintain the air-fuel ratio in the narrowest possible range. Cordierite ceramics were used for the honeycomb substrates for the catalyst due to their excellent durability at high temperatures [1].

It is thus clear from the trend in advances in reciprocating engines that the change in social conditions accelerates the increased usage of ceramics in automobile engines to meet requirements for finer engine control through utilizing the exceptional properties of ceramics compared with metals and polymers. Because the more advanced engine to be built in the next generation is currently being developed to meet the needs reflected the more recent and prospective social conditions, any further application of ceramics to engines should take immediate and consequential account of ecological issues.

CERAMICS FOR THE NEXT GENERATION

No predictions can readily be made as to when the next generation of application ceramics will be realized. Table 3 showing the four stages in the application of ceramics to automobile, however, indicates that each new stage of application appears at a rate of once a decade. It is noteworthy that the fundamental ceramic technologies were already developed prior to the each stage of application although the changes in the social condition at the time directly enhanced the application. For example, the first stage application of alumina to spark plug insulators in the 1950s was enhanced by the extended use of tetraethyl lead addition to gasoline although the production of high-alumina content insulators had been possible since 1932. Alumina as the second stage application to the mechanical seal in the 1960s would have been enhanced by the application to a thread guide and a ceramic tool. The third stage application of oxygen sensors and catalyst substrates was based on the development in sintering technologies of special oxide ceramics in the 1960s. The fourth stage application of silicon nitride in the 1980s, which was led by customer needs for amenities, was based on the extensive work done on non-oxide ceramics and fracture mechanics in the 1970s principally for gas turbine use. These trends

TABLE 3. History of automobile ceramics application

Years	Major ceramics application
1950s	High-alumina content spark plug insulator
1960s	Alumina mechanical seal
1970s	Cordierite and zirconia
1980s	Silicon nitride

seem to suggest the next stage that will appear in the 1990s. Specifically, ceramic composites might be used as the major ceramic materials in the next generation because they were developed in the 1980s, if they can be made to provide a solution to pressing ecological problems.

The typical reciprocating engine components of ceramics being considered for development are given in Table 4. The application of ceramics is generally thought to be advantageous in its high-temperature durability, wear resistance and light weight. Applications utilizing high-temperature durability are considered mainly to the combustion chamber of the diesel engine involving such parts as the piston cap, cylinder head plate, exhaust valve and valve sheet, although the heat insulation without an energy recovery system results in increasing the thermal energy of the exhaust gas [34]. The focus of development is also on the application of the outstanding wear resistance properties to such valve train components as the valve guide, valve spring retainer, tappet and cam lobe, and on lightweight applications such as the piston, piston pin and exhaust valve for the gasoline engine.

In order to decrease the CO_2 emission, automobile makers are making tremendous strides to improve engine fuel consumption. The diesel engine is still a hopeful candidate as the future engine due to its inherently good fuel economy, even though the difficulties in reducing the other emissions such as NO_x and carbon particulates as well as quieting mechanical noise remain. Therefore, a study will be emphasized to further improve fuel consumption and to produce cleaner emissions. Heat insulation of the combustion chamber in the diesel engine achieved by using ceramics having low thermal conductivity is capable of realizing a high thermal efficient engine as already shown by R. Kamo *et al.*[35, 36] if an energy recovery system such as a turbocompound from the high-temperature exhaust gas can be established.

The particulate trap, of course, is a major practical candidate for the cleaner emissions produced by diesel engines [37, 38]. A ceramic port liner [39] and a thin-wall catalyst substrate may be candidates for decreasing emissions from gasoline engines before warming. The key to replacing a component with a ceramic part for enhanced wear resistance and lightweight application would be cost reduction unless a great advantage that clearly compensated for the cost increase were to be discovered.

TABLE 4. Typical reciprocating engine components of ceramics being considered for development

Ceramics advantages	Possible application
High-temperature durability	Piston cap, cylinder head plate, exhaust valve, Valve sheet
Wear resistance	Valve guide, valve spring retainer, tappet, cam lobe
Light weight	Piston, piston pin

It is noteworthy that further investigations undertaken to expand ceramics application should be focused on the following two fields. The first is to ensure the potential of ceramics application to the combustion chamber of the diesel engine, with a special focus on ameliorating ecological problems. The second is development of novel ceramics processing that remarkably reduces production cost. Production cost reduction remains a very important approach to expanding the practical use of ceramics because advantages in applying ceramics to wear-resistant parts as well as lightweight parts have already been confirmed. However, application of ceramics to the combustion chamber, especially of direct-injection diesel engines, requires extensive work on an energy recovery system and on efficient combustion including lowering emissions. It also requires a new processing technology that remarkably lowers the production cost because ceramic components for heat insulation are generally large in size.

CONCLUDING REMARKS

The research program on the ceramic gas turbine started from the early 1970s [40] has been well known to have accelerated the application of structural ceramics to reciprocating engines. The program was aimed at confirming the high thermal efficiency of the gas turbine engine running at high temperatures around 1350°C through the use of high-performance ceramics such as silicon nitride and silicon carbide. Sintering aids for non-oxide ceramics, a process for forming complex shapes, and a mechanism for toughening ceramics have been studied during the decades of research. A new concept for advanced diesel engines has also been proposed. An adiabatic diesel engine capable of eliminating the need for a cooling system has been the focus of increasing thermal efficiency by producing a range of parts with ceramics having low thermal conductivity, although developing a system capable of recovering energy such as a turbocompound from the high-temperature exhaust gas has continued to be important for achieving high thermal efficiency [35, 36].

In the early 1980s, several ceramic diesel engine demonstrations were given in Japan [41], and ceramic engine components using silicon nitride subsequently began to be used. The initial component was the ceramic glow plug introduced in 1981, followed by the swirl chamber, rocker arm pad and turbocharger rotor. The purpose of these applications, however, has been different from that for gas turbine and adiabatic diesel engines. The advantages of using ceramics at the time were not only high-temperature durability but also achieving significant wear resistance and lighter weight. Furthermore, these applications were limited to the practical substitution of engine components, and no engines with high thermal efficiency were produced and marketed using ceramics. From the viewpoint of thermal efficiency, however, improvement in high-temperature engines is preferable such that the use of ceramic materials will prove to be indispensable to more advanced engines as a result of their high-temperature durability.

Several key points should be emphasized to enhance the application of ceramics to automobile engines. Ascertaining the advantages in using ceramics especially on ecological problems is most important for generating any new application. Development of a novel ceramics processing that can be done at a lower production cost is indispensable for expanding the practical use of ceramics. Regarding low reliability in ceramics which had generally been thought to be one of the major barriers to development, this problem is becoming more manageable as a result of the recent fracture mechanics-based work promoted to enable ceramics to be more fully applied to automobile engines. In addition, we must emphasize again that the enhanced application of ceramics in the near future can be achieved through the extremely close cooperation fostered among all fields of research from the academic to the industrial, all related engineers and researchers, and companies manufacturing and employing ceramics.

Acknowledgments
The authors thank Drs. Y. Nakajima, Y. Matsui, S. Kimura and H. Kuroda for their helpful comments and discussions, and NGK Spark Plugs Inc. and NGK insulators Inc. for their useful information on ceramics application.

REFERENCES

1. Taguchi, M., Application of high-technology ceramics in Japanese automobiles. Adv. Ceram. Mater., 1987, **2**, 754-62.

2. Owens, J. S., Hinton, J. W., Insley, R. H. and Poland, M. E., Development of ceramic insulators for spark plugs. Amer. Ceram. Soc. Bull., 1977, **56**, 437-9, 443.

3. Nishio, K., Spark plugs. Serammikusu (*in Japanese*), 1982, **17**, 15-19.

4. Nishio, K., Effect of resistor spark plug on noise suppression. Nainen-Kikan (*in Japanese*), 1973, **12**, [10] 31-40 .

5. Nishio, K., Trend in resistor spark plugs. J. Soc. Automotive Engineers Jpn. (*in Japanese*), 1982, **36**, 1082-87.

6. Fleming, W. J., Howarth, D. S. and Eddy, D. S., Sensor for on-vehicle detection of engine exhaust gas composition. SAE Paper, 1973, No. 730575.

7. Dueker, H., Friese, K. H. and Haecker, W. D., Ceramic aspect of the Bosch Lambda-sensor, SAE Paper, 1974, No. 750223.

8. Hamann, E., Manger, H. and Steinke, L., Lamda-sensor with Y_2O_3-stabilized ZrO_2-ceramic for application in automobile emission control system. SAE Paper, 1977, No. 770401.

9. Tien, T. Y., Stadler, H. L., Gibbons, E. F. and Zacmanidis, P. J., TiO_2 as an air-to-fuel ratio sensor for automobile exhausts. Amer. Ceram. Soc. Bull., 1975, **54** 280-82, 85.

10. Esper, M. J., Logothetis, E. M. and Chu, J. C., Titania exhaust gas sensor for automobile applications. SAE Paper, 1979, No. 790140.

11. Takami, A., Matsuura, T., Sekita, T., Okawa, T. and Watanabe, Y., Progress in lead tolerant titania exhaust gas oxygen sensors. SAE Paper, 1985, No. 850381.

12. Takami, A., Matsuura, T., Miyata, S., Furusaki, K. and Watanabe, Y., Effort of precious metal catalyst on TiO_2 thick film HEGO sensor with multi-layer alumina substrates. SAE Paper, 1987, No. 870290.

13. Kamo, T., Chujo, Y., Akatsuka, A., Nakano, J. and Suzuki, M., Lean mixture sensor. SAE Paper, 1985, No. 850380.

14. Schwochert, H. W., Performance of a catalytic converter on nonleaded fuel. SAE Paper, 1969, No. 690503.

15. Howitt, J. S., Thin wall ceramics as monolithic catalyst supports. SAE Paper, 1980, No. 800082.

16. Arrigoni, V., Calvi, G. F., Gaetani, B., Giavazzi, F. and Zanoni, G. F., Recent advances in the detection of knock in S.I. engines. SAE Paper, 1978, No. 780153.

17. Kraus, B. J., Godici, P. E. and King, W. H., Reduction of octane requirement by knock sensor spark retard system. SAE Paper, 1978, No. 780155.

18. Oota, J. and Kamo, T., Knock sensor. Serammikusu (*in Japanese*), 1982, **17** 30-33.

19. Ueda, A., Knock control system. J. Soc. Automotive Engineers Jpn. (*in Japanese*), 1982, **36**, 1074-81.

20. Ishikawa, K. and Kino, K., in Fine Ceramic Handbook (*in Japanese*), ed. Hamano, K., Asakura Shoten, Tokyo, 1984, pp.950.

21. Oki, M., Nara, A. and Hori, M., PTC heater for automobiles. Erekutoroniku Seramikusu (*in Japanese*), 1988, 67-71.

22. Kawamura, H. and Yamamoto, S., Improvement of diesel engine startability by ceramic glow plug start system. SAE Paper, 1983, No. 830580.

23. Matsuoka, H., Kawamura, H. and Toeda, S., Development of ceramic precombustion chamber for the automobile diesel engines. SAE Paper, 1984, No. 840426.

24. Kamiya, S., Murachi, M., Kawamoto, H., Sato, S., Kawakami, S. and Suzuki, S., Silicon nitride swirl chamber for high power turbocharged diesel engines. SAE Paper, 1985, No. 850523.

25. Sakurai, S., and Matsuoka, T., Development of low particulate engine with ceramic swirl chamber. SAE Paper, 1986, No. 861407.

26. Tanimoto, I., Kano, M. and Abe, M., Trend in wear-resistant materials for valve train. J. Soc. Automotive Engineers Jpn. (*in Japanese*), 1988, **42** 711-18.

27. Ogawa, Y., Machida, M., Miyamura, N., Tashiro, K. and Sugano, M., Ceramic rocker arm insert for internal combustion engines. SAE Paper, 1986, No. 860397.

28. Katayama, K., Watanabe, T., Matoba, K. and Katoh, N., Development of Nissan high response ceramic turbocharger rotor. SAE Paper, 1986, No. 861128.

29. Matsuo, I. and Nishiguchi, F., The development of second generation ceramic turbocharger. SAE Paper, 1988, No. 880703.

30. New materials technologies in Nissan. Nissan Koho Shiryo (in Japanese), Nissan Motor Co., Ltd., Tokyo, October 1985.

31. Kimura, S., Matsui, Y. and Ito, T., Effects of combustion chamber insulation on the heat rejection and thermal efficiency of diesel engines. To be presented in 1992 SAE International Congress.

32. Gagliardi, J. C. and Ghannam, F. E., Effects of tetraethyl lead concentration on exhaust emissions in customer type vehicle operation. SAE Paper, 1969, No. 690015.

33. Weaver, E. E., Effects of tetraethyl lead on catalyst life and efficiency in customer type vehicle operation. SAE Paper, 1969, No. 690016.

34. Kosuge, H., Ito, T. and Ishii, M., Effects of insulation of combustion chamber on performance of a diesel engine. J. Soc. Automotive Engineers Jpn. (*in Japanese*), 1986, **33** 25-32.

35. Kamo, R. and Bryzik, W., Adiabatic turbocompound engine performance prediction. SAE Paper, 1978, No. 780068.

36. Bryzik, W. and Kamo, R., TACOM/Cummins adiabatic engine programs. SAE Paper, 1983, No. 830314.

37. Howitt, J. S. and Montierth, M. R., Cellular ceramic diesel particulate filter. SAE Paper, 1981, No. 810114.

38. Wade, W. R., White, J. E. and Florek, J. J., Diesel particulate trap regeneration techniques. SAE Paper, 1981, No. 810118.

39. Hensler, P., Kirchdorffer, G. and Jaksch, A., The Porsche 944 turbo engine. MZT Motertechnische Zeitschrift, 1985, **46** 39-42, 45-47.

40. McLean, A. F., Ceramics in small vehicular gas turbines. in Ceramics for High-Performance Applications, ed. Burke, J. J., Gorum, A. E. and Katz, R. N., Brook Hill, 1974, pp. 9-36.

41. Samejima, Y., Kyocera Research Laboratory. Gas Turbine Soc. J. (*in Japanese*), 1988, **15** 129-31.

MATERIAL AND LUBRICANT RELATIONSHIPS IN THE TRIBOLOGY OF INTERNAL COMBUSTION ENGINES*

C. S. Yust
Metals and Ceramics Division
Oak Ridge National Laboratory
Oak Ridge, TN 37831-6063 USA

ABSTRACT

Ceramics are being introduced into production engines for both wear resistance and inertial response, as well as for thermal barrier and elevated temperature capability. The selection of both materials and lubricants for future generations of engines will be influenced by the demands for efficient, environmentally sound, and minimally serviced but long-lived operation. The long-term tribological response of the selected materials is a significant issue which does not generally receive sufficient attention. An experimental approach to the evaluation of the long term-wear response of engine candidate ceramics, based on an earlier wear mode transition diagram, is discussed in this paper. Data are presented for initial studies of a silicon carbide whisker-silicon nitride composite.

INTRODUCTION

The internal combustion (IC) gasoline engine has evolved in the past century from simple beginnings to the sophisticated, complex, highly controlled entity it is today. Fundamentally the engine is very much unchanged, continuing to be a device lubricated primarily by petroleum-based liquids, driven by the combustion of gasoline and air mixtures within an iron-alloy confinement. The system incorporates a wide variety of

*Research sponsored by the U.S. Department of Energy, Assistant Secretary for Conservation and Renewable Energy, Office of Transportation Materials, Tribology Program, under contract DE-AC05-84OR21400 with Martin Marietta Energy Systems, Inc.

sliding/rolling tribocontacts, as, for example, the piston ring-cylinder wall, cams, valve guides, bearings, wrist pins, and oil pump. The refinement of materials and lubricants that has led to the very successful configurations of those tribocontacts in the modern engine is the result of a lengthy evolution. Significant changes in the both the social and physical environment in which the internal combustion engine operates now dictate important modifications of the engine.

The impending changes in internal combustion engines are the response to two related, compelling challenges; development of an environmentally acceptable vehicle, and the reduction of energy consumption. Environmental acceptability will require lower levels of pollution from the fuels and lubricants used, as well as from the system exhaust. Alternative fuels may be introduced as an anti-pollution measure, serving to preserve petroleum resources, but also introducing new corrosion and lubrication issues. Lubricant viscosity may be reduced in the interest of lower engine friction losses, resulting in thinner films at critical contacts. Lowered heat rejection, consequently higher operating temperatures demanding more heat and corrosion resistant system materials, will be an element in future designs. Lower system mass, smaller inertial forces, and greater wear resistance are also features desired of coming generations of engines. In sum, while the necessity for significant modifications in materials and lubricants in engines is clear, the selection of the optimal material and lubricant systems will be the subject of extended discussions, some of which constitute the proceedings of this meeting. This paper discusses some of the tribological features of the internal combustion reciprocating engine and some issues influencing the future selection of materials and lubricants for engines.

ACCOMPLISHMENTS TO DATE

Materials

The initial concepts for advanced IC engines included substantial increases in cylinder operating temperatures. As a consequence, ceramics became prime candidate materials [1]. In addition to high-temperature capability, ceramics were perceived to be chemically inert and highly wear resistant, even without lubrication, because of their hardness.

Unlubricated sliding wear experiments soon demonstrated the fallacy of this belief [2]. More recent results of lubricated sliding wear tests of ceramics simulating the reciprocating motion in a piston ring-cylinder wall contact have shown that friction and wear performance comparable to that in current diesel engines can be achieved [3]. In addition, ceramic components are being used experimentally in many applications and gradually being introduced in production engines. Examples of both phases of use were reported at past meetings of this symposium [4,5]. Nissan Motor Company has described the use of a silicon nitride rocker arm pad in an LPG-fueled taxi engine [6]. The engines in these vehicles operate at idle speed for up to 70% of the total operating time, conditions which result in minimal lubrication of the pad. Replacement of the sintered iron-based alloy pad by silicon nitride has resulted in greatly improved service of this component.

A table illustrating the history of the adoption of ceramic components into IC engines has been prepared by Nissan, Table I. A slow but steady rate of increase in the use of ceramic components is indicated by this compilation. Both silicon nitride and partially stabilized zirconia are being evaluated for use as cam follower rollers in engine simulation tests, with promising results. Wear resistant ceramic coatings for cylinder walls are also being developed by Caterpillar Inc. and by Cummins Engine Company, Inc.[7,8]. The objective of these activities is the development of coatings for piston ring and cylinder liner components to provide low wear, low friction, high thermal-shock resistance, good adherence, and reproducible coatings. Friction and wear test results indicate that successful coatings will result from both projects [7,8]. Results from the Cummins tests indicate that the soot content of the engine oil can strongly influence the wear response of the coating [9]. It is evident that the numerous efforts underway will ultimately lead to the development of a variety of high-temperature materials and coatings for tribological applications in advanced engines.

Lubricants

As engine temperatures become more elevated, new lubricants will be required. Traditionally, engine lubricants have been petroleum-based liquids, but many new concepts are under consideration. Synthetic base stocks with high-temperature capability are an alternative for liquid lubricant development. Solid lubricants may be applicable

TABLE I
Ceramic component use in production engines.

Component	Material	Manufacturer	Initial Use
Glow Plug	Si_3N_4	Isuzu Mitsubishi Nissan	1981 1983 1985
Hot Plug	Si_3N_4	Isuzu Toyota Mazda	1983 1984 1986
Rocker Arm Pad	Si_3N_4	Mitsubishi Nissan	1984 1987
Turbocharger Rotor	Si_3N_4	Nissan Isuzu Toyota	1985 1988 1989
Fuel Injector	Si_3N_4	Cummins	1989
Port Liner	Al_2TiO_5	Porsche Lamborghini	1985 1989

in some circumstances, as in the use of plasma-sprayed coatings containing lubricant phases [10]. Vapor transport of lubricants to the tribocontact has also been shown to be successful [11]. Continuous lubrication of ceramic surfaces at high temperatures has been demonstrated by catalytic transformation of an organic gas to form a lubricating carbonaceous layer [12]. In addition, solid lubricants can be incorporated as a constituent of the microstructure of the solid surfaces.[13,14].

Some discussions of proposed liquid lubricant properties have indicated that future oil viscosities will be lower, and film thicknesses thinner. As a result, either new materials or surface modification of existing materials will be required to provide sufficient wear resistance with the new lubricants. Interesting lubrication concepts discussed elsewhere in this volume include the use of soft metallic coatings in conjunction with a synthetic liquid [15], and tribopolymerization of monomers included in lubricants to form protective surface films [16]. The long term solution to advanced engine lubrication may include entirely new ideas, but in the near-term engines will continue to use enhanced liquids.

Ceramic Tribology

The proposal to use ceramics as components in engines generated many questions about the strength, fracture toughness, and wear resistance of these materials. Each of these issues has been investigated intensively, and many improvements in ceramic properties have been the result. While still incomplete, a body of data is accumulating related to the tribological behavior of a variety of ceramics. The tribological studies at Oak Ridge National Laboratory have emphasized the significance of the mild-to-severe wear transition in wear applications, and the necessity of avoiding that transition during the projected wear life of a component [17]. The dimensional changes, increase in surface roughness, and the debris production associated with severe wear would be intolerable in almost any machine configuration. It should also be recognized that machine components, including those in engines, are generally exposed to cyclic stresses, and some consideration should be given to the cumulative effect of cyclic stress. Wear tests, however, are generally short-term tests, as compared to realistic component service lifetimes, and the effects of accumulated cycles of stress may be overlooked. Interest in fatigue processes in ceramics is increasing, and many studies are presently in progress[18,19,20].

The importance of wear mode transitions in the wear of metal alloys has been recognized for some years, and a wear mode transition diagram has been developed by the International Research Group of the Organization for Economic Cooperation and Development (IRG-OECD) [21]. The diagram shows the range of applied force and velocity for which a particular wear mode may be expected for a specific lubricant and material pair, Fig. 1(a). The limit of mild wear is indicated by the transition line. Fig. 1(b) illustrates the experimental arrangement for evaluating the wear mode as a function of applied force and velocity. The wear mode transition is determined by sudden variations in friction coefficient μ or wear factor k as load is increased in small increments at a given velocity, Fig. 1(c), and by the corresponding change in dimension and appearance of the wear flat on the ball. Plotting the mild (open symbols) and severe (filled symbols) on Fig. 1(a) indicates the position of the transition line. A designer can refer to this diagram to assess the suitability of a given material for service at a particular

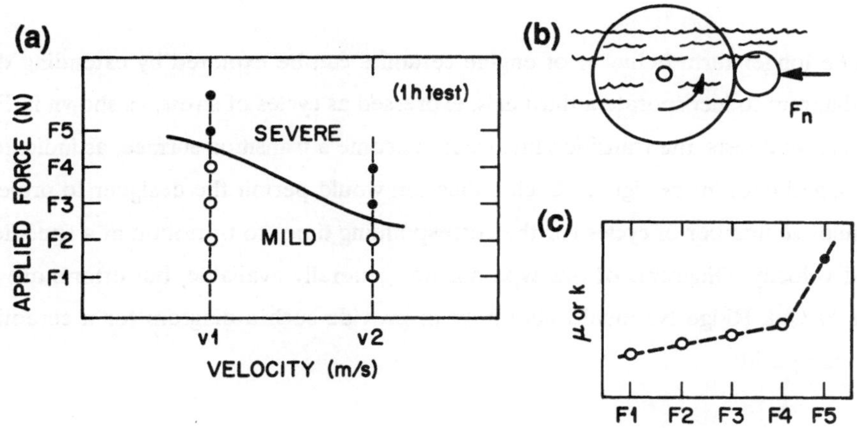

Fig. 1. (a) Schematic OECD wear mode transition diagram. The open symbols indicate mild wear, and filled symbols indicate severe wear. The transition curve separates mild and severe wear regimes. (b) Test configuration for evaluation the wear mode. (c) A sudden increase in friction coefficient, μ, or wear factor, k, indicate the load at which the transition occurs.

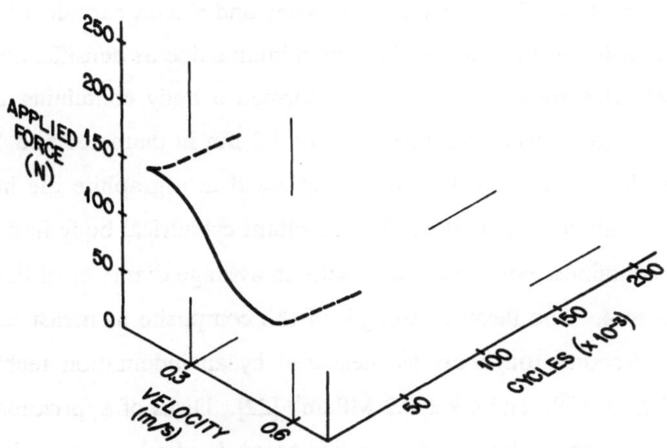

Fig. 2. The OECD wear transition diagram can be adapted to incorporate the effects of cyclic stress. The transition line extends to form a wear transition surface.

design condition. One limitation of the OECD diagram is the restriction of test duration in the determination of the diagram to 1 hour.

The longer term behavior of engine ceramics can be explored by extending the OECD diagram to incorporate a third axis, expressed as cycles of stress, as shown in Fig 2. For extended tests, the transition line would become a transition surface, as indicated by the dashed lines in the figure. Such a diagram would permit the designer to project the cumulative number of cycles (or the corresponding time) to transition at a specified load and velocity. Diagrams of this type are not generally available, but efforts are in progress at Oak Ridge National Laboratory to provide such a diagram for a ceramic-ceramic composite.

WEAR MODE DIAGRAM DEVELOPMENT AT ORNL

Materials and Methods

The work in progress at ORNL is exploring the wear mode transition behavior of a silicon nitride-silicon carbide whisker composite. The ceramic composite was prepared by hot pressing a mixture of silicon nitride powder and silicon carbide whiskers. The powder contained 6% yttrium oxide and 2% aluminum oxide as densification aids. The proportion of whiskers added to the powder formed a body containing 20 vol % of whiskers. The average whisker dimensions were 1.2 μm in diameter and 10-50 μm in length. The powder-whisker mixture was hot-pressed in a graphite die in a nitrogen atmosphere for 60 minutes at 2050 K. The resultant cylindrical body had a density of 3.22 Mg/m^3 and contained elongated grains with an average diameter of 0.5 μm and an aspect ratio of 3 to 4. The flexural strength of the composite as measured in 4-point bending and the fracture toughness as measured by an indentation technique were, respectively, 771 ± 39 MPa and 6.9 ± 0.2 $MPam^{1/2}$ [22]. Disks of approximately 40 mm in diameter were prepared by cutting the hot-pressed cylinder perpendicular to the pressing direction, the cylindrical axis.

Polished surfaces were prepared on the disk specimens to facilitate the detection and measurement of very small amounts of wear. Use of polished surfaces also allowed the measurement of the inherent friction between the solids and evaluation of the wear response in a surface containing the minimum of flaws. The disk surfaces were prepared by grinding on a series of diamond-impregnated metal wheels, the final-stage wheel having a 600-grit diamond content. After grinding, the specimens were polished on vibratory polishers, first using a metal mesh with 6-μm diamond particles in water, then a low-nap cloth with 3-μm diamond in water, and finally nylon cloth with 1 μm diamond in water. The surface of the polished disks had an average roughness value, R_a, of 0.02 μm, as determined by surface profilometry.

The surfaces of the 9.35-mm diameter silicon nitride spheres used as the counterface pins were finished to grade 10 ball bearing specifications, which corresponds to a surface roughness (R_a) of 0.02 μm. The density of the spheres was 3.20 Mg/m^3. The spheres were obtained from a commercial source.*

The specimens were ultrasonically cleaned for 10 minutes in acetone and trichlorotrifluoroethane before testing. The specimens were then dried and positioned in the test system. Specimens were cleaned after lubricated tests by washing in a soap solution, followed by the cleaning procedure indicated above.

The linear reciprocating test system used a sphere sliding on the composite flat as the test configuration. The stroke of the reciprocating motion was 10 mm in length and oscillation frequencies of 15 Hz and 30 Hz, producing average velocities of 0.3 and 0.6 m/s, respectively, were used. Room temperature tests were of varied durations up to 200,000 cycles of stress, and the normal force was adjusted to evaluate the wear mode transition load at a given velocity. The tests were performed in a white mineral oil (Saybolt viscosity 150 maximum).

Results

Figure 3 illustrates the results obtained at a sliding velocity of 0.6 m/s. At a normal force of 30 N, wear of both the pin and disc surfaces is negligible for 1.5 x 10^5 cycles, whereas at a normal force value of 200 N, severe wear is observed at 50 x 10^3 stress cycles. In

* Norton Company, Northboro, MA

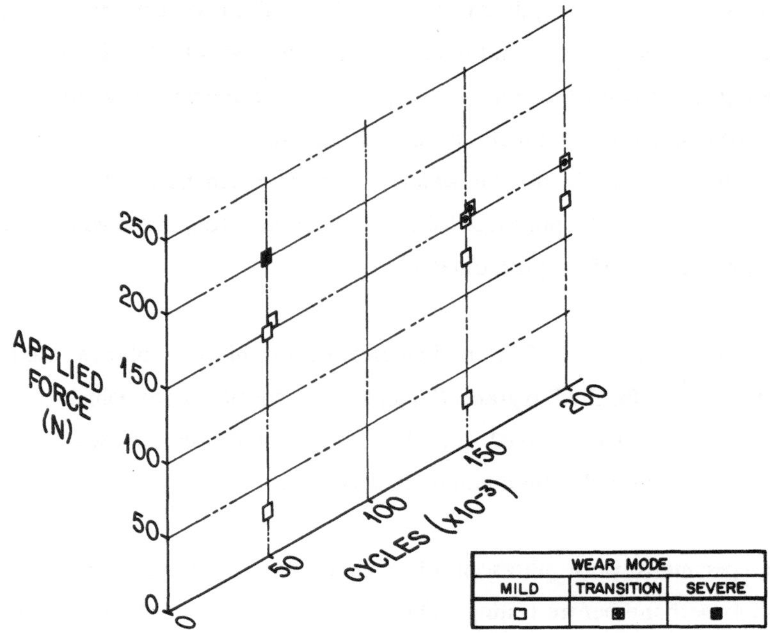

Fig. 3. The wear mode results at an average sliding velocity of 0.6 m/s. Open symbols indicate mild wear, the filled symbol denotes severe wear, and symbols containing a dot represent the start of the transition from mild to severe wear. Severe wear occurs at an applied force of 200 N, but at 150 N, mild wear is observed at short test durations and the transition condition is noted at 150 x 10^3 cycles of stress.

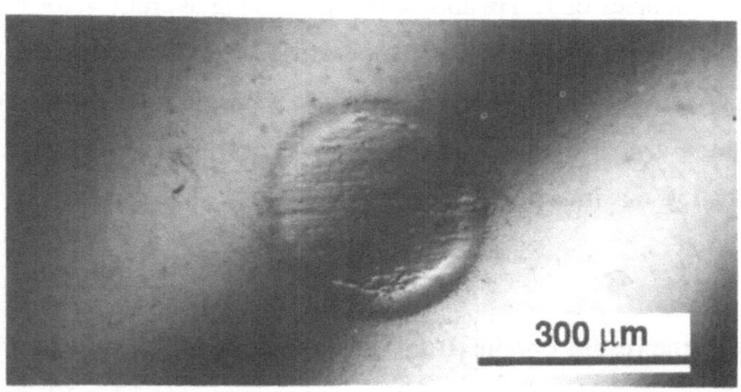

Fig. 4. Mild wear on a pin tip after 150 x 10^3 cycles at an applied force of 150 N at a velocity of 0.3 m/s. Light abrasion marks, caused by sliding on the flat formed by elastic deformation under load, are observed. On removal of the load, the tip returns to a spherical shape. Reflected light photo.

this context, severe wear corresponds to a readily visible loss of surface material and a related wear factor of 10^{-6} mm^3/N-m or greater; mild wear is characterized by a very limited, almost invisible, surface loss of material and a wear factor of 10^{-8} mm^3/N-m or less; and the transition condition is identified by the start of visible wear damage and wear factors of about 10^{-6} mm^3/N-m. The result at 200 N applied force is in contrast with the behavior at 150 N where mild wear is noted at 50 x 10^3 cycles, but the start of the transition to severe wear is evident at 1.5 x 10^5 cycles. This latter result is indicative of the susceptibility of the wear surface to time and cycle dependent processes which promote a transition from mild to severe wear. The data of Fig. 3 are shown in an isometric projection since the total data set is subsequently compared in a three-dimensional plot.

Figure 4 illustrates the condition of the pin tip characterized as mild wear. The surface of the tip is lightly abraded, but contact profilometry reveals that the tip is essentially unworn. Although a wear flat appears to have formed, the surface of the pin is spherical and only a slight interruption of the curvature of the sphere, corresponding to the annular ring seen in the figure, is observed. In contrast, Fig. 5 illustrates the pin tip condition as severe wear begins. The material at the tip is fractured and the profilometer trace of the tip shows the formation of a flat containing an irregular profile. A residue of debris remains around the circular contact surface.

The condition of the wear paths on the composite disk were found to be consistent with the wear mode evaluations made on the basis of the observed pin tip condition. For those pin tips indicating mild wear, the disk wear paths showed only light abrasion marks and were free of damage. For experiments of sufficient duration for the transition to severe wear to have started on the pin, the formation of pits was observed on the disk wear path. Such a condition is exemplified by the initiation of surface pitting on a specimen in which the transition to severe wear has started at 1.5 x 10^5 cycles of stress, Fig 6.

The experiments performed to date in the region adjacent to the transition surface indicate the limits of mild wear for the silicon nitride composite in mineral oil, Fig. 7. Results are shown in this figure for experiments at both sliding velocities, and it is

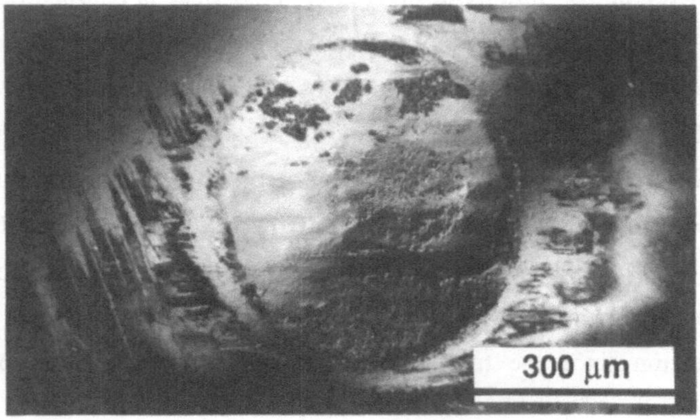

Fig. 5. A pin tip showing the start of the transition to severe wear after 150 x 10^3 cycles at 150 N load and velocity of 0.6 m/s. The contact area is flat and patches of fracture surface are present. Reflected light photo.

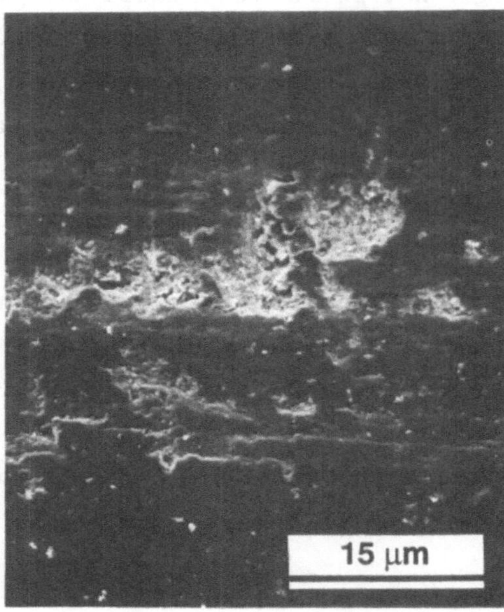

Fig. 6. Pit formation on a disk wear path after 150 x 10^3 cycles at a load of 150 N and velocity of 0.6 m/s. The pits contain a fine-scale fracture surface and enlarge as sliding continues.

evident that at 0.3 m/s severe wear is incurred at an applied force of 250 N, but not at 150 N. At 0.6 m/s, however, 200 N causes severe wear at only 50 x 10^3 cycles, and evidence of the start of the transition is observed at 150 N and 150 x 10^3 cycles of stress. Additional data are now being obtained to fully define the position of the transition surface emerging from these results.

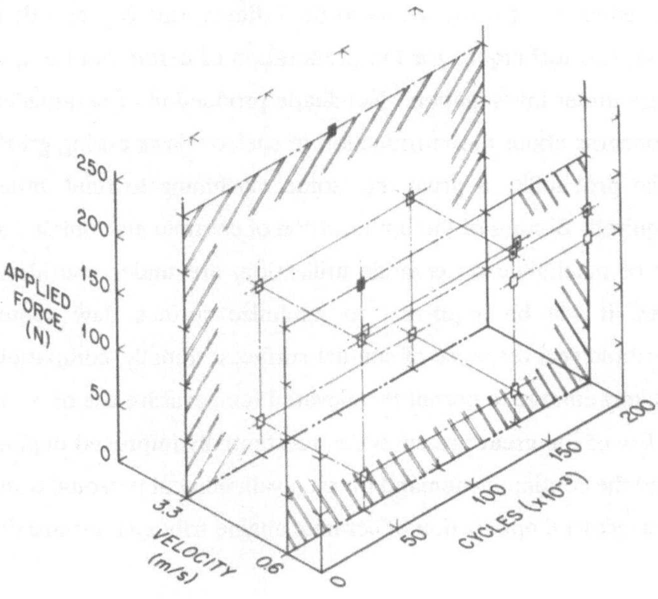

Fig. 7. Data accumulated to date at sliding velocities of 0.3 and 0.6 m/s. At 150 N applied force, mild wear is retained at 0.3 m/s for the range of testing investigated, but at 0.6 m/s the start of the transition is detected at 150 x 10^3 cycles.

CONCLUDING REMARKS

Wear mode transition diagrams will vary as the material pairs and lubricants are varied. Diagrams could be developed for materials and lubricants in common use to assist in the selection of materials for specific designs. Introduction of wear response as a function of time of exposure to stress is an important element of this approach. Generalized wear mode transition information will assure effective, long-service life designs for ceramics.

With collection of sufficient data, in conjunction with the appropriate lubricants, the limiting conditions for the use of ceramics as machine components should be realized. Performance testing in full-scale systems will always be an element of the tribological evaluation process, but the definition of the general limits should permit designers to match material capability and design requirements. Although the development of processes for the reduction of components to final dimensions is presently an issue of lively discussion, several techniques for the preparation of ceramics in net, or near net shape condition are under investigation. Net-shape production of ceramic components would alleviate concerns about the introduction of surface flaws during grinding and/or machining, but the probability is great that some machining to final dimensions will continue to be required. Studies of the optimization of ceramic machining, including the economic impact of machining on ceramic utilization, are under consideration. For machined surfaces, it will be important to minimize surface flaw content and to characterize the tribological response of contact surfaces. Finally, compatible lubricant systems will very likely emerge to permit the elevated temperature use of ceramic engine components. In view of the great pressures for substantially improved engine efficiency and cleanliness and the continuing demand for the availability of personal transportation, the prospects for successful application of ceramic engine tribosystems are encouraging.

REFERENCES

1. R. Kamo, M. Woods, and P. Sutor, Development of Tribological System and Advanced High-Temperature In-cylinder Components for Advanced High-Temperature Diesel Engines, in Proc. 1987 Coatings for Adv. Heat Engines Workshop, July 27-30, Castine , ME, USDOE, Conf-870762.

2. J. Breznak, E. Breval, and N. H. Macmillan, "Sliding Friction and Wear of Structural Ceramics," J. Mater. Sci. 20 (1985) 4657-4680.

3. P. A. Gaydos and K. F. Dufrane, "Studies of Dynamic Contact of Ceramics and Alloys for Advanced Heat Engines," in Proc. 27th Automotive Technology Development Contractor's Coordination Meeting, Oct. 23-26, 1989, SAE, Warrendale, PA, 1990, pp 149-153.

4. R. R. Wills and R. E. Southam, "Ceramic Engine Valves," in Ceramic Materials and Components for Engines, Nov. 27-30, 1988, American Ceramic Society, 1989, pp. 1429-1437.

5. J. C. Bentz, T. M. Yonushonis, T. Aoba and Y. Fujimoto, "Silicon Nitride Ceramic Wear Resistant Machine Elements," ibid., pp. 1489-1494.

6. M. Kano and I. Tanimoto, Wear Mechanism of High Wear-Resistant Materials for Automotive Valve Trains, in Wear of Materials - 1991, ed K. C. Ludema and R. G. Bayer, ASME, New York, 1991, pp. 83-89.

7. C. D. Weiss, Wear Resistant Coatings, in Ceramic Technology for Advanced Heat Engines Project Semiannual Progress Report, ORNL/TM-11719, pp. 194-199.

8. M. G. S. Naylor, Development of Wear Resistant Ceramic Coatings for Diesel Engine Components, Ibid, pp. 200-227.

9. M. G. S. Naylor, "Wear Resistant Ceramic Coatings," in Proc. Annual Automotive Technology Development Contractors' Meeting, Soc. Auto. Engrs., Warrendale, PA, 1991, pp. 273-281.

10. H. E. Sliney and C. DellaCorte, "Tribological Prperties of PM212: A High Temperature, Self-Lubricating Powder Metallurgy Composite," Lubr. Engrg., 47(4) 298-303, 1991.

11. E. E. Klaus, A Study of Tricresyl Phosphate as a Vapor Delivered Lubricant, Lubr. Eng., 45(11) 717-723, 1989.

12. J. L. Lauer, Continuous High-Temperature Lubrication of Ceramics by Carbon Generated Catalytically, in Material Research Society Symposium Proceedings, Vol 140 (New Material Approaches to Tribology), 1989, pp. 363-368.

13. C. R. Blanchard and R. A. Page, Effect of Silicon Carbide Whisker and Titanium Carbide Particulate Additions on the Friction and Wear Behavior of Silicon Nitride, J. Am. Ceram. Soc., 73(11) 3442-52, 1990.

14. C. R. Blanchard and R. A. Page, Effect of Particulate Additions on the Contact Damage Resistance of Hot-Pressed Si_3N_4, J. Mater. Sci., 23, 946-957, 1988.

15. O. O. Ajayi, A. Erdimir, J.-H. Hsieh, R. A. Erck, and F. A. Nichols, Boundary Lubrication of Ceramic Materials by Soft Metallic Coating and Synthetic Oil, this volume.

16. M. J. Furey and C. Kajdas, Tribopolymerization as a Novel Approach to Ceramic Lubrication, this volume.

17. C. S. Yust, Wear Transition Surfaces for Long-Term Wear Effects, in Tribological Modeling for Mechanical Designers, ed. K. C. Ludema and R. G. Bayer, ASTM-STP 1105, 1991, pp 153-161.

18. A. P. Nikkila and T. A. Mantyla, "Cyclic Fatigue of Silicon Nitride," Ceram. Eng. Sci. Proc., 10(7-8) 646-656, 1989.

19. R. H. Dauskardt, D. B. Marshall, and R. O. Ritchie, "Cyclic Fatigue-Crack Propagation in Magnesia-Partially-Stabilized Zirconia Ceramics," J. Amer. Cer. Soc., 73(4) 893-903, 1990.

20. S. Suresh and R. O. Ritchie, "Propagation of Short Fatigue Cracks," Intl. Met. Rev., 29(6) 445-453, 1984.

21. A. W. J. deGee, A. Begelinger, and G. Salomon, "Lubricated Wear of Steel Point Contacts - Application of the Transition Diagram," in Wear of Naterials - 1983, ed K. C. Ludema, ASME, New York, 1983, pp. 534-540.

22. T. N. Tiegs, J. W. Geer, P. D. Tennis, and S. M. Leahy, Ceramic Technology for Advanced Heat Engines Project Semiannual Progress Report for April Through September 1988, Oak Ridge National Laboratory, ORNL/TM-11116, pp 92-97, 1989.

MICROSTRUCTURE OF SILICON CARBIDE AFTER PLASMA SINTERING

Kazunori KIJIMA
Kyoto Institute of Technology
Matsugasaki, Kyoto, JAPAN

ABSTRACT

Plasma sintered SiC-C composites were observed by transmission electron microscopy in order to investigate the existing state of carbon, interface structure between SiC and C, and grain boundary structure in SiC. The present observation was helpful in discussing the sintering mechanism and in explaining the high hardness and high toughness of plasma sintered SiC-C composites. The results obtained are summarized as follows.

1. Graphite in plasma sintered SiC-C composites existed as (1) fine polycrystals, (2)fibrous crystals and (3) single crystals. The fine polycrystals and fibrous crystals had no orientation around a pore, but some orientation between SiC grains because of the compressive stress generated by the difference in the thermal expansion coefficient between SiC and graphite. Single crystals of graphite were observed at grain boundaries between SiC crystals. In the graphite, cracks were formed parallel to the basal plane because of the thermal stress.
2. High hardness and high toughness of plasma sintered SiC-C composites are presumably due to the residual compressive stress caused by the difference in the thermal expansion coefficient. It is also considered that the toughness was improved by cracks in graphite grains which inhibited the crack propagation.
3. At the interface between SiC and graphite, there existed the structures that the basal plane of graphite was parallel to the interface, that SiC was transformed near the interface and that SiC had step structures in an amorphous layer formed by reaction on graphite. These structures might be formed to reduce high energy at the interface. A second phase was not observed at grain boundaries in SiC, however, structure relaxation might occur within a thickness of 1nm.

INTRODUCTION

SiC is difficult to sinter, but it can be sintered to a high density when treated by plasma. SiC containing an excess quantity of carbon is generally impossible to sinter by conventional methods, but can be easily sintered by the plasma method to produce an SiC-C composite material. The SiC-C composites have a high hardness and high toughness, and is expected to a material for sliding parts and seals for rotary parts because carbon in the composite acts as a lubricant.

In this study, the microstructures of plasma-sintered SiC-C composites containing B and C, were observed by transmission electron microscopy (TEM) to clarify the existing state of carbon, the interface structures between SiC and C and grain boundary structure in SiC, and thereby to discuss the sintering mechanism, and the origin of high hardness and high toughness of plasma sintered SiC.

MATERIALS AND METHODS

α-SiC(average grain size: 0.49μm), amorphous boron and phenol resin for carbon source were mixed in the ethanol by using a ball mill for 16h. The mixture, after drying, was formed into a shape, rectangular bar with 5x5x40mm^3 in size, by pressing at 300kg/cm^2 and then by an isostatic pressing at 2000kg/cm^2 .It was then calcined at 1270K for 60min. in an Ar atmosphere to carbonize the phenol.

Plasma was generated by a high-frequency generator (frequency: 4MHz, output: 15kW). Low-temperature plasma was formed in a water-cooled, double-tube type quartz furnace evacuated to 0.3torr to produce an egg shaped RF-thermal plasma of about 10000K in an Ar atmosphere. The specimen was sintered by introducing into the plasma flame gradually.

Two types of specimens were prepared, A and B containing carbon of 3.0 and 7.0wt%, respectively. The conditions under which the specimens were prepared are given in Table 1.

TEM observation was performed using a JEOL JEM 4000 FX and the specimen was made by ion thinning.

Table 1. Specimen preparation conditions for sample A and sample B.

Sample	B (wt.%)	C (wt.%)	Anode Voltage (kV)	Ar gas Flow (SCCM)	Pressure (torr)	Sintering Time (s)	Density (g/cm³)	Relative Density (%)
A	1.0	3.0	7.0	80	50	300	2.969	92.5
B	2.0	7.0	7.5	40	50	120	3.056	95.2

RESULTS AND DISCUSSIONS

Sintering Conditions and Structures

Figures 1 (a) and (b) show the microstructures of specimens A and B observed by TEM, where the white parts represent pores. As is shown, specimen A, which was lower in density, was more porous. The average grain size of the specimens A and B was about 1 m and 1.5 m respectively.

Carbon was present around the pores and between the SiC grains in both specimens but more frequently between the grains in specimen B.

The existing state of carbon hoases are discussed in detail in the subsequent section.

Sintering density generally decreases as carbon contentincreases, if specimens is sintered under the same conditions. However, specimen B containing more carbon was apparently higher in density than specimen A conceivably because the former was sintered at a higher anode voltage and lower Ar flow rate, and hence at a higher temperature.

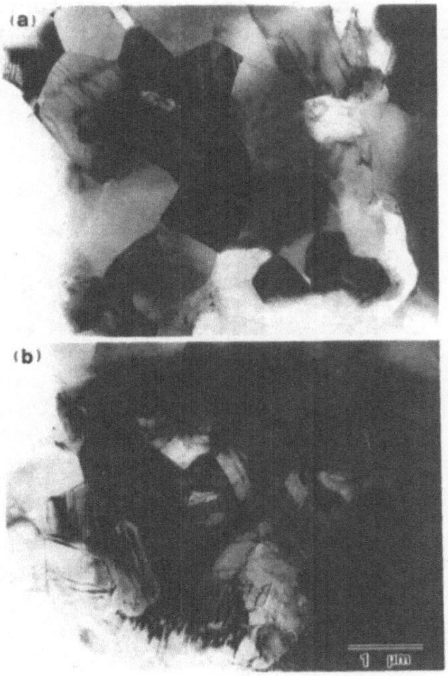

Fig. 1. TEM micrographs of (a) sample A(3% carbon) and (b) sample B(7% carbon).

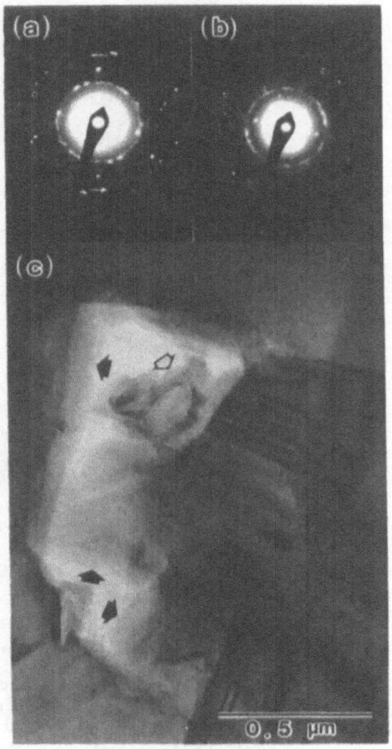

Fig. 2. TEM photographs of a region around a pore for sample A(3% carbon). (a) Diffraction pattern at a fibrous graphite and (b) at a fine polycrystalline graphite, and (c) TEM image of a fibrous graphite (black arrow) and fine polycrystals (white arrows) around a pore.

Carbon Phases

As shown in Fig. 2 (c), each pore in specimens A or B was surrounded by fibrous (white arrow) or fine polycrystalline (black arrow) graphite. Figures 2 (a) and (b) show the selected area diffraction patterns of the fibrous and polycrystalline graphite. Each diffraction ring was almost circular, irrespective of the diffraction conditions, indicating that the fibrous and polycrystalline graphite around the pore had no orientation. Since graphite was always found around the pores, as mentioned above, it is possible to postulate that carbon existed in the grain boundaries of SiC was diffused to the pores when the grains were rearranged during the initial stage of plasma sintering.

Figure 3 shows typical graphite grains existing at the grain boundary of SiC (white arrow). These graphite grains are single crystals having a relatively large size of 0.4 to 0.7 m, and are characterized by cracks running in parallel to the basal plane.

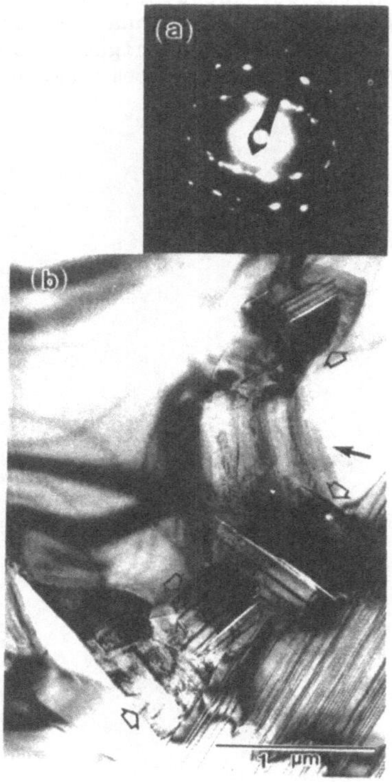

Fig. 3. TEM photographs of a region between SiC grains for sample B(7% carbon). (a) Diffraction pattern at a region including a particle indicated by the black arrow, and (b) TEM image of single crystals of graphite (white arrows) and fine polycrystal or fibrous crystal of graphite (black arrows).

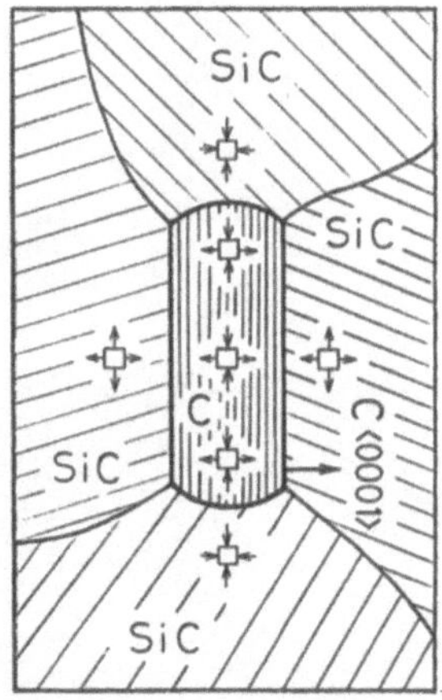

Fig. 4. An illustration of stress distribution in SiC-C composite material.

Thermal expansion coefficients along the a- and c-axis of graphite are 9.5x10 /K and 2.8x10 /K, whereas that of SiC is 4.0x10 /K. Therefore a stress distribution within the specimens after it is cooled becomes an illustration shown in Fig. 4. The Young's modulus along the a- and c-axis of graphite are 1020GPa and 36.4GPa, whereas that of SiC is 400GPa. Presuming that the specimen is heated to 2800K and the strain produced at the interfaces changes continuously, a compressive stress of 4070MPa along the a-axis and a tensile stress of 2730MPa along the c-axis should be produced in the specimen cooled to room temperature. These stresses will produce the cracks running in parallel to the basal plane. Since the minimum stress required to flay the basal plane is about 0.4MPa, the above stresses are sufficient to produce the cracks. A black contrast, considered to be a stress-induced contrast, appears in the interface between graphite and SiC (Fig.3), indicating that there is a residual stress after cracking.

Figures 3 (a) and (b) show the diffraction patterns of the region containing a grain, indicated by the black arrow. The pattern is not a circle but an ellipse. Some spots had high intensity, which corresponded to the before-mentioned relatively large graphite grains. The other continuous ellipse is due to fine polycrystalline or fibrous graphite, corresponding to the substance indicated by the black arrow. The diffraction ellipse gradually changed to a circle by shortening its longitudinal axias, depending on diffraction conditions, which means that the grain has some orientation. The orientation direction is indicated by the black arrow in the figure. The polycrystalline or fibrous graphite grain has the c-axis oriented basically in the direction indicated by the arrow, rotating another direction around the c-axis. The difference in thermal expansion coefficient between graphite and SiC produced a compressive stress in parallel to the basal plane of the graphite, causing the polycrystalline or fibrous graphite grains to be rearranged in such a way that their basal plane are overlapped one another.

Figure 5 shows the graphite single crystal sandwiched by the SiC grain. The photograph also shows the cracks running along the basal plane, considered to be produced in similar manner as those described earlier. The black contrast in the SiC grain around the graphite grain represents a stress region.

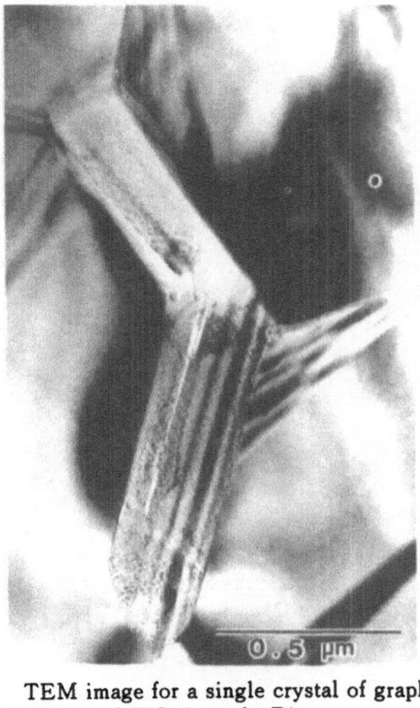

Fig. 5. TEM image for a single crystal of graphite in grain boundary of SiC (sample B).

As discussed above, the graphite grain present in plasma-sintered SiC-C composite was in the form of (a) a fine polycrystal, (b) a fibrous crystal, or (c) a single crystal. The polycrystalline or fibrous graphite grains were distributed randomly when present around the pores, whereas they were oriented when present between the SiC grains. The single crystals were found between the SiC grains, where cracks were formed by compressive and tensile stresses.

Plasma-sintered SiC has a higher hardness and higher toughness than SiC sintered by the conventional method, conceivably resulting from the effects of compressive stress generated by the difference in thermal expansion coefficients between SiC and graphite. A stress distribution, shown in Fig. 4, is produced around the graphite grains in SiC, where cracks run in parallel to the basal plane to cause tensile stress to be significantly lower than compressive stress, with the result that the latter stress plays a more important role within the sintered body. Another possible reason for the high toughness is the contribution of cracks in the graphite to inhibit the crack propagation through the body.

SiC-Graphite Interface

Figure 6 shows a high-resolution TEM image of the interface between SiC and Graphite (C), showing that the basal plane of graphite is parallel to the interface. The interface energy will not be high, because the bonding over the basal plane, being characterized by bonding has low anisotropy. As discussed earlier, however, a tensile stress is generated on the SiC side and a compressive stress on the graphite side, along the interface, because of the difference of the thermal expansion coefficient between a-axis of graphite and SiC. The figure shows the black contrasts along the interface, reflecting the residual strain. Thus the interface parallel to the basal plane of graphite was frequently observed in this study.

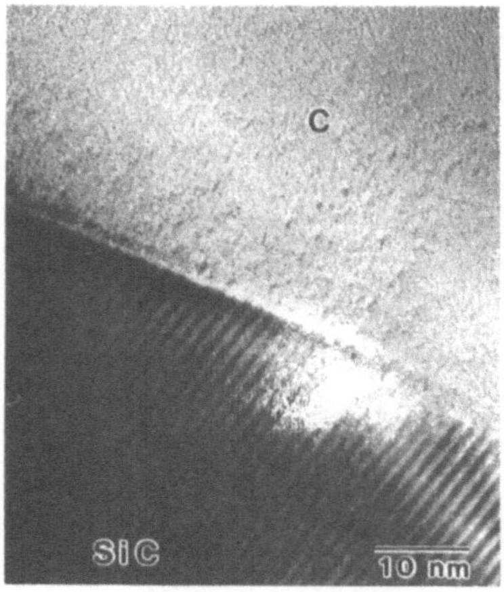

Fig. 6. HR-TEM image of the interface between SiC and graphite, indicating that the basal plane of graphite was parallel to the interface (sample A).

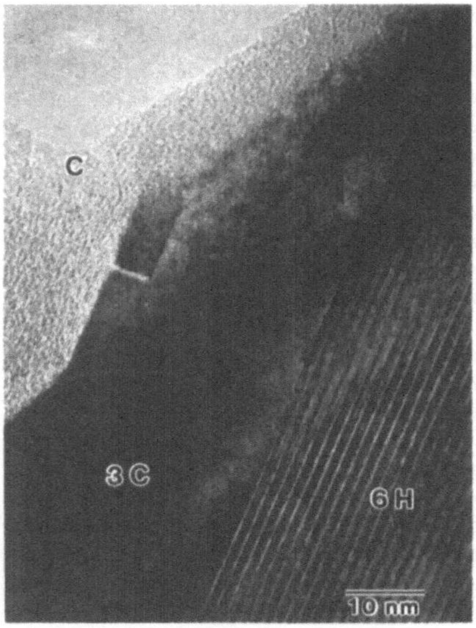

Fig. 7. HR-TEM image of the interface between SiC and graphite, indicating that SiC was transformed from 6H to 3C near the interface (sample B).

Figure 7 shows the transformation of SiC from 6H to 3C in the vicinity of the interface between SiC and graphite. Such a phenomenon is sometimes observed in the grain boundaries of SiC sintered by the conventional method. The transformation is considered to have the effect of reducing interface energy. It is also observed in this case that the basal plane of graphite is in parallel to the interface between 3C-SiC and graphite, Figure 7 also shows the microcracks on the SiC side, conceivably resulting from tensile stress along the interface, as shown in Fig. 4.

Figure 8 shows an example that the basal plane of graphite is not in parallel to the interface between SiC and graphite. In such a case, SiC frequently has step structure, as seen in the photograph. The height of step is about 1.5nm, roughly equivalent to the c-axis height of the 6H-SiC unit cell. It was also found that there is about a 3nm thick layer of low contrast, which may represent the amorphous phase, because no lattice image was observed for this phase. It is considered to be the product of the reaction between SiC and graphite. In such a case, the interface will be affected by the bond, and will have a high interface energy because of the high anisotropy of bonding. The amorphous reaction layer may be formed to reduce the high interface energy.

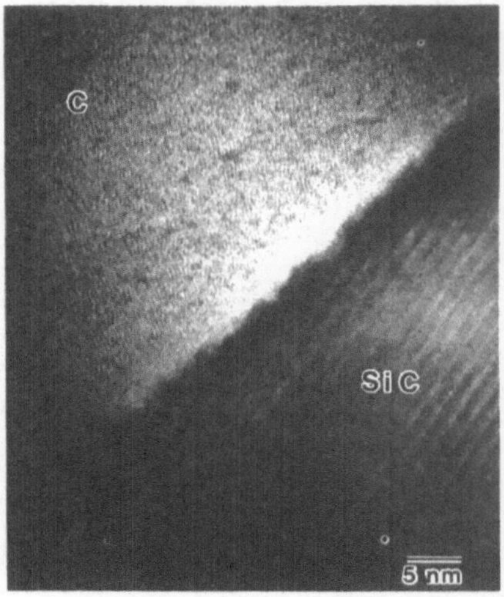

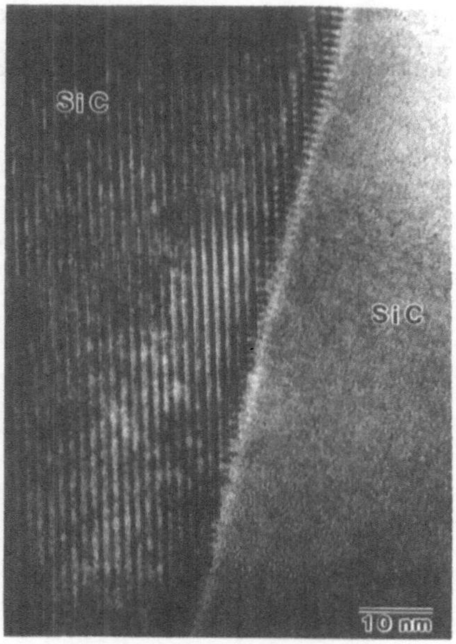

Fig. 9. HR-TEM image of grain boundary in SiC, indicating that no second phase existed at the boundary (sample A).

SiC-SiC Grain Boundary

Figure 9 is the high-resolution TEM photograph of the SiC-SiC grain boundaries. This photograph shows no secondary phase. However, structure relaxation might occur within a thickness of 1nm, judging from the contrasts in the grain boundaries.

REFERENCES

1) K.Kijima, "Proceedings of The 7th Inter. Symp. on Plasma Chemistry," IUPAC, Eindhoven, Vol.2,662(1985).
2) K.Kijima, 6th High-Temperature Structural Materials Basic Seminar Proceedings Abstract, pp.73-78(1986).
3) Handbook of Chemistry, third edition, edited by Chemical Society of Japan, Maruzen, II-22(1984).
4) H.Abe et al., "Engineering Ceramics," Giho-do, pp.16-20(1984).
5) T.Oku, "Introduction to Carbonaceous Materials, revised edition," edited by the Carbonaceous Materials Society of Japan, p.63-65(1984).
6) Victor A. Drits, "Electron Diffraction and High-Resolution Electron Microscopy of Mineral Structures," Springer-Verlag, Berlin(1987) pp.149-154.
7) K.R.Kinsman and S.Shinozaki, Proc. Int. Conf. Solid Phase Transform. 605-9(1082)

OXYNITRIDE GLASS SYSTEMS AND SUBSEQUENT GLASS-CERAMIC HEAT TREATMENTS

S HAMPSHIRE, R FLYNN, J LONERGAN AND A O'RIORDAN
Materials Research Centre, University of Limerick
Plassey Technological Park, Limerick, Ireland

ABSTRACT

Silicon nitride based ceramics contain oxynitride glass phases at the grain boundaries which can impair subsequent high temperature properties. Studies of the effect of nitrogen on properties of these glasses show that nitrogen increases viscosity, hardness and glass transition temperature. Heat treatment of the ceramics to crystallise these glasses is well established but further improvements are possible if glass-ceramic processes using two-stage heat treatments are introduced. This paper reports on the optimisation of glass-ceramic heat treatments in various oxynitride systems (notably RE-M-Si-O-N systems where (RE = Nd, Sm, M=Mg, Al) and the types of crystalline phases produced. One of the major crystalline phases is apatite, $MRE_4Si_3O_{12}N$, which exists as a range of solid solution. The use of the glass-ceramic heat treatment results in improved hardness and this is used to optimise the heat treatment processes.

INTRODUCTION

The occurrence of oxynitride glasses as grain-boundary phases in silicon nitride based ceramics and the desire to understand their nature has resulted in a number of investigations on oxynitride glass formation and properties [1-5]. Studies of the effect of nitrogen on properties of these glasses show that small concentrations of nitrogen in silicate glasses increase their glass transition temperature, viscosity and hardness [3-5]. Empirical investigations of crystallisation of selected oxynitride glasses, mainly in the Y-Si-Al-O-N system, have been carried out but the aim of that work was to complement more extensive studies of phase equilibria in M-Si-Al-O-N systems and the effects of vitreous phases on high-temperature mechanical properties of silicon nitride based ceramics [1,6,7].

Glass-ceramics have been developed because of their combination of mechanical strength with thermal, optical or electrical properties which arise from the presence of crystalline phases formed by two consecutive heat treatments of the parent glass to produce (a) nucleation and (b) growth of crystals. These thermal treatments are carried out at temperatures corresponding to those at which the maximum nucleation and growth rates for the crystals would be observed. The subsequent properties of the glass-ceramics depend on the composition of the parent glass, the extent of crystallisation which has taken place and the interactions of the vitreous and crystalline phases.

So far, the only systematic study of the two-stage glass-ceramic process as applied to oxynitride glasses is that reported by Morrissey et al. [8]. This paper reports on the further optimisation of process parameters controlling the preparation of oxynitride glass-ceramics.

MATERIALS AND METHODS

Glass Preparation

Parent glasses were prepared from mixtures of silicon nitride powder (Starck LC12) together with high purity (99.9%) oxides (Nd_2O_3, Sm_2O_3, Y_2O_3, Al_2O_3, MgO, SiO_2) to give the required chemical composition, in equivalent percent (e/o) cations/anions, the equivalent concentration of nitrogen for any composition given by 3[N]/(2[O]+3[N]), where [O] and [N] are, respectively, the atomic concentrations of oxygen and nitrogen. By using constant cation ratios, the effects of nitrogen-oxygen ratio on glass properties such as viscosity, hardness, glass transition temperature can be clearly demonstrated [5].

The powders were blended in isopropanol for 10 min using a Janke and Kunkel Ultra-Turrax T25 homogeniser. The alcohol was then evaporated off and the powder mixture dry-mixed for a further 2 min before being pressed into pellets. Large batches (60g) were melted in a boron nitride lined graphite crucible under 0.1MPa nitrogen at 1700°C for 1 h in a vertical tube furnace after which the crucible was withdrawn rapidly from the hot zone, lifted from the furnace and poured into a pre-heated graphite mould at 900°C and annealed for 1 h at 850-900°C prior to slow furnace cooling.

Examination of the Glasses

Simultaneous differential thermal analysis/thermogravimetry was carried out using a Stanton-Redcroft STA-780 simultaneous TG/DTA analyser to determine glass transition (T_g) and crystallisation (T_c) temperatures. Hardness tests were carried out using a Wilson Tukon microhardness tester with an indentation load of 200 g applied for 15 s. Viscosity was determined using three point beam bending of glass test bars under an applied load and measurement of strain rate. Assessment of Young's modulus was carried out using an ultrasonic technique [9] at ENSCI, Limoges, France.

Glass-Ceramics Heat Treatments

The conventional glass-ceramic process involves heating the as-prepared glass to an initial heat-treatment temperature at which internal nucleation occurs followed by heating to a second higher temperature to allow crystal growth of the nuclei. In the absence of any information about annealing temperatures or viscosities, initial heat treatments were carried out for 2.5 h at temperatures (°C) in the range T_g to (T_g + 90) based on results from TGA. For the second heat treatments, temperatures (°C) from (T_c - 40) to T_c were employed.

The glass-ceramics were characterised by X-ray diffraction, SEM and microhardness, as for the parent glasses. From the hardness results, values of indentation fracture toughness were calculated [10].

RESULTS AND DISCUSSION

Properties of glasses

Figure 1 shows d.t.a. traces for a series of Sm-Si-Al-O-N glasses (28:56:16 composition) with a range of N contents from 0 to 25 e/o. Similar traces are obtained for glasses in the Y-Si-Al-O-N and Mg-Nd-Si-O-N systems [8]. The glass transition temperature, T_g, is given by the small endothermic changes observed in the range 850 - 970°C. T_g increases almost linearly with N content.

Figure 2 shows the variation of glass transition and crystallisation temperatures with nitrogen content as obtained from the maximum point of the exothermic peak(s) of the d.t.a. traces for various RE-Si-Al-O-N glasses (RE=Y,Nd,Sm) T_c increases initially with N content and there appears to be a levelling effect at higher nitrogen contents.

The variation of viscosity with temperature for a glass is important at different stages of manufacturing, particularly in annealing and the heat treatments necessary for glass-ceramic production. Comparative plots of viscosity at different N contents for the Nd-Si-Al-O-N glasses are given in Figure 3. Clearly, nitrogen increases the viscosity of a glass of constant cation ratio. The viscosity changes by three orders of magnitude over 100 degrees (K) in the temperature range 850-950°C. Similar behaviour is observed for other glasses investigated. As inferred by Figure 2, for RE-Si-Al-O-N glasses, viscosity changes in the order Sm<Nd<Y.

Figure 4 shows the variation in Young's modulus with nitrogen content for two different Nd-Mg-Si-O-N glass compositions studied. For the 12:24:64 (Nd-Mg-Si) compositions, E increases substantially with N content while for the 24:12:64 composition, an increase is also observed up to 17 e/o N but the high N sample has a similar value to the oxide glass.

Optimisation of glass-ceramic process for Nd-Mg-Si-O-N glasses.
Increase in microhardness following heat treatment was used as the main criterion for assessing increases in strength due to crystallisation.' MORRISSEY et al **[9]** showed that heat treatment at a single temperature resulted in a small increase in hardness for the 12:24:64 composition but two stage heat treatments resulted in much higher increases. Further work has concentrated on optimising the glass-ceramic process for the whole range of glasses within this system and results for the 12:24:64 composition are given as representative.

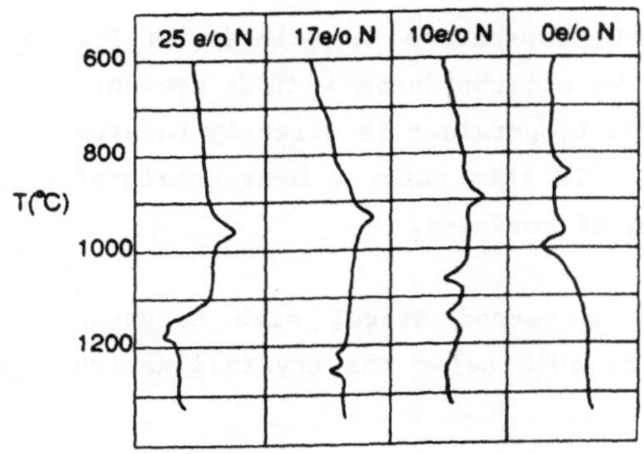

Figure 1: d.t.a. traces for Sm-Si-Al-O-N glasses with 28:56:16 cation composition (e/o) showing the effects of nitrogen.

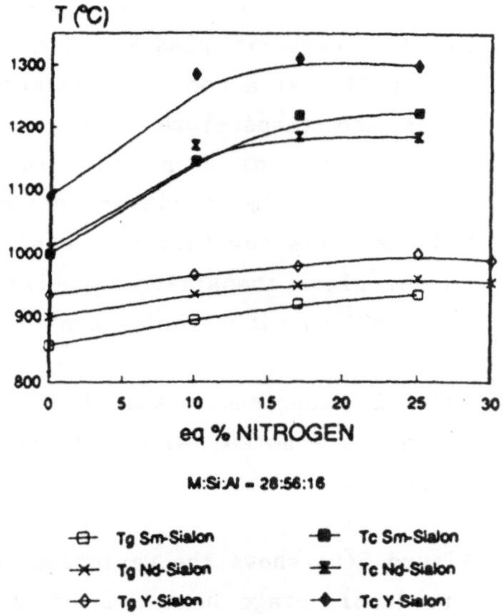

Figure 2: Effects of nitrogen on glass transition temperature, T_g, and crystallization temperature, T_c, for various oxynitride glasses.

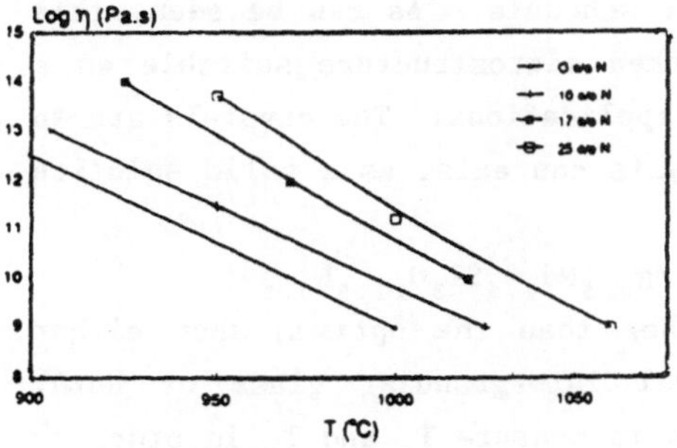

Figure 3: Variation of viscosity with temperature for Nd-Si-Al-O-N glasses with 0 and 17 e/o N.

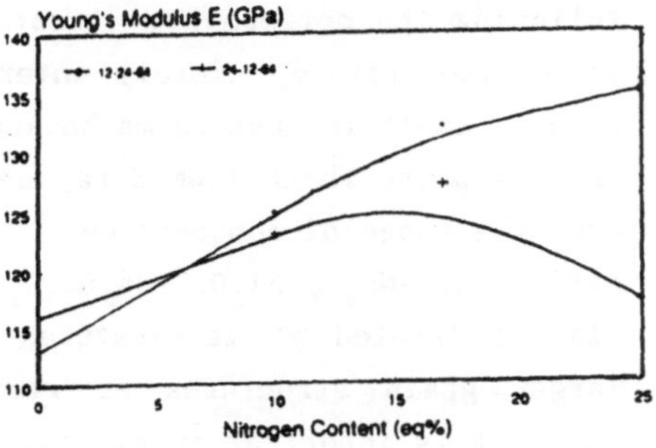

Figure 4: Variation in Young's modulus with nitrogen content for Nd:Mg:Si:O:N glasses.

Figure 5(a) shows the effect of single heat treatments on microhardness of glasses after 2.5 hours at 900°C, T_g+50(°C) and T_g+90(°C). At a constant temperature of 900°C, the trend is not clear since the temperature relative to T_g will be different at each N content, whereas when heat treatment temperature is related to T_g, there is an almost linear increase in microhardness with N content. This confirms the fact that nucleation temperature is directly related to the glass transition temperature. In this case, a heat-treatment of T_g+90(°C) gives the highest values of hardness.

Initial experiments showed that in a second stage, high hardness values were noted after treatment at 40°C below the crystallization temperature.

Figure 5(b) shows the variation in microhardness with nitrogen for all the double-stage heat-treated samples. For the processes where the nucleation stage is directly related to T_g and the crystal growth stage to T_c, there is an almost linear increase in hardness as nitrogen content increases, The optimum treatment giving the highest hardness values is at T_g+90(°C) followed by T_c-40(°C) K_{Ic} increases with nitrogen content for all the glass-ceramics produced.

Figure 6 shows a scanning electron micrograph of 12:24:64 composition following the optimum heat treatment schedule. As can be seen, this is a fine-grained, closely interlocked microstructure suitable as a glass-ceramic for use in mechanical applications. The crystals are an apatite phase which from X-ray analysis can exist as a solid solution over the range of composition:

$MgNd_4Si_3O_{13}$-$Nd_{4.67}Si_3O_{13}$-$Nd_5Si_3O_{12}N$-$Mg_{0.5}Nd_{4.5}Si_3O_{12.5}N_{0.5}$

Glasses treated at temperatures other than the optimum show either larger grain structures or residual intergranular glass or both. Thus, it is important in all systems to measure T_g and T_c in order to determine the appropriate heat treatment parameters for the glass-ceramic process.

(a)

(b)

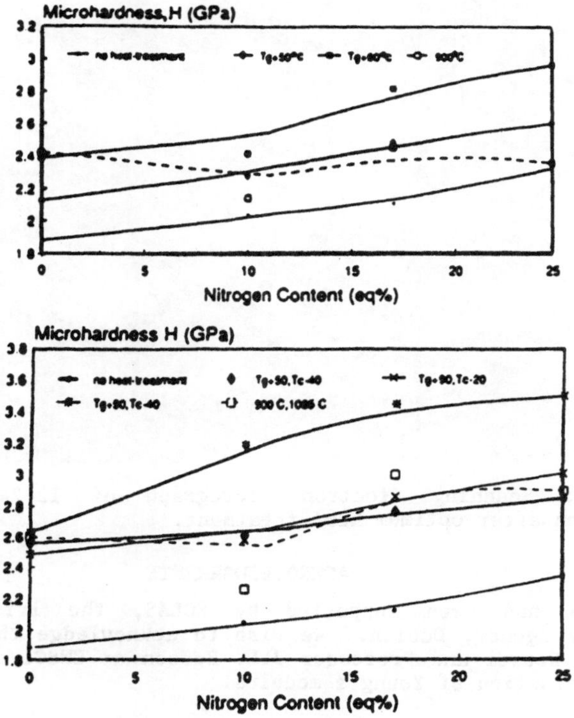

Figure 5: Variation of microhardness with nitrogen content for 12:24:64 (Nd:Mg:Si) composition after (a) single heat treatments, (b) double-stage heat-treatments.

CONCLUSIONS

Glasses have been prepared in various RE-Si-Al-O-N (RE = Y, Nd, Sm) systems and in the Nd-Mg-Si-O-N system with up to 25e/o N by melting appropriate mixtures of oxides and nitrides in nitrogen at 1700°C. When the cation composition is kept constant and only the oxygen/nitrogen ratio is varied, properties such as T_g, microhardness, viscosity and Young's modulus increase with increasing nitrogen content.

From results of differential thermal analysis, T_g and T_c for the glasses have been measured. Heat treatments have been carried out on glasses for 2.5 h at various temperatures involving also two-stage treatments. The optimum schedule as evidenced by increases in microhardness is 2.5h each at (T_g+90) and (T_c-40). This leads to formation of fine-grained glass-ceramics containing a N-apatite phase.

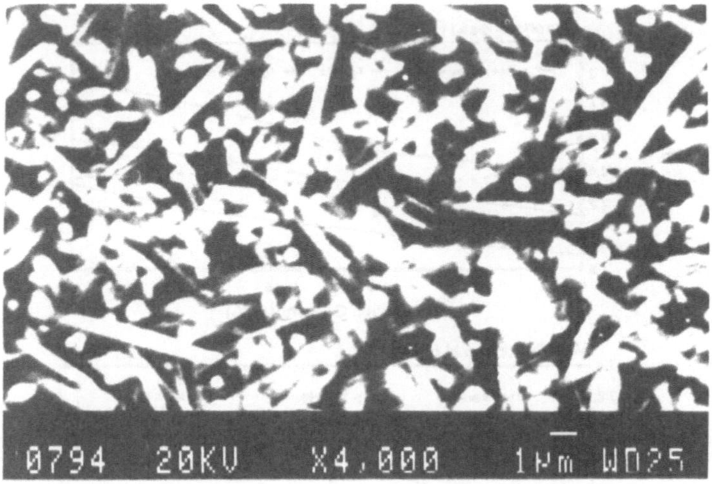

Figure 6: Scanning Electron Micrograph of 12:24:64 (Nd:Mg:Si) composition after optimum heat treatment.

ACKNOWLEDGEMENTS

This work has been supported by EOLAS, the Irish Science and Technology Agency, Dublin. We wish to acknowledge the assistance of Dr Tanguy Rouxel and Professor J.L. Besson at ENSCI, Limoges, France for determination of Young's modulus.

REFERENCES

1. Jack, K.H., Sialons-A Study in Materials Development. In Non-Oxide Technical and Engineering Ceramics, ed. S. Hampshire, Elsevier-Applied Science Publishers, London 1986, pp. 1-30.
2. Jack, K.H., Sialon Glasses. In Nitrogen Ceramics. ed. F.L. Riley, Noordhoff, The Hague, 1977, pp. 257-262.
3. Loehman, R.E., Preparation of Yttrium-Silicon-Aluminium-Oxynitride Glasses, J. Amer. Ceram. Soc., 1979, 62, 491-494.
4. Drew, R.A.L., Hampshire, S. and Jack, K.H., Nitrogen Glasses. In Special Ceramics 7, ed. P. Popper, Brit. Ceram. Proc., 1981, 31, 119-132.
5. Hampshire, S., Drew, R.A.L. and Jack, K.H. Oxynitride Glasses. Phys. Chem. Glass., 1985, 182-186.
6. Lewis, M.H., Leng-Ward, G. and Mason, S., Microstructural Design of High-Temperature Ceramics. In Engineering with Ceramics 2, eds. R. Freer, S. Newsham and G. Syers, Brit. Ceram. Proc., 1987, 39, 1-14.
7. Lewis, M.H., Mason, S. and Szweda, A., Syalon Ceramic for Application at High Temperature and Stress. In Non-Oxide Technical and Engineering Ceramics, ed. S. Hampshire, Elsevier-Applied Science Publishers, London, 1986, pp. 175-190.
8. Morrissey, V., Lonergan, J., Pomeroy, M.J. and Hampshire, S., Crystallization Treatments for Neodymia-Containing Oxynitride Glasses and Glass-Ceramics, In Fabrication Technology, eds. R.W. Davidge and D.P. Thompson, Brit. Ceram. Proc., 1990, 45, 23-32.
9. Rouxel, T., Besson, J.-L., Gault, C., Goursat, P., Leigh, M. and Hampshire, S., Viscosity and Young's Modulus of an Oxynitride Glass. J. Mater. Sci. Lett., 1989, 8, 1158-1160.
10. Evans, A.G. and Charles, E.A., J. Am. Ceram. Soc., Discussions and Notes, 1976, 59 **[7-8]**.

TEM STUDIES OF SRBSN WITH UNSTABILIZED ZrO_2 ADDITIONS

H.-J. Kleebe[1], W. Braue[2], W. Luxem[2], M. Rühle[1]

(1) Max-Planck Institut für Metallforschung, Institut für Werkstoffwissenschaft, D-7000 Stuttgart 1, FRG

(2) Deutsche Forschungsanstalt für Luft- und Raumfahrt (DLR), Institut für Werkstoff-Forschung, D-5000 Köln 90, FRG.

ABSTRACT

We report on a correlation of the densification behavior and microstructural development of ZrO_2-fluxed SRBSN as monitored by dilatometer experiments and TEM studies. Two distinct densification events are resolved by means of highly sensitive dilatometry. In this material ZrO_2 only acts as an effective sintering aid because of the increased sintering temperatures and high N_2-overpressue during the second dilatometer maximum which promote the active participation of ZrO_2 in the liquid phase formation process.

I. INTRODUCTION

Silicon nitride materials cannot be fully densified by classical solid-state sintering techniques due to its high covalent bonding character and low self-diffusivity, respectively. The addition of sintering aids is a prerequisite to promote liquid phase sintering in order to achieve nearly complete densification during processing. The Si_3N_4 bodies investigated were prepared via RBSN-route and subsequent gas-pressure sintering with unstabilized ZrO_2 additions to ensure liquid phase formation. Unexpectedly, the SRBSN grades could be fully densified during post sintering using only 5 wt.% of the sintering aid. The overall microstucture of the dense bodies is characterized by large elongated β-Si_3N_4 grains embedded in a fine-grained β-Si_3N_4 matrix. The crystalline secondary phase located at triple grain junctions is homogeneously distributed in the material.

As unstabilized m-ZrO_2 was added to the powder blends prior to nitridation and subsequent sintering and, moreover, m-ZrO_2 was detected as the predominant crystalline secondary phase after densification, the densification mechanism remained unclear. Therefore, this study focusses on the correlation of microstructural features observed by means of HREM and AEM with distinct densification events as monitored by a unique dilatometer design of improved sensitivity. The following outline gives a comprehencive summary of the project; a full draft is published elsewere [1].

II. EXPERIMENTAL AND INSTRUMENTATION

The SRBSN material was processed via a two-step gas-pressure sintering cycle with increased N_2-pressure as described in [2]. In the dilatometer experiment two distinct densification events are resolved at 1730-1750°C and at 1900-1920°C depending on the heating rate of the sintering furnace. The dilatometer measurements revealed densification rates unique to the ZrO_2-fluxed system compared with any other sintering additive. It is important to note that the first dilatometer maximum was very low with 0.5 µm/min compared to the second, which was strongly pronounced with 70 µm/min (see Fig. 1).

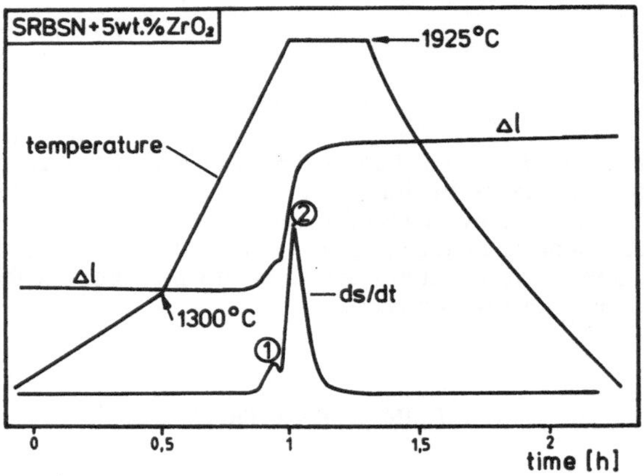

Figure 1. Heating rate, shrinkage and densification rate of the dilatometer experiment revealing two dinstinc densification maxima 1 and 2.

Microstructural development of the material correlating to the first and second dilatometer maximum, respectively, was analysed by HREM, small probe micro-analysis (EDS, EELS), and CBED. TEM foils of the corresponding materials were prepared by standard ceramographic techniques.

III. RESULTS AND DISCUSSION

The specimen of the **first dilatometer maximum** showed distinct differences compared with the densified material. The low-temperature sintered body consists of homogeneously fine-grained β-Si_3N_4 particles with a relatively high amount of residual porosity of approximately 15 vol%. No elongated large β-Si_3N_4 grains were present at this stage of sintering. Some extended aggregates of the crystalline secondary phase (up to 5 µm) presumably due to agglomeration during powder processing were observed (see Fig.2). It is important to note that these agglomerates are only present under the conditions of this low-temperature sintering event.

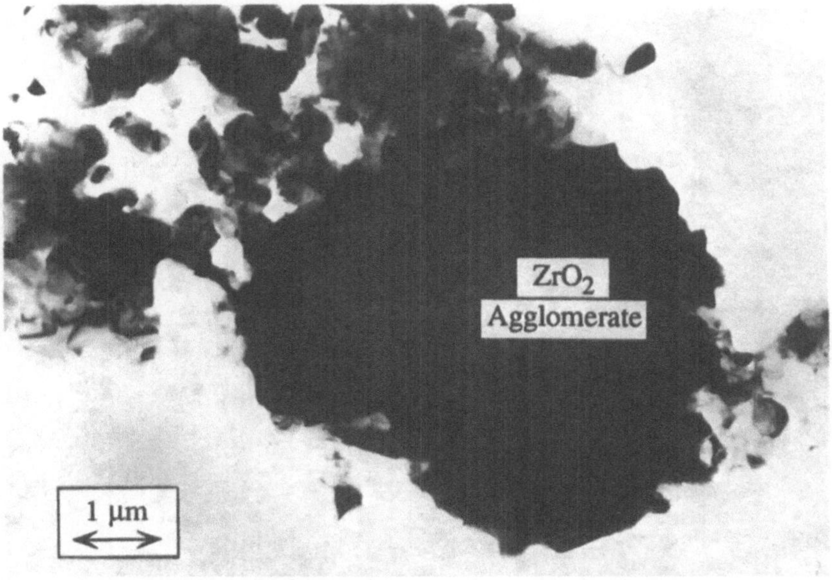

Figure 2. Low-magnification TEM micrograph depicting a large agglomerate of the crystalline secondary phase surrounded by a fine grained β-Si_3N_4 matrix which shows residual porosity (1st dilatometer maximum).

The Zr-oxinitride phases reported in the literature [3] were not confirmed by x-ray and electron diffraction. Apart from Zr and O EDS and EELS analysis of the secondary phases revealed a low amount of Si, but no N could be detected on the basis of a 5 at.% detection limit. Moreover, due to thermodynamic calculations in the ZrO_2-ZrN system [4] most Zr-oxinitride phases are shown to decompose irreversibly to ZrO_2, ZrN and N_2 at temperatures above 1200°C. ZrN was only present on the specimen surface as a thin coating but could not be detected in the bulk sample. During sintering Si_3N_4 could react with ZrO_2 on the specimen surface to form ZrN according to [4]

$$4\,Si_3N_4 + 6\,ZrO_2 \rightarrow 12\,SiO_{(g)} + 6\,ZrN + 5\,N_{2(g)} \tag{1}$$

where the gaseous constituents can leave the system forming a ZrN cover on the specimen surface. However, above 1400°C following reactions (2) and (3) can take place

$$2\,SiO_2 + ZrN \leftrightarrow ZrO_2 + 2\,SiO_{(g)} + 0.5\,N_{2(g)} \tag{2}$$

$$3\,SiO_2 + 2\,ZrN \leftrightarrow Si_2N_2O + 2\,ZrO_2 + SiO_{(g)} \tag{3}$$

retransforming ZrN back to ZrO_2 mainly in the bulk sample. The x-ray data from this material reveal a combination of t- and m-ZrO_2 while HREM and CBED showed heavily twinned m-ZrO_2 as the prominent secondary phase constituent (Fig. 3).

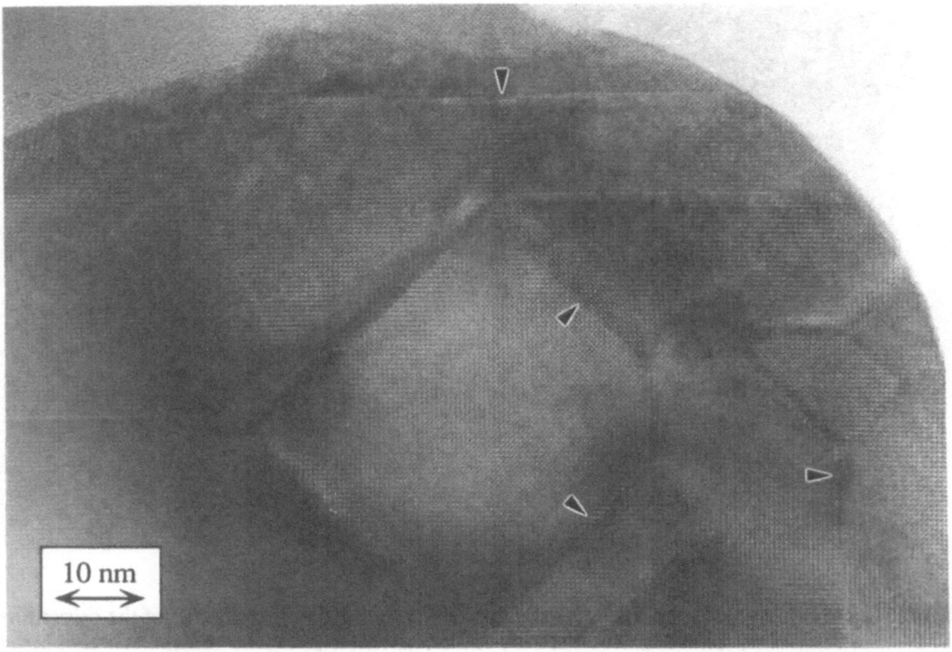

Figure 3. HREM image of the heavily twinned m-ZrO_2 phase from the first low-temperature sintering event (twin boundaries are marked by arrows).

Apart from the Zr-containing second phase Si_2N_2O was observed in these specimens. The heavily faulted Si_2N_2O contained spherical inclusions which were identified as α-Si_3N_4 by means of HREM and CBED. This is consistent with former observations of Braue [5] indicating that small α-Si_3N_4 particles may act as nuclei during growth of Si_2N_2O in SiO_2-Si_3N_4 materials. However, in the ZrO_2-fluxed SRBSN material an eutectic SiO_2-rich liquid is formed at approximately 1680°C and, as the α-β-Si_3N_4 transformation is not yet completed, Si_2N_2O is formed as a transient phase surrounding the α-Si_3N_4 particles which is in accordance with phase relationships observed by Gauckler et al. [6].

Under the conditions of the **second dilatometer maximum** both densification behavior and microstructure of the ZrO_2-fluxed SRBSN changed dramatically. The material was fully densified and no more large secondary phase agglomerates were observed. This indicates that ZrO_2 is actively involved in the liquid phase formation process. Si_2N_2O is no longer present in the material. Huang et al. [7] reported on the thermal degradation of Si_2N_2O above 1830°C releasing additional SiO_2 and Si_3N_4. Hence, both Si_2N_2O and ZrO_2, participating in the liquid phase formation, give rise to the strongly pronounced sintering rate observed in the second dilatometer maximum.

With increasing sintering temperature the large agglomerates of the sintering aid, still present at the first dilatometer maximum, start now to melt thus forming a heterogeneous liquid phase distribution within the material. Therefore, locally enhanced solution-diffusion and reprecipitation of Si_3N_4 will occur resulting in the formation of large elongated β-Si_3N_4 grains (see Fig. 4).

Figure 4. Bright field image of the densified ZrO_2-fluxed material showing large elongated β-Si_3N_4 grains embedded in a fine grained β-Si_3N_4 matrix. The crystalline secondary phase is homogenously distributed at triple grain junctions (no large agglomerates are observed).

Due to the formation of such in situ-grown elongated β-Si_3N_4 particles an improved fracture resistance of 10 MPa(m)$^{1/2}$ was measured in this material [8] compared to commercial SSN-grades with an average toughness of 4-5 MPa(m)$^{1/2}$. The main toughening mechanisms are crack bridging and crack deflection.

The crystalline secondary phase of the material corresponding to the second dilatometer maximum is homogeneously distributed at Si_3N_4 three- and four-grain junctions. CBED confirmed m-ZrO_2 as the apparent modification with typical (110) and (100) twins observed by means of TEM. An interesting feature along the Si_3N_4/ZrO_2 phase boundary are strong strain contours. This can be due to the martensitic t->m phase transformation upon cooling with a volume increase of 3-5%. On the other hand it can be related to a characteristic microstructural feature of the dense material: HREM studies of the ZrO_2-fluxed Si_3N_4 system revealed unexpectedly thin phase and grain boundary films (see Fig. 5).

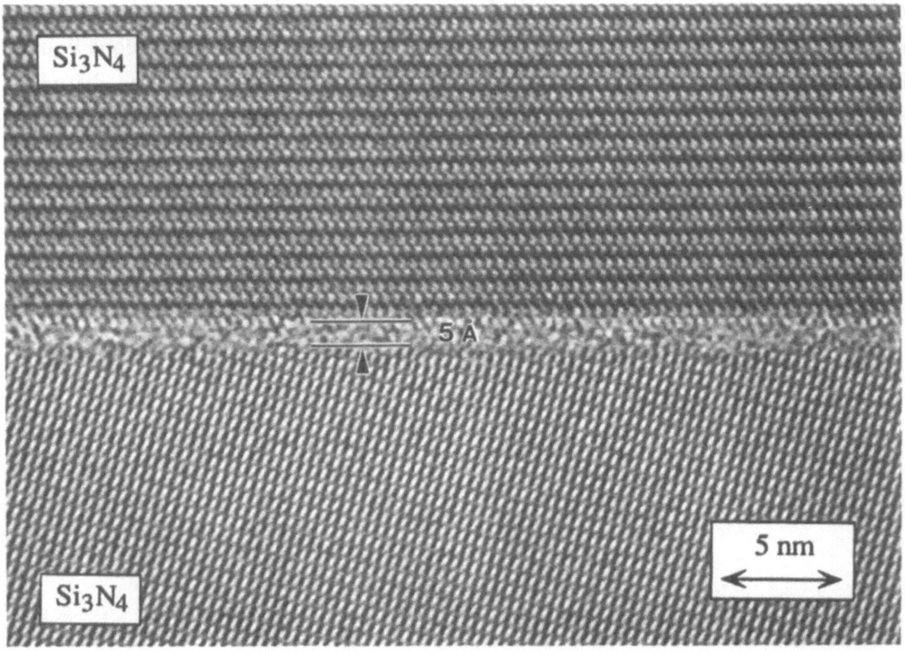

Figure 5. HREM image of a Si_3N_4/Si_3N_4 grain boundary showing an extremely thin amorphous grain boundary film along the interface.

The film thickness was up to one order of magnitude less compared to Y_2O_3 and other RE-doped SSN systems! Hence, such thin phase and grain boundary films cannot release internal strain due to thermal expansion mismatch (and/or volume change during phase transformation). Additional chemical analysis of the grain boundary films using a dedicated STEM with a probe size of 0.8 nm revealed excess O but no Zr could be detected in the amorphous film. This is consistent with the ZrO_2-SiO_2 phase diagram and the formation of a SiO_2-rich and highly viscous eutectic liquid during sintering of ZrO_2-fluxed SRBSN. Apart from the thinness of these films the chemical composition of the grain boundary films governs the softening temperature of these glassy structures. Such a silica-rich intergranular film probably shows a T_g which is higher than the t->m transformation temperature indicating that residual stresses cannot be released via viscous flow of the amorphous phase or rearrangement of the surrounding grains.

IV. SUMMARY

In conclusion it can be stated that the two dilatometer maxima are correlated unequivocally with the microstructural development observed by means of HREM and AEM. The relatively little pronounced first dilatometer maximum is due to the formation of a highly viscous silica-rich liquid phase. During this stage of sintering ZrO_2 is not involved in the liquid formation process. The second maximum with a pronounced densification rate is attributed to a radically change in secondary phase composition incorporating the sintering aid as monitored by the observations of corresponding TEM foils. The results suggest a good performance of this material under a large variety of service conditions due to (i) the formation of in-situ grown large β-Si_3N_4 grains, (ii) a completely crystallized secondary phase (m-ZrO_2), and (iii) surprisingly thin amorphous phase and grain boundary films. The latter aspect is pursued in ongoing investigations.

REFERENCES

1. Kleebe, H.-J., Braue, W. and Luxem, W., Densification of SRBSN with unstabilized zirconia: dilatometry and electron microscopy, submitted to J. Mater. Sci. (1991).

2. Kleebe, H.-J., Ziegler, G., Influence of crystalline secondary phases on densification behavior of reaction-bonded silicon nitride during post-sintering under increased nitrogen pressure, J. Am. Ceram. Soc. 72 (1989) 2314-2317.

3. Ikeda, S., Yagi, T., Ishizawa, N., Mizutani, N. and Kato, M., A new face-centered cubic phase in the ZrO_2-ZrN system, J. Solid State Chem. 73 (1988) 52-56.

4. Weiss, J., Gauckler, L.J., Lukas, H.L. and Petzow, G., Determination of phase equilibria in the system Si-Al-Zr/N-O byexperiment and thermodynamic calculation, J. Mater. Sci. 16 (1981) 2997-3005.

5. Braue, W., Konvergente Elektronenbeugung in der analytischen Elektronenmikroskopie keramischer Werkstoffe - eine Anleitung für die Praxis, Mat.-wiss. u. Werkstofftech. 21 (1990) 72-84.

6. Gauckler, L.J., Weiss, J. and Petzow, G., Stability of Si_3N_4 and SiC based materials containing ZrO_2, in: Proc. of 4th Int. Meeting on Modern Ceramics Technologies, Saint Vincent, Italy 28-31 May (1979), ed. P. Vincenzini, Elsevier Sci. Publ. Co. (1980), pp. 671-679.

7. Huang, Z.K., Greil, P. and Petzow, G., Formation of Si_2N_2O from Si_3N_4 and SiO_2 in the presence of Al_2O_3, Ceram. Int. 10 (1984) 14-17.

8. Kleebe, H.-J., Evans, A.G., The microstructure and toughness of in-situ reinforced silicon nitride, submitted to J. Am. Ceram. Soc. (1991).

HIGH-TEMPERATURE REACTIONS IN THE B-Al-Si-N-C SYSTEM

R.J.Oscroft and D.P.Thompson
Wolfson Laboratory, Materials Division,
Department of Mechanical,Materials & Manufacturing Engineering,
University of Newcastle upon Tyne, U.K.

ABSTRACT

Boron, in its tetragonal and rhombohedral forms, is characterized by an ability to incorporate a wide spectrum of metals (e.g Si,Y,Zr,Ln,U) and non-metals (C,O,P,As) into the crystal structure whilst still retaining the basic structure type. The present work has extended this by studying the incorporation of nitrogen into boron carbide,and at temperatures in the range 800 -1100°C, the solubility is ≈3.8 w/o. At these and higher temperatures, there is a competing reaction whereby the B_4C structure is broken down and an amorphous B-N-C product is formed. This resists crystallization even at temperatures as high as 1500°C.

Oxygen is always present as a surface layer on carbide and nitride powders used in high-temperature reactions. Its role in the densification of nitrogen ceramics is well known; the present work describes its role in the formation and densification of Al_4SiC_4 ceramics.

INTRODUCTION

An important feature of boron crystal chemistry is the extensive range of other elements which can be incorporated in structures containing B_{12} units; some of these elements stabilize the tetragonal boron arrangement (e.g. Be in BeB_{12}), but most stabilize the rhombohedral boron arrangement. It is interesting that rhombohedral boron itself, the unit cell of which consists of three B_{12} icosahedra arranged in a rhombohedral stacking sequence, melts at ≈2300°C (showing strong covalent bonding between adjacent B_{12} units), whereas boron carbide, which has a similar unit cell consisting of three $B_{11}C$ icosahedra but linked by C-B-C chains, exhibits only slightly stronger bonding with a melting point of ≈2500°C. Hydrogen treatment easily strips

TABLE 1. Unit cell dimensions (Å) for some rhombohedral boron structures

MB_{12}			$B_{13}X_2$			B_4X		
M	a	c	X	a	c	X	a	c
Zr	5.238	12.83	C	5.66	12.28	O	5.37	12.35
Lu	5.278	12.93	Si	6.33	12.75	P	5.934	11.75
U	5.284	12.95				As	6.142	11.88
Y	5.303	12.99	Rhombohedral boron:					
Dy	5.304	12.99	a = 4.908, c = 12.56					

carbon atoms from B_4C up to the composition $B_{10}C$ and in many ways the carbon atoms in B_4C can be treated as interstitials. This is more obviously the case for the $B_{50}C_2$ carbide [3]. Table 1 summarizes a selection of MB_{12}, $B_{13}X_2$ and B_4X compounds which exhibit rhombohedral boron structure types. This similarity in structure can in some cases lead to benefits in materials processing. Thus, for example, the similarity between B_4Si and B_4C results in a small (≈2.5 a/o) solubility of Si in B_4C as reported by Telle [2], which assists sintering because of the low melting (Si,B,C) eutectic; the silicon is then subsequently incorporated into the final B_4C structure. In contrast, aluminium does not form a stable AlB_4 compound of the B_4C type, and Al solubility in B_4C is therefore low.

The main limitation of B_4C materials is the relatively poor oxidation resistance because of easy loss of the "interstitial" carbon as CO and conversion of the boron residue to low-melting B_2O_3. Despite the existence of the $B_{13}O_2$ oxide [1], this is a relatively unstable intermediate in which the oxygen atoms are again interstitial in character and easily displaced in mildly oxidizing or reducing environments, so that a $B_{13}O_2$ surface on a B_4C material would not offer satisfactory oxidation protection at temperatures above ≈1000°C.

The incorporation of nitrogen into B_4C has not been explored in any depth in previous work. Will and Kossobutzki [4] described the formation of a $B_{50}N_2$ compound, but again this is relatively unstable in both oxidizing and reducing environments. The present work is part of a more general investigation of the B-Al-Si-N-C system, and describes the preparation of nitrogen-containing B_4C materials.

The formation of dense materials in these and other high-temperature carbide and nitride systems relies on additives which promote transfer of material by either solid-state or liquid-phase processes. Previous work on the formation of Al_4SiC_4 by reaction between Al_4C_3 and SiC [5] has shown that dense materials can be produced by hot-pressing without additives, whereas pressureless sintering gives very little densification. Oxygen is an important factor in the reactions taking place, and this paper describes further work on its role in the densification process.

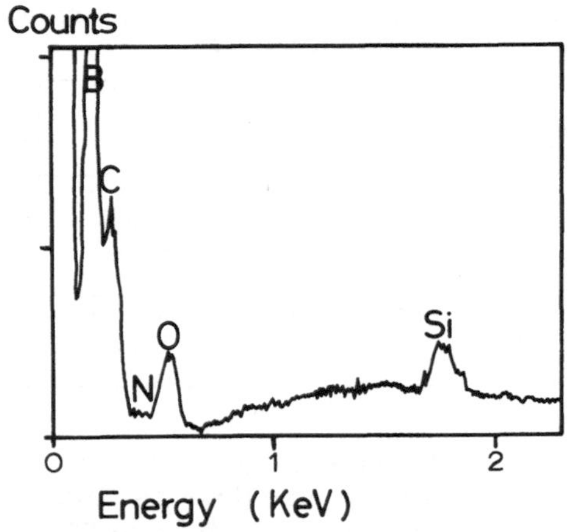

FIGURE 1. EDX spectrum of as-received Starck HS B_4C powder.

EXPERIMENTAL METHODS

The investigation of nitrogen incorporation into B_4C used as-received B_4C powder (H.C.Starck-Berlin, Grade HS, FSSS 0.78 μm, 75.80% B, 22.08% C, 1.30% O, 0.45% N, 0.03% Fe, 0.08% Si, 0.005% Al, see EDX spectrum shown in Figure 1), which was heated in flowing ammonia (flow rate 300 cc min^{-1}) at 100°C intervals between 900 and 1100°C. Phase identification was carried out using Hägg-Guinier X-ray diffraction photographs and monochromatic $CuK\alpha_1$ radiation. Microanalysis was carried out on powders using windowless EDX (eXL model, Link Systems, U.K.) in conjunction with a Camscan S4-80DV scanning electron microscope (Camscan U.K.). Nitrogen contents of the nitrided B_4C powders were measured using a combustion analyzer (Carlo Erba, Model 1106).

Reactions involving Al_4C_3 and SiC were carried out on mixed powders prepared by ball milling equimolar amounts of Al_4C_3 (H.C. Starck-Berlin, special grade, FSSS 2μm, 72.44% Al, 24.5% C, 2.2% O, 0.8% N and 0.06% Fe) and SiC (H.C. Starck-Berlin, Grade A10, FSSS 0.59 μm, 30.19% C, 0.72% O, 0.013% Fe, 0.02% Al, 0.003% Ca). The main impurity element was oxygen, present in significant amounts in the Al_4C_3 powder and further increased by subsequent processing. It is well known that Al_4C_3 is unstable in moist air, resulting in the formation of a surface layer of amorphous alumina. However, once some hydrolysis has occurred, the surface layer slows down further hydrolysis. Cold uniaxially and isostatically compacted powder samples were embedded in BN powder in graphite crucibles and fired in a graphite-element resistance furnace under argon. Firing was carried out at 100°C intervals between 1000-1700°C and at 50°C intervals between 1700 and 2100°C for 1 hour using a heating rate of 30Kmin^{-1} and a natural cooling rate of ≈30Kmin^{-1}.

TABLE 2. Unit cell parameters and nitrogen content of nitrogen containing boron carbide

Sintering conditions T(°C)/time/gas	B_4C unit cell parameters a(Å)	c(Å)	V(Å^3)	w/o nitrogen (±0.15)
As received B_4C	5.6087	12.0829	329.17	0.40
800/12h/NH_3	5.6071	12.0865	329.08	2.69
900/12h/NH_3	5.6050	12.0871	328.84	3.48
1000/12h/NH_3	5.6062	12.0819	328.85	3.76
1100/12h/NH_3	5.6040	12.0919	328.87	3.53

RESULTS AND DISCUSSION

The nitriding of boron carbide

Table 2 shows the results of nitriding B_4C powder in fast-flowing ammonia at 800-1100°C. X-ray diffraction photographs of the product showed B_4C as the only crystalline phase, but careful measurements of unit cell dimensions showed that both **a** and **c** dimensions had systematically decreased with increasing nitriding temperature, corresponding to an increase in nitrogen content (Figure 2). EDX spectra (Figure 3) confirmed the presence of

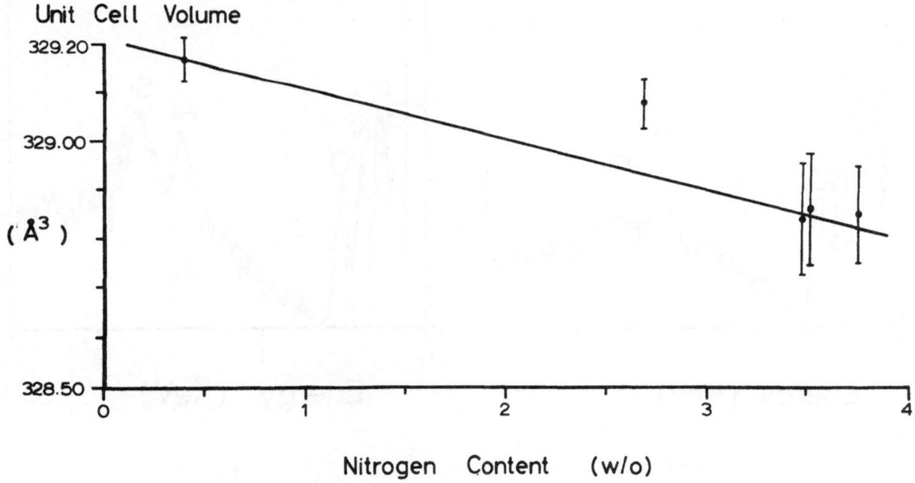

FIGURE 2. Unit cell dimensions of B_4C as a function of nitrogen content.

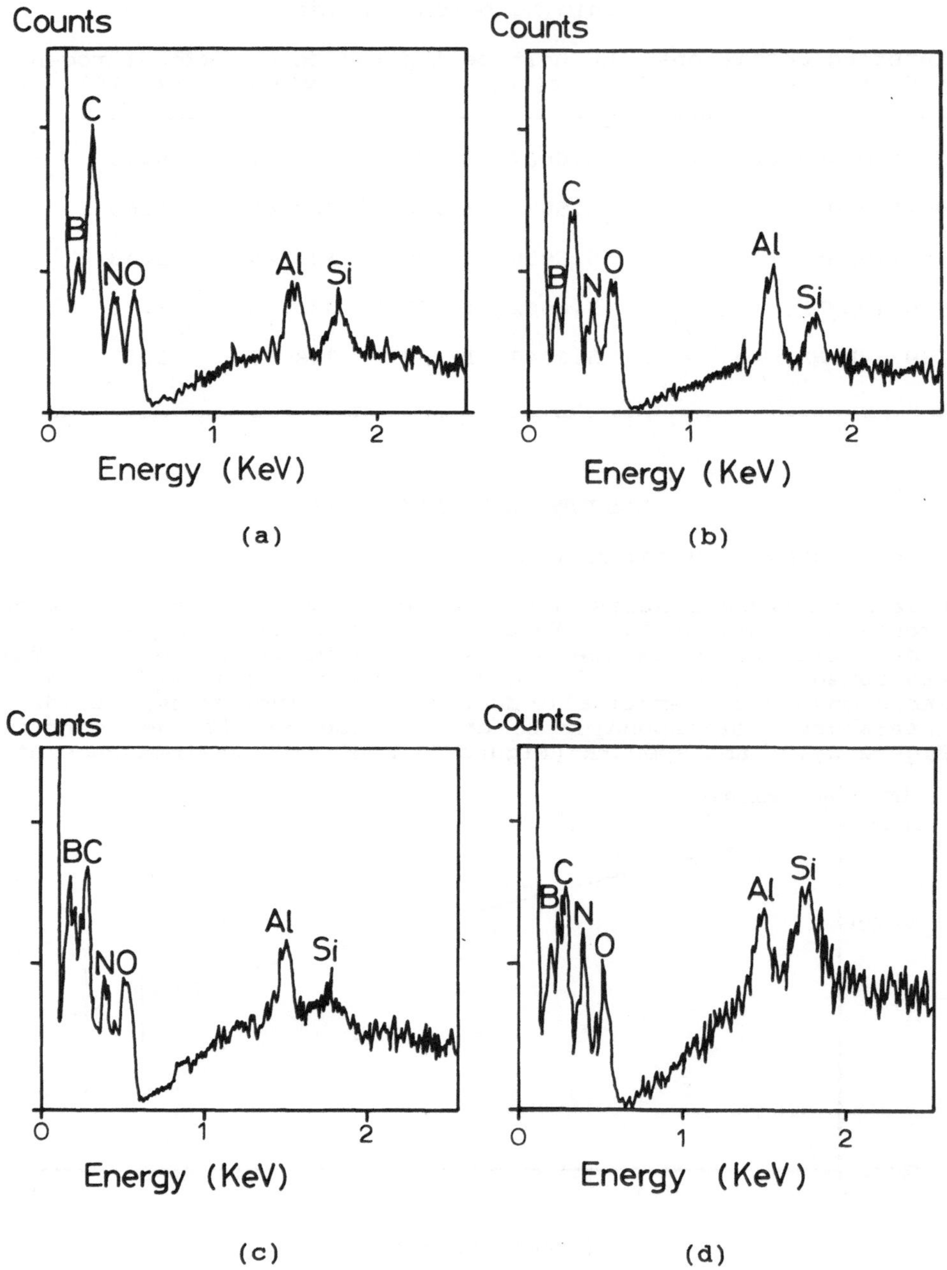

FIGURE 3. EDX spectra of B_4C powders nitrided in ammonia at (a) 800°C, (b) 900°C, (c) 1000°C and (d) 1100°C for 24 hours.

nitrogen; precise quantitative estimation from this data is difficult. The Si and O peaks in these spectra are from impurities present in the starting B_4C powder, and the aluminium peak is from the sample holder. The decrease in unit cell dimensions resulting from nitrogen incorporation can be attributed either to substitution of C by N in chain sites, or to the nitrogen atoms occupying similar sites to those occupied by aluminium in C_4AlB_{24}, i.e. bonded to the central B atom in the C-B-C chain [6]. Under more aggressive nitriding conditions, the B_4C structure is destroyed and an amorphous product remains. A sample nitrided at 1000°C and showing an amorphous X-ray diffraction pattern, had a nitrogen content of 35.4% and a carbon content of 6.7%. Heat-treatment of this material at 1500°C failed to achieve crystallization; further work is in progress to identify the nature of this amorphous phase, and to determine whether new crystalline B-N-C phases can be produced in this way.

Densification of Al_4SiC_4

Densification of carbon-based ceramics is not a simple process, and alternative methods of achieving fully dense products by pressureless sintering at temperatures below 2000°C are of great interest. In the current work, a detailed study of the densification behaviour of Al_4SiC_4 has been carried out.

The reaction sequence for the formation of Al_4SiC_4 is shown schematically in Figure 4. Starting phases were 9R Al_4C_3 and a mixture of 6H- and 4H-SiC. Reaction starts at ≈1500°C with the formation of Al_2OC by reaction between the surface layer of alumina present on Al_4C_3 grains and the Al_4C_3 itself; this is consistent with the Al_2O_3 - Al_4C_3 diagram given by Foster et al. [7]. As the temperature increases, the amount of Al_2OC maximizes at ≈20% at 1600°C and then decreases because of reaction with SiC to produce Al_4SiC_4 plus small amounts of an aluminosilicate-liquid rich in aluminium. The Al_4SiC_4 content increases to a maximum at ≈1750°C, assisted by solution precipitation via the newly-formed liquid phase. At 1775°C, some $Al_4Si_2C_5$ also starts to form because of nitridation of the original Al_4C_3, which on

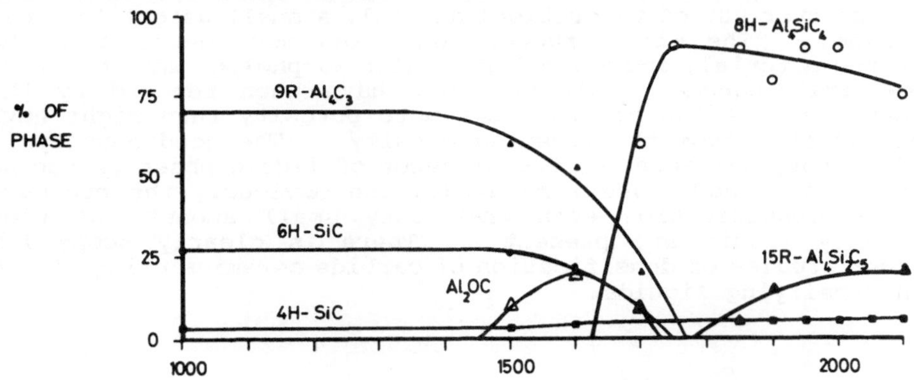

FIGURE 4. Reaction sequence for the formation of Al_4SiC_4 from mixed Al_4C_3/SiC powders.

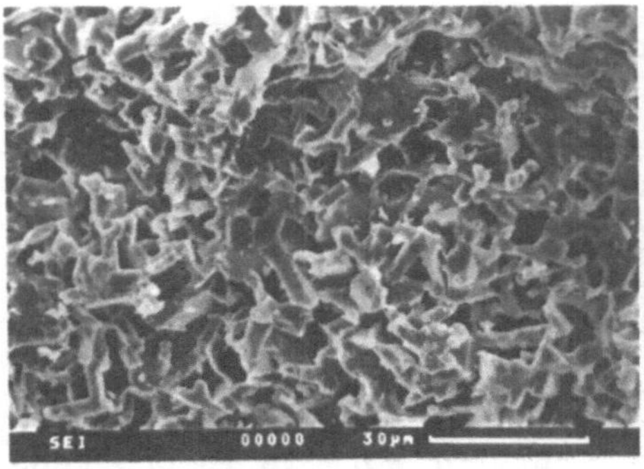

FIGURE 5. Microstructure of Al_4SiC_4, pressureless sintered at 1800°C for 1h.

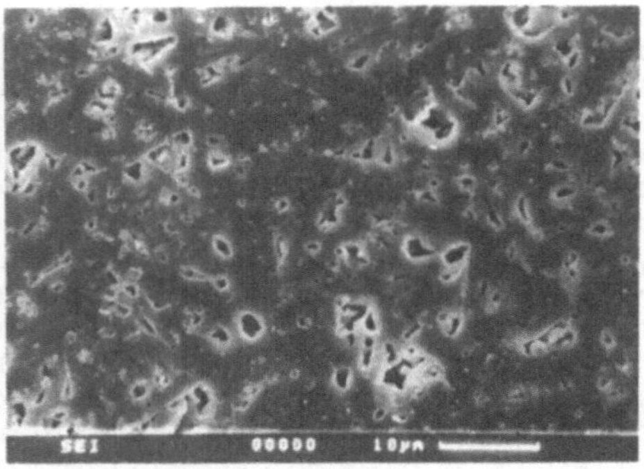

FIGURE 6. Microstructure of a polished and etched surface of Al_4SiC_4, HP 1900°C for 1h.

reaction with silicon carbide, moves the overall composition towards an M_5X_6-type $nSiC.Al_4C_3.(2-n)AlN$ composition. The original 4H SiC remains constant throughout the reaction; this is thought to be due to stabilization by aluminium, which is a well-known stabilizer of the 4H polytype [8,9]. Densities of pressureless-sintered samples are very low (typically 50-60% of theoretical) and weight losses of 10% are observed, probably attributable to dissociation of the liquid at high temperature. Fracture surfaces show little evidence of liquid phase sintering, perhaps due to the small amount of liquid phase, its high viscosity and also, at these temperatures, its dissociation. Figure 5 shows a typical microstructure of an Al_4SiC_4 composition pressureless-sintered at 1800°C for 1 hour. The fracture surface is very porous and shows large tabular grains of Al_4SiC_4.

Hot-pressing was used to improve the density of the products. Figure 6 shows a polished and etched microstructure of an Al_4SiC_4 product, hot-pressed at 1900°C for 1 hour. The measured density was just over 98% of theoretical and only a small weight loss was observed. The microstructure was also more dense than the sintered material, and a similar tabular morphology was observed. Some grain-boundary material must have been removed by the etchant because the sample shows more porosity than might have been expected from the measured density. The good density is confirmatory evidence for the presence of liquid phase (since no densification additives were used), and obviously the pressure assists densification even when only small amounts of high viscosity liquid are present. There is clearly scope for further studies of densification of carbide ceramics using Al-O-C rich densifying liquids.

CONCLUSIONS

≈3.8 w/o of nitrogen can be incorporated into boron carbide at 1000°C. This level can probably be increased using careful preparative techniques at lower temperatures (700-900°C). At higher temperatures, the B_4C structure breaks down and an amorphous B-C-N phase is produced. Further studies will be directed towards identifying the sites occupied by nitrogen in the B_4C structure, and also determining how the nitrogen solubility changes in the presence of other elements (Al,Si).

Whereas silica-rich aluminosilicate liquids have high viscosities and low carbon solubilities, these are not useful for densifying carbide-based materials, whereas aluminium-rich liquids based on the Al_2OC composition have lower viscosities and higher carbon solubilities and do achieve some densification in the case of Al_4SiC_4 type materials. Hot-pressing is still needed, however, to obtain full densities in these materials.

Acknowledgements

The authors gratefully acknowledge the assistance of Mr D. Dunbar, Department of Chemistry, University of Newcastle upon Tyne, for carrying out the nitrogen analyses. One of us (R.J.O.) gratefully acknowledges support from the Science and Engineering Research Council for this work.

REFERENCES

1. Matkovich, V.I., Giese, R.F.Jr. and Economy, J., Z. Krist. 1965, **122**, 116-30.
2. Telle, R. in The Physics of Carbides, Nitrides and Borides, ed. R. Freer, Kluwer Academic Publishers, Dordrecht, 1990, pp. 249-68.
3. Will, G. and Kossobutzki, K.H., Z. Krist., 1975, **142**, 384-97.
4. Will, G. and Kossobutzki, K.H., J. Less Comm. Met., 1976, **47**, 33-8.
5. Oscroft, R.J., Korgul, P. and Thompson, D.P., in Proc. 3rd Int. Symp. 'Ceramic Materials and Components for Engines', ed. V.J. Tennery, pub. The American Ceramic Society, 1989, pp. 661-670.
6. Perrotta, A.J., Townes, W.D. and Potenza, J.A., Acta Cryst., **B25**, 1223-1229.
7. Foster, L.M., Long, G. and Hunter, M.S., J. Amer. Ceram. Soc., 1956, **39**, 1-11.
8. Schwetz, K.A. and Lipp, A, Sci. Ceram., 1980, **10**, 149-158.
9. Mitomo, M., Inomata, Y. and Kumonomido, M., Yogyo Kyokai Shi, 1970, **78**, 224-228.

CHARACTERIZATION OF PRESSURELESS SINTERED ALPHA SiC BY ADVANCED TECHNIQUES OF SCANNING ELECTRON MICROSCOPY AND X-RAY DIFFRACTION

A. CAMANZI, B.A. DE ANGELIS, G. GIUNTA, S. LORETI, C. RIZZO and P. ALESSANDRINI.
ENIRICERCHE S.p.A., 00015 Monterotondo, Rome, Italy.

ABSTRACT

Sintered alpha-SiC has been investigated by Scanning Electron Microscopy (SEM) and X-Ray diffraction (XRD). By using channeling contrast effects produced when SEM is operating with low beam energy and high current density, grain size and shape, boron and carbon inclusions, porosity are identified on the same surface without chemical or thermal etching of the samples. From XRD data, by applying the Rietveld method, the weight percentage of SiC polytypes has been evaluated without using standard powder mixtures of alpha-SiC. The percentage of the 4H polytype has been studied as a function of the sintering temperature and has been tentatively related to the presence of exaggerately grown grains.

INTRODUCTION

The microstructure of pressureless sintered silicon carbide has been widely investigated and a large amount of data, obtained with a variety experimental techniques, has been reported (1-4). However, there is still room for a more detailed study of some aspects, particularly those regarding the role of the sintering aids and the mechanism of their action in promoting densification (5).
While the data on grain size and shape, grain boundaries and inclusions (and their relationship to the processing conditions) are abundant, considerably less attention has been dedicated to the determination of the polytype distribution. This kind of investigation has been carried out mainly on samples obtained from beta-SiC starting powders (6,7). The data on alpha-SiC are scanty (8).
In this work we report on the microstructure of alpha-SiC sintered at different temperatures, as studied by scanning electron microscopy (SEM) at low beam energy and high current density. For the same samples the polytype composition has been obtained from X-ray diffraction data by using the

Rietveld method (9,10).

MATERIALS AND METHODS

Samples Preparation
A commercially available alpha-SiC was used: UF-10 (Lonza), BET 10.5 m^2/g. The sintering aids were amorphous boron (H.C. Starck) and carbon black (Degussa). The samples were sintered in argon atmosphere in a graphite furnace. Relative density of the samples sintered at 2100 °C was 97-98%.
The polished pellets were either examined as such or after chemical etching (boiling Murakami solution) or thermal etching (30 minutes, 1550 °C, Ar atmosphere).

SEM Techniques
To obtain orientation informations from small areas and with enough contrast a selected area electron channeling pattern (SACP) technique was used. In order to apply SACP techniques complicated methods are generally used, but resolution in grain size and contrast sensitivity are usually very limited. By equipping a standard SEM with a high brightness gun (LaB_6 cathode) and a solid state backscattered electron detector having high collection efficiency and by using a suitable take-off angle and low electron beam energy it was possible to identify in backscattered electron channeling mode grains smaller than 0.5 um and inclusions of B and C smaller than 0.1 um thanks to the high sensitivity in contrast obtained (11).

X-Ray Techniques
In order to evaluate the phase concentration of polytypes in alpha-SiC, X-Ray diffraction data were elaborated by using a modified Rietveld method. Each polytype was quantitatively determined without using standard powder mixtures and with an error less than 1%. Moreover, crystallographic texture effects were reduced because the fitting between the experimantal diffraction pattern and the theoretical one was performed on the whole diffraction spectrum taken on 2-theta angles between 30° and 80°. The method, in order to obtain the theoretical diffraction pattern, requires the knowledge of the crystallographic parameters of each polytype. Suitable computer programs were developed.

RESULTS

The backscattered channeling image of a polished surface of sintered SiC (2100 °C, 1 hour) is reported in Fig.1. The minimum detectable grain size is about 0.5 um. The same surface has been observed in secondary electron contrast after thermal etching (Fig.2). Grains, inclusions and pores are clearly visible in both micrographs.
Fig.3a shows the secondary electron image of a complex defect on the unetched surface of a sample sintered at 2100 °C for 2 hours. Fig.3b, 3c and 3d give the X-ray maps of carbon, boron and silicon, respectively. These maps are obtained by using a low electron beam energy (6 kV) and a windowless detector.
Regions containing boron, carbon and boron, carbon and silicon can be easily distinguished. The black area on the top of the

Figure 1. Backscattered channeling micrograph of a polished surface of sintered SiC. Bar=5um

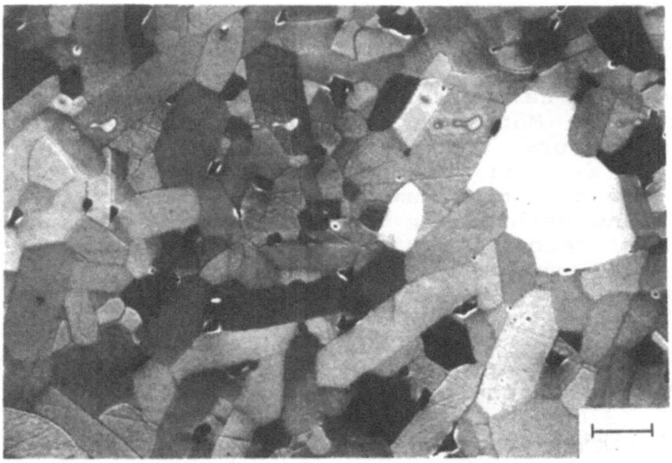

Figure 2. Secondary electron image of the same surface of Fig.1 after thermal etching. Bar=5um

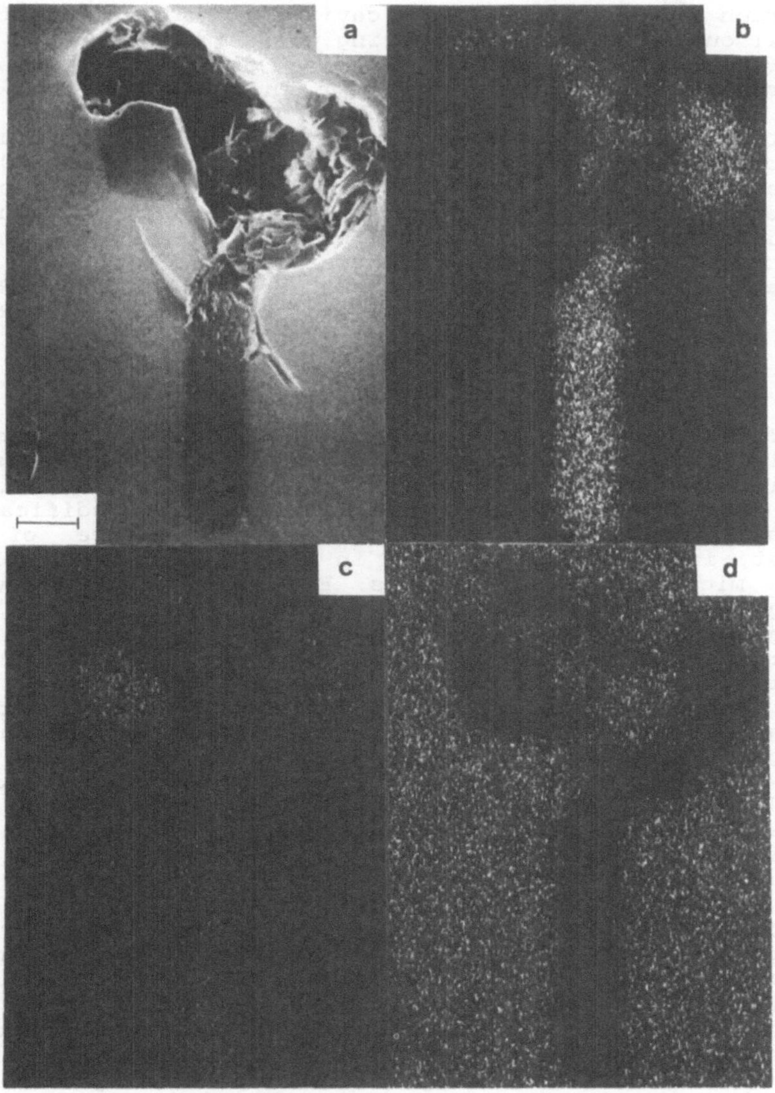

Figure 3. Secondary image of a complex defect on unetched SiC (a) and X-ray maps of carbon (b), boron (c) and silicon (d). Bar=1um.

micrograph (Fig.3a) is an empty cavity, probably originated by the pull out of an inclusion during polishing.
Micrographs of two samples with a different content of exaggerately grown grains are shown in Fig.4 and Fig.5. The sample of Fig.4 was sintered at 2100 °C with no holding time, while that of Fig.5 was held 1 hour at 2180 °C. The amount of added carbon black was also different (4g and 2g for 100 g of SiC, respectively). In Fig.6 and Fig.7 the XRD spectrum of the same samples of Fig.4 and Fig.5 are reported. The calculated X-ray diffraction patterns of the 6H, 15R and 4H polytypes are also given. The polytype concentrations, as wt%, have been calculated for the starting alpha-SiC powder (6H=80%, 15R=15%, 4H=6%) and for the sintered samples of Fig.4 (6H=83%, 15R=11%, 4H=6%) and Fig.6 (6H=75%, 15R=10%, 4H=15%).

DISCUSSION

The SACP technique permits the observation of grains, inclusions and pores on polished and unetched surfaces. By observing the surface at different orientation angles relative to the beam the contrast is changed. This tilt modification, besides indicating the crystallographic nature of the contrast, is an additional tool in the identification of the various microstructural features. For example, it allows to distinguish carbon inclusions from pores, which always have black-gray contrast. In general, etching is a perturbation of the surface. It can eliminate some features, like inclusions, and introduce artifacts that are difficult to interpret. This is particularly true for high resolution pictures and for chemical etching, which is usually more damaging than thermal etching.
An interesting feature of Fig.3a in the boron inclusion of about 1 um diameter observed in the secondary electron image of the unetched sample. The contrast is due mainly to electrical conductivity differences. The amount of boron added to all samples we have prepared was 0.4 g/100 g SiC and this is not much different from the solubility limit of B in SiC. Nevertheless, this amount is sufficient to promote densification and give inclusions containing B or B and C. This would seem to indicate that the quantity of boron effectively acting as a sintering aid is rather low.
Our data on polytype concentration show that the number of exaggerately grown grains and the amount of the 4H phase both grow as the sintering temperature and holding time are increased. We have analyzed twenty samples of sintered alpha-SiC. The growth of 4H starts at 2100 °C.

CONCLUSIONS

The microstructure of pressureless sintered alpha-SiC has been studied by using the channeling contrast effects produced in a SEM apparatus operated at low beam energy and high current density. Grain size and shape, boron and carbon inclusions and porosity are identified on the same surface without chemical or thermal etching. The Rietveld method has been applied to the determination of the polytype distribution. An increasing percentage of 4H has been observed as the sintering

Figure 4. Backscattered electron channeling image of a sample sintered a 2180 °C for 1 hour. Bar=20um.

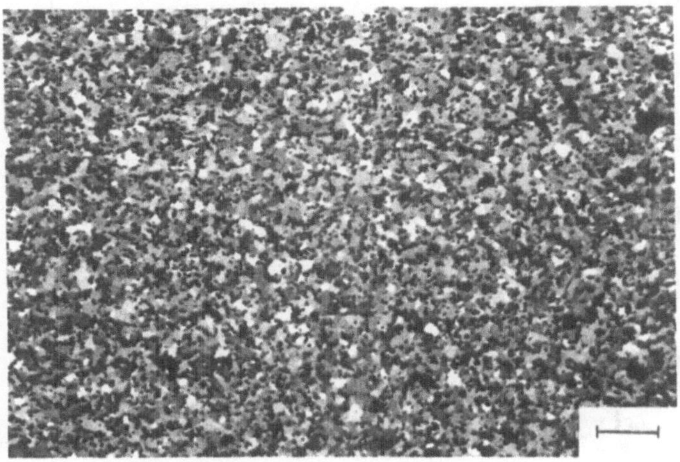

Figure 5. Backscattered electron channeling image of a sample sintered at 2100 °C, with no holding time. Bar=20um

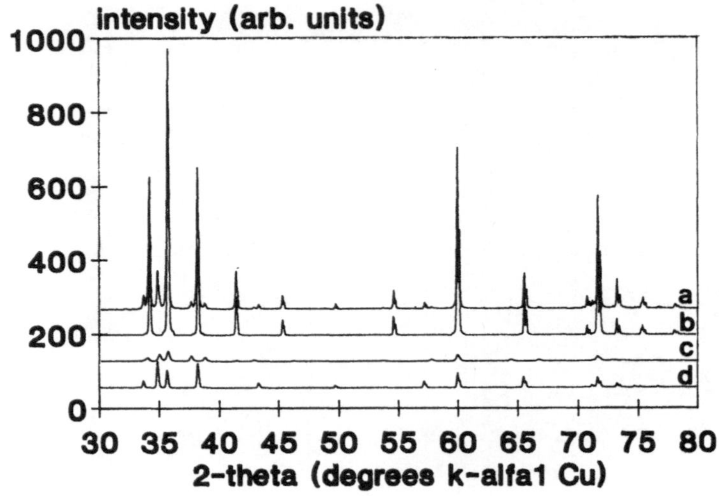

Figure 6. XRD pattern of the sample of Fig. 4. (a) is the experimental spectrum, (b), (c), (d) are the calculated spectra of 6H, 15R, 4H, respectively.

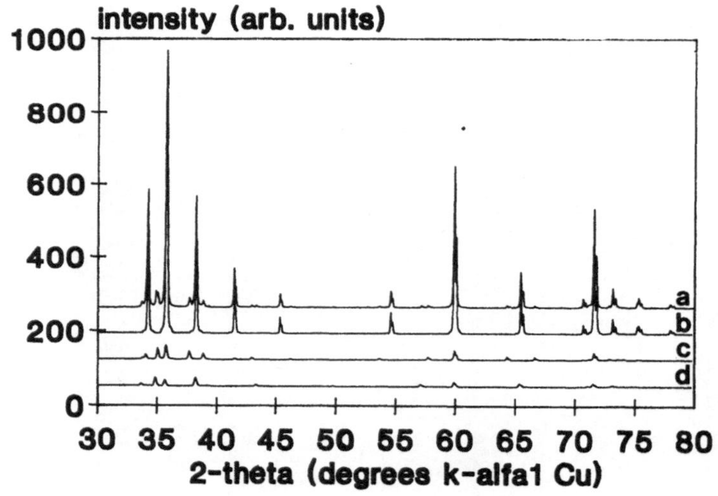

Figure 7. XRD pattern of the sample of Fig. 5. (a) is the experimental spectrum, (b), (c), (d) are the calculated spectra of 6H, 15R, 4H, respectively.

temperature and holding time increased. The amount of the 4H polytype has been tentatively related to the presence of exaggerately grown grains.

ACKNOWLEDGEMENTS

Technical help in sample preparation by Mr. E. Ferretti is acknowledged. This work was supported by ENI.

REFERENCES

1. Prochazka, S., The sintering process for silicon carbide : a review. General Electric Report No. 81CRD314, 1981.

2. Bocker, W. and Hausner, H., The influence of boron and carbon additions on the microstructure of sintered alpha silicon carbide. Powder Metall, Int.,1978, 10, 87-89.

3. Hamminger, R., Grathwohl, G. and Thummler, F., Microanalytical Investigations of sintered SiC. Part 1. Bulk material and inclusions. J. Mater. Sci., 1983, 18, 353-364.

4. Browning, R., Smialek, J.L. and Jacobs N.S., Multielement mapping of a-SiC by scanning Auger microscopy. Adv. Ceram. Mat.,1987, 2, 773-779.

5. Maddrell, E.R., Pressureless sintering of silicon carbide. J. Mater. Sci. Lett., 1987, 6, 486-488.

6. Johnson, C.A. and Prochazka, S., Microstructure of sintered SiC. In Ceramic Microstructures '76, eds. R.M. Fulrath and J.A. Pask, Westview Press, Boulder, 1977, pp.366-377.

7. Tajima, Y. and Kingery, W.D., Solid solubility of aluminum and boron in silicon carbide. J. Am. Ceram. Soc., 1982, 65, C27-C29.

8. Shinozaki, S.S., Hangas, J., Maeda, K. and Soeta, A., Enhanced formation of 4H polytype in silicon carbide materials. In Silicon Carbide '87, eds. J.D. Cawley and C.E. Semler, Ceramic Transactions, 1989, 2, 113-120.

9. Rietveld, H.M., Profile refinement method for nuclear and magnetic structures. J. Appl. Cryst., 1969, 2, 65-71.

10. Bish, D.L. and Howard, S.A., Quantitative phase analysis using the Rietveld method. J. Appl. Cryst., 1988, 21, 86-91.

11. Camanzi, A., Giunta, G., Parretta, A. and Vittori, V., Characterization of ceramic coatings by channeling effects in scanning electron microscopy. In Inst. Phys. Conf. Ser. No. 93, Vol.2, eds. P.J. Goodhew and H.G. Dickinson, 1988, pp. 525-526.

MICROSTRUCTURE AND PROPERTIES OF MIXED α'+β'-SIALONS

G.Z. Cao*, R. Metselaar* and G. Ziegler#
* Eindhoven University of Technology, the Netherlands
# Bayreuth University, Bayreuth, F.R. Germany

ABSTRACT

The microstructure and mechanical properties of fully dense yttrium-containing α'+β'-sialon ceramics prepared by gas pressure sintering were studied. The typical microstructure of mixed α'+β'-sialon ceramics consists of needle-like grains of the β'-phase (the ratio of length to width is about 7) dispersed in a matrix of equi-axed α'-sialon grains (mean grain size is about 3 μm) with a small fraction of an yttrium-rich amorphous phase (atom ratio Y:Si:Al =0.8:1.0:0.5) at grain boundaries. Vickers hardness (Hv0.5) and fracture toughness (K_{IC}) were measured by using the indentation technique and the biaxial bending strength (σ_{bi}) was determined with the "ball-on-ring" test. The relationships between the composition, microstructure and mechanical properties are discussed.

INTRODUCTION

One of the aims of development of sialon ceramics is to minimize the amount of the secondary amorphous and/or crystalline phases because it deteriorates the mechanical properties of the final products. A significant difference between sialons and other nitride ceramics is that the former can incorporate the constituents of the secondary phase into the crystal lattice [1,2]. Two types of sialons based on the silicon nitride structure have been developed. β'-sialon ceramics possess a good strength and have been widely investigated, and some commercial products are available now. α'-sialon offers excellent hardness and thermal shock resistance but is still in the very early stage of development [2,3].

Mixed two-phase sialon ceramics offer possibilities of tailoring the composition, microstructure and consequent properties of the final products, as many of the properties of a two-phase ceramics are additive. It has been realized that mixed α'+β'-sialon ceramics can yield benefits compared with monolithic α'- and/or ß'-sialon ceramics in engineering applications, since these materials can combine a good strength of the ß'-sialons with a significantly higher hardness of the α'-phase. It also has been found that the formation of the α'-sialons starts at a lower temperature than that of the ß'-phase [3,4], which is probably beneficial to the formation of elongated ß'-grains. These needle-like grains consequently will influence the fracture toughness and strength of the final products. One good example is that the metal cutting tools made of mixed α'+ß'-sialon ceramics demonstrate a considerable increase of the lifetime and improved wear resistance in comparison with that of ß'-sialons [5-6].

This paper describes the microstructural analysis of mixed α'+ß'-sialon ceramics and their mechanical properties. The relationships between the composition, microstructure and the mechanical properties are discussed.

EXPERIMENTAL PROCEDURES

Raw materials used in the present investigation were silicon nitride (LC-12, Starck, with 1.57 wt% Oxygen), aluminium nitride (Grade C, Starck, with 1.80 wt% Oxygen), alumina (CR 10, Baikalox, 99.99%), and yttria (Ventron, 99.99%). The composition of the starting mixtures and the green densities of the compacts are presented in Table 1. The composition of the mixtures is restricted to the region of α'- and ß'-sialon phases as shown in Figure 2. The oxygen impurity in the nitrides was not taken into account. The powders were mixed by using ball milling. After drying, the mixtures were pressed uniaxially and then isostatically under 250 MPa. Specimens were 25 mm in diameter and 7 mm in thickness.

The specimens, embedded in a mixture of 75 wt% silicon nitride, 20 wt% aluminium nitride and 5 wt% yttria, were sintered in a gas pressure furnace first at 1800°C for 60 min under a nitrogen pressure of 0.5 MPa and followed by 1900°C for 30 min under 10 MPa N_2. Fully dense α'+ß'-sialon ceramics (relative density not less than 99.8%) were obtained and the overall weight loss was about 1 wt%. The details of sintering are given elsewhere [7].

The microstructural analysis was performed mainly by means of SEM and TEM (with EDS) observation. For the SEM observation polished specimens were etched by using Plasma Enhanced CVD under the following conditions: Temperature 300°C,

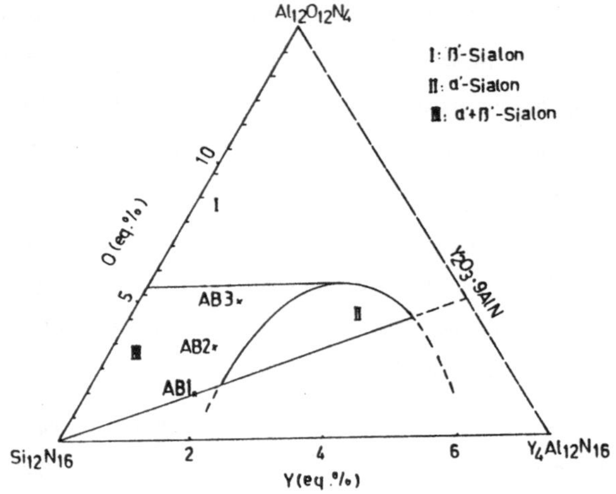

Figure 1. Phase diagram of the concentration plane of $Si_{12}N_{16}$-$Si_4Al_8O_8N_8$-$Y_4Al_8N_{12}$.

pressure 600 mtorr, 300 kHz, 100 W for 10 min with a gas of mixed CF_4/O_2 (5 vol%) with a flow rate of 200 ml/min. In addition, also fracture surfaces were studied with the SEM. The TEM specimens were cut to very thin foils of about 20 μm and then ion milled.

TABLE 1
The Composition and the Relative Densities of the Green Compacts

Samples	Composition (wt%)				Density
	Si_3N_4	AlN	Al_2O_3	Y_2O_3	(%)
AB1	87.37	7.83	0	4.80	64
AB2	83.88	8.83	2.50	2.50	62
AB3	80.62	9.76	4.83	4.79	62

The measurements of the Vickers hardness Hv0.5 (under a load of 0.5 N) and fracture toughness K_{IC} (under a load of 15 N) were conducted by using the indentation technique. The biaxial bending strength was measured with a "ball-on-ring" test [8-9]. The measurements were conducted at ambient temperature (about 20°C) under 0.1 MPa nitrogen at a dew point of -35°C (relative humidity 1 %). A crosshead speed of 85 mm/h was used. The specimens were 20 mm in diameter and 1 mm in thickness. The surface roughness of both sides was less than 0.2 μm. Not less than 12 measurements were performed for each composition.

RESULTS AND DISCUSSION

MICROSTRUCTURE

SEM observations reveal that the typical microstructure of mixed α'+β'-sialons consists of needle-like grains of the β'-phase dispersed in a matrix of equi-axed α'-sialon grains with a small amount of secondary phase remaining at grain boundaries as demonstrated in the fracture surface photograph of Figure 2. Figure 3, a back-scattered SEM picture of a polished surface, shows that there are black bars of the β'-phase dispersed in the grey matrix phase of the α'-sialon, also with a small amount of white spots of the secondary phase. The α'-sialon always appears in equi-axed grains, with a mean grain size of about 3 μm. For the β'-phase, an aspect ratio of about 7 was observed, with an average length of 14 μm.

The TEM observation, together with EDS analysis and the selected-area electron diffraction technique, reveals that in mixed α'+β'-sialon ceramics, the secondary phase is an yttrium-rich amorphous phase. It is present not only at the intersections of three grains (triple junctions) and four grains, but also at the intersections of two grains. That means that all crystal grains both of α'- and β'-sialons are covered with a thin film of amorphous material. Figure 4 gives a TEM picture together with the electron diffraction pattern and the EDS analysis result of the secondary phase. The composition of the

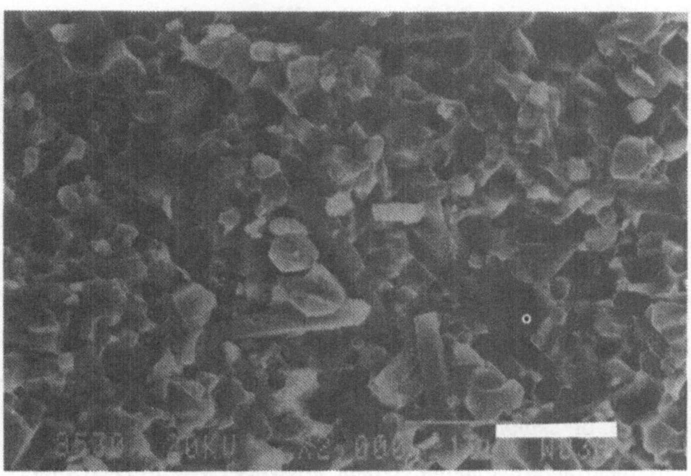

Figure 2. SEM fractography of sample AB1. The bar is 10 μm.

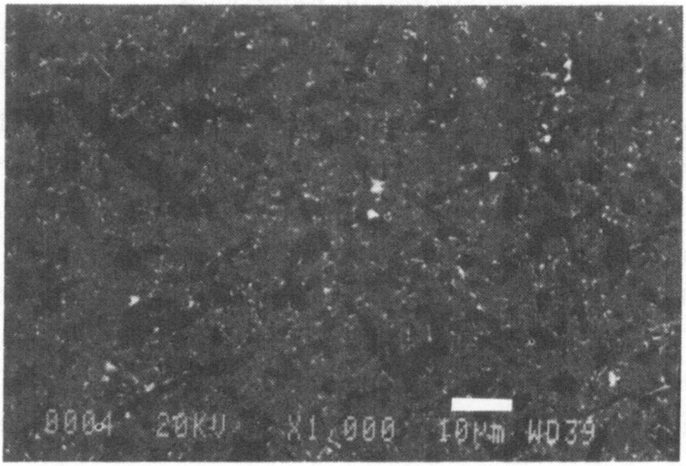

Figure 3. Back-scattered SEM microphotography of sample AB1. The bar is 10 μm.

amorphous phase was observed to be Y:Si:Al =0.8:1.0:0.5 in atom ratio, in which the yttrium content is rather higher than the overall composition of mixed α'+ß'-sialons. Approximately 3.5 vol % amorphous phase was observed.

The increase of the alumina content in the starting mixtures was seen to result in an increase of the ß'-phase content from 50 vol% in sample AB1, to 60 vol% in AB2 and then to 65 vol% in AB3. It was found that the aspect ratio of the ß'-grains decreases from 7 to about 4 with increasing Al_2O_3 content. Whereas the mean grain size of the α'-sialon remains approximately the same. The amount of the amorphous phase, remaining at grain boundaries, increases slightly, even though the alumina theoretically is totally soluble into the structure of α'- and ß'-sialons.

MECHANICAL PROPERTIES

Table 2 gives the Vickers hardness (HV0.5), fracture toughness (K_{IC}) and biaxial bending strength (σ_{bi}) of mixed α'+ß'-sialon ceramics, together with the standard deviations of the mechanical properties. It can be seen that the best mechanical properties are achieved in the case of sample AB1, without using alumina powder in the starting mixture. Although the fracture toughness data are very close to each other, the Vickers hardness and the biaxial bending strength are seen to decrease with the increasing amount of alumina powder used in the starting mixtures (c.f. Table 1).

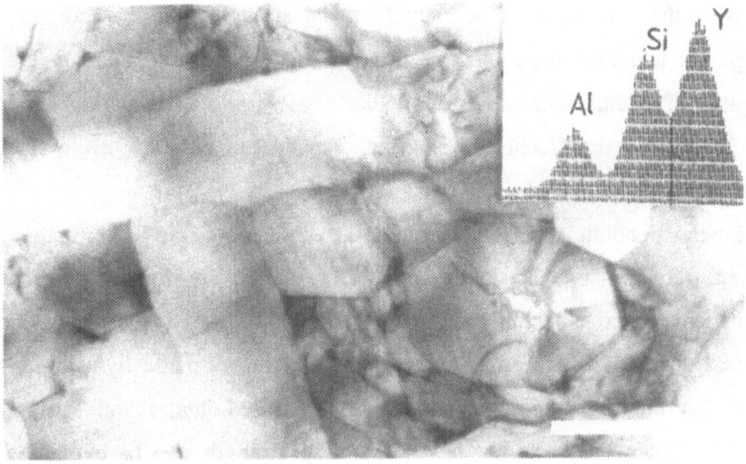

Figure 4. TEM picture of sample AB1 with the EDS analysis results of the grain boundary phase. The bar is 2.5 μm.

TABLE 2
The Properties of Mixed α'+ß'-Sialon Ceramics with the Standard Deviation

Samples	Density (g/cm^3)	Hv0.5 (GPa)	K_{IC} ($MPa.m^{1/2}$)	σ_{bi} (MPa)
AB1	3.23	17.2±1.5	6.0±0.7	734±36
AB2	3.22	16.8±1.0	5.8±0.6	680±51
AB3	3.23	14.5±1.0	5.7±0.7	600±74

Dortmans and de With [9] have reported that the three point bending strength (for specimens of 50 x 4.5 x 3.5 mm) appears to be about 75 % of the biaxial bending strength under certain conditions. With this method, the highest three point bending strength of mixed α' +ß'-sialon ceramics is calculated to be about 550 MPa.

DISCUSSION

The Vickers hardness of mixed α'+ß'-sialon ceramics is additive as reported in literature [5,6]. α'-sialons have a very high hardness of above 21 GPa, whereas ß'-sialon ceramics have about 14 GPa. The Vickers hardness of AB1 and AB2 therefore correspond to the composition well. However, the hardness of sample AB3 is lower than what we expected.

However, the fracture toughness and biaxial strength are not additive. The high fracture toughness achieved for mixed α'+ß'-sialon ceramics is explained by the presence of the elongated ß'-grains. The variation of the aspect ratio of the ß'-grains has shown very little influence on the fracture toughness, which decreases slightly with decreasing aspect ratio.

The biaxial bending strength of mixed α'+ß'-sialon ceramics is higher than that of both monolithic α'- and ß'-sialon ceramics reported in literature. For α'-sialon ceramics the typical biaxial bending strength is about 520 MPa [7], whereas for ß'-sialon it is slightly less than 500 MPa [10]. All these materials were made by using gas pressure sintering and the biaxial bending strength was measured under the same conditions described. This enhancement of the biaxial bending strength can be explained in terms of the microstructure of the mixed α'+ß'-sialon ceramics. Firstly, it is seen that the pull-out of ß'-grains, with a high aspect ratio, results in increasing the fracture energy, and also the mean grain size of the α'-phase in the case of mixed α'+ß'-sialons is somewhat smaller than in monolithic α'-sialon ceramics. Secondly, the amount of the amorphous phase remaining at grain boundaries is much smaller in mixed α'+ß'-sialons than that in monolithic sialon ceramics (generally not less than 8 vol%) [7], which probably results in a reduction of the size of the flaws.

Besides, the biaxial bending strength of mixed α'+ß'-sialon ceramics decreases in sequence of AB1 > AB2 > AB3, in corresponding to the increasing amount of alumina used in the starting composition. This is very similar to the behaviour of ß'-sialon ceramics [11].

CONCLUSIONS

In summary, the typical microstructure of mixed α'+ß'-sialon ceramics consists of needle-like grains of the ß'-sialon dispersed in an equi-axed α'-sialon grain matrix with a small amount of yttrium-rich amorphous phase at grain boundaries. The best mechanical properties achieved are HV0.5 17.2 GPa, K_{IC} 6.0 $MPa \cdot m^{1/2}$ and σ_{bi} 734 MPa.

The mechanical properties of mixed α' + ß'-sialon ceramics are improved by tailoring the microstructure of the final products.

Addition of alumina powder to the starting mixtures results in an increase of the ß'-sialon content and a decrease of the aspect ratio of the ß'-grains. It is also seen that the mechanical properties of mixed $\alpha'+\beta'$-sialon ceramics decrease with increasing amount of alumina powder used in the starting mixtures.

ACKNOWLEDGEMENT

The authors acknowledge the assistance of Ing. J.P.G.M. van Eijk and Drs. H.F. Scholten. Thanks are also due to J.W. Feng in the Shanghai Institute of Ceramics, for the TEM observation.

REFERENCES

1. Jack, K.H., Review: Sialons and Related Nitrogen Ceramics, J. Mater. Sci., 11(1976)1135.
2. Cao, G.Z. and Metselaar, R., α'-Sialon Ceramics: A Review, Chemistry of Materials, 3(1991)242.
3. Thompson, D.P., Sun, W.Y. and Walls, P.A., α'- β' and O'- β'-Sialon Ceramics, Ceramic Materials and Components for Engines, Eds. Bunk, W. and Hausner, H., DKG, Berlin, 1986, p643.
4. Cao, G.Z., Metselaar, R. and Ziegler, G., Gas Pressure Sintering of α'-Sialon Ceramics, Euro-Ceramics, I, eds. de With, G., Terpstra, R.A. and Metselaar, R., Elsevier Science Publ., London, 1989, p346.
5. Ingelström, N. and Ekström, T., Relation between Composition, Microstructure and Cutting Tool Performance of Alpha-Beta Sialons, J. Phys. (Paris), 47(1986)C1-347.
6. Ekström, T., Fabrication and Properties of Yttrium Sialon Ceramics, Mater. Sci. Forum, 34/36(1988)605.
7. Cao, G.Z., Preparation and Characterization of α'-Sialon Ceramics, Ph.D Thesis, Eindhoven University of Technology, April, 1991.
8. de With, G. and Wagemans, H.H.M., Ball-on-ring Test Revisited, J. Am. Ceram. Soc., 72(1989)1538.
9. Dortmans, L.J.M.G. and de With, G., Weakest-link Failure Predictions for Ceramics Using Finite Element Post-processing, to be published in J. Euro-Ceram. Soc.
10. Kokmeijer, E., Sintering Behaviour and Properties of β'-$Si_3Al_3O_3N_5$ Ceramics, Ph.D Thesis, Eindhoven University of Technology, Oct., 1990.
11. Weiss, J., Silicon Nitride Ceramics: Composition, Fabrication Parameters, and Properties, Ann. Rev. Mater. Sci., 11(1981)381.

HIGH-PRESSURE CARBOTHERMAL PREPARATION OF Si_3N_4. INFLUENCE OF TEMPERATURE PROGRAM AND ADDITIVES.

Ekelund M. and Forslund B.
Department of Inorganic Chemistry, Arrhenius Laboratory, Stockholm University, S-10691 Stockholm, Sweden

ABSTRACT

High-pressure nitridation of $C+SiO_2$ powder mixtures for the preparation of Si_3N_4 powder offers some advantages over the ordinary atmospheric pressure process. For one thing, the useful temperature range can be much expanded upwards without formation of unwanted SiC, which starts at about 1500°C at 0.1 MPa (1 bar).

Here we present and discuss the results of a recent series of runs in the pressure range 1-6 MPa, T = 1830°C, with the samples characterized by XRD, SEM and chemical analyses.

Stoichiometry (C/O), α-phase content and particle size are important powder characteristics that we have studied as functions of heating rate, nitrogen pressure, gas flow rate, starting material types and additives.

It was found that large surface area silica (50 m^2/g) produced large Si_3N_4 particles (hexagonal prisms, ~ 5μm in size), while small surface area quartz (1 m^2/g) yielded smaller Si_3N_4 particles (~ 1μm). Control of the powder grain size below 1μm was achieved by "seeding", *i.e.* addition of ~5wt% of Si_3N_4 to the starting mixture. The effect was obtained when large Si_3N_4 particles (~ 5μm) and submicron powder were used as "seeds".

A high α-phase content is promoted by reactive starting materials, and by moderately low gas flow rates.

INTRODUCTION

Carbothermal reduction and simultaneous nitridation ("CTN") of SiO_2 is a potentially important production process for high-quality Si_3N_4 powder [1]. A material prepared by this method from C + SiO_2 powder mixtures generally has a high α-phase content and a narrow Si_3N_4 grain size distribution [2,3].

The influence of process parameters (temperature program, starting material, gas flow rate, etc.) on conversion rate and powder characteristics such as α-content, residual oxygen content and particle size has been studied in flowing nitrogen at atmospheric pressure [2-5] and at elevated pressure [6,7]. One advantage of the high pressure carbothermal procedure is that high temperatures can be used without inducing competitive SiC formation [8], which starts at about 1500°C for $p(N_2)$ = 0.1 MPa (1 bar) [9].

The present study of CTN at 1830°C and $p(N_2)$ = 1.5-9MPa, comprising a number of nitridation experiments with small charges (0.4 g) of C + SiO_2 mixtures, has been undertaken with the primary purpose of relating the starting material type (SiO_2 surface area), gas flow rate and heating rate to the resulting α-content and Si_3N_4 grain size. The effect on the powder grain size of "seeding" the starting mixtures was also studied.

Control of the morphology of the Si_3N_4 powder is of great importance for the final properties of the sintered products. Si_3N_4 usually crystallizes as whiskers or hexagonal prisms, as reported by most authors. The size of the prisms has been found to vary from <0.5 µm up to ~10 µm. The morphology of the starting material (*eg.* SiO_2 and C) was reported [2,11] to have no influence on the morphology of the final Si_3N_4 powder. This suggests that the main reaction path proceeds *via* the gas phase. The usually preferred type of Si_3N_4 powder, for use in high performance ceramics, has a particle size < 1µm and a narrow size distribution.

EXPERIMENTAL

The experimental set-up used in this investigation has been described earlier [7]. It consists of a vertical graphite furnace inside a pressure vessel with in- and outlets for the gas. All gas introduced in to the autoclave is forced through the charge and then the shortest way out *via* a reducing valve and flowmeter. The reaction was followed by continuously monitoring the carbon monoxide content in the outgoing gas by means of an IR-detector.

Starting mixtures with different surface areas of silica ("Aerosil", Degussa A.G., Frankfurt, Germany) and quartz ("pure natural", provided by Gunnar Hällgren, Kema Nord Industrikemi, Ljungaverk, Sweden) as well as different seeding materials have been used, and are listed in Table 1. The mixtures were homogenized by rolling carbon (Nordisk Philblack Co., Malmö, Sweden) and silica/quartz powders,in an ethanol slurry with SIALON milling balls, in stochiometric proportions according to the intended reaction:

$$3SiO_{2(s)} + 6C_{(s)} + 2N_{2(g)} \longrightarrow Si_3N_{4(s)} + 6CO_{(g)} \quad (1)$$

In some cases "seeding material" was added to the starting material The mixtures were granulated to macaroni shaped pellets (Ø = 1.5 mm, length 1-10 mm), dried and then carefully charged as a thin layer in the reactor. The nitrogen gas ("Nitrogen Plus", AGA Co., Lidingö Sweden) was used without further purification.

The temperature program was controlled by a PID regulator. The column "T-prog" in Table 2 shows "heating rate / temperature", which was followed by annealing for one hour at the final plateau temperature. The temperature was measured with a W, W-Re thermocouple, calibrated to ± 5°C.

The samples were characterized by XRD, SEM and chemical analyses.

Table 1.
Starting mixtures and raw materials.

Symbol	Starting mixture	Grain size	BET surface	seed material
A	Carbon Silica	20 nm 40 nm	115 m^2/g 50 m^2/g	
B	Carbon Silica	20 nm 40 nm	115 m^2/g 50 m^2/g	15 w% Si_3N_4, size <1μm
C	Carbon Quartz	20 nm 34 μm	115 m^2/g 1 m^2/g	
D	Carbon Quartz	20 nm 34 μm	115 m^2/g 1 m^2/g	3w% amorphous, Si_3N_4 *
E	Carbon Silica	20 nm 40 nm	115 m^2/g 50 m^2/g	5w% Si_3N_4, size ~5μm
F	Carbon Silica	20 nm 40 nm	115 m^2/g 50 m^2/g	5w% BN

* submicron powder, provided by Dr Stephen Danforth, Rutgers University N.Y., USA

RESULTS

The results are presented in Table 2. In all runs the conversion to Si_3N_4 was almost complete, implying a carbon content of ~ 3% and an oxygen conent of ~1% in the produced Si_3N_4 powder. The departure of the product from the original stoichiometry (C/SiO_2=2) is due to irreversible loss of $SiO_{(g)}$ from the sample [7].

The presence of CO in the feed gas generally decreases the nitridation rate, the more the lower the temperature [10]. At the high temperature used in this study, however, 1% CO in # 9 did not induce any retarding effect compared to # 8 (with N: 37.5%, O: 0.94%, C: 3.6%, *cf.* #9 with N: 37.8%, O: 0.91%, C: 3.7%)

The minimum and maximum grain size was observed in the SEM and is recorded in Table 2. In all cases the resulting Si_3N_4 occured as hexagonal prisms. The α content of each sample was checked and is included in Table 2.

DISCUSSION

Inoue *et al.* [12] reported that the size and shape of the final powder could be controlled by seeding the starting mixture with Si_3N_4. It was also found that the conversion rate of the starting powder was promoted, and it has furthermore been reported that the particle size of the Si_3N_4 increases with decreasing particle size of the reactants [3].

In this study we found that powder mixtures with small SiO_2 particles produced Si_3N_4 with particle sizes in the range ~ 4 - 7 μm (Fig 1). Larger SiO_2 particles however gave smaller Si_3N_4 crystals, ~ 1 - 2 μm (Fig 2). The size of the large particles decreased to < 1μm when the starting mixture

Table 2.
Experimental parameters and results (A':C/SiO_2 = 2.1, all others: C/SiO_2 = 2.0)

#	Mix	p [MPa]	Φ [l·min⁻¹]	T-prog [K·min⁻¹/°C]	α [%]	Grain size
1	A'	25	1.8	10/1520 2/1830	100	~3-5 μm
2	A'	25	20	10/1520 2/1830	82	~3-4 μm
3	B	25	1.9	2/1520 10/1830	90	~0.1-0.3 μm
4	A	25	20	10/1520 2/1830		~4-7 μm
5	A	25	20	10/1830	80	~4-7 μm
6	A	65	20	10/1520 2/1830	100	
7	C	26	1.9	10/1830	81	~1-2 μm
8	C	15	11	10/1830	51	~1-2 μm
9	C	15	11 1.9	10/1680 10/1830	59	1%CO in N_2 ~1-2 μm
10	C	25	1.8	2/1830	73	~1-2 μm
11	D	25	2.0	10/1830	81	~0.5-1 μm
12	E	25	1.9	10/1830	100	~0.5-1 μm
13	F	25	20	10/1520 2/1830	100	~3-5 μm
14	A	90 20	2 20	10/1520 2/1830	87	~3-5 μm

was seeded with Si_3N_4 (Fig 3), in accordance with the findings of Inoue *et al.* [12]. In contrast to Inoue *et al.* we did not observe any increase in the conversion rate when the C + SiO_2 powder mixture was seeded with Si_3N_4.

The type of Si_3N_4 seeding material did not seem to be of importance; α-Si_3N_4 powders of various fineness, as well as an amorphous Si_3N_4 powder with small particle size, were tested, and they all gave the same result: Si_3N_4 particles in the ~ 1μm range (Fig 4). This is not in accordance with the findings of Shanker *et al.* [13], who investigated the relation between the particle size of the seeding material and of the Si_3N_4 formed. They reported that the grain size of the produced powder was roughly the same as that of the seeding powder.

Neither was the heating rate critical for the Si_3N_4 grain size, as seen in Table 2; 2 or 10 K/min produced same type of powder. Nor did a gas flow rate between 2 and 20 l/min influence the grain size or shape.

Cannon & Zhang [3] explain the narrow particle size distribution,

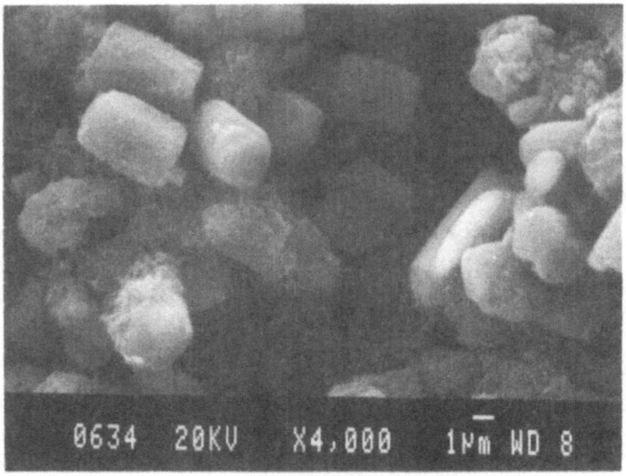

Figure 1. Si_3N_4 particles formed from a starting mixture with small SiO_2 particles (BET area 50 m^2/g). Run # 5.

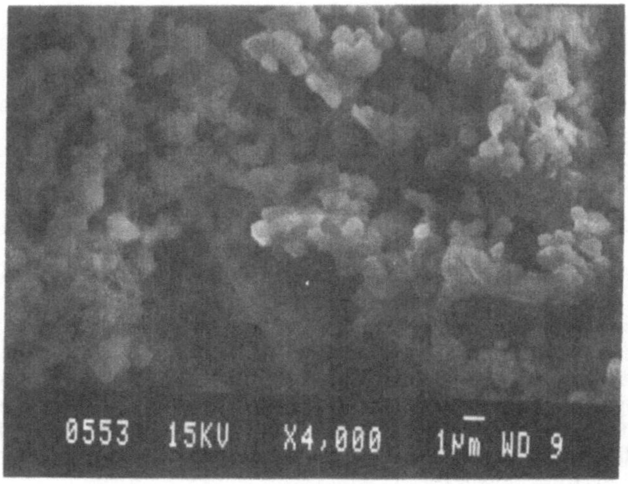

Figure 2. Si_3N_4 particles formed from a starting mixture with large SiO_2 particles (BET area ~1 m^2/g). Run # 7.

Figure 3. Si_3N_4 particles formed from a starting mixture seeded with small Si_3N_4 particles (d < 1μm). Run # 3.

Figure 4. Si_3N_4 particles formed from a starting mixture seeded with large Si_3N_4 particles (d ~5 mm). Run # 12. A seed crystal is seen in the center of the picture.

generally obtained in the CTN of SiO_2, by suggesting that when the initial nucleation has occurred, the concentration of the reactant $SiO_{(g)}$ is so low that only growth continues. This means that there is a burst of nucleation followed by growth, so that all particles reach approximately the same size.

It has also been reported that a large carbon-to-silica ratio in the starting mixture gives small Si_3N_4 particles. An explanation put forward for this is that the carbon particles act as "spacers" [11]. This could explain the somewhat smaller grain size observed in runs # 1 and 2 than in # 4 and 5. However, our attempts to seed the starting mixture with inert BN powder as spacer, (#13) resulted in large Si_3N_4 particles. That the particle size decreases both with excess of carbon and with Si_3N_4 additives indicates that the surface of the "seeding material" (C or Si_3N_4) is not critical; instead the activity of some gaseous species may be important.

The formation of α-Si_3N_4 could possibly be due to a high supersaturation of p(SiO), resulting in α-Si_3N_4 formation for kinetic reasons [10.14]. It has been found that a low gas flow rate favours the α-Si_3N_4 formation, while the amount of β-Si_3N_4 is larger in runs with higher gas flow rates [10]. A lower gas flow rate would imply a higher p(SiO) and thus a high α-Si_3N_4 crystal concentration, as the sample space is not drained of the $SiO_{(g)}$ as efficiently as with a faster flow. The present results support this idea (*cf.*#1 and 2, #7 and 8).

Run #14 in Table 2 was an attempt to supersaturate the gas phase during the heating by using a high pressure and then to initiate nucleation from many sites on a pressure release, in order to obtain a fine-grained powder. Evidently this procedure had a negligible effect on the particle size.

Previously we have also found that the α-formation is promoted when the carbothermal synthesis is performed at higher temperatures [10]. This could also be explained by the supersaturation of the p(SiO), as both the rate of SiO evaporation and p(SiO) at equilibrium are favoured by an increase in temperature. Possibly, such a supersturation effect could explain the somewhat higher α-content obtained in #7, at a heating rate of = 10 K/min, compared with #10 (2 K/min).

CONCLUSION

* The carbothermal synthesis of Si_3N_4 powder from SiO_2 in pressurized nitrogen (1.5-6MPA) goes virtually to completion within 1 hour at 1830°C, leaving 0.9% oxygen in the produced Si_3N_4 material.
* The grain size of the powder can partly be controlled by:
  - choice of the SiO_2 grain size
  - stochiometry (C/SiO_2) of the starting mixture
  - addition of a few % Si_3N_4 to the starting mixture

The heating rate (between 2 and 10 K/min) was found to have no effect on the Si_3N_4 grain size, nor did dilution of the starting mixture by addition of an indifferent powder (BN)

* The formation of α-Si_3N_4 in the carbothermal process is promoted by:
  - reactive starting materials
  - a moderatly low gas flow rate
  - a high heating rate

The α-content was found not to be affected by addition of Si_3N_4 to the starting mixture.

AKNOWLEDGEMENTS

The authors are grateful to Professor Lars Kihlborg for comments on the manuscript. This work has recieved support from the Swedish Natural Science Research Council (NFR) and the Swedish Board for Technical Development (STU).

REFERENCES

1. Natansohn S., p. 27 in *Ceramic Materials and Components for Engines.* Edited by V.J. Tennery, The American Ceramic Society, Inc., USA (1989).
2. Rahman I. A. & Riley F.L., *J. Eur. Ceram. Soc.*, **5**, 11-22 (1989).
3. Cannon W.R. & Zhang S.-C., p. 86 in *Ceramic Materials and Components for Engines.* Edited by V.J. Tennery, The Am. Ceram. Soc. Inc., USA (1989).
4. Komeya K. & Inoue H., *J. Mater. Sci.*, **10**, 1243-46 (1975).
5. Zhang S.-C. & Cannon W. R., *J. Am. Ceram. Soc.*, **67**, 691-5 (1984).
6. Ekelund M., Forslund B. and Johansson T., p.101 in *Ceramic Materials and Components for Engines.* Edited by V.J. Tennery, The American Ceramic Society, Inc.,USA (1989).
7. Ekelund, M. & Forslund, B., p.337 in *Ceramic Powder Science III.* Edited by G.L. Messing, The American Ceramic Society, Inc., USA (1990).
8. Ekelund M., Forslund B., Niklewski T., Eriksson G., Hällgren G., Johansson T. and Hatcher M., No.392, p.567 in *Extended Abstract Catalogue 86-1,* The Electrochemical Society's 169th Meeting, May 4-9, 1986, Boston, Ma, USA (1986).
9. Lee J. G. & Cutler I.B., p.175 in *Nitrogen Ceramics.* Edited by F.L. Riley, Noordhoff, Leyden, The Netherlands, (1977).
10. Ekelund M. & Forslund B., Carbothermal prepatation of silicon nitride. Influence of starting material and synthesis parameters., accepted for publication in *J. Am. Ceram. Soc.*
11. Durham S. J. P., Shanker K. and Drew R. A. L., p. 313 in *Ceramic Powder Processing Science,* (Eds:. Hausner, Messing, Hirano), Proceedings of the Second International Conference, Berchtesgaden, FRG, Oct 12-14, (1988).
12. Inoue H., Komeya K. and Tzuge A., *J. Am. Ceram. Soc.*, **65**, C-205 (1982).
13. Shanker K., Grenier S. and Drew R. A. L, p. 321 in *Ceramic Powder Science III.* Edited by G.L. Messing, The American Ceramic Society, Inc.,USA (1989).
14. Ekelund M. & Forslund B., Reactions within quartz-carbon mixtures in a nitrogen atmosphere.,accepted for publication in *J. Eur. Ceram. Soc.*

THERMAL EXPANSION OF DIBORIDE SOLID SOLUTIONS

E. Fendler[1], O. Babushkin[2], T. Lindbäck[2], R. Telle[1] and G. Petzow[1]
[1]Max-Planck-Institut für Metallforschung, Institut für Werkstoffwissenschaft, PML
Heisenbergstr. 5, D-7000 Stuttgart 80, Germany
[2]Luleå University of Technology, Department of Engineering Materials
S-95187 Luleå, Sweden

ABSTRACT

The thermal expansion of TiB_2, CrB_2 and W_2B_5 is re-investigated with a high temperature x-ray diffractometer up to 2000 K. The results are compared with published data [1]. The thermal expansion behaviour of TiB_2 is identical in both works. The extraordinary anisotropy of CrB_2 is confirmed but the thermal expansion is lower in the present work. W_2B_5 seems to have a different thermal expansion behaviour at higher temperatures. The thermal expansion of $(Ti,Cr)B_2$ and $(Ti,W)B_2$ solid solutions are investigated. The thermal expansion coefficients of $(Ti,Cr)B_2$ do not obbey the rule of mixture. Especially in the crystallographical c-direction there is a high increase of the lattice parameters with temperature compared to the expected data. For the $(Ti,W)B_2$ solid solutions, there is a decrease of the thermal expansion at lower temperatures compared with TiB_2.

INTRODUCTION

The attractive properties of the transition metal diborides such as high hardness and melting point, good thermal and electrical conductivity and sufficient resistance against many corrosive media and oxidation, offer a wide range of applications. Diborides as abrasive and protective materials are used for cutting tools and coatings. They show considerable promise as structural materials in high-temperature environments. Additionally, diborides are added to hard metals and ceramic composites.

The high-temperature use of the diborides requires the knowledge of the thermophysical properties. The thermal expansion behaviour is of special interest if the diborides are used as coatings or as composite components in combination with other materials. The mismatch

in thermal expansion limits the use with respect to the materials and the temperature range. TiB_2 is an example of a diboride with a large thermal expansion and a pronounced anisotropy caused by different expansions of the crystallographic axis during heating. The more than 30 % anisotropy of TiB_2 is disadvantageous for high temperature applications for example as an electrode for the aluminium electrolysis. A variation of the thermal expansion behaviour may be reached by the formation of solid solution because the different diborides have a different thermal expansion behaviour. The precondition for improving properties by the formation of solid solutions is a sufficient range of solubilities between the diborides and a precise knowledge of the thermal expansion behaviour of the diborides.

Lönnberg [1] investigated the thermal expansion of the group IV-VII transition metal diborides up to 1500 K and discussed the correlation between thermal expansion and bonding strength. Due to the anisotropy of the bonding strength between the crystallographical a- and c-direction it is well-known, that the thermal expansion of the diborides is larger in the c-direction than in the a-direction. So for transition metal diborides the dependance of the thermal expansion on the bonding strength is confirmed by the fact that the mean thermal expansion increase linearly with their reciprocal melting point.

The anisotropy of the thermal expansion of CrB_2 exhibit an opposite behaviour. The value in the a-direction is considerably larger than the values for other diborides. In the c-direction the thermal expansion is correspondingly low compensating the high α_a value. Spin fluctuations, which explain the high experimental value of the specific heat compared with the value determined by band structure calculations, explain this effect.

From literature it is known that the diboride solid solutions are formed by a simple substitution of the metal atoms [2]. Therefore the bonding forces of a solid solution should have the sum of the fractional parts of the bonding forces of the pure diborides. This means that the properties of the solid solututions can be correlated with their composition. Using the extraordinary thermal expansion behaviour of CrB_2, a solid solution of TiB_2 and CrB_2 may diminish the strong anisotropy of TiB_2. A solid solution exhibiting no anisotropy of thermal expansion in a certain temperature range is theoretically possible. A reduction of the thermal expansion of TiB_2 may be obtained by the formation of solid solutions with a diboride possessing lower thermal expansion, for example W_2B_5.

TiB_2 and CrB_2 exhibit a complete solid solubility and the solubility of "WB_2" in TiB_2 is nearly 50 mole-% at 2270 K so that a wide range of variation is possible. Up to now, no published data are available about the thermal properties of diboride solid solutions. The present work investigates the thermal expansion behaviour of $(Ti,Cr)B_2$ and $(Ti,W)B_2$ solid solutions and whether the values can be controlled by composition or not. The thermal expansion of the pure diborides TiB_2, CrB_2 and W_2B_5 are re-examined up to 2000 K and compared with the data of Lönnberg [1].

MATERIALS AND METHODS

Sample Preparation

The starting materials were TiB_2, CrB_2 and W_2B_5 supplied by H.C. Starck, Berlin, having a purity of 98-99 wt.-% according to the producer specification. Powder mixtures of the following compositions were produced:

TABLE 1
Powder compositions in mole-% . In the brackets are the compositions of the sintered samples determined by EDX.

Sample Nr.	TiB_2	"WB_2"	CrB_2
1	100.0		
2		100.0	
3			100.0
4	41.4 (42.7)		58.6 (57.1)
5	63.5 (59.3)		36.5 (40.6)
6	80.9 (81.1)		19.1 (18.7)
7	76.9 (81.1)	23.1 (18.4)	
8	50.0 (53.0)	50.0 (47.0)	

All samples were hot-pressed in hexagonal BN-coated dies in Ar-atmosphere under a pressure of 20 MPa. The pure diborides CrB_2, W_2B_5 and TiB_2 were sintered for 15 min at 2170 K, 2270 K, and 2370 K, respectively. The powder mixtures were sintered at a temperature of 2370 K for 60 min.. An additional heat treatment of 6 h at 2270 K was nessecary to get homogeneous solid solutions. The compositions of the phases were verified with an EDX analysis (tab. 1).

X-ray Experiments

Investigation of thermal expansion was performed using a high-temperature X-Ray diffractometer (HT-XRD) in a vacuum (10^{-3} Pa). The main features of the HT-XRD include a conventional Philips powder diffractometer, in which the 2Θ slewing motor is under direct computer control, and the sample furnace, which is controlled by a microprocessor and computer system. The standard PAAR-ANTON high-temperature attachment was reconstructed to enable 2270 K to be reached in inert atmosphere and vacuum. The overall system has many peripheric devices which serve to store and display data. The HT-XRD system was carefully aligned in order to produce reliable data with good counting statistics. Temperature was measured with a WRe5-WRe26 thermocouple placed adjacent to the sample. The temperature measurement was adjusted by observing the melting of platinum and nickel. To avoid a temperature gradient within the sample and reduce the power consumption the furnace was heat-protected by a Mo-shield packed with carbon fibres.

The temperature range of the HT-measurements was between 1000 and 2100 K. The samples were scanned between 20° and 65° at room-temperature and at additional five temperature points. The scanning time was limited by the maximum running time of the HT-XRD.

RESULTS

Lattice constants were calculated with a program based on least-squares refinement. The room-temperature lattice constants of the diborides are summarized in table 2.

TABLE 2
Some crystallographical data of the investigated samples.

Sample	a [Å]	c [Å]	V [Å^3]	c/a
1	3.0317±0.005	3.2312±0.005	25.72	1.066
2	2.9860±0.005	13.894±0.005	107.29	4.653
3	2.9743±0.005	3.0710±0.005	23.53	1.033
4	2.9977±0.005	3.1318±0.005	24.37	1.045
5	3.0067±0.005	3.1545±0.005	24.70	1.049
6	3.0180±0.005	3.1962±0.005	25.21	1.059
7	3.0355±0.005	3.2124±0.005	25.63	1.058
8	3.0351±0.005	3.1921±0.005	25.47	1.052

The lattice parameters versus temperature were fitted using second order polynomials. Figures 1-3 show the temperature dependance of the lattice parameters of the pure diborides.

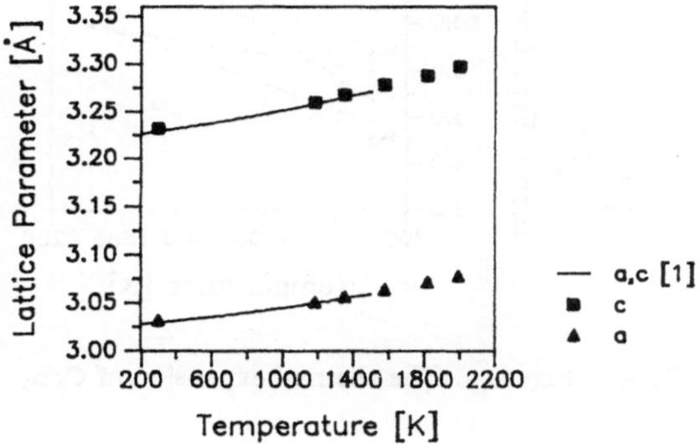

Figure 1. The lattice parameters of TiB_2 versus temperature.

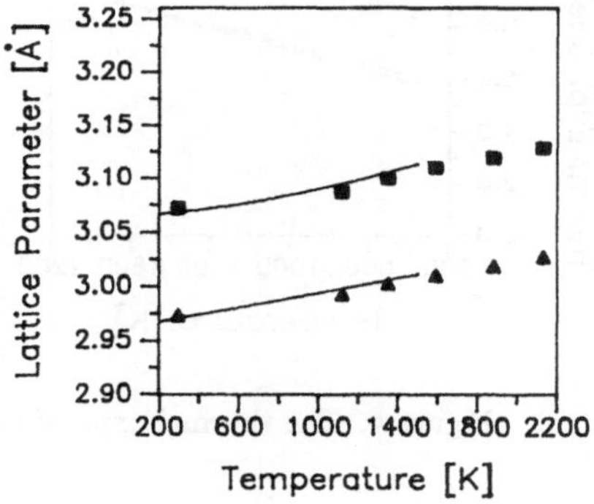

Figure 2. The lattice parameters of CrB_2 versus temperature.

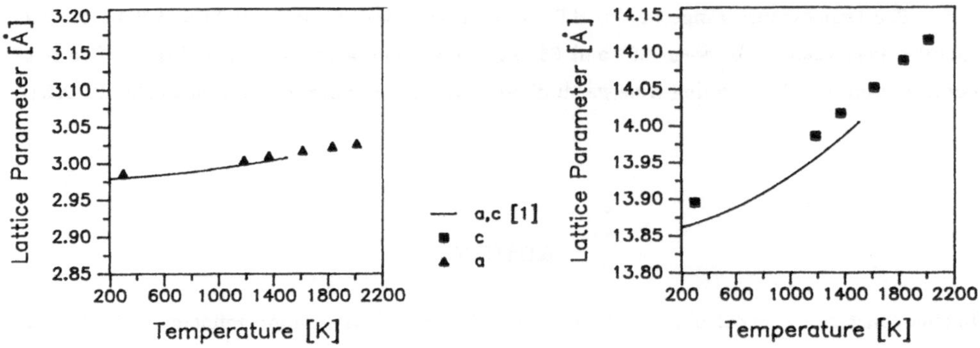

Figure 3.The lattice parameters of W_2B_5 versus temperature.

The linear thermal expansion coefficients were calculated using the following equation:

$$\alpha_a = \frac{a - a_0}{a_0\,(T - T_0)},$$

where T_0 is room temperature (298 K) and a_0 the lattice parameter at T_0. The values of α_c are attained using the c-parameter instead of a. Figures 4-9 show the thermal expansion coefficients versus temperature. The results of Lönnberg [1] are added in the plots of the pure diborides.

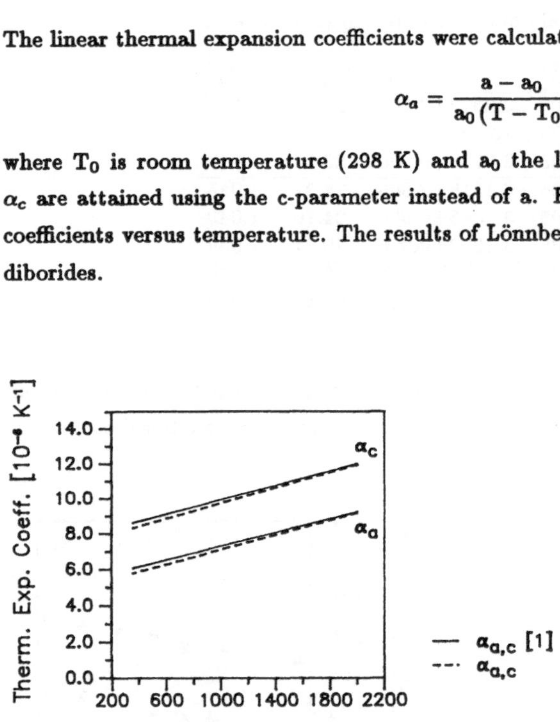

Figure 4. The thermal expansion of TiB_2.

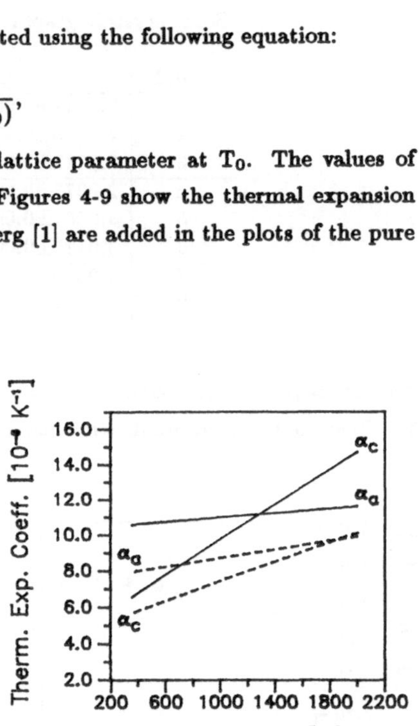

Figure 5. The thermal expansion of CrB_2.

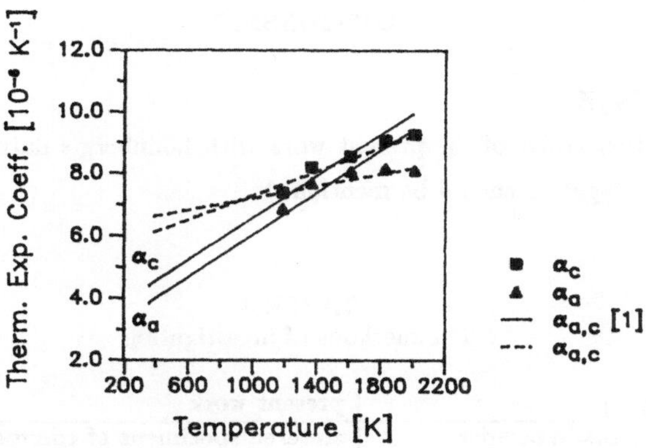

Figure 6. The thermal expansion of W_2B_5.

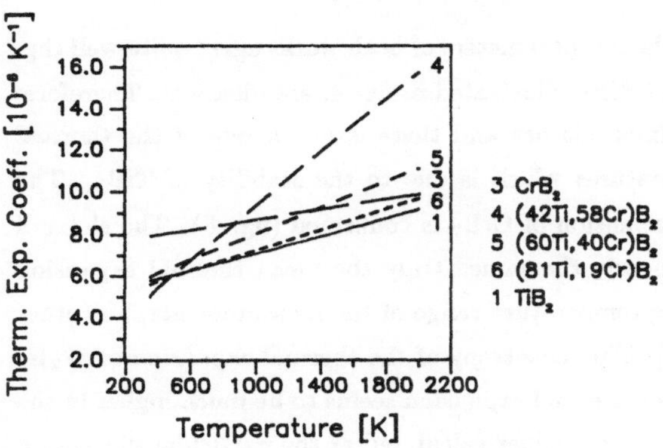

Figure 7. The thermal expansion of $(Ti,Cr)B_2$ solid solutions in a-direction.

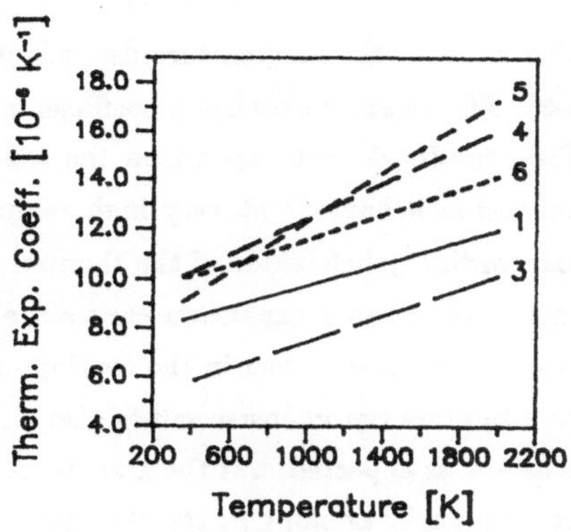

Figure 8. The thermal expansion of $(Ti,Cr)B_2$ solid solutions in c-direction.

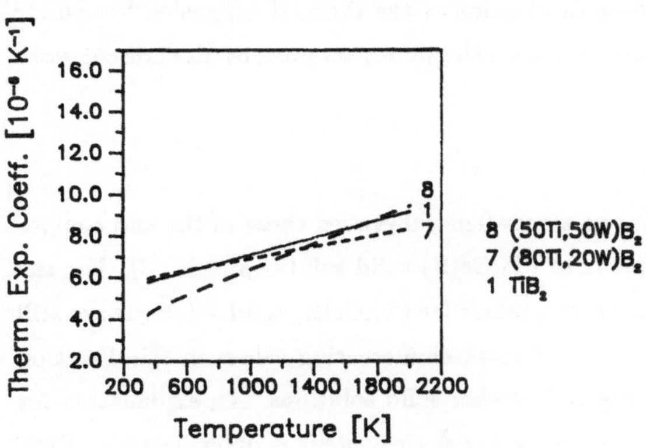

Figure 9. The thermal expansion of $(Ti,W)B_2$ solid solutions in a-direction.

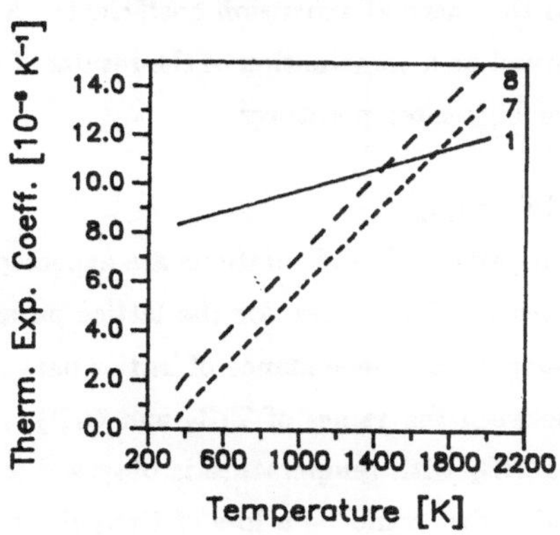

Figure 10. The thermal expansion of $(Ti,W)B_2$ solid solutions in c-direction.

DISCUSSION

TiB_2, CrB_2 and W_2B_5

Before comparing the results of the present work with Lönnberg's data [1], the differences in the methods of investigation should be mentioned:

TABLE 3
The methods of investigation

Lönnberg [1]	present work
laboratory-made powder	sintered specimens of commercial powder
temperature range: 298-1500 K	1000-2000 K

The values of the temperature-dependent lattice parameters of both works agree quite well (fig. 1-3). The thermal expansion coefficients of TiB_2, illustrated in fig. 4, are identical. Therefore, TiB_2 results do not depend on the specimen history and there is no change of the thermal expansion behaviour at very high temperatures which is due to the stability of TiB_2. The extraordinary behaviour of the thermal expansion of CrB_2 is confirmed (fig. 5). The slope of the mean thermal expansion curves are nearly the same. Only the mean thermal expansion is somewhat lower even in the overlapping temperature range of the measurements, but other investigators report lower values also [3,4]. The anisotropy of the thermal expansion of W_2B_5 is as low as expected, but the amount of the thermal expansion seems to be much higher in the present work. Looking at the thermal expansion values calculated at the particular measuring temperatures (fig. 6), it is obvious that there is a change of the thermal expansion behaviour at higher temperatures. The crossing of the α_a and α_c curves is also an effect of the linear fit to the thermal expansion coefficients. A good description of the thermal expansion behaviour would be a combination of the results of Lönnberg with the values attained by the present work for higher temperatures.

$(Ti,Cr)B_2$

Properties of solid solutions are expected to be somewhere inbetween those of the end compositions. This is true for the lattice parameters of $(Ti,Cr)B_2$ solid solutions (tab. 2). For the temperature dependance of lattice parameters, the values for $(Ti,Cr)B_2$ solid solutions are still between the values of TiB_2 and CrB_2, but a trend towards increasing values in the direction of TiB_2 with temperature is observed for the CrB_2-richer solid solutions. An explanation for this effect could be a loss of CrB_2 during heating. A lower Cr-content is determined by EDX after the high temperature measurement. The lattice parameters of sample 4 and 5 are slightly higher after heat treatment than before. Although this effect suggest a higher expansion it does not explain the high thermal expansion of the $(Ti,Cr)B_2$ solid solutions in the c-direction (fig. 8). On sample 6 no decomposition could be detected but the α_c-value is nevertheless higher than the values of TiB_2 and CrB_2. An increase of the slopes and the amount of the mean thermal expansion is observed with increasing CrB_2 content. Concerning the anisotropy of the solid solutions no decrease is observed. For all solid solutions the thermal expansion is higher in the c-direction than in the a-direction. Since the increase in the lattice parameter with temperature is steeper in the a-direction there is a decrease in the anisotropy at higher temperature.

(Ti,W)B_2

The temperature dependance of the lattice parameters of (Ti,W)B_2 solid solutions can only be compared with TiB_2 because the "WB_2" crystallizes in the W_2B_5 structure type which can be described as a stacking variation of the AlB_2-type structure in which TiB_2 and CrB_2 crystalise [5]. The a-lattice parameters of (Ti,W)B_2 solid solutions are very similar to the values for TiB_2. In the c-direction much lower values are measured for the (Ti,W)B_2 solid solutions than for TiB_2 (tab. 2) . The thermal expansion in the a-direction of sample 7 and 8 are very close to the behaviour of TiB_2 (fig. 9). Thus, the properties in the a-direction of (Ti,W)B_2-solid solutions are not influenced by an increasing W content. This is in agreement with observations of Lönnberg [1], that the thermal expansion of diborides do not change much with the size of the metal radius. An opposite behaviour is observed in the c-direction (fig. 10). At low temperature the thermal expansion is very small and there is a large increase with temperature. It should be pointed out that the thermal expansion behaviour at lower temperature is not measured in the present work but extrapolated from the high temperature behaviour. A slightly higher thermal expansion at lower temperatures is therefore possible. The low thermal expansion of W_2B_5 in the c-direction is explained by the high bonding forces due to puckered boron layers resulting in shorter layer distances. This is concerned with the special crystal structure of W_2B_5. Therefore it was not expected that (Ti,W)B_2 solid solutions possess such low thermal expansion coefficients.

CONCLUSIONS

The thermal expansion of the diborides TiB_2, CrB_2 and W_2B_5 are measured up to 2000 K and the results are presented. Comparing with published data most results agree quite well. A different thermal expansion was found for the CrB_2. The lower values are good for this materials if used combination with other materials possessing a low thermal expansion. The re-examination was necessary to obtain standards for the solid solutions because they only can be produced by sintering.

The thermal expansion behaviour of the $(Ti,Cr)B_2$ solid solutions was not expected. Vegard's rule, which states that the properties of a solid solution are in accordance with the average of the fractional properties of the end compositions, is not obeyed. Because of the very high thermal expansion of $(Ti,Cr)B_2$ solid solutions especially in the c-direction, the bonding forces seem to be low. This agrees with the observed decomposion of the $(Ti,Cr)B_2$ solid solutions at temperatures above 1500 K.

For crystalchemical explanations of the thermal expansion behaviour of diboride solid solutions further investigations are necessary. The present investigations did not make measurements up to 1000 K due to the measuring time limit. Investigations in this temperature range are more easily studied.

REFERENCES

1. Lönnberg, B., Thermal expansion studies on the group IV-VII transition metal diborides, *University of Uppsala, internal report, UUIC-B18-182*, 1987.

2. Post, B., Glaser, F.W. and Moskowitz, D., Transition metal diborides, *Acta Metall.*, 1954, 2, 20-25.

3. Castaing, J. and Costa, P., Properties and uses of diborides, In *V.I. Matkovich (ed.), Boron and refractory borides*, Springer, Berlin, 1977, pp.390-412.

4. Post, B., In *R.M. Adams (ed.), Boron, metallo-boron compounds and boranes*, Wiley, New York, 1964, pp. 301-371.

5. Kiessling, R., The borides of some transition elements, *Acta Chemica Scand.*, 1950, 4, 209-227.

SYNTHESIS OF AlN POWDER BY CARBOTHERMAL NITRIDATION OF Al_2O_3 AT ELEVATED N_2 PRESSURES.

B. Forslund and J. Zheng
Stockholm University, Department of Inorganic Chemistry,
Arrhenius Laboratory, S-106 91 Stockholm, Sweden.

ABSTRACT

AlN is a potentially important ceramic material for use in heat engines. In this paper we report some results of synthesizing AlN powder by carbothermal reduction of Al_2O_3 at elevated nitrogen pressures. Small charges of C + Al_2O_3 powder mixtures were heat-treated (1300 - 1700°C) in a reactor permitting controlled flows of pressurized N_2 (0.1 - 5 MPa) through the sample space. The samples were characterized by XRD, SEM and chemical analyses (N,O).

The rate of conversion to AlN was found to be a function of temperature, gas flow rate, C/Al_2O_3 ratio and also, to some extent, of nitrogen pressure.

The residual oxygen content obtained after a 1-hour run at 1600°C and 1 MPa was considerably lower (0.4 wt-%) than the corresponding value obtained at atmospheric pressure (3%), using starting mixtures of carbon black (80 m^2/g) and α-Al_2O_3 (0.6 m^2/g).

The AlN particle size (in the range 0.5 - 3 μm) and the degree of agglomeration were also found to be dependent on gas flow rate and pressure, as well as on raw-material type.

INTROCUCTION

AlN has some properties that makes this material attractive for use by the electronics industry [1]: high electrical resistivity, high thermal conductivity, relatively high strength and low thermal expansion (compared with Al_2O_3 and BeO). Due to a combination of mechanical properties, superior to that of most other oxide and non-oxide materials, such as low density, high thermal conductivity and low thermal expansion, AlN is also potentially useful in heat-engine applications [2].

Current research worldwide on AlN powder synthesis is directed towards minimizing production cost, controlling the grain size and reducing the impurity levels. In particular, the residual oxygen content has been found to be decisive for the thermal conductivity of sintered AlN bodies.

AlN can be synthesized utilizing a variety of reactions. Among them the direct nitridation of Al and carbothermal reduction of Al_2O_3 in the presence of N_2 are probably the most commonly used methods for commercial powder production. A merit quoted for the latter, carbothermal nitridation:

$$Al_2O_3 + 3C + N_2(g) \longrightarrow 2AlN + 3CO(g) \quad (1)$$

is the possibility of preparing homogeneously sized and non-aggregated powders from relatively cheap raw materials.

We report here some results of a recent study of AlN powder synthesis by the carbothermal route in *pressurized* nitrogen. Important powder characteristics, such as residual oxygen level and grain size and shape, have been found to depend on the nitrogen pressure, as well as on other process parameters.

EXPERIMENTAL

Three solid reagents were used: coarse α-Al_2O_3, fine γ-Al_2O_3 and carbon black. Their properties are listed in Table I. The nitrogen gas (total oxygen < 10ppm) was used without further purification.

Table I.

Reagent	Grain size	BET surface [m^2/g]	Source, grade
α-Al_2O_3	1-10μm	0.55	Fisher Scientific Co, "A-591"
γ-Al_2O_3	0.1-1μm	138	made in-house
C	c:a 30nm	82	Nordisk Philblack Co, "N330"

The starting materials were three mixtures of alumina and carbon black powders, in the following denoted by letters A,B,C (Table II).

Table II

Symbol	wt ratio	mol ratio	$-\Delta m/m_o$ for complete conversion (%)
A	α-Al_2O_3/C= 3	0.353	38.8
B	α-Al_2O_3/C= 2/3	0.078	22.3
C	γ-Al_2O_3/C= 2/3	0.078	22.3

Every composition was milled in a plastic bottle with sialon balls in an ethanol medium for 96 hours. The slurries were subsequently dried to a plastic mass, then pelletized to granules with Ø= 1.5mm, l= 1-6mm. These were further dried for removal of the alcohol prior to charging into the autoclave.

The carbothermal reduction at elevated N_2 pressure was performed in a gas autoclave with a vertical graphite furnace (Fig.1). The furnace tube and sample holder (Fig.2) were made of a pure graphite material and designed so that the nitrogen gas introduced into the autoclave was forced through the charge and then the shortest way out *via* a reducing valve and a flowmeter. The nitrogen flow rate (Φ) was in the range 0.5 to 25 l(stp.)/min. The CO content in the outlet gas was continuously measured by means of an IR-detector.

About 0.7 g of charge was used in every run and carefully spread on a perforated graphite foil in the sample holder. A standardized temperature program, controlled by a PID regulator, was followed in all runs; rapid heating at a rate of 20 K/min to a constant temperature, holding at this level and then cooling at 30 K/min. Normally the gas outlet valve was opened at 1100°C on heating and closed at 1400°C on cooling.

The temperature was measured with a W-Re thermocouple, calibrated against the melting points of Au,Pd and Pt. The reading was found to be accurate to within +/- 5K, while a constant temperature could be kept within +/- 1K.

The experimental products were characterized by XRD (Guinier-Hägg focusing cameras) and SEM (JEOL 820). The elemental (O,N) analyses were performed by an independent laboratory, using a fusion-gas analysis method. Carbon contents were measured by thermogravimetry in an oxygen atmosphere. The powder surface area and particle size were determined (by the BET method and SEM respectively) on samples decarburized by heating in air at 700°C.

RESULTS AND DISCUSSION

The relative weight change, $\Delta m/m_o$ (m_o is the charge weight), determined in each run, was used as a measure of conversion to AlN in the present study.

As expected, the rate of conversion increased with temperature. *E.g.* for four-hour-runs with starting mixture "A" at 0.5 *M*Pa nitrogen pressure and a gas flow rate of 1 l(stp)/min, $-\Delta m/m_o$ was found to be 1.9 wt% at 1300°C, 8.0 wt% at 1400°C, 30.0 wt% at 1500°C and 38.0 wt% at 1600°C (38.8% corresponds to complete conversion).

The rate of conversion was higher with starting mixtures "B" and "C" than with "A", the former two having an Al_2O_3/C ratio = 0.67 by weight; the latter, 3.0 (the stoichiometric ratio for reaction (1) is 2.83). This is in accordance with earlier work [3,4] stating that the nitridation rate is generally enhanced by an increased carbon/alumina ratio.

For all compositions AlN, Al_2O_3 and C were the only product phases obtained according to XRD and TG.

As has also been found in earlier work [5], we noted a dependence of conversion on the gas flow rate. The diagrams in Figs. 3-5 show the resulting degrees of conversion in a number of nitridation runs in this study, each with one hour's annealing time, plotted against gas flow rate and nitrogen pressure.

Our data in Figs. 3 and 4 (obtained under conditions giving a somewhat reduced nitridation rate, *i.e.* run at low temperature or with a non-ideal C/Al_2O_3 ratio) indicate that flows about 5 l/min are needed with a 1 g charge in order to eliminate the influence of gas flow rate on the conversion rate, in the present experimental set-up at $p(N_2)$=0.1-1 MPa. The much reduced $-\Delta m/m_o$ for 1 l/min at 5 MPa compared with 0.5 MPa suggests that the crucial parameter is in fact the superficial gas velocity past the granules. With regard to the formulation of reaction (1), a plausible explanation for the flow-dependence of conversion is that the rate of CO(g) removal is due to the velocity of the fresh nitrogen gas venting the sample space.

Comparing one-hour-runs at 1600°C with $\Phi(N_2)$ = 5 l/min but different $p(N_2)$, a pressure dependence of the conversion of mixture "A" is noted for p < 1 MPa (Fig. 4). Thus, in spite of the relatively higher superficial gas velocity in the sample holder at atmospheric pressure, the $-\Delta m/m_o$ just reached about 3/4 of the value obtained at 1 MPa. Even with the more "active" starting mixture "B" at 1600°C, an increase of the conversion rate with increased pressure from 0.1 to 0.5 MPa was noted for runs with $\Phi(N_2)$ = 5 l/min (Fig. 5).

Obviously, the maximum value of $-\Delta m/m_o$ was attained relatively rapidly for samples with carbon excess (within 1 hour at 1600°C), provided sufficiently rapid gas flows and a nitrogen pressure 2 or 3 times the atmospheric pressure were used.

A more sensitive and useful measure of conversion is the residual oxygen content of the sample. This quantity was determined for some "B" samples run for 1 hour at 1600°C, $\Phi(N_2)$ = 5 l/min, as shown in Fig. 6; and it was found to depend on $p(N_2)$, thus reflecting the trend for $-\Delta m/m_o$. While about 3% oxygen by weight remained after the one-hour run at atmospheric pressure, the oxygen contents of the samples run at 0.5 and 1.0 MPa were just 0.37 and 0.42%. The residual carbon contents of these samples rated to 58%.

Four-hour runs at 1600°C, $\Phi(N_2)$ = 1-5 l/min, $p(N_2)$ = 0.2, 0.5 and 5.0 MPa all produced AlN powders containing 0.32% oxygen. This might therefore be close to the equilibrium value under the prevaling experimental conditions. The figure recalculated to correspond to a decarburized powder is 0.76% oxygen.

The total surface area (BET) of a "B" sample, run for 4 hours at 1600°C, 0.5 MPa, 5 l/min, and subsequently decarburized, was determined to be 2.0 m^2/g.

It turned out that quite different powders, with respect to particle size and shape, could be produced by varying the process parameters $\Phi(N_2)$, $p(N_2)$ and the starting mixture type.

Naturally, mixture "A", having an Al_2O_3 excess, or poorly reacted samples of "B" and "C" displayed irregularly shaped particles with a wide size distribution. However, as seen in the SEM pictures, Figs. 7a-f, also the appearances of completely nitrided (and decarburized) samples could differ a lot, due to different synthesis conditions. Thus, while highly agglomerated powders were produced from "B" mixtures run at 1600°C and a low $p(N_2)$, discrete particles formed at $p(N_2)$ > 1 MPa (Figs. 7a,b,c). With "C" as the starting mixture, discrete particles formed at all values of $p(N_2)$ and $\Phi(N_2)$ tried so far (Figs. 7d,e,f). As the discrete AlN particles observed in the SEM do not seem to bear any resemblance to the Al_2O_3 particles in the starting mixtures, we believe that a crucial step in the formation of AlN is a gas phase reaction. This would also explain why the AlN grain size is dependent on $p(N_2)$ and $\Phi(N_2)$.

As a comparison of Figs. 7c and f shows, a larger mean grain size resulted when high-surface-area γ-Al_2O_3 was the reagent (in "C") instead of the coarse α-Al_2O_3 in "B". This might be due to a higher feed rate to the gas, in the former case, of some Al-containing species needed for growth of AlN.

The difference in grain size between the runs in Figs. 7e and f, both with mixture "C", possibly indicates that a more efficient removal of CO(g) from the sample space when a high gas flow is used, causes nucleation at a lower temperature, produceing a fine-grained powder. Still, a comparison between Figs. 7d and e proves that a higher superficial gas velocity does not generally imply a reduction of crystallite size, but that the $p(N_2)$ and $\Phi(N_2)$ parameters act independently.

A more exhaustive study of carbothermal synthesis of AlN, including a number of different mixture types, and also computer calculations of pertinent equilibrium compositions in the Al-C-O-N system at different p and T, is now in progress at this Laboratory.

CONCLUSIONS

* The rate of conversion to AlN in a powder mixture of C + Al_2O_3, reacting in flowing nitrogen, is dependent on:

- temperature
- nitrogen gas flow rate
- C/Al_2O_3 ratio
- nitrogen pressure (0.1 - 1 MPa)

* A mixture (m_C/m_{Al2O3} = 1.5) of carbon black (BET surface area 82 m^2/g) and Al_2O_3 (0.6 m^2/g) can be converted to AlN powder (2 m^2/g, 0.32 % O, 58 % C) within 4 hours at 1600°C, $p(N_2)$ = 1 MPa, $\Phi(N_2)$ = 1 $l_{(stp)}$/min.

* By varying the process parameters the size and shape of the AlN particles can be controlled, to some extent.

- A highly agglomerated powder results when low-surface area Al_2O_3 (0.6 m^2/g) is run at low $p(N_2)$, while discrete particles are obtained from high-surface area Al_2O_3 (138 m^2/g) at $p(N_2)$ > 0.2 MPa.

- The size of the discrete AlN particles can be minimized to about 1 μm by proper choice of nitrogen pressure and gas flow rate (2 MPa, 5 l/min).

ACKNOWLEDGEMENTS

The authors are grateful to Prof. Lars Kihlborg for valuable comments on this manuscript and to Dr Sven Westman, who made the language revision. Thanks are also due to Mr Bengt Gruvin at Kema Nord Industrikemi AB, Ljungaverk, for help with the surface area determinations of the powders. This work has received support from the Swedish Natural Science Research Council (NFR) and the Swedish Board for Technical Development. (STU).

REFERENCES

1) Sheppard, L.M.; "AlN: a Versatile but Challenging Material", *Ceramic Bull.* **69(11)**, 1801-12 (1990).

2) Blakely, K.A., Schorr, J.R. and Shaffer, P.T.; "AlN Ceramics, a Potential Material for Heat Management in Internal Combustion Engines", p. 1515 *in "Ceramic Materials and Components for Engines"*, ed.: V.J. Tennery, The American Ceramic Society (1989).

3) Shanker, K., Grenier, S. and Drew, R.A.L.; "Synthesis of Nitride Ceramics by Carbothermal Reduction", p. 321 *in "Ceramic Powder Science III"*, eds.: G.L. Messing, S-I. Hirano and H. Hausner, The American Ceramic Society (1990).

4) Hirai, S., Miwa, T., Iwata, T. and Katayama, H.G.; "Some Examination on the Formation of AlN by Simultaneous Reduction- Nitriding of Al_2O_3", *JJIM* **54(2)**, 181-5 (1990).

5) Kosolapova, T.Ya.; Yakovleva, D.S.; Oleinik, G.S.; Bartnitskaya, T.S.; Tel'nikova, N.P. and Timofeeva, I.I.; "Some Features of the Formation of AlN During the Reduction-Nitriding of Ultrafinely Divided Alumina", *Poroshk. Metall.* **11(263)**, 14-9(1984)

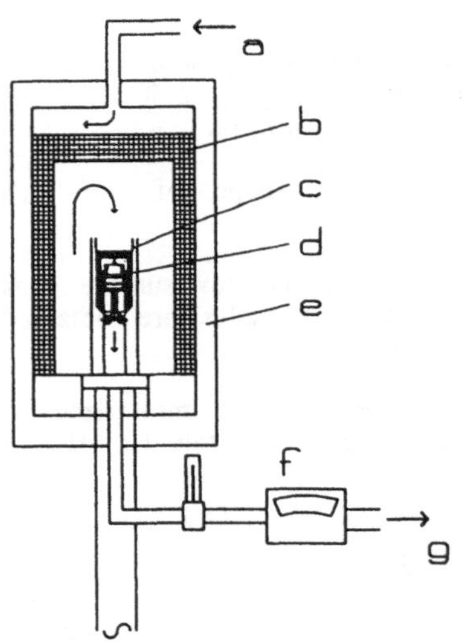

Fig. 1 The autoclave with the graphite furnace.
a) gas inlet b) thermal insulation c) furnace tube d) sample holder e) pressure vessel f) IR detector (CO) g) gas outlet

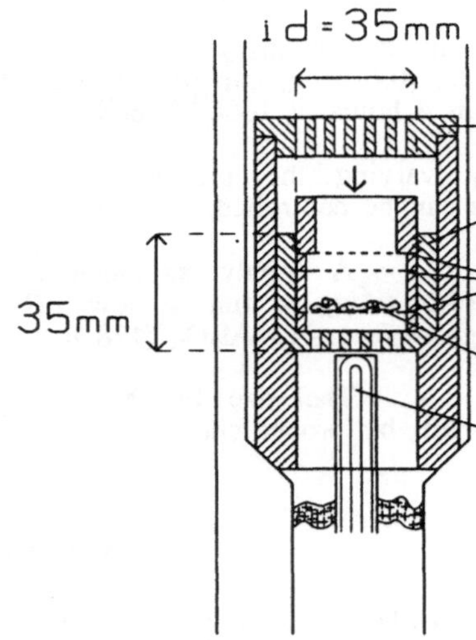

Fig. 2 The sample holder (graphite) a) perforated lid b) sample holder with screw ring c) perforated graphite foils and spacer rings d) sample e) Al_2O_3 and graphite thermocouple shielding tubes

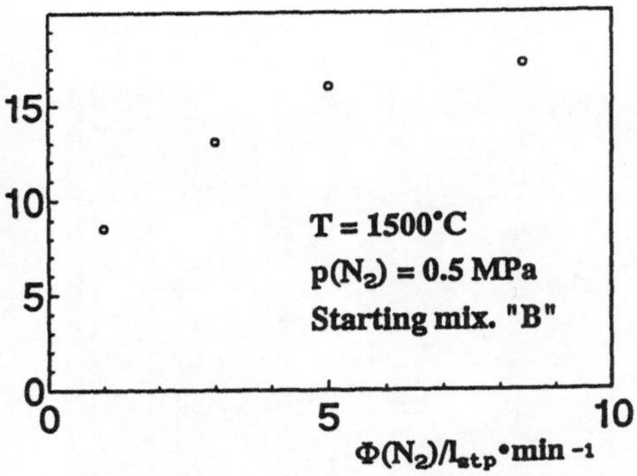

Fig. 3 Conversion as a function of gas flow rate for 1 hour runs at 1500°C with starting mixture "B". Full conversion corresponds to $-\Delta m/m_o$ = 22.3 %.

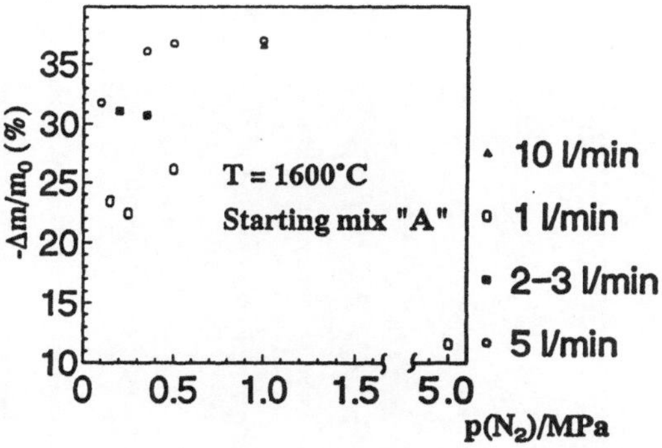

Fig. 4 Conversion as a function of nitrogen pressure for 1 hour runs at 1600°C and different gas flow rates, with starting mixture "A". Full conversion corresponds to $-\Delta m/m_o$ = 38.8 %

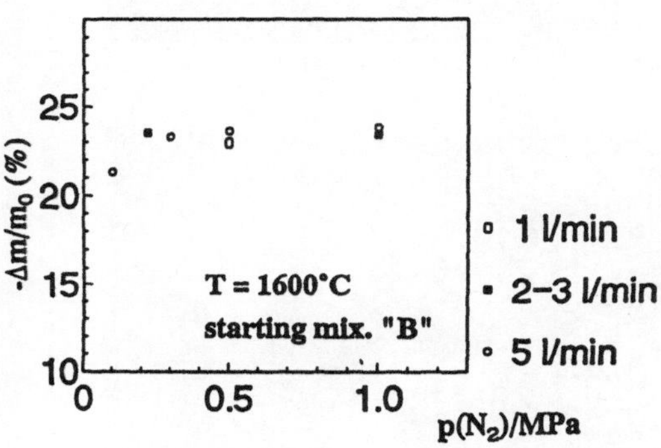

Fig. 5 Conversion as a function of nitrogen pressure for 1 hour runs at 1600°C and different gas flow rates, with starting mixture "B".

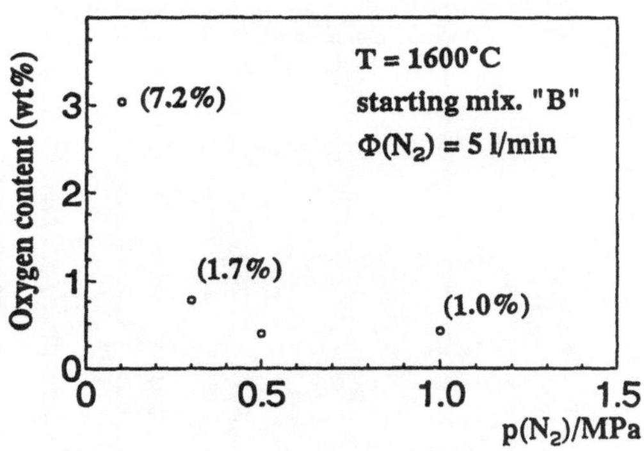

Fig. 6 Residual oxyygen contents in samples of starting composition "B", run for 1 hour at 1600°C at different $p(N_2)$ values. The residual carbon contents were 58%. Figures in parentheses are the oxygen values calculated for decarburized samples.

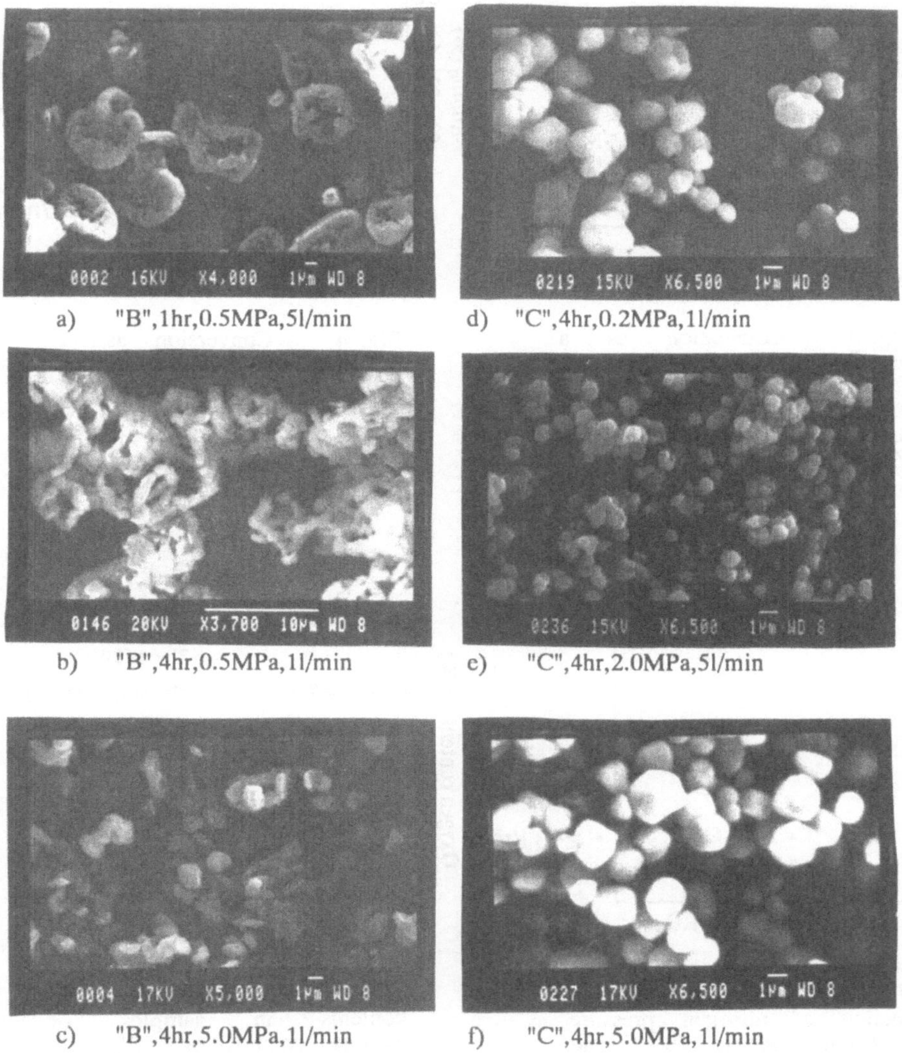

a) "B",1hr,0.5MPa,5l/min

b) "B",4hr,0.5MPa,1l/min

c) "B",4hr,5.0MPa,1l/min

d) "C",4hr,0.2MPa,1l/min

e) "C",4hr,2.0MPa,5l/min

f) "C",4hr,5.0MPa,1l/min

Fig. 7 SEM pictures of "B" and "C" samples, run at 1600°C, different $p(N_2)$ and $\Phi(N_2)$ and subsequently decarburized by heating in air at 700°C.

HIGH TEMPERATURE PROPERTIES AND MICROSTRUCTURES OF Si_3N_4 WITH AZ-TYPE NON-TOXIC, NON-OXIDE ADDITIVES

Ge Changchun+, Xia Yuanluo, Chen Limin, Yuan Yi
Laboratory of Special Ceramics and Powder Metallurgy (LSCPM)
University of Science and Technology Beijing
Beijing 100083, China

ABSTRACT

A new type of Si_3N_4 ceramics (ZAN) is developed in our laboratory. Densification of ZAN is promoted by non-toxic, non-oxide AZ-type additives. In this work high temperature (HT) properties and microstructures of ZAN are investigated.

INTRODUCTION

Silicon nitride is one of the most important candidate materials for various high temperature (HT), high-stressed components of engines. Due to the low self-diffusivity, Si_3N_4 ceramics are generally densified by the oxide additives such as Al_2O_3, MgO, Y_2O_3 etc. which react with silica on the surface of Si_3N_4 powders to form liquid phase during sintering and promote densification. But the glassy phase in grain boundaries formed after sintering usually causes series degradation of HT strength.

The amount and characteristics of the intergranular glassy phase are closely related to the oxide content of the Si_3N_4-based ceramics. Utilization of non-oxide additives instead of oxide additives can significantly reduce the amount of glassy phase and raise its softening temperature, that favour the improvement of its HT strength but make it difficult to be sintered. Prochazka and Greskovich used $BeSiN_2$ as additive [1], which led to high densification and good HT properties. But $BeSiN_2$ is toxic and the fracture toughness of the ceramic is low. Highly dense Si_3N_4 ceramics with additives of metal nitrides were fabricated by expensive high pressure hot-pressing under 3 GPa at

---

+ Former research fellow of Alexander von Humboldt Foundation

1500~1800°C or HIP at 1650~1800°C under 150GPa by Shimada, N. Uchida and M. Koizumi, with no HT bending strength reported [2]. Recently O. Abe reported relatively high strength was maintained below 1300°C on hot-pressed Si_3N_4 with alkaline-earth nitrides. One problem for practical application is the chemical unstability of alkaline - earth nitrides in air[3].

A new type of Si_3N_4 ceramics (ZAN) is developed in our laboratory [4] [5], densification of which is promoted by non-toxic, non-oxide proprietary AZ-type additives either with gas-pressure sintering or conventional hot - pressing . In this work HT properties and microstructures of ZAN are investigated in comparison with a conventional Si_3N_4 ceramic with spinel as oxide additive.

MATERIALS AND METHODS

Commercial Si_3N_4 powders (Shanghai Electroceramics Factory, BET-specific surface area Sg 5.8 m^2/g) and self-made AZ-type powders (Sg 5.4 m^2/g) are used as raw materials. The oxygen contents of Si_3N_4 powders and AZ-type powders are 2.25 and 2.45 % respectively. The metallic impurities content of Si_3N_4 powders is < 4590ppm and that of AZ-type powder is < 1285ppm .

Si_3N_4 with $MgO \cdot Al_2O_3$ as oxide additive (MAO) is prepared as reference material for comparison.

Si_3N_4 powders with different additives are mixed and ball-milled with ethanol as medium for 96h, vacuum dried, sieved through -200 mesh with Sg of 12.5 m^2/g and then hot-pressed in a small hot-pressing furnace under N_2 atmosphere with 10 MPa at different temperatures.

Dimensions of specimens for 3-pt bending strength are 3×4×40 mm with span of 30 mm; loading rate is 0.5mm/min; specimens (5 replicates) are tested with the loading direction coincident with that of hot pressing.

The microstructures are observed with SEM, TEM with EDX and HREM.

RESULTS AND DISCUSSIONS

1. Oxidation Behavior at High Temperature

Specimens are oxidized at 1400°C for 110h. The weight increase -- time relations are shown in Fig.1. The oxidation behavior in the first stage of oxidation can be expressed as:

$$(\Delta W/A)^2 = K \cdot t$$

ΔW --- weight increase during oxidation in kg;
t --- oxidation duration in sec;
A --- surface area of specimen in m^2;
K --- oxidation rate constant in $kg^2 \cdot m^{-4} \cdot sec^{-1}$;
K (ZAN) = 3.5×10^{-11} $kg^2 \cdot m^{-4} \cdot sec^{-1}$, is two orders lower than
K (MAO) = 1.15×10^{-9} $kg^2 \cdot m^{-4} \cdot sec^{-1}$.

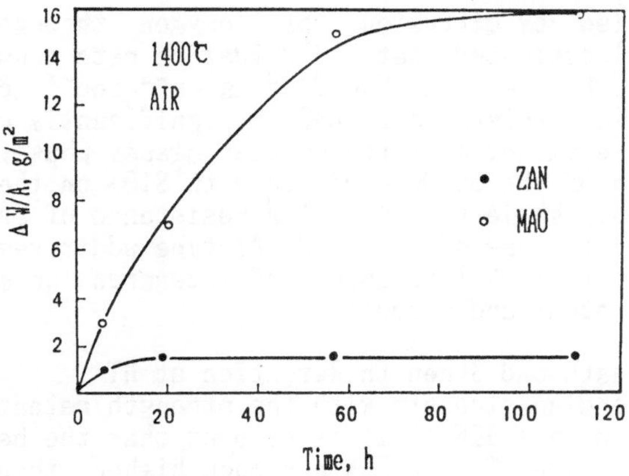

Fig. 1 Dependence of weight increase on oxidation time.

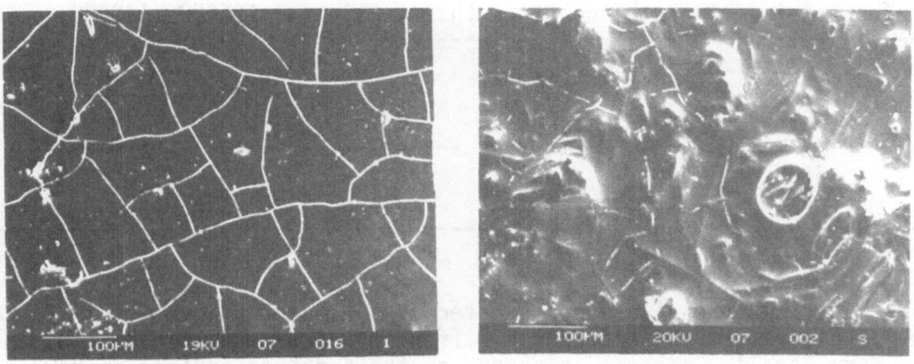

(a) ZAN (b) MAO

Fig. 2 SEM micrographs of oxidized surfaces of (a) ZAN and (b) MAO.

SEM micrographs of oxidized specimens (Fig. 2) show that many protrusions appear on the porous oxidation films on specimens MAO, while rather smooth oxidation films adhere on specimens ZAN, indicating better oxidation resistance.

Passive oxidation of Si_3N_4 proceeds according to the reaction:

$$Si_3N_4 + 3O_2 = 3SiO_2 + 2N_2$$

It is controlled by diffusion of oxygen through surface SiO_2. Greskovich et al. estimated that the oxidation rate constant of CVD-Si_3N_4 at 1400°C should be less than 5×10^{-13} $Kg^2 \cdot m^{-4} \cdot sec^{-1}$ [6].

The oxidation resistance of MAO is significantly reduced because of the considerable amount of intergranular glassy phase, which is formed through reaction of the oxide additive with SiO_2 on the surface of Si_3N_4 powder particles. While the oxidation resistance of ZAN is much better than MAO through the use of non-oxide AZ-type additives instead of oxide additives, due to much less amount of intergranular glassy phase with much higher N content and viscosity.

2. Bending Strength and Strength Retention at HT

The RT and HT bending strength with the strength retention at 1250°C and 1400°C are listed in TABLE 1. It is evident that the bending strength at RT, 1250°C and 1400°C for ZAN are much higher than MAO. The more important fact is that the strength retentions at 1250°C and 1400°C of ZAN reach 90.2% and 71.0%, while those of MAO are only 61% and 41.5%.

TABLE 1.
The bending strength and strength retention at 1250°C, 1400°C.

Specimen	σ, MPa			$\frac{\sigma_{1250°C}}{\sigma_{RT}}$ %	$\frac{\sigma_{1400°C}}{\sigma_{RT}}$ %
	RT	1250°C	1400°C		
ZAN	686	619	484	90.2	71.0
MAO	544	332	226	61.0	41.5

SEM fractographs of specimens after bending strength tests at 1400°C shown in Fig. 3, further prove the different characteristics of ZAN and MAO. Many small "spheres" appeared on the fracture surface of MAO at 1400°C, which formed as liquid phase at 1400°C under the action of surface tension.

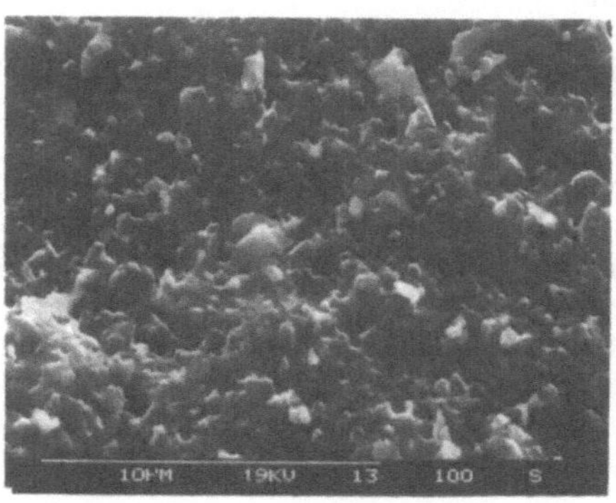

(a) ZAN

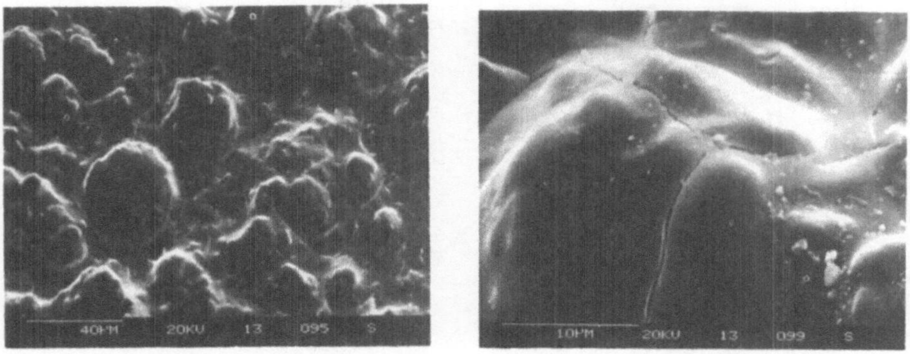

(b) MAO

Fig.3 SEM fractographs after bending strength test at 1400°C

Cracks between "spheres" and matrix developed after a certain amount of plastic deformation through sliding along the surfaces of these "spheres". While the fracture surface of the bending strength specimen ZAN at 1400°C keep the brittle characteristics like that of specimen broken at room temperature.

3. Microstructure Features of Si_3N_4 with AZ-Type Non-Oxide Additive

Superior RT and HT mechanical properties and oxidation behavior of ZAN are closely related to the microstructures. Hexagonal of elongated β'-ss grains of $<1\mu$ are shown in specimen hot pressed at 1800°C under TEM. Many ZrN particles dispersed in matrix and secondary crystalline phase distributed in triple regions of grain boundaries are observed.(Fig.4,5)

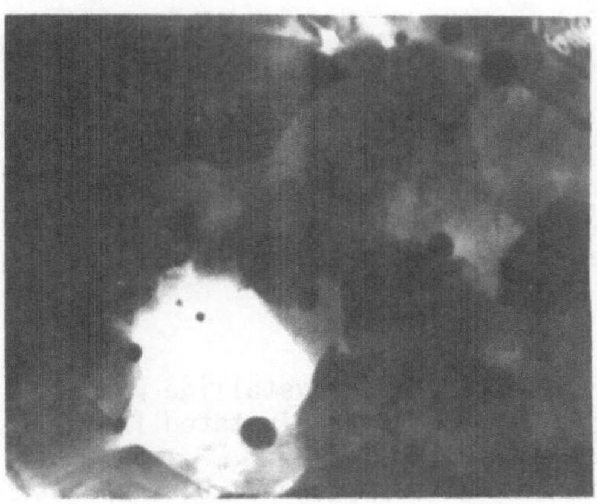

(a)

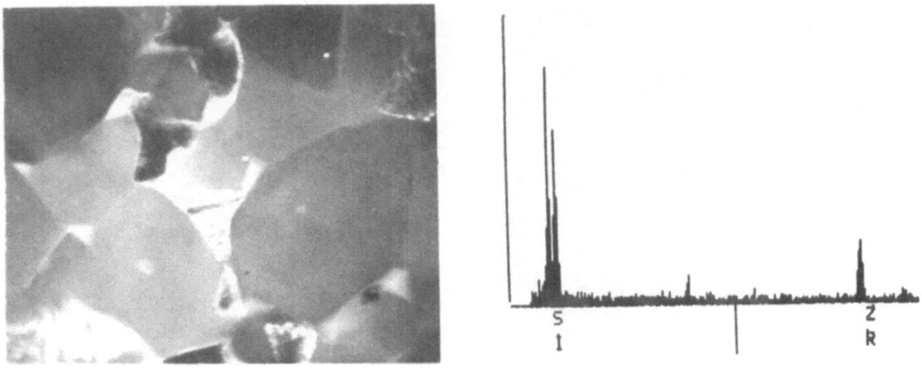

(b)

Fig.4 TEM of ZAN showing secondary crystalline phase in triple regions.
(a) ZrN particles dispersed in matrix, 60,000×
(b) EDX analysis of the compositions of secondary crystalline phase, 60,000× : atom ratio of Zr/Si=1/3

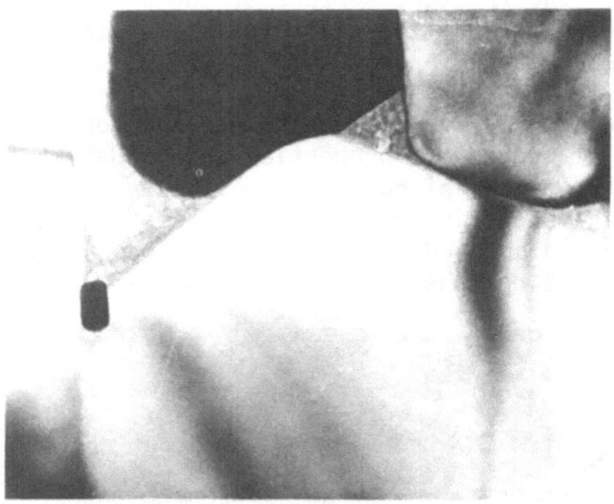

Fig.5 Intergranular glassy phase in the triple regions of β'-Si_3N_4 matrix. 120,000×

Zr and Si are identified in this crystalline phase with EDX which might be oxynitrides of Si and Zr precipitated from liquid phase during cooling after hot-pressing.

Minor intergranular glassy phase is observed at few triple regions, but is not found in most of grain boundaries as show in lattice images in Fig. 6.

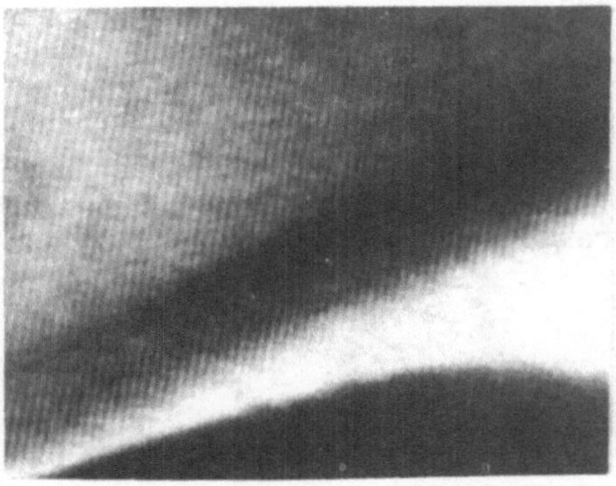

(a)

(b)

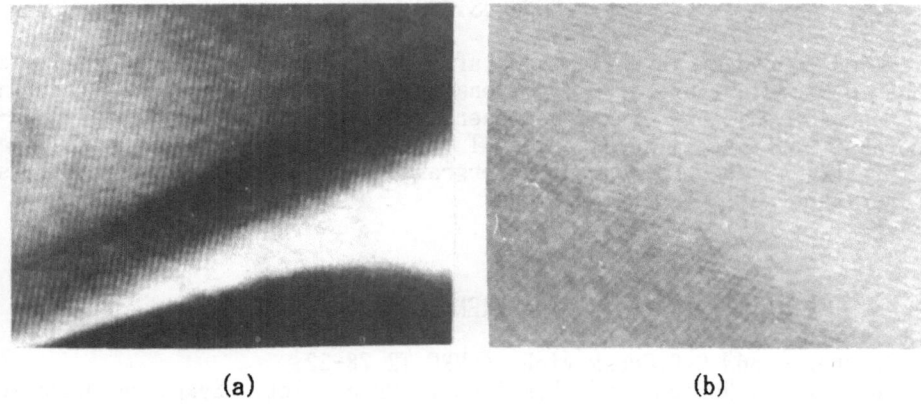

Fig.6 Lattice images of ZAN
(a) Intergranular glassy phase in triple regions of β'-Si_3N_4 grains
(b) No apparent glassy phase found between two adjacent grains
1.5mm=7.608Å

In specimens ZAN hot-pressed at 1650°C, different superlattice structures in α'-Si_3N_4 are found (Fig. 7). It is suggested that Ca^{2+} are located interstitially in α'-Si_3N_4 cells and $Ca_y(Si_{1-x}Al_x)_3N_4$ type compound might form [7], which distributes according to certain periodicity and leads to the formation of superlattice structures.

Fig.7 Superlattice structure found in α'-Si_3N_4 of ZAN
1.5mm=7.608Å

CONCLUSIONS

The microstructure investigations afford convincing evidence for the favorable combination of RT strength, HT strength and oxidation resistance of this new Si_3N_4-based ceramic with AZ-type non-toxic, non-oxide as additive. The superior HT properties are attributed to the significantly reduced amount of intergranular phase and increase of its viscosity and softening temperature.

REFERENCES

1. S.Prochazka and C.D.Greskovich, AMMRC TR 78-32.
2. M.Shimada, N.Uchida and M.Koizumi, Proc. Int. Symp. on Ceramic Components for Engines, Japan, 1983.
3. O. Abe, Ceramics Int., 1990, 16, 53-60.
4. Ge Changchun, Xia Yuanluo, Tan Yiqin and Shao Guoqiang, Int. Conf. of P/M, Dusseldorf, 1986.
5. Ge Changchun, Xia Yuanluo, Chen Limin, EURO-CERAMICS V.1, pp.426, Elsevier Appl. Science.
6. S. C. Singhal, J. Mat. Sci. 1976, 11, 500.
7. Wen Suling, J. of Inorganic Materials, Shanghai, 1987, 3, 199-206

ACKNOWLEDGEMENTS

This work is supported by National Natural Science Foundation of China, National Education Committee of China and Invention Association of Beijing.

Nd-Ce-TZP POWDERS AND CERAMICS: HYDROTHERMAL PREPARATION OF HIGH QUALITY POWDERS FROM SULPHATE PRECURSORS

LAILA GRAHL-MADSEN[1], NANNA PETERSEN[1], KATHRYN WARNER[1], JESPER SAND DAMTOFT[2], and JOHN ENGELL[1]

[1]Technical University of Denmark, Building 204, DK-2800 Lyngby, Denmark
[2]Aalborg Portland, Post Box 165, DK-9100 Aalborg, Denmark

ABSTRACT

High quality powders of tetragonal zirconia of the composition $Nd_{0.01}Ce_{0.08}Zr_{0.91}O_{2-x}$ have been prepared hydrothermally by in-situ crystallization (best 225°C/2.4 MPa, 7-8 h, pH 10-11) of co-precipitated sulphate derived gels. The best powders contain 0.2 wt% residual SO_3. Uniaxially pressed tablets of the as prepared freeze dried powders sinter to >95% TD in 1 h at 1350°C, whereas tablets of calcined and milled powders sinter to >99% TD in 1 h at 1320°C. These ceramics have an average grain size around 1 μm, a 4-point bending strength of 450 MPa, and show a pseudo-plastic behaviour. The toughness is too high to be measured by Vickers indentation. Sintering to 1400°C for 2 h results in sufficient grain growth to induce a slow spontaneous transformation of tetragonal→monoclinic zirconia upon cooling.

INTRODUCTION

Preparation of tetragonal zirconia powders normally involves heat treatment of co-precipitated hydroxy gels made from chloride or nitrate precursors. The use of sulphate precursors is normally avoided, as the removal of residual sulphates from co-precipitated hydroxy gels requires higher calcination temperatures, i.e. ca. 1000°C [1-3]. The present work on the use of sulphate precursors is part of an assessment study of eudialyte, an acid soluble Zr-Y-REE-Nb-silicate, as an alternative raw material for zirconia based engineering ceramics [1-2].

Hydrothermal processing offers a low temperature alternative to conventional precipitation-calcination-milling techniques for the production of oxide powders. The literature of hydrothermal synthesis of zirconia powders has recently been reviewed [4-5]. Using chloride or nitrate precursors, the dissolution-precipitation formation of equilibrium phases, as deduced from phase diagrams, is favoured under acid conditions (pH < 3, [4-5]). At intermediate and basic conditions hydrothermal crystallization occurs by in-situ transformation or mixed processes respectively, under these conditions Ostwald's Rule of Successive Transformation applies, and the observed sequence of crystallization is c→t→m zirconia [4]. Li et al. [6] have recently studied the hydrothermal crystallization of zirconia from solutions of sulphuric acid and zirconium carbonate, as well as from similar solutions containing $MgSO_4$. Crystallization at 250°C results in the formation of metastable zirconium oxysulphate and acicular particles of monoclinic zirconia. At intermediate and high H_2SO_4 concentration treatment times in excess of 48 hours are necessary in order to eliminate the zirconium oxysulphate phase.

The present work concerns hydrothermal crystallization of Nd-Ce-Zr hydroxy gels derived from sulphate precursors. A low residual sulphate content (<2 wt% SO_3) in such hydroxy gels requires precipitation at pH >10 and an elaborated washing scheme [1]. However, these gels have a high specific surface area (>160 m^2/g), and calcined powders prepared from such gels contain more residual SO_3, than similarly prepared powders derived from high sulphate hydroxy gels precipitated at lower pH, and having a lower specific surface area [1].

Previous work has shown that using inexpensive mixtures of Nd-Ce-oxides it is possible to produce a variety of TZP ceramics optimized either according to strength, toughness, or having improved ageing behaviour in aqueous solutions [2]. One effect of the Nd is to reduce grain growth during sintering. Ceramics prepared from the composition selected for the present work ($Nd_{0.01}Ce_{0.08}Zr_{0.91}O_{2-x}$) show a high fracture toughness but a poor ageing behaviour in aqueous solutions at low temperature (20°-90°C, [2]).

EXPERIMENTAL

The powders were synthesized as outlined in Figure 1, using Zirconium Basic Sulphate (Magnesium Electron), $Ce_2(SO_4)_3$ (Fluka AG, purum), Nd_2O_3 (96%, Union Molycorp), Ammonia Water and H_2SO_4 of technical grade. Teflon containers were used in the autoclave, and polyethylene or polypropylene bottles for all other handling.

Homogeneous solutions were prepared by dissolving the stabilizers in stock solutions of Zr-sulphate (ca. 4 wt% ZrO_2). Co-precipitation was induced by addition of aqueous ammonia

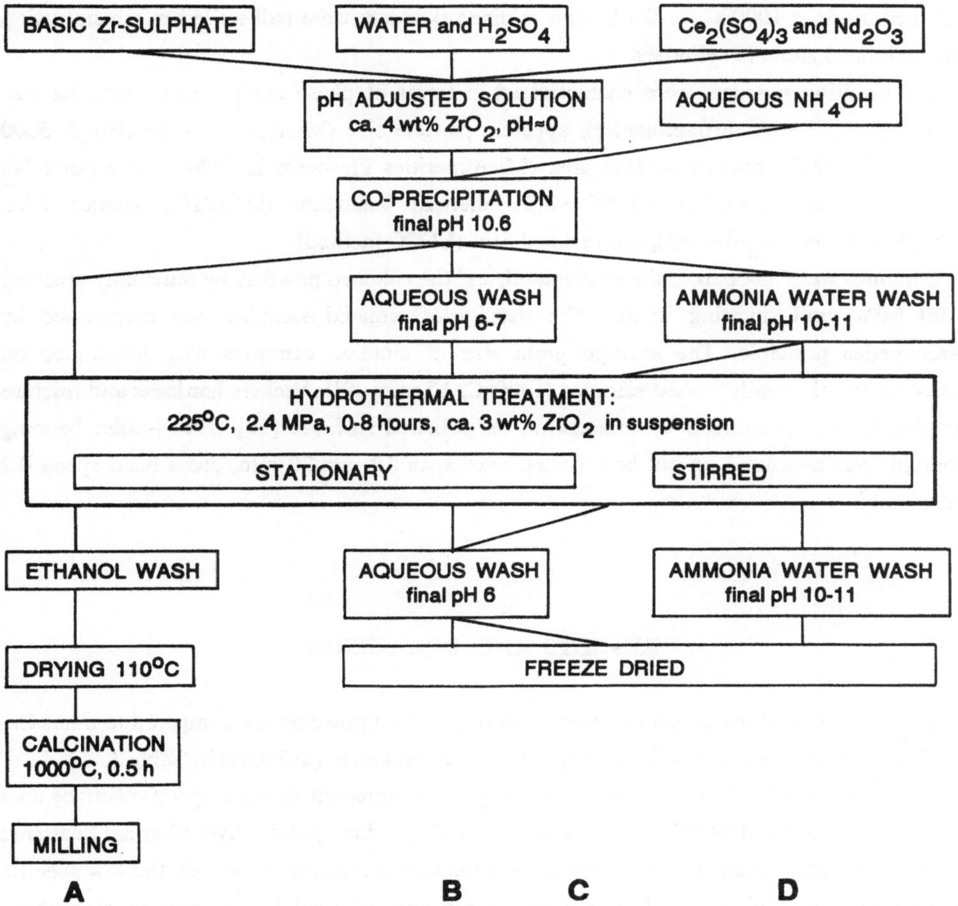

Figure 1. Flow diagram showing the four routes used for the preparation of TZP powders. All washing was done using a continuous tangential flow filter and 10 liter solvent per mol ZrO_2.

until pH=10.6 while stirring vigorously (high speed mixer). The resulting ammonium-sulphate containing hydroxy gel was aged at least 24 h before decanting the surplus liquid. Different washing schemes (Figure 1) were employed in order to investigate the influence of residual sulphate on the hydrothermal reaction. The hydrothermal treatment was done at 225±5°C and 2.4 MPa under stationary conditions (Route A, B & C), or while stirring (Route D). Freeze drying was done in a Heto freeze dryer. Some of the dried gels (Figure 1) were calcined in air at 1000°C for 0.5 h. The calcined powders were ball milled in ethanol using Mg-stabilized zirconia cylinders.

The resulting powders were characterized in terms of phase content and crystallite size (XRD, [7-8], Philips diffractometer), apparent particle size (Micromeritics SediGraph 5000 ET; STDV $\pm 2\%$), specific surface area (Micromeritics Flowsorp II 2300, single point N_2 adsorption and desorption, STDV <2%), thermal behaviour (DTA/TG, Mettler TA1, 10°C/min in dry CO_2-free air), and residual SO_3 (LECO method).

Ceramics were made from the as prepared, and the calcined powders by uniaxially pressing (100 MPa) and sintering in air. The density of sintered samples was determined by Archimedes principle. The average grain size of sintered ceramics was determined on polished and thermally etched samples (1300°C, 15 min; [9]). Vickers hardness and fracture toughness was determined by indentation on polished surfaces [10]. The 4-point bending strength was measured on cut bars (inner/outer span 10 mm/20 mm, cross head speed 0.2 mm/min).

RESULTS AND DISCUSSION

The properties of the present hydrothermally prepared powders are compared to other Ce-TZP powders derived from sulphate [1] or chloride precursors (TOSOH) in Table 1.

Upon drying at 110°C the as prepared amorphous co-precipitate has a specific surface area of 123 m^2/g and residual SO_3 >15 wt% (Route A, 0 h, Table 1). After hydrothermal treatment for 1.2 h the only crystalline phase detected is tetragonal zirconia. However, the low specific surface area and high weight loss indicate the presence of a sulphate-containing amorphous phase. Essentially the same results are obtained using a treatment time of 7.3 h (Table 1). Using Route A (Figure 1) calcination at 1000°C for 0.5 h is necessary in order to obtain high quality TZP powders (Table 1).

Nearly identical high quality TZP powders were prepared by the three other routes used in the present work (Route B, C & D, Figure 1 & Table 1). The average apparent particle size of the freeze dried powders is in the range 2-4 μm. The crystallite sizes calculated from XRD and BET are identical showing that the particles formed are built of an open network of 7 nm crystallites (Table 1). The residual SO_3 content decreases from 1 wt% for powders prepared by Route B, to a minimum of 0.2 wt% for powders prepared by Route D. These powders show a total weight loss of only ca. 4 wt% upon heating to 1000°C. It is perhaps significant that this minimum SO_3 content is similar to the content in the calcined powders prepared by Route A (Table 1).

TABLE 1.

Comparison of $Nd_{0.01}Ce_{0.08}Zr_{0.91}O_{2-x}$ powders prepared in this work and other differently prepared Ce-TZP powders.

POWDER PREP. ROUTE	Time 225°C h	Calc. 1000°C 0.5 h	SPECIFIC SURFACE AREA m^2/g	PHASES M vol%	PHASES T vol%	CRYSTALLITE SIZE (T) XRD nm	CRYSTALLITE SIZE (T) BET nm	APPARENT PARTICLE SIZE $d_{50\%}$ µm	WEIGTH LOSS <1000°C wt%	SO_3 wt%
A	0		123 ±2	Amorph.				3.5	49.6	>15
	1.2		42 ±1		100	6	23	6.6	29.2	
	7.3		47 ±1		100	7	21	2.8	20.5	
B	7		141 ±3	<5	>95	7	7	3.4	4.0	1.0
C	7		137 ±3		100	7	7	3.2	4.1	0.8
D	8		143 ±3		100	7	7	2.5	3.8	0.2
A	0	+	34.5±0.7	17	93	25	29	1.6		0.2
	1.2	+	17.5±0.4		100	21	57			
	7.3	+	18.0±0.5		100	20	56	6.2		
Damtoft et al. [1]										
$Ce_{0.08}Zr_{0.92}O_2$		+	20	85	15	20	50			0.2
$Ce_{0.12}Zr_{0.88}O_2$		+	25	60	40	28	40			0.4
TOSOH TZ-12Ce*			9-10	7-43	93-57	21	100	0.2	0.4	

* Contain minor cubic $Ce_{0.75}Zr_{0.25}O_{2-x}$.

TABLE 2.

Characterization of $Nd_{0.01}Ce_{0.08}Zr_{0.91}O_{2-x}$ ceramics prepared from the present powders (Table 1) and by a co-precipitation/calcination/milling route [2].

POWDER PREP.	Time 225°C h	Calc. 1000°C 0.5 h	SINTERING °C	SINTERING h	%TD	AV. GRAIN SIZE µm	MECHANICAL PROPERTIES Vickers Hardness GPa	Fracture Toughness $MPa{\cdot}m^{0.5}$	4-Point Bending Strength MPa	m	n
A	0	+	1320	1	100			+++			
	7.3	+	1320	1	99.5		8.4		450	5	14
B	7		1350	1	96.6	1.1	8.9	+++			
C	7		1350	1	94.9	0.9	6.5	+++	344	5	14
Damtoft et al. [2]			1350	2	-	1.0	8.2	+++			

+++: Pseudo-plastic behaviour.

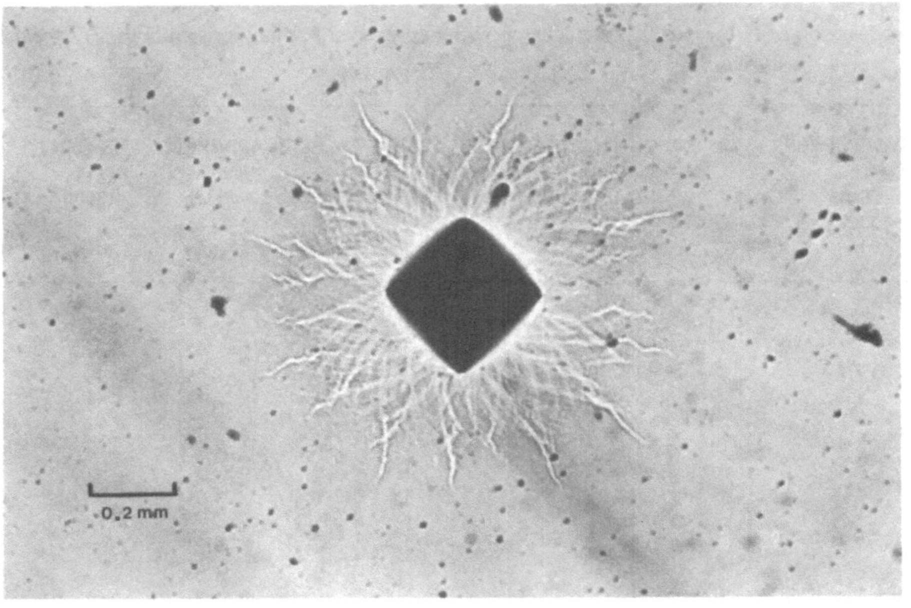

Figure 2. Pseudo-plastic deformation in indented $Nd_{0.01}Ce_{0.08}Zr_{0.91}O_{2-x}$ ceramic sintered at 1350°C for 1 h (B, Table 2).

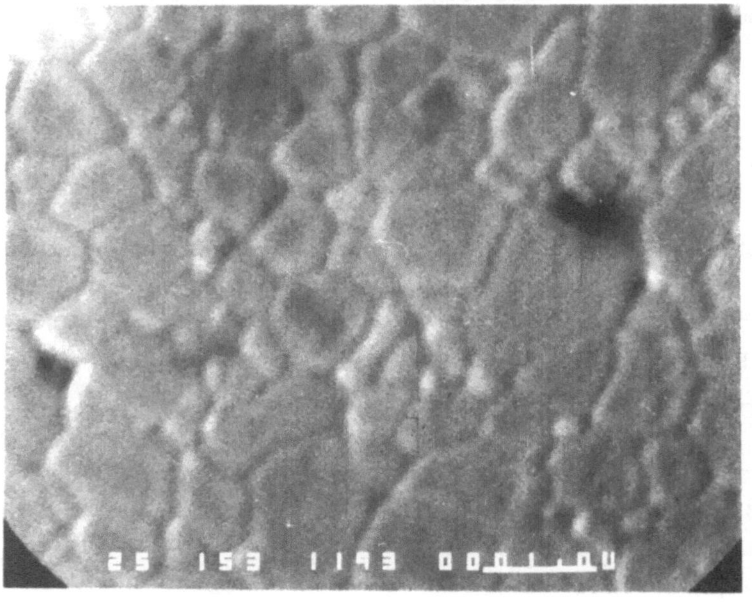

Figure 3. SEM picture of microstructure in the TZP ceramic shown in Figure 2.

The unit cell dimensions of the tetragonal $Nd_{0.01}Ce_{0.08}Zr_{0.91}O_{2-x}$ in the calcined powders are a = $0.3617_{0\pm8}$ nm and c = $0.521_{2\pm1}$ nm as calculated from XRD data. From this the theoretical density of this TZP material has been calculated to be 6.21 g/cm^3.

The properties of TZP ceramics prepared from the present powders are given in Table 2. Uniaxially pressed tablets made from the as prepared hydrothermal powders sinter to 95-96% TD in 1 h at 1350°C (B & C, Table 2). Using the calcined and milled powders near theoretical density ceramics have been obtained after sintering at 1320°C for 1 h. Sintering to 1400°C for 2 h results in sufficient grain growth to induce a slow spontaneous transformation of t→m zirconia upon cooling.

The prepared ceramics show a pseudo-plastic behaviour upon indentation, and the fracture toughness is too high to be determined by this method (Figure 2). XRD examination of the sintered surfaces shows tetragonal zirconia only. The ceramics sintered at 1350°C have an granular microstructure with an average grain size of approximately 1 μm (Figure 3, Table 2). This is close to the critical grain size of tetragonal zirconia stabilized with 8 mol% CeO_2 and 1 mol% $NdO_{1.5}$, and the ceramic has a high fracture toughness but an inferior ageing behaviour in aqueous media at low temperature (20°-90°C). The 4-point bending strength of the fully dense material is 450 MPa.

CONCLUSION

The present work shows that it is possible to prepare high quality TZP powders by hydrothermal processing of sulphate derived Nd-Ce-Zr hydroxy gels. The best results was obtained using Route D (Figure 1). Powders prepared in this way contain only 0.2 wt% residual SO_3 and show a total weight loss of less than 4 wt% upon heating to 1000°C. High density TZP ceramics can be made directly from the as prepared hydrothermally crystallized powders.

TZP ceramics of the composition $Nd_{0.01}Ce_{0.08}Zr_{0.91}O_{2-x}$ and with an average grain size of 1 μm have a reasonable bending strength (450 MPa), a high toughness, and show pseudo-plastic deformation behaviour.

ACKNOWLEDGEMENTS

The financial support from EEC through EURAM contract no. MA1E/0019/C and from the Danish Centre for Advanced Ceramic is gratefully acknowledged.

REFERENCES

1. Damtoft, J.S., Engell, J. & Frederiksen, J., Stabilized Zirconia Powder Derived from Sulphate Solutions. In New Materials and Processes: Proceedings of the 5th Scandinavian Symposium on Materials Science, eds. I.L.H. Hansson & H. Lilholt, 1989, pp.221-228.

2. Damtoft, J.S., Engell, J., Frederiksen, J., Garcia-Coronado, N., Gilbart, E., Grahl-Madsen, L. & Brook, R.J. (in prep), Alternative Stabilizers for Zirconia Ceramics: Powders Derived by Processing of Eudialyte. Extended Synthesis Report, EURAM contract no. MA1E/0019/C.

3. Smith, A. & Baumard, J.-F., Sinterability of Tetragonal ZrO_2 Powders. Am. Ceram. Soc. Bull., 1987, 66, 1144-1148.

4. Grahl-Madsen, L., Engell, J. & Riman, R., Hydrothermal Preparation of Stabilized Zirconia Powder. In Ceramic Powder Science III, eds. G.L. Messing, S.-I. Hirano & H. Hausner, Am. Ceram. Soc., 1990, pp.33-40.

5. Denkewicz, R.P., TenHuisen, K.S. & Adair, J.H., Hydrothermal Crystallization Kinetics of m-ZrO_2 and t-ZrO_2. J. Mater. Res., 1990, 5, 2698-2705.

6. Li, C., Yamai, I., Musrase, Y. & Kato, E., Formation of Aciculat Monoclinic Zirconia Particles under Hydrothermal Conditions. J. Am. Ceram. Soc., 1989, 72, 1479-1482.

7. Klug, H.P. & Alexander, L.E.: X-ray Diffraction Procedures for Polycrystalline and Amorphous Materials, John Wiley & Sons, 2.ed. 1974, pp.1-966.

8. Toraya, H., Yoshimura, M. & Somiya, S., Calibration Curve for Quantitative Analysis of the Monoclinic-Tetragonal ZrO_2 System by X-ray Diffraction. J. Am. Ceram. Soc., 1984, C119-C121.

9. Mendelson, M.I., Average Grain Size in Polycrystalline Ceramics. J. Am. Ceram. Soc., 1969, 52, 443-46.

10. Anstis, G.R., Chantikul, P., Lawn, B.R. & Marshall, D.B., A Critical Evaluation of Indentation Techniques for Measuring Fracture Toughness: I, Direct Crack Measurements. J. Am. Ceram. Soc., 1981, 64, 533-543.

MICROSTRUCTURE AND MECHANICAL PROPERTIES OF SINTERED REACTION BONDED SILICON NITRIDE (SRBSN)

M. HERRMANN, S. KESSLER, Ch. Taut
Institute of Ceramic Technology and Materials Science in the Central Institute of Solid-State Physics and Materials Research, O-8020 Dresden, Winterbergstr.28, Germany
G. A. GOGOTSI, V. P. ZAVADA
Institute of Problems of Strength, 252014 Kiev, Timiryasevskaya str 2, USSR

ABSTRACT

The distribution of grain size and aspect ratio of the β-Si_3N_4-grains in SRBSN were estimated. These microstructural parameters and the amount of glassy phase were correlated with mechanical properties of the materials at room and high temperatures.

INTRODUCTION

During the last years, silicon nitride ceramics attracted increasing interest. This development has been caused by the potential applications and by the actual progress in the field of materials technology and microstructural design .
Previous work [1, 2] indicated that the high fracture toughness of Si_3N_4-materials is due to its needle like grain morphology. But there exist only some studies of the influence of the amount and properties of the glassy phase on the mechanical properties of silicon nitride.
The aim of this work is to correlate the mechanical properties of sintered reaction bonded silicon nitride (SRBSN) at room and elevated temperatures with the amount of the glassy phase.

MATERIALS AND METHODS

The starting materials were prepared from a high purity silicon powder (surface area: 3.2 m^2/g; oxygen content: 0.73%), Al_2O_3 (ALCOA A 16) and a mixture of rare earths (64

% Nd_2O_3; 17% Pr_2O_3; 14% La_2O_3;1% Sm_2O_3; content of all rare earths >99.98%; surface area: 5.3 m/g). The molar ratio of Al_2O_3 and rare earth oxide R_2O_3 was 1:1 (Tab. 1). Milling was carried out in ethanol in a planetary ball mill. The jars were made from agate. The resulting suspensions were dried, and the dried powder mixture pressed at 200 MPa to green bodies of an approximate size of 5x6x60 mm^3. These samples were then nitrided in a nitrogen/hydrogen atmosphere with stepwise increasing of the temperature up to 1380°C in 22 h. Some characteristics of the RBSN materials are given in Tab. 1. The phase content has been determined by X-ray diffraction analysis, density and porosity by the Archimedes technique. Sintering was carried out in a two step regime at 1850°C and a nitrogen pressure up to 50 atm (furnace: FPW 150/200 KCE). The results of sintering experiments are given in Table 2.

TABLE 1
Properties of the samples (RBSN-state)

Number	Composition			Phase content α/β Si_3N_4	Porosity
	Si	Al_2O_3 w/o	R_2O_3	%	%
1	88.87	2.59	8.54	90 : 10	29.0
2	86.13	3.22	10.65	90 : 10	29.1
3	82.40	4.1	13.50	90 : 10	29.5

TABLE 2
Microstructure and properties of materials sintered at 1850°C (Sintering time 90 min; A_m - maximum measured aspect ratio, E - dynamic elastic modulus)

Material	Density ρ (g/cm³)	Thickness d (μm)	A_m	Part of grains ratio larger than 3.5 (%)	5.5 (%)	E GPa
1	3.27	0.58	10	69	26	306
2	3.32	0.57	9	-	-	303
3	3.31	0.58	8	60	15	291

The microstructure of the sintered materials was characterized by analysis of SEM photomicrographs of etched polished samples. For all grains the length and thickness were measured and the apparent aspect ratio was calculated. The distribution of the aspect ratio determined in this way differs from the real distribution of aspect ratio in the material, since the measured aspect ratio depends on the orientation of the grain in relation to the polished section. In order to determine the distribution of the real aspect ratio of the grains in the bulk a mathematical model has been developed [3]. The results of analysis are shown in Table 2.

For the determination of four point bending strength (distances between the supports 40/20 mm) ground samples (3.5*4*50mm) were used. The investigations were carried out at a special precision loading equipment, which allows to minimize the error associated with torsion and constraction [4]. The equipment is completed with an measuring device for registration of the real deformation of the sample.

This allows the determination of the load-deformation diagram even at low loads to be performed with higher accuracy than the registration of the cross head displacement only. The details of the determination of the R-curve behaviour are given in [5]. The investigations of the high-temperature properties were carried out in air in a furnace with silicon carbide heating elements. The investigations were carried out at two cross head velocities of 0.5 and 0.008 mm/min. The fracture toughness (K_{Ic}) was determined by the SENB-method (thickness of the notch 0,15 and 0,2 mm) and additionally by the direct crack measurement after indentation.

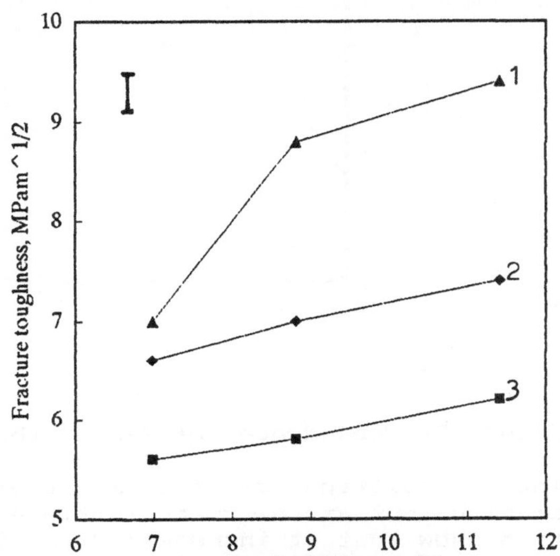

Figure 1: Dependence of the K_{Ic} on the amount of sintering additives (1,2- SENB,1- r=0,2, 2-r=0,15mm; 3-DCM)

RESULTS AND DISCUSSION

As shown in Table 2 the microstructure of the materials 1 - 3 is very similar. With increasing content of additives the amount of grains with high aspect ratio decreases slightly. The main difference between the three materials is the different amount of the additives producing a different content of the grain boundary glassy phase.
The fracture toughness measured by the SENB and by the indentation methods increases with increasing amount of additives. (Fig.1). For the material SRBSN-3 an R-curve behaviour was observed (Fig.2). The fracture surface and the cracks initiated by the Vickers indentations were observed (Fig. 3,4) to find the reason for this phenomenon. The analysis of micrographs of the fracture surface shows a high percentage of intergranular fracture and extended crack branching (Fig.3).

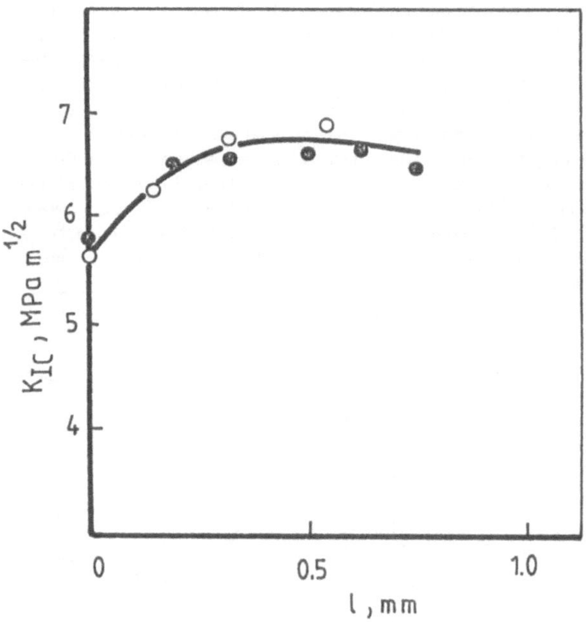

Figure 2: Dependence of K_{Ic} on the crack length (l)

The calculation of the amount of intergranular fracture carried out at the crack initiated by the Vickers indentation show that it increases with rising amount of the glassy phase. The part of intergranular fracture for the material SRBSN-1 was 85% and for the material SRBSN-3 90 %. With increasing amount of intergranular fracture the crack deflection and grain interlocking mechanisms can be worked more effectively and the fracture toughness increases.

Another reason for increasing K_{Ic} is the intensification of the crack branching. This phenomenon is correlated with different strength of the grain boundaries due to slightly different amounts of glassy phases in the grain boundary (Fig.3)

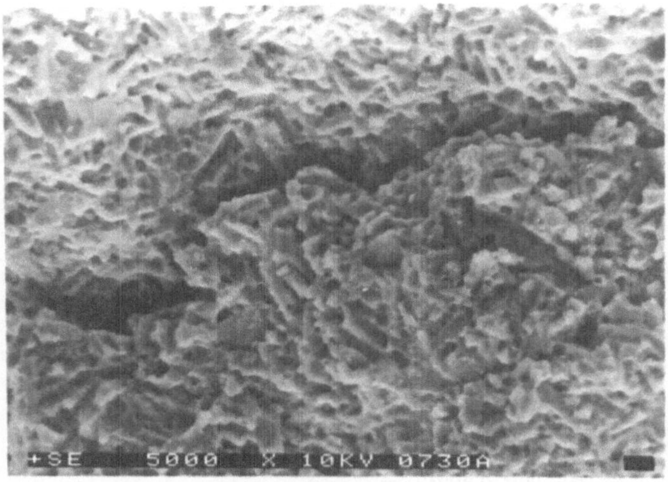

10 µm

Figure 3: SEM micrograph of the fracture surface of SRBSN 3

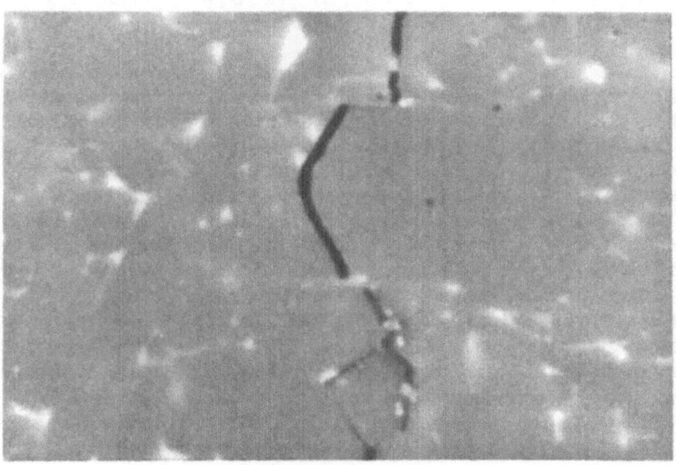

2 µm

Fig. 4: Crack initiated by the the Vikers indentation

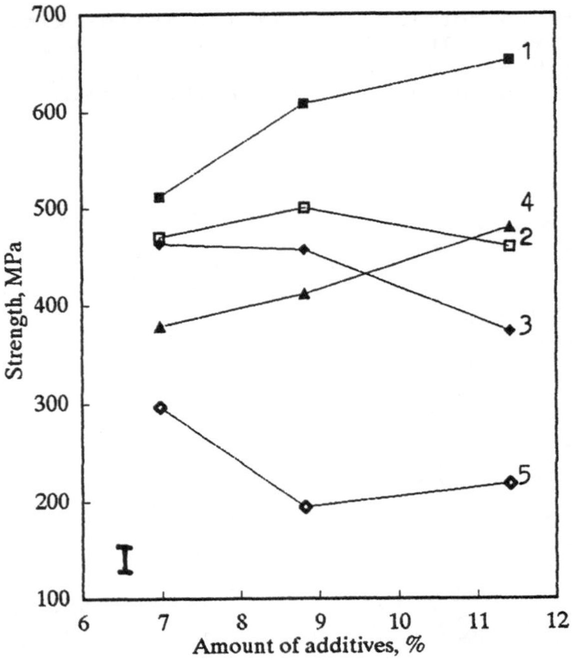

Figure 5: Four-point-bending strength as a function of the amount of additives at room temperature (1), 900°C (2,3) and 1200 °C (4,5) (1,3,4 - loading rate v=0,5mm/min; 2,5 - v=0,008mm/min)

The strength of the materials at different temperatures and loading rates are shown in Fig. 5. At room temperatures the loading rate has no significant influence on the strength. This had been expected, because silicon nitride materials have a high resistance against subcritical crack growth at room temperature. The lower strength of the SRBSN-1 cannot be explainad with the lower K_{Ic}. It is due to a higher defect size of the the material, because the amount of the sintering additives is not high enough to eliminate the defects in the same manner as in the materials SRBSN-2 and -3. At 900°C the strength of the materials at lower loading rates is higher than that at a high cross head velocity. Near the transformation point of the glassy grain boundary phase the resistance against subcritical crack growth is high enough, that this mechanism does not determine the strength. But the viscosity of the glassy phase is low enough, that relaxation processes can take place at low loading rates. Such an explanation is supported by the fact, that at 900°C and low loading rates some pull out of elongated grains takes place.

Above 1000 °C subcritical crack growth is the main mechanism which determin the strength. At the high loading rate the material with the higher room temperature strength (lower defect size) has also the higher strength at 1200°C. At low loading rates the strength degradation increases with increasing amount of glassy phase.

CONCLUSIONS

The results of in vestigations of three SRBSN-materials with nearly constant grain size and shape but different amount of glassy phase allows to draw the following conclusions:

1. The fracture toughness at room temperature increases with rising amount of glassy phase.
2. At 900°C the strength at low loading rates is higher than at high loading rates. The reason for this is the pull out of elongated grains.
3. At high temperatures subcritical crack growth is the most important mechanism of destruction of the materials. With increasing amount of glassy phase the subcritical crack growth increases.

REFERENCES

1. G. Ziegler, J. Heinrich, G. Wotting, Rev. Relationships between Processing, Microstructure and Properties of Dense and Reaction Bonded Silicon Nitride, J. mat. Sci.,22, (1987), 304
2. M. Herrmann, Chr. Schubert, P. Obenaus, W. Kreher, S. Hess, W. Hermel in: Technische keramische Werkstoffe Editor: J. Kriegesmann, DWD, Koln, (1989), 5.1.2.1,
3. P. Obenaus, M. Herrmann, Methode zur quantitativen Charakterisierung von Stengelkristalliten in Siliciumnitridkeramik, Praktische Metallographie, 27, (1990), 503
4. G. Gogotsy L. Grushevsky V. P. Zavada, Attestaziya Keramiki po mechanicheskim svoistvam, Ogneupory ,9, (1988), 33
5. G. A. Gogotsy, Deformation behaviour of ceramics, Journ. of the Eur. Cer. Soc., in press
6. P. L. Swanson, C. J. Fairbanks B. R. Lawn at al J. Am. Ceram. Soc., 70, (1987), 279

EFFECT OF INTERGRANULAR MICROSTRUCTURE ON PROPERTIES OF Si_3N_4 BASED COMPOSITE CERAMIC MATERIALS

E. MARIA KNUTSON-WEDEL, LENA K.L. FALK and THOMMY EKSTRÖM*
Department of Physics, Chalmers University of Technology
S-412 96 Göteborg, Sweden
* AB Sandvik Hard Materials, S-126 80 Stockholm, Sweden

ABSTRACT

The microstructures of Si_3N_4 based ceramic materials containing ZrO_2 have been characterized by analytical electron microscopy and x-ray diffractometry. The ceramics were fabricated by hot isostatic pressing (HIP) and contained different additions of Al_2O_3, AlN and Y_2O_3 as sintering additives. Ceramics formed without the separate addition of Y_2O_3 were densified in the presence of a transient liquid phase; these microstructures contained an extremely small volume fraction of residual glass. Increased volumes of intergranular glass formed in materials containing larger additions of separately added Y_2O_3. These larger glass volumes had a negative influence on hardness. Formation of prismatic β-Si_3N_4 grains was promoted by additions of Y_2O_3, either separately or in the form of Y_2O_3 stabilized ZrO_2 starting powder. This morphology of the Si_3N_4, as well as larger additions of ZrO_2, increased toughness.

INTRODUCTION

The intrinsic properties of the compound Si_3N_4 make Si_3N_4-based ceramic materials potential candidates for high temperature structural applications. The strongly covalent interatomic bond in Si_3N_4 results, however, also in a very low self diffusivity which inhibits sintering of pure Si_3N_4; generally, densification of Si_3N_4 powder compacts to full theoretical density requires an addition of metal oxide sintering aids. During densification, the metal oxides react with the surface silica present on the Si_3N_4 powder particles and also some of the Si_3N_4 to form an oxynitride liquid which promotes densification [1,2]. The kinetics of the densification process may be interpreted by the three stages of Kingery's model for liquid phase sintering [3]; particle rearrangement in the liquid, solution of α-Si_3N_4 followed by precipitation of β-Si_3N_4 and coalescence.

The morphology of the Si_3N_4 grains is dependent upon the properties of the oxynitride liquid phase sintering medium, e.g. composition and viscosity. Larger volumes of an isotropic liquid environment will allow growth of prismatic β-Si_3N_4 grains [4]. During processing, secondary crystalline phases may also form from the oxynitride liquid phase, and remaining liquid will solidify as a residual intergranular glass phase present as thin intergranular films merging into pockets at multigrain junctions. The number, chemistry and amount of secondary phases will be dependent upon starting powder composition and densification process, e.g. temperature/time programme.

Grain boundary phases present in the microstructure of Si_3N_4 ceramics have a conclusive influence on mechanical and chemical properties. In general, the intergranular glassy phase will have a detrimental effect on high temperature properties, and it becomes important to minimize the amount of residual glass. Densification by hot isostatic pressing (HIP) makes it possible to reduce the amount of metal oxide sintering additives, and such materials will have a comparatively low volume fraction of residual glass. By careful selection of metal oxide system, it is also possible to control the intergranular microstructure by formation of secondary crystalline phases and/or formation of solid solutions. As demonstrated for Si_3N_4 ceramics formed in the Y-Si-Al-O-N system [5,6], post-densification heat treatments may result in a substantial crystallisation of the residual glass present in the microstructure after sintering. Formation of crystalline solid solutions, e.g. sialons, will consume constituents of the liquid phase sintering medium, and hence reduce the amount of residual glass.

The present paper concerns a microstructural investigation of ceramics based on Si_3N_4 and ZrO_2, containing additions of Y_2O_3, Al_2O_3 or AlN as sintering additives and formed by HIP. The results have been related to room temperature hardness and indentation fracture toughness, and special attention was paid to the intergranular microstructure.

MATERIALS AND METHODS

The Si_3N_4 ceramic materials in this investigation were densified by hot isostatic pressing, HIP, at 1750 °C for 1 hour using a pressure of 200 MPa. Prior to HIP, the green bodies were glass encapsulated. TABLE 1 gives the different compositions of the characterized ceramics. The Si_3N_4 starting powder was H.C. Starck-Berlin, LC1, which contains 94-95 wt% α-Si_3N_4 and has an oxygen content corresponding to 2.9 wt% surface silica. The additives were ZrO_2 (Chema Tex, SC16) and ZrO_2 stabilized with 3 mol% Y_2O_3 (TOSOH, TZ-3Y), Al_2O_3 (Alcoa, A16SG), Y_2O_3 (H.C. Starck, Finest) and AlN (H.C. Starck-Berlin, Grade D). Materials containing 5 wt% ZrO_2 and 2 wt% Y_2O_3 were formed both with Al_2O_3 additions and with AlN. The large (20 wt%) additions of ZrO_2 were stabilized with 3 mol% Y_2O_3. Processing prior to HIP is described in more detail in reference [7].

The microstructures were characterized by analytical electron microscopy (SEM, TEM, STEM, EDX). Scanning electron microscopy (SEM) was performed on polished surfaces using a CamScan S4-80 DV scanning electron microscope. Transmission electron microscopy was carried out using a Jeol 2000-FX TEM/STEM/SEM instrument equipped with a Link energy dispersive x-ray (EDX) system for quantitative elemental analysis. As Si_3N_4 is an insulating material, carbon was evaporated onto all samples in order to avoid charging in the microscope.

Phase compositions were determined by x-ray diffractometry (XRD) of polished sections. Room temperature hardness was determined by Vickers indentation using a 10 kg load, and indentation fracture toughness was calculated from the indentation marks according to the equation of Anstis [8].

TABLE 1
Additives (wt%), phase compositions and EDX analysis of the samples in the present investigation.

Al_2O_3	AlN	Y_2O_3	ZrO_2	Phase composition	Calculated; Y_2O_3/ZrO_2 [mol%]	EDX results; Y_2O_3/ZrO_2 [mol%]	Al in Si_3N_4 [at%]	Glass
/	/	/	2	α,β, SNO, t	/	/	/	/
2	/	/	2	α,β, SNO, t	/	/	2.5	/
2	/	2	5	β, SNO, t	17.9	4.5	1.1	g
/	2	2	5	β, SNO, t	17.9	3.9	2.0	g
4	/	2	5	α,β, SNO, t	17.9	4.0	1.6	g
/	4	2	5	β, t	17.9	3.0	3.3	g
/	/	6	5	β, c/t	39.6	10.6	/	g
4	/	6	5	β, c/t	39.6	9.6	–	g
/	/	2	10	β, SNO, t	9.9	7.6	/	/
2	/	2	10	β, SNO, t	9.9	10.2	–	g
/	/	/	20*	β, SNO, t	3.0	2.8	/	/
2	/	/	20*	β, SNO, t	3.0	2.9	–	/

Legend: *=prereacted with 3 mol% Y_2O_3; –=not measured; g=glassy pockets analyzed;
α=α-Si_3N_4; β=β-Si_3N_4; SNO=Si_2N_2O; t, c=tetragonal, cubic ZrO_2

RESULTS AND DISCUSSION

General

The general microstructure of these ceramics consisted of β–Si_3N_4, ZrO_2 and a residual intergranular glassy phase, see TABLE 1. Samples formed with only small amounts of oxide additives also contained some retained α-Si_3N_4. All compositions but three resulted in a formation of Si_2N_2O; materials formed with 6 wt% Y_2O_3 and the material formed with 4 wt% AlN did not contain Si_2N_2O. The volume fractions of the different phases and the morphology of the Si_3N_4 and ZrO_2 grains were strongly dependent upon starting powder composition.

The effect of ZrO_2

Previous work has indicated that additions of ZrO_2 participate in the formation of the oxynitride liquid phase sintering medium and hence promote densification [9]. This has also been demonstrated by the ceramics in the present study; an addition of 2 wt% ZrO_2 resulted in a fully dense body, and a significant transformation of α to β-Si_3N_4 was achieved during densification [10]. In the fully densified body, the ZrO_2 was present as a secondary crystalline phase, and only extremely thin intergranular glass films were present in the microstructure. This implies that addition of ZrO_2 is one way to achieve a transient liquid phase sintering process.

The Si_3N_4 grains in the material formed with 2 wt% ZrO_2, or 2 wt% ZrO_2 together with 2 wt% Al_2O_3, had an equiaxed shape which indicates that liquid volume and chemistry did not promote growth of the β-Si_3N_4 grains which formed during densification. Larger (20 wt%) additions of ZrO_2 powder, which had been stabilized with 3 mol% Y_2O_3, resulted in a different morphology of the Si_3N_4 matrix. A full conversion of α to β-Si_3N_4 was obtained, and

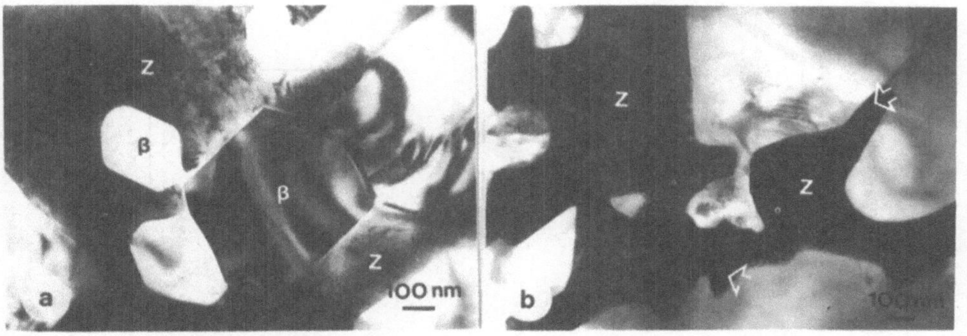

Figure 1. Bright field TEM images of samples formed with 2 wt% Y_2O_3 and 5 wt% ZrO_2 together with *a)* 2 wt% Al_2O_3, *b)* 4 wt% AlN. The ZrO_2 grains (z) are highly irregular in shape filling out space between β-Si_3N_4 grains (β) and the grain shape is more dendritic (arrowed) in the sample formed with AlN.

prismatic β-Si_3N_4 grains characteristic of growth in an isotropic liquid phase environment were present in the microstructure. Also these materials contained an extremely small volume fraction of residual glass and the ZrO_2 grains, which had retained the Y_2O_3, had an irregular shape filling out space between adjacent Si_3N_4 grains. Development of this fibrous matrix microstructure could be due to a larger liquid volume present in the compact during densification as well as to the presence of Y_2O_3. Small additions of Y_2O_3 have been shown to provide a liquid phase which promotes growth of prismatic β-Si_3N_4 grains [10,11].

Y was detected in the ZrO_2 grains by EDX in the STEM also when the Y_2O_3 was separately added. This implies that the separate additions of ZrO_2 and Y_2O_3 reacted via the oxynitride liquid phase sintering medium. The amount of Y_2O_3 incorporated into the ZrO_2 structure was sufficient for a stabilization of tetragonal and cubic structures, see TABLE 1.

Previous work has demonstrated that also N may enter the ZrO_2 structure during densification of Si_3N_4/ZrO_2 ceramics [12,13]. It has been suggested, however, that the presence of Y_2O_3 may affect the N uptake, and that an increased Y_2O_3 concentration in the ZrO_2 would be associated with a reduced N content [13]. The present work demonstrated that the addition of AlN instead of Al_2O_3 resulted in a reduced amount of incorporated Y_2O_3 in the ZrO_2 structure, TABLE 1. This could possibly be explained by an increased simultaneous incorporation of N into the anion lattice from the more N rich liquid. In the AlN containing materials, some of the ZrO_2 grains had grown in a more dendritic fashion than in the Al_2O_3 containing materials, see Fig. 1. This indicates the presence of larger local volumes of liquid during densification; N has been reported to lower the eutectic temperature in these systems [6].

Al_2O_3 and AlN additions

Additions of Al_2O_3 and AlN to Si_3N_4 ceramics enables formation of solid solutions in the Si-Al-O-N system. Particularly, Si and N in β-Si_3N_4 can be replaced by Al and O to form a β-sialon [4]. Considering charge balance, the composition is given by the formula $Si_{6-z}Al_zO_zN_{8-z}$ where z is not greater than 4.2 [14]. Adding Al_2O_3 and/or AlN is thus a powerful way to enhance densification by increasing the amount of liquid present at the sintering temperature but at the same time reduce the amount of residual glass by solid solution.

EDX analysis in the STEM showed that both Al_2O_3 and AlN additions resulted in an

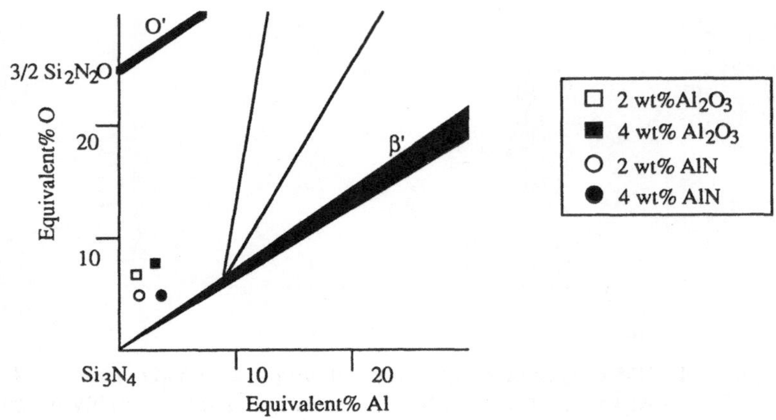

Figure 2. The Si_3N_4 corner of the SiO_2-Al_2O_3-Si_3N_4-AlN behaviour diagram. The compositions of the Al_2O_3 and AlN containing samples are indicated.

incorporation of Al into the β-Si_3N_4 structure. A comparison between materials with the same additions of Y_2O_3 and ZrO_2 showed that the dissolution of Al was significantly higher when adding the same amount (wt%) of AlN instead of Al_2O_3, see TABLE 1. This may also be expected from the SiO_2-Al_2O_3-Si_3N_4-AlN behaviour diagram; the AlN containing materials are more Al rich compositions closer to the β-sialon solid solution, see Fig. 2. The higher N content of the material formed with 4 wt% AlN resulted in a composition too close to the β' for formation of Si_2N_2O.

Residual Intergranular Glass Phase

The incorporation of Al into the β-Si_3N_4 structure as well as the formation of a crystalline ZrO_2 phase, which contained dissolved Y and possibly also N, would consume constituents of the liquid phase sintering medium and hence reduce the volume fraction residual glass. However, separate additions of Y_2O_3 resulted in increased amounts of residual glass, Fig. 3. When as much as 6 wt% Y_2O_3 had been added in combination with 0 or 4 wt% Al_2O_3 and 5 wt % ZrO_2, larger glassy "zones" formed in the microstructure, see Figs. 3 c and 4. EDX analysis in the

TABLE 2

EDX analysis of cation contents (at%) in glassy pockets of the samples in the present study.

Al_2O_3	AlN	Y_2O_3 (wt%)	ZrO_2	Si	Al	Y	Zr
2	/	2	5	50-87	4-19	5-30	≈2
/	2	2	5	72-90	5-50	5-16	≈2
4	/	2	5	63-93	5-19	5-30	≈2
/	4	2	5	72-90	5-50	5-16	≈2
/	/	6	5	63-94	/	5-34	≈2
4	/	6	5	50-87	2-24	4-26	≈2
2	/	2	10	54-75	≈0	24-40	≈3

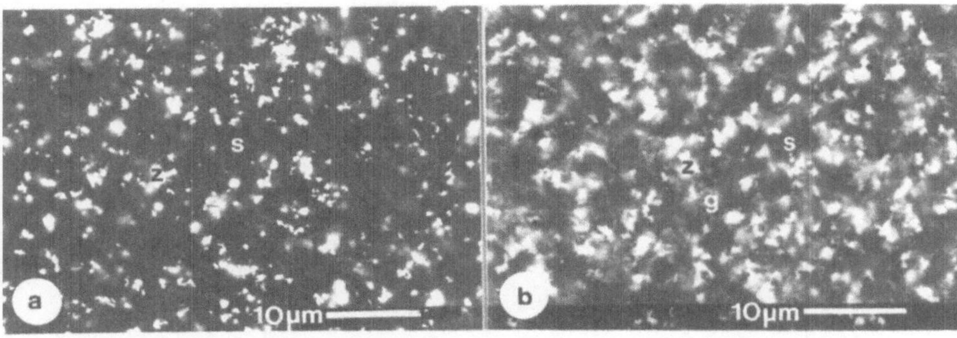

Figure 3. Back-scattered SEM images of polished surfaces. The black areas are Si_3N_4 or Si_2N_2O (s) the white ZrO_2 (z) and the greyish areas intergranular glass (g). The samples are formed with addditions of *a)* 2 wt% ZrO_2 and 2 wt% Al_2O_3, *b)* 10 wt% ZrO_2, 2 wt% Al_2O_3 and 2 wt% Y_2O_3 and *c)* 5 wt% ZrO_2 and 6 wt% Y_2O_3. A clear increase in the amount of residual glass with the Y_2O_3 addition can be noticed.

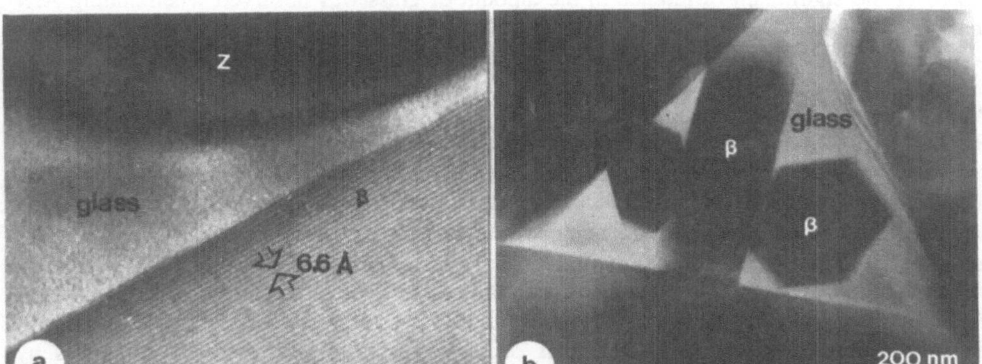

Figure 4. TEM images of samples formed with additions of *a)* 10 wt% ZrO_2, 2 wt% Al_2O_3 and 2 wt% Y_2O_3 and *b)* 5 wt% ZrO_2, 4 wt% Al_2O_3 and 6 wt% Y_2O_3. *a)* The intergranular glassy thin films between adjacent grains merge into pockets at multigrain junctions as shown in this HREM image. *b)* Increasing the Y_2O_3 addition to 6 wt% resulted in glassy "zones" containing faceted β-Si_3N_4 grains as shown in this dark field TEM image.

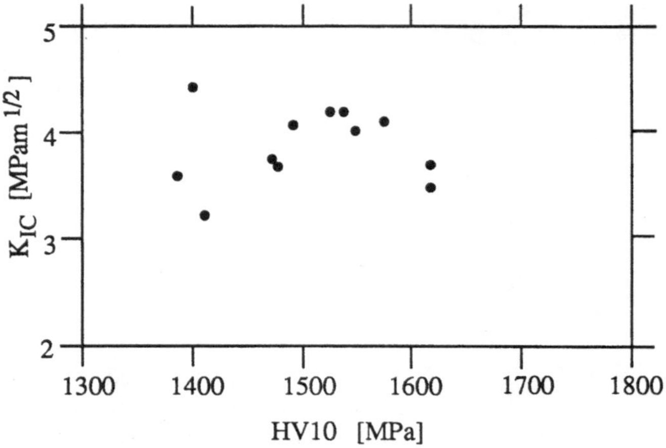

Figure 5. Hardness and indentation fracture toughness at room temperature of the samples in this investigation.

STEM showed that the composition of the residual glass varied significantly between individual glass pockets in these materials, TABLE 2. Additions of separately added Y_2O_3 promoted also formation of prismatic β-Si_3N_4 grains, see Fig. 4.

The presence of N from AlN additions could be anticipated to result in the formation of larger volumes of oxynitride liquid phase sintering medium, and consequently in a larger amount of residual glass, since N is reported to lower the eutectic temperature of the system [6]. However, no significant difference in the amount of glass was observed when adding AlN instead of Al_2O_3.

Mechanical Properties

Room temperature hardness and fracture toughness are shown in Fig. 5. Materials containing only extremely small volume fractions of residual glass showed comparatively high hardness. The glassy phase had, when present in larger volume fractions, an adverse effect on hardness.

Higher values on fracture toughness was promoted by addition of Y_2O_3 and/or larger additions of ZrO_2. Y_2O_3 additions resulted in the development of prismatic β-Si_3N_4 grains, which may increase toughness by e.g. crack deflection. The equiaxed morphology of the β-Si_3N_4 grains in material formed without Y_2O_3 additions was reflected in a reduced toughness. ZrO_2 additions resulting in a tetragonal ZrO_2 phase could contribute to toughness by mechanisms such as transformation toughening, ferroelasticity and particle reinforcement.

CONCLUDING REMARKS

Additions of ZrO_2, either pure or partially stabilized with Y_2O_3, makes it possible to produce dense Si_3N_4 based ceramic materials which contain only extremely thin intergranular films of residual glass. Additions of Al_2O_3 or AlN, which could be anticipated to increase the volume of oxynitride liquid phase during densification, did not increase the amount of glass due to β-sialon solid solution. Separate additions of Y_2O_3 to the Si_3N_4/ZrO_2 ceramics resulted in an increased volume fraction of Y-rich residual glass and also in a partial stabilization of the ZrO_2 with Y_2O_3.

Additions of Al_2O_3, AlN and Y_2O_3 to Si_3N_4/ZrO_2 ceramics makes it possible to control Si_3N_4 grain morphology and intergranular microstructure, which both have a pronounced influence on mechanical properties.

ACKNOWLEDGEMENT

This project was supported by the Swedish National Board for Technical Development (STU).

REFERENCES

1. M.H. Lewis and R.J. Lumby, "Nitrogen ceramics: liquid phase sintering". Powder. Met., 1983, **26**, 73.
2. F.F. Lange, "Dense Si_3N_4: interrelation between phase equilibria, microstructure and mechanical properties". In Nitrogen Ceramics, ed. F.L. Riley, Noordhoff, Leyden, 1977, pp. 491.
3. W.D. Kingery, "Densification during sintering in the presence of a liquid phase". J. Appl. Phys., 1959, **30**, 301.
4. M.H. Lewis, B.D. Powell, P. Drew, R.J. Lumby, B. North and A.J. Taylor, "The formation of single-phase Si-Al-O-N ceramics". J. Mater. Sci., 1977, **12**, 61.
5. L.K.L. Falk and G.L. Dunlop, "Crystallisation of the Glassy Phase in a Si_3N_4 Material by Post-Sintering Heat Treatments". J. Mater. Sci., 1987, **22**, 4369
6. R.A.L. Drew, S. Hampshire and K.H. Jack, "Nitrogen glasses". In Special Ceramics, ed. D.Taylor and P. Popper, Proc. Brit. Ceram. Soc., 1981, **31**, 119.
7. T. Ekström, L.K.L. Falk and E.M. Knutson-Wedel,, "Si_3N_4-ZrO_2 Composites with Small Al_2O_3 and Y_2O_3 Additions-Prepared by HIP". Accepted for publication in J. Mater. Sci.
8. G.R. Anstis, P. Chantikul, B.R. Lawn and D.B. Marshall, "A Critical Evaluation of Indentation Techniques for Measuring Fracture Toughness: I, Direct Crack Measurements". J. Amer. Cer. Soc., 1981, **64**, 533.
9. L.K.L. Falk and M. Holmström, "Microstructural Development During Processing of a Si_3N_4/ZrO_2 Material". In Euroceramics, ed. G. de With, R.A. Terpstra and R. Metselaar, Elsevier Applied Science, London, 1989, pp. 1.373.
10. E. M. Knutson, L.K.L. Falk, H. Björklund and T. Ekström, "Si_3N_4 Ceramics Formed by HIP using Different Oxide Additions - Relation between Microstructure and Properties". Accepted for publication in J. Mater. Sci.
11. E. M. Knutson, L.K.L. Falk and T. Ekström, "Microstructures of Si_3N_4 Ceramics Formed by HIP". In Euroceramics, ed. G. de With, R.A. Terpstra and R. Metselaar, Elsevier Applied Science, London, 1989, pp. 1.416.
12. F.F. Lange, "Compressive Surface Stresses Developed in Ceramics by an Oxidation-Induced Phase Change". J. Amer. Cer. Soc., 1980, **63**, 38.
13. F.F. Lange, L.K.L.Falk and B.I. Davis, "Structural ceramic composites based on Si_3N_4-$ZrO_2(+Y_2O_3)$ compositions". J. Mater. Res.,1987, **2**, 66.
14. M.H. Lewis, A.R. Bhatti, R.J. Lumby and B. North, "The microstructure of sintered Si-Al-O-N ceramics". J. Mater. Sci., 1980, **15**, 103.

THE MICROSTRUCTURE AND THE OXIDATION BEHAVIOUR OF Si_2N_2O/ZrO_2 COMPOSITES

COLETTE O'MEARA
Department of Physics, Chalmers University of Technology,Göteborg S-412 96, Sweden

ABSTRACT

The microstructure and oxidation behaviour of Si_2N_2O/ZrO_2 composite ceramic materials have been examined. Based on the reaction sintering of Si_3N_4 and $ZrSiO_4$, high density composites were formed by the NPS technique from Si and $ZrSiO_4$ with Y_2O_3/Al_2O_3 or $CaMg(CO_3)_2$ as stabilisation/sintering aid. The composites consisted of a submicron grain size Si_2N_2O phase of fibrous morphology, micron size irregularly shaped ZrO_2 grains and an amorphous phase. XRD and TEM examination revealed that in the Y_2O_3/Al_2O_3 stabilised materials the tetragonal phase occured mainly as the non-transformable *t'*-ZrO_2 polymorph while in the $CaMg(CO_3)_2$ stabilised materials XRD indicated the transformable *t*-ZrO_2 polymorph to be present. Significant oxidation of the materials occurred only at temperatures >1000^oC with the formation of porous scales containing SiO_2 and $ZrSiO_4$. At temperatures above 1250^oC severe degradation of the materials occured following oxidation.

INTRODUCTION

Much attention is now being devoted to the development of zirconia toughened nitrogen ceramics composites using oxide additives (e.g. Y_2O_3, MgO) whose function is both to stabilise the ZrO_2 phase and serve as a sintering aid. Such materials are thought to combine the refractoriness of the nitride ceramic matrix with an increased toughness which arises from a ZrO_2 transformation mechanism. The composites are considered to have good potential for high temperture applications. The toughness of doped zirconia polycrystalline material results from a stress induced martensitic transformation of the metastable tetragonal phase into monoclinic symmetry (*t-m*). Work on several zirconia toughened Si_3N_4 and sialon ceramic systems where Y_2O_3 was used as sintering aid/stabiliser has shown that no significant toughening occurs in these composites due to the formation of the non-transformble t'-ZrO_2 phase [1-2]. It has been suggested that the tetragonal phase formed in these composites is stabilised by nitrogen during sintering [3]. However some increase in toughness may be acheived, even at high temperatures, through a ferroelastic domain-switching mechanism [4].

Additionally, the application of these composites may be limited by commercial considerations such as high raw material and production costs, excessive shrinkage during sintering and the formation in the microstructures of deleterious zirconia oxynitride phases which oxidise readily at intermediate temperatures (600-800^oC) with a significant volume increase and a consequent reduction in strength of the composite [5].

This paper examines the microstructure and oxidation resistance of Si_2N_2O/ZrO_2 composites in the temperature range 700-1350^oC. The materials were fabricated by the NPS technique [6] using inexpensive raw materials, Si and $ZrSiO_4$, and with Y_2O_3/Al_2O_3 or $CaMg(CO_3)_2$ as sintering aid/stabiliser.

EXPERIMENTAL

Material Fabrication

Sub-micron powder mixtures of $ZrSiO_4$ (Ventron), Si (KemaNord) and Y_2O_3/Al_2O_3 (H. C. Starck/Sumitomo AKP 30) or $CaMg(CO_3)_2$ as sintering aid /stabiliser were prepared by milling in plastic bottles for 70 hours in ethanol using Si_3N_4 milling balls. The volume (wt%) of Y_2O_3/Al_2O_3 and $CaMg(CO_3)_2$ sintering additives used was 5:6.5, and 25 and the materials are hereafter referred to as Material Y and C respectively. A stoichiometric $ZrSiO_4$/Si ratio was used. The powder compacts were uniaxially pre-pressed into rectangular bars of dimensions 7x7x40 mm followed by cold isostatic pressing (CIP) at 280 MPa. The green bodies had a density of 55% of theoretical. Nitridation was carried out in a graphite furnace in N_2 at 0.1MPa. A two-step nitridation program was used with the following heating cycle (temperature/holding time): 1150^oC/0.5h-1250^oC/1.5h. The nitrided density was 67-68% of theoretical. Pressureless sintering was carried out in a graphite oven in a Si_3N_4 powder bed (preoxidation:1200^oC/0.5h) in N_2 at 0.1MPa. The sintering program for Material Y was 1600-1640^oC with 1-4h holding time and 2h at 1350^oC for Material C. The samples had a final sintered density of 98-99% of theoretical.

Oxidation and microstructural analysis

The materials were oxidised in air for 50h at temperatures between 700-1350^oC as is detailed in Table 1. The phase composition of the as-sintered materials was characterised by x-ray diffractometry (XRD) of bulk and powdered samples. For Material Y the microstructure was characterised by analytical transmission electron microscopy (TEM/STEM/EDX) in a Jeol 2000 FX TEM/STEM instrument with a Link Systems AN 10,000 EDX spectrometer while for Material C analytical scanning electron microscopy (SEM/EDX) of polished surfaces was used. The microstructure of the oxidised material was characterised by a combination of XRD and analytical scanning electron microscopy in surface and cross section using a CAM Scan S-4 80DV instrument equipped with a Link eXL System.

RESULTS AND DISCUSSION

General microstructure of as-sintered composites

Electron microscopy examination and XRD revealed that the general microstructure of the materials consisted of a Si_2N_2O matrix in which micron size ZrO_2 grains were homogeneously distributed and a significant volume of amorphous phase residual from the sintering liquid (Fig.1). The Si_2N_2O phase, identified by XRD as the main phase present in the microstructures had a fibrous, interweaving morphology with grains of high aspect ratio. The ZrO_2 phase had a micron size, irregular, intragranular morphology with ellipsoidal protrusions extending into the glassy phase and between the Si_2N_2O crystals (Fig. 2) indicating that this phase grew in the available space in the microstructure left over after the formation of the Si_2N_2O network. The amorphous phase was continuous throughout the microstructures.

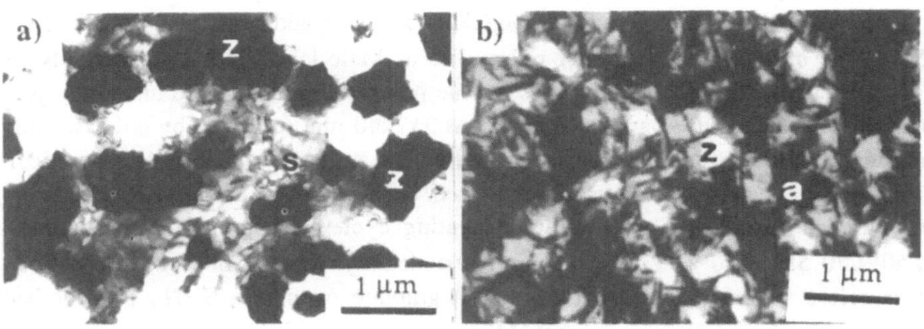

Figure 1. (a) TEM micrograph showing the general microstructure of Material Y, micron size ZrO_2 grains (Z) are seen within a Si_2N_2O (S) matrix. (b) SEM backscattered electron image of a polised section of Material C showing the the distribution of the Si_2N_2O (S) phase, ZrO_2 grains (Z) and amorphous phase (a).

XRD of the as-sintered materials revealed that Material Y contained mostly *t* and *c*-ZrO_2 with a small amount of *m*-ZrO_2 (<10 vol %). Materials C contained only *t* and *m*-ZrO_2 indicating that no cubic ZrO_2 phase formed at the low sintering temperature used (1350°C). In this material XRD of the bulk as-received materials showed the *t*-ZrO_2 polymorph to be the main zirconia phase present in the microstructure. However, following mechaninal grinding, XRD of the powders revealed that a large volume of the tetragonal phase had transformed to *m*-ZrO_2 indicating that in this composite the transformable *t* and not the *t'* phase is present.

For the Y_2O_3 stabilised material, the *t*-ZrO_2 polymorph in the microstructures was identified by selected area diffraction and dark field imaging of different {112} reflections, forbidden in the cubic flourite structure (Fig. 2). Using this technique different sets of complementary tetragonal laths could be imaged showing some grains to be completly tetragonal. As is shown in Fig. 2 the presence of anti-phase domain boundaries (APB's) was diagnostic for the presence of the non-transformable *t'*-ZrO_2 variant and the majority of the tetragonal grains examined were found to be of this type . The absence of the transformable tetragonal phase in Material Y indicates that improvement in the fracture toughness of these materials cannot be achieved through the martensitic transformation toughening mechanism. However, some toughening may be achieved, even at higher temperatures through a ferro-elastic domain swithching mechanism [4].

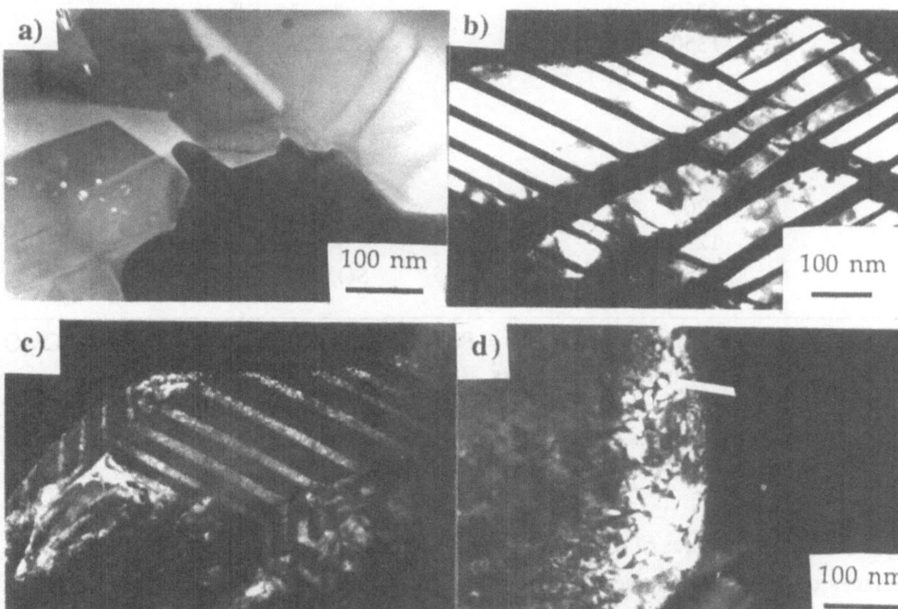

Figure 2 (a-d). (a) TEM micrograph of Material Y showing the irregular morphology and intragranular protrusions of the ZrO_2 phase in the microstructure.(b-d) TEM centered dark field images of a *t*-ZrO_2 grain formed by different {112} reflections. APB's (arrowed) can be seen in (d).

Oxidation behaviour

Both materials exhibited good oxidation resistance at temperatures up to 1250°C for the experimental exposure time of 50h. The phase composition of the oxide scales, as determined by XRD, are shown in Table 1.

TABLE 1

Oxidation temperatures and phase composition of the oxide scales of Materials Y and C following 50h exposure in air.

Material	Oxidation Temperature (°C)	Phase composition of oxide scale
Y	700	AM, SO
	900	" "
	1000	AM, SO, ZSO
	1250	" " "
	1350	SO, ZSO, YS, UN
C	700	AM, SO
	900	" "
	1000	AM, SO, ZSO, CMO
	1200	" " " "
	1250	" " " "
	1350	SO, ZSO, CMO

Legend: AM = amorphous phase, SO = SiO_2, ZSO = $ZrSiO_4$, YS = $Y_2Si_2O_7$ CMO =$CaMg(SiO_3)_2$, UN = unidentified peaks.

At temperatures of < 1000°C very little oxidation reaction could be detected. The surfaces of the materials were covered in a thin (≤ 5μm), predominantly amorphous oxide films although XRD indicated that a small volume of cristobalite had formed in the scales. No cracking or spalling of the materials was observed.

In the temperature range 1000-1250°C, in addition to an amorphous phase, $ZrSiO_4$ and SiO_2 were also found in the scales. The volume fraction of crystalline phase in the scales increased with increasing temperature. In Material Y no yttrium or aluminium silicates were detected while, in Material C, a significant volume of diopside ($CaMg(SiO_3)_2$) was present in the scales at all temperatures. XRD of the bulk material following oxidation indicated that in Material C a significant transformation of the t-ZrO_2 phase to the m-ZrO_2 polymorph had occured.

SEM examination of the oxidised materials in cross section revealed that scale thicknesses varied between 5-30mm in different parts of the samples. No spalling or pitting of the scales was observed however, as is shown in Fig. 3 at 1200 and 1250°C the scales of both Material Y and C contained a considerable volume of pores which most probably arose from nirogen gas evolution. EDX analysis indicated amorphous phase in the scales to be rich in Si and also to contain additive cations of the respective sintering additives of Materials Y and C.

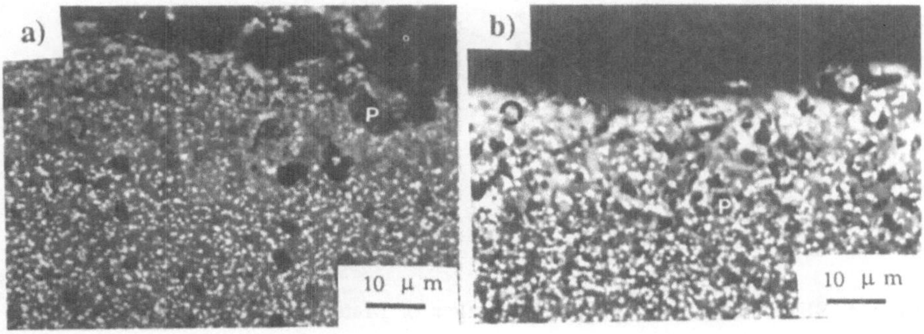

Figure 3. SEM backscattered electron images of cross sections of: (a) Material Y and (b) Material C following oxidation at 1250°C. The oxide scales (O) are seen to be porous (P) and contain a considerable volume of crystalline phases.

At 1350°C catastrophic oxidation of both Materials Y and C occured indicating a temperature of ≈ 1200°C to be the maximum operating temperature for these composites. As is shown in Fig. 4 the scales were several hundred microns thick, predominantly crystalline and very porous. Cracking occured both within the scale and at the scale matrix interface.

CONCLUDING REMARKS

The microstructures of Si_2N_2O/ZrO_2 composite ceramics fabricated by the NPS technique consist of a major Si_2N_2O phase of submicron grain size, a homogeneous distribution of micron sized, irregularly shaped, ZrO_2 grains and significant volume of amorphous phase. For materials fabricatedwith Y_2O_3/Al_2O_3 as sintering aid/stabiliser

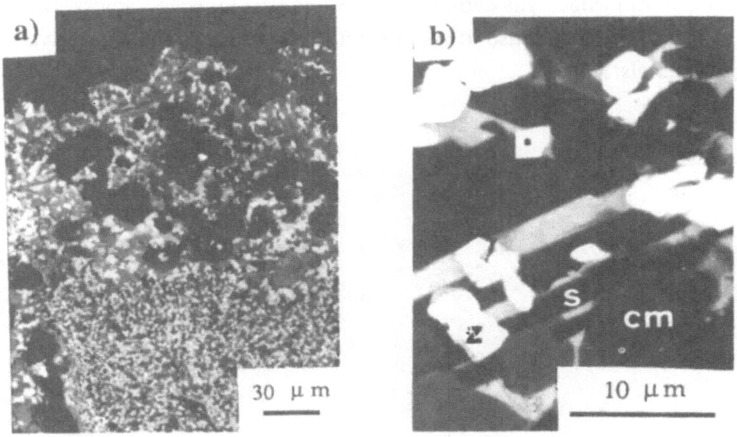

Figure 4. SEM backscattered electron images of Material C following oxidation at 1350°C showing (a) the (a) The scale consists of crystalline SiO_2 (S), $ZrSiO_4$ (Z), $CaMg(CO_3)_2$ (CM) and a small volume of amorphous phase (A).

the tetragonal phase in the microstructure was found to be the t'-ZrO_2 variant. The absence of the transformable tetragonal phase in this material indicates that improvement in the fracture toughness of these materials cannot be achieved through the martensitic transformation toughening mechanism. However some toughening may be acheived through a ferroelastic domain swithching mechanism. Results so far indicate the transformable t-ZrO_2 polymorph to be present in the $CaMg(CO_3)_2$ stabilised materials . Both materials showed good oxidation resistance at temperatures up to 1250°C for the experimental exposure times of 50h. No cracking or spalling of the materials occured but at temperatures ≥ 1000°C the scales were quite porous and contained a considerable volume of crystalline phases. Oxidation at 1350°C resulted in severe degradation of the samples.

ACKNOWLEDGEMENTS

Financial support from the Swedish Board for Technical Development (STU) is gratefully acknowledged. Doc. R. Pompe and Dr. K Rundgren of the Swedish Ceramic Institute (SCI) are thanked for fabricating and providing the experimental materials.

REFERENCES

1. Rundgren, K., Ph.D thesis, Chalmers University of Technology, Göteborg, 1988,ISBN 91-7900-623-X.

2. Falk, L.K.L. and Holmström, M., Microstructural development during processing of a Si_3N_4/ZrO_2 material, In Euro-Ceramics. Volume 1. Processing of Ceramics,eds. G. de With, R.A. Terpstra and R. Metselaar, Elsevier Applied Science Publishers, London, 1989, pp. 1.373-1.377.

3. Cheng, Y and Thompson, D.P., Nitrogen containing tetragonal zirconia, J. Am. Ceram. Soc., 1991, 74.

4. Virkar, A.V. and Matsumoto, R.L.K., Ferroelastic domain-switching as a toughening mechanism in tetragonal zirconia, J. Am. Ceram. Soc., 1986, 69,C-244-C-226.

5. Lange, F.F., Compressive surface stresses developed in ceramics by an oxidation - induced phase change, J. Am. Ceram. Soc., 1982, 63, 38-41.

6. Pompe, R., Hermansson, L. and Carlsson, R., Development of commercially advantageous techniques for fabrication of low shrinkage Si_3N_4-based materials, Sprechsaal, 1982, 115,1098-1101.

SURFACE ACID-BASE, AND ELECTROKINETIC PROPERTIES OF $Si_3N_4, ZrO_2, Al_2O_3, Y_2O_3$, AND TiO_2 POWDERS

Alf B. A. Pettersson*[,1], Hedvig Byman-Fagerholm [1,2], Jarl B. Rosenholm[1]

[1])Department of Physical Chemistry, Åbo Akademi University, Porthansgatan 3-5, SF-20500 Åbo, Finland
[2])Swedish Ceramic Institute, Box 5403, S-402 29 Göteborg, Sweden

ABSTRACT

The surface acid-base properties of $Si_3N_4, ZrO_2, Al_2O_3, Y_2O_3$, and TiO_2 powders have been estimated by the nonaqueous titration method using a series of Hammet indicators, and correlated with electrophoretic measurements. All powders are components in slips for different casted ceramics. Pure and yttrium stabilized zirconia as well as rutile and anatase forms of titania were analyzed. The results show good correlation between the rank orders of the iso electric points (IEP:s) and the acid-base properties of the powder surfaces studied.

INTRODUCTION

Slip casting offers a cheap method for providing complicated geometries to sintered high tech ceramics. High quality demands of the final product requires accurate processing when preparing the slip [1].

Well characterized powder surfaces contribute to the understanding of the action of dispersants upon different slip properties. Furthermore, improved knowledge of the surfactant interaction facilitates the efficient use of these additives and to meet the demand of producing high quality slip casting products.

The powders studied are typical components in sintered ceramics such as Si_3N_4, Si_3N_4/ZrO_2, Al_2O_3/ZrO_2, and Al_2O_3/TiO_2. The Al_2O_3 and Y_2O_3 powders are used as added sintering agents in the first two products. The necessary stabilizing additives (e.g. SiO_2, MgO, Fe_2O_3 and other oxides) in aluminium titanate materials are however not dealt with in this study [2].

Particle electrophoresis studies have proved to be useful in many practical situations where colloid stability is involved The present electrophoretic mobility measurents concerns dilute powder dispersions (0.05 wt%). Of special interest is to relate the slip stability with the surface acidity/basicity and the surface charge, for which the connections are less known. Stability problems in slip casting are known to appear when the slip consists of different powders with different properties such as density, size and surface charge, resulting in preferential settling of one or some powder species, or heterocoagulation of powders with opposite charges [1,2,3,4].

EXPERIMENTALS

Materials and methods.

The Si_3N_4 used was denoted E 10 by UBE Industries. The yttrium stabilized zirconia (YSZ) used was denoted TZ-3YS (5.25 wt% Y_2O_3, of slip casting grade) and supplied by Tosoh Corporation, Japan. The pure zirconia was of the type TZ-O by Toyo Soda Co., Japan. The Y_2O_3 used was denoted "grade fine" by HC Starck, Germany. The Al_2O_3 used was of the type AKP-30, i.e. a $\alpha - Al_2O_3$ powder, and supplied by Sumitomo Chemicals, Japan. The titanias were delivered by Kemira Oy, Pori, Finland. They were produced by the sulfate process and were taken directly from the calcining furnace without after-treatments. The rutile form contained Al^{3+} as doping agent. Interpretation of X-ray diffraction patterns of the substrates , using Ni-filtered copper Kα radiation, indicated that the samples consisted of pure rutile and anatase, respectively. All powders were used as-received. The titanias were also washed with water in order to free the powder surfaces of soluble salts originating from the manufacturing processes. After the washing the titanias were dried in vacuum at 80 °C, and therefore the powders must be considered to be at least partly hydrated. The electrophoretic mobility measurements were carried out on the pre-washed titanias and the surface acid-base experiments on both types of samples.

The water used was double distilled, and subsequently passed through a Milli-Q water purification system. The minimum resistivity of the purified water was $10^7 \Omega \cdot cm$, and it was used in all experiments.

The BET areas of the powders, as determined by the standard N_2 gas adsorption method, and mean particle sizes, measured using laser light scattering, are shown in table 1. The average size of each powder were measured at pH=11 of a water dispersion, except Al_2O_3 for which pH was 6.4, in order to maintain maximum dispersivity.

TABLE 1
BET-areas and particle sizes of the powders used.

Powder	BET-area / m^2g^{-1}	Average grain size / μm
Si_3N_4	10.2	0.482
ZrO_2 (TZ-3YS)	8	0.416
ZrO_2 (TZ-0)	15	0.368
Y_2O_3	13.0	0.836
Al_2O_3	7.5	0.427
TiO_2 (rutile)	8.73	0.269
TiO_2 (anatase)	7.13	0.390

The acidity (the number of acid sites per area) at different acid strength (the ability of acid sites to convert an adsorbed basic Hammet indicator into its conjugate acid) of the acid sites on the powders were estimated by the widely used titration method described by Johnson [5] and Benesi [6]. The titrator was a 0.1 N solution of n-butylamine (Fluka, puriss.) in the apolar solvent cyclohexane (Fluka, puriss. p.a.). When the indicator used undergoes a colour change, the value of the Hammet acidity function H_0 (measure of the acid strength of the sites) is equal to or lower than the pK_a of the indicator. The amount of acid sites of the powder added is the number of mole n-butylamine consumed in the titrator. The change of colour was visually recorded. The basicity was measured using the method described by Yamanaka and Tanabe [7,8], and gives the basicity on a common scale using the same basic indicators as for mesuring acidic properties. The titrator was a 0.1 N solution of trichloroacetic acid (Merck, p.a. grade) in cyclohexane. The value of the H_0 (measure of the basic strength of the sites) now being equal to or greater than the pK_a of the indicator. The basic indicators used with respective pK_a values were: Nile blue A (+9.5, Fluka), neutral red (+6.8, Merck), methyl red (+4.8, Merck), phenylazonaphthylamine (+4.0, Aldrich), benzeneazodiphenylamine (+1.5, Aldrich) and crystal violet (+0.8, Merck) [9,10].

There has been a lot of debate about whether the method gives the sum of Brønsted and Lewis acid sites or only Brønsted surface acid groups, because of steric hindrance of the indicator molecules to adsorb onto the sites. The requirements when recording the colour change visually is that the colour of the indicator in the acid form masks the colour of the basic form and vice versa. There are also reports that surface sites other than the acidic or basic are able to react with the indicators. These are serious drawbacks to the method. It may, however, be useful as an easy and quick way of screening surface acidities and basicities, when the surface is amphoteric [11,12].

Yamanaka and Tanabe introduced the concept of $H_{0,max}$, as the H_0 -value of the strongest acid groups of the solid surface (lowest H_0-value). Approximately it is also the H_0-value of the strongest base groups of the surface (highest H_0-value). $H_{0,max}$ is received graphically from figures 1&2 at the H_0-value when the amounts of surface acid and base sites are zero. A low $H_{0,max}$ meaning strong acid sites and weak base sites. Vice versa a high $H_{0,max}$ means strong base sites and weak acid sites [7,8].

The electrophoretic mobilities and particle size analysis respectively of the powders were carried out using laser light scattering equipments (Malvern Zetasizer IIC and Malvern system 4700C, both by Malvern Instruments, UK). The dispersions for the electrophoretic mobility measurements were prepared by adding powder to solutions with pH values ranging from 3 to 11, but constant ionic strength (0.001 M, using NaCl as background electrolyte). The pH were adjusted with HCl or NaOH (Merck titrisols). The powder amount was 0.05g/l dispersion (the exact powder amounts added were recorded). The dispersions were processed in ultrasonic bath, and allowed to equilibrate for 24 hours before the measurements. The pH:s of the dispersions were measured prior to or during the electrophoretic mobility measurements.

RESULTS AND DISCUSSION

Surface acidity-basicity.

Figure 1 shows the acidity (the number of acid sites) and basicity (the number of basic sites) at various acid-base strengths of the as-received powders. The results indicates that all the powder surfaces are amphoteric, having both acidic and basic sites. The existence of very weak acid groups on Al_2O_3 is somewhat uncertain. According to literature some indicators may give faint acid colors with samples of activated aluminas [11]. The $H_{0,max}$ values are given in table 2.

Silicon nitride, the only non-oxide powder studied in this work is known to have acidic silanol ($Si-OH$) groups and basic secondary amine (Si_2-NH) groups on the powder surface. Figure 1 shows that the Si_3N_4 used has strong acidic surface groups and weak base groups, the latter sites outnumbering the former, resulting in a IEP of 6.3. The IEP of Si_3N_4 usually varies between 4.2 and 7.6, depending on the ratio of acidic and basic surface sites [13,14].

Zirconia is more basic than titania (see fig. 1). The addition of stabililizing yttrium in the form of Y_2O_3 to pure zirconia renders the surface more basic.

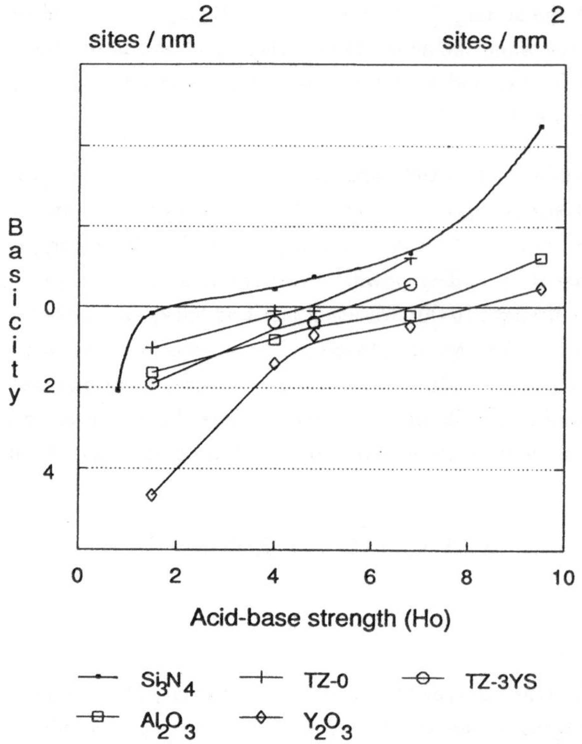

FIGURE 1. Acid -base strength distribution of Si_3N_4, ZrO_2 (TZ-0), ZrO_2 (TZ-3YS), Al_2O_3, and Y_2O_3.

The influence of washing titania with water on the acid-base properties are shown in figure 2. As shown the as-received rutile is more acidic as as-received anatase (rutile having lower $H_{0,max}$). The washing reverses this order, making the anatase more acidic and reducing the acidity of rutile. This is considered to be due to that some of the hydration water dissociate and is thence bound as hydroxo ligands to surface four- or fivefold coordinated cations such as Lewis acidic Ti^{4+}. The mild drying process leaves the surface partly hydrated which causes the acidity of anatase to increase during the water treatment. These results may also support the theory by Benesi and Winquist [11], that the indicator method does not measure Lewis acidity correcly, but mainly Brønsted acidity. Moreover, high calcining temperature makes the titania surface less acidic. The rutile looses acidity when soluble acidic ions from the manufacturing process leave the surface [15,16].

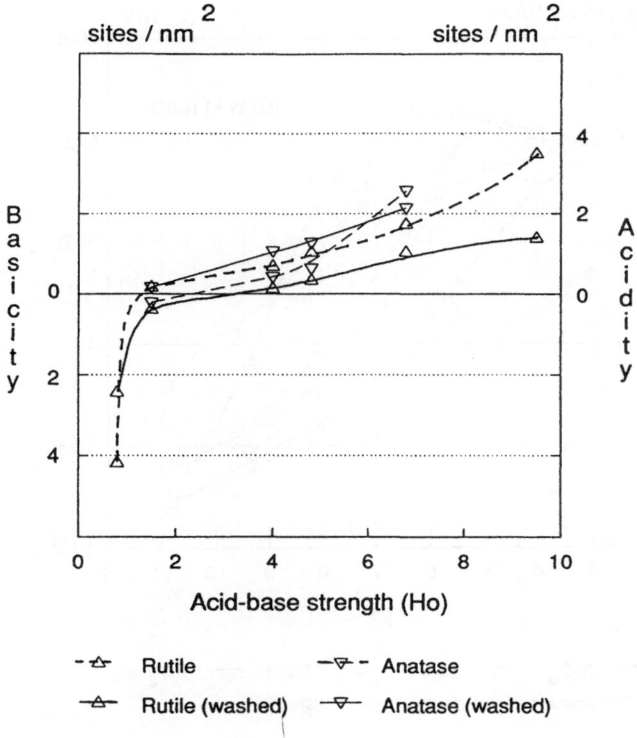

FIGURE 2. The influence of washing with water on the acid-base strength distribution of rutile and anatase surfaces.

TABLE 2
IEP and $H_{0,max}$ values of the powder surfaces.

Powder	IEP / pH	$H_{0,max}$
Y_2O_3	10	8
Al_2O_3	9.2	7.5
ZrO_2 (TZ-3YS)	8	5.5
ZrO_2 (TZ-0)	7.2	5.0
TiO_2 (rutile)	6.5	3
Si_3N_4	6.3	2
TiO_2 (anatase)	6	1.5
TiO_2 (rutile, as-received)		1.5
TiO_2 (anatase, as-received)		2.5

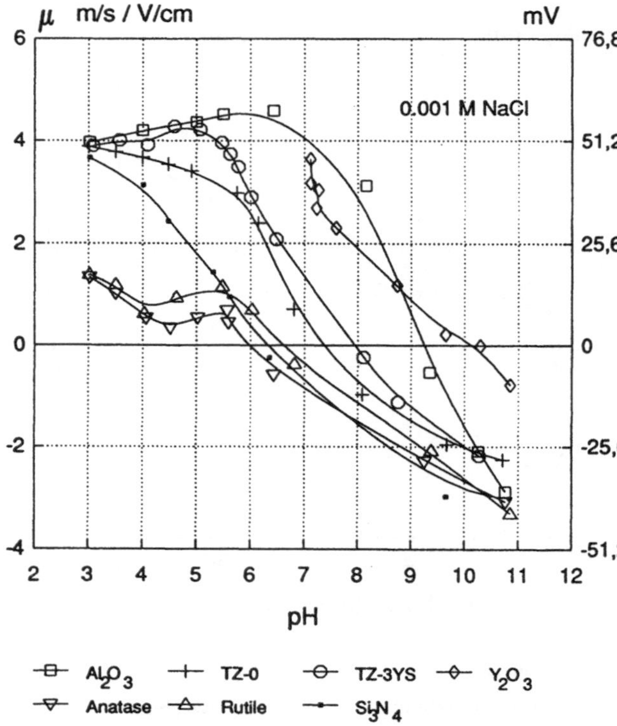

FIGURE 3. Electrophoretic mobilities and ζ-potentials versus pH of different dispersed particles in water at 298 K.

Electroforetic mobilities and ζ-potentials.

The results from the electrophoretic mobility experiments are shown in figure 3. The IEP:s of the powders are reported in table 2.

As shown in figure 1 the YSZ is more basic than pure zirconia, resulting in a shift of the IEP to a higher value (figure 3). According to literature YSZ with 2 mole% Y_2O_3 (TZ-2Y, Tosoh Corporation, 3.53 wt% Y_2O_3), has an IEP at pH 7.4 [17].

The pH:s of the Y_2O_3 dispersions equilibrated 24 hours remained above pH 7, probably due to the solubility of yttrium ions in the acidic region.

The ζ-potentials of the particle surfaces were evaluated from the electrophoretic mobility data, using the Helmholtz-Smoluchowski equation (the particle sizes in table 1 and the electrolyte concentration lie close to the the validity limit of the equation). The ζ-potentials are reported in figure 3. ζ-potential measurements show the importance of electrostatic stabilization [18].

The correlation of the acid-base properties of the powder surfaces to the electrophoretic mobility measurements are good, because the succession of IEP and $H_{0,max}$ values for the powder surfaces are the same (table 2). Concequently the surface acid and base sites produce charges in aqueous dispersions predominantly through hydration reactions. However dissociation of ions from the surface probably contributes to the charge formation. Of the studied powders this may hold at least for yttrium and yttrium stabilized zirconias.

ACKNOWLEDGEMENTS

The authors wish to thank the research foundation Neste Oy:n Säätiö for the financial support to H. Byman-Fagerholm, and the Swedish Ceramic Institute for supplying most of the powders. The Kemira Oy, Finland is also acknowledged for the financial support to A. B. A. Pettersson and for supplying the TiO_2 samples.

REFERENCES

1. Persson M., "Slip Casting and Pressing of Ceramics based on Colloidal Processing Techniques", Thesis, Chalmers University of Technology, Göteborg, Sweden, (1989).
2. Peuckert D., Hausselt J., and Kriechbaum G. W., Proc. International Symposium on Ceramic Materials & Components for Engines, Las Vegas, (1988).
3. Wohlfromm H., Pena P., Moya J. S., and Requena J., Proc. 7th Cimtec-World Ceramic Congress, Montecatini, Italy, June (1990).
4. Rehnfeld G., Staudt Th., and Zografou D., Proc. 1st International Conference on Ceramic Powder Processing Science, Volume 2, Orlando, (1987).
5. Johnson O., J. Phys. Chem., **59**, 827, (1955).
6. Benesi H. A., J Phys. Chem., **61**, 970, (1957).
7. Yamanaka T., Tanabe K., J. Phys. Chem., **79**, 2409, (1975).
8. Yamanaka T., Tanabe K., J. Phys. Chem., **80**, 1723, (1976).
9. Tanabe K., "Solid Acids and Bases", Academic Press, New York, (1970).
10. Tanabe K., in Anderson J. R., Boudart M. (eds), "Catalysis - Science and Technology", Volume 2, Springer-Verlag, Heidelberg, pp. 231-273, (1981).
11. Benesi H. A., Winquist B. H. C., Adv. Catal., Volume 27, pp. 97-182, (1978).

12. Hodgkin J. H., Hawthorne D. G., Swift J. D., Solomon D. H., Commonwealth Scientific and Industrial Research Organization, U.S. Patent 3,834,923, (September 10, 1974)

13. Bergström L., Bostedt E., Colloids and Surfaces, **49**, 183, (1990).

14. Bergström L., Pugh R. J., J. Am. Ceram. Soc., **72**, 103, (1989).

15. Solomon D. H., Hawthorne D. G., "Chemistry of Pigments and Fillers", John Wiley & Sons, New York, (1983).

16. Jones P., Hockey J. A., J. Chem. Soc. Faraday Trans. 1, **67**, 2679, (1971).

17. DeLiso E. M., van Rijswijk W., Cannon W. R., Colloids and Surfaces, **53**, 383, (1991).

18. Hiemenz P. C., "Principles of Colloid and Surface Chemistry", Marcel Dekker Inc., New York, (1986).

STRUCTURAL DETAILS OF SIALONS - THEIR INFLUENCE ON PHASE STABILITY

JÖRGEN SJÖBERG & ROBERT POMPE
Department of Inorganic Chemistry, Chalmers University of Technology and University of Göteborg, and Swedish Ceramic Institute, Göteborg, Sweden

ABSTRACT

In view of the importance for phase relationships, preparation and properties, details of the structures of O'-sialons and β'-sialons have been studied. In addition to conventional powder x-ray diffraction, more recently developed techniques have been used: Neutron powder diffraction for determination of site occupancies; MAS NMR and EXAFS to study local atomic arrangements. The results are compared to literature data from related work. It is proposed that there exists a certain degree of local ordering in O'- and β'-sialon structures, and that this sets limits on the ranges of solid solubility for the two phases. This may also affect the strength of sialon grains in the matrix.

INTRODUCTION

The sialon materials represent an interesting example of ceramic alloying [1]. It originates from the effort to facilitate the sintering of silicon nitride based materials by the addition of a sintering aid ($Al_2O_3 \cdot AlN$) and yet obtain a material with no secondary crystalline phase and very little glassy phase at the grain boundaries. This is achieved by the formation of a single crystalline phase, β'-sialon, with the same structure as β-Si_3N_4, but with the general formula $Si_{6-z}Al_zN_{8-z}O_z$. As much as 2/3 of the silicon atoms can be replaced by aluminium, but the mechanical properties of the materials produced remain similar to those of Si_3N_4 based materials. The chemical properties, on the other hand, are more "alumina like". The situation is similar for Si_2N_2O, which together with Al_2O_3 forms O'-sialon, $Si_{2-x}Al_xN_{2-x}O_{1+x}$, but here, a smaller fraction of the silicon atoms can be replaced. The sialon polytypoids also form ranges of solid solution. The concept of ceramic alloying can be extended to include a second oxide additive, *e.g.* Y_2O_3, whereby the formation of a suitable secondary phase can be controlled.

Local structural details, such as domain formation, may be expected to affect the ratio of intergranular to transgranular fracture and thus the strength and toughness. However, more detailed structural studies mainly on β'-sialons have been restricted to x-ray work to determine lattice parameters [2].Few neutron diffraction studies to determine site occupancies [3] as well as investigation of local atomic arrangements by ^{15}N, ^{27}Al and ^{29}Si MAS NMR have been done [4,5]. The present work is concerned with the local structure of sialons. More data - mainly on O'-sialons - based on x-ray diffraction, neutron diffraction and MAS NMR are presented, along with results obtained from an EXAFS study on β'-sialons. A general interpretation of the results is proposed.

MATERIALS AND METHODS

The compositions, starting powders and sintering conditions of the samples used in this study are given in Table 1.

TABLE 1
Description of the samples

Sample	Composition	Starting powders	Sintering conditions	ref
O1	O': x=0.06, 5%YO	SN,SO,AO,YO	1650°C,3h,N_2	[6]
O2	O': x=0.16, 5%YO			
O3	O': x=0.30, 5%YO			
O4	O': x=0.40, 5%YO			
O5	O': x=0.04,	SN,SO,AO	1820°C,0.5h,N_2	[6]
O6	O': x=0.10			
O7	O': x=0.24			
O8	O': x=0.40			
O9	O': x=0.10,^{15}N,5%YO	SN15,SO,AO,YO	1650°C,3h,N_2	[7]
O10	O': x=0.24,^{15}N,	SN15,SO,AO	1820°C,3h,N_2	[7]
O11	O': x=0.28,^{15}N,5%YO	SN15,SO,AO,YO	1650°C,3h,N_2	[7]
B1	β': z=0.25	*	1850°C,2h,N_2,	[2]
B2	β': z=1.0		10MPa	
B3	β': z=2.7			
B4	β': z=0.6,^{15}N	SN15,SO,AO	1820°C,3h,N_2	[7]
B5	β': z=0.9,^{15}N			

Legend: % = wt%; O'=$Si_{2-x}Al_xN_{2-x}O_{1+x}$; β'=$Si_{6-z}Al_zN_{8-z}O_z$; ^{15}N=enriched in ^{15}N; YO=Y_2O_3, H.C. Starck; SN=Si_3N_4, UBE SN-E-10; SN15 ^{15}N-enriched Si_3N_4 AO=Al_2O_3, APK 30 HP, Sumitomo; SO=SiO_2, spray dried silica sol; *=prepared by AB Sandvik Hard Materials.

The ^{15}N enriched Si_3N_4 (denoted SN15) used in some samples was prepared from Si and α-Si_3N_4 according to a procedure developed by Harris et al.[5]. A sub-set of samples is used for each of the experiments, as detailed in Table 2.

TABLE 2
The selection made for the different experiments

Experiment	Samples
X-ray diffraction	O1 - O8
Neutron diffraction	O3, O4, O6, O7
MAS NMR	O5 - O7, O9 - O11, B1, B3 - B5
EXAFS	B2, B3

X-ray diffraction
Monochromatic CuKα_1 x-ray radiation from a sealed x-ray tube, which was focused in a bent quartz crystal, was used to obtain high resolution powder diffraction films by the Hägg-Guinier technique. The films were scanned with an optical line scanner which allowed phase identification and refinement of the lattice parameters in the least squares program *Pirum*. The lattice parameters were used to observe changes in unit cell dimensions as a function of composition of the starting powder, thereby determining the range of solid solution and the associated structural changes [6].

Neutron diffraction
Samples selected for neutron diffraction were mounted on the medium resolution powder diffractometer *Polaris* at ISIS, Rutherford Appleton Laboratory, U.K. Time-of-flight neutrons in the range 2000-20000 µs with a total flight path of 12 m, corresponding to a range in d(Å) of 0.2-3.2 with a resolution of about $5 \cdot 10^{-3}$ were used. Recorded diffraction data were focused, normalized and corrected for attenuation. Crystal structure information (including coordinates, isotropic temperature factors and site occupancies) was extracted by the Rietveld profile refinement technique, using the program *TF15LS*, developed at ISIS [6].

MAS NMR
For ^{15}N and ^{29}Si magic angle spinning (MAS) NMR a Varian VXR300 spectrometer was used. It operated at frequencies of 30.4 MHz and 59.6 MHz, respectively. Spinnings speeds of 4-6 kHz were used during acquisition. The ^{27}Al NMR spectra were recorded with a Bruker AM-500 instrument operating at 130.2 MHz with spinning speeds of 5-17 kHz. The high field and high spinning speeds were used for the ^{27}Al spectra to reduce the second order quadrupolar interactions, which cause severe line broadening [7].

EXAFS
The extended x-ray absorption fine structure (EXAFS) spectra were recorded on the soft EXAFS station (3.4) at the Synchrotron Radiation Source at Daresbury Laboratories, U.K. The samples were mounted on flat sample holder of stainless steel and exposed to monochrome x-ray radiation at both the Al and Si K-edges. The experiments were carried out in vacuum to reduce attenuation of the low energy x-rays and detection was made by an Auger yield detector. Data analysis was carried out on normalized and background subtracted data in k-space, using the program *EXCURVE90* with k^3 weights applied [8].

RESULTS AND DISCUSSION

All O'-sialon samples contained β'-sialon and α-Si_3N4, while samples O4, O7 and O8 also contained some X-phase ($Si_3Al_6O_{12}N_2$). All β'-sialons were single phase except B3, which contained some 15R ($SiAl_4O_2N_4$). The amounts of secondary phases are low, typically less than 5 %, the notable exceptions being samples O1, with only some Al_2O_3 added as a sintering aid, and O8, in which considerable amounts of x-phase and β'-sialon was present. Some glassy phase is expected to have been formed.

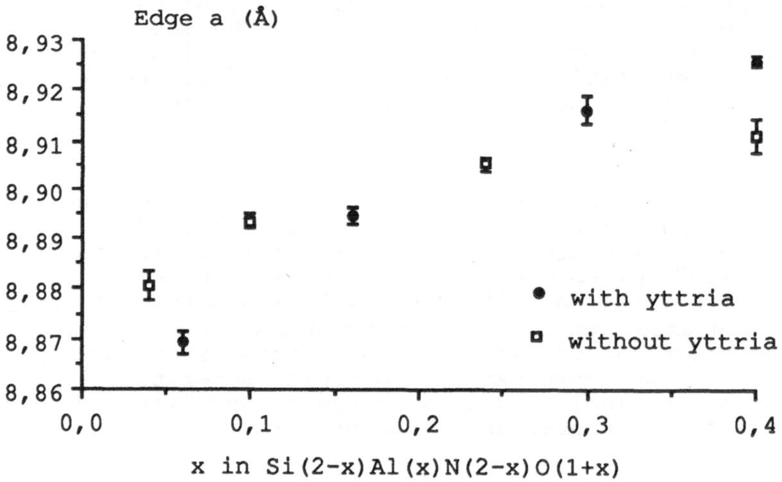

Figure 1. Cell edge *a* in O'-sialon as a function of composition.

O'-sialon compositional range
The only lattice parameter to change noticeably with increased amount of aluminium content in the O'-sialons is the unit cell edge *a*. In Figure 1 the calculated *a*-values including standard deviations obtained from the least squares refinements are shown as a function of the composition. It is clear that the unit cell dimension does not change beyond $x \approx 0.2$ in the general formula for O'-sialons for samples containing no yttrium. This is in

TABLE 3
Site occupancies (Si) calculated from neutron diffraction data

Sample	O6	O7	O3	O4
Fraction (Si)	0.92(3)	0.81(3)	0.80(4)	0.72(4)
Given as x	0.17(6)	0.38(6)	0.40(8)	0.56(8)

perfect agreement with results reported by Trigg & Jack [9]. For yttrium containing samples, the unit cell edge *a* continues to increase at least up to x=0.4. The effect of Y_2O_3 is not known, but according to a TEM investigation [10], a small content (<1.5 wt%) of yttrium is found in the O'-sialon crystals.

This is rather low yttrium level was not expected to contribute significantly to neutron diffraction intensities, and were not included in the calculations of site occupancies found, Table 3, which show a trend similar to that in Figure 1. It should be noted that the site occupancies obtained from the refinements are not very precise on an absolute scale, probably due to inadequacies in the model used in the Rietveld refinement. However, it is believed that the present procedure using a uniform treatment of all data sets provides good results on a relative scale. The only inter-atomic distance to change with composition is the Si/Al - O bond perpendicular to the puckered layers in the b-c plane.This is plotted in Figure 2 as a function of composition as calculated from the powder compositions used for synthesis. It is compared to the correlation between bond length and composition in layered silicates. The agreement is indeed good, and this is a strong indication that the trend in site occupancies is correct. Thus, from the x-ray and neutron diffraction data it seems clear that the presence of yttria in the system does increase the aluminium and oxygen content in the O'-sialon phase formed. The small amount of yttrium found in the crystals suggests that the phase is now slightly out of plane of the sialon phase system. The most probable sites for the Y^{3+} cations are between the O atoms bridging the Si/Al - N layers.

The local arrangements in sialons

The NMR results are shown in Table 4. Only the chemical shifts which can be correlated to the main phases are included. No changes in the shifts for ^{15}N and ^{29}Si occur within the compositional ranges of O'-sialon and β'-sialons studied. The ^{27}Al NMR spectra for the β'-sialons, on the other hand, are composition dependent. The shift at about 110 ppm is assigned to an AlN_4 environment, while that at ≈60 ppm is believed to represent AlO_4 units. It has been proposed from similar data [4] that silicon is only present in SiN_4 units in β'-sialons and aluminium is present mainly in AlO_4 units, especially for high aluminium sialons. Extending this idea to the O'-sialons one would assume silicon to be present only in $SiON_3$ tetrahedra. However, it is also possible that the shielding of the studied nucleus (^{15}N or ^{29}Si) is not changed throughout each phase region

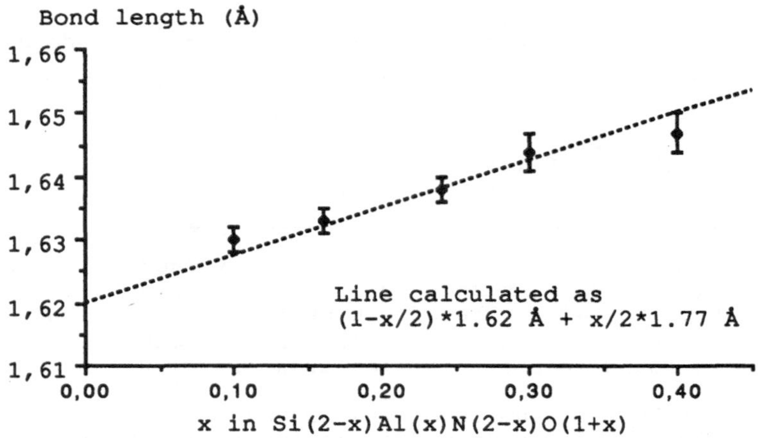

Figure 2. The bond length Si/Al-O of O'-sialon as a function of composition with values for layered silicates indicated [11].

TABLE 4
The NMR results

Sample	^{15}N-shift ppm	^{29}Si-shift ppm	^{27}Al-shift ppm
O5		-61.5	
O6		-61.0	
O7		-61.3	64.7
O9	-312.6	-60.9	
O10	-312.3		
O11	-312.3	-60.9	
B1			110
B3	-301.0/-284.3	-48.4	
B4	-301.2/-284.4	-48.0	
B5		-47.9	69.1

regardless the degree of substitution. This is possible, as the shielding depends on the electron densities of the valence electrons, and the substitution does not change the crystal structure and probably not the electronic structure. Thus, these shifts do not provide any conclusive evidence of local ordering.

The results of the EXAFS study, Table 5, show that the average bond length of the silicon tetrahedra in β'-sialons is reduced as compared to that in Si_3N_4, almost by the value expected for oxygen and nitrogen in random distribution over the non metal sites. It is therefore assumed that silicon is not exclusively present in SiN_4 units in β'-sialons. By analogy to the β'-sialons, silicon containing tetrahedra than $SiON_3$ are

TABLE 5
Results from EXAFS

Sample	B2	B3	SO	AO	SN	AN
Bonds (Å):						
Si-N/O	1.69	1.68	1.61		1.71	
Al-N/O	1.77	1.76		1.74		1.83

Legend: SO=SiO_2(crist.); AO=Zeolite NaA; SN=Si_3N_4, UBE; AN=AlN, H.C. Stark.

believed to be present in the O'-sialons. This would also be consistent with neutron diffraction data on β'-sialons [3], and removes the implication that Loewenstein's rule [12] for distribution of AlO_4 tetrahedra should be violated in β'-sialons [4]. The EXAFS results on the Al K-edge, on the other hand, seem to imply some preference for the formation of Al-O bonds, rather than a random distribution. However, the precision of these calculations is rather poor, due to the short Al edge available for analysis. The ^{27}Al NMR also indicate a preference for AlO_4 units. Here, line broadening leads to uncertainty of the results which therefore are not conclusive evidence for the preference of AlO_4 units.

CONCLUSIONS ON THE STRUCTURES AND PHASE STABILITY

The regions of solid solubility in the system SiO_2-Si_3N_4-AlN-Al_2O_3 exist because of the similarity between the two isoelectronic groups Al-O and Si-N. When aluminium replaces silicon, an adjacent nitrogen atom is also likely to be replaced so that charge neutrality is maintained. It is possible that the Al-O group can adapt to the electronic configuration of the phase structure, which is not necessarily energetically favourable to the Al-O bond and it may thus have a destabilizing effect. According to Pauling's second rule for crystal structures, Al-O and Si-N bonds are more likely to form than Al-N and Si-O, the present experimental data are consistent with this. The formation of small domains in the crystals having a stoichiometry different from the average for the whole structure is therefore likely. This may have two implication on the phase relationships in the system. (1) With increasing degree of substitution the driving force for the formation of new phases in which the aluminium is octahedrally coordinated by oxygen increases. This is known to be favourable to phase structures rich in these two elements. (2) Structural inhomogenieties of the sialon crystal grains may cause strain and thereby reduce their mechanical strength.

It is not surprising that the O'-sialons display a much shorter range of solid solution. The higher amount of oxygen, which is evenly distributed, with at least one oxygen atom in each tetrahedron, makes formation of domains with Al-O bonds difficult. The effect of yttrium could be to modify the electronic structure, which would enable more aluminium to enter the structure. However, the amount of yttrium found in the crystals seems to be too low, and the additional replacement of Si by Al to compensate for the possible presence of Y^{3+} ions in the structure is not alone enough to explain the lattice expansion. The effect of yttrium may also be to lower the viscosity of the melt during sintering, thereby increasing mobility and facilitating equilibrium to be established. Hence, the observed maximum value of x would be dependent on synthesis conditions and not necessarily reflect true equilibrium. The effect of temperature may then be significant.

REFERENCES

1. Jack, K.H., Silicon Nitride, Sialons and Related Ceramics. Ceramics and Civilisation, 1987, Vol III, 259-88.

2. Ekström, T., Käll, P.-O., Nygren, M. & Olsson, P.-O., Dense Single-Phase β-sialon Ceramics by Glass-Encapsulated Hot Isostatic Pressing. J.Mater.Sci., 1989, **24**, 1853-61.

3. van Dijen, F.K., Metselaar, R. & Helmholdt, R.B., Neutron Diffraction of β'-sialon. J.Mater.Sci.Lett., 1987, **6**, 1101-2.

4. Dupree, R., Lewis, M.H. & Smith, M.E., Structural Characterization of Ceramic Phases with High Resolution ^{27}Al NMR. J. Appl.Cryst., 1988, **21**, 109-16.

5. Harris, R.K., Leach, M.J. & Thompson, D.P., Synthesis & MAS NMR of ^{15}N-Enriched Silicon Nitride. Chem.Mater., **2**, 320-23.

6. Lindqvist, O., Sjöberg, J., Hull, S. & Pompe, R., Structural Changes in O'-sialons, $Si_{2-x}Al_xN_{2-x}O_{1+x}$, $0.04 \leq x \leq 0.40$. Acta Cryst. B, 1991, in press.

7 Sjöberg, J., Harris, R.K. & Apperley, D., A ^{29}Si,^{27}Al and ^{15}N MAS NMR Study of O'-sialons and Some Related Phases. J.Mater.Chem., 1991, submitted for publication.

8. Sjöberg, J.,Ericsson, T. & Lindqvist, O., The Local Structure of β'-sialons - an EXAFS Study. J.Mater.Sci., 1991, in preparation.

9. Trigg, M.B. & Jack, K.H., Solubility of Aluminium in Silicon Oxynitride. J.Mater.Sci.Lett., 1987, **6**, 407-8.

10. Sjöberg, J., O'Meara, C. & Pompe, R. J.Eur.Ceram.Soc., 1991, in preparation.

11. Smith, J.V. & Bailey, S.W. Acta Cryst., 1963, **16**, 801-811.

12. Loewenstein, W., The distribution of aluminium in the tetrahedra of silicates and aluminates. Am.Miner., 1954, **39**, 92-96.

THE MICROSTRUCTURE CONTROL IN THE SOL-GEL DERIVED ALUMINA-ZIRCONIA COMPOSITE

V. Srdić and L. Radonjić
Faculty of Technology, Novi Sad, Yugoslavia

ABSTRACT

Alumina-zirconia composites were prepared by two different sol-gel routes: 1. mixing of two sols and 2, peptizing of alumina by zirconium-salt. Originaly different gel composite microstructures have great influence on the microstructural changes under heating. Transformation temperatures to alpha-alumina, t-zirconia and t - m are somewhat higher in the route 2 than in 1. Final alumina and zirconia grain size are higher too in the samples after the route 2 than 1.

INTRODUCTION

It has been well demonstrated that the tetragonal to monoclinic (t - m) phase transformation of dispresed zirconia particles can be utilized to increase the fracture toughness of alumina ceramics. Several toughening mechanisms based on this transformation, such as stress-induced transformation toughening, microcrack toughening, crack deflection, formation of compressive surface stresses etc. [1-4] enable additional energy dissipation at crack tipes. The type and magnitude of toughening is controlled by the microstructural parameters, such as: volume fraction of zirconia, its size, the distribution and amount of tetragonal/monoclinic zirconia.

Preparation of alumina-zirconia composites through sol-gel methods have attracted considerable attention in recent years [5-9], because in it is possible to alter many variables which can designe the composite microstructure using sol-gel methods.

Two different sol-gel routes, determined by some of sol-gel processing variables, were used to produce two different gel microstructures. The main objective of this study is to determine the relationship between the original gel microstructure and the final heat treated composite microstructure. This relationship is still not well understood, but it is the key to the application of sol-gel processing in the alumina-zirconia composite.

EXPERIMENTAL PROCEDURE

The starting alumina-zirconia sols have been prepared by two different sol-gel routes.

In the first one (with sample notation 1-AZ-X, where X is a partion of zirconia in composite - wt.% ZrO_2), two sols were separately formed starting from aluminium-sec-butoxide, $Al(OC_4H_9)_3$, and zirconium-n-propoxide, $Zr(OC_3H_7)_4$. The alumina sol was prepared by hydrolysis of $Al(OC_4H_9)_3$ with 100 mol of water per mol of alkoxide and peptizing with nitric acid under reflux and vigorous stirring at t = 90°C. Simultaneously a zirconia sol was prepared by dissolving $Zr(OC_3H_7)_4$ in anhydrous ethanol and by hydrolising and peptization of alkoxide with water-acid mixture at ambient temperature. Finally, both prepared sols were mixed together and resultant clear solution was placed in an oven at 70°C until the gel was formed.

In the second sol-gel route (with sample notation 2-AZ-X) $Al(OC_4H_9)_3$ and zirconium-ohychloride, $ZrOCl_2 \cdot 8H_2O$, were used as a precursors. Aluminium-hydroxide prepared in the same manner as in the first method, was peptized by appropriate ratio of $ZrOCl_2 \cdot 8H_2O$. Further treatment of the composite sols were exact the same as in the first method.

Drying of the gels were carried out carefully for a few days up to 120°C. After drying gel fragments were heat treated up to 1450°C and held for 60 min.

X-ray diffraction (XRD) analysis were conducted on a computer-controlled Philips APD 1700 diffractometer. The ratio of tetragonal to monoclinic ZrO_2 was determined using the integrated intensity of the tetragonal (111) and the monoclinic (111) and ($\bar{1}$11) peaks [10]. The crystal size of the heat treated samples was determined by the Scherrer formula ($B_{(2\theta)}$ = 0.9 /L·cosθ) [11]. Microstructural development was analyzed from SEM micrographs of fractured surfaces made on SEM JEOL 35.

RESULTS AND DISCUSSION

Resultant gel microstructures in the method 1-AZ and 2-AZ are different.

The X-ray diffraction patterns presented in Figure 1 show that both types of the composite gels have poor crystalline of boehmite like structure, but without any traces of zirconia. The gels have different crystalline size: being higher for composite 1-AZ-20 then for 2-AZ-20, because of different: temperature of hydrolysis, dilution, precursor type (in the 2-AZ ionic being) and peptization agents (in the 2-AZ zirconium salt itself).

SEM micrographs of the dried composite gels are shown in Figure 2. These gels show homogenous, fine particulate, uniform size distributed microstructure in which it is not possible to distinguish presence of two phases - alumina and zirconia. SEM micrograph of the 2-AZ-20 composite gel (Figure 2b) shows the difficult to distinguish particles because of its poor crystallinity, confirming X-ray results that the composite gel 1-AZ-20 has a coarser gel microstructure than 2-AZ-20.

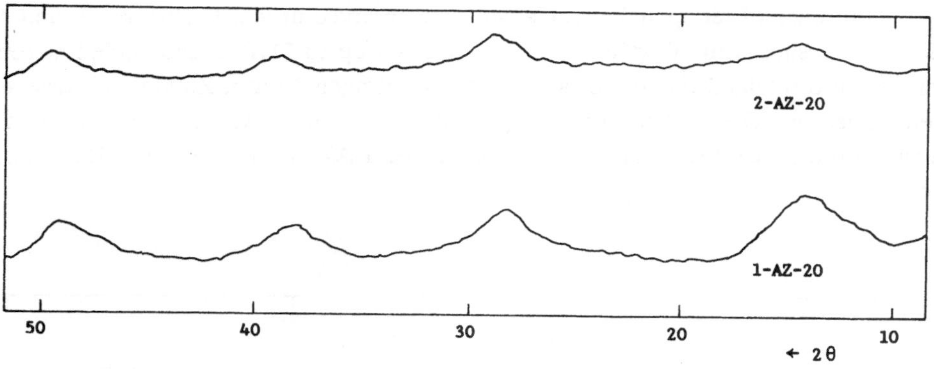

Figure 1. X-ray difraction patterns of 1-AZ and 2-AZ dried composite gels

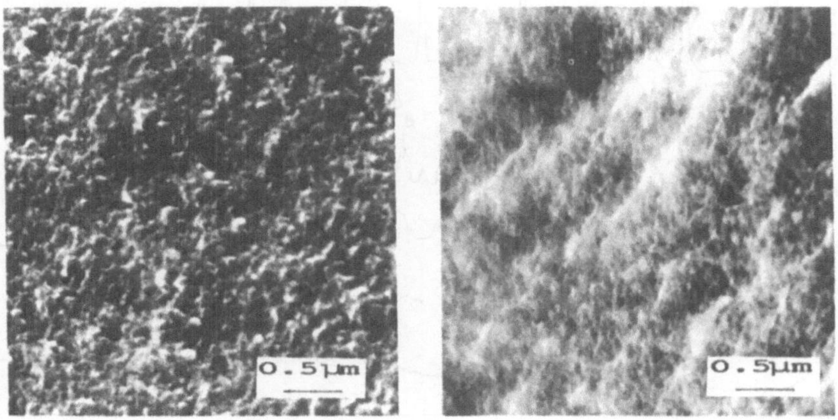

Figure 2. SEM micrographs of the dried composite gels; a) 1-AZ-20 and b) 2-AZ-20

On the heating boehmite transforms topotacticaly to transitional phases of alumina, and in the final stage transformation to stable alpha-alumina occurs by a nucleation and growth process [12-13]. The phase transformations with temperature according to the XRD results for the 1-AZ-20 and 2-AZ-20 samples are presented in Figures 3. and 4. respectively. Comparing with results previously reported for pure sol-gel derived alumina [14,15] it is obvious that the presence of zirconia greatly inhibits the transformation to alpha-alumina. This transformation in presence of 20 wt% ZrO_2 beginies at 1170 °C for the 1-AZ composite (Figure 3.) and at somewhat higher temperature for the 2-AZ composite (Figure 4.), while the pure alumina prepared by the same route is already tcompletely trasformed to alpha-alumina at 1100°C [14,15].

Presence of tetragonal zirconia was first observed at 900°C (Figures 3. and 4.). Both composites with 20 wt% ZrO_2 heat treated up to 1170°C and cooled at room temperature retained the almost all zirconia in tetragonal form. Zirconia particle size determined by Scherrer line broadening (Table 1.) increases with temperature. In this temperature range crystalline size is larger in the 1-AZ-20 than in the 2-AZ-20 composite.

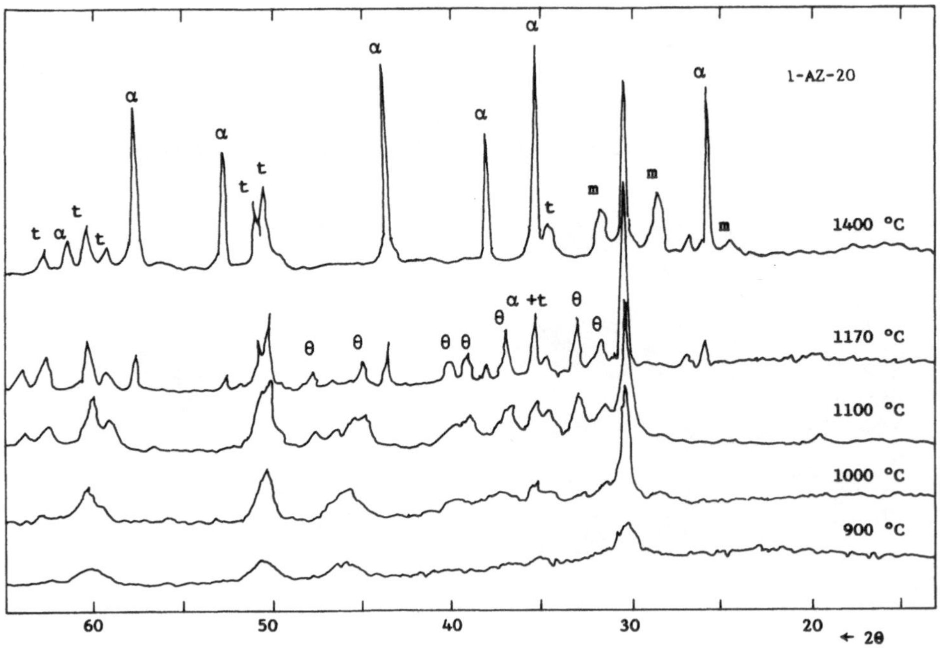

Figure 3. X-Ray Diffraction Patterns of Heat Treated 1-AZ samples

TABLE 1.
Particle size broadening determined by the Scherrer equation ($B_{(2\Theta)}=0{,}9\lambda/L\cos\Theta$)*

Temperature [°C]	Size of ZrO_2 measured from tetragonal (111) line [nm]	
	1 - AZ - 20	2 - AZ - 20
900	7,5	-
1000	16,1	9,3
1100	18,2	15,8
1170	34,2	25,7

* λ =0,15418 nm, L - width at half height in radians, Θ - Bragg angle

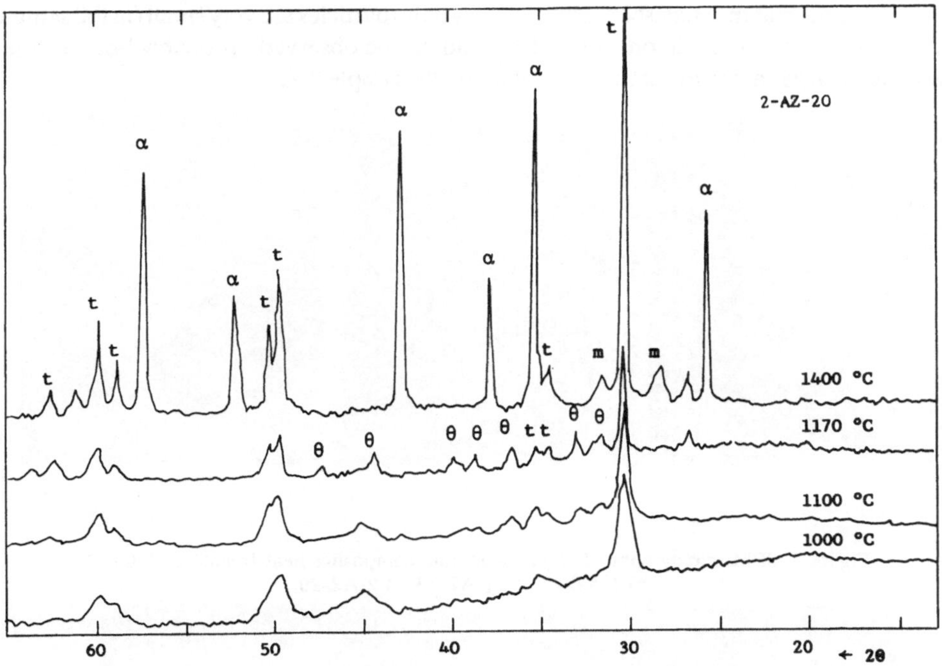

Figyre 4. X-ray diffraction patterns of heat treated 2-AZ samples

Different starting gel microstructures determined behaviour of matrix and zirconia in the temperature range between 900°C and 1170°C. The first, alpha-alumina nucleation from the transitional forms of aluminas is inhibited and shifted towards the higher temperatures (in respect to the pure alumina), and the second, crystalline forms of zirconia appear at higher temperatures than in pure zirconia and being almost all tetragonal.

On the heating the composite gels in the temperature range between 1200°C and 1450°C, alpha-alumina increase in size (Figure 5.). Beside the temperature, processing route and zirconia content have influence on the grain size of alumina. By increasing zirconia content alpha-alumina grains decrease (Figure 5.) and grain size of alumina in the composite 2-AZ-20 is larger than in composite 1-AZ-20 (Figure 5.).

The XRD patterns of gels heat treated at 1400°C (Figures 3. and 4.) show that different t/m ZrO_2 ratio are present in the 1-AZ-20 and 2-AZ-20 composites. The sample 2-AZ-20 containes more tetragonal zirconia (90,8 vol%) than the sample 1-AZ-20 heat treated at 1400°C (67,7 vol%). But unfortunately this transformation t - m could not be explained by grain size of zirconia. Zirconia grain size (Figure 5.) in both samples is very small and mainly distributed in side the large alpha-alumina grains. Zirconia grains at 1400°C are larger in the composite 2-AZ-20 than in the composite 1-AZ-20, ev, probably because of coalescence of very fine zirconia particles in system 2..

It seems that the coalescence appear only if the particles are very small (in the sample 1-AZ-20 coalescence of zirconia particles could not be observed, probably because they were much larger at lower temperature than in the sample 2-AZ-20).

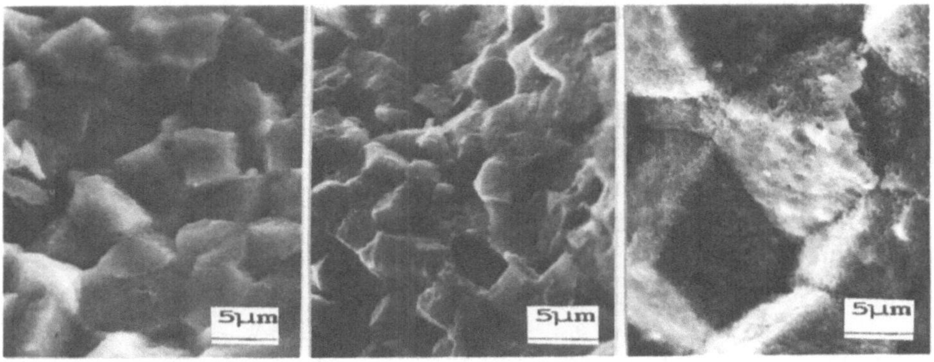

Figure 5. SEM micrographs of alumina-zirconia composites heat treated of 1400°C; a) 1-AZ-20 , b) 1-AZ-33, c) 2-AZ-20

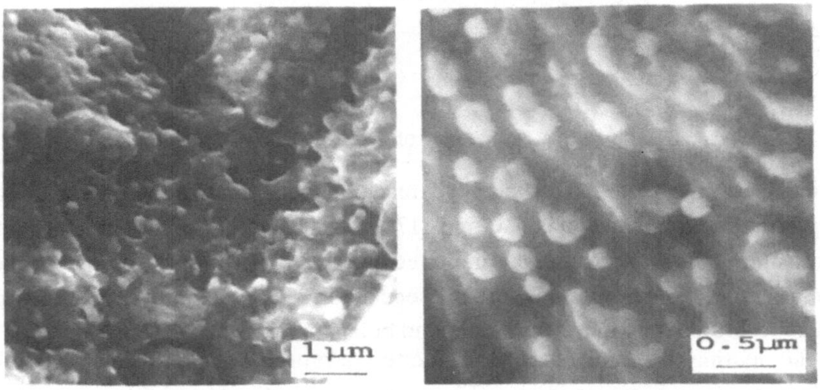

Figure 6. SEM micrographs of alumina-zirconia composites heat treated at 1400°C; a) 1-AZ-20 and b) 2-AZ-20

Experimental results obviously show the high importance of starting composite gel microstructure because of its great influence on the alumina crystallization, zirconia crystallization and grain growth of both the zirconia and the alumina at high temperature.

ACKNOWLEDGEMENT: We are deeply greatful to Milos T. Bokorov from Institute of Biology, PMF, University of Novi Sad, for its intelligent help in SEM results,

REFERENCES

[1] Claussen, N. and Ruhle, M., Design of transformation-toughened ceramics. In Advances in Ceramics Vol 3, Science and Technology of Zirconia, ed. A.H. Heuer and W.L. Hobbs, The American Ceramic Society, Columbus, Ohio, 1981, pp. 137-63.

[2] Lange, F,F, Transformation taughening. J. Mater. Sci., 1981, 17, 225-63.

[3] McMeeking, R.H. and Evans, A.G., Mechanics of transformation-toughening in brittle materials. J. Am. Ceram. Soc., 1982, 65, 242-6.

[4] Wang, J. and Stevens, R., Review-Zirconia-toughened alumina (ZTA) ceramics. J. Mater. Sci., 1989, 24, 3421-40.

[5] Messing, G.L. and Kumagai, M., Low-temperature sintering of seeded sol-gel derived, ZrO_2-toughened Al_2O_3 composites. J. Am. Ceram. Soc., 1989, 72, 40-4.

[6] Debsikdar, J.C., Influence of synthesis chemistry on alumina-zirconia powder characteristics. J. Mater. Sci., 1987, 22, 2237-47.

[7] Low, I.M. and McPherson, R., Crystallization of gel-derived alumina and alumina-zirconia ceramics. J. Mater. Sci., 1989, 24, 892-898.

[8] Bach, J.P. and Therenot, F., Fabricaation and characterization of zirconia-toughened alumina obtained by inorganic and organic precursors. J. Mater. Sci., 1989, 24, 2711-21.

[9] Pugar, E.A. and Morgan, P.E.D., Coupled grain growth effects in Al_2O_3/10 vol% ZrO_2. J. Am. Ceram. Soc., 1986, 69, C-120-3.

[10] Toraya, H., Yoshimura, M. and Somiya, S., Calibration curve for quantitative analysis of the monoclinic-tetragonal Zr_2O system by X-ray diffraction. J. Am. Ceram. Soc., 1984, 67, C-119-21.

[11] Warren, B.E., X-Ray Diffraction, Reading, Massachusetts, 1989, pp. 251-4.

[12] Kumagai, M. and Messing, G.L., Controlled transformation and sintering of a boehmite sol-gel by alpha-alumina seeding. J. Am. Ceram. Soc., 1985, 68, 500-5.

[13] Dynys, F.W. and Halloran, J.W., Alpha alumina formation in Al_2O_3 gels. In Ultrastructure Processing of Ceramics, Glasses, and Composites, ed. L.L. Hench and D.R. Ulrich, Wiley, New York, 1984, pp. 142-51.

[14] Radonjić, L., Srdić,V., and Nikolić, L., Boehmite gels microstructure relationship to their transformation to alpha-alumina. Mater. Res. Bull., in print.

[15] Radonjić, L., Srdić, V., and Nikolić, L., The microstructural transformations of alumina gels under heating. In Structural Ceramics, Processing Microstructure and Properties, ed. J.J. Bentzen, Bilde-Sorensen, N. Christiansen, A. Horsewell and B. Ralph, RISO National Laboratory, Roskilde, Denmark, 1990, pp. 471-6.

INTERFACES IN CERAMIC MATRIX COMPOSITES FOLLOWING HIGH TEMPERATURE DEFORMATION

HÅKAN A. SWAN, COLETTE O'MEARA AND GORDON L. DUNLOP*
Department of Physics, Chalmers University of Technology, S-412 96 Göteborg, Sweden
*Department of Mining & Metallurgical Engineering, University of Queensland, St. Lucia, QLD 4067, Australia

ABSTRACT

The role of whisker interfaces during high temperature deformation of ceramic matrix composites has been studied. The effect of the whiskers depends on the actual CMC system used and in particular on the presence or absence of an intergranular amorphous phase. SiC_w/Si_3N_4 and SiC_w/Al_2O_3 have been creep tested at elevated temperatures and reasons for the difference in creep behaviour of the two composites are discussed. It is concluded that the effect of the whiskers upon the high temperature properties of the CMC material is strongly dependent upon the microstructure and properties of the matrix.

INTRODUCTION

Ceramic matrix composite (CMC) materials involve the incorporation of fibres and/or whiskers into a ceramic matrix material in order to improve both the room temperature and high temperature properties. These materials are of special interest for high temperature use since they can retain their mechanical strength at temperatures considerably in excess of 1000°C.

The most studied systems at present include SiC_f reinforced ceramics with Si_3N_4, Al_2O_3 or SiC matrix material, and SiC_w reinforced Al_2O_3 and Si_3N_4. As yet, SiC fibres are not stable at temperatures in excess of ~1100°C, thus making them less suitable for long term applications at elevated temperatures than SiC whiskers. Another advantage of whisker reinforced ceramics is that they can be fabricated using the same processing techniques used for non-reinforced ceramics.

It has been shown that the whisker/matrix interface properties of CMC materials are of paramount importance for their room temperature fracture toughness because the interface has to be sufficiently weak to allow debonding during crack propagation [1]. The role of the interfaces at elevated temperatures has as yet not been fully ascertained. The effect of the whiskers depends on the actual CMC system used and in particular on the presence or absence of an intergranular amorphous phase. In the present work SiC_w/Si_3N_4 and SiC_w/Al_2O_3 have been tested at elevated temperatures in order to study the effect of the whiskers and especially the effect of the interfaces on the high temperature deformation properties of the material. The effect of the whiskers and the whisker/matrix interfaces will be discussed in general terms, followed by a closer examination of the two CMC systems investigated in this work.

INTERFACES IN CMC MATERIALS

The microstructure of high performance ceramic materials such as Si_3N_4 or Al_2O_3 consists of grains of the major phase surrounded by an amorphous phase. Secondary crystalline phase(s) may also be present depending on the amount and type of sintering additives and the processing conditions used. The volume fraction of amorphous phase also depends upon the amount of sintering additives used and the impurity concentration. However, a thin amorphous film is virtually always present at the grain boundaries. This film is also present at whisker/matrix interfaces and the properties of this glassy phase have a strong effect on the creep and crack propagation characteristics of CMC materials.

At room temperature the role of whiskers in a CMC is to increase the fracture toughness of the material. If the interface is weakly bonded, a number of fracture toughening mechanisms can lead to a non-linear stress-strain curve and an increase in fracture toughness. Possible toughening mechanisms include deflection, pullout, crack bridging, debonding and microcracking. The properties of the interfacial glassy film are thus of paramount importance. For example, a silica rich film provides a strong bond whereas a graphite film at the interface results in a loose bonding between the whisker and the matrix.

At elevated temperatures the role of whiskers is somewhat different. The aim in this case is to reduce the creep rate by the introduction of whiskers, and to improve the creep rupture properties by slowing down the process of crack growth.

The effect of whiskers is dependent upon several factors. Firstly, structural aspects such as volume fraction, homogeneity, whisker alignment and whisker dimensions are all of importance. Secondly, mechanical aspects such as the whisker/matrix interaction during creep, the crack arrest mechanisms during creep fracture and the role of interfaces in these mechanisms must be considered. Another important factor in the fabrication of CMC's is the chemical compatibility between the whisker/fibre and the matrix. Degradation can be enhanced by impurities that are remnant at the whisker/fibre interfaces. In the SiC_w/Si_3N_4 and SiC_w/Al_2O_3 CMC's investigated here the whiskers are not chemically inert with respect to the two matrix materials. However, degradation can be minimized by careful choice and control of the processing parameters.

Finally, the oxidation resistance of the composite is an important factor in high temperature applications. SiC is normally quite oxidation resistant due to the formation of a passivating oxide layer, however this passivating layer is not always formed in bulk CMC materials. The oxidation rate in SiC_w/Al_2O_3 is an order of magnitude greater than for monolithic SiC due to the effect of the surrounding matrix [2]. In SiC_w/Si_3N_4 the oxygen diffuses through the amorphous glassy phase and the effect of oxidation becomes dramatic above ~1350°C which is the lowest eutectic temperature for the Si_3N_4 composites investigated here.

SiC WHISKER REINFORCED Al_2O_3

Commercially available single phase alumina materials normally have a very low glassy phase content and a grain size of 1-5 μm. Small additions of MgO or CaO as sintering aids are common, and these result in thin intergranular glassy films. Since alumina is an oxide, the corrosion properties are excellent, however the creep resistance of oxides is generally poor. The creep deformation mechanism of the material is considered to be grain boundary sliding, accomodated by diffusion of oxygen and aluminium ions. The lattice diffusion rate for oxygen is much lower than for aluminium, but grain boundary diffusion of oxygen is faster, thus resulting in the diffusion of aluminium cations as the rate controlling mechanism. One therefore has lattice or grain boundary cation diffusion control resulting in a creep exponent, n, of 1 or interface controlled diffusional creep with n = 2 [3].

The introduction of SiC whiskers into alumina was thought to provide a solution to the problem of creep strengthening alumina and several creep tests of SiC_w/Al_2O_3 have all shown a reduced creep rate of from one up to three orders of magnitude, depending on the temperatures and loads used [4-6]. The stress exponent was generally observed to change from 1-2 to ~4-6 with the introduction of whiskers, thus indicating a change in deformation mechanism.

In the present work, compressive creep tests of SiC whisker reinforced Al_2O_3 have been performed in air at temperatures ranging from 1200°C to 1350°C. The experimental material was a commercially available CMC containing 30 vol% SiC whiskers†. A more detailed description of the experiments and results is given elsewhere [7]. A stress exponent of ~1.5 was obtained, and the activation energy was calculated to be 370 ± 50 kJ/mol. The value of the stress exponent suggests interface controlled grain boundary sliding as the rate determining mechanism, just as for non-reinforced alumina. The creep rates were only decreased by a factor of ~20, see Fig. 1. The activation energy is similar to, but slightly lower than, the activation energy for cation diffusion in alumina. The activation energy for Al^{3+} lattice diffusion is 478 kJ/mol [8] and for grain boundary diffusion is 418 kJ/mol [9], however this latter value is known to be dependent upon the chemistry of the grain boundaries.

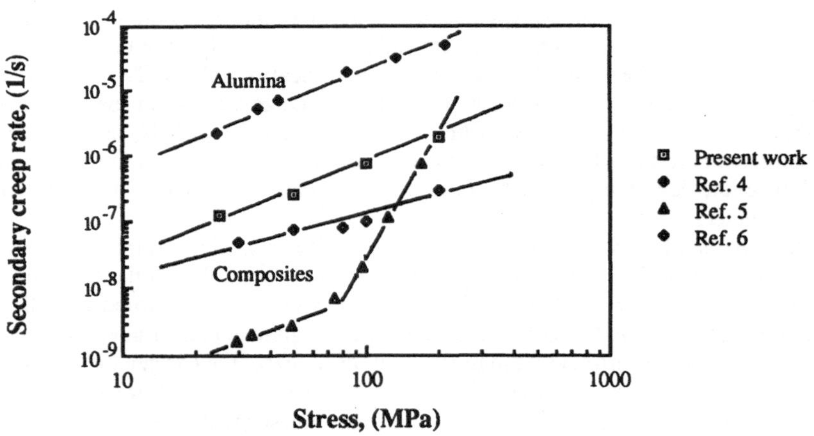

Figure 1. Relationship between secondary creep rate and stress for the SiC_w/Al_2O_3 composite tested at 1300°C (Present work). The results are compared with a nonreinforced alumina material [4], and with the results for similar CMC's from refs 5 (4-point bending) and 6 (compressive creep in argon environment).

---

†ARtuff CC7000, ACMC, Greer, SC.

TEM specimens were cut from the specimens tested at 1300°C with up to 10% compressive strain. These were assumed to be representative for all temperatures and loads since the stress exponent and activation energies remained approximately the same. SEM specimens were also prepared from the same specimens. Investigations of the as-received material showed that it appeared to have been fabricated from spray dried granule compacts which were hot pressed. The microstructural investigation of the crept samples showed that extensive cavitation had taken place as a result of non-accomodated grain boundary sliding. The cavitation was predominantly observed at the whisker/matrix and whisker/whisker interfaces and at whisker deficient regions. Many of the cavities have coalesced into larger microcracks (Fig. 2). Fig. 3 is a TEM micrograph showing a whisker which has debonded along the interface. No change in the alumina grain size could be detected in the whisker rich regions (within the granules) after creep deformation. The whiskers thus prohibit grain growth during creep testing.

The results of the microstructural analysis suggest that the rate determining creep deformation mechanism was diffusion accommodated grain boundary sliding but in many cases the rate of diffusion was too slow to accomodate the high rates of grain boundary shear resulting in cavitation. The introduction of long, rigid whiskers into the matrix slows down the rate of deformation by increasing the amount of grain shape accommodation required for diffusion accommodated GBS. The mean diffusion lengths are thus increased by the presence of whiskers resulting in slower creep rates and an increased amount of non-accommodated GBS. Dislocation activity has been observed in some Al_2O_3 grains, however the extent was too small to be rate-determining.

Oxidation of the whiskers during creep resulted in the formation of a silica rich glass film along the whisker interface, slowly spreading intergranularly into the alumina matrix. However this effect was only observed in the vicinity of the specimen surface and along cracks penetrating into the bulk of the specimen. The effect of environment has previously been examined by other investigators [6] showing an increased creep rate for specimens subjected to creep in air as compared to argon (Fig. 1).

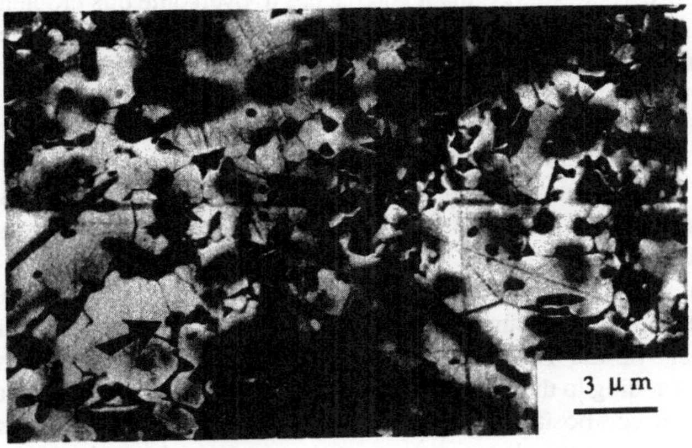

Figure 2. SEM secondary electron micrograph of the specimen tested at 1300°C showing cavities that have coalesced into larger microcracks. The surface was slightly etched in ortophosphoric acid for better contrast.

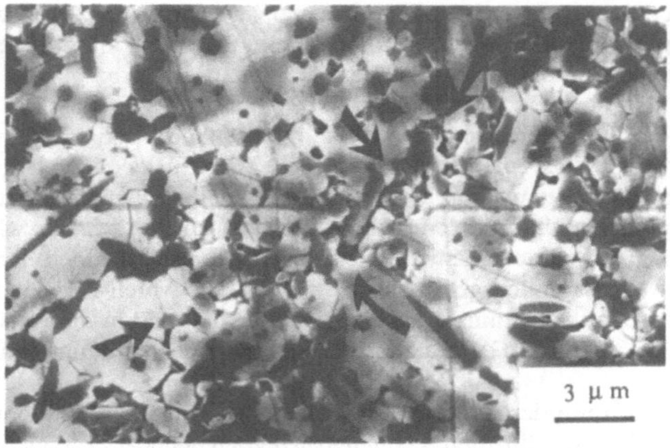

Figure 3. TEM bright field micrograph showing debonding at a whisker interface in the same specimen as Fig. 2.

SiC WHISKER REINFORCED Si_3N_4

Hot pressed Si_3N_4 ceramic materials normally contain Y_2O_3 and/or Al_2O_3 as sintering aids, resulting in the presence of a liquid phase during sintering. This liquid phase is formed at a eutectic temperature of ~1350°C in a reaction between the sintering additives, surface silica on the Si_3N_4 grains and Si_3N_4 itself. The liquid solidifies upon cooling, resulting in a glassy phase surrounding the grains. This amorphous phase begins to soften at ~1000°C and this severely affects the creep rate at elevated temperatures, where the dominant creep mechanism is that of grain boundary sliding and results in extensive cavitation [10]. At lower temperatures, creep rates rapidly decrease due to a strain hardening effect [10]. The intergranular phase also acts as a path of increased diffusivity in the material, thus further limiting the use of these materials at elevated temperatures. There are two main approaches to overcome this problem, either through the elimination/minimization of the intergranular glassy phase, or through the introduction of SiC whiskers to the material. The former approach includes hot isostatic pressing without additives, or crystallization of the secondary phase.

Although evidence exists for an increase in the room temperature fracture toughness with the introduction of whiskers [11], no improvement of the creep resistance at elevated temperatures has as yet been observed [12,13]. Stress vs strain measurements performed at elevated temperatures have indicated that the creep properties of hot pressed Si_3N_4 were improved by the incorporation of SiC whiskers [14], and creep experiments on SiC whisker reinforced Si_3N_4 materials were performed using 4-point bending in air at 1200°C to 1375°C [15]. The deflection was measured directly on the loading rod, and thus included the deformation effects of the HIP SiC rollers and the total loading system in the strain measurements, thus adding to the experimental errors. The experimental material was a nitrided pressureless sintered composite with large additions of Y_2O_3 and Al_2O_3 as sintering aids. Extensive crystallization of the glassy phase was observed, leaving only thin amorphous films at the interfaces.

A primary creep region characterized by a decreasing creep rate, strain hardening behaviour, was obtained and no steady state creep region was observed. An increase of the load shifted the creep curve slightly due to the elastic strain of the specimen and the testing rig and the settlement of the loading rollers into the specimen. No consistent change of strain rate due to the increased load could be detected since the strain hardening contribution dominated the creep curve behaviour. The traditional creep parameters such as the creep exponent, n, and the creep activation energy, Q, cannot be determined directly from a creep curve of this type. Another method of creep characterisation has been developed by Fett et al.[16] for this type of behaviour. However the testing rig proved to be inadequate to obtain sufficiently accurate data for quantitative analysis according to the above mentioned method. No improvement in creep resistance due to the incorporation of SiC whiskers was detected for the Si_3N_4 material.

When the temperature during creep testing was raised to 1375°C the appearance of the creep curve changed to that of steady state behaviour. However, this creep region was not investigated further due to specimen failure. The strength of the material was considerably decreased at this temperature. These results indicated that the test temperature was above the eutectic temperature for the material composition used, therefore changing the creep mechanism as well as the specimen strength. A eutectic in the region of 1350°C to 1375°C has also been recorded in other studies. The strength above this temperature decreases significantly while the oxidation rate increases, thus limiting the application of these materials to temperatures below ~1350°C.

Fig. 4 shows the microstructure of the material below the oxidation scale in the region of maximum strain (~1%). Debonding and microcracking/cavitation at the whisker/matrix interface was commonly observed (Fig. 4a), but also along the interfaces of the Si_3N_4 grains (Fig. 4b). Furthermore, a number of large cracks of up to ~50 μm were observed. The microstructural analysis suggests that microcracking and the formation of larger cracks was the rate determining mechanism for the loads and temperatures used. Whisker surface impurities had a strong influence on the degradation properties of the whiskers during processing. Some whiskers had oxidized with the formation of a degradation zone adjacent to the remains of the whisker.

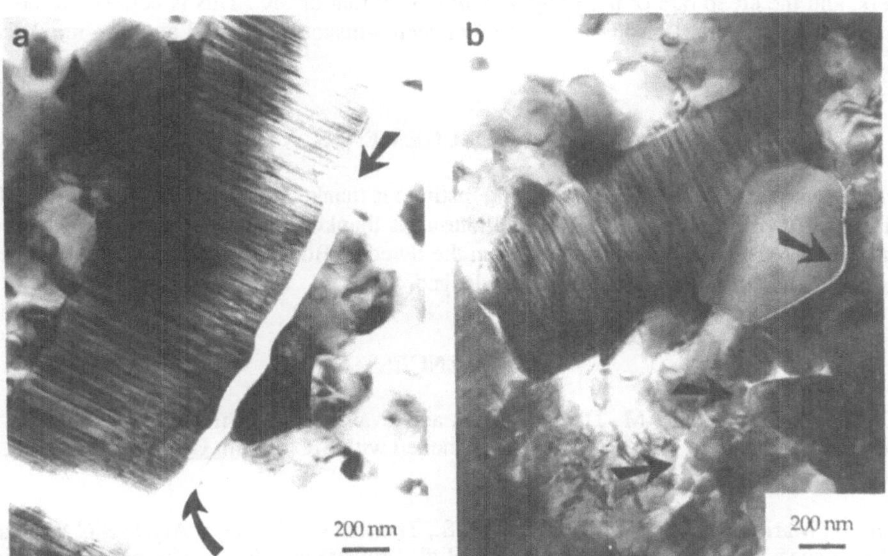

Figure 4. Bright field TEM micrograph of the SiC_w/Si_3N_4 material after creep testing at 1375°C showing a) debonding along the whisker/matrix interface and b) intergranular microcracks (arrowed).

CONCLUDING DISCUSSION

The results of these experiments show that the effect of whiskers upon the high temperature properties of the CMC materials is strongly dependent upon the microstructure and properties of the matrix. The creep properties were found to improve with the addition of whiskers in SiC_w/Al_2O_3, whereas no change could be observed in a SiC_w/Si_3N_4 material. Some general conclusions on the behaviour CMC materials at elevated temperatures can be made from these observations.

The impurity content at the fibre/whisker interfaces is a problem during fabrication and high temperature application of CMC materials. Furthermore the processing of CMC's is more complex than for non-reinforced matrix materials with an increased probability of porosity and inhomogeneities such as whisker agglomerates. These factors also have an influence on the creep properties of the material. For example Porter observed a lower creep rate for SiC_w/Al_2O_3 when the whiskers were homogeneously dispersed because of a different processing method [17]. In SiC_w/Si_3N_4, the whiskers had to be milled to a low aspect ratio in order to obtain fully dense materials, resulting in a reduced effect of the whisker addition.

The role of the fibres/whiskers during creep is dependent upon the matrix microstructure and properties. The presence and amount of interfacial glassy phase is the most important factor for the high temperature properties of CMC's. Since grain boundary sliding is the predominant deformation mechanism in the materials considered, the presence and viscosity of the glassy phase has a strong influence on the creep properties. Accomodation and/or cavitation is also greatly simplified by the glassy phase. Furthermore, the effect of the whiskers is also influenced by the morphology and properties of the main matrix phase. In the ionically bonded alumina material, the grains were of low aspect ratio and have a substantially higher creep rate than SiC. However the β–Si_3N_4 grains had a high aspect ratio with lengths comparable to the milled whiskers, and the creep rate of pure Si_3N_4 is similar to that of SiC. This is believed to be the major reason for the similarity in creep rate between whisker reinforced and non-reinforced Si_3N_4.

ACKNOWLEDGEMENTS

Dr. Robert Lundberg of the Swedish Ceramic Institute is thanked for supplying the SiC_w/Si_3N_4 material. Prof. Mike Swain of CSIRO, Melbourne, is thanked for his assistance in the creep testing of SiC_w/Al_2O_3. Financial support from the Swedish Board for Technical Development and the Swedish Institute is gratefully acknowledged.

REFERENCES

1. Campbell, G.H., Rühle, M., Dalgleish, B.J. and Evans, A.G., Whisker toughening: A comparison between Al_2O_3 and Si_3N_4 toughened with SiC. J. Am. Cer. Soc., 1989, **73**, pp. 521-530.

2. Lin, F., Marieb, T., Morrone, A. and Nutt, S., Thermal Oxidation of Al_2O_3-SiC Whisker Composites: Mechanisms and Kinetics. In High Temperature/High Performance Ceramics, ed. F.D. Lemkey, S.F. Fishman, A.G. Evans and J.R. Strife, Materials Research Society, Pittsburgh, Pa, 1988, **120**, pp. 323-332.

3. Cannon, R.M., Rhodes, W.H. and Heuer, A.H., Plastic Deformation of Fine-Grained

Alumina (Al_2O_3): I, Interface-Controlled Diffusional Creep. J. Am. Ceram. Soc., 1980, **63**, pp. 46-53.

4. Chokshi, A.H. and Porter, J.R., Creep Deformation of an Alumina Matrix Composite Reinforced with Silicon Carbide Whiskers. J. Am. Ceram. Soc., 1985, **68**, pp. C144-C145.

5. Lipetzky, P., Nutt, S.R. and Becher, P.F., Creep Behavior of an Al_2O_3 - SiC Composite. In High Temperature/High Performance Ceramics, ed. F.D. Lemkey, S.F. Fishman, A.G. Evans and J.R. Strife, Materials Research Society, Pittsburgh, 1988, **120**, pp. 271-277.

6. Arellano-López, A.R., Cumbrera, F.L., Domínguez-Rodríguez, A., Goretta, K.C. and Routbort, J.L., Compressive Creep of SiC-Whisker-Reinforced Al_2O_3. J. Am. Ceram. Soc., 1990, **73**, pp. 1297-1300.

7. Swan, A.H., Swain, M.V. and Dunlop, G.L., Compressive Creep of SiC Whisker Reinforced Alumina. To be published.

8. Paladino, A.E. and Kingery, W.D., Aluminium Ion Diffusion in Aluminium Oxide. J. Chem. Phys., 1962, **37**, pp. 957-962.

9. Cannon, R.M. and Coble, R.L., In Deformation of Ceramic Materials, ed. R.C. Bradt and R.E. Tressler, Plenum, New York, 1975, pp. 61-100.

10. Clarke, D.R., The Microstructure of Nitrogen Ceramics. In Progress in Nitrogen Ceramics, ed. F.L. Riley, Martinus Nijhoff, The Hague, 1983, pp. 341-358.

11. Becher, P.F., Hsueh, C.H., Angelini, P. and Tiegs, T.N. , Toughening Behaviour in Whisker-Reinforced Ceramic Matrix Composites. J. Am. Ceram. Soc., 1988, **71**, pp. 1050-1061.

12. Porter, J.R., Lange, F.F. and Chokshi, A.H., Processing and Creep Performance of Silicon Carbide Whisker-Reinforced Silicon Nitride. In Mat. Res. Soc. Symp. Proc., Materials Research Society, 1987, **78**, pp. 289-294.

13. Backhaus-Ricoult, M., Castaing, J. and Routbort, J.L., Creep of SiC-Whisker Reinforced Si_3N_4. Revue Phys. Appl., 1988, **23**, pp. 239-249.

14. Swan, A.H., Olsson, E., Linde, K.M. and Lundberg, R., The influence of Microstructure on the Mechanical Properties of SiC Whisker Reinforced Si_3N_4. In New Materials and Processes, ed. I.L.H. Hansson and H. Lilholt, Danish Society for Materials Testing and Research, Copenhagen, 1989, pp. 655-662.

15. Swan, A.H. and O'Meara, C., Creep Properties and Interfacial Microstructure of SiC Whisker Reinforced Si_3N_4. In Interfaces in Composites, ed. C.G Pantano and E.J.H. Chen, Materials Research Society, Pittsburgh, PA, 1990, **170**, pp. 223-228.

16. Fett, T., Keller, K. and Munz, D., An Analysis of the Creep of Hot Pressed Silicon Nitride in Bending. J. Mater. Sci., 1988, **23**, pp. 467-474.

17. Porter, J.R., Dispersion Processing of Creep Resistant Whisker-Reinforced Ceramic Matrix Composites. Mater. Sci. Eng., 1989, **A107**, pp. 127-132.

PHASE RELATIONSHIPS OF α', ß'-SIALONS IN THE Y,SI,AL/N,O SYSTEM AND DESIGN OF SIALON COMPOSITE MATERIALS

W.Y.SUN, L.J.TONG and T.S.YEN
Shanghai Institute of Ceramics, Chinese Academy of Sciences, Shanghai 200050, China

ABSTRACT

A brief account of the single α'-Sialon phase region and the α'-Sialon + ß'-Sialon (ß-Si_3N_4) region on the α'-Sialon plane in the Y,Si,Al/N,O system is presented based on our previous investigation [1,2]. Two compositions within the ß-Si_3N_4-α'-Sialon compatibility region were designed and fabricated into samples. In general, they do show the additivity characteristics of the better strength and fracture toughness of elongated ß-Si_3N_4 crystals plus the higher hardness of equiaxed α'-Sialon grains. By making intelligent use of the information provided by the phase relationships of the Y,Si,Al/N,O system, promising materials can be tailored along with processing studies.

INTRODUCTION

ß-Si_3N_4, α' and ß'-Sialons have been recognized to form materials with superior high-temperature strength(ß or ß'-solid solutions) or better hardness and possibly higher thermal shock resistance (α'-solid solutions). Recent thorough study of the phase relationships of the Y,Si,Al/N,O system combined with a critical assessment of the existing information obtained in our laboratory and in collaboration with the laboratories of the University of

Michigan and University of Newcastle Upon Tyne[1,2] made clear the phase boundaries of α'-Sialon region and an α'+ β'-Sialons two phase region within the system. The objectives of this research are (1) to present the relevant results of α'-Sialon and α'+ β'-Sialon regions and their compatibility relationships with other phases and (2) to design and fabricate materials with composite α'+ β'(or α'+ β-Si_3N_4) phases with an aim to tailor the properties of the multi-phased materials. It is intended to see how the higher hardness of α'-Sialon may incorporate with the better strength of β-Si_3N_4 to give more attractive properties. A microstructure with elongated prismatic β-Si_3N_4 grains of high aspect ratio will also be expected to show an in situ strengthening effect.

EXPERIMENTAL

Si_3N_4(Starck LC-12), Y_2O_3 and La_2O_3(99.9% purity) and AlN(32% nitrogen) were used as the starting materials. YN will easily undergo hydrolysis, so whenever YN is necessary to be used as a component, it was carefully prepared under dry flowing nitrogen in the laboratory and was used for specimen preparation and sintering on the same day.

Two compositions of β-Si_3N_4 plus α'-Sialon composite compositions situated essentially on the Si_3N_4-Al_2O_3:AlN-YN:3AlN plane(α'-Sialon plane) were designed(with nominal compositions of 80:20 and 60:40 respectively) as in Fig.1(c). Gas pressure sintering (GPS) up to 1.5MPa N_2 pressure was used to process these specimens with N-rich contents to achieve full densification. Even under such sintering conditions, a small amount of La_2O_3(3-4wt%) as an additive was incorporated to enhance densification. It has been found that La_2O_3 will not enter into the α'-Sialon structure and yet forms a refractory grain boundary glassy phase which will not deteriorate the high-temperature properties of the material[5]. Part of the specimens were given a post-sintering heat treatment at 1400^oC for 10h. The phase compositions of the specimens were semi-quantitatively determined by x-ray diffraction. Their microstructures were observed by TEM and SEM techniques. Their mechanical properties were determined from room temperature to 1300°C by 3-point bending with a cross-head speed of 0.5mm/min and the fracture toughness values by the indentation method using a load of 10kg.

RESULTS AND DISCUSSION

Fig.1 shows the α'-Sialon plane within the Y,Si,Al/N,O prism. The solubility limits of α'-Sialon solid solutions along the Si_3N_4-YN:3AlN join in the nitride system Si_3N_4-YN-AlN has been determined[4,6]. The general formula of α'-Sialon can be represented as $Y_{m/3}Si_{12-(m+n)}Al_{m+n}N_{16-n}O_n$, and the end members of α'-Sialon single phase at the pure nitride side range from m=1.3 to m=2.4. When it extends to the O-containing side, the m value gradually decreases until m=1.0. This side of α'-Sialon phase region is continuously compatible with ß-Si_3N_4. The maximum oxygen content of α'-Sialon is approximately at n=1.7 and at this composition it is compatible with ß'-Sialon from z=0-0.8. This information is shown in Fig.1.

The two compositions designed (ß:α'=80:20 and 60:40 respectively) are essentially along the Si_3N_4-Y_2O_3:9AlN join, the intersect of two planes: the Si_3N_4-AlN-Y_2O_3 plane and the α'-Sialon plane. They fall on the ß-Si_3N_4 + α'-Sialon two phase region. In principle, all the oxide phases added except La_2O_3 could be absorbed into the α'-Sialon solid solution. However, after annealing at 1400°C for 10h, a small amount of J-phase($2La_2O_3.Si_2N_2O$) and H-phase($La_{10}(SiO_4)_6N_2$) were detected at the grain boundaries which are believed to be Y-containing solid solutions[7], since the complete absorption of Y_2O_3 into the α'-Sialon solid solution is kinetically not likely.

The main phases of ß-Si_3N_4 and α'-Sialon were determined semi-quantitatively to be 80:20 and 60:40 respectively, which are close

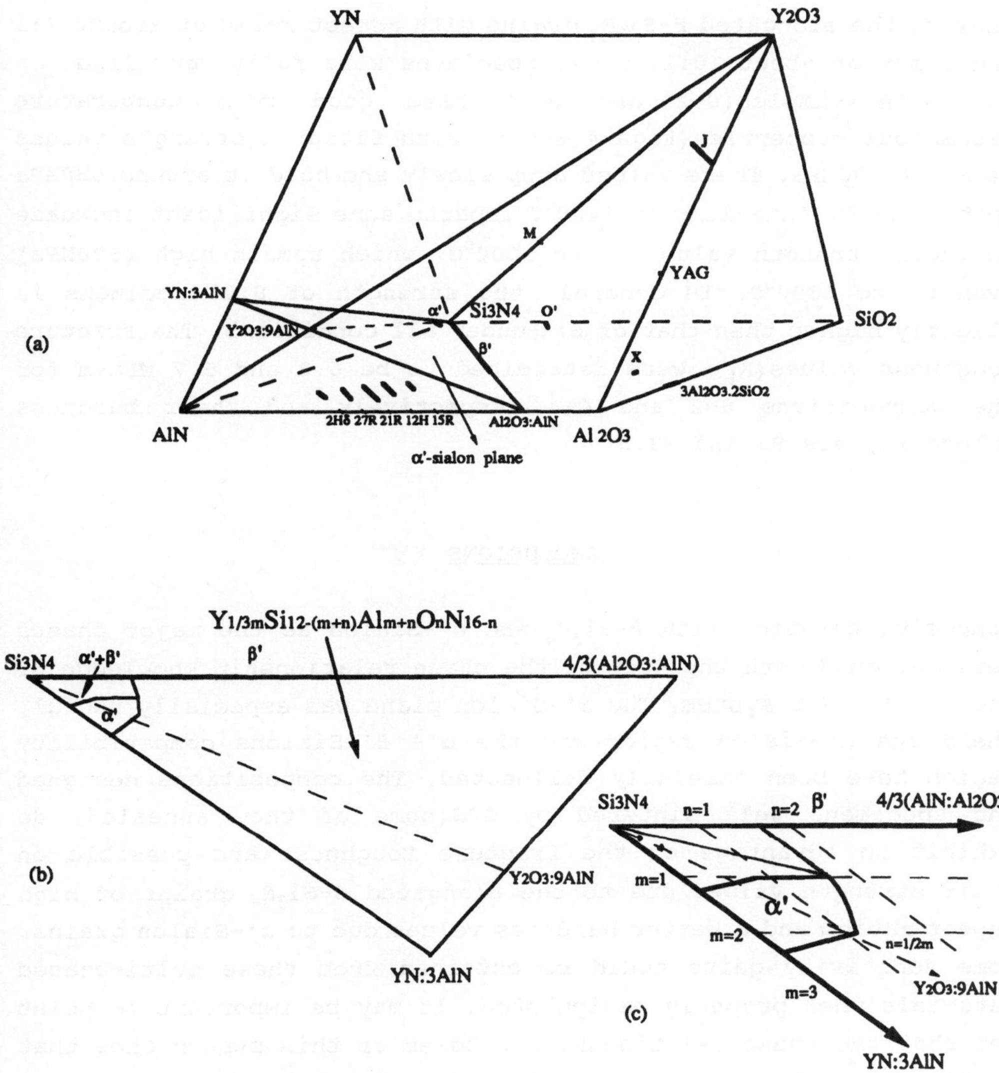

Figure. 1(a) Representation of α'-Sialon plane in Y,Si,Al/N,O system; (b) α'-Sialon plane; (c) an enlarged corner of α'-Sialon plane showing α', α' + β-Si_3N_4 and α' + β' regions and the positions of the compositions studied[3,4].

to the expected compositional values. Fig.2-3 show the typical TEM replica photomicrographs of 8-2(80:20) and 6-4(60:40) specimens showing the elongated β-Si_3N_4 grains with aspect ratio of around 7:1 and a max of about 10:1. These specimens were fully densified.

Both samples(8-2 and 6-4) have good room temperature mechanical properties(Figs 4 and 5) with flexural strength values of about 700MPa. These values drop slowly and hold at around 550MPa up to 1300°C. Annealing at 1400°C imparts some significant increase to their strength values up to 1000°C, which remain high (600MPa) even up to 1300°C. In general, the strength of 8:2 specimens is slightly higher than that of 6:4 under all conditions. The fracture toughness values(K_{1c}) were determined to be 6.8 and 5.7 MPa$\sqrt{m}$ for the compositions 8-2 and 6-4 respectively and their hardness values(Ra) are 93 and 93.5.

CONCLUSIONS

Composite ceramics with β-Si_3N_4 and α'-Sialon as the major phases were designed with the help of the phase relationship knowledge of the Si,Al,Y/N,O system. The α'-Sialon plane was especially useful, where the α'-Sialon region and the α'+ β'-Sialons compatibility region have been carefully delineated. The compositions designed and specimens well sintered by GPS(some of them annealed) do exhibit an advantage on the fracture toughness and possible on their strength values due to the elongated β-Si_3N_4 grains of high aspect ratios and a better hardness values due to α'-Sialon grains. Some additivity gains could be obtained from these multi-phased materials when properly manipulated. It may be important to point out that the phase relationships studies of this system show that α'-Sialon, β'-Sialon are compatible with YAG, AlN and its polytypes and other phases forming complicated compatibility tetrahedra[8,9]. Promising materials could be tailored by carefully designing their compositions within this system along with proper processing studies.

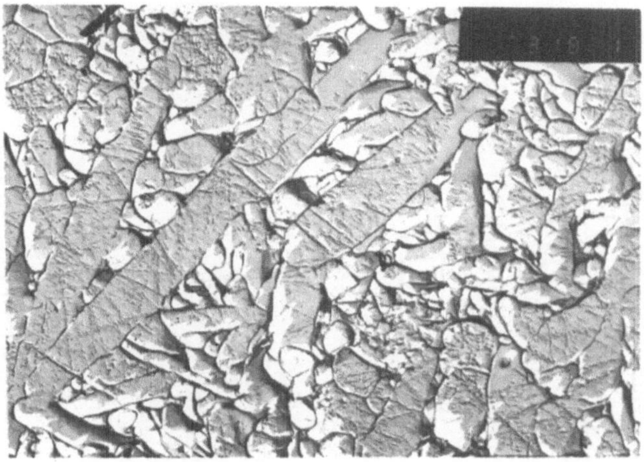

Figure 2. TEM micrograph of the specimen 8-2

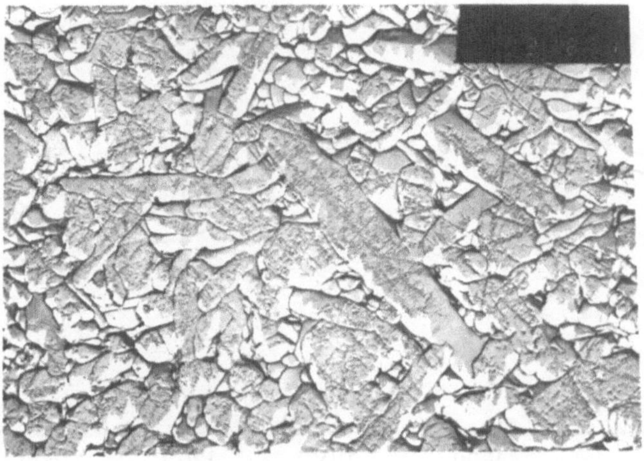

Figure 3. TEM micrograph of the specimen 6-4

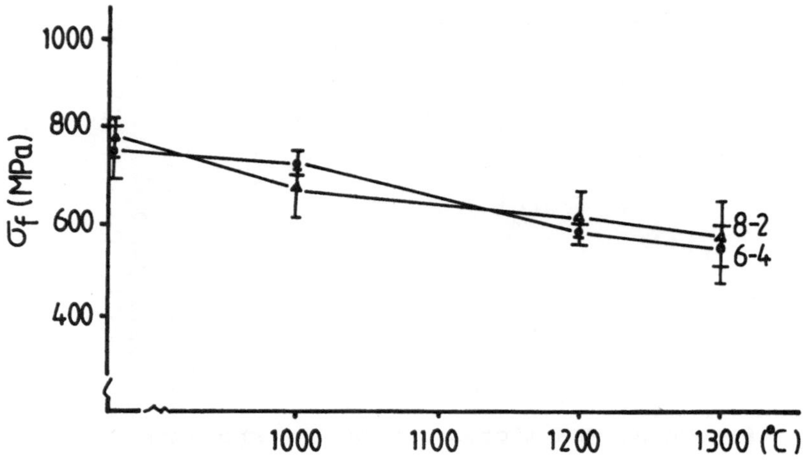

Figure 4. Flexural strength vs temperature for as-fired specimens

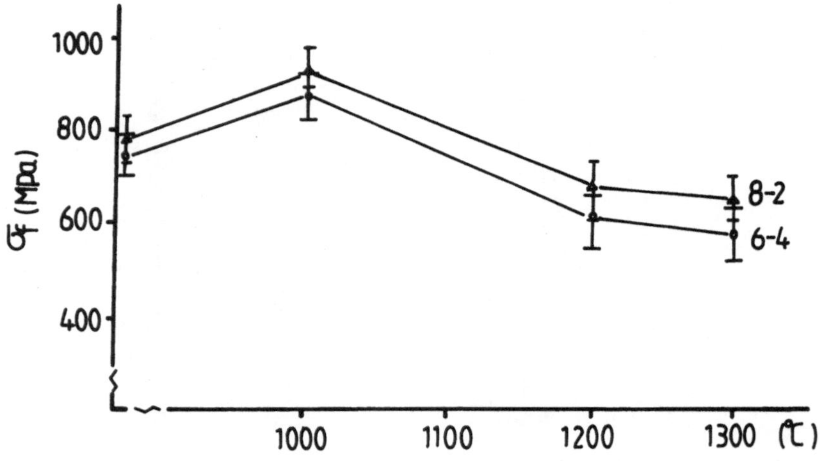

Figure 5. Flexural strength vs temperature for specimen after annealing at 1400°C for 10h

REFERENCES

1. Sun, W.Y., Yen, T.S. and Tien, T.Y., Graphical representation of the subsolidus phase relationships in the reciprocal salt system Si,Al,Y/N,O, I: Si_3N_4-$ß_{60}$-Al_2O_3-SiO_2-Y_2O_3, to be published in Scientia Sinica.

2. Sun, W.Y., Yen, T.S. and Tien, T.Y., Graphical representation of the subsolidus phase relationships in the reciprocal salt system Si,Al,Y/N,O, II: Si_3N_4-AlN-YN-Al_2O_3-Y_2O_3, to be published in Scientia Sinica.

3. Slasor, S and Thompson, D.P., Preparation and characterization of Yttrium α'-Sialon, in Proceeding of International Conference: Non-oxide Technical and Engineering Ceramics, ed. S.Hampshire, Elsevier, Amsterdam, 1986, pp.223-30.

4. Sun, W.Y., Tien, T.Y. and Yen,T.S., Solid solubility limits of single phase region of the α'-Sialon solid solutions and their compatibility relationships in the system Si,Al,Y/N,O, accepted to be published in J. Am. Ceram. Soc.

5. Xu, Y.R., Huang, L.P., Fu, X.R. and Yen, T.S., Hot-pressed silicon nitride ceramics with rare-earth oxide additives, Scientia Sinica(Series A), 1985, 28, 556-60.

6. Slasor, S. Doctoral Dissertation, University of Newcasrle Upon Tyne, 1986.

7. Cao, G.Z., Huang, Z.K. and Yan, D.S.(Yen,T.S.), Phase relationships in the Si3N4-Y2O3-La2O3 system, Science in China(Series A), 1989, 32, 429-33.

8. Huang, Z.K., Tien, T.Y. and Yen, T.S., Subsolidus phase relationships in Si_3N_4-AlN-rare-earth oxide systems, J. Am. Ceram. Soc., 1986, 69, C241-42.

9. Sun, W.Y., Tien, T.Y. and Yen, T.S., Subsolidus phase relationships in part of the system Si,Al,Y/N,O, II: the system Si_3N_4-AlN-YN-Al_2O_3-Y_2O_3, accepted to be published in J. Am. Ceram. Soc.

EFFECT OF ZrO_2 ON THE CRYSTALLIZATION OF GRAIN BOUNDARY GLASSY PHASE IN SINTERED SILICON NITRIDE

J.ZHAO, L.WANG, G.PENG and J.G.WU
Department of Materials Science and Engineering
Tsinghua Univerity, Beijing 100084, P.R. of China

ABSTRACT

Effect of ZrO_2 as nucleating agent on the crystallization of grain boundary glassy phase in sintered silicon nitride materials with Y_2O_3, La_2O_3 and Al_2O_3 additives were studied. It was found that the nucleator ZrO_2 could reduce the period of nucleation and prompt the crystallization of glassy phase. The composition of crystalized phase after different heat treatment time was investigated by X–RAY diffraction. High temperature bending strength of the Si_3N_4 materials doped suitable amount of ZrO_2 after 30 hour heat treatment increased obviously and could keep up to 1300℃.

INTRODUTION

Because of the good high temperature properies, Si_3N_4 ceramics are one of the most promising structural materials for application in structural components of heat engine. Si_3N_4 material is very difficult to sinter to full density due to its low atomic diffusion coeffcient. Sintering additives are generally used to prompt desification by liquid–sintering mechanism. But the liquid phase resides as grain boundary glassy phase after cooling and degrades the high temperature mechanical properies of the materials. The improvement of grain boundary glassy phase becomes the key to increase the high temperature properties of Si_3N_4 materials. The common used metheds to improve the glassy phase are summed up as following: 1) using advance sintering technique and decreasing the amount of glassy phase(eg. GPS, HIP). 2) increasing the refractoriness of the grain boundary by using high purity and sintering activity of Si_3N_4 powder or selecting appropriate additives.3) decreasing the amount of grain boundary glassy phase by transition phase sintering(eg. β–sialon, α–sialon etc.). 4) crystallizition of glassy phase by heat treatment and increasing the strength of grain boundary.

Researchers have already done many works on above subjects. [1–6] Some experimental results also have been reported about crystallization of glassy phase for system of Si–Y–Al–O–N and Mg–Al–Si–O–N. [7–8] But there are few research reports about the effect of nucleating agent on the crystallization of glassy phase in sintered Si_3N_4. The purpose of the present study was to investigate the effect of ZrO_2 as nucleating agent on the crystallization of glassy phase and to improve the mechanical properties of pressureless sintered Si_3N_4 with La_2O_3, Y_2O_3 and Al_2O_3 additives.

EXPERIMENT

The Si_3N_4 powder (94%α phase, oxygen content 1.19wt%) with concurrent addition of La_2O_3, Y_2O_3, Al_2O_3 and ZrO_2 (purity all >99%) were mixed according to the selected composition shown in table 1. The mixtures were milled for 24h using Si_3N_4 mill ball and ethanol agent and the powder slurry was then dried. The compacts were shaped by press and isostatic press and then sintered in a graphite furnace at 1750–1850℃ in N_2 atmosphere. The relative density of as-sintered compacts were more than 97%. The samples were heat treated at 1350℃ in N_2 atomosphere and holding time was 2h,30h respectively.

The samples were ground and then polished with a 7μm diamond paste and the final size of samples was 3x4x35mm. The phase composition were determined by X-ray diffraction analysis. The phase contents were quanlitatively evaluated by diffraction peak. The three-point flexural strength was measureed at room temperature, 1200℃ and 1300℃ in air with a span of 30mm and a crosshead speed of 0.04mm / min.

Table 1
Composition of the Samples

NO	Composition(wt%)				
	Si_3N_4	Y_2O_3	La_2O_3	Al_2O_3	ZrO_2
I	76.5	17.25	2.75	3.5	0
II	76.5	17.25	2.75	3.5	1.5
III	76.5	17.25	2.5	3.5	5

RESULTS

<u>Composition of Crystallized Phases after Different time Heat Treatment Process</u>

The experimental results are shown in Table2,3,4, respectively. Before heat treatment it is found that the main crystal phase is β–Si_3N_4 in all compositions but there exists small amount of 5–1–1 in No II, III samples. After 2 hour heat treatment at 1350℃, there appears new crystallized phases(H,K phase) and trace of $LaYO_3$ phase in all compositions. H and $LaYO_3$ crystall phases are advantageous to the improvement of grain boundary. Especially due to its high melt point (~ 1750℃) the existence of $LaYO_3$ phase might increase the refractory of grain boundary and improve high temperature strength of Si_3N_4 ceramics. After longer heat treatment(30hx1350℃), there are some change in the kind of crystallized phase compared with short time heat treatment and obvious change of relative amount of all crystallized phases especially the increase of the amount of K phase. In No III samples there only appears K phase after longer time heat treatment. The experimental results show that the longer time heat treatment may prompt the crystallization of K phase.

Table 2
Phase Composition of as–Sintered Samples

NO	β'–Si_3N_4	5–1–1*	H	K	$LaYO_3$	Glassy
I	VS**	—	—	—	—	S
II	VS	S	—	—	—	W
III	VS	S	—	—	—	S

* 5–1–1 = $5Y_2O_3Si_3N_4Al_2O_3$ H = $YSiO_2N$ K = $Y_5N(SiO_4)_3$
** VS = Very Strong MS = Middle Strong S = Strong
VW = Very Weak MW = Middle Weak W = Weak

Table 3
Composition of Crystallized Phase after 2 Hour Heat Treatment

NO	β'–Si_3N_4	5–1–1	H	K	$LaYO_3$	Glassy
I	VS	—	MS	W	VW	W
II	VS	W	S	W	VW	W
III	VS	—	S	—	VW	W

Table 4
Compositon of Crystallized Phases after 30 Hour Heat tratment

NO	β'–Si_3N_4	5–1–1	H	K	$LaYO_3$	Glassy
I	VS	—	MS	S	VW	MW
II	VS	—	MS	S	VW	MW
III	VS	—	MS	S	VW	MW

Effect of Nucleating Agent on High Temperature Properties

Fig1 shows the flexural strength of the samples added different amount of ZrO_2. Before heat treatment the maxium bending strength was obtained on No II samples with 1.5wt% content of ZrO_2 and its high temperature flexural strength(at 1200℃) is 545MPa. After 2 hour heat treatment, the flexural strength of No I samples at 1200℃ increased obviously($\sigma_{1200℃}$ = 668Mpa) and there are small increase in strength of No II, III samples. After longer time heat treatment, $\sigma_{1200℃}$ Of No II samples is up to 720Mpa and the flexiual strength of No I, III samples decreased. These results show that the flexural strength is inflenced obviously by the content of ZrO_2 in samples.

The shrinkage and weight loss of all sapmles are shown in fig2. Fig2 shows the weight loss get larger with the increase of amount of ZrO_2. At the surface of as–sintered No Ⅱ, Ⅲ samples, ZrN phase is determined by XRD. The forming of ZrN may be due to the following chemical reaction:

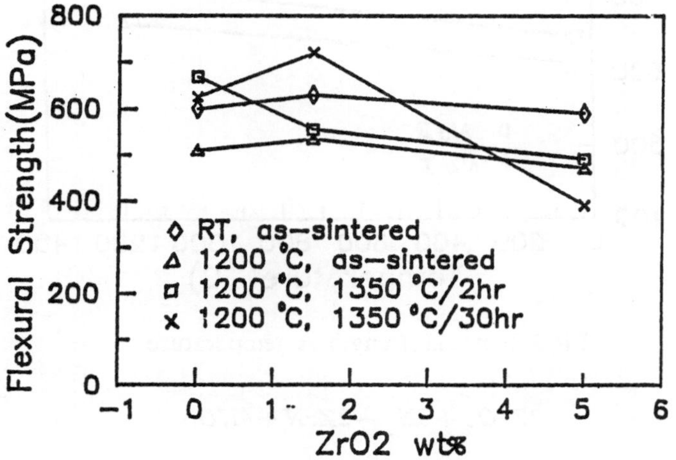

Fig 1 Effect of ZrO_2 on flexural strength

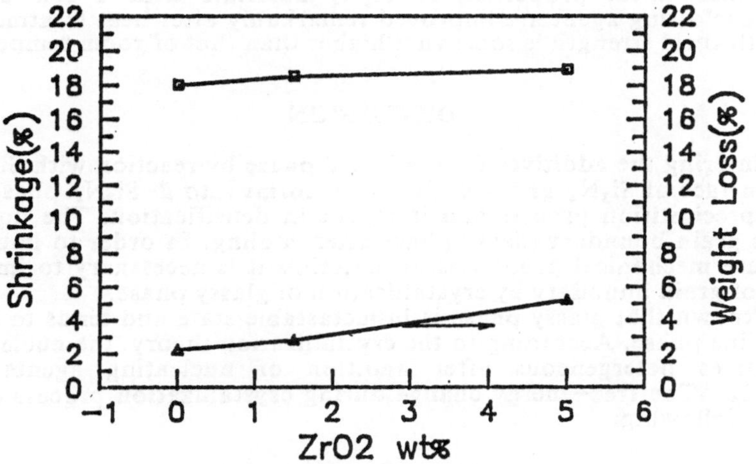

Fig 2 Shrinkage and weight loss curve of sintered samples

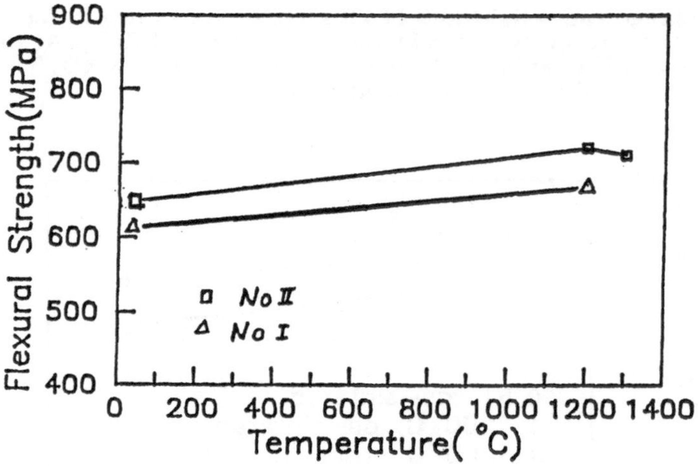

Fig 3 Flexural strength vs. temperature

$$2ZrO_2 + 3N_2 \rightarrow 2ZrN + 4NO$$

$$6ZrO_2 + 4Si_3N_4 \rightarrow 6ZrN + 12SiO + 5N_2$$

The change of flexural strength with temperature for No II samples after 30 hour heat treatment at 1350℃ is shown in fig3. The results show that high temperature mechanical properties of Si_3N_4 materials with a few amount of ZrO_2 as nucleating agent are improved remarkably after heat treatment and its 1300℃ ° flexural strength is somewhat higher than that of room temperature.

DISCUSSION

During sintering the additives form a liquid phase by reaction with SiO_2 on the particle surface of Si_3N_4 and $\alpha-Si_3N_4$ transforms into $\beta-Si_3N_4$ phase via a solution–reprecipitition process and it results in densification. The liquid phase resides as grain boundary glassy phase after cooling. In order to improve high temperature mechanical properties of materials it is necessnary to improve the property of grain boundary by crystallization of glassy phase.

It is known that glassy phase is in metastable state and tends to transform to crystalline phase. According to the crystallization theory, the nucleation process becomes hetergeneous after addition of nucleating agents such as ZrO_2, TiO_2. The free–energy change during crystallization process can be estimated as following:

$$\Delta G_k = S\gamma - V\Delta g_v$$

Where $\Delta g_v = \Delta H \Delta T / T_e$

S——Surface area of new phase
γ——Interfacial energy
V——Volume of new phase

Δg——Chemical energy per unit volume
T_e——Equilibium temperature

The heterogeneous nucleation rate may be writen:

$$I_s = K_s exp\left(\frac{-\Delta G'_k}{KT}\right)$$

Where $\Delta G'_k = \Delta G_k f(\theta)$

$\Delta G'_k$——Free–energy change of hetergeneous nucleation
$f(\theta)$——Function of contact angle

After a stable nuclus has been formed, the rate of growth is determined by the condition of temperature and the degree of supersaturation. Because of the addition of ZrO_2 nucleating agent, ZrO_2–new phase interface energy($\gamma_{Np-ZrO2}$) is less than new phase–liquid interface energy(γ_{Np-L}), and it results in reduction of the barriers to nucleation and decrease of $\Delta G'_k$. If the same nucleation rate is kept, the undercoolings ΔT could be decreased slightly and the peak of nucleation rate is closer to that of growth rate. It makes crystallization easier due to the increase of crystallized region. The results of Table 2 show that there begin crystallization in No Ⅱ, Ⅲ samples before heat treatment and there is no crystaliization in No Ⅰ samples without addition of nucleation agent.

In research of microcrystal glass Zdnieski [8] found out that the phaes transformation process in which ZrO_2 as nucleating agent follows the classical theroy of transfomation kinetics. In present study it is found that addition of ZrO_2 could shorten the breeding period of nucleation and increases the rate of nucleation. ZrO_2 crystal is not found in all samples before and after heat treatment. W.Brawe [1] discovered trace of ZrO_2 crystal phase only after 50 hour heat treatment at 1400℃ temperature. It implies that ZrO_2 might exist in grian boundary as colloid precipitate.

Crystallization occurs in as–sintered samples with addition of ZrO_2 nucleating agent. The crystallization of glassy phase results in increase of viscosity of resided glassy phase and high temperature strength of as–sintered Si_3N_4 materials is improved without heat treatment. After 2 hour heat treatment, the high temperature strength of materials is further enhanced due to increase of amount of crystallized phase. After 30 hour heat treatment, the high temperature strength of No Ⅱ sample is improved obviously. These results show that ZrO_2 nucleating agent could effectively prompt nucleation and crystal growth, and then improve the high temperature mechanical properties of Si_3N_4 matreials. It is consistent with the conclusion of W.Brawe's research works.

CONCLUSION

Crystallization of grain boundary glassy phase in sintered Si_3N_4 ceramics with ZrO_2 as nucleating agent was studied. It was found that ZrO_2 nucleating agent could shorten the period of nucleation and prompt crystallization of glassy phase. With addition of ZrO_2 the crystallized phase composition are different after various time heat treatment and especially the amount of K phase increased after longer time heat treatment. High temperature flexural strength of Si_3N_4 materials doped with suitable amount of ZrO_2 after 30 hour heat treatment at 1350℃ increases obviously and could keep up to 1300℃.

REFERENCE

1. W.Brawe, G.Wotting, and G.Zigler, Ceramic Material Components for Engines Proceedings of Second International Symposium, ed. W.Bunk, H.Hausner, Lubeck, FRG.,1986 pp 503
2. W.A.Sanders, Am. Ceram. Soc. Bull., 1985, 4, pp 304–09
3. A.Tsuge et al., J. Am. Ceram. Soc., 1975, 58, pp 323
4. K.H.Jack, J. Mater Sci., 1976, 11, pp 1135–1158
5. N.Hirosaki, A.Okads and M.Maoba, J. Am. Ceram. Soc., 1988, 71, C–144–147
6. A.B.Steven and K.R.Karssek, J. Mater. Sci., 1987, 6, pp 791
7. R.R.Wills, J. Am. Ceram. Soc., 1975, 58, pp 335
8. W.A.Zdnieski, J. Am. Ceram. Soc., 1978, 61, PP 799

THE INFLUENCE OF PRE-OXIDATION ON THE PROCESSING OF SILICON POWDER

R.G. STEPHEN AND F.L. RILEY*
Division of Ceramics, School of Materials,
University of Leeds, Leeds LS2 9JT, UK.

ABSTRACT

Silicon powder is readily oxidised during milling in aqueous media at pH > 5 to yield colloidal silicon dioxide which during drying deposits as a thin film on silicon particle surfaces. Interaction between the films causes strong agglomeration in compacted powders, leading to inhomogeneity in the silicon particle packing which is retained during the subsequent argon sintering and nitriding stages of silicon processing to reaction bonded silicon nitride.

INTRODUCTION

Silicon powder reacts with water during milling at a rate controlled by pH to produce amorphous silicon dioxide (SiO_2) and hydrogen [1-4].

$$Si_{(s)} + 2H_2O_{(l)} = SiO_{2(s)} + 2H_{2(g)} \quad (1)$$

The rate of oxidation in propan-2-ol (IPA) is too slow to be readily measured. During drying, the SiO_2 deposits as a film of ~ 50 nm nodules on silicon particle surfaces, and has a marked influence on subsequent powder compaction behaviour. Compacted powder strength can be high, but the formation of strong agglomerates can also lead to the development of microstructural inhomogeneity in the compacted powder which is ultimately undesirable for the mechanical properties of the derived reaction bonded silicon nitride [5]. We report here the results of studies of the argon sintering and nitriding behaviour of silicon powders containing different amounts of SiO_2 coating produced by milling in IPA, and water. The object of this programme was to assess the influence of the SiO_2 coating on microstructure at the successive stages of processing to reaction bonded silicon nitride.

*To whom all correspondence should be addressed.

EXPERIMENTAL

The silicon powder (KemaNord Sicomill IIC) had a median sedimentation particle size of 7.8 μm ("Sedigraph 5000ET", Micromeritics). 25 g batches of powder were gently milled for 7 or 14 ks using 450 g of stabilised zirconia media in a polypropene flask, by the vibratory action of a laboratory flask shaker, under 75 cm^3 IPA, or distilled water of pH between 5 and 10 obtained using aqueous hydrochloric acid or ammonium hydroxide. The amount of SiO_2 formed was calculated from the volume of hydrogen collected over water and on the basis of equation (1) [4,5]. Powders were dried at ~ 90 °C, and characterised using specific surface area (a_s), transmission electron microscopy (TEM) and Fourier transform infra-red spectroscopy (FTIR). Water-milled powders were passed through a 250 μm mesh sieve before compaction.

Discs of powder ~ 17 mm in diameter and ~ 3 mm in thickness were uniaxially pressed in a steel die, and then iso-pressed at ~ 225-250 MPa to a constant density of 1.4 Mg m^{-3}. Discs were argon sintered at 1200 °C, and nitrided at 1370 °C using flowing N_2/5% H_2, in a vertical tube furnace with continuous measurement of weight changes (Mk. 3 Microforce balance, CI Electronics, Salisbury, UK). Samples were lowered over 5 s into the furnace hot-zone at the required temperature and under the appropriate gaseous atmosphere. Fracture surfaces of broken discs were examined by scanning electron microscopy (SEM). Porous materials were impregnated with methyl methacrylate monomer and after curing polished to 1 μm for light microscopy (LM).

RESULTS

The specific surface area of milled powder increases as a function of hydrogen yield as shown in Figure 1 for powder milled under different conditions for 7 ks. TEM micrographs of silicon milled for 14 ks in IPA (silicon (IPA)) and water of pH 10 (silicon (PH10)) are shown in Figure 2(a) and (b). FTIR showed the characteristic SiO stretching absorption at 1120 cm^{-1}. The calculated SiO_2 yields for silicon powders (IPA), (PH5), (silicon milled in water of pH 5) and silicon (PH10) milled for 14 ks were 0, 1.1 and 16.0% by mass respectively. Mass losses occurred during sintering under argon, the loss being related to the extent of pre-oxidation at the milling stage (Figure 3). Compacted and argon sintered silicon (IPA) had a homogeneous microstructure (Figure 4a) and smooth fracture surface (Figure 4b). Silicon (PH5) was broadly similar in behaviour to silicon (IPA), although slight localised particle sintering and microstructural coarsening occurred (Figure 5a and 5b). Silicon (PH10) showed marked inhomogeneities of dimension ~ 100 μm (Figure 6a) and fracture faces were rough (Figure 6b).

Isothermal nitridation rates, as determined from mass gains, were strongly influenced by the SiO_2 content; nitridation was fastest for silicon (IPA), and slowest for silicon (PH10) in which the large volume SiO_2 initially inhibited reaction (Figure 7). No attempt was made to achieve complete nitridation, but the LM micrographs show the relative homogeneity of the microstructure of nitrided silicon (IPA), compared with the relative inhomogeneity of microstructure of silicon (PH10) (Figure 8a and 8b). Fracture surfaces of nitrided silicon (IPA) (Figure 9a) were very smooth in comparison to those of silicon (PH10) (Figure 9b).

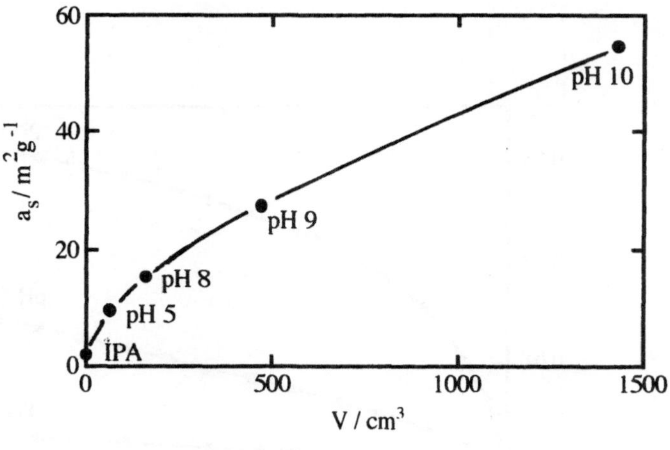

Figure 1. Specific surface area (a_s) as a function of hydrogen evolved (V) for silicons (IPA), (PH5), (PH8), (PH9) and (PH10) milled for 7 ks.

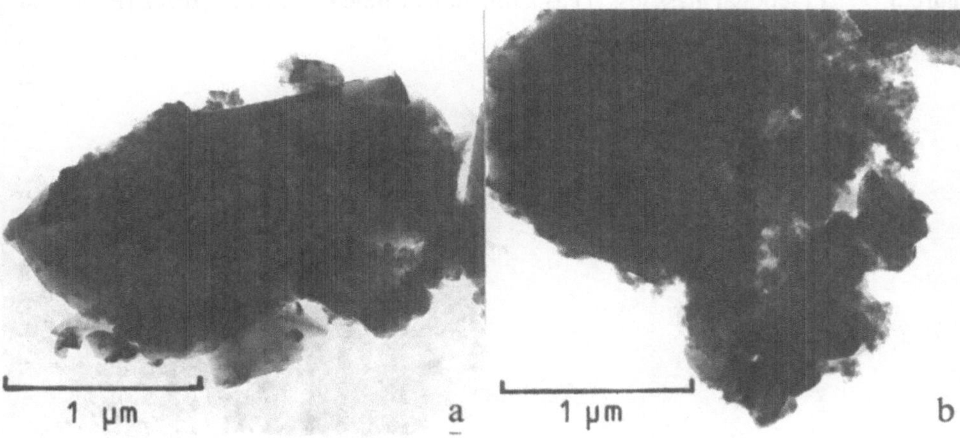

Figure 2. TEM micrograph of silicon particles milled for 14 ks in (a) IPA and (b) water of pH 10.

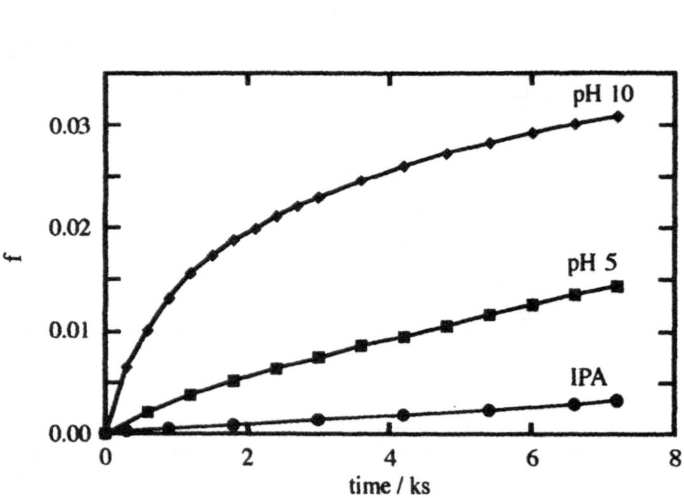

Figure 3. Fractional mass loss (f) as a function of time for silicons (IPA), (PH5) and (PH10).

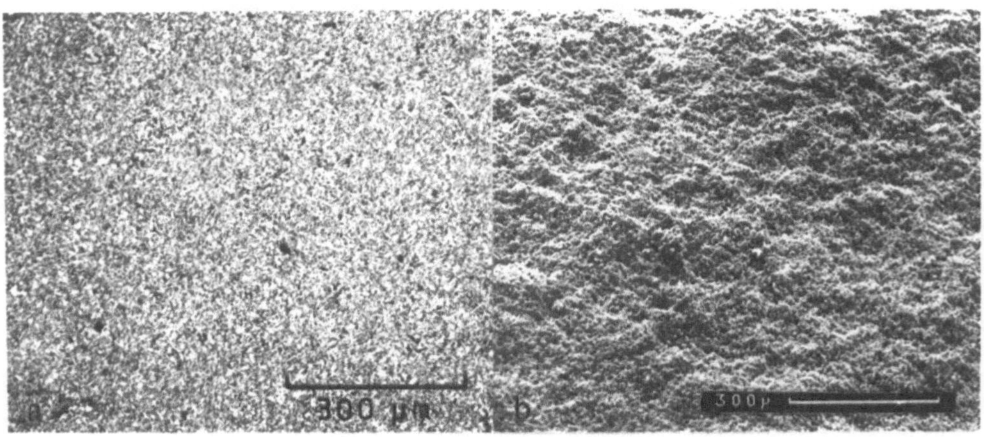

Figure 4. LM and SEM fracture face micrographs of silicon (IPA) argon sintered at 1200 °C for 7 ks.

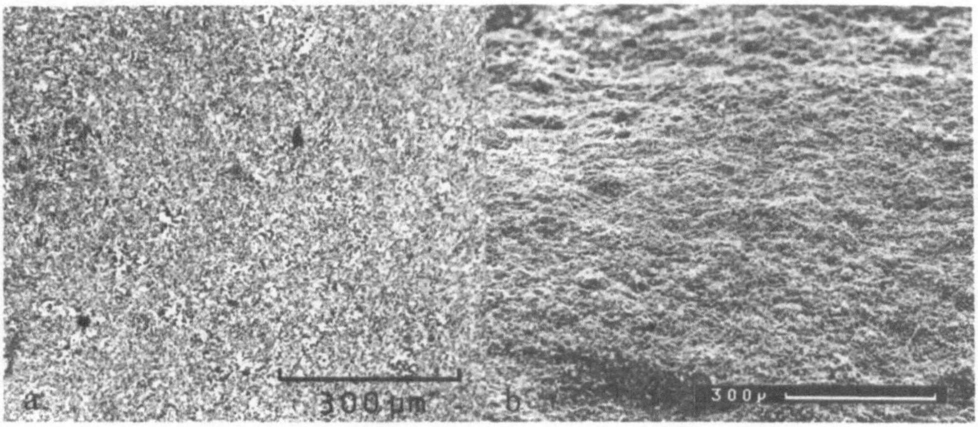

Figure 5. LM and SEM fracture face micrographs of silicon (PH5) argon sintered at 1200 °C for 7 ks.

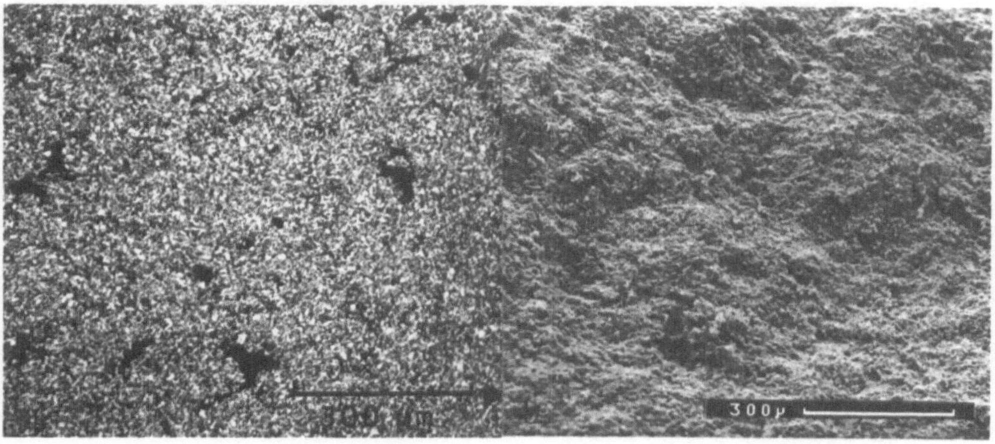

Figure 6. LM and SEM fracture face micrographs of silicon (PH10) argon sintered at 1200 °C for 7 ks.

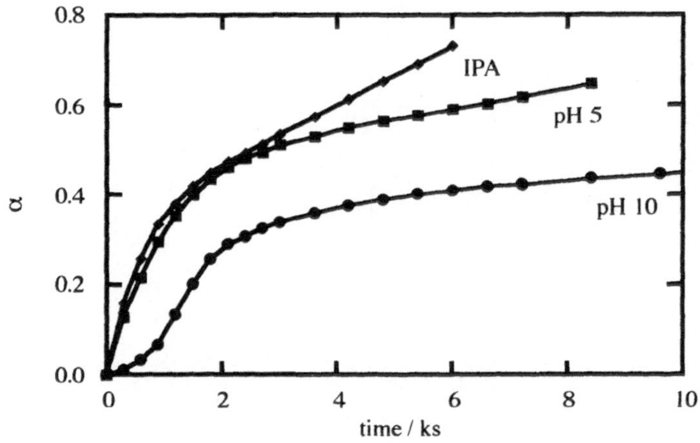

Figure 7. Fraction (α) of Si_3N_4 formed as a function of time for silicons (IPA), (PH5) and (PH10).

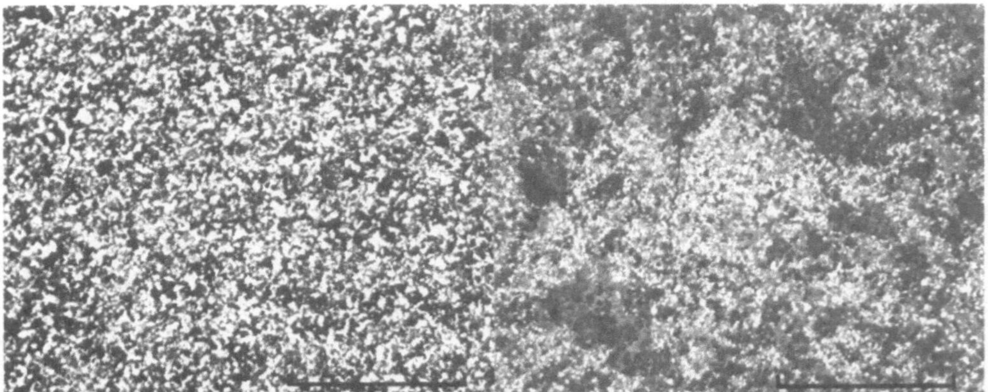

Figure 8. LM micrographs of polished sections of silicon nitrided at 1370 °C; (a) silicon (IPA), (b) silicon (PH10).

Figure 9. SEM micrographs of fracture faces of silicon nitrided at 1370 °C; (a) silicon (IPA), (b) silicon (PH10).

DISCUSSION

It is clear that the surface properties of silicon powders are determined by the wet milling treatments to an extent which is not the result of simple particle fracture. It is assumed that deposition of the colloidal SiO_2 formed by reaction (1) occurs during drying, with some agglomeration to yield the ~ 50 nm nodules seen by TEM. Interaction between these coated particles then gives high strength on compaction, and is likely to be responsible for microstructural inhomogeneity in compacted powder after argon sintering. Similar effects have been reported after the milling of alumina powders [6]. Some loss of SiO_2, presumably as SiO, occurs at the argon sintering stage;

$$SiO_{2(s)} + Si_{(s)} = 2SiO_{(g)} \qquad (2)$$

at low hydroxyl ion concentration there appears to be some associated influence on silicon particle sintering behaviour (Figure 5). In the case of silicon (PH10) the mass loss at 1200°C, and assumed to be predominantly that of SiO_2, is ~ 3% over 7 ks; ~ 80% (13/16) of the SiO_2 thus remains to enter the nitridation stage. The existence of agglomerates in silicon (PH10) is shown by the irregular fracture face of the argon sintered material, which has a strong resemblance to that of the compacted powder. In comparison sintered silicon (IPA) and silicon (PH5) have much more homogeneous microstructures and give smoother fracture faces.

The nitridation kinetics of argon sintered silicon (PH10) suggested that a significant amount of SiO_2 remained to inhibit nitridation [7,8], although it would be expected that complete nitridation would be achieved by a second, higher temperature, stage. The partially nitrided microstructure revealed in the fracture face shows that inhomogeneities present in the earlier processing stages can not be removed.

Extreme conditions were used deliberately to produce silicon powder containing large amounts of SiO_2 for this study, and as a consequence differences in the microstructure at all stages of powder compaction, argon sintering and nitridation were marked. It is clear, however, that the nature of the silicon particle surfaces, caused for example by milling, must be carefully controlled if compacted silicon powder of maximum homogeneity and thus giving reaction bonded silicon nitride of maximum homogeneity and minimum pore size, is to be obtained. The best powder in these respects was obtained by non-aqueous milling.

CONCLUSIONS

Colloidal SiO_2 forms readily on silicon particle surfaces as a result of wet processing; such coatings are detrimental to microstructural homogeneity. No homogenisation of microstructure occurs during the subsequent argon sintering and nitriding stages. Good homogeneity at the silicon powder compaction stage is achieved by preventing surface oxidation through control of pH. Compacted silicon powder of good homogeneity and small maximum void size has been obtained after milling in IPA.

ACKNOWLEDGMENTS

This work was supported by an SERC CASE Studentship in association with T & N Technology Ltd., Rugby, UK. The assistance of A.P. Bromley with transmission electron microscopy is acknowledged.

REFERENCES

1. Kolbanev, I.V. and Butyagin, P.Yu., "Mechano-chemical reaction of silicon with water", Kinet. Katal., 1982, 23 (2), 327-33.
2. Montenyohl, V.I. and Olson, C.M., "Silica from finely divided silicon", U.S. Patent No. 2 614 993, 1952.
3. Tachivana, A., Koizumi, M., Teramane, H. and Yamabe, T., "Tunnelling reaction path for the interaction of silicon atoms and water", J. Am. Chem. Soc., 1987, 109
4. Stephen, R.G. and Riley, F.L., "Oxidation of silicon in water", J. Eur. Ceram. Soc., 1989, 5, 219-22.
5. Stephen, R.G. and Riley, F.L., "The compaction behaviour of oxidised silicon powder", to be published.
6. Niesz, D.E. and Bennett, R.B., "Structure and Properties of Aggomerates", in Ceramic Processing Before Firing, ed. G.Y. Onoda and L.L. Hench, John Wiley, New York, 1978, pp 61-73.
7. Campos-Loriz, D, and Riley, F.L., "Effect of silica on the nitridation of silicon", J. Mat. Sci. Lett., 1976, 11, 195-198.
8. Dervisbegovic, H. and Riley, F.L., "The role of H_2 in the nitridation of silicon powder compacts", J. Mat. Sci., 1978, 13, 1945-55.

AN OXIDATION RESISTANT Al_2O_3-TiO_2 SINTERED SILICON NITRIDE

D.N. OVREBOE[+], S.K. BISWAS, F. CASTRO[*] AND F.L. RILEY[1]
Division of Ceramics, School of Materials,
University of Leeds, Leeds, LS2 9JT, U.K.
[+]Norsk Hydro, Porsgrunn, Norway.
[*]CEIT, 20009, San Sebastian, Spain.

ABSTRACT

Silicon nitride has been hot-pressed and pressureless sintered using densification aids based on the Al_2O_3-SiO_2-TiO_2 system. Hot-pressing can be achieved satisfactorily at 1700 °C with an Al_2O_3-TiO_2 ratio of 1:1; sintering at 1850 °C requires liquid phase compositions richer in Al_2O_3 and giving a transient ternary liquid at temperatures in the region of 1500 °C. Dense materials prepared in this system show good resistance to oxidation at temperatures up to 1500 °C.

INTRODUCTION

In order to sinter silicon nitride powder to high density, oxide additives forming liquids at high temperature must be used [1]. The additive reacts with the natural silicon dioxide film on the silicon nitride particles to form silicates (or aluminosilicates when aluminium oxide is present) which, on cooling, generate the silicon nitride intergranular glass, or series of crystalline phases. In many cases the subsequent high temperature oxidation rate of the multi-phase silicon nitride is strongly influenced by the intergranular phase [2-4]. This is because the ability of silicon nitride to withstand oxidation depends on the formation of a protective silicon dioxide film of low permeability towards oxygen [5].

$$Si_3N_{4(c)} + 3O_{2(g)} = 3SiO_{2(c)} + 2N_{2(g)} \qquad (1)$$

---

[1] To whom all correspondence should be addressed.

Silicates tend to be less effective in this respect. If the intergranular phase metal oxide is not in thermodynamic equilibrium with SiO_2 an outwards flow of cations (together with charge counterbalancing anions) down a chemical potential gradient from the intergranular phase to the surface oxide film can occur, with subsequent reaction with the silicon dioxide to give the related silicate or aluminosilicate. Kinetic factors may also be important: intergranular Mg^{2+} is a mobile cation; Y^{3+} is less mobile and silicon nitride materials densified with yttrium oxide generally appear to have better oxidation resistance than those based on magnesium oxide provided the formation of oxidisable intergranular quaternary silicon yttrium nitride oxides is avoided by careful control of the Y_2O_3/SiO_2 ratio [6,7].

The ideal densification aid from the point of view of oxidation behaviour would be based on metal oxides thermodynamically stable in contact with silicon dioxide at the temperature of application of the component, and thus showing no tendency to form silicates. This would minimise the outwards flow of metal oxide from the intergranular phase with consequent loss of protection provided by the silicon dioxide surface film. Ti(IV) oxide (TiO_2) largely meets this criterion in forming no binary compound with SiO_2, or ternary compound in the Al_2O_3-SiO_2-TiO_2 system) [8,9], although at high temperature there is a region of limited solid solution. In the SiO_2-TiO_2 binary system the lowest eutectic is at 1540 °C, but because this liquid contains ~90% SiO_2 its viscosity appears to be too high for it to be an effective densifying liquid for silicon nitride. In the Al_2O_3-SiO_2-TiO_2 system the SiO_2-rich ternary eutectic is at 1490 °C, and the liquidus surface occurs at temperatures below 1700 °C away from SiO_2 in the mullite-TiO_2 solid solubility region [9]. The Al_2O_3-TiO_2 combination has been examined previously as a hot-pressing densification aid for silicon nitride [10]; it was selected for study in this programme as one potentially providing also oxidation resistant materials.

EXPERIMENTAL

Silicon nitride materials were made by sintering, and by hot-pressing, Starck LC12N silicon nitride powder (specific surface area 23 $m^2\,g^{-1}$; oxygen content 1.6% by weight, or the equivalent of 2.9% SiO_2). Initially aluminium oxide was added as 0.4 μm powder (Baco BAX 36 R grade) or aluminium nitrate; titanium dioxide was added as 0.5 μm rutile (Tioxide NP89/228) or titanium isobutoxide. Components were mixed by extended milling with dense silicon nitride media under propan-2-ol (IPA), followed by removal of the IPA, and calcination at 500 °C for 2h to eliminate bonded water, anions, and organic groups. The oxides were also incorporated, with markedly improved microstructural homogeneity, as surface coatings by controlled heterogeneous precipitation from an IPA solution of aluminium isopropoxide and titanium isobutoxide, by the slow addition at 25 °C of a solution of water in IPA, followed at completion of hydrolysis by drying and calcination at 500 °C.

Hot-pressing was carried out at 1700 °C and 20 MPa in boron nitride powder coated graphite dies. Alternatively discs of powder were iso-pressed at 200 MPa to a green density of ~52%, and sintered in a powder bed under 3.5 bar nitrogen at 1850 °C.

Dense materials were surface ground and cut into 5 mm cubes for oxidation supported by a Pt cradle in still air at 1400 - 1500 °C. Surfaces of polished materials were examined by light microscopy (LM), scanning electron microscopy (SEM), and X-ray diffraction (XRD); small flakes of material were polished and then ion-beam thinned for transmission electron microscope (TEM) examination with an X-ray micro-analysis (XMA) attachment.

RESULTS

Typical $Al_20_3/Ti0_2$ coated silicon nitride particles after calcination at 500 °C are shown in Figure 1.

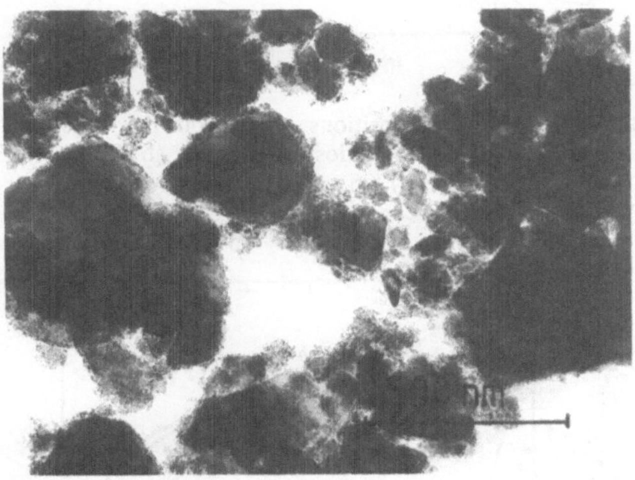

Figure 1. TEM micrographs of 10 weight % $Al_20_3/Ti0_2$ (9:1 ratio) coated silicon nitride powder, after calcination at 500 °C.

At a constant 10% total weight the proportions of $Al_20_3/Ti0_2$ added to the silicon nitride were varied within the range 0:10 to 10:0. Additive starting compositions for the $Al_20_3/Ti0_2$ ratio of 1:9, 2:8, 3:7, 4:6, 5:5 are shown on the Al_20_3-$Si0_2$-$Ti0_2$ ternary diagram in Figure 2; the presence of 2.9% $Si0_2$ in the silicon nitride powder was assumed. Hot-pressing densification data are expressed in terms of instantaneous densification rate as a function of density in Figure 3. The effectiveness as densification aids of different compositions at increasing hot-pressed density is shown in Figure 4, in which densification rate at a specified density is plotted as a function of composition expressed in terms of the $Al_20_3/(Al_20_3 + Ti0_2)$ ratio. Final densities of 3.06 Mg m^{-3} or better were obtained for the compositional range $Al_20_3/Ti0_2$ 4:6 to 10:0, with the best value for composition 9:1. This composition was chosen for further pressureless sintering tests after it was shown that the 5:5 composition, although adequate for hot-pressing, was

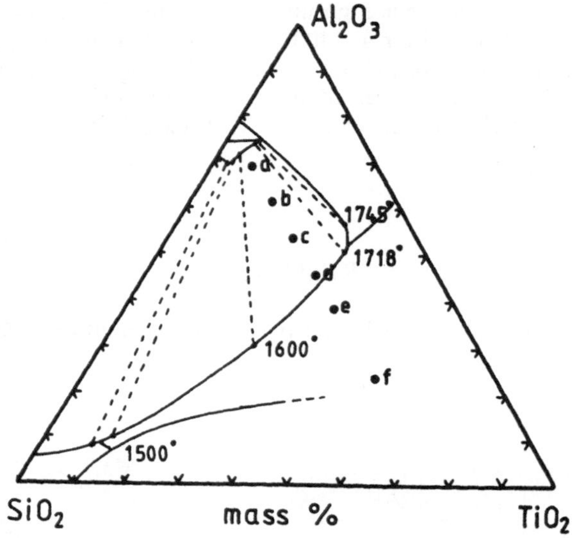

Figure 2: Additive starting compositions shown on the Al_2O_3-SiO_2-TiO_2 ternary diagram: The Al_2O_3/TiO_2 ratios are (a) 9:1, (b) 8:2, (c) 7:3, (d) 6:4, (e) 5:5, (f) 3:7.

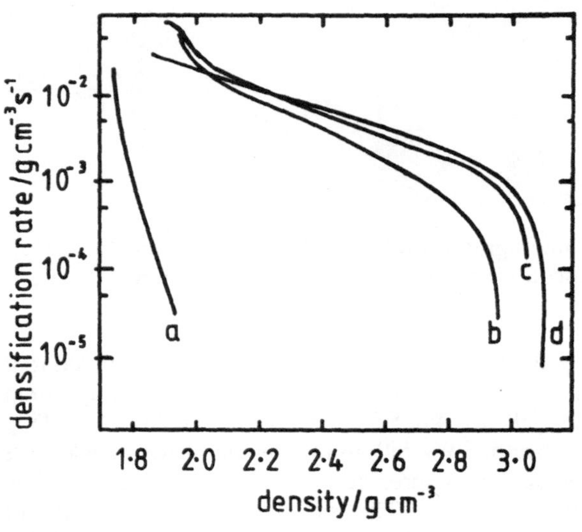

Figure 3. Hot-pressing instantaneous densification rates as a function of density, at 1700 °C and 20 MPa: The Al_2O_3/TiO_2 ratios are (a) 0:10, (b) 3:7, (c) 5:5, (d) 9:1.

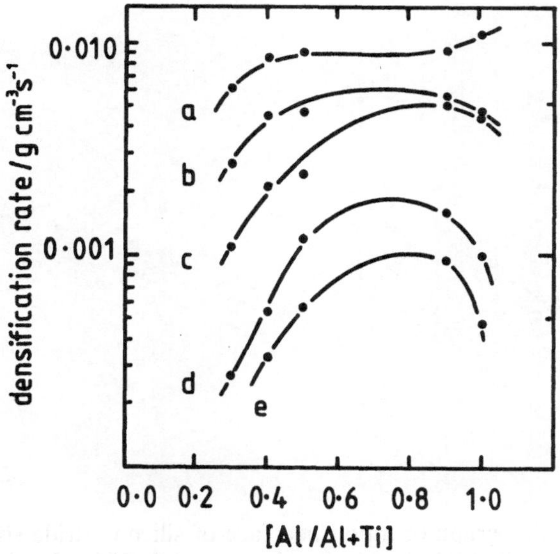

Figure 4. Hot-pressing instantaneous densification rate, at 1700 °C and 20 MPa, for specified densities and additive compositions: (a) 2.3 g cm^{-3}, (b) 2.5 g cm^{-3}, (c) 2.7 g cm^{-3}, (d) 2.9 g cm^{-3}, (e) 3.0 g cm^{-3}.

unsatisfactory as a sintering additive in giving too slow a densification rate. Densities of 98% of theoretical (which was determined using crushed powder by a standard pycnometer method) were obtained by sintering at 1850 °C with a 2h hold at temperature. There was no significant improvement in densification behaviour when the amount of additive was increased to 15% at an Al_2O_3/TiO_2 ratio of 9:1. The low residual void volume and homogeneity of microstructure obtained with mixing the sintering additive using a solution phase shown by the LM micrograph of Figure 5. Microstructural finer details are shown in the TEM micrograph of Figure 6. XMA analysis of a number of clearly defined grains at the edge of the ion-beam thinned section indicated the approximate mean atomic proportions: Si 85%, Al 11%, Ti 4%. XRD examination showed ß-Si_3N_4 to be the major phase, with silicon oxynitride (Si_2N_2O) and titanium nitride (TiN) as minor constituents. The α-Si_3N_4 to ß-Si_3N_4 conversion was therefore complete. Oxidation at 1400 °C gave a mass gain of 1.4 g m^{-2} after 30h; at 1500 ° the mass gain after 20h was 2.5 g m^{-2}. After longer times at 1500 °C (to 44h) small mass losses were measured (~5 g m^{-2}) as shown schematically in Figure 7. Repeated Vickers micro-indentation measurements on polished surfaces gave mean values for K_{1c} of 5.3 MPa $m^{3/2}$, and for H_v of 15.6 GPa.

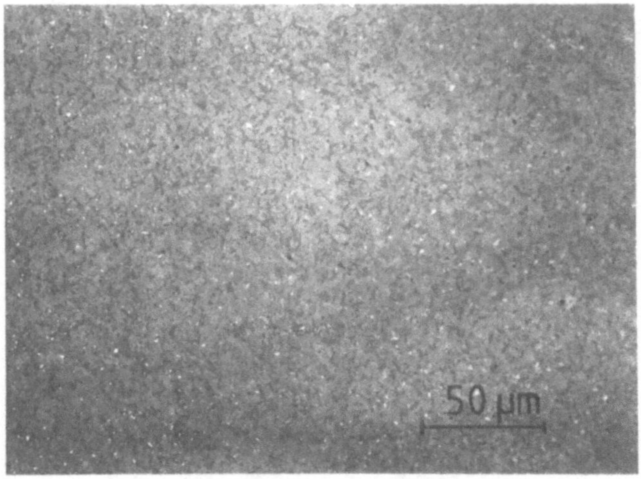

Figure 5. LM micrograph of a polished face of silicon nitride sintered at 1850 °C for 2h with 10% chemically blended Al_2O_3/TiO_2 of ratio 9:1.

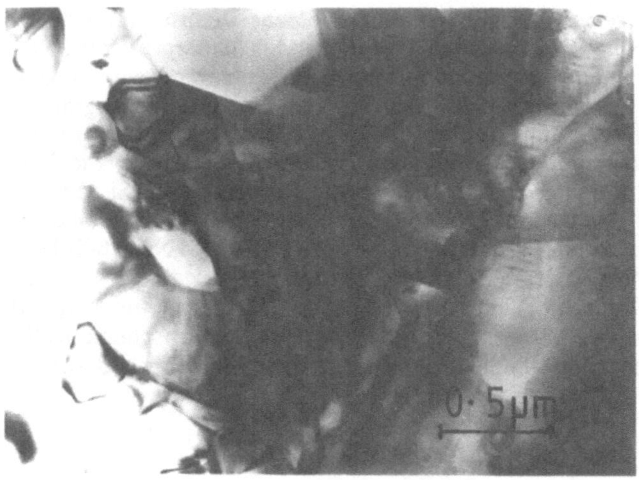

Figure 6. TEM micrograph of silicon nitride sintered at 1850 °C for 2h, with 10% Al_2O_3/TiO_2 of ratio 9:1.

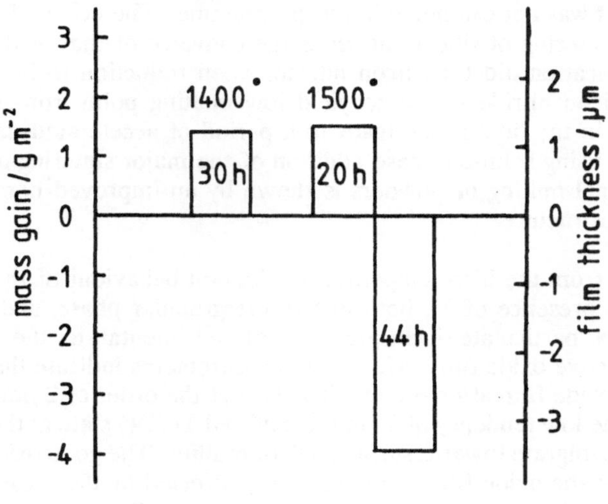

Figure 7. Oxidation behaviour of sintered silicon nitride, in air at 1400 °C, and 1500 °C.

DISCUSSION

The Al_2O_3-SiO_2-TiO_2 system at the 10 weight % addition level permits satisfactory sintering of sub-micrometre silicon nitride powder. While the Al_2O_3-SiO_2-TiO_2 phase diagram provides a guide to the liquid forming behaviour of this system, it cannot be applied too rigorously because of the reduction of TiO_2 to Ti_3O_5 and TiO, under the low oxygen potential conditions of the sintering environment, and reaction with Si_3N_4 to form TiN and Si_2N_2O [11]

On the basis of experience concerning the carbothermal reduction of TiO_2 to TiN, the lifetime of TiO_2 would be expected to be short at sintering temperatures [12]. Simultaneously the Al_2O_3 is largely incorporated within the ß-silicon nitride structure, forming a ß'-sialon of z-value ~0.7, on the basis of the XMA analysis of grains imaged in the TEM, and which is to be compared with a value of ~0.5 calculated on the assumption of complete reaction of the 9% Al_2O_3 with the silicon nitride. It is of interest that a significant quantity of (~4 atom % of metallic elements) of Ti appears also be absorbed into the ß'-sialon structure; such sialon solid solution phases have not previously been reported, though seen in separate studies on the use of TiO_2 as a densification aid in ß'-sialon systems [13]. The presence of aluminum titanate as an intergranular phase in the sintered material was looked for, but not detected by XRD. The Al_2O_3-SiO_2-TiO_2 ternary liquid generated is thus a transient liquid, and it would be expected that the high temperature mechanical properties of the material would be good, though this aspect was not examined in this programme. The effect of the reduction of TiO_2 during the sintering of silicon nitride is the converse of that of the similar use of Fe_2O_3 as a densification aid for silicon nitride, when reduction to Fe and subsequent reaction with silicon nitride occurs to yield low melting point iron silicides. Silicon nitride solubility in the liquid then leads to a period of accelerated densification [14]. An advantage of using solution phase addition of the major sintering additives instead of the conventional milling of powders is shown by an improved homogeneity of the final dense microstructure.

It is clear from the high temperature oxidation behaviour of the dense sintered material that the presence of Ti, both in the intergranular phase, and in the ß'-sialon grains or as TiN particulate inclusions, is not detrimental to the stability of the developing protective oxide film. Mass gain measurements indicate that the maximum thickness of the oxide film after 44h at 1500 °C is of the order of 2 μm. This would be consistent with the low tendency of Ti in the oxidised Ti (IV) state at the silicon nitride-oxide interface to migrate towards surface SiO_2 or mullite. The good oxidation resistance indicates also that the oxide film is not adversely affected by the presence of ~10% of mullite. TiO_2 and Al_2TiO_5 are both stable in contact with SiO_2 at temperatures below 1490 °C, and at 1400 °C no interaction with the surface film would therefore be expected. At 1500 °C a ternary liquid should start to appear in the surface oxide, but on the basis of the extent of oxidation observed it is still a good barrier of low permeability towards oxygen. Small mass losses seen at 1500 °C for longer oxidtion times are believed to be primarily the result of slight sticking of the now oxidised melting phase to the platinum cradle and loss of material on separation

CONCLUSIONS

Al_2O_3-TiO_2 is an effective sintering additive for silicon nitride, and gives materials of good resistance to oxidation at temperatures up to 1500 °C. This system provides a transient liquid in that the Al_2O_3 is largely incorporated with the ß'-sialon grains, and the TiO_2 is reduced and nitrided to the stable TiN. Ti also appears able to substitute to a small extent for Al in the ß'-sialon crystal lattice.

ACKNOWLEDGMENTS

This work was part-funded by the Commission of the European Communities through Award MA1E-0037-C(A). Financial support was also provided by the British Council.

References

1. Jack, K.H., The significance of structure and phase equilibria in the development of silicon nitride and sialon ceramics. In Science of Ceramics, ed. R. Carlsson and S. Karlsson, Swedish Ceramic Society, 1981, 11, 125-42.

2. Singhal, S.C., Oxidation of silicon nitride and related materials. In Nitrogen Ceramics, ed. F.L. Riley, Noordhoff, Leiden, The Netherlands, 1977, 607-26.

3. Billy, M., The kinetics of gas-solid reactions and environmental degradation of nitrogen ceramics. In Progress in Nitrogen Ceramics, ed. F.L. Riley, Martinus Nijhoff, The Hague, The Netherlands, 1983, 403-419.

4. Billy, M. and Desmaison, J.G., High temperature oxidation of silicon-based ceramics, High Temperature Tech., 4, (3) 1986, 131-139.

5. Lamkin, M.A., and Riley, F.L., Oxygen diffusion in silicon dioxide and silicates: A Review, to be published.

6. Andrews, P. and Riley, F.L., The microstructure and composition of oxide films formed during high temperature oxidation of a sintered silicon nitride, J. Europ. Ceram. Soc., 5, 1989, 245-256.

7. Lange, F.F., Singhal, S.C. and Kuznicki, R.C., Phase relations and stability studies in the Si_3N_4-$Si0_2$-Y_20_3 pseudo binary system, J. Amer. Ceram. Soc., 60 [5-6] 1977, 249-52.

8. DeVries, R.C., Roy, R. and E.F. Osborn, The System $Ti0_2$-$Si0_2$, Trans. Brit. Ceram. Soc., 53 [8] 1954, 524-540.

9. Agamawi, Y.N. and White, J., The system Al_20_3-$Si0_2$-$Ti0_2$, Trans. Brit. Ceram. Soc., 51 (1951-52, 293-325.

10. Kishi, K. and Umebayashi, S., Some mechanical properties of hot-pressed ß'-sialon-TiN sintered body prepared from α-Si_3N_4, aluminium iso-propoxide and titanium iso-propoxide, J. Ceram. Soc. Japan, Inter Ed, 96, 1988, 710-714.

11. Trigg, M.B. and McCartney, Reactions in the system silicon-titanium-oxygen-nitrogen, J. Amer. Ceram. Soc., 63 [1-2] 1980, 103-4.

12. Li, W.Y. and Riley, F.L., J. Eur. Ceram. Soc., to be published.

13. Patel, M., Riley, F.L., Castro, F., LeDoussal, H. and Cales, B., ß'-sialon from a low-cost siliceous waste material, to be published.

14. Stalios, A.D., Luyten, J., Hemsley, C.D., Riley, F.L. and Fordham, R.J., The interaction of iron during the hot-pressing of silicon nitride, J. Eur. Ceram. Soc., 7 [2], 1991, 75-81.

A BRAZED CERAMIC-TO-METAL JOINT IN A CAR ENGINE TAPPET

I. A. BUCKLOW, S B DUNKERTON, The Welding Institute, Abington, UK; W. G. HALL, Johnson Matthey Technology Centre, Sonning Common, UK; B. CHARDON, Peugeot, Velizy, France

ABSTRACT

The achievements of a EURAM programme to make a brazed ceramic faced metal tappet are outlined. A patented corrugated washer for accommodating differential contraction strains is described, and progress is reported on a production technology for rapidly-solidified tape of fine microstructure.

1. INTRODUCTION

The programme described in this paper was chiefly performed at The Welding Institute (TWI), and Johnson Matthey Technology Centre (JMT), UK, in conjunction with Peugeot (PSA) France. The work was part-funded by the UK and French governments under the EURAM initiative of the CEC.

The specific objectives of the programme were to join a ceramic disc to the head of a steel tappet body and to develop a technology for the production of braze-foils. The ceramic-faced tappet was part of a programme to improve the overall performance of an automotive internal combustion engine. Performance improvement would be effected by: firstly, reducing the wear of both the tappet head and a contacting rotating cam, secondly, by reducing the friction between the two, and thirdly by reducing the overall weight of the tappet. The braze-foil production technology was part of a wider programme to develop multilayer systems designed to assist in the manufacture of a wide range of ceramic-to-metal joints. But in addition to these specific aims, a larger benefit would accrue to the partners as a result of their appreciation of the way in which ceramics could be more widely used in internal combustion engines, and of the technical and production problems involved in their use.

The overall design of the tappet was such that it was not possible to employ mechanical fixing to hold the ceramic onto the steel. Furthermore, the component was required to operate in oil between temperature extremes of -20°C and +150°C, thus it was necessary to use brazing to join the two materials. The principal tasks to be accomplished were, therefore: to devise a ceramic-to-metal brazing technique (TWI); to develop and produce suitable braze-filler metals in a convenient form (JMT); and to run the tappets in a test engine (PSA).

The paper will concentrate on the first and second of these three tasks.

2. CERAMIC-TO-METAL JOINT

2.1 The Problems

The two major problems that were present at the outset of the work were believed to be representative of those involving most metal/ceramic joints introduced as substitutes for all-metal components in existing designs.

The first problem was that of the difference in thermal expansion (a factor of 4 in the present case) between the steel tappet body and the Syalon 101 disc which constituted the ceramic head. A suitable alloy for brazing ceramics for the expected service conditions would have a solidus temperature above 600°C, and so the thermal contraction mis-match between the two materials, as they cooled from the brazing temperature, would impose considerable strains on the joint. The conventional way of overcoming this problem of contraction mis-match is to interpose a compliant layer between the two materials such that the contraction of the layer is intermediate between those of the ceramic and the metal, or to use a layer of suitably graded composition. In either case the solid compliant layer usually requires to be of a depth that is a significant fraction of its diameter.

In the particular case of this tappet, however, the component had to fit into an existing design and the resultant constraints would not allow for the necessary length of solid compliant layer to be inserted. The ceramic disc was 23mm in diameter, Fig 1, and both the tappet length and total weight were fixed. In addition, substantial compressive and shear forces were exerted on the ceramic by the rotating cam.

The problem of the compliant layer had, therefore, to be solved primarily by design rather than by material. This was confirmed by finite element analysis which showed that a solid intermediate layer of allowable thickness would place excessive stresses on the component, Fig 2. The resultant doming distortion of the ceramic disk gave a simple means of assessing those stresses.

2.2 The Solution

2.2.1 <u>The compliant layer:</u> The compliant layer had to have a relatively low radial stiffness to allow thermal contraction and subsequent cycling to take place without imposing too great a stress on the ceramic, but it was also required to withstand the shear forces imposed in service. Furthermore, the layer had to be sufficiently strong to withstand the vertical service stresses.

The first attempt to fulfil these conditions employed a metal honeycomb with the cells aligned parallel to the length of the tappet, but the brazed contact area between the ends of the cell walls and the flat surface on either side was too small to give the required shear strength.

Various other designs were explored but the one which was finally adopted was a circularly corrugated washer made by pressing or by electroforming. Electroforming has the advantage of being versatile and relatively cheap for prototypes because matching dies are not required. Pressing, on the other hand, is ideal for mass production once the pattern has been optimised and the dies made to match.

An early design of washer, Fig 3, shows the essential features of a basically very simple device, and it can be seen that many variations on this theme are possible. This doubly-rigid washer was made by electroforming and was deposited onto an oxidised stainless steel mandrel.

Mock tappets made for shear testing, Fig 4, consisted of a simple steel thimble with the corrugated washer sunk into a recess and two discs of 50μm thick Ag/Cu/Ti braze alloy tape to braze the washer to the steel on one side, and to the ceramic on the other. The tape was made by JMT by a rapid-solidification process which resulted in a very fine micro-structre within the tape. The specimens were tested to destruction, or to the point where no further deformation could take place, and were subsequently sectioned, Fig 5; it can be seen that failure occurred by tearing of the corrugation walls. The shear load applied to initiate failure was 190kg.

In other cases, failure began by a tearing of the corrugations followed by failure in the ceramic where the braze/ceramic interface remained intact but material was torn out of the ceramic.

2.2.2 Design of the washer: There are many factors which affect the radial stiffness and vertical strength of the washer. For example, the material itself and its thickness, the number and spacing of corrugations, the height and angle of the corrugation walls, and the area of the flats on the top and the bottom. All these must be considered in relation to the dimensions of the main body of the tappet. Finite element analysis was used to model a number of these variations, Fig 6, and to highlight the areas where maximum distortion would occur as the component cooled from an assumed solidus of 600°C to room temperature. The full black line again shows the outline of the original position of the various elements, plus the location of the area of maximum distortion in this particular arrangement. The analysis also predicted ceramic doming figures in the range of 1 - 2 microns for the best washer designs, and these figures were borne out in practice.

The eventual design was a compromise which employed pure iron as the material to form a single circular corrugation. The washer was recessed into the tappet head so that the resulting shoulders gave some support to the ceramic against the vertical forces encountered in service.

A trial batch of 80 tappets has been made for engine tests which are currently being run at a PSA plant near Paris.

The principles behind the corrugated washer illustrated in this paper can be applied to many different configurations of joints, and a patent to cover these ideas, plus certain uses of metallic honeycombs, has been applied for.

3. BRAZE-FOIL PRODUCTION TECHNOLOGY

3.1 The Brazing Alloy

Many ceramics are successfully wet by "active" brazes containing titanium or zirconium. One formulation which appeared to be appropriate to the present work is based on the Ag/Cu eutectic plus some 1.5% Ti and 12% In with a solidus at about 600°C and a liquidus at about 715°C. It is commercially available as a foil produced by conventional casting and rolling, but this route results in the formation of large and complex Ti intermetallics which can sometimes occupy the full thickness of the tape. Much of the ceramic-wetting Ti is tied up in these intermetallics which must first be dissolved before Ti becomes available for reaction, hence large intermetallic particles slow down the brazing reaction times and may influence wetting. An alternative foil manufacturing process - tape casting - should alleviate, or possibly eliminate, the problem, and this was the technique chosen by JMT to develop as a braze-tape production process.

3.2 Tape Casting

3.2.1 The process: Tape casting, Fig 7, is one form of the rapid solidification processes which, in effect, squirt liquid metal onto a rapidly moving chilled surface to leave a thin, non-adherent layer on that surface. If the layer is very thin and the quench-rates are sufficiently high (approximately 10^6 °C/sec) virtually amorphous alloys result, but braze tapes are usually required to be between 25 and 100µm thick and so the concomitant quench rates are too low to achieve completely amorphous structures. The microstructures are, however, generally very much finer than those from conventional casting and rolling production.

3.2.2 The equipment: The first small-scale experiments were carried out with 200g of the liquid alloy held in a quartz crucible, the bottom of which formed a nozzle which was sealed during melting by an external plug. The gap between the nozzle and the circumference of the water-cooled casting wheel was fixed. A protective atmosphere of flowing argon containing less than 100ppm of oxygen was maintained around the nozzle area. Clean and ductile tape was produced but dimensional control was difficult because molten metal wetted the bottom of the quartz nozzle and gave an uncontrolled spread of braze-alloy at the casting point. Various designs of casting nozzle were explored but the problem was eliminated only when a coating of boron nitride (which was not wet by the braze alloy) was sprayed onto the nozzle.

The process was then scaled up in two stages to cast a charge of 4kg through a graphite nozzle - again coated with boron nitride. Graphite had the advantage of being easily machined to high tolerances, and so complex nozzle shapes could be produced more simply than in quartz. During a casting run, both the crucible and the rotating wheel expanded, thus changing the gap between the two and so influencing tape thickness. An automatic gap control system was therefore installed, which, via a proximity sensor, maintained the gap at any desired setting. The external stopper of the original equipment was replaced by an internal plug which contained the thermocouple used to monitor and control the melt temperature.

3.2.3 The product: Braze tapes some 25mm wide by 40µm thick are produced, although thinner tapes can be made; experience shows that tape thickness is relatively insensitive to the level of liquid metal in the charge pot. The tape is sufficiently ductile to be rolled if required, and the microstructure as-cast, Fig 8, shows that the Ti intermetallic particles are very fine and evenly dispersed. The fine microstructure is retained on heating virtually up to the solidus, and so tape can be annealed after stamping or cutting to give a ductile braze pre-form without springback and without losing its rapid reaction times.

The work has now reached the stage where significant progress has been made towards the development of the manufacturing technology for the production of active brazing alloy strip by tape casting. The product has been shown to possess properties superior to those of conventionally processed materials which are currently commercially available.

4. CONCLUSIONS

The combination of the rapidly-solidified tape and the corrugated washer insert holds considerable promise as a versatile system for joining ceramics to metal without incurring dimensional penalties in the component.

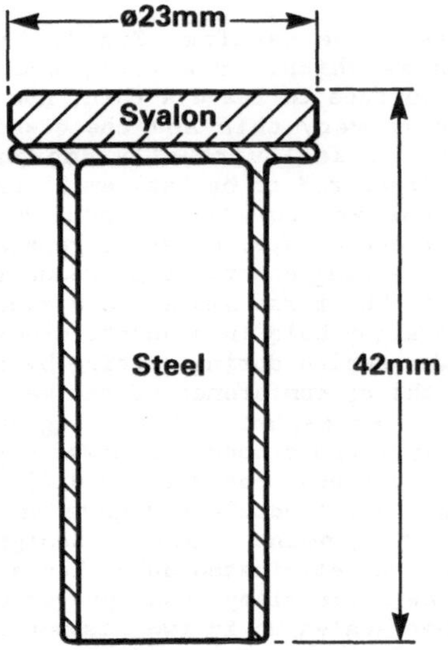

Figure 1. The ceramic-faced tappet

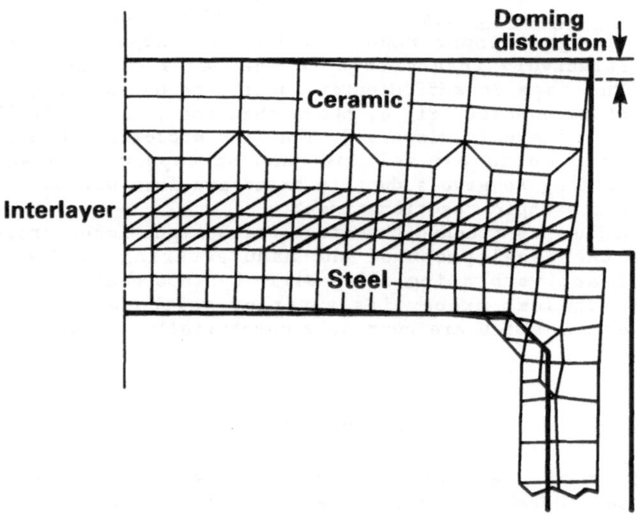

Figure 2. Distortion with a solid intermediate layer

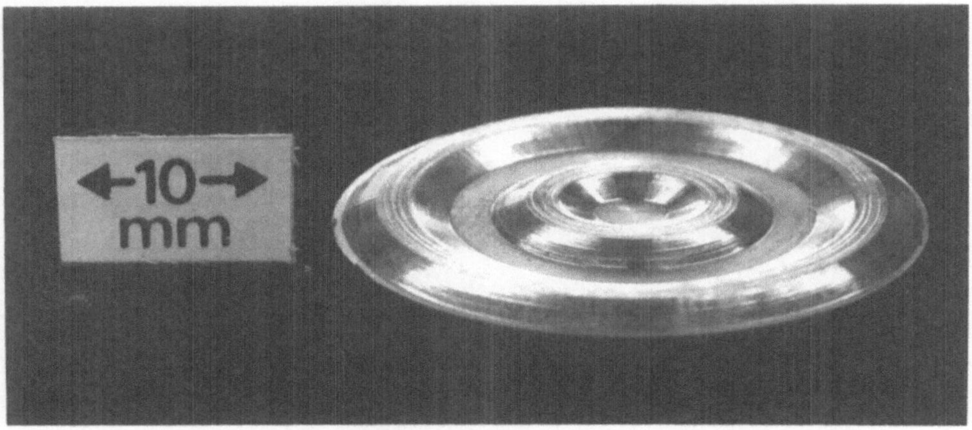

Figure 3. Corrugated washer

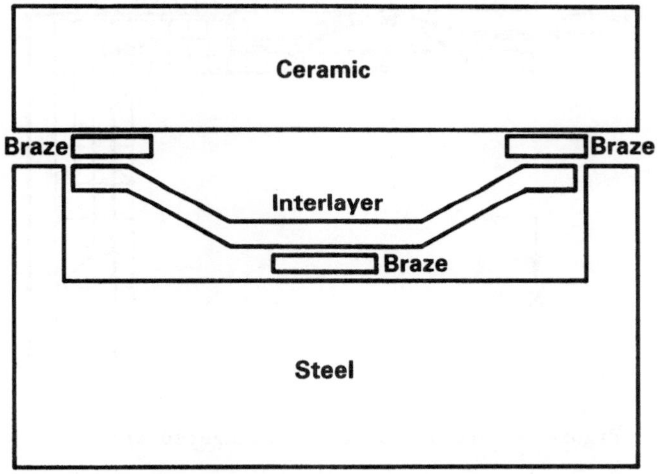

Figure 4. Joint design principle

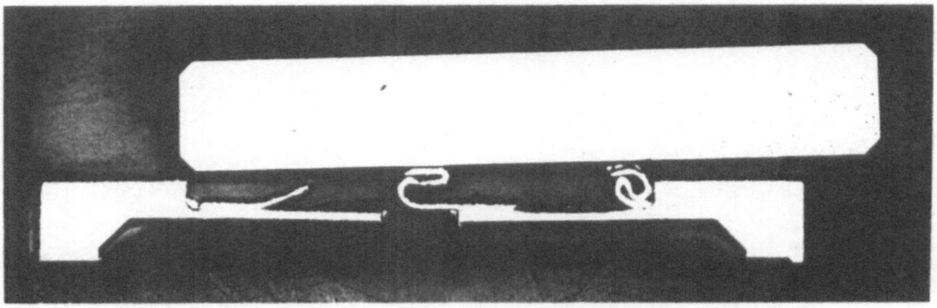

Figure 5. Sectioned shear-test specimen, x5

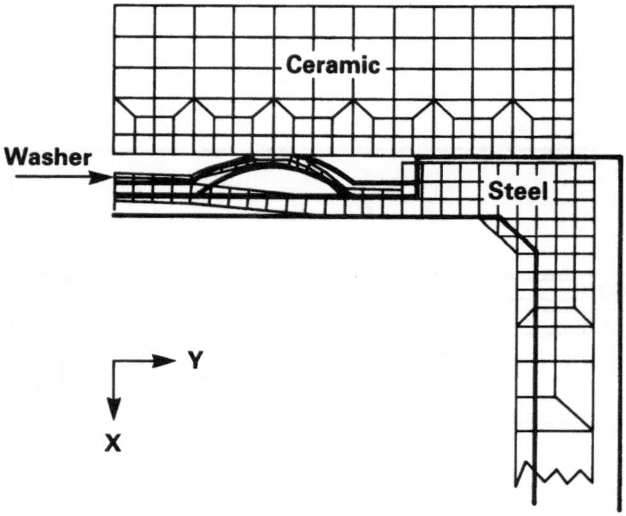

Figure 6. Distortion with corrugated washer

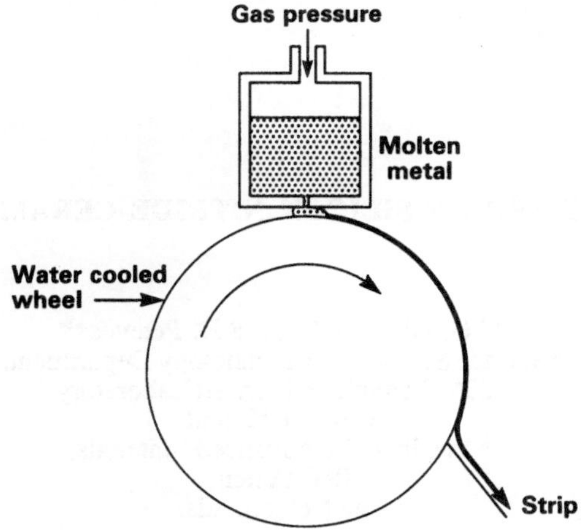

Figure 7. Tape casting

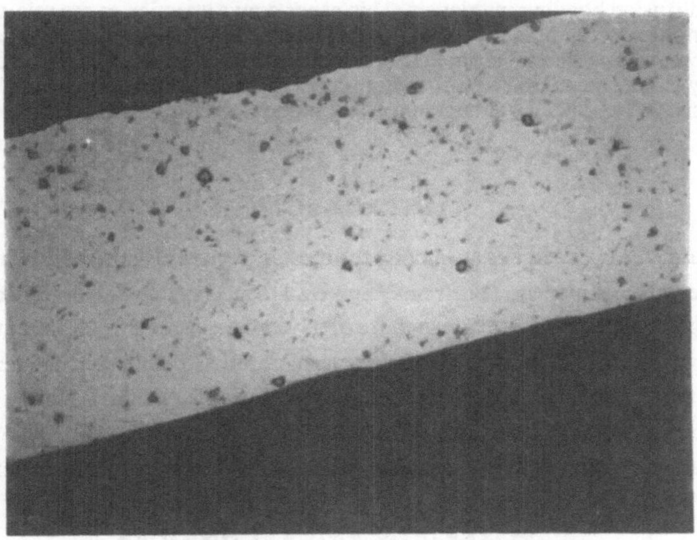

Figure 8. Microstructure of JMT rapidly-solidified braze tape, x500

JOINING OF SILICON NITRIDE CERAMICS

M.G. Nicholas* and S.D. Peteves**
*Surface Science and Technology Department,
AEA Technology Harwell Laboratory
Oxon, UK, and
**Institute for Advanced Materials,
JRC Petten,
The Netherlands.

ABSTRACT

The joining of silicon nitride ceramics can be promoted by the presence of reactive elements in brazes or diffusion bonding interlayers. This paper draws attention to the role of Cr and Ti in forming bridging compounds at ceramic-metal interfaces when Ag, Cu or Ni based brazes or Ni diffusion bonding interlayers are used. Laboratory trials showed that bond quality was affected by the concentration of Cr or Ti in brazes, by coating the ceramics with Cr or Ti, and by process variables such as temperature, environment and surface finish.

INTRODUCTION

Silicon nitride, Si_3N_4 ceramics are among the most important of those now being developed for use as engine components. Their high temperature mechanical properties are excellent, with strength/weight ratios better than those of nickel base super alloys. Effective exploitation of these properties can depend on the development of reliable joining techniques, and complementary programmes at our Laboratories have examined the usefulness of brazing and diffusion bonding. Particular attention was paid to the effects of controlling the chemistry of joint formation by introducing the reactive metals Cr or Ti into the joining materials.

MATERIALS AND TECHNIQUES

Commercial quality Si_3N_4 containing Al_2O_3 and Y_2O_3 or La_2O_3 sintering aids was used in this work; Ceralloy from Ceradyne, Tosnite from Toshiba, Ekasin - S from ESK, and two grades of sialon, Sialon 101 and Sialon 201 from Lucas-Cookson . For some experiments, ceramic surfaces were coated with Cr or Ti prior to bonding.Brazes from various sources were used, 99.9% Cu from Goodfellows, Ag-28%Cu and Ag-27%Cu-2%Ti from

GTE Wesgo, and laboratory made Cu-Ti alloys. Ni-11%P, Ni-11%P-14%Cr and Ni-11%P-23%Cr and Ni-5%Si-3%B-7%Cr brazes given by Wall Colmonoy, while 125 μm Ni-20%Cr foil from Goodfellows was used as a diffusion bonding interlayer.Unless stated otherwise, compositions are quoted as weight percentages.

Brazing and wettability trials were performed at temperatures in furnaces evacuated to less than $1x10^{-5}$ mbar, while diffusion bonded samples were produced in a pressing chamber evacuated to $1x10^{-6}$ mbar or evacuated and then backfilled with 1 bar of Ar whose N_2 content was 50 ppm. The ceramic samples used in the brazing and wettability trials were 15x15x3 mm, with surfaces finished to an R_a roughness of 0.3 μm. The diffusion bonded ceramic samples were 15x15x15 mm, and their surfaces were usually finished to an R_a of 0.1 μm. The wetting behaviour of the brazes was assessed using the sessile drop technique to measure contact angles (1).The quality of both brazed and diffusion bonded Si_3N_4-Si_3N_4 sandwich joints was evaluated by microstructural examination and shear or bend strength testing.

RESULTS

Ti promoted brazing

Sessile drop tests showed that Cu did not wet Ekasin S or the sialon ceramics even at 1150°C, but could be induced to do so by the addition of 5 to 10% Ti, Figure 1. The Ag-27%Cu-2%Ti braze alloy also wetted well at 900°C, just above it's liquidus temperature.

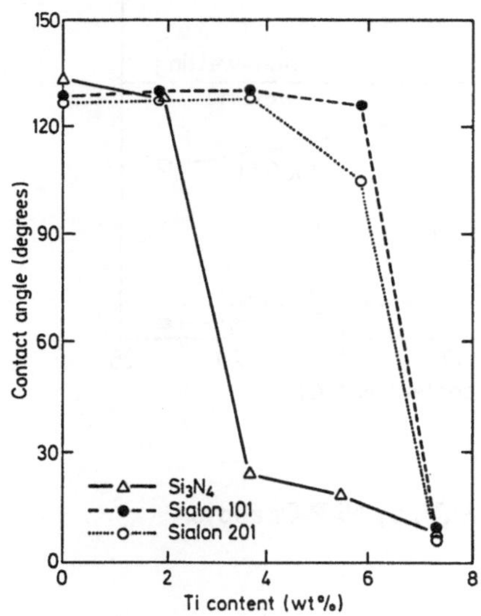

Figure 1. Wettability of Si_3N_4 ceramics by Cu-Ti alloys at 1150°C.

The Cu-Ti alloys reacted vigorously to form duplex layers 10 to 30 μm thick in 20 minutes at temperature. The thicker layers were cracked, and evidence of penetration into occasional ceramic voids indicated that this cracking occurred while the brazes were still molten.Electron Probe Microanalysis, EPMA, surveys showed the layers to be rich in Ti and N or Si, in proportions approximating to TiN next to the ceramic and Ti_5Si_3 next to the braze. Similar layers formed by the Ag-Cu-Ti braze were typically 2 μm thick and uncracked with room temperature shear strengths of 54 MPa.

Sialon ceramics were coated with Ti by vapour deposition as 60 nm thick layers or by sputter ion plating as 1 μm thick layers. The sputtered layers enabled the Ti free Ag-28%Cu braze to wet and form bonds with shear strengths of about 95 MPa, but the vapour deposited layers did not induce wetting or bonding.

Cr promoted brazing

The Ni-Si-B-Cr braze did not wet Sialon 101, the contact angle was 140° at 1075°C in vacuum. Ti sputter coated Sialon 101 had a contact angle of 19° and shear strengths of about 55-60 MPa. Work using Ceralloy and Ni-P-Cr brazes revealed a systematic variation of wettability with Cr content at 1000 and 1200°C, Figure 2. The interfaces formed by the Ni-11%P brazes contained many voids, but those formed by the other brazes were sound. Some erosion of the ceramic was caused at 1200°C, but there was no optical evidence of reaction.

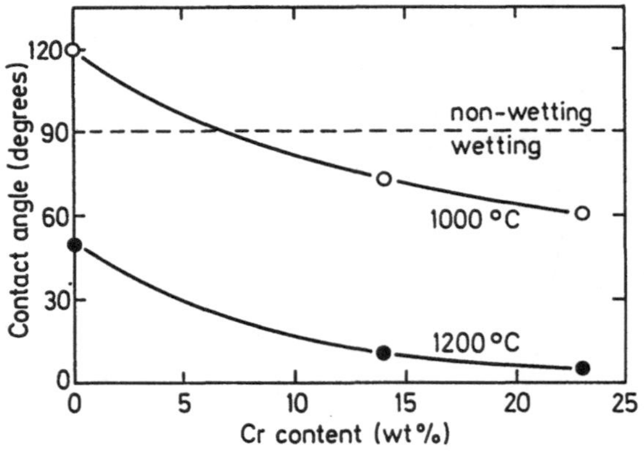

Figure 2. Wettability of Ceralloy Si_3N_4 by Ni-P-Cr alloys.

Diffusion bonding uncoated Si_3N_4

Si_3N_4 can be bonded using Ni-20%Cr interlayers at 1050°C and above (2), and in this work a bonding temperature of 1130°C was used as a norm with a pressure of 100 MPa being applied for 60 minutes in Ar. The joints produced with Ceralloy Si_3N_4 were soundly bonded by a reaction product layer typically 8-10 μm thick, Figure 3. EPMA surveys showed these layers to be rich in Cr and N as were precipitates produced in grain boundaries within the foils some distance from the interfaces. Spot analyses Table 1, showed the reaction products to be impure CrN and the precipitates to be Cr_2N. The reaction products contained some Al and Y derived from the oxide sintering aids, but neither were present in the precipitates.

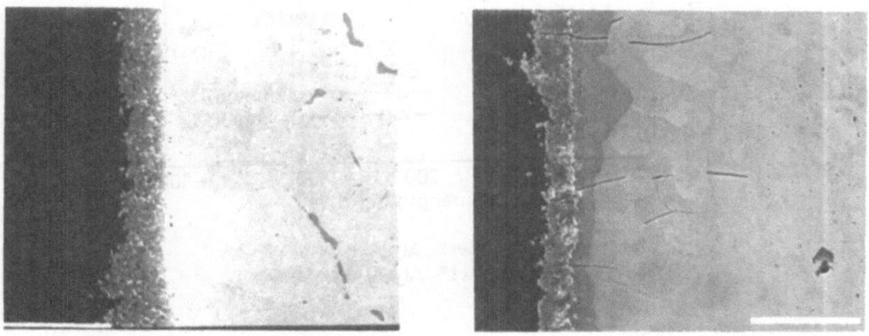

Figure 3. Microstructures of Ceralloy Si_3N_4/Ni-20%Cr interfaces diffusion bonded, left, in Ar at 1130°C and, right, in a vacuum at 1200°C.20μm markers.

Table 1
Bond product chemistries.

Area	Elemental concentration, atom % Ni	Cr	Si	N	Y	Al
Reaction layer	6.2	41.3	5.1	45.2	1.3	0.9
Precipitates	9	62.1	0.6	28.3	-	-

In addition to these microstructural changes, the bonding process caused depletion of Cr to about 8 atom% and ingress of Si to produce concentrations of about 14 atom% in the alloy adjacent to the reaction product layer. Concentration gradients of Cr and Si extended about 30μm into the foils, and could be fitted to error function diffusion profiles (3) to yield coefficients of 7.7×10^{-14} m^2s^{-1} for Cr and 4.9×10^{-14} m^2s^{-1} for Si in good accord with published data for volume diffusion (4,5).

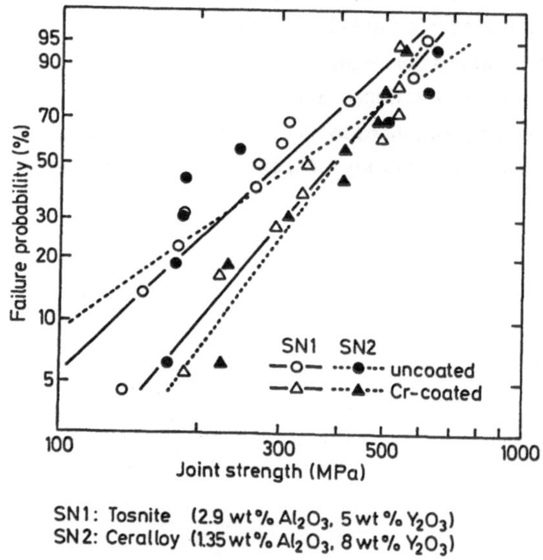

Figure 4. Weibull plot of bend strength data for uncoated and coated Si_3N_4 diffusion bonded using Ni-20%Cr interlayers

The average room temperature bend strengths of the bonded Ceralloy Si_3N_4 samples was 309 MPa, but individual strength values were scattered as shown in Figure 4, and the Weibull modulus was only 2.3. This scatter was considered to be excessive and attention was focussed on means of diminishing it by coating and varying fabrication parameters.

Table 2

Effect of Cr coating on the strength parameters* of diffusion bonded Si_3N_4

Silicon nitride	Uncoated ceramic joints	Coated ceramic joints
Tosnite	Σ=342, m=1.8	Σ=390, m=3.4
Ceralloy	Σ=309, m=2.3	Σ=384, m=2.9

*Σ, average joint 4-point bend strength in MPa; m, Weibull modulus.

Diffusion bonding coated Si_3N_4

Tosnite and Ceralloy Si_3N_4 were sputtered with 0.1 μm of Cr before being bonded using the Ni-20%Cr foil interlayers. The microstructures and microchemistries of these samples were very similar to those of uncoated ceramics. The reaction product layer was 8-10 μm of impure CrN with precipitation of Cr_2N in grain boundaries. However,strength characteristics were significantly different. The average strengths were higher and the individual strengths were less scattered as illustrated in Figure 4 and Table 2.

Diffusion bonding process parameters

The most dramatic effects of process parameters were produced by increasing the bonding temperature to above 1150°C. Si_3N_4 can dissociate at high temperatures, particularly in low N activity environments, to liberate N_2 and elemental Si. Diffusion from the Si enriched surface of the ceramic produces high local concentrations in the alloy foil, and even compositions close to those of some commercial nickel brazes that are molten at 1150°C . The effects of this excessive ingress of Si is illustrated in Figure 3 for a sample bonded in vacuum at 1200°C. A thick,20 μm, N lean reaction product layer of impure CrN (49 atom% Cr, 38 atom% N, 4 atom% Ni and 7 atom% Si) was formed and the foil melted close to the ceramic, converting on cooling to a two phase structure with a cracked matrix of Ni_3Si. The lighter islands in this structure were undissolved foil material containing slightly less than than the 14 atom% solubility limit of Si in Ni.

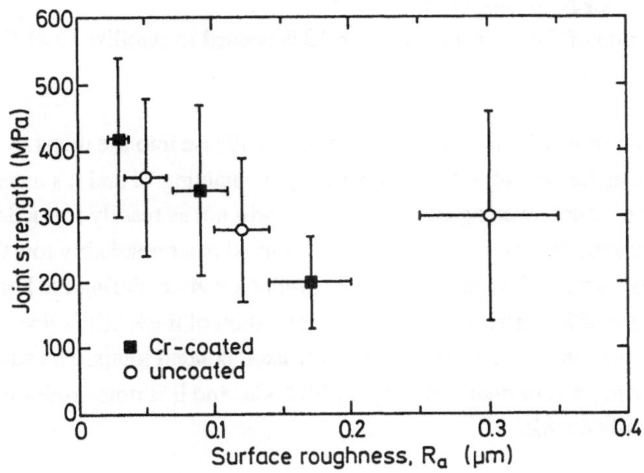

Figure 5. Effects of surface roughness on diffusion bonded Ceralloy Si_3N_4 strengths.

In contrast, polishing the ceramic to a smoother finish improved the strengths of both uncoated and coated samples, Figure 5. These improvements were not associated with discernible changes in microstructure and it is assumed that they were due to a decrease in the frequency of surface flaws caused by earlier stages of the preparation process. Finally, it was noticed that variations in the strengths of samples produced using ceramics from different sources could be related to the bulk strengths of the ceramics.

DISCUSSION

This work has shown that small changes in the chemistry or processing conditions can have major effects on the microstructures of joints formed by liquid and solid phase bonded Si_3N_4 ceramics.

Cr and Ti promoted bonding by forming nitride bridging layers adjacent to the ceramic at both brazed and diffusion bonded interfaces, and this may be a reflection of the much lower solubility of N than Si in Ni solvents (6) and it's neglible solubility in Ag and Cu. Contact with alloys lean in Si and N will cause some dissociation of Si_3N_4. Si and N will be released in the ratio of 3 to 4, and subsequent diffusion into the metal could cause rapid localised saturation by N and formation of nitride reaction products adjacent to the ceramic. The most stable nitrides when elemental metals are present are TiN and Cr_2N, but the nitride actually formed at diffusion bonded interfaces is impure CrN. This is believed to be a degradation product

$$Cr_2N \text{ --------->} \; CrN + Cr$$

caused by depletion of the Cr level below the 12% needed to stabilise Cr_2N (7)

As the interactions proceed, ultimately enough Si can diffuse into the metal to produce silicides. Ti_5Si_3 was formed adjacent to braze alloys containing Ti and it's metallic nature may account for their excellent wetting behaviour. Ni_3Si will not as readily formed in diffusion bonded joints because the activity coefficient of Si in Ni is a remarkably low 0.02 (8), but ultimately it does form, as the matrix of a thick two phase zone, during the high temperature bonding in vacuum of Si enriched ceramic.The formation of thick brittle layers was mechanically detrimental for both brazed and diffusion bonded joints. The strongest joints were produced using a very ductile braze, Ag-28% Cu, and it is noteworthy that Ni-20% Cr is also exceptionally ductile.

This work has shown that strong joints can be produced by both brazing and diffusion bonding. The strengths and microstructures actually achieved in practice depend on both chemical and kinetic factors. Reactive brazing is already being used to produce Si_3N_4 components for engines (9), but this work suggests diffusion bonding may be the more attractive technique in the long term because of the greater control it offers. The chemistry of the joining materials can be varied more readily than for brazes because wetting is no longer a criterion for selection. Similarly, reaction kinetics can be varied by trading off pressure against temperature as means of achieving ceramic-metal contact.The next few years could see diffusion bonding of ceramic-metal components emerge as a major fabrication technique.

REFERENCES

(1) Iida, T. and Guthrie, R.I.L., The Physical Properties of Liquid Metals. Oxford Science Publications, Oxford, 1988, pp 114-116

(2) Peteves, S.D. and Nakamura, M., Solid-state bonding of Si_3N_4 ceramics with Ni-Cr interlayers, J Amer. Ceram. Soc., 1990, **73**, pp 1221-1227

(3) Shewmon P., Diffusion in solids. McGraw-Hill,, New York, 1963, pp 12-13

(4) Smithells Metals Reference Book, ed. Brandes, E.A., 6TH edition, Butterworths, London, 1983, p 13/72

(5) Delauny, D., Huntz, A.M.,and Lacombe, P., Influence de l'yttrium sur la diffusion volume et aux joints du Ni et les alliages Ni-Cr 80/20., Scripta Met., 1979, **13**, pp 419-424

(6) Massalski, T.B., Binary Alloy Phase Diagrams. ASM, Menlo Park,1986, pp 1653 & 1756

(7) Wang, L. and Pehlke, R.D., High temperature thermodynamics of the $Cr\text{-}Cr_2N\text{-}N_2$ system, Metal. Trans. B. 1988,**19B**, pp 471-476

(8) Schwerdtfeger, K., and Engell, H-J., Activity measurements in nickel-silicon melts in the range 1480-1610°C, Trans. Met. Soc. AIME. 1965,**233**, pp 1327-1332

(9) Suga, T., Designing Interfaces For Technological Applications. ed. Peteves, S.D, Elsevier Applied Science,London,1989, pp 247-265.

DIFFUSION BONDING OF Si_3N_4/TiN AND Si_3N_4/TiB_2 COMPOSITES TO INCOLOY 909

Richard Larker [1*], Liu-Ying Wei [1], Mikael Olsson [2] and Bengt Loberg [1]
[1] Division of Engineering Materials, Luleå University of Technology, Sweden
[2] Department of Materials Science, Uppsala University, Sweden

ABSTRACT

Reduction of thermo-mechanical stresses in ideal-elastic diffusion joints between silicon nitride and superalloys can be accomplished by ceramic joint pieces of graded compositions made of Si_3N_4 and a supplementary ceramic phase.
In this work, TiN and TiB_2 were chosen as supplementary ceramic phases in particulate composites with Si_3N_4 due to their suitable properties and the availability of such materials. Diffusion bonding of the composites to Incoloy ® 909 was performed by Hot Isostatic Pressing (HIP) at 1200 K (927°C) and 200 MPa for 4 h. The influences of the TiN phase or TiB_2 phase on diffusion reactions between the Si_3N_4 phase and the superalloy constituents (Fe, Ni, Co, Nb, Ti, Si) were compared by examination with TEM/STEM/EDS.

INTRODUCTION

The introduction of structural ceramics such as silicon nitride in heat engines is today usually restricted by the lack of efficient joining methods to superalloys. Silicon nitride should due to manufacturing limitations and costs be applied only in parts where its properties can be utilized efficiently, chiefly in load-bearing components facing very high temperatures and/or aggressive environment. Superalloys are preferred, when their properties are sufficient. The possibilities to design with an appropriate combination of these materials are strongly related to the durability of the joint. Joining methods [1] for structures used at elevated temperatures must deal with two major restrictions, namely the **high thermo-mechanical stresses** and **excessive reaction layers** that can result in the bonded region during joining and use.

Very high stresses occur due to the large mismatch in Coefficient of Thermal Expansion (CTE) between the joined materials (normally $\alpha_{metal} >> \alpha_{ceramic}$), and can cause fracture in the joint or at some distance into the ceramic, either during cooling from joining temperature or during thermal cycling in use. Reaction layers usually have inferior properties compared with the joined materials, thereby reducing the possible stability and strength of the joint. It is particularly important that the joining reactions does not proceed during use at high temperatures.

The possible bonding methods for joints used at elevated temperatures (500-700°C) are mechanical attachment (shrink fit), active brazing or diffusion bonding. In the ceramic gas turbine, the hub of the ceramic turbine wheel must be attached to a metal shaft. A proper design based on a shrink fit combined with diffusion bonding could be used at high temperatures, due to the absence of wide reaction zones with low melting points and limited chemical stability.

Hot Isostatic Pressing is mainly considered to be an efficient sintering method for densifying powders of metals or ceramics to fully dense materials [2] but does also offer some advantages for joining of dissimilar materials. In diffusion bonding, the most crucial aspect is the degree of alignment between the joined surfaces. The high isostatic pressure working on the encapsulated joint forces the metal to plasticize and accommodate to the surface of the ceramic. The pressurized encapsulation prevents both unwanted reactions with the furnace atmosphere and the formation of voids due to released gases, e.g. nitrogen during joining of silicon nitride. These advantages permit reduction of bonding temperatures, resulting in lower residual stresses and a very thin interface, if chemical stability can be obtained.

The number of suitable materials for the interlayers needed between Si_3N_4 and superalloys are limited. The stress levels can to some extent be reduced by selection of a superalloy with relatively low thermal expansion, such as the Incoloy 909, but the coefficient increases rapidly for this austenitic alloy above the ferro- to paramagnetic transition at ≈400°C, confirmed by dilatometric measurement performed in the interval 20-1000°C, see figure 1:

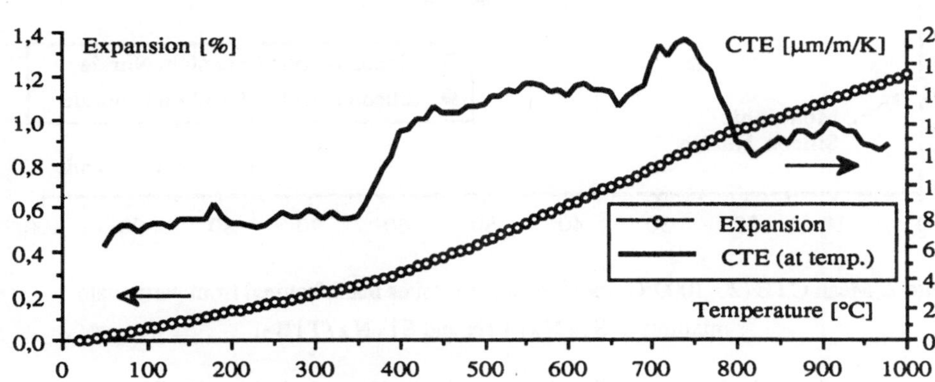

Figure 1. Thermal expansion and momentary CTE for Incoloy 909.

This behaviour results in a mean value >12 × 10^{-6} K^{-1} in a typical temperature interval for joining (20-900°C), being similar to mean values for ferritic steels.

Refractory metals having CTE values close to Si_3N_4 ($\alpha = 3.5 \times 10^{-6}$ [K^{-1}]), such as W or Mo ($\alpha = 4.0 - 6.2 \times 10^{-6}$ [K^{-1}]), might reduce the residual stresses in the ceramic, but these metals are very unstable at high temperatures in the oxidizing environments usually present in engines. Soft interlayers of ductile metals with low yield stress and considerably higher CTE, such as Ni or Cu, can relieve stresses through plastic deformation on cooling from joining temperature, but they may be susceptible to fatigue during thermal cycling in use.

The possibility to **gradually increase the CTE in the ceramic** towards the interface could increase the performance of the joint, allowing an ideal-elastic joint to be formed by diffusion bonding. For this purpose, diffusion reactions between particulate composites containing Si_3N_4 and a supplementary ceramic phase and superalloys have been analysed. Requirements for this supplementary ceramic phase are:

- intermediate CTE ($\alpha_{Si_3N_4} < \alpha_{supplementary\ ceramic} < \alpha_{superalloy}$);
- sufficient stability against reactions with Si_3N_4 (during sintering);
- useful properties of the particulate composite formed with Si_3N_4;
- controlled reactivity against the superalloy (during joining and use).

In this work, TiN and TiB_2 were chosen as supplementary ceramic phases with Si_3N_4 due to their suitable properties and the availability of a variety of such composites, produced by ABB Cerama AB, Sweden, using glass-encapsulated HIP at 1600°C. The mean CTEs were determined in the interval 20-1000°C by dilatometric measurements and presented in figure 2:

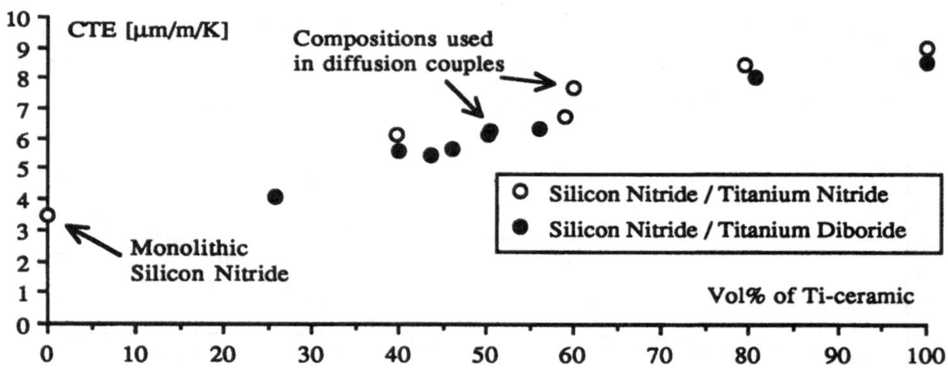

Figure 2. Mean CTE (20-1000°C) for HIPed composites manufactured from particulate mixtures of Si_3N_4/TiN and Si_3N_4/TiB_2.

The steepest change in the mean CTE takes place in the interval 40-60 vol% TiN or TiB_2, where both Si_3N_4 and the supplementary phase are continuous. The value at 80 vol% TiN could reduce the CTE mismatch against Incoloy 909 to $\Delta\alpha = 3.5 \times 10^{-6}$ [K^{-1}].

MATERIALS AND METHODS

The two ceramics used in diffusion couples in this study were composites densified by HIP from powder mixtures of Si_3N_4 and either 60 vol% TiN or 50 vol% TiB_2.
The superalloy used was Incoloy® 909 (42.1 wt% Fe, 38.3 wt% Ni, 12.9 wt% Co, 4.7 wt% Nb, 1.5 wt% Ti, 0.4 wt% Si), heat treated for optimum tensile properties by solution annealing at 980°C/1h/AC, followed by ageing at 720°C/8h/FC + 620°C/8h/AC. The joining temperature at 1200 K (927°C) was chosen considering results from thermo-dynamic calculations of possible reactions [3-4] and with the aim to retain a useful microstructure in the superalloy after joining [5].

The reaction zones between diffusion bonded bulk samples of silicon nitride and Incoloy 909 have earlier been shown [6] to be very thin, being on the limit of the element resolution for SEM/EDS. Recent work in TEM [7-9] have shown the possibilities to resolve the details of reaction products between Si_3N_4 and metals. Therefore diffusion couples, adapted for the preparation of TEM specimens, were fabricated according to the following route:

- Ultrasonic machining of Si_3N_4/TiN & Si_3N_4/TiB_2 and electro-discharge machining of Incoloy 909 samples to the dimensions Ød=2.35 mm and length=4.0 mm.
- Grinding to obtain a half cylinder with a plane surface (4.0 × 2.35 mm) on each sample.
- Polishing of the plane surfaces using slurries down to 1 μm diamonds.
- Encapsulation of paired samples in a stainless steel tube ($Ød_o$=3.0 mm, $Ød_i$=2.35 mm).
- Joining by HIP at 1200 K (927°C) and 200 MPa for 4 h.
- Cutting of thin slices (0.4 mm) of the joined diffusion couples surrounded by the capsule.
- Planar grinding and polishing to approximately 70 μm thickness.
- Dimple grinding to approximately 7 μm thickness, (measured by optical microscopy).
- Selectively ion beam thinning for approximately 2 h, (perpendicular to the joint).

The influences of the TiN phase or TiB_2 phase on diffusion bonding reactions between Si_3N_4 and the superalloy constituents (Fe, Ni, Co, Nb, Ti, Si) were then compared by examination with TEM / STEM / EDS; (the light elements N, B and O were omitted in the measurements due to the beryllium window on the EDS detector). The interface is consistently displayed vertically, with the ceramic on the left side and the superalloy on the right side.

RESULTS

The reaction zones were in general very thin, usually less than 1 μm in total thickness. The most evident reaction phase in the joint was due to the enrichments of Ni, Nb and Si proposed to be **Laves phase**, formed in the superalloy at the interface. Laves is a grain-boundary phase in the Incoloy 909, preventing grain growth and grain-boundary sliding at high temperatures.

At the joint interface between Si_3N_4/TiB_2 and Incoloy 909, the **Laves phase** was continuous along the interface, with a thickness varying between 100-500 nm.
Fine **titanium nitride** crystals were also detected by SAED in a Ti-rich layer between Si_3N_4 grains and Laves phase. The crystal size was below 50 nm and the layer thickness approximately 100 nm.
Silicide clusters (rich in Ni, Fe, Co and Si) appeared occasionally, often between Si_3N_4 and TiB_2 grains near the interface. Their size varied from a few hundred nm to one μm.
The appearance of these phases at the interface is shown at high magnification in the TEM micrograph in figure 3:

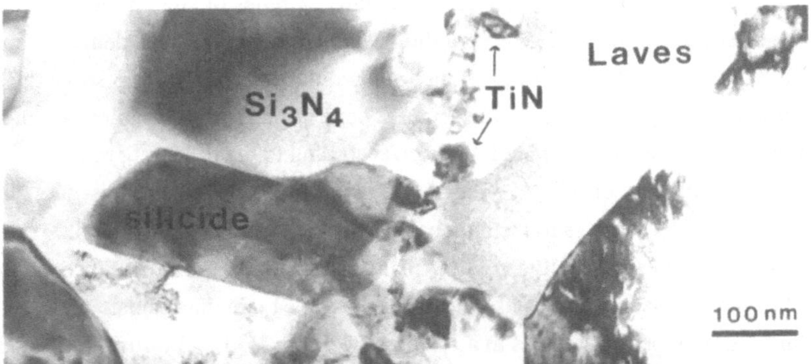

Figure 3. TEM micrograph from an interface between Si_3N_4/TiB_2 and Incoloy 909

Diffusion profiles perpendicular to the two kinds of ceramic-metal interfaces (Si_3N_4-Incoloy 909 and TiB_2-Incoloy 909) were obtained by EDS point measurements with 25 nm spacing. The results from ZAF calculations at each point are presented in figure 4:

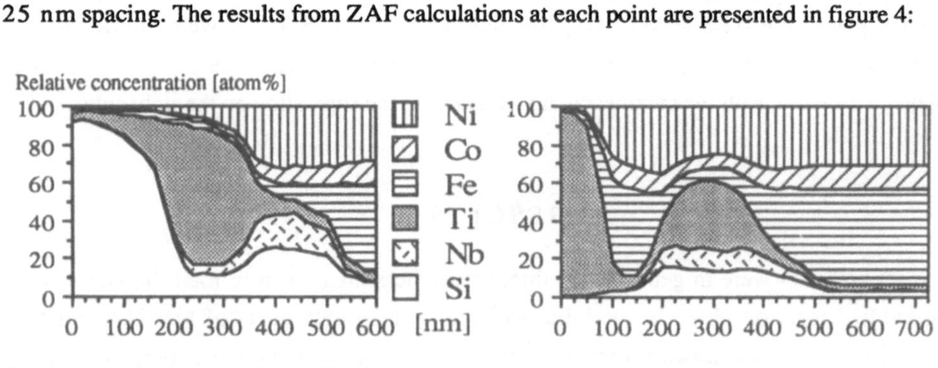

Figure 4. Diffusion profiles by EDS (25 nm spacing) at interfaces between Si_3N_4 - Incoloy 909 (left) and TiB_2 - Incoloy 909 (right) in a Si_3N_4/TiB_2 - Incoloy 909 joint.

The appearance of the continuous Laves phase along the interface between Si_3N_4/TiB_2 and Incoloy 909 can be seen in the STEM micrograph (left) with corresponding EDS maps for Nb and Ti close to the interface (right) in figure 5. The brighter areas are polycrystalline Si_3N_4 surrounded by larger grains of TiB_2. A silicide cluster, rich in Ni, Fe, Co and Si, was formed between the large upper TiB_2 grain and the polycrystalline Si_3N_4 area. A grain-boundary in the superalloy was intersected by the ceramic/metal interface, thus acting as a Nb supply for Laves formation, as shown by the EDS map for Nb. The thin layer containing TiN between the Si_3N_4 area and the Laves phase can be detected in the EDS map for Ti.

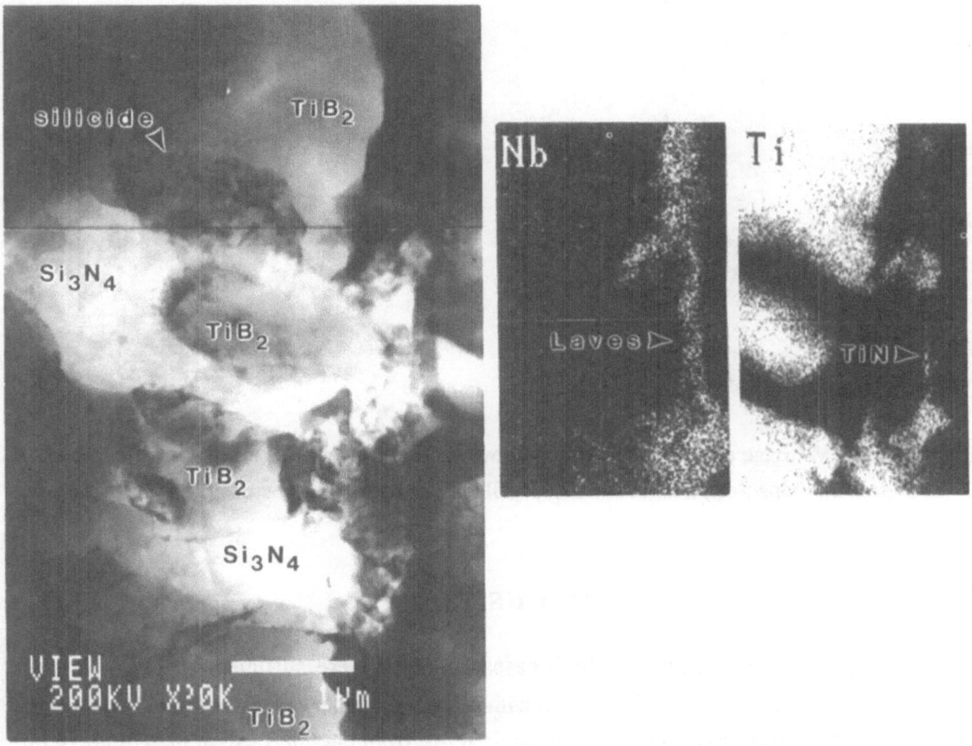

Figure 5. STEM micrograph from an interface between Si_3N_4/TiB_2 and Incoloy 909 (left) with corresponding EDS maps for Nb and Ti (right).

At the joint interface between Si_3N_4/TiN and Incoloy 909, only the **Laves phase** layer, (rich in Ni, Nb and Si) was detected. The layer was in this case semi-continuous, and the occurrence did not seem to depend on the phase of the adjacent ceramic grains (Si_3N_4 or TiN). The thickness of the layer varied between 100-500 nm.

The interface is shown in a STEM image with corresponding EDS element maps (left) and in three diffusion profiles by EDS with 25 nm point spacing across the joint (right) in figure 6:

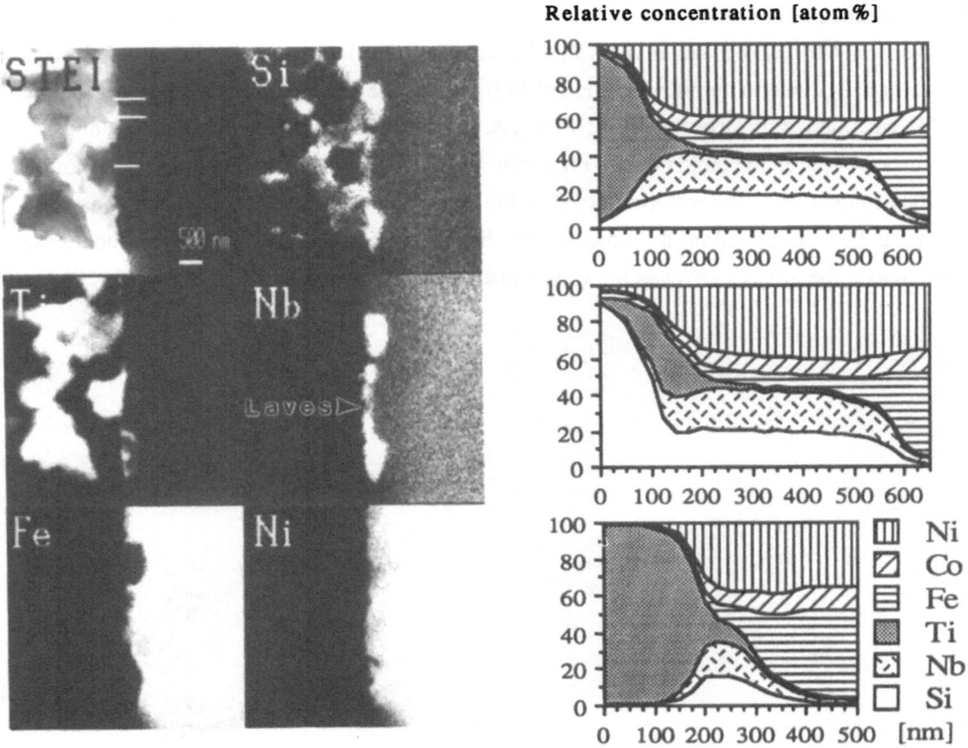

Figure 6. STEM image of the joint interface between Si_3N_4/TiN and Incoloy 909 with EDS element maps (left) and three EDS profiles (25 nm spacing) across the joint (right).

DISCUSSION

The reactions were very limited in both cases, but they were sufficient to give a strong adhesion. This was seen in larger butt joints with 10 mm diameter, where convex cracks in the composites formed during cooling at a few hundred μm from the interface, in the volume with the highest residual stresses. The behaviour is similar to butt joints of monolithic Si_3N_4 to Incoloy 909. The reduced CTE mismatch for the joints with the composites ($\Delta\alpha = 4.4 - 5.9 \times 10^{-6}$ [K^{-1}]) compared to joints with monolithic Si_3N_4 ($\Delta\alpha = 8.8 \times 10^{-6}$ [K^{-1}]) resulted, however, in considerable reduction of cracking problems during preparation of the TEM specimens, in spite of the higher intrinsic strength of monolithic Si_3N_4. This is promising, since the necessary reduction of the residual stresses formed in joints with application dimensions could be obtained by applying a joint piece with compositional gradients, using higher amounts of TiN or TiB_2 close to the superalloy. Concepts based on graded joint pieces could thus make it feasible to manufacture ideal-elastic joints by diffusion bonding with HIP.

CONCLUSIONS

- Laves phase occurred in the first 100-500 nm of metal for both composite / superalloy joints. The formation seems to be promoted by the increased access to silicon at the interface compared to that in Incoloy 909 (0.4 wt% Si).
- The TiB_2 phase seems to be less inert compared with the TiN phase as a supplementary phase with Si_3N_4, due to the more pronounced formation of nitrides and silicides.
- TiN is therefore preferred as supplementary phase in future work on composite joint pieces with compositions graded from monolithic Si_3N_4 to 80-90 vol% TiN.
- Cermets composed of TiN and ferritic steel might together with optimized joint design lower the residual stress level further in ideal-elastic joints formed between Si_3N_4 and superalloys using diffusion bonding by HIP.

REFERENCES

1. Elssner, G. and Petzow, G., Metal-Ceramic Joining. ISIJ International, 1990, **30** [12] pp. 1011-32.
2. Adlerborn, J., Burström, M., Hermansson, L. and Larker, H.T., Development of High Temperature High Strength Silicon Nitride by Glass Encapsulated Hot Isostatic Pressing. Mater. & Design, 1987, **8** [4] pp. 229-32.
3. Schuster, J.C., Weitzer, F., Bauer, J. and Nowotny, H., Joining of Silicon Nitride Ceramics to Metals: The Phase Diagram Base. Mater. Sci Eng., 1988, **A105/106**, pp. 201-6.
4. Loehman, R.E., Tomsia, A.P., Pask, J.A. and Johnson, S.M., Bonding Mechanisms in Silicon Nitride Brazing. J. Am. Ceram. Soc., 1990, **73** [3] pp. 552-8.
5. Heck, K.A., Smith, D.F., Smith, J.S., Wells, D.A. and Holderby, M.A., The Physical Metallurgy of a Silicon-Containing Low Expansion Superalloy. In Superalloys 1988, ed. S. Reichman, D.N. Duhl, G. Maurer, S. Antolovich and C. Lund, The Metallurgical Society, 1988, pp. 151-60.
6. Larker, R., Loberg, B. and Johansson, T., Diffusion Bonding Reactions between Silicon Nitride, Silicon Oxynitride and Incoloy 909 by Hot Isostatic Pressing. In 3rd Int. Symp. on Ceramic Materials & Components for Engines, Las Vegas, 1988. Ed. V.J. Tennery, American Ceramic Society, 1989, pp. 503-12.
7. Brito, M.E., Yokohama, Y., Hirotsu, Y. and Mutoh, Y., Si_3N_4-Ni Bonding and its Reactions. In MRS International Meeting on Advanced Materials, Tokyo 1988. Vol. 8., Metal-Ceramic Joints. MRS, Pittsburgh (1989).
8. Suganuma, K., Niihara, K. and Fujita, T., Solid State Bonding of Silicon Nitride with a Nickel Interlayer. J. Less-Common Metals, 1990, **158**, pp. 59-69.
9. Ishikawa, T., Brito, M.E., Inoue, Y., Hirotsu, Y. and Miyamoto, A., Interfacial Structure and Mechanical Strength of β-sialon-Ni Bonded System. ISIJ International, 1990, **30** [12] pp. 1071-7.

APPLICATION OF CERAMIC-METAL EUTECTICS FOR SOLID-STATE BONDING BETWEEN CERAMICS

STANISLAW SERKOWSKI
Institute of Material Science and Engineering
Silesian Technical University, Katowice, Poland

ABSTRACT

A new concept for solid-state bonding between ceramics was developed. An interlayer, which forms as a result of the ceramic-metal eutectic works as a bonding agent.
It has been found that the $(Al,Cr)_2O_3$-Cr eutectic gives possibility to obtain an excellent solid-state join between ceramic elements with the bend strength over 200 MPa for pressureless sintering.
On the way of controlled phase transformation the interlayer eutectic can be obtained as a fully oxide phase. As a result of phase transformation the melt temperature of the interlayer increases up to that of bonded materials; however formation of the join takes place at much lower temperature.

INTRODUCTION

Glass and ceramic seals are used in a wide variety of industrial and consumer goods, where they provide hermeticity, environmental protection, mechanical and dielectric strengths, and thermal stability. During the course of the research on bonding of ceramic-ceramic a new bonding method in which a ceramic-metal eutectic forms the bonding layer has been developed. In this method the interlayer is an eutectic material containing a joining oxide. The advantages of this method result from the unique behaviours of ceramic-metal eutectics.

CHARACTERIZATION OF OXIDE-METAL EUTECTICS

The refractory metals such as W, Mo, Ta and Cr form eutectics with some oxides. Up to now, about 40 oxide-metal eutectics are known - Tab.1 - but the most of these eutectics are created by very rare lanthanides elements.

Table 1.Summary of known oxide-metal eutectics [1-2].

SYSTEM	SYSTEM	SYSTEM
$(Al,Cr)_2O_3$-Cr	$Gd_2O_3(CeO_2)$-W	$Nd_2O_3(CeO_2)$-Mo
$(Al,Cr)_2O_3$-Mo	GeO_2-Ge	$Nd_2O_3(CeO_2)$-W
$(Al,Cr)_2O_3$-W	$HfO_2(CeO_2)$-W	Sm_2O_3-W
CeO_2-Mo	$HfO_2(Y_2O_3)$-W	$Sm_2O_3(CeO_2)$-W
CeO_2-W	HfO_2-Ta	Ta_2O_5-Ta
Cr_2O_3-Cr	Ho_2O_3-Mo	UO_2-W
Cr_2O_3-Mo	La_2O_3-Mo	UO_2-Nb
Cr_2O_3-Nb	La_2O_3-W	UO_2-Ta
Cr_2O_3-Re	$La_2O_3(CeO_2)$-W	$Y_2O_3(CeO_2)$-Ta
Cr_2O_3-Ta	$La_2O_3(CeO_2)$-Mo	Y_2O_3-Mo
Cr_2O_3-V	$LaCrO_3$-Cr	$YCrO_3$-Mo
Cr_2O_3-W	$LaCrO_3$-Mo	$YCrO_3$-Cr
$Er_2O_3(CeO_2)$-Mo	$LaCrO_3$-W	ZrO_2-Ta
$Gd_2O_3(CeO_2)$-Mo	MgO-W	ZrO_2-W

Briggs and Hart [1] tried to establish criteria for predicting oxide-metal system which exhibit eutectics structure. The possible criteria which have been considered are stoichiometry of the oxide, thermodynamics, ions field strength, crystal structure of the metal and oxide, and electronic structures of the metal and oxide. Up to now, none of these criteria have been successfully used to explain observed behaviours.

Now, the oxide-metal eutectics are used to fabricate a new class of ceramic matrix composites by the way of directional solidification [3-5]. There are potential advantages of these structures such as the uniform fiber spacing, low porosity, if any, high thermodynamic stability and each square centimeter of these materials contains many millions of metalic fibres with diameters of less than 1 μm. The behaviour of liquid oxide-metal eutectics which was observed during the investigation of directional solidification as well as phase equilibrium in oxide-metal systems gives the possibility to create a new conception of joining ceramics. It has been found that the oxide-metal liquid eutectics wet very well, both, metal and oxide, and are able to solve a large quantity of solid components. The eutectic phase is physically compatible to both eutectic components and a stable chemical equilibrium exists also in this system up to the melting temperature. All of this give a good chance to produce a strong bonding.

THE CONCEPT OF BONDING CERAMICS WITH OXIDE-METAL INTERLAYER

The concept of application of oxide-metal eutectics for bonding of ceramic parts will be presented on the example of joining alumina samples. Alumina, the most popular ceramic material does not form eutectic with refractory metals by itself. Addition of Cr_2O_3 to Al_2O_3 is the reason that metals can dissolve and eutectic is formed - Fig. 1 [6].

If the Cr_2O_3-Cr eutectic is in contact with alumina at the temperature over 1660°C, alumina will dissolve in the liquid eutectic phase. But the coexistence of metal and oxides in the eutectic liquid requires the adequate oxygen content. It can be maintained by controlling the oxygen activity in the atmosphere over the melt. For Cr_2O_3-Cr and $(Al,Cr)_2O_3$-Cr eutectics the equilibric oxygen partial pressure in function of temperature is shown in Fig.2. In this condition the metal and oxide can coexist in the melt without changing metal/oxide ratio. But if the eutectic raw material is melted in atmosphere with higher oxygen partial pressure, the metal will be oxidized. As a result of the changing of metal/oxide ratio the melt will crystallize. If the eutectic raw material ($[Al,Cr]_2O_3$-Cr) - marked as A in Fig.1 - is melted under a bit higher oxygen partial pressure than the required equilibric pressure, the process of metal oxidizing is accelerated and the fully ceramic material is obtained. This phase transformation produces

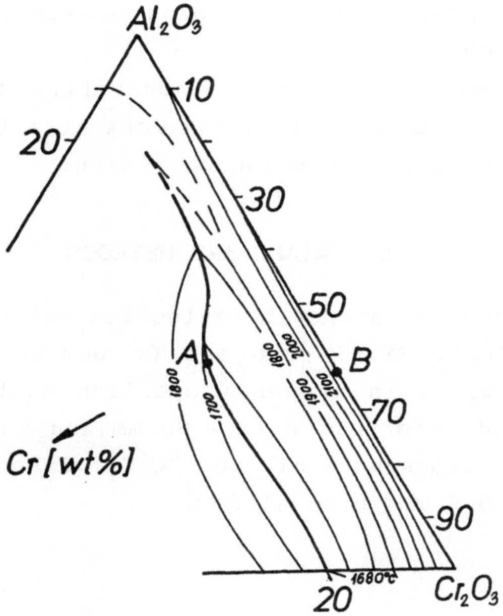

Figure 1. Al_2O_3-Cr_2O_3-Cr system

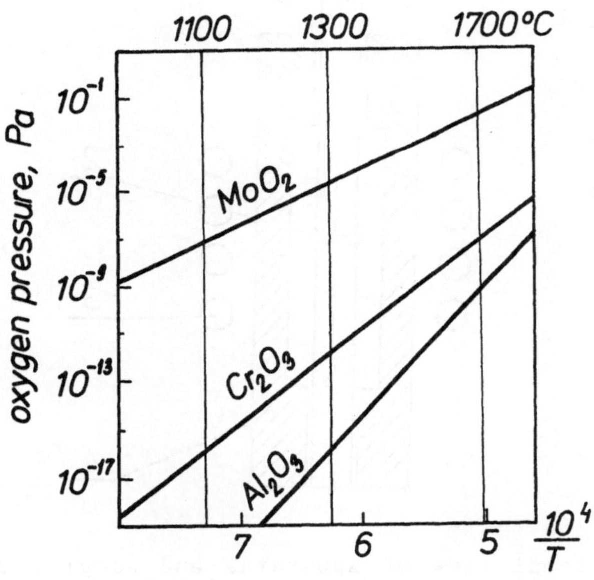

Figure 2. Metal-oxide equilibria vs partial oxygen pressure for various temperatures [7].

new material - marked as B in Fig. 1 - with the melting temperature over 2100°C.

The foregoing process gives possibility to obtain joining between alumina samples, which can work at a temperature higher than the temperature of bonding operation.

MATERIALS AND METHODS

The powders used for preparing eutectics materials were : Al_2O_3 99.8% pure, Cr_2O_3 99.9% pure and Cr powder 99.4% pure. This powders were mixed in proper proportion with ethylene glycol binder, pressed into rods Φ 10 × 80 mm and then sintered in an argon/hydrogen atmosphere at 1500 °C for 2 hours. Sintered rods were cut into 0.5 mm thick slides.

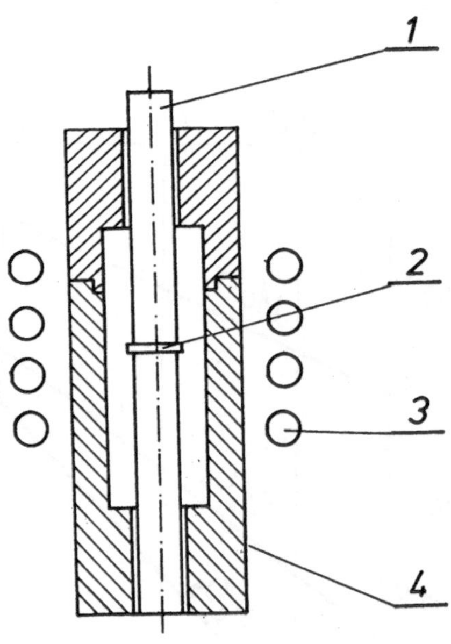

Figure 3.Sectional view of apparatus and samples used in the joining process (1-alumina sample, 2-eutectic, 3-rf coil, 4-preheater)

The alumina rods ϕ 10 x 50 mm were placed in molybdenum-made support as it is shown on Fig.3. This support worked as an induction preheater.

The atmosphere was prepared from argon and hydrogen mixture in order to maintain oxygen partial pressure of 10^{-5}Pa (10^{-10}atm.) at 1700 ^{o}C. According to Fig.2 the oxygen content was much higher than it is required for coexistence of Cr and Cr_2O_3 in the melt state. The mechanical strength was determined in 3-point bending test. Some of joined specimens were sectioned, mounted in epoxy and polished for examination in a scanning electron microscope (SEM).

RESULTS

The bend strength, determined on 6 joined samples, was in range of 190 - 225 MPa. Although, the samples were held at the processing temperature for different periods of time (from 2 to 10 min.), it did not influence on the mechanical strength. In order to test the thermal shock resistance, short alumina bars, joined in the earlier described way, were heated to a tempera-

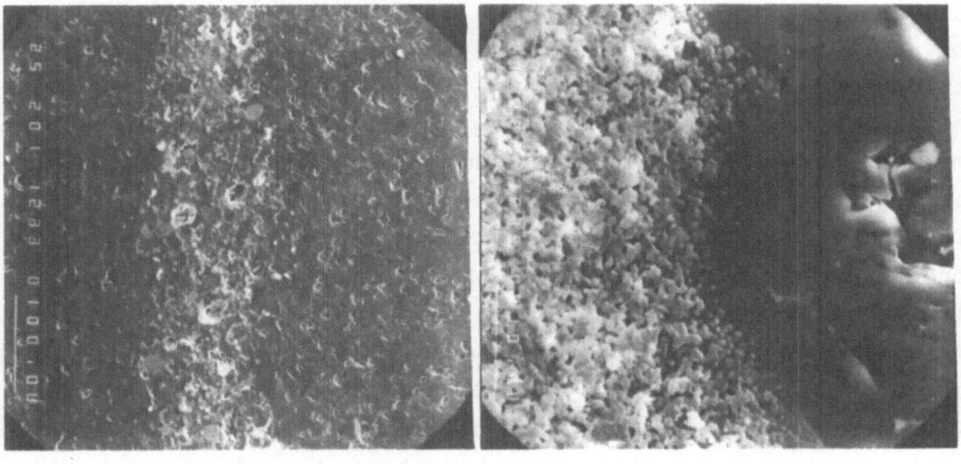

a) b)

Figure 4.Transverse section of joined alumina samples (a-interlayer, b-interlayer - sample boundary)

ture of 1500 ^{o}C on air and than cooled rapidly in water. Though, some of alumina bars failed, no cracks across bonding layer were observed. Microstructure of eutectic interlayer in bonded alumina-alumina couple is presented in Fig.4.
The interlayer, which was obtained as the result of phase transformation of oxide-metal eutectic is slightly visible. The boundary between alumina specimen and eutectic interlayer can be indicated under higher magnification only. In the interlayer there is no metallic phase. The eutectic, which was used for joining alumina samples in the melted state, dissolved the joined material and - in reaction with oxygen - changed its phase composition. These processes increase the melting temperature and the liquid crystallizes. As a result, the fine crystalline microstructure of interlayer is obtained, producing the good strength of joining.

CONCLUSIONS

The results which were obtained confirm that the concept of application of ceramic - metal eutectics for bonding ceramics gives a new possibility of solving this technological problems. The most important advantage of this method is composition change of interlayer material during the joining process. Oxidation of metallic component increases the melting temperature and leads to densification of interlayer. As a result, a satisfactory strength of bonding is obtained as well as a high melting point of the resulted interlayer, higher than the temperature of bonding operation.
The described method gives good effects without using of pressure, while other methods of joining need a high pressure to improve wettability of bonded phases.
This application of ceramic-metal eutectics for bonding ceramics is not the only way of using this specific material. These eutectics can be applied to joining ceramic and metal as well. Of course in this case the phase transformation, which needs adequate oxygen partial pressure can not be performed.
The application of ceramic-metal eutectics for bonding ceramics and metals is a new way of practical realization of

K.Suganuma's method for solid-state bonding between ceramics and metals [8]. The thermal expansion coefficient of this composite interlayer is intermediate between those of constituents of the couple. The excellent wettability of both ceramic and metal provided by oxide-metal eutectic liquid gives the possibility of reducing high pressure during bonding operation. In Suganuma's method producing of the composite interlayer between ceramic and metal needs very high pressure (3 GPa).

To sum up, many ceramics, and ceramic and metal couples can be easily bonded using the oxide-metal eutectic as the interlayer material. It is expected that materials bonded by this method will have high strength and thermal shock resistance.

REFERENCES

1.Brigs J., Hart P.E., Refractory oxide-metal eutectics., J.Am.Cer.Soc.,59,1976,530-531

2.Serkowski S.,Production and properties of directionally solidified oxide-metal eutectics.(polish),Zeszyty Nauk.Pol.Sl. Z.34,Gliwice 1989

3.Ashbrook R.L.,Directionally solidified ceramic eutectics., J.Am.Cer.Soc., 60,1977,428-435

4.Serkowski S.,Ceramic metal eutectics of fibrous microstructure., Ceram.Int., 13,1987,105-107

5.Banik G.,Wruss W.,Vendl A.,Entwicklunten auf dem Gebiet faserverstarkter sonderkeramischer Werkstoffe., Sprechsaal, 113,1980, 261-265

6.Serkowski S.,The Al_2O_3-Cr_2O_3-Cr phase system.(polish), Inzynieria Materialowa, 3, 1989, 89-93

7.Trzczalowski T.,Plants for heat treatment in special atmospheres.(polish), ed.WNT, Warszawa, 1975

8.Suganuma K.,Okamoto T.,Shimada M.,Koizumi M.,New method for solid-state bonding between ceramics and metals., J.Am.Cer. Soc., 1983, C-117

Fe-TiC/N CERAMIC-METAL COMPOSITES

I.W.M.BROWN* & G.V.WHITE
DSIR Chemistry, Lower Hutt, New Zealand
& G.L.DUNLOP
University of Queensland, Brisbane, Australia

ABSTRACT

A fine powder consisting of iron metal and titanium carbide or nitride can be obtained by reduction of the mineral ilmenite ($FeTiO_3$) with carbon under controlled atmospheres. Hot-pressing of Fe-TiC powder causes transformation of ≈80% of the α-Fe to Fe_3C, whereas for Fe-TiN powder the extent of transformation to Fe_3C is very minor. The densification process is controlled by the extent to which the ceramic is wetted by the metal matrix and has been found to be very different between TiN and TiC-based systems. Microstructural examination of these specimens shows retention of sharp (cubic) TiN grain boundaries in Fe-TiN whereas the TiC grains in Fe-TiC are heavily rounded. Physical property measurements for hot-pressed specimens show that density and hardness increase with temperature while porosity reduces to zero by ≈1330°C for Fe-TiN and ≈1530°C for Fe-TiC. Vickers hardness values of ≈1100 and ≈1380 are obtained for fully dense specimens of Fe-TiN and Fe-TiC respectively.

INTRODUCTION

Ceramic-metal composites, often called "cermets", may be considered as dispersions of ceramic grains in a continuous metallic matrix. In some cases, such as WC-based "hardmetals", there may be a relatively low volume fraction of the metal matrix, resulting in only a thin intergranular layer. It is a feature of most cermet systems that the two key physical properties of ceramics and metals, that is their hardness and toughness respectively, are found to be either complementary or additive.

Composites of iron metal and titanium carbonitride offer the opportunity to use relatively cheap and abundant raw materials, reducing the dependence on strategic materials such as WC, Co and Ni. There was an early period of exploration of Fe-TiC compositions for hard metal substitution in the late 1930's and again in the 1950's [1]. A recent resurgence of interest in these materials has been focussed at what has been perceived as a "gap" in toughness and hardness behaviour between high speed steels and cemented carbides, such as WC-Co. Some of the earlier materials had limited success due to the ceramic grain size being too large, but more recent developments, including the use of pre-formed Fe-Ti alloy powders as starting materials, have permitted reduction in ceramic grain size with a consequent improvement in properties and performance [2,3].

The material described in this paper is sourced from the mineral ilmenite ($FeTiO_3$). The principle is simple: since ilmenite contains iron and titanium mixed approximately equally on an atomic level, careful processing should enable the formation of Fe-TiC/N composites with very fine grain sizes. This paper describes preliminary synthesis and densification studies of both Fe-TiC and Fe-TiN composites derived directly from beneficiated ilmenite ore.

EXPERIMENTAL

Ilmenite powder was ball-milled in ethanol using steel milling media to achieve a mean particle size of ≈3 μm. Typically, 250g. ilmenite was milled for 72 hours. Two sources of ilmenite powder were used. The first was from the West Coast of the South Island of New Zealand and the second was an Australian East Coast ilmenite. The New Zealand material contained garnet impurities occluded in the ilmenite grains at the micron and sub-micron level, which made beneficiation difficult, as can be seen in the XRF analyses in table 1. The Australian ilmenite was more pure, but still contained residual aluminosilicate impurities.

TABLE 1
XRF Analyses for New Zealand and Australian Ilmenites.

	New Zealand	Australian
TiO_2	45.17	50.16
Fe_2O_3	4.16	16.62
FeO	36.61	28.18
SiO_2	6.23	1.14
Al_2O_3	2.62	0.42
MnO	1.70	1.10
MgO	0.14	0.92
CaO	1.26	0.02
Na_2O	0.58	0.00
K_2O	0.26	0.00
P_2O_5	0.24	0.02
LOI	0.57	0.66

The fine ilmenite powder was blended with carbon in ethanol for 3 hours using steel milling media. Two carbons were used: BDH Carbon (activated, acid washed) as a general purpose reagent grade material and Degussa Lampblack when a higher purity, low ash carbon was required. The carbon was added to the ilmenite so as to give a stoichiometric excess of up to 30% according to the following reaction schemes:

$$FeTiO_3 + 3C + 0.5N_2 \rightarrow Fe + TiN + 3CO \quad (1)$$
$$FeTiO_3 + 4C \rightarrow Fe + TiC + 3CO \quad (2)$$

In reaction (1) the ilmenite/carbon blend was heated under flowing N_2 whereas in (2) the reaction atmosphere was vacuum or flowing argon. Typical powder synthesis conditions were 2-4 hours at 1350-1400°C.

Disc-shaped specimens of 22 mm diameter suitable for hot-pressing were prepared by uniaxially pre-pressing 3g. of Fe-TiC/N powder blended with 5wt% oleic acid as a pressing aid, using ethanol as a dispersing agent. The organic component was removed by heating the pressed discs in flowing N_2 at 0.5°C/min to 400°C, followed by a 2 hr. soak at this temperature. Densities of the unfired discs were typically 3.47 for Fe-TiN and 3.42 for Fe-TiC. These represent ≈58% of final density. Hot-pressing was carried out in a graphite die with an Astro HP20 hot-press (Thermal Technology Ltd., Santa Barbara, USA) using flowing N_2 for Fe-TiN and vacuum for Fe-TiC, respectively. A layer of boron nitride powder was used at the graphite/specimen interfaces to prevent damage to the die. The temperature was thermocouple controlled, but each experiment was independently monitored with an optical pyrometer. The applied force was controlled at 1000kg. on the 25mm diameter graphite ram, but frictional losses reduced the effective force on the hot-pressed specimens.

RESULTS

Physical Properties of Hot-Pressed Fe-TiC Specimens.
Full densification of Fe-TiC composites was not achieved for the given applied pressure unless the temperature was above 1530°C (see Figure 1). This is a relatively high temperature and probably indicates that densification in these specimens is achieved only when the melting point of iron metal (1538°C) is reached. The powder synthesis conditions have clearly permitted little, if any, free carbon to remain available for Fe-C eutectic formation, which otherwise might have lowered the melting temperature of the metallic phase. The iron-carbon system can achieve a minimum eutectic melting temperature of ≈1150°C for a composition of 4.3wt% carbon in iron. There is some evidence from the specimens hot pressed below 1400°C that a small amount of metallic segregation occurred, but this appears to involve only a tiny fraction of the total iron in the specimens and did not contribute to the densification process at the temperatures and pressures used in this study. To take advantage of the Fe-C eutectic chemistry, additional excess carbon could be used either in the powder synthesis or blended into the Fe-TiC powder prior to pre-pressing. Such experiments are proceeding. The calculated density for an Fe-TiC body derived from Australian ilmenite is 5.91, in good agreement with the maximum density of 5.90 achieved for these specimens.

XRD experiments on the Fe-TiC powder used for composite synthesis show two phases only: α-Fe with a unit cell of 2.8665Å and TiC with a unit cell of 4.315Å. Although the TiC unit cell is found to vary in the third decimal place according to the exact temperature/time schedule of the synthesis conditions, the Fe cell is almost always within measurement error of the unit cell for pure α-Fe (2.8664Å, JCPDS 6-696). The result is consistent with there being virtually no dissolution of carbon in iron under the conditions used for the Fe-TiC powder synthesis. The unit cell for pure TiC (JCPDS#32-1383) is 4.3274Å, so the TiC prepared here is either non-stoichiometric or contains a small quantity of residual oxygen.

XRD of fully dense hot-pressed discs shows that by the time melting has occurred, at or above ≈1530°C, substantial modification to the iron crystal structure has taken place. There is a continuous rise in the proportion of secondary iron phases as full densification is approached, with a gradual reduction in the level of α-Fe to ≤20% of that in the source powder. Fe_3C predominates at the highest firing temperature. This change is coincident with

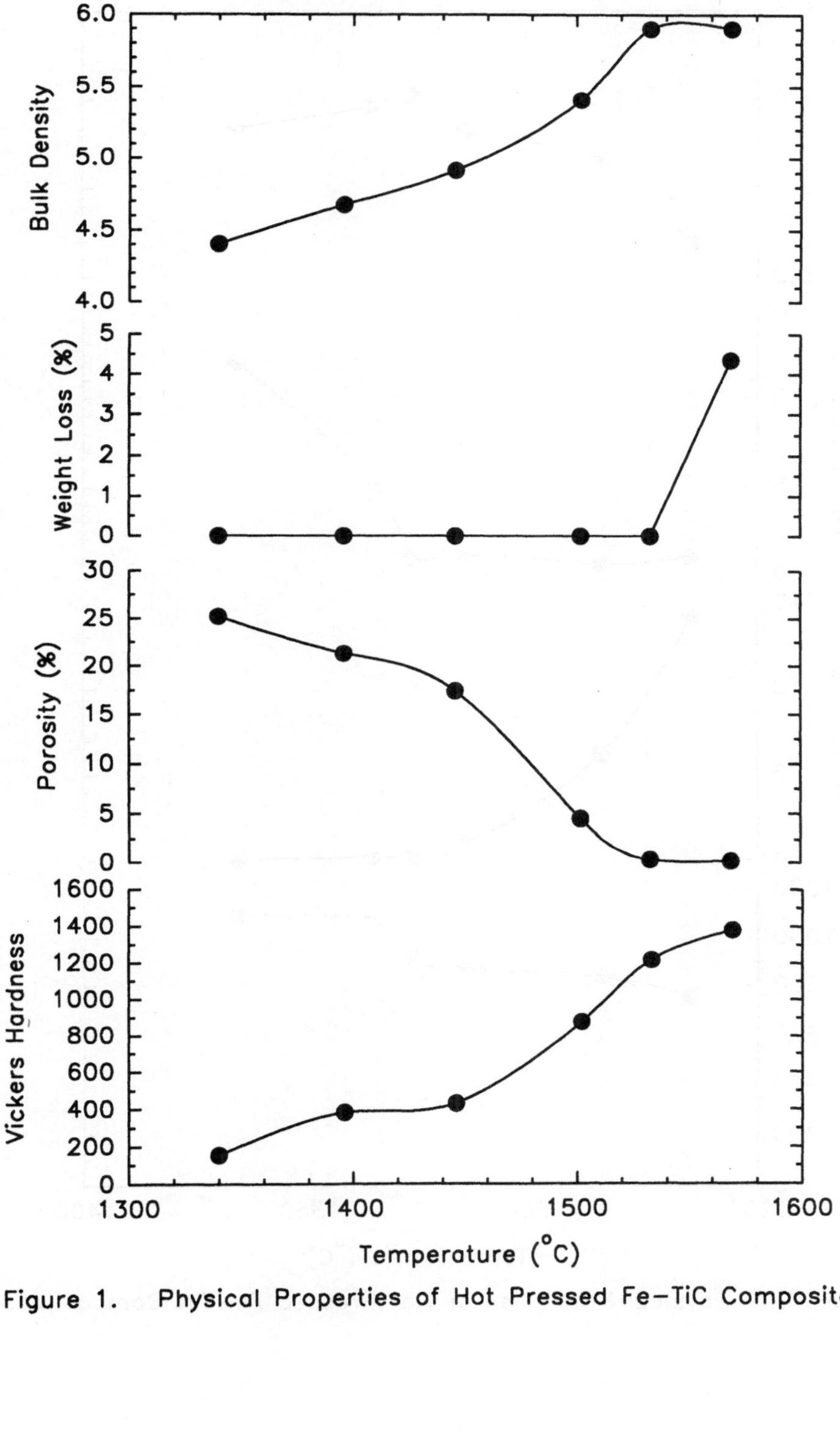

Figure 1. Physical Properties of Hot Pressed Fe–TiC Composites.

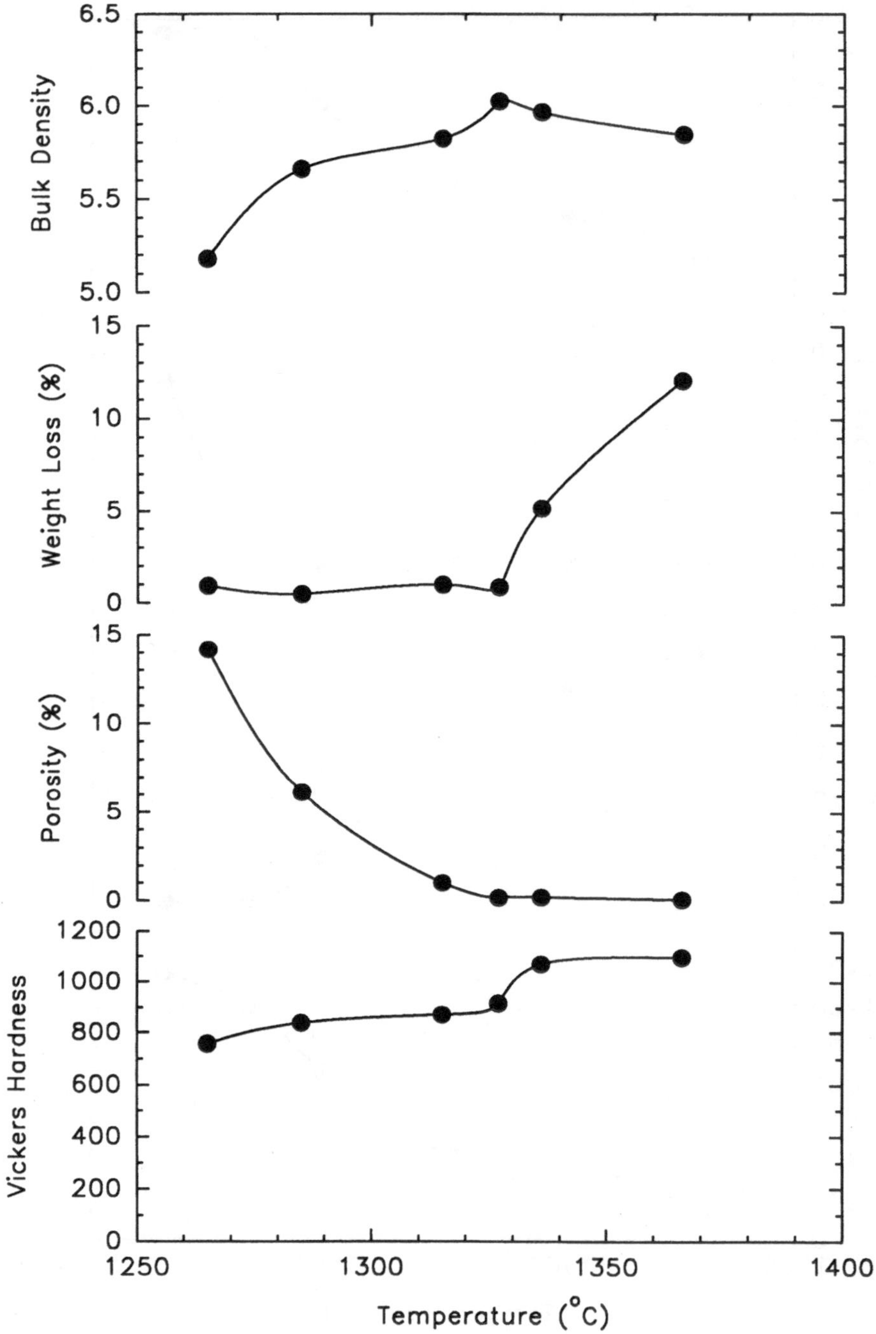

Figure 2. Physical Properties of Hot Pressed Fe–TiN Composites.

a change in TiC microstructure (discussed below) and may suggest that the TiC is the source of the carbon for the iron carbide phases. The TiC unit cell is ≈4.318Å in the highest density sample with a line width >2 times that of specimens hot-pressed below 1500°C, indicating considerable disorder (and probably a loss of stoichiometry). Coincident with the increase in sample density, the specimen porosity decreases to nil when full density is achieved (see Figure 1). Evidence of the limited extent of sintering achieved below 1500°C is indicated by very high open porosity values (17-25%) for these specimens. Once full density has been achieved, there is a tendency for the hot-pressing to squeeze out "excess" iron from the composite, as evidenced by the weight loss data in Figure 1. Interestingly, although ≈4wt% has been lost from the specimen pressed at the highest temperature (1569°C), this has not yet registered as a density reduction in the specimen, which remains stable at 5.90. This may indicate that the optimum density lies within the temperature range 1533-1569°C. Further evidence that this may be the case can be seen by examination of the Vickers hardness data in Figure 1 (for a 30 kg. load). The values for HV_{30} rise continuously as the hot pressing temperature increases, reaching ≈1380 by 1569°C.

Physical Properties of Hot-Pressed Fe-TiN Specimens.

The physical property data for Fe-TiN composites reported here in Figure 2 are in sharp contrast to the results for Fe-TiC. Firstly, Fe-TiN composites achieve full density by ≈1330°C, a full 200°C lower than for Fe-TiC. In both cases the effect of increasing the hot pressing temperature is a gradual increase in bulk density to the point of achieving zero porosity. In the case of the Fe-TiN composites this maximum density value (≈6.03) is slightly higher than the comparable value for Fe-TiC (≈5.90) and can be compared to a calculated value of 6.07. The difference may indicate some residual closed porosity in the Fe-TiN specimens. The Fe-TiN composites have an obvious maximum in their bulk density data which coincides with achieving zero open porosity. Thereafter, weight is lost from the composites as excess iron is squeezed out from the specimens and is forced up in between the walls of the graphite die. This effect is much more pronounced in the Fe-TiN composites, which tend to lose iron more readily than their Fe-TiC counterparts (see figures 1 and 2).

XRD data for the Fe-TiN source powder shows that the Fe unit cell is 2.8675Å, slightly larger than that for pure α-Fe. The TiN unit cell is 4.2600Å, higher than the expected value for pure TiN of 4.2417Å (JCPDS#38-1420), indicating a composition of approximately $TiC_{0.25}N_{0.75}$. Fully dense hot-pressed Fe-TiN specimens still contain α-Fe as the principal iron phase, although a small but continuous increase in Fe_3C, similar to that formed in Fe-TiC composites, can be observed as the pressing temperature and density rise. In Fe-TiN composites, the proportion of Fe_3C is vastly smaller than in the Fe-TiC composites, in support of the microstructural evidence presented below. The Ti(C,N) cell reduces to 4.253(2)Å on hot-pressing ie. becomes more nitrogen rich than the source powder.

Microstructure of Hot-Pressed Fe-TiC/N.

Figures 3 and 4 show typical microstructural development for Fe-TiC hot-pressed at 1569°C and 1000kg and for Fe-TiN at 1340°C and 800kg, respectively. Several features are common to both micrographs. In particular, the ceramic grain size is similar in both specimens (1-3μm) and the distribution of ceramic and metal matrix is similar. In both cases there is some evidence of partial aggregation of the ceramic grains, a likely sign of imperfect wetting. The principal difference between the materials is the retention of the sharp (cubic) grain boundaries in the TiN-based composite (figure 4) and the rounded, almost globular appearance of the TiC grains (figure 3). Almost certainly the additional 200°C required to densify the Fe-TiC

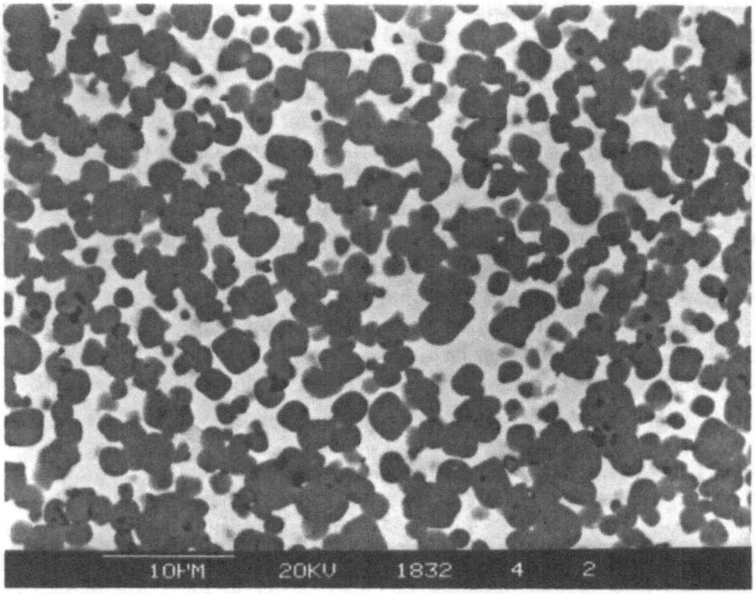

Figure 3. Electron micrograph of composite prepared by hot-pressing Fe-TiC powder.

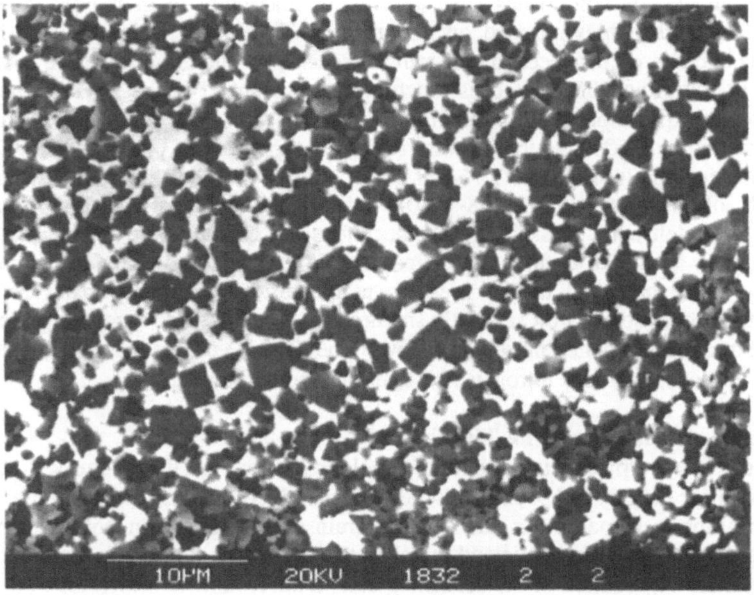

Figure 4. Electron micrograph of composite prepared by hot-pressing Fe-TiN powder.

composite has been a dominant factor here.

It is essential to relate these observations to the grain size distribution in the source Fe-TiC/N powders. Although micrographs are not reproduced here, both Fe-TiN and Fe-TiC show similar characteristics: all powder material greater than ≈5μm (ie. that which can be polished readily) shows a dispersion of blocky (cubic) TiN or TiC grains averaging 1-3μm dispersed in an α-Fe matrix. The microstructure of both powder materials is almost indistinguishable from that shown in figure 4 for hot-pressed Fe-TiN, implying that no further grain growth occurs on hot-pressing the Fe-TiN powder, while the blocky structure that is carried over suggests that there is only very limited interfacial reaction between the iron matrix and the TiN grains. ie. wetting of the TiN is poor. In contrast, the TiC grains are heavily rounded upon hot-pressing, indicating much greater interfacial reaction between the matrix and TiC and much better wetting.

The wetting of ceramic by metal is also affected by other factors. Specimens hot-pressed using powder derived from the less pure ilmenite (Table 1) can be fully densified without difficulty, but examination of the microstructure reveals that the aluminosilicate impurities become the active wetting agents and the glassy phase so formed often interferes with or prevents the desired iron to TiC/N contact. The result is dense bodies with a very inhomogeneous microstructure that contain too much brittle phase between the TiC/N grains and insufficient ductile (tough) iron phase.

In summary, the grain size distribution of the TiC/N is closely controlled by the characteristics of the Fe-TiC/N source powder. Production of finer TiC/N grain sizes will require more careful control of the powder process chemistry, rather than any substantial modification to the hot-pressing or densification method.

CONCLUSIONS.

1. Carbothermal reduction of ilmenite ($FeTiO_3$) under N_2, Ar or vacuum gives Fe-TiC and Fe-TiN powders with 1-3μm sized TiC/N grains suitable for the manufacture of ceramic-metal composites.
2. Fully dense composites can be prepared by hot-pressing Fe-TiN powders at ≥1330°C and Fe-TiC powders at ≥1530°C, giving HV_{30} values of 1100 and 1380, respectively.
3. Imperfect wetting is apparent in both composites above the minimum densification temperature, more particularly so in the case of the Fe-TiN system.
4. The microstructure of Fe-TiN composites retains the sharp (cubic) TiN grain boundaries evident in the source powder, whereas the Fe-TiC composites display completely rounded TiC grains. In both cases the ceramic grain size is 1-3μm.

REFERENCES.

1. Ellis, J.L., Carbides that are weldable, machinable and heat treatable. The Tool Engineer, April 1957, 103-105.
2. Lehuy, H., Cliche, G. and Dallaire, S., Synthesis and characterisation of Ti(C,N)-Fe cermets produced by direct reaction. Materials Science and Engineering, 1990, A125, L11-L14.
3. Von Holst, J.P. and Oskarsson, R.G., Compound body and method of making the same. US Patent 4,618,540, 1986.

WETTING AND INTERACTIONS BETWEEN SELECTED CERAMICS AND Ni-BASE BRAZE ALLOYS

M. HOLMSTRÖM, L. LJUNGBERG, A. WARREN, R. WARREN
Department of Engineering Metals,
Chalmers University of Technology,
S-412 96 Sweden

ABSTRACT

Wetting tests were performed with two commercial Ni-base braze alloys on a selection of ceramic materials in vacuum between 1000 and 1150 °C. The two alloys were Ni-5Si-3B and Ni-7Cr-5Si-3Fe-3B(wt%). The ceramic substrates included Al_2O_3, Si_3N_4, TiC/TiN coated Si_3N_4, SiC, a SiC/SiC CVI composite and a TiC/TiN coated SiC/SiC CVI composite. In one experiment a thin foil of Ti was placed between the braze alloy and the ceramic in an attempt to promote wetting. With Al_2O_3, Si_3N_4 and TiC/TiN coated Si_3N_4 there were no observable reactions between braze and substrate, the wetting was poor (contact angles $\theta \geq 110^o$) and the braze drops detached themselves after solidifying. With a thin foil of Ti between the braze alloy and the Si_3N_4 ceramic a restricted reaction was noticed and wetting occurred ($\theta \approx 70^o$). The braze detached itself but in this case this was due to fracture through a brittle reaction product. The bulk SiC substrate exhibited a strong chemical reaction with the braze. The SiC/SiC composite exhibited "good" wetting but this could be explained by the open porosity of the material and a penetration of the braze alloy. On the two SiC/SiC materials the braze alloy adhered to the ceramic and reaction could be observed. Wetting tests were also performed on the superalloys, Hastelloy X and Incoloy 909; both exhibited satisfactory wetting behaviour.

INTRODUCTION

The joining of ceramics to metals is becoming of growing significance with the increasing use of high performance ceramics. One of the most promising methods of joining is brazing which involves using an intermediate material (e.g. a braze alloy). This has a lower melting point than the materials to be joined and is brought to the molten state during the process. A weakness of the method is that the temperature capability of the joint is limited by the melting point of the braze alloy. Thus in joining high temperature materials such as ceramics there is a need to find braze alloys with high melting points. Other requirements of a braze alloy are satisfactory wetting and bonding without excessive chemical reaction with the substrate. The main objective of this work was to investigate the wetting and interactions between Ni-base alloys and selected ceramics and Ni-based superalloys substrates of relevance to high temperature applications. For a preliminary study, two commercially avaiable Ni-base brazes were chosen. These have been developed primarily for the brazing of superalloys (1) and consequently little is known of their interactions with ceramics. Their recommended brazing temperatures are around 1050°C and they are suitable for service up to about 800°C. Nicholas (2)has shown that for certain combinations of ceramics and Ni-base brazes satisfactory wetting can be achieved.

EXPERIMENTAL PROCEDURE

Materials

The two braze alloys were supplied in the form of powder by Lucas- Milhaupt, Inc.,USA. Their compositions are given in Table 1.

TABLE 1
Composition of braze alloys

braze alloy	contents (wt%)				
	B	Cr	Fe	Ni	Si
AMS 4778 A	3	-	-	Bal.	5
AMS 4777	3	7	3	Bal.	5

As well as a variety of ceramic substrates, two superalloy substrates were investigated namely Hastelloy X and Incoloy 909. Hastelloy X is a nickel based, solid solution strengthened superalloy with the composition given in Table 2. The material was heat-treated at 1170°C, 10 min in air followed by air cooling. Incoloy 909 is a low thermal expansion superalloy. It is a precipitation hardening alloy, the hardenability being achieved by additions of niobium and titanium. Its nominal chemical composition is given in table 2. All tests were carried out on material heat treated for optimum strengthening as follows: 1) 980°C for 1 hour and then air cooled to RT 2) 720°C for 8 hours and then cooled in furnace to 620°C 3) 620°C for 8 hours and then air cooled to RT.

TABLE 2
Composition of substrate alloys

substrate	contents (wt%)											
	Al	C	Co	Cr	Fe	Mn	Mo	Nb	Ni	Si	Ti	W
Hastelloy X	-	0.1	1.5	22	18.5	≤1	9	-	Bal.	≤1	-	0.6
Incoloy 909	0.03	0.01	13	-	42	-	-	4.7	38	0.4	1.5	-

The ceramic substrates were as follows:
Al_2O_3 - Standard α-alumina "AL 23" (supplied by Friedrichfeld GmBH, Germany); purity ≥99.5 % and density between 3.7 to 3.95 g/cm^3. Si_3N_4 - hot isostatically pressed at ABB Cerama, Sweden to full theoretical density from silicon nitride powder type " S-95", Kema Nord, Sweden. 1% of Y_2O_3 was added as sintering aid. About 1% of WC and Co can also be expected from the milling process of the powder. TiC/TiN coated Si_3N_4 - A thin double CVD layer of TiC/TiN prepared at Sandvik Coromant AB, Sweden on a substrate of Si_3N_4. The coating consists of a base layer of about 5 μm TiC with a top coating of about 3.5 μm TiN produced in a reaction chamber at about 1000 °C. The coating was applied to inhibit the expected chemical reaction between SiC and Ni while at the same time producing a wettable surface. SiC - hot isostatically pressed to full density at ABB Cerama , Sweden. The raw material was SiC-powder type undoped Lonza VF 15. CVI SiC/SiC fibre composite - A composition formed by the chemical vapour infiltration of SiC yarn fibres with a SiC-matrix (supplied by SEP, France). The fibres themselves are prepared in the form of a multifilament yarn by the pyrolysis of a

polycarbosilane to yield a microstructure consisting of a mixture of SiC, silica and free C. In this case the composite consisted of 51 vol% amorphous SiC matrix, 9 vol% porosity and 40 vol% woven fibre with 0°/90° directions stacked to give a 3 mm thick composite plate. TiC/TiN coated CVI SiC/SiC fibre composite - The CVD layers were prepared at Tixon AB, Sweden with the same standard treatment as for TiC/TiN coated Si_3N_4.

Methods

The wetting tests were carried out in a horizontal Mo-wound alumina resistance furnace. The temperature is measured and controlled to about +/-5°C with a Pt/Pt-Rh thermocouple placed about 10 mm from the sample. The vaccum achieved by an oil diffusion pump system is about 10^{-6} mbar at room temperature and 10^{-5}mbar at 1200°C. The sample in the furnace is photographed during the test.

All the substrates were ground with 600 grit SiC paper except for the CVI SiC/SiC-fibre composite samples, which were used as-received. All samples were ultrasonically cleaned in acetone before the wetting experiment. The braze alloy powder was mixed with heptane to form a pellet and then placed on the substrate.

Photographs were taken of the drop profile after 5 min holds at 1050, 1100, 1150 °C. The wetting angles were evaluated from the photographs normally to an accuracy of around 10 %.

Cross-sections of the samples from the wetting experiments were polished for microstructural analysis, which were performed with a JEOL 733 scanning electron microscope.

RESULTS

The wetting behaviour on the various substrates is presented in Table 3 and Figure 1. With the Al_2O_3, Si_3N_4 and TiC/TiN coated Si_3N_4 there was no observable reaction between braze and substrate; the wetting was poor (contact angles $\theta \geq 110$ °) and the braze drops detached themselves after solidifying. However small satellite drops (originating from powder particles detached from the braze pellet) remained attached to the substrates and on the TiC/TiN coated Si_3N_4 some of these small drops exhibited better wetting than the main braze drop.

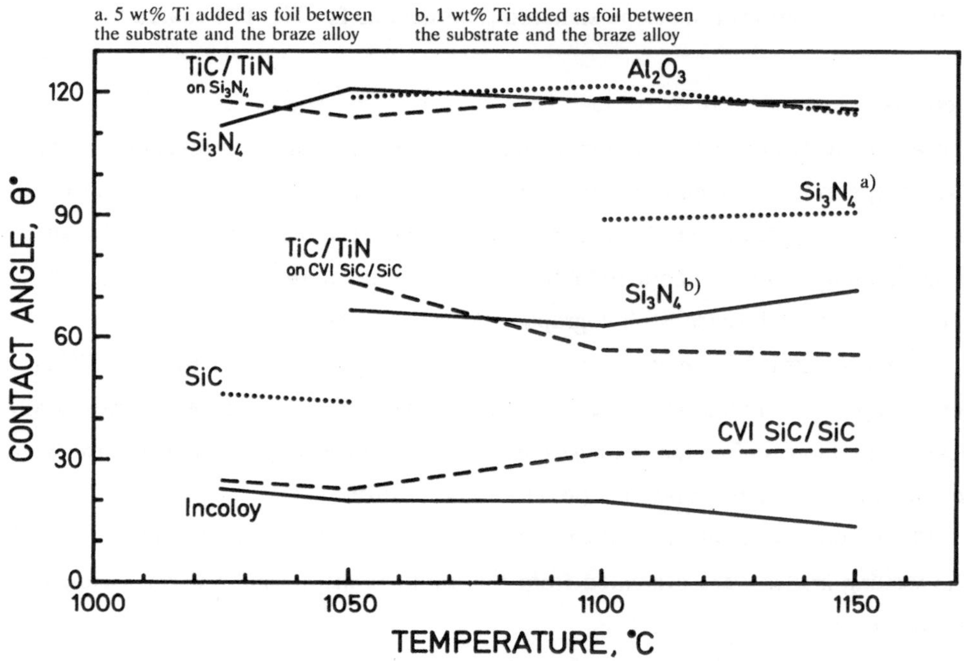

Figure 1. Wetting angles for Ni-7Cr-5Si-3Fe-3B on selected ceramic substrates and superalloys.

With a thin foil of Ti between the braze alloy and the Si_3N_4 ceramic a restricted reaction was observed at the interface and wetting occurred ($\theta \approx 70°$). The reaction product was gold-coloured and is assumed to be TiN. The braze drop detached from the substrate but with substrate fragments attached indicating fracture through the brittle reaction product or the Si_3N_4.

TABLE 3
Wetting angles

substrate	wetting behaviour (wetting angle at 1150°C) Ni-5Si-3B	Ni-7Cr-5Si-3Fe-3B
Al_3O_4	128° / detached	115° / detached
Si_3N_4	-----	118° / detached
	-----	73° / detached (+ 1 % Ti)
	-----	91° / detached (+ 5 % Ti)
TiC/TiN on Si_3N_4	129° / detached	116° / detached
SiC	-----	44° / reaction, adhered
SiC/SiC composite	33° / reaction, adhered	42° / reaction, adhered
TiC/TiN on SiC/SiC	-----	55° / reaction, adhered
Incoloy 909	-----	14° / adhered
Hastelloy X	16° / adhered	-----

The bulk SiC-substrate exhibited a strong chemical reaction with the braze. A 25 μm thick reaction product layer was produced. Ni, Fe and Si were detected in this layer by electron probe micro analysis. Cracks had formed both in the reaction product layer and in the ceramic substrate. It is probable that a crack initiates in the reaction layer and continues into the ceramic substrate (see Figure 2). In spite of the reaction the wetting was only moderate ($\theta \approx 45°$).

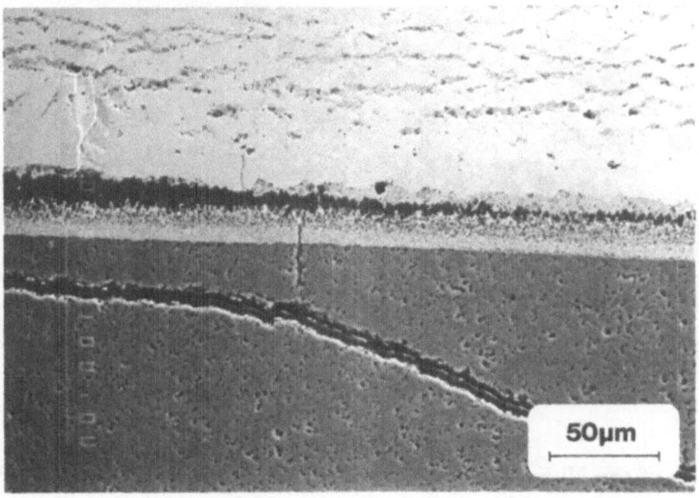

Figure 2. Interfacial region between Ni-7Cr-5Si-3Fe-3B and bulk SiC.

Both the coated and uncoated SiC/SiC composite exhibited apparently better wetting than the monolithic SiC but this could be explained by penetration of the open porosity of the material by the molten braze alloy. The braze also adhered to both substrates. An extensive reaction zone in which both the matrix and the fibres had reacted was found between the braze and the substrates (Fig 3). The presence of the TiC/TiN coating in the coated composite was evidently ineffective as a barrier to the reaction. The marked reactions between the SiC-containing substrates and the braze alloys are probably dominated by the reactions of the type:

$$SiC + Ni \rightarrow \text{Ni-silicides} + C$$

Electron probe microanalysis indicated that the reaction zones including the reacted fibres contained both Ni and Si. The substrate surface collapsed under the braze with the result that cavities were formed.

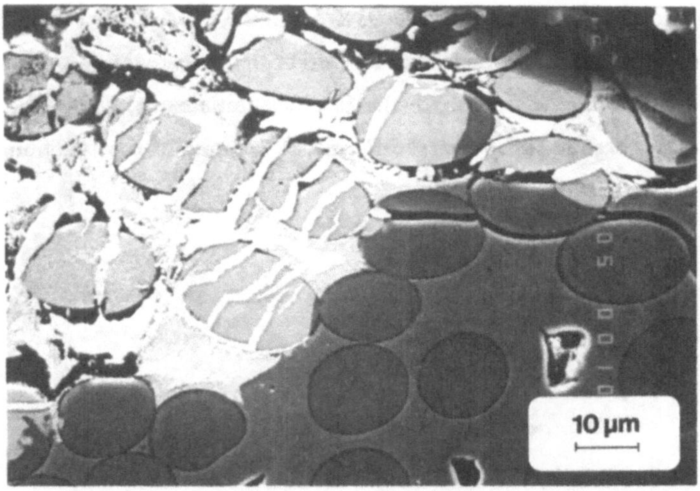

Figure 3. Position of reaction zone between Ni-7Cr-5Si-3Fe-3B braze alloy and coated SiC/SiC composite.

The wetting of the Ni-base alloy substrates by the brazes was satisfactory as was to be expected.

DISCUSSION AND CONCLUSIONS

The investigated Ni-braze alloys do not perform satisfactorily as braze alloys on the range of ceramics investigated. Either the wetting is poor with wetting angles greater than 90° or excessive reactions occur at the interface. The observed interactions are consistent with known phase relationships between the ceramic compound and metallic nickel. Thus the braze reacts with SiC substrates (3) but is stable with respect to silicon nitride and alumina (4,5). The alloying additions in the braze do not appear to have a significant effect on the reactions and neither do they lead to a significant improvement of wetting - i.e. they are not active with respect to these substrates. Through unstable active additions Ni-base brazes may have potential as brazes for Al_2O_3 and Si_3N_4.

ACKNOWLEDGEMENTS

We would like to thank Dr. L.Pejryd and Dr. R.Lundberg both of Volvo Flygmotor AB, Trollhättan, Sweden for valuable help and advice. The work was supported financially by the Swedish Board of Techical Development (STU).

REFERENCES

1. Knotek, O., Lugscheider, E., Brazing Filler Metals Based on Reacting Ni-Cr-B-Si Alloys. Welding Journal, 1976, **10**, 314s-318s

2. Nicholas, M.G., Peteves, S.D., this conference

3. Warren, R., Andersson, C.H., Silicon Carbide Fibres and Their Potential Use in Composite Materials, Part 2, Composites, 1984, **2**, 101-111

4. Weitzer, F., Schuster, J.C., Phase Diagrams of the Ternary Systems Mn, Fe, Co, Ni-Si-N, J. Solid State Chemistry, 1978, 178-184

5. Rühle, M., Mader, W., Structure and Chemistry of Metal/Ceramic Interfaces, In Designing Interfaces for Technological Applications, ed. S.D.Peteves, Elsevier Applied Science, London, 1986, pp 145-195

CMZP - A NEW HIGH TEMPERATURE THERMAL BARRIER MATERIAL

D.A.HIRSCHFELD, D.M.LIU, and J.J.BROWN
CIT Center for Advanced Ceramic Materials
Virginia Polytechnic Institute and State University
Blacksburg, VA, 24061-0256 USA

ABSTRACT

New $(Ca_{1-x},Mg_x)Zr_4(PO_4)_6$ ceramics in the ultra low expansion NZP family not only have near zero thermal expansion but also have thermal conductivities lower than those of conventional thermal barrier materials such as ZrO_2. In addition, CMZP is less dense than ZrO_2 with a theoretical density of approximately 3.2 gm/cm^3 versus 5.8 gm/cm^3. Techniques have been developed to fabricate lightweight ceramics with both open and closed pore structures with relative densities ranging from 0.2 to 0.8. The tensile, compressive, and flexure strengths are shown to compare favorably with those of ZrO_2. Furthermore, no strength loss following air quenching from temperatures up to 1500°C was observed indicative of good thermal shock resistance.

INTRODUCTION

A new class of ceramics in the [NZP] structural group have been identified as having an ultra-low coefficient of thermal expansion.(1,2,3,4) One particular member of this group, $(Ca_{1-x},Mg_x)Zr_4(PO_4)_6$ (CMZP), is chemically stable to 1500°C and also possesses a thermal conductivity lower than that of conventional thermal barrier materials while maintaining good thermal shock resistance. The objective of this research is to demonstrate the potential application of lightweight CMZP by comparing its properties with those of ZrO_2, a common high temperature insulator.

The thermal expansion of [NZP] can be explained in terms of the rotation of polyhedra which make up the lattice (3). The crystal structure of [NZP] is hexagonal with $R\bar{3}C$ space symmetry built up of PO_4 tetrahedra and ZrO_6 octahedra which are linked by corners to form a three-dimensional network. The bonds between the polyhedra are strong and can bend and rotate without breaking the structure. The interstitial sites can be left vacant or allow a variety of ionic substitutions without changing the crystal structure(5,6). In the case of CMZP, Ca^{+2} and Mg^{+2} are substituted for Na^{+} in the [NZP] structure to create a new, low-thermal-expansion, low-thermal-conductivity material. Axial expansion measurements have shown that the **a** axis contracts while the **c** axis expands upon heating which results in an overall near zero thermal expansion(7,8). It has also been shown that as the Mg content is increased from 0.0 to 0.4, the expansion anisotropy is reduced yielding a bulk coefficient of thermal expansion over the temperature range of 25°C to 750°C of -4.4 x 10^{-7}/°C for $CaZr_4(PO_4)_6$ and +4.8 x 10^{-7}/°C for $(Ca_{0.6},Mg_{0.4})Zr_4(PO_4)_6$(7).

The thermal conductivity of dense CMZP at high temperatures is less than half that of ZrO_2 (Figure 1). The conductivity of CMZP was determined from the thermal diffusivity which was measured using the laser flash technique (7). The data for ZrO_2 are from Hassellman, et al (8). Furthermore, the thermal conductivity tends to decrease with increasing Mg content. To reduce the thermal conductivity even more, CMZP can be made into a porous lightweight ceramic.

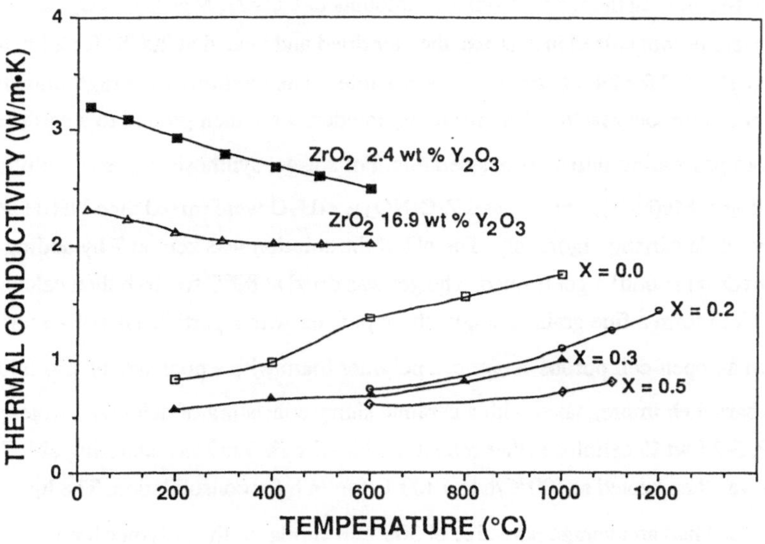

Figure 1. Thermal Conductivity of CMZP and ZrO_2

Porous lightweight ceramics may be formed with either an open-cell or a closed-cell structure. Typically, open-cell ceramics are formed by coating a reticulated polymer substrate with a ceramic slurry then sintering. The main advantages of this technique are first, that the pore size and interconnectivity of the lightweight ceramic is controlled by the morphology of the polymer substrate and secondly, that products having complex shapes are easily formed (9). Closed-cell structures can be formed by one of two techniques: sintering of hollow spheres (11) or by mixing a polymer powder with a ceramic powder then sintering the resulting compact. The tensile strength, compressive strength, flexure strength, and thermal shock resistance of lightweight ceramics made by both processing techniques were investigated.

EXPERIMENTAL PROCEDURE

First, CMZP powders were synthesized by both solid-state reaction and sol-gel techniques. In the solid-state reaction method, stoichiometric amounts of $CaCO_3$, $MgCO_3$, ZrO_2, and $NH_4H_2PO_4$ were homogenized in acetone, then air dried and heated at 200°C for 24 h, at 600°C for 8 h, and at 1250°C for 24 h to drive off the volatiles. The mixtures were reground after each heat treatment to promote reaction. The resulting powders were then ground to a particle size of 3-5 μm before processing into a porous ceramic. For sol-gel synthesis, aqueous solutions of $Ca(NO_3)_2 \cdot 4H_2O$, $Mg(NO_3)_2 \cdot 6H_2O$, and $ZrO(NO_3)_2 \cdot xH_2O$ were mixed then $NH_4H_2PO_4$ was added slowly while stirring vigorously. The pH of the solution was kept at 7 by adding ammonium hydroxide until a gel formed. The gel was dried at 80°C for 16 h then calcined at 650°C for 20 h to form a fine grained single phase powder with a particle size of 45 nm.

To form an open-cell porous ceramic, a polymer foam with a pore size of 250-300 μm was cut into bars then impregnated with a ceramic slurry consisting of deionized water, 10-25 wt % CMZP, 5-25 wt % cellulose ether binder, and 0 - 5 wt% ZnO as a sintering aid. The coated foam was then heated at 100°C/h to 1300°C for 24 h for consolidation. The lightweight ceramic produced had an average pore size of 300 μm similar to the polymer foam.

Lightweight ceramics having primarily a closed pore structure were produced using the polymer powder method. This was accomplished by mixing CMZP with poly(vinyl chloride) powder having an average particle size of 25 μm and 0 - 5 wt% ZnO. The mixture was then consolidated into a bar using a hydraulic press and stainless steel die. The bars were then heated using the same firing conditions as the lightweight ceramic made using the polymer foam method. The resulting lightweight ceramics exhibit a range of pore sizes less than 100 μm.

The theoretical density of CMZP is 3.2 gm/cm^3 (1). The polymer foam method yielded lightweight ceramics with relative densities of less than 0.35 while the polymer powder method was used to produce CMZP with relative densities greater than 0.35. The densitites were determined by the ASTM boiling water method designated C 20-87 which used the Archemedes method with water as the fluid medium.

For comparison, ZrO_2 was made using the polymer powder method described previously. Zirconia, 5-25 wt % poly(vinyl chloride), and 4.2 wt % MgO were pressed into bars then slowly heated (100°C/h) to 1500°C and held for 6 h. The theoretical density of magnesia stabilized zirconia is 5.9 gm/cm^3.

Tensile, compressive, and flexure strengths of the lightweight ceramics were measured at room temperature using an Instron Model 4204 universal testing machine with a load capacity of 4500 N (12). Tensile strength was determined by loading specimens formed using solid state reaction derived powders and having dimensions 6mm x 10mm x 90mm at a rate of 0.5mm/min. The tensile specimens were attached to aluminum grip adaptors with an epoxy adhesive. The compression specimens, also made using powders derived by solid state reaction, were 12.5mm x 50mm x 50mm loaded at a rate of 0.6mm/min (ASTM C 165-83). The flexure strength for CMZP synthesized by both solid-state reaction and sol-gel methods was measured using 3-pt. bending (ASTM C 203-85) for CMZP synthesized by both solid-state reaction and sol-gel methods.

The thermal shock resistance of lightweight CMZP was studied by air quenching specimens from temperatures up to 1500°C at room temperature then determining the flexure strength. Specimens were made using the polymer powder method with and without ZnO sintering aid.

RESULTS AND DISCUSSION

The typical microstructure of lightweight CMZP with an open-cell structure formed using the polymer foam method is shown in Figure 2, and with a closed-cell structure formed by the polymer powder method with a relative density of 0.67 in Figure 3. As the relative density of ceramics made by the polymer powder method decreases, the pores become more interconnected.

The flexure strength of lightweight CMZP is independent of composition and processing technique, however when formed using sol-gel derived powders the MOR is two to three times greater than that of ceramics formed using coarse solid state reaction powders as shown in Figure 4 for relative densities of less than 0.5. This behavior appears to be due to the observed presence of holes within the struts between the pores in specimens prepared with powders

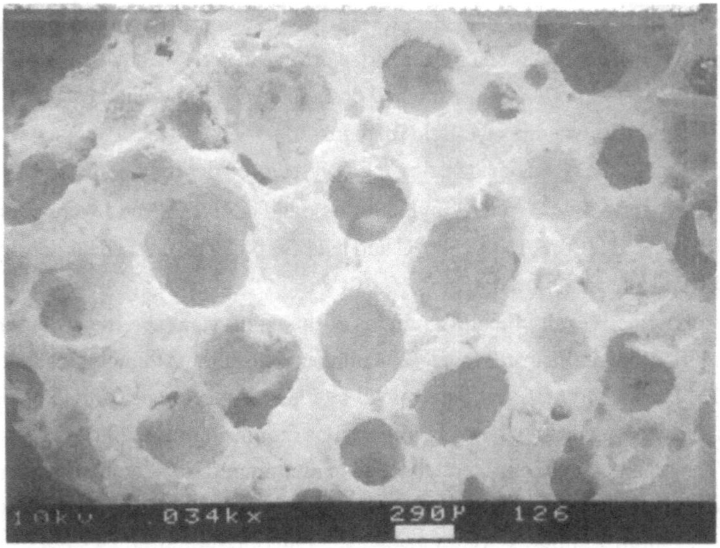

Figure 2. Open-cell structure of lightweight CMZP

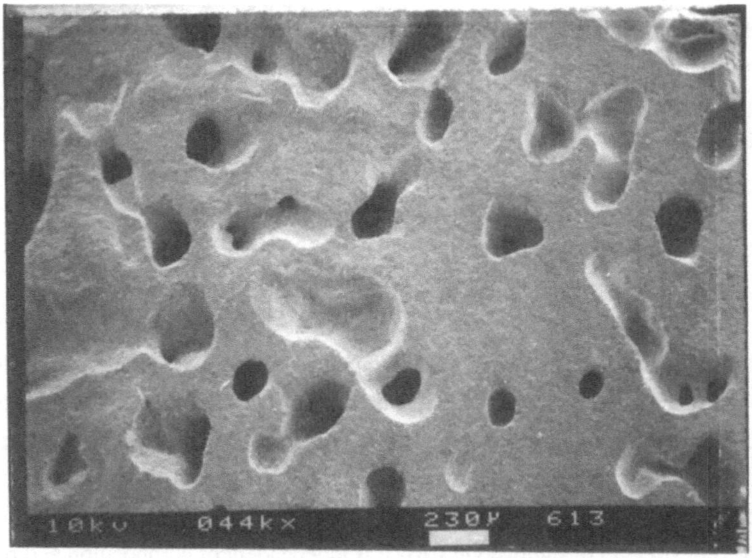

Figure 3. Closed-cell structure of lightweight CMZP

synthesized by solid-state reaction. The struts in the sol-gel synthesized powders appear dense with no defects. Furthermore, the MOR of lightweight CMZP compares favorably with that of ZrO_2.

The tensile and compressive strengths also compare favorably with that of ZrO_2 on the basis of bulk density as shown in Figures 5 and 6, respectively. These strengths are independent of CMZP composition. The compressive strength is dependent on the processing method with the specimens produced by the polymer foam method exhibiting a lower compressive strength than those produced by the polymer powder method. This decrease in strength appears to be related to the small volume of ceramic between pores (Figure 2).

The thermal shock resistance of lightweight CMZP made by the polymer powder method using ceramic powders synthesized by solid state reaction is shown in Figure 7. Lightweight ceramics with relative densities of 0.87 and 0.54 exhibited no significant strength loss following air quenching from any temperature up to 1500°C. The addition of ZnO, a sintering aid, increased the modulus of rupture and did not appear to affect the thermal shock resistance.

CONCLUSIONS

Lightweight CMZP ceramics have been produced with either an open-cell or closed-cell structure. The tensile, compressive, and flexure strengths of CMZP compares favorably with those of lightweight ZrO_2. The flexure strength of lightweight CMZP produced using powders synthesized by sol-gel techniques is significantly greater than that produced using coarse solid-state reaction powders due to the reduction of defects in the solid material between the pores. CMZP exhibits good thermal shock resistance.

ACKNOWLEDGEMENTS

This work was supported by the U.S.Department of Energy, Ceramic Technology for Advanced Heat Engines Project, Subcontract 86X22049C, and the Virginia Center for Innovative Technology. The research was conducted at the Center for Advanced Ceramic Materials at Virginia Polytechnic Institute and State University, Blacksburg, VA.

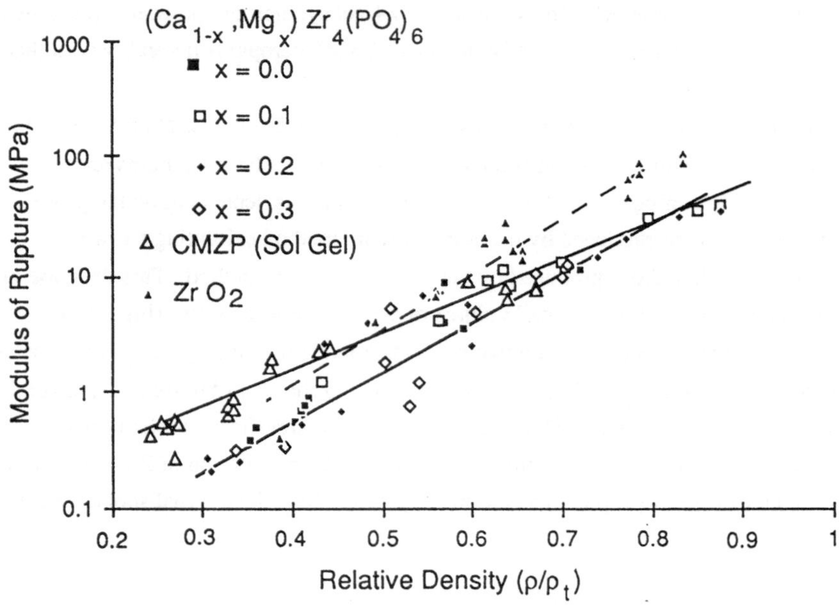

Figure 4. Flexure strength of CMZP and ZrO_2

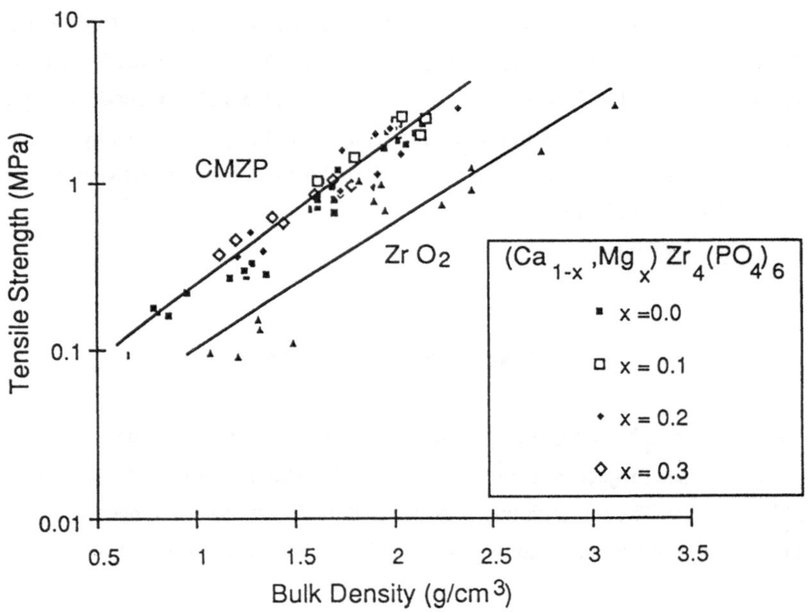

Figure 5. Tensile strength of CMZP and ZrO_2

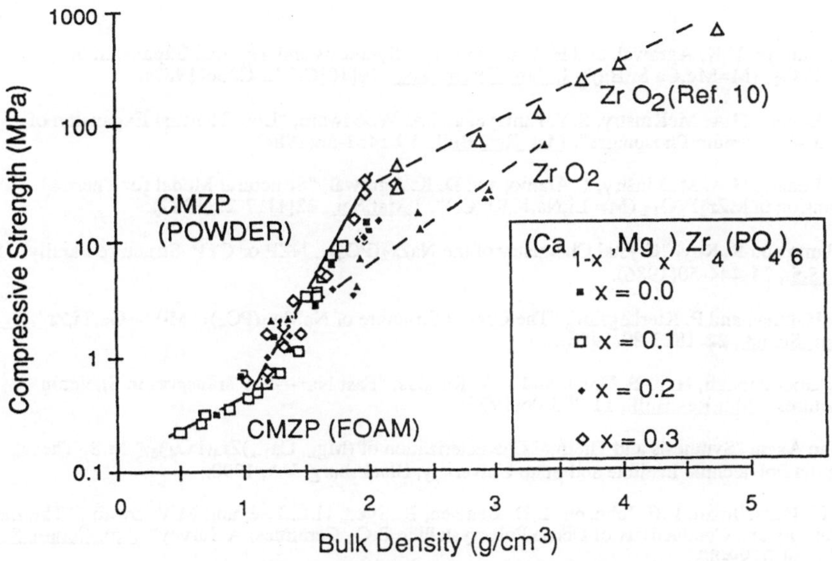

Figure 6. Compressive strength of CMZP and ZrO_2

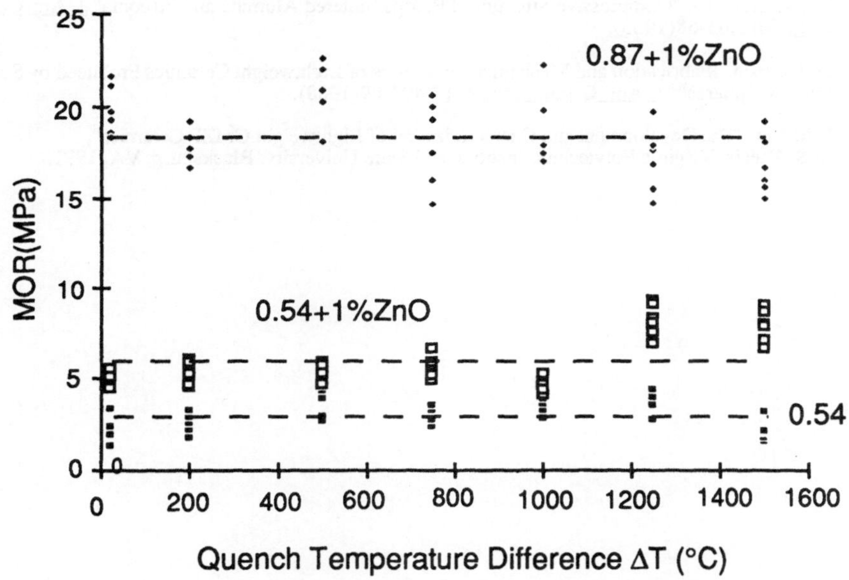

Figure 7. Thermal shock resistance of CMZP

REFERENCES

1. S.Y. Limaye, D.K. Agrawal, and H.A. McKinstry, "Synthesis and Thermal Expansion of $MZr_4P_6O_{24}$ (M=Mg,Ca,Sr,Ba)", J. Am. Ceram. Soc., **70**[10]C 232-C236(1987).

2. G.E. Lenain, H.A. McKinstry, S.Y. Limaye, and A. Woodward, "Low Thermal Expansion of Alkali---Zirconium Phosphates", Mat. Res. Bull., **19** 1451-56(1984).

3. G.E. Lenain, H.A. McKinstry, J. Alamo, and D. K. Agrawal, "Structural Model for Thermal Expansion in $MZr_2P_3O_{12}$ (M= Li,Na,K,Rb,Cs)", J.Mat.Sci., **22**[1]17-22(1987).

4. J. Alamo and R. Roy. "Crystal Chemistry of the $NaZr_2(PO_4)_3$, NZP or CTP, Structure Family", J. Mat. Sci., **21** 444-50(1986).

5. L.O. Hagman and P. Kierkegaard, "The Crystal Structure of $NaM^{iv}_2(PO_4)_3$, M^{IV} = Ge,Ti,Zr", Acta. Chem. Scand., **22** 1822-89(1986).

6. J. B. Goodenough, H.Y.-P. Hong, and J. A. Kafalas, "Fast Na^+---Ion Transport in Skeleton Structures", Mat.Res.Bull., **11** 203-06(1976).

7. S. Van Aken, "Synthesis and Thermal Characterization of $(Mg_x, Ca_{1-x})Zr_4(PO_4)_6$," M.S. Thesis, Virginia Polytechnic Institute and State University, Blacksburg, VA, 1990.

8. D.P.H. Hassellman, L.F. Johnson, L.D. Bentsen, R. Syed, H.L. Lee, and M.V. Swain, "Thermal Diffusivity and Conductivity of Dense Polycrystalline ZrO_2 Ceramics: A Survey", Am. Ceram. Soc. Bull., 66[5]799-806.

9. F.F. Lange, and K. T. Miller, "Open-Cell, Low-Density Ceramics Fabricated from Reticulated Polymer Substrates", Adv. Ceram. Mat., **2**[4]827-31(1987).

10. E. Ryshkewitch, "Compressive Strength of Porous Sintered Alumina and Zirconia" J. Am. Ceram. Soc., **36**[2]65-68(1953).

11. D. J. Green, "Fabrication and Mechanical Properties of Lightweight Ceramics Produced by Sinter of Hollow Spheres", J. Am. Ceram. Soc., **68**[7]403-09(1988).

12. D.M.Liu, "The Development and Characterization of Lightweight CMZP Ceramics", M.S. Thesis, Virginia Polytechnic Institute and State University, Blacksburg, VA, 1991.

MICROSTRUCTURE AND MECHANICAL PROPERTIES OF ZrO_2 TOUGHENED MULLITE SYNTHESIZED BY SOL-GEL METHOD

M.G.M.U.Ismail, H.Shiga, K.Katayama, Z.Nakai and T.Akiba,
Chichibu Cement Co.Ltd., Kumagaya, Saitama, Japan
* S.Sōmiya , The Nishi Tokyo University, Uenohara, Yamanashi, Japan

ABSTRACT

Zirconia dispersed mullite composites were prepared by adopting the sol-gel method. The maximum amount of ZrO_2 that could be dispersed in mullite without causing a decrease in strength was 15 vol%. The flexural strength and fracture toughness of 15 vol% ZrO_2 dispersed mullite-zirconia ceramic at room temperature were 500 MPa and 4.3 $MPa.m^{1/2}$, respectively. The flexural strength at 1400°C of the same composition was 300 MPa. The ZrO_2 in these composites occupy both inter and intra granular positions. Morphology of mullite grains changed from equiax to needle like on sintering at temperatures above 1670°C. The mechanical properties were deteiorated with the change of microstructure.

INTRODUCTION

The high-temperature strength, creep resistance and good thermal stability are the main characteristics of pure mullite ceramics. The low fracture toughness is the main barrier to overcome, and Claussen and John (1) showed, by dispersing zirconia in mullite, that the fracture toughness could be increased. The method adopted for dispersion of zirconia in mullite were reaction of zircon ($ZrSiO_4$) with Al_2O_3 (2), (3), or mechanical mixing of mullite powder with zirconia (4). In the present investigation sol-gel method was adopted to get a homogeneous dispersion of zirconia in muulite. The mechanical properties of composites containing different proportions of ZrO_2 were determined . It was found that the maximum amount of ZrO_2 that could be dispersed to obtain a high flexural strength was 15 vol%.

EXPERIMENTAL

Mullite sol was prepared by mixing boehmite and silica sols. The Al_2O_3/SiO_2 mole ratio was 1.5 and kept constant through out this investigation. Zirconyl chloride solution of 1M concentration was added to the mullite sol and mixed to form a homogeneous dispersion. The composite sol was gelled within few minutes on leaving at room temperature. The gel was dried, ball milled and calcined at temperatures between 900°C and 1400°C for 1 hour. The calcined gel was ball milled in aqueous medium for 15 hours using TZP grinding media. The slurry was dried at 120°C and used for sintering and other property determinations. Powder compacts were cold isostatically pressed under 2 tons/cm^2 , and sintered at temperatures ranging from 1550°C to 1680°C for 3 hours. Flexural strength was determined by three-point bend method. Fracture toughness was determined by single edge pre-cracked beam (SEPB) method . The t-ZrO_2 content in the powder and sintered specimens were determined by the method described by Garvie and Nicholson (5). The microstructure of the sintered specimens were observed under SEM. The grain size was determined by linear intercept method.

RESULTS AND DISCUSSION

The differential thermal analysis (DTA) paterns of pure and 15 vol.% ZrO_2 dispersed mullite gels were taken upto 1400°C and showed the addition of ZrO_2 favoured to form mullite at a lower temperature. The mullite formation temperature decreases slightly with the increase of ZrO_2 content. The XRD patterns of composite gel calcined at different temperatures showed the crystallisation of tetragonal zirconia (t-ZrO_2) at temperatures low as 600°C. Mullite was formed at temperatures above 1250°C. The TEM microphotograph of the composite gel powder containing 15 vol.% ZrO_2 calcined at 1200°C and 1400°C for 1 hour are given in figure 1. The ZrO_2 particles were embedded in mullite particles and were mainly of spherical morphology. The zirconia particles were mainly of tetragonal form. The ZrO_2 particles are

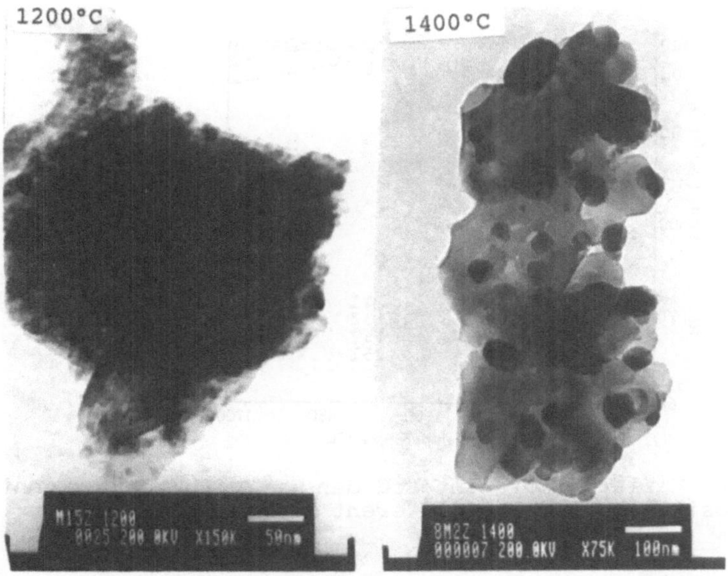

Figure 1. TEM microphotographs of Mullite/ ZrO_2 15 vol.% powder calcined at 1200°C and 1400°C for 1 hour

in the order of 40 nm in the powder calcined at 1200°C and increased to 80 nm on calcination at 1400°C.

The composite powders containing 10, 15 and 25 vol.% of ZrO_2 were calcined at 1400°C and ball milled to reduce the average particle size to 1.2 μm . These powders were sintered at temperatures between 1500°C and 1700°C for 3 hours and the variation of sintered density with temperature is given in figure 2. As shown in figure 2 the temperature required for full densification decreases with the increase of ZrO_2 fraction. Pure mullite of same particle size requires temperatures in the region of 1650°C for full densification. The SEM microphotographs of 5, 10 and 15 vol.% ZrO_2 containing composites are given in figure 3. The bright particles are ZrO_2 and were mainly occupying intergranular positions. Mullite grain size decreased from 1.2 μm

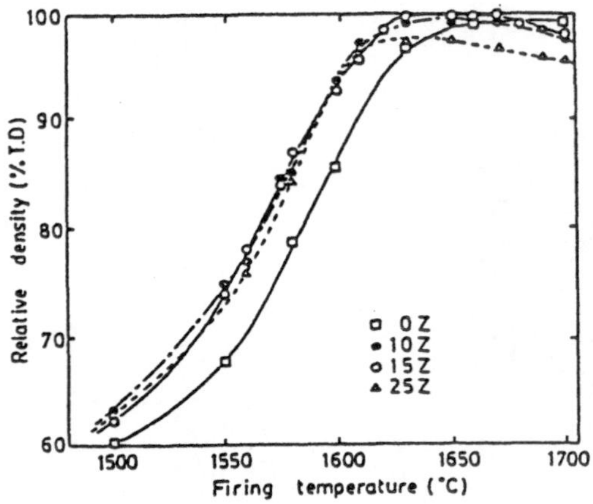

Figure 2. Variation of sintered density with temperature of composites containing different fractions of ZrO_2

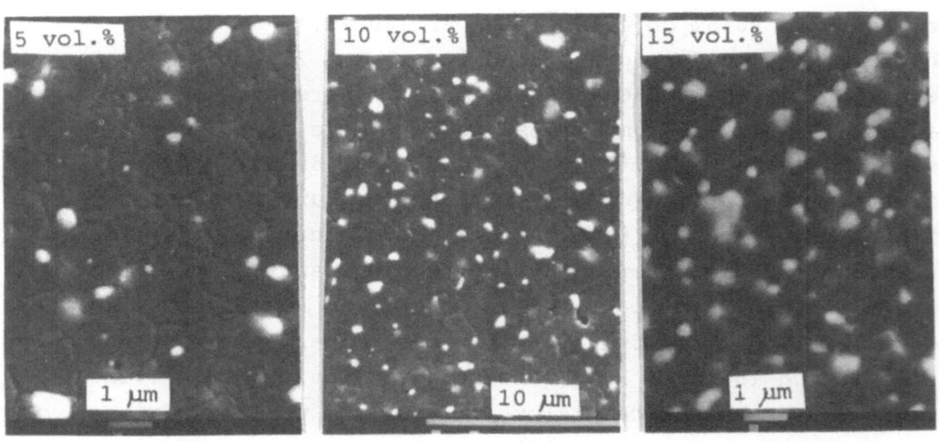

Figure 3. SEM microphotographs of 5, 10 and 15 vol.% ZrO_2 composites sintered at 1600°C for 2.5 hours.

in the composite containing 5 vol.% of ZrO_2 to 0.8 µm in 15 vol.% ZrO_2 composite. The mullite grain growth was more predominant in composites containing low fraction of ZrO_2.

The effect of total zirconia content on the flexural strength and fracture toughness was studied earlier (6) and it was observed that the flexural strength increases with increasing total ZrO_2 content and decreases above 15 vol.% ZrO_2. The maximum flexural strength observed (500 MPa) is for a composite containing 15 vol.% ZrO_2, whereas the fracture toughness increases with increasing ZrO_2 content due to the microcrack toughening observed in composite containing high percentages of ZrO_2 , as a result of spontaneous t -> m transformation on cooling.

The effect of calcination temperature on sintering of 15 vol.% ZrO_2 composite is given in figure 4. The composite powders were calcined at 900°C, 1200°C and 1300°C for 1 hour and ball milled to reduce the particle size to 1.2 µm. The t-ZrO_2 fraction of these powders after ball milling were 60%, 50% and 45% for powders calcined at 900°C, 1200°C and 1300°C respectively. The powder calcined at 900°C sinter to a low density even at 1640°C compared

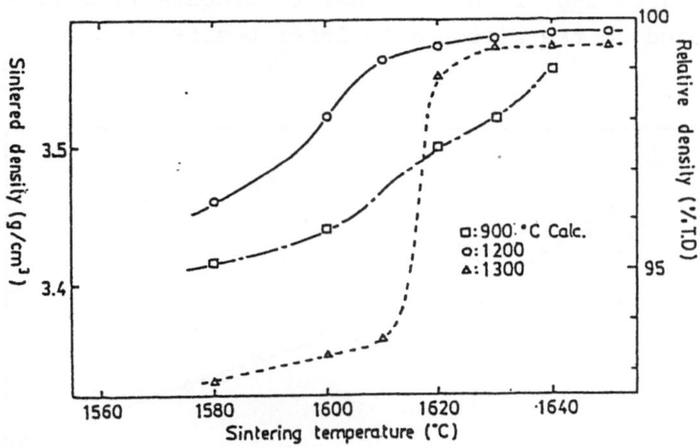

Figure 4. Sintering curves of 15 vol.% ZrO_2 dispersed composites calcined at 900°C, 1200°C and 1300°C for 1 hour and sintered at the given temperatures for 3 hours

to the other two powders. The powder calcined at 900°C contains an appreciable amount of spinel phase γ-Al_2O_3 and during ball milling a part of these transforms to γ-$Al_2O_3.H_2O$, which leaves pores on de-hydration while sintering and hence the densification is retarded. The powder calcined at 1200°C sinter to a higher density at a lower temperature compared to powder calcined at 1300°C. The powder calcined at 1200°C is not completely crystallised to mullite. Presence of δ-Al_2O_3 was observed. This could have promoted the densification at a lower temperature. It was also observed that the increase of calcination temperature shifted the temperature required for full densification to a higher region. The t-ZrO_2 fraction in the fired composite also decreased.

The variation of t-ZrO_2 with sintering temperature for composite containing 15 vol.% ZrO_2 is given in figure 5. As shown in the figure the variation could be split into two portions, one between 1500°C and 1570°C , the other between 1580 and 1700°C. This shows the presence of ZrO_2 in two different positions in the matrix. At first the inter-granular zirconia coarsen at low temperature and undergoes transformation to monoclinic form. Whereas the coarsening of intra-granular zirconia is a slow process compared to the zirconia in inter positions.

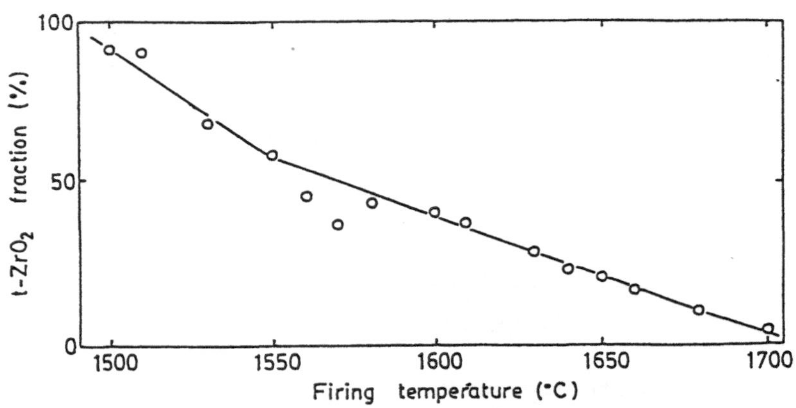

Figure 5. Effect of sintering temperature on t-ZrO_2 fraction in Mullite/ZrO_2 15 vol.% composite (Duration 3 hours)

The morphology of zirconia in intra-granular positions were spherical and tranformation to monoclinic form was a slower process compared to irregular morphology zirconia grains in inter-granular positions.
The variation of flexural strength of composite containing 15 vol.% ZrO_2 sintered at 1630°C for 3 hours is given in figure 6. The room temperature strength of 500 MPa decreases with the increase of temperature. At 1000°C the flexural strength was 400 MPa and decreased to 300 Mpa at 1400°C. In the case of pure mullite the strength at 1400°C was 300 MPa. In this case the presence of zirconia has not deteiorated the high temperature properties of mullite.
The wear properties were determined using an alumina disk. The wear rate of mullite/ZrO_2 15 vol.% composite ceramic was lower than that of mullite and TZP ceramics. The fracture toughness of this composite was 4.3 MPa.$m^{1/2}$.

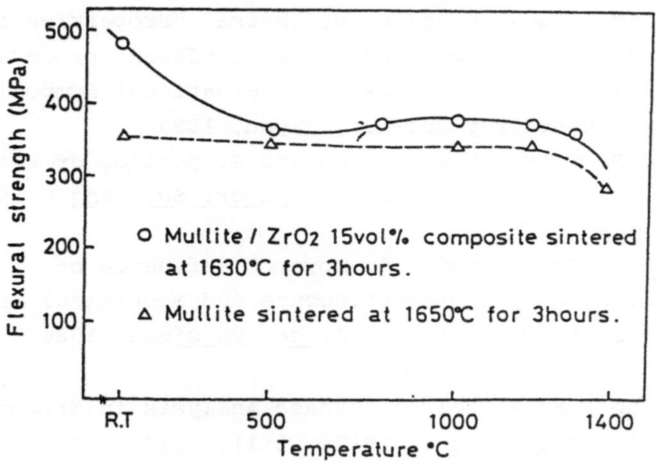

Figure 6. Variation of flexural strength with temperature for 15 vol.% ZrO_2 dispersed mullite composite.

SUMMARY

The maximum amount of ZrO_2 that could be dispersed in mullite by adopting sol-gel method without causing a decrease in strength was 15 vol.%. Mullite grain growth was hindered in the presence of ZrO_2. But at higher temperatures (1670°C) the morphology of mullite grains changed from equiaxed to elongated needle like. This is due to exsolution of SiO_2 from mullite to form high alumina mullite leaving a liquid phase along grain boundaries. The flexural strength and fracture toughness of 15 vol.% ZrO_2 dispersed mullite-zirconia ceramic at room temperature were 500 MPa and 4.3 $MPa.m^{1/2}$ respectively. The high temperature strength of this composite was same as that of pure mullite .

REFERENCES

1. N. Claussen and J. Jahn, " Mechanical properties of sintered in situ-reacted mullite zirconia composites", J. Am. Ceram. Soc. 1980, 63(3-4), 228-29
2. J.S. Wallace, N. Claussen and S. Prochazka, "Properties of Mullite-Zirconia composites prepared by different processing routes", pp642-49 in Proceedings of International Symposium on ceramic components for Engines, Japan, 1983
3. P. Boch and J.P. Giry, "Preparation and Properties of Reaction Sintered Mullite-Zirconia Ceramics", Mater. Sci. Eng., 1985, 71, 39-48
4. P. Miranzo, M.I. Osendi and J.S. Moya, " Influence of Processing Method on the Microstructure and Mechanical Properties of Mullite/Zirconia Composites", J. de. Physique, 1986, 47, C1-417
5. R.C. Garvie and P.S. Nicholson, "Phase Analysis in Zirconia Systems", J. Am. Ceram. Soc., 1982, 65(11), C190-C191
6. M.G.M.U. Ismail, Z. Nakai and S. Sōmiya , "Properties of Zirconia-Toughened Mullite Synthesized by the Sol-Gel Method", pp 119 - 125, in The Advances in Ceramics , Volume 24, 1988

SINTERING AND CHARACTERISTICS OF ELECTROCONDUCTIVE Al_2O_3-BASED COMPOSITES

ALIDA BELLOSI, GIAN NICOLA BABINI
CNR-IRTEC, Research Institute for Ceramics Technology
Via Granarolo, 64, 48018 Faenza, Italy

ABSTRACT

Al_2O_3-based composites with the addition of 30 vol% of TiN, TiC, TiB_2 have been produced by hot pressing and by gas-pressure sintering. Mechanical, electrical and thermal properties have been correlated to the microstructure and to the composition. Mechanical properties of the hot pressed samples reveal an increase in fracture toughness of 30-70% and in flexural strength of ~70%, in comparison with the Al_2O_3 matrix. Electroconductive ($\rho \sim 10^{-3}-10^{-5} \Omega$.cm), toughened and reinforced composites useful for manufacturing complex shapes by electrical discharge machining have been produced, with the addition of secondary phases in an amount greater than 20-30 vol%.

INTRODUCTION

Alumina ceramics are strong and hard, but brittle. In order to improve strength and fracture toughness, composites can be prepared by dispersing a second phase in an Al_2O_3 matrix.

Metallic borides, carbides and nitrides have recently been attracting considerable attention as new engineering ceramics. TiB_2 in particular, but also TiN and TiC, have many outstanding properties such as high melting points, high hardness and electrical conductivity (1).

The addition of these phases to Si_3N_4 or SiC (2,3) or to Al_2O_3 (4) can increase strength and toughness; moreover, if their amount is higher than 20-30 vol%, they lower drastically the electrical resistivity of the composites to about $10^{-2}-10^{-5} \Omega$.cm. These electroconductive toughened ceramics can be electrical discharge machined (2,3,5) to manufacture complex shapes and are suitable for high temperature heaters, glow-plug heaters, and heaters with a controllable specific electrical resistance, which therefore can be used for low voltage, i.e. for 12 Volt automotive batteries, heat exchangers, igniters. Moreover, they would also find application as wear resistant materials and cutting tools.

In this study, Al_2O_3-based composites with the addition of 30 vol% of TiC, TiN, TiB_2 have been hot pressed and characterized. Electrical and mechanical properties, thermal expansion, oxidation resistance have been correlated to the composition and microstructure, and compared with the Al_2O_3 matrix characteristics.

EXPERIMENTAL METHODS

Al_2O_3 powders have been homogenized with 30 vol% of the secondary phase in plastic jars and alumina balls with isobuthyl alcohol. The composites have been hot pressed in vacuum at 1600-1650°C at a pressure of 30MPa. Microstructural characteristics have been studied through X-Ray Diffraction, Scanning Electron Microscopy and Microanalysis.

Young's modulus (E) has been calculated using Resonance Frequency method. The modulus of rupture (σ) has been measured on samples 3x3x30mm, in a 4-pt bending fixture, 26mm as outer span and 13mm as inner span, with a crosshead displacement of 0.5 mm/min. Fracture toughness (K_{IC}) has been evaluated using the Direct Crack Measurement method with a load of 98 N. Hardness (H) has been measured on polished surfaces with 0.5 kg load.

Electrical resistivity was measured by a four linear probes method, with a distance among the contact points of 3.78 mm.

Thermal expansion has been measured up to 1000°C, with a heating rate of 5°C/min. Oxidation resistance in air has been evaluated through the continuous weight gain, detected by T.G apparatus, during isothermal runs (30 hours) at temperatures in the range of 600-1100°C. Electrical discharge machining tests have been performed on 8 mm thick samples using the wire technique and the machined surfaces have been characterized.

RESULTS AND DISCUSSION

Densification and microstructure

The densities of the hot pressed materials (Table 1 and Fig. 1) indicate the good densification behaviour of all the mixes in comparison with Al_2O_3. The presence of TiB_2 and TiC lowers the temperature at which densification starts (~1150°C and 1180°C respectively). By contrast, the sample with TiN starts to densify above 1280°C. The presence of an inert second phase in Al_2O_3, such as TiC, TiN, TiB_2, which has limited sinterability, decreases the densification rate by forming an interconnected framework which interferes with the driving forces necessary for bulk diffusion around the inert particles, but their presence has no effect on ultimate shrinkage.

Porosity is virtually absent in the samples containing

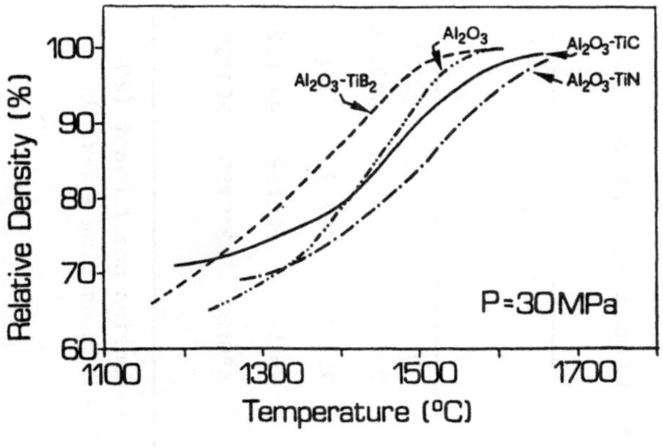

Figure 1. Densification behaviour during hot pressing.

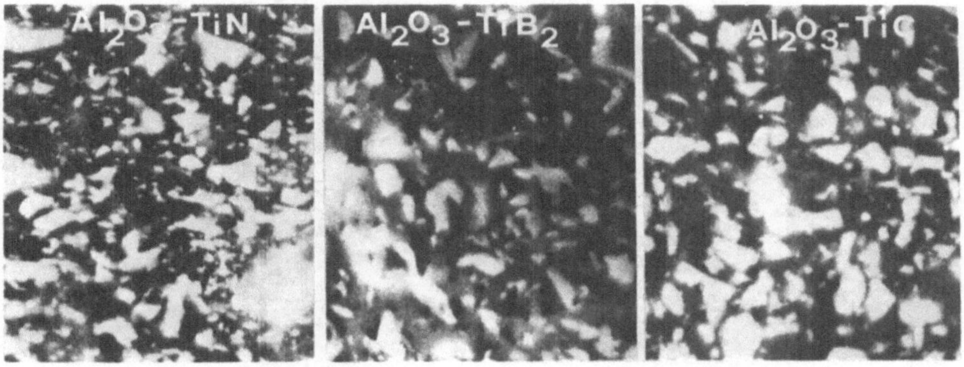

Figure 2. Microstructure of polished surfaces (3000X)

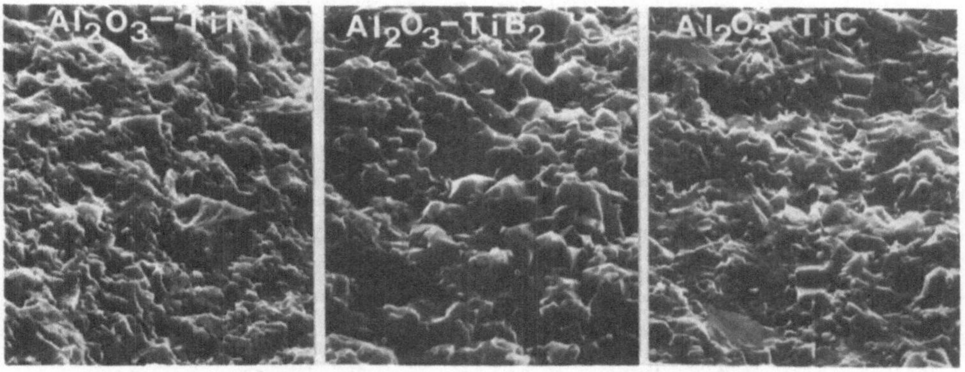

Figure 3. Microstructure of fracture surfaces (3000X)

TABLE 1

Density, microstructural characteristics and properties of the hot pressed composites containing 30 vol.% of secondary phases

Sample	r.d.	$\bar{l}$	ρ	α	E	H_v	K_{IC}	σ		
	%	um	Ω.cm	°C^{-1}	GPa	GPa	$MPa.m^{-1/2}$	MPa		
								R.T.	700°C	800°C
Al_2O_3-TiN	99.3	0.8	$6.94\ 10^{-4}$	8.11	415	19.0±.6	5.2±.4	729±53	521±44	448±16
Al_2O_3-TiB_2	98.6	1.3	$1.14\ 10^{-3}$	7.82	424	19.2±.9	5.7±.6	711±9	487±23	437±14
Al_2O_3-TiC	99.4	0.9	$6.15\ 10^{-3}$	7.94	410	20.7±.9	4.3±.5	785±86	565±52	491±12
Al_2O_3	100		/	8.01	396	18.7±.6	3.2±.2	436±35	346±35	364±19

r.d. = relative density; $\bar{l}$ = mean grain size; ρ= electrical resistivity; α= thermal expansion coefficient (20-1000°C); E = Young's modulus; H_V = **Vickers** hardness; K_{IC} = fracture toughness; σ= 4-points bending strength.

TiN and TiC and is very scarce in the sample with TiB_2 (fig. 2a-c). Fracture surfaces (fig. 3a-c) reveal different morphology and grain size of the composites in comparison with the hot pressed Al_2O_3 matrix. The second phase effectively inhibited grain growth of Al_2O_3, slowing down movement of the grain boundaries. A chemical reaction between the second phase and the matrix has been observed only in the sample with TiB_2, given that small amounts of aluminium borate have been detected in X-Ray Diffraction analyses.

Properties

With reference to the results presented in Table 1, the first important point is the noticeable increase in the fracture toughness and in the modulus of rupture in respect to the base-line Al_2O_3. The operative toughening and reinforcing mechanisms can mainly be related to crack interactions with the hard second phase particles. These interactions may include crack deflection and crack bridging, with associated stress redistribution at the crack tip when the particles are encontered. A progressive decrease in strength is observed at high temperatures. The strength is linked to the defect population and it is known that this population changes unpredictably with the temperature. In these composites the second phases are very reactive to oxidation, generating in this way new potential weak points at high temperature, owing to a large volume expansion that occurs during oxidation of TiN, TiC, TiB_2 to TiO_2. It may cause cracking at the matrix-dispersoids interface. Young's modulus and hardness values increase in the composites in comparison to the base-line Al_2O_3, the second phase being stiffer and harder than the matrix.

The thermal expansion coefficient of the composite is related to the volume fraction of each phase; the dilatometric curves have no hysteresis and suggest the absence of microcracks in the sintered body.

The values of the electrical resistivity point to a slight difference in the three composites. Considering that the resistivity of these second phases is about the same, the lower resistivity of Al_2O_3-TiN could be associated with the finer particle size of the second phase, as has previously been observed (3). The probability of forming a threefold network required for percolation conduction is higher when smaller grains are used. Moreover, in the sample with TiB_2 the presence of aluminium borate, with unidentified electrical characteristics and physical properties, could affect the electrical properties.

Oxidation resistance

As TiN, TiC, TiB_2 are reactive to oxidation, their use as secondary phases strongly affects the thermal stability of the composites. The weight gain as a function of temperature after 1000 min of isothermal oxidation is shown in fig. 4. Al_2O_3-TiN and Al_2O_3-TiB_2 start to oxidize at about 700°C, Al_2O_3-TiC at about 800°C. Two distinct behaviours are then observed: the oxidation of the composites containing TiC and TiN increases substantially with the increase in temperature; the sample

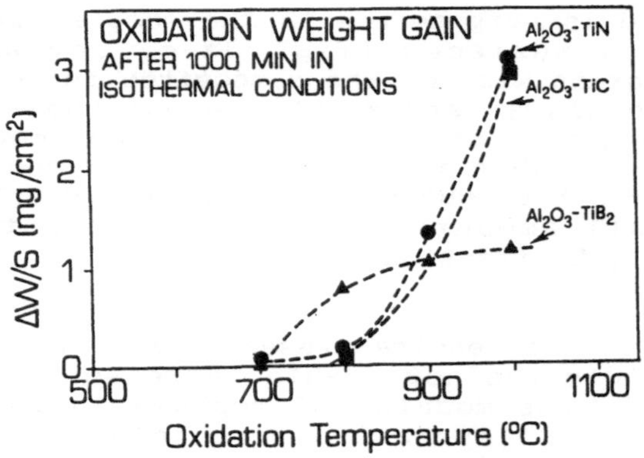

Figure 4. Oxidation weight gain vs temperature

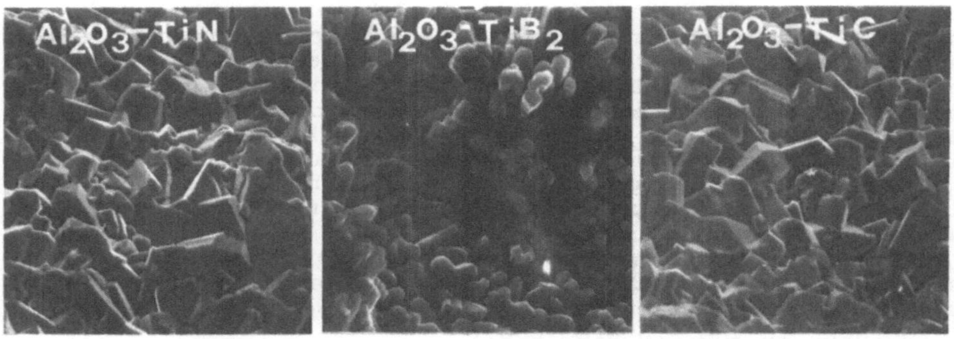

Figure 5. Morphologies of the oxidized surfaces after 30h at 1000^oC (600X)

with TiB_2 after a high oxidation at about 800^oC, shows a progressive stability up to 1000^oC. This behaviour can be explained on the basis of the structure (Fig. 5) and composition of the oxide scale and of the kinetics of the process which are reported elsewhere (6). Briefly, on the sample with TiN and TiC, the oxide scale is mainly composed of TiO_2, which may be non protective owing to cracking of the oxide layer resulting in an increase in the active area for oxidation.

In the case of Al_2O_3-TiB_2 at $T > \sim 800^oC$, in addition to TiO_2, $Al_4B_2O_9$ has been detected, which could help to protect the bulk from the massive oxidation of TiB_2 particles.

Electrical discharge machining (EDM)

For all the tested composites the electrical conductivity level was good enough for EDM. The material removal rate (Fig.

6) measured during the machining of 8 mm thick samples under the same experimental conditions using brass wire and water as dielectric, and the characteristics of the as-cut surfaces suggest an easier machinability of the composites containing TiB_2 and TiN samples. For the sample with TiC the material removal rate is twice or four times lower than that for Al_2O_3-TiB_2 and Al_2O_3-TiN respectively. Both material removal rate and surface roughness (therefore the damaged layer thickness) are directly related to the electrical resistivity of the samples: the higher the resistivity, the higher the surface roughness and the lower the cutting rate.

Figs. 7a-c compare the microstructure of the machined surfaces

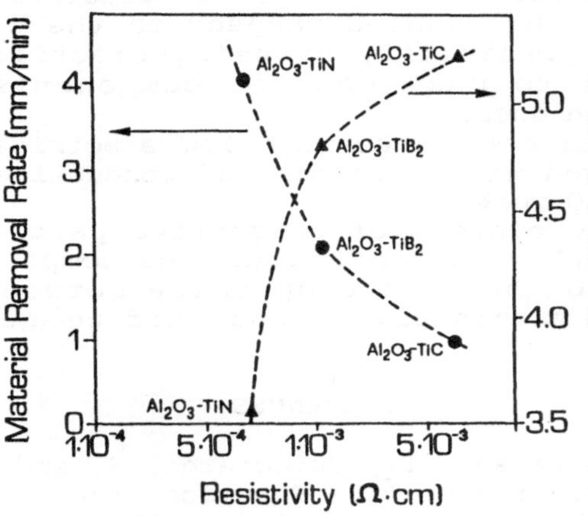

Figure 6. Material removal rate and surface roughness vs the resistivity of the composites during EDM

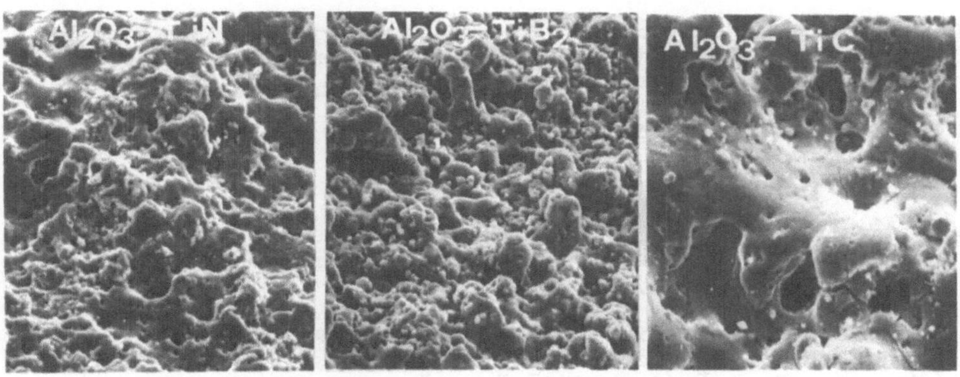

Figure 7. Morphologies of EDM machined surfaces (600X)

Discharges create craters and the overall surface appears pitted with rounded molten looking bumps with overlapping craters of varying diameter and morphology. The damage and crater size increase with the decrease in cutting speed. Al_2O_3-TiC samples show significant surface damage with microcracking. Small droplets are resolidified brass from wire (white areas in the back scattered electron image of Fig. 7).

CONCLUSIONS

New electroconductive Al_2O_3-based composites can be produced by particle reinforcement of the matrix mainly with TiN, TiB_2 or TiC additions. The most attractive properties of this new class of advanced ceramics are improved fracture toughness and strength that can be doubled respect to the baseline Al_2O_3 matrix, and the possibility of using electrical discharge machining (EDM) to make complex components instead of conventional techniques.

High strength composites with low electrical resistivity can be manufactured with a volume of conductive particles in the range of 20-40 vol%.

The size and content of conductive particle additions have to be carefully controlled and the composition must be adjusted to find the best compromise between mechanical properties, EDM machinability and surface quality of the composites.

REFERENCES

1. Bellosi, A., Graziani, T., Guicciardi, S. and Tampieri, A., Characteristics of TiB_2 Ceramics. Special Ceramics 9, London 18-20 Dec. 1990, in print on the Proceeding.

2. Mc Murtry, C.H., Boecker, W.D.G., Seshadri, S., Zanghi, J.S. and Garnier, J.E., Microstructure and Material Properties of SiC-TiB_2 Particulate Composites. Am. Ceram. Soc. Bull., 1987, **60**, 325.

3. Bellosi, A., Guicciardi, S. and Tampieri, A., Development and characterization of electroconductive Si_3N_4-TiN composites. In print on J. Europ. Ceram. Soc..

4. Matsushita, J., Shinsuke, H. and Saito, H., Pressureless sintering of TiB_2-Al_2O_3. J. Ceram. Soc. Jpn. Inter. Ed., 1989, **97**, 1200-1205.

5. Martin, C., Cales, B., Vivier, P. and Mathieu, P., Electrical Discharge Machinable Ceramics Composites. Mat.Sci. Eng., 1989, **A109**, 351-356.

6. Tampieri, A., Biasini, V. and Bellosi, A., Oxidation of monolithic TiB_2 and of Al_2O_3-TiB_2 Composites. European Ceramic Society Second Conference, Augsburg, Sep. 11-14, 1991.

FABRICATION AND CHARACTERIZATION OF SLIP-CAST LAYERED Al_2O_3-ZrO_2 COMPOSITES

Raymond A. Cutler and Charla B. Brinkpeter
Ceramatec, Inc.
Salt Lake City, Utah 84119, USA

and

Anil V. Virkar and Dinesh K. Shetty
University of Utah
Salt Lake City, Utah 84112, USA

ABSTRACT

Monolithic and three-layered Al_2O_3-15 vol. % ZrO_2 composites were fabricated by slip casting aqueous slurries. The outer and inner layers of three-layer composites contained unstabilized and partially stabilized ZrO_2, respectively. Transformation of part of the unstabilized ZrO_2 led to surface compressive stresses in the outer layers. Strain gage, x-ray, indentation crack length, and strength measurements were used to determine the magnitude of residual stresses in the composites. The strength of the three-layer composites (1.1 to 1.2 GPa) was 500-700 MPa higher than that of the monolithic outer layer composites at room temperature and 350 MPa higher at 750°C. The strength differential decreased rapidly above the monoclinic to tetragonal transformation temperature. Three-layered composites showed higher Weibull modulus and excellent damage resistance. Cam follower rollers were fabricated to demonstrate the applicability of this technique for making automotive components.

INTRODUCTION

Transformation of ZrO_2 from the partially stabilized tetragonal polymorph to the equilibrium monoclinic form has been successfully used to toughen a variety of ceramic matrices[1]. Transformation of the metastable tetragonal ZrO_2 in the near surface regions of ceramics has also been used to produce compressive residual stresses and, thus, strengthen ceramics containing ZrO_2[2,3]. The techniques developed to date for this purpose, such as grinding[2] or partial removal of the stabilizer[3], produce only modest thickness (≈30 μm) of the compressive stress zones. These ceramics do not sustain the improved strength under severe damage conditions that may produce a flaw larger than the compressive stress zone. The compressive stress produced by grinding can also be irreversibly lost by a high-temperature exposure and reverse transformation of the monoclinic ZrO_2 to the tetragonal form.

Virkar et al.[4] recently developed a technique by which significant compressive residual stress can be introduced in ceramics with surface compression zones of the order of 300 to 1000 μm. The technique involves fabrication of three-layer composite ceramics consisting of outer layers that contain unstabilized ZrO_2 in an oxide matrix, and an inner layer that contains ZrO_2, partially stabilized with an oxide additive such as Y_2O_3, dispersed in the same oxide matrix. On cooling from the fabrication temperature, a large fraction of the ZrO_2 in the outer layer transforms to the monoclinic form, while nearly all of the ZrO_2 in the inner layer is retained metastably in the tetragonal form. This selective transformation of the ZrO_2 (with the accompanying volume expansion) in the outer layers and the constraint of the bulk inner material leads to significant compressive stress in the outer layers and balancing tensile stress in the bulk. The residual stress will not decrease with temperature until the monoclinic to tetragonal transformation temperature is reached, since monoclinic and tetragonal ZrO_2 polymorphs have nearly the same coefficients of thermal expansion.

Cutler et al.[5,6] have successfully applied the three-layer technique to Al_2O_3-ZrO_2 composites. Using dry pressing to form the sandwich composites, a compressive stress of 400 MPa was produced in the outer layers of Al_2O_3-15 vol. % ZrO_2 composites with an outer layer thickness of 375 μm. A significant fraction of this residual stress (≈200 MPa) was retained at 750°C. Substitution of HfO_2 for ZrO_2 increased the temperature to which residual stresses were retained. The purpose of this paper is to report progress made in optimizing three-layer Al_2O_3-15 vol. % ZrO_2 composites for heat engine applications using slip casting as a forming technique.

EXPERIMENTAL PROCEDURES

Al_2O_3-15 vol. % ZrO_2 composites were prepared by dry pressing using techniques discussed previously[5-7] and by slip casting using Al_2O_3 (ERC-DBM, Reynolds Metal Co., Bauxite, AR.) and ZrO_2 (DK-1, Daiichi Kaguku Kogyo Co. Ltd, Osaka, Japan) as starting materials for the outer layer monolithic material. The inner layer monolithic material used the same source of Al_2O_3 but partially stabilized (3.0 mol. % Y_2O_3) ZrO_2 (HSY-3.0, Daiichi) was used in place of unstabilized zirconia. The two slips were dispersed using 0.5 wt. % citric acid and 2.0 wt. % Darvan C (R. T. Vanderbilt, Norwalk, CT.) in an aqueous slip at 70 wt. % (35 vol. %) solids. The slips were vibratory milled 16 hours with ZrO_2(3.0 mol. % Y_2O_3) media and degassed prior to slip casting. Three-layer composites were slip cast in plaster molds by first casting the outer layer slip for a given time period and then pouring out the outer layer slip and quickly pouring in the inner layer slip. The inner layer slip was allowed to remain in the mold when making solid plates or cylinders. When hollow tubes for cam followers were fabricated, the inner layer slip was drained followed by introduction of the outer layer slip for a second time. The thickness of the layers was controlled by the slip casting times of the outer layers. The slip cast parts were dried under controlled conditions and sintered at 1587°C for 30 minutes. The sintered parts were subsequently HIPed in 200 MPa Ar at 1550°C for 30 minutes.

Characterization for slip cast materials was similar to that used previously for dry-pressed samples[4-8]. Strength testing was generally performed on bars (see Figure 1) in four-point bending. Thermal shock testing was performed on 6.5 mm diameter by 50 mm long rods. The samples were heated to various temperatures prior to quenching in ice water (0°C). Strength of thermally-shocked rods was measured in three-point bending at room temperature.

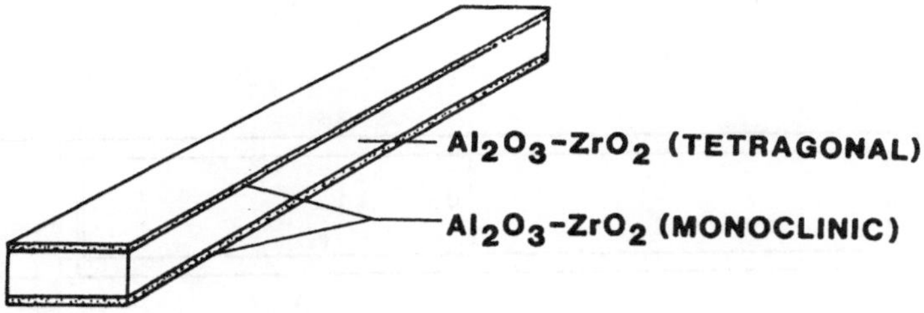

Figure 1. Schematic of three-layer Al_2O_3-15 vol. % ZrO_2 composites with unstabilized ZrO_2 in outer layers and partially stabilized ZrO_2 in inner layer.

RESULTS AND DISCUSSION

Strength Testing

A schematic of the three-layer composites is shown in Figure 1. Both the outer and the inner layers consist of Al_2O_3-15 vol. % ZrO_2 so that thermal expansion coefficients are similar. The main difference is that the ZrO_2 in the outer layers is unstabilized and the ZrO_2 in the inner layer has been coprecipitated with 3 mol. % Y_2O_3. Upon cooling from sintering temperatures (≈1600°C) where ZrO_2 in both outer and inner layers is tetragonal, most of the unstabilized ZrO_2 in the outer layers transforms to the monoclinic phase with an accompanying volume expansion. The constraint of the inner layer puts the outer layers under compression and the inner layer in tension (see Figure 2). Assuming a square wave stress distribution the accompanying residual compressive stress, σ_1, in the outer layers is

$$\sigma_1 = -\Delta\varepsilon_o E d_2/(1-v)d \quad (1)$$

where $\Delta\varepsilon_o$ is the unconstrained strain in the outer layers from the transformation of ZrO_2, E is Young's modulus, d is thickness, v is Poisson's ratio and the subscripts 1 and 2 refer to the outer and inner layers, respectively. Correspondingly, the residual tensile stress, σ_2, in the inner layer is

$$\sigma_2 = 2\Delta\varepsilon_o E d_1/(1-v)d \quad (2)$$

Based on fracture of three-layer composites from within the outer layers, as would be expected in flexure, the failure strength, σ_f, is

$$\sigma_f = \sigma_o + \Delta\varepsilon_o E d_2/(1-v)d \quad (3)$$

where σ_o is the failure strength of the outer layers in the absence of residual stress. A plot of strength as a function of normalized inner layer thickness (d_2/d) would be expected to follow a linear relationship with slope equal to $\Delta\varepsilon_o E/(1-v)$ and intercept equal to σ_o. Experimental verification of Equation (3) has been demonstrated for samples with flaw populations typical of "as-sintered" samples[5], as well as for samples with well characterized indentation flaws[6,7]. The value of $\Delta\varepsilon_o$ determined from these measurements agrees with estimates from x-ray measurements[5].

Virkar[4,8] has developed a technique for determining the residual stress using inexpensive strain gages. This technique can be applied to determine the residual stress profile in sintered ceramics. A strain gage is attached to one side of a three-layer ceramic which initially has outer layers of equal thickness. One side is then incrementally ground off (see Figure 3) and the strain (ε) is measured as a function of thickness removed (δ). Considering a symmetric stress profile ($\sigma_{xx}=\sigma_{yy}$) so that residual stress is a function of z (thickness direction of a three-layer composite) only and can be denoted by $\sigma(z)$, the

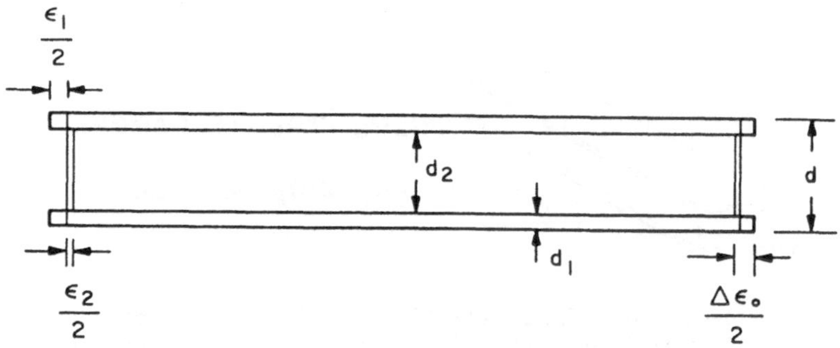

Figure 2. Schematic of three-layer sample showing unconstrained outer and inner layers as well as constrained length of sample.

measured strain, $\varepsilon_M(\delta)$, vs δ data can be used to determine the residual stress profile. Virkar[9] has used simple beam theory to show that

$$\varepsilon_M(\delta) = (1-v)/E(d-\delta)^2 \int_0^{d-\delta}[6z-4(d-\delta)]\sigma(z)dz \tag{4}$$

It is possible to predict the shape of the strain vs thickness removed[8,9] plot for a three-layer composite. Verification of this predicted response is shown in Figure 4 where the data points represent experimental measurements and the solid lines are the predictions based on Equation (4) and constant stress profiles in the layers. As expected, the monolithic outer and inner layer composites show no change in strain as a function of thickness of material removed, while the three-layer composites show significant residual compressive stress in the outer layers. The discontinuity in the slope of the $\varepsilon_M(\delta)$ vs δ plot occurs at $\delta=d_1$ (the interface between outer and inner layers). The magnitude of $\Delta\varepsilon_o$ was measured to be 1.2×10^{-3}, giving a calculated stress in the outer layers of $\approx$-520 MPa[8].

Strength measurements as a function of temperature are shown in Figure 5 for monolithic outer, inner and three-layer samples. At room temperature the three-layer composites have a strength of 1150 MPa as compared to a strength of 660 MPa for the outer layer monolithic bars. The strength differential between the three-layer and outer layer monolithic bars, as shown in Figure 6, is due to the compressive residual stress in the outer layers of the three-layer bars. The experimentally observed value of -490 MPa from strength testing is close to the value of -520 MPa from strain gage measurements.

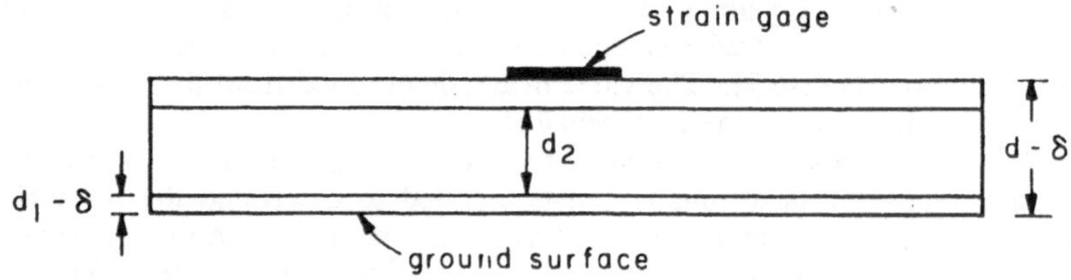

Figure 3. Schematic showing the three-layer sample with a strain gage mounted on one face. The other face is incrementally ground off and strain is recorded as a function of the thickness removed.

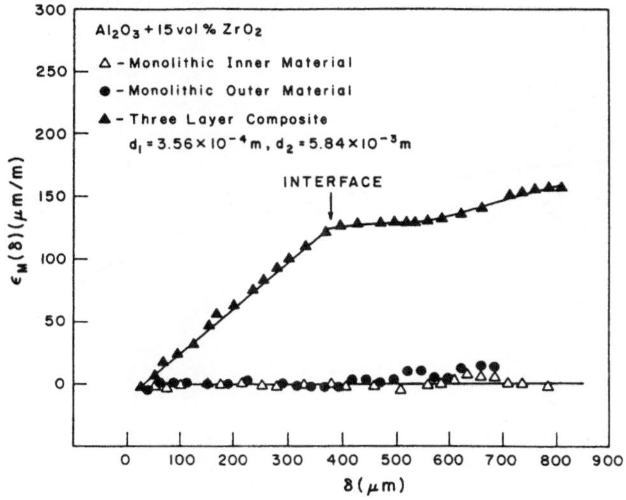

Figure 4. Measured strain vs thickness removed for monolithic and three-layer Al_2O_3-15 vol. % ZrO_2 samples. Monolithic samples show no residual stress while three-layer sample shows σ_c=-520 MPa[8].

The residual stresses are effective in strengthening the three-layer composites until the monoclinic to tetragonal (m--->t) transformation is completed at a temperature above

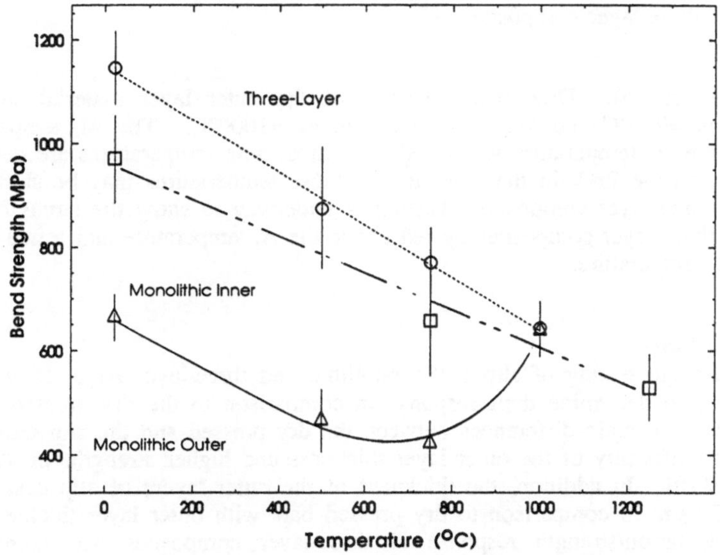

Figure 5. Strength of three-layer Al_2O_3-15 vol. % ZrO_2 composites as a function of temperature, in comparison to monolithic inner (Al_2O_3-15 vol. % ZrO_2(3 mol. % Y_2O_3)) and outer (Al_2O_3-15 vol. % ZrO_2) layer materials.

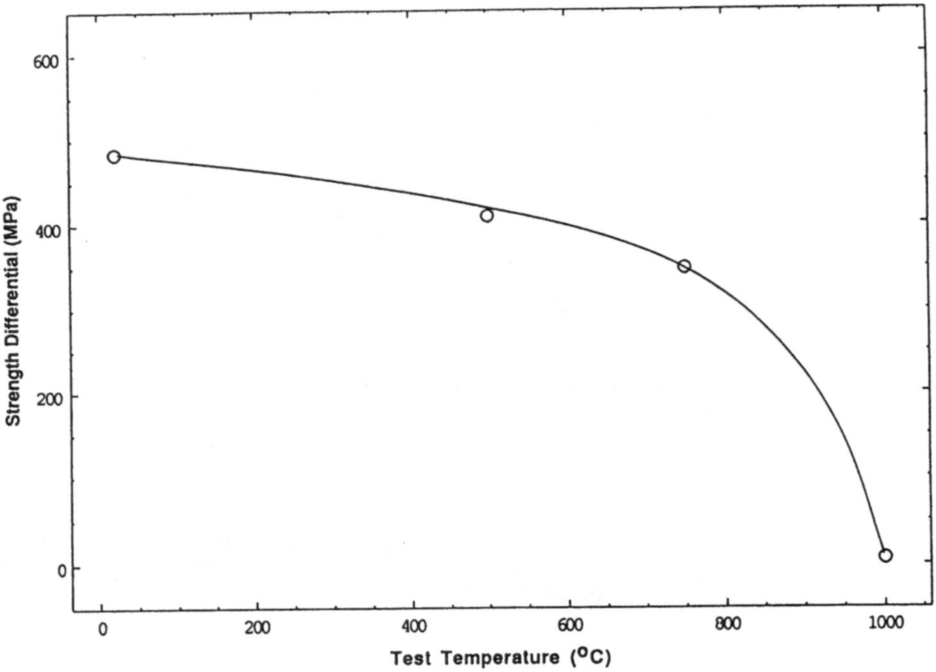

Figure 6. Strength differential between three-layer and outer layer materials. Strength differential is due to compressive residual stress in the outer layers of the three-layer composite.

750°C (see Figure 6). Dilatometric studies of the outer layer material show the A_s temperature as ≈900°C and the A_f temperature as ≈1000°C. The M_s temperature was ≈600°C and the M_f temperature was ≈500°C. Since these temperatures are dependent on the constraint on the ZrO_2 in the outer layers, these temperatures may be shifted slightly lower in the three-layer composite. Testing is underway to show the strength hysteresis expected for three-layer composites cycled above the A_f temperature and tested on cooling above the M_s temperature.

Damage Resistance

Indentation/strength testing of slip cast monolithic and three-layer Al_2O_3-15 vol. % ZrO_2 was performed to determine their response in comparison to the dry pressed bars tested previously[7]. The main differences between the dry pressed and the slip cast bars were the improved uniformity of the outer layer thickness and higher strengths of all three slip cast materials[10]. In addition, the thickness of the outer layers of slip cast three-layer bars was ≈250 μm in comparison to dry pressed bars with outer layer thickness of ≈375 μm. The indentation/strength response of three-layer composites was compared with theoretical expectation based on superposition of stress. It can easily be shown[7] that the strength expected for three-layer composites which have been indented with a Vickers indenter of load P is given by

$$\sigma_f = CK_{Ic}^{4/3}/((E/H)^{1/6}P^{1/3}) + \Delta\varepsilon_o E d_o/(1-v)d \qquad (5)$$

where C is a constant equal to 2.02[7], K_{Ic} is fracture toughness and H is hardness. According to Equation (5), a plot of σ_f vs $P^{-1/3}$, as shown in Figure 7, should yield a straight line with a slope related to K_{Ic} and an intercept giving the compressive residual stress, σ_c. Taking values of E=340 GPa, H=17 GPa, linear regression of the data for monolithic specimens gave slopes corresponding to K_{Ic} values of 5.35 and 5.03 MPa·m$^{1/2}$ for outer and inner materials, respectively. Both materials had intercepts near zero, showing that they were free of residual stress (see Figure 7). Linear regression of the data for the three-layer composites gave a slope corresponding to a fracture toughness of 5.75 MPa·m$^{1/2}$, a value similar to the monolithics as expected from theory[7]. It is interesting to note, however, that there was very little decrease in strength at high indentation loads (greater than 125 N) suggesting that the three-layer material has even better damage resistance than predicted by Equation (5). The intercept from the linear regression gave a value of -588 MPa for the compressive stress. This value of σ_c is higher than the difference in strength of -497 MPa obtained from the strengths of the unindented bars. These data confirm the superior damage resistance of materials made using the three-layer concept and show that improved resistance to contact damage can be expected for ceramic components made using this technique.

In order to show the practical extension of this technology, totally encapsulated ≈5 mm diameter rods were fabricated by slip casting such that pits on the order of 50-250 μm were prevalent on the surface but were rarely present in the bulk. The outer layer of the

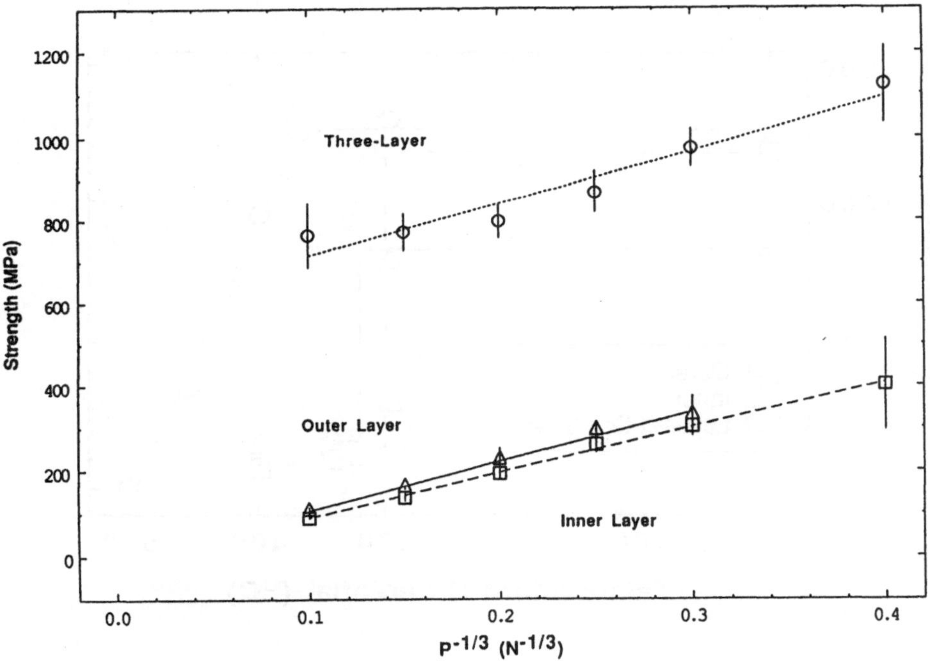

Figure 7. Fracture stress (σ_f) versus inverse cube root load ($P^{-1/3}$) plots for the slip cast three-layer and monolithic ceramics.

rods was ≈425 μm thick and surface compressive stress was on the order of 400 MPa. The strength (5 rods broken in 4-point bending using an inner span of 20 mm and an outer span of 40 mm) of the "as-HIPed" material was 908±116 MPa, as compared to strength of 1,211±123 MPa for three-layer rods which were ground to a 30μm surface finish. In contrast, monolithic outer layer rods had strength of 476±84 MPa in the "as-HIPed" state and 830±27 MPa in the ground state. The increase in strength of the layered composite over that of the monolith was improved at larger flaw sizes. This means that as long as the outer layer thickness of the "three-layer" composite is sufficiently larger than the surface flaws, strength improvements consistent with the indentation/strength data can be expected in components. To further explore the use of three-layer composites, monolithic and three-layer composite cam followers were fabricated and have been sent to an automotive manufacturer for testing.

Thermal Shock Testing
Fracture strengths of monoliths and three-layer composites after thermal shocking over different temperature ranges are plotted in Figure 8 for monolithic outer, inner and "three-layer" rods. The as-HIPed strengths were 830±27 MPa, 1185±109 MPa, and 1206±36 MPa for outer, inner and layered composite rods, respectively. These strengths are similar to those reported above for bars in four-point bending. The inner layer rods had a ΔT of slightly less than 300°C, the outer layer rods had a ΔT of ≈325 and the layered composite rods had a ΔT of ≈425. The exposed ends of the layered composite rods were the regions most susceptible to thermal shock and totally encapsulated rods would likely have resulted in a higher ΔT for this material. Individual layered composite rods had strengths greater than 1200 MPa at temperatures up to 425°C. Since thermal shock under severe cooling

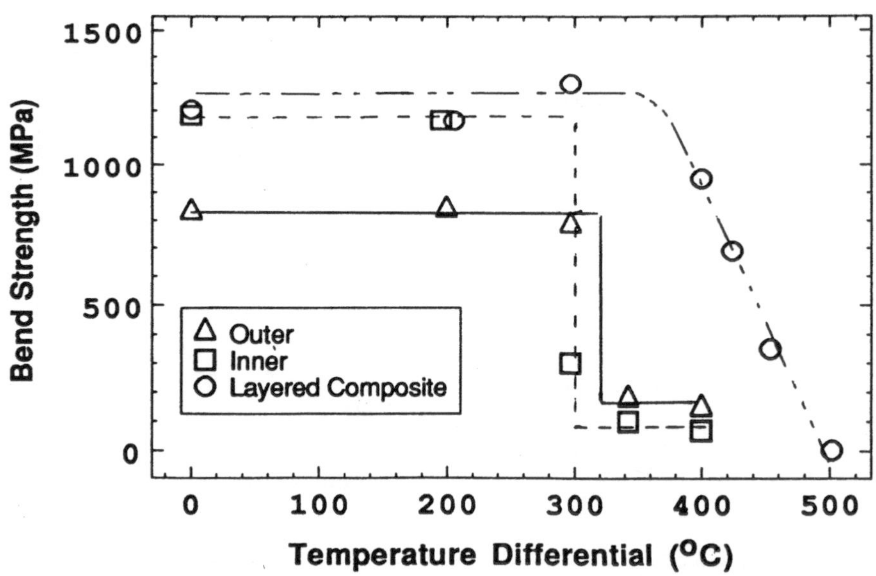

Figure 8. Thermal shock behavior of monolithic and layered composite Al_2O_3-15 vol. % ZrO_2 quenched in ice water. Layered rods with surface compressive stress of ≈400 MPa have critical ΔT 100°C higher than monolithic rods.

conditions generally initiates from the outer surface of monolithic components, the increase in thermal shock resistance of the three-layer rods over the outer layer rods is of interest. This increase of 100°C can be compared with what would be expected for the strength improvement due to the surface compression. The compressive stress, σ_1, in the outer layer and the balancing tensile stress, σ_2, in the inner core can be easily calculated assuming a square-wave stress distribution. These stresses are approximately given by

$$\sigma_1 = -(A_2/A)E\Delta\varepsilon_o/(1-v) \quad (6)$$

and

$$\sigma_2 = (A_1/A)E\Delta\varepsilon_o/(1-v) \quad (7)$$

where A is the cross-sectional area of the rods. The outer case thickness was approximately 0.375 mm or 1/12 the diameter of the rods. Taking values of 4.86×10^{-6} m^2, 1.10×10^{-5} m^2 and 1.59×10^{-5} m^2 for A_1, A_2, and A, respectively, E of 340 GPa, $\Delta\varepsilon_o$ of 1.3×10^{-3} and Poisson's ratio of 0.25 results in a compressive residual stress of -409 MPa in the outer layer and a residual tensile stress of 180 MPa in the core. The calculated compressive stress is in good agreement with the increase in room temperature strength between the outer and the layered composite rods of 376 MPa.

Based on the early work of Hasselman[11,12] it is possible to explain the improved thermal shock resistance of the Al_2O_3-15 vol. % ZrO_2 ceramics having substantial compressive surface stresses as compared to the monolithic ceramics of similar composition, modulus and thermal expansion. The expected increase in ΔT_c, the critical temperature difference to which the rods are subjected in order to initiate crack growth and decrease strength, for a material with compressive residual stress is given by

$$\Delta T_c = \Delta T_c^{\circ} + [\sigma_c(1-v)/\alpha E]f(k/ha) \quad (8)$$

where ΔT_c° is the temperature differential in the absence of residual stress given by $[\sigma_f(1-v)/\alpha E]f(k/ah)$, where σ_f is the strength of unquenched rods, α is linear coefficient of thermal expansion, k is the thermal conductivity, h is the surface heat-transfer coefficient and a is the characteristic heat transfer length. The work of Becher et al.[13] investigating the effect of sample size on the thermal shock resistance of ceramics shows the need to take the Biot modulus (ah/k) into account in calculating ΔT. For a conservative prediction, a relatively high heat transfer coefficient, h=10 W/cm$^{2\circ}$C was assumed. With a=0.225 cm and k=0.25 W/cm°C, the Biot modulus equals 9. For this value of ß, f(k/ah)=0.433. From Equation (8), the predicted increase in thermal shock resistance for three-layer rods is 243°C for σ_c=-376 MPa. The increase obtained in experiments was 120°C. The reason for this discrepancy is likely related to the fact that the ends of the rods were not encapsulated. The reason for the low values for ΔT (300-425°C) measured in the present study in contrast to the high values ($\Delta T > 800$°C) reported by Becher[14] for monolithic Al_2O_3-ZrO_2 composites is believed to be the more vigorous quench of the ice water as compared to boiling water[15] and the difference in sample thickness between the two studies.

Improved thermal shock resistance of the three-layer composites is expected due to the presence of residual compressive stresses. These results show that residual compressive stress of substantial depth in layered composites not only increases strength, apparent toughness[5] and damage resistance, but also makes the materials more resistant to thermal shock.

Improved Reliability

Superposition of temperature stress (due to difference in thermal expansion mismatch) on transformation-induced stress (due to volume expansion differences between monoclinic and tetragonal ZrO_2) was demonstrated using ZrO_2(3 mol. % Y_2O_3)-40 vol. % Al_2O_3 as the inner layer material in three-layer slip cast composites. The temperature-induced stress enhances the strength of three-layer composites at low temperatures.

The expected compressive residual stress in the outer layer of the bars is the combination of transformation-induced and temperature-induced stresses. The transformation-induced compressive stress, σ_c, in the outer layer, assuming a square wave stress distribution, is

$$\sigma_c = \frac{-(E_1 E_2 d_2 \Delta\varepsilon_o)}{[(1-v)(2E_1 d_1 + E_2 d_2)]} \tag{9}$$

in an analogous manner to Equation (1). The temperature induced stress, σ_T, in the outer layer, also assuming a square wave distribution, is given as

$$\sigma_T = \frac{-[(E_1 E_2 d_2 \Delta T)(\alpha_2 - \alpha_1)]}{[(1-v)(2E_1 d_1 + E_2 d_2)]} \tag{10}$$

where ΔT is the temperature range over which the stress builds up and α is the coefficient of linear thermal expansion. By superposition, the expected residual stress in the outer layer is

$$\sigma_1 = \frac{-[(E_1 E_2 d_2)(\Delta\varepsilon_o + \Delta T(\alpha_2 - \alpha_1))]}{[(1-v)(2E_1 d_1 + E_2 d_2)]} \tag{11}$$

Taking E_1 as 340 GPa, E_2 as 275 GPa, $\Delta\varepsilon_o$ as 1.3×10^{-3}, ΔT as 1000°C, α_2-α_1 as 1×10^{-6}/°C, v as 0.25, d_1 as 500 μm and d_2 as 5 mm, the expected compressive residual stress in the outer layer is slightly over -1 GPa. The corresponding residual tensile stress in the inner layer is approximately 200 MPa. Strain gage measurements resulted in a measured compressive stress of -1.1 GPa, in excellent agreement with prediction.

Room temperature fracture strength of three-layer composites were 1,275 MPa as compared to a strength of 549 MPa for the outer layer bars. The compressive stress of

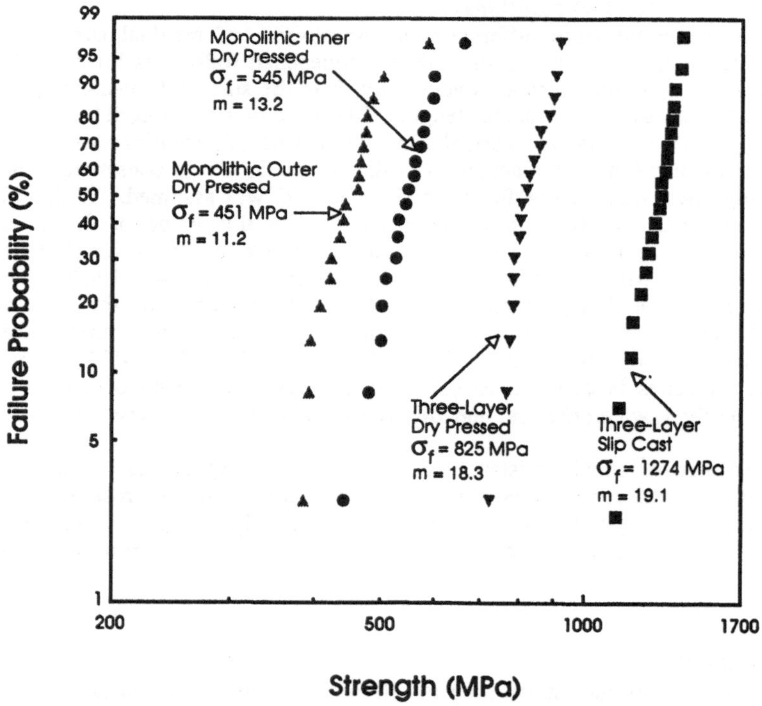

Figure 9. Linearized Weibull plots of fracture stress of monolithic and three-layer composites. Slip cast composites have both transformation-induced and temperature-induced residual stress. Note higher modulus of three-layer composites compared to monolithics, indicating enhanced reliability.

-725 (difference between inner and three-layer bars) is 70% of the predicted value. More importantly, it shows that high strengths can be achieved in layered ceramic composites using a combination of transformation and temperature-induced stresses, as expected.

Figure 9 shows linearized Weibull plots of fracture stresses of monolithic outer, monolithic inner and three-layer composites fabricated by dry pressing as well as three-layer composites fabricated by slip casting. It is noted that the three-layer composites exhibit both increased strengths as well as improved Weibull moduli relative to the monolithic ceramics. Improvements in uniformity in the thickness of the layers by slip casting is the main reason for the improved strength of the slip cast composites. The improvement in Weibull modulus is due to superposition of stress[5]. Residual compression deliberately introduced on the surface of structural ceramics can be a viable approach to increase Weibull modulus and reliability.

SUMMARY AND CONCLUSIONS

Three-layer oxide ceramics with compressive residual stress ranging between 300 and 600 MPa in the outer layers were fabricated using dry pressing and slip casting. The outer layer thickness was controlled in the green state. The outer layer protected against damage due to surface flaws or sliding contact.

Transformation-induced stresses were present at temperatures in excess of 750°C. Stress due to the thermal expansion mismatch between inner and outer layers were superimposed to give strength increases greater than 500 MPa at room temperature.

Slip casting was used to improve the uniformity of the interface between layers to allow composites with outer layer thicknesses of 200-300μm to be fabricated. The room temperature strength of Al_2O_3-15 vol. % ZrO_2 composites increased from 825 MPa to 1150 MPa and the strength at 1000°C increased from 320 MPa to 640 MPa. Strength in excess of 1200 MPa at room temperature was achieved by superimposing temperature stress on transformation-induced stress.

In addition to strength measurements, residual stresses could be detected by strain gage measurements, characterization of monoclinic content as a function of temperature by x-ray diffraction, indentation/ strength measurements and thermal shock testing. The use of layered composites to enhance damage resistance and enhance reliability will be tested in cam followers for automotive applications.

REFERENCES

1. Evans, A. G. and Cannon, R. M., Toughening of Brittle Solids by Martensitic Transformations. Acta Metall., 1986, **34**, 761-800.

2. Swain, M. V., Grinding-Induced Tempering of Ceramics Containing Metastable Zirconia. J. Mater. Sci. Lett., 1980, **15**, 1577-79.

3. Green, D. J., A Technique for Introducing Surface Compression into Zirconia Ceramics. J. Am. Ceram. Soc., 1983, **66**[9], C-178-C-179.

4. Virkar, A. V., Huang, J. L. and Cutler, R. A., Strengthening of Oxide Ceramics by Transformation Induced Stresses. J. Am. Ceram. Soc., 1987, **70**[3], 164-70.

5. Cutler, R. A., Hansen, J. J., Virkar, A. V., Shetty, D. K. and Winterton, R. C., Strength Improvement in Transformation Toughened Ceramics using Compressive Residual Surface Stresses. In Advanced Structural Ceramics Vol. 78, ed. by P.F. Becher, M. V. Swain, and S. Somiya, Materials Research Society, Pittsburgh, PA, 1987, pp. 155-63.

6. Cutler, R. A., Bright, J. D., Virkar, A. V. and Shetty, D. K. Strength Improvement in Transformation-Toughened Alumina by Selective Phase Transformation. J. Am. Ceram. Soc., 1987, **70**[10], 713-18.

7. Hansen, J. J., Cutler, R. A., Shetty D. K. and Virkar A. V., Indentation Fracture Response and Damage Resistance of Al_2O_3-ZrO_2 Composites Strengthened by Transformation-Induced Residual Stresses. J. Am. Ceram. Soc., 1988, **71**[12], C-501-5.

8. Virkar, A. V., Jue, J. F., Hansen, J. J. and Cutler, R. A., Measurement of Residual Stresses in Oxide-ZrO_2 Three-Layer Composites. J. Am. Ceram. Soc., 1988, **71**[3], C-148-51.

9. A. V. Virkar, "Determination of Residual Stress Profile Using a Strain Gage Technique," J. Am. Ceram.Soc., **73**[7] 2100-02 (1990).

10. Cutler, R. A., Brinkpeter, C. B., Bruner, S. L., Prouse, D. W., Virkar, A. V., and Shetty, D. K., Transformation-Toughened Ceramics with Strength Retention to High Temperatures. In Proc. 27th Automotive Tech. CCM, SAE, Warrendale, PA, 1990, pp.155-63.

11. Hasselman, D. P. H., Unified Theory of Thermal Shock Fracture Initiation and Crack Propagation in Brittle Ceramics. J. Am. Ceram. Soc., 1969, **52**[11] 600-04.

12. Hasselman, D. P. H., Figures of Merit for the Thermal Stress Resistance of High-Temperature Brittle Materials. Ceramurgia Int., 1979, **4**[4] 147-50.

13. Becher, P. F., Lewis, D., Carman, K. R. and Gonzalez, A. G., Thermal Shock Resistance of Ceramics: Size and Geometry Effects in Quench Tests. Am. Ceram. Soc. Bull., 1980, **59**[5] 542-48.

14. Becher, P. F., Transient Thermal Stress Behavior in ZrO_2-Toughened Al_2O_3. J. Am. Ceram. Soc., 1981, **64**[1], C-37-9.

15. Becher, P. F., Effect of Water Bath Temperature on the Thermal Shock of Al_2O_3. J. Am. Ceram. Soc., 1981, **64**[1], C-17-18.

LASER ASSISTED PLASMA COATING FOR CARBON MATRIX COMPOSITES

S.SASAKI*, H.SHIMURA*, K.HASEGAWA** and K.HIRANO***

* Mechanical Engineering Laboratory, Tsukuba, Ibaraki 305 Japan
** Kawasaki Steel Co., Kawasaki-cho, Chiba 260, Japan
*** Nippon Steel Co., Nakahara-ku, Kanagawa 211, Japan

ABSTRACT

To improve the environmental resistance of C/C composite materials under high temperature conditions, we have designed a multi-functional layer coating system. This laser and plasma hybrid spraying method, which couples a high power CO_2 laser with low pressure plasma spraying equipment, was developed to synthesize a highly dense and highly adhesive coating layer. Using this hybrid method, we succeeded in synthesizing a multi-functional coating system of Mo/W/WC - $MoSi_2$ - ZrO_2 on monolithic carbon in experimentally. The synthesized multi-layer coatings were stable at 1700°C in inert atmosphere. In oxidative conditions, the coatings were confirmed to be stable up to 1000°C minimum and they also survived five heat cycle tests of 500 to 1500°C under vacuum.

INTRODUCTION

Carbon fiber reinforced carbon matrix (C/C) composite materials are being applied to, or considered for, a growing range of engineering uses. Initially developed as thermally stable replacements for plastic-matrix composites or as tougher forms of carbon for short- term use in hot structural parts of rockets and missiles[1][2], C/C composites materials are now are found in aircraft brake disks, nuclear reactor systems, gas turbines, metal-forming tools, and under study for biomedical applications,as pistons in diesel engines, and electric commutators[3]. All of these applications take advantage of the excellent physical properties of C/C composite materials such as low density, high rigidity, low thermal expansion, high wear resistance and high temperature strength. The specific tensile strength of C/C composite materials compares with high temperature metal alloys and various ceramics. For example, in Fig.1. C/C composite materials provide superior specific strength and significantly enhanced temperature capabilities.

Based on their mechanical performance, C/C composite materials have been identified as attractive structural materials for ceramics gas turbine applications. However, these applications involve extended periods of operation in oxidizing environments under erosive and corrosive conditions. As a result, the development of reliable oxidation protection is crucial to utilizing the full potential of C/C composite materials. In this paper, we outline a multi-functional layer coating system and a laser and plasma hybrid spraying method. Actual multi-layer coatings synthesized on carbon materials and some of their thermal properties are also described.

Concept of environmental resistant coating

To improve the environmental resistance of C/C composite materials under high temperature conditions, we designed a multi-functional layer coating system, which consisted of a maximum of five different functional layers, as shown in Fig. 2. The characteristics of each layer are described below :

(1) 1st layer coating

The function of the 1st layer coating is to increase the adhesion between the matrix and the following coating layers. The following properties are required of the 1st layer coating materials:

(a) The materials must possess moderate reactivity with carbon; the carbide generation temperature should be high and the carbide generation speed should be slow in the temperature range of 1500 to 2000°C.
(b) The melting point must be higher than 2000°C.
(c) That the materials must not diffuse elements which deteriorate the carbon substrate; for example, oxide ceramics are not available.
(d) That the materials must have low thermal expansion coefficients.

Figure 3 shows the free energy of various metals to form carbides as a function of temperature. The thermal expansion coefficients of high melting point materials are shown in Table 1. Judging based on those properties, it is considered that the materials which satisfy the above conditions are Mo, W and WC.

(2) 2nd layer coating

The function of the 2nd layer is to trap the oxygen which permeats through the upper layers by a self-sacrificing oxidation resistant mechanism. This sacrificing oxidation mechanism means that the material reacts with oxygen and forms oxides, which then permits no further oxygen diffusion. Materials which are considered to satisfy those demands are $MoSi_2$ and WSi_2. $MoSi_2$ forms SiO_2 and Mo_5Si_3 in the following reaction:

$$5MoSi_2 + 7O_2 \rightarrow Mo_5Si_3 + 7SiO_2$$

(3) 3rd layer coating

The function of the 3rd layer is to inhibit the diffusion of oxygen to the underlying layers. The 3rd layer is considered necessary for extending the life of the 2nd layer, that is, the self-sacrificing oxygen barrier layer. Ir and Pt, which do not react with oxygen, are considered suitable as the coating materials. This layer is also required to be dense without pinholes.

(4) 4th layer coating

The 4th layer is a thermal barrier coating (TBC). This layer acts to control the thermal gradient and to protect the multi-layers from thermal shock and thermal stress. These coating materials are required to possess low thermal conductivity and stability under an oxidizing atmosphere. Porous ZrO_2 is considered as satisfying those requirements.

(5) 5th layer coating

The outer surfaces are exposed to high temperature gas which flows at high speed and which contains small particles and corrosive substances. The function of the 5th layer is to protect the coating on the lower levels from erosion and corrosion under high temperature conditions. A denser and stronger ZrO_2 coating than the 4th layer is considered suitable.

EXPERIMENTAL

Laser and plasma hybrid spraying system

The key concern in synthesized multi-functional layer coating performance is considered to depend on the quality of the 1st layer coating. Thus, this laser and plasma hybrid spraying system was developed to produce a high quality 1st layer coating on carbon matrix composites.

There are many coating methods to produce coating films: plasma spraying, sputtering, electron beam evaporation, chemical vapor deposition (CVD), pack cementation, ion implantation, sol gel techniques, etc. The CVD method is the most popular method for ceramics coatings on C/C composite materials[4]. This method can produce a strong adhesive coating layer on carbon materials, but it is difficult to increase film thickness or coat large components. It is easy to produce thick coatings using the ordinary low pressure plasma spraying method, but plasma sprayed coatings typically have low adhesion, low density and high porosity-stressed permeable structures. To improve these properties, the substrate surface must be heated to at least 2000°C, and this is a difficult task using only the plasma spraying.

Consequently, we developed a laser and plasma hybrid spraying method, which couples a high power CO_2 laser with low pressure plasma spraying equipment. In this method, it is possible to easily heat a coated material surface to temperatures higher than 3000°C by controlling laser power: thus, when the laser power is 3 kW and the laser beam dia. is 5 mm, the irradiation power density is approximately 4×10^7 W/m^2. Figure 4 shows a schematic of the laser and plasma hybrid spraying system we used. This system mainly consists of a low pressure plasma spraying system having a vacuum chamber with an inside dia. of 2 m and a length of 3 m, two high power CO_2 lasers, an NC controlled spray robot, a super fine particle production device and an infrared temperature measurement and analysis system.

Spraying experiments

As a preliminary study, several basic spraying experiments were conducted using laser and plasma hybrid spraying on monolithic carbon and 2D-C/C composite materials. In this case, the multi-functional layer coating system was composed of 3 layers: Mo/W/WC - $MoSi_2$ - ZrO_2. The dimensions of the spraying test piece was 17x14x70 mm. Using the spray robot to handle the test pieces, it was impossible to coat all surfaces in one operation. Therefore, in this experiment, only five faces were coated. Prior to spraying, the test pieces were preheated with the laser for 3 minutes under vacuum. The spraying conditions are shown in Table 2. The test piece was fed at a 2 mm pitch at a scanning speed of 2000 mm/min. Only the 1st layer coating was synthesized using the laser and plasma hybrid spraying method. The 2nd layer coating, $MoSi_2$, was synthesized by low pressure plasma spraying and the 3rd layer coating, ZrO_2, was synthesized by normal plasma spraying.

RESULTS

SEM micrographs of the coating layers are shown in Fig. 5. The results of the heating tests are summarized in Table 3. The WC coating layer was confirmed stable up to 2000°C in Ar. After five heat cycle tests of 500 to 1500°C in vacuum, no cracks or delamination were observed in any of the two layers coating systems of Mo/W/WC - $MoSi_2$ on monolithic carbon. The WC-MoSi2 coating system on monolithic carbon was also stable under oxidizing conditions even at 1300°C. Further, the WC-MoSi2 coating layer on C/C composite materials was found to be stable up to 1300°C.

CONCLUSIONS

1. To improve the oxidation and corrosion resistance of C/C composite materials, we have designed the concept of a multi-functional coating system. Selection of materials which satisfy the function of each coating layer are discussed.

2. To synthesize multi-functional layer coatings, a laser and plasma hybrid coating method was devised and a hybrid spraying technique was developed.

3. Using devised laser and plasma hybrid spraying, it was determined that a thick coating layer on carbon matrix materials, which had been considered a difficult problem, was possible to be formed. Multi-functional coating systems have been synthesized using monolithic carbon and C/C composite materials.

4. The stabilities of the synthesized muti-layer coatings were tested under high temperature conditions. Two layer coatings of Mo/W/WC - $MoSi_2$ were stable up to 1700°C in Ar and they survived five cycle heat testing of 500 to 1500C° under vacuum. Three layer coatings of WC - $MoSi_2$ - ZrO_2 were also determined to be stable up to 1500°C in Ar. In air, WC - $MoSi_2$ on carbon showed stability at 1300°C.

5. Further works are needed to synthesize multi-functional coatings stable up to 2000°C. In particular, the concept of functionally gradient materials (FGM) must be applied to multi-functional layer coating.

REFERENCES

[1] Schmidt, Donald L.: "Carbon/Carbon Composites," SAMPE J., May/June 1972, P.9.

[2] Jortner.J: "Thermostructural Behavior of Carbon-Carbon Composites," edited by Jortner.J, Booklet AD-11 ASME 1986, P.1.

[3] Fitzer,E. and Gkogkidis,A.: "Carbon-Fiber-Reinforced Carbon Composites Fabricated by Liquid Impregnation,"In Petroleum-Derived Carbons, edited by J.D.Bacha, ACS Symposium Series 303, ACS., 1986, P.346.

[4] Witold, Kowbel: " Graded-Codeposited ZrC-BN Coating for The Thermal Protection of Carbon-Carbon Composites," Proc. of Ceramic Materials & Components for Engines, Edited by V.J.Tennery, 1988, P.290.

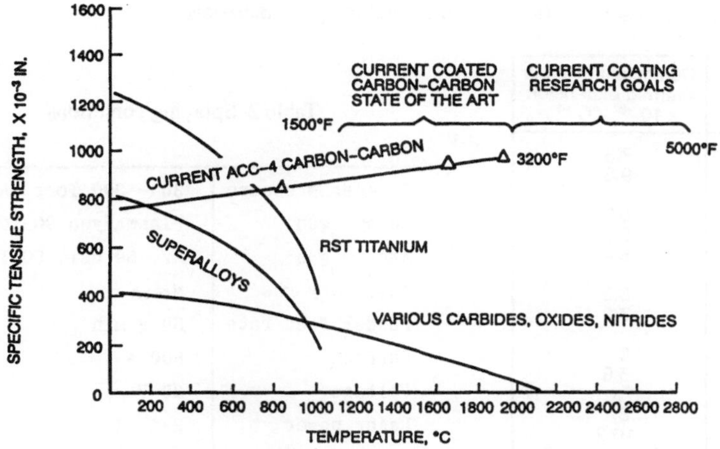

Fig.1 Specific tensile strength of C/C composite materials compared with high temperature materials.

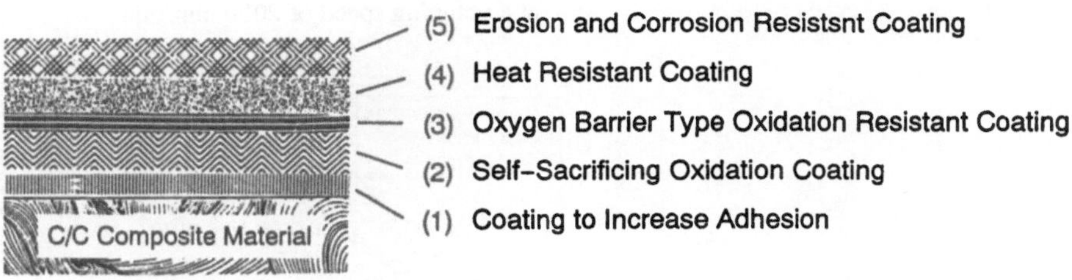

Fig.2 Concept of environmental resistant coating system

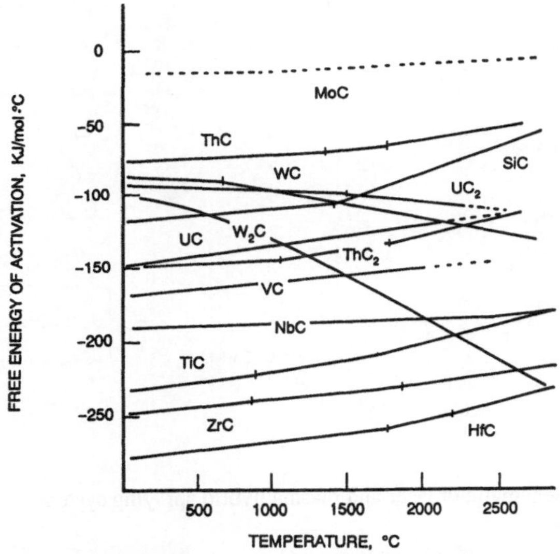

Fig.3 Free energy of metals to form carbides as a function of temperature

Table 1 Thermal expansion coefficients of high melting point materials

Materials	Coefficient of linear thermal expansion $\times 10^{-6}$ °C^{-1}
Mo	7.3
Mo_2C	9.3
W	5.7
WC	3.8
W_2C	6.4
$MoSi_2$	8.3
WSi_2	7.9
Al_2O_3	8
ZrO_2	5.6
Ir	8
Pt	10.2
graphite	5
C/C	<±1

Table 2 Spraying conditions

Chamber pressure	30 - 300 Torr (Ar)
Spray gun	Plasmadyne SG-100
Arc gas	Ar, 50 psi, 50 l/min
Auxiliary gas	None
Powder feed rate	50 g/min
Current	800 A
Voltage	33 V
Laser power	2-4 kW
Focus dia.	5 mm (approximately)

The test piece was fed at a 2 mm pitch at a scanning speed of 2000 mm/min

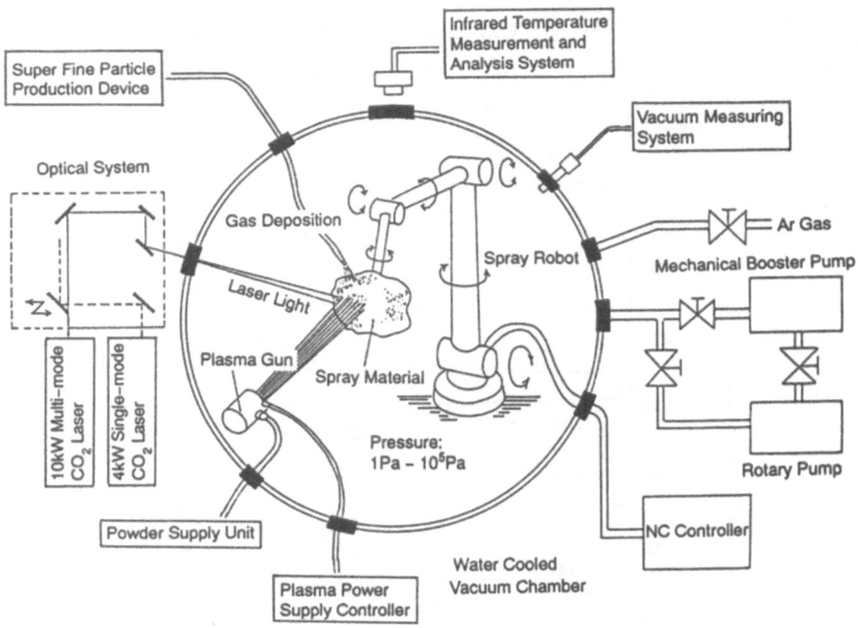

Fig.4 Schematic of laser and plasma hybrid spraying system

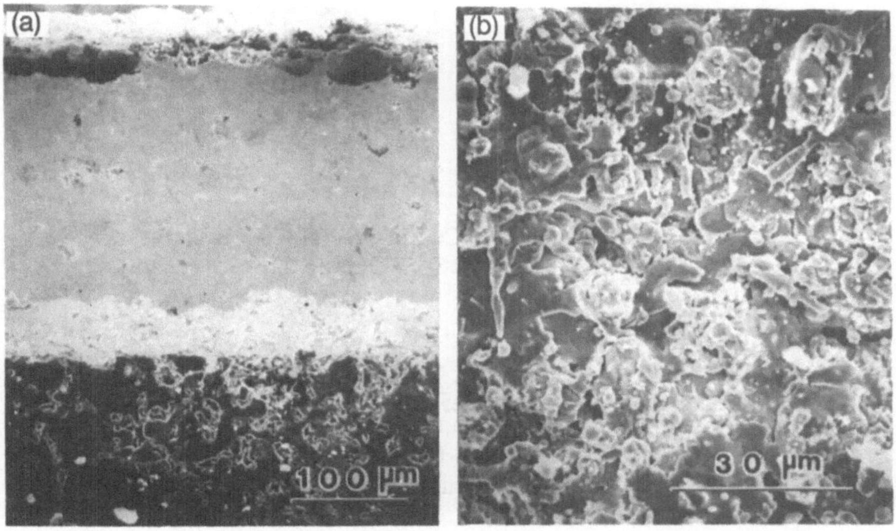

Fig.5 SEM micrograohs of coating layers.
(a) Cross sectional view of WC–$MoSi_2$ coating on monolithic carbon
(b) $MoSi_2$ sprayed surface

Table 3 Results of heating tests

Coating System	in Ar *(1)			in Air *(2)			Heat Cycle Tests *(3)
	1000°C	1500°C	2000°C	1000°C	1300°C	1500°C	500 to 1500C°x 5 times
(1) Mo-Base Coating							
1-1 Mo	O	O	×	–	–	–	–
1-2 Mo-$MoSi_2$	O	O	O (1700°C) ×	–	–	–	O
1-3 Mo-$MoSi_2$-PSZ	O	Δ	×	–	–	–	–
(2) W-Base Coating							
2-1 W	O	O	×	–	–	–	–
2-2 W-$MoSi_2$	O	O	O (1700°C) –	O	Δ	–	O
2-3 W-$MoSi_2$-PSZ	O	Δ	–	–	–	–	–
(3) WC-Base Coating							
3-1 WC	O	O	O	–	–	–	O
3-2 WC-$MoSi_2$	O	O	O (1700°C) –	O	O	–	O
3-3 WC-$MoSi_2$-PSZ	O	O	–	–	–	–	–
(4) on C/C							
4-1 W-$MoSi_2$	O	O (1300°C)	–	–	–	–	–

O No delamination × Delamination or Crack

*(1) rate of heating: 6°C/min, rate of cooling: 20°C/min, maintaining a definite temperature for 2 hours.
*(2) rate of heating: 6°C/min, rate of cooling: uncontrolled maintaining a definite temperature for 2 hours.
*(3) rate of heating and cooling: 20°C/min.

$SiC-Si_3N_4$ GRADIENT COMPOSITE CERAMICS BY SPECIAL HIP PROCESSING

Jiang Dongliang, She Jihong, Tan Shouhong, Guo Jingkun
Shanghai Institute of Ceramics, Academia Sinica

Peter Greil
Technische Universitat Hamburg-Harburg(TUH), FRG

ABSTRACT

α-SiC fine powder clad with special glass were sintered by HIP processing under 200 MPa at 1800 °C. The density of material obtained is close to 97.5% of theoretical density, bending strength reached 582 MPa, which is about 50% higher than that of common pressureless sintered silicon carbide. After post-HIP treatment, density can be further increased to above 98.5% of theoretical density. The material with different composition can also be got in various atmospheres. According to thermodynamic calculation, SiC can be transformed into Si_3N_4 under optimum temperature and high N_2-pressure. Experimental work was successful in fabricating $SiC-Si_3N_4$ gradient composite material. This kind of new $SiC-Si_3N_4$ gradient composite material were also identified by X-ray diffraction analysis. The strength and fracture toughness of this material is 900 ± 100 MPa and 8.4 $MPa.m^{1/2}$ respectively, which are about one times higher than those of hot-pressed SiC material. This study offered a new HIP processing technology for preparing high performance $SiC-Si_3N_4$ gradient composite ceramics directly from pure SiC powder.

INTRODUCTION

SiC and Si_3N_4 are the most attractive advanced structural ceramics because of their excellent high-temperature mechanical properties. They have been widely used as various kinds of engineering components, such as cutting tools, mechanical sealing rings, bearings, heat exchangers and components in heat engines. However, the intrinsic brittleness of these materials is still considered as a major limitation and drawback for structural applications. Therefore, the studies on multiphase ceramics have been carried out and some results obtained have shown that the mechanical properties of monolithic ceramics can be enhanced through the composite approach. For example, the development of the SiC-TiC [1-2], the $SiC-Al_2O_3$ [3] and the Al_2O_3-TiC [4] composite ceramics has demonstrated that fracture toughness or strength and high-temperature properties can significantly be improved by particle reinforcement, thus many new materials and some new applications have been exploited. Similarly, the Si_3N_4-SiC composites have been investigated for many years. It has been reported that the fracture toughness can be increased in silicon nitride systems using SiC as the dispersed particle [5]. Peter Greil et al. [6] have obtained Si_3N_4/SiC-particulate composite materials by HIPing, and they have founded that when the content of SiC particles is higher, a distinct improvement of strength had been achieved, but toughness hadn't been improved. Yoshio AKIMUNE [7] have reported the best results of Si_3N_4-SiC composites fabricated by using ultrafine starting powders and applying pressureless-sintering together with hot isostatic pressing, the maximum strength was up to 1226 MPa and the fracture toughness was 5.4 $MPa.m^{1/2}$. Another study on SiC-Si_3N_4 composites has also been reported by Hiroshi NAKAMURA et al. [8], the bending strength of 920 MPa was obtained by using submicro-meter SiC and Si_3N_4 starting powders and utilizing the hot-pressing technique. Generally, the expensive Si_3N_4 powders are used as the matrix of SiC-Si_3N_4 composites. Thermodynamic evaluation of the reaction between SiC and N_2 indicates that the transformation of SiC and Si_3N_4 is feasible under given conditions (temperature, pressure and atmosphere). Up to now, there has been no report that the SiC-Si_3N_4 composites were directly fabricated from the SiC starting materials. Consequently, the investigation are well worth developing that the SiC-Si_3N_4 composites can be directly fabricated by making use of the plentiful and cheap SiC materials. Based on thermodynamic evaluation, the research reported in this paper explores the possibilities of fabricating Si_3N_4-SiC functionally gradient material (FGM) by the hot isostatic pressing technique. A new technique was provided for the development of new materials.

EXPERIMENTAL PROCEDURE

The starting material was α-SiC powder having an average particle size of 0.8μm. Al_2O_3 was used as sintering additive. SiC powders and Al_2O_3 additives were mixed with ethanol in a polyethylene ball mill for 48 hours. The mixture was then dried, sieved and cold isostatically pressed with 200 MPa. Subsequently, the powder compacts were dewaxed in a vacuum heated furnace and encapsulated with a special glass. Hot isostatic pressing was carried out at 1850 °C under 200 MPa pressure. HIPed specimens were machined into 3mm x 4mm x 36mm bars for measuring three-point bending strength, and 2.5mm x 5mm x 36mm bars for measuring fracture toughness. Both bending strength and fracture toughness measurements were performed using 1195 Instron testing machine at a crosshead speed of 0.5mm/min. Microstructure characterization of fracture surfaces was made by SEM and TEM. Reaction products formed on the surface were identified by X-ray diffraction.

EXPERIMENTAL RESULTS AND DISCUSSION

I. Thermodynamic Analysis

Fig.1 shows the possibility of the reaction between SiC and N_2 under given temperature and N_2-pressure, which was drawn according to the following thermodynamic calculation

$$\Delta G^0_{298} = \Delta H^0_{298} - T\cdot\Delta S^0_{298} - RT\cdot Ln(P)$$

where $\blacktriangle H^0_{298}$ is the standard enthalpy change, $\blacktriangle S^0_{298}$ is the standard entropy change, T is the absolute temperature, P is the partial pressure of N_2 at temperature T, $\blacktriangle G^0_{T,P}$ is the overall free energy change for a reaction under given conditions (pressure P and temperature T).

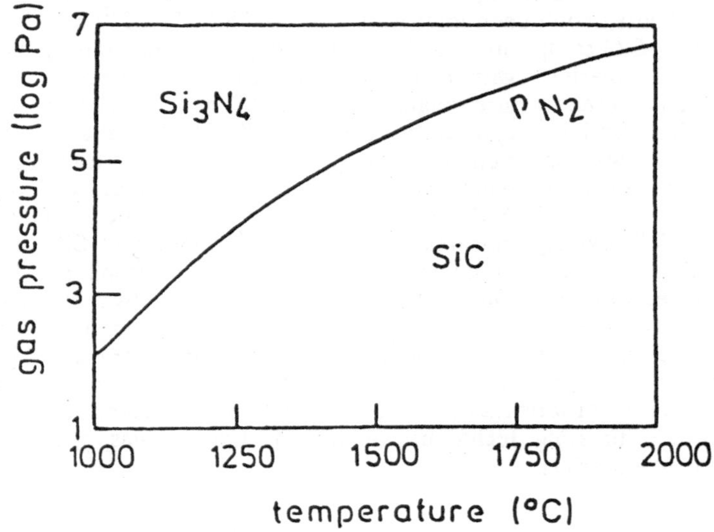

Fig.1. Partial pressure of N_2 at 1000-2000°C for the reaction between SiC and N_2.

The thermodynamic data is from the Thermochemical Properties of Inorganic Substances [9]. The chemical reaction between SiC and N_2 is given by

	3SiC	+	$2N_2$	=	Si_3N_4	+	3C
$\blacktriangle H^0_{298}$	-17.5		0		-178		0
S^0_{298}	3.97		45.77		27		1.37

It can be seen in Fig.1 that at high temperature, SiC can be transformed into Si_3N_4 only when the N_2-partial pressure is high enough. For instance, when the temperature is 1850 °C, the N_2-pressure must be higher than 80 MPa. At low temperature, the N_2-pressure which causes SiC to be transformed into Si_3N_4 can be reduced, but the densification of presintered ceramics can not be occurred. In the present study, because of the simultaneous application of high temperature and high pressure, not only was near-theoretical density obtained, but also SiC was partially transformed into Si_3N_4, resulting in the formation of SiC-Si_3N_4 composite. This result was identified by X-ray diffraction (the diffraction peak of ß-Si_3N_4 was appeared in the low-angle X-ray diffraction zone).

II. Possibility of Densification by HIPing at Lower Temperature

It is well known that SiC is a strong covalent bonded compound which is difficult to be fully densified by conventional sintering techniques. Even if submicrometer powders and suitable sintering aids are used, only 93-95% of the theoretical density can be obtained by pressureless-sintering. In the case of the hot-pressing, SiC ceramics with near-theoretical density can be gained, but the strength which depends on sintering aids and processing is only 500-700 MPa, and the fracture toughness is just 4MPa.m$^{1/2}$ or so. Densification by pressureless-sintering or hot-pressing requires very high temperature (usually higher than 2000 °C), therefore, grain growth can hardly be avoided and the further increase of strength is difficult to be achieved. In order to investigate the possibility of sintering SiC ceramics at lower temperature, the green formed samples were encapsulated using special glass, hot isostatic pressing was carried out under 200 MPa in Ar at 1850 °C. The HIPed SiC ceramics had a relative density of 97.5% and a bending strength of 582 MPa, which is obviously better than the pressureless-sintered silicon carbide and corresponds to the hot-pressed SiC with near-theoretical density. Because the HIPed SiC ceramics exhibited fine-grained structure and no coarse grain growth, they had higher strength in spite of the porosity of 2-3%. It is estimated that the further improvement of strength should be obtained if the density can be further increased and the same fine structure can be kept. Consequently, the HIPed SiC was post-HIPed after the container was removed. The result showed that the density was increased to 98.5% of theoretical density and the strength was up to 668 MPa.

III. Effect of Atmosphere on Structure and Properties of SiC

Under the same post-HIPing conditions, the results will be completely different if different atmospheres are used. When argon was used as the pressure transmitting medium for post-HIPing, the HIPed SiC was further densified, but the improvement of properties was not obvious. However, the considerable improvement of properties was achieved when post-HIPing was carried out in N_2. Table 1 listed the properties of SiC ceramics post-HIPed using two kinds of gases.

TABLE 1
Properties of SiC ceramics post-HIPed in various gases.

	Density (g/cm^3)	Strength (MPa)	Fracture toughness (MPa·m$^{1/2}$)
HIP	3.13±0.02 [10]	582±78 [5]	5.7±1.2 [5]
POST-HIP(Ar)	3.16±0.01 [6]	668±72 [5]	
POST-HIP(N_2)	3.20±0.01 [8]	907±110 [7]	8.4±1.2 [4]

* Number of specimens is given in brackets.
** HIPing and post-HIPing conditions: 1850 °C, 200 MPa, 60 min dwell time.

The difference in properties was investigated by X-ray analysis. Under high temperature and high N_2-pressure, the reaction between SiC and N_2 occurred, which brought about the formation of Si_3N_4-SiC composite with a fine and densified structure. This reaction did not result in the change of macro volume of SiC ceramics. Contrarily, because the expansion coefficient of Si_3N_4 is smaller than that of SiC, the material was in condensed stress during cooling, leading to the distinct improvement of strength. Fig.2 showed the microstructure of sample before and after post-HIPing.

(A) (B)

Fig.2. SEM micrographs of fracture surfaces of SiC before (A) and after (B) post-HIPing in N_2 (1850 °C, 200 MPa, 1 h).

Since the density of HIPed silicon carbide was up to above 96%, the reaction between SiC and N_2 might begin with surfaces. Once the pores are closed by Si_3N_4 products, the further penetration of N_2 will be prevented, resulting in the end of the reaction. Therefore, the inside of sintered bodies is still silicon carbide. In order to verify this idea, the strength was determined after the surfaces were removed by 0.2-0.3 mm thickness, which corresponded to that of SiC ceramics post-HIPed in argon. Furthermore, X-ray analysis revealed the disappearance of Si_3N_4. Thus it can be seen that SiC monolithic ceramics can be transformed into Si_3N_4-SiC composites by post-HIPing under high N_2-pressure, forming the gradient composite ceramics with different Si_3N_4/SiC ratios from the outside to the inside. In addition, Auger Electron Spectroscopy (AES) analysis also identified the change of Si_3N_4/SiC ratio from the surface to inside. Fig.3 shows N and C weight fraction varying with depth of sample. Because the diameter of electron ray is only 0.1-0.5 μm, it is very hard for whole sample to be quantitatively described. The statistical method or other technology are needed to get more information. Fig.4 was the TEM picture of sample. There is a non-crystalline secondary phase located at triple junction, which consist of Al_2O_3 and SiO_2.

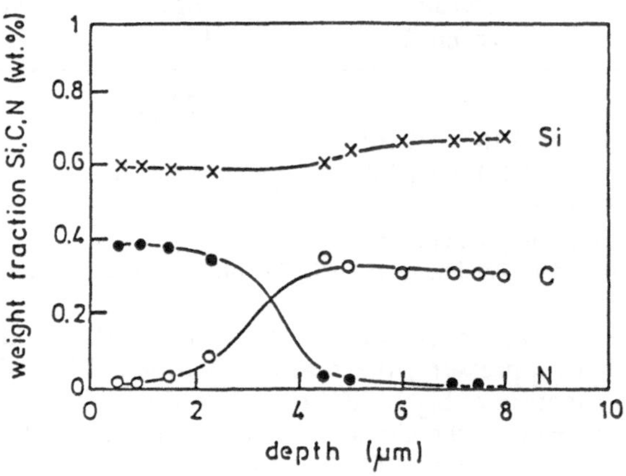

Fig.3. Elemental distribution from the surface to the bulk of the surface-nitrided SiC (1850 °C, 200 MPa, 1 h) obtained by AES.

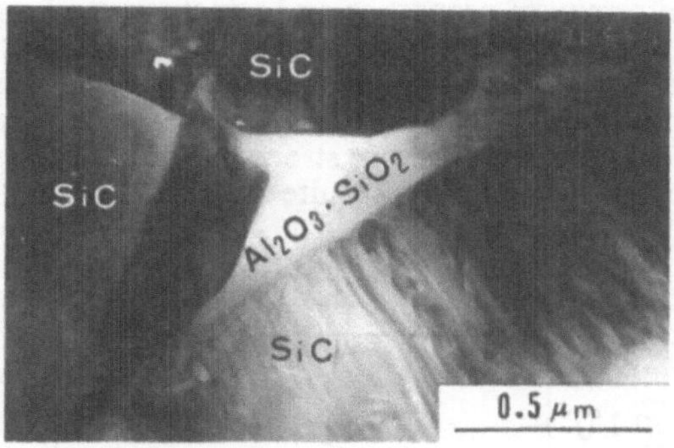

Fig.4. Bright field TEM micrograph of amorphous intergranular phase in Al_2O_3-doped SiC HIPed at 1850 °C under 200 MPa for 1 h.

IV. Effect of Ratio of Si_3N_4/SiC on Strength

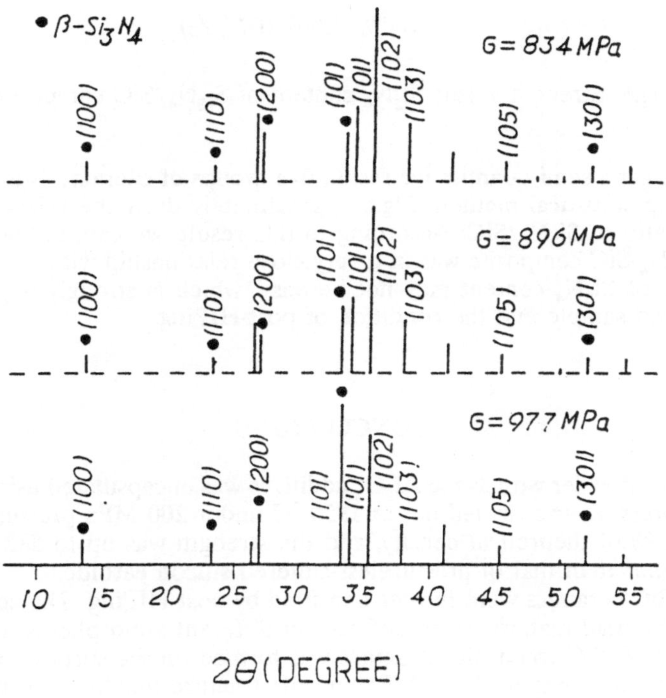

Fig.5. X-ray differaction spectrum of SiC ceramics post-HIPed under 200 MPa in N_2 at 1850 °C.

In common, more uniformity of properties and structure can be got by HIPing process than the other processing. Nevertheless, in this work we found that the strength deviation was very big. In order to explain this phenomenon, after the strength of each bar was tested, they were examined by X-ray diffraction. Fig.5 gave the X-ray diffraction spectrum of three samples which exhibited different strength value. In this figure, dark spots are ß-Si_3N_4 characteristic spectrum, which is quite different for the sample with different strength.

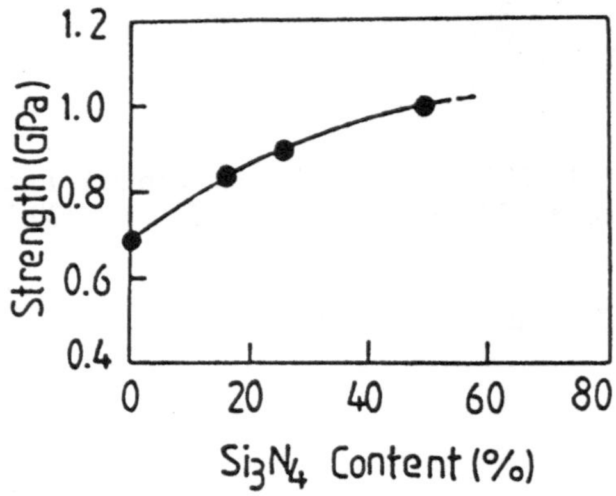

Fig.6. Strength versus Si_3N_4 content of Si_3N_4/SiC composite.

In order to get a semi-quantitative result, five groups of characteristic spectrum were treated by using statistical method. Fig.6 approximately drew the relationship between strength and ratio of Si_3N_4/SiC. According to this result, we can, at least, say that the strength of Si_3N_4-SiC composite was in a very close relationship with Si_3N_4 content. But the controlling of Si_3N_4 content may not be easy, which is strongly dependent on the density of HIPed sample and the condition of post-HIPing.

CONCLUSIONS

1. The α-SiC powder with 3wt% Al_2O_3 additives was encapsulated using special glass. Hot isostatic pressing was carried out at 1850 °C under 200 MPa pressure. The density was above 97.5% of theoretical density, and the strength was up to 582 MPa, which is about 50% higher than that of pressureless-sintered silicon carbide.

2. HIPed SiC ceramics were further densified by post-HIPing. The composition and properties of the final material were different in different atmospheres. Under high N_2-pressure, the Si_3N_4-SiC composite structure was formed on the surface of SiC ceramics, which exhibited the strength of 900 MPa and the fracture toughness of 8.4 MPa.m$^{1/2}$.

3. The gradient Si_3N_4-SiC composite ceramics were directly made from the SiC monolithic ceramics by a new hot isostatic pressing technology.

ACKNOWLEDGEMENT

Financial support from the Natural National Science Foundation of China (NNSFC) and the Germany Science Foundation (DFG) is gratefully acknowledged.

REFERENCES

[1] Wei, G.C. and Becher, P.F., Improvements in Mechanical Properties in SiC by the Addition of TiC Particles. J. Am. Ceram. Soc., 1984, **67**, 571-4.

[2] Jiang, D.L., Wang, J.H., Li, Y.L. and Ma, L.T., Studies on the strengthening of silicon carbide-based multiphase ceramics. I. The SiC-TiC system. Mater. Sci. Eng., 1989, **A109**, 401-6.

[3] Katsumura, Y. and Kobayashi, M., Some properties of Al_2O_3-SiC sintered compact. Journal of the Japan Society of Powder and Powder Metallurgy, 1988, **35**, 137-41.

[4] Wahi, R.P. and Ilschner, B., Fracture behaviour of composites based on Al_2O_3-TiC. J. Mater. Sci., 1980, **15**, 875-85.

[5] Baljan, S.T., Baldoni, J.G. and Huchabee, M.L., Si_3N_4-SiC Composites. Am. Ceram. Soc. Bull., 1987, **66**, 347-52.

[6] Greil, P. and Petzow, G., Sintering and HIPping of Silicon Nitride-Silicon Carbide Composite Materials. Ceramics International, 1987, **13**, 19-25.

[7] Akimune, Y., High strength SiC-particle/Si_3N_4 Composites. Nippon Seramikkusu Kyokai Gakujutsu Ronbunshi, 1990, **98**, 424-38.

[8] Nakamura, H., Umebayashi, S. and Kishi, K., Mechanical properties of hot-pressed silicon nitride-silicon carbide composites. Nippon Seramikkusu Kyokai Gakujutsu Ronbunshi, 1989, **97**, 1517-20.

[9] Barin, L. and Knacke, O., Thermochemical properties of inorganic substances, 1973.

NOVEL α'/β' SIALON CERAMICS

C.A. JASPER+ AND M.H. LEWIS*
+Vesuvius Zyalons Midlands Ltd., Solihull, UK.
*Centre for Advanced Materials Technology,
University of Warwick, Coventry, UK.
(+Now at British Gas plc, Solihull, UK.)

ABSTRACT

New mixed-phase α'/β' Sialon ceramics have been fabricated by transient liquid-phase sintering resulting in minor glass residues. The glass has been completely devitrified to YAG because of the receptiveness of the combined α' and β' phases in accommodating non stoichiometric components. The resulting ceramics have distinctive microstructures of isolated, non-faceted, YAG in an α'/β' phase mixture, typically in 70/30 ratio.

Hardness, oxidation and creep-resistance for these ceramics are superior to those of earlier β' Sialons.

INTRODUCTION

Silicon nitride has long been recognised as a potential candidate for use in high temperature engineering applications. It exhibits the necessary combination of intrinsic properties; high creep and oxidation resistance, good thermal and chemical stability, and high elastic modulus to specific gravity ratio, which make it attractive for gas turbine components. However, the strong, covalent, Si-N bonding, fundamental to these properties, results in a high activation energy for self diffusion which inhibits solid-state sintering below the temperature at which decomposition occurs. Fabrication to full density without the application of exceptional pressure/temperature conditions requires the addition of one or more oxide agents to catalyse the liquid-sintering reaction.

Following the early identification of effective sintering additives [1] and the mechanism of solution-reprecipitation within silicate-based liquid systems [2], subsequent development of the liquid sintering chemistry has proceeded largely via empiricism. Much effort has been expended in modifying the volume and composition of the liquid residues to achieve the best compromise between sinterability and retention of the most desirable properties. A key step in the development of commercial materials was the 'alloying' of silicon nitride with Al and 0 from the additives to generate a new range of materials known generically as 'sialons' [3,4]. Recent work has concentrated on further refinement of the liquid composition to leave a residue which can be crystallised by a post-sintering heat treatment to enhance high temperature performance [5].

This paper seeks to review the current status of sialon liquid-sintering chemistry, to examine the evolution and influence of the secondary phases, and illustrate how the microstructure may be modified further to generate novel materials with improved properties.

Evolution of Sialon microstructures

The ability to accommodate Al and 0 in the silicon nitride structure stems from the similarity between the Si-N and Al-0 bond lengths, 0.174nm and 0.175nm respectively. In practice, substitution is achieved by a complex series of reactions involving atomic transport in a silicate related eutectic liquid. Silica derived from the surface contaminant on the Si_3N_4 starting powder (usually the α-Si_3N_4 form) initially combines with the added oxides (Y_20_3, Al_20_3) to form a liquid at the ternary eutectic composition. As the sintering temperature is increased progressive solution of the bulk nitride species occurs to give an oxynitride liquid from which β'-sialon then precipitates. The reaction rate is governed by the volume and viscosity of the liquid which in turn is dependent upon the increment in sintering temperature above the eutectic liquidus. The addition of nitrogen increases the liquid viscosity and the amount dissolved in conjunction with the concentration of cationic species present (Y/Al) determines the crystal form of sialon evolved, α'or β', and the nature of the final residual glass.

The crystal structure of the α'-sialon derivative requires a modifier cation (from the oxide additive) for stability. The range of homogeneity is smaller than for β' and pure α'-sialon materials free from residual glass and unreacted powder cannot be prepared by simple fabrication routes because of the acute constraints upon composition.

The ability to fully crystallise a liquid residue depends on the average ceramic composition in relation to the bulk Si_3N_4-based species and the potential crystalline matrix phase. In its simplest form the average composition, C, is on a binary tie-line between β' and the matrix phase which crystallises from the glass at constant composition (see Janecke prism representation Fig. 1). This is rarely possible but an example is provided by the Nd-Sialon system [6]. More frequently it is necessary that as the matrix phase composition moves out of the glass forming region to the terminal oxide phase the excess components diffuse into the existing β' or precipitate as more β' but of a modified composition [7]. This flexibility in variation of the β' composition thus provides an extra degree of freedom in the selection of a suitable matrix phase. In practice complete crystallisation is rarely achieved. For example in β'/YAG ceramics (Syalon 201) excess yttrium cannot be accommodated in the β' Sialon phase and may remain within a very small glass residue.

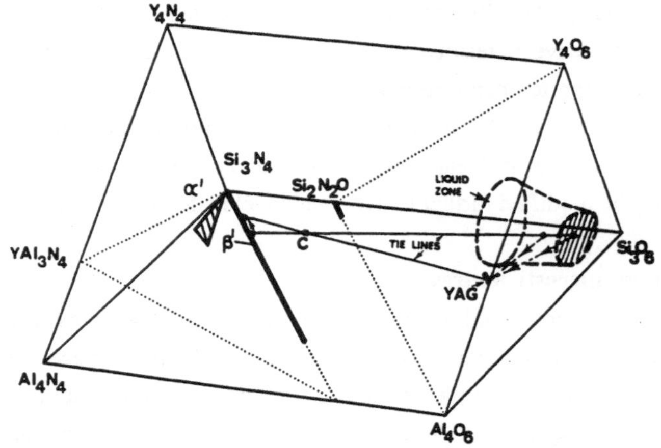

Figure 1. Janecke prism, illustrating the relation between ceramic components, major phases α' and β' sintering liquid and glass crystallisation products

Pressureless-sintered α'/β' sialon ceramics

Even though the evolution of α' offers the potential for solid-solution of all sintering additive elements within the silicon nitride structure, pressureless sintered α'-sialon containing materials have not received as much attention as the β'-variants because of the constraints on compositional control. However, α'/β'/glass materials have reached a stage of commercial availability, for example Kyon 3000* cutting tips where additional hardness afforded by inclusion of the α' species leads to increased wear resistance. The benefits of complete crystallisation of a third (oxide) phase, utilising the flexible partitioning of residual elements within α' and β' solid solutions, have not been explored. This study was undertaken with a view to the parallel improvement in both low and high temperature properties of sialon ceramics.

PREPARATION AND SINTERING

Densification studies were carried out on powder mixtures of Si_3N_4 (Grade 1002, Cookson UK), 21R Polytypoid (Cookson UK), and Y_2O_3 (99.99%, Meldform Metals) compacted isostatically (~138 MPa) and sintered isothermally for 5 hours at temperatures up to 1800°C. Compositions were prepared slightly above the α'/β' equilibrium phase field and along the α'/β'-liquid tie-line to ensure a slight excess of liquid during sintering. Final materials and components were fired using an additional step incorporating a 2 hour hold at an intermediate temperature to improve homogeneity and develop the appropriate α'/β' ratio prior to full densification. Post-sintering cycles were 2-stage involving a long initial period for crystal nucleation followed by a higher temperature hold for completion of crystallisation.

MICROSTRUCTURE OF THE α'/β' MATERIALS

'As-sintered' Materials

The sintering reaction sequence of the materials under observation was similar to that for pure α'/β' materials reported previously[8] except that α' precipitation did not begin until 1450°C (cf. 1400°C) after substantial β' reprecipitation had occurred. The final amount of residual glass, (3-4 vol% Figure 2), compared favourably with 10-12 vol% normally present in β'/YAG materials prior to devitrification. The α'-sialon phase constituted approximately 70% of the microstructure. Grains were generally larger (1-15μm) than in β' ceramics with less faceted hexagonal morphology.

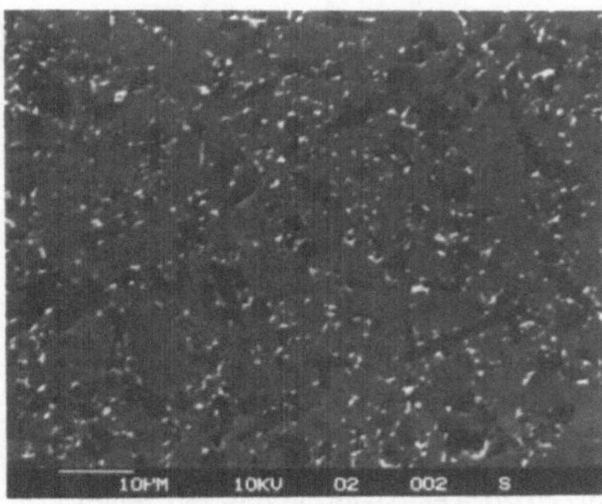

Figure 2. Microstructure of α'/β'ceramics in the 'as-sintered' state.

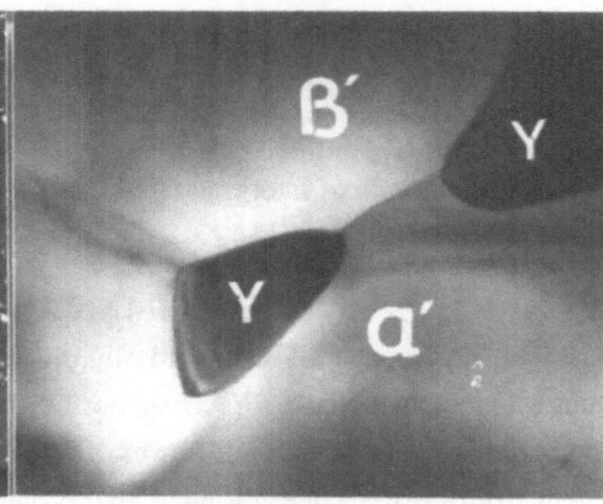

Figure 3. Microstructure of the α'/β'ceramics after post-sintering heat treatment.

Crystallised materials

Upon heat treating, the small volume of residual glass was crystallised to YAG which existed as small isolated equiaxed grains, with curved boundaries rather than straight facets, at some of the triple points (Fig.3). YAG crystals in close proximity had similar crystallographic orientations. High resolution lattice imaging showed that there was no obvious film of residual glass sandwiched between adjacent grain interfaces (Fig. 4).

Similar microstructures have been reported [7] in the surface of β'/YAG materials following extensive creep testing and exposure in oxidising environments. Here it was concluded that the excess intergranular metallic ions had diffused out to the surface silica oxidation layer during creep deformation. Removal of the grain boundary residues clearly resulted in increased solid/solid contact and the requirement to balance the interfacial energy isotropy was satisfied by diffusive rearrangement of the YAG/β' interfaces. In the present materials transformation was throughout the bulk. The α' crystals have acted as an effective 'sink' for the non-stoichiometric elements (Fig.5). Triple grain junctions had dihedral angles of 120° which is indicative of an equivalence between α'/β', α'/YAG and β'/YAG interfacial free energies.

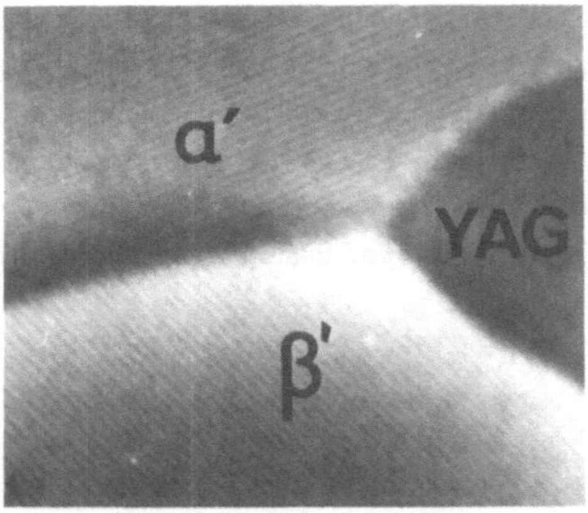

Figure 4. TEM lattice image showing an absence of intergranular glass.

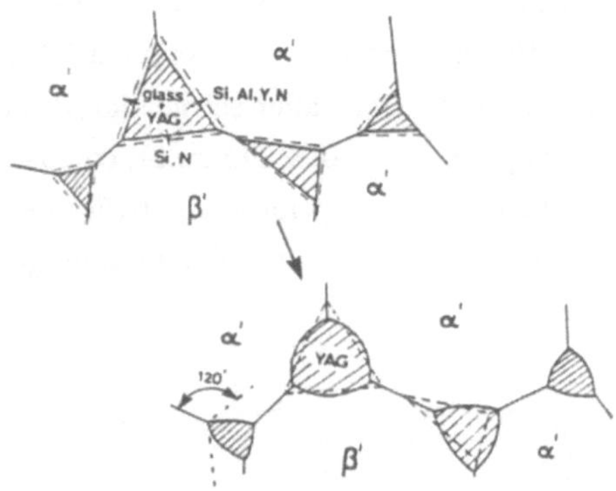

Figure 5. Illustrating the transport of non-stoichiometric elements during YAG crystallisation.

MECHANICAL PROPERTIES

The α'/β'/YAG materials exhibited slightly higher room temperature modulus of rupture values than conventional β'/YAG sialons (800 MPa cf. 725 MPa). Strength retention was improved at temperatures in excess of 1300°C (620 MPa cf. 500MPa at 1400°C) believed to stem from greater cohesive bond energy between adjacent sialon grains and removal of the glass residues at triple points which have a tendency for the creation of internal flaws through cavity formation at high stresses.

Hardness was increased typically from 1600 kg mm^{-2} to 2010 kg mm^{-2} influenced mainly by the high α' content and believed to originate from the higher Peierls Stress for dislocation motion consistent with an increase in the c-axis Burgers vector. (In the α' variant the c unit cell dimension is almost twice that for β').

Fracture toughness (SENB technique) was increased from 5.5 MPa $m^{1/2}$ to 6.5 MPa $m^{1/2}$. The small improvement in K_{1c} may result from enhanced grain boundary cohesion coupled with the pull-out and crack bridging influence of the anisotropically-shaped β' phase normally effective in β' sialons. In addition, in β'/YAG materials, there is a minor volume change as the glass is devitrified to YAG giving rise to internal stresses which may lower the fracture toughness by contributing to crack propagation. In the present materials devitrification occurs within a smaller glass volume.

In pressureless sintered β' materials high temperature properties are dictated by the nature and chemistry of the matrix phase. At low stress, prior to matrix crystallisation, creep is controlled by viscous flow processes. At high stress, or in materials with incomplete crystallisation, grain boundary shear causes hydrostatic tension in the glass residues and cavity nucleation. In fully crystalline materials creep is dominated by interfacial or grain boundary diffusion mechanisms (Coble creep).

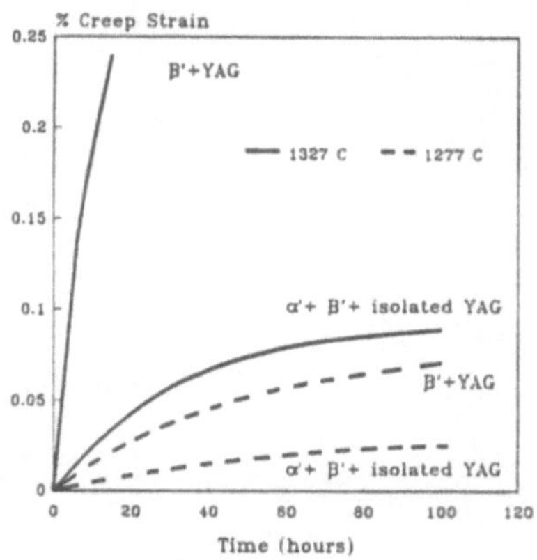

Figure 6 Comparison of creep curves for α'/β' and β' ceramics.

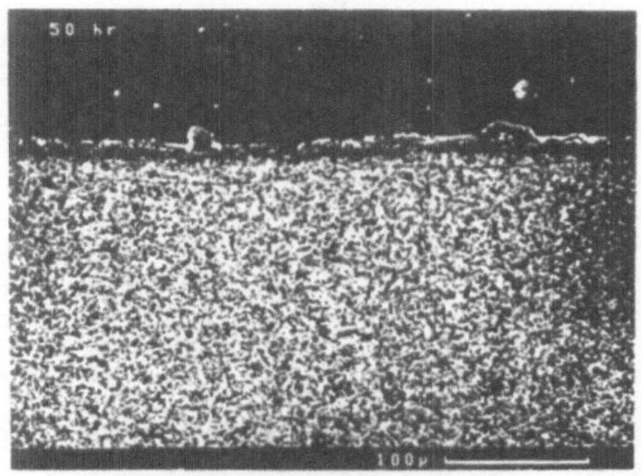

Figure 7 SEM of a SiO_2-rich oxidation layer found on α'/β' ceramics at 1300°C in 50 hours.

Comparison between the creep properties of β'/YAG and α'/β'/YAG materials showed minor differences at 1277°C (Fig.6). Both materials exhibited behaviour characteristic of largely crystalline ceramics - a primary viscoelastic stage rapidly tending towards a 'pseudo-steady state' strain rate characteristic of Coble creep. The β'/YAG curve showed a longer transient period consistent with small intergranular residues of glass being recrystallised or diffusing out to the surface as deformation progressed. The α'-containing material showed a lower steady state creep rate due to the increase in grain size (strain rate is proportional to $1/d^3$ where d is average grain size) and reduced grain boundary diffusion rate. At 1327°C creep and failure in the β'/YAG material became catastrophic. Creep deformation is enhanced by oxidation because the surface diffusion couple between SiO_2 and YAG regenerates a ternary eutectic liquid formation. In the α'/β'/YAG material there is less YAG, and it exists in an isolated morphology such that liquid formation is inhibited. The reduced metallic ion content of grain boundaries in α'/β'/YAG ceramics and slower grain boundary diffusion also inhibits oxidation. After 50 hours at 1300°C in air a 10μm oxidation layer is formed (Fig.7) compared to 40μm for β'/YAG ceramics.

COMPONENT PERFORMANCE

Trials were conducted on wire drawing dies. Conventional sialon materials have had difficulty becoming established in this area because performance is only comparable to that of tungsten carbide. Brass rod (DIN CuZn36 Pb1.5) was drawn through supported dies (Fig. 8) at a rate of 0.35 ms^{-1} with a 12% reduction in cross-sectional area. Wear was monitored until the diameter was increased to the upper tolerance specification. Improvements of 50% in total pull quantity were regularly achieved (Fig.9) and the products had a superior surface finish.

Figure 8. An α'/β' Sialon ceramic wire-drawing die.

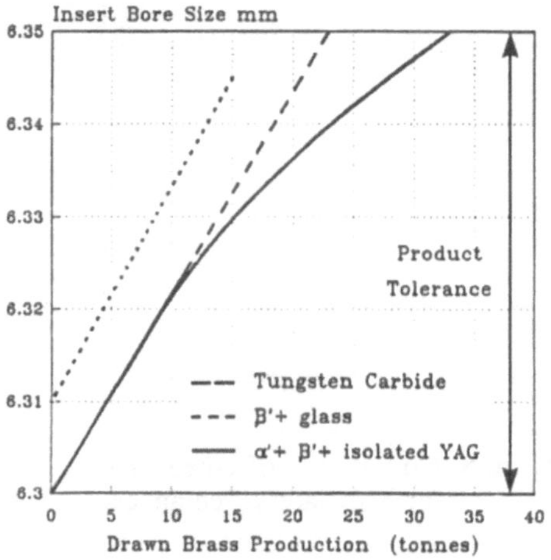

Figure 9. Wear performance during brass wire-drawing.

CONCLUSION

By careful control of the starting composition and manipulation of microstructure using a complex sintering/heat-treatment cycle novel α'/β'-sialon materials with improved properties can be prepared. Fabrication involves nearly complete transient liquid phase sintering and the additives which facilitate easy preparation are absorbed into the sialon microstructure and have less detrimental influence at temperature. The ability to achieve full crystallisation is a major advantage and may be decisive in allowing these ceramics to be used to their full potential.

---

* Kyon 3000 is marketed by Kennametal Inc. USA.

REFERENCES

1. Deeley, G., Herbert, J.M. and Moore, N.C., Powder Met., 8, 145, (1961).

2. Drew, P., and Lewis, M.H., J. Mat. Sci., 9, 1833, (1974).

3. Oyama, Y. and Kamigaito, O., Jap. J. App. Phys., 10, 1637 (1971).

4. Jack, K.H., and Wilson, W.I., Nature, 238, 28-29, (1972).

5. Lewis, M.H., Bhatti, A.R., Lumby, R.J., and North, B., J. Mat. Sci., 15, 103, (1980).

6. Lewis, M.H., Leng.Ward, G. and Jasper, C., in Ceramic Transactions; Ceramic Powder Science II, ed. G.L. Messing, E.R. Fuller and H. Hausner (Amer. Ceram. Soc. 1988) 1019.

7. Lewis, M.H., and Lumby, R.J., Powder Met. 26, No.2, 73 (1983).

8. Walls P.A., Ph.D. Thesis, University of Newcastle Upon Tyne, (1986)

YTTRIA DOPED Si_2N_2O CERAMICS

T. EKSTRÖM, M. HOLMSTRÖM and P.-O. OLSSON+
Dept. Inorg. Chemistry, University of Stockholm, S-106 91 Stockholm, Sweden
+National Defense Research Establishment, S-172 90 Sundbyberg, Sweden

ABSTRACT

Silicon oxynitride ceramics have been prepared from Si_3N_4 and SiO_2 by pressureless sintering at 1600 and 1775°C and by hot isostatic pressing (HIP) at 1600, 1750 and 1900°C, with Y_2O_3 (0-10 wt%) as a sintering aid. Pressurelessly sintered samples were in general porous, whereas the HIPed samples were dense at all temperatures. The reaction to form Si_2N_2O was very sluggish in the absence of Y_2O_3, and the samples mostly consisted of unreacted starting material, whereas addition of Y_2O_3 and high temperatures greatly favoured the formation of Si_2N_2O. Dense Si_2N_2O materials, prepared by HIP at 1900°C, were hard (HV10 = 1550 $kgmm^{-2}$) and brittle (K_{1C} = 3.3 $MPam^{1/2}$) at room temperature for low Y_2O_3 additions. The hardness decreased slightly, but the fracture toughness remained the same with increasing yttria content.

INTRODUCTION

The Si_2N_2O based ceramics are of special interest as engineering materials because of the expected good oxidation resistance at high temperatures, compared to silicon nitride based ceramics. These materials are considerably more difficult to prepare in pure form, however, than the corresponding silicon nitride-based ceramics. The formation of Si_2N_2O from equimolar mixtures of high-purity Si_3N_4 and SiO_2 heated at high temperatures is known to be very sluggish due to kinetic hindrance. It has been shown in a number of reports that additions of various sintering aids, which form a liquid phase at the sintering temperature, greatly facilitate the formation of silicon oxynitride (1-3).

Ceric oxide, CeO_2, has been used to prepare Si_2N_2O ceramics because of its high eutectic temperature with silica, but also because it is among the least expensive of all rare earth oxides (4-6). Otherwise the most effective sintering aids are mixtures of Y_2O_3 and Al_2O_3, which make pressureless sintering possible (7-9). In the presence of alumina the Si_2N_2O phase forms a narrow solid solution range represented by the formula $Si_{2-x}Al_xO_{1+x}N_{2-x}$, where x varies from zero to 0.2 (8). Solidification of materials along this line, with only Al_2O_3/AlN added, gives fully dense materials by pressureless sintering at temperatures about 1800°C, but with addition of Y_2O_3 the solidification process takes place even at 1600°C (9).

For Si_2N_2O materials to be used at high temperatures, addition of metal oxides like Y_2O_3 alone, which forms a refractory Y-Si-O-N glass, is a most attractive alternative. This report will describe the effects of yttria additions on the preparation and properties of Si_2N_2O ceramics made from equimolar mixtures of silicon nitride and silica.

EXPERIMENTAL

The selected compositions for this study have been made of equimolar mixtures of Si_3N_4 and SiO_2 as the parent, with additions of 0, 1, 5 or 10 wt% Y_2O_3. The source materials used were silicon nitride (H.C. Starck-Berlin, grade LC1), silicon dioxide (Anal. purity) and yttrium oxide (H.C. Stark-Berlin, grade Finest). The starting powders were carefully weighed in a total batch size of 500 g for each composition, mixed in water-free propanol and milled in a vibratory mill for 17 hours with sialon milling medium. After drying, the powder mixes were dry-pressed (125 MPa) into compacts of size 16x16x6 mm. The samples were pressurelessly sintered at 1600 and 1775°C for 2 hours, or glass encapsulated and hot isostatically pressed at 1600, 1750 and 1900^oC for two hours with 200 MPa of argon. Density measurements using Archimedes principle were made on the as-sintered samples. Hardness (HV 10) and indentation fracture toughness (K_{1C}) at room temperature were obtained by a Vickers diamond indenter using 98 N (10 kg) load. The fracture toughness was evaluated by assuming a value of 290 GPa for Young's modulus, and the precision of repeated measurements on the same sample was $\pm$ 0.2. The hot hardness was measured with a load of 9.8N (HV1). The phase analysis was based on X-ray powder patterns recorded by Guinier-Hägg cameras. Scanning electron microscopy was performed on carbon-coated materials, using a Jeol JSM 820 instrument equipped with a Link AN 10000 EDS analyzer.

RESULTS

Pressureless sintering at the temperature 1600°C gave highly porous materials, but HIP can be used to obtain dense materials also at this low temperature, see Figure 1. The HIP materials prepared at the different temperatures were also examined by optical microscopy on polished cross-sections and were found to be fully dense. Pressureless sintering at 1775°C of

$Si_3N_4 + SiO_2$ mixtures, however, gave materials with fairly good density (98-99% TD).

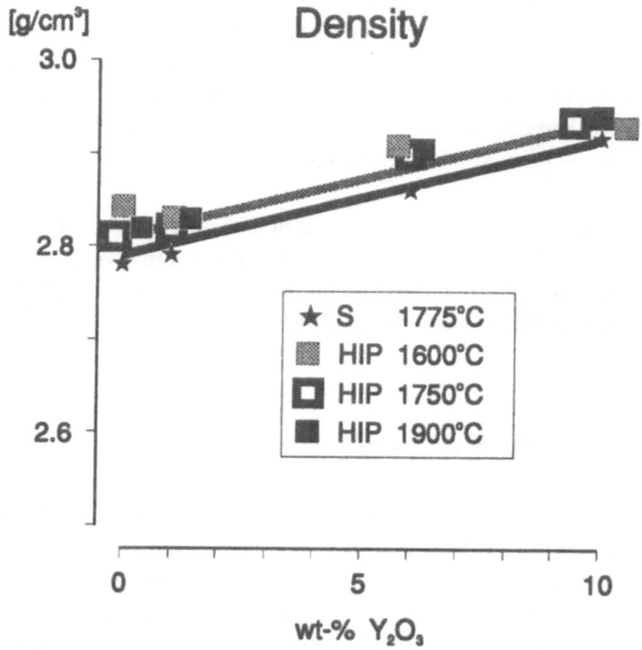

Figure 1. The recorded density as a function of Y_2O_3 content.

The effect of the Y_2O_3 additions and the temperature on the formation of Si_2N_2O from the $Si_3N_4 + SiO_2$ mixtures was examined by X-ray diffraction. Without yttria addition, only small amounts of the oxynitride phase were found besides the initially added Si_3N_4., Addition of yttria favoured the Si_2N_2O formation in most cases. Increasing the temperature also promoted the reaction. The results are summarized in Figure 2 for materials either pressurelessly sintered or HIPed at different temperatures. The calculated orthorhombic unit cell parameters of Si_2N_2O were within a= 0.8875 ± 0.0005 nm, b= 0.5496 ± 0.0005 nm and c= 0.4853 ± 0.0003 nm in all preparations, regardless of the level of Y_2O_3-addition.

The microstructures of fully dense HIP materials were carefully examined by SEM. Typical appearances for different yttria additions at high temperatures are illustrated in Figures 3a-c, recorded from the preparations made by HIP at 1900°C. Without yttria added, the product consisted mainly of unreacted starting materials with a grain diametre about 1 um, and the image contrast was poor. With addition of 1 wt% yttria, elongated Si_2N_2O grains appeared in the microstructure, see Figure 3b. The yttrium, with a bright contrast, was found somewhat irregularly distributed at multi-grain junctions as an intergranular glassy phase. At 6 wt% addition or higher, the intergranular glassy phase was evenly distributed and the elongated Si_2N_2O crystals had reached diametres around one micron and aspect ratio about eight, see Figure 3c. The Si_2N_2O grains were observed to have a somewhat irregular contrast, giving lighter areas in the middle of the crystals. This will be more carefully analyzed by TEM. EDS-analysis of the larger Si_2N_2O grains found in the preparations doped with 6 and 10 wt% Y_2O_3 showed, however, only silicon present, whereas the intergranular phase besides Si contained all added Y.

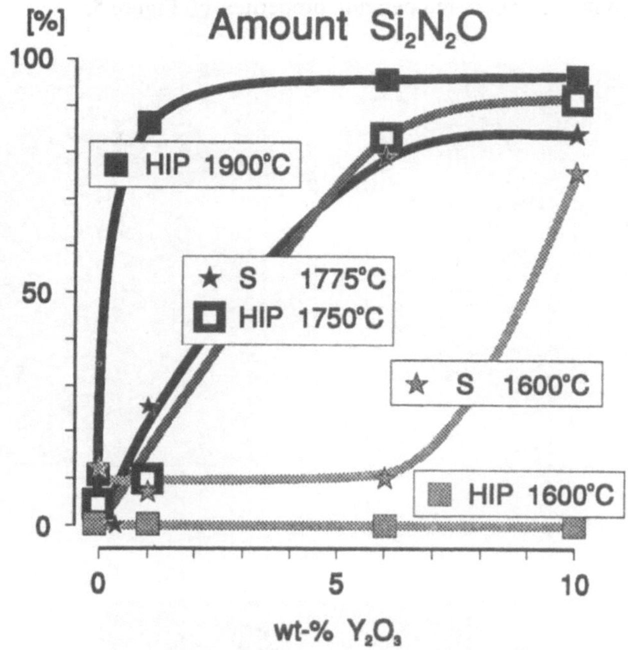

Figure 2. The amount of Si_2N_2O formed from mixtures of Si_3N_4 and SiO_2, as a function of the amount of added Y_2O_3 and of preparation temperature and method. Note the low amount of silicon oxynitride formed in the series sintered by HIP at 1600°C.

Since only small amounts of Si_2N_2O were formed in the samples without sintering aid, it is difficult to determine the mechanical properties of monophase Si_2N_2O based on the results obtained in this study. A good approximation is, however, a Vickers hardness, HV10, of 1550 $kgmm^{-2}$ and a fracture toughness around 3.3 $MPam^{1/2}$, obtained from measurements on materials HIPed at 1900°C (with only 1 wt% of yttria), which mainly consisted of crystalline Si_2N_2O. Addition of yttria to silicon oxynitride was found not to increase the fracture toughness significantly, see Figure 4. The hardness was higher for materials prepared at high temperature, containing more silicon oxynitride, but decreased for the 1900°-preparations with yttria addition.

Finally, it should be mentioned that the hot hardness of a sample prepared by HIP at 1900°C and of overall composition "Si_2N_2O" is preserved at temperatures at least up to 1200°C, the highest temperature used in this study. This sample consisted mainly of unreacted starting material and only partly of the phase Si_2N_2O, cf. Figure 2. The addition of a small amount of Y_2O_3 (1 wt%) had a slight influence on the overall hardness, but not on the hot hardness behaviour. It is thus evident that the small amount of glassy Y-Si-O-N phase formed in this sample does not deteriorate the hot-hardness properties. In a sample having the nominal composition $Si_{1.9}Al_{0.1}N_{1.9}O_{2.1}$ + 1 wt% Y_2O_3 , however, the small amount of Al present was enough to change the composition of the glassy phase and this had a pronouncedly negative effect, giving poor high-temperature properties, cf. Figure 5.

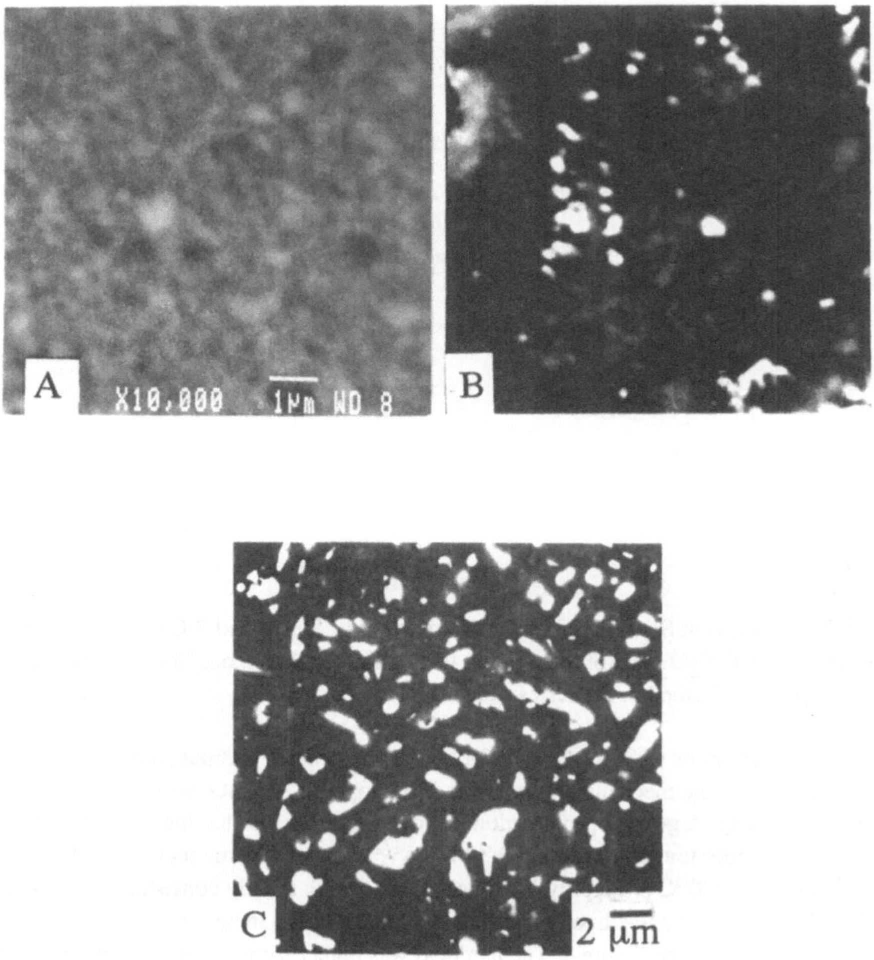

Figure 3. The microstructures of Si_2N_2O ceramics prepared by HIP at 1900°C with additions of a) 0, b) 1 and c) 6 wt% Y_2O_3. Note that the magnification of c) is different from the other two micrographs.

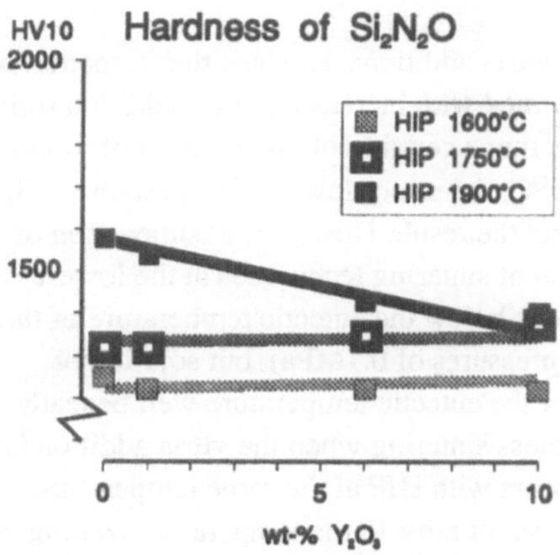

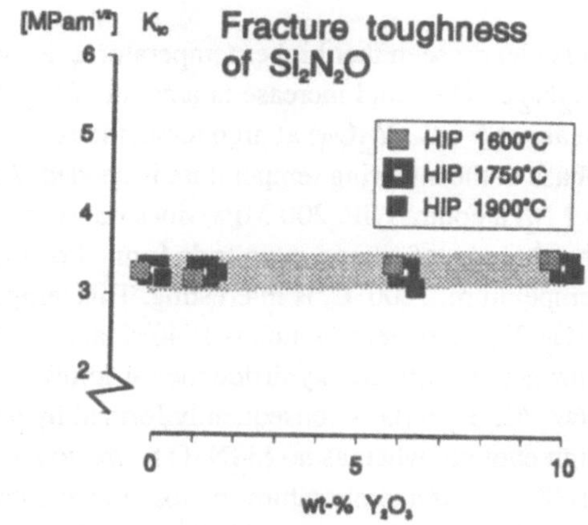

Figure 4. The hardness and indentation fracture toughness of fully dense materials prepared by HIP and with different amounts of yttria added.

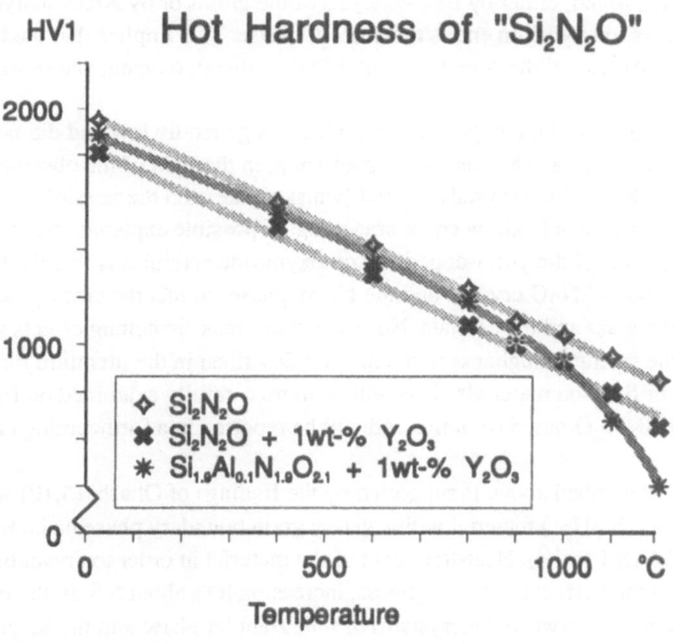

Figure 5. The hot hardness of silicon oxynitride ceramics with small additions of sintering aids. The sample of overall composition "Si_2N_2O" (prepared by HIP at 1900°C) consisted mainly of Si_3N_4 and only partly of the phase Si_2N_2O, cf. text and Figure 2.

DISCUSSION

It is clearly seen that higher temperatures, as well as yttria additions, enhance the formation of Si_2N_2O. The rapid increase in amount of Si_2N_2O formed with increasing yttria addition (up to about 5 wt% Y_2O_3) at high temperatures indicates that a certain volume fraction of liquid phase at the sintering temperature is needed. At 1775°C, the use of low or high pressure (PS, 0.1 MPa contra HIP, 200 MPa) does not seem to affect the result. However, a comparison of the amounts of silicon oxynitride formed by the different sintering techniques at the lowest temperature, 1600°C, is interesting. This temperature is below the eutectic temperature of the SiO_2-Y_2O_3 system (which is 1640°C at a "normal" pressures of 0.1MPa), but solution of nitrogen to form an oxynitride melt is known to lower the eutectic temperature well beneath 1600°C. Si_2N_2O is consequently formed by pressureless sintering when the yttria addition is high enough, whereas no Si_2N_2O formation at all occurs with HIP at the same temperature (1600°C). At a temperature so close to the eutectic, a small error in the temperature reading of the HIP may be responsible for the effect, or the high pressure may have raised the eutectic temperature in the "SiO_2-Y_2O_3" system above 1600°C, possible because of a lower solubility of nitride in the oxide melt.

No evidence was found, either by EDS-analysis of the grains or by XRD-analysis of the lattice parameters, that yttrium enters the Si_2N_2O phase. This implies that the intergranular glassy phase contained all the added yttrium, besides silicon, oxygen, and nitrogen.

The fracture toughness of the oxynitride ceramics was generally low and did not improve with the addition of yttria. This may seem surprising, in the light of the observed very elongated grain shape of the crystals formed (whisker-like) and the possibility that the glassy phase, when present, would allow crack branching. A possible explanation for this poor toughness behaviour of the yttria-doped silicon oxynitride ceramics is that the bonding is too strong between the Si_2N_2O crystals and the glassy phase, so that the crack proceeds straight through both the glass and the crystals. No pull-out or crack-branching effects will therefore contribute to the fracture toughness in the manner described in the literature for yttria-doped silicon nitride or β sialon materials. This will be more carefully examined on fractured surfaces of the Si_2N_2O based ceramics and will be reported in a forthcoming communication.

The hypothesis described above is supported by the findings of Ohashi (5,10) who reported that a ceria-doped Si_2N_2O material with a glassy grain boundary phase had a fracture toughness of 2.5 ($MPam^{1/2}$). Heat-treatment of the material in order to crystallize the glassy phase had a dramatic effect on the toughness, increasing it to about 5.5. In the heat-treated material, the bonding between the crystallized intergranular phase and the Si_2N_2O crystals was much weaker. The hypothesis also finds support in the recent findings by Kishi (11), who prepared single phase β sialon ceramics with either equi-axed or elongated grains. The Vickers hardness (HV 10) and indentation fracture toughness values were similar for the two materials, being around 16 GPa and 3.3 $MPam^{1/2}$, respectively. The elongated grains thus had no effect on the fracture toughness and it was concluded that β sialon with elongated grains can not be toughened without the existence of a weakly bonded glassy grain boundary phase which allows crack branching.

CONCLUSIONS

* Additions of Y_2O_3 and high temperatures greatly enhance the formation of Si_2N_2O.
* Additions of Y_2O_3 give rise to a glassy intergranular phase and elongated Si_2N_2O crystals in the microstructure.
* Additions of Y_2O_3 have only minor effects on the fracture toughness and slightly lower the hardness.
* Additions of Y_2O_3 have no negative effect on the hot-hardness at temperatures below 1200°C.

REFERENCES

1. Bergman, B. and Heping, H., The Influence of Different Oxides on the Formation of Si_2N_2O from SiO_2 ands Si_3N_4, J. Europ. Ceram. Soc., 1990, **6**, 3-8.

2. Barta, J., Manela, M. and Fischer, R., Si_3N_4 and Si_2N_2O for High Performance Radomes, Mater. Sci. Eng., 1985, **71** 265-72.

3. Huang, Z.K., Greil, P. and Petzow, G., Formation of Silicon Oxynitride from Si_3N_4 and SiO_2 in the Presence of Al_2O_3, Ceram. Int., 1984, **1** 14-17.

4. Ohashi, M., Tabata, H. and Kanzaki, S., High-Temperature Flexure Strength of Hot-Pressed Silicon Oxynitride, J. Mater. Sci. Letters, 1988, **7**, 339-40.

5. Ohashi, M., Kanzaki, S. and Tabata, H., Reaction Sintering Process and Mechanical Properties of Silicon Oxynitride, Yogyo Kyokai Shi, 1988, **96**, 1073-80.

6. Ohashi, M., Kanzaki, S. and Tabata, H., Influence of Seeding on the Reaction Sintering of Silicon Oxynitride, Yogyo Kyokai Shi 1989, **97**, 559-65.

7. Lewis, M.H., Reed, C.J. and Butler, N.D., Pressureless-sintered Ceramics Based on the Compound Si_2N_2O, Mater. Sci. Eng. 1985, **71**, 87-94.

8. Trigg, M.B. and Jack, K.H., Solubility of Aluminium in Silicon Oxynitride, J. Mater. Sci. Letters 1987, **6**, 407-408.

9. Trigg, M.B. and Jack, K.H., The Fabrication of O'-Sialon Ceramics by Pressureless Sintering, J. Mater. Sci. 1988, **23**, 481-7.

10. Ohashi, M., Kanzaki, S. and Tabata, H., Processing, Mechanical Properties and Oxidation Behaviour of Silicon Oxynitride Ceramics, J. Am. Ceram. Soc. 1991, **74**, 109-14.

11. Kishi, K.,Umebayashi, S. and Tani, E., Influence of Microstructure on Strength and Fracture Toughness of β-Sialon, J. Mater. Sci. 1990, **25**, 2780-84.

MECHANICAL PROPERTIES OF GAS PRESSURE SINTERED SILICON NITRIDE

Cornelia Boberski and Rainer Hamminger
Hoechst AG, 6230 Frankfurt 80, FRG

ABSTRACT

The densification behavior of Si_3N_4-powders during gas pressure sintering with additions of Y_2O_3 and Al_2O_3 was studied by dilatometer measurements. High densities of 99.5% of the theoretical value were achieved by 2-step gas pressure sintering while the density of pressureless sintered material was 1-2% less. The reliability of the sintered ceramic material was improved by applying gas pressure after the main densification step. The Weibull modulus of pressureless sintered materials of m=8-11 increased to m=20-25 by applying gas pressure sintering. The average bending strength (4-point) was 800-900 MPa for pressureless as well as gas pressure sintered materials.

INTRODUCTION

In automotive applications high interest is focused on strength and rupture properties of a low cost ceramic with sufficient reliability in the intermediate-temperature range, between 600 and 1000° C. Pressureless sintering of silicon nitride is the most cost-effective process in comparison to hot isostatic pressing or gas pressure sintering. To fabricate high-density materials by pressureless sintering, large amounts of additives are necessary [1,2]. However, pressureless sintered Si_3N_4-ceramics yield low reliability of the Weibull modulus between m=8 to 11 due to residual pores in the material. Gas pressure sintering has enabled densification of Si_3N_4 to be carried out at higher temperatures then pressureless sintered materials due to the applied nitrogen pressure which counteracts the tendency of Si_3N_4 to decompose [3-5]. Surpression of the decomposition of silicon nitride and the physical effect of the applied nitrogen pressure during gas pressure sintering can eliminate pores as well as hindering the formation of pores during sintering [6]. Therefore, gas pressure sintering was applied to silicon nitride ceramics with relative high amounts of the sintering aids Y_2O_3 and Al_2O_3 in order to improve the materials reliability. Furthermore, the influence of gas pressure sintering on the final microstructure was studied by taking into account the powder characteristics [7-12].

EXPERIMENTAL PROCEDURE

The Si_3N_4-powder (synthesized by the diimid route) was mixed with 10 Wt.% Y_2O_3 and 2.2 Wt.% Al_2O_3. This powder was then mixed and milled with Si_3N_4 balls in water, dryed, sieved and then formed into test bars by cold isostatic pressing (300 bar). The sintering runs were carried out in a graphite resistance furnace under nitrogen pressure ranging from 1 to 90 bar. The temperature was obtained from an optical pyrometer reading as well as by thermocouple measurements (W/Rh). The sintering procedure starts with heating up the samples under 3 bar pressure of nitrogen up to 1750° C. Temperature and nitrogen pressure were varied between 1800 and 1900° C and 3 to 90 bar, respectively. The cooling down procedure occured under applied nitrogen pressure.

The sintering behavior was studied simultaneously with a dilatometer inside the graphite furnace. Small zylinders of 10 mm in diameter and 10 mm hight were used. Relative density, weight loss and shrinkage of the samples were measured at unload. The sintered samples were cut and polished into test bars of 3x4x42 mm for measuring the 4-point bending strength at room temperature. The fracture toughness was measured by the singel-notch-etch method (with 120-140 μm notch of 0.7 to 1.3 mm depth which yield a reproducibility of +/- 0.5 MPa). The absolute values, however, are higher than determined by indentation method.

The microstructures of the sintered materials were observed by scanning electron microscopy. The samples were prepared by plasma etching [13].

RESULTS AND DISCUSSION

Sintering Behavior

Shrinkage and densification rates are shown in Fig. 1 for increasing pressure steps during gas pressure sintering at 1750 and 1800° C. Rearrangement starts at 1400° C and is indicated by the beginning of shrinkage. Maxima of the shrinkage rate occur at 1450 and 1600° C (Fig. 1a, Tab. 1). These processes can be explained by solution and reprecipitation processes during liquid phase sintering [7,14]. The processes observed at 3 bar nitrogen pressure are similar to those obersved during pressureless sintering for the same additive system and equivalent Si_3N_4-powders [15]. Increasing the pressure from 3 to 60 bars can cause a significant decrease of the shrinkage rate and final density when applied too early in relation to the densification stage of the sample (Fig. 1 a,b; Tab.1). Here, the pressure increase from 3 to 60 bars was set as soon as 50 % of the final shrinkage for low pressure was reached (Fig. 1b). This sample contains numerous pores. An additional densification of 1 to 2 % up to totally 98.5 to 99.5 % of the theroretical density can be achieved by applying higher pressure after 80-90% of the sample shrinkage. This sinter-hip-effect [8] was systematically observed when the pressure was increased from 3 to 30 and 90 bars, respectively, after the main densification step (Fig. 2).

Microstructure and Mechanical Properties

A wide range of different types of microstructures were observed by varying the sintering steps, the amorphous grain boundary phase is seen as a white phase (Fig. 3). The silicon nitride grains were described by grain numbers per area as grain size parameter and aspect ratios for the shape of the grains [7,16]. At low pressure and low temperature sintering conditions, random microstructures of small needle-like grains of silicon nitride with aspect ratios of 9.6 were observed. Small equiaxial Si_3N_4-grains were observed in the interstitial space between larger elongated grains. At higher sintering temperatures (Fig. 3b) grain growth occurs, indicated by less grains per area and more homogeneous grain distributions. The aspect ratio of 4,5 of this needle-like Si_3N_4 is very small and an interstitial filling of very small grains cannot be observed. The grain size distribution seems to be smaller than in the sample discussed before. A detailed image analysis will be carried out to prove these observations in the near future.

The fracture toughness and the bending strength of 900 MPa indicate that the microstructure shown in Fig. 3a is the most favourable microstructure, what is in good agreement with the model discussed [7]. Further investigations are planned in order to describe the different microstructures in detail and to correlate them to crack propagation behavior.

Reliability

By pressureless sintering, Si_3N_4-ceramics with numerous small pores (1-5 μm) are produced. An analyses of the fracture surfaces is given in Fig. 4. Pores of different sizes as fracture origins can be detected. The elimination of large pores by gas pressure sintering is assessed in Fig. 5. The reliability is probably improved by liquid filling of the large pores [6]. The analysis of the fracture surfaces indicate clearly that concentrations of the glass-phase become an important fracture origin. Further improvement of the mechanical properties is anticipated from better distribution of the glassphase or by the reduction of the total amount of sintering additives. However, the results prove that gas pressure sintering can be applied for improving the reliability of Si_3N_4-ceramics, even in case when high amounts of additives are used.

CONCLUSIONS

Gas pressure sintering with an increase of pressure after a presintering phase at low nitrogen pressure can be successfully applied to increase the final density, reduce the weight loss, modify the microstructure and improve the reliability of high performance Si_3N_4-ceramics. Finally, the reliability of technical engine parts of Si_3N_4 is strongly dependent on processing steps like milling and mixing, whereas their microstructural occurrence is mainly controlled by the properties of the primary powders and by the sintering conditions (Fig. 6.).

ACKNOWLEDGEMENT

This work would not have been possible without the collaboration of A. Bärreiter, K.-U. Dickopf, S. Fischer, J. Häfner-Schmidt, V. Herglotz, V. Olt, R. Rat and J. Steuper.

REFERENCES

[1] Loehman, R.E. and Rowcliffe, D.J., Sintering of Si_3N_4-Y_2O_3-Al_2O_3, J. Am. Ceram. Soc., 1980, 63, p. 144.

[2] Crosbie, G. M., Nicholson, J. M. and Stiles, E.D., Sintering Factors for a Dry-Milled Silicon Nitride-Yttria-Alumina Composition, Ceram. Bull., 1989, 68, p. 1202.

[3] Mitomo, M., Pressure Sintering of Si_3N_4, J. Mater. Sci., 1976, 11, p. 1103.

[4] Galasso F. and Veltri R., Sintering of Si_3N_4-Y_2O_3 Using Nitrogen Pressure, Powder Met . Int., 1982, 14, p. 217.

[5] Mitomo, M. and Mizuno K., Ceramic Materials and Components for Engines, ed by W. Bunk and H. Hausner, German Ceram. Soc., Bad Honnef, Germany 1986, p. 263.

[6] Kuang, S.-J., Greil, P., Mitomo, M. and Moon, J.-H., Elimination of Large Pores During Gas-Pressure Sintering of ß-Sialon, J. Am. Ceram. Soc., 1989, 72, p. 1166.

[7] Wötting, G., Kanka, B. and Ziegler, G., Microstructural Development, Microstructural Characterization and Relation to Mechanical Properties of Dense Silicon Nitride, Proc. Intern. Conf. "Non-Oxide Technical and Engineering Ceramics", Limerick/Ireland, ed. by S. Hamphire, Elsevier Appl. Sc. Publ., London, 1986, p. 83.

[8] Franz, G., Laubach, B., Wickel, U., Wötting G. and Gugel, E., Optimization of a High Purity Silicon Nitride Powder Concerning Sintering and Mechanical Properties, in "Ceramic Materials and Components for Engines", Las Vegas, 1988, ed. V.J. Tennery, Am. Ceram. Soc., p. 1.

[9] Mitomo, M., In Situ Microstructure Control in Silicon Nitride Based Ceramics, Adv. Ceram. Proc. Lect. Meet., 1988, 2, p. 147.

[10] Mitomo, M., Tsutsumi, M., Tanaka, H., Uenosono, S. and Saito, F., Grain Growth During Gas-Pressure Sintering of ß-Silicon Nitride, J. Am. Ceram. Soc., 1990, 73, p. 2441.

[11] Wötting, G. and Ziegler, G., Powder Characteristics and Sintering Behaviour of Si_3N_4, Powder Met. Int., 1986, 18, p. 25.

[12] Mitomo, M., Yang, N., Kishi, Y. and Bando, Y., Influence of Powder Characteristics on Gas Pressure Sintering of Si_3N_4, J. Mater. Sci., 1988, 23, p. 3413.

[13] Mitomo, M., Sato, Y.-I., Ayuzawa, N. and Yashima, I., Plasma Etching of α-Sialon Ceramics, J. Am. Ceram. Soc., 1991, 74, p. 856.

[14] Kingery, W.D., Densification During Sintering in the Presence of a Liquid Phase, J. Appl. Phys., 1959, 30, p. 301.

[15] Boberski, C., Bestgen, H. and Hamminger, R., Microstructural Development During Liquid Phase Sintering of Si_3N_4-Ceramics, J. Europ. Ceram. Soc., 1991, submitted.

[16] Obenaus, P. and Herrmann, M., Method of Quantitatively Characterising Columnar Crystals in Silicon Nitride Ceramics, Pract. Met., 1990, 27, p. 505.

Table 1:

Sintering behavior of gas pressure sintered Si_3N_4-ceramics

Run	1	2	3	4
Temperature (°C)	1750	1800	1800	1800
Pressure (bar)	3	60	60	60
Duration (min)	150	125	125	190
Increase of Pressure (min)	0	60	65	120
Green Density (g/cm3)	1.90	1.78	1.77	1.86
Density (%th. D.)	98.5	93.9	99.0	98.5
Weigth Loss (%)	0.80	0.78	1.16	1.75
Shrinkage (%)	18.6	15.8	19.0	17.4
1. Maximum of Shrinkage Rate (μm/min)	18	---	20	12
2. Maximum of Shrinkage Rate (μm/min)	235	185	250	220

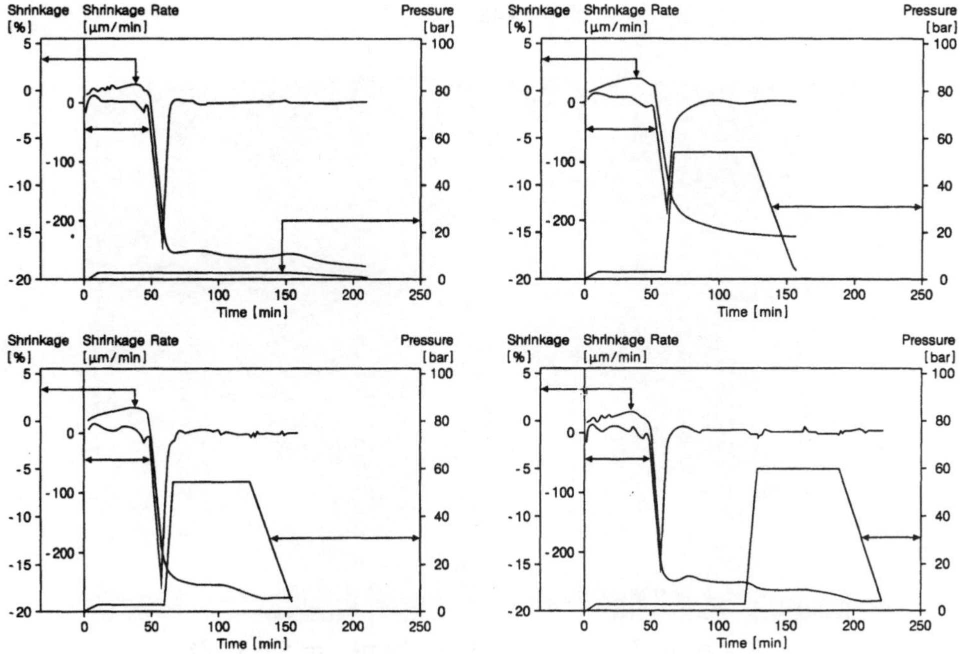

Fig. 1.:
Sintering behavior of gas pressure sintered Si_3N_4-ceramics with Y_2O_3 and Al_2O_3 additives. Linear shrinkage, shrinkage rate is shown for different pressure-increase-cycles (with maximum temperature at 1800°C). Data are listed in table 1.

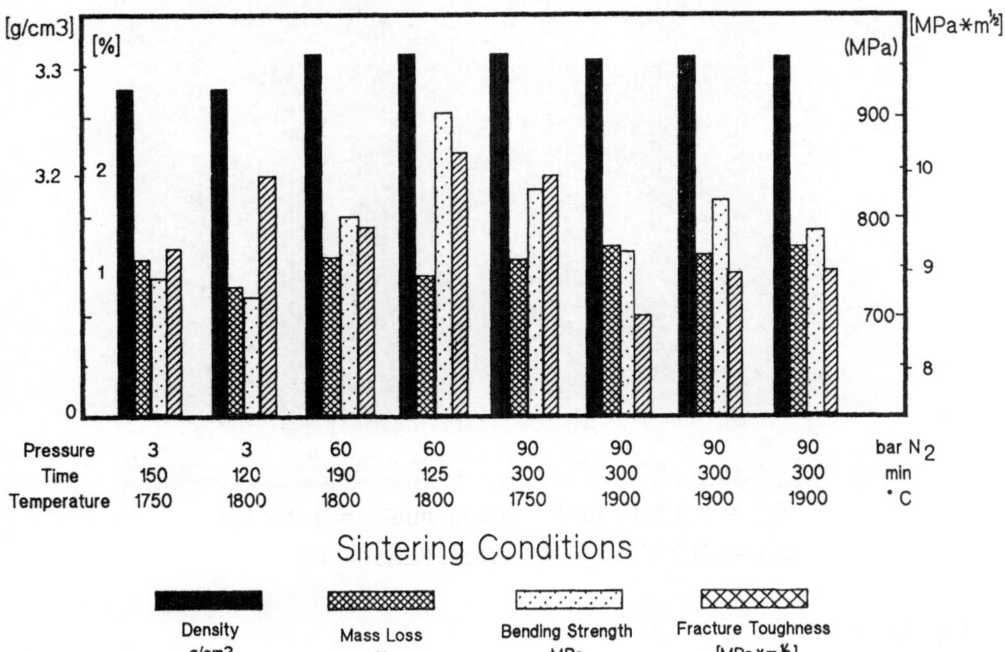

Fig. 2.:
Properties of gas pressure sintered Si_3N_4-materials.

a)

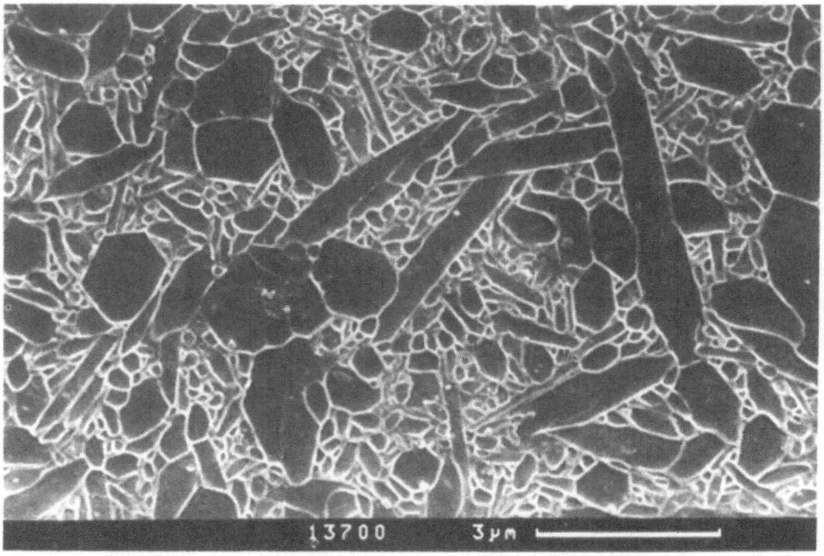

K_{IC} = 10.3 MPa.m$^{1/2}$ Grains/mm^2 = 5.6 x 10^6
σ_{RT} = 900 MPa Aspect Ratio = 9.6

b)

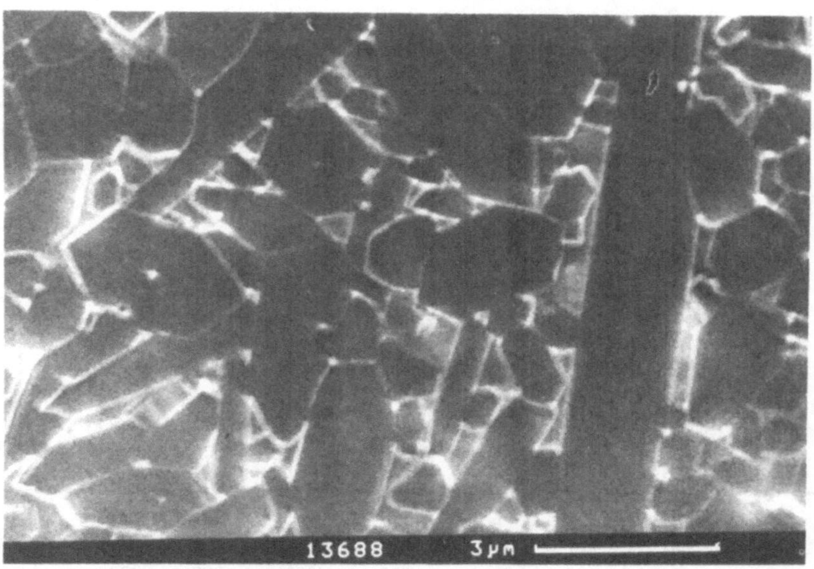

K_{IC} = 8.9 MPa.m$^{1/2}$ Grains/mm^2 = 1.2 x 10^6
σ_{RT} = 813 MPa Aspect Ratio = 4.5

Fig. 3.:
Microstructures and mechanical properties achieved by different gas pressure sintering conditions a) 1800° C, 60 bar N_2, 125 min and b) 1900° C, 90 bar N_2, 300 min.

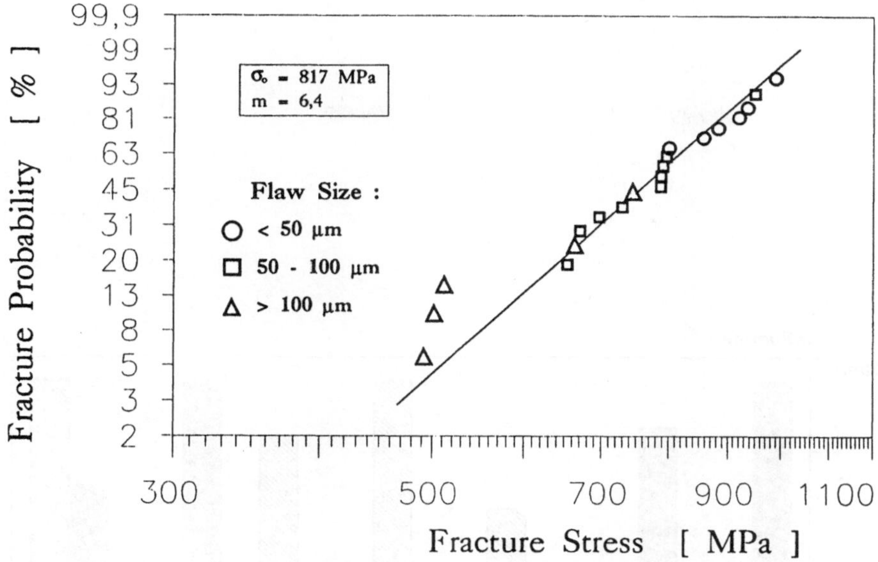

Fig. 4.:
Strength and fracture origins of pressureless sintered Si_3N_4-materials.

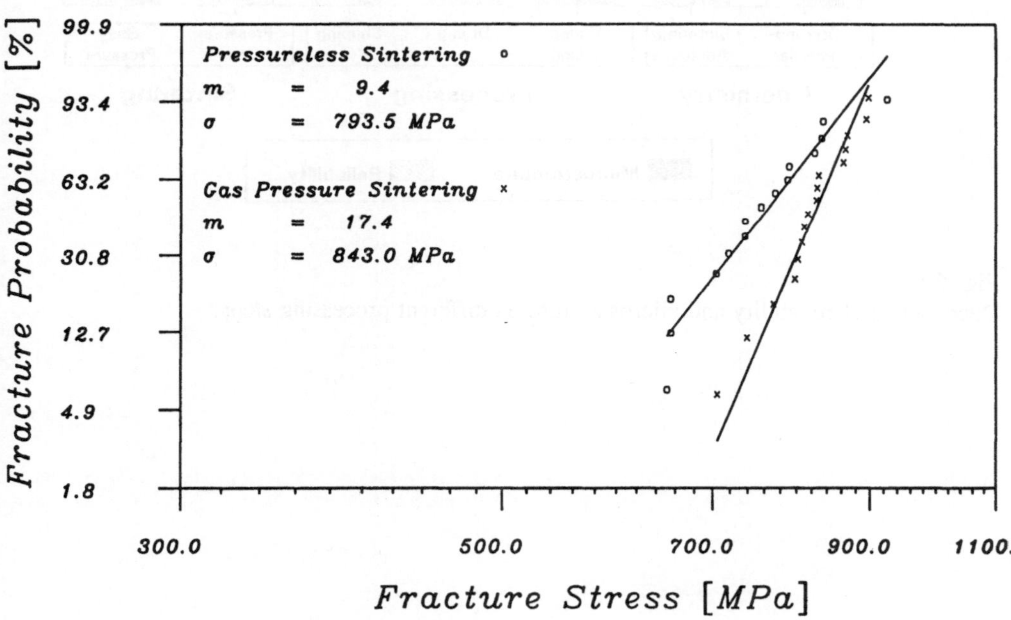

Fig. 5.:
Increase of reliability by gas pressure sintering.

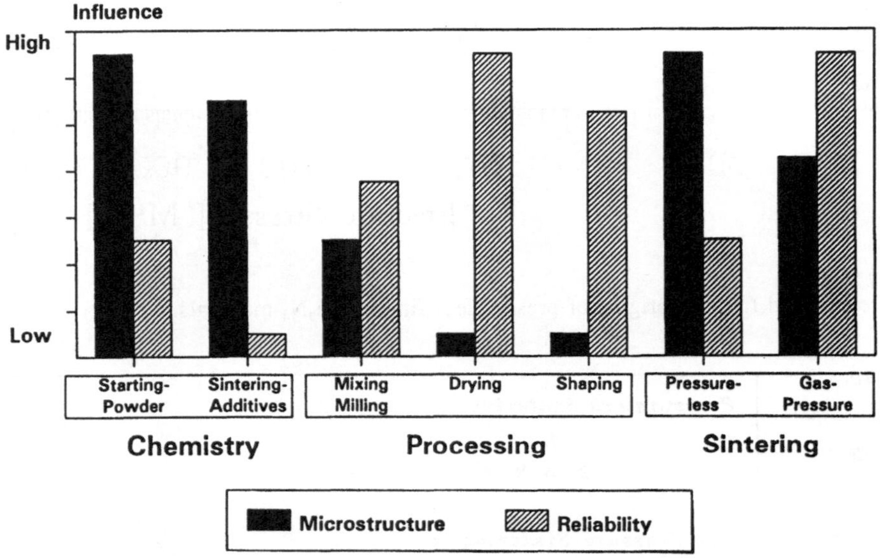

Fig. 6.:
Dependency of reliability and microstructure on different processing steps.

STATIC AND CYCLIC FATIGUE BEHAVIOR OF GAS PRESSURE SINTERED SILICON NITRIDE

SATOSHI IIO, TOMONORI NIWA, YO TAJIMA AND MASAKAZU WATANABE
Research and Development Center
NTK Technical Ceramics
NGK Spark Plug Co., Ltd.
2808 Iwasaki, Komaki, Aichi 485, Japan

ABSTRACT

Crack propagation behavior in a gas pressure sintered silicon nitride was investigated under static and cyclic loading at room temperature to clarify the cycling effect. All the testing was conducted in 3-point bending for specimens with artificial flaws introduced by indentation on the center of the mirror finished tensile surfaces. Crack propagation occurred under cyclic loading with maximum applied stress below the applied stress with which crack propagation stagnated under static loading, showing clearly that crack propagation velocity under cyclic loading was higher than that under static loading. Although intergranular crack propagation with crack-bridging was observed in both cases under static and cyclic loading, crack-bridgings, that were observed after static loading, were found to be fractured by the subsequent cyclic loading. It was thought that crack shielding force by crack-bridging, which is effective for restraining crack propagation under static loading, was decreased by cyclic loading and consequently crack propagated faster under cyclic loading than under static loading.

INTRODUCTION

In the recent several years, silicon nitride ceramics have been actively developed for engine parts such as turbo charger rotors and industrial parts [1]. For such structural applications, it is important to understand a fatigue behavior when considering the reliability and durability. There are many studies on the fatigue behavior of silicon nitride ceramics for various shapes of specimens under various testing conditions [2-10]. In the case of specimens without artificial flaws, time dependent fatigue behavior was observed by the present authors [2], whereas an effect of stress ratio was reported by Masuda et al. [3,4]. On the other hand, accelerated slow crack growth rate due to cycling was observed for pre-cracked specimens [5-10]. Although studies on fatigue behavior have been conducted for various silicon nitride ceramics fabricated under various conditions such as sintering method, additives and porosity, the mechanism for the fatigue behavior is not fully understood.

We have already reported on the static, cyclic and dynamic fatigue behavior of gas pressure sintered silicon nitride (GPSSN) specimens without artificial flaws, and it is suggested that all the result can be correlated in terms of a single subcritical crack growth (SCG) parameter and that there is no significant enhancement of the SCG due to cycling at room temperature [2]. In the present work, fatigue and crack propagation behavior in GPSSN was investigated under static and cyclic loading at room temperature for specimens with pre-cracks to clarify the cycling effect.

EXPERIMENTAL PROCEDURE

The materials studied was a GPSSN with the additions of alumina and yttria. It contains virtually no pores. Some properties of GPSSN at room temperature are summarized in Table 1. Specimens were cut from blocks of the material and ground into flexure bars (approximate dimensions 3 by 4 by 40 mm) with a 140-grit diamond wheel. Pre-cracks (2c, about 180μm) were introduced by Vickers indentation with indentation load of 49 N on the center of tensile surfaces (4 by 40 mm) of specimens.

All the testing was conducted in 3-point bending under static and cyclic loading with a span of 30 mm at room temperature in ambient air atmosphere. In the cyclic tests, the specimens were subject to a sinusoidal cyclic stress with a stress ratio (R) of 0.1 or 0.5 and a frequency of 1 or 10 Hz. Tensile surfaces of some samples were mirror finished before introducing pre-cracks, and an iterative process of loading and measuring surface clack length by optical microscope of 400 magnifications was conducted to estimate the crack propagation velocity. The surface crack after static and cyclic loading was observed by scanning electron microscope (SEM).

Table 1
Properties of gas pressure sintered silicon nitride (GPSSN).

Density (g/cm^3)	Young's Modulus (GPa)	Vickers Hardness (GPa)	Fracture Toughness by SEPB (MPa·m$^{1/2}$)	Thermal Expansion Coefficient (/K)	Thermal Conductivity (W/m·K)
3.23	320	14.3	6.0	2.8×10^{-6} (r.t.-800°C)	27

RESULTS AND DISCUSSION

Fatigue Behavior

The results of the static fatigue test and of the cyclic fatigue tests under different loading conditions are shown in Fig. 1. All the data are plotted as a function of effective time to failure, to relate the static and cyclic fatigue data directly [10]. The effective time to failure was calculated by using the fatigue parameter obtained by the static fatigue tests. It is clearly seen that the life time of cyclic fatigue was shorter than that of static fatigue, in contrast with the time dependent fatigue behavior of specimens without artificial flaws [2].

The applied stress - time to failure diagram (S-T diagram) obtained by cyclic fatigue tests shows an inverted S shape, trend being the same as the results reported by Kawakubo et al. [6]. The effective time to failure was shorter at low stress ratio and/or at high frequency. However, cyclic fatigue data at both 1 and 10 Hz at stress ratio of 0.1 plotted as a function of number of cycles seemed to show no significant difference as shown in Fig. 2, and the cyclic fatigue behavior seems to be cycle dependent. From these results, it was found that the fatigue behaviors of GPSSN samples with and without artificial flaws were completely different. Therefore, measurement of crack propagation velocity, which affects the life time, and observation of crack extension were attempted to clarify the cyclic effect on crack propagation in pre-cracked samples.

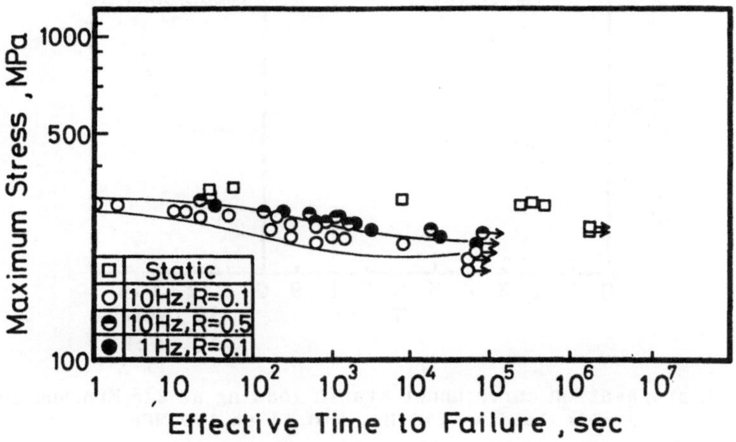

Fig. 1. Results of static and cyclic fatigue tests for pre-cracked specimens under various loading conditions.

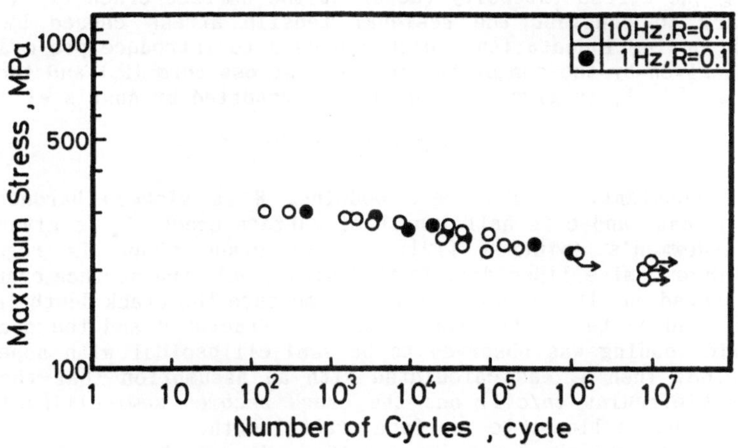

Fig. 2. Results of cyclic fatigue test for pre-cracked specimens.

Crack Propagation Behavior

Fig. 3 shows the surface crack length (2c) as a function of time (t) or cycles (N) for the sample with pre-crack, tested under static loading with applied stress of 245 MPa and subsequent cyclic loading (10 Hz, R=0.1) with maximum applied stress of 199 MPa. Crack propagation was stagnated at around surface crack length of 400μm under static loading. However crack propagation occurred under cyclic loading with maximum applied stress of 199 MPa, below the applied stress under static loading, showing clearly that crack propagation velocity under cyclic loading was higher than that under static loading. The above mentioned trend is same as the result obtained by Horibe [9].

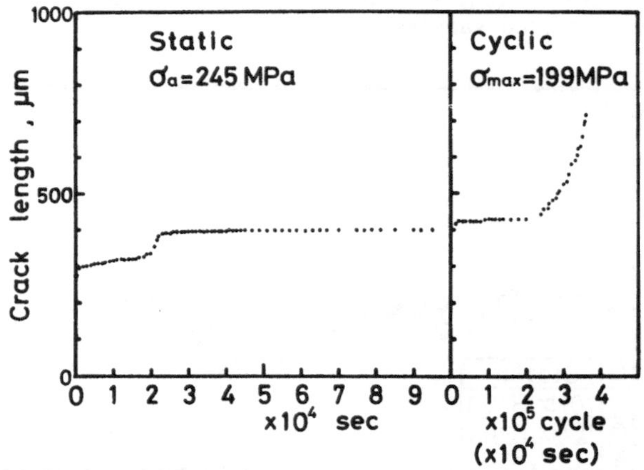

Fig. 3. Crack propagation curve under static loading at 245 MPa and subsequent cyclic loading (10 Hz, R=0.1) at 199 MPa.

The crack propagation velocity (dc/dt or dc/dN) is given by a differential of the curve obtained by smoothing these data points [9]. For calculating the stress intensity factor at the surface crack tip (K_{Imax}), it is necessary to consider the residual tensile stress caused by plastic deformation during indentation, which was used to introduce pre-cracks, and the K_{Imax} is given by the sum of the residual stress term (K_r) and the applied stress term (K_s). K_r is given by Eq.(1), as reported by Anstis et al. [11].

$$K_r = X(E/H)^{1/2}P/c^{3/2} \tag{1}$$

where X is constant, E is Young's modulus, H is Vickers hardness, P is indentation load, and c is half length of surface crack. K_s is given by well known Raju-Newman's solution [12], if the crack shape is regarded as semicircular or semi-ellipsoidal. In this study, only the surface crack length (c) was measured and it was not possible to measure the crack depth (a). After the cyclic loading test, the sample was fast fractured and the crack shape after cyclic loading was observed to be semi-ellipsoidal with aspect ratio (a/c) of 0.68. Then K_s was calculated with an assumption that the initial crack is semicircular (a/c=1) and the crack becomes semi-ellipsoidal with aspect ratio change linear to surface crack length.

According to the above mentioned method, K_{Imax}-V diagrams for the sample under cyclic loading (10 Hz, R=0.1) at 228 MPa and that for the sample under static loading at 287 MPa were obtained independently and shown in Fig. 4. X value in Eq. (1) was selected in such a way that a single line is obtained for the result under cyclic loading, and the value was 0.011, a little different from the value X=0.016 reported by Anstis et al. [11]. The crack propagation velocity under cyclic loading based on number of cycles was converted simply into that based on time, regarding that 10 cycles are 1 second.

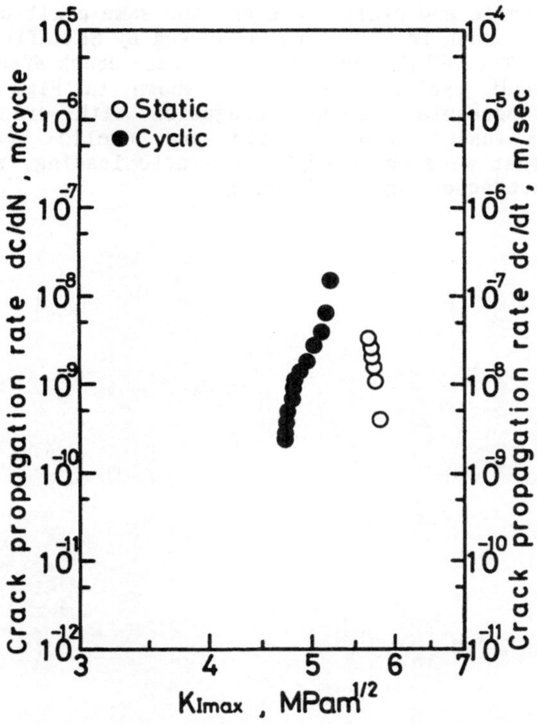

Fig. 4. K_{Imax}-V diagrams under cyclic and static loading.

The K_{Imax}-V diagram under cyclic loading shows an inverted S shape and suggests that the threshold of K_{Imax} (K_{Ith}) for crack propagation exists, the trend being similar to the S-T diagram obtained by the rupture testing as shown in Fig. 1. Although the approaches were different, these results were thought to be in a good agreement. The method of measuring the crack propagation behavior in this study has an advantage that the fatigue behavior is characterized with fewer specimens in shorter time compared with the rupture testing.

The K_{Imax} under cyclic loading was lower than that under static loading for the same V value, indicating that the crack propagation velocity under cyclic loading was faster than that under static loading. In addition, the effective time for maximum applied stress under cyclic loading was shorter than the actual loading time, and thus the cycling effect was thought to be more significant.

On the other hand, crack propagation velocity was calculated to decrease with increasing K_{Imax} accompanied by crack extension under static loading in this study.

From these results, crack propagation mechanisms for the static loading and that for the cyclic loading were thought to be different. Although several mechanisms are proposed for the cycling effect [13-16], such as asperity contact [13], intrinsic residual stress [14], and non-linear process zone [15], evidence for each mechanism was not clearly obtained in our previous study [10]. In an attempt to clarify the difference of crack propagation mechanisms under static and cyclic loading, the same position of the surface crack of the sample shown in Fig. 3 was observed by SEM after static loading and cyclic loading. The SEM pictures of the surface crack after static loading and after subsequent cyclic loading are shown in Fig. 5 (A) and (B) respectively. Intergranular crack propagation with crack-bridgings was observed in both cases under static and cyclic loading. However, crack-bridgings, that were observed after static loading, were found to be fractured by the subsequent cyclic loading.

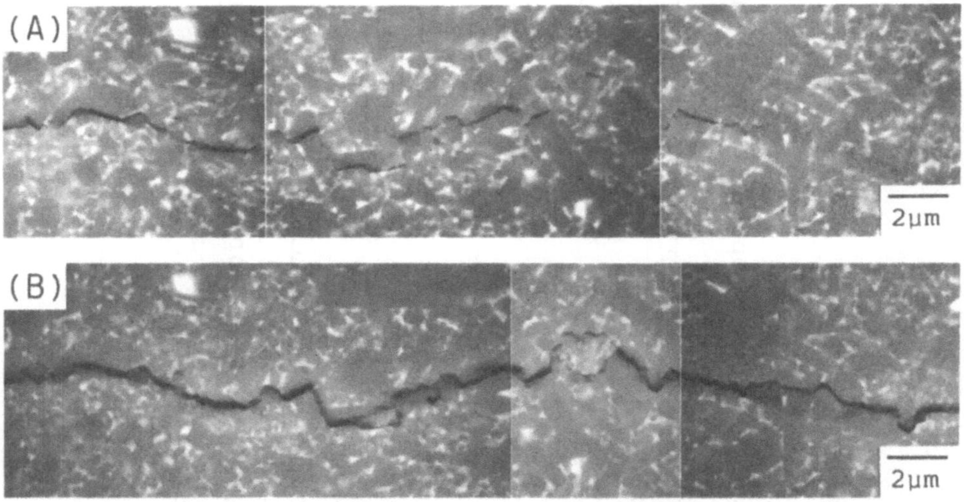

Fig. 5. SEM pictures of surface crack after static loading and subsequent cyclic loading.

From the foregoing results, the difference of crack propagation mechanisms under static loading and cyclic loading was considered and schematically shown in Fig. 6. In the case of static loading, stress intensity factor (K_{static}) is equal to applied stress term (K_{app}) unless another force exists. However, the crack-bridging, which is effective for restraining crack propagation, was observed by SEM. Therefore the K_{static} was diminished by crack shielding force (ΔK_{static}) and therefore $K_{static} = K_{app} - \Delta K_{static}$. On the other hand, crack propagation accompanied by fracture of crack-bridgings was observed under cyclic loading. It was thought that bonding of grains at crack-bridging was reduced by cyclic loading and consequently the crack shielding force by crack-bridging became smaller than that under static loading ($\Delta K_{cyclic} << \Delta K_{static}$). Therefore, crack intensity factor under cyclic loading was larger than that under static loading, and consequently crack propagated faster under cyclic loading than under static loading.

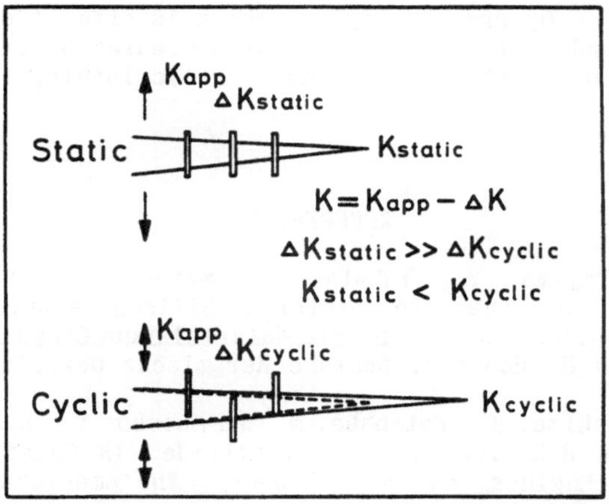

Fig. 6. Schematic of difference of crack propagation mechanisms under static and cyclic loadings.

Maniette et al. measured the R-curve behavior by utilizing a compact tension specimens of GPSSN, and reported that rising R-curve behavior was observed with transition crack length of about 1 mm [17]. Although samples used in this study were different in shape and crack geometry, rising R-curve behavior is thought to exist as well, because the crack-bridgings were observed even 80μm behind the crack tip under static loading. And it is suggested that R-curve behaviors under static and cyclic loading are unequal, and rising of crack growth resistance is larger under static loading than under cyclic loading. Because K_{Imax}-V diagram obtained under static loading, with assumption that the effect of residual stress caused by indentation is same under static and cyclic loading, shows that crack propagation velocity decreased with crack propagation. No cycling effect for the specimens without artificial flaw is thought to be related with the R-curve behavior; negligible effect of crack-bridgings for these specimens due to very small initial flaw size and high applied stress.

From the foregoing results, it is concluded that the fatigue behavior of GPSSN with artificial flaw was affected by the loading condition, and crack propagated faster under cyclic loading than static loading because of the difference of crack shielding force by crack-bridging.

CONCLUSIONS

Crack propagation behavior in a gas pressure sintered silicon nitride was investigated for pre-cracked specimens under static and cyclic loading by 3-point bending at room temperature to clarify the cycling effect. Crack propagation occurred under cyclic loading with maximum applied stress below the applied stress with which crack propagation stagnated under static loading, showing clearly that crack propagation velocity under cyclic loading was higher than that under static loading. Although intergranular crack propagation with crack-bridging was observed in both cases under static and cyclic loading, crack-bridgings, that were observed after static loading, were found to be fractured by the subsequent cyclic loading. It was thought that crack shielding force by crack-bridging, which is effective for restraining crack propagation under static loading, was decreased by cyclic loading and consequently crack propagated faster under cyclic loading than under static loading.

REFERENCES

1. Hattori, Y., Tajima, Y., Yabuta, K., Matsuo, Y., Kawamura, M. and Watanabe, T., Gas Pressure Sintered Silicon Nitride Ceramics for Turbocharger Application. In Ceramic Materials and Components for Engines, ed. W. Bunk and H. Hausner, Deutshe Keramische Gesellschaft, 1986, pp. 165-72.
2. Tajima, Y., Urashima, K., Watanabe, M. and Matsuo, Y., Static, Cyclic, and Dynamic Fatigue Behavior of Silicon Nitride. IN Ceramic Materials and Components for Engines, ed. V.J. Tennery, The American Ceramic Society Inc., 1989, pp. 719-28.
3. Masuda, M., Soma, T., Matsui, M. and Oda, I., Fatigue of Ceramics (Part 1). Nippon Seramikkusu Kyokai Gakujutu Ronbunshi, 1988, **96**, 277-83.
4. Masuda, M., Yamada, N., Soma, T., Matsui, M. and Oda, I., Fatigue of Ceramics (Part 2). Nippon Seramikkusu Kyokai Gakujutu Ronbunshi, 1989, **97**, 520-24.
5. Yamauchi, Y., Sakai, S., Ito, M., Ohji, T., Kanematsu, W. and Ito, S., Fatigue Test for HP-Si_3N_4 Ceramics Indented with Knoop Pyramid. Nippon Seramikkusu Kyokai Gakujutu Ronbunshi, 1986, **94**, 631-35.
6. Kawakubo, T. and Komeya, K., Static and Cyclic Fatigue Behavior of a Sintered Silicon Nitride at Room Temperature. J. Am. Ceram. Soc., 1987, **70**, 400-05.
7. Kishimoto, H., Ueno, A. and Kawamoto, H., Crack Propagation Characteristics of Sintered Si_3N_4 under Static and Cyclic Loads. J. Soc. Mat. Sci., Japan, 1987, **36**, 1122-27.
8. Kishimoto, H., Ueno, A., Kawamoto, H. and Fujii, Y., Influence of Wave Form and Compressive Loads on the Crack Propagation Behavior of Sintered Si_3N_4 under Cyclic Loads. J. Mat. Sci., Japan, 1989, **38**, 1212-17.
9. Horibe, S., Cyclic Fatigue Crack Growth from Indentation Flaw in Si_3N_4. J. Mater Sci. Lett., 1988, **7**, 725-27.
10. Niwa, T., Urashima, K., Tajima, Y. and Watanabe, M., Effects of Crack Size on Fatigue Behavior in Silicon Nitride. Nippon Seramikkusu Kyokai Gakujutu Ronbunshi, 1991, **99**, 296-99.
11. Anstis, G.R., Chantikul, P., Lawn, B.R. and Marshall, D.B., A Critical Evaluation of Indentation Techniques for Measuring Fracture Toughness: I, Direct Crack Measurements. J. Am. Ceram. Soc., 1981, **64**, 533-38.
12. Raju, I.S. and Newman Jr., J.C., Engng. Fract. Mech., Vol. 11, 1979, pp. 817-29.
13. Evans, A.G., Fatigue in Ceramics. Int. Journ. of Fracture, 1980, **16**, 485-98.
14. Lewis, D. and Rice, R.W., Comparison of Static, Cyclic, and Thermal-Shock Fatigue in Ceramic Composites. Ceram. Eng. Sci. Proc., 1982, **3**, 714-21.
15. Kobayashi, H. and Kawakubo, T., Fatigue - Difference between Ceramics and Metal. Bulletin of Japan Institute of Metals, 1988, **27**, 757-65.
16. Horibe, S., Cyclic Fatigue of Ceramic Materials. Tetsu-To-Hagane, 1989, **75**, 578-86.
17. Maniette, Y., Inagaki, M. and Sakai, M., Fracture Toughness and Crack Bridging of a Silicon Nitride Ceramic. J. Europ. Ceram. Soc. 1991, **7**, 255-63.

MECHANICAL PROPERTIES AND MICROSTRUCTURE OF PRESSURELESS SINTERED DUOPHASE SIALON

Ran-Rong Lee, Bruce E. Novich, George Franks, Debbie Ouellette
Ceramics Process Systems Corporation
Milford, MA 01757, USA
Mattison K. Ferber, Camden R. Hubbard, and Karren More
High Temperature Materials Laboratory
Oak Ridge National Laboratory
Oak Ridge, TN 37831, USA

ABSTRACT

Duophase (α'/β') sialon is being developed for ceramic engine applications by using the Quickset™ injection molding process, followed by pressureless sintering and a thermal treatment. The sialon had an average four-point flexural strength of 670 MPa at room temperature and 490 MPa at 1370°C. It survived the flexural stress rupture test at 1300°C and 340 MPa for 190 hours. X-ray diffraction (XRD) and transmission electron microscopy (TEM) characterization showed that crystallization of the grain boundary phase improved the high temperature flexural strength of this sialon material. The creep behavior was also found to be affected by the crystallized grain boundary phases. The formation of a yttrium aluminum garnet (YAG) phase and elongated grains yielded better creep resistance. The correlation between mechanical properties and microstructure is discussed.

INTRODUCTION

Silicon nitride (Si_3N_4) is one of the prime candidate materials for high temperature ceramic turbine engine applications.[1] Most of the high performance silicon nitrides used in the Advanced Turbine Technology Applications Project (ATTAP) require hot-isostatic-pressing (HIP) to achieve high density and good mechanical properties. Two important issues related to the HIPped silicon nitride need to be addressed: 1) HIP is a very expensive process, and is thus not a cost-effective manufacturing process for ceramic engine components and 2) the mechanical properties of silicon nitride with an as-HIPped surface are worse than that with an as-machined surface,[2] probably due to the glass reaction layer on the as-HIPped surface resulting from the glass encapsulation process. In comparison, pressureless sintering is a lower cost manufacturing process and does not require glass-encapsulation. However, to achieve full density, greater amounts of sintering aids must be employed for pressureless sintered ceramics. Typically, the high temperature mechanical properties and oxidation resistance of pressureless-sintered silicon nitride are greatly reduced by the increased amounts of grain boundary phases generated form the sintering aids.

Several techniques involving the modification of the grain boundary phases have been used to improve the high temperature mechanical properties. They can be classified into the following approaches: (1) Decrease the amount of the grain boundary phase by using less or no sintering aid. However, this method requires glass-encapsulation HIP or hot-pressing to achieve high density.[3] (2) Increase the viscosity of the glassy grain boundary phase by using more refractory sintering aids such as yttria, instead of magnesia (3) Crystallize the grain boundary phase.[4] (4) Diffuse some of the elements within the grain boundary phase into the matrix grains during densification.[5] For the pressureless-sintered material examined in this study, the focus was on approaches (2), (3) and (4).

Beta silicon aluminum oxynitride (β'-sialon), formed from the liquid phase sintering of silicon nitride with the addition of yttria and alumina, has been under development for more than a decade for cutting tool, extrusion die and other engineering applications. However, the application of β' sialon in the ceramic turbine engine was limited by the degradation of the grain boundary phase at temperatures exceeding 1200°C, similar to other types of pressureless-sintered silicon nitride materials. Recently, the duophase (α'/β') sialon material was found to perform very well in cutting tool applications.[6] The purpose of this study is to develop the duophase (α'/β') sialon for ceramic engine applications. This paper reports the room temperature and high temperature mechanical properties of the duophase sialon and their correlation with microstructure and processing. The focus is on grain boundary engineering for improving the high temperature mechanical properties through the post-sintering treatment.

MATERIALS AND METHODS

Materials Preparation

The duophase sialon was prepared from silicon nitride powder with optimum amounts of yttria (Y_2O_3) and 21R ($SiAl_6O_2N_6$) as sintering aids. The green billets having dimensions of 25 mm x 50 mm x 9.4 mm were produced using the Quickset™ injection process, which is the forming process for engine component fabrication at Ceramics Process Systems Corporation. The details of the forming process are described elsewhere.[7,8] The green parts were sintered in a graphite furnace in a flowing nitrogen atmosphere. The nitrogen pressure was only slightly higher than ambient pressure. The post-sintering thermal treatment was carried out in the same furnace, but at lower temperatures. Three different batches of samples, #G, #H and #I, having the same composition, but different firing cycles, are evaluated in this paper.

Mechanical Testing

The flexural strength was measured by using a four-point loading fixture following MIL-STD-1942(MR) specification. The "A" bar configuration, with a specimen size of 1.5 x 2 x 30 mm, was used for measuring the room temperature flexural strength and the "B" bar configuration, with a size of 3 x 4 x 50 mm, was used for the high temperature flexural strength and stress rupture testing. The loading rate for the room temperature flexural testing was 0.2 mm/min and that for the high temperature flexural testing was 5 lb/s. The room temperature fracture toughness was measured by using the controlled-flaw strength method.[9]

The flexural creep tests were conducted using a SiC four-point fixture having inner and outer spans of 20 and 40 mm, respectively. During the testing, an LVDT tracked the downward displacement of the load ram. A computer monitored this displacement as well as the load on each specimen and provided necessary adjustments to maintain the desired stress level.

The calculation of the tensile creep strain from the flexural strain data was based on the formulation of Hollenburg.[10] The creep behavior generated from the flexure test are described by the equation,[11]

$$d\varepsilon_s/dt = A_o(\sigma_a/\sigma_o)^n$$

where $d\varepsilon_s/dt$ is the steady-state creep rate, A_o is a pre-exponential factor, σ_a is the applied stress, σ_o is a normalizing parameter (=1 MPa), and n is the creep exponent.

TEM/EDS Characterization

Transmission electron microscopy (TEM) was used to identify the secondary phases and characterize the morphology of the grain boundary phases in each sample. TEM specimens were cut from the samples before and after creep deformation. The specimens were mechanically thinned to ~75 μm in thickness, dimpled down to ~25 μm in thickness and ion milled at 6 kV and 1 amp at 15°. The analytical electron microscopy (AEM) was performed at 200 kV in a JOEL 2000FX equipped with a Kevex Quantum ultra-thin window energy dispersive spectrometer (EDS).

X-ray diffraction (XRD)

X-ray diffraction was used to identify the intergrangular phases and α'/β' contents in the duophase sialons. The XRD system consisted of a Cu X-ray tube operated at 45 kV and 40 mA, 230 mm radius goniometer, and a Ge solid-state detector. Patterns were collected at 0.1°/min scanning rate. Phase identification was based on reference patterns contained in Sets 1-40 of the Powder Diffraction Files.

RESULTS

Mechanical/Thermal Properties

Flexural strength The four-point flexural strengths of the duophase sialon are plotted in Figure 1. The strength of the as-sintered material decreased gradually from 745 MPa at 25°C to 550 MPa at 1200°C, and then dropped to 350 MPa at 1300°C and 1370°C. In comparison, the duophase sialon after post-sintering thermal treatment had a lower room temperature flexural strength of 670 MPa, but maintained relatively good strength at high temperatures, having a flexural strength of 490 MPa at 1300°C and 1370°C. These results demonstrate the importance of the post-sintering thermal treatment on the mechanical properties of the duophase sialon.

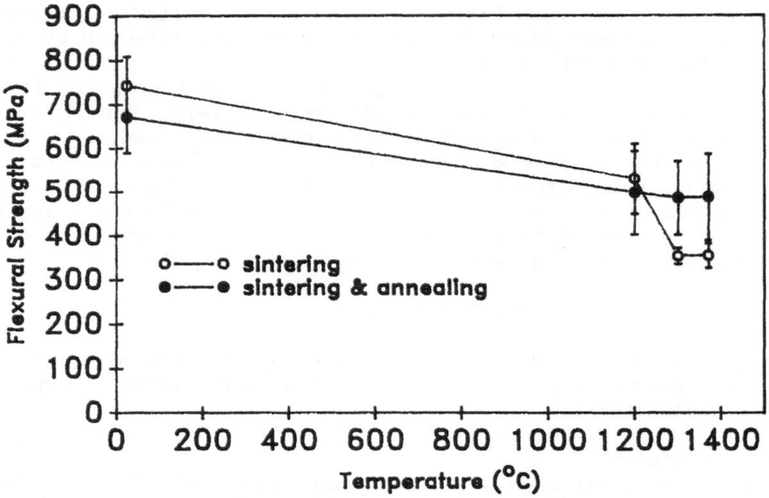

Figure 1. Comparison of the flexural strength of the duophase sialons before and after a post-sintering thermal treatment.

Flexural stress rupture The flexural stress rupture was tested at stresses of 240 MPa and 340 MPa at 1300°C. The strain verse time curves of the samples, #G, #H and #I, are plotted in Figure 2. At 240 MPa, samples #H and #I had better creep resistance than sample #G. When the stress was increased from 240 MPa to 340 MPa, #I showed better creep resistance than #H, indicating that sample #H was more sensitive to the applied stress. The secondary creep rates of the these samples tested at different conditions are summarized in Figure 3. Notice that the creep rate of sample #I was not as sensitive to the applied stress. From these data, the creep exponents (n) of #H and #I were calculated to be 2.7 and 0.7, respectively.

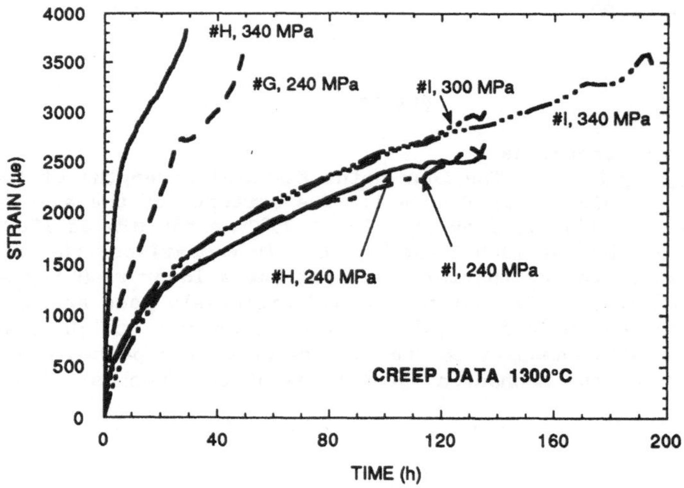

Figure 2. Creep behavior of samples #G, #H, and #I at 1300°C and several stress conditions

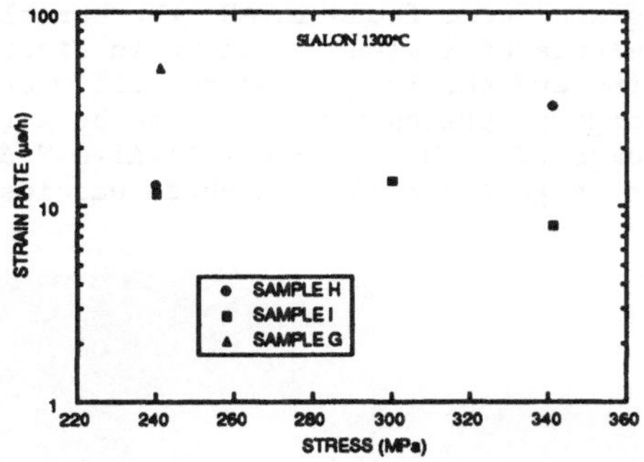

Figure 3. Secondary steady-state creep rates of samples #G, #H and #I under various conditions

Hardness The hardness, measured using a Vickers indenter, was 20.5 GPa, which was higher than the hardness of pure β' sialon of 17 GPa.[12]

Microstructure/Phase Characterization

The microstructure and phases of the sialon materials were characterized using XRD and AEM. In the as-sintered samples, the sialon had an amorphous grain boundary phase, as determined by electron diffraction. After a post-sintering heat treatment, the grain boundary phases were crystallized. In addition to equiaxed grains of both α' and β', elongated β' grains were observed, as shown in Figure 4. Several crystalline grain boundary pockets are arrowed in this figure.

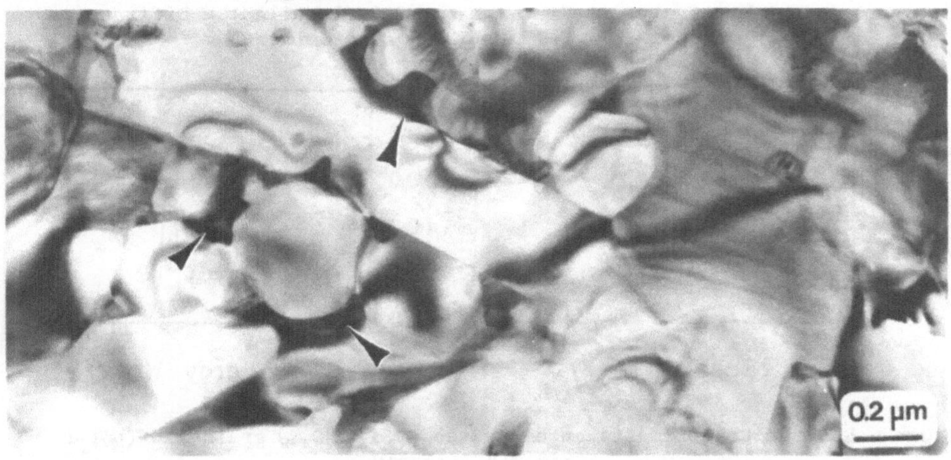

Figure 4. TEM image showing typical microstructure of α'/β' sialon material (sample #I). Several crystalline grain boundary pockets are arrowed.

There were several microstructural/phase differences observed between the three batches of samples, #G, #H and #I, corresponding to the differences in creep behavior of these three samples. From XRD, the major difference between the #G, #H and #I was the α' and β' contents. #G had much less α' sialon (~19 %) than #H and #I (~37 %), and #G also was found to have fewer elongated β' grains than #H and #I. Among these three samples, #H had the highest amount of elongated grains. Two different types of elongated grains were found in #H, the normal elongated β' grains and highly faulted elongated grains, as shown in Figure 5 (see morphology of the labelled grains and the inset electron diffraction pattern). From EDS, the faulted elongated grains were found to be a (high-Al)-Si-O-N (Figure 6(a)), as compared with the (high-Si)-Al-O-N β' sialon grains (Figure 6(b)). This (high-Al)-Si-Al-O-N phase was identified as ζ-$Si_3Al_7O_3N_9$[13] by XRD.

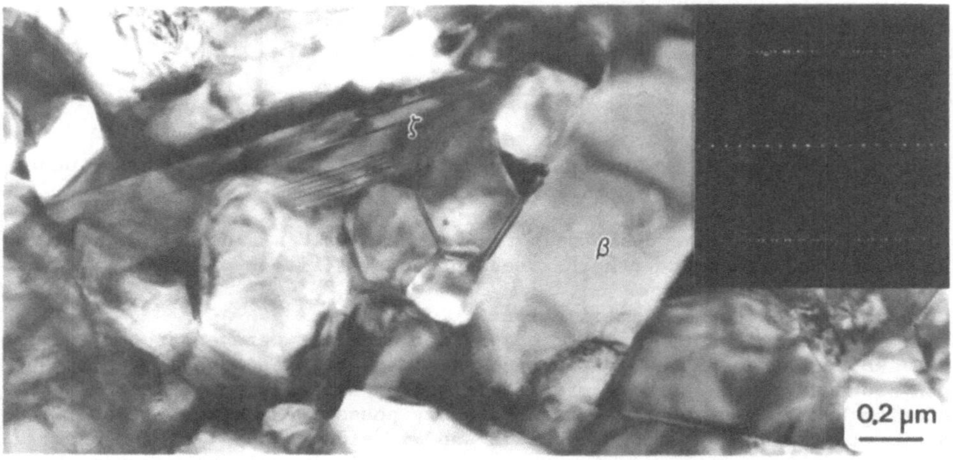

Figure 5. TEM image of sample #H showing elongated β' grains as well as elongated and highly faulted ζ-$Si_3Al_7O_3N_9$ grains (labelled). Diffraction pattern of a ζ-$Si_3Al_7O_3N_9$ grain is inset.

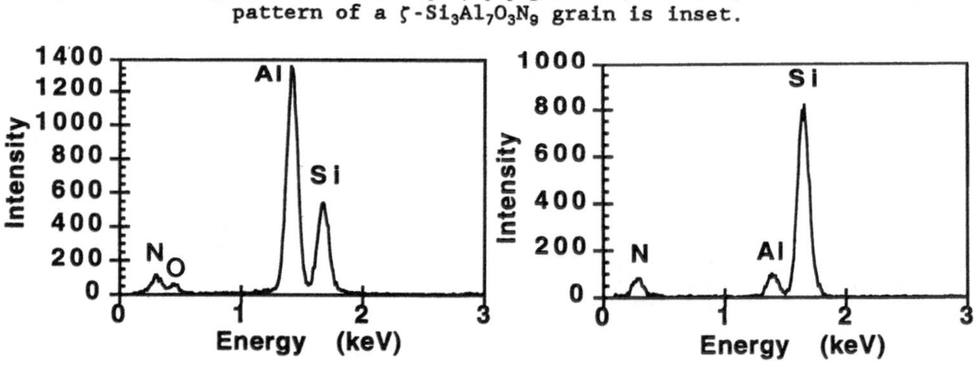

(a) (b)

Figure 6 EDS spectra from (a) an elongated ζ-$Si_3Al_7O_3N_9$ grain and (b) an elongated β' grain

The major grain boundary phase found in these three samples was a yttrium aluminum garnet (YAG) phase as identified by XRD. From EDS characterization, a small amount of silicon (Si) was found in the YAG phase ($Y_3Al_5O_{12}$), forming a Si-diffused YAG phase. Other minor grain boundary phases identified by XRD were K-phase ($YSiO_2N$) and melilite ($Y_2Si_3O_3N_4$). Sample #G appeared to have more of the Y-Si-O-N (K-phase or melilite) from AEM characterization.

Following creep testing, TEM samples were prepared from both the tensile and compressive faces of the test bars. Sample #G had many cracks at the tensile side of the specimen, with the cracks running perpendicular to the tensile face, as shown in Figure 7. In fact, TEM samples were extremely difficult to prepare from this sample as the surface repeatedly crumbled during the grinding and polishing steps. Samples #H and #I showed no obvious microstructural differences following creep. There were no microcracks and no cavitation was observed in either sample. The phases remained unchanged in each sample after creep at 1300°C, as determined by XRD.

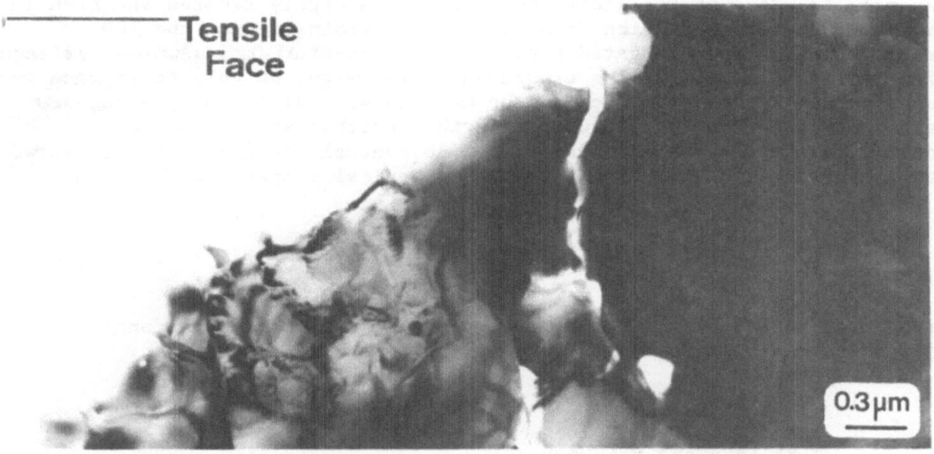

Figure 7. TEM image of #G after creep testing at 240 MPa and 1300°C, showing cracks running perpendicular to the tensile surface.

DISCUSSION

The improvement of the high temperature flexural strength of the duophase sialon after the post-sintering thermal treatment can be related to the crystallization of the grain boundary phases. At the same time, the crystallized grain boundary phases also decreased the room temperature flexural strength, perhaps due to residual stresses resulting from the thermal expansion mismatch between the crystallized grain boundary phases and the sialon grains.

The creep behavior of the duophase sialon was also affected by the microstructure and grain boundary phases. Samples #G, #H, and #I, having the same formulation but different firing cycles, had quite different creep behaviors. Sample #G had the worst creep resistance and the lowest α' content. The smaller amount of α' phase in #G indicates that more yttrium was left in the grain boundary phases. Thus, more yttrium-containing grain boundary phases formed. The melilite and K-phases found in #G appeared to be more than that in #H and #I. Therefore, the creep resistance was worse than that of sample #H and #I. The elongated grains in sample #H and #I are also believed to improve the creep resistance. The creep resistances of #H and #I were close to each other at the lower stress (240 MPa), but #I material had better creep resistance at the higher stress (340 MPa). Further evaluation on the differences between these two samples is in progress.

The creep exponent (n) could be calculated using the equation $\dot{\varepsilon}_1/\dot{\varepsilon}_2 = (\sigma_1/\sigma_2)^n$, where $\dot{\varepsilon}_1$ and $\dot{\varepsilon}_2$ are the strain rates and the σ_1 and σ_2 are the applied stresses. From the data in Figure 3, the stress exponents of samples #H and #I were calculated to be 2.7 and 0.7, respectively. Sample #H had a stress exponent larger than one, indicating that some creep mechanism, in addition to Coble diffusional creep, played an important role in the creep behavior of this material. Sample #I had a stress exponent close to one, corresponding to its good creep resistance.

In addition to the relationship between the grain boundary phase and the high temperature mechanical properties of the duophase sialon, the development of elongated grains was also of interest for the following reasons: (1) Some of the elongated grains were highly faulted and rich in aluminum, not normal Si-rich Si-Al-O-N sialon grains. From the XRD results, the faulted elongated grains are a zeta-$Si_3Al_7O_3N_9$ aluminum silicon oxide nitride phase. (2) The elongated grains might be able to improve the toughness and creep resistance of the materials. (3) Too many elongated grains might degrade the flexure strength. Further experiments are underway to explore the new phase, ways to control the growth of elongated grains, and their correlation with the mechanical properties.

CONCLUSIONS

A duophase (α'/β') sialon for high temperature ceramic turbine engine application has been developed to have the following properties.

Density	3.25 g/cm^3
Hardness	20.5 MPa
Fracture Toughness	5.5 MPa-m$^{1/2}$
4-pt Flexural Strength	
25°C	670 MPa
1200°C	510 MPa
1300°C	490 MPa
1370°C	490 MPa
Flexural Stress Rupture	
1300°C, 240 MPa	No failure after 190 hrs
Strain rate	0.0010%/hr
1300°C, 340 MPa	No failure after 190 hrs
Strain rate	0.0014%/hr

The grain boundary phases played an important role in controlling the high temperature mechanical properties of this duophase sialon. Optimum post-sintering thermal treatment improved the flexural strength from 350 MPa to 490 MPa at 1300°C and 1370°C. This was due to the formation of isolated and crystallized yttrium-rich refractory grain boundary phases.

The creep behavior of the duophase sialon was affected by the crystallized grain boundary phase, α'/β' content and grain morphology. The formation of a YAG grain boundary phase and elongated grains yielded better creep resistance.

ACKNOWLEDGEMENTS

This work was carried out under contract to Allison Gas Turbine Division of General Motors Corp. sponsored by the U.S. Department of Energy. The authors acknowledge Mr. Gilbert Rancoule of Vesuvius Crucible and Dr. Ian Wilson of Vesuvius Zyalon for their helpful technical discussion. Research sponsored in part by the U.S. Department of Energy, Assistant Secretary for Conservation and Renewable Energy, Office of Transportation Technologies, as part of the High Temperature Materials Laboratory User Program, under contract DE-AC05-OR21400 with Martin Marietta Energy Systems, Inc.

REFERENCE

1. Proceeding of the Annual Automotive Technology Development Contractors' Meeting, Society of Automotive Engineers, Inc. Dearborn, Michigan, October 22-25, 1990

2. Helms H.E., Haley P.J., Groseclose L.E., Hilpisch S.J. and Bell A.H.III., Advanced Turbine Technology Applications Project (ATTAP). in Proceeding of the Annual Automotive Technology Development Contractors' Meeting, Society of Automotive Engineers, Inc. Dearborn, Michigan, October 22-25, 1990, pp.89-103.

3. Prochazka S. and Rocco W.A., High Pressure Hot Pressing of Silicon Nitride Powders. in Nitrogen Ceramics, edited by Riley F.L., Noordhoff. Netherlands, 1981

4. Clarke D.R., Lange F.F. and Shnitgrund G.D., Strengthening of Sintered Silicon Nitride by Post-fabrication Heat Treatment. J. Am. Ceram. Soc. 1982, 4, C51-C52

5. Jack K.H., Crystal Chemistry of SiAlONs and Related Nitrogen Ceramics. in Nitrogen Ceramics. Edited by F.L. Riley, Noordhoff, Netherlands, 1977

6. Aucote J. and Foster S.R., Performance of sialon Cutting Tools When Machining Nickel-Base Aerospace Alloys. Met. Sci. Technol., 1986, 2, 700-708.

7. Novich B.E., Lee R.-R., Franks G., and Ouellette D., Quickset™ Injection Molding of High Temperature Gas Turbine Engine Components. in Proceeding of the Annual Automotive Technology Development Contractors' Meeting, Society of Automotive Engineers, Inc. Dearborn, Michigan, October 23-26, 1989, pp.311-318.

8. Novich B.E., Lee R.-R., Franks G., Ouellette D., and Groseclose L.E., Ceramic Engine Components Fabricated by Using the Quickset Injection Molding Process. in 4th International Symposium on Ceramic Materials & Components for Engines., Goteborg, Sweden, June 10-12, 1991, published in this issue.

9. Chantikul P., Anstis G.R., Lawn B.R. and Marshall D.B., A Critical Evaluation of Indention techniques for Measuring Fracture Toughness: II Strength Method. J. Am. Ceram. Soc., vol.64, 9, 1981, PP.539-543.

10. Hollenburg, G.W., Terwilliger G.R., and Gordon R.S., Calculation of Stresses and Strain in Four-point Bending Creep Tests. J. Am. Ceram. Soc., vol.54, 4, 1971, pp 196-99

11. Riedel H., Fracture at High Temperature, Springer-Verlag, Berlin, Heidelburgh, 1987

12. Lee R.-R., unpublished work.

13. Land P.L., Wimmer J.M. Burns, R.W., and Chourhury N.S., Compounds and Properties of the Systems Si-Al-O-N, J. Am. Ceram. Soc., vol.61, 1-2, 1987, pp 56

CYCLIC FATIGUE OF SILICON NITRIDES AND A WHISKER-REINFORCED SILICON NITRIDE

T.A. MÄNTYLÄ[1], A.-P. NIKKILÄ[1], P.T. VUORINEN[1],
M. ENOKI[2], and T.KISHI[2]
1) Tampere University of Technology, Institute of Materials Science, Tampere, Finland
2) The University of Tokyo, Research Center for Advanced Science and Technology,Tokyo, Japan

ABSTRACT

Cyclic stresses derived from mechanical loads or temperature changes are common in engines. Compression-tension cycling is recently found to degradate the mechanical properties of common engineering ceramic materials. In this study, the cyclic fatigue behaviour of sintered and hot-pressed silicon nitride is compared to the behaviour of SiC-whisker-reinforced silicon nitride. Alternating cyclic testing, R=-1, was done by using modified four-point bending arrangement. Cycling resulted in clear degradation of mechanical properties in all materials studied. The composite structure had different behaviour as compared to the monolithic materials showing a bimodal behaviour with little degradation at low cycle area but increased degradation at high cycle area. The results of the test are presented and the behaviour is interpreted and discussed based on the results of fractographic analysis.

INTRODUCTION

Successful use of structural ceramics in load bearing application requires reliable information on material behaviour under the appropriate loading type. In many applications alternating cyclic stresses exist and material response on such loads should be known. The mechanical properties of metals and polymers are susceptible to degradation under repeated loading, i.e. cyclic fatigue. Such fatigue has not been considered to be effective in ceramics due to the absence of appreciable crack tip plasticity [1]. Recently it has been found that cyclic loading causes clear degradation of the mechanical properties of ceramics [2-6] and it has been found that alternating tension-compression loading is more effective in reducing residual strength or time to failure than cycling only with tension-tension or compression-compression loads [7-9]. Although at this moment there is no unanimous understanding of the mechanism of cyclic fatigue, since it is difficult to separate the role of slow crack growth from the real cyclic fatigue mechanisms. Several possible mechanisms have been proposed, but their experimental verification is difficult.

In designing more advanced microstructures to ceramics to increase the toughness or high temperature stability little attention has been paid to the fatigue behaviour. However, it seems that some of these new structures are more prone to fatigue than basic monolithic structures [3]. Often these materials are aimed to demanding mechanically loaded applications and it is very important to know their fatigue behaviour. In this study we have compared the response of different silicon nitride materials to alternating cyclic stresses by using recently developed simple testing procedure [8].

MATERIALS AND METHODS

Three different commercial silicon nitrides and an experimental whisker-reinforced silicon nitride were used in this study. In order to determine the role of the whisker-reinforcement, the composite and the reference material, which had the same composition as the matrix of the composite, were both made with the same densification additives and processing steps. Two of the commercial materials were sintered silicon nitrides and the third was hot-pressed. The composite had 10 vol% of SiC whiskers . The composite and reference material contained 10 wt% of Y_2O_3 and 5 wt% of Al_2O_3 as densification additives. Both materials were hot-pressed, the composite at 2053 K and the reference material at 1973 K [10].

The strength of each material was measured by four-point bending . The specimens in strength and fatigue tests were rectangular bars and their size was 3x4x40 mm. In the hot-pressed materials the longest dimension was perpendicular to hot-pressing direction. The surface roughness of each specimen was determined by profilometer. The loading span was 10 mm and the supporting span 30 mm. The cross-head speed was 0.5 mm/min. The testing was done in air at room temperature. The number of specimens for commercial materials was 10 and for composite and reference material 5. The fracture toughness of the materials was measured by SENB (single edge notched beam) for commercial materials and by SEPB (single edge precracked beam) [11] for composite and reference material.

The cyclic fatigue tests were carried out by a specially designed bending jig with 8 loading rolls in four-point bending geometry allowing cyclic tension-compression loading [8] . The loading mode was sinusoidal. Loading ratio R (minimum stress/ maximum stress) was in all tests -1. The cycling was done with two different units having the frequencies of 10 Hz and 25 Hz. All the specimens were tested as non-indented.

The crystal structure of the materials was characterized by XRD and the microstructures were characterized by SEM with EDX. The composite and the reference material was also studied by TEM. Fractography for each specimen was done by SEM with EDX.

RESULTS AND DISCUSSION

The strength values of the materials are given in Table 1. The materials A and C had high Weibull modulus, the material B and the experimental materials had much higher scatter in strength values which is indicated by low values of Weibull moduli. Since the number of specimens was small the values of Weibull moduli are quite inaccurate, but due to the limited availibility of the materials, it was not possible to make more extensive analysis.

All commercial materials had β- Si_3N_4 structure with rod-like grains which had the following dimensions (length/diameter in micrometers) 4/1 for A, 3/1 for B, and 3/1 for C. The matrix of the composite had also the β-structure, but in the reference material some α-Si_3N_4 was found. In the composite the whiskers were mainly oriented into the hot-pressing plane. The length of the whiskers determined by SEM and TEM in composites varied from few micrometers up to 20...30 μm.

TABLE 1
The mechanical properties of tested materials

	Material				
	SSN A	SSN B	HPSN C	Si_3N_4-SiC	REF.HPSN
Strength, N/mm^2	734	652	870	895	747
- max	768	765	970	1014	833
- min	671	593	833	807	640
Weibull modulus	28 [1)]	9 [1)]	16 [1)]	11 [2)]	10 [2)]
- 90% confidence limits	18-44	6-14	10-25	6-22	6-20
K_{Ic}, $MN/m^{-3/2}$	5.9 [3)]	5.1 [3)]	6.4 [3)]	7.0 [4)]	5.6 [4)]
Surface roughness					
R_a(∥), μm	0.07	0.08	0.05	0.06	0.05
R_a(⊥), μm	0.07	0.08	0.04	0.12	0.04
Producer	ESK [5)]	Kyocera [6)]	ESK [7)]		

1) 10 specimen, calculated by Maximum Likehood method, 2) 6 specimen, calculated by Maximum Likehood method, 3) SENB with 5 specimen, 4) SEPB with 5 specimen, 5) EKasin S, 6) SN-220, 6) Ekasin D

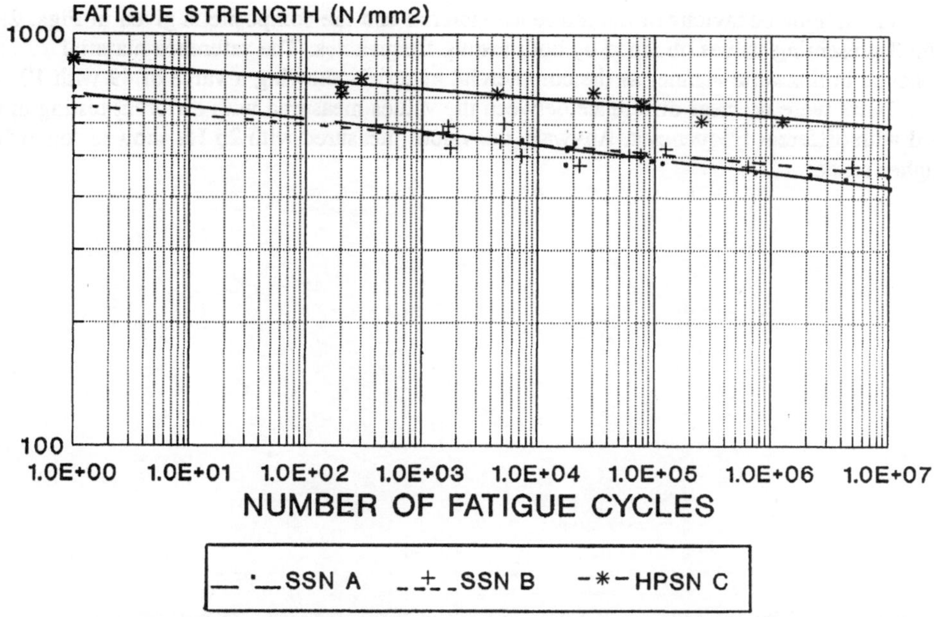

Figure 1. Fatigue strength of studied silicon nitride materials as a function of the number of cycles. Testing was done by 25 Hz at room temperature and in air.

The fatigue strength of the commercial materials as a function of the number of cycles is given in Fig. 1. All of them showed same kind of degradation of the strength as function of the cycles. The severity of this degradation can be described with parameter n which can be determined according to the equation

$$\sigma^n N = \text{constant}$$

where σ is the fatigue strength and N is the number of cycles. Following values for parameter n were obtained from experimental data: $n_{material\ A} = 31$, $n_{material\ B} = 41$, and $n_{material\ C} = 43$. The scatter of measured fatigue strength values for these materials was quite small and it seemed to decrease when the number of cycles increased. The decrease in strength as the number of cycles increased was similar for hot-pressed material C and for sintered material B. Although the material A had higher bend strength than material B, it was more prone to fatigue than material B, which is indicated by higher n value. No clear fatigue limit was observed for these commercial materials when testing was done up to a few million cycles. However, based on fractographic observations and by measuring the residual strength of some surviors which did not break up to several million cycles we have estimated that fatigue limit for these materials is about 2/3 of the average bending strength. This is of the same order of magnitude as reported for silicon nitrides in literature. Masuda et. al. [9] showed with different silicon nitrides that their specimens survived up to 10^{10} cycles of alternating cyclic loading when the fatigue stress amplitude was lower than 60% of the initial strength.

The fatigue behaviour of the reference material and the composite is given in Figs. 2. and 3 presenting values obtained by both testing frequencies. The reference material behaved in the same manner as the commercial materials showing n value of 49 with 10 Hz. There were no clear difference between the values measured with different testing units and with different frequencies. Anyway, the values measured with 25 Hz showed somewhat higher scatter.

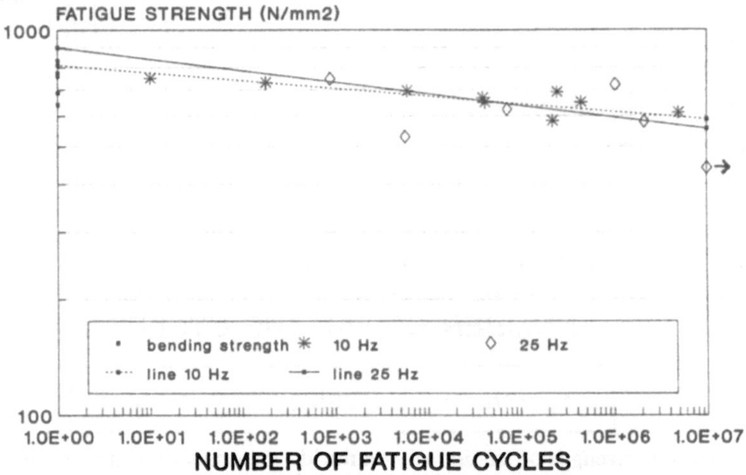

Figure 2. Cyclic fatigue behaviour of the reference material as a function of the cycles tested with two frequencies at room temperature in air.

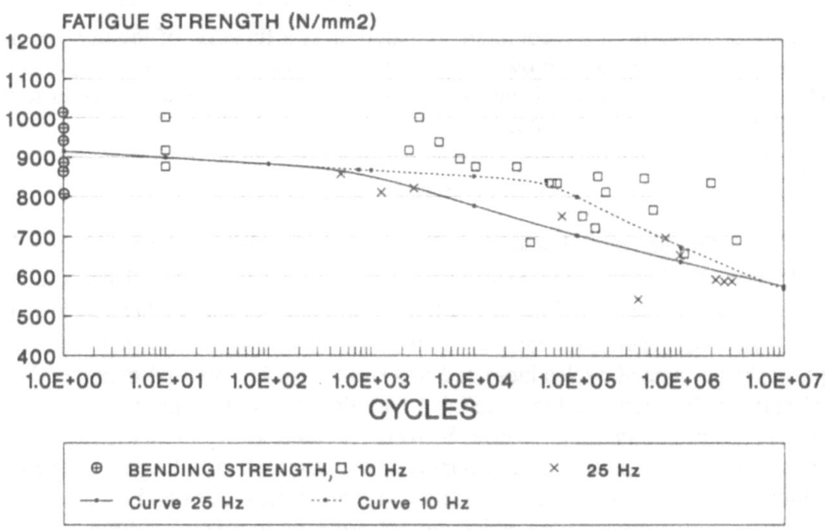

Figure 3. Cyclic fatigue behaviour of the Si_3N_4-SiC_w . Fatigue strength is given as a function of the cycles. Testing was done with two frequencies at room temperature in air.

The fatigue behaviour of the composite was quite different showing an apparent bimodal behaviour. This is especially clear in values measured by the lower frequency, 10 Hz. In the low cycle range (number of cycles < 10000) relatively little degradation was observed which gave the n value of 128. At high cycle area the degradation was rapid showing n value of 13. In this area n value was 23 when 25 Hz was used. The scatter in the values in this range was large.

The fractoraphic analysis did not reveal any clear evidence of specific cyclic degradation mechanism. Fracture originated typically from internal defects like pores or pore agglomerates. A sligth smoothening of pore edges on some fatigued specimens as compered to bend tested specimens was observed. This is caused by slow crack growth before the final fracture. Cracks typically propagated intragranually. Our earlier results showed [8] that the crack growth is faster in the case of alternating cyclic loading than pulsating cyclic loading by using the same maximum tensile stress amplitude.

In the composites and the reference material fracture originated either from pores or iron containing impurity inclusions. The source of this iron contamination is not clear at this moment. In the fracture surfaces of composites some debonding of whisker-matrix interface was observed. Some examples of empty whisker sites are given in Fig.4. This observation may explain the observed bimodal behaviour in fatigue strength. The cyclic alternating loading degradates gradually the strength of the interface and after certain number of cycles total debonding happens. When this happens in several whiskers close to each other large internal defects are generated and they grow rapidly to critical size. The length of this final stage depends on the local microstructure, eg. the whisker/whisker contacts, whisker orientation or the distance between neighbouring whiskers and causes the scatter to the results.

The other possibility is that the whiskers partly bridge the subcritical cracks up to the point where whisker cracking is generated by repeated tension-compression loading or the strength of the interface is lost and whisker is pulled out from the matrix. Some holes of pulled out whiskers were found on the fracture surfaces. The detaching and pull-out of whiskers were already observed in specimens which broke at low cycle area, see Fig 4. This may indicate the weakness of the interface, which causes the failure at lower stresses after certain number of cycles. Anyway, very few out reaching whiskers were found on fracture surfaces. However, the detaching surfaces may cut them during the final fracture, since the breaking surfaces may touch each other. This kind of fracture behaviour may explain the slightly lower fatigue strength values which were obtained by higher frequency since higher frequency may lead to more rapid softening of the glassy boundary phase due to thermal effects than lower frequency. Masuda et. al. [9] showed that frequency does not affect to the fatigue strength of sintered silicon nitrides. Our results also support this when monolithic materials are considered, but in composite structures the situation is different due to second phase whiskers, different interface stresses and high amount of glassy phase. The fracture of bridging whisker may also form asperites or trapped particles into the growing fracture surface which creates tensile stresses at crack tip during compression cycle.

a) b)

Figure 4. Examples of the debonded whiskers at the fracture surfaces of the composite specimens: a) Cycled by 810 N/mm^2 and 25 Hz 1260 cycles and b) Cycled by 650 N/mm^2 and 25 Hz 10 cycles. In the latter case the fracture has originated from large whisker agglomerate where the orientation of the whiskers has also been inferior

CONCLUSIONS

Cyclic fatigue testing by using alternating stresses with R=-1 was done in four-point bending geometry for different silicon nitrides and a whisker-reinforced silicon nitride (10 vol% of SiC_w).

1) The fatigue behaviour of HPSN was similar as SSN materials. The fatigue limit for the tested commercial monolithic materials was about 3/5 of the four-point strength.

2) The whisker-reinforced composite showed a bimodal behaviour with little degradation at low cycle area (number of cycles < 10000) and rapid degradation at high cycle area. The fatigue strength also showed slight dependency on frequency so that the strength was somewhat lower with higher frequency.

3) No clear cyclic fatigue mechanism could be found in fractoraghic analysis for these materials. In composites some clear whisker-matrix interface failures were found. Based on these observations at least two possible fatigue mechanisms for the composite could be proposed: a) cyclic loading leads to the failure in whisker matrix interface, which accumulates local defects approaching the critical defects size, and b) bridging whiskers detach from the matrix due to softening of the glassy phase which leads to particle trapping and the wedging effect. The latter also explains the frequency dependency found in experiments. It must be pointed out that sofar these mechanisms are only speculative and need more experimental verification.

ACKNOWLEDGEMENTS

The authors thank Dr. Y. Akimune of Nissan Motor Co. for fabricating the composite and the reference material. The help of Mrs. T. Stenberg and Mr. J. Laurila in XRD analysis and transmission electron microscopy is gratefully acknowledged. The Academy of Finland and the Technology Development Centre is thanked for partial funding of this work.

REFERENCES

1. Evans,A.G., Fatigue in ceramics, Int. J. of Fracture, 1980, **16**, 485-98.

2. Kawakubo, T. and Komeya, K., Static and Cyclic Fatigue Behavior of a Sintered Silicon Nitride at Room Temperature, J. Am. Ceram. Soc., 1987, **70**, 400-405.

3. Dauskardt,R.H., Yu,W., and Ritchie,R.O., Fatigue Crack Propagation in Transformation-Toughened Zirconia Ceramic, J. Am. Ceram. Soc., 1987, **70**, C248-52.

4. Reece,M.J., Guiu,F., and Sammur,M.F.R., Cyclic Fatigue Crack Propagation in Alumina under Direct Tension-Compression Loading, J. Am. Ceram. Soc., 1989, **72**, 348-52.

5. Gratwohl,G., Ermudung von Keramik unter Schwingbeanspruchung, Mat.-wiss. u. Werkstofftech., 1988, **19**, 113-24.

6. Horibe,S., Fatigue of Silicon Nitride Ceramics under Cyclic Loading, J. Europ. Ceram. Soc., 1990, **6**, 89-95.

7. Masuda,M., Soma,T., Matsui,M., and Oda,I., Cyclic Fatigue of Sintered Si_3N_4, Ceram. Eng. Proc., 1988, **9**, 1371-82.

8. Nikkilä, A.-P. and Mäntylä, T.A., Cyclic Fatigue of Silicon Nitrides. Ceram. Eng. Sci. Proc., 1989, **10**, 646-56.

9. Masuda,M., Soma,T., and Matsui,M., Cyclic Fatigue Behavior of Si_3N_4 Ceramics, J. Europ. Ceram. Soc., 1990, **6**, 253-58.

10. Akimune, Y., Katano,Y., and Matoba,K., Spherical-Impact Damage and Strength Degradation in Silicon Carbide Whisker/Silicon Nitride Composite, J. Am. Ceram. Soc., 1989, **72**, 791-98.

11. Nose,T. and Fujii,T., Evaluation of Fracture Toughness for Ceramic Materials by a Single-Edge-Precracked-Beam Method, J. Am. Ceram. Soc., 1988, **71**, 328-33.

MECHANICAL PROPERTIES OF Ce-TZP/Al_2O_3 CERAMICS WITH SMALL TiO_2 ADDITIONS

Senfeng CHEN, YiYuan WU, JianDong YE, ShaoTang ZHAO
South China University of Technology
Dept. of Inorganic Materials
Wu Shan, GuangZhou 510641, P.R.China

ABSTRACT

Ce-TZP has been prepared utilizing ZrO_2, CeO_2, Al_2O_3 and TiO_2 as starting raw materials. As expected, the CeO_2 affected the bending strength and fracture toughness. The additions of Al_2O_3 and TiO_2, over the range studied, had little effect on strength. Increasing additions of TiO_2 appeared to lower the toughness. The use of experimental design to optimize mechanical properties, for the system studied, is discussed.

Key words: Zirconia; alumina; transformation toughening; Ce-TZP

INTRODUCTION

Phase transformations often produce internal stress in ceramic matrices, and are sometimes regarded as unfavourable in certain ceramic systems. This is not the case for transformation toughened ceramics. In 1975, R.C.Garvie et al.[1] showed that ZrO_2 could be transformation toughened successfully. This toughness improvement relies on the retention of metastable tetragonal phase ZrO_2 in the ceramic matrix. When ceramic matrix is subjected to external tension stress, tetragonal phase ZrO_2 (t-ZrO_2) transforms to monoclinic phase (m-ZrO_2) with an accompanying volume expansion surrounding the crack tip of 3-4%. This volume expansion produces pressure on the crack and thus crack extension is retarded. As a result, the toughness and strength of transformation toughened ceramics are raised.

CeO_2 containing tetragonal zirconia polycrystalline (Ce-TZP) has better fracture toughness than other transformation toughened ceramics. It is not as sensitive to lower temperature degradation in moist environments as Y-TZP[2], and its critical grain size for phase transformation is bigger. Ce-TZP with good mechanical properties has been prepared using technical ZrO_2 and CeO_2 as raw materials. It is well recognized that the addition of Al_2O_3 to Y- and Ce-TZP results in increased strength, presumably due to alumina acting as a grain growth inhibition. The increase in strength is generally accompanied by a decrease in fracture toughness due to finer ZrO_2 grains being less susceptible to transformation.Additionally, Al_2O_3 additions increase Young's modulus and decrease the transformation zone size and limit or eliminate auto-catalysis.It is also well known that the amount or stabilizer should be controlled if optimum toughness is desired since too little stabilizer results in monoclinic ZrO_2 and too much stabilizer results in cubic ZrO_2. In the fabrication of Ce-TZP, therefore, the contents of stabilizer (CeO_2) and additives (Al_2O_3,TiO_2) affect the bending strength and fracture toughness remarkably. Excellent performance can be achieved by controlling the contents of CeO_2, Al_2O_3 and TiO_2 in TZP appropriately.[3]

The purpose of this work was to optimize the Ce-TZP with Al_2O_3 and TiO_2. Factorial experiments were carried out varying the contents of CeO_2, Al_2O_3, and TiO_2, as well as sintering temperature in order to gain optimum processing parameters. The effect of additives on the σ_f and K_c is discussed.

EXPERIMENTAL PROCEDURE

This study was performed using multi-varible orthogonal testing design for the contents of CeO_2, Al_2O_3, TiO_2 and sintering temperature.[4] Three levels have been set for each factor, as shown in Table 1.

The samples were made up with $L_9(3^4)$ test table, according to the formula: (100-X) mol% ZrO_2 + X mol% CeO_2 + Y wt% Al_2O_3 + Z wt% TiO_2. ZrO_2 powder was mixed with CeO_2 by vibratory milling for 24 hrs in alcohol using Y-TZP milling media. The mixed powder was calcined at 1250°C for 1 hr and then was vibratorily milled again for 20 hrs. After milling, the powder was isostatically cold-pressed into test bar at 200 MPa. The green bodies were sintered in air at different temperatures, as listed in the test table, for 2 hrs.

Three-point bending strength measurements were performed using 30-mm span and a cross-head speed 0.4mm/min. The fracture toughness was evaluated by an indentation method and K_c values were calculated using the equation given by Niihara.[5] The Vickers hardness Hv of samples was measured. The fraction of monoclinic phase on the ground and polished surface was determined using the X-ray intensity ratio of the monoclinic (111) and (111) peaks to the tetragonal (111) peak.[6] Scanning electronic microscopy was used to examine the microstructure of samples.

TABLE 1
Factors and levels test table

factor / level	CeO_2 mol%	Al_2O_3 wt%	TiO_2 wt%	Sintering Temp. °C
1	10	9	1	1510
2	12	11	2	1530
3	14	13	3	1550

EXPERIMENTAL RESULTS

The experimental results are listed in Table 2.

The factorial design was analyzed using integrated software[7]. The maximum differences are shown in Fig.1 and Fig.2.As can be seen in the two figures, the average K_c value of Ce-TZP is sensitive to the small variations in Al_2O_3 and TiO_2 contents used in this experimental design. There appears to be no advantage to adding TiO_2 if environments of mecchanical properties are desired. Table 2 shows that the best properties

TABLE 2
Test conditions and results

#	CeO_2 m/o	Al_2O_3 w/o	TiO_2 w/o	Temp °C	Density g/cm³	Strength MPa	Kc MPa.√m	Hardness GPa	$m\text{-}ZrO_2$ v/o
1	10	9	1	1510	4.94	240	4.0	3.17	86.7
2	10	11	2	1530	5.48	206	3.9	4.36	87.5
3	10	13	3	1550	5.51	236	3.5	4.50	91.4
4	12	9	2	1550	5.78	662	7.0	7.84	5.4
5	12	11	3	1510	5.61	639	6.9	8.38	4.7
6	12	13	1	1530	5.50	746	7.8	9.23	8.6
7	14	9	3	1530	5.77	523	5.3	8.62	1.0
8	14	11	1	1550	5.67	539	6.4	6.99	2.1
9	14	13	2	1510	5.49	485	6.0	7.64	0.9

		CeO_2	Al_2O_3	TiO_2	Temp
σ_f	I	227	475	508	455
	II	683	461	492	492
	III	516	466	479	479
	R	456	57	37	37
Kc	I	3.78	5.42	6.05	5.61
	II	7.24	5.69	5.65	5.67
	III	5.89	5.80	5.22	5.64
	R	3.46	0.38	0.83	0.03

Prediction Equation:

$$S=-156882 + 1937.08C - 77.71C^2 - 110.17A + 5.167A^2 - 165.83T + 36.167T^2 + 190.58Temp - 6.2083Temp^2$$

Kc = ...

D = ...

Hv = ...

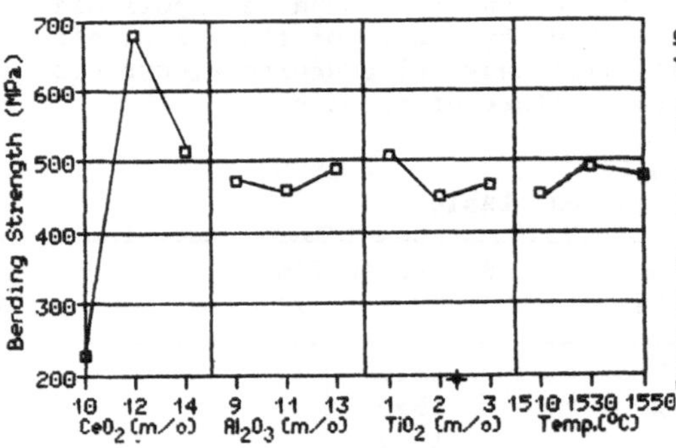

Fig. 1 Processing Parameters Effects on σ_f

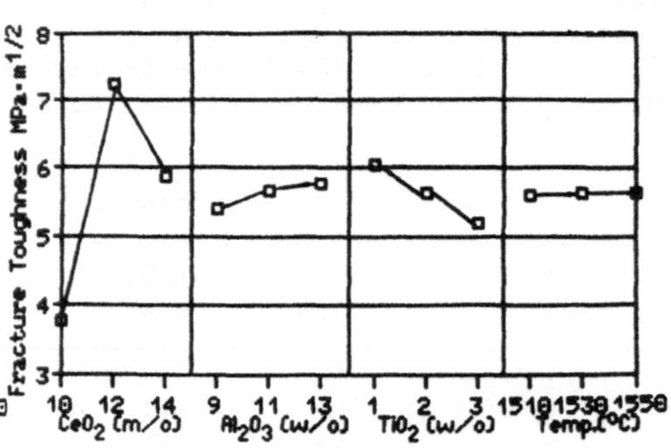

Fig. 2 Processing Parameters Effects on Kc

were measured in the case of containing 12 mol% CeO_2, 13 wt% Al_2O_3, 1 wt% TiO_2 and sintering at 1530°C, or No.6 test condition in testing design table. The bending strength was 746 MPa and fracture toughness was 7.8 MPa.√m. Moreover, by comparing the results of maximum differences, it is obvious that the CeO_2 content has the most effect on the bending strength and fracture toughness, as expected. The influence of sintering temperature on the bending strength and fracture toughness within the temperature range of 1510-1550°C is not evident, as expected, due to the grain growth inhibition of ZrO_2 by Al_2O_3.

A quadratic mathematic model was utilized to fit the relationship between processing parameters and properties, as can be seen in Fig.3[6]. When the quantities of the various factors are put into the empirical equation, the mechanical properties could be forecasted. For example, strength can be calculated with the following equation:

$$\begin{aligned} S = -156882 &+ 1937.08 \times [CeO_2] - 77.7083 \times [CeO_2]^2 \\ &- 110.167 \times [Al_2O_3] + 5.16667 \times [Al_2O_3]^2 \\ &- 165.833 \times [TiO_2] + 36.1667 \times [TiO_2]^2 \\ &+ 190.583 \times [Temp] - 6.20833 \times 10^{-2} \times [Temp]^2 \end{aligned}$$

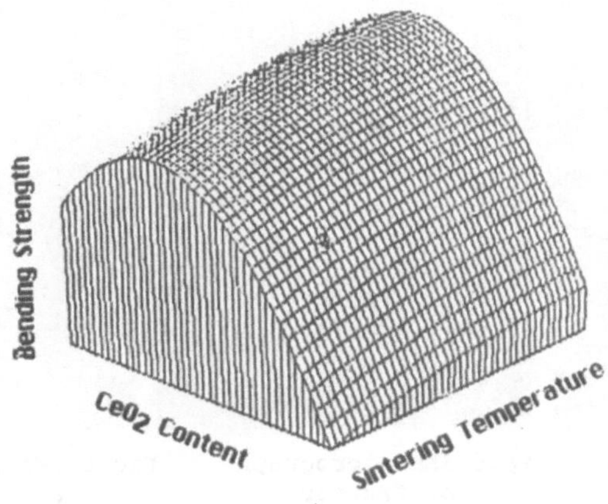

Fig. 3 Influence of CeO_2 Content and Sintering Temperature on Bending Strength

According to this equation, the maximum strength value in this experiment was forecast to be 764 MPa when $[CeO_2]$=12.46 mol%, $[Al_2O_3]$=13 wt%, $[TiO_2]$=1 wt%, and [Temp]=1535°C. Under the same test condition as above,the sample strength was 759 MPa and toughness was 7.9 MPa√m, in excellent agreement with predicted results.

DISCUSSIONS AND ANALYSES

1. Effect of CeO_2 content on K_c and σ_f of Ce-TZP

The optimum CeO_2 content is 12 mol% in this experiment and 12.45 mol% predicted by the forecast equation. From XRD phase analysis, the average m-ZrO_2 fraction at room temperature was 88.5% when CeO_2 content was 10 mol%. As expected, most of the t-ZrO_2 transformed to m-ZrO_2 during the cooling process, so the phase transformation failed to play a substantial role in increasing toughness. Due to the high monoclinic content, the strength and hardness were relatively low. For the sample containing 12 mol% CeO_2, the average m-ZrO_2 fraction was 6.2%. As expected, the higher stabilizer content allowed more metastable t-ZrO_2 to be retained at room temperature. When samples were subjected to external stress, a part of metastable t-ZrO_2 could transform to m-ZrO_2, which produced a transformation zone limiting crack extension, so the toughness and strength of ceramics were enhanced. For the sample containing 14 mol% CeO_2, the average m-ZrO_2 fraction only was 1.5%. The resistance of phase transformation increases with CeO_2 content.

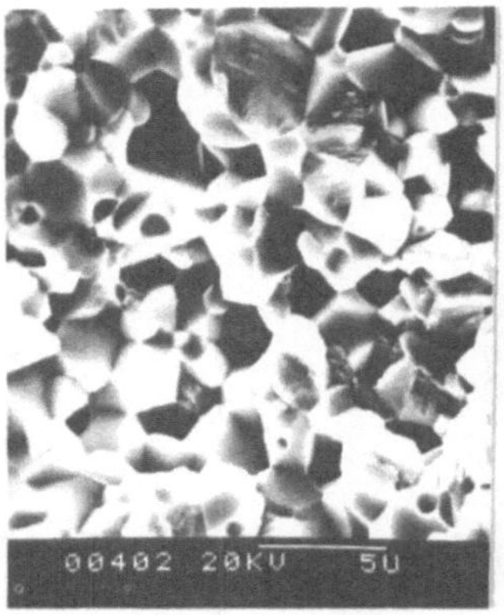

Fig.4 SEM micrograph of fractured surface of 10 mol.% CeO_2

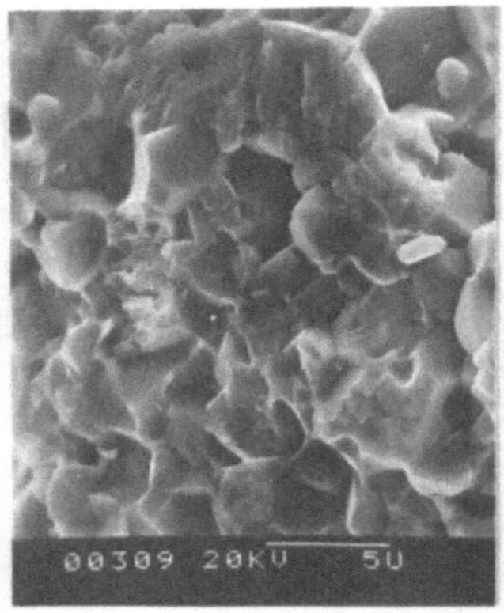

Fig.5 SEM micrograph of fractured surface of 12 mol.% CeO_2

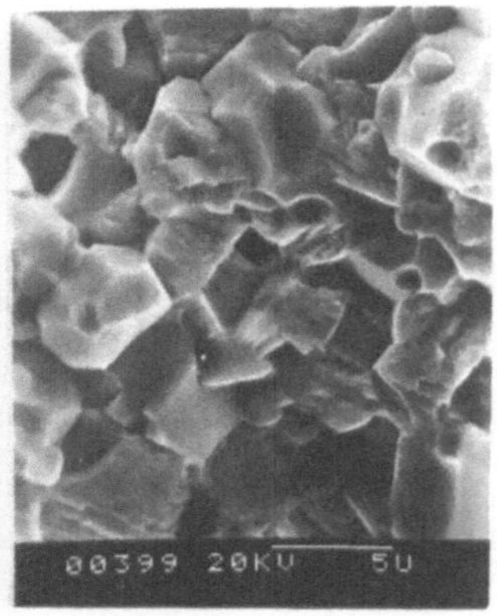

Fig.6 SEM micrograph of fractured surface of 14 mol.% CeO_2

Fig. 4, Fig. 5, and Fig. 6 are SEM micrographs of fracture surfaces of the samples which contained different CeO_2 content and were sintered at 1530°C for 2 hrs. From these figures, it can be found that intergranular fracture was main fracture mode for CeO_2 content being 10 mol% with transgranular fracture increasing with increasing CeO_2 content. For the samples containing 14 mol.% CeO_2, the proportion of transgranula fracture was less.

2. Effect of Al_2O_3 content on K_c and σ_f of Ce-TZP

The experimental results show no conclusive trends for the effect of Al_2O_3 on strength or toughness over the range studied, although it is well known that Al_2O_3 has a substantial effect when varied over larger ranges. When Al_2O_3 is uniformly distributed in t-ZrO_2 matrix, the Al_2O_3 grains are under slight compression. A.G. Evans [9] proposed that the two mechanisms, phase transformation toughening and crack deflection and branching toughening could occur simultaneously.It is generally recognized that alumina additions to TZP matrices increase strength and decrease toughness due to their effect on ZrO_2 grain size. It is, however, also possible to increase both the strength and toughness of Ce-TZP with Al_2O_3 additions [10]. The increase in toughness is again related to ZrO_2 grain size since there is an optimum grain size for transformation. The crack deflection and branching results in the increase of fracture surface energy and the volume fraction of transforming t-ZrO_2. For this reason, the fracture toughness of Ce-TZP increases.

3. Effect of TiO_2 content on K_c and σ_f of Ce-TZP

The sintering behavior of Ce-TZP can be improved by addition of small amount of TiO_2 which form a grain boundary liquid phase with Al_2O_3 and SiO_2 (a impurity existed in raw materials) during sintering and this liquid phase speeds up diffusing and sintering process. But with the increase of TiO_2 content the K_c of Ce-TZP appeared to decrease. The chemical valence of Ti element is varible (trivalent or tetravalent). As a result, under high temperature and a reducing atmosphere (inside the ZrO_2 body), TiO_2 is reduced to Ti_2O_3 according to the following fomula:

$$TiO_2 \longrightarrow (Ti^{+4}{}_{1-2x}\ \ Ti^{+3}{}_{2x})\ O_{2-x}V_x + x/2\ O_2$$

or $$2TiO_2 \longrightarrow Ti_2O_3 + 1/2\ O_2$$

V_x = oxygen ion vacancy

The phenomenon of oxygen losing from TiO_2 is clear when the temperature increase above 1380°C. The higher the temperature is, the more the oxygen losses. For this reason, the TiO_2 content should be minimized. Small amount of TiO_2 additions is needed for improvement of sintering.

CONCLUSIONS

In this experimental scope, CeO_2 10-14 mol%, Al_2O_3 9-13 wt%, TiO_2 1-3 wt%, and sintering temperature 1510-1550°C:

1. As expected, CeO_2 contents near 12 mol.% resulted in high toughness and strength, whereas 10 mol.% and 14 mol.% CeO_2 materials had less of tendency to transform.

2. The variation in strength and toughness for small changes in Al_2O_3 content (9-13 wt.%) were insignificant, as expected.

3. When TiO_2 is added to Ce-TZP, sintering may be enhanced, but there was no improvement seen in mechanical properties. Due to the possibility of TiO_2 reduction or aluminium titanate formation, excessive additions of TiO_2 are not recommended.

4. There is an optimum CeO_2 content for its influence on σ_f and K_c, being 12 mol% in this experiment and 12.45 mol% according to the forecasting equation.

5. The experimental results show that Ce-TZP with good mechanical performance can be prepared by using technical ZrO_2. The maximum K_c value and σ_f value are 746 MPa and 7.8 MPa$\sqrt{m}$ respectively when CeO_2=12 mol%, Al_2O_3=13 wt%, TiO_2=1 wt% and sintering temperature=1530°C.

6. On the basis of forecasting equation, for the experimental design used, the optimum processing parameters are:

CeO_2	= 12.45 mol%
Al_2O_3	= 13 wt%
TiO_2	= 1 wt%
sintering temperature	= 1535°C

REFERENCES

1. R.C. Gravie, R.H.J. Hannink and R.T.Pascoe, "Ceramic Steel?", Nature, 1975, 258 [12] 703-704.

2. Toyo Soda, pp.155-163 in Advanced Structural Ceramics, Vol. 78, edited by P. F. Becher, M. V. Swain and S. Somiya, MRS (Pittsburgh, PA),1987.

3. S. G. Yang and K. Chen, "Preparation of Ce-TZP and Its Mechanical Behaviour", pp. 904-911 in Proceedings of the 3rd Intl. Symp. on Ceramic Materials and Components for Engines, edited by V. J. Tennery, Las Vegas, NV, 1988.

4. Liangchao Jin, Orthogonal Test Design and Multi-Index Analysis (Chinese press), The Railways Press, Beijing, 1983.

5. K. Niihara, R. Morena and D.P.H. Hasselman, "Evaluation of K_{IC} of Brittle Solids by the Indentation Method with Low Crack-to-Indent Ratios", J. Mater. Sci. Lett., 1982, 1 [1] 13-16.

6. R.C. Gravie and P.S. Nicholson, "Phase Analysis in Zirconia Systems", J. Am. Ceram. Soc., 1972, 55 [6] 303-305.

7. Shao-Tang ZHAO, " Advanced Orthogonal Test Design Data Manipulation", Intern. Conf. on EPMESC III, Aug. 1990, Macao. Proceedings Vol. 6, pp. 1251-1260.

8. Shao-Tang ZHAO, " Application of 3-Dimension Graphics Softwares in Ceramic Materials Research ", Proceedings on Structural Ceramics 90' Annual Conference, pp. 299-300, Oct 1990, Xianyang, P.R. China.

9. A.G. Evans, p. 193 in Advanced in Ceramics Vol. 12, Ed. N. Claussen, M. Ruhle and A.H. Heuer, Am. Ceram. Soc., Columbus, OH, 1984.

10. T. Li, Z. Shen and Z. Ding, "The Mechanical properties of ZrO_2-Al_2O_3 Ceramic Composites", J. Chinese Ceram. Soc., 1990, 18 [1] 39-46.

SUPERPLASTIC DEFORMATION OF α'-SiAlON

J. Gerretsen, J.P.G.M. van Eijk, Centre for Technical Ceramics-TNO,
P.O.Box 595, 5600 AN Eindhoven, The Netherlands.
P.R. Put, Institute of Higher Professional Education Eindhoven, The Netherlands

ABSTRACT

Recent developments show that covalent ceramic materials such as silicon nitride may have superplastic forming (SPF) properties [1,2]. SPF allows large strains to be achieved under a low flow stress, with great potential as an industrial shaping technique. α'-Sialon is a promising material for SPF because of the presence of a transient liquid phase during sintering and its relatively stable grain size. In this paper, the results of an experimental investigation of the deformation behaviour of α'-sialon in compression will be presented. Samples are prepared by reaction sintering of a mixture of Al_2O_3, AlN, Si_3N_4 and Y_2O_3 at 1600 °C, in a 0.1 MPa nitrogen atmosphere, giving a density of 3.15×10^3 kg/m^3 (96 % TD). Samples are deformed in a graphite-based hot press, under a constant uniaxial stress of 30 to 100 MPa, at temperatures ranging from 1450 to 1600 °C. Upon deformation the material shows a moderate degree of work hardening with a strain-rate proportional to $e^{-0.37\pm0.06}$. Variation in the degree of work hardening from sample to sample results in a large spread in the stress sensitivity. Assuming a power law type constitutive creep equation and taking the strain-rate at 0.5 % strain, the stress sensitivity (n) was found to be 0.94 at 1500 °C, 0.56 at 1550 °C and at 1600 °C, n = 0.2. The activation energy for creep is 900 kJ/mol and correlates with the activation energy for Y^{3+} diffusion which may imply that the pressure solution accommodation mechanism is controlled by the transport of this cation. SEM observations indicate show that virtually no grain growth occurs and that the grain shape remains equiaxed.

INTRODUCTION

Superplastic deformation is the ability of a material to undergo high degrees of straining without mechanical failure. Metals and alloys have been known for a long time to behave superplastically and superplastic forming (SPF) as well as superplastic diffusion bonding are nowadays used as an industrial manufacturing technique for a range of applications as for instance in the aerospace industry. Ceramic materials are in general considered to be very resistant to plastic deformation up to very high temperatures. However, recent investigations have shown that under certain conditions oxide ceramics such as alumina, zirconia [3] and even covalent materials such as silicon nitride-silicon carbide composites [1] and sialon [2] can be deformed up to large strains. SPF offers the advantage that components can be made to near net shape, employing various forming methods such as extrusion, bulging, bending, blowing and shell forming. Provided high deformation-rates can be achieved at relatively low temperatures, SPF offers a flexible shaping technique combined with high productivity, low energy consumption and low tooling costs [4], with the scope for large scale implementation as a production technique in the ceramics industry.

The progress in experimental investigations of superplasticity of ceramics has been recently reviewed in a number of publications [2,5] and the material properties and microstructural requirements for superplastic behavior of ceramics are found to be of a similar nature as for metals and alloys: small grain size (< 1μm), equiaxed grains, no extensive grain growth, a high strain-rate sensitivity ($0.5<m<1$) and high diffusivity (lattice diffusion, grain boundary diffusion with or without the presence of a liquid phase). Most of these prerequisites for SPF are met by α'-sialon. Of special interest to SPF is the transient nature of the grain boundary phase present in the a'-sialons. The formation of an intermediate oxynitride liquid which is gradually absorbed into the crystal lattice of the final product not only promises better high temperature properties after fabrication, but the transient liquid may also provide a fast diffusion path for material transport along the grain boundaries during SPF.

In this paper, the results of an experimental investigation into the deformation behavior of α'-sialon in compression creep experiments will be presented, the deformation behaviour will be discussed in terms of a power law type, constitutive creep equation and the microstructural development will be illustrated by SEM micrographs of fracture surfaces.

MATERIAL AND METHODS

Cylindrically shaped α'-Sialon samples were prepared by uniaxial pre-pressing followed by isostatic pressing of a mixture of 72.01 wt% silicon nitride (UBE SNE-10), 2.55 wt% alumina (AKP30), 16.01 wt% aluminum nitride (Starck, Grade C) and 9.43 wt% yttria (Starck Finest Grade) + processing additives. Samples, embedded in a BN powder mixture were sintered in a graphite based, gas-pressure sinter furnace in a nitrogen atmosphere ($P = 0.5$ MPa) using a heating and cooling rate of 10 °C/min up to a maximum temperature of 1600 °C. The shrinkage behavior was measured by dilatometry. After sintering samples were obtained with an average density of 3.15×10^3 kg/m^3 (Archimedes method). Weight loss was limited to an average of 0.06 g. The sintered materials (see Figure 1) consisted of equiaxed grains with hexagonal cross sections and an average diameter of 0.5 μm. Fairly large pores of approximately equivalent size as the grains have been developed. In order to investigate the effect of density on the creep behaviour, samples were also sintered with a 10 MPa nitrogen gas pressure and a soak of 30 minutes. The density obtained for these samples is 3.3×10^3 kg/m^3. However, for most of the experiments reported here, preference was given to samples with a short heat-treatment in order to ensure the presence of a liquid grain boundary phase thought to be required for SPF.

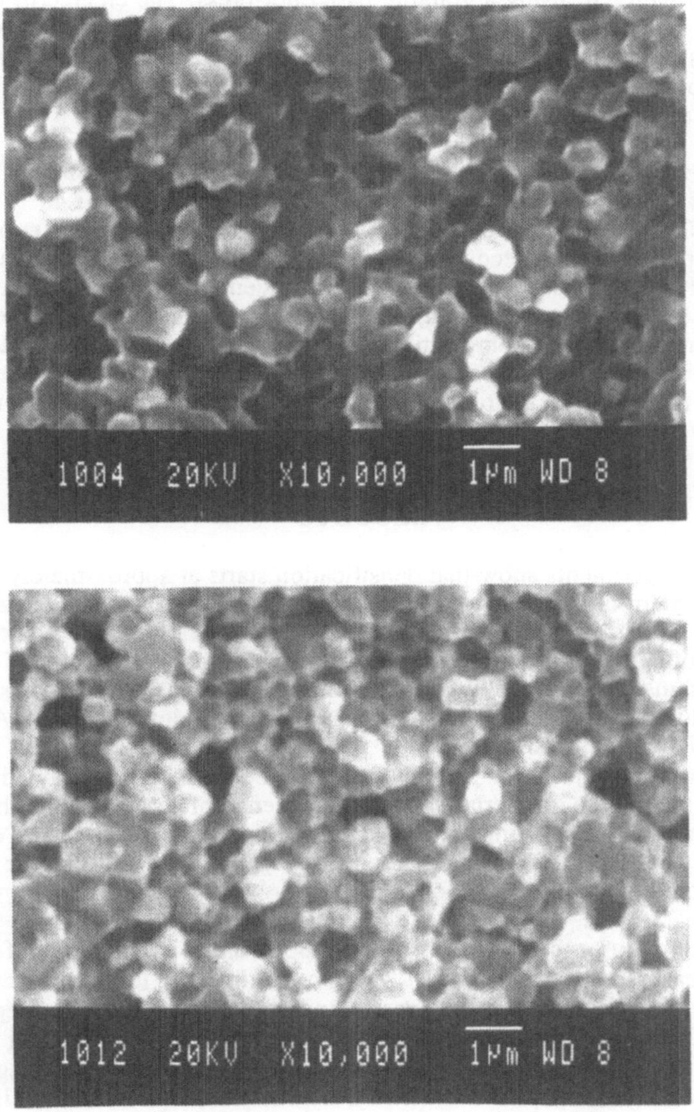

Figure 1. SEM micrographs of a fracture surface of α'-sialon a) sintered (top); b) deformed at 1500 °C and 40 MPa (bottom).

Uniaxial compression creep deformation experiments were carried out in a 0.1 MPa nitrogen gas, using a graphite based hot-press. The change in length of the sample during the creep experiments was obtained from an externally mounted displacement transducer. The temperature was controlled with a B/BC thermocouple. Load, temperature and displacement were simultaneously recorded as a function of time.

Samples were deformed at temperatures ranging from 1450 to 1600 °C (1450, 1475, 1500, 1550,1600 °C), under a stress ranging from 40 to 80 MPa (30,40,60,80,100 MPa). One sample was deformed in a stress, temperature stepping experiment as a reconnaissance run. True strain and strain-rates were calculated from digitized recordings of displacement vs. time curves. After recovery of the samples the length, diameter, weight and density in the relaxed state were measured. The composition and microstructure of the as-sintered and of two deformed samples were analyzed by XRD on the content of the starting substances and various sialon phases and by SEM on grain size and grain morphology respectively.

RESULTS

Dilatometry experiments show that densification starts at approximately 1250 °C. A first maximum in the sintering rate, probably related to rearrangement occurs at 1300 °C. Most of the densification occurs around 1500 °C, related to the dissolution of silicon nitride into the oxide eutectic liquid and precipitation of α'-sialon. This maximum is followed by a third and slightly lower maximum at 1562 °C, which is probably related to a maximum in the intermediate YAG phase ($3Y_2O_3.5Al_2O_3$) concentration. These observations are in agreement with the results from Cao [6] for α'-sialon of similar composition. X-ray diffraction analysis of the sintered material shows the presence of the α'-sialon phase only and no traces of the starting materials or intermediate YAG phase could be detected. Within the error of this method (± 5%) it can be said that the material consists of predominantly α'- sialon.

At temperatures below 1500 °C no appreciable amount of deformation could be measured during loading at 80 MPa for times up to 20 minutes (strain-rate << 5×10^{-7} s^{-1}). For temperatures at and above 1500 °C, Ln(creep rate) is calculated from the true strain ($\ln(1+ \partial l/Lo)$) vs time curves obtained from the creep experiments (see Figure 2). During deformation, not only the length of the specimens changed but also the density, as shown in Table 1. No significant weight loss was observed. From Table 1 it can be seen that at high stress (80 MPa) and high temperature (1550 °C and 1600 °C) the density decreases whereas in all other deformation experiments the density increases. Cao [9] investigated the isothermal shrinkage behaviour of α'-sialon without an externally applied differential stress and found for material of a similar composition that densification can be described by the equation $\Delta l/lo = A.\ t^b$, with b = 0.09. This implies that any contribution from the 'non-stressed' densification mechanisms operating (i.e. surface energy driven) can be neglected and that the observed changes in density are predominantly due to the externally applied stress.

TABLE 1
Relative density change in % as a result of deformation.

Temp [°C]	40 MPa	60 MPa	80 MPa
1500	2.75	0.77	-0.21
1550	1.93	0.74	-0.79
1600	1.56	1.75	+1.47

From Figure 2, it follows that the creep-rate is strongly temperature dependent and that variations in the stress do not lead to large changes in the creep-rate. Furthermore, it can be observed that all samples show a more or less similar degree of work hardening. From Figure 2 it can be seen that the strain-rate initially remains approximately constant. After an incubation time (t_i) the strain-rate starts to decrease linearly with log(t), following the relationship de/dt = t^m. The incubation time was found to be independent of the stress, but it decreases with increasing temperature (t_i(1500°C) = 5000 s, at 1550 t_i = 400 s and at 1600 t_i = 250 s). The power law exponent m ranges from -0.2 to -0.38, with an average of -0.27±0.05, m being slightly lower at low stress and low temperature.

In order to obtain an impression of the stress sensitivity of the deformation behaviour, the phenomenon of work hardening should be taken into account. Following [7] and assuming a power law-type creep equation the stress sensitivity can be derived from the integrated expression of the strain-rate:

$$\dot{\varepsilon}(t) = \frac{B}{p+1} . \sigma^{\frac{n}{p+1}} . t^{\frac{-p}{p+1}} \qquad (1)$$

From Figure 2, using equation 1, it follows that m = -p/p+1= -0.27, or p = 0.37. This implies that the stress sensitivity (n) is a factor 1.37 times larger than the stress exponent obtained from Figure 2. However, as the rate of work hardening varies during the experiment and also varies from one experiment to another the strain-rate does not appear to follow a straight forward power law dependence to the applied stress. At 1600°C for instance, when increasing the stress from 60 to 80 MPa the strain-rate increases, whereas from 40 to 60 MPa the strain-rate decreases again, implying a negative stress sensitivity. The maximum stress sensitivity (n) calculated from Figure 2 is 1.8 (at 1550 °C and maximum strain). This complicated creep behaviour may be due to a competition between more than one deformation mechanism. This problem will be addressed in more detail in the discussion. Taking the strain-rate at 0.5 % strain the stress sensitivity (n) was found to be 0.94 at 1500 °C, 0.56 at 1550 °C and at 1600 °C n = 0.2. This very low stress sensitivity is confirmed by additional creep experiments at a temperature of 1600 °C on fully densified samples over a wider stress range from 30 to 100 MPa. As a result of deformation the density of these samples increased by 0.1 % to a value of 3.30 x 10^{+3} kg/m^3.

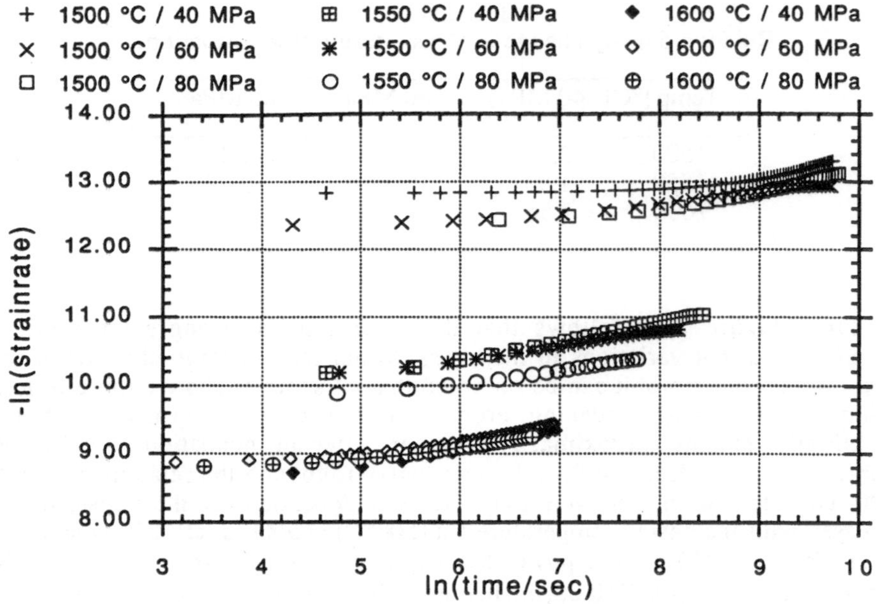

Figure 2. Ln(creep rate) curves of α'-sialon as a function of Ln(time) at various temperatures and stresses as indicated.

From a log(strain-rate) vs 1/T plot the activation energy was calculated of approximately 1200 kJ/mol at low T and 800 kJ/mol at the high temperature interval from 1550 to 1600 °C. However, considering the complex relationship between applied stress and strain-rate and the limited number of data points, such an interpretation should be made with some precaution. The average activation energy calculated from all data points give an activation energy of the order of 900 kJ.

Figure 1b shows the microstructure of a deformed samples as obtained from scanning electron microscopy of a fractured sample. From this micrographs it can be seen that in comparison with the undeformed material (Figure 1a) no significant modifications have occurred. All SEM observations show submicron, more or less equiaxed grains, and cavities of the same order as the sialon grains. From other SEM observations it appears however, that the grain size distribution in the high temperature sample is somewhat larger then compared with the low T deformed sample. Furthermore, the grain boundaries at high T are more diffuse as compared to the low T sample. However, it can be concluded that the changes are minor and more detailed analysis would be required to elucidate any substantial changes. X-ray analysis shows that in the as-sintered material no YAG phase is present, at least not in a crystalline form. After deformation at 1500 °C, the YAG phase has increased. The XRD analysis of the sample deformed at 1600 °C does show some evidence for the reappearance of the YAG phase but the data is not conclusive.

DISCUSSION

Within the range of the values of temperature and stress investigated, the α'-sialon material shows a complex deformation behaviour. Both the stress sensitivity and the temperature dependence change considerably with applied stress and strain. However, combining the measured density changes with the strain-rate in relation to the applied stress and temperature for all experiments, it becomes apparent that those samples which have been deformed at high stress, particularly at high temperature, deform at a lower strain rate then expected on the basis of the lower stress experiments and, that these conditions of applied load and temperature can also be correlated with a decrease in density. It can thus be concluded that creep by cavitation is active and that the deformation behaviour reflects a combination of stable grain boundary sliding (GBS) and cavitation creep. When cavitation occurs, part of the work done by shortening the sample (W = stress x strain x V_0) is used for the expansion of the sample (stress x dV) and the sample can be expected to deform at a lower rate. The complete set of experimental results thus appears to reflect the partitioning of two deformation mechanisms controlled by the presence or absence of cavitation.

So far, no data for the deformation behaviour of α'-sialon in this temperature range has been published and the results cannot be compared directly with other work. Karutnaratne and Lewis [8] investigated the creep behaviour of β'-sialon with MgO and Mn_3O_4 as sintering aids in the range from 1200 to 1400 °C and found a non-integral stress exponent of 1.5 for the MgO/Mn_3O_4 doped material and 1 for the MgO doped material. Cavitation was observed to occur only in the MgO/Mn_3O_4 doped material. The activation energy for the non-cavitating Mg-sialon at 1400 °C was found to be 830 kJ/mol. Of the same order as was found here for the temperature dependence of the α'-sialon creep behavior. Sato et al. [9] investigated the high temperature oxidation of silicon nitride based ceramics by water vapor. On the basis of their results they argue that the oxidation rate, characterized by an activation energy of 800 kJ/mol, was controlled by the rate of outward diffusion of cations such as Mg^{2+} and Y^{3+} through the unoxidized substrate layer into the oxide film. In the case of SPF of α'-sialon, in the presence of a grain boundary liquid phase, the rate controlling shape accommodation mechanism is very likely to be pressure solution. This mechanism involves (1) dissolution of the α'-sialon into the grain boundary liquid phase, either in the form of sialon or in dissociated form ($3Y_2O_3.5Al_2O_3$, AlN and Si_3N_4), (2) transport of the solutes through the grain boundary followed by (3) deposition of the α'-sialon in unstressed regions. The correlation of the activation energy for creep and Y^{3+} transport may imply that the pressure solution accommodation mechanism is controlled by the transport of this cation. This hypothesis is further supported by the observation that the intermediate YAG phase, which after sintering was not present as a crystalline phase and probably taken up in the α'-sialon, returned as a result of deformation.

CONCLUSIONS

The results of the experiments on the compressive creep behaviour of α'-sialon at high temperature show that high strain-rates up to $10^{-4}\ s^{-1}$ can be obtained at relatively low stresses. The low stress sensitivity ($0.2 < n < 1$) in combination with a moderate degree of work hardening predict a high deformation stability. Virtually no grain growth was observed and the grain shape remains equiaxed and superplastic forming capabilities may be expected provided cavitation can be prevented. The high value for the activation energy (900 kJ/mol) for creep implies that the material can be given shape relatively easy at high temperature, whereas at lower temperatures (< 1500 °C) plasticity is reduced and hence the load bearing capacity increased. The observations suggest that the YAG phase does not necessarily have to be present as a grain boundary phase prior to deformation but may be formed as a result of the applied stress during SPF due to dissociation of the α'-sialon. At present, experiments are conducted to investigate the creep behavior of α'-sialon with a higher density, over a wider stress range. Future experimental work will be directed towards extrusion of α'-sialon components with a relatively simple geometry and attention will be given to the nature of the grain boundary phase, influence of grain size, alternative additives and stress states with the objective to increase the maximum obtainable strain-rate.

REFERENCES

1. Wakai, F., Komoda, Y., Skaguchi, S. Murayama, N., Izaki, K. and Niihara, K. A superplastic covalent crystal composite. Nature, 1990, 334, 421-3.

2. Chen, I-W and Xue, L.A. Development of Superplastic Structural Ceramics. 1990, J. Am. Ceram. Soc., 73, 2585-2609.

3. Carry, C. High ductilities, superplastic behaviors and associated mechanisms in fine grained ceramics. In Superplasticity, eds. Kobayashi, M and Wakai, F. Proc. Int. Meet. Adv. Mat., Mat .Res. Soc 1989, 7,199-215.

4. Wu, X and Chen, I-W. Superplastic bulging of fine grained zirconia. J. Am. Ceram. Soc., 1990, 73, 746-49.

5. Maehara, Y. and Langdon, T.G. Review: Superplasticity in Ceramics. J. Mat. Sc. 1990, 25, 2275-2286.

6. Cao, G-Z. Preparation and Characterization of α'-sialon ceramics. PhD Thesis, Eindhoven University of Technology, The Netherlands, 1991.

7. Munz, D. and Fett, T. Mechanisches Verhalten keramischer Wirkstoffe. Werkstoff-Forschung und Technik 8, 1989. Springer-Verlag, Berlin.

8. Karutnaratne, B.S.B. and Lewis, M.H. High -temperature fracture and diffusional defromation mechanisms in Si-Al-O-N ceramics. J. Mat. Sc., 1980, 15, 449-62.

9. Sato, T. Haryu, K., Endo, T. and Shimada, M. High-temperature oxidation of silicon nitride-based ceramics by water vapor. J. Mat. Sc., 1987, 22, 2635-40.

DETERMINATION OF SUBCRITICAL CRACK GROWTH THRESHOLD IN CERAMICS WITH INDENTATION FLAWS

J.H.Gong and Z.D.Guan
Department of Materials Science and Engineering
Tsinghua University, Beijing 100084, P. R. China

ABSTRACT

A technique for determining high temperature subcritical crack growth threshold in ceramics with indentation flaws is presented. This technique requires prior knowledge of the residual stress field about the indentation. Detailed analysis shows that the residual stress is the only driving force for crack growth of indentation flaws in absence of any applied stress. Under certain conditions, it can cause a spontaneous extension or healing of the indentation flaw. Making the duration from just indented to measurement of crack dimension long enough, an equilibrium configuration for the indentation flaw system may be obtained and the threshold can be calculated with parameters which are used to characterize this configuration. This technique is then applied to a pressureless sintered Si_3N_4. It is shown that this technique may be a simple and reliable method for the determination of subcritical crack growth threshold of ceramics at elevated temperatures.

INTRODUCTION

One of the properties of ceramics that limit their use as structural materials at high temperatures is the steady extension of preexisting flaws, i.e., subcritical crack growth which (SCG) occurs at stress intensity less than the critical value, K_{IC}. The dependence of crack velocity, v, on the stress intensity, K_I, has been established. It is generally noted that the crack velocity will tend to zero when the stress intensity drops to a certain value, which is described as subcritical crack growth threshold [1-4]. The engineering signifinance of the SCG threshold is recognized as a design criterion for high performance ceramics since it allows the calculation of the applied stress below which the material do not exhibit subcritical crack growth [2]. So it is necessary to devise a simple method for rapid and accurate evaluation of SCG threshold.

The present study explores the possibility of a technique for estimating the SCG threshold in ceramics by means of crack dimension measurements during the spontaneous extension or healing of indentation flaws at various temperatures. The theoretical basis for the technique is discussed in detail. A pressureless sintered silicon nitride, some of whose indentation fracture mechanical parameters have been measured in our previous work [5], is chosen as the material for study.

THEORETICAL CONSIDERATION

By means of microhardness testing, a permanent indentation is introduced into the surface of a mechanical test specimen. Meanwhile, an effective wedge opening force generated by the indenter acts at the deformation zone and results in the formation of a small surface flaw beneath the indentation. This indentation flaw is usually used as an artificial flaw to simulate the natural flaws preexisting in materials.

As shown by Marshall and Lawn [6], the stress field about a indentation flaw is elastic–plastic and a residual stress, which results from mismatch between the plastic zone beneath indentation and the surrounding matrix, remains after indenter removal. With the indentation fracture mechanical analysis, the residual stress intensity, K_r, can be expressed by [7]:

$$K_r = \frac{\chi_r P}{c^{3/2}} \tag{1}$$

where P is indentation load; c is crack half–length; χ_r is a constant dependent on indenter geometry and material properties but assumed independent of indentation load.

The residual stress is the only driving force for the growth of an indentation flaw in absence of any applied loads. Under certain conditions, it can cause an extension of indentation flaw or impede the crack healing.

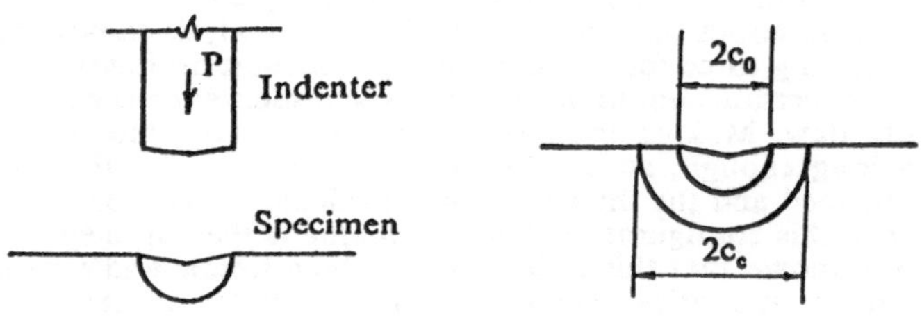

(A) (B)

Fig.1 Schematic of Knoop Indentation Flaw System:
(A) Initial State; (B) A Spontaneous Extension Occurs at R.T.

Extension of Indentation Flaws at Room Temperature

Figure 1 shows a schematic diagram of the crack evolution during a Knoop indentation test. In stage (A), a Knoop indenter with peak load, P, introduces a surface flaw into specimen. According to Eq.(1), the residual stress intensity of this surface crack, K_{r0}, can be written as:

$$K_{r0} = \left(\frac{\chi_0 P}{c_0^2}\right)\sqrt{c_0} \tag{2}$$

where c_0 is initial crack dimension after indentation.

Although the diagonal of the Knoop indentation and crack length of the Knoop indentation–induced crack result from different physical processes, it has been reported by Petrovic [7] that actual crack dimension is close to the diagonal length only happened at a lower indentation load range (<70N). In our previous works [8, 9], a constant value of P/c^2, instead of $P/c^{3/2}$, was also obtained for Knoop cracks at the similar load range (<50N). It seems to say that the Knoop diagonal is approximately equal to crack length and can be used in the lower indentation load range.

Since P/c_0^2 can be treated as a constant, Eq.(2) shows that K_{r0} increases with $\sqrt{c_0}$, thereby with the indentation load. So it can be expected that a critical indentation load exists above which K_{r0} will be larger than SCG threshold, K_{th}, then a spontaneous extension of indentation flaw occurs. This case is shown as Figure 1(B).

Such a spontaneous extension causes the indentation flaw to extend, then the residual stress intensity, K_r, diminishes, in accordance with Eq.(1). When K_r drops to a mininum, K_{th}, the crack extension will stop and crack system shown in Figure 1(B) attains to a postindentation equilibrium configuration. Denoting $c=c_e$ as ultimate crack dimension appropriate to this stable state, we obtain:

$$K_{th} = K_{re} = \frac{\chi_r P}{c_e^{3/2}} \qquad (3)$$

This is our basic equation for the determination of SCG threshold.

It is clear that there are two prerequests for the use of Eq.(3). First, the indentation flaw must have a spontaneous extension. If there is no extension of indentation flaw after indenter removal, Eq.(3) is no longer valid. In this case, the crack dimension may be close to the diagonal of indentation, i.e., it is the value of P/c_0^2 which is constant. The second prerequest for using Eq.(3) is that the value of $P/c_e^{3/2}$ for indentation flaws produced with different loads must be a constant. In general, this requirement can be satisfied by making the duration between completion of indentation test and measurement of crack dimension long enough, i.e., making sure that the velocity of extension has decreased to zero.

Healing of Indentation Flaw at Elevated Temperatures

A recent explanation for SCG threshold suggests that the net driving force for crack growth is more properly described by the sum of a force driving crack growth and a force impeding growth [3]. When the aiding and impeding influences become comparable and the net driving force for crack growth reduces to zero, the crack velocity will tend to zero and it reveals an apparent threshold. The force impeding crack growth arises from the surface interacting across the barely separated crack walls near the crack tip [3] and transmission electron microscopy observations has indicated that a strong closure force exists between crack surfaces separated by small distance and causes crack healing in a variety of materials [10].

In continuation of this idea, let us consider the equilibrium of the crack system shown as Figure 1 at elevated temperatures.

In preliminary experiments, crack healing is observed for indentation flaws in Si_3N_4 when temperature rises. This fact seems to indicate that the net driving force for crack growth is less than zero in this case. As a result of crack healing, however, the diminish in crack dimension causes a increase in K_r and a decrease in the driving force for crack healing. In other words, the net driving force for crack growth is increasing during crack healing. Thus, if keeping an indentation flaw at a given temperature for enough time, a new equilibrium of indentation crack system may be expected to attain. With the measurement of crack dimension appropriate to this equilibrium, we can calculate the SCG threshold using Eq.(3).

It can be seen easily from the discussion above that there are also two prerequests for the use of Eq.(3) at elevated temperatures: a constant value of $P/c_0^{3/2}$ and the completion of a spontaneous healing of indentation flaws.

EXPERIMENTAL PROCEDURE

A pressureless sintered Si_3N_4 is used as test material in this study. Specimens are in slab form with flat, parallel machined surface. A careful polishing is done, ultimately with M0.5# diamond paste, on one surface of specimens to produce an optical finish. Finally, all specimens are subjected to air annealing at 1100℃ for 30 min in order to eliminate the residual stress generated by machining.

By varying the indentation load (2.65 to 4.65Kg), several flaws of varying size are introduced in the polished surface of the specimen by Knoop indentation. The diagonal length of indentation on the surface of the specimen is measured carefully by optical microscopy at once and after 24h. at room temperature. The specimen containing indentation flaws is then heated to temperatures of 1100℃ and 1200℃, respectively, and maintained for durations of 30 min and 7h, respectively. Changes in crack dimensions are measured by optical microscopy after the specimen cools to room temperature. Before measurement, careful polishing is done with M0.5# diamond paste on the surface of specimen in order to remove the oxide layer formed at high temperatures.

RESULTS AND DISCUSSIONS

Indentation Load–Crack Length Relationship for Various Temperatures

The dimension of indentation flaws caused by different indentation loads at room temperature is given in Table I. No changes in crack dimension in delayed measurement are observed. That means the final crack configuration produced by Knoop indentation in Si_3N_4 can be maintained for at least 24 hour in air at room temperature. But this configuration is not an equilibrium state described as Eq.(3), for the results show an increase in $P/c^{3/2}$ with indentation load. A constant value of P/c^2 indicates that these indentation flaws have no extension or healing after indenter removal. So it is imposible to calculate the SCG threshold at room temperature using Eq.(3).

TABLE I
Knoop Indentation Flaw Dimensions Caused by Different Loads at Room Temperature

P(Kg)	2c(μm)		P/c^2 (MPa)	$P/c^{3/2}$ (MPa$\sqrt{m}$)
	Just Unloading	After 24 hours		
2.65	190.8±3.7	190.6±3.6	2853.48	27.87
3.15	208.0±2.8	208.6±3.2	2854.11	29.11
3.65	225.2±1.9	225.0±1.8	2821.25	29.94
4.15	237.8±0.8	237.4±0.8	2876.81	31.37
4.65	252.0±2.5	252.4±1.5	2870.37	32.22

Marshall has studied the relationship between indentation load and crack size for Knoop indentation flaws in Si_3N_4[11]. It is shown that a constant value of $P/c^{3/2}$ exists at higher load range (>200N) while an increasing tendency of $P/c^{3/2}$ at lower range. This is consistent with the present study.

Changes in crack dimension in delayed measurements at elevated temperatures are given in Table II. The spontaneous healing of indentation flaws is evident, for the crack dimension decreases and the value of $P/c^{3/2}$ increases with time. A duration of 7 hour for crack healing seems to be a sufficient time for indentation crack system to attain an equilibrium configuration since the results show that a constant $P/c^{3/2}$ exists which indicates that the indentation crack system is in stable state as described in section 2.

TABLE II
Crack Dimension as a Result of Healing at Elevated Temperature

Temp.	P(Kg)	30min after Indentation		7 hour later	
		2c(μm)	$P/c^{3/2}$(MPa$\sqrt{m}$)	2c(μm)	$P/c^{3/2}$(MPa$\sqrt{m}$)
1100℃	2.65	177.0±1.9	31.93	157.5±1.9	37.16
	3.15	195.3±2.1	31.98	176.5±1.9	37.24
	3.65	213.3±1.9	32.47	195.3±2.3	37.06
	4.15	229.2±2.6	33.16	211.0±1.6	37.54
	4.65	238.8±2.0	34.92	229.5±2.0	37.02
1200℃	2.65	180.4±2.3	30.12	136.6±2.3	46.01
	3.15	188.4±3.4	33.76	155.2±2.8	45.16
	3.65	202.8±1.6	38.24	169.8±1.7	45.73
	4.15	208.4±1.5	38.24	184.2±2.0	46.01
	4.65	216.4±2.1	40.49	202.2±1.8	44.90

Determination of SCG Threshold at Elevated Temperatures

A method for measuring residual stress parameter, χ_r, by means of strength test have been presented in our previous works [5, 9]. The value of χ_r, for the Knoop indentation flaw in the material used in the present study, is determined to be 0.0425 at 1100℃ and 0.0309 at 1200℃ [5]. The discussions above show that two prerequests for the use of Eq.(3) are statisfied at elevated temperatures. So the SCG threshold can be calculated with the experimental data listed in Table II. Table III gives the results.

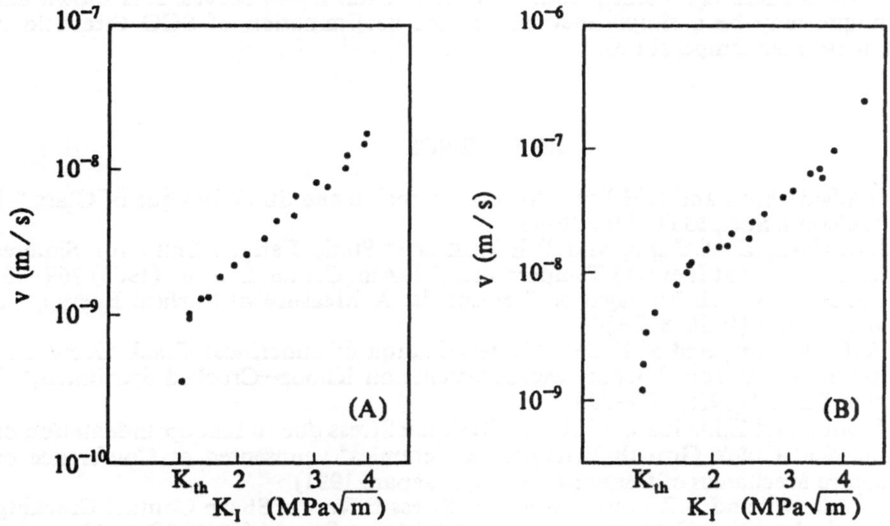

Figure 2 $K_I \sim v$ curves for SSN at (A)1100℃ and (B)1200℃

TABLE III
Comparison between Experimental Results of K_{th} $(MPa\sqrt{m})$

Temperature	This Study	From $K_I \sim v$ Curves [5]
1100℃	1.58	1.52
1200℃	1.41	1.39

The elevated temperatures SCG behavior of the present material has been studied by extending indentation flaws in specimens loaded in 3–point bending [5] and measured average velocity is plotted as a function of average stress intensity in Figure 2. From these experimental $K_I \sim v$ curves, the SCG threshold can be estimated directly and the results are also listed in Table III for comparison. It is seen that the technique presented in this study is an effective method for determining the SCG threshold at elevated temperatures, for the two series of experimental data show a close agreement.

SUMMARY

A technique for estimating SCG threshold in ceramics is presented. Detailed analysis shows that the residual stress about indentation is the only driving force for crack growth in absence of any applied stress. Under certain conditions, it can cause a spontaneous extension or healing of indentation flaw. Then a constant value of $P/c^{3/2}$ is obtained for indentation flaws produced by different loads and the threshold can be calculated.

This technique is applied to a pressureless sintered Si_3N_4 at various temperatures. No stable state existe for an indentation crack system at room temperature while a constant $P/c^{3/2}$ is evident for each flaw when temperature is elevated. The threshold is determined to be $1.58MPa\sqrt{m}$ at 1100℃ and $1.41MPa\sqrt{m}$ at 1200℃. These results are identical with the results obtained from the experimental $K_I \sim v$ curves. It is shown that this technique may be a simple method for the determination of SCG threshold in ceramics at elevated temperatures.

REFERENCE

〔1〕 S.M.Wiederhorn and L.H.Bolz, "Stress Corrosion and Static Fatigue of Glass," J. Am. Ceram. Soc., 53 (1970) 543–48

〔2〕 E.J.Minford, D.M.Kupp, and R.E.Tressler, "Static Fatigue Limit for Sintered Silicon Carbide at Elevated Temperatures," J. Am. Ceram. Soc., 66 (1983) 769–73

〔3〕 R.F.Cook, "Crack Propagation Thresholds: A Measure of Surface Energy," J. Mater. Res., 1 (1986) 852–60

〔4〕 T.Fett, D.Munz, and K.Keller, "Determination of Subcritical Crack Growth on Glass in Water from Lifetime Measurements on Knoop–Cracked Specimens," J. Mater. Sci., 23 (1988) 798–803

〔5〕 J.H.Gong and Z.D.Guan, "Effect of Residual Stress due to Knoop Indentation on Subcritical Crack Growth Behavior in Ceramics"; presented at Conference on Fracture Mechanics of Ceramics (Nagoya, Japan, 1991)

〔6〕 D.B.Marshall and B.R.Lawn, "Residual Stress Effects in Sharp Contact Cracking: Part I .Indentation Fracture Mechanics," J. Mater. Sci., 14 (1979) 2001–12

〔7〕 J.J.Petrovic, "Effect of Indenter Geometry on Controlled–Surface–Flaw Fracture Toughness," J. Am. Ceram. Soc., 66 (1983) 277–83

〔8〕 J.H.Gong and Z.D.Guan, "A Modified Indented Bending Beam Method for the

Determination of Fracture Toughness of Ceramics," J. Chinese Ceram. Soc., 19 (1991) 234–41

[9] Z.D.Guan, J.H.Gong, Y.Y.Zhao and X.F.Zhao, "The Characteristic Stress Between Two Kinds of Delayed Failure for Ceramics at Elevated Temperature"; this volume (B84).

[10] B.J.Hockey; pp.637–58 in Fracture Mechanics of Ceramics, Vol.5. Edited by R.C.Bradt, A.G.Evans, D.P.H.Hasselman, and F.F.Lange. Plenum, New York, 1983

[11] D.B.Marshall, "Controlled Flaws in Ceramics: A Comparison of Knoop and Vickers Indentation," J. Am. Ceram. Soc., 66 (1983)127–31

DELAYED FAILURE OF HOT ISOSTATIC PRESSED SILICON NITRIDE AT HIGH TEMPERATURES

ANDERS MICSKI & BILL BERGMAN
Dept of Physical Metallurgy and Ceramics
Royal Institute of Technology
100 44 STOCKHOLM, SWEDEN

ABSTRACT

The static fatigue behaviour of three series of HIP Si_3N_4 materials has been studied. Two series were produced from powder supplied by KemaNord Industrikemi (Ljungaverk, Sweden), one was alloyed with 2.5 w/o Y_2O_3, and the other one with 2.5 w/o Y_2O_3 + 1 w/o Al_2O_3. A third serie, produced from Si_3N_4 powder supplied by UBE Industries (Tokyo, Japan), was alloyed with 4 w/o Y_2O_3. Static fatigue testing was performed at temperatures from 1100°C to 1400°C, using stress levels ranging from 20 to 75% of the appropriate fast fracture high temperature strength. The creep rate was found to be significantly higher for the KemaNord material containing alumina. The susceptibility for slow crack growth was also much larger for this material compared to the materials without alumina additions.

INTRODUCTION

Silicon nitride is primarily being studied for use as a high-performance material, especially for applications where oxidation resistance, high strength and creep resistance are required. Numerous investigations have shown that the high-temperature behaviour of Si_3N_4-based materials are highly dependent on the amount and properties of the residual glass phase, formed by a reaction between sintering aid(s) and impurities in the starting powders. Hot isostatic pressing is a means of producing dense materials with low amounts of sintering aids, thus decreasing the amount of glass phase in the final product.

The high-temperature strength (1, 2) and oxidation behaviour (3) has previously been reported for HIP-Si_3N_4. This paper reports on the static fatigue behaviour of three different HIP-Si_3N_4 materials.

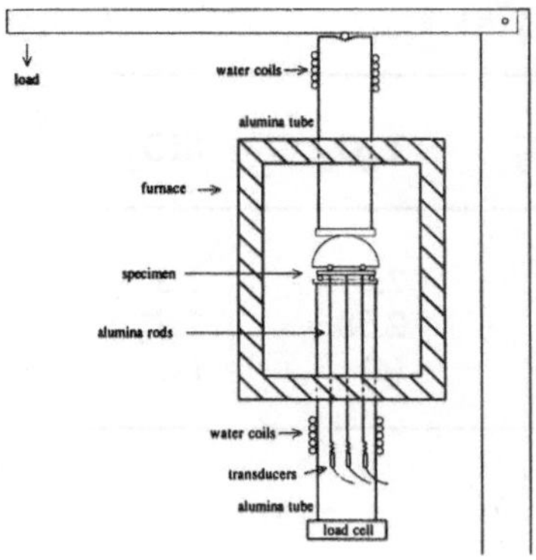

Figure 1. Experimental setup for the static fatigue experiments.

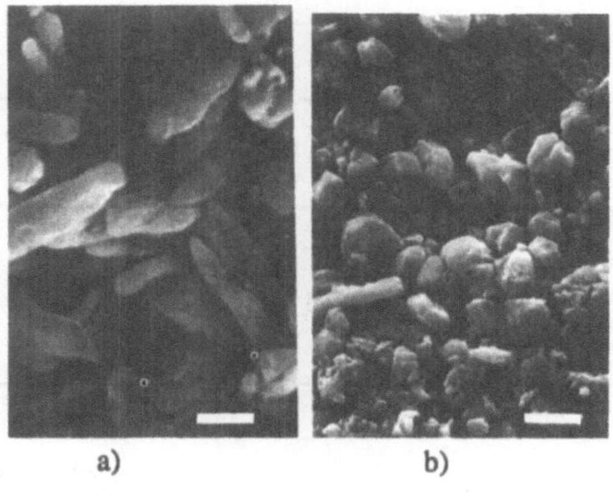

Figure 2. Microstructure of etched a) KN-material and b) UBE-material. The bars are 1μm.

MATERIALS AND METHODS

The static fatigue behaviour of three series of Si_3N_4 has been studied. Two series were produced from powder supplied by KemaNord Industrikemi (Ljungaverk, Sweden) (grade P95J), one was alloyed with $2.5\,^{w}/_{o}$ Y_2O_3 (denoted KN0), and the other one with $2.5\,^{w}/_{o}$ Y_2O_3 + $1\,^{w}/_{o}$ Al_2O_3 (denoted KN1). A third serie, produced from Si_3N_4 supplied by UBE Industries (Tokyo, Japan) (grades SN-E 10 (25 %) + SN-E 03 (75 %)), was alloyed with $4\,^{w}/_{o}$ Y_2O_3 (denoted U0).

Bend bars with the dimensions 3.5 x 4.5 x 50 mm^3 and 3.5 x 4.5 x 65 mm^3, respectively, for the KN-materials and the UBE material, respectively, were supplied by ABB Cerama in as-cut condition. The edges were chamfered in order to avoid fracture initiation from edge flaws introduced during cutting.

The static fatigue experiments were performed in an alumina four-point bending fixture (inner/outer span 19/38 mm, respectively). Dead weight loads ranging from 200 to 750 N were applied via a lever arm. The loads were continuously monitored by a load cell at the bottom of the system, see Fig. 1.

Hollenberger et al (4) were the first to present an analysis in which stresses and strains in creep deformed beam specimens could be calculated from bend test data. In the present study stresses and strains were calculated assuming a creep stress exponent of unity.

To avoid including creep of the alumina bend fixtures and tubes to the creep of the specimens, the deflection at the center of the beam was measured relative the deflection at the two inner load rolls, see Fig. 1. This deflection will be referred to as midpoint deflection. The strain can also be calculated by only measuring the deflection at either of the two inner load rolls, and will be referred to as load-point deflection. The deflection was measured by using three alumina rods in contact with the specimen, with one rod below each load roll and with one under the center of the specimen. The displacements were monitored by using three water-cooled linear voltage displacement transducers (LVDT), placed at the other end of the rods. When the creep specimens failed, the lever arm automatically switched of the furnace.

The fracure surface were examined in SEM (JEOL 840) and light optical microscope (LOM) in order to identify the fracture initiating flaw, and, when appropriate, the creep mechanism.

TABLE 1
Results from chemical analysis (w/o)

HIP material	Al	Y	O	Al_2O_3	Y_2O_3	SiO_2
KN0	-	1.98	2.24	-	2.51	3.2
KN1	0.58	2.03	2.59	1.10	2.58	2.87
U0 (estimated)			(2.3)		(4)	(4.3)

RESULTS

Chemical Composition and Phases

The results from X-ray diffraction of the KN-materials has previously been reported (2). The materials contain mainly β-Si_3N_4 with traces of Si_2N_2O. KN0 also contained traces of $Y_2Si_2O_7$. The UBE material, however, contained substantial amounts of α-Si_3N_4 together with β-Si_3N_4. The amount of α-Si_3N_4 was estimated to 37 w/o, following the procedure given by Käll (5), where the intensities from given α- and β-Si_3N_4 reflection from the Guinier-Hägg films are used. No corrections for absorption in the sample was made, resulting in an estimated relative error of 6%.

A chemical analysis of the Al, Y and O content of the KN-materials (2) has been made, see Table 1. All Al and Y are assumed to have combined with the oxygen to form Al_2O_3 and Y_2O_3, respectively, and the remaining oxygen is assumed to be present as SiO_2. This gives an estimate of the amount of amorphous intergranular phase present in the material. The amount of oxygen in the UBE material was estimated by using previous knowledge of Si_3N_4-materials of similar composition (6), see Table 1.

In order to characterize the microstructure the materials were etched in a mixture of molten NaOH and KOH for 30 s. The mean grain diameter for the KN-materials is 0.9 µm with an aspect ratio of 4, see Fig. 2 a. The mean grain diameter of the UBE-material was determined to 0.6 µm, and the majority of the grains found to be equiaxed, see Fig. 2 b.

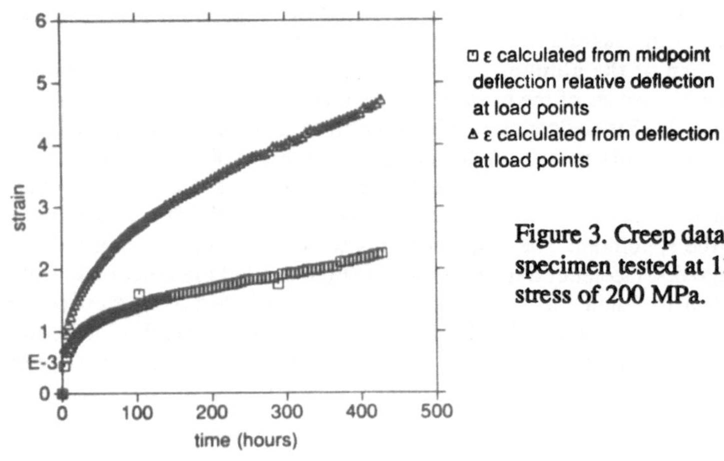

Figure 3. Creep data collected from a KN0 specimen tested at 1300°C and an applied stress of 200 MPa.

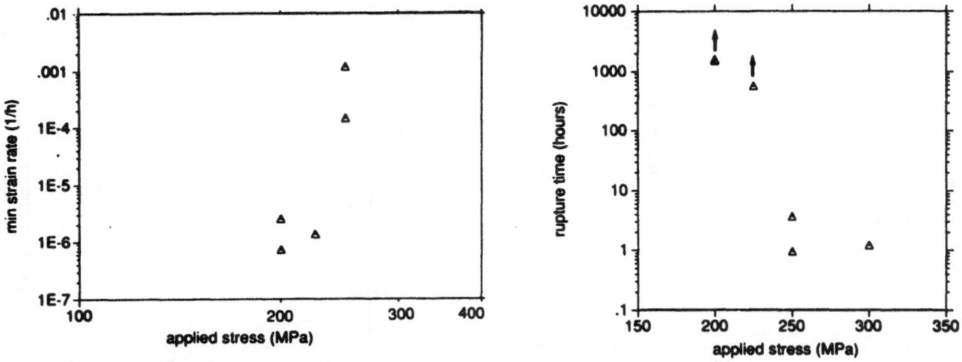

a) b)

Figure 4. Creep data for KN0 tested at 1300°C. a) $\dot{\varepsilon}_{min}$ vs. σ and b) t_f vs. σ.

Creep data

An example of creep data obtained in the present experiments is shown in Fig. 3. For comparision the creep curve obtained for the same specimen but where strain is calculated based on load point deflection only, is included in the same diagram. As expected, the curve calculated from midpoint deflection gives lower strain values compared to the curve calculated from load point deflection, in this case the difference being a factor two. This difference is attributed to creep of the test rig and creep of the part of the specimen situated between the inner and outer loading roll. The creep data presented in the following has all been evaluated by using the midpoint deflection.

Many specimens failed before steady-state conditions was reached, and for these the minimum strain rate is presented rather than the steady-state creep rate. Failure always occured without any indications of change in creep rate, suggesting that rapid crack growth abruptly inhibits the normal creep processes in these specimens, i.e. Stage 3 creep was never observed.

Considerable run-to-run scatter was observed, see Fig. 4-6 where minimum strain rate ($\dot{\varepsilon}_{min}$) and rupture time (t_f) are plotted as a function of applied stress (σ), respectively. This is

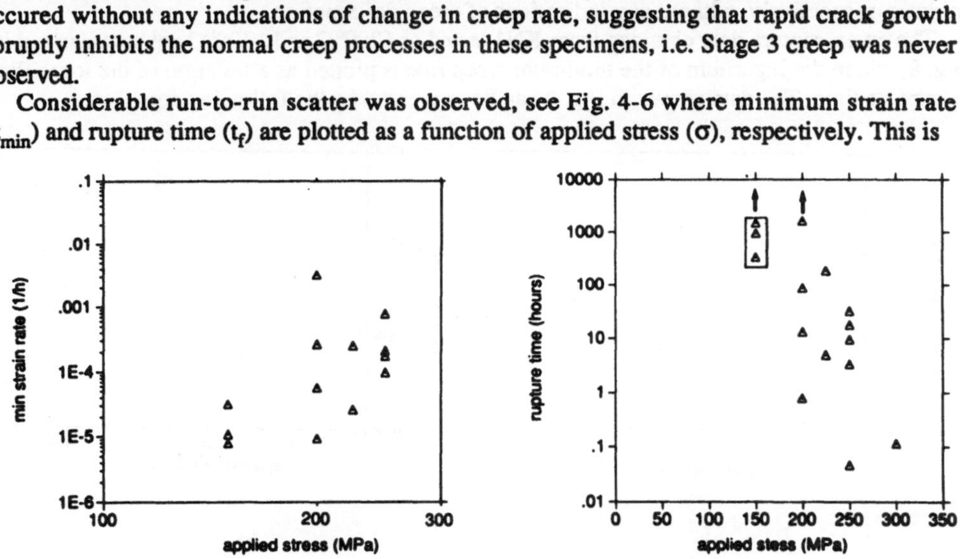

a) b)

Figure 5. Creep data for KN1 tested at 1200°C. a) $\dot{\varepsilon}_{min}$ vs. σ and b) t_f vs. σ.

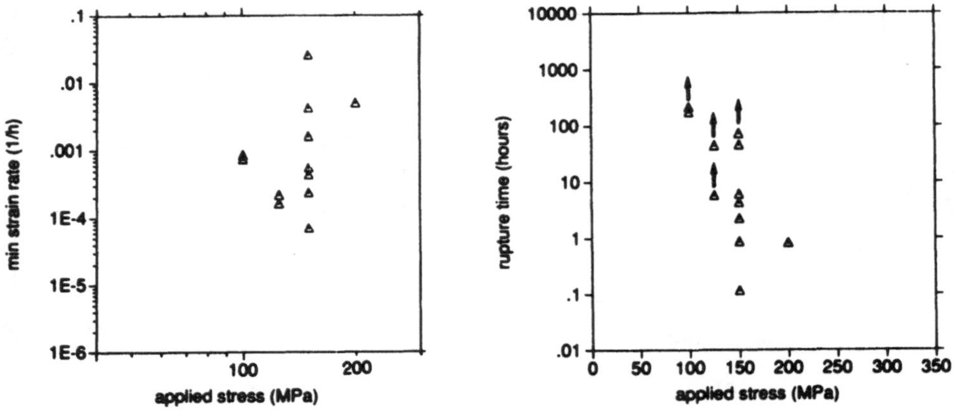

a) b)

Figure 6. Creep data for KN1 tested at 1300°C. a) $\dot{\varepsilon}_{min}$ vs. σ and b) t_f vs. σ.

done for KN0 at 1300°C, and KN1 at 1200°C and 1300°C. Specimens which did not fail during the testing period are marked with arrows. Only a few specimens failed during testing of U0 specimens why it is only appropriate to present $\dot{\varepsilon}_{min}$ plotted as a function of σ. Testing was performed at 1220°C and 1350°C, see Fig. 7 a and b.

Three KN0 specimens were creep tested at 1200°C, at 150, 250 and 300 MPa, respectively. They all survived throughout the testing period, which was between 292 and 394 h. However, in neither case was it possible to determine any creep. Furthermore, two KN0 specimens were creep tested at 1400°C, using a stress of 150 MPa. These specimens failed after 3 and 15 minutes, respectively, without any indications of creep deformation.

The creep-rupture data obtained from KN1 tested at 1200°C and 1300°C are summarized in Fig. 8, where the logarithm of the minimum creep rate is plotted as a function of the logarithm of rupture time. The data shown in Fig. 8 can be expressed by the following equation:

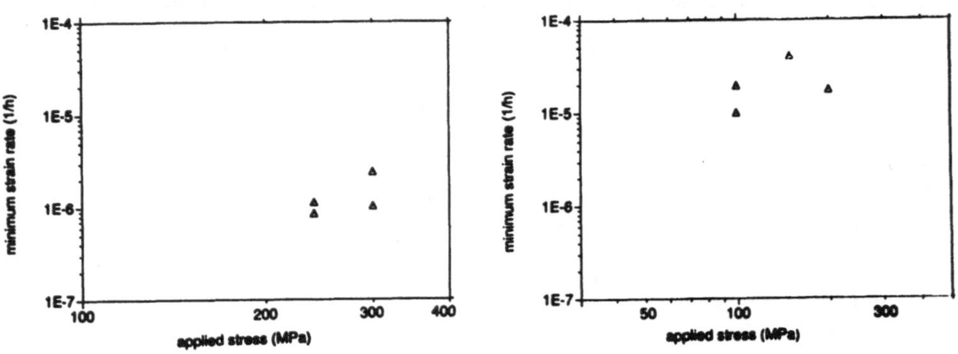

a) b)

Figure 7. $\dot{\varepsilon}_{min}$ vs. σ for U0 tested at a) 1220°C and b) 1350°C.

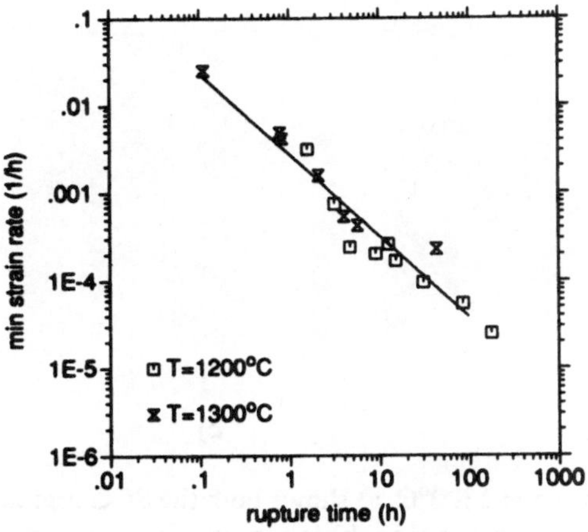

Figure 8. Creep-rupture data fitted to a Monkman-Grant type of plot for KN1.

$$t_f \dot{\varepsilon}^a_{min} = b \qquad (1)$$

where the constants a = 1.07 and ln (b) = - 6.3, were obtained by a linear regression analysis of ln t_f upon ln $\dot{\varepsilon}_{min}$ for 17 sets of data, including stresses in the range 150 MPa to 300 MPa and temperatures of 1200°C and 1300°C. Equation (1) is the Monkman-Grant (7) relationship, originally used to describe the creep fracture of metals. The exponent, a, in this equation is not significantly different from 1, the value usually used to describe creep fracture of metals.

The rupture strain was found to increase with decreasing applied stress, for all materials.

Fractography was made on all failed specimens. Furthermore, the tensile side and side surfaces of survived specimens were examined for indication of e.g. slow crack growth (SCG).

Of the four KN0 specimens, which failed at 1300°C, three fractured from intrinsic flaws, i.e. surface defects or near-surface flaws. The fourth specimen failed due to SCG, starting from a volume flaw, where the SCG region covered only a very small portion of the fracture surface. The two specimens tested at 1400°C failed due to SCG, initiated from a corner. Even though the testing time was relative short, less than 15 minutes, the SCG region covered about 20 % of the fracture surface.

The KN1 material showed a substantial susceptibility to SCG. Fig. 9 a shows both the SCG region (lower part of the micrograph) and the fast fracture region of a KN1 specimen tested at 1300°C. The rough appearance of the SCG region is a result of the crack propagating mainly in the intergranular phase, following the grain boundaries. Due to the high testing temperature, a layer of glass is formed on the fracture surface, making fractography difficult. After etching this layer away it can be seen that the crack has propagated mainly intergranularly in the SCG region (see Fig. 9 b). The crack grows stably to a critical size, followed by fast fracture (see Fig. 9 c), where it propagated mainly transgranularly. Of the 17 specimens tested at 1200°C and 1300°C, 16 failed due to SCG, and one failed due to the combined effect of SCG and a large pore. An interesting feature was that the SCG region covered 20 to 25% of the fracture surface for specimens tested at 1200°C but 40 to 50% for specimens tested at 1300°C. This indicates that the cracks developed at 1300°C are less severe than corresponding cracks at 1200°C, probably a consequence of rapid oxidation forming SiO_2 at the crack tip which blunts the crack tip. Examination of survived specimens showed that the tensile surface and side surfaces of two specimens tested at a stress of 125 MPa at 1300°C were severily cracked and the specimens would probably not have survived very much longer.

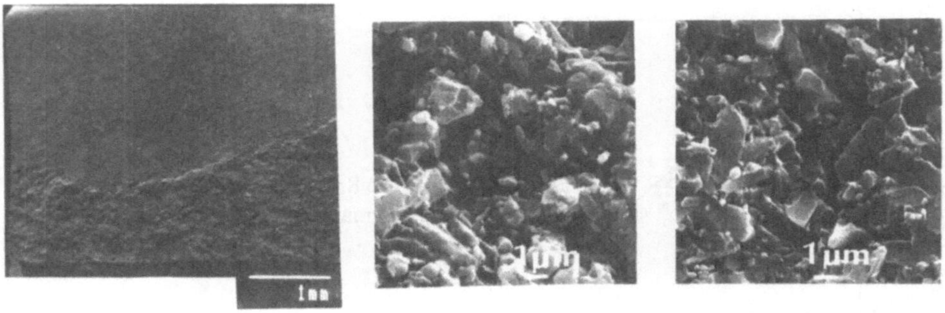

a) b) c)

Figure 9. Fracture surface of a KN1 specimen tested at 1300°C. a) shows both the SCG region (lower part of the fracture surface) and the fast fracture region, b) is a magnification of the SCG region and c) is a magnification of the fast fracture region.

Only four U0 specimens out of 14 specimens failed, one after testing at 1100°C, the rest after testing at 1220°C. No signs of SCG was found in these specimens, for which fracture seems to have initiated from flaws situated in the near vicinity of the tensile surface. However, for the two specimens which survived testing at 1350°C, using stresses of 150 MPa and 200 MPa, respectively, there were faint but evident signs of SCG formation.

DISCUSSION

To the authors' knowledge, very few, if any, studies have been done on the static fatigue behaviour of HIP-Si_3N_4. The static fatigue behaviour of hot pressed silicon nitride (HPSN), however, is well characterized and certain grades of HPSN are used as model structural materials vulnerable to static fatigue.

It is well established (8,9) that static fatigue failure in HPSN can be caused by a variety of phenomena, e.g. SCG and creep fracture. In SCG, the deformation occurs in the immediate vicinity of preexisting crack tips, and leads to crack propagation. Creep fracture occurs when there is a significant deformation throughout the material leading to the formation of extensive creep microcrack networks, which coalesce, eventually leading to fracure. High temperature and low stress promote the latter.

The temperature influence on the susceptibility to SCG is clearly shown for KN0 where at 1200°C none of the three specimens failed and no creep deformation was detected. At 1400°C, however, the two tested specimens failed after only a short period of time, without showing any creep deformation, while at the intermediate temperature of 1300°C, creep deformation was detected and one failure was due to SCG. The existence of a threshold stress, below which mainly creep deformation occurs, is indicated in Figs. 4 b, and seems to be around 250 MPa.

As already has been observed by e.g. Smith and Quackenbush (10) and the present authors (2), an alumina addition to the intergranular glass phase will promote the development of

SCG, and, indeed, SCG was responsible for all failures for the KN1 material. The scatter in $\dot{\varepsilon}_{min}$ and t_f are very large. This may probably be explained by the microstructure of the material. The amount of additions are kept low, and is known that the main amount of amorphous phase is found at triple grain junctions (11). Therefore, in a microscale, the material is very inhomogeneous, and whether the tip of the most severe crack is in a region rich in intergranular phase or not, will have a great impact on the development of SCG, on lifetime and, consequently, on strain to rupture.

Since the UBE material was tested at different temperatures compared to the KN-material, a comparision is not straight-forward. By comparing Figs 7 a and b with 4 a, the creep rate seems comparable to that of KN0. The higher level of yttria in U0 therefore seems to have no effect on the susceptibility to SCG. This is not surprising, since a higher yttria content will result in a grain boundary phase with a higher viscosity.

CONCLUSIONS

The static fatigue behaviour of three series of HIP-Si_3N_4 have been studied. The run-to-run scatter in minimum strain rate and time to failure was found to be large. It was found that the rupture strain and time to failure increased with decreasing applied stress for all materials. The alumina containing material was found to have the highest susceptibility to slow crack growth (SCG). The susceptibility to SCG was found to be very low for the UBE-material. A threshold stress, above which SCG dominates over creep deformation, seems to exist at about 250 MPa and 1300°C for the KemaNord material not containing alumina. A Monkman-Grant type of relationship exists between rupture time and minimum creep rate for the alumina containing grade of KemaNord Si_3N_4.

REFERENCES

1. Micski, A.L. and Bergman, B., High temperature fracture of hot isostatically pressed Si_3N_4. In Br. Ceram. Proc., No. 29, (1987) pp. 45-54.

2. Micski, A.L., and Bergman B., High temperature strength of silicon nitride HIP-ed with low amounts of yttria ot yttria/alumina. J. Eur. Ceram. Soc., 1990, 6, pp. 291-301.

3. Micski, A.L., and Bergman, B., Oxidation of HIP-Si_3N_4 with low amounts of sintering aids. In Br. Ceram. Proc., No. 46, (1990) pp. 271-86.

4. Hollenberger, G.W., Terwilliger, G.R. and Gordon, R.S., Calculation of stresses and strains in four-point bending creep tests. J. Am. Ceram. Soc., 1971, 54, pp. 196-9.

5. Käll, P-O., Quantitative phase analysis of Si_3N_4-based materials. Chemica Scripta, 1988, 28, pp. 439-46.

6. Adlerborn, J., ABB Cerama, Private communication 1990.

7. Evans, H.E.,Mechanisms of creep fracture, Elsevier Applied Science Publishers, London, 1984, pp. 18-22

8. Quinn, D.G., Static fatigue in high-performance ceramics. Proceedings of ASTM Symposium on Methods for Assessing the Structural Reliability of Brittle Materials, San Fransisco, December 1982, pp. 177-93.

9. Grathwohl, G., Creep and fracture of hot-pressed silicon nitride with natural and artificial flaws. Proceedings from the Conference of Creep and Fracture, 1984, pp. 565-77.

10. Smith J.T., and Quackenbush, C.L., Phase effects in Si_3N_4 containing Y_2O_3 or CeO_2. I, Strength. Am. Ceram. Soc. Bull., 59 (1980) pp. 529-32.

11. Falk, L.K.L., Development of microstructure in high performance ceramics. PhD Thesis, Department of Physics, Göteborg, 1986.

COMPARISON BETWEEN IMPACT AND HERTZIAN INDENTATION ON GLASS

Jonas Persson, David J. Rowcliffe and Kristin Breder
Royal Institute of Technology
Department of Physical Metallurgy and Ceramics
S - 100 44 Stockholm

ABSTRACT

As a preliminary study of impact phenomena in ceramic materials, experiments have been conducted on glass with small steel ball indenters. Both dynamic impact experiments as well as quasi-static Hertzian indentations were performed. We show that different damage mechanisms act in the glass in the two different cases. The impact experiments have been performed with a small calibre gas gun with velocities up to 225 m/s. Comparison between impact and quasi-static experiments show that the theories that relate a maximum impact load to an impact velocity overestimate the impact load. Hertz theory can be used only as long as the target responds with tensile stress damage, that is the low velocity region. Hertzian indentations show excellent agreement with theory where the predicted contact zone radii are in very good agreement with the measured ones. The radius of the largest ring crack formed in the experiments is approximately 1.25 times the theoretical contact zone radius.

INTRODUCTION

Hertz's theory for the contact between two solids, developed over 100 years ago, has proved to be accurate in determining the contact area. It is derived with the assumption that both materials behave elastically [1, 2]. The theory has been applied for both quasi-static and dynamic experiments [3]. The main equation is

$$a^3 = \frac{3PR}{4E^*} \qquad (1)$$

where a is the contact zone radius, P is the load, R is the radius of the indenter and

$$\frac{1}{E^*} = \left(\frac{1-\nu^2}{E} + \frac{1-\nu'^2}{E'}\right) \qquad (2)$$

where ν and E are Poisson's ratio and Young's modulus (primes refer to the indenter material).

MATERIALS AND METHODS

The materials used in this study were slabs of soda-lime glass (E=70 GPa, ν=0.25), approximately 50 mm * 50 mm * 15 mm. These were impacted or indented with 3 mm diameter hardened steel spheres (E=200 GPa, ν=0.3).

The gas-gun consists of a 570 mm long barrel with inner diameter 3 mm, loaded with compressed nitrogen. Gas is released by letting a spring loaded needle puncture a brass diaphragm. By choosing a combination of gas pressure, ranging between 0.5 and 4 MPa, and diaphragm thickness, 0.025 or 0.050 mm, the velocity (ranging from 15 to 225 m/s) can be selected within approximately ±5 m/s (less error at low velocities). The velocity of the ball is measured by a commercial system (WEINLICH GmbH & Co, Reilingen, Germany) that is designed to measure the velocity of projectiles from different weapons. It works on the principle of time of flight between two lightbeams.

The indentations were performed in a computer controlled universal testing machine (Instron, High Wycombe, G.B.) run with a controlled loading rate and constant monitoring of load and displacement. The maximum load was varied from 100 N up to 4500 N with loading rates in the range from 0.050 N/s up to 500 N/s. The advantage of doing experiments in this machine is that both the maximum load and the loading rate can be controlled simultaneously. In the impact experiments these two are dependent.

The impacts and the indentations were performed on one of the large flat faces on the glass slabs. One impact or up to nine indentations were made on each glass slab.

RESULTS

Impact

The results from the impacts show that certain velocity regions can be defined where different types of cracks are formed. These velocity regions depend on the projectile and target materials. At velocities below 20 m/s no cracks can be seen. When the velocity of the ball is in the region 20 to 30 m/s ring and cone cracks are formed. At velocities in the range 30 to 40 m/s radial and lateral cracks are also formed. Hertz's theory for the contact zone can still be used up till this point. At velocities above 45 m/s the central contact zone begins to be crushed. When crushing occurs no ring cracks can clearly be defined and therefore nothing can be said about the contact zone size. When the velocity is raised further median cracks penetrating deep into the material can also be seen and material removal by chipping begins.

Indentation

Almost all indentations seem to follow Hertz's theory. At loads up to 1800 N there is good agreement between observed and calculated contact zones, see Table 1. At loads above 2300 N the centre of the indentation begins to be crushed and Hertz's theory does not seem to hold at these very high loads.

The damage that occurs during indentation was examined after each experiment and seemed to follow the maximum load encountered during the experiment. The crack types can be ranged relative to increasing maximum load. The cracks that occur at the indentations are: ring, cone, radial and lateral cracks in this sequence according to increasing maximum load. Cook and Pharr have recently given a good description of the different crack systems that occur during indentation of a brittle material (glass or ceramic) with a sharp indenter [4]. These crack systems are also formed when glass is indented with a spherical indenter. Figures 1 to 4 illustrate the similarities and differences between the impact and the indentation damage. The different crack systems are pointed out in the figures. Cone cracks can be seen below the surface. The median cracks cannot be seen in these figures since they extend into the glass perpendicular to the surface.

Table 1.

Measured and theoretical radii of contact zone and ring cracks in the quasi-static indentations. The ring crack radii are divided by 1.25 which give good agreement with the calculated values. The measured contact radii are measured on coated specimens indented at 10 N/s. All radii are in micrometers.

Load (N)	Radius of ring cracks/1.25				Radius of the measured contact area	Calculated contact radius from eq. (1)
	Loading rate (N/s)					
	5	10	100	500		
100						126.4
200	158.4	161.2				159.2
300	226	178.8			187.5	182.3
400	204	195.6			205	200.6
500	206	207.2	194	199.6	211	216.1
600	223.6	221.2	222	224.4	207	229.6
700	238	242	296		224.5	241.8
800	248.8	262	252		250	252.8
900	262.8	258.4	252	264.4	217.5	262.9
1000	276	285.6	307.2	282	253	272.3
1100	291.2	318.4	316.8	295.6	260	281.1
1200	308.8	291.2	299.2	292	271	289.3
1300	303.2	312.4	300	306	295	297.2
1400	314	300	377.2	322.4	311.5	304.6
1500	332	333.2	302.4	338.4	319	311.7
1600	331.2	327.2	315.2	332	334	318.5
1700	342	330	311.2	336	332	325.0
1800	352	350.8	348	344	312.5	331.2

DISCUSSION

Although derived with the assumption that no cracks are formed in either material, Hertz's theory seem to be able to predict accurate contact zones even when the indented material responds to the loading by crack formation. This can be explained if the stress field under the spherical indenter is taken into account. Zeng et al. [5] have calculated the stress field and show that it is tensile outside the contact zone while it remains compressive in a volume below the indenter. When the damage of the indented glass pieces is related to the stress field it can be seen that cracking occurs in volumes where the largest stresses are tensile (outside the contact zone) while crushing occurs right below the indenter where the stresses are compressive. The shape of the compressive zone is near spherical and coincides well with the crushed zone. Therefore it is probable that cracks are formed by tensile stresses while crushing is compressive damage.

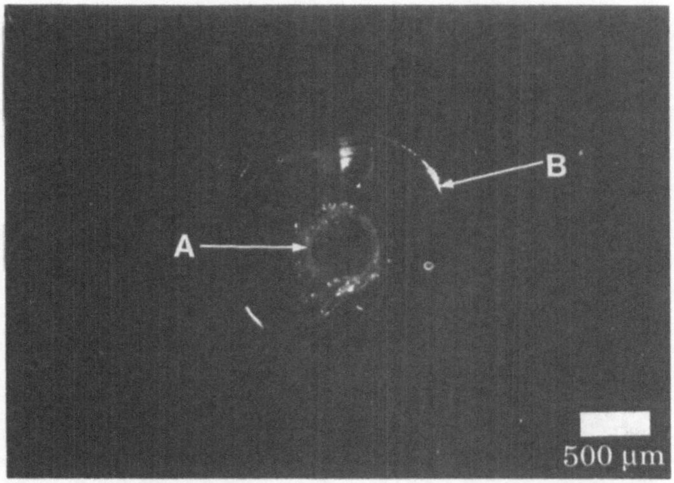

Figure 1a. Impact damage at 30.9 m/s. Close spaced ring cracks (A), one cone crack (B), the central area is undamaged, no crushing.

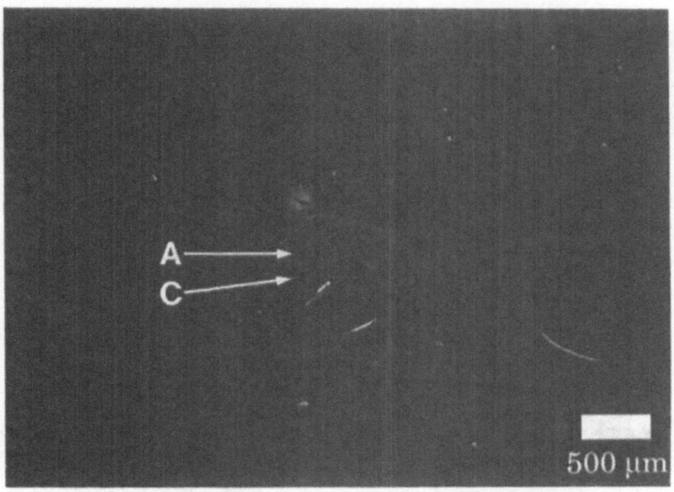

Figure 1b. Indentation damage at maximum load 505 N, loading rate 5 N/s. One ring crack (A), short radial cracks (C).

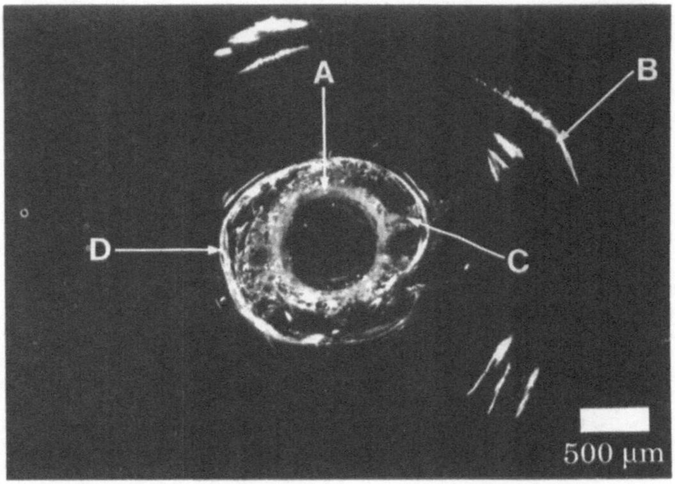

Figure 2a. Impact damage at 39.6 m/s. Close spaced ring cracks (A), one cone crack (B), radial cracks (C), lateral cracks (D).

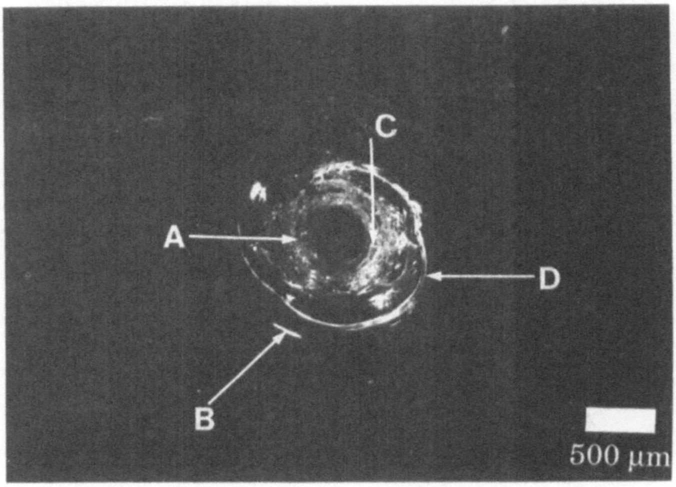

Figure 2b. Indentation damage at maximum load 1804 N, loading rate 5 N/s. Several ring cracks (A), cone crack (B), radial cracks (C), lateral cracks (D).

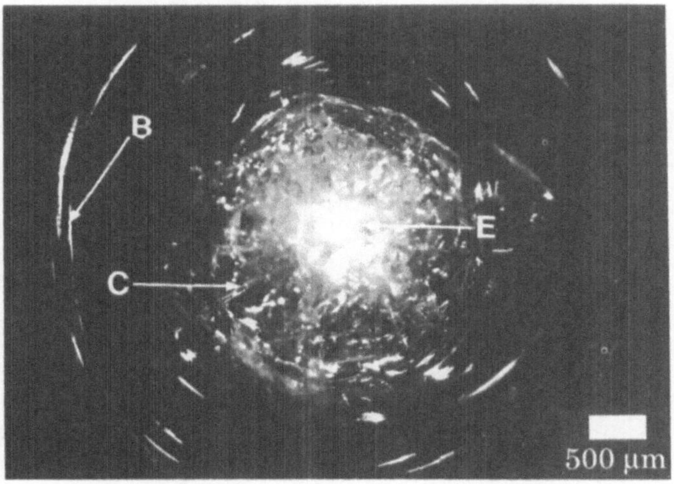

Figure 3a. Impact damage at 42.4 m/s. One cone crack (B), short radial cracks (C), crushing (E) begins.

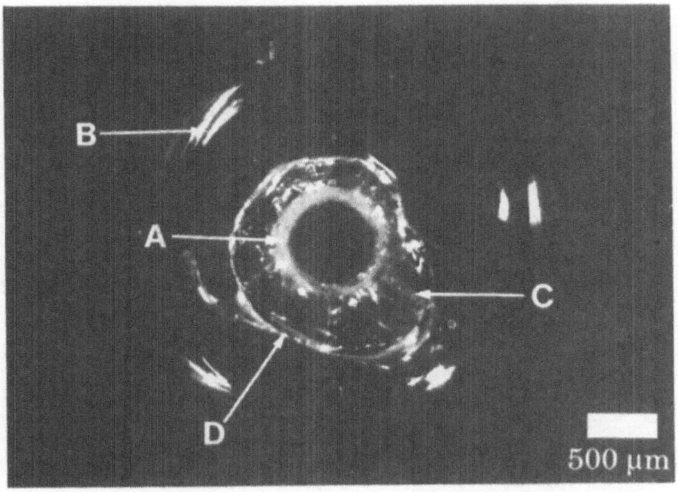

Figure 3b. Indentation damage at maximum load 1795 N, loading rate 500 N/s. Several close spaced ring cracks (A), cone crack (B), radial cracks (C), lateral cracks (D).

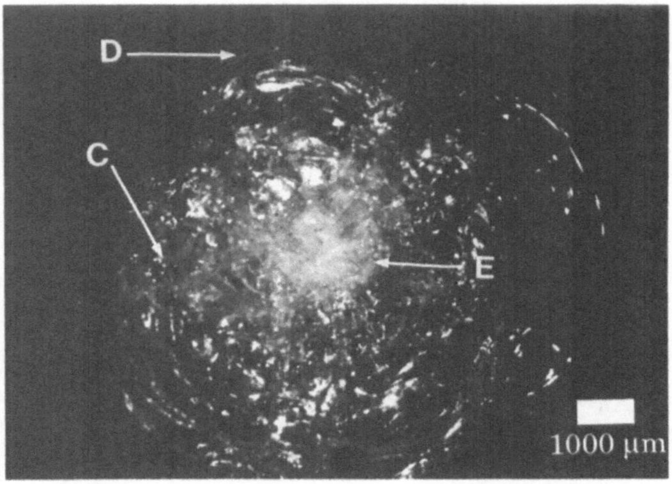

Figure 4a. Impact damage at 222.5 m/s. Several large radial cracks (C), lateral cracks (D), median cracks, extensive crushing (E).

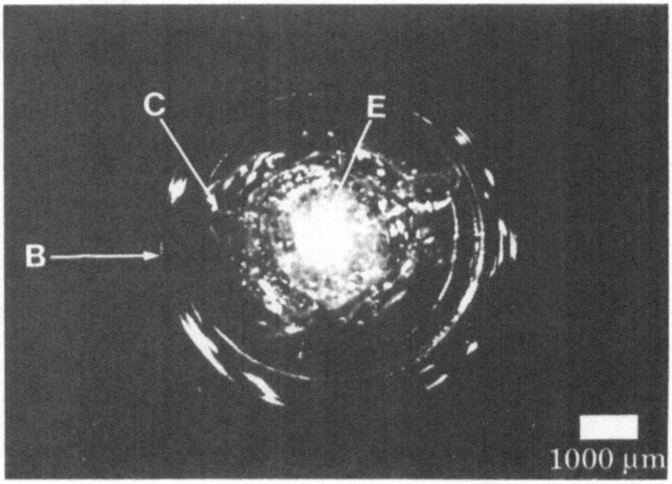

Figure 4b. Indentation damage maximum load at 2440 N, loading rate 10 N/s. Cone crack (B), radial cracks (C), crushing (E).

A difference between the dynamic impacts and the quasi-static indentations is that no median cracks have been encountered in the quasi-static case. In the impact case median cracks occur at higher velocities than is needed to cause crushing. Since Hertz theory cannot be used at these high velocities, the presence of median cracks in the impact experiments is an indication of the deviations from Hertz theory at high velocities. This can also be compared with indentation with a Vickers diamond where median cracks are associated with a plastic zone beneath the indentation.

This comparison between dynamic and quasi-static methods has been made to see if it is possible to simulate impacts with indentations since indentations are easy to control and follow closely during the whole process. These results show that impacts made at low velocities can be simulated by indentations but as the velocity increases, the deviations between the impacts and the quasi-static indentations grow.

CONCLUSIONS

Hertz's theory can be used to predict contact zones between two solids as long as the damage of the indented or impacted material is caused only by tensile stresses. Indentation techniques can only be used to simulate low velocity impacts.

REFERENCES

1. Johnson, K.L., One Hundred Years of Hertz Contact, Proc. Instn. Mech. Engrs., 1982, 196, 363 - 78
2. Frank, F.C. and Lawn, B.R., On the Theory of Hertzian Fracture, Proc. Roy. Soc. A, 1967, 299, 291 - 306
3. Wiederhorn, S.M. and Lawn, B.R., Strength Degradation of Glass Resulting from Impact with Spheres, J. Am. Ceram. Soc., 1977, 60, 451 - 58
4. Cook, R.F. and Pharr, G.M., Direct Observation and Analysis of Indentation Cracking in Glasses and Ceramics, J. Am. Ceram. Soc., 1990, 73, 787 - 817
5. Zeng, K., Breder, K. and Rowcliffe, D.J., The Hertzian Stress Field and Formation of Cone Cracks. Part I - Theoretical Approach, submitted for publication, 1991

ACKNOWLEDGEMENT

We are grateful to Mr. Bengt Möllerberg, Department of Solid Mechanics at the Royal Institute of Technology, who skilfully built the gas gun system.

TESTING OF ENGINEERING CERAMICS UNDER BIAXIAL LOADING

Ewald Staskewitsch, Klemens Stiebler
and Andreas Burblies
Fraunhofer-Institut für Angewandte Materialforschung
2820 Bremen 77, FRG
Remanuel Arone
Israel Institute of Metals, Technion City, Haifa, Israel

ABSTRACT

A new test procedure was developed to proof the strength of engineering ceramics under biaxial loading. This test procedure enables a simultaneous loading of thin-walled tubular specimen by using internal hydraulic pressure and an axial force. This leads to a wide range of biaxial stress states of combined tension/compression stresses. First test results of silicon carbide show the usefulness of the test setup. It will be used to perform comprehensive investigations of engineering ceramics. The evaluated biaxial strength behaviour and an additional investigation of the fracture toughness of engineering ceramics should lead to a probabilistic modeling of ceramic failure under multiaxial loading.

INTRODUCTION

The majority of potential engineering ceramic components are intend to serve under multiaxial loading. This lends extreme importance to the experimental determination of ceramic fracture conditions under multiaxial loading and to the dedevelopment of adequate models for failure prediction. By contrast, most experimental data on the strength of engineering ceramics relate to uniaxial compressive or tensile loading resp. bending. The lack of reliable data on the mechanical behaviour of the engineering ceramics under multiaxial loading requires more effort to this problem.

Therefore, the aims of a joint research work between the Fraunhofer-Institut für Angewandte Materialforschung and the Israel Institute of Metals are:

a) The development of experimental procedure for the biaxial testing of engineering ceramics.
b) Experimental determination of the strength and fracture mechanisms of engineering ceramics under biaxial stresses.
c) Further development of probabilistic failure models and numerical procedures intended for the strength and reliability under multiaxial loads.

In this paper the development of a test procedure for biaxial testing of engineering ceramics is reported.

DEVELOPMENT OF A BIAXIAL TEST PROCEDURE

The existing methods of engineering ceramics testing under multiaxial loading can be subdivided into three main groups:

a) Testing of thin-walled tubular specimens under combined action of internal or external hydraulic pressure and axial loading [1-4].
b) A combination of torsion-tension or torsion-compression test of rods or tubular specimens [5,6].
c) Flexure of disk specimens concentrically supported at their periphery [7,8].

The advantage of the tubular specimen test combining pressuration and axial loading is that significant stress gradients through the specimen thickness can be avoided and that a wide range of biaxial stress states can be achieved with one and the same specimen geometry. Therefore a test system with combined internal hydraulic pressure and axial loading was built up, Figure 1. The principles of the system operation are as follows: The tubular ceramic specimen is loaded by the universal testing machine which procedures compression stresses by the force F. Simultaneously the internal pressure p_i from a second universal testing machine is applied to the specimen through the pressure multiplier, producing circumferential tension stresses. The two above mentioned universal testing machines are controlled by a com-

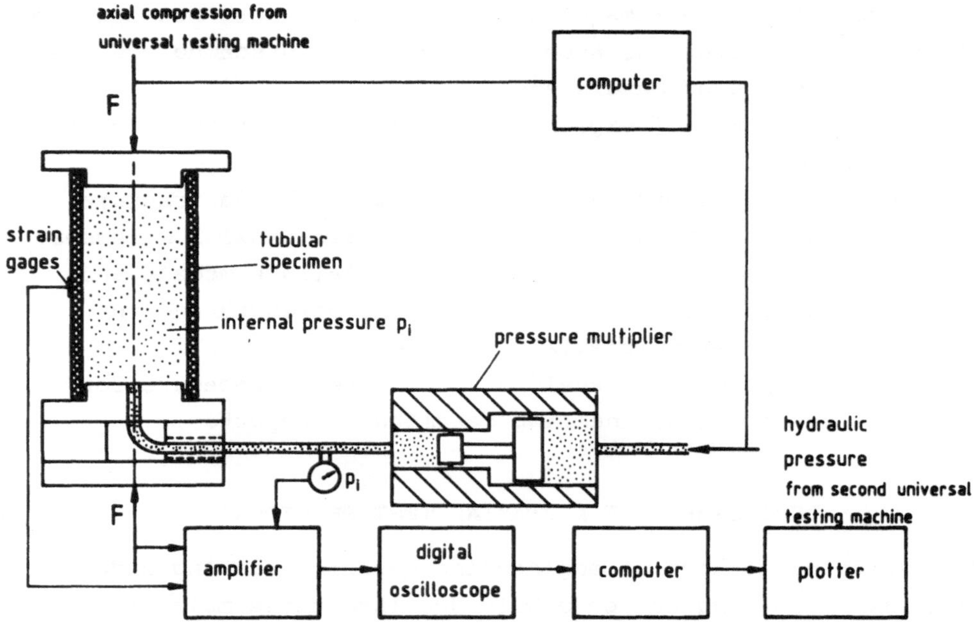

Figure 1. Schematic representation of the biaxial test system

puter to start simultaneously and to get desired values of principal stress ratios. The four following signals: (1) the compression load F, (2) the internal pressure p_i, the strains in (3) axial and (4) circumferential direction measured with strain gages, are amplifiered and stored in a digital oscilloscope. Afterwards, the data are transmitted to a main computer for data acquisition, processing and display.

After fabrication of the necessary elements first tests were carried out to prove the test system. The tubular specimens were machined by grinding in a shape with a strong reduced cross section at the test section by having the wall thickness at the ends three times thicker, Figure 2. To ensure uniform loading over the circumference the ends of the specimen were of high parallelism. A remained small deviation has been eliminated through an alignment plate. Alignment is controlled by three strain gages attached to the test section

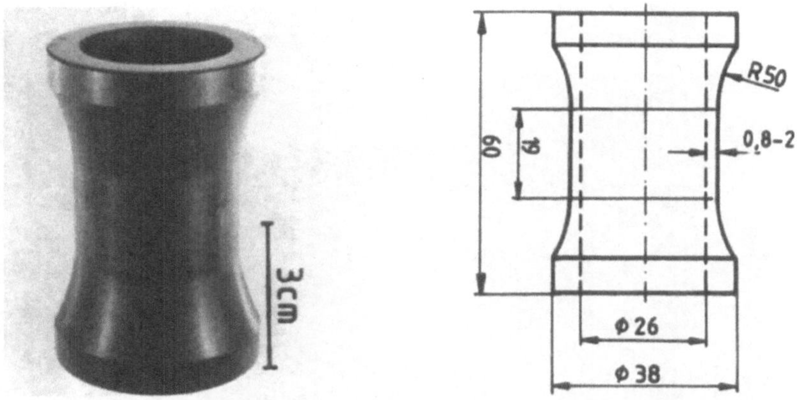

Figure 2. Shape and dimensions of the thin-walled tubular ceramic specimen

in distances of 120° respectively. The whole test system is shown in Figure 3 and the test fixture containing the tubular specimen in Figure 4.

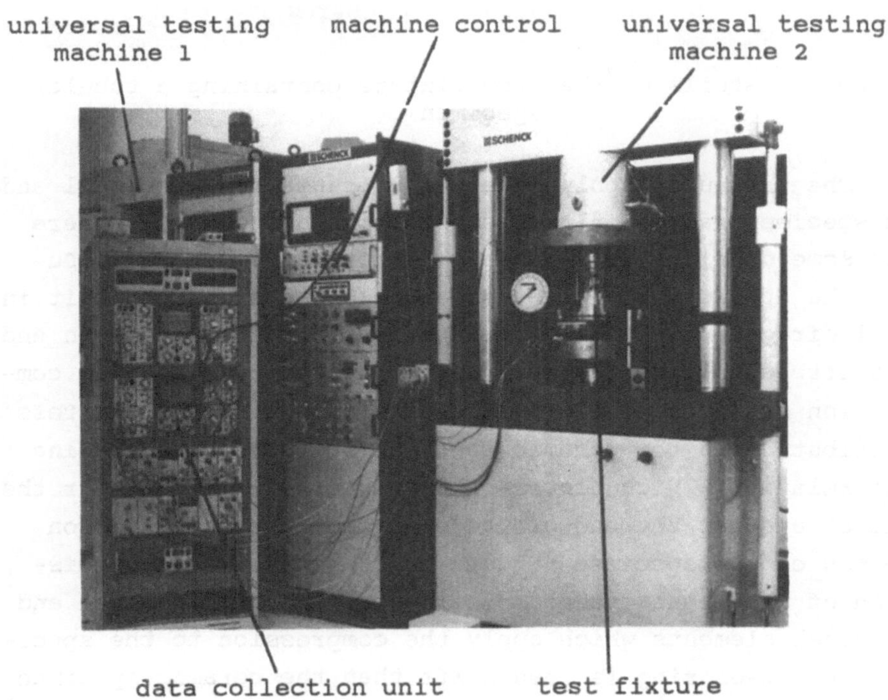

Figure 3. Photograph of the entire biaxial test system.

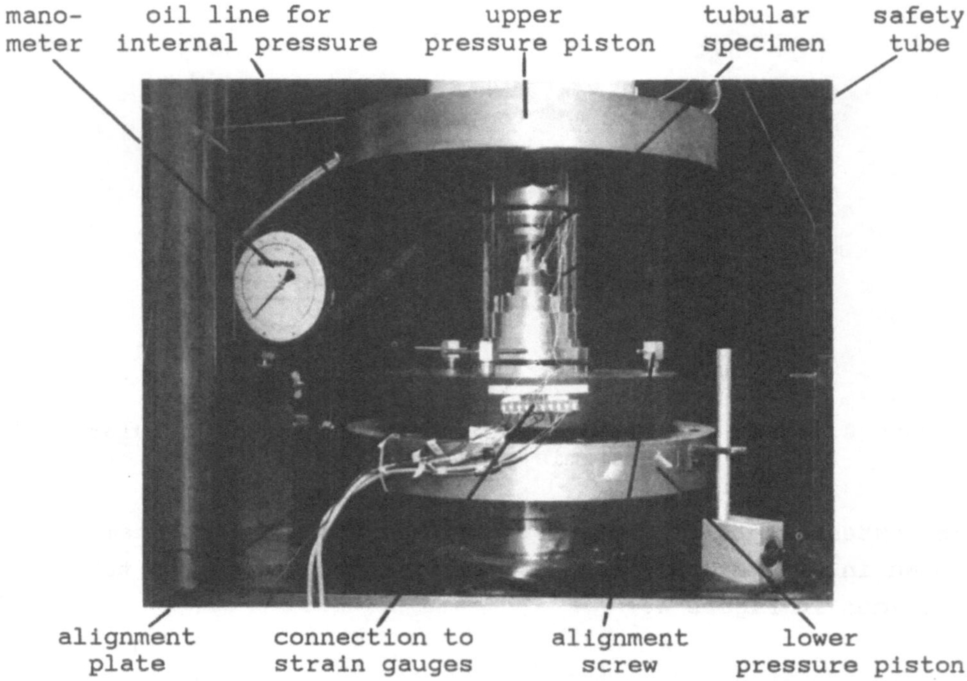

Figure 4. Details of the test fixture containing a tubular specimen

The test using only internal pressure ran very well and the specimen failed in the test section as expected. There were some difficulties with high compressive stresses. During the application of the axial load the specimen split in axial direction. This splitting starts from the specimen ends at a stress value of approximately 50 % of the expected compression strength. A Finite Elemente analysis of the stress distribution in the ceramic specimen was performed showing that relatively high tensile hoop stresses appeared near the contact area at the end of the specimen, where the piston presses on the specimen, Figure 5. The reason was the mismatch of the transverse compliance between the specimen and the steel elements which apply the compression to the specimen. The steel ring is less stiff than the ceramic specimen and has a higher lateral displacement because of the different elastic constants. Under compression load this mismatch in the lateral displacements causes tension stresses in the ceramic in the area of the interface because a large friction coefficient obviates sliding in the contact zone.

Further analyses showed that a stiffer ring of hard cutting metal (Young's modulus = 550 GPa) was more suitable to

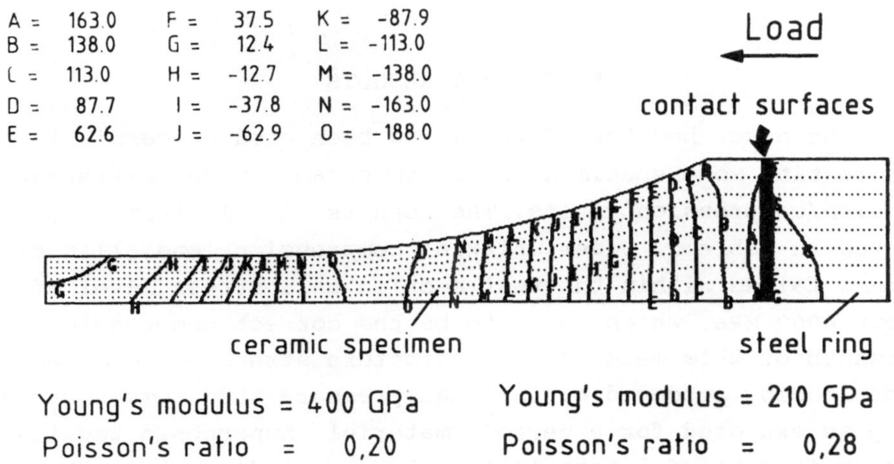

Figure 5. Tension hoop stress distribution for a compression test using steel rings between load piston and ceramic specimen (values in MPa).

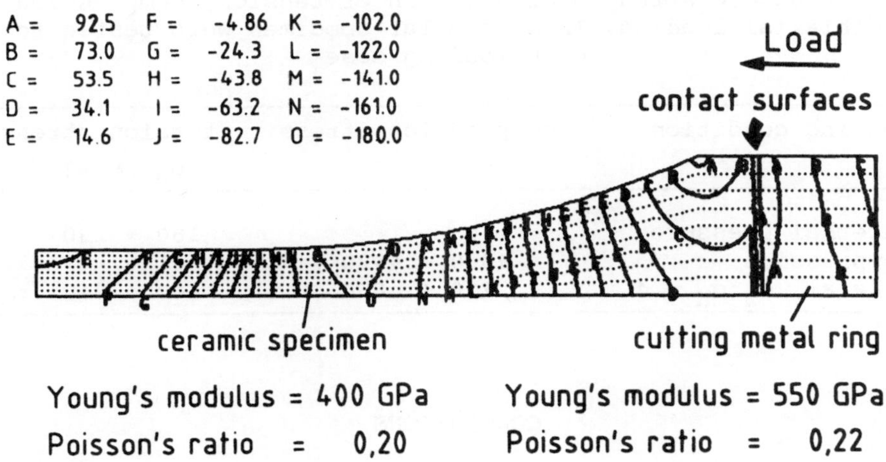

Figure 6. Tension hoop stress distribution for a compression test using cutting metal rings between load piston and ceramic specimen (values in MPa).

apply the axial compression load to the ceramic specimen. The lateral expansion of the ceramic specimen was reduced and this resulted in tension stresses approximately half of that appearing in the case of a steel rings, as can be seen by comparing the stress distribution in Figure 6 with that in Figure 5.

FIRST TEST RESULTS

The described test fixture has been used successfully for testing some specimens of SiSiC-ceramic under different tension/compression ratios. The results of this tests are listed in Table 1. Under uniaxial compression load all specimens failed within the test section on a stress level of about 4000 MPa, which seems to be the correct compression strength of this material. The fracture strength under tension loading produced by internal pressure show some scattering as expected for a ceramic material. Superposed tension stress in a biaxial test lowers the compression strength dramatically.

TABLE 1

Fracture strength of SiSiC under tension, compression and biaxial loading. Three tubular specimen were tested for each loading case

loading condition	compression stress σ_Z [MPa]	tension stress σ_H [MPa]
internal pressure	-	180 + 230
compression	4030 + 4420	-
biaxial, $\sigma_Z:\sigma_H \sim 4$	720 + 850	160 + 240

CONCLUSIONS

To perform proper tests under biaxial loading of ceramics a new test system was developed and successfully proved. The results show that the new experimental procedure works well and that the planned test series for the determination of the strength of ceramics can be done as a part of the further development of probablistic failure models and numerical procedures to asses the strength and reliability of engineering ceramics under multiaxial loads.

ACKNOWLEDGEMENT

The authors acknowledge the German Ministery of Research and Technology (BMFT) for the support of this research programs.

REFERENCES

1. Broutman, L.J., Krishnakumar, S.M., and Mallik, P.K., Effects of Combined Stresses on Fracture of Alumina and Graphite. J. Amer. Ceram. Soc., 1970, 53, 649-654.

2. Ely, R.E., Strength of Titania and Alumina Silicate Under Combined Stresses. J. Amer. Ceram. Soc., 1972, 55, 347-350

3. Adams, M., and Sines, G., Determination of Biaxial Compressive Strength of Sintered Alumina Ceramics. J. Amer. Ceram. Soc., 1976, 59, 300-304.

4. Tappin, G., Davidge, R.W., McLaren, J.R., The Strength of Ceramics under Biaxial Stresses. 3. Int. Symp. on Fracture Mech. of Ceramics, Flaws and Testing, ed. R.C. Bradt, D.P.H. Hasselman and F.F. Lange, Plenum Press, New York - London, 1978, Vol. 3, pp. 435-449.

5. Petrovic, J.J. and Stout, M.G., Fracture of Al_2O_3 in Combined Tension/Torsion: I. Experiment, J. Amer. Ceram. Soc., 1981, 64, 656-660.

6. Petrovic, J.J. and Stout, M.G., Fracture of Al_2O_3 in Combined Tension/Torsion: II. Weibull Theory. Ibid, 661-686.

7. Shetty, D.K., Rosenfield, A.R., Duckworth, W.H., and Held, P.R., A Biaxial-Flexure Test for Evaluating Ceramic Strength. J. Amer. Ceram. Soc., 1983, 66, 36-42.

8. Godfrey, D.J., and John, S., Disc Flexure for the Evaluation of Ceramic Strength. In Ceramic Materials and Components for Engines. Proceedings of the Second International Symposium, Luebeck-Travemuende, FRG, Verlag Deutsche Keramische Gesellschaft, 1986, pp. 657-665.

TOUGHENING INVESTIGATION OF MODE Ⅲ CRACK BY HOLOGRAPHIC INTERFERENCE TECHNIQUE

SUN Qing–Ping*, HUANG Yong*, LIN Shu–Tian** and SI Wen–Jie*
* Department of Materials Science and Engineering,
** Department of Engineering Mechanics,
Tsinghua University, Beijing 100084, P.R.China

ABSTRACT

The transformation toughening effect in mode Ⅲ crack of Ce–TZP ceramics and the crack growth process of the specimen are investigated by the holographic interference technique. The crack growth path under applied loading and the corresponding displacement field of the specimen were clearly recorded by the holographic interferometer. The experimental results showed that there is an obvious stable crack growth stage before final failure and there is obvious toughening effect (R–curve effect) for mode Ⅲ crack. It is also shown that the crack does not grow along the original crack plane, but deflects with an angle of about 60°.

INTRODUCTION

Transformable ceramics such as Mg–PSZ and Ce–TZP have very important applications due to their fine properties. During the last decade, a lot of investigations have been done on the transformation toughening and fracture behavior of such transformation toughened structural ceramics[1], and most of the research has been concentrated on mode I crack. For the cracks under mode Ⅲ loadings (tearing mode), which is also very important in practice, the corresponding investigation on toughening and fracture behavior has been rarely seen in the literature. The objective of the present paper is to investigate the crack growth process and the transformation toughening effect of mode Ⅲ crack in Ce–TZP ceramics by the holographic interference technique. Based on the authors' previous theroetical investigation [2], the experimental results are analyzed and discussed.

MATERIALS AND METHODS

The material used is the ceria–stabilized tetragonal zirconia polycrystalline ceramics (10mol%–Ce–TZP) which is sintered at 1400℃ for 3 hours and is furnace–cooled to the room temperature. The material then is cut and ground into a plate of size 50mm × 60mm × 10mm. An edge crack with length of 30mm and width of 0.2mm is cut by a diamond grinding wheel. To perform mode Ⅲ loading, the specimen is assembled on the special loading equipment as shown in figure 1. The crack growth path during loading and the corresponding surface off–plane displacement field of the specimen are recorded by the holographic interferometer as shown in figure 2.

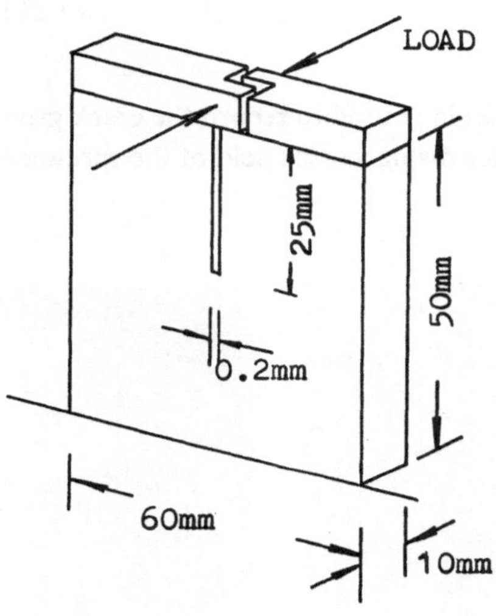

Figure 1. Specimen under mode Ⅲ loading.

RESULTS

Figure 3 (a) and (b) show the photographs of the measured iso–displacement interference fringe pattern of the off–plane surface displacement field of the specimen corresponding to the applied load of 8kg and 14kg respectively. When load is increased to 37kg and 47kg, the crack has grown some distance, the measured surface displacement fringe patterns in the near crack tip region are shown in figure 4(a) and (b), where the amount of crack growth can be determined from the surface displacement discontinuity.The measured load~crack length curve is shown in figure 5.

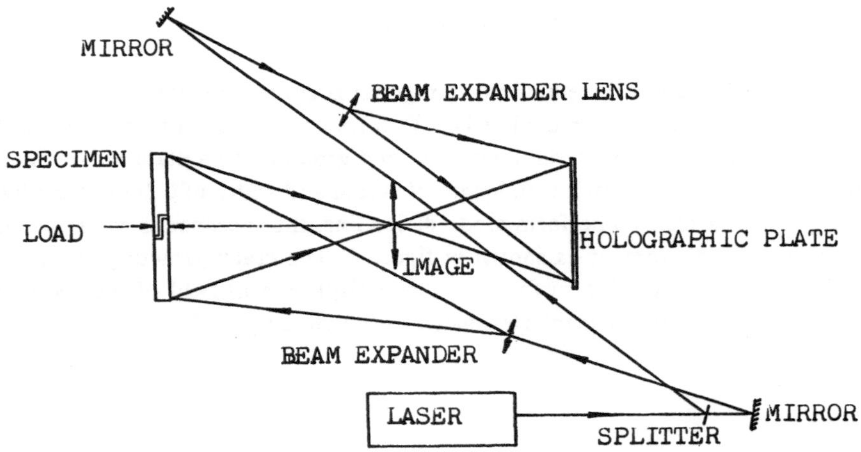

Figure 2. Holographic technique used to record the crack growth path and the surface displacement field of the specimen.

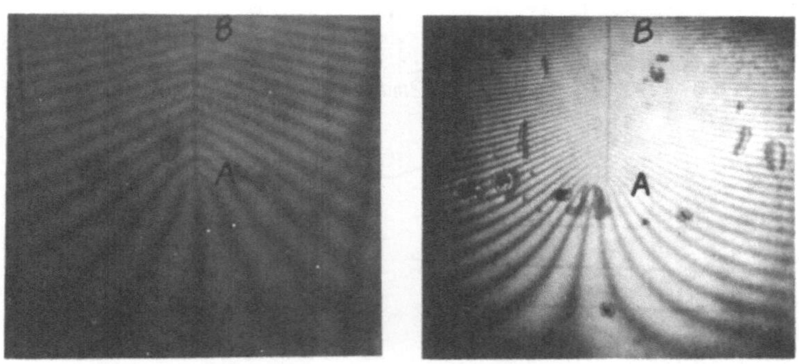

AB------Original Crack
A------Crack Tip

(a) (b)

Figure 3. The surface displacement fields of the stationary mode III crack specimen corresponding to the loads of (a) 8kg and (b) 14kg, respectively.

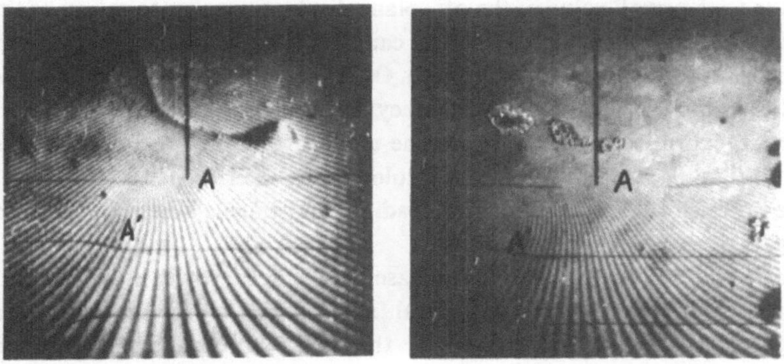

(a) (b)

Figure 4. The surface displacement fringe patterns of the specimen with loads of 37kg (a) and 47kg (b), respectively, showing the crack growth and deflection.

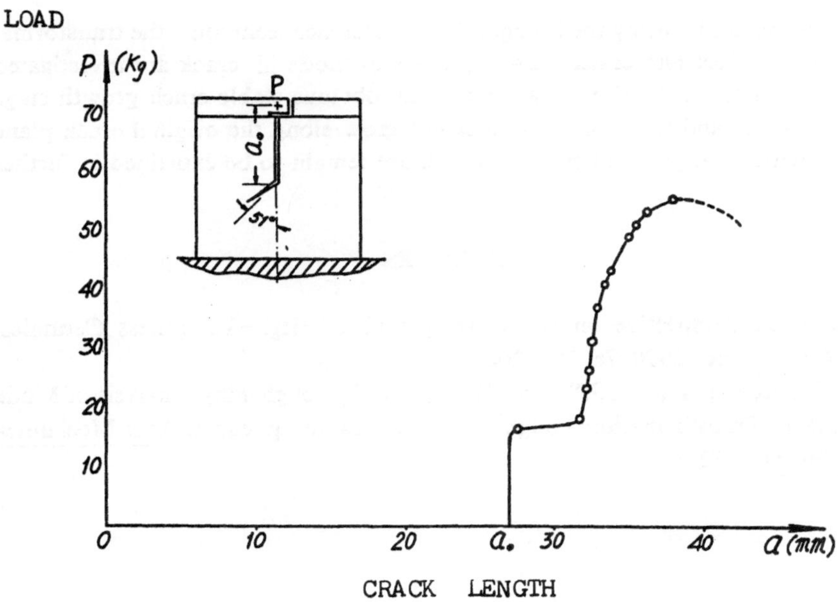

Figure 5. The measured load~crack length curve.

DISCUSSION

Compared with the mode Ⅰ crack, the observation of mode Ⅲ crack is much more difficult due to the small value of the off–plane displacement, uncertainty of the crack growth direction and the crack close effect caused by the volume dilation of stress–induced transformation during crack growth. On the other hand, The advantages of the holographic technique such as high accuracy, high resolution, total field and in–situ measurement, can in some extent overcome the above difficulty. From the results in figure 3 and figure 4, we can see that the holographic interference method can clearly record the crack growth process during loading, and thus is a very potential method to investigate the mode Ⅲ crack problem.

From figure 5 an obvious load increase with crack growth is observed, that is there exists stable crack growth before final failure (point A) for this brittle ceramics. This toughening phenomenon caused by the stress–induced transformation is in agreement with the authors′ previous theoretical analysis [2].

Figure 4 and figure 6 show that the crack does not grow along the original crack plane, but deflects with an angle of about 60 ° . Further research work in both theoretical calculation and microstructural analysis is needed to explain such asymetric crack growth phenomenon.

CONCLUSIONS

In the present paper, by using the holographic interference technique, the transformation toughening effect and crack growth process of mode Ⅲ crack are investigated experimentally. The results show that there is an obvious stable crack growth stage before final failure and that the crack does not grow along the original crack plane but deflects with an angle of about 60 ° which are remain to be explained in further research.

REFERENCES

1. Evans, A.G., Perspective on the Development of High–Toughness Ceramics. J.Am. Ceram. Soc., 1990, 76, 187–206.
2. Sun, Q.P., Huang, Y., Yu, S.W. and Hwang, K.C., Toughening Analysis of Mode Ⅲ Crack in Transformation Toughened Ceramics. to appear in Acta Mechanica Solida Sinica, 1991.

MECHANICAL PROPERTIES AND MICROSTRUCTURE OF ZrO_2/ Si_3N_4 COMPOSITE

BAOQING ZHANG, JIEMO TIAN, XIAOHUA TONG,
LIMIN DONG, LINGLING WANG
Beijing Fine Ceramic Laboratory of Institute of Nuclear
Energy Technology Tsinghua University
P.O.Box 1021, Beijing 102201, China

ABSTRACT

The Si_3N_4 ceramic reinforced with 30 wt% 3YZrO₂(ZrO_2 containing 3 mol% Y_2O_3) has been fabricated by two kinds of processes (hot press and pressureless sintering). The ZrO_2/Si_3N_4 composites exhibited a high fracture toughness and flexural strength, and lower thermal conductivity than Si_3N_4 matrix material. The ZrO_2/Si_3N_4 composite is suggested to be a very hopeful candidate material as component of heat engine applications. In this paper, the microstructure and phase composition of the composites are studied by SEM and XRD. The toughening mechanism of ZrO_2 in the ZrO_2/Si_3N_4 composites is discussed.

INTRODUCTION

As it is well known that, in order to develop a ceramic composite with high strength and high toughness, the dispersion of ZrO_2 particles in a ceramic matrix and utilization of the stress-induced martensitic transformation of ZrO_2 from the tetragonal to monoclinic phase have been investigated comprehensively, and in the oxide system, significant effectiveness have been obtained. For example, ZrO_2 toughened Al_2O_3 achieves the fracture toughness of 12 $MPa \cdot m^{1/2}$ and flexural strength of 1200 MPa[1]. In some non-oxide ceramics including Si_3N_4 and SiC, silicon nitride is a very hopeful candidate material to be used as high temperature structural components in heat engine, because it possesses high strength at high temperature and high hot shock resistance. But in general, the fracture toughness of Si_3N_4 is less than ZrO_2, and the thermal conductivity of Si_3N_4 is higher than ZrO_2. So in the present study, we hope to increase the fracture toughness and decrease the thermal conductivity by the dispersion of ZrO_2 particles in Si_3N_4 matrix.

In last years, some researchers have already reported on some experimental results of ZrO_2/Si_3N_4 composites. N. Claussen, et al[2] found the flexural strength and fracture toughness of ZrO_2/Si_3N_4 composites increases with increasing content of ZrO_2, reaching a maximum with 20-25 Vol% ZrO_2, the fracture toughness increases from 5 MPa. $m^{1/2}$ to 7 MPa · $m^{1/2}$ and flexural strenght is near 600 MPa in the pressureless sintering condition, while in hot pressed material the fracture toughness increases From 5.5 MPa · $m^{1/2}$ to 8.5 MPa · $m^{1/2}$, and flexural strength is near 1000 MPa. But in the meantime, the reports of ZrO_2/Si_3N_4 composites about high temperature mechanical properties are not found.

EXPERIMENTAL PROCEDURE

The ZrO_2/Si_3N_4 composites were prepared with two kinds of sintering processes. One is hot press sintering with a axial pressure of 30 MPa at 1600℃ under argon atmosphere, another one is pressureless sintering at 1800℃ under <0.1 MPa nitrogen atmosphere by protective powder bed. In the pressureless sintering process, the power mixtures were added a little Al_2O_3 to inhibit the formation of ZrN[3-4],and the powder mixtures were preformed to green compacts by cold isostatic pressing. The microstructure and phase composition of ZrO_2/Si_3N_4 composites were studied by SEM and XRD.The flexural strength and fracture toughness of the composites measurements were performed using a four point device with a inner span of 20 mm and outer span of 40 mm, and fracture toughness was measured by the chevron notched method. The size of all specimens is 45 mm × 4 mm × 3 mm, and they are finished with a D 120 diamond wheel, the chevron notches are cut with a diamond wheel of 0.1 mm thickness. In this paper, the thermal conductivity of the composites was also measured.

RESULTS AND DISCUSSION

In the present paper, the composites were based on Si_3N_4 matrix toughened by dispersion of 30 wt% $3YZrO_2$(ZrO_2 containing 3 mol% Y_2O_3). Experimental results showed the composite possessed lower thermal conductivity than Si_3N_4 and a high flexural strength and fracture toughness. The results are showed as follawing :

For hot pressed composites:

Thermal conductivity = 18.1±0.5 W/m·℃ at 400℃

σ_b = 870±50 MPa; Kic = 8±1 MPa·$m^{1/2}$ at 25℃

σ_b = 780±20 MPa; Kic = 6±1 MPa·$m^{1/2}$ at 850℃

For pressureless sintered composites:

Thermal conductivity = 12.5±0.5 W/m·℃ at 400℃

σ_b = 610±50 MPa; Kic = 7±1 MPa·$m^{1/2}$ at 25℃

σ_b = 530±20 MPa; Kic = 7±1 MPa·$m^{1/2}$ at 850℃

By XRD analysis, before fracture at 25℃, the monoclinic phase content of ZrO_2 (m-ZrO_2) is near zero, but after fracture, about 5wt% m-ZrO_2 phase was found on the fracture surface (Figure 1). Obviously, these m-ZrO_2 phase results from the stress induced martensitic transformation of ZrO_2.

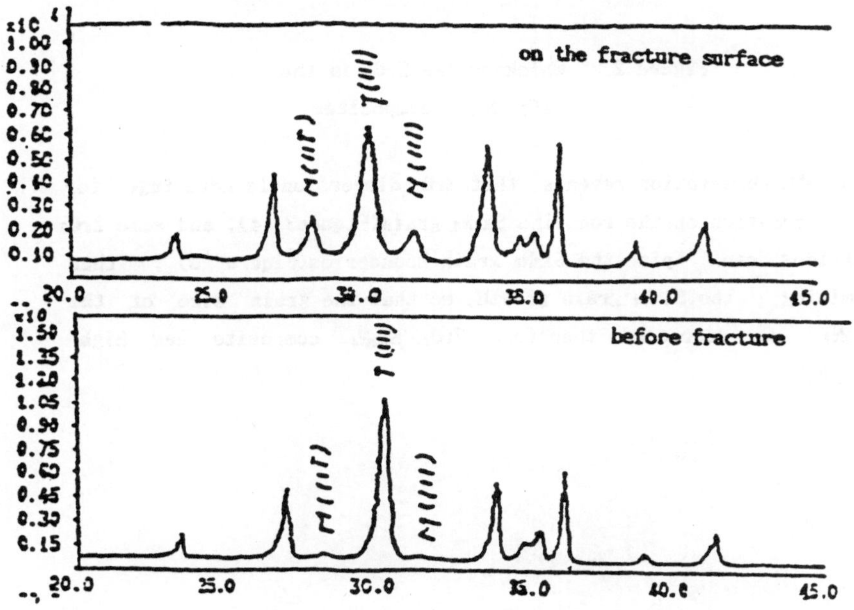

Figure 1 XRD Patterns of room temperature fractured specimens of ZrO_2/Si_3N_4 Composites

When external stress induced phase transform from the tetragonal phase to the monoclinic phase, microcracks are generated with the increase in volume of ZrO_2 particles alloyed with Y_2O_3. These microcracks can make crack energy loss, so ZrO_2/Si_3N_4 composite gave high toughness. Besides, some whiskerlike ZrO_2 grains inthe pressureless sintered composites observed by SEM can also contribute to the increase of the toughness(Figure 2).

Figure 2 Whisker-like ZrO_2 in the ZrO_2/Si_3N_4 composites

The SEM observation reveals that ZrO_2 dispersion is advantage for the formation of the rod-like Si_3N_4 grain(Fiqure3, 4), and some ZrO_2 partides can pin the Si_3N_4 grain bounderies(Fiqure 5), thus limiting the Si_3N_4 grain growth, so that the grain size of the Si_3N_4 was decreased, therefore, ZrO_2/Si_3N_4 composite has high strength.

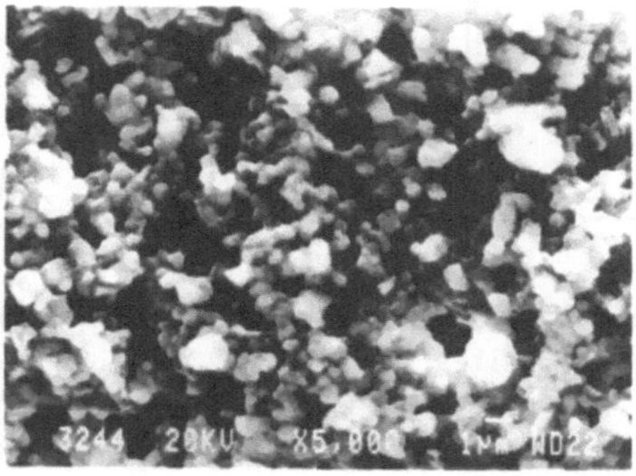

Figure 3 SEM images of the microstructure of Si_3N_4 matrix material

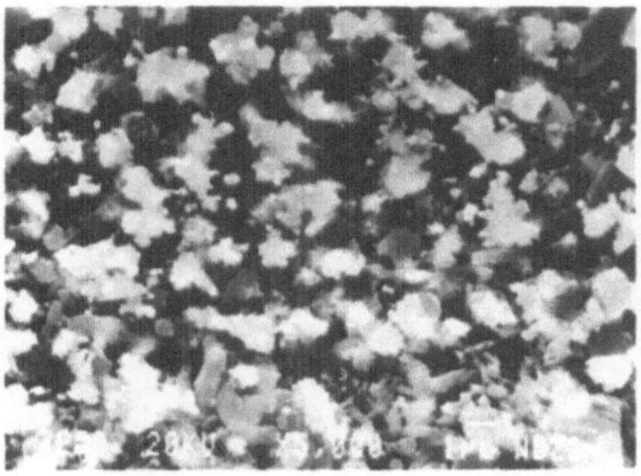

Figure 4 Rod-like Si_3N_4 grains in the ZrO_2/Si_3N_4 composites

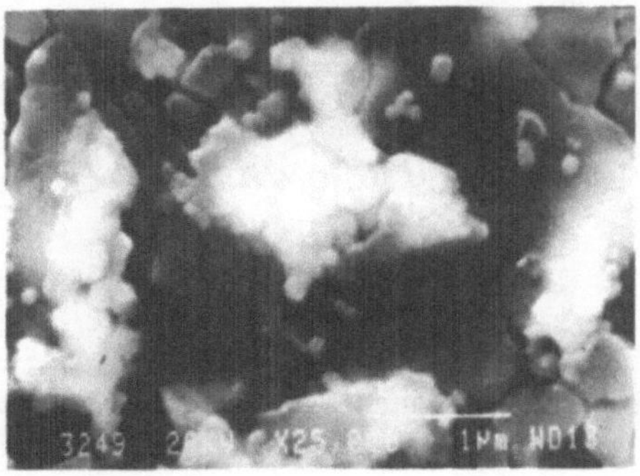

Figure 5 Some ZrO_2 particles pining the Si_3N_4 grain bounderies in the ZrO_2/Si_3N_4 composites

SEM observation of the fracture surfaces of the composites indicates that they are intercrystalline and transcrystalline fracture(Figure6,7). For high temperature (850℃) fracture surface, a little intergranular glassy phase is observed.

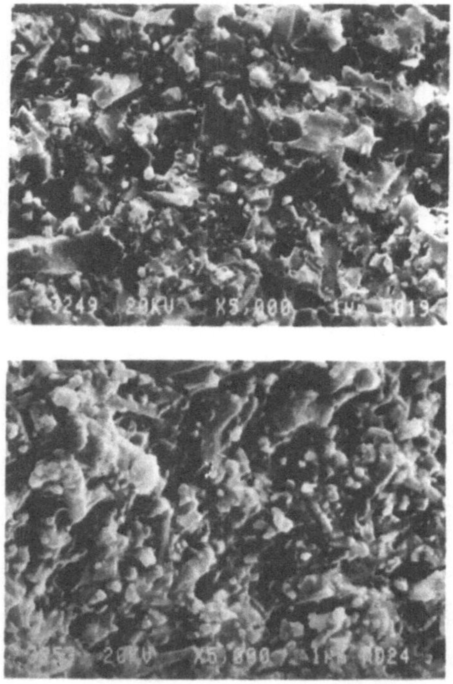

Room temperature specimen

850℃ specimen

Fiqure 6 SEM images of fracture surfaces

CONCLUSIONS

1. The ZrO_2/Si_3N_4 composites possessed lower thermal conductivity than Si_3N_4, It exhibited a high flexural strength and fracture toughness. So it is suggested to be used as component of heat engine.
2. Acoording XRD analysis, the Zr-O-N and ZrN phases are not found in the composites by two kinds of sintering processes.
3. It is suggested that toughening mechanism of the ZrO_2/Si_3N_4 composites is mainly the stress-induced martensitic transformation and microcracking.
4. The SEM observation reveals that the ZrO_2 particle dispersion promotes the formation of rod-like Si_3N_4 grains, which are advantageous for composite strength.

REFERENCES

[1] F.F.Lange, J.Mat. Sci 17(1982)247-254

[2] N.Claussen and J.Jahn, J.Am. Ceram. Soc. 61[1-2](1978)94

[3] A.K.Tjernlund, et al, 《British Ceramic proceedings》 No.37 Oct 1986, 29-34

[4] J.Weiss, L.J. Gauckler, H.L.Lukas, et al, J. Mat. Sci., 16[1] (1981)2997

THE CHARACTERISTIC STRESS BETWEEN TWO KINDS OF DELAYED FAILURE FOR CERAMICS AT ELEVATED TEMPERATURE *

Z.D.Guan, J.H.Gong, Y.Y.Zhao, and X.F.Zhao
Department of Materials Science and Engineering
Tsinghua University, Beijing 100084, P. R. China

ABSTRACT

The delayed failure behavior of ceramics is characterized with stress rupture and crack extension experiments. A linear relationship is evident between stress rupture data in higher applied stress level, denoting that the failure of material in this stress level is controlled by SCG. A evident deviation appears when extrapolating the straight line from higher stress level to lower stress level, implying a change in mechanism for failure. The characteristic stress value, σ_0 , for discerning between different stress regions in high temperature rupture can be calculated with SCG threshold, K_{th} . Calculated result is consistent with the measured result of stress rupture. Finally, a criterion for distinguishing between two mechanisms for delayed failure, SCG and creep rupture, with an intrinsic parameter of material, K_{th}, is obtained.

INTRODUCTION

In most studies on lifetime prediction for ceramics, stress rupture testing is commonly used as a simple and most reproducible test technique to observe delayed failure phenomenon, with which the lifetime of ceramics under a certain applied stress can be obtained directly. Since the most reliable data for lifetime prediction may come from stress rupture testing [1] , the delayed failure behavior of ceramics under this loading condition must be evaluated over a wide range of failure time.

In the 1970s and early 1980s, only subcritical crack growth (SCG) is recognized as the main source of delayed failure in ceramics and the stress rupture data are usually treated as a power relationship between crack velocity and applied stress intensity [2, 3] . More recently, much attention has been focused on creep rupture. For example, Grathwohl has reported the fact that growing cracks in hot-pressed silicon nitride, under static fatigue condition, can be stopped due to creep if a stress is applied to the specimen that is below a characteristic value; under these conditions, creep rupture is observed after the critical creep strain is reached [4] . Similar phenomena have been observed by Quinn in other Si_3N_4 [5] and a commercial Al_2O_3 [1] . These inverstigations show that delayed failure in ceramics may be governed by different mechanisms.

The work presented in this paper will add to these investigations and presents

* This paper is done under the cooperative research contract with Hitachi Chemical Co., Ltd.

additional experimental data to distinguish more clearly between creep rupture and subcritical crack growth in ceramics under steady loading at elevated temperature.

EXPERIMENTAL PRECEDURES

Material and Specimen Preparation

A pressureless sintered Si_3N_4 is used in the present study. Test specimens are of dimensions 4mm wide × 3mm thick × 36mm long approximately. The tensile surface of specimens is polished carefully using M0.5[#] diamond paste and the edges are chamfered. Then all specimens are subjected to air annealing at 1000℃ for 1 hour to eliminate the residual stress generated by machining.

Determination of Residual Stress Magnitude Parameter of Indentation

An Knoop indented bending beam method, which has been studied in detail in our earlier works [6, 7], is applied to determine the residual stress magnitude parameter, χ_r, of indentation for the test material. Specimens are precracked using a Knoop microhardness indenter. The flaw is carefully placed in the center of the tensile surface of specimens in an orientation perpendicular to the tensile stress direction whose dimension and geometry is controlled by the choice of indentation load. Then, specimens are tested in three point bending with a span of 30mm and a crosshead speed of 0.05mm / min. All tests are conducted in air at 1200℃. Specimens are held at test temperature for about 5 min before loading is started.

Specimens without precracks are also tested in same condition, but with a crosshead speed of 0.5mm / min, to determine the strength of the test material.

Indirect Measurements of SCG parameters by Stress Rupture test

In order to observe the delayed failure phenomenon of the test material, stress rupture testing is conducted in air at 1200℃ in three point bending with a span of 30mm. A mechanical test furnace, in which loading is done by a deadweight level arm system which transmits a force into the furnace and on to the fixtures, is used to perform these tests. The fluctuation in temperature of this test furnace has been calibrated within ± 1℃ and the errors in stress from loading system are kept to less than ± 1%. Specimens are held at test temperature for about 5 min before full load is applied. During preheating of the furnace, the specimen is subjected to a small stress of about 25Mpa to prevent crack healing at elevated temperature.

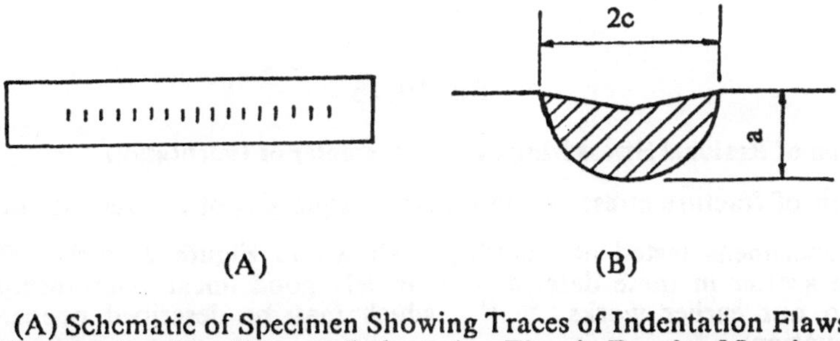

Fig. 1 (A) Schematic of Specimen Showing Traces of Indentation Flaws and (B) Fracture Profile of Knoop Indentation Flaw in Depth of Specimen

Complete SCG Behavior Observation by Direct Crack Extension Measurements

In order to obtain the SCG velocity data as a function of stress intensity, the indentation–induced–flaw (IIF) technique is applied to study the characteristics of SCG for the test material during fracture process at 1200℃. This technique has been studied in detail by Mendiratta et.al [8], Kawai et.al. [9], and the present authors [10], respectively.

Several indentation flaws of same size are introduced into the tensile surface of specimen by Knoop indentation. These flaws are oriented prependicular to the longitudinal axis of specimens and at least 2mm apart. Figure 1(A) shows schematicly the traces of flaws placed on the tensile surface of specimen and Figure 1(B) shows the flaw profile on the fractur surface of specimen.

After measuring the initial dimension, indentation flaws are extended by subjecting the specimen to constant bending stresses at 1200℃ for durations of 10s to 7 hours, respectively. All experiments are conducted in air and specimens are held at test temperature for about 5min before full load is applied. During preheating the furnace, a preload of about 25MPa is applied on test specimen to prevent from crack healing. The crack extension on the surface of specimen is measured by optical microscopy after unloaded and cooled to room temperature. In order to demonstrate a clear crack tip, a thin surface oxide layer formed at high temperature is removed by slightly polishing the surface before measurements.

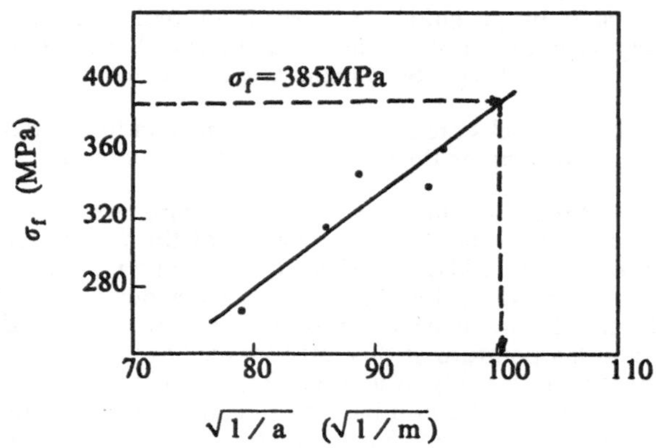

Fig.2 $\sigma_f \sim \sqrt{\frac{1}{a}}$ Relation for SSN

RESULTS

Determination of Residual Stress Magnitude Parameter of Indentation

The variation of fracture stress, σ_f, with inverse square root of crack depth, $\sqrt{\frac{1}{a}}$, for precracked specimens tested at 1200℃, is shown in Figure 2. Although there is a considerable scatter in these data, a reasonablely good linear relationship is evident, according to our earlier works [6, 7], which may be described properly with the following equation:

$$\sigma_f + \frac{\chi_r}{Y_a}\left(\frac{P}{c^2}\right)\sqrt{\frac{c}{a}} = \frac{K_{IC}}{Y_a}\sqrt{\frac{1}{a}} \qquad (1)$$

where a and c are crack depth and half–length; P is indentation load; χ_r is a dimensionless constant which characterizes the magnitude of residual stress; Y_a is geometric factor of semi–elliptical surface defect.

Table I gives some measured results of flexural strength tests of precracked specimens. The fact that the value of a / c, P / c^2, and Y_a can be treated as constants in an appropriate range of indentation load [7] is confirmed again with these data.

TABLE I
Measured Results of Flexural Strength Test of Precracked Specimens at 1200℃

Indentation Load (g)	a / c	P / c^2 (MPa)	Y_a	K_{IC} ($MPa\sqrt{m}$)	χ_r
2700～4700	0.852±0.018	2880.8±111.2	1.258	6.45	0.0533

Using linear regression, the straight line shown in Figure 2 can be described as:

$$\sigma_f + 130.12 = 5.11\sqrt{\frac{1}{a}} \quad (r = 0.9240) \qquad (2)$$

Now the values of K_{IC} and χ_r can be obtained conveniently with Eq.(1) and (2). Calculated results are also listed in Table I .

Following the works of Govila [11–13], extrapolation of the data in Figure 2 to the mean fracture stress of uncracked specimens indicates that the inherent flaws in Si_3N_4 can be treated by analogy of a semi–elliptical surface flaw placed at the tensile surface perpendicular to longitudinal axis located at middle point of bending span with a mean depth of about 98μm.

Stress Rupture Test Results

Figure 3 presents the results of the stress rupture test trails. Twenty–eight specimens are subjected to applied stresses from 190 to 310 Mpa and a delayed failure phenomenon is evident. The larger scatter in time to failure may be due to the inherent material variability.

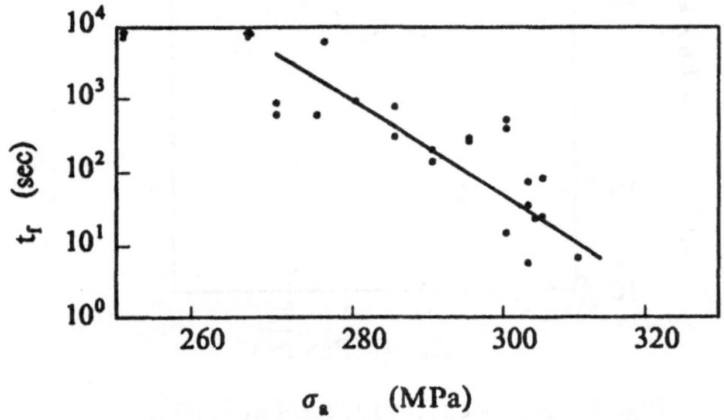

Fig.3 Results of Stress Rupture Testing at 1200℃

A linear relationship exists on the stress rupture data by a log stress ~ log time graph in the higher applied stress range (>265MPa). Following the work of Evans and Wiederhorn [14], this linear relationship can be explained with the SCG mechanism. Using the power relation between crack velocity, v, and stress intensity, K_I, $v=\frac{da}{dt}=AK_I^N$; the time to failure, t_f, for specimen subjected to a higher applied stress, σ_a, can be given by:

$$t_f = B\sigma_a^{-N} \tag{3}$$

Taking logarithms of Eq.(3) gives:

$$\log t_f = -N\log\sigma_a + \log B \tag{4}$$

where logB is a constant.

Eq.(4) shows that a linear relationship exists between $\log\sigma_a$ and $\log t_f$ and the negative of the slope on such a relation would be the SCG exponent N. For the present material, N can be determined to be 33.16.

Extrapolating the straight line from higher stress region in Figure 3 to a lower stress level, one can expect that the specimen which is subjected to an applied stress of 250MPa will not survive for more than 7 hours. However, experiments show that a specimen do not failure at this stress level at time of more than 100 hours in fact. Meanwhile, specimens subjected to rather lower stresses also maintained a longer time duration than that expected with SCG mechanism. On the other hand, an evident creep deformation exists in these specimens. These phenomena seem to indicate that a characteristic stress exists here, above which the failure of ceramics is controlled by SCG mechanism and below which controlled by another mechanism.

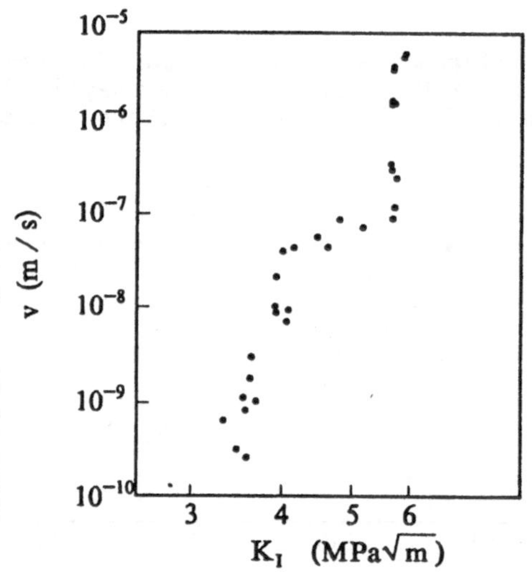

Fig. 4 K_I~v curves for SSN at 1200℃

SCG Behavior by Direct Crack Extension Measurements

Using the IIF technique and the method for K_I~v calculation outlined in Ref.[10], the K_I ~v relation at 1200℃ for same material is shown in Figure 4. There are three region in this diagram and the power law relation between crack velocity and stress intensity can be applied to described each region with different values of A and N (Table II). Specially, a distinct, relatively abrupt, subcritical crack growth threshold, K_{th} , is observed from Figure 4 at about $K_I = 3.3 MPa\sqrt{m}$, implying that no SCG occurs when stress intensity about the tip of flaw drops below the threshold value.

TABLE II
Experimental Results of SCG Parameters

Region	Range of K_I ($MPa\sqrt{m}$)	A	N
I	3.3~4.2	2.51×10^{-22}	22.9
II	4.2~5.6	2.19×10^{-10}	3.6
III	5.6~6.5	1.51×10^{-37}	40.9

It is noted that a difference exists between the values of N obtained with stress rupture and crack extension experiments, respectively. Silimar result has been reported by other authors [15] . A direct explaination for this result may be that the power relationship between crack velocity and stress intensity is used assuming constant values of A and N in stress rupture while different values in crack extension.

DISCUSSION

Recently, it is suggested that delayed failure in ceramics at elevated temperature can occur by several mechanisms including SCG and creep rupture [4, 5] . A characteristic stress level, σ_0 , is proposed by Grathwohl [4] to discern between different stress regions in high temperature rupture. In the present study, a similar conclusion can be obtained with the results of stress rupture in Figure 3. At higher applied stress level, a linear relationship is evident between the log stress and the log time to failure and it is suggested that failure is controlled by SCG in this region. In the lower stress level, however, material can survive for a longer time than that expected by SCG mechanism, implying a change in the criterion for failure. It seems to indicate that creep rupture may govern the failure in this region, for a relative larger creep deformation is observed. The characteristic stress, which is proposed by Grathwohl, for the present material can be determined approximately to be 265MPa from Figure 3 at 1200℃.

In fact the value of this characteristic stress can also be calculated with SCG threshold, K_{th} , in certain environment temperature, for K_{th} is a intrinsic parameter for describing the SCG behavior below which no SCG occurs. Using the relationship among flaw size, a , applied stress, σ_a , and stress intensity, K_I,

$$K_I = Y\sigma_a\sqrt{a} \qquad (6)$$

(where Y is a constant depending on specimen and crack geometry) we obtain:

$$\sigma_0 = \frac{K_{th}}{Y\sqrt{a}} \qquad (7)$$

Since the inherent flaws in the present material can be equivalent to a semi–elliptical surface flaw with a mean depth of about 98μm, the value of Y for the surface flaw produced by Knoop indentation, =1.26, can be used approximately. Taking $K_{th}=3.3MPa\sqrt{m}$, it is shown with Eq.(7) that the mean value of σ_0 is about 264MPa, which is consistent with the result obtained with stress rupture test.

Now, we can propose another criterion for distinguishing between the two mechanisms for delayed failure, SCG and creep rupture, with an intrinsic parameter, K_{th}: if the stress intensity at the tip of the most serious flaw in material due to applied stress, K_I, is larger than the SCG threshold, K_{th}, the delayed failure of materials under steady loading is controlled by SCG; otherwise, creep rupture is the dominating process.

CONCLUSIONS

Stress rupture and crack extension experiments are performed at 1200℃ with a pressureless sintered Si_3N_4 to characterize the delayed failure behavior. The following conclusions are obtained:

(1) A linear relationship is evident between stress rupture data in higher applied stress region, denoting that the failure of material is controlled by SCG in this region.

(2) A clear deviation appears when extrapolating the straight line from higher stress region to lower stress region, implying a change in mechanism for failure.

(3) The characteristic stress, σ_0, for discerning between different stress regions in high temperature rupture, can be calculated with SCG threshold, K_{th}. Calculated result is consistent with the measured result of stress rupture.

(4) If the stress intensity at the tip of flaw due to applied stress is larger than SCG threshold, the failure of material under steady loading is controlled by SCG; otherwise, creep rupture is the dominating process.

REFERENCE

〔1〕 G.D.Quinn, "Delayed Failure of a Commercial Vitreous Bonded Alumina," J. Mater. Sci., 22 (1987) 2309–18

〔2〕 G.G.Trantina, "Strength and Life Prediction for Hot–Pressed Silicon Nitride," J. Am. Ceram. Soc., 62 (1979) 377–80

〔3〕 G.D.Quinn and J.B.Quinn, "Slow Crack Growth in Hot–Pressed Silicon Nitride"; pp.603–36 in " Fracture Mechanics of Ceramics 6" . Edited by R.C.Bradt, A.G.Evans, et.al. (Plenum, New York, 1983)

〔4〕 G.Grathwohl, "Regimes of Creep and Slow Crack Growth in High–Temperature Rupture of Hot–Pressed Silicon Nitride"; pp.573–86 in "Deformation of Ceramics 2". Edited by R.E.Tressler and R.C.Bradt (Plenum, New York, 1984)

〔5〕 G.D.Quinn, " Static Fatigue Resistance of Hot–Pressed Silicon Nitride" ; pp.319–32 in " Fracture Mechanics of Ceramics 8" . Edited by R.C.Bradt, A.G.Evans, et.al. (Plenum, New York, 1983)

〔6〕 Z.D.Guan and J.H.Gong, "Investigation of Test Reliability in K_{IC} Determination with Knoop Indented 3–PT Bending Method"; pp.696–700 in "Proceedings of the Third International Symposium on Ceramic Materials and Components for Engines". Edited by V.J.Tennery (The Am. Ceram. Soc. Inc., 1989)

〔7〕 J.H.Gong and Z.D.Guan, "A Modified Indented Bending Beam Method for the Determination of Fracture Toughness of Ceramics," J. Chinese Ceram. Soc., 19 (1991) 234–41

〔8〕 M.G.Mendiratta and J.J.Petrovic, "Slow Crack Growth from Controlled Surface Flaws in Hot–Pressed Si_3N_4," J. Am. Ceram. Soc., 61 (1978) 226–30

〔9〕 M.Kawai, H.Abe, and J.Nakayama, " Indentation–Induced Flaw Method for Measuring Crack Velocity in Sintered Si_3N_4"; pp.587–601 in "Fracture Mechanics of Ceramics 6". Edited by R.C.Bradt, A.G.Evans, et.al. (Plenum, New York, 1983)

〔10〕 J.H.Gong and Z.D.Guan, "Effect of Residual Stress due to Knoop Indentation on Subcritical Crack Growth Behavior in Ceramics" ; presented at the 5th International Symposium on Fracture Mechanics of Ceramics (Nagoya, Japan, July 15–17, 1991)

〔11〕 R.K.Govila, "Material Parameters for Life Prediction in Ceramics"; pp.536–67 in " Ceramics for High–Performance Application 3" . Edited by E.M.Lenon, R.N.Katz, and J.J.Burke (Plenum, New York, 1983)

〔12〕 R.K.Govila, "Phenomenology of Fracture in Sintered Alpha Silicon Carbide," J. Mater. Sci., 19 (1984) 2111–20

〔13〕 R.K.Govila, "Fracture Phenomenology of a Sintered Silicon Nitride Containing Oxide Additives," J. Mater. Sci., 23 (1988) 1141–50

〔14〕 A.G.Evans and S.M.Wiederhorn, "Proof Testing of Ceramic Materials —— An Analysical Basis for Failure Prediction, " Int. J. Fract. Mech., 10 (1974) 379–92

〔15〕 B.J.Pletka and S.M.Wiederhorn, " A Comparison of Failure Predictions by Strength and Fracture Mechanics Techniques," J. Mater. Sci., 12 (1982)

FAST CATALYTIC DEBINDING OF INJECTION MOULDED PARTS

J.H.H. TER MAAT, J. EBENHÖCH and H.-J. STERZEL
BASF AG, 6700 Ludwigshafen, Germany

ABSTRACT

In a new approach to powder injection moulding, BASF has developed a new, polyacetal-based binder system. The rheological properties of the binder enable part dimensions and shapes that were previously impossible to mould. Debinding is achieved by a new catalytic process: The polyacetal binder is depolymerised in the presence of an acidic gas and volatilises as formaldehyde. Debinding proceeds from the surface inward with a speed of up to 2 mm/h, which is more than ten times faster than the conventional debinding technology allows. The low debinding temperature excludes warping, sagging or slumping.

INTRODUCTION

The reduction of costs by the integration of several functions in one part, miniaturising and net shape forming is a most important objective in the manufacturing industry. Among the production routes to complex precision parts by sintering a powder preform, powder injection moulding is the most attractive alternative: It achieves a very high production rate, combines a near-net shape capability with a good surface finish, and is easy to automate.

In the course of the last twenty years, intensive research has been conducted to develop a commercially feasible powder injection moulding process. For a review, the reader is referred to Edirisinghe and Evans (1,2). The initially used conventional thermoplastic polymers like polyethylene, polypropylene, polystyrene, have now largely been abandoned due to the painfully slow pyrolytical debinding operation (several days for 3 mm thickness). Presently a Wiech-type process (3) is favored. The binder is essentially a solution of a polymer, usually polypropylene in oil or a low-melting wax.

Debinding is a combination of evaporation and pyrolysis (3-5). Higher removal rates are claimed by pressurised debinding (6) or by extraction of the solvent with supercritical carbon dioxide (7).
The debinding cycle length in a conventional Wiech-type process is reduced to 1/2-1 day for a part with 3 mm wall thickness. This has been sufficient to find some niche applications for powder injection moulding.

MARKET

The powder injection moulding route is accessible for two important classes of structural materials, metals and ceramics. Metal Injection Moulding (MIM) has always had a substantial lead on Ceramics Injection Moulding (CIM) in terms of market penetration. This is due to the higher flaw tolerance of metals, and also due to the relative coarseness of metal powders, which alleviates some of the debinding problems. MIM has since the seventies been pursued mainly by small specialist firms, and has found a market in precision parts, for example parts for office equipment, watches and medical tools. After years of steady, rather slow growth the MIM market is now rapidly expanding at the expense of die casting. The turnover in MIM is still modest, but the annual growth rate is becoming quite impressive (Table 1).

Table 1: Market estimates and forecast for powder injection moulding worldwide (1990 - 2000)

	Year	MIM	CIM
Market volume	1990	500 - 800 t/a	200 - 300 t/a
Turnover	1990	80 million US$	100 million US$
Market growth	1990 - 1995	45 % p.a.	10 % p.a.
Market growth	1995 - 2000	25 % p.a.	-
Turnover	2000	1 billion US$	-

In CIM thread guides have been the first application. After considerable efforts, Japanese firms have succeeded in introducing injection moulded turbocharger rotors and diesel engine precombustion chambers in passenger cars. In spite of these successes, CIM is not on the verge of a breakthrough. The CIM turnover figure is prejudiced by the fact that CIM is pushed by few large companies with a large captive use; the market growth rate is not more than the corresponding figure for technical ceramics, i.e. CIM has not been able to reduce the dominance of the simple, but laborious powder pressing method. Forecasts beyond 1995 for CIM are very volatile, because they depend heavily on action rather than plans to introduce ceramics in car engines, and these figures are therefore not given in Table 1.

Powder injection moulding has needed a long time for penetrating the sintered metals business, and it is still stagnating in the ceramics sector. The cause for this hesitating acceptance are the limitations of the conventional powder injection moulding formulations; these limitations originate in the innate properties of the conventional binder.

FLAWS OF THE CONVENTIONAL BINDER

The efforts to shorten the debinding cycle led to a steady increase in the fraction of volatiles in the conventional binder, and ultimately reached about 80 % (4,5,8,9) as a compromise between shape retention of the as-moulded parts and speed of binder removal. Unfortunately, the high fraction of low-melting compounds has unpleasant consequences in mouldability, debinding and sintered part integrity.

The low viscosity of the molten binder phase makes the conventional powder injection moulding formulations prone to powder particle segregation. Segregation can manifest itself in several ways, depending on the cause: Shear gradients or differential acceleration.
Shear-induced particle migration can cause seizure of the screw in an injection moulding machine when the volume fraction of solids is high and appropriate internal lubrication is lacking. In piston extruders this type of segregation is noticed as the squeezing-out of binder (10). At a solids content which is not quite so close to the limit for immediate dilatancy, the effect will be noticed as a pressure limitation for injecting into the cavity. Generally, the conventional binders do not allow higher pressures than 500-600 bar, and this puts rather severe restrictions on part geometry (maximum length: diameter ratio approximately 10). Wall shear-induced segregation is the cause for the occurrence of weld lines. This is particularly pronounced when a solution of a higher-molecular weight polymer in oil (or wax) is used as the binder. The binder itself can segregate at the interface by bleeding out the oil, which aggravates the already occurring powder particle segregation. These weld lines do not cure in the mould, and persist throughout debinding and sintering, giving rise to a weak plane or cracks.
Segregation by differential acceleration can occur when the injected melt is forced to change its direction of flow. This is quite common practice to avoid jetting into the mould cavity. During the change of direction, the binder and the powder particles experience, due to the difference in specific gravity, a different centrifugal force. The difference in acceleration causes an increase in binder content on the gate side of the moulded part. This leads to warping and sink marks.
Both types of segregation will show a linear dependence of the extent of segregation on the viscosity of the binder phase. This explains the shear sensitivity of conventional powder injection moulding formulations.

A further consequence of the presence of a substantial volume fraction of low-melting compounds in conventional binders is that the moulded part will deform during the debinding operation by viscous flow. To avoid sagging and slumping the moulded part is supported in a powder bed (5,10,11) or on a porous ceramic having the contours of the moulded part (9). This incurs handling problems and additional costs.

In spite of the improvements made in debinding, powder injection moulding is currently not considered for parts with a wall thickness above eight millimeters. This sets another size limit in part geometry.

From this overview the intrinsic properties of conventional binders emerge as the source of most, if not all practical obstacles for producing sintered metal or ceramic parts economically by powder injection moulding. Exactly here the BASF binder system offers considerable improvements.

THE BASF BINDER SYSTEM

The research at BASF has been aimed at developing a binder with improved mouldability, which can be removed quickly from the moulded part without leaving a char residue. For this purpose a preselection was made of polymers that decompose mainly by unzipping into monomer units, for example, polymethylmethacrylate and polyacetal. The clean and relatively quick pyrolysis of polyacetals had already drawn attention (12). At an early stage in our work it was discovered, that the depolymerisation of polyacetals is catalysed in the presence of an acid. This proved to be an immense advantage.

Normal, commercially available polyacetals were found to be not very well suited for powder injection moulding:

- Polyacetals are highly crystalline. This invokes substantial shrinkage during cooling in the mould. Near-net shaping is difficult; thicker parts exhibit shrinkage voids in the center.

- Polyacetals are shear sensitive. Upon exposure to high shear stresses during compounding and moulding they tend to chain rupture and degradation. The evolution of formaldehyde causes pores and entails a poor reproducibility of the volume fraction of binder.

Much of our work has been devoted to the development of polyacetals that meet the numerous processing requirements. The metal or ceramic powders are dispersed in these modified polyacetals by a proprietary, mildly shearing compounding process. The resulting granulate is ready for moulding, and contains 50-60 Vol.-% powder. The BASF binder system has demonstrated to work well with standard submicron ceramic powders (Si_3N_4, SiC, Al_2O_3, Y_2O_3 and ZrO_2), and with stainless steel, iron and nickel alloy powders.

Moulding can be performed with a screw injection moulding machine, preferably equipped with an abrasion resistant screw, cylinder liner and check ring. The cylinder is heated to 160 - 190 °C, depending on feedstock grade and part size. Injection pressure may be taken up to the machine limit, usually 1500 - 2000 bar. The mould temperature must be optimised for a given part; a suitable range is 110 - 150 °C, using a mould cycle time of 10 - 60 seconds. Under optimised moulding conditions we have observed no significant binder segregation. We attribute this in part to the higher viscosity of the BASF binder system compared with conventional binders. A second contribution originates in the polar character of polyacetals; this provides for an improved adhesion between the binder and the powder surface.

Debinding of the moulded green parts is achieved by depolymerisation of the polyacetal-based binder. The depolymerisation of polyacetals into formaldehyde is catalysed by introducing a gaseous acid in the atmosphere. Good

results are obtained with HNO_3, HCl and BF_3.
We must immediately and strongly advise against the use of HCl as a catalyst. In this case, the symmetrical dichlorodimethylether may be formed in the gas phase; this substance is known to be highly carcinogenic.
The catalytic debinding treatment proceeds rapidly at temperatures above 100 °C, which is well over the boiling point of formaldehyde. Through the action of the catalyst (preferrably HNO_3), the polyacetal binder unzips rapidly and vaporises off as formaldehyde gas (Figure 1).

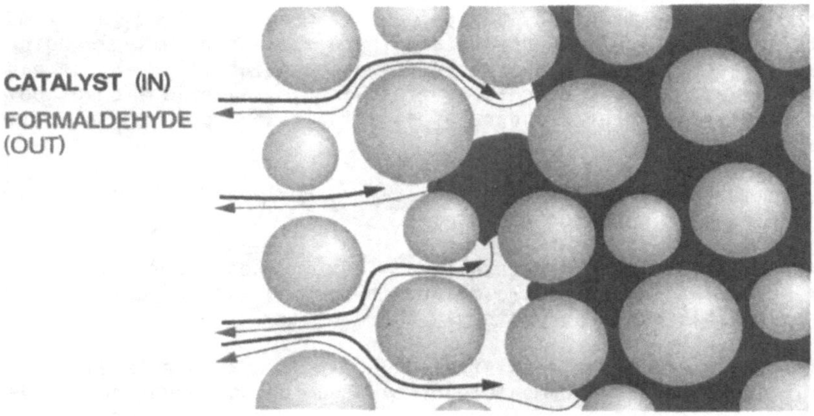

Figure 1: Schematic representation of the debinding mechanism

By choosing a debinding temperature below the softening point of the polyacetal-based binder (150-160 °C), some crucial advantages over conventional binders are secured:

- Deformation of the green part due to gravitational influence or stress relaxation is excluded.
- Depolymerisation takes place as a direct solid-gas transition. Capillary forces play no role.
- Depolymerisation takes place only at the gas-binder interface. Pyrolytic decomposition inside the binder phase is impossible.
- Debinding is easy to control by temperature and catalyst concentration.
- Debinding is much faster than in conventional systems. The cycle time is shortened by a factor of 10 or more.

For the debinding operation the green parts are transferred into an oven. Suitable is a simple, gas tight oven with an inner casing of stainless steel (see Figure 2).

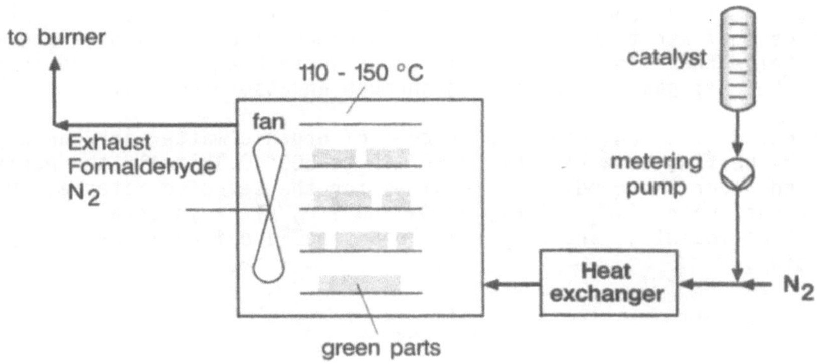

Figure 2: Flow sheet of the debinding process

The oven is heated to 110 - 150 °C and purged with preheated nitrogen gas to avoid an explosive atmosphere during the actual treatment, at the same time avoiding excessive surface oxidation of sensitive powders. The debinding process is not disturbed by minor variations in temperature (± 5 °C), so inexpensive equipment can be used. After reaching the process temperature 0,1 - 2 Vol.-% of gaseous catalyst is introduced into the hot purge gas. To achieve thorough mixing and a proper heat transfer a circulating fan is commendable.
By virtue of the fact, that gas exchange is limited to the already debinded, porous zones, debinding proceeds according to a shrinking core model (see Figure 3). The interface travels inward with a speed of up to 1-2 mm/h, depending on powder size and stage of debinding. Above 10 mm part thickness, diffusion through the porous zone starts to be the rate-limiting process in the final stage of debinding.

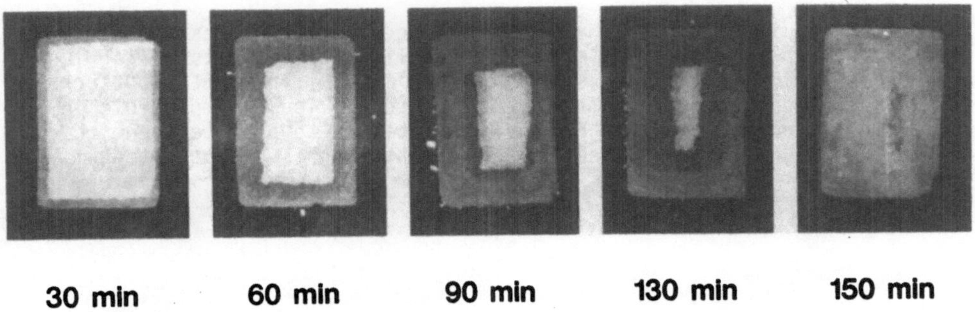

Figure 3: Stages in debinding of a green bar (4 x 6 mm), moulded from a feedstock, based on a 0,7 um Al_2O_3 powder. The dark coloured zones have been penetrated by a dye solution, which occasionally damaged the fragile debinded zone.

The exit gas, mainly formaldehyde in the carrier gas nitrogen, is burnt with an excess of air to carbon dioxide and water. The traces of NO_x-gas stay well below the allowed limits. In cases where BF_3 is preferred as a catalyst, the exit gas must be passed through an absorbing unit.

After debinding, the few remaining percent of organic matter are rapidly pyrolysed by heating at a rate of 50 °C/min to 500 °C. Finally the parts are sintered under the conditions required for the selected material. We have experience with Si_3N_4 (HPSN), Al_2O_3 and ZrO_2 (TZP) in ceramics injection moulding, and with stainless steel and iron in metal injection moulding.

OUTLOOK

The prospects for powder injection moulding have much improved with the newly developed BASF binder system. A comparison with a conventional MIM-feedstock is presented in Table 2, demonstrating the much higher flexibility with respect to shape, the improvements in near-net shaping and the higher rate of binder removal for a process, based on the BASF feedstock.

Table 2: Comparison of MIM-feedstocks (57 Vol.-% of 4 um Fe powder) on some essential processing features.

Feature	Conventional	BASF
Injection pressure	< 600 bar	< 2000 bar
Size ratio (L/d)	< 10	< 100
Segregation	yes	no
Deformation	yes	no
Debinding rate	0.1 mm/h	1-2 mm/h
Part thickness	< 8 mm	< 40 mm

The comparison is made for MIM-feedstocks, because the processing parameters are better known in this area. Our results with ceramics so far indicate, that similar improvements are obtained in CIM. The BASF feedstocks allow part dimensions and geometries that were previously impossible in powder injection moulding. In Figure 4 the ability to produce long, thick or large parts by moulding, debinding and sintering is demonstrated.

Figure 4: Various sintered shapes, demonstrating the flexibility of the BASF MIM-feedstocks. The length bar indicates 10 mm.

It is the intention of BASF to market <u>feedstocks</u>, rather than the binder or finished parts. The feedstocks will be based on well-established sintering compositions, e.g. 93 % Si_3N_4 - 5 % Y_2O_3 - 2 % Al_2O_3 for gas pressure-sintered silicon nitride (HPSN). Currently we are working on the development of several standard feedstock grades. For CIM, this includes Al_2O_3, Si_3N_4 (SSN and HPSN), ZrO_2 (TZP) and SiC (SSiC). In MIM, Fe-Ni alloy (2-8 % Ni), stainless steel and pure Fe feedstocks are under active development. Customers will be supported in determining the optimum moulding conditions and in establishing the debinding technology. Market introduction will be done with pilot plant quantities, collaborating with few innovative companies. The production capacities will follow the market needs.

With the production of the feedstock BASF is taking care of the most difficult part in the powder injection moulding technology. The access to BASF-feedstocks will enable smaller firms to enter an entirely new field of activities, armed with a better technology, and requiring relatively low investments. This will increase competitiveness in a market which is controlled by a few companies, which in turn will push out established technologies like die casting and cold isostatic pressing quicker. We expect therefore, that the BASF binder system will help powder injection moulding to its long-anticipated breakthrough.

REFERENCES

1. M.J. Edirisinghe and J.R.G. Evans, Int. J. High Technology Ceramics <u>2</u> (1986) 1-31
2. M.J. Edirisinghe and J.R.G. Evans, Int. J. High Technology Ceramics <u>2</u> (1986) 249-278
3. R.E. Wiech, US Patent 4 404 166 (1983)
4. S. Ito and T. Mizuno, European Patent 0 125 912 (1987)
5. K. Kato, M. Shirai and A. Nagoya, German Patent 36 30 690 (1989)
6. M. Inoue, Y. Kihara and Y. Arakida, Interceram. <u>2</u> (1989) 53-57
7. T. Miyashita, Y. Ueno, H. Nishio and S. Kubodera, European Patent 0 202 852 (1990)
8. S. Sasaki, US Patent 4 701 106 (1987)
9. T. Matsuhisa, S. Sasaki, German Patent 35 07 804 (1989)
10. C.L. Quackenbush, K. French and J.T. Neil, Ceram. Eng. Sci. Proc. <u>3</u> (1982) 20-34
11. J.A. Mangels and W. Trela, Adv. Ceramics <u>9</u> (1984) 220-233
12. G. Farrow and A.B. Conciatori, European Patent 0 114 746 (1989)

ADVANCED PROCESSING OF SILICON NITRIDE CERAMICS

J.T. NEIL, S. NATANSOHN, AND A.E. PASTO
Materials Science Laboratory
GTE Laboratories Incorporated
40 Sylvan Road, Waltham, MA 02254, USA

ABSTRACT

The results of a study designed to improve the mechanical strength and reliability of structural silicon nitride components are reported. The objective of the study was to eliminate the frequently found metallic inclusions that originate from the abrasive wear that occurs in processing this hard material in metallic equipment and which adversely affects the material properties. The feasibility of replacing metals with polymeric materials in equipment where substantive wear occurs was addressed. The data showed that it is possible to successfully compound silicon nitride with injection molding binders in a three-way oval orbital mixing chamber, realizing melt index viscosities required for shaping of complex components. Furthermore, engineering plastics were identified which have the requisite combination of properties (such as abrasion and erosion resistance, thermal stability, and cost) to make them suitable for components of an injection molding machine. Initial evaluation results showed excellent wear resistance and greatly reduced metallic inclusions in the molded components.

INTRODUCTION

The industrial utilization of silicon nitride ceramics, materials whose unique properties allow new and innovative solutions to component design problems, is hampered by the low reliability of these brittle solids. Silicon nitride ceramics are characterized by a strength distribution which is related to the material's fracture toughness and its flaw concentration and characteristics. Several approaches can be taken to improve its strength reproducibility (Weibull modulus), and these were discussed previously [1]. Recently, GTE Laboratories has been engaged in a program* designed to improve the reliability of silicon nitride ceramics; its

*Research sponsored by the U. S. Department of Energy, Assistant Secretary for Conservation and Renewable Energy, Office of Transportation Systems, as part of the Ceramic Technology for Advanced Heat Engines Project of the Advanced Materials Development Program under contract DE-AC05-840R21400 with Martin-Marietta Energy Systems, Inc., Work Breakdown Structure Subelement 1.1.4.3.

emphasis is on identifying strength limiting factors and minimizing or eliminating them through improved processing and process control techniques. The overall project objective is to develop processing techniques which would result in the reproducible fabrication of silicon nitride ceramics with the following properties:

- Average tensile strength of 900 MPa at 25°C
- Average tensile strength of 500 MPa at 1370°C
- Weibull modulus of 20 in both instances

These challenging goals require significant technical advances. It is noteworthy that the program objectives call for a substantial increase in both the tensile strength and Weibull modulus in a two-pronged thrust to enhance ceramic component reliability and thus reduce the probability of component failure. This is being accomplished by determining the source(s) of failure-causing defects and by modifying and controlling the manufacturing process to minimize their occurrence. All potential sources of defects are evaluated, from raw materials through individual powder processing and densification steps and finally through machining and surface finishing of the test specimen. The details of the overall program and its progress in the various phases have been discussed elsewhere [1–3]; this report reviews the progress of efforts to eliminate metallic inclusions in the ceramic parts.

RESULTS AND DISCUSSION

The material selected by GTE Laboratories for this program was a commercial silicon nitride powder of high purity with 6 w/o of yttria as sintering aid (GTE Labs' PY6 composition). Injection molding is being used to form the components which are densified by hot isostatic pressing. The composition and fabrication process were selected because they have been shown to yield ceramics capable of meeting the challenging goals of this program. The initial results obtained in the evaluation of the tensile strength of the test specimens (1.1 cc volume under stress) were, at 455 MPa, lower than expected. The monitoring of the samples in both green and densified states by x-ray microfocus radiography has revealed the pervasive presence of high density inclusions (HDI), which were then identified as being of metallic origin. Their largest dimension (length) is in the range between 50 and 200 mm. The tensile rods failed at these inclusions, and analysis of the fracture origins by SEM/EDX showed the presence of Fe, Ni, Cr, indicative of steel, and Zn, Cu, which implies contamination from brass. X-ray dot maps and backscatter images indicate that these contaminants are not present as discrete metallic flaws but are dispersed within the silicon nitride matrix in a localized region at the inclusion site. The composition of the inclusions identifies erosion of the processing equipment (steel from the injection molding machines and steel as well as brass from the compounder) as the major contamination source. The need to eliminate contact between the abrasive ceramic formulation and the metal surfaces of the processing machinery is compelling because most tensile bars fail at identifiable metallic inclusions. Thus, any attempts to improve ceramic performance through powder, formulation, mixing, or machining modification are overshadowed by the inclusion effects.

There are several approaches to eliminating this contamination problem. The currently practiced one is the hardfacing of the metallic equipment; however, such coatings will eventually be breached, and this process is not applicable to some of the equipment. Another alternative is to replace some of the metallic components with ceramic ones, but densification of large ceramic parts is difficult, and machining of such components to the required tolerances is expensive. In the present approach, the feasibility of using engineering plastics was evaluated.

Nylon Compounder

In the initial step, nylon was used to construct a noncontaminating compounding system. A single silicon nitride mixing ball was used for the dispersion of the powder and injection molding binder formulation. This small mixer proved the feasibility of the concept in that it was capable of compounding 20-g batches of PY6 using a preheated mixing chamber and ball. Melt index measurements indicated that a 10-min orbital mixing time resulted in melt index values comparable to those observed on PY6 batches compounded in standard metallic sigma blade mixers.

Subsequently, a larger compounder, capable of compounding 600 g of PY6 per batch, was built. The mixing chamber was again constructed of nylon with an oval internal configuration. The system utilizes a three-way orbital mixing approach and a single mixing ball. Several large-diameter mixing balls were evaluated:

- Isopressed and HIPed PY6 (density = 3.2 g/cc)
- Isopressed and sintered zirconia (density = 6.0 g/cc)
- Nylon capped steel ball (composite density = 3.5 g/cc)
- Nylon capped WC ball (composite density = 6.0 g/cc)
- Nylon capped WC ball (composite density = 10.0 g/cc)

In compounding the formulation, the powder, binder, and mixing ball were loaded into the nylon mixing container and preheated to the compounding temperature. The sealed mixing chamber was then loaded into the orbital mixer and compounded for the desired period (typically 10 min); little heat is lost during the mixing period. The compounded mix was then removed, cooled and granulated. As anticipated, no metallic inclusions were found in the material compounded in the nylon mixer.

The suitability of the compounded materials prepared by this technique for injection molding was evaluated by melt index viscosity measurements for four processing conditions:

- PY6 mixing ball and warm mixer
- PY6 mixing ball and warm mixer, followed by twin screw compounding
- Nylon capped WC ball and warm mixer
- Nylon capped WC ball and cooler mixer

The resulting viscosity curves are shown in Figure 1.

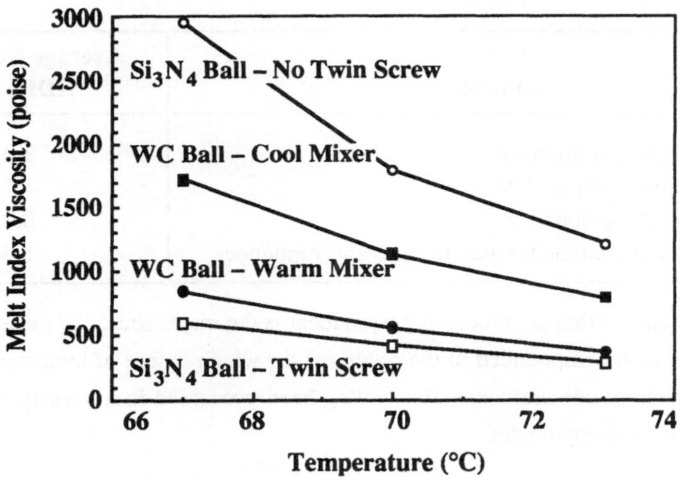

Figure 1. Melt index viscosities of PY6/binder formulations compounded in nylon orbital mixer.

The material mixed with the lower density PY6 ball has a high viscosity, but it is still within the range suitable for injection molding. Use of twin screw compounding after this mixing results in a substantial decrease in viscosity. It was also observed that the use of a higher density mixing ball showed a major improvement in compounding, and that a higher temperature preheat is desirable for best mixing quality. The conditions of a high density mixing ball along with a higher mixing temperature resulted in a mix viscosity comparable to that realized by processing the material in a high shear twin screw compounder. To date, there has been no observable wear of the nylon mixing container or the capped balls in repeated compounding experiments using this system.

Based on these experimental findings, the larger mixer was utilized to produce sufficient compounded mix for tensile specimen injection molding trials. Half of the prepared mix was used direct from the plastic compounder; the remaining material was passed through the twin screw compounder prior to granulation. Other batches were prepared with the standard large Abbe sigma-blade compounder to serve as a baseline. Material from these batches has been molded into tensile specimens, dewaxed, and HIPed. They are being machined into tensile test specimens. Green-state NDE proved that less metallic contamination was incurred utilizing the noncontaminating mixer, as expected (Table 1).

TABLE 1

Effect of compounding equipment on the concentration of metallic inclusions

Equipment	Average Number of HDI's/cm^3
Large ABBE compounder	3.0
Small ABBE compounder	1.7
Nylon orbital compunder	0.0
Nylon orbital compunder with twin screw compunder	1.0

The data confirm that the processing equipment is the major source of metallic inclusions because the material compounded in the nylon mixer, which is free of inclusions, does show the presence of these contaminants after having been processed in a subsequent step in the metallic twin-screw compounder.

Plastic Processing Equipment

In an extension of this effort, a study was initiated to evaluate engineering plastics for potential use in silicon nitride processing equipment. The goal of the activity was to replace metallic equipment components prone to wear with ones made from engineering plastics [4-6], thus decreasing or eliminating metallic inclusion defects in injection molded silicon nitride. This study was designed to evaluate a range of plastic materials for room-temperature erosion and abrasion resistance, taking into consideration the thermal properties of the materials [7, 8] and their consequent suitability for operating at injection molding temperatures, typically 70–100°C. The cost of the material [9] as well as its availability in the requisite sizes was also considered. Generally, engineering plastics which are stable at higher temperatures are substantially higher in cost. The list of the plastics evaluated and their abbreviations is given in Table 2.

The abrasion resistance of the materials was measured by calculating the sample volume loss when forced against a 30-μm metal-bonded diamond polishing wheel with a 1-kg load.

The weight losses on 1.27 × 1.27 × 0.635 cm samples were measured at 1-min intervals (5-min total test time) and converted to volume loss using the sample density. The linear slope of volume loss versus abrasion time was calculated to determine abrasion rate [10].

Erosion resistance was measured by grit blasting the samples at 5.5 kPa using 50-μm Al_2O_3 powder. The sample was placed 13 mm from the blaster nozzle, and sample volume loss was calculated after 1 minute of blasting [10].

TABLE 2
List of evaluated plastics and their abbreviations

Abbreviation	Name
UHMW PE	Ultra-high molecular weight polyethylene
Nylon 6/6	Polyamide
PEEK	Polyetheretherketone (Victrex 450G)
PE	Polyethylene
PP	Polypropylene
PI	Polyimide (unfilled: Vespel SP-1, filled: Vespel SP-21)
PEI	Polyether imide (Ultem 1000)
PAI	Polyamide imide (Torlon 4301)
PTFE	Polytetrafluoroethylene

Tests such as these are useful in a general way in ranking the performance of different materials, but neither test accurately duplicates the actual wear mechanism. It is expected that the flow of a submicron Si_3N_4 powder in a heated fluid binder over a surface will produce wear mostly by abrasion, but with some erosive contribution as well.

The results of this study are summarized in Table 3, which also includes information on maximum use temperature (temperature where deflection occurs under a load of 18.2 kPa) and bulk raw material cost where available. Data obtained under the same conditions on a hardened tool steel sample (Crucible 95) are included for comparison purposes.

The measured wear rates in the study for both tests span two orders of magnitude, showing the wide performance range of available plastics. While many of the plastics tested showed poor wear resistance relative to tool steel, several plastics showed comparable or superior wear rates. For room-temperature applications, UHMW PE would be desirable. This material showed minimal wear in both tests, but it is only usable at low temperatures. Nylon would be a good choice at intermediate temperatures (50–90°C); its wear was comparable to tool steel, and it is readily available at low cost and in large sizes. For high-temperature use, or moderate-temperature/high-stress applications, polyetheretherketone (PEEK) appears to be the first choice. This material is, however, expensive and somewhat limited in sizes of stock available. A second possible choice in this temperature range is polyetherimide (PEI). PEI had a medium wear rate, but is readily available in large sizes at 20% of the cost of PEEK. It was also noted in the study that materials with good abrasion resistance generally had good erosion resistance as well. This is shown graphically in Figure 2.

TABLE 3
Results of wear study

Material	Heat Deflection Temp (°C)	Bulk Cost ¢/cm³	$\times 10^{-3}$ Erosive Wear Rate (cc/min)	$\times 10^{-3}$ Abrasive Wear Rate (cc/min)
UHMW PE	48	0.20	0.3	0.5
Nylon 6/6	88	0.69	3.9	2.0
Tool Steel	—	—	8.1	2.0
PEEK	160	6.67	4.4	2.4
Teflon (PTFE)	56	2.93	19.2	3.7
Low Pressure PE	38	0.10	6.4	4.1
PP	54	0.09	2.1	4.8
PI (Unfilled)	349	5.80	30.9	5.0
PEI	200	1.31	13.8	8.8
PI (Filled)	360	5.92	56.1	13.0
PAI	282	5.86	49.2	23.6
LCP (Filled)	204	6.10	44.4	24.2

To evaluate the performance of an actual plastic component, an injection screw for a 15-ton Boy injection molding machine was fabricated from PEEK. In designing replacement components from this material, it is necessary to provide for the difference in the thermal coefficient of expansion between PEEK and steel, which is approximately three times larger for the former [11]. This screw was used to injection mold about 5 kg of PY6 silicon nitride into turbocharger-sized test samples. Inspection of the screw after this test revealed some polishing of the screw, but no significant wear.

Based on this successful result, four screw segments for a twin screw compounder were made from PEEK. The segments produced were for the feed end of the compounder, where conditions are believed to be most abrasive. The twin screw compounder using the four PEEK screw segments was also successfully tested. About 40 kg of precompounded PY6 (powder plus binder) was passed through the compounder. The barrel was removed after the test, and no noticeable wear of the PEEK segments was observed. Construction of an entire pair of PEEK screws for this compounder is now planned. In addition, a PEEK injection screw has been designed for the 200-ton injection molding machine. This component failed in torsion on its second molding run and is currently being redesigned. Thus, plastic components are being evaluated in a variety of processing operations to establish the viability of their continued use for injection molding of silicon nitride. Concomitantly, the tensile strength of ceramics made by these techniques is being evaluated in order to demonstrate the anticipated improvement in mechanical properties of inclusion-free specimens.

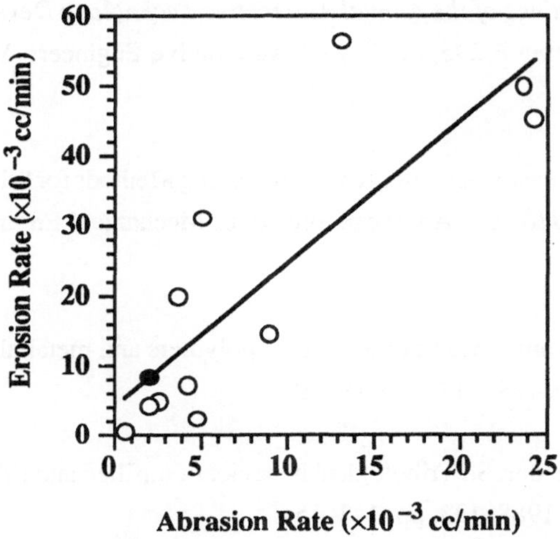

Figure 2. Relationship between abrasion and erosion rates (solid point is tool steel).

CONCLUSIONS

The results of this study indicate the feasibility of using plastic components in silicon nitride processing equipment, and thus reducing significantly the concentration of metallic inclusions. Successful compounding of silicon nitride formulations has been realized in orbital mixers made from nylon. A comprehensive evaluation of a variety of engineering plastics has identified materials that could be used in injection molding machinery as well as conditions for their use. At room temperatures, UHMW PE is a suitable material; at intermediate temperatures, nylon is useful; while at more stressful conditions, PEEK is the material of choice. Test results to date in actual processing equipment show these materials to have superior wear resistance. It is anticipated that ceramics fabricated under these conditions will have greatly improved strength, commensurate with the programs' objectives.

REFERENCES

1. Pasto, A.E., and Natansohn, S., Advanced Processing of High-Performance Silicon Nitride Ceramics. In *Proceedings of the 27th Automotive Technology Development Contactors'*

Coordination Meeting P-230, Society of Automotive Engineers, Warrendale, PA, 1990, pp. 197–206.

2. Pasto, A. E., and Natansohn, S., Improved Processing Methods for High Reliability Silicon Nitride. In *Proceedings of the Annual Automotive Technology Development Contractors' Coordination Meeting P-243,* Society of Automotive Engineers, Warrendale, PA, 1991, pp. 143-160.

3. Natansohn, S., and Pasto, A.E., Improved Processing Methods for Silicon Nitride Ceramics. *Paper No. 91-GT-316,* The American Society of Mechanical Engineers, New York, NY, 1991.

4. Madsen, B.W., A comparison of the wear of polymers and metal alloys in laboratory and field slurries. *Wear,* 1989, 134, pp. 59–79.

5. Jain, V.K., and Bahadur, S., Tribological behavior of unfilled and filled poly(amide-imide) copolymer. *Wear,* 1988, 123, pp. 143–154.

6. Cadillac Plastic and Chemical Company, *CADCO® Engineering Plastics,* 1989.

7. Miller, B., Resins to turn to when the heat is on. Plastics *World,* August 1990, pp. 40–45.

8. Anon., Plastics. *Machine Design,* October 1990, pp. 61–149.

9. Sherman, L., Pricing update. *Plastics Technology,* January 1991, pp. 79–88.

10. Baldoni, J.G., Wayne, S.F., and Buljan, S.T., Cutting tool materials: Mechanical properties — wear resistance relationships. *ASLE Transactions,* 1986, 29, pp. 347–352.

11. Farrow, G.J., Wostenholm, G.H., Darby, M.I., and Yates, B., Thermal expansion of PEEK between 80 and 470 K. *J. Mater. Sci. Letters,* 1990, 9, pp. 743–744.

CERAMIZED INJECTION MOULDING MACHINE FOR CONTAMINATION FREE PREFORMING

J.O.H. Nilsson* and H.T. Larker
ABB Cerama AB, S-915 00 Robertsfors, Sweden

ABSTRACT

Ceramic injection moulding is normally performed in conventional moulding machines. The machine parts subject to wear are made of wear resistant alloys in order to achieve acceptable machine life. Even so, the worn off material is enough to contaminate the ceramic compound and generate critical defects in the ceramic end product. Such contamination will limit the strength and reliability of ceramic products.

Defects caused by worn off material were virtually eliminated by substituting the steel parts of mixing and moulding machines in contact with the moulded material with ceramic parts. The screw, cylinder, valve and sprue bushing were made of silicon nitride and the inlet parts of polymer materials.

The design of the machine is described. An evaluation of the machine was made by comparing tensile test results of moulded rods from the conventional machine and the ceramized machine.

INTRODUCTION

Injection moulding shape making and glass encapsulated hot isostatic pressing produce excellent intricately shaped ceramic parts with close dimensional accuracy and surface finish. That fact has since two years been utilized in a running production of wear resistant parts for the textile industry. [1]

The process route is as follows:

* Powder is milled and mixed with sintering additives. The milling is normally performed in ball mills with ceramic balls.

* The powder is mixed with an organic binder in a Z-mixer.

* The mix is injected into a mould.

* The binder is removed from the moulded part by slow heating in vacuum.

* The part surfaces are covered with glass particles, loaded into a hot isostatic press and HIPed.

* Encapsulation glass is removed.

For parts requireing **high tensile strength**, especially in large loaded volumes, two steps of the injection moulding process have to be carefully considered. These process steps are the mixing of powder and binder and the moulding itself. In both these steps, especially the mixing step, the ceramic powder and binder compound are in intimate contact with the surrounding equipment under high pressure and high shear. There is an obvious risk of contamination of the ceramic material with worn off particles from the equipment.

The aim of the presented work was to minimize the risk for such contamination when processing silicon nitride based ceramics and to evaluate the result by analyzing the fracture initiation points in tensile tests with large loaded volume.

EQUIPMENT

The newly developed equipment was designed for processing of silicon nitride and virtually all parts of the mixing and moulding equipment in contact with ceramic powder were made of silicon nitride.

Mixer
A Z-type mixer with housing and paddles of silicon nitride was made. The parts of the mixer are shown in Figure 1. They were preformed with cold isostatic pressing and green machined. They were densified by using the Cerama glass encapsulation technique. Only mating surfaces and the outer diameter of the paddles were machined by diamond grinding after densification. The housing was mounted in an outer steel mantle and heated by blowing hot air inbetween the mantle and the housing. The paddles were connected to separate electrical motors gears with different speeds of revolution. The capacity of the mixer is about 1 - 1.5 kgs of silicon nitride powder per batch and a typical mixing time is eight hours.

By lowering the temperature at the end of the mixing cycle the compound can be crushed to particles small enough to be fed into a moulding machine.

Injection moulding machine
The parts of a screw-type injection moulding machine coming in contact with the moulded material are the cylinder with nozzle and inlet and the screw with valve assembly, Figure 2.

Figure 1. Ceramic Z-mixer

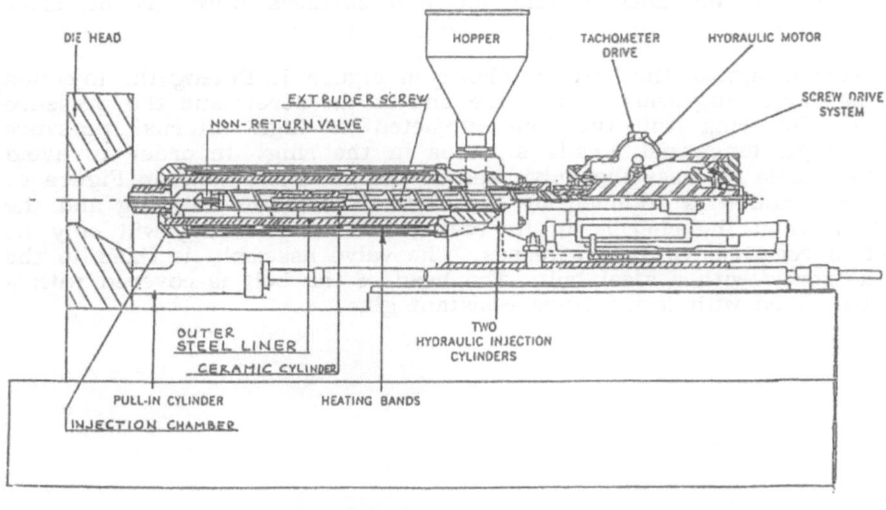

Figure 2. Injection moulding machine

The granulated compound is moved forward, compressed and passed through the valve by rotation of the screw during retraction. During injection the screw is moved forward without rotation and the compound is extruded through the nozzle into the mould. The maximum injection pressure is 100 MPa.

The cylinder was built up of a supporting steel cylinder with silicon nitride liner. The liner was made in four segments. The front segment, the only part subjected to high internal pressure, was pre-stressed in the steel cylinder. The pre-stress is high enough to avoid tangential tensile stresses in the liner at maximum internal pressure and working temperature. The other liner segments were fit into the steel cylinder with a close fit allowing axial movement relative to the cylinder when the working temperature is varied. The liner segments are axially kept together by ten pre-stressed rods. The rods acts as springs and compensate for the difference in thermal expansion when the temperature of the whole assembly varies. The inlet part of the cylinder was lined with nylon.

During the injection stroke a varying part of the front liner segment is subjected to high internal pressure. This generates axial bending of the liner wall and axial tensile stresses. The outer diameter of the liner was choosen to keep these stresses to a level lower than 100 MPa.

The nozzle was made of two silicon nitride parts pre-stressed in an outer steel part.

The cylinder and nozzle are heated with four individually controlled heating zones.

The screw was built up of four hollow segments. The segments are kept together axially by a central pre-stressed steel rod. They are aligned by centering bushings and the torque is transmitted by ceramic pins.

The screw segments were preformed by radial pressing and green machining. The outer diameter and the end surfaces were ground after HIPing.

The standard design of the valve is shown in Figure 3. During the injection stroke, the valve ring seals against the end of the screw and the pressure builds up. The ring will then be subjected to high internal pressure generating high tangential tensile stresses in the ring. In order to avoid these high tensile stresses the valve was re-designed as shown in Figure 4. The sealing function is obtained by the close fit between the ring and the central bolt when the ring is in its rear position. The ring will only be subjected to compressive axial stresses. The valve assembly is fixed to the end of the screw with a steel bolt. The head of the bolt is covered with a ceramic disc fixed with temperature resistant glue.

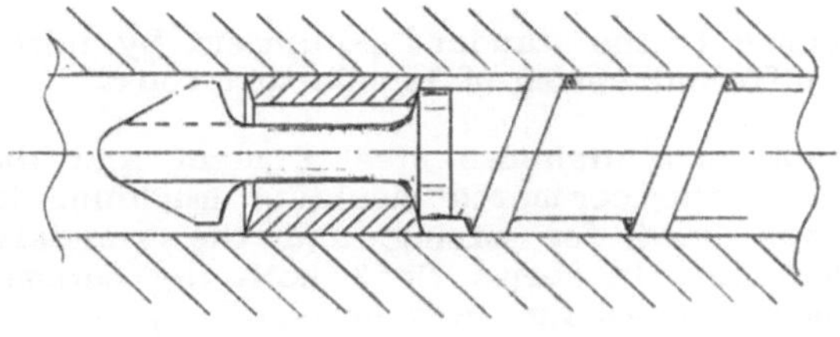

Figure 3. Standard valve

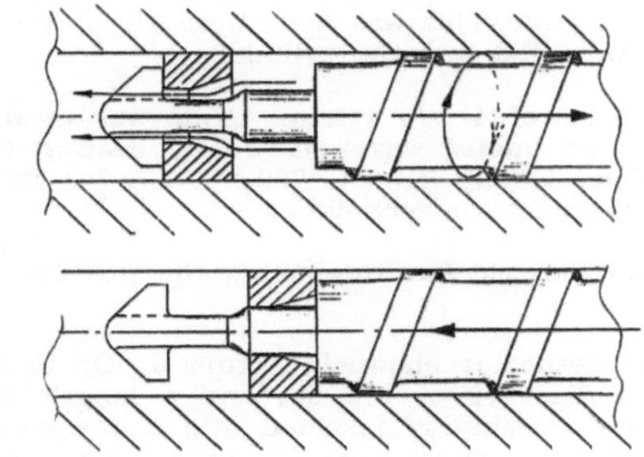

Figure 4. Ceramic valve design

Figure 5 shows the ceramic screw with the mounted valve and a segment of the ceramic cylinder liner.

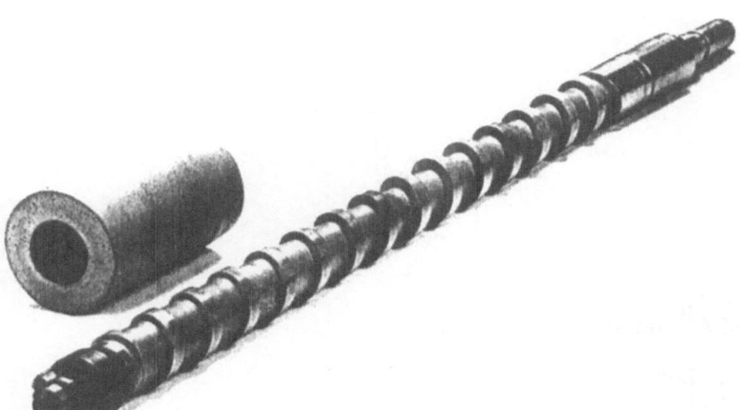

Figure 5. Ceramic components of injection moulding machine.

Experimental
The ceramized equipment was compared to the standard equipment by injection moulding and HIPing of three different series of tensile test rods.

In series No 1 the mixing was made in a stainless steel Z-mixer and the injection moulding was carried out in the ceramized moulding machine. In series No 2 the ceramic mixer was used for mixing and the standard machine was used for injection moulding. In series No 3 both the ceramic mixer and the ceramized injection moulding machine was used.

There was no significant difference in function of the ceramized equipment compared to the standard equipment. Even so, special care was taken not to overload the torque with the ceramic screw. This was done by a slight preheating of the granules before they were fed into the cylinder.

In all series silicon nitride with addition of 4 w/o yttrium oxide was used. The binder content was 16 w/o and the binder was removed by heating to 600 °C in vacuum. The rods were densified by HIP at 1750 °C and 200 MPa during 1 h using the ABB Cerama encapsulation technique.

The rods were tested by the Swedish Ceramic Institute using the ASCERA^R tensile test method (Table 1).

The principle for the ASCERA^R test method is shown in Figure 6. On each end of the ceramic test rod with a diameter of 9.5 mm and a length of 120 mm a steel piston is glued. The assembly is inserted into a pressure vessel with high pressure seals positioned against each of the pistons. The pressure between the pistons is increased until the test rod is pulled off. The ultimate strength can be calculated from the recorded maximum pressure at rupture.

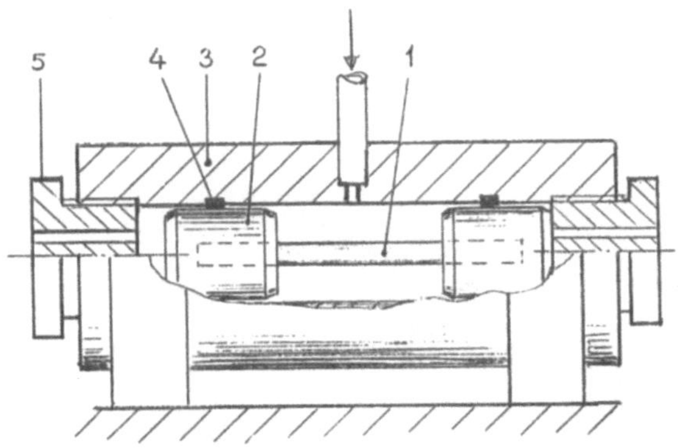

Figure 6. ASCERA^R tensile test method.

1. Test bar
2. Piston
3. Pressure chamber
4. High pressure seal
5. End closure

TABLE 1
Tensile test results

Series No	Number of rods	Mean strength MPa	S_n
1	8	261	49
2	31	331	48
3	12 (11)	393 (413)	102 (63)

Note: Values within paranthesis in Series No 3 if one test rod fractured at a large surface defect is omitted.

The cause of the fracture is of more importance than the strength values for the evaluation of the equipment. Therefore the fracture initiation points were analyzed. This work was also made by the Swedish Ceramic Institute.

Series No 1: C, Mo, Cr, Fe and Ni were detected in all (7) bulk initiated fractures.

Series No 2: C, Fe, Cr, Br were detected in 3 out of 12 investigated bulk initiations. In one, Ni was also detected.

Series No 3: There were three bulk initiations with detectable impurities dominated by C.

Most significant is the result from series No 1. Detected impurities are all alloying elements in the stainless steel mixer used for the mixing. Figure 7 shows a microphoto of a typical fracture initiation point and Figure 8 the microanalysis of the fracture initiation.

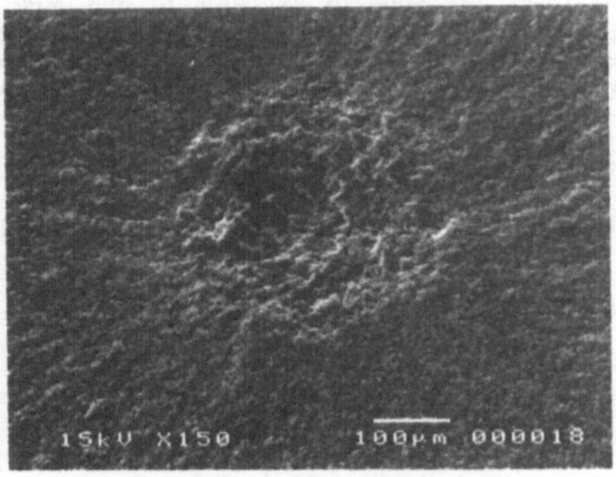

Figure 7. Fracture initiation

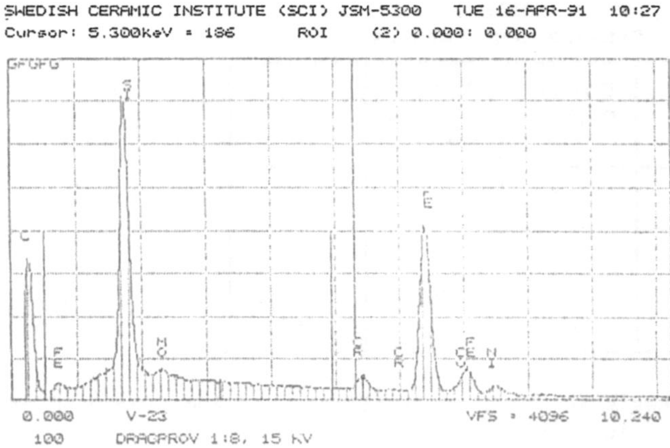

Figure 8. Micro analysis of fracture iniation

The impurities found in series No 2 correspond well with the composition of the screw and the cylinder material in the standard injection moulding machine.

Carbon inclusions found in series No 3 are probably binder residue or worn off nylon from the cylinders inlet and could probably be avoided by a modified binder removal cycle.

Conclusions

A Z-mixer and a ceramized injection moulding equipment have been designed and manufactured. The equipment has been tested and can be used for prototype work.

Analysis of the fracture origins in the tensile test rods confirms that metallic contamination from the mixing and moulding process is avoided.

The strength level was increased from 291 MPa in average to 393 MPa. Further improvements in strength should be possible if the remaining carbon contaminations from binder residue or the nylon line inlet of the screw machine could be eliminated. This is probably possible with an improved binder removal cycle.

ACKNOWLEDGEMENTS

The project was supported by the Swedish National Board for Technical Development (STU).

REFERENCES

1. Larker, H.T. and Karlsson, E., Glass Encapsulated HIP - A Production Reality, 5th International Fine Ceramics Workshop, March 15-16, 1990, Nagoya, Japan.

2. Nilsson, J. and Mattsson, B., A New Tensile Test Method for Ceramic Materials, Ceramic Materials and Components for Engines, ed. W. Bunk and H. Hausner, Deutsche Keramische Gesellschaft, Bad Honnef, 1986, pp. 651-6.

THE CAUSES AND PREVENTION OF INJECTION MOULDING DEFECTS IN LARGE CERAMIC SECTIONS

T. ZHANG* and J.R.G. EVANS
Department of Materials Technology,
Brunel University, Uxbridge, Middlesex, UB8 3PH, UK.

ABSTRACT

During the solidification stage, a range of defects appears in large ceramic sections in conventional injection moulding. These include shrinkage voids and cracks which may appear after storage or reheating. The origins of these defects are being explored experimentally in terms of material properties and machine conditions and quantitatively by computer modelling. Several interventions are described which have in common the objective of eliminating voids without resort to high hold pressures. These include the use of an enlarged sprue, a heated sprue and modulated hold pressure.

INTRODUCTION

The objectives of the work of which this forms a part are to observe the incidence of defects which originate during solidification of large sections in ceramic injection moulding, to understand their causes qualitatively, to model the origin of defects quantitatively and to predict machine interventions or combinations of material and machine parameters which will prevent such defects.

Solidification-induced defects have been thoroughly described (1) and there is already considerable progress in quantitative modelling of their origins (2,3). It is important for the practitioner to recognize that these defects are not inconvenient accidents but have causes which are inherent in the non-uniform cooling and pressure decay of the injection moulding process. Furthermore, some defects, which only become apparent during pyrolysis of the organic binder, have their origins in the solidification stage. Correct diagnosis of such defects is desirable. In ceramic fabrication, the logical strategy is to address problems which are inherent in the shaping operation in order of descending size of defect; that is the policy followed in this laboratory.

The experimental work uses an alumina suspension in polypropylene. It is not presented as an ideal injection moulding binder. It is sufficient that it has a formal similarity to thermoplastic suspensions in general and that its thermal properties and equation of state are known (4,5). The economical acquisition of such properties for diverse suspensions by the application of volume fraction dependencies and group contribution has been described (4,5).

The voids which are described below result from the decay of pressure in the molten core of the moulding after the sprue has solidified. This pressure can be calculated if the temperature distributions in the body are known at all times by numerical computation and if the equation of state is known (2). The stresses in mouldings which can give rise to cracks are the result of two related effects. The solidification of successive layers generates a centre-tensile surface-compressive distribution which depends on the rate of cooling in the cavity but is independent of pressure and can be calculated for simple shapes (3). The decay of pressure during solidification perturbs the stress in layers which are already solid and this leads to an additional effect of pressure on stress distribution which can only be approximated by calculation at present (3).

EXPERIMENTAL DETAILS

The ceramic powder was RA6 kindly donated by Alcan Chemicals Ltd. The organic binder was atactic polypropylene, isotactic polypropylene and stearic acid in the weight ratios 4:4:1. The mixing procedure has been described previously (6) and the ceramic volume fraction was 0.56.

A Negri-Bossi NB90 reciprocating screw machine adapted for ceramics use was employed to mould 40mm diameter cylinders 60mm in length. Mould temperatures were 20°C and 80°C. The barrel temperatures were 190-200-210-210°C feed to nozzle.

The modulated pressure device has been described previously (7) as has the heated sprue (8).

RESULTS AND DISCUSSION

Conventional Moulding

In conventional moulding, the incidence of voids in 40mm diameter cylinders is a function of hold pressure. The exact hold pressure at which voids cease to be formed depends on geometry, injection and solidification temperatures and thermal properties as well as hold pressure. In any injection moulding suspension with positive coefficients of thermal expansion in the liquid and solid states there will be a regime within which such voids appear. They are clearly seen in the sections shown in figure 1. Table 1 summarizes the incidence of defects in the conventional moulding operations.

TABLE 1
Defects in conventional moulding

Mould temperature 20°C:		Mould temperature 80°C:	
Hold pressure/MPa	Defect type	Hold pressure/MPa	Defect Type
65	large voids	32	large voids
87	" "	76	" "
108	voids	108	crack, voids
119	small void	130	crack
130	crack	151	crack
141	"		
151	"		

The thermal properties of the suspension are well known (4) and so the temperature distributions inside the moulding can be calculated by a finite difference method. The method allows the variation in thermal diffusivity with temperature and the melting enthalpy to be incorporated. The surface temperature was measured as a function of time in order to avoid making assumptions about heat transfer coefficient (9). The results are shown in figure 2.

In order to predict the injection moulding parameters for void-free moulding, the time necessary for the pressure in the molten core to decay to zero is required. This was obtained by a numerical method incorporating the temperature distributions, the thermal expansion and the equation of state (10). The sprue solidification time, which has a pronounced influence in this calculation, was also calculated by a finite difference method but the supercooling and pressure-dependence effects on solidification temperature were neglected. These effects tend to cancel each other in the range of pressures of interest (1). The sprue solidification times for the 10mm diameter sprue were 18s at 80°C and 13s at 20°C. The times taken for the centre of the mouldings to reach zero pressure are shown in figure 3.

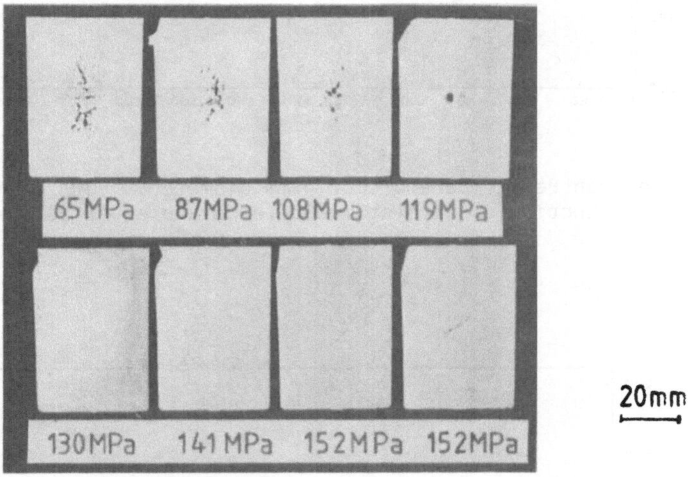

Fig. 1. Defects in 40mm diameter cylindrical mouldings made under various hold pressures.

By comparing figures 2 and 3, the conditions for void-free moulding can be predicted. The results agree quite well with the experimental results in table 1 and figure 1. If the time taken for the centre of the moulding to reach the solidification temperature of 145°C in figure 2 exceeds the time taken for the pressure to fall to zero in the molten core, then a void should form. The transition occurs for a hold pressure of 105 MPa for the 80°C mould temperature and between 120 and 150 MPa at 20°C mould temperature.

Effect of Sprue Diameter

There are clearly a number of material parameters which would confer lower solidification temperatures and lower injection moulding temperatures as well as low coefficients of expansion. However, materials selection is restricted by the demands of the subsequent binder removal operation. The purpose of this exploration is to evolve machine interventions. Clearly, the simplest action is to increase mould temperature. There is, however, a

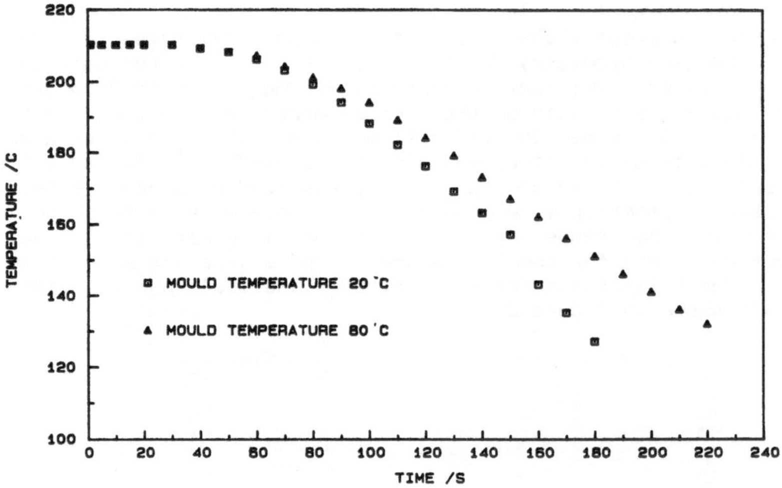

Fig. 2. The centre temperature of 40mm diameter x 60mm cylinders as a function of time at mould temperatures of 20°C and 80°C.

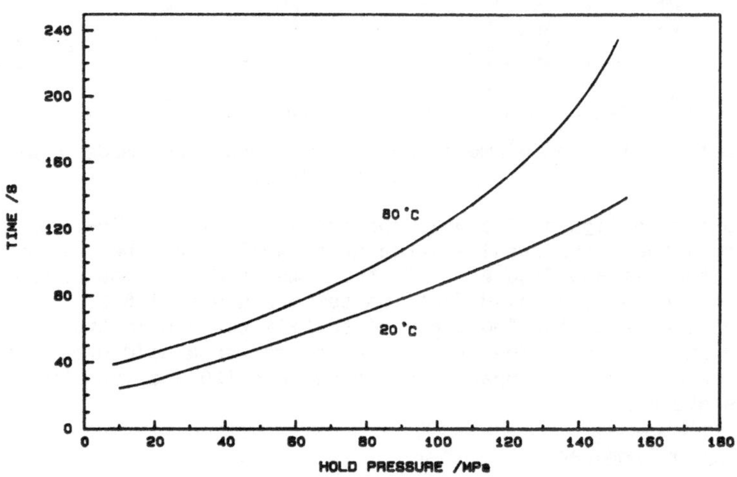

Fig. 3. The time required for the pressure in the molten core to fall to zero as a function of hold pressure for 40mm cylinders.

limit to mould temperatures for all moulding compositions above which the moulding cannot be ejected without damage. Increasing the hold pressure also removes the voids, but, because it results in a large drop in pressure from the beginning to the end of solidification, high stresses are established and the moulding cracks (figure 1).

Another simple solution is to employ a large diameter sprue. It should be remarked that all sections of the sprue, runner and gate, must give this extended time. The practical limitation is that to use sprue diameter as an intervention, the sprue size must approach that of the largest section of the moulding demanding subsequent machining. This intervention will not prevent defects in mould cavities in which large sections are separated from the sprue by narrow channels.

Effect of Modulated Pressure

The use of modulated hold pressure has been described previously (7). The effect of frequency is shown in figure 4. For any combination of material and mould cavity there is an optimum frequency at which the maximum power is supplied to the sprue as viscous losses in oscillatory flow. At low frequencies, time is wasted at top and bottom dead centre of the piston travel. At high frequencies, oscillatory flow in the sprue is incomplete. At a fixed pressure amplitude, the effect of frequency on sprue solidification time is considerable (figure 4).

The modulated pressure amplitude also has a strong influence on sprue solidification time. Figure 5 shows this extension to sprue solidification time in a rather dramatic way; at pressures above 98 MPa the sprue solidification time actually exceeds the time taken for the centre of the moulding to solidify. This means that modulated pressure is supplying a heat source to the core of the moulding as well as to the sprue. However, it is in the pressure region at, and just above 98 MPa that mouldings free from both voids and cracks are obtained. Figure 6 shows radiographs of a moulding made just below this region which contain voids and one made at just above this region which is free from shrinkage-related defects. The advantage of the modulated pressure technique is that it can deal with cavities in which a large section is remote from the gate and separated by a narrow channel. The main disadvantage is that in single gated modulated pressure moulding, the volume of material engaged in oscillatory flow is restricted by the cavity size and liquid compressibility. In current work, this oscillating volume is calculated as a function of time and estimates are made of the various heat sources. In multigated modulated pressure moulding, this restriction is removed (11).

Effect of a Heated Sprue

The sprue is designed with both heating and cooling facilities (8) so that the weakness of conventional hot runner moulding is avoided. Thus a complete sprue can be ejected and the first material to enter the cavity on the next shot is at the nozzle temperature. Mouldings can be made with much lower hold pressures which are free from both voids and cracks. In fact, hold pressures below 20 MPa can be used without the appearance of voids in 40mm cylinders. Cracks do not appear because there is a small pressure drop during the course of solidification. In these mouldings, an initial high injection pressure is used to fill the cavity but the machine switches to low hold pressure at the earliest opportunity. Like the large sprue, the heated sprue does not address the problems of large moulded sections separated from the gate by a narrow channel.

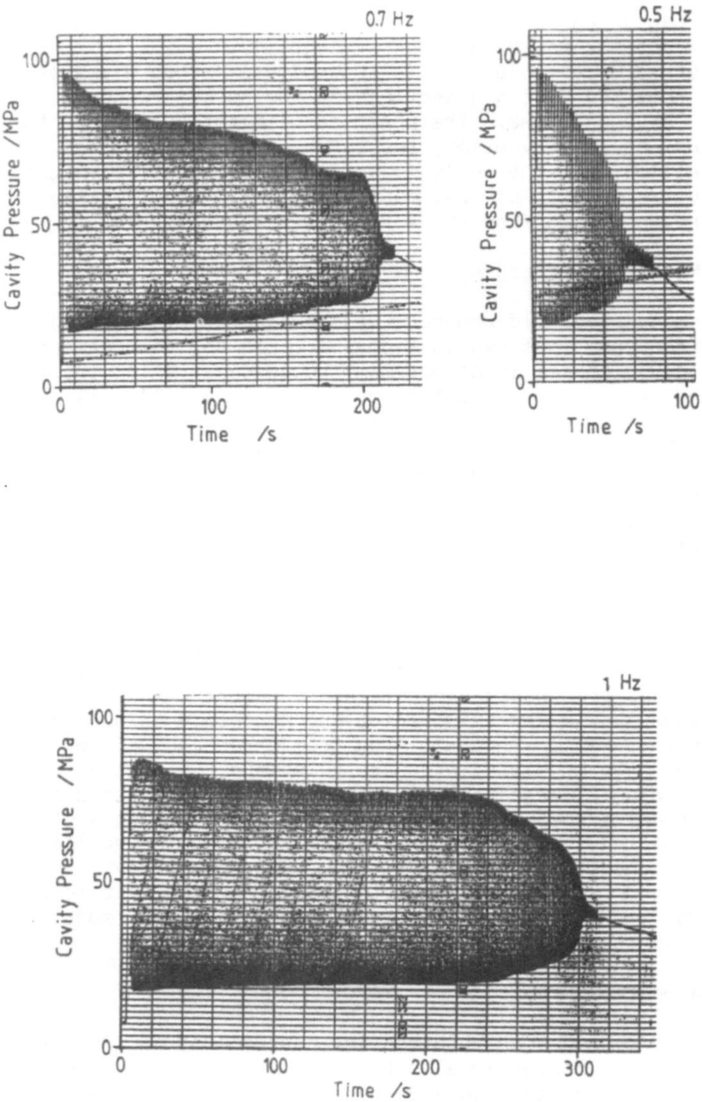

Fig. 4. The effect of oscillating pressure frequency on sprue solidification time at a mould temperature of 80°C for 40mm diameter cylinders.

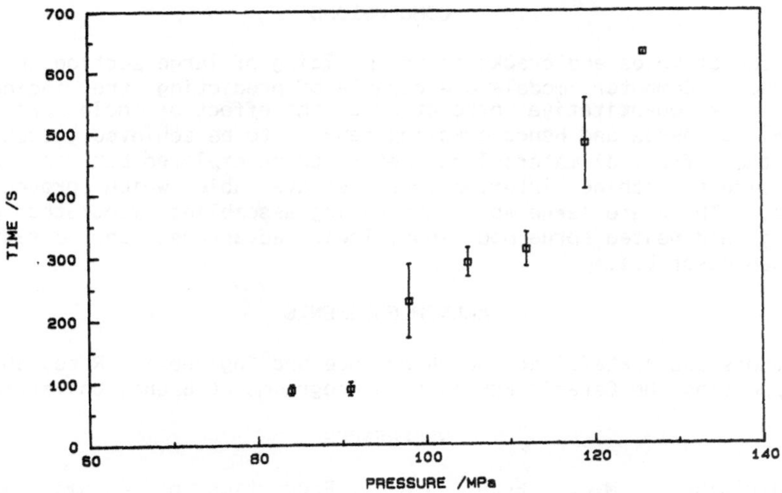

Fig. 5. The effect of oscillating pressure amplitude on sprue solidification time at a mould temperature of 80°C for 40mm diameter cylinders.

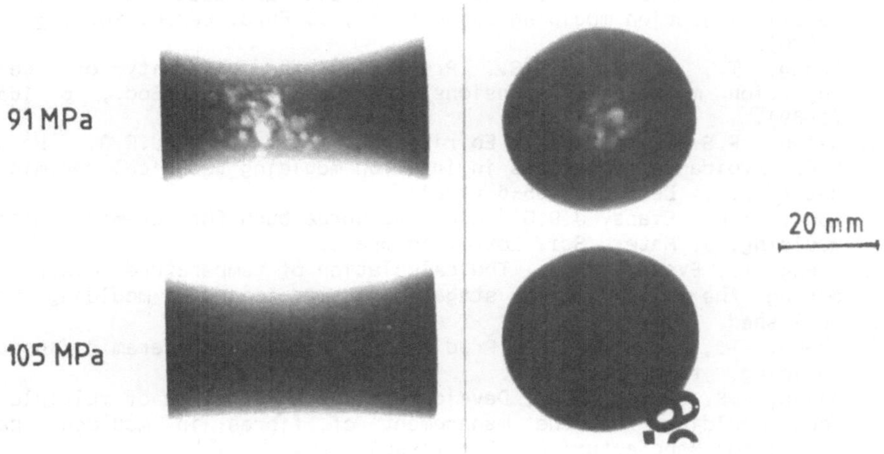

Fig. 6. Radiographs of 40mm cylinders made under oscillating pressure (MPa) conditions at 1Hz frequency and 80°C mould temperature.

CONCLUSIONS

The causes of voids and cracks in the moulding of large sections have been described. Computer models are capable of predicting the incidence of voids. The quantitative prediction of the effect of hold pressure on residual stresses and hence cracking remains to be achieved. Such models allow the effect of material parameters to be explored but in addition, three direct machine interventions are available which prevent these defects. These are large sprue and gating assemblies, modulated pressure moulding and heated sprue mouldings. Their advantages and disadvantages have been described.

ACKNOWLEDGEMENTS

The authors are grateful to the UK Science and Engineering Research Council for supporting the Ceramic Fabrication Programme at Brunel University.

REFERENCES

1. Edirisinghe, M.J., Evans, J.R.G., Properties of ceramic injection moulding formulations I. Melt rheology, J. Mater. Sci., 22 2267-73 (1987).
2. Hunt, K.N., Evans, J.R.G., Woodthorpe, J., Computer modelling of the origin of defects in ceramic injection moulding I. Shrinkage voids, J. Mater. Sci., 26 292-300 (1991).
3. Hunt, K.N., Evans, J.R.G., Mills, N.J., Woodthorpe, J., Computer modelling of the origin of defects in ceramic injection moulding IV. Residual stresses, J. Mater. Sci., in press.
4. Zhang, T., Evans, J.R.G., Dutta, K., Thermal properties of ceramic injection moulding suspensions in the liquid and solid state, J. Euro. Ceram. Soc., 5 303-9 (1989).
5. Zhang, T., Evans, J.R.G., Thermal expansion and equation of state for ceramic injection moulding suspensions, J. Euro. Ceram. Soc., 6 15-21 (1990).
6. Zhang, T., Evans, J.R.G., Predicting the viscosity of ceramic injection moulding suspensions, J. Euro. Ceram. Soc., 5 165-172 (1989).
7. Allan, P.S., Bevis, M.J., Edirisinghe, M.J., Evans, J.R.G., Hornsby, P.R., Avoidance of defects in injetion moulding technical ceramics, J. Mater. Sci., Lett., 6 165-6 (1987).
8. Hunt, K.N., Evans, J.R.G., A heated sprue bush for ceramic injection moulding, J. Mater. Sci. Lett., in press.
9. Zhang, T., Evans, J.R.G., The calculation of temperature distribution during the solidification stage in ceramic injection moulding, to be published.
10. Zhang, T., Evans, J.R.G., Predicting the voids in ceramic injection moulding, to be published.
11. Allan, P.S., Bevis, M.J., Development and application of multiple live feed moulding for the management of fibres in moulded parts, Composites Manufacturing 1 79- (1990).

THE EFFECT OF CLEANROOM PROCESSING ON THE PROPERTIES OF SILICON NITRIDE

H. Schubert

Max-Planck-Institut für Metallforschung
Institut für Werkstoffwissenschaft
Pulvermetallurgisches Laboratorium

Heisenbergstraße 5
W-7000 Stuttgart 80

Introduction

High-tech ceramics offer a spectrum of mechanical properties which cannot be achieved by any other class of materials. This has led to a number of applications as structural materials in motor and machine engineering. A wider application has however been hindered by the large scatter of the mechanical properties. In the past few years, different strengthening concepts have been applied, aiming to increase the damage tolerance[1,2]. The success of property improvement was remarkable and the understanding of the microstructure property relation developed. Nevertheless, it must be recognized that high-tech ceramics are still brittle materials and that their strength is predominantly determined by processing related defects.

The reliability of ceramics is linked with the defect-size distribution[3]. The aim to prepare a high strength ceramic with low scatter requires the limitation of defects. There is a large number of possibilities for defect formation during a preparation process. After bend test, however, only three major types of defects can be detected on the fracture surface: a) foreign particles, b) pores, and c), surface defects. Foreign particles are not very common, but they might be extraordinarily large. Pores may be a result of pressing defects or of hard granulates. Their size and frequency is a function of many process parameters (spray parameters, amount of pressing aids, residual humidity, etc.). Surface defects are created during surface grinding or during mechanical load and they depend on the microstructure.

The use of clean room technology aims especially at the diminishing of foreign particle defects. Once being introduced metallic and ceramic particles remain in the material throughout the whole process. Organic particles burn out leaving pores behind. Neither can be removed (repaired) by process modifications, i.e. they have to be avoided completely from the beginning.

At the Max-Planck-Institut für Metallforschung a clean room facility has been installed which was constructed for the purposes of ceramic processes. The processing of small defect size Si_3N_4 and SiC ceramics is part of the BMFT project 03M20128 and has been carried out experimentally since 1989. This paper reports the experience in finding possible ways to handle ceramic powders and to prepare small defect size ceramics.

Cleanroom Technology

The central unit of the clean room installation at MPI is a class 100 (100 particles per cubic foot; n=3) work station installed within the clean room for powder handling and processing. Ceramic processing requires the use of solvents and acids; both are strictly prohibited for recycling air. The usual construction for handling these liquids would be an exhaust hood. The flow direction would be upwards, which is contradictory to the laminary flow concept of clean rooms. The chosen construction is a down-flow clean bench with air passing the experiment and being exhausted through a perforated counter. The part of the air being contaminated with solvent vapor is waste air while the air in the personnel way is recycled (Fig. 1). The air velocity in the clean bench is higher, allowing a class 100 purity (class 100 is defined: 100 paricles / foot ^{3} according to Federal Standard 209b with 0.5 μm reference particle size, respectively 1000 particles / m ^{3} according to VDI Richtlinie 2083; n = 3 for 1 μm reference particle size) compared to class 1000 in the personnel area.

Two Service rooms with class 10000 installations are located on both sides of the clean room. All together, there are 3 clean rooms and 4 service rooms available. The work stations are equipped with high purity gases (Ar, N_2, O_2, H_2), high purity water, fresh water, cooling water, a vacuum line and electrical power. The waste air, powders and liquids are separated in the work stations.

The cleanliness of the facilitiy is measured continuously by counting particles in the size range of 0.3 to 10 μm using a 6-channel laser counting device.

Powder Handling in the Cleanroom

The processing of ceramic powders in the cleanroom entails several basic problems which had to be solved in terms of a working concept.

Cleanroom technology has been applied in other types of industry (electronics, pharmacy, and medicine) because particulate contaminants - even if they are small - are dangerous for the reliability and the functionality of the products (substrate, layered structures, integrated circuits) or because they destroy the sterility of the products. Basically, any airborne particle is considered to be a contaminant. This concept, however, is not tolerable for ceramic powders. An amount of 25 kg Si_3N_4 represents an amount of 10^{16} powder particles of 0.5 μm mean size. Preliminary experiments in the clean room work stations showed that the filling and refilling of these powders evolve particle concentrations of 100 000 and more in the environment. Handling of ceramic powders is a natural contradiction to the needs of particle freedom which had to be taken into account in the working concept.

For further investigations, it was necessary to handle batches up to 10 kg in size. This, however, made it impossible to install the whole processing line in the clean room. Therefore, a concept had to be developed which solves both main problems

a) particle emission by powder itself and
b) handling of larger batches.

The Working Concept

Ideally, the powder would be handled throughout the whole process under the protection of the highest clean room class in the work station. This, however, is not possible because of the need of large batche sizes. Fig. 2 shows a materials flow diagram for Si_3N_4 processing which solves the requirements in a satisfactory way. The different clean room classes are noted and the path of the material is drawn as a line.

The commercial material is delivered in barrels which have to be cleaned before they are transferred into the clean room area ("clean-lab" in Fig. 2, class 10 000). The powder is then stored in the clean room (class 100/1000). The first processing steps of weight in, and the addition of water, deflocculant and sintering aid (Al_2O_3 / Y_2O_3) carry the highest risk of contamination and are therefore carried out in the work station. The further processing steps, namely milling, spray drying and isostatic pressing, cannot proceed in the class 100 protection because the equipment is much too large and it could not be modified to an extent which allows installation in class 100. Therefore, the open handling of powder is restricted to the area of the work station. Only there are the powders weighed and put into closed containers (henceforth referred as transport system) and taken to the further stations (service room, class 10 000). The transport system is of great importance. Its path is depicted in Fig. 2 as a broad line. It allows a further protection of the material for any treatment outside of class 100. The transport system is conserving the purity of class 100 and transports it to the further stations. After a green body has been formed, no further particles can be introduced into the body and destroy the homogeneity. Thus, further processing is done outside the clean room.

Experience with processing Si_3N_4

The investigations were carried out with an Si_3N_4 material containing 6 wt.% Y_2O_3 and 2 wt.% Al_2O_3 because there is information in the literature about microstructure and achievable properties for this composition [4,5]. A standard commercial powder (LC12S, H.C. Starck, Gosslar, D) was used because this material has fairly low variations in powder characteristics from batch to batch.

Generally, the ceramic process could be carried out applying the working concept as described above. The working techniques and the equipment had to be checked for their feasability in the clean room. In some cases technical modifications were necessary. The process allows the use of commercial equipment and offers the protection throughout the whole procedure. It can be carried out with the same number of personnel as the standard process.

Homogeneous contaminations

Sampling and decomposed (PTFE vessels) [6] carried out in the clean room. The determination of the contents was done using a ICP (Jobin Yvon JY 70+, France). The O-content was determined using hot gas extraction (Leco TC 436, USA). Both a pure Si_3N_4 batch (batch A) and a material with additives (batch B) were processed. The results are given in Tab. 1.

The Al-, Mg- and Y-content are a result of wear from the milling balls and the lining of the mill. Their values increase which correlates with the weight loss of the milling media. The Fe-content could be attributed to wear of a ring near the rotor bearing. This ring has since been replaced by a SiC piece. The determination of the Mg and Fe content in batch B shows similar numbers. The B content does not vary which is an indication that no leaching effect from the glass occurrs.

The O, C and H_2O content increase linearly with milling cycles due to the increase in surface area. Again the value of batch A and B were comparable.

Particluate Contaminants

The use of the transport system gave a complete protection of the material during dispersion and milling. The main source for particulate contaminants was therefore the spray dryer. After sieving the different contaminants were found. In every case the contamination source could be identified and removed. The main contaminations were Fe or Fe/Cr particles. During sintering a reaction with Si_3N_4 matrix took place and Fe silicides were formed. A corrosion of the heating elements was identified as main source for this type of particles. As a consequence, a stainless steel hot gas filter was built into the gas supply which stopped the particle contamination.

Properties

The strength of Si_3N_4 made by a conventional unprotected process is governed by the superposition of three defect distributions:

- particulate incusions
- pores as a result of hard granulates or pressing voids
- surface defects.

This type of ceramics had strength value which scattered between 300 and 800 MPa with all three defect types being present. The use of clean room technology and the application of the working concept has limited the influence of particulate contaminants to such an extent that they do not limit the strength of the material. Surface defects were found to be the only strength controlling defect type (Fig. 3). Values up to 930 MPa was measured. From the literature it is known that the strength values of directly HIPed material of the same composition is in the order of 924 MPa [4] up to 1080 MPa [5]. The difference is attributed to the less effective sintering technique applied in this work.

Final Remarks

Experiments under clean room protection require a special education of the personnel. The use of complex process technology forces the discussion of materials handling directly with the technical personnel working in the clean room. Since the technical staff will work 99% of the time alone, the clean room concept must be developed and accepted from all personnel involved in the operation. Due to the extensive, time-consuming processes necessary for working in proper clean rooms, all experimental steps must also be planned from the beginning in great detail.

Acknowledgement

This work was supported by the Federal Minister of Science and Technology (BMFT) under the contract 03 M 2012 8. The analytical work was supported by DFG under the contract Schu 660/1-1. The author thanks R. Kröner, J. Stettler and G. Kaiser for analytical work, furthermore Prof. Dr. Dr. h.c.mult. G. Petzow and all colleagues in this project for helpful discussion.

Literature

1) A.G. Evans
"Perspective on the Development of High-Toughness Ceramics"
J. o. Am. Ceram. Soc., Vol. 73, No 2, (1990) p187

2) H. Schubert und G. Petzow
"What makes Ceramics Tougher?" in J.O.Materials Education,
Vol. 10 No. 6 (1988) 601-620

3) V.D. Frechette and J.R. Varner (edt.)
Fractography of Glasses and Ceramics, Advances in Ceramics Vol. 22
The American Ceramic Society, Westerville, Ohio, USA, 1986

4) C.L. Quackenbush and J.T. Smith
"GTE Sintered Silicon Nitride"

5) D. Sordelet, J. Neil, M. Mahoney and A. Hecker
"Development of Injection-Molded Gas Turbine Components: Investigation Into Toughening GTE PY6 Si_3N_4", Presented at the International Gas Turbine and Aeroengine Congress and Exposition, Orlando, FL June 3-6, 1991, paper 91-GT-65.

6) G. Kaiser, W. Wörner, K. Schulze, H. Schubert, G. Petzow
"Associated Analysis of Contamination of Si_3N_4 Clean Room Processing"
to be published in Microchemica Acta

Tables

	Batch A Milling Cycles			Batch B Milling Cycles		
	0	2	4	0	2	4
Fe (ppm)	64 ± 5	151 ± 2	209 ± 5	69 ± 2	130 ± 4	174 ± 6
Mg (ppm)	52 ± 2	71 ± 2	81 ± 2	57 ± 5	75 ± 2	85 ± 3
Al (ppm)	371 ± 7	416 ± 8	436 ± 9			
Y (ppm)	1,4 ± 0,9	59 ±10	100 ± 9			
C (%)	0,2 ±0,01	0,3 ±0,01	0,37 ±0,01	0,2 ±0,01	0,32 ±0,01	0,33 ±0,01
H_2O (%)	0,88 ±0,02	0,96 ±0,02	1,05 ±0,02	0,9 ±0,01	0,99 ±0,02	1,11 ±0,01
O (%)	1,89 ±0,02	2,06 ±0,05	2,14 ±0,03	4,05 ±0,06	4,20 ±0,04	4,30 ±0,01

Tab. 1: Analytical results for Batch A (pure Si_3N_4) compared with Batch B (Si_3N_4 + 6 wt.% Y_2O_3 + 2 wt.% Al_2O_3).

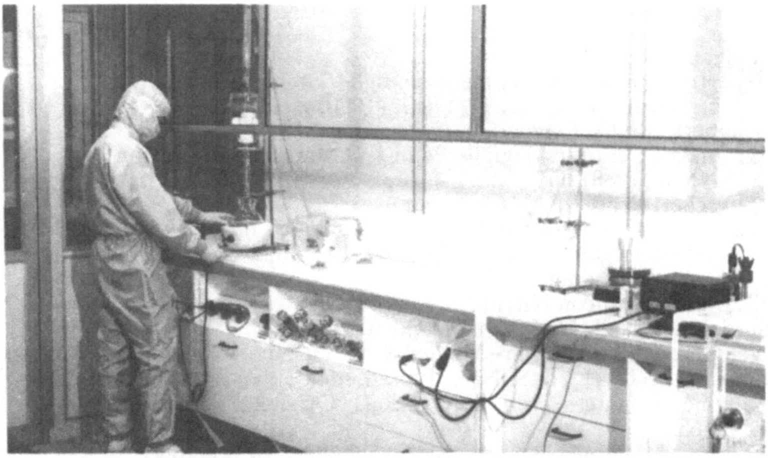

Fig. 1: Work station for Si_3N_4 processing (class 100).

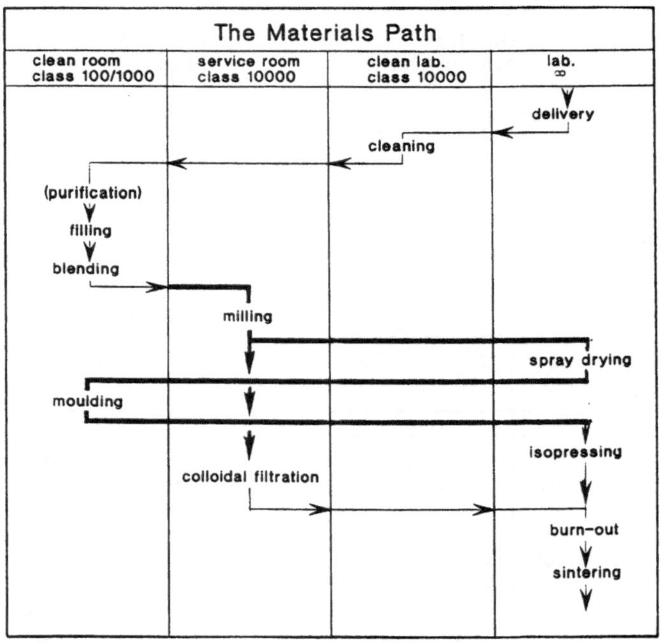

Fig. 2: Materials path for Si_3N_4 processing. The use of the transport system is noted as a broad line.

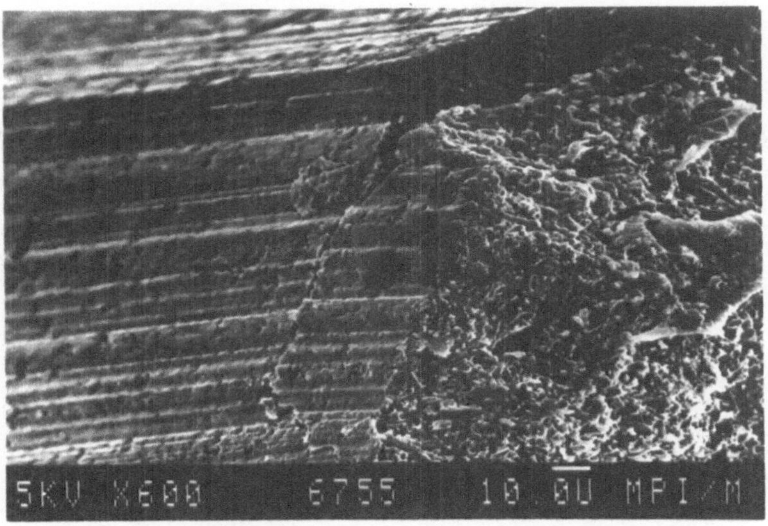

Fig. 3: Surface defects as the only strength controlling origin in clean room processed Si_3N_4 (930 MPa).

SELF-REINFORCED SILICON NITRIDE - A NEW MICROENGINEERED CERAMIC

ALEKSANDER J. PYZIK, DANIEL F. CARROLL, C. JAMES HWANG, and ARTHUR R. PRUNIER

The Dow Chemical Company
Central Research Advanced Ceramics Laboratory
Midland, Michigan 48674, USA

ABSTRACT

A low cost, high fracture toughness silicon nitride material can be obtained through the in-situ growth of whisker-like elongated grains. This ability to form a reinforced material in-situ during densification avoids the processing difficulties and potential health hazards generally associated with whisker reinforced ceramic composites. The in-situ growth of elongated silicon nitride grains is controlled by the glass phase chemistry and processing conditions. A new class of self-reinforced silicon nitride materials have been developed based upon the Si_3N_4-Y_2O_3-MgO-CaO system, where yttrium can be replaced by seven other elements, magnesium can be replaced by six other elements and calcium with 19 other elements. These compositions have been found to yield fine grained, high aspect ratio silicon nitride ceramics with a unique combination of high flexure strength and fracture toughness.

INTRODUCTION

The formation of elongated grains in silicon nitride has been observed for a long period of time [1]. Lange [2] was one of the first investigators that correlated an improvement in mechanical properties with the observance of elongated grains in silicon nitride. Work conducted by Wotting et al. [3] indicated that the aspect ratio of the silicon nitride grains could be changed in a Y_2O_3/Al_2O_3 based system by varying the ratio of the sintering additives. Following the work of Wotting, other studies have investigated the formation of elongated grains in the Si_3N_4-Y_2O_3-Al_2O_3 system, resulting in materials with improved fracture toughness [4,5]. However, the application of these materials above 1200°C has generally been limited by excessive deformation of the glassy phase and its devitrification products [6].

The objective of this paper is to demonstrate that there are several other silicon nitride based systems that can exhibit controlled formation of these elongated beta silicon nitride grains. This technology is based upon adjusting the chemistry of the starting powders to control the nucleation and growth of elongated grains and crystallization of the glassy phase. This ability to control the formation of elongated grains has resulted in the production of silicon nitride materials with unique combinations of high fracture toughness and flexure strength. Due to the compositional flexibility with these systems, a class of self-reinforced silicon nitrides can be developed for a broad range of applications at low and high temperatures.

EXPERIMENTAL

The self-reinforced silicon nitride materials were made using a silicon nitride powder that was composed of more than 95% of the alpha phase with a surface area of 8.2 m^2/g, 1.3 wt% oxygen, < 30 ppm Fe, < 100 ppm Cl and <50 ppm Ca and Al. The densification additives were selected based upon their purity and particle size. All compositions were attrited for 1 hour in methanol, dried, passed through a 60 mesh screen and hot pressed at 1700°C to 1850°C for 1 to 6 hours in a nitrogen atmosphere. The microstructure of each material was examined using scanning electron microscopy to determine the relationship between glass chemistry and grain morphology.

Selected billets were sectioned for fracture toughness and flexure strength evaluation. Approximately twelve specimens (3 x 4 x 45 mm) were broken in four point bending to determine the room temperature flexure strength. The flexure strengths measured at elevated temperatures were also tested in four point bending in an air environment. The fracture toughness of these materials were determined by fracturing five to seven chevron-notched bend beam (CNB) specimens in three point-bending. The dimensions of the CNB specimens were b=3.0 mm, w=4.0 mm, $a_0/w=0.42$, $a_1/w=1.0$ and s/w= 10. The fracture toughness was calculated using the method developed by Xian [7]. Extreme care was taken to ensure stable crack growth throughout the CNB test.

The post-densification heat-treatments were conducted at 1250 to 1600°C for various times in nitrogen atmosphere. The crystallized products were identified by X-ray diffraction. The fracture toughness and flexure strength of the heat treated materials were measured with the previously mentioned techniques.

The metal milling tests were performed by Metcut Research Associates (Cincinnati, Ohio). Self-reinforced silicon nitride blanks were ground to an insert geometry of SNG434T00630. Each milling test used one insert held at a 15° lead angle and -5° axial and radial rake angles. Billets of grade 30 gray cast iron 0.102 m wide and 0.61 m long were center milled using the following conditions: depth of cut = 1.524 mm, feed per revolution = 0.33 mm, cutting speed = 15.2 m sec^{-1} and radius of rotation = 0.152 m. The flank wear of the insert was measured periodically during the milling and testing was stopped just after 0.26 mm of wear. The relative performance of various compositions was rated by comparing how much cast iron had been removed at 0.254 mm of flank wear.

RESULTS AND DISCUSSION

Chemistry of Self-Reinforced Silicon Nitrides

At densification temperatures, the glass phase becomes a major factor in controlling the rate of dissolution, nucleation, alpha to beta transformation and grain growth. It was found [8] that if the chemistry of silicon nitride powder and glass are controlled, silicon nitride grains can be formed with different sizes and aspect ratios. The morphology of the beta grains can be tailored by adjusting the ratio of densification aid (e.g. MgO) to transformation aid (e.g. Y_2O_3) in presence of a whisker growth agent (e.g. CaO). The phenomenon of controlled growth of elongated grains was confirmed not only for Y_2O_3-MgO-CaO based glasses [9,10] but also for several other systems including seven substitutions for Y_2O_3, six substitutions for MgO and 19 for CaO [11]. The main densification aids are selected from elements in group II of the periodic table while the conversion aids are selected mainly from the rare earth compounds and the whisker enhancing compounds from groups II, III and IV elements. The formation of elongated silicon nitride grains can be promoted by either the oxide or non-oxide derivatives of these elements.

The Effect of Glass Chemistry on Microstructure

The development of silicon nitride grain morphology is dependent upon mass transport through the glass phase and ability of the glass to dissolve the alpha silicon nitride. Both group of additives, the densification aids and the transformation aids, affect the dissolution and transport properties differently at high temperatures [12]. With MgO, the glass viscosity is low and the transport of material is rapid. In this system, the kinetics of densification is generally faster than the kinetics of the dissolution process (alpha to beta transformation). With Y_2O_3, the glass viscosity is higher and the transport of material through the glass phase is slow. In this system, the alpha to beta transition is generally finished before complete densification occurs. It has been shown [8] that in a low viscosity systems the grain shape changes from high to low aspect ratio with increasing saturation of the glass phase. In high viscosity systems, the grain shape also changes from low to high aspect ratio with increasing saturation. By changing the ratio of densification aid to conversion aid, the glass phase characteristics can be optimized to enhance the formation of elongated silicon nitride grains. The optimum ratio for enhanced elongated grain formation depends on chemistry of the silicon nitride powder, processing conditions and type of additives. In general, the ratio of densification aid to transformation aid ranges from 10:1 to 1:1. For Y_2O_3-MgO based glasses, the optimum range was found between 3:1 and 1:1.

The addition of whisker-growth agents, such as CaO, facilitates the formation of elongated grains. Aspect ratio of silicon nitride grains were found to increase from 1-2 to 7-11 in presence of only 0.1 to 0.2 % of CaO. This dependence of crystal morphology on the presence and concentration of impurities has been investigated for several systems [13,14]. However, it is still not clear how or why the growth rate varies with crystal orientation in the presence of impurities. One possible hypothesis is that the rate difference arises from changes in the number of growth sites and local solute diffusivity [15].

The densification aids such as MgO, conversion aids such as Y_2O_3 and whisker-growth agents allows the supersaturation condition of the glass to be altered and, as a result, changes the nucleation and grain growth processes. By controlling the glass chemistry together with the processing conditions, microstructures with tailored grain sizes, aspect ratios and grain quantities can be formed. Figure 1 shows examples of the two materials with controlled microstructures that are characterized by small grains (average diameter ~0.35μm) and large grains (average diameter >1μm).

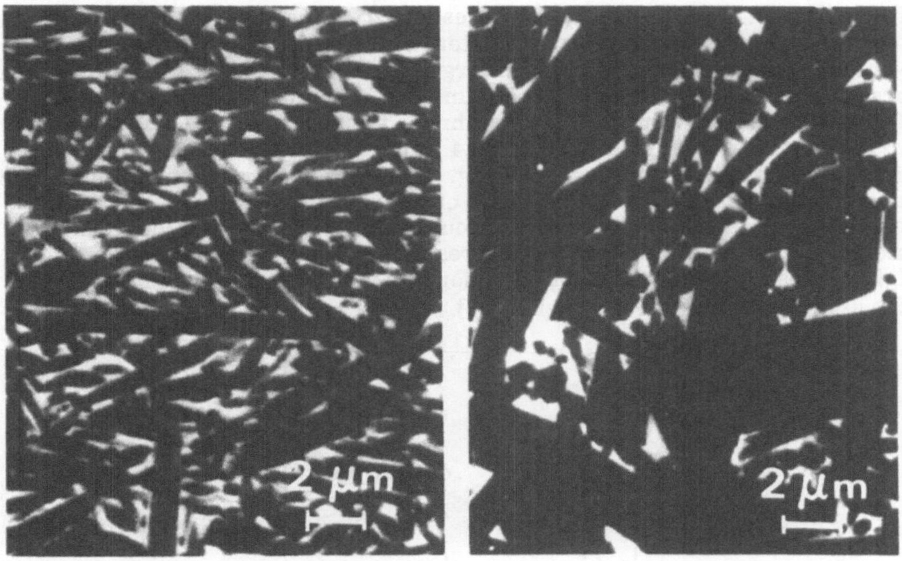

Fig.1. The microstructure of Si_3N_4-Y_2O_3-MgO-CaO materials with small and large elongated beta silicon nitride grains.

The Effect of Grain Morphology on Mechanical Properties

Fracture toughness is a material property which can benefit most from the formation of elongated grains with high aspect ratios. Figure 2 is a graph from Becher [16] modified by addition of our data. The graph illustrates a correlation between fracture toughness, flexure strength and grain diameter in self-reinforced silicon nitride. For the Si_3N_4-Y_2O_3-Al_2O_3 system (B), the fracture toughness is found to increase with increasing grain diameter. However, this increase in fracture toughness is at the expense of a lower flexure strength. The effect of grain size on the mechanical properties is a major difference between our data and previously reported results. Our materials based on the Si_3N_4-Y_2O_3-MgO-(whisker-growth agent) systems (A) have much smaller grain diameters and, as a result, exhibit different mechanical behavior. They possess a combination of high flexure strength and high fracture toughness.

The fracture toughnesses of the Si_3N_4-Y_2O_3-MgO-(whisker-growth agent) systems are strongly dependent upon the diameter of the elongated grains. Figure 3 shows that the fracture toughness of a selected self-reinforced silicon nitride composition can be increased from ~7.0 MPa $m^{1/2}$ for an average grain diameter of 0.18 μm to ~9.0 MPa $m^{1/2}$ for an average grain diameter of 0.44μm. At larger diameters, the fracture toughness decreased reaching a value of approximately 6.6 MPa $m^{1/2}$ for an average grain diameter of 1.15 μm. This maximum in fracture toughness is unusual for silicon nitride-based materials. Past investigations have shown that the fracture toughness of silicon nitride generally increases with grain size due to an increase in the amount of crack deflection and bridging [16,17]. The reason for the unusual behavior in our materials can be seen on the fracture surfaces of the broken CNB specimens. Figure 4a, and b represents fracture surfaces of the materials with an average grain diameters of 0.44 and 1.15 μm, respectively. The fracture mode in Figure 5a is intergranular. When intergranular failure occurs, the main toughening mechanisms are crack deflection and crack bridging. The increase in fracture toughness for materials with an average grain diameter ranging from 0.19 to 0.44 μm is due to an increase in the amount of crack deflection and bridging. The decrease in fracture toughness at larger grain sizes is due to the onset of transgranular failure (figure 5b). When the crack starts to propagate through grains instead of along the grain boundaries, the enhanced toughening resulting from crack deflection and bridging is lost. A similar type of behavior has been reported in Si_3N_4-Y_2O_3-Al_2O_3 and Si_3N_4-cordierite systems [18].

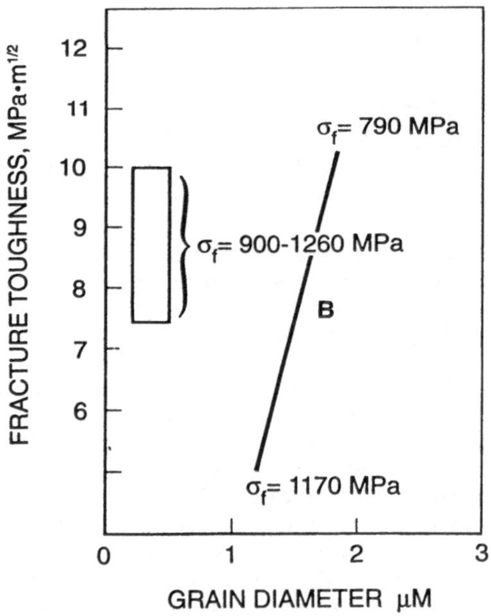

Fig.2. Mechanical property summary of various self-reinforced silicon nitrides. (A) Si_3N_4-Y_2O_3-MgO-(whisker-growth agent) systems.
(B) Si_3N_4-Y_2O_3-Al_2O_3 system [16].

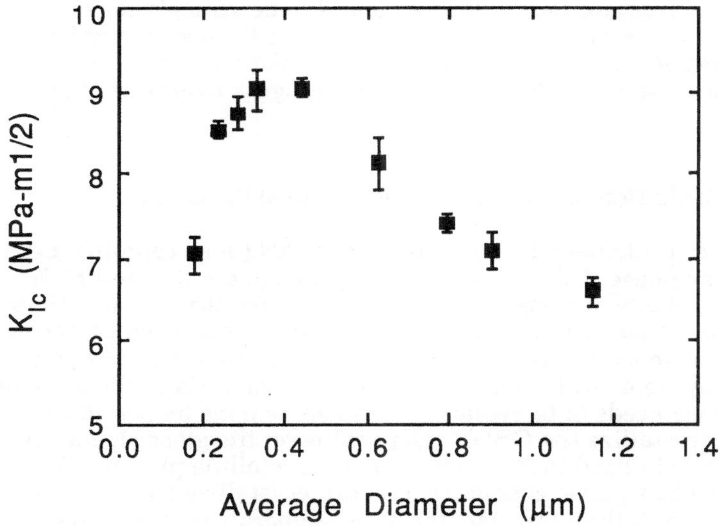

Fig.3. Fracture toughness as a function of average grain diameter for a self-reinforced silicon nitride composition.

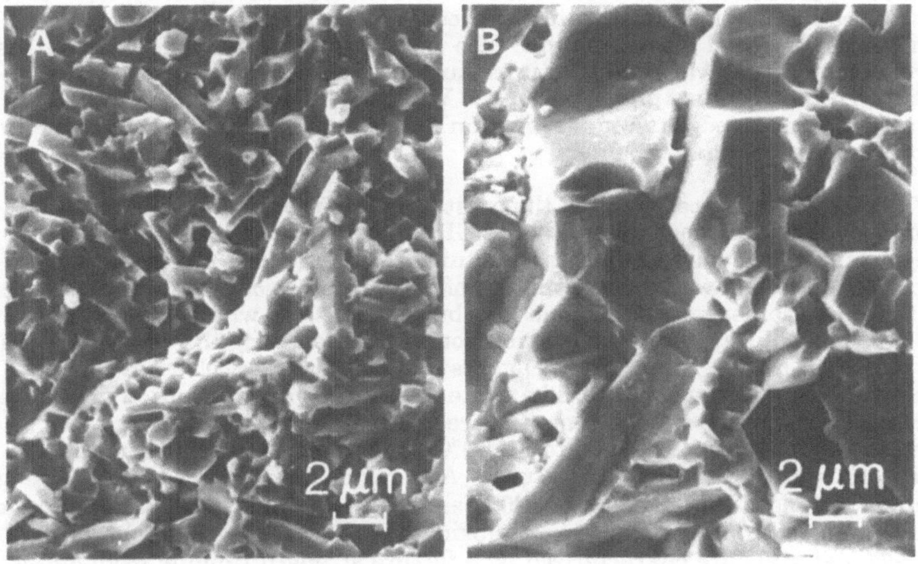

Fig. 4. Fracture surfaces of self-reinforced silicon nitride with average grain diameters of 0.44 and 1.15 µm, respectively.

Even though the fracture toughness was found to be strongly dependent upon grain size for this material, the room temperature flexure strengths were found to be relatively independent of grain size. For materials with average grain diameters less than 1μm, the flexure strengths averaged approximately 1100 MPa.

The Effect of Grain Boundary Phase on Mechanical Properties

As in most silicon nitrides, the self-reinforced Si_3N_4 also contains a grain boundary glassy phase (GBGP). Its presence affects the low and high-temperature mechanical properties. At room temperature, the optimum fracture toughness and strength for most of the self-reinforced Si_3N_4 systems is obtained in presence of 4 to 10 volume % of glass. However, in order to improve the high temperature properties in these materials, the amount of amorphous phase needs to be reduced. This can be done by post-heat treatments to crystallize the GBGP. Depending on the glass chemistry, the GBGP can be crystallized to form one or more crystalline phases. By controlling the glass phase chemistry, one can crystallize the glass phase to form desirable crystalline products for high temperature properties.

In materials based on the Si_3N_4-Y_2O_3-MgO-(whisker-growth agent) system, the crystallization nature of the glass phase is complex and, as a result, difficult to control. The Y_2O_3-MgO based system tends to experience a glass phase separation at temperatures of 1200 to 1600°C. Consequently, after post-heat treatment, the crystallization process is not complete and a tangible amount of residual glass remains. The crystallized products from this glass system consists of various structures, such as garnet, apatite and pyroxene. The high-temperature properties of the Y_2O_3-MgO based materials are not markedly improved after crystallization and therefore, are recommended for applications less than 1200°C.

When MgO is replaced by SrO, the crystallization process becomes more controlled. The composition of the GBGP can be easily adjusted to crystallize a secondary phase with an apatite structure. The structural flexibility of the apatite phase allows for the formation of an extensive solid solution with many different cation species [19]. This flexability allows the whisker-growth agents to be changed in the starting composition without altering the crystallization process. In fact, whisker-growth agents like Ca, Ti, Al and Na have been shown to exist in the crystallized apatite phase. The crystallization of apatite occurs over a wide range of temperatures. The apatite structure appears as low as 1000°C and is stable up to ~1600°C. Moreover, it remains compatible with the Si_3N_4 at these temperatures. As a result, the four point bend strength can be increased by 5 to 30% at 1375°C as the amount of crystallization product increases. A similar increase in the oxidation resistance is also observed. These materials are characterized by a flexure strength ranging from 400 to 500 MPa at 1375°C and fracture toughness between 6.5 and 8.0 MPa $m^{1/2}$.

The Effect of Grain Morphology and Glass Chemistry on Wear Behavior in Milling of Gray Cast Iron

One very promising application for self-reinforced Si_3N_4 is in metal machining. It is generally recognized that cutting tools based on Si_3N_4 exhibit outstanding performance compared to carbide or other ceramic materials in the machining of gray cast irons [20]. Of the various tool failure mechanisms observed during metal machining [21], gradual wear by abrasion predominates in the machining of grey cast iron [22]. Since abrasive wear resistance of most cutting tool materials is proportional to $K_{IC}^{3/4}$ $H^{1/2}$, any changes in microstructure that enhance fracture toughness and hardness should improve the cutting performance [22]. An example of this effect can be seen in the two hot pressed silicon nitride materials shown in figure 5. Both materials are based on similar glass phase chemistries. The only difference is that for the material in figure 5A, the ratio of transformation aid to whisker growth agent was adjusted to achieve a self-reinforced microstructure. This material removed 0.0105 m^3 of gray cast iron in a standard milling test as opposed to only 0.0061 m^3 for the material in figure 5B, a 72% improvement. Numerous milling tests with various glass phase compositions have consistently shown an improved milling performance when an optimum self-reinforced microstructure was achieved.

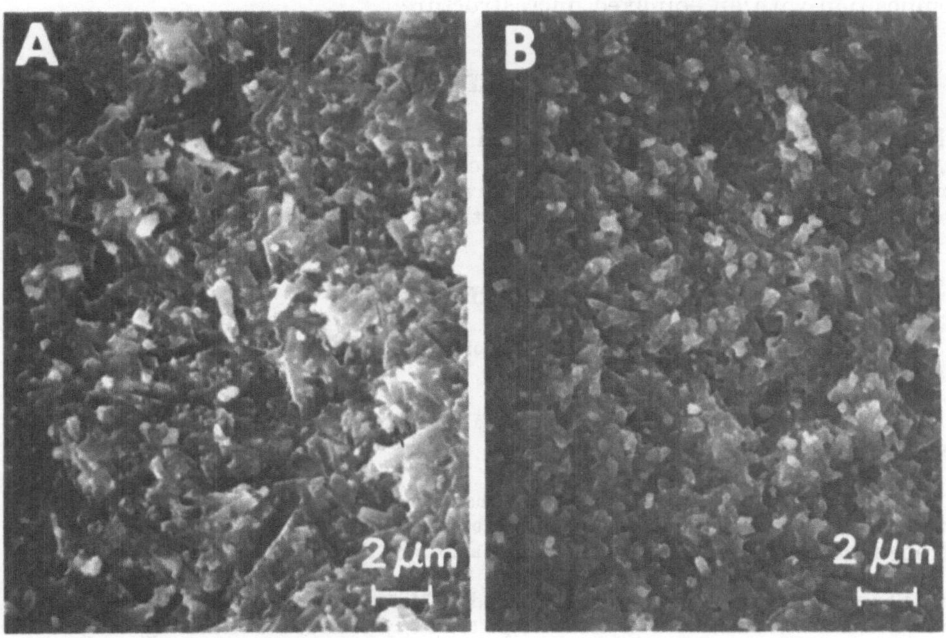

Fig. 5. Fracture surfaces of two silicon nitrides tested for gray cast iron machining performance. The material having the self-reinforced microstructure (A) strongly outperformed the material with an equiaxed microstructure (B).

SUMMARY

1. A low cost, high fracture toughness silicon nitride materials can be obtained through the in-situ growth of whisker-like elongated grains. This ability to form a reinforced material in-situ during densification avoids the processing difficulties and potential health hazards generally associated with whisker-reinforced composites. This new class of self-reinforced silicon nitride is based upon the Si_3N_4-Y_2O_3-MgO-CaO system where yttrium can be replaced by seven other elements, magnesium can be replaced by six other elements and calcium by 19 other elements.

2. By controlling the morphology of the elongated grains through glass phase chemistry, self-reinforced silicon nitrides can be made with room temperature fracture strengths ranging from 800 to 1250 MPa and fracture toughnesses ranging from 8 to 11 MPa $m^{1/2}$. With the controlled crystallization of the glassy phase, these materials can have fracture strengths approaching 500 MPa at 1375°C.

3. The compositional flexibility of the self-reinforced silicon nitride systems allows the microstructures to be tailored for specific applications. In milling tests on grey cast iron, a self-reinforced nitride cutting tool exhibited approximately 72% improvement in cutting performance over a similar composition with an equiaxed microstructure.

ACKNOWLEDGEMENTS

The authors wish to thank S. Piper and H. Rossow from Central Research and T. Allen from Intertec Design for their contributions in sample preparation and characterization. Special thanks are due to A. Hart for supporting this research. The authors also appreciate the discussions with D. Beaman.

REFERENCES

1. Drew P., Lewis M.H., The microstructures of silicon nitride ceramics during hot pressing transformations, J. Mat. Sci.,1974, **9**, pp. 261-69.

2. Lange F.F., Fracture toughness of Si_3N_4 as a function of the initial α-phse content, J. Amer. Ceram. Soc., 1979, **62**, pp. 428-30.

3. Wotting G., Kanka B.,and Ziegler G., Microstructural development, microstructural characterization and relation to mechanical properties of dense silicon nitride. In Non-Oxide Technical and Engineering Ceramics, ed. S. Hampshire, Elsevier, London, 1986,pp. 83-96.

4. Selkregg K.R.,More K.L., Seshadri S.G.,and McMurty C.H., Microstructural characterization of silicon nitride ceramics processed by pressureless sintering, overpressure sintering, and sinter hip, Ceram. Eng. Sci. Proc. 1990, 11, 7-8, pp. 603-15.

5. Lee R.R., Nowich B.E., Franks G., Quellette D.,Milford M.A.,Ferber M.K., Hubbard C.R., and More K., Duophase sialon for ceramic engine component application, presented at the 93rd American Ceram. Soc. Mtg., Cincinnati, OH, April 29, 1991.

6. Quinn G. D. and Braue W.R., Secondary phase devitrification effects upon the static fatigue resistance of sintered silicon nitride, Ceram. Eng. Proc., 1990, **11**, 7-8, pp. 616-632.

7. Shang-Xian, Wu, Compliance and stress-intensity factor of chevron-notched three-point bend specimen, Chevorn-Notched Specimens: Testing and Stress Analysis, ASTM STP 855, ed. S. W. Frieman and F. I. Baratta, ASTM, Philadelphia, 1984, pp. 177-192.

8. Pyzik J.A., Beaman D.R., Self-reinforced silicon nitride, J. Amer. Ceram. Soc., to be published.

9. Pyzik A.J., Dubensky W.J.,Schwarz D.B., and Beaman D.R., U.S. Pat. No. 4,883,776, 1989.

10. Pyzik A.J., Schwarz D.B., Dubensky W.J., and Beaman D.R., U.S. Pat. No. 4,919,689, 1990.

11. Pyzik A.J., Schwarz D.B., Beaman D.R. and Dubensky W. J., U.S. Pat. No. 5,021,372, 1991.

12. Vincenzini P. and Babini G.N., The influence of secondary phases on densification, microstructure and properties of hot pressed silicon nitride, In Sintered Metal-Ceramic Composites, ed. by G.S. Upadhyaya, Elsevier Sci. Publish., Amsterdam, Niderlands,1984, pp. 425-54.

13. Brice J.C., The growth of crystals from liquids, ed. E.P. Wohlfarth, Selected Topics In Solid State Physics, v. 12, Nort Holland/American Elsevier, London, 1973, pp. 227-45.

14. Khamskii E.V., Crystallization From Solutions, Consultants Bureau, New York-London, 1969, pp. 47-8.

15. Hwang C.J and Tien T.Y., Microstructural development in silicon nitride ceramics, Material Science Forum, 1989, **47**, pp. 84-109.

16. Becher P.F., Microstructural design of toughened ceramics, J. Amer. Ceram. Soc., 1991, 74, **2**, pp. 255-69.

17. Li, C. W. and Yamanis, J., Super-tough silicon nitride with R-curve behavior, Ceram. Eng. Sci. Proc., **10** , (7-8), 1990, pp. 632-645.

18. Hwang C.J. and Tien T.Y., A critical evaluation of the microstructural effect on the fracture toughness of silicon nitride, preseneted at the 90th American Ceram. Soc. Mtg., Cincinnati, OH, 1988.

19. Ito J., Silicate Apatites and Oxyapatites, Am. Mineralogist, **53**, 1968, pp. 890-907.

20. Samanta S.K., and Subramanian K., Hot pressed Si_3N_4 as a high performance cutting tool material, Proc. 13th N. Amer. Manufact. Res. Conf., SME, Dearborn Michigan, 1985, pp.402-407.

21. Kramer B.M., On tool materials for high speed machining, ASME Prod. Eng. Div., 1984, **12**, pp.127-40.

22. Baldoni J.G., Wayne F., and Buljan S.T., Cutting tool materials: mechanical properties - wear-resistance relationships, ASLE Trans, 1985, **29**, 3, pp. 347-52.

DEVELOPMENT OF ß-Si_3N_4 FOR SELF-REINFORCED COMPOSITES

Dale E. Wittmer*, Dilip Doshi and Thomas E. Paulson
Department of Mechanical Engineering, Southern Illinois University,
Carbondale, IL 62901 USA

ABSTRACT

ß-Si_3N_4 was obtained by the doping of a high-purity, commercial Si_3N_4 powder with 2 wt.% Y_2O_3. With additional processing the ß-Si_3N_4 was used as seed material for the development of self-reinforcing microstructures in a Si_3N_4 composite, containing Al_2O_3 and Y_2O_3 sintering aids.

A Si_3N_4 composition (A2Y6), containing a 5 wt.% addition of ß seed, was processed by turbomilling and pressure casting. Discs of the baseline Si_3N_4 (composition without seeds) and the seeded Si_3N_4 were sintered to 99-100% of theoretical density at 1880°C for 4 h at 43.5 KPa (300 psig) N_2. Based on limited data, 4-pt flexural strength was found to be 925-1025 MPa for the baseline Si_3N_4 and 750-950 MPa for the seeded Si_3N_4. Fracture toughness, using the Cook and Lawn modified indentation method, was found to be 8-10 MPa-m$^{1/2}$ for the baseline Si_3N_4 and 10-13 MPa-m$^{1/2}$ for the seeded Si_3N_4. SEM results show that the seeded Si_3N_4 developed a more mature ß-Si_3N_4 grain structure than the baseline composition.

INTRODUCTION

It has long been recognized that toughened monolithic ceramics can be produced through microstructural engineering of grain boundary phases, grain size and morphology. Over the last 20 years considerable research effort has been devoted to Si_3N_4 and Sialon ceramics because of the ability to effect the mechanical properties through the control of type and amount of grain boundary phases and microstructure. Depending on the processing conditions, elongated rod-like ß-Si_3N_4 grains may form during sintering. Often the microstructure produced contain interlocking networks of these ß-Si_3N_4 grains which impart significant increases in fracture toughness. However, reduction in strength is usually observed to accompany these increases in fracture toughness because the large grains that are produced can also act as flaw origins.

Recently the potential for improving the toughness, through the controlled formation of the elongated ß phase in-situ during sintering, of silicon nitride (Si_3N_4) has been re-emphasized.[1-6] Because the toughening is produced through the in-situ formation of an elongated and potentially oriented microstructure, a new term "self-reinforcing" has been coined. Improvements in toughening by self-reinforcing have been reported for Si_3N_4 by controlling the in-situ development of the ß phase through modification of the composition and amount of the sintering aids.[1,3] These self-reinforced Si_3N_4 compositions contain a significant volume of liquid phase and rely primarily on the solution-precipitation mechanism for the growth of the elongated ß phase.

It is generally thought that the $\alpha \rightarrow \beta$ transformation in Si_3N_4 is required to produce elongated instead of equiaxed ß morphology. Also, it has long been postulated that this transformation only occurs in the presence of a liquid phase by a solution-precipitation mechanism. In addition, it has been suggested that both the presence of a liquid phase and a high α-Si_3N_4 content are required in the starting powders in order to produce strong, dense Si_3N_4 ceramics.[7-9]

Recently, Wittmer[10] patented a process for the high-temperature formation of pure β-Si_3N_4 whiskers from Si_3N_4 substrates of slip cast, high-purity, >98% α-Si_3N_4 powder, with and without sintering aids. Wittmer and Ranganathan[11] characterized these whiskers and determined that they were most likely formed by a vapor-condensation mechanism and not by solution-precipitation. Additionally, Ranganathan and Wittmer[12] investigated the $\alpha \rightarrow \beta$-Si_3N_4 transformation in the absence of liquid phase additions by using pure α-Si_3N_4 whiskers as the starting material. The transformation was studied over a temperature range of 1650 to 1850°C with a N_2 overpressure of up to 100 atm. to control volatilization of the Si_3N_4. The results proved that the $\alpha \rightarrow \beta$ transformation can occur entirely in the absence of any liquid phase, and depending on the conversion conditions, large diameter, well-facetted, rod-like β-Si_3N_4 may be produced. The proposed mechanism for this conversion was vaporization-condensation.

Because conversion to β-Si_3N_4 was observed for the pure α phase in the absence of a liquid phase, it was thought that similar conversion could occur for Si_3N_4 powders with high starting α content. Therefore, similar conversion trials were attempted with commercially available, high-purity α-Si_3N_4 powders with less than satisfactory results. Doping of the commercial Si_3N_4 powders with rare earth oxides was found to enhance the nucleation and growth of the β-Si_3N_4.

The concept of controlling and enhancing the formation of specific conversion products by "seeding" is well recognized in the chemical processing, crystal growing and ceramic raw material industries. Seeding is accomplished though the addition of high-purity particulate which act as sites for growth of the preferred phase. For powder processes it is generally desired to obtain a high degree of dispersion of high surface area seed material in order to control the size and conversion rate of the product. For other processes, such as crystal growth, it is usually preferred to have larger seeds with well-facetted surfaces so that growth in a preferred orientation can be initiated and maintained.

This investigation explores the use of β-Si_3N_4 seed for the enhanced microstructural development of sintered Si_3N_4.

PROCEDURE

Preparation of ß-Si_3N_4 Seed

A 2 wt.% addition of Y_2O_3[1] was made to 500 g UBE E-10 Si_3N_4[2] by processing in a 9 cm diameter turbomill[3], containing 1 kg of 4 mm diameter YTZ[4] milling media and 625 cc distilled water. After the turbomill was loaded, the pH was adjusted to 8.4 by the addition of ammonium hydroxide to obtain heterocoagulation of the Si_3N_4 and Y_2O_3. Following turbomilling for 1 h at 1200 r.p.m., the resulting slurry was discharged through a 60/200 mesh screen stack. The screened slurry was then pressure cast, dried, and granulated to -150 mesh. The granulated powder was then converted in a graphite crucible at 1800°C for 4 h at an N_2 overpressure of 29 KPa (200 psig). The morphology of the transformation product was observed by SEM and the phases identified by standard x-ray diffraction methods[17,18]. The ß-Si_3N_4 seed was then prepared by turbomilling the transformation product, using the YTZ media, in 500 cc distilled water for 1 h at 1200 r.p.m.. The morphology of the ß-seed was then observed by SEM prior to the incorporation of the seed in a selected Si_3N_4 matrix.

Preparation of ß-Si_3N_4 Baseline and Seeded Compositions

A2Y6, Si_3N_4[5] containing 2 wt.% Al_2O_3[6] and 6 wt.% Y_2O_3[1], was selected as the baseline composition for this study. The raw materials were batched in the appropriate amounts for 0.5 kg lots and processed in the turbomill, containing 2 kg of the YTZ milling media and 500 cc distilled water. After the turbomill was loaded, the pH was adjusted to about 10 by the addition of ammonium hydroxide. Following turbomilling for 1 h at 1200 r.p.m., the resulting slurry was discharged through a 60/200 mesh screen stack. The slurry was then pressure cast (using N_2 gas gradually increased from 700 Pa (5 psig) up to 14.5 KPa (40 psig)) into green discs 7.6 cm in diameter by ≈ 2.2 cm thick, using a commercial filter press[7], and

---

1 Y_2O_3 - High Purity Grade, Hermann C. Starck, W. Germany
2 Si_3N_4 - Grade E-10, UBE Industries America, NY, NY
3 The turbomill is a high shear mixing and milling apparatus designed for processing advance ceramics and composites. The turbomill is built by the end user and is not commercially manufactured.[13-16]
4 YTZ - Tohso, America, Atlanta, GA
5 Si_3N_4 - LC-10N Grade, Hermann C. Starck, W. Germany
6 Al_2O_3 - Ceralox HPA-0.5 AF Grade, Ceralox Corp., Tuscon AZ
7 N.L. Baroid, Dallas, TX

then dried under vacuum at 100°C for 24 h. Following drying, the discs were isopressed at 310 MPa to obtain a green density of about 60-65% of theoretical.

A similar procedure was used for the addition of the ß-Si_3N_4 seed to the baseline A2Y6 composition. After turbomilling the baseline composition for 2 h, the appropriate amount of ß-Si_3N_4 was added to obtain a 5 wt.% addition. Turbomilling was then continued for an additional 30 min. to disperse the seed. The resulting slurry was then pressure cast, dried and isopressed for the same conditions as the baseline composition without ß seed.

The baseline and seeded compositions were overpressure sintered in graphite crucibles using A2Y6 setter powder to tightly pack the discs. Sintering conditions were 1880°C for 4 h at an N_2 overpressure of 43.5 KPa (300 psig).

Standard x-ray diffraction methods were used to determine the phases present, while SEM was used to observe the microstructure and determine the morphology of the ß-Si_3N_4.

For flexural strength[19] and fracture toughness[20] measurements, test bars were machined into 3X4X45 mm test bars with 45° edge chamfers. The density and porosity of each test bar was measured by Archimedes' method in distilled, deionized water. 4-pt. loading for both measurements used an internal span of 20 mm and an outside span of 40 mm, and a loading rate of 0.05 cm/min. Flexural strength was measured using the as-machined tensile surface. For fracture toughness measurements, the tensile surface of each bar was diamond polished to 1 μm, prior to making 5 Vicker's indentations about 1 mm apart in the center and along the long axis of the bar, using a 5 kg load.

RESULTS AND DISCUSSION

ß-Si_3N_4 Seed

XRD results confirmed complete transformation of the Si_3N_4 doped with 2 wt.% Y_2O_3, after firing to 1800°C for 4 h under N_2 overpressure. The conversion product used as the source of ß-Si_3N_4 seed before and after turbomilling is shown in Figs. 1 and 2, respectively. As seen in Fig. 1, the fired Y_2O_3 doped Si_3N_4 powder formed a very uniform, high aspect ratio ß-Si_3N_4 approximately 1-2 μm in diameter. As shown in Fig. 2, turbomilling for 1 h produced nearly all particulate ß-Si_3N_4 seed approximately 0.25-1 μm in diameter and very few elongated grains. Assuming a ß-Si_3N_4 seed with a diameter of 0.5 μm and a green density of 65 % of theoretical, a green disk with a 5 wt.% seed addition would contain about 5×10^{11} ß-Si_3N_4 seeds per cc of green volume.

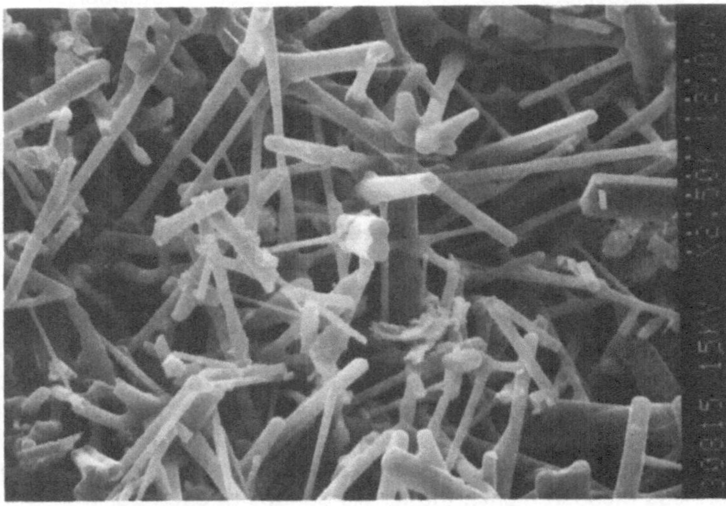

Figure 1. SEM micrograph of Si_3N_4 with 2 wt.% Y_2O_3 after firing to 1850°C for 4 h.

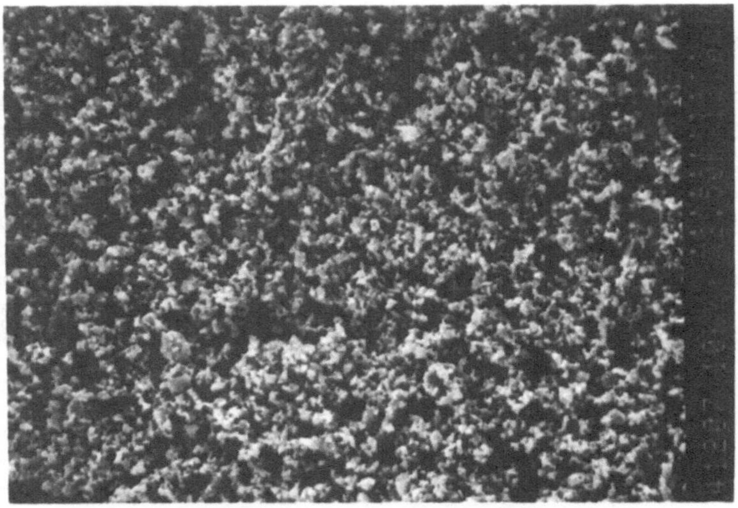

Figure 2. SEM micrograph of β-Si_3N_4 seed after turbomilling.

A2Y6-Si_3N_4 Matrix

The A2Y6 matrix sintered to >99% of theoretical density at 1880°C for 4 h under 300 psig N_2 overpressure. Based on 4 test bars each, the 4-pt flexural strength was 925-1025 MPa and the fracture toughness was 8-10 MPa-$m^{1/2}$. XRD results indicated complete conversion to ß-Si_3N_4. The microstructure of theetched A2Y6 matrix showed a small amount of uniformly distributed porosity about 6 μm in diameter, some fine grained material, and some elongated ß-Si_3N_4 approximately 1-2 μm in diameter by 4-6 μm long. The microstructure given in Fig. 3 is characteristic of the A2Y6 matrix after sintering and thermal etching.

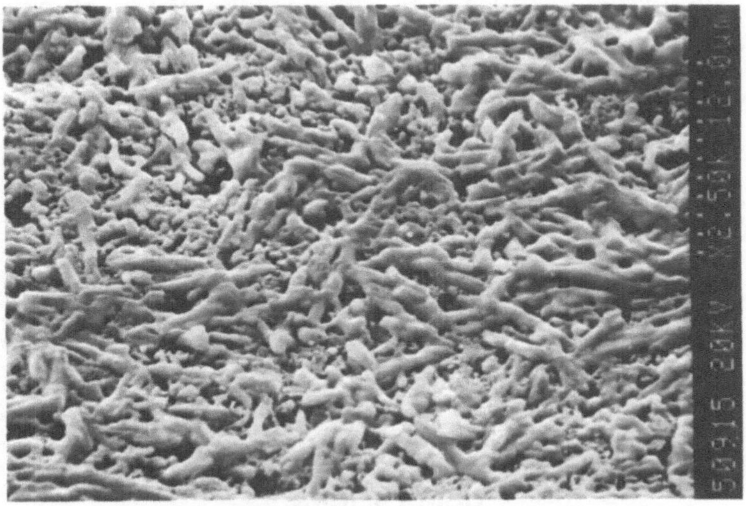

Figure 3. Thermally etched surface of sintered A2Y6 matrix.

A2Y6-Si_3N_4 with 5 wt.% ß-Si_3N_4 Seed

The seeded A2Y6 also sintered to >99% of theoretical density at 1880°C for 4 h under 300 psig N_2 overpressure. Based on 4 test bars each, the 4-pt flexural strength was 750-950 MPa and the fracture toughness was 10-13 MPa-$m^{1/2}$. As with the matrix, XRD results indicated that the seeded composition completely converted to ß-Si_3N_4. The microstructure of the seeded Si_3N_4 (Fig. 4) showed a small amount of uniformly distributed porosity about 6 μm in diameter, however there was much less fine grained material, and a greater amount of elongated ß-Si_3N_4 than observed in the A2Y6 matrix. Also, the ß-Si_3N_4 appears to be larger in diameter (2-3 μm) and slightly longer (6-8 μm) for the seeded A2Y6. The reduction in the fine grained material would be expected to lower the flexural strength, while the increased volume of the elongated ß phase would likely increase the toughness, as was observed. Based on these limited results, it appears that ß seeding of Si_3N_4 compositions is a means of altering the sintered microstructure to increase the fracture toughness.

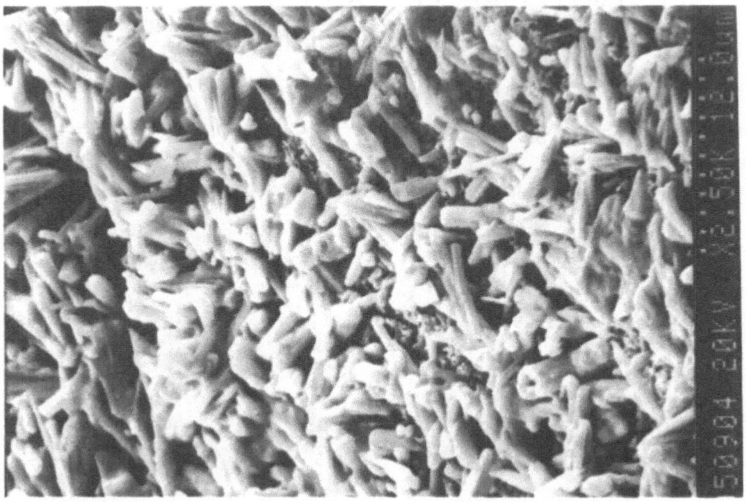

Figure 4. Thermally etched surface of sintered A2Y6, containing 5 wt.% ß seed.

CONCLUSIONS

Although the data is limited to just a few test bars, it would appear that there is enough evidence to establish that ß-Si_3N_4 seed can be used to increase both the amount and size of the elongated ß-Si_3N_4 phase in Si_3N_4 containing Al_2O_3 and Y_2O_3 sintering aids. This microstructural enhancement was found to increase the fracture toughness with a slight decrease in the flexural strength, which is consistent with expectations.

ACKNOWLEDGEMENTS

This research was sponsored by the U.S. Dept. of Energy, Assistant Secretary for Conservation and Renewable Energy, Office of Transportation Systems, as part of the Ceramic Technology for Advanced Heat Engines Project of the Advanced Materials Development Program, under contract DE-AC05-840R21400 with Martin Marietta Energy Systems, Inc. The authors would like to thank D.R. Johnson, Manager Ceramic Technology for Advanced Heat Engine Project, ORNL for his support and encouragement, and T.N. Tiegs, ORNL for his many useful discussions. The authors would also like to express their appreciation to Ceralox Corp., Tuscon, AZ for providing gratis samples of Al_2O_3.

REFERENCES

1. E. Tani, S. Umebayashi, K. Kishi, K. Kobayashi, and M. Nishijima, "Gas-Pressure Sintering of Si_3N_4 with Concurrent Addition of Al_2O_3 and 5 wt% Rare Earth Oxide: High Fracture Toughness Si_3N_4 with Fiber-Like Structure," Am. Ceram. Soc. Bull., 65 [9] 1311-1315 (1986).

2. C. Li and J. Yamanis, "Super-Tough Silicon Nitride with R-Curve Behavior," Ceram. Eng. Sci. Proc., 10 [7-8] 632-45 (1989).

3. A.J. Pyzik and D.R. Beaman, "Processing and Microstructure of Self-Reinforcing Silicon Nitride," 14th Annual Conf. of Composites and Adv. Ceramics, Cocoa Beach, FL, Jan. 14-17, 1990 (to be published in Ceram. Eng. Sci. Proc. 11 (1990)

4. K. White and A.J. Pyzik, "R-Curve Behavior of Self-Reinforcing Si_3N_4," ibid.

5. D.F. Carroll and A.J. Pyzik, "Microstructural/Mechanical Property Relationships in Self-Reinforced Silicon Nitride," ibid.

6. A.J. Pyzik and D.R. Beaman, "Processing, Microstructure, and Properties of Self-Reinforced Silicon Nitride," 92nd Annual Am. Ceram. Soc. Meeting Abstracts, Dallas, TX, April 22-26, 1990, pp. 74.

7. C. Greskovich and S. Prochazka, "Observations on the $\alpha \rightarrow \beta$-Si_3N_4 Transformation," J. Am. Ceram. Soc. 60, 471-72 (1977).

8. L.J. Bowen, R.J. Weston, T.G. Carruthers, and R.J. Brooks, "Hot-Pressing and the α-β Transformation in Silicon Nitride," J. Mat. Sci. 13, 341-350 (1978).

9. F.F. Lange, "Silicon Nitride Polyphase Systems: Fabrication, Microstructure, and Properties," Int. Metals. Rev. 25, 1-20 (1980).

10. D.E. Wittmer, "Process for Producing Beta Silicon Nitride Fibers," U.S. Patent No. 4,717,693, Jan. 5, 1988.

11. D.E. Wittmer and P. Ranganathan, "Characterization of Pure ß-Si_3N_4 Whiskers," 91st Annual Am. Ceram. Soc. Meeting, Indianapolis, IN,

12. P. Ranganathan and D.E. Wittmer, "$\alpha \rightarrow \beta$ Transformation in Pure Si_3N_4 Whiskers," 92nd Annual Am. Ceram. Soc. Meeting Abstracts, Dallas, TX, April 22-26, 1990, pp. 69.

13. D.E. Wittmer, "Alternative Processing Through Turbomilling," American Ceramic Society Bulletin, Vol.67, No. 10, pp. 1670-72 (1988).

14. D.E. Wittmer, "Improved Dispersion Technique for Ceramic Whisker-Ceramic Matrix Composites," Ceramic Engineering and Science Proceedings, Vol. 9, No. 7-8, pp. 735-40 (1988).

15. D.E. Wittmer and W. Trimble, "Ceramic Composites Processed by Turbomilling," 3rd International Symposium on Ceramic Materials and Components for Engines, November 27-30, 1988, Las Vegas, NV, V.J. Tennary, Ed., Am. Ceram. Soc. Pub., pp. 273-289 (1989).

16. D.E. Wittmer, W. Trimble, and T. Paulson, "Processing of Ceramic Matrix Composites by Turbomilling," Proc. of the Vacuum Industries Symp. on High-Temperature Materials, Boston, MA, Oct. 18-20, 1989.

17. Methods and Practices in X-Ray Powder Diffraction, R. Jenkins, Ed., Center for X-Ray Diffraction Studies, JCPDS-ICDD, Swarthmore, PA (1989).

18. C.P. Gazzara and D.M. Messier, "Determination of Phase Content of Si_3N_4 by X-Ray Diffraction Analysis," Ceram. Bull., Vol. 56, No. 9, pp. 777-780 (1977).

19. "Test Method for Flexural Strength of Advanced Ceramics at Ambient Temperature," ASTM Vol. 15.01, C 1161-90 (1991).

20. R.F. Cook and B.R. Lawn, "A Modified Indentation Toughness Technique," J. Am. Ceram. Soc., Vol. 66, No. 11, C 200-201 (1983).

SINTER-CANNING FOR HIP-PROCESSING OF SILICON NITRIDE

HORST WEDEMEYER, HANS-JOACHIM RITZHAUPT-KLEISSL
AND ELMAR GÜNTHER
Kernforschungszentrum Karlsruhe GmbH
Institut für Materialforschung, IMF III
POB. 3640, D-7500 Karlsruhe, Fed. Rep. of Germany

ABSTRACT

A new sinter-canning method has been developed for the coating of silicon nitride parts prior to hot-isostatic pressing. This method is based on thick-film technology. Using sol-gel methods different shapes of silicon nitride have been coated with an alumina or zirconia gel. After drying and sintering a dense layer of the oxides are formed, suitable as a ceramic canning material. Following the HIP-process, dense silicon nitride is achieved with a surface layer containing the aluminium or zirconium ions in a crystalline phase. This method of sinter-canning prevents the formation of glassy phases and can be used for a variety of ceramics to be densified by hot-isostatic pressing.

INTRODUCTION

Since several years oxide and non-oxide materials are developed to be used as engineering ceramics. Those materials are especially distinguished by high melting points, high mechanical strengths and/or high corrosion resistance against agressive atmospheres. Pressing and sintering of the powdered ceramic materials usually is used for the fabrication of different shapes and bodies, whereas the covalent bonding often prevents a total densification. High densities are achieved with those ceramics by hot-isostatic pressing (HIP) using temperatures and pressures of up to 2000 °C and 3000 bar, resp. But, a pre-densification to closed porosity, i.e. about 95 % th.d., is necessary for the success of HIP-processing. If this predensification is impossible by sintering the pressed powder compacts have to be encapsulated prior to hot-isostatic pressing. Metals with high melting points, such as tantalum or niobium, and also quartz [1] is used as such canning materials, whereas metallic capsules may be used for simple shaped bodies only. Metallic contaminations or even the formation of second phases in the surface areas of the bodies are observed after hot-pressing by diffusion or impregnation of the capsule materials due to the strong contact between capsule and pressed body. Those surface effects may influence the mechanical properties of the dense ceramic materials especially if the formation of glassy phases are observed.

The disadvantages of metallic or quartz capsules for the HIP-process, as pointed out above, can be avoided by using ceramics as canning materials [2-3] . The ceramic encapsulation can be achieved using high sinterable ceramic powders obtained from sol-gel processes together with the so called thick-film technology [4-5] , as pointed out within this paper, for the total densification of pressed compacts of silicon nitride, Si_3N_4.

MATERIALS PROCESSING AND CERAMIC ENCAPSULATION

Compacts of silicon nitride, Si_3N_4, were prepared from a commercial powder by cold-isostatic pressing (CIP) using a dry-bag equipment or by usual axial pressing. The obtained green densities were about 45 % th.d. or about 30 % th.d., resp. The commercial Si_3N_4 powder (Fa. UBE) was processed from silicon tetrachloride, $SiCl_4$, following the diamide process. The particel sizes of the powder are about one micron, mostly in form of small agglomerates. The powders were milled prior to CIP-compaction or used as processed for axial pressing.

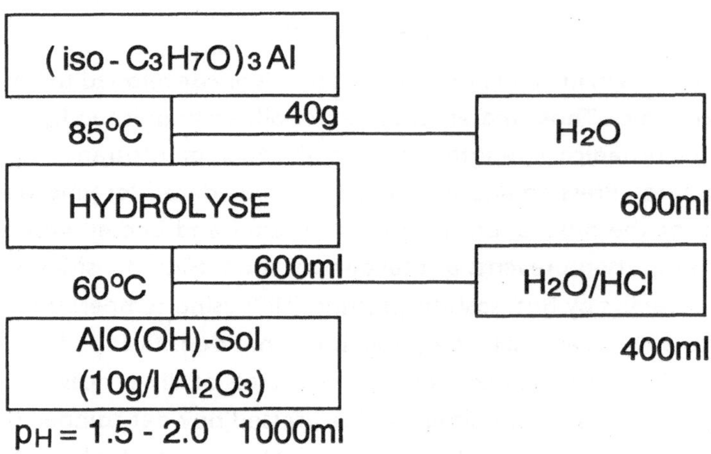

Figure 1. Preparation of alumina sols

Sols of alumina or zirconia (monoclinic or yttria stabilized tetragonal) were prepared from aluminium isopropoxide, $(iso\text{-}C_3H_7O)_3Al$, or zirconium n-propoxide $(n\text{-}C_3H_7O)_4Zr$, as shown in figure 1 and 2.

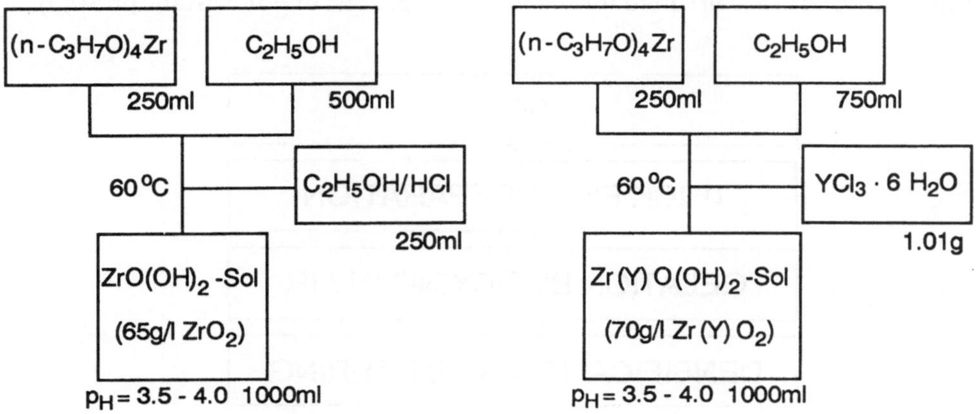

Figure 2. Preparation of zirconia sols
(left: monoclinic zirconia, right: yttria stabilized zirconia)

The alumina sols were easily prepared by hydrolysis of the isopropoxide suspended in water and slow addition of hydrochloric acid to achieve a p_H-value of 1.5 to 2.0, which is important to obtain a stable sol. Stable zirconia sols were prepared by dissolution of the n-propoxide dissolved in ethanol without further addition of water for hydrolysis. To prepare monoclinic zirconic coatings the p_H-value was kept to 3.5 to 4.0 by slow addition of hydrochlorid acid dissolved in ethanol. To prepare yttria stabilized coatings yttrium chloride, YCl_3 6 H_2O, was added to the sol instead of hydrochloric acid.

Prior to coating the compacts were carefully dried using different methods: drying in normal air, inert atmospheres (argon, nitrogen) or in an ammonia atmosphere at about 900 °C within some hours. After cooling the compacts were impregnated with methanol to refuse the different gases enclosed in the compacts. The coated compacts were densified by hot-isostatic pressing using a sinter-HIP process with temperatures of up to 1950 °C and pressures of up to 2000 bar.

The pressed compacts of silicon nitride than were coated with the high-concentrated sols of alumina or zirconia (see figure 3). After gelation in air dense layers or thick-films in the range of 10 to 100 microns were obtained. These coatings can be densified by a normal sintering process due to high sinterability of the extremely fine-grained gel powders with grain-sizes in the range of some ten nanometers forming a ceramic capsule. As it is well known that those sintered ceramics show a high plasticity behaviour at elevated temperatures [6-7] , the ceramic capsules are well suited for the following process of hot-isostatic pressing.

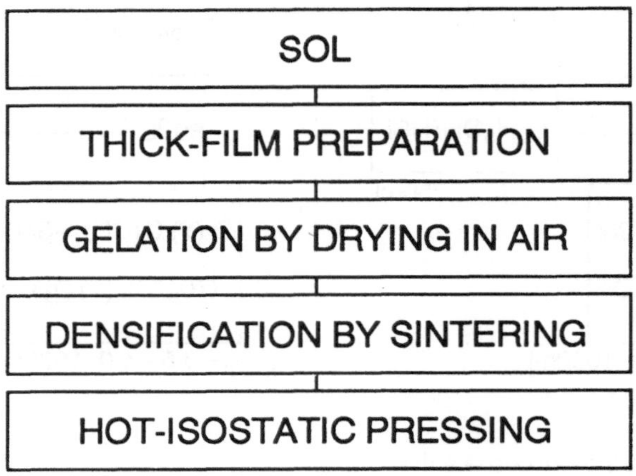

Figure 3. Ceramic encapsulation of pressed compacts

EXPERIMENTAL RESULTS

The alumina coated silicon nitride compacts were completely densified during hot-isostatic pressing. Metallographic examination showed a dense single-phase structure at the inner parts of the body with the α-Si_3N_4 particles transformed into the β-Si_3N_4 phase. A two-phase structure was observed at the surface area. Further analysis by X-ray diffraction showed a well crystallized phase, which could be identified to be an alumina containing phase, $Si_2Al(N,O)$ (SIALON). Pure alumina, which was assumed to be separated in a further phase, could not be detected. Scanning electron microscopy was used for examination of the structure and the grain-boundaries. As a result, no glassy phases could be identified. Single pores were observed, mostly less than one micron in diameter.

It is assumed that the formation of the alumina containing phase worked as a sintering aid during the HIP-process. Thus, the outer shell of the compacted body might be densified in an early stage of the sinter-HIP process, the open porosity closed, and an ideal ceramic capsule formed for the high-pressure densification.

The zirconia coated silicon nitride compacts were nearly completely densified during hot-isostatic pressing. The metallographic examination showed some porosity with decreasing amount towards the inner parts of the bodies. A two-phase structure was observed at a broader surface area as compared to the alumina coated compacts. The X-ray diffraction analysis resulted in a two-phase structure formed by well crystallized β-Si_3N_4 and monoclinic or tetragonal zirconia. No further phases could be detected. Scanning electron microscopy again was used to examine the structure and the grain boundaries. As a result, no glassy phases could be identified.

The results lead to the assumption that the zirconia had not worked as a sintering aid. The assumed super-plastic behaviour of the sol-gel derived zirconia might allow the thin zirconia shell to be pressed deeper into the compact structure as compared to the alumina canned silicon nitride samples, from which a larger amount of porosity is observed in the outer parts of the densified bodies. A thicker zirconia layer or a repeated coating procedure may help to overcome this problem.

CONCLUSION

Ceramic coatings, prepared from sol-gel derived materials are suited to prepare a sinter-canning prior to hot-isostatic pressing. This method of sinter-canning prevents the formation of glassy phases and can be used for a variety of ceramics, included non-oxide ceramics, to be densified by HIP- or sinter-HIP processes. As a further advantage of this sinter-canning process, dense and single-phase bodies can be processed using ceramic canning materials with the same chemical composition, which is of special interest for the fabrication of single-phase parts of silicon nitride, aluminium nitride, boron nitride, boron carbide, etc. The only precondition is a method for the preparation of the corresponding stable sols, which easily tend to form gels. A variety of possible procedures are given in the literature [8-10] , from which oxide as well as non-oxide sol-gel derived ceramics may be processed.

REFERENCES

1. Okada, H., Homma, K., Fujikawa, T. and Kanda, T., Fabrication of dense Si_3N_4 by hot-isostatic pressing, Proc. 6th Intern. World Congr. High Tech Ceramics, Milan, Italy 1986. In High Tech Ceramics, Part A, ed. P. Vincencini, Elsevier, Amsterdam, Oxford, New-York, Tokyo 1987, pp. 1023-32.

2. Heinrich, J. and Böhmer, M., Werkstoff- und Bauteileentwicklung durch heiß-isostatisches Pressen am Beispiel von Siliciumnitrid. In Keramische Komponenten für Fahrzeug-Gasturbinen III, Springer, Berlin 1984, pp. 159-65.

3. Heinrich, J. and Böhmer, M., Comparison of sinter-HIP and canning HIP of RBSN and Si_3N_4 powder compacts. In Ceramic Materials and Components for Engines, eds. Bunk, W. and Hausner, H., Deutsche Keramische Gesellschaft 1986, pp. 243-53.

4. Budd, K.D. and Payne, D.A., Ceramic processing of thin-layer capacitors by sol-gel methods, Proc. 2nd Intern. Conf. Ceramic Powder Processing Science, Berchtesgaden, Fed. Rep. of Germany, 1988. In Ceramic Powder Processing Science, eds. Hausner, H., Messing, G.L. and Hirano, S., Deutsche Keramische Gesellschaft 1989, pp. 513-20.

5. Abell, J.S., Wellhofer, F. and Shields, T.C., Textured superconducting thick-films of $YBa_2Cu_3O_{7-x}$, Proc. 1st European Ceramic Soc. Conf., Maastricht, The Netherlands, 1989. In Euro-Ceramics, eds. de With, G., Terpstra, R.A. and Metselaar, R., Elsevier Applied Science, London, New York 1989, Vol. 2, pp. 476-80.

6. Wakai, F., Kodama, Y., Sakaguchi, S., Murayama, N., Izaki, K. and Nihara, K., A superplastic covalent crystal composite, Nature (London), 1990, **344**, 421-3.

7. Chen, I-W., Xue, L.A., Development of superplastic structural ceramics, J. Am. Ceram. Soc., 1990, **73**, 2585-609.

8. Pouskouleli, G., Metallorganic compounds as preceramic materials I. Non-oxide ceramics, Ceramic International, 1989, **15**, 213-29.

9. Pouskouleli, G., Metallorganic compounds as preceramic materials II. Oxide ceramics, Ceramic International, 1989, **15**, 255-70.

10. Szweda, A., Hendry, A. and Jack, K.H., The preparation of silicon nitride from silica by sol-gel processing, Proc. British Ceramic Soc., 1981, **7**, 107-13.

SIC DISPERSOID - REINFORCED Si_3N_4 COMPOSITES.

S.M.KETCHION, G. LENG - WARD and M.H.LEWIS
Centre for Advanced Materials Technology, University of Warwick, Coventry, UK.

ABSTRACT

A series of composites were prepared by dispersing SiC whiskers and platelets into a Si_3N_4 matrix, intended for high temperature applications, and fully densified by HIP. The effect of dispersoid additions on the mechanical properties was evaluated and related to interfacial microstructure and fracture surface studies; the latter giving an insight into the toughening mechanisms taking place. Creep resistance and stress rupture data revealed information on the high temperature deformation behaviour of selected compositions and illustrated the importance of the crystallisation behaviour of the intergranular region.

INTRODUCTION

A potential method of improving the thermo - mechanical properties of brittle ceramic materials is by incorporating high temperature resistant SiC whiskers in which bridging and pullout toughening mechanisms can take place. Previous work on SiC whisker reinforced Al_2O_3 showed significantly improved mechanical properties and creep resistance at elevated temperatures over the monolith [1,2]. However, work on SiC reinforced Si_3N_4 composites have been met with varying success [3-6], the major problems being with the interface characteristics and the crystallisation behaviour of the intergranular region. In this study two types of SiC dispersoid were used; a new source of SiC whiskers which have relatively large diameters and smooth surfaces, and a recent commercially available source of platelets which have been shown to improve fracture toughness [7] and reduce the health hazards associated with whiskers. These were dispersed into a tailored Si_3N_4 matrix composition which has been shown to give desirable crystallisation products and high temperature properties [8].

MATERIALS AND EXPERIMENTAL PROCEDURE.

The dispersoids used were cubic β – SiC whiskers†, with an average diameter of 1.2μm, and hexagonal α –SiC platelets‡, with an average particle size of 16μm and a thickness of between 2 and 2.5μm. Compositions were prepared by aqueous shear blending of pre-milled matrix material; a composition of 94wt.% Si_3N_4*, 5wt.% Y_2O_3 and 1%wt.% SiO_2, with the required weight of dispersoid. The flocculation conditions were controlled by adjusting the pH, then the slurry was degassed and slip cast into plaster of Paris moulds to produce green billets of approximately 25mm thickness. These were CIPed at 150 MPa, fired at 1050°C for 4 hours in N_2 atmosphere, then HIPed§ at 1725°C, 160 MPa for 1 hour.

The mechanical test samples were rectangular bars with a cross-section of 3.0 x 3.0mm, machined from the HIPed billets with the tensile face and chamfered edges given a final polish using a 1μm diamond paste. The tensile surface was that parallel and closest to the bottom surface of the plaster of Paris mould. Modulus of rupture and indentation initiated fracture toughness tests were performed on 15 test bars of each composition, at room temperature in four-point loading, with inner and outer spans of 12 and 20mm respectively. Creep testing was performed in four-point bending and creep rates were determined from the load point displacement [9]. The specimens were left at temperature for 24 hours before the test load was applied.

MICROSTRUCTURE.

The density of the HIPed composites containing at least 30wt.% whiskers and 40wt.% platelets were very close to theoretical density. Optical micrographs showed good homogeneity and preferential alignment of the dispersoids parallel and close to the bottom surface of the billets [10]. X-Ray diffraction indicates that there was near complete transformation from α - to β – Si_3N_4 and the intergranular region crystallised to α – $Y_2Si_2O_7$ and Si_2N_2O, although samples were further heat treated at 1280°C for 24 hours to ensure crystallisation before microscopy and mechanical testing. SEM micrographs also show that the intergranular phase was finely distributed throughout the composites [10]. TEM micrographs show that the whiskers and platelets are very stable after elevated processing temperatures and there was no degradation of their structure. The matrix microstructure of both types of composite and the monolith, do not differ substantially. The Si_3N_4 had a broadly bimodal grain structure, consisting of fine equiaxed grains, mostly in the lower part of the 0.1 - 1.0 μm range, and elongated grains with dimensions in the 1 - 5 μm range.

† SNIA FIBRE, Cesano Maderno, Italy.
‡ C - Axis Technology, ALCAN Ltd, Jonquiere, Quebec, Canada.
* SN E-10, UBE Industries Ltd, Tokyo, Japan.
§ ABB Cerama, Robertsfors, Sweden.

High resolution microscopy of the whisker composite, Fig.1, shows a crystalline triple junction which electron diffraction confirmed was α - $Y_2Si_2O_7$. EDAX of these regions occasionally shows metal impurities such as calcium in whisker composites and aluminium in the platelet composites. These impurities originate from the dispersoids and have been incorporated into the intergranular region when the minor phase was formed by reaction of Y_2O_3 with SiO_2 on the surface of the Si_3N_4 and SiC [10]. However, lattice imaging shows there was a very thin interfacial layer of approximate 2 nm thickness between the disilcate and the Si_3N_4 grains, and between the Si_3N_4 grains and the dispersoid. It was likely that the progressive crystallisation of the interganular phase occurred until the disilicate composition was exhausted and the residual material remained as a thin layer of Y-Si-O glass at the boundary.

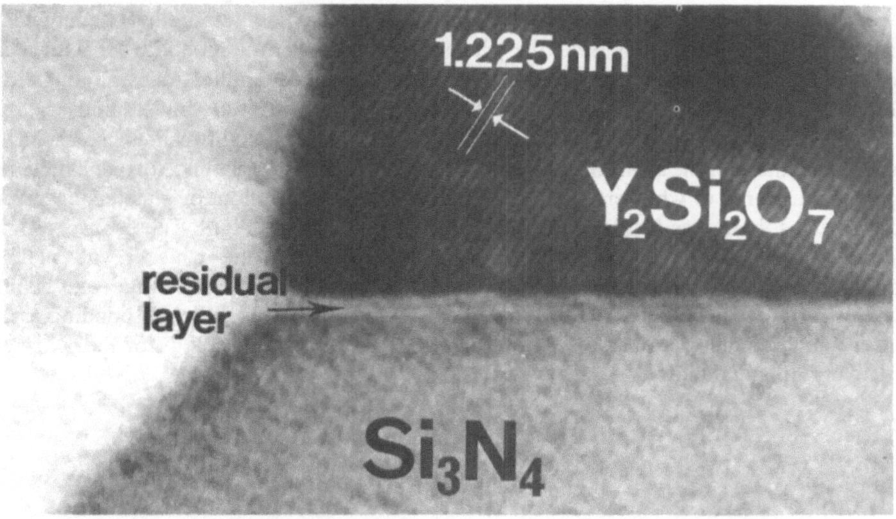

Fig.1. TEM micrograph showing crystalline intergranular triple junction and thin residual layer.

MECHANICAL PROPERTIES

Room temperature mechanical properties of representative compositions, Table 1, show that whereas the whisker composite maintains the fracture strength of the monolith, there was a significant decrease with the platelet composite. This was probably because of the relatively large size of the platelets compared to the matrix grains, and possibly thermal expansion mismatch, which will introduce microcracking. Fracture surfaces of the test bars show that they generally fail from near the corner of the tensile face from no obvious origin, indicating machining related flaws, although some of the platelet containing samples indicate failure near small clusters of platelets.

Material	M.O.R. MPa	Weibull Modulus	K_{1c} MPa.m$^{1/2}$	Vickers Hardness GPa
Monolith	700	14.9	4.6	14.0
30 wt.% Whiskers	710	15.1	5.6	17.4
30 wt.% Platelets	480	10.7	6.0	15.7

Table.1 Room Temperature Mechanical Properties of Selected Compositions.

The Weibull modulus of the whisker composite and monolith give similar values, but the platelet composite has a lower reliability due to a wider strength reducing flaw distribution. There was a increase in the Vickers Hardness over the monolith for both types of composite which could give benefits in wear resistance applications, although the differences in value for the whisker and platelet materials was not fully understood.

For both types of composite there was a small but not significant increase in the fracture toughness. For a similar volume fraction of dispersoid, platelets enhance the toughness compared to whiskers. Fracture surfaces reveal that pullout of platelets occurred, Fig.2, whereas there was little evidence of protruding whiskers, Fig.3. These observations suggest differences in the interfacial characteristics between the whisker and platelet composites because these are extremely important in allowing crack wake toughening mechanisms to occur in the brittle reinforcement of ceramics. This indicates bonding at the whisker / matrix interface was too strong and/or the surface morphology of the whiskers restricts bridging and pullout .

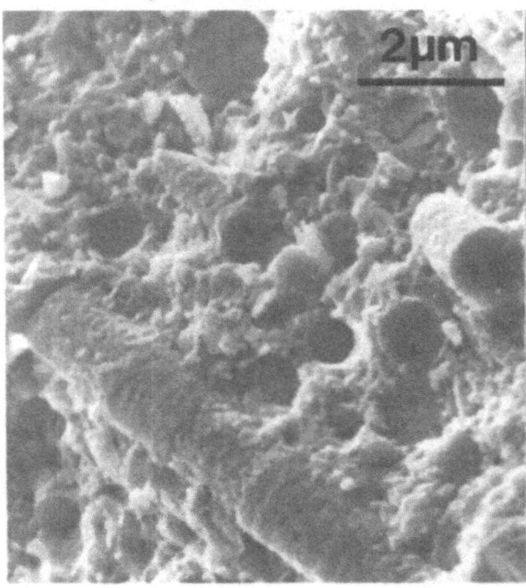

Fig.2. Fracture surface of whisker composite

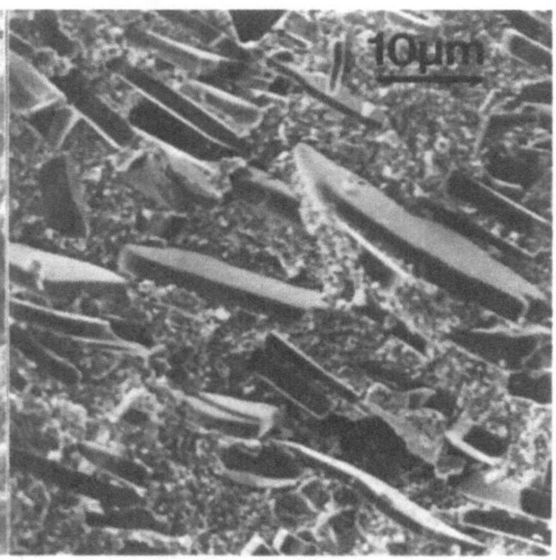

Fig.3. Fracture surface of platelet composite

Studies of the microstructural features of toughening in a TEM sample, Fig.4, shows the typical fracture path in an accidently damaged platelet composite sample. There was clearly preference for fracture and debonding along the platelet / matrix interface. As a result intact ligaments can be retained which bridge the propagating crack and when the platelet was pulled out there would be an increase in frictional work . However, pullout distances are relatively short, so fracture toughness increases are small. There was little or no debonding at the whisker / matrix interface resulting only in some crack deflection and transgranular fracture of the whiskers. This behaviour has been attributed to the thin residual layer between whiskers and the matrix which promotes development of a strong bond and some mechanical interlocking because of the microscopically irregular whisker surface.

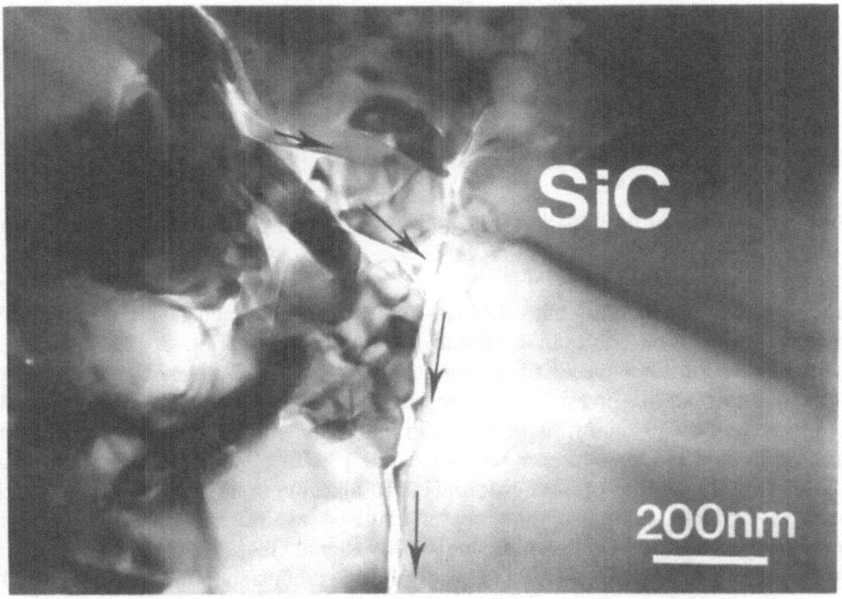

Fig. 4. TEM micrograph showing crack propagation up the platelet /matrix interface.

HIGH TEMPERATURE DEFORMATION

In the use of ceramics as high temperature structural materials, creep resistance and stress rupture properties are the main thermo-mechanical criteria . A plot of the strain against deformation time at a constant stress of 150 MPa, Fig.5, shows time dependent deformation becomes increasingly important as the temperature was raised. At 1300°C creep rates were minimal and virtually identical for the two materials, but at 1400°C differences in creep properties were noticeable. After initial elastic strain there were short primary creep stages . There is some controversy over the existence of true steady state

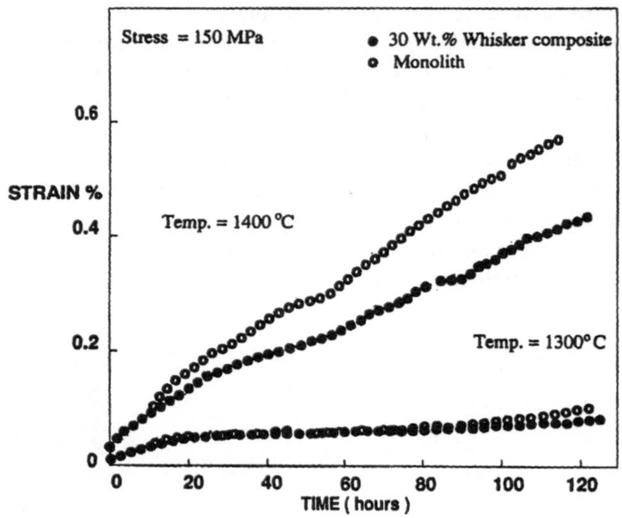

Fig.5. Comparison of the Creep Resistance of Monolith and Whisker Materials

creep for Si_3N_4 based ceramics, but the data indicates a near constant creep rate after approximately 0.2% strain for both monolith and whisker material. A 30wt.% platelet material fractured soon after the test load was applied at 1400°C. It can be seen that the whisker containing material had a better creep resistance than the monolith, presumably because of increased diffusional flow distances in materials which have Coble creep as the dominant deformation mode. Whiskers will also inhibit grain boundary sliding in a matrix which has, essentially, similar properties to the monolith. In comparison, previous work [6] showed that a whisker composite had not crystallised as much as the monolith and consequently there was increased grain boundary diffusional creep.

Steady state creep rates over a range of stress values at a constant temperature of 1400°C for the selected materials are shown in Fig.6. Since the reduced creep rate of the whisker composite did not result in a significant change in the value of the stress exponent n, found using the relationship $\dot{\varepsilon} \propto \sigma^n$, it appears that the same mechanisms may control the deformation properties of both materials.

Stress rupture results in Fig.7 show that fracture was relatively rapid at stresses above 275 MPa. However, there appears to be a threshold stress of about 250 MPa at which pre-existing flaws do not lead to failure during a substantial increase in deformation time. Both long term and short-term failures occur at the same stress level which can be attributed to a statistical distribution of defect sizes where the larger defects are subjected to a stress intensity above the threshold value and the smaller defects subjected to one below the threshold value. Although relatively few specimens were tested and for deformation times shorter than one would wish, the results indicate that this plateau in stress rupture behaviour, by both types of material, was caused by crack blunting by diffusional creep,with no obvious signs of slow crack growth, common in glassy materials.

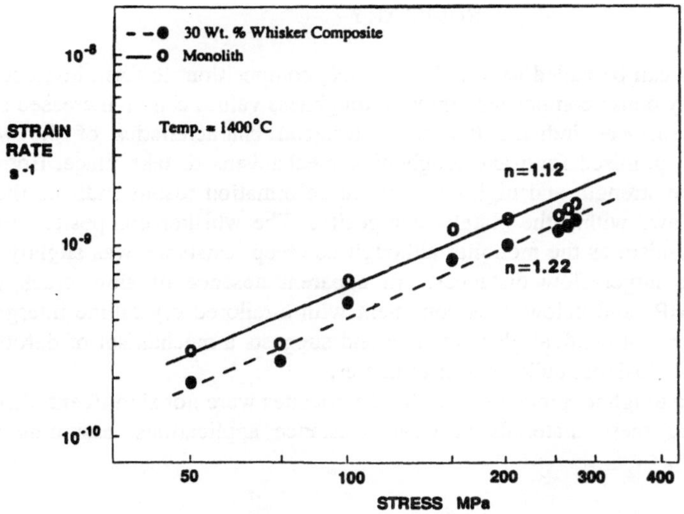

Fig.6. Comparison of the Creep Rates over a range of Deformation Stress

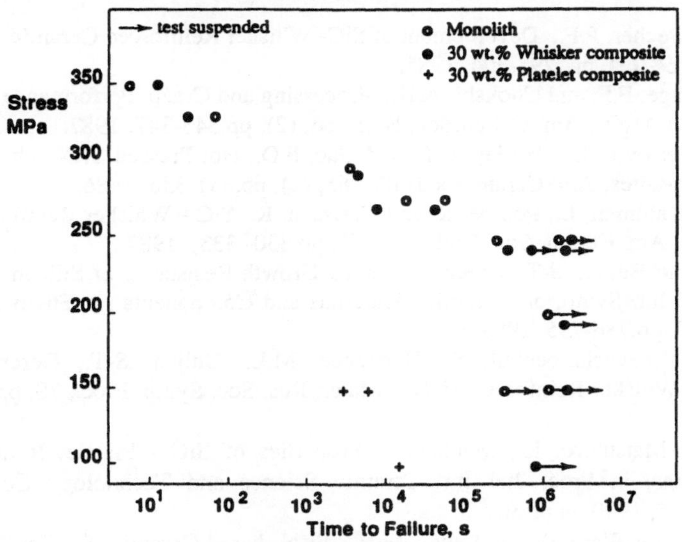

Fig.7. Stress Rupture values at 1400°C

TEM micrographs of a whisker composite material which had undergone substantial creep strain showed similar microstructures to an untested specimen. There was no evidence of the formation of triple point voids, nor apparently along grain boundaries. This is characteristic of microstructures in which creep cavitation is suppressed because of a sub-critical residual glass volume for nucleation of cavities at these stress levels. This confirms why there was no evidence of slow crack growth at stresses of 250 MPa and below, in the limited available data.

SUMMARY

SiC dispersoids can be added to a tailored Si_3N_4 composition, to form homogeneous, anisotropic and fully dense composites. Fracture toughness values can be increased slightly although fracture surfaces indicate that the interfacial characteristics of the whisker composite are not optimised for wake toughening mechanisms to take place. Both room temperature fracture strength and high temperature deformation results indicate there are strength limiting flaws within the platelet composites. The whisker composite showed a similar creep mechanism as the monolith although its creep resistance was slightly better, presumably due to larger flow distances. An apparent absence of slow crack growth behaviour at 250 MPa and below, was consistent with a tailored crystalline intergranular phase with a sub-critical residual glass volume and suggests a mechanism of deformation by grain boundary diffusional bulk transfer of matter.

Although fracture toughness increases for the composites were not significant, there may be potential use of these materials in wear resistance applications, assuming it was financially viable.

REFERENCES

[1] Wei, G.C. and Becher, P.F., Development of SiC- Whisker Reinforced Ceramics , Am. Ceram.Soc. Bull., 64, (2), pp. 298-304, 1985.

[2] Porter, J.R., Lange, F.F. and Chokshi, A.H., Processing and Creep Performance of SiC Whisker Reinforced Al_2O_3, Am. Ceram.Soc. Bull., 66, (2), pp.343-347, 1987.

[3] Shalek, P.D., Petrovic, J.J., Hurley, G.F. and Gac, F.D., Hot Pressed SiC- Whisker/ Si_3N_4 Matrix Composites, Am. Ceram.Soc.Bull., 65, (2), pp.351-356, 1986.

[4] Lundberg, R., Kahlman, L., Pompe, R. and Carlsson, R., SiC - Whisker Reinforced Si_3N_4 Composites, Am. Ceram. Soc. Bull., 66, (2), pp.330- 333, 1987.

[5] Baldoni, J.G. and Buljan, S-T., Creep and Crack Growth Resistance of Silicon Nitride Composites, in 3rd Int. Symp. on Ceramic Materials and Components for Engines, Las Vegas, Nov. 27-30, pp.786-795, 1988.

[6] Nixon, R.D., Chevacharoenkul, S., Huckabee, M.L., Buljan, S-T., Deformation Behavior of SiC Whisker Reinforced Si_3N_4, Mater. Res. Soc. Symp. Proc., 78, pp. 295 - 302, 1987.

[7] Sakai, H. and Matsuhiro, K., Mechanical Properties of SiC - Platelet Reinforced Ceramic Composites, in Proc. 2nd Int. Ceramic Science and Technology Congress, Orlando, Nov.12 - 15, 1990, in print.

[8] Tuersley, I.P., Leng-Ward, G. and Lewis, M.H., Si_3N_4 based Ceramics for Gas Turbine Applications, in 3rd Int. Symp. on Ceramic Materials and Components for Gas Turbine Applications, Las Vegas, Nov. 27-30, pp.856 - 870, 1988.

[9] Hollenberg, G.W., Terwilliger, G.R. and Gordon, R.S., Calculation of Stresses and Strains in Four Point Bending Creep Tests, J.Am. Ceram. Soc., 54, (4), pp. 196-199, 1971.

[10] Ketchion, S.M., Leng-Ward, G. and Lewis, M.H., SiC Dispersoid Reinforced Si_3N_4 Composites, in Proc. 2nd Int. Ceramic Science and Technology Congress, Orlando, Nov.12-15, 1990, in print.

INVESTIGATION ON THE SYNTHESIS PROCESS AND CHARACTERISTICS OF Si_3N_4 POWDERS AND WHISKERS MADE FROM IMIDE

Ge Changchun+, Xia Yuanluo, Xie Jiangang, Huang Xiangdong
University of Science and Technology Beijing
Laboratory of Special Ceramics and Powder Metallurgy
Beijing 100083, China

ABSTRACT

A modified imide process for making Si_3N_4 powders and whiskers is developed. Thermal decomposition of $Si(NH)_2$ under H_2 or N_2 atmosphere at different tempertures and the characteristics of Si_3N_4 powders are investigated with DTA, XRD, SEM and XPS analysis.

INTRODUCTION

Silicon nitride is recognized as one of the most prospective high temperature (HT) structural ceramics due to its superior combination of properties. The development of high quality Si_3N_4 powders plays an important role for making high performance Si_3N_4 ceramics. Among many processes, the thermal decomposition of $Si(NH)_2$ is a suitable process for preparation of high quality Si_3N_4 powders with high productivity.

A modified synthesis process of Si_3N_4 powders by thermal decomposition of imide is developed in our laboratory. In this article thermal decomposition under various atmospheres and the characteristics of Si_3N_4 powders are investigated.

EXPERIMENTAL PROCEDURES

1. Synthesis and Thermal Decomposition of $Si(NH)_2$

The main features of our modified synthesis process of imide consist of: (1) N_2 as carrier gas of $SiCl_4$ vapor to react with NH_3 gas at room temperature; (2) Special design of a series of powder gathering containers with a number of barrier plates to prolong the precipitation

---

+ Former research fellow of Alexonder von Humboldt Foundation.

time of powders and avoid the plugging problem usually occurred in ammonolysis with gas phase reactions; (3) much simpler apparatus and prevention of organic liquid, refrigerator and carbon contamination in powder products.

The mole ratio of $SiCl_4$ to NH_3 used is 1:4.5~5.

The mixture of $Si(NH)_2$ and NH_4Cl is calcined (in gaseous ammonia atmosphere at 900 °C, and than thermal decomposed under N_2 or H_2 atmosphere.

The formation of amorphous Si_3N_4 and phase transformation are traced with DTA, Fourier Transform infrared (FTIR) spectroscopy, and XRD.

2. Characterization of Powders and Whiskers

Powders and whiskers morphology are examined with SEM. Specific surfaces are measured with BET method, and agglomerated-particles are measured with light transmission method.

XPS experiments are performed with a VG-SCIENTIFIC multiple electron spectrometer with MgK (1253.6ev) radiation and 240W (12KV 20mA) power.

RESULTS AND DISCUSSIONS

1. Formation of Crystalline Si_3N_4 and its Phase Compositions under Different Atmospheres

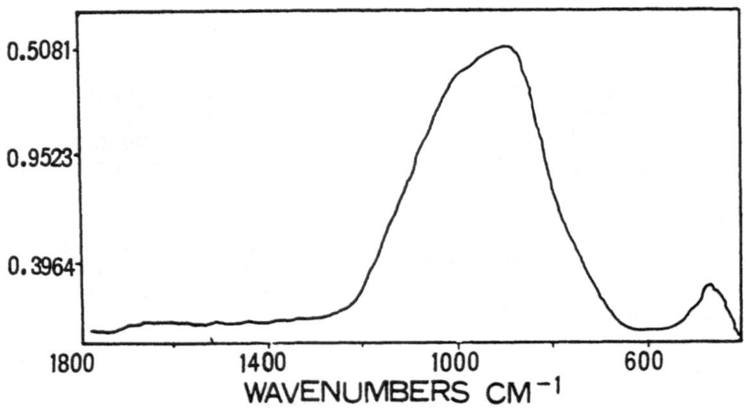

Fig. 1 Infrared spectrum of amorphous Si_3N_4 powders obtained from thermal decomposition at 1000 °C under H_2 atmosphere

Fig.1 shows the infrared spectrum of thermal decomposed product obtained at 1000 °C under H_2 atmosphere. Absorption bands appear at 850~1100cm^{-1} and 468cm^{-1}. which are assigned to unsymmetrical and symmetrical stretching vibrations of Si-N bonds respectively, agreeable to the characteristic spectrum of amorphous Si_3N_4,(1)(2) indicating that the thermal decomposition of $Si(NH)_2$ has completed.

Quantitative XRD results for thermal decomposed products obtained at 1000~1500 °C under H_2 atmosphere are given in TABLE 1.

TABLE 1.
Phase compositions of thermal decomposed products obtained at different temperatures under H_2 atmosphere

TEMPERATURE	PHASE COMPOSITION wt%			
°C	AMORPHOUS	α-Si_3N_4	β-Si_3N_4	Si
1000	100	0	0	
1100	57	36	7	
1200	23	70	7	
1300	0	96.3	3.7	
1350	0	96.7	3.3	
1400	0	96.2	3.8	
1450	0	96.7	3.3	ND
1500	0	85.4	14.6	ND
1550	0	75.2	24.8	ND
1600	0	37.3	45.1	17.6

The amorphous Si_3N_4 begins to crystallize at 1000~1100 °C and completes at 1200~1300 °C. It is unexpected that a small portion of β-Si_3N_4 simultaneously is produced with α-Si_3N_4 during crystallization. As shown

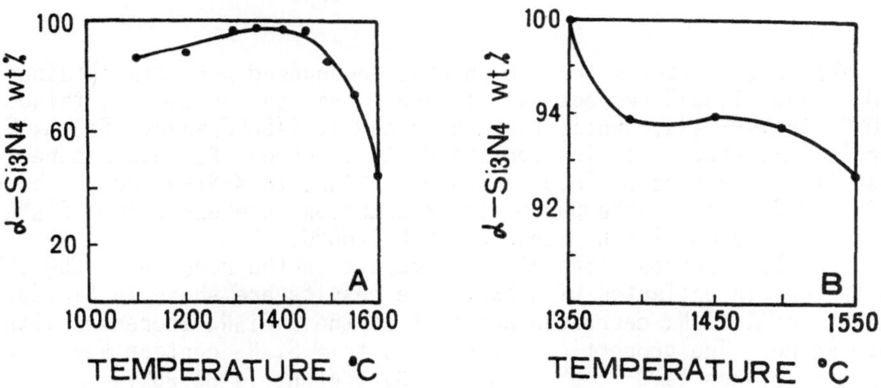

Fig.2 Phase compositions of crystallized Si_3N_4 versus thermal decomposition temperature
A: under H_2 atmosphere B: under N_2 atmosphere

in Fig. 2-A, the content of α-Si_3N_4 increases with temperature in the range of 1100~1300 °C and keeps constant at ~96% from 1300°C to 1450°C. Above 1500 °C, the content of β-phase significantly increases and at 1600 °C free Si content as high as 17% is noted due to dissociation of Si_3N_4.

The DTA-curve for the thermal decomposition of $Si(NH)_2$ under N_2 atmosphere shows two endothemic peaks: the first peak is located between 1310~1350°C, and the second is in the range of 1390~1460°C. (Fig. 3)

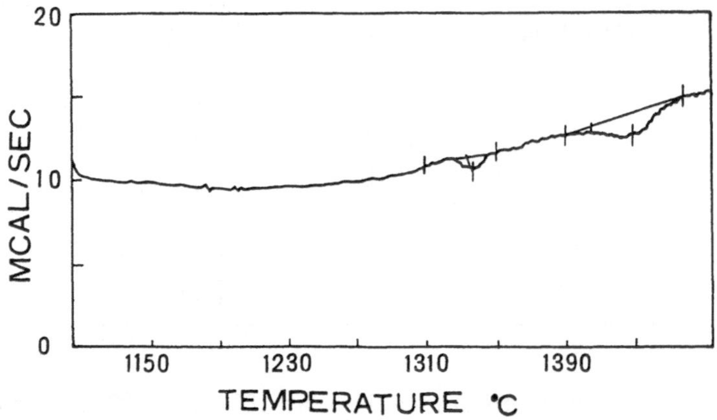

Fig. 3 DTA-Curve of thermal decomposition of $Si(NH)_2$ under N_2 atomsphere

The XRD patterns for the thermal decomposed products obtained at 1300°C and 1450°C respectively indicate that the product obtained at 1350°C is α-Si_3N_4, while that obtained at 1450°C apart from α-Si_3N_4 emerges β-Si_3N_4. It is confirmed that under N_2 atmosphere the crystallization process from amorphous Si_3N_4 to α-Si_3N_4 occurs between 1310~1350°C, and the phase transformation process from α-Si_3N_4 to β-Si_3N_4 proceeds in the range of 1350~1460°C.

The phase compositions of the products in the range of 1350~1550°C are further investigated with XRD, the results are shown in Fig.2B. The content of α-Si_3N_4 decreases and that of the β-Si_3N_4 increases with the temperature. The proportions of α-Si_3N_4 to β-Si_3N_4 content keep nearly constant between 1400~1500°C, as α-Si_3N_4 content is 95~96%.

It is suggested from the above mentioned results that H_2 molecules not only favor the reconstruction of Si-N bonds and cracks of N-H bonds during the transformation of $Si(NH)_2$ to Si_3N_4, showing lower formation temperature of amorphous Si_3N_4, but also promote the crystallization process and transformation process to proceed at lower temperature in contrast to N_2 molecules.

2. Characteristics of Powders and Whiskers

2.1 Morphology of powders and whiskers

SEM micrographs of thermal decomposed products obtained at 1000°C, 1300°C under H_2 atmosphere are given in Fig. 4-A and B.

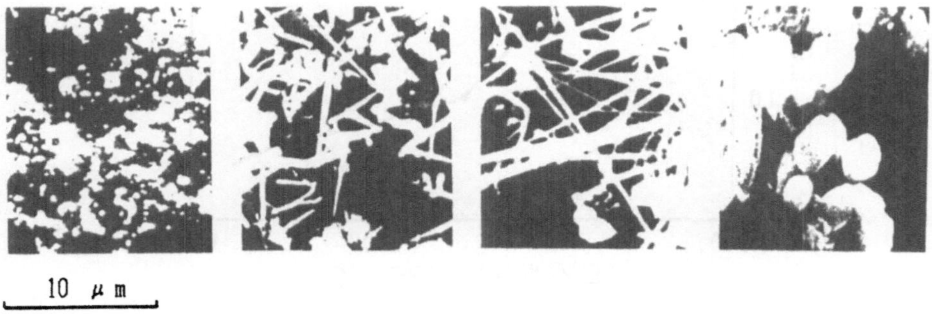

A B C D

Fig. 4 SEM micrographs of Si3N4 powders and whiskers obtained A: at 1000°C under H_2 atmosphere, B: at 1300°C under H_2 atmosphere, C: at 1500°C under N_2 atmosphere, D: at 1450°C under N_2 atmosphere for compacted amorphous powders.

As shown in TABLE 1, product obtained at 1000°C is amorphous; that of 1300°C is completely crystallized Si_3N_4 (α-Si_3N_4~96.3%, β-Si_3N_4~3.7%). Fig. 4-C shows the SEM micrograph of whiskers obtained at 1500°C under N_2 atmosphere. (α-Si_3N_4~95.4%, β-Si_3N_4~4.6%).

As shown in Fig. 4-A,B amorphous Si_3N_4 exists in the form of very fine spherical powders, while crystallized Si_3N_4 (mainly α-Si_3N_4) is basically in the form of whiskers. It is suggested that stronger growth driving force along the c-axis of hexagonal cells of crystallized Si_3N_4 is attributed to the formation of whiskers. When a certain amount of nuclei forms and much space exists around them, free growth along c-axis occurs and needle-like whiskers form.

According to the above suggestions, spherical powders are produced from compacted amorphous Si_3N_4 powders treated at 1450°C for 1h under N_2 atmosphere, due to limited space for free growth of nuclei, as shown in Fig. 4-D.

2.2 Specific surfaces (Sg) of powders

Log Sg in relation to thermal decomposition temperature under H_2 atmosphere is shown in Fig. 5. The specific surface of the amorphous Si_3N_4 powders as high as ~350 m^2/g is measured. Dramatic decrease of Sg due to transformation from amorphous powders to whiskers is noted. Specific surfaces between 28~5 m^2/g through variation of thermal decomposition parameters are obtained. Powders with specific surface of 12 m^2/g are obtained after thermal decomposition of imide at 1300°C for 30min under H_2 atmosphere. The average size of agglomerated powders measured with light transmission method are around 2μ.

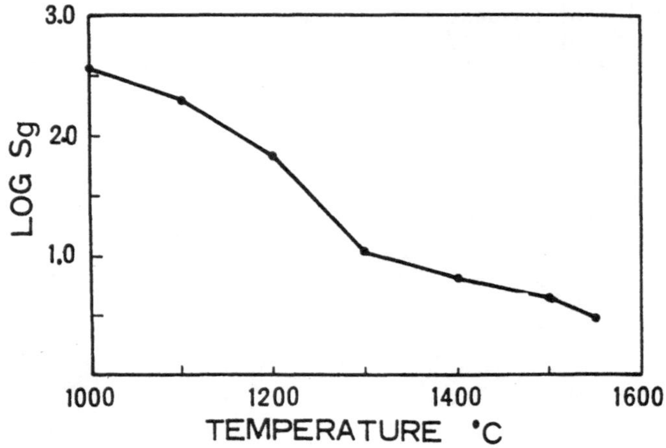

Fig. 5 Specific surfarce of powders versus thermal decomposition temperature under H_2 atmosphere

2.3 Chemical purity of Si_3N_4 powders
The impurities of Si_3N_4 powders made in this work are as follows: Oxygen < 2%. Chlorine ~20ppm. Metallic impurities < 300ppm.

3. XPS Studies of Si_3N_4 Powders
Samples of Si_3N_4 powders for XPS investigation were made at different thermal decomposition temperatures under N_2 with our modified imide process: S1 - 1350°C; S2 - 1450°C; S3 - 1500°C; S4 - 1550°C.

Samples were exposed in air except S2, which was kept in a closed vessel.

Two other samples were used for comparison: S5 - made from imide one year before; C1 - commercial Si_3N_4 powder from Si nitridation, made in Shanghai Electroceramics Factory.

3.1 The overall spectrum analysis and the existing forms of oxygen on Si_3N_4 powders surfaces
Fig. 6 shows the overall spectra for S4 and C1, in which strong peaks labelled Si_{2p}, Si_{2s}, N_{1s}, O_{1s} are detected.

The electron binding energy (B.E.) of elements and the shifts of characteristic peaks of elements are shown in TABLE 2. The Si_{2p} peaks are located near 102ev, which approaches the Si_{2p} peak of Si_3N_4 (101.9 ev) and is away from Si_{2p} of SiO_2 (103.4 ev), indicating the chemical environments of Si in surface regions of samples differ from that of Si in SiO_2. The N_{1s} peaks of all these samples lie near the N_{1s} peak of Si_3N_4 (397.70 ev). The B.E of Si_{2p} and N_{1s} peaks in these samples shifts within 0.5 ev, showing that the surface compositions change in small degrees as the thermal decomposition temperature increases.

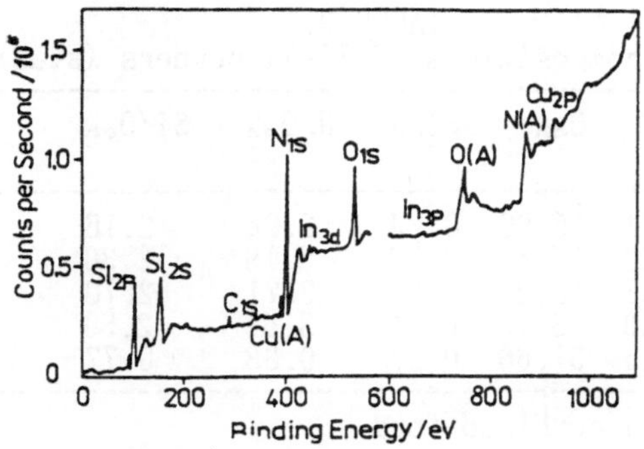

Fig. 6 Overall XPS spectrum of S2 (made from imide,decomposed at 1450°C)

TABLE 2.
Electron B.E. of elements in samples (ev)

Sample	Si_{2p}	N_{1s}	O_{1s}		
S1	102.22	397.88	532.41	530.21	------
S2	101.91	397.60	532.08	------	------
S3	101.83	397.32	531.90	529.33	------
S4	101.85	397.58	532.11	529.59	------
C1	102.30	397.83	532.59	530.55	528.71

Most of the samples made from imide have double O_{1s} peaks, indicating two different existing forms of oxygen --- combined form (531.9~532.4 ev) and absorbed form (530.2~529.3 ev). For S2 (kept in closed vessel), no absorbed peak appears. For S5 (kept in air for one year) and C1 (commercial powder from Si nitridation), one peak for combined form and two peaks for absorbed form with a B.E. difference of 1.86~2.55 ev are found in the XPS spectra. It is thought that the second small absorbed peak shows a week-absorbed form of oxygen on the surface of strongly absorbed oxygen layer after exposure in air for a long time.

3.2 Quantitative XPS analysis of surface compositions of Si_3N_4 powders

Surface compositions of Si_3N_4 powders calculated from quantitative XPS analysis are given in TABLE 3, from which the following results are drawn:

The atom ratios Si/O_{ch} N/O_{ch} and N/Si are much higher on powder surfaces in samples S1~S4 than C1.

TABLE 3.
Surface compositions of Si_3N_4 powders (atm%)

Sample	Si	N	O_{ch}	Si/N	N/O_{ch}	Si/O_{ch}	SixNyO y	x
S1	34.29	46.74	15.90	0.73	2.94	2.16	2.94	2.15
S2	35.39	49.20	15.41	0.72	3.19	2.30	3.19	2.30
S3	35.10	48.33	13.01	0.73	3.71	2.70	3.71	2.70
S4	34.38	48.29	13.76	0.71	3.51	2.50	3.51	2.50
C1	24.33	26.15	31.66	0.93	0.83	0.77	0.83	0.77

O_{ch} -- oxygen in combined form

The surface compositions of S1 and S4 are significantly different from those of C1: for S1~S4, the chemical formula can be written as $Si_{\sim 2.7}N_{2.9\sim 3.7}O$; while for C1 it is $Si_{0.8}N_{0.8}O$.

CONCLUSIONS

1. Under H_2 atmosphere, formation of amorphous Si_3N_4 and its transformation to crystallized Si_3N_4 complete at 1000°C and 1300°C respectively, while under N_2 atmosphere, crystallization occurs between 1310~1350°C.

2. Under controlled conditions, either whiskers or spherical powders with α-phase content higher than 95% has been obtained.

3. The specific surface (Sg) of the amorphous Si_3N_4 powders as high as ~350 m^2/g is measured, while Sg of the crystallized powders can be adjusted between 28~5 m^2/g through variation of decomposition parameters.

4. Powders of high purity with amount of metallic impurities < 300ppm, oxygen < 2% and chlorine ~20ppm are obtained and C contamination is prevented with this process.

5. Much lower oxygen content and higher atom ratio of N/O_{ch}, Si/O_{ch} and N/Si are found on Si_3N_4 powders made with this process than that of powders from Si nitridation.

6. It is confirmed that no SiO_2 surface layer exists on Si_3N_4 powders made with modified imide process and those made from Si nitridation. XPS quantitative analysis indicates that $Si_{2.2\sim 2.7}N_{2.9\sim 3.7}O$ exists on surfaces of Si_3N_4 powder made with modified imide process in contrast to $Si_{0.8}N_{0.8}O$ on surfaces of commercial Si_3N_4 powders from Si nitridation.

REFERENCES

1. T.L. Chu et al, J. Eltrochem.Soc., 1967, 114(7), 717-22.
2. N.Wada et al, J. Non-cryst. Solids., 1981, 43, 7-15 .

FABRICATION OF SILICON NITRIDE / OXYNITRIDE BY REACTION BONDING AND POST SINTERING

A. Bartek#, T. Johansson#, D.E. Niesz*, T. Lindbäck# and B.Q. Lei#
Luleå University of Technology, Luleå, Sweden
* Rutgers University, Piscataway, New Jersey, USA

ABSTRACT

Reaction bonding and post sintering of silicon nitride / oxynitride was investigated as a route to fabricate a material with good oxidation resistance and good high temperature strength. Silicon powders, mixed with 0, 2, 4, 8 and 16 wt% silica, were CIPed at 150 MPa and nitrided using the nitrogen demand principle. Four different nitriding gas compositions were used, consisting of different amounts of hydrogen, argon and helium mixed with nitrogen. Post sintering at atmospheric pressure and by HIP at 160 MPa were investigated to densify the materials. Samples were characterized by X-ray diffraction, SEM and Hg-porosimetry in both the nitrided and sintered state.

The α/β silicon nitride ratio after nitriding varied with gas composition and decreased with increasing silica content. During nitriding, some of the silica was transformed to silicon oxynitride, and the porosity of the nitrided samples was 25-35 %. Nitridation in the presence of hydrogen and helium, respectively, resulted in a larger amount of residual silicon. Fracture surfaces show a submicron grain size with grains in clusters. The porosity becomes coarser with increasing silica percentage, and the clusters become more obvious. This has been confirmed by Hg-porosimetry measurements. Post sintering at atmospheric pressure at 1750 °C for two hours does not increase the bulk density, but increases the density locally in the clusters, increases the pore size and decreases the α/β silicon nitride ratio. HIPing at 1750 °C and 160 MPa for two hours resulted in complete transformation to β and a considerable increase in density.

INTRODUCTION

Reaction bonding and post sintering of reaction bonded silicon nitride as a route for fabrication of components for structural applications has been a topic of research for some time. This is mainly because of RBSN's good high temperature properties and low shrinkage upon post sintering. Several review papers on the formation of RBSN have been published [1, 2, 3, 4, 5], and factors such as nitriding gas composition, presence of impurities, temperature and time influence on the formation of silicon nitride were discussed. It is important to understand how the microstructure of RBSN can be affected to produce the most suitable structure for post sintering.

α silicon nitride is mainly proposed to form from vapour phase and β mainly from liquid phase. For the vapour phase reaction volatilazation of silicon is essential and is known to be hindered by surface silica . For post sintering purposes a high α/β ratio is of great importance as the transformation of α to β silicon nitride aids densification, and the effect of silica therefore must be taken into consideration.

The effect of hydrogen in the nitriding gas has been investigated [6] as means for reducing this problem. Hydrogen seemes to increase the volitalization of the silica layer in the initial stage of nitridation. This volitalization of the silica to SiO gas leaves the silicon surface free for reaction and thereby increases the reaction rate and formation of α. The influence of other gases on the reaction bonding process has also been reported [7, 8]. Additions of helium and hydrogen to the nitridation gas increases the thermal conductivity of the gas mixture leading to lower temperature gradients. Argon and helium increase the viscosity of the nitriding gas and thereby control the flow rate of nitrogen, so that the risk of local overheating in the compact is decreased.

Other factors that have been found to be important in increasing the α content are low nitriding temperatures and slow reaction rates. Due to more nucleation a finer structure is also achieved. Impurities on the other hand can lead to formation of a liquid phase below 1200 °C leading to the formation of more β silicon nitride and a coarser structure. Commercial silicon, for example, contains Fe, Ca and Al. Iron is known to aid the disrupting of the oxide layer on silicon and is therefore often a desired impurity.

For post sintering, sintering aids are necessary. They can be added to the silicon before nitriding or infiltrated into the RBSN material. In the first case the additives affect the nitridation. An investigation has been made on nitriding with different oxides present [9]. A silicon / 8% silica mixture was nitrided in a gas mixture of nitrogen and 5% hydrogen at 1250 °C. The silica did not seem to affect the nitridation to any great extent even though some decrease in conversion to silicon nitride was observed. Silicon oxynitride was also formed.

Post sintering [10, 11] and post HIPing [12, 13, 14] has mainly been performed at temperatures >1800 °C and 1700-2000 °C respectively. The sintering aids were mainly magnesia, yttria and alumina. HIPing produced materials close to theoretical density while less dense materials were obtained by sintering.

The aim of this investigation is to achieve an understanding of how RBSN, containing different amounts of silica, should be produced for successful post sintering to a dense Si_3N_4 / Si_2N_2O material.

MATERIALS AND METHODS

Five mixtures of silicon with additives of silica and binder, from KemaNord, were used: 0, 2, 4, 8 and 16 w% SiO_2. Powders used for the mixtures were Sicomill 4D and Aerosil silica with 5 w% PEG. The powders were compacted by CIPing in silicon rubber molds to sizes 35x25x15 mm at a pressure of 150 MPa. Binder was removed by stepwise firing up to a temperature of 500 °C in hydrogen atmosphere. Nitriding was performed using the nitrogen demand principle up to 1450 °C with four different gas compositions:

1. 50 mbar Ar + 950 mbar N_2
2. 50 mbar Ar + 20 mbar H_2 + 930 mbar N_2
3. 20 mbar H_2 + 980 mbar N_2
4. 50 mbar He + 950 mbar N_2

The first nitridation was run at a slower rate than the three other. The first and third runs were interrupted and restarted, due to equipment malfunction.

One sample of each composition from the first and second nitridation run was sintered at atmospheric pressure in 1750 °C for two hours in a silicon nitride powder bed. This sintering was performed at SCI in Gothenburg, Sweden. An identical set of samples was HIPed at 160 MPa in 1750 °C for two hours at ABB Cerama in Robertsfors, Sweden.

All reaction bonded and post sintered samples were analyzed by X-ray diffraction for identification of present phases. Densities were measured by the water immersion technique for all reaction bonded and sintered materials. Microstructures were studied by SEM for sintered and reaction bonded samples from the first and second nitridation run. Hg-porosimetry was done on a number of samples.

RESULTS

Phases

X-ray diffraction data show that α and β silicon nitride is formed during nitridation. For increasing additions of silica also silicon oxynitride is formed in increasing amounts. At an addition of 16% silica, the amount of silicon oxynitride formed, is approximately 12-15 %.

The ratio of α to β silicon nitride varies with nitridation gas composition and silica content according to fig 1a. In this figure it is seen that small additions of silica increase the α ratio for all conditions while larger additions decrease the same. Other significant trends shown in this figure are that samples nitrided slowly in argon have the highest α ratio, between 0.6 and 0.8 and that those nitrided in argon+hydrogen and helium have the lowest ratio, between 0.4 and 0.6. Besides affecting the α to β ratio, the gas composition and silica addition affects the amount of unreacted silicon. For all compsitions nitrided in Ar and Ar+H the unreacted silicon was only a few percent. Nitridation in H gave higher amounts of residual silicon for all samples, with the largest amounts at 0 and 8 percent silica. Nitridation in helium resulted in higher amounts of unreacted silicon in the sample without silica addition and almost no reaction at all took place in the sample with 16 percent silica.

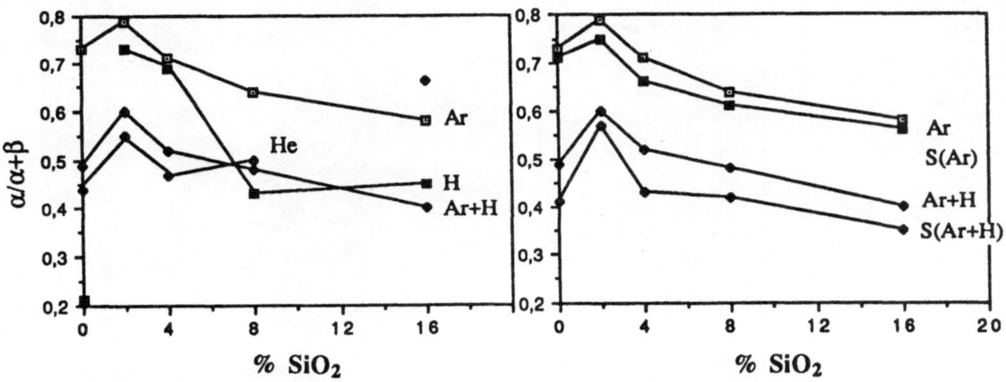

a) b)

Figure 1. a) α to β ratio for materials nitrided in different atmospheres. b) α to β ratio for nitrided and post sintered materials where S=sintered.

Figure 1b shows the α to β ratio for materials post sintered at atmospheric pressure after nitriding in Ar and Ar+H, respectively. It is seen that the trends shown for nitrided and post sintered materials are similar. It is however noted that some of the α silicon nitride transforms to β and therefore the ratios are reduced. This transformation is larger in samples with more β silicon nitride. When HIPing is used all α silicon nitride transforms to β.

Microstructure
Fracture surfaces of the nitrided and sintered samples from the first (Ar) and second (Ar+H) nitridation run were investigated. For the nitrided samples it appears that the porosity grows coarser with increases in silica addition and that submicron grains are present in denser clusters, see fig 2a-b. At high SiO_2 contents there are also some large grains present. The main difference between the two nitridation runs seems to be that the first nitridation run mainly has smaller pores, more evenly distributed, while the pores are coarser in the second nitridation run with denser areas present,see fig 2c. After post sintering denser areas tend to become even more dense and the large pores grow. The grain size increases somewhat but still submicron grains are present, see fig 2d.

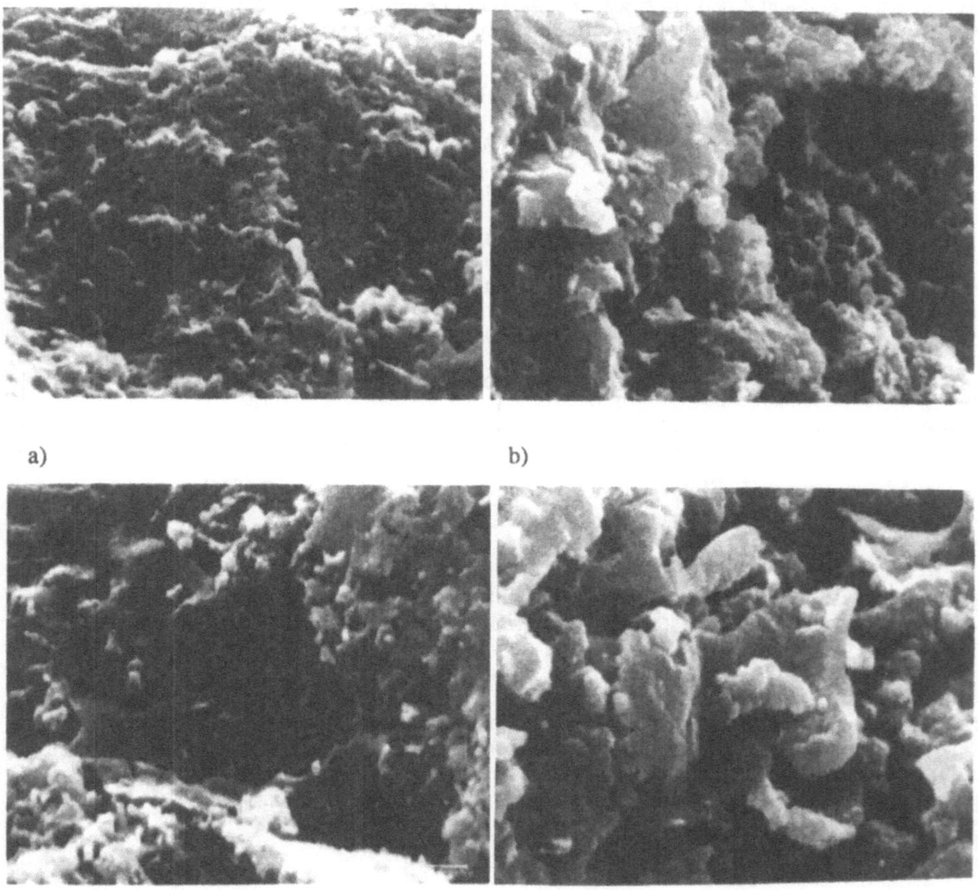

a) b) c) d)

Figure 2. Fracture surfaces at the magnification of 10 000x where ⌴ equals 1 μm.
a) 2% SiO_2 nitrided in Ar. b) 16% SiO_2 nitrided in Ar. c) 2% SiO_2 nitrided in Ar+H. d) 16% SiO_2 nitrided in Ar and post sintered at 1750 °C.

Density and porosity
The density trends for the different nitriding runs seem to be similar. With increasing silica addition a decrease in density can be observed, see fig 3a. Highest densities are obtained for samples from the argon and helium runs. After post sintering at atmospheric pressure the densities change only slightly in most cases, which can be seen in fig 3b, while a larger increase in density occurs after post HIPing.

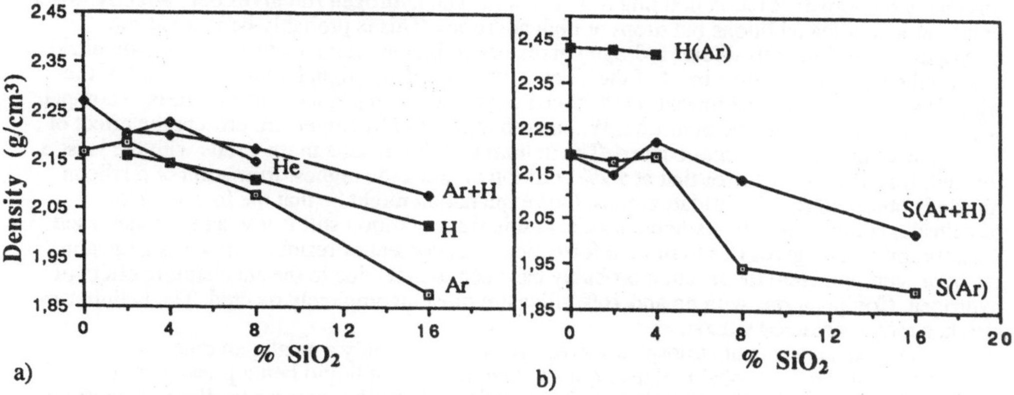

Figure 3. Density variations with silica addition for materials a) nitrided in different gas mixtures b) post sintered and post HIPed materials nitrided in Ar and Ar+H, S= sintered and H=HIPed.

The open porosity is high, 25-35 % in the as nitrided samples, and increases after post sintering in all samples but the 4% SiO_2 composition where it is constant for the argon gas containing run and decreases for the argon+hydrogen run. Post HIPing decreases the amount of open porosity in all cases. Porosimetry confirms the observed microstructures by showing an increase in pore size with silica addition, see fig 4a. This behavior has been found for other oxide additives earlier [15]. There is also a difference between the first and second nitriding run. In samples with no addition this difference is larger than at 16% silica addition. After post sintering both pore size and the percentage of pores increases, see fig 4b.

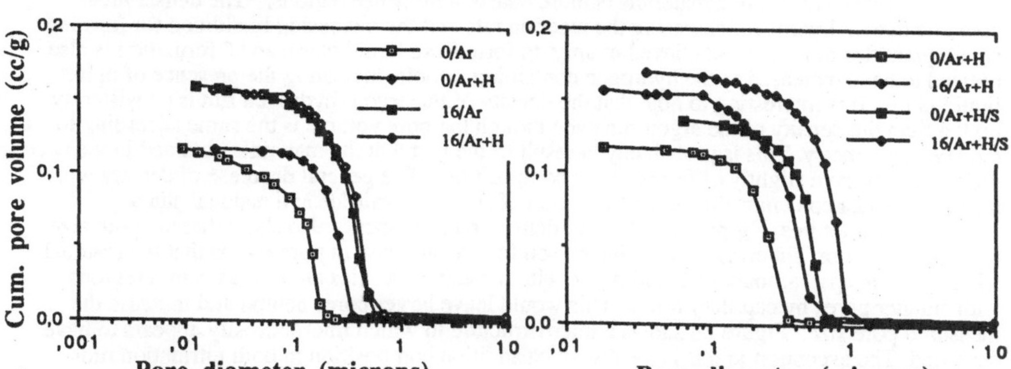

Figure 4. Cummulative pore volume plotted against pore diameter for materials with 0 and 16% silica addition a) nitrided in Ar and Ar+H b) nitrided in Ar+H and post sintered at 1750 °C.

DISCUSSION

Both the influence of different gases and the presence of silica on the nitriding behaviour described in the literature seems to agree with this investigation. The importance of the different mechanisms, however, varies somewhat.

The explanation for the highest α silicon nitride content observed in the argon run is obviously the longer reaction bonding time allowed. The hydrogen run gives only slightly lower values at low silica additions but drops at high additions. This is probably because of the hydrogen´s reaction with silica. Hydrogen gives rise to increased nucleation of α silicon nitride at low silica additions, when most of the silica is volatilized. At higher additions not all of the silica is volatilized, and the formation of silicon oxynitride and β silicon nitride starts to compete with the formation of α silicon nitride. When both argon and hydrogen are present the effect of hydrogen seems to be less pronounced. The helium addition results in materials with the least α formed. It is interesting to see that at a 2% addition of silica the highest formation of α silicon nitride is observed for all nitridation runs. One explanation might be that the formation of oxynitride, competing with α silicon nitride, at this silica addition still is low and the increased evaporation of SiO gives rise to more α formation. The content of residual silicon is largest in the hydrogen run. Here the reaction probably has been too fast due to the accelerating effect of hydrogen. Compositions with no and 16% silica are most incompletely reacted. The helium run also has more unreacted silicon.

After post sintering at atmospheric pressure at 1750 °C only a small percentage of α transforms to β. This is probably due to only a small amount of liquid being present at the sintering temperature leading to an incomplete reaction. Most silica reacted to silicon oxynitride in the nitriding process but a small amount could have stayed unreacted, forming a liquid with impurities. Unreacted silicon would also form a liquid. Small additions of silicon have been seen to increase the transformation of α to β [17]. The greater transformation of α to β in samples with more β could be a consequence of the already formed β acting as growth sites [16]. After HIPing all α transformed to β. In the HIPing process the pressure increased the diffusion rate and total transformation was possible.

The investigation of fracture surfaces showed that the materials from the first nitridation run (Ar) have a more homogeneous structure with smaller pores and that materials from the second run (Ar+H) have denser areas with larger pores in between. According to the micrographs and Hg-porosimetry measurements, this difference is most noticable at low silica additions while the difference is much smaller at the 16% addition. The finer structure at low silica additions is likely due to the slow reaction rate of the Ar run while for the high silica additions the formation of SiO gas is increased and gas composition is less important. The structure of these high silica materials is more coarse with dense regions. The denser areas could be due to hydrogen increasing the reaction rate and the increasing likelihood for forming small quantities of melt. β is believed mainly to form from liquid phase and β formation is also noticed to have increased in the hydrogen containing sample, indicating the presence of more liquid phase. It is interesting to note that the density of the argon+hydrogen run is consistently higher than the density of the argon run even though the pore volume is the same according to the Hg-porosimetry. This is probabably a result of the fact that the materials obtained in the two different runs have slightly different phase compositions. The general decrease of density with increasing silica content is due to the formation of silicon oxynitride and residual silica.

Microstructures, Hg-porosimetry and density measurements also show that the pore size increases with post sintering. A possible reason for this increase in pore size is that the residual silica forms a small amount of liquid phase with present impurities and is drawn into regions with smaller pores by capillary force. This would leave larger pores behind and increase the measured pore size. Figure 2d shows a microstructure in which microporosity appears to have densified. The exception to this is the 4% silica addition composition in both nitridation runs where density increases. Here the percentage of liquid is probably too low to enlarge the larger pores but high enough to allow some densification in the smallest pore areas. Due to the higher diffusion rate, caused by applied pressure, post HIPing can densify the materials to a higher degree.

CONCLUSIONS

It has been shown that the effect of hydrogen additions to the nitriding gas, on α formation, is decreased with larger additions of silica in the samples. The silica additions increase the pore size in both nitrided and post sintered (atm pressure) samples. Neither post sintering (atm pressure) nor post HIPing at 1750 °C was able to densify the samples, but post HIPing increased the density. The highest percentage α silicon nitride was formed by using a slower reaction rate. Similar amounts of α silicon nitride were obtained when nitriding low silica samples with hydrogen at a faster rate, but larger amounts of unreacted silicon remained. Nitriding samples containing silica resulted in silicon oxynitride formation. This together with the pore coarsening makes silica unsuitable as sintering aid for post sintering. Small silica additions on the other hand seem to promote the formation of α silicon nitride and could be successfully utilized in the reaction bonding process.

ACKNOWLEDGEMENTS

The authors would like to acknowledge the financial support to this project by The Swedish Board for Technical Development. We would also like to thank Dr. Daniel Ashkin for help in making the Hg-porosimetry measurements and discussing the results.

REFERENCES

1. Moulson, A.J., Review Reaction-bonded silicon nitride: its formation and properties. J. Mater. Sci., 1979, 14, 1017-1051.

2. Riley, F.L., Silicon Nitridation. In Progress in Nitrogen Ceramics, ed. F.L. Riley, Martinus Nijhoff Publishers, Boston/The Hague/Dordrecht/Lancaster, 1983, pp.121-133.

3. Jennings, H.M., Review On reactions between silicon and nitrogen. Part 1 Mechanisms. J. Mater.Sci., 1983, 18, 951-967.

4. Ziegler, G. and Wötting, G., Post-treatment of Presintered Silicon Nitride by Hot Isostatic Pressing. Int. J. High Technology Ceramics, 1985, 1, 31-58.

5. Riley, F.L., Reaction Bonded Silicon Nitride. Material Science Forum 1989, 47, 70-83.

6. Barsoum, M., Kangutar, P. and Koczak, Nitridation Mechanisms of Silicon Powder compacts. Ceram. Eng. Sci. Proc. 1989, 10, 794-806.

7. Jennings, H.M., Dalgleish, B.J. and Pratt, P.L., Reactions between silicon and nitrogen. J. Mater. Sci. 1988, 23, 2573-2583.

8. Washburn, M.E. and Coblenz, W.S., Reaction-Formed Ceramics. Ceram. Bull. 1988, 67, 356-363.

9. Antona, P.L., Giachello, A. and Martinengo, P.C., Nitridation of Silicon in Presence of Oxides. In Ceramic Powders, P. Vincenzini, Elsevier Scientific Publishing Company, Amsterdam, 1983, pp. 753-766.

10. Mangels, J.A., Sintering of Reaction Bonded Silicon Nitride. In Progress in Nitrogen Ceramics, ed. F.L. Riley, Martinus Nijhoff Publishers, Boston/The Hague/Dordrecht/Lancaster, 1983, pp. 231-236.

11. Popper, P., Sintering of Silicon Nitride, A Review. In Progress in Nitrogen Ceramics, ed. F.L. Riley, Martinus Nijhoff Publishers, Boston/The Hague/Dordrecht/Lancaster, 1983, pp. 187-210.

12. Richerson, D.W. and Wimmer, J.M., Properties of Isostatically Hot-Pressed Silicon Nitride. J. Am. Ceram. Soc., 1983, 66, C173-176.

13. Ziegler, G., Heinrich, J. and Wötting, G., Review Relationships between processing, microstructure and properties of dense and reaction-bonded silicon nitride. J. Mater. Sci., 1987, 22, 3041-3086.

14. Heinrich, J., Backer, E and Böhmer, M., Hot Isostatic Pressing of Si_3N_4 Powder Compacts and Reaction-Bonded Si_3N_4. J. Am. Ceram. Soc., 1988, 71, C28-31.

15. Kleebe, H.J., Wötting, G. and Ziegler, G., Influence of RBSN-Characteristics on the Microstructural Development of Gas-Pressure Sintered Silicon Nitride. In Science of Ceramics 14, ed. D. Taylor, The Institute of Ceramics, Shelton, UK, 1988, pp. 407-412.

16. Ekström, T., Ingelström, N. Brage, R., Hatcher, M. and Johansson, T., α–β Sialon Ceramics Made from Different Silicon Nitride Powders. J. Am. Ceram. Soc., 1988, 71, 1164-1170.

17. Park, J.Y., Kim, J.R. and Kim, C.H., Effects of Free Silicon on the α to β Phase Transformation in Silicon Nitride. J. Am. Ceram. Soc., 1987, 70, C240-242.

4th INTERNATIONAL SYMPOSIUM ON CERAMIC MATERIALS AND COMPONENTS FOR ENGINES

Proceedings of the 4th International Symposium on Ceramic Materials and Components for Engines, organized by the Swedish Ceramic Society and held at Göteborg, Sweden, 10–12 June 1991

Co-sponsored by

American Ceramic Society
Australian Ceramic Society
Ceramic Society of Japan
European Ceramic Society
International Ceramic Federation

4th INTERNATIONAL SYMPOSIUM ON CERAMIC MATERIALS AND COMPONENTS FOR ENGINES

Edited by

R. CARLSSON

T. JOHANSSON

and

L. KAHLMAN

Swedish Ceramic Society,
Gotebörg, Sweden

SPRINGER-SCIENCE+BUSINESS MEDIA, B.V.

WITH 174 TABLES AND 932 ILLUSTRATIONS

© 1992 Springer Science+Business Media Dordrecht
Originally published by ELSEVIER SCIENCE PUBLISHERS LTD in 1992
Softcover reprint of the hardcover 1st edition 1992

British Library Cataloguing in Publication Data

Ceramic Materials and Components for
Engines: 4th
International Symposium on Ceramic
Materials and Components for Engines
I. Title. II. Carlsson, R. III. Johansson, T.
IV. Kahlman, L.
666

ISBN 978-94-010-5280-1 ISBN ISBN 978-94-011-2882-7 (eBook)
DOI 10.1007/978-94-011-2882-7

Library of Congress CIP data applied for

No responsibility is assumed by the Publisher for any injury and/or damage to persons or property as a matter of products liability, negligence or otherwise, or from any use or operation of any methods, products, instructions or ideas contained in the material herein.

Special regulations for readers in the USA

This publication has been registered with the Copyright Clearance Center Inc. (CCC), Salem, Massachusetts. Information can be obtained from the CCC about conditions under which photocopies of parts of this publication may be made in the USA. All other copyright questions, including photocopying outside the USA, should be referred to the publisher.

All rights reserved. No part of this publication may be reproduced, stored in a retrieval system, or transmitted in any form or by any means, electronic, mechanical, photocopying, recording, or otherwise, without the prior written permission of the publisher.

PREFACE

The 4th International Symposium on Ceramic Materials and Components for Engines, organized by the Swedish Ceramic Society in Göteborg on 10–12 June 1991, continues the series of conferences which started in 1983 at Hakone, Japan, followed by Travemünde, Germany, in 1986 and Las Vegas, USA, in 1988. The aim of these conferences is to bring together engineers, scientists and students working on ceramics for engine applications to discuss the state-of-the-art and recent developments. Three hundred and twenty-five participants from 23 countries attended the conference, which indicates the worldwide interest in the subject.

The success of a conference depends primarily on the quality of the papers presented and on the participants. At the 4th symposium 173 papers were presented, divided into 12 invited papers, 91 oral presentations in three concurrent sessions and 70 posters. This book contains the written version of 142 of these papers. Every paper has been refereed by two experts. About one-third of the papers come from the industry and two-thirds from universities, institutes and government laboratories.

The Local Organizing Committee would like to thank all the authors for the effort put into their presentations at the conference and during the preparation of their manuscripts. We also acknowledge the International Advisory Committee for their continuous support and the co-sponsoring organizations for their marketing efforts. Special thanks go to all the members of the Committee of Referees for their very important work in assessing and checking all the manuscripts. Finally, we wish to express our gratitude to Ulla-Britt Jigholm and Margareta Jansson of the Swedish Ceramic Institute for their outstanding job in dealing with all practical details before, during and after the conference.

The 5th symposium will be organized in China in 1994 and the 6th symposium is provisionally planned to take place in Japan. The new President of the International Advisory Committee is Professor T. Yen of Academia Sinica, China.

Roger Carlsson
Thomas Johansson
Lars Kahlman

CONTENTS

Production and Properties

Applications

INTERNATIONAL ADVISORY COMMITTEE

T. Johansson	Höganäs Eldfast AB/Luleå University of Technology, Sweden (*President*)
G. N. Babini	NRC-Research Institute for Ceramics Technology, Italy
B. Bertrand	GIE.PSA Etudes et Recherches, DRAS, France
C. Bonnet	SEP, France
D. Broussaud	Rhône-Poulenc Centre de Recherches, France
K. Esaki	Japan Fine Ceramics Center, Japan
M. K. Ferber	Oak Ridge National Laboratory, USA
X. Fu	Shanghai Institute of Ceramics, China
H. Hausner	Technische Universität Berlin, Germany
W. P. Holbrook	American Ceramic Society, USA
M. H. Lewis	University of Warwick, UK
R. Metselaar	Eindhoven University of Technology, The Netherlands
M. Morita	Japan Fine Ceramics Center, Japan
R. Neumann	Preussag AG, Germany
D. E. Niesz	Rutgers University, USA
G. Petzow	Max-Planck-Institut für Metallforschung, Germany
H. B. Probst	NASA Lewis Research Center, USA
S. Saito	Kanagawa Academy of Science and Technology, Japan
V. Shevchenko	Academy of Sciences, Russia
S. Sōmiya	The Nishi Tokyo University, Japan
C. Sorrell	Sydney University of Technology, Australia
R. M. Spriggs	Alfred University, USA
V. J. Tennery	Oak Ridge National Laboratory, USA
E. Tiefenbacher	Daimler Benz, Germany
M. H. Van de Voorde	CEC Joint Research Centre, The Netherlands
P. Vincenzini	NRC-Research Institute for Ceramics Technology, Italy
T. Yen	Academia Sinica, China

LOCAL ORGANIZING COMMITTEE

T. Johansson	Höganäs Eldfast AB/Luleå University of Technology, Sweden (*President*)
M. Bergström	Scania AB, Sweden
R. Carlsson	Swedish Ceramic Institute, Sweden (*Secretary*)
L. Kahlman	Swedish Ceramic Institute, Sweden (*Assistant Secretary*)
H. T. Larker	ABB Cerama AB, Sweden
L. Pejryd	Volvo Flygmotor AB, Sweden
J. Rehn	United Turbine AB, Sweden
D. J. Rowcliffe	Royal Institute of Technology, Sweden
A. Wendel	AB Volvo, Sweden

COMMITTEE OF REFEREES

Babini, Gian N.	CNR-IRTEC, Faenza, Italy
Bellosi, Alida	CNR-IRTEC, Faenza, Italy
Bergström, Magnus	Saab-Scania AB, Södertälje, Sweden
Bönsch, Christof	Fraunhofer Institute, IPT, Aachen, Germany
Brandt, Gunnar	AB Sandvik Coromant, Stockholm, Sweden
Breder, Kristin	Royal Institute of Technology, Stockholm, Sweden

Brinkman, Charles	Martin Marietta Energy Systems Inc., Oak Ridge, TN, USA
Broussaud, Daniel	Rhône-Poulenc Recherches, Aubervilliers, France
Brown, Ian	DSIR, Chemistry, Lower Hutt, New Zealand
Burström, Martin	Institute for Production Engineering Research, Luleå, Sweden
Cales, Bernard	Céramiques Techniques Desmarquest, Evreux, France
Cannon, Roger	Rutgers University, Center for Ceramic Research, Piscataway, NJ, USA
Carlsson, Lennart	Swedish National Testing Laboratory, Borås, Sweden
Carlsson, Roger	Swedish Ceramic Institute, Göteborg, Sweden
Clegg, William	ICI Solid State Science Group, Runcorn, UK
Collin, Marianne	AB Sandvik Coromant, Stockholm, Sweden
Cutler, Raymond	Ceramatec Inc., Salt Lake City, UT, USA
Demaestri, Pier Paolo	Centro Ricerche Fiat, Orbassano, Italy
Dunlop, Gordon	University of Queensland, Brisbane, Australia
Dworak, Ulf	Elektroschmelzwerk Kempten GmbH, Kempten, Germany
Ekberg, Inga-Lill	Swedish Ceramic Institute, Göteborg, Sweden
Ekström, Thommy	AB Sandvik Hard Materials, Stockholm, Sweden
Engell, John	Technical University of Denmark, Copenhagen, Denmark
Engström, Håkan	AB Sandvik Hard Materials, Stockholm, Sweden
Ericsson, Torsten	Linköping University of Technology, Linköping, Sweden
Ernstsson, Marie	Institute for Surface Chemistry, Stockholm, Sweden
Fabbri, Luciano	Eniricerche SpA, Monterotondo, Italy
Falk, Lena	Chalmers University of Technology, Göteborg, Sweden
Fu, Xiren	Shanghai Institute of Ceramics, Shanghai, China
Funatani, Kiyoshi	Japan Fine Ceramics Center, Nagoya, Japan
Furey, Michael	Virginia Polytechnic Institute and State University, Blacksburg, VA, USA
Ge, Changchun	University of Science and Technology, Beijing, China
Gugel, Ernst	Cremer Forschungsinstitut GmbH & Co. KG, Rödental, Germany
Hampshire, Stuart	University of Limerick, Limerick, Ireland
Hausner, Hans	Technische Universität Berlin, Berlin, Germany
Heinrich, Jürgen G.	Hoechst CeramTec AG, Selb, Germany
Hermansson, Leif	Doxa Certex AB, Uppsala, Sweden
Hirschfeld, Deirdre	Virginia Polytechnic Institute and State University, Blacksburg, VA, USA
Jack, Kenneth	The Cookson Group plc, Wallsend, UK
Jiang, Dongliang	Shanghai Institute of Ceramics, Shanghai, China
Johansen, Knut	Elkem A/S, Keramer, Kristiansand, Norway
Johansson, Thomas	Höganäs Eldfast AB, Höganäs, Sweden
Kahlman, Lars	Swedish Ceramic Institute, Göteborg, Sweden
Kamo, Roy	Adiabatics Inc., Columbus, IN, USA
Karlsson, Sven	Swedish Ceramic Institute, Göteborg, Sweden
Knutson-Wedel, Maria	Chalmers University of Technology, Göteborg, Sweden
Larker, Hans T.	ABB Cerama AB, Robertsfors, Sweden
Leuchs, Martin	MAN Technologie AG, München, Germany
Lewis, M. H.	University of Warwick, Coventry, UK
Linde, Kerstin	AB Sandvik Hard Materials, Stockholm, Sweden
Lundberg, Robert	Volvo Flygmotor AB, Trollhättan, Sweden
McLaren, Malcolm	Rutgers University, Center for Ceramic Research, Piscataway, NJ, USA
Mäntylä, Tapio	Tampere University of Technology, Tampere, Finland
Natansohn, Samuel	GTE Laboratories Inc., Waltham, MA, USA
Niesz, Dale	Rutgers University, Center for Ceramic Research, Piscataway, NJ, USA

Nygren, Mats	University of Stockholm, Stockholm, Sweden
Okada, Akira	Nissan Motor Co., Central Eng. Lab., Yokosuka, Japan
Olsson, Per-Olof	Defence Research, Stockholm, Sweden
Pejryd, Lars	Volvo Flygmotor AB, Trollhättan, Sweden
Persson, Åke	Dynamec Research AB, Södertälje, Sweden
Persson, Michael	Eka Nobel AB, Bohus, Sweden
Petzow, Günther	Max-Planck-Institute for Metals Research, Stuttgart, Germany
Rowcliffe, David	Royal Institute of Technology, Stockholm, Sweden
Rundgren, Kent	Swedish Ceramic Institute, Göteborg, Sweden
Schulz, Robert	US Department of Energy, Washington, DC, USA
Sjöberg, Jörgen	Chalmers University of Technology, Göteborg, Sweden
Sōmiya, Shigeyuki	The Nishi Tokyo University, Tokyo, Japan
Spriggs, Richard M.	Alfred University, Center for Advanced Ceramic Technology, Alfred, NY, USA
Swab, Jeffrey J.	US Army Materials Technology Lab., Watertown, MA, USA
Tennary, Victor	Oak Ridge National Laboratory, Oak Ridge, TN, USA
Van de Voorde, Marcel	CEC Joint Research Centre, Petten, The Netherlands
Warren, Richard	Chalmers University of Technology, Göteborg, Sweden
Wendel, Agneta	AB Volvo, Technical Development, Göteborg, Sweden
Yen, T. S.	Shanghai Institute of Ceramics, Shanghai, China
Yust, Charles S.	Oak Ridge National Laboratory, Oak Ridge, TN, USA

OPTIMIZATION OF THE SILICON NITRIDATION PROCESS

B.Q. Lei, D. Ashkin, T. Johansson and T. Lindbäck
Department of Engineering Materials, Luleå University of Technology,
S-95187 Luleå, Sweden

ABSTRACT

By investigating the nitridation processes using a large mass of silicon under different atmospheres, it was found that the nature of heat conduction and mass transport were influenced by the size of the silicon powder compact and/or the number of compacts. Two main reaction maxima were observed and both accelerating as well as retarding influences on the reaction rate could be identified. The extent and starting temperature of the first reaction maximum was found to be greatly influenced by gas composition. Also investigated was the influence of different nitridation atmospheres (N_2+Ar, N_2+H_2, N_2+H_2+Ar and N_2+He) on microstructure of large silicon powder compacts. A homogeneous and complete reaction could be achieved by tailoring the gas composition.

INTRODUCTION

Reaction Bonded Silicon Nitride (RBSN) has been studied for decades because of its moderate cost of production, good reliability at high temperature and the fact that the overall dimensions of the powder compact do not change significantly during nitriding (1-3). The nitriding reaction, in its simplest form, can be written as

$$3Si(s) + 2N_2(g) \rightarrow Si_3N_4(s) \qquad \Delta G_f^\circ = -723 + 0.315\,T \ [\text{kJ mol}^{-1}]$$

Due to contamination and the exothermic nature of this reaction the practical nitridation process is more complex than this reaction formula. To minimize the complexity of this process much of the RBSN research has been performed on a laboratory scale. When these results have been applied to industrial scale, however, limitations concerning powders, mass of reactant, billet size, temperature, time and similar factors are experienced. Many of these limitations are due to the fact that the nitridation reaction is so strongly exothermic and that the thermal conductivity of the RBSN is low. This can lead to both temperature and pressure gradients in large samples. To combat these problems careful control of the temperature and use of auxiliary gases, such as H and He, is needed. Generally, however, there is a deficient public knowledge of the effect of these gases on large masses of reactants. Previous investigations have included studies of both processing parameters and raw materials (4). In this investigation, the nitridation process has been performed on a pilot-plant scale according to the nitrogen demand method. Using this method both temperature and gas pressure is controlled by a computer interactive procedure. In addition the effect of auxiliary gases can be studied.

EXPERIMENTAL

The silicon powders were produced by KemaNord Industrikemi AB, Table 1. They were cold isostatically pressed at 300 MPa to the dimensions of 50x33x13 mm. The green density was approximately 60% of theoretical. The green compacts containing PEG were de-waxed in a alumina tube furnace with an appropriate heating schedule up to 500°C under flowing hydrogen (0.5 l/min). The residual binder was determined to be less than 0.3% (wt).

<table>
<tr><th>powder</th><th>d_{50} (μm)</th><th>Apparent density (g/cm^3)</th><th>Surface area BET(m^2/g)</th><th>Impurity (wt.%)</th></tr>
<tr><td>4C</td><td>7</td><td>0.7</td><td>1.2</td><td rowspan="3">Fe = 0.07
Al = 0.07
Ca = 0.01
C = 0.06
O = 0.2-0.7*</td></tr>
<tr><td>4D</td><td>5</td><td>0.6</td><td>2.1</td></tr>
<tr><td>4E</td><td>3</td><td>0.5</td><td>3.5</td></tr>
</table>

* The oxygen content is dependent on the particle size distribution

After de-waxing the green compacts were placed in graphite crucibles and embedded in a powder bed containing Si_3N_4+ Si 6%(wt) . The mass of silicon in each running was about 1500 g. The nitridation was carried out in a cold wall vacuum/pressure furnace, with a hot zone volume of 15 dm^3, specially designed for RBSN and SSN. The upper and lower pressure limits were set to be 900 mbar and 870 mbar respectively and was programmed such that when the total pressure was reduced to the lower limit only pure nitrogen was filled. This kept the partial pressure of the auxiliary gases, H_2, Ar and He constant. The heating rate was programmed to be 2.4 °C/h, and was reduced to zero when the nitrogen consumption exceeded a preset value (to avoid overheating). The conversion fraction was calculated by weight gain and the relative abundances of difference phases was determined by the method of Gazzara and Messier (5). From these XRD determinations of phases the conversion fraction could be compared.

RESULTS

The nitridation process was studied by recording the change in nitrogen pressure per unit time with different auxiliary gases. These measurements were performed for both the whole nitridation run, Fig. 1, and for the initial stages of the nitridation runs, Fig. 2. Figure 1 and 2 also show the temperatures at which heating rate was arrested to prevent overheating in the compact. These are characterized by hold times at the temperature at which rapid nitrogen consumption occurs.

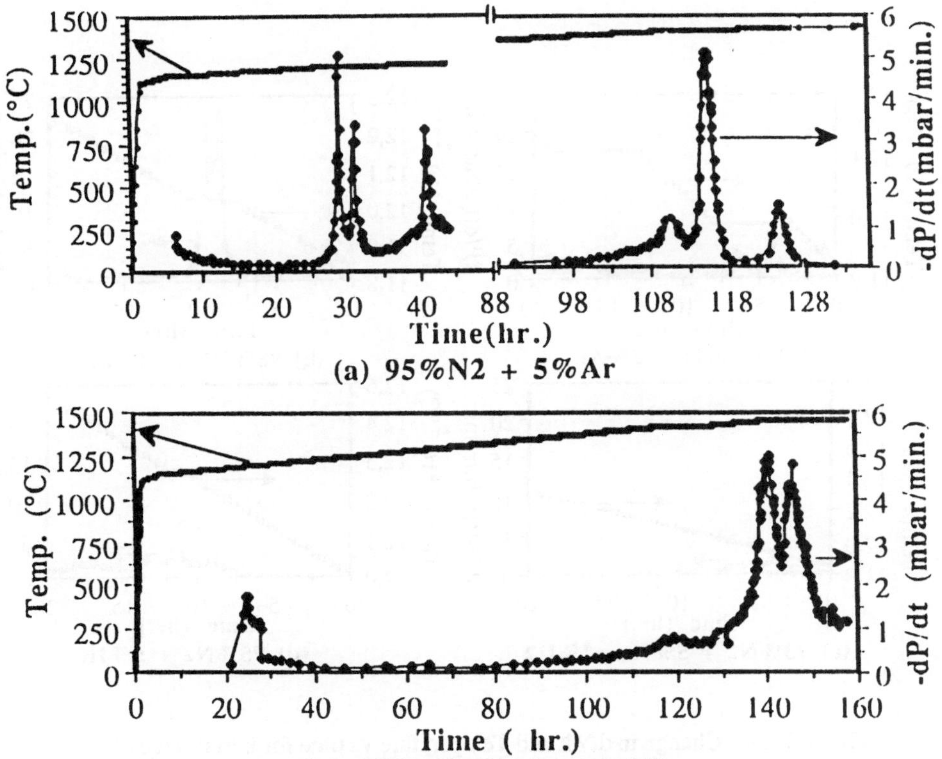

Fig 1(a,b) The change in dP/dt and temperature vs. time for two different gas compositions.

From these measurements it is apparent that nitridation starts as soon as nitrogen is admitted at 1150 °C and that there are two distinct maxima in nitrogen consumption, the first occurring around 1200 °C and the second arround 1400 °C. It is also noticed that each stage consists of several peaks and that the height vary considerably with gas composition. The temperature at which the first nitridation maximum occurs is additionally found to be influenced by the atmosphere. For nitridation with an atmosphere of 95%(vol.)N_2 + 5% Ar, Fig. 2A, the maximum occurs in the range 1200 - 1211 °C, while for nitridation with H additions (Fig. 2B-C) the maximum decreases to 1190 - 1205 °C and for additions of He the maximum occurs in the range of 1220 - 1240 °C.

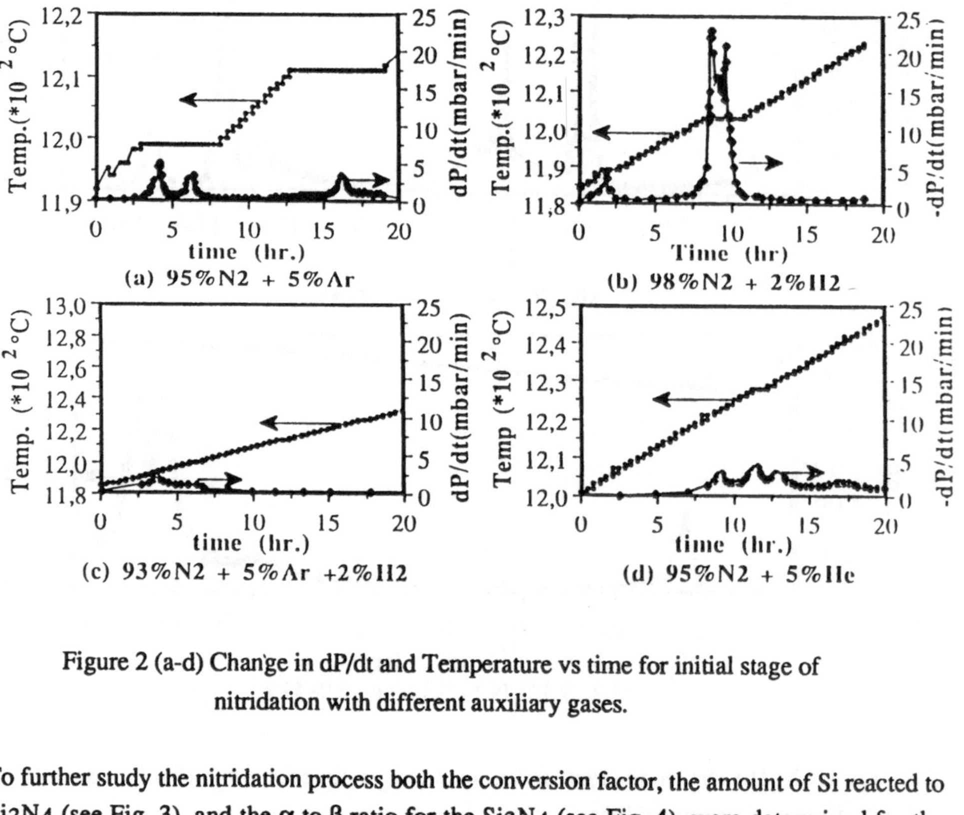

Figure 2 (a-d) Change in dP/dt and Temperature vs time for initial stage of nitridation with different auxiliary gases.

To further study the nitridation process both the conversion factor, the amount of Si reacted to Si_3N_4 (see Fig. 3), and the α to β ratio for the Si_3N_4 (see Fig. 4), were determined for the different powders and gas compositions. It is seen that samples nitrided with Ar as an auxiliary gas have the lowest conversion factors and samples nitrided with He have the highest conversion factor and α to β ratios. Samples nitrided in He, it should be noted, also had the fastest and most even nitridation according to the dP/dT diagrams.

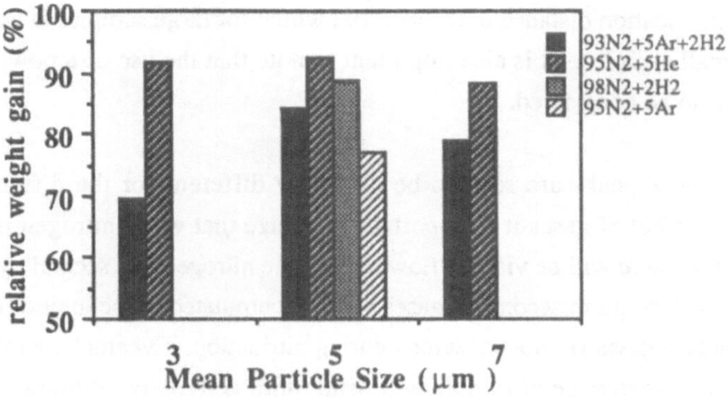

Fig.3 Relative weight gain varied under different atmospheres

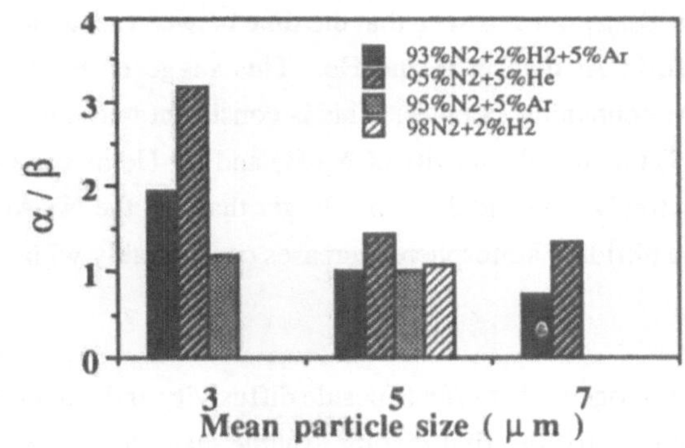

Fig.4 Ratio of α to β under different atmospheres

Discussion of Results

From the results it is seen that there are two distinct maxima in nitrogen consumption and that each stage associated with these maxima consist of several peaks. The presence of several peaks for these maxima is different from that of nitridation results for small masses of silicon, which are usually characterized by distinct single peaks (4,6,7). The probable explanation for these multiple peaks, see fig 2a-d, in large samples is that as the nitrogen is consumed, gas concentration and pressure gradients are built up within the material. In simple terms it takes time for the nitrogen to diffuse and flow into the compacts and the reaction is restricted until the concentration of nitrogen is high enough. The size of these gradients are related to the diffusion and permeation distance in the material which for large samples is naturally greater than that of small samples. It is also important to note that the use of a powder bed and its thickness needs to be considered.

The nature of these peaks are seen to be distinctly different for the different gases. In considering the effect of gases it is important to realize that when nitrogen is consumed in large samples that there will be viscous flow of both the nitrogen and secondary gases into the compact. This will result in secondary gases being accumulated in the interior of the compact since the secondary gases are not consumed during nitridation. Eventually a reverse diffusion of the secondary gas to edge of sample will result until activity equilibration. As such this process is quite complex, but it is significant to note that the effect of secondary gases is greater for larger samples.

The first effect that will be discussed is between Ar, fig. 2A, and that of H, fig. 2B, and He, fig. 2D. In contrasting their behaviour it is seen that the time between separate peaks is much longer for samples nitrided in Ar than in H and He. This suggests that the transport of nitrogen is lower for the Ar containing samples. This is consistent with the calculations by H. Kim and C. H. Kim (8,9) that the diffusivity of N_2-H_2 and N_2-He mixtures, in the range 1200 to 1700 K, are respectively 2.65 and 3.3 times larger than for the N_2-Ar combination and that the viscosity of the nitriding atmosphere increases considerably with the addition of Ar.

For samples nitrided with hydrogen other effects beside diffusivity and gas viscosity have to be considered. The mechanism for the first distinguishable stage has been suggested, by A.G.Ravesz (10) and Boyer et al. (11), to be due to Fe devitrifying the amorphous silica film and causing to its disruption. The presence of exposed silicon is thought to result in a drastic increase in nitriding reaction rate, but an explanation for the onset temperature of the first reaction maximum being affected by the atmospheres has not presented. Hydrogen, however, when added to the nitriding gas is known to decrease the partial pressure of oxygen and accelerate the nitridation process (1,6). This effect is particularly evident for the run shown in fig. 2b. Here only a small addition of hydrogen produced sharp and narrow dP/dt peaks that were approximately four times as high as peaks in other runs. This suggests that the accumulative effect of the secondary gas discussed earlier is particularly important for this run. It should also be remembered that the nitridation reaction is strongly exothermic and therefore a rapid reaction will cause both a large increase in temperature of the compact and a temperature gradient within the sample and furnace. Since the temperature that is recorded is the furnace temperature this can explain the low temperature for the first maxima in this run and run shown in fig. 2c.

For the He run both the nature of the peaks and the temperature of the first maxima in fig. 2d is distinctly different from the H and Ar runs. The temperature of onset is quite high, between 1220 and 1240 °C and the nitridation process is characterized by a rapid and stable nitridation. This likely due to the fact that He unlike H does not speed the nitridation process by reacting with silica and that compared to Ar the addition of He increases both the diffusion and thermal conductivity of the atmosphere. He, also, like Ar acts to dilute the nitrogen and decrease the nitridation rate. This reduces the risk of overheating in these materials. Another possible effect that He has on the nitriding process is that He can initiate the formation of atomic N as discussed by Kaiser (12) and Patel et al (13). As shown in Fig.4, the presence of atomic N is thought to lead to the increased formation of α phase (2).

The conversion fractions shown in figure 3 give some measure of the relative importance of different factors in nitriding large samples. As expected from the dP/dt plots the He samples have the highest conversion factors. The Ar samples, however, have the lowest conversion factor despite having a reasonably controlled reaction rate. This suggests that the decreased diffusivity and increased viscosity of Ar is more deleterious to the nitridation process than increased reaction rate caused by H.

SUMMARY

The nitridation of RBSN was investigated by the nitrogen demand principle for large masses of Si. It was found that two distinct reaction stages occurred at temperatures of around 1200 and 1400 °C and that each stage consisted of several speed-up and slow-down processes. The cause of these changes in reaction rate are believed to be due to competition between the consumption and feed rate of nitrogen and were found to be affected by the use of auxiliary gases. The use of He as an auxiliary gas was found to be most beneficial in controlling the nitridation process. This was evident from the dP/dt plots and calculated conversion factors for the nitridation reaction. Other gases studied were H and Ar and beside dP/dt plots the onset temperature of first reaction maxima and α to β ratio were considered for all gas compositions.

ACKNOWLEDGMENTS

The authors would like to acknowledge the financial support of STU.

REFERENCES

1. A.J.Moulson, J. of Mater. Sci., Vol.14, (1979), 1017-1051
2 H.M.Jennings, J. of Mater. Sci., Vol.18, (1983), 951-967
3. F.L.Riley, Mater. Sci. Forum., Vol. 47, (1989), 70-83
4. R.Pompe, L.Hermansson, T.Johansson, E.Djurl and M.E.Hatcher, Mater. sci. and Eng., Vol.71, (1985), 355-362
5. C.P.Gazzara and D.R. Messier, Bulletin of Am. Ceramic Soc., Vol.56(9), (1977), 777-780
6. M.Barsonm, P.Kangutar and M.J.Koczak, Ceram. Eng. Sci. Proc., Vol.10(7-8), (1989), 795-806
7. W.Mustel and D.Broussaud, Sci. of Ceram. Vol.12, (1982), 131-137
8. H. Kim and C.H.Kim, J. of Mater. Sci., Vol.20, (1985), 149-156
9. H.Kim and C.H.Kim, J. of Mater. Sci., Vol.20,(1985), 141-148
10. A.G. Ravesz, J. Non-Crystalline solids, Vol.11, (1973), p309
11. S.M.Boyer, D.Sang and A.J.Moulson, Nitrogen Ceramics, Ed.by F.L.Riley, Noordhoff, Leyden, (1977), p 297
12. W.Kaiser and C.D.Thurmond, J.of Appl.Phys., Vol.30(3), (1959), 427-431
13. C.K.Patel, P.K.Tien and J.H.McFee, Appl. Phys. Letter, Vol.7(11), (1965), 290-292

GAS PHASE PRODUCTION OF SILICON NITRIDE USING A DC PLASMA.

P.D. Harmsworth*, A.G. Jones, T.A. Egerton, S.R. Blackburn
Tioxide Technical Division, Central Laboratories
Stockton-on-Tees, Cleveland, UK. TS18 2NQ

ABSTRACT

Silicon nitride powder has been prepared by the gas phase reaction of the metal chloride with ammonia using a nitrogen DC plasma as the heat source. A wide range of particle sizes from 0.005 - 0.2μm can be prepared by control of the gas recirculation in the reaction zone.

This synthesis method results in a high purity ultra fine white powder with a spheroidal morphology. Either amorphous or crystalline forms of silicon nitride can be produced dependent upon reaction temperature. The product is associated with a surface oxygen layer as shown by XPS surface analysis. However fine tuning of the process to grow the particles can result in bulk oxygen contents of less than 1.5 % O, whilst retaining a specific surface area greater than 25 m^2/g.

Sintering studies using relatively low levels of common additives indicate that high density and high α - ß conversion can be achieved at low temperatures. The sintered product has a sub-micron grain size and exhibits high strength.

INTRODUCTION

The use of gas phase synthesis for ceramic powders has been found to offer advantages over conventional nitridation/carbothermal preparation routes, such as finer particle size, higher purity and low agglomeration [1]. The gas phase methods that have been tried, eg, laser, furnace and plasma have predominantly been only at laboratory scale. Plasma systems have however, been used for several years by Tioxide for the manufacture of pigmentary TiO_2 at up to 50,000 tpa scale. This technology has been adapted for the production of oxide ceramic powders [2] and Titanium Nitride [3]. In this work, development of a route to manufacture silicon nitride powder using a nitrogen DC plasma is described.

METHOD

A flow diagram of the reaction sequence is shown in Figure 1.

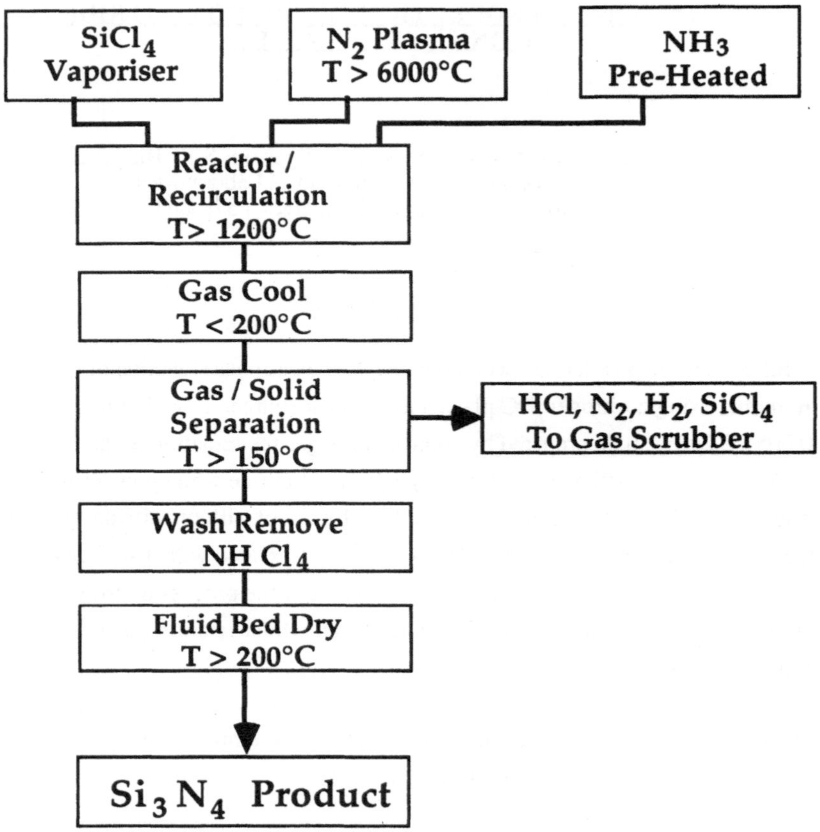

Figure 1. Process Flow Chart for Tioxide Silicon Nitride Powder and Important Temperatures.

Silicon tetrachloride and ammonia are vaporised and then fed into a reaction chamber, into which the plasma torch fires. Reaction occurs in this chamber at temperatures in excess of 1200°C. The recirculation of the powders in this chamber causes them to stick together at temperatures above the Tamman temperature. The use of recirculation to control particle size in a plasma system has been described in greater detail in Ref [3] for Titanium Nitride. Controlling this process effectively allows regulation of the particle size of the product. Once the ceramic particles have formed, they pass into a gas cooling section and on into a filtration system. The solids are then taken out with the remaining gases, passing through to a gas scrubbing unit to remove HCl and any unreacted $SiCl_4$.

The powder was characterised by ^{29}Si MAS NMR, Transmission electron microscopy, X-ray photoelectron spectroscopy and X-ray diffraction. Bulk oxygen was analysed by electrofusion with carbon, separating the derived gases by chromatography and infra-red detection of the separated CO_2, using a LECO TC136.

RESULTS AND DISCUSSION

The vapour phase reaction of silicon tetrachloride with ammonia, results in a fine (<0.1μm) white powder with a spheroidal morphology, Figure 2.

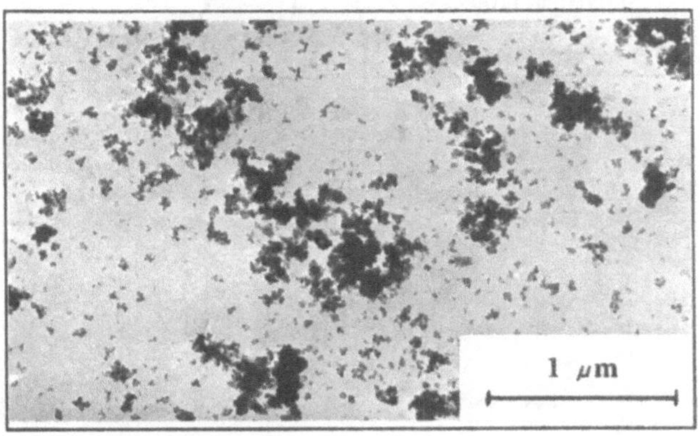

Figure 2. TEM micrograph of Tioxide Silicon Nitride Powder.

The powder has been produced in either an amorphous or crystalline form dependent upon the temperature of reaction and degree of recirculation. Analysis of the amorphous product using ^{29}Si MAS NMR and comparing the peaks with silicon imide (Figure 3a), shows that the immediate product is not the imide. However, heating the imide to 1100°C gives a spectrum similar to that of the Tioxide product (Figure 3b) and both of these have chemical shifts similar to that of crystalline silicon nitride (Figure 3c). It is therefore concluded that an amorphous silicon nitride is produced directly.

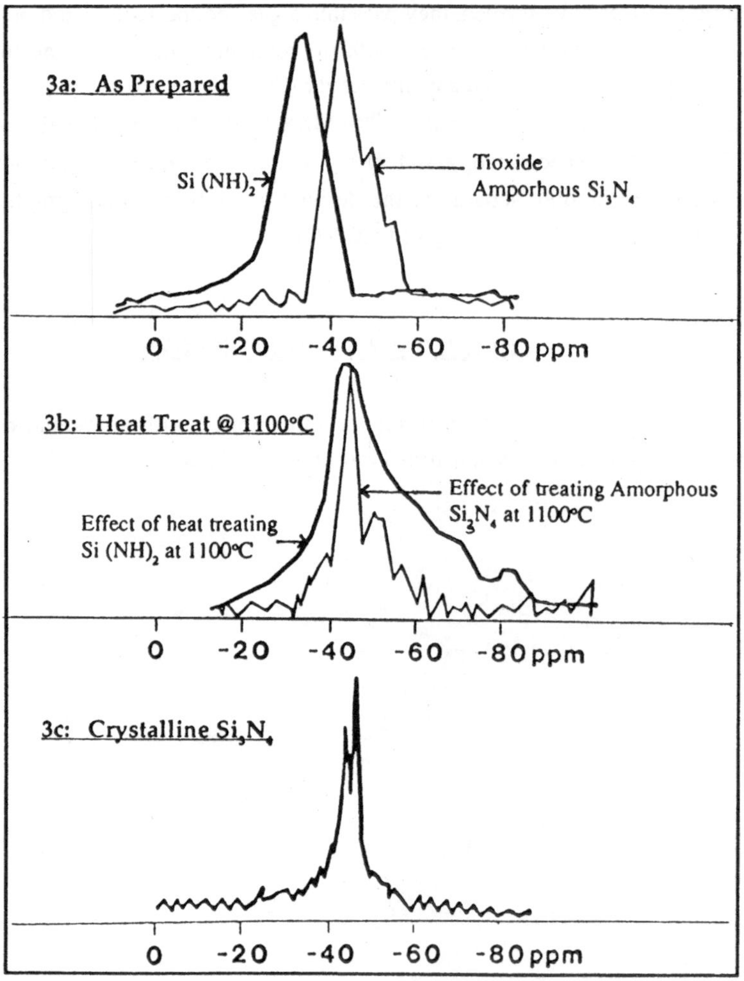

Figure 3. ^{29}Si NMR spectra of Tioxide Silicon Nitride Powder.

3a Traces of as prepared Tioxide amorphous silicon nitride & silicon imide
3b Effect of heat treating Tioxide material and the imide at 1100°C.
3c Trace for crystalline Tioxide silicon nitride

Quantitative XRD analysis of the crystalline product has shown the α phase content to be greater than 80%.

The powder has been produced in a range of particle sizes from 0.01 - 0.1 μm (SSA 70-15 m^2/g) by controlling the gas recirculation in the reaction zone. The silicon nitride powder reacts with oxygen on exposure to atmosphere. Comparison of surface oxygen analysis by XPS with bulk analysis using a LECO, has shown that all of the oxygen is accommodated within the outer surface of the particles. It is useful to compare the results from this work with the model of Greil et al [4]. They found typical surface compositions for commercial silicon nitride powders of 41% oxygen in a surface depth of 0.42nm. This average composition represents an SiO_2/SiON surface. Based on this, a relationship between oxygen content and surface area has been calculated. This is the lower line in Figure 4. The maximum surface composition found by Greil represents silicon nitride powder with a SiO_2 surface layer. Using this basis, the upper line of Figure 4 was calculated.

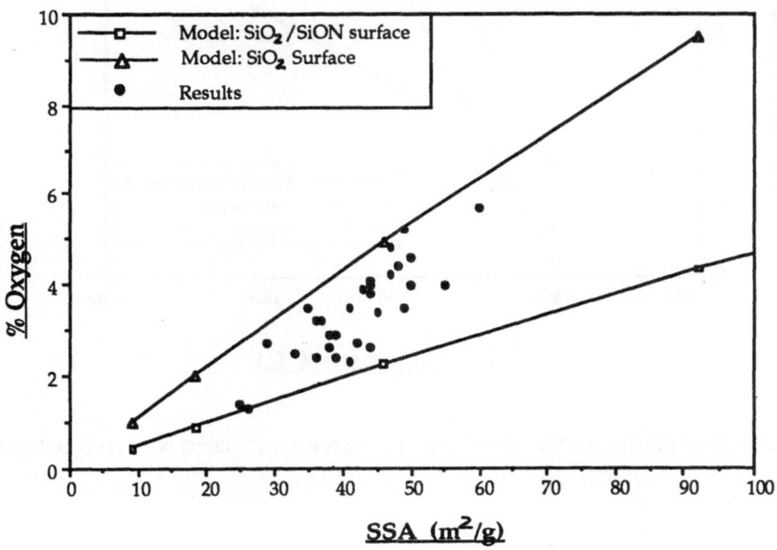

Figure 4. Relationship between oxygen content and Specific Surface Area for Si_3N_4 Powder.

The present results for the plasma derived powders fall between the calcluated lines and clearly demonstrate the increase in oxygen content with increasing surface area. This relationship supports the XPS analysis, that the oxygen in the gas phase powders is only present on the outer surface of the particles. Growing the particles in the reaction chamber can, in effect, be used to control the oxygen content of the material by varying the surface area of the powder. Bulk oxygen contents as low as 1.5% O have been achieved whilst retaining a specific surface area of 25 m^2/g.

Preliminary pressureless sintering studies under nitrogen using a standard sintering additive of 6 wt% Y_2O_3 / 3 wt% Al_2O_3 with a 1 hour dwell, have been undertaken. These have shown that the gas phase powders sinter to a higher density at lower temperatures than powders produced by the more common nitridation and carbothermal routes, Figure 5.

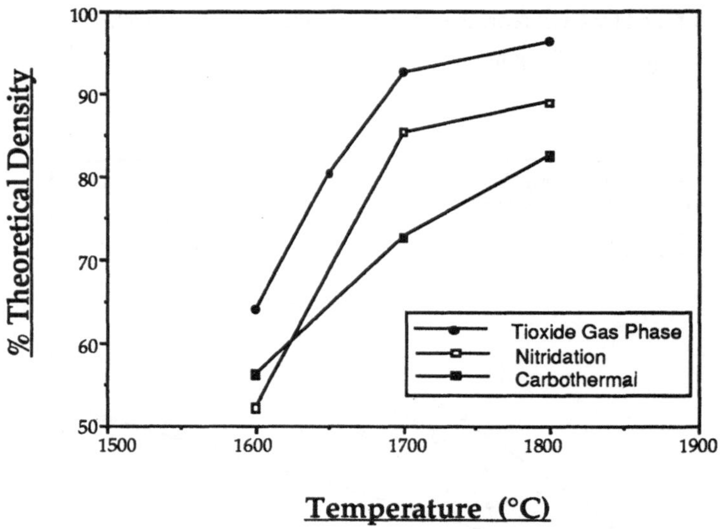

Figure 5. Densification curves for Tioxide gas phase route compared with conventional routes.

This is believed primarily to be due to the much finer particle size of the gas phase material, which has a SSA of 25-40m^2/g compared with typical values of less than 15 m^2/g for powders prepared by nitridation or carbothermal routes.

The sintered bodies have a sub-micron grain size, as seen from the fracture surface shown in Figure 6. XRD analysis of the surface shows it to have a high ß-phase content. Mechanical testing of sintered articles is underway, some initial results have indicated that strengths in excess of 800 MPa are achieveable.

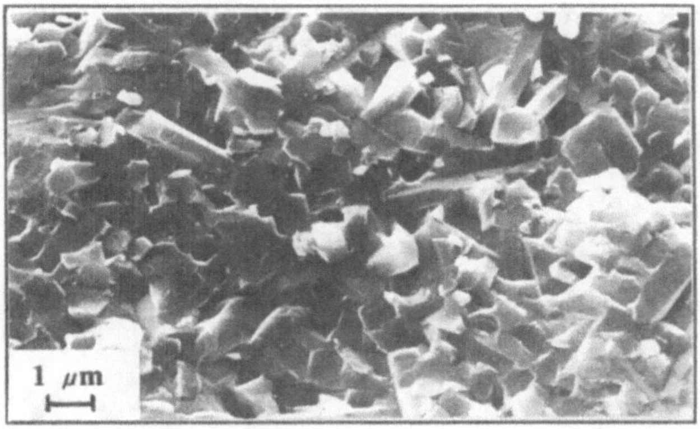

Figure 6. SEM fracture surface of Tioxide material showing sub-micron grain size.

CONCLUSIONS

A gas phase synthesis method has been developed for the production of silicon nitride powder. The process can be controlled to vary the particle size, oxygen content and crystal form by altering the degree of recirculation in the reaction zone. The powder has been found to offer advantages over powders prepared by conventional routes such as:- low temperature sintering,, fine grain size and high strength. These properties can be atributed to the high purity and ultra fine crystal size of the powders, which arises from plasma synthesis.

ACKNOWLEDGEMENT

The authors gratefully acknowledge Tioxide Group Plc for their kind permission to publish this paper.

REFERENCES

1 Cannon, W.R., Danforth, S.C., Haggerty, J.S., Marra, R.A., Sinterable ceramic powders from laser-driven reactions: II powder characteristics and process variables. J.Am.Ceram Soc., 1982, vol.65, no7, 330-335.

2 Dransfield, G.P., Fothergill, K.A., Egerton, T.A., The use of plasma synthesis and pigment coating technology to produce an yttria stabilised zirconia having superior properties. In Euro Ceramics, vol 1, ed. G. de With, R.A. Terpstra, R.M Metsalaar, Elsevier Applied Science Publishers, London, 1989, pp. 275-279.

3 Blackburn, S.R., Egerton,T.A., Jones, A.G., "Vapour phase synthesis of nitride ceramic powders using a DC plasma" to be published in Proc.Br.Ceram Soc, Fine ceramic powders.

4 Pleukert, M., Greil, P., Oxygen distribution in silicon nitride powders J.Mater.Sci., 1987, **22**, 3717-3720.

PRESSURELESS SINTERING IN THE Si-Y-Mg-O-N SYSTEM

F.M. Mahoney, M.J. Hoffmann, G. Petzow
Max Planck Institute for Materials Research
Powder Metallurgy Laboratory
Stuttgart, Germany

C. Boberski
Materials Research Department
Hoechst AG
Frankfurt am Main, Germany

ABSTRACT

The mechanical properties of pressureless sintered silicon nitride are often limited at temperatures above 1000°C due to the softening of the grain boundary phase which forms from the sintering additives necessary for liquid phase sintering. Improvements in the high temperature properties may therefore be possible by minimizing the amount of sintering additives needed for complete densification. However, to minimize the total amount of additives, a better understanding of the liquid phase properties and phase relations within the ceramic system is necessary. In this study, the densification behavior of pressureless sintered silicon nitride containing constant volume fractions of oxide sintering additives of varying composition was investigated. The densification behavior was analyzed in terms of the additive composition, the liquid phase properties, and the effects of phase relations within the ceramic system.

INTRODUCTION

Pressureless sintering of silicon nitride ceramics is recognized as a more cost-effective process for densification compared to processes such as hot-isostatic-pressing or gas-pressure-sintering. Without the aid of pressure, however, pressureless sintered Si_3N_4 generally requires a correspondingly higher fraction of densification aids which, in turn, lead to more severe degradation of the mechanical properties at high temperatures.

It follows that an optimization of pressureless sintered Si_3N_4 ceramics may be possible by reducing the amount of additives to the minimum necessary for complete densification. Due to the fact that different sintering additives have different liquid phase and thermochemical properties, however, there cannot exist a universal minimum amount of densification aid for sintered silicon nitride. For example, it has been shown that silicon nitride can be completely densified by pressureless sintering using only 5w/o MgO, whereas difficulty in densification is found for silicon nitride doped with up to 7w/o Y_2O_3 [1]. On the other hand, it has been long recognized that low softening temperatures of the grain boundary phases eliminate single additive MgO-doped Si_3N_4 materials from serious consideration for high temperature application [2], whereas Y_2O_3 has been shown to produce more favorable grain boundary phases due, at least in part, to its higher softening temperature [3,4]. Since there does not appear to be any single additive available which yields both adequate densification and acceptable high temperature mechanical properties, alloying of additives is necessary.

Previous Sintering Studies

The pressureless sintering of Si_3N_4 doped with combinations of Y_2O_3 and MgO additives has been previously investigated [5,6,7,8]. In general, similar trends have been reported. Giachello, et al [5] found that whereas full density could not be achieved using only Y_2O_3 as a sintering additive, a small addition of MgO allowed complete densification. It was reasoned that the MgO had an appreciable effect on the densification by lowering the viscosity of the liquid phase, although it should be noted that the addition of MgO would also increase the volume of liquid phase during sintering, additionally enhancing densification.

Pomeroy, et al [6] studied the pressureless sintering of Si_3N_4 containing constant 10w/o fractions of densification aids composed of different ratios of Y_2O_3 and MgO. Starting with 10w/o Y_2O_3, MgO was substituted incrementally for Y_2O_3 to 10w/o MgO. Improved densification was reported with increasing fractions of MgO. Similar to Giachello, the improved densification was explained in part by a decrease in the liquid phase viscosity with increasing MgO content. It was also pointed out that, due to the different crystalline densities between Y_2O_3 (5.01 g/cc) and MgO (3.58 g/cc), increasing <u>weight</u> fractions of MgO also lead to increasing volume fractions of liquid phase.

Thus, in general, it can be stated that a larger volume of lower viscosity liquid is beneficial for densification [6]. It is unclear, however, if the improved densification observed in these studies is controlled more significantly by the decrease in the liquid phase viscosity or by the increase in the volume of the liquid phase. In an effort to minimize the amount and optimize the composition of sintering additives, it would therefore be advantageous to investigate the pressureless sintering behavior of Si_3N_4 containing <u>constant volume fractions</u> of liquid phase. The requirements and limitations of this concept are discussed in the following section.

REQUIREMENTS AND LIMITATIONS OF CONSTANT VOLUME CALCULATIONS

Based on the results of previous studies, it is desirable to separate the influence of two important parameters on the densification behavior of pressureless sintered silicon nitride: liquid phase volume and liquid phase viscosity. A limitation to applying constant volume fractions of liquid phase, however, arises from the variation of nitrogen solubility as a function of the liquid composition. The higher the solubility of nitrogen in the liquid phase, the more Si_3N_4 is dissolved and thus the greater the liquid phase volume during sintering. Nitrogen solubilities have not been determined quantitatively for many ceramic oxide liquids, although it has been reported that yttrium-containing glasses have a higher nitrogen solubility than magnesium-containing glasses [9]. Thus, although the use of constant volume liquid phase fractions is desirable, the unknown relationship of the nitrogen solubility as a function of composition prevents its application. The calculation of compositions must therefore be limited to constant volume fractions of total sintering additives.

An important requirement for calculating constant volume fractions of additives arises from the SiO_2 content in silicon nitride powders. All silicon nitride powders contain SiO_2 in the form of a surface layer on the Si_3N_4 grains, and since SiO_2 is a metal oxide and melts to form part of the liquid phase during sintering, it must be properly considered as a sintering additive. The amount of SiO_2 can be calculated from the oxygen content in the silicon nitride powder assuming one mole of oxygen forms one-half mole of SiO_2. The amount of SiO_2 resulting from the silicon nitride powder contained in a sintering composition depends therefore on the oxygen content of the silicon nitride powder and the volume fraction of sintering additives chosen. Consequently, the influence of the SiO_2 present in the silicon nitride powder becomes increasingly significant as the chosen additive volume fraction is reduced. Figure 1 shows the percentage of the total additive volume which is made up of SiO_2 as a function of the chosen volume% of additive. The different curves shown represent different oxygen contents in the silicon nitride powder. The effect of the SiO_2 present in the silicon nitride powder is shown by the dotted lines for a powder with 2w/o oxygen. For example, in choosing 20v/o total sintering additives, the SiO_2 present in the silicon nitride powder constitutes 23% of the total additive volume. Reducing the total additive volume to 10v/o, SiO_2 occupies 51% of the additive volume. At 6v/o, the SiO_2 present in the silicon nitride powder will compose 89% of the additive volume - by far the most significant additive in the sintering composition. Thus, properly considering the role of the SiO_2 present in the silicon nitride powder as a sintering additive, it becomes obvious that by minimizing the total additive fraction in Si_3N_4 sintering compositions, the influence of SiO_2 on the densification behavior becomes increasingly significant.

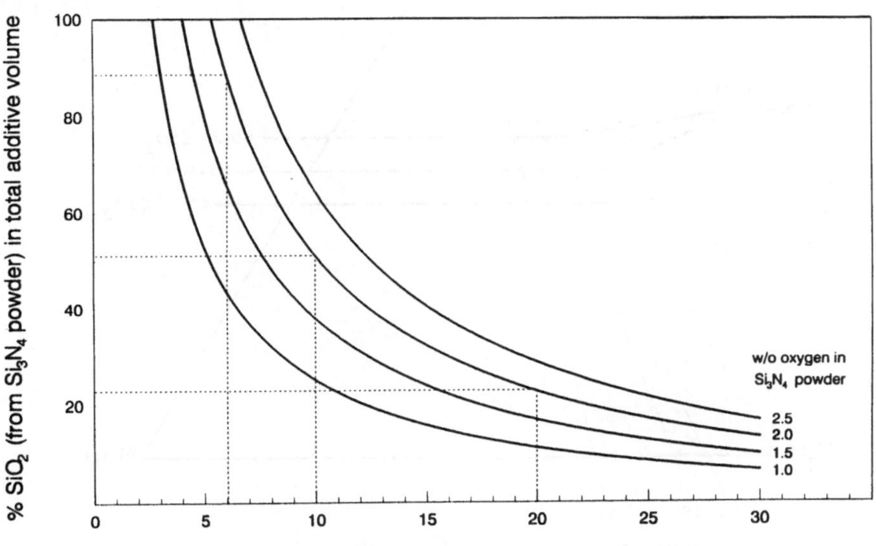

Figure 1. % of total sintering additive volume composed of SiO_2 (from the silicon nitride powder) as a function of the total additive volume fraction and oxygen content of the silicon nitride powder

EXPERIMENTAL PROCEDURE

Sintering compositions in this study consisted of Si_3N_4 (Stark LC-12S, 1.94w/o O) doped with 10v/o total additives. The nine different additive compositions are represented on the pseudoternary SiO_2-Y_2O_3-MgO oxide system shown in Fig. 2 and cover a range of compositions based on the results of previous studies. From calculations in the previous section, the minimum possible volume fraction of SiO_2 in the additive compositions was calculated as 47% (10v/o total additives, 1.94w/o oxygen, see Figure 1). Additional amounts of SiO_2 were added to study higher fractions of SiO_2, and three different volume ratios of MgO/Y_2O_3 were investigated.

Additive compositions were attrition-milled in isopropyl alcohol with silicon nitride powder for 4 hours using Si_3N_4 milling media. Cylindrical test samples (1.5 cm diameter, 1.5 cm height) were cold-pressed and subsequently pressureless sintered in a silicon nitride powder bed in N_2 atmosphere. Sintering was conducted at 1800°C for 1, 2, and 4 hours.

After sintering, the relative densities were measured by Archimedes method using a glycerol solution as the suspending medium [10]. Measured densities were confirmed qualitatively by SEM micrographs of polished and plasma-etched samples [11]. X-ray diffraction was also conducted on various samples to determine the α-Si_3N_4 content remaining after sintering. To relate the trends in densification to variations in liquid phase viscosity, data from the Research Center of the Belgian Ceramic Industry were used to estimate the viscosity of the oxide liquid phase compositions at the sintering temperature.

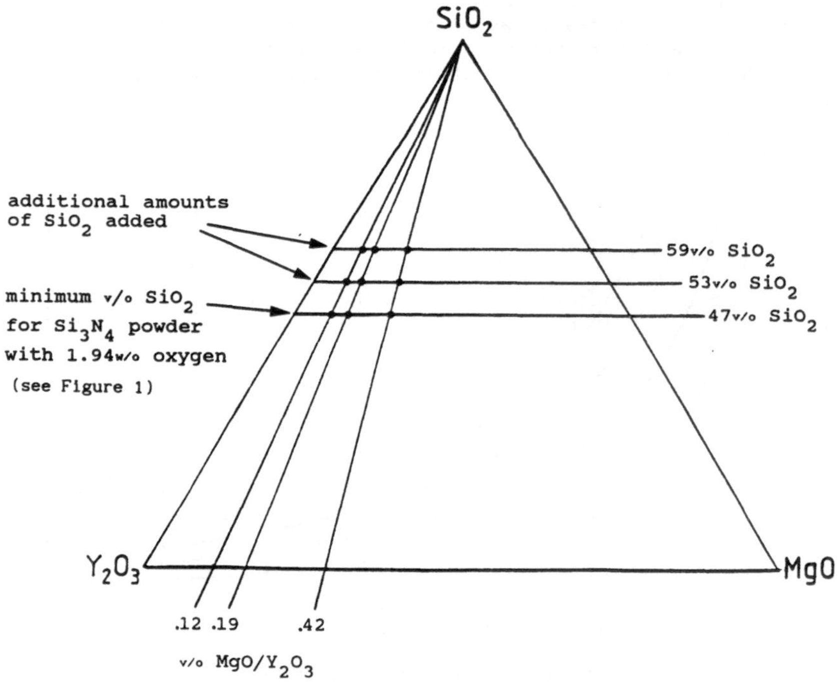

Figure 2. Sintering additive compositions chosen to study the densification of Si_3N_4 containing 10v/o total additives

RESULTS AND DISCUSSION

Figure 3 shows schematically the relative densities of samples sintered at 1800°C for 1 hour as a function of the additive composition. The arrows indicate direction of improved densification. In general, similar trends were observed for all three sintering times and can be summarized as follows. For the minimum volume fraction of SiO_2 (corresponding to SiO_2 resulting solely from the silicon nitride powder), densification improved with increasing fractions of MgO. This result supports the trends observed in the previous sintering studies where densification increased with increasing amounts of MgO [5,6,7]. With additional amounts of SiO_2, changes in the densification behavior depended upon the volume ratio between MgO and Y_2O_3. For the highest MgO/Y_2O_3 volume ratio, densities decreased with increasing SiO_2 content; for lower MgO/Y_2O_3 ratios densities increased with increasing SiO_2 content.

It is interesting to note here that the trends in densification as a function of increasing SiO_2 content were not observable in previous studies since the concept of investigating increased fractions of SiO_2 was not evident. By properly considering the influence of the SiO_2 present in the silicon nitride powder, it became obvious that sintering additive compositions could be evaluated in the entire SiO_2-Y_2O_3-MgO ternary system instead of merely within the Y_2O_3-MgO binary system.

Figure 4 gives the estimates of oxide liquid phase viscosities at 1800°C as a function of additive composition. Arrows indicate direction of decreasing viscosity. To calculate these values, an experimental model developed for estimating the viscosity of complex melts was used [12]. It is important to note that the viscosity values listed are for oxide melts and thus do not include the effect of nitrogen which would be present in such liquids during sintering. The viscosity of actual oxynitride liquids would therefore be higher than those listed here [9]; however, the trends in viscosity related to metal cation composition are clear. A more exact analysis of the influence of liquid phase viscosity on the densification behavior would only be possible with a more complete quantitative description of the nitrogen solubility as a function of composition and temperature.

Using this data, the argument of improved densification for decreasing liquid phase viscosity was analyzed. Considering first the additive compositions with minimum fractions of SiO_2, it can be seen that the relative sintered densities increase with decreasing liquid phase viscosity (compare Figure 3 with Figure 4). Also for the highest MgO/Y_2O_3 volume ratio investigated, densities increase with decreasing viscosity. These trends support the argument of improved densification due to decreased liquid phase viscosity. Considering lower MgO/Y_2O_3 ratios, however, increased densities are observed for <u>increasing</u> viscosities indicating that the variation in the liquid phase viscosity as a function of metal cation composition cannot solely explain the variation in the densification behavior. Other parameters must also have a significant affect on the densification.

X-ray diffraction was conducted on a number of samples after sintering. Based on the relative intensities of the α-Si_3N_4 and β-Si_3N_4 peaks, the mass% of each component can be calculated [13]. Figure 5 shows the mass% α-Si_3N_4 in samples sintered at 1800°C for 1 hour as a function of the sintering additive composition. The relative position of the Mg-Y apatite phase is also schematically represented. The results reveal that additive compositions nearest the apatite composition exhibit incomplete (thus slower) transformation of α-Si_3N_4 to β-Si_3N_4 after 1 hour at 1800°C. The

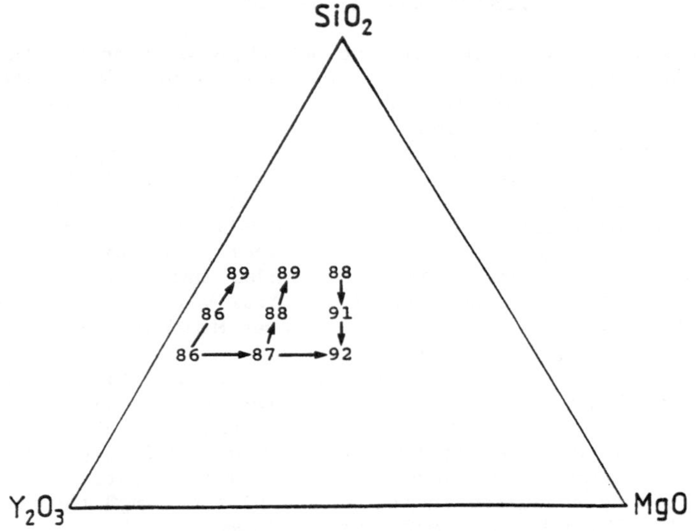

Figure 3. % theoretical density of sintered Si_3N_4 samples containing 10v/o sintering additives as a function of additive composition after 1 hour at 1800°C (arrows indicate direction of increasing density)

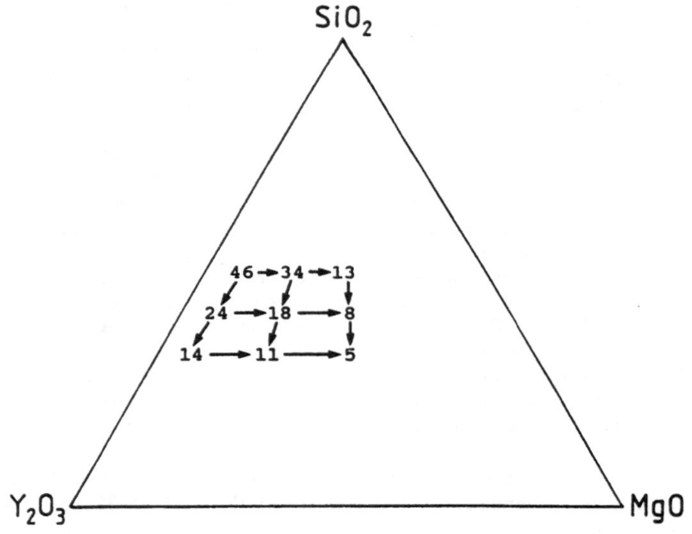

Figure 4. Viscosity (in poise) of oxide melts as a function of composition at 1800°C (Courtesy of Dr. F. Cambier)
(arrows indicate direction of decreasing viscosity)

samples which showed incomplete transformation also exhibited lower sintered densities (Fig. 3) suggesting that phase relations involving the apatite phase may have a decided effect on the densification behavior. The formation of a crystalline phase such as the apatite phase during the sintering process, for example, would reduce the amount of liquid phase available for densification.

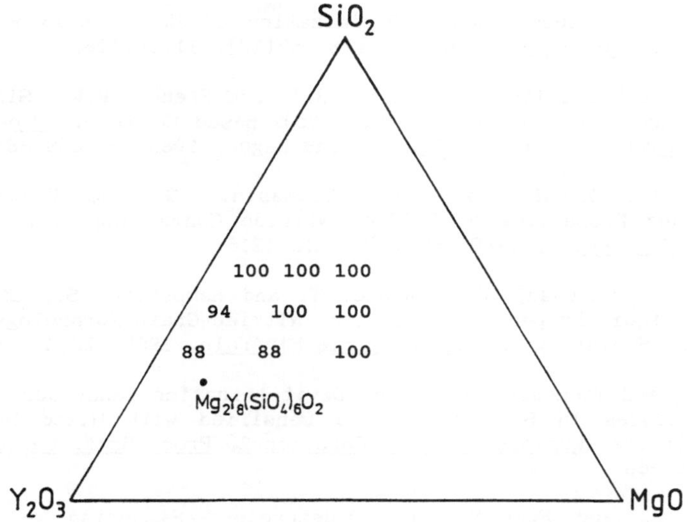

Figure 5. %β-Si_3N_4 content as a function of additive composition for samples sintered for 1 hour at 1800°C
(Composition of Mg-Y apatite phase is shown schematically)

CONCLUSIONS

The pressureless sintering of Si_3N_4 containing constant volume fractions of oxide sintering additives from the pseudoternary SiO_2-Y_2O_3-MgO system has been studied. An important requirement for the application of constant volume fractions of total additives is the proper consideration of the SiO_2 present in the silicon nitride powder as a sintering additive. The relative densities of samples sintered for different times at 1800°C show that the oxide additive composition has a significant effect on the densification behavior. For minimum fractions of SiO_2, densification improved with increasing fractions of MgO. This trend supports results from previous studies in the Y_2O_3-MgO additive system. For increasing fractions of SiO_2, trends in the densification behavior depended on the volume ratio MgO/Y_2O_3 where increased densification with additional fractions of SiO_2 were found only for low MgO/Y_2O_3 volume ratios.

The argument of improved densification with decreasing liquid phase viscosity was shown to explain trends in the densification behavior only for certain additive compositions (minimum SiO_2 content, high MgO/Y_2O_3 ratios). For a more complete description of the sintering behavior, the variation of the nitrogen solubility as a function of liquid phase composition must be understood.

REFERENCES

1. Hampshire, S. and Jack, K.H., The Kinetics of Densification and Phase Transformation of Nitrogen Ceramics. Special Ceramics 7, Proc. Brit. Ceram. Soc., 1981, **31**, 37-49.

2. Gazza, G.E., Hot-Pressed Si_3N_4. J. Am. Ceram. Soc., 1973, **56**[12], 662.

3. Weaver, G.Q. and Lucek, J.W., Optimization of Hot-Pressed Si_3N_4-Y_2O_3 Materials. Am. Ceram. Soc. Bull., 1978, **57**[12], 1131-1136.

4. Quackenbush, C.L., Smith, J.T., Neil, J.T. and French, K.W., Sintering, Properties and Fabrication of Si_3N_4 + Y_2O_3 Based Ceramics. Progress in Nitrogen Ceramics, Martinus Nijhoff, The Hague, 1983, pp.669-682.

5. Giachello, A., Martinengo, P.C., Tommasini, G. and Popper, P., Sintering and Properties of Silicon Nitride Containing Y_2O_3 and MgO. Am. Ceram. Soc. Bull., 1980, **59**[12], 1212-1215.

6. Pomeroy, M.J., Saruhan, B., McArdle, T. and Hampshire, S., Effect of Oxynitride Liquid Properties on Silicon Nitride Grain Morphology During Pressureless Sintering. Revue de Chimie Minérale, 1985, **22**[4], 555-563.

7. Patel, J.K. and Thompson, D.P., The Devitrification Behaviour of Grain Boundary Glasses in Silicon Nitride Densified with Mixed Magnesia, Yttria Additions. Engineering with Ceramics 2, Proc. Brit. Ceram. Soc., 1987, **39**, 61-68.

8. Hampshire, S. and Pomeroy, M., Pressureless Sintering of Silicon Nitride with Mixed Oxide Additives. Ann. Chim. Fr., 1985, **10**, 65-72.

9. Drew, R.A.L., Hampshire, S. and Jack, K.H., Nitrogen Glasses. Special Ceramics 7, Proc. Brit. Ceram. Soc., 1981, **31**, 119-132.

10. Pennings, E.C.M. and Grellner, W., Precise Nondestructive Determination of the Density of Porous Ceramics. J. Am. Ceram. Soc., 1989, **72**[7], 1268-1270.

11. Täffner, U., Hoffmann, M.J., and Krämer, M., A Comparison of Different Physical-Chemical Methods of Etching for Silicon Nitride Ceramics. Practical Metallography, 1990, **27**, 385-390.

12. Urbain, G., Cambier, F., Deletter, M. and Anseau, M.R., Viscosity of Silicate Melts. Trans. J. Brit. Ceram. Soc., 1981, **80**, 139-141.

13. Gazzara, C.P. and Messier, D.R., Determination of Phase Content of Si_3N_4 by X-ray Diffraction Analysis. Am. Ceram. Soc. Bull., 1977, **56**[9], 777-780.

SOME FEATURES AND LIMITATIONS OF ZIRCONIA TOUGHENED NITROGEN CERAMICS

Y.Cheng[§], C.o'Meara[*], S.Slasor[+], R.Pompe[£], D.P.Thompson[§]
[§]University of Newcastle upon Tyne (U.K.)
[+]Cookson Group plc. (U.K.)
[*]Chalmers University of Technology (Sweden)
[£]Swedish Ceramic Institute (Sweden)

ABSTRACT

It has previously been assumed that good toughness can be achieved in silicon nitride based ceramics up to moderate temperatures through the addition of small amounts of zirconia particles to the ceramic matrix, providing the crystalline structure of the zirconia is partially stabilized. However, in all reactions involving the partially stabilized zirconia and silicon nitride in a nitrogen atmosphere, the non-transformable tetragonal zirconia (t') phase forms by dissolution of small amounts of nitrogen and simultaneous introduction of oxygen vacancies in the crystal structure. The increasing numbers of vacant sites in the anion lattice produces additional stabilization to the zirconia and consequently $t \rightarrow m$-ZrO_2 transformation is prevented. The effect of nitrogen dissolution in zirconia is clearly evidenced in zirconia containing O' and O'-ß' composites. As a result, no significant improvement in toughness is observed although the tetragonal zirconia phase is present in the samples.

INTRODUCTION

Improving the fracture toughness of engineering ceramics has been an important focus for research effort in recent years. Among the various mechanisms available, toughening by incorporating a dispersion of second phase particles offers advantages of easy fabrication and low cost. Early success in the development of zirconia-toughened alumina and mullite ceramics stimulated further research activity aimed at achieving similar increases in toughness for zirconia-containing silicon nitride and sialon ceramics, based on the assumption that both crack deflection and stress-induced martensitic transformation toughening mechanisms could operate simultaneously in these composites if a dispersion of metastable tetragonal zirconia particles could be retained within the matrix.

Claussen and Jahn [1] studied Si_3N_4-ZrO_2 composites containing up to 30 v/o *m*-ZrO_2 and found some improvement in fracture toughness, which was attributed to microcrack toughening generated by spontaneous $t \rightarrow m$ transformation during cooling. However, when Y_2O_3 was added to Si_3N_4-ZrO_2 mixes to stabilize the zirconia in a metastable tetragonal form and also to prevent zirconium nitride formation, no toughening was observed. Ekström et al [2,3] found that in both pressureless sintered and hot-pressed Si_3N_4-ZrO_2 composites containing up to 20 w/o TZ-3Y ZrO_2 powder (TOSOH, Japan), the fracture toughness remained unchanged. Rundgren et al (4) obtained toughness values in the range of 6 - 8 $MPam^{½}$ for composites of Si_3N_4-30w/o ZrO_2(TZ-3Y) which had a microstructure free from glassy grain boundary phase except for a thin amorphous film between the grains. This toughness level was similar to what is regularly attained for Si_3N_4 specimens with a glassy grain boundary phase thus pinpointing some role of zirconia. Although tetragonal zirconia was observed in the final materials, there was no transformation after mechanical grinding.

Clearly when tetragonal zirconia particles are present in a Si_3N_4 (or ß'-sialon) matrix with the correct grain size to operate a transformation toughening mechanism, no increment in toughness is observed which can be attributed to this process. The present study extends these observations to O'-ZrO_2 and O'-ß'-ZrO_2 composites and gives an explanation for these phenomena in terms of the role of nitrogen in these materials.

EXPERIMENTAL

Sub-micron powder mixtures of $ZrSiO_4$ (Ventron) and Si (KemaNord) with Y_2O_3/Al_2O_3 (H C Stark/Sumitomo AKP 30) as sintering aids/stabilisers were prepared by milling in plastic bottles for 70 hours in ethanol using Si_3N_4 milling balls. A stoichiometric $ZrSiO_4$/Si ratio was used and a sintering additive ratio (w/o) of 5:6.5. The powder compacts were uniaxially pre-pressed into rectangular bars of dimensions 7x7x40 mm followed by cold isostatic pressing (CIP) at 280 MPa. The green bodies had a density of 55% of theoretical. Nitridation was carried out in graphite furnace in N_2 at 0.1 MPa. A two-step nitridation program was used with the following heating cycle (temperature/holding time): 1150 °C/0.5h-1250 °C/1.5h. The nitrided density was 67-68% of theoretical. XRD analysis showed the green bodies to contain $ZrSiO_4$ partially disproportionated to *m* and *t*-ZrO_2. No other crystalline phases were detected. Pressureless sintering was carried out in a graphite oven in a Si_3N_4 powder bed (preoxidation:1200 °C/0.5h) in N_2 at 0.1 MPa. The sintering program was 1600-1640 °C with 1-4h holding time. The samples had a final sintered density of 98-99% of theoretical.

The microstructure of the zirconia phase in the composites was characterised by a combination of X-ray diffractometry (XRD) of powdered samples, and analytical transmission electron microscopy (STEM/EDX) of thin foils in a JEOL 2000 FX TEM/STEM instrument with a Link Systems AN 10,000 EDX spectrometer.

Four additional compositions were chosen in order to understand the behaviour of zirconia in mixed O'-ß' sialon composites (Table 1). These compositions were designed to give varying proportions of O' and ß' sialon in the final products. Samples were placed in BN-lined carbon crucibles and sintered at 1500 °C for half an hour followed by a one hour soak at 1700 °C in a carbon resistance furnace in nitrogen.

Starting powders used to explore possible Si_3N_4/ZrO_2 and N_2/ZrO_2 reactions were TZ-3Y $ZrO_2^{\$}$, ZrN^{*}, $Si_3N_4^{+}$, $Al_2O_3^{¶}$, $Y_2O_3^{§}$ and $SiO_2^{¥}$. Calculated compositions were mixed in water-free propanol in either an agate mortar or a rubber-lined mill using silicon nitride balls. All samples were placed in carbon crucibles lined with BN and sintered in a carbon furnace in flowing nitrogen. X-ray analysis was carried out using a Hägg-Guinier focusing camera with Si as a reference standard. The fracture toughness K_c of the samples was evaluated by indentation using a 10 Kg load.

RESULTS AND DISCUSSION

O'-ZrO_2 composites

The general microstructure of the composite consisted of a Si_2N_2O matrix, zirconia grains ($\approx$30 v/o) and a considerable volume of intergranular amorphous phase. XRD indicated *t* and *c*-ZrO_2 to be the main zirconia phases present in the microstructure with only a small volume of *m*-ZrO_2 (<10 v/o). The zirconia grains in the microstructure were observed to have an irregular, intragranular morphology with ellipsoidal protrusions extending into the glassy phase and between the Si_2N_2O crystals (Fig.1). The distribution of this phase in the matrix was homogeneous and a grain size distribution of 0.5-3 μm was observed. The overall impression from the microstructures is that the ZrO_2 phase grew in the available space in the microstructure left over after the formation of the Si_2N_2O platelet network.

For the identification of *t*-ZrO_2 in the microstructures, selected area diffraction was used whereby on tilting the thin foil to a <111> zone axis different {112} reflections, forbidden in the cubic fluorite structure, can be used for dark field imaging. Using this technique different sets of complementary tetragonal laths could be imaged showing the grains to be completely tetragonal (Fig.2). The presence of anti-phase domain boundaries (APB's) when imaging {112} fluorite forbidden reflections in dark field is diagnostic for the presence of the non-transformable *t'*-ZrO_2 variant [5], and most of the tetragonal grains examined were found to be of this type (Fig.2). This observation is also consistent with the measured yttrium content of the grains, as determined by EDX analysis, which showed the yttrium content to be >5 w/o and relatively constant from grain to grain.

Fig.3 shows a "tweed" structure which was frequently observed in thin parts of the ZrO_2 grains. Within these grains fine 5-10 nm precipitates could be observed in dark field. The "tweed" structure is known to be present in *c*-ZrO_2 grains and is usually associated with homogeneous precipitation of *t*-ZrO_2 within the *c*- matrix. However work by Lanteri et al. [5] indicates that the small particles may obscure possible APB contrast and therefore even in these grains the *c*- matrix may have undergone transformation to the *t'* symmetry.

$ TOSOH, Japan; * PFALTZ & BAUER, Inc., U.S.A.; + LC10, Starck Berlin; ¶ A17, ALCOA; § Rare Earth Products; ¥ BDH Products

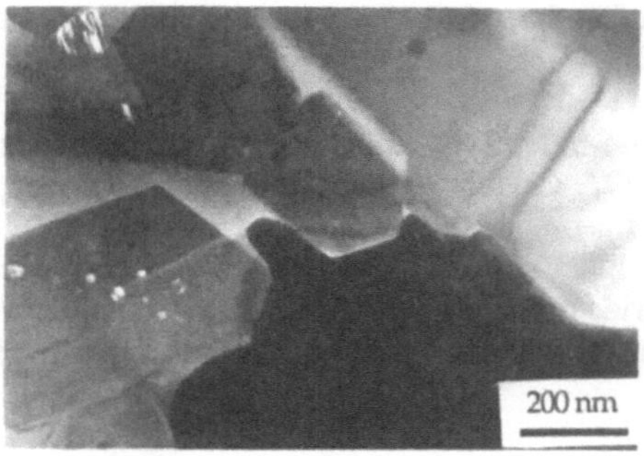

Fig.1 TEM micrograph of O'/ZrO_2 material showing the ZrO_2 phase to have an irregular, intergranular morphology with ellipsoidal protrusions extending into the glassy phase and between the Si_2N_2O crystals.

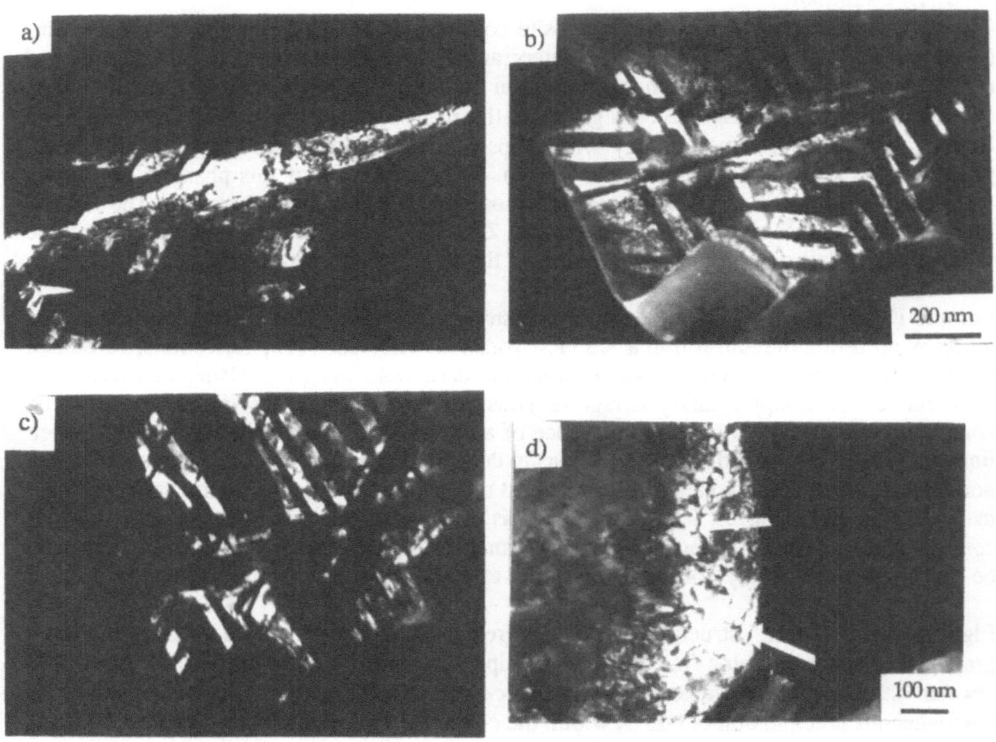

Fig.2 TEM centred dark field images of a t-ZrO_2 grain in the O'/ZrO_2 material after tilting to a <111> zone axis orientation showing: (a-c) tetragonal laths using different (112) reflections; (d) anti-phase domain boundaries (APB's) which indicate the presence of the t'-ZrO_2 variant (arrowed).

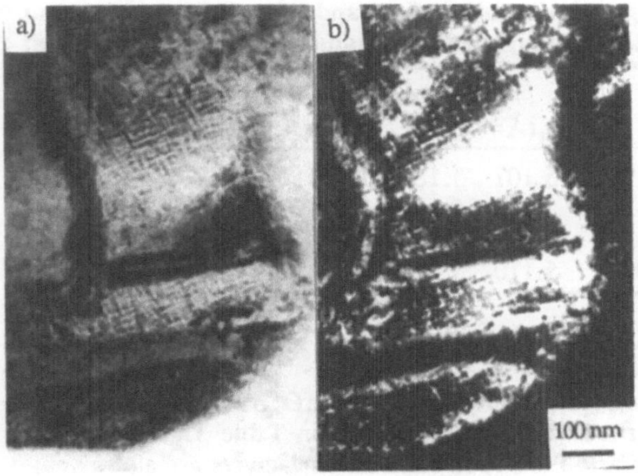

Fig. 3 TEM micrograph of O'/ZrO_2 material (a) bright field and (b) dark field, showing a "tweed"-type structure usually associated with the homogeneous precipitation of *t*-ZrO_2 material within the *c*-matrix.

The absence of the transformable tetragonal phase in these O'-ZrO_2 composites indicates that improvement in the fracture toughness of these materials cannot be achieved through the martensitic transformation toughening mechanism. In this respect O'-ZrO_2 materials behave in a comparable way to Si_3N_4-ZrO_2 and ß'sialon-ZrO_2 ceramics.

O'-ß'-zirconia reaction

Reactions involving O', ß' sialon and zirconia were examined for four compositions, in which the proportions of O' and ß' were varied (Table 1). X-ray results of powder samples showed only small amounts of *m*-ZrO_2 in the O'-ß' composites, but the transformation from tetragonal to monoclinic was completely suppressed if no O' was present in the matrix. Tetragonal zirconia was observed in all four samples, however, the *c*/*a* ratio in all samples gradually decreased as the composition approached the ß' side even though the Y_2O_3:ZrO_2 ratio of 4:25 was kept constant in all samples, indicating an increasing stability of the tetragonal phase with increasing ß' content (Table 2). One of the major differences between the O' and ß' phases is that the nitrogen content in ß' is much higher than that of O'. Therefore, the cell dimension change of zirconia in O'-ß' composites can be at least partly related to the fact that there is more nitrogen to react with the zirconia when more ß' phase is introduced into the matrix. The absence of transformable tetragonal phase in these composites clearly prevents any improvement in fracture toughness (Table 2).

Table 1. Compositions and X-ray results for zirconia-containing O'-ß' composites

	Starting composition (w/o)				X-ray results				
Samples	O'	ß'	ZrO_2	Y_2O_3	O'	ß'	*t(t')*	*m*	ZrN
BOZ8	44	31	25	4	ms	ms	s	vw	
BOZ9	30	45	25	4	m	s	s	vw	
BOZ10	10	65	25	4		vs	s		
BOZ11		75	25	4		vs	s		mw

Table 2. Cell dimensions of tetragonal zirconia before and after heat-treatment at 800°C for 2h in air and hardness (H_v) and toughness (K_c) values before heat-treatment

Samples	Before Oxidation a_t (Å)	c_t (Å)	c_t/a_t	After oxidation a_t (Å)	c_t(Å)	c_t/a_t	H_v (GPa)	K_c (MPam$^{1/2}$)
BOZ8	5.108	5.189	1.0159	5.107	5.184	1.0151	12.2	4.4
BOZ9	5.111	5.184	1.0143	5.108	5.184	1.0149	12.8	4.5
BOZ10	5.116	5.189	1.0143	5.110	5.187	1.0151	12.4	4.1
BOZ11	5.121	5.172	1.0100	5.110	5.178	1.0133	12.8	4.2

Reaction Between Zirconia and Nitrogen

Selected results of sintering 1 g pellets of partially stabilized ZrO_2(3Y) with or without additional ZrN at different temperatures in nitrogen are shown in Table 3. The nitrogen content of the products analyzed by a Carlo Erba 1106 combustion analyzer are also given.

Table 3. Product characterization of samples sintered in nitrogen for 2 hours at various temperatures

	Temperature (°C)	Phases	Cell dimension of ZrO_2 (Å) a_t	c_t	a_c	c_t/a_t	N w/o
	ZrO_2(3Y)						
	unsintered	*t*	5.102	5.178		1.0149	
1	1600	*t*'(vs), *m*(vw)	5.109	5.156		1.0092	0.44
2	1700	*c*(vs), *t*'(mw)			5.121		1.05
3	1800	*c*(vs)			5.118		1.50
	ZrO_2(3Y)+10 w/o ZrN						
4	1600	*c*(vs), *m*(w)			5.116		1.54
5	1700	*c*(vs), *m*(w)			5.115		1.82
6	1800	*c*(vs), *m*(w), ZrN(w)			5.116		1.96

The original unsintered ZrO_2(3Y) powder shows a tetragonal unit cell with a significant difference between the *a* and *c* dimensions. Even without additional ZrN, the sintering of zirconia in a nitrogen atmosphere clearly enables nitrogen to enter the zirconia lattice, providing an additional stabilization of the crystal structure. As a result, the tetragonal zirconia becomes increasingly stable and in some cases transformed to the cubic form. Moreover, martensitic transformation could not be initiated by mechanical grinding although tetragonal zirconia was observed on the X-ray photographs. Clearly, the original metastable tetragonal ZrO_2 became non-transformable after sintering in nitrogen. TEM observations on the zirconia particles in ZrO_2-ß'sialon composites showed that the tetragonal zirconia grains in the materials exhibited a fine scale lath-like intragranular microstructure, exactly as in the

t' phase [6-8]. In the ZrO_2-Y_2O_3 system, the phase transition from $m \rightarrow t \rightarrow t' \rightarrow c$ zirconia is promoted by continuous introduction of a stabilizing cation (e.g. Y^{3+}) into the zirconia lattice, resulting in a gradual reduction in the c/a ratio [9]. It is believed that the oxygen vacancies created by the dissolved nitrogen are disordered and behave in a similar way to the stabilization achieved by cation addition. The pre-existing vacant sites in ZrO_2(3Y) are essential for preventing further ordering of the nitrogen related vacancies because the reaction between m-ZrO_2 and nitrogen produces neither tetragonal nor cubic zirconia but a zirconium oxynitride phase, $Zr_7O_{11}N_2$ [7,10].

The incorporation of an additional 10 w/o of ZrN into the ZrO_2(3Y) powder encourages more nitrogen to dissolve in the zirconia lattice. For complete reaction, 10 w/o of ZrN can introduce 1.34 w/o of nitrogen into the samples, which, together with nitrogen provided by the sintering atmosphere, significantly increases the total nitrogen content of the zirconia. This is proved by the fact that nitrogen containing cubic zirconia forms at 1600 °C when ZrO_2(3Y) reacts with both ZrN and N_2, whereas zirconia forms as tetragonal (t') if the ZrO_2(3Y) reacts merely with nitrogen gas (Table 3). Further increase in the sintering temperature up to 1800 °C leads to the formation of ZrN in the sample, and in contrast, molecular nitrogen can only stabilize the cubic phase at this temperature. The appearance of ZrN in the sample indicates a limit to the nitrogen solubility in zirconia before the fluorite structure breaks down. These results suggest that the reaction between atomic nitrogen and zirconia exhibits a more negative free energy than that between molecular nitrogen and zirconia, with nitrogen atoms diffusing more easily in the former case. In the same way, it is clear that the same type of reaction occurs in compositions containing both zirconium nitride and partially stabilized zirconia. Obviously the driving force for the reaction depends on the type of nitride, the sintering temperature and the severity of the reducing environment. Trace amounts of m-ZrO_2 are consistently found in all samples originally containing ZrN (Samples 4-6); this clearly does not arise from martensitic transformation during post-grinding for X-ray sample preparation since there is only cubic zirconia in the samples; phase equilibria in the Zr-Y-O-N system need more detailed study to understand this phenomenon.

Further evidence for nitrogen dissolution in the zirconia phase is provided by oxidation results at an intermediate temperature. It was found that after heat-treatment at 800 °C for 2 hours in air, the c/a ratio of what was previously t' phase increased significantly, showing a reduction in non-transformability. It is assumed that the change in zirconia lattice dimensions during oxidation at 800 °C is merely due to partial substitution of nitrogen by oxygen since the oxidation of nitrogen-containing zirconia starts at 700 °C [8].

Discussion
In nitride based composites, zirconia grains are normally surrounded by both a nitrogen-containing matrix and a nitrogen sintering atmosphere and therefore the reaction between zirconia and nitrogen is unavoidable although it could be prevented in the gas phase [11] by controlling the oxygen potential at the sintering temperature. However this reaction has not been fully studied particularly when Y_2O_3 is also present in the zirconia phase. It has been reported that nitrogen stabilizes zirconia in the cubic form [12], by partial substitution of nitrogen for oxygen in the anion lattice with simultaneous creation of oxygen vacancies, exactly similar to the case when lower-valence cations are added. However, the related conclusion that nitrogen can also make the tetragonal phase non-transformable, does not appear to have been deduced, and yet this is fundamental to all attempts at achieving

transformation toughening in silicon nitride based materials.

Partially stabilized zirconia can dissolve a certain amount of nitrogen without deviating from its original fluorite structure when this phase reacts with either atmospheric nitrogen or a solid nitride at high temperature. As a result, the zirconia becomes martensitically non-transformable even though it may still possesses tetragonal symmetry. In a nitride based ceramic composite (e.g. sialons), the reaction between partially stabilized zirconia and a solid nitride phase can be accelerated because solid-state reaction offers the easiest route for partial substitution of oxygen by nitrogen. This is particularly the case when the nitride phase accommodates large amounts of nitrogen (e.g. ß'-sialons); it is therefore, not possible for zirconia to toughen ß'-sialon ceramics by a transformation toughening mechanism.

CONCLUSIONS

Dissolution of nitrogen in partially stabilized zirconia results in additional stabilization of the zirconia which prevents the martensitic transformation from $t \rightarrow m$ under mechanical stress. As a result, the expected advantage of transformation toughening arising from the inclusion of zirconia particles in O'-ZrO_2 composites (fabricated using the NPS technique) and O'-ß'-ZrO_2 composites (prepared by normal powder routes) is neutralized. Therefore, more careful chemical processing and other toughening mechanisms must be considered in an attempt to improve toughness of zirconia-containing silicon nitride based composites.

REFERENCES

[1]. N.Claussen and J.Jahn, *J. Am. Ceram. Soc.*, 61(1-2)94,1978
[2]. T.Ekström, L.K.L.Falk and *J. Mater. Sci. Lett.*, 9,823,1990
[3]. T.Ekström, *J. Mater. Sci.*, private communication 1990
[4]. K.Rundgren, *Ph.D thesis*, Chalmers Univ. Techn., Göteborg, 1988
[5]. V.Lanteri, R.Chaim and A.H.Heuer, *J. Am. Ceram. Soc.*, 69,C-258,1986
[6]. L.K.L.Falk & M.Holmström, *Euro-Ceramics*, Vol.1, ed. by G. de With et al., p.373,1989
[7]. F.F.Lange, L.L.L.Falk & B.I.Davis, *J. Mater. Res.*, 2(1)66,1987
[8]. Y.Cheng and D.P.Thompson, *J. Am. Ceram. Soc.*, 74(5)1991
[9]. Y.Cheng and D.P.Thompson, *J. Mater. Sci. Lett.*, 9,24,1990
[10]. Y.Cheng and D.P.Thompson, *Special Ceramics 9*, to be published
[11] A.K.Tjernlund, R.Pompe, M.Holmström and R.Carlsson, *Special Ceramics 8*, Brit. Ceram. Proc. 37, 29, 1986
[12]. N.Claussen et al., *J. Am. Ceram. Soc.*, 61(7-8)369,1978

FABRICATION OF SILICON NITRIDE CERAMICS BY MICROWAVE HEATING*

T. N. Tiegs and J. O. Kiggans
Oak Ridge National Laboratory, Oak Ridge, TN 37831-6069

ABSTRACT

Microwave processing of silicon nitride-based ceramics is currently being examined to identify aspects that may accelerate densification or produce unique microstructures. Because the microwaves preferentially couple with the sintering additives and the intergranular phases, relatively high additive contents are required to produce samples that will densify and are crack-free. Improved densification of green powder compacts and enhanced weight losses are observed during microwave heating.

INTRODUCTION

Numerous applications exist for silicon nitride-based ceramics in advanced heat engines for such items as turbocharger rotors, valves, and cam roller followers.[1] Conventional processing of silicon nitride uses high α-phase Si_3N_4 starting powders mixed with sintering aids, such as MgO, Y_2O_3, and Al_2O_3 (either as powders or other chemical precursors). The mixtures are consolidated into a shape and then heated to temperatures of 1700-1900°C to densify in nitrogen at one atmosphere pressure or greater (≥0.1 MPa). The sintering temperature is dependent on the type and quantity of sintering additives utilized. The nitrogen pressure, in turn, is dictated by the temperature and must be sufficiently high to suppress decomposition of the silicon nitride, which is significant at temperatures >1800°C.[2] Alternatively, silicon powder, mixed with the sintering additives, is initially nitrided (at temperatures of 1200-1450°C) and then heated to elevated temperatures to densify in a manner similar to that described for the starting α-Si_3N_4 powders.

Microwave heating is a unique method for volumetric thermal heating of ceramic materials.[3] Differences in microwave and conventional heating techniques have been reviewed by Sutton[3] and Binner.[4] Previous studies have shown that silicon nitride-based materials are relatively low loss materials and microwave heating occurs by coupling to the sintering additives or grain boundary phases.[5-7] Improvements in densification, accelerated nitridation of silicon, and improved high-temperature mechanical properties with thermal annealing have all been reported for microwave processed silicon nitrides in earlier investigations.[8-10] In the present research, fabrication of silicon nitride ceramics was further examined to optimize the sintering behavior for certain compositions in the microwave.

---

* Research sponsored by the U. S. Department of Energy, Assistant Secretary for Conservation and Renewable Energy, Office of Transportation Technologies, as part of the Ceramic Technology for Advanced Heat Engines Project of the Advanced Materials Development Program, under contract DE-AC05-84OR21400 with Martin Marietta Energy Systems, Inc.

EXPERIMENTAL

Appropriate quantities of Si_3N_4,* Y_2O_3,# Al_2O_3,† SiAlON,†† SiC,## or Si** were milled together in isopropanol for 2-6 h and then dried with constant stirring. Samples for the sintering studies were fabricated by isopressing ~100 g of the powder mixture into cylinders with 20 MPa pressure.

All densities after firing were determined by the Archimedes method. Microwave sintering was performed in either a 2.45-GHz or a 28-GHz furnace with the samples packed in Si_3N_4*** powder (with 2 wt. % Y_2O_3-2 wt. % SiC) in an insulation package previously described.[8] Temperatures were monitored by a thermocouple surrounded by the specimens. Conventional sintering and annealing was done in a graphite element furnace with the specimens packed in the same powder mixture as the microwave heated materials. All tests were performed at 0.1 MPa (1 atmosphere) nitrogen.

RESULTS AND DISCUSSION

A summary of the results from the sintering of silicon nitride powder compacts is given in Fig. 1. As shown, higher densities were achieved by the microwave sintered materials as compared to those fired in a conventional furnace. Conventionally, silicon nitride is densified by liquid phase sintering, and this is also true for the microwave heated materials. While the microwave sintered materials showed improved densification, they also showed a greater susceptability to cracking, especially at the lower sintering additive contents. This is a result of the heating of the sintering additives and subsequent heating of the surrounding silicon nitride grains. In a powder compact with low thermal conductivity non-uniform heating can lead to a sequence of "hot-spots," followed by differential shrinkage and finally cracking. At the higher additive contents the heating is more uniform and cracking is less of a problem.

One method to improve the uniform heating of powder compacts is to add particulates into the samples that couple well with the microwave energy.[3,9] Silicon carbide, SiAlON, and silicon powders all couple well with microwaves. The silicon has an additional advantage in that it reacts in-situ to form silicon nitride. Shown in Fig. 2 are the results of sintering with the addition of coupling additives. Improved densification and reduced cracking were observed. In the case of the compacts made completely from the SiAlON powders, the coupling was so uniform that no cracking was observed, and the samples easily achieved full density.

Another method to decrease the heating variability is to improve the microwave field uniformity by utilizing higher frequencies. This has been demonstrated to improve the sintering and reduce the cracking of zirconia powder compacts.[11] The results of sintering silicon nitride-based materials at 28-GHz as compared to 2.45-GHz, shown in Fig. 2, illustrate the increase in densification with the higher microwave frequency.

---

* Ube, Japan; Grade E-10
# Molycorp, White Plains, NY ;Grade 5600
† Reynolds, Malakoff, TX ;Grade RC-HP DBM
†† Vesuvius Crucible, Pittsburgh, PA; Grade AA, Sinterable
Starck, Berlin, Germany;Grade A-10
** Elkem, Buffalo, NY ; Grade Silicon HQ
*** Elkem, Buffalo, NY ; Grade Silicon Nitride HQ

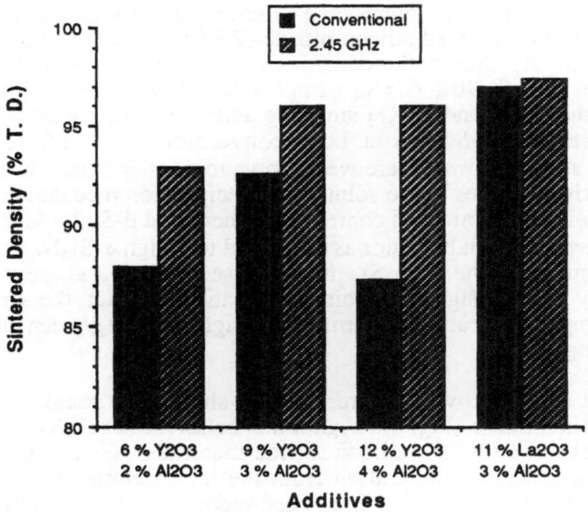

Figure 1. Summary of results on sintering of silicon nitride with various additive contents. All samples sintered at 1750°C for 1 hour. Theoretical densities for the samples are 3.26 g/cm^3, 3.33 g/cm^3, 3.38 g/cm^3, and 3.38 g/cm^3 (as viewed from left-to-right).

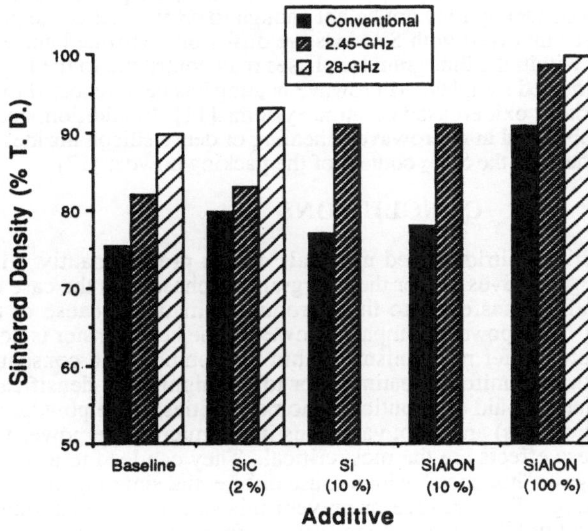

Figure 2. Summary of results on the sintering of silicon nitride with various additives to aid in coupling in the microwave furnace. The baseline material is Si_3N_4-6% Y_2O_3-2% Al_2O_3. The SiC, Si, and SiAlON additives are based on weight percent addition to the baseline material. The SiAlON powder is a sinterable grade containing Y_2O_3. All samples sintered at 1750°C for 1 hour. Theoretical densities for the samples are 3.26 g/cm^3, 3.25 g/cm^3, 3.26 g/cm^3, 3.26 g/cm^3, and 3.19 g/cm^3 (as viewed from left-to-right).

Because silicon powder couples well in the microwave furnace, an expanded study was made to investigate sintered reaction-bonded silicon nitride (RBSN). The results of that study are summarized in Fig. 3. As shown, the densities were dependent on the initial sample composition with the 10% Si-90% Si_3N_4 samples showing the best sintering behavior. In the presence of the Y_2O_3 and Al_2O_3 sintering additives, nitridation of Si is known to produce predominantly ß-Si_3N_4 in both conventional and microwave heating.[12,13] High α-Si_3N_4 starting powders are well known to densify better than high ß-Si_3N_4 powders due to the driving forces in the solution-reprecipitation mechanism.[14] Consequently, in the samples with large initial Si contents, the increased ß-Si_3N_4 levels are believed to have reduced the densification behavior as compared to a high α-Si_3N_4 starting powder. For the samples containing the 10% Si, the increased ß-Si_3N_4 content after nitridation was not sufficiently high enough to inhibit densification. In fact, the samples may have benefited from the improved heating uniformity and higher starting green density afforded by the reaction bonding.

Comparisons of conventional and microwave sintering also showed increased weight losses during sintering for the latter case (Fig. 4). Again this behavior is believed due to the manner in which heat is absorbed by the silicon nitride materials. Conventionally, weight losses are associated with SiO volatilization from the intergranular phases and decomposition of the silicon nitride.[2,15] It has long been recognized that to minimize weight losses and achieve good densification during sintering, the materials need to be packed in powder beds to provide a protective environment.[16] In the case of microwave sintering, this appears to be of critical importance and will be an area of continued study. The high weight losses can be attributed to a couple of factors. In the microwave, the samples are the hottest part in the system and there is typically a steep thermal gradient existing from the sample outward through the insulation. Thus, the powder beds are at a lower temperature and consequently may not be hot enough to provide adequate protection. Additionally, the processes involved with SiO loss are diffusion controlled and enhanced diffusion of Si-O species within the intergranular phases may contribute to the high weight losses for the microwave heated samples. Microwave heating has been reported to increase diffusional processes in other oxide-based ceramic systems.[11] In addition, accelerated weight losses have been observed in microwave annealing of dense silicon nitrides and was found to be highly dependent on the SiO_2 content of the packing powder.[17]

CONCLUSIONS

Microwave heating of silicon nitride-based materials occurs predominantly via power absorbtion by the sintering additives and/or the intergranular phases. In the case of dense materials, the heat is readily transferred to the surrounding matrix because of the high thermal conductivity.[18] With powder compacts, however, the heat transfer is dominated by convection and radiant transfer mechanisms within the compact and consequently is very slow. Problems with non-uniform heating prior to any significant densification can occur due to imperfect sintering aid distribution, inhomogeneities in the powder compact (such as density or agglomerates) and also, variations in the microwave power field. At low additive contents, these effects are the most critical. They can lead to a sequence of localized thermal runaways or "hot-spots" which cause differential sintering and shrinkage, and finally sample cracking. To minimize or prevent this series of events, one can (1) increase the additive content or (2) increase the microwave frequency. In the present study, we observed improved heating of powder compacts with additions of Si, SiC, and SiAlON particulates. The silicon additions were nitrided to silicon nitride prior to sintering. Increasing the sintering additive contents (in this case, Y_2O_3 and Al_2O_3) also exhibited improved heating and densification. By increasing the microwave frequency from 2.45 to

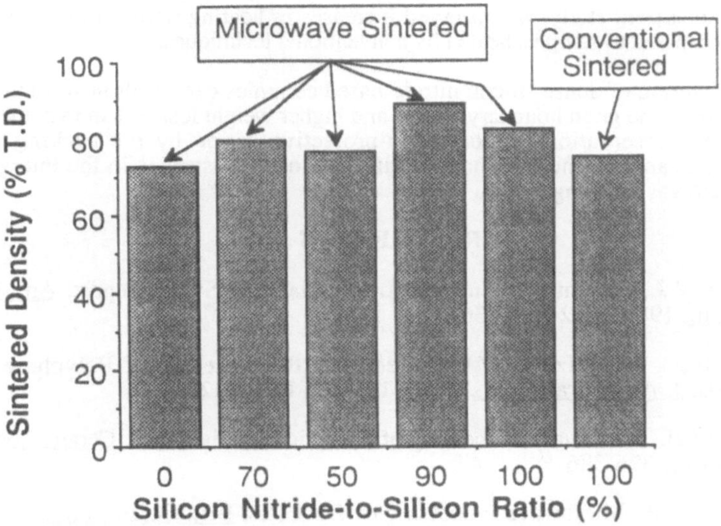

Figure 3. Summary of results on microwave sintering of reaction-bonded silicon nitride. All samples containing Si were initially nitrided to >85% reaction in a 24 h cycle to 1350°C followed by a sintering step to 1750°C for 1 h. The samples containing no Si were just subjected to the 1750°C for 1 h. All samples were fabricated to contain 6% Y_2O_3-2% Al_2O_3 as the final composition. Theoretical densities for the samples are 3.26 g/cm^3.

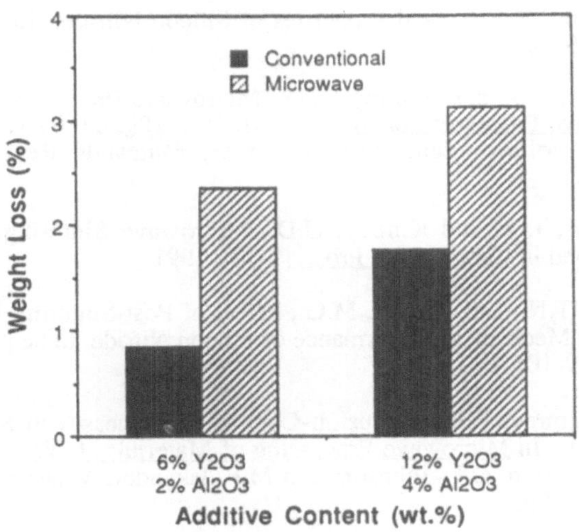

Figure 4. Comparison of weight losses during sintering of silicon nitrides with different additives. Samples sintered at 1750°C for 1 h. Microwave heating at 2.45-GHz.

28-GHz, in the current study, we observed increases in heating uniformity and sintering behavior as compared to samples heated by conventional techniques.

Microwave sintering of dense silicon nitride-based ceramics can result in substantial SiO volatilization from the grain boundary phases and higher weight losses than in conventional heating. These observations indicate non-protective action by the packing powder surrounding the samples and/or enhanced diffusion of Si-O species in the intergranular phases by microwave heating.

REFERENCES

1. Tennery, V.J., Ceramics in Engines-An International Status Report, Am. Ceram. Soc. Bull., 1989, **68**[2], 362-365

2. Lange, F.F., Volatilization Associated with the Sintering of Polyphase Si_3N_4 Materials, J. Am. Ceram. Soc., **65**[8], 1982, C-120-C-121.

3. Sutton, W.H., Microwave Processing of Ceramic Materials, Am. Ceram. Soc. Bull., 1989, **68**[2], 376-386.

4. Binner, J. G.P., Microwave Processing of Ceramics, Brit. Ceram. Proc., 1990, **45**, 97-108.

5. Clarke, D.R. and Ho, W.W., Effect of Intergranular Phases on the High-Frequency Dielectric Losses of Silicon Nitride Ceramics. In Additives and Interfaces in Electronic Ceramics, Advances in Ceramics, Vol. 7, ed. M.F. Yan and A.H. Heuer, Amer. Ceram. Soc., Westerville, OH, 1983, pp. 246-252.

6. Ho, W.W., High-Temperature Dielectric Properties of Polycrystalline Ceramics. In Microwave Processing of Materials, Vol. 124, eds. W.H. Sutton, M.H. Brooks, and I.J. Chabinsky, Materials Research Society, Pittsburgh, PA, 1988, pp. 137-148.

7. Ohno, H. and Katano, Y., Electrical Properties of Silicon Nitride. In Mater. Sci. Forum, Vol. 47, 1989, pp. 215-227.

8. Tiegs, T.N., Kiggans, J.O., and Kimrey, H.D., Microwave Processing of Silicon Nitride. In Microwave Processing of Materials-II, Vol. 189, ed. by W.B. Snyder, W.H. Sutton, D.L. Johnson, and M.F. Iskander, Materials Research Soc., Pittsburgh, PA 1991.

9. Tiegs, T.N., Kiggans, J.O., and Kimrey, H.D., Microwave Sintering of Silicon Nitride, to be published in Ceram. Eng. Proc., [9-10], 1991.

10. Ferber, M.K., Tiegs, T.N., and Jenkins, M.G., Effect of Post-Sintering Microwave Treatments Upon the Mechanical Performance of Silicon Nitride, to be published in Ceram. Eng. Proc., [9-10], 1991.

11. Janney, M.A. and Kimrey, H.D., Diffusion-Controlled Processes in Microwave-Fired Oxide Ceramics. In Microwave Processing of Materials-II, Vol. 189, ed. by W.B. Snyder, W.H. Sutton, D.L. Johnson, and M.F. Iskander, Materials Research Soc., Pittsburgh, PA, 1991.

12. Antona, P.L., Giachello, A., and Martinengo, P.C., Nitridation of Silicon in Presence of Oxides. In Ceramic Powders, ed. P. Vincenzini, Elsevier Sci. Pub., Amsterdam, 1983, pp. 753-766.

13. Kiggans, J. O., Hubbard, C.R., Steele, R.R., Kimrey, H.D., Holcombe, C.E., and Tiegs, T.N., Characterization of Silicon Nitride Synthesized by Microwave Heating, to be published in Ceramic Transactions, Microwaves: Theory and Application in Materials Processing, ed. D.E. Clark, F.D. Gac, and W.H. Sutton, American Ceramic Society, Westerville, OH, 1988.

14. Lewis, M.H., Leng-Ward, G. and Jasper, C., Sintering Additive Chemistry in Controlling Properties of Nitride Ceramics. In Ceramic Transactions, Ceramic Powder Science, II, B., ed. G.L. Messing, E.R. Fuller, Jr., and H. Hausner, American Ceramic Society, Westerville, OH, 1988, pp. 1019-1033.

15. Hirosaki, N. and Okada, A., Change in Oxygen Content of Y_2O_3-Nd_2O_3-Doped Silicon Nitride During Firing, J. Am. Ceram. Soc., **72**[12], 1989, 2359-2360.

16. Pompe, R. and Carlsson, R., Sintering of Si_3N_4-Based Materials Using the Powder Bed Method. In Progress in Nitrogen Ceramics, ed. F.L. Riley, Martinus Nijhoff Pub., Netherlands, 1983, pp. 219-224.

17. Tiegs, T.N., Ferber, M.K., Kiggans, J.O., More, K.L., Hubbard, C.M., and Coffey, D.W., Microstructure Development During Microwave Annealing of Dense Silicon Nitride, to be published in Ceramic Transactions, Microwaves: Theory and Application in Materials Processing, ed. D.E. Clark, F.D. Gac, and W.H. Sutton, American Ceramic Society, Westerville, OH, 1988.

18. Johnson, D.L., Microwave Heating of Grain Boundaries in Ceramics, J. Am. Ceram. Soc., **74**[4], 1991, 849-850.

PREPARATION OF Ln-α-SIALON POWDERS AND SINTERING BEHAVIOR

Yamada,T., Nakayasu,T., Takahashi,T., Yamao,T. and Kohtoku,Y.
Corporate Research & Development, UBE Industries, Ltd., Ube, Japan

ABSTRACT

Various Ln-α-Sialon (where Ln= yttrium and lanthanide elements) powders were prepared by reacting mixtures of amorphous Si_3N_4, AlN, Al_2O_3 and Ln_2O_3 powders at high temperatures. The x-value in the chemical formula of $Ln_x(Si,Al)_{12}(O,N)_{16}$ was determined by the Rietveld analysis of the X-ray diffraction pattern and was found to increase with decreasing radii of Ln^{3+} ions. Pressureless sintering of the mixtures of Ln-α-Sialon and α-Si_3N_4 powders was performed and highly dense Ln-SiAlON-based ceramics were obtained. Bending strength of Ln-Sialon-based ceramics was increased with decreasing radii of Ln^{3+} ions. It was concluded that thermo-mechanical properties of Sialon-based materials could be much improved by the selection of Ln elements.

INTRODUCTION

Silicon Nitride (Si_3N_4) and related Sialon-based ceramics have attracted attention in the field of structural applications at high temperatures because of their outstanding properties such as high strength and toughness. α-Sialon is the solid solution with expanded α-Si_3N_4 crystal structure. The chemical compositon is generally formulated as $M_x(Si,Al)_{12}(O,N)_{16}$, where M is a "modifying cation" such as Li, Mg, Ca, Y or lanthanide (Ln) elements except La and Ce, and $0< x \leqq 2$ [1-4]. Polyphase Sialon-based ceramics comprising both α-Sialon and β-Sialon phases represented by $Si_{6-z}Al_zO_zN_{8-z}$ $(0< z \leqq 4.2)$ have been reported to exhibit better mechanical properties, especially better thermal shock resistance and high-temperature strength [5-7].

Although yttrium is commonly used as a modifying cation M, Y-containing Sialon-based ceramics show low oxidation resistance at high temperatures [8,9], which limits the application of these materials for high-temperature engine components. Therefore, it is important to improve the oxidation resistance of the Sialon-based ceramics by optimizing the microstructure and composition of the grain boundary phase.

It is our purpose to report the experimental results on the preparation of Ln-containing α-Sialons (referred to as Ln-α-Sialon hereinafter) powder from amorphous Si_3N_4, their sintering behavior and the characteristics of resultant Ln-Sialon-based ceramics made from mixtures of Ln-α-Sialon and α-Si_3N_4 powders. The effect of ionic radii of Ln elements on the thermo-mechanical properties of Ln-Sialon-based ceramics will be discussed.

EXPERIMENTAL PROCEDURE

Powder Preparation and Characterization

Si_3N_4 powders used in this experiment were produced by UBE's own imide decomposition process [10,11]. In this process, amorphous Si_3N_4 or intermediate compounds between Si_3N_4 and silicon diimide ($Si(NH)_2$) [12,13], either of which was referred to as amorphous Si_3N_4 hereinafter, were produced by calcining $Si(NH)_2$ at about 1100℃, and then converted to α-Si_3N_4 powder by heating at a higher temperature. A flow diagram of the process for producing α-Si_3N_4 and Ln-α-Sialon powders is shown in Fig. 1.

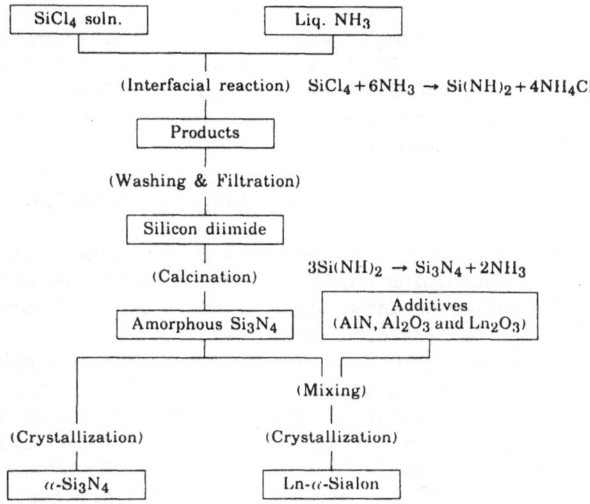

FIGURE 1 Flow diagram of the process for producing α-Si_3N_4 and Ln-α-Sialon powders by UBE'S imide decomposition method

Amorphous Si_3N_4, AlN (Tokuyama Soda Co.), Al_2O_3 (AKP-30 from Sumitomo Chemical Co.) and Ln_2O_3 (Shin-Etsu Chemical Co.) powders were used as the starting materials for the preparation of Ln-α-Sialon powders. After mixing these powders with intended Ln-α-Sialon compositions by vibration milling, the powder mixtures were heated in nitrogen atmosphere up to 1600℃ to crystallize [14,15].

Nitrogen and oxygen contents of Si_3N_4 and Ln-α-Sialon powders were determined by alkali fusion and inert gas fusion methods, respectively. The contents of cations or metal impurities of the powders were analyzed by inductively coupled plasma (ICP) emission spectroscopy after the samples were decomposed by the digestion with hydrofluoric acid in a teflon pressure vessel (UNISEAL equipment) and dissolved in hydrochloric acid solution. The single point BET method was used to measure the specific surface area of the powders.

Phase identification of the powders was performed by X-ray powder diffractometry with Ni-filtered CuKα radiation. The lattice constant and crystal structural parameter of $Ln_x(Si,Al)_{12}(O,N)_{16}$ are determined by the application of Rietveld analysis of their X-ray powder diffraction data [16-18].

Sintering of Ln-Sialon and Evaluation

Powders of Ln-α-Sialon and α-Si_3N_4 (SN-E10 grade) in the weight ratio of 40 to 60 were mixed by ball-milling in ethanol for 48 hours, and then the slurry was dried to obtain the mixture. The mixed powder was die-pressed at 20 MPa followed by cold isostatic pressing at 150 MPa in a rubber bag. The green compacts were embedded in the powder composed of Si_3N_4, Al_2O_3, Ln_2O_3 and h-BN in a crucible and sintered at 1780 ℃ for 2 hours in nitrogen atmosphere at normal pressure.

Bulk density of the sintered Ln-Sialon was measured by Archimedes method and the theoretical density was calculated by the rule of mixtures. The phase of the sintered Ln-Sialon was identified by X-ray powder diffraction. The microstructure of the fracture surface of the sintered Ln-Sialon was observed by scanning electron microscopy (SEM).

Bending strength was measured on a 4-point loading fixture at room temperature and 1200℃ according to JIS-R1601 method. As a test of the oxidation resistance of the Ln-Sialon-based ceramics, the samples were heated in air at 1350℃ for 48 hours and the weight gain of the samples caused by oxidation was measured.

RESULTS AND DISCUSSION

The amorphous Si_3N_4 powder was so active that it easily reacted with the compounds containing Ln, Al, O and N elements to crystallize into Ln-α-Sialon. Characteristics of Ln-α-Sialon powders obtained were compared with those of α-Si_3N_4 powder (SN-E10) in Table 1.

The formation of Ln-α-Sialon powder was much influenced by the ionic radii of Ln elements which were accommodated in the interstices of the expanded α-Si_3N_4 crystal lattice. The relationship between the phase composition of the reaction products and the ionic radii of Ln elements is shown in Fig. 2. Various Sialon powders which were mainly composed of Ln-α-Sialon were prepared at 1600℃. However, La_2O_3 did not lead to the stable formation of an α-Sialon structure. These powders also contained small amounts of α-Si_3N_4 and β-Sialon phases with either phase sharing less than 10% by weight.

The x-value in the formula of $Ln_x Si_{12-4.5x} Al_{4.5x} O_{1.5x} N_{16-1.5x}$ and the lattice parameter of Ln-α-Sialon can be determined by Rietveld analysis of their X-ray powder diffraction patterns. The effects of ionic radii of Ln elements on the x-value and the lattice constant of Ln-α-Sialon powders are shown in Fig. 3. The lattice constant of Ln-α-Sialon was increased with increasing radii of Ln^{3+} ions. In contrast with this, the x-value of Ln-α-Sialon was decreased with increasing ionic radii from 0.86 to 0.89Å for Ln=

TABLE 1 Typical characteristics of Si_3N_4 and Ln-α-Sialon powders produced by UBE process

Grade	Si_3N_4 (SN-E10)	Ln-α-Sialon (Under research)
Nominal formula	α-Si_3N_4	$Ln_{0.5}Si_{9.75}Al_{2.25}O_{0.75+Z}N_{15.25-Z}$
Chemical analysis		
Si (wt%)	60.0	(44.2 ~ 40.1)*
Al (wt%)	< 0.005	(9.8 ~ 11.8)*
Ln (wt%)	–	(7.2 ~ 12.7)*
N (wt%)	38.7	(33.8 ~ 31.3)*
O (wt%)	1.3	(4.5 ~ 5.0)*
		(Excess amount of oxygen 3.0)
C (wt%)	0.15	0.15
Fe (ppm)	< 50	< 50
Ca (ppm)	< 20	< 20
Grain morphology	Equiaxed	Equiaxed or lump
Specific surface area (m^2/g)	11	~ 2
Crystal Phase (wt%)	α-phase > 97	α-Sialon > 80

* Depending on the atomic weight of Ln elements

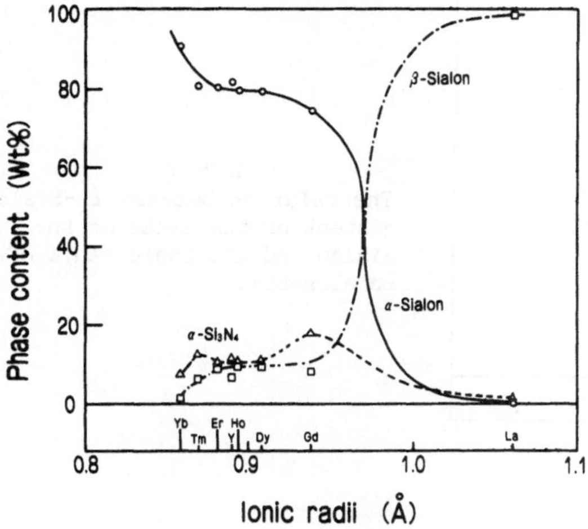

FIGURE 2
The relation between phase composition of the reaction products and the ionic radii of Ln elements.

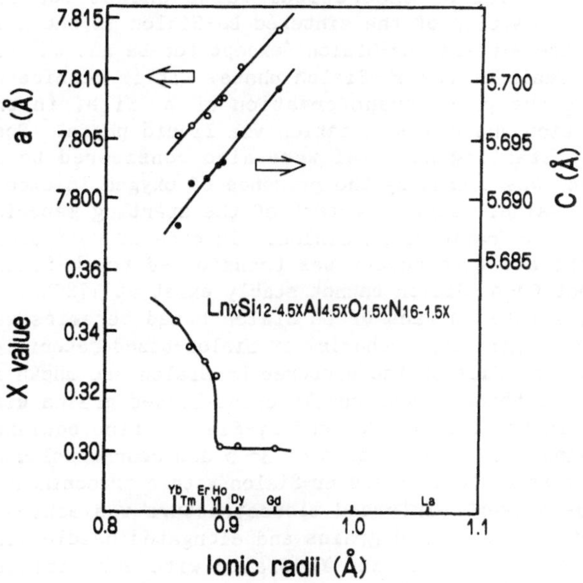

FIGURE 3
Effects of ionic radii of Ln elements on the x-value and lattice constant of Ln-α-Sialon powders.

Y, Er, Tm and Yb. In the case of Ho-, Dy- and Gd-Sialon, the x-value was not dependent on ionic radii. It was the constant value of 0.30 which almost agreed with the lower limit of solid solubility of α-Sialon reported by Izumi et al. [17,18]. As the Ln^{3+} ions occupy the large closed interstices of the (Si,Al)-(N,O) network, Ln^{3+} ions with smaller radii can be solid-solved more easily in the α-Sialon lattice.

Densification and phase transformation of Sialon-based ceramics were examined by the pressureless sintering of the mixture of 40 wt% of Ln-α-Sialon and 60 wt% of SN-E10 without any other additives. The said starting powders were so sinteractive that Ln-Sialon-based ceramics were consolidated to fully

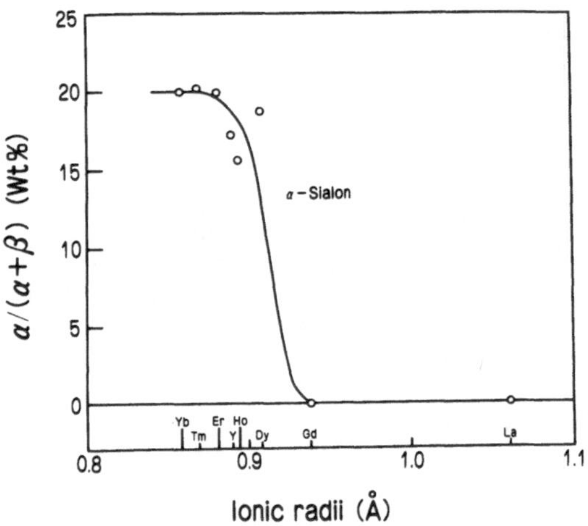

FIGURE 4
The relation between α-Sialon content of the sintered Ln-Sialon and the ionic radii of Ln elements.

dense bodies for all the Ln elements excluding lanthanum. Bulk density of the sintered Ln-Sialon was 98.0~98.8% of theoretical value except for La-containing material. Phase composition of the sintered Ln-Sialon is shown in Fig. 4. α-Sialon content of the sintered Ln-Sialon (except for La and Gd) was about 20% by weight and the remainder was β-Sialon phase. It is considered that β-Sialon was formed by the phase transformation of α-Si_3N_4 in the starting material by the solution and reprecipitation via liquid phase. Some parts of Ln-α-Sialon in the starting material were also considered to be transformed to β-Sialon, which was caused by the presence of oxygen in excess amounts to the stoichiometric value. Oxygen content of the starting material had much influence on the α to β ratio of Ln-Sialon. In case of Gd-Sialon, all the α-Sialon phase in the starting powder was transformed to β-Sialon after sintering. It seems that Gd-α-Sialon cannot stably exist at 1780℃.

Microstructure of the fracture surface of Ln-Sialon-based ceramics was observed to speculate on the strengthening mechanism of Sialon-based materials. SEM photographs of the fracture surface of the sintered Ln-Sialon are shown in Fig. 5. There were observed the three-dimensionally cross-linked grains with different shapes and sizes. In the case of Gd- and Dy-Sialon, fine equiaxed grains, small needle-like grains with aspect ratio of 3~5 and coarse columnar grains were observed. In the case of Ho-, Y- and Er-Sialon with a predominantly homogeneous texture, in-situ whisker reinforced microstructure was achieved owing to the cross-linking of fine equiaxed grains and elongated needle-like grains with aspect ratio of 6~8. In Tm- and Yb-Sialon with more complex microstructure, the grains were finer equiaxed, needle-like and differently sized. However elongated growth of needle-like grains was not enough and their aspect ratio was about 5. It is considered that these complicated microstructures, containing two types of Sialon phases and various kinds of grains with different morphology, create the multitoughening effects such as microcracking, crack-deflection and crack-bridging, which result in the improvement of the mechanical properties of Sialon-based ceramics, as mentioned in the following paragraph.

Engineering properties of Sialon-based ceramics can be improved by controlling the physical and chemical properties of grain boundary phase which is mainly composed of lanthanide alumino-silicates. Bending strength and

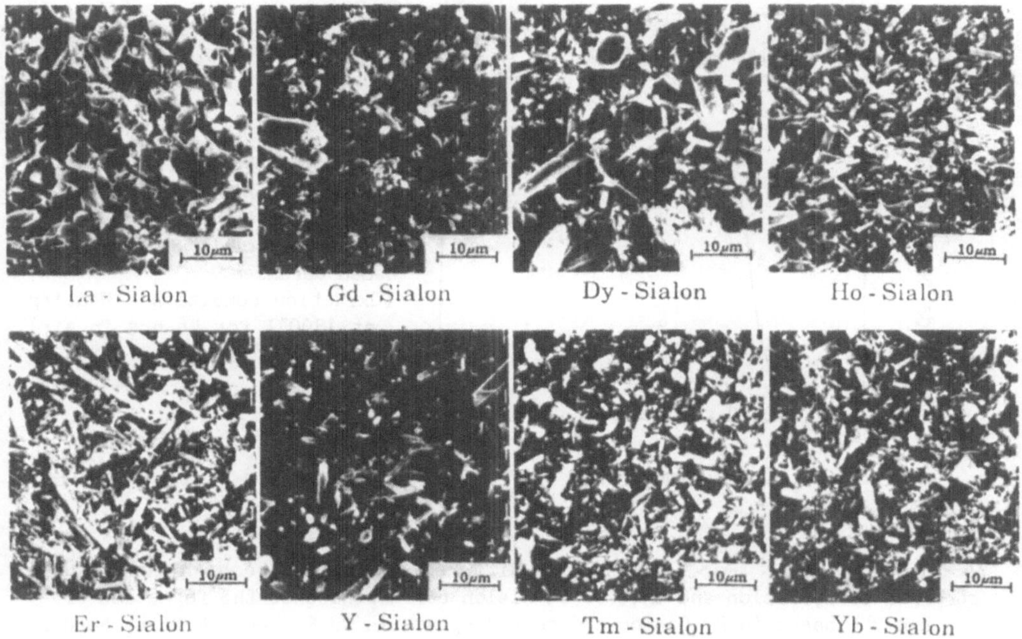

FIGURE 5 SEM photographs of the fracture surface of the sintered Ln-Sialon.

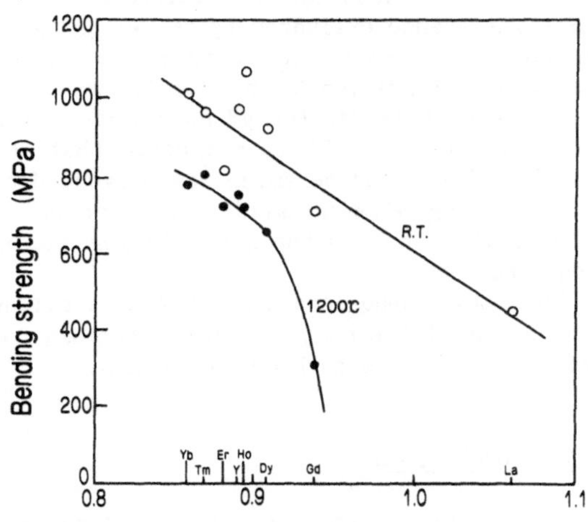

FIGURE 6
Effect of ionic radii of Ln elements on the bending strength of Ln-Sialon-based ceramics at room temperature and 1200℃.

oxidation resistance of Ln-Sialon-based ceramics were evaluated in order to clarify the effects of ionic radii of Ln elements on the thermo-mechanical properties of Sialon materials, as shown in Figs. 6 and 7. Bending strength of the sintered Ln-Sialon at both room temperature and 1200℃ was substantially increased with decreasing radii of Ln^{3+} ions. Especially, the strength was much raised by the formation of polyphase Sialon which comprised both α - and

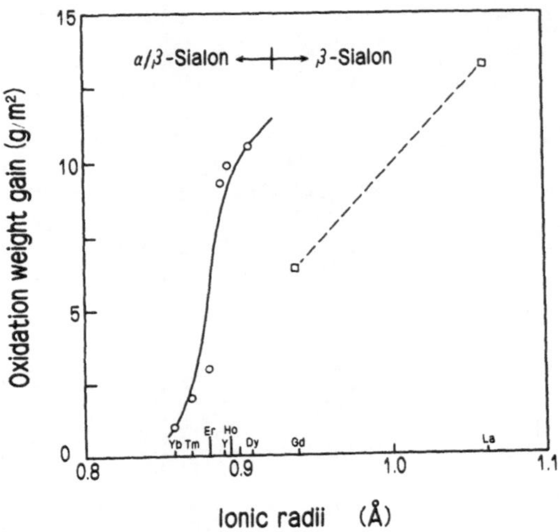

FIGURE 7
Effect of ionic radii of Ln elements on the oxidation weight gain of Ln-Sialon-based ceramics.
(Oxidation condition : Heating at 1300℃ for 48 hrs in air)

β-Sialon phases. For example, the difference between the high-temperature strength of Dy-Sialon and that of Gd-Sialon corresponded to the formation of α- and β-phases in Dy-Sialon and only β-phase in Gd-Sialon. 4-Point bending strength of Tm-Sialon-based ceramics was 960 MPa at room temperature and 810 MPa at 1200℃, which was the maximum high-temperature strength in this experiment.

Regarding the oxidation resistance of Ln-Sialon-based ceramics, the effect of ionic radii of Ln elements was more striking. That is, the smaller the ionic radius became, the better the oxidation resistance evaluated by the weight gain was obtained. Much difference in the oxidation weight gain was observed between Y-Sialon and Er-Sialon. Minimum weight gain, obtained for the sintered Yb-Sialon, was 1.0 g/m^2 after heating in air at 1350℃ for 48 hours, which was nearly equal to the data for SiC ceramics [19,20]. It is suggested that the progress of oxidation at the surface of Sialon-based ceramics was prevented by the formation of a protective Yb-containing alumino-silicate layer which restrained the oxygen permeability or mobility due to the high binding energy between O^{2-} and Yb^{3+} with smallest ionic radius.

Accordingly, it is evident that Yb-Sialon-based ceramics exhibit excellent thermo-mechanical properties, which offer one of the most feasible materials for high-temperature structural applications such as gas turbine components.

SUMMARY AND CONCLUSIONS

Various Ln-α-Sialon (where Ln= Y and lanthanide elements) powders were prepared from the imide-derived amorphous Si_3N_4, then, their sintering behavior was studied. The following results were made clear.

(1) The x-value of the Ln-Sialon ($Ln_x(Si,Al)_{12}(O,N)_{16}$) powders was determined by the Rietveld analysis of their X-ray diffraction patterns and the amount of Ln_2O_3 solid-solved in α-Sialon lattice was found to incease with decreasing radii of Ln^{3+} ions.

(2) Highly dense Ln-Sialon-based ceramics were fabricated by the pressureless

sintering of mixtures of Ln-α-Sialon and α-Si_3N_4 powders (in the weight ratio of 40/60) without any other additives. It was an effective way to use the mixtures of Ln-α-Sialon and SN-E10 as the starting materials for the consolidation of high performance ceramics.

(3) Bending strength, especially high-temperature strength, of Ln-Sialon-based ceramics was much raised by the formation of both β-Sialon and α-Sialon which accommodate the additives in α-Sialon lattice. Maximum high-temperature strength, obtained for the sintered Tm-Sialon, was 810 MPa at 1200℃.

(4) The smaller the radius of the Ln^{3+} ion, the better the oxidation resistance of the Ln-Sialon-based ceramics. Minimum oxidation weight gain, obtained for the sintered Yb-Sialon, was 1.0 g/m^2 at 1350℃ for 48 hours.

As a result of this study on the preparation of Ln-Sialon powders and their sintering behavior, it was demonstrated that thermo-mechanical properties of Ln-Sialon-based ceramics could be much improved by the selection of lanthanide elements. It was also proved that Yb-Sialon-based ceramics exhibit excellent oxidation resistance, which is most suitable for the application to advanced ceramic components for engines.

REFERENCES

1) Hampshire, S., Park, H.K., Thompson, D.P. and Jack, K.H., Nature, 1978, 274, 880-82.
2) Grand, G., Demit, J., Ruste, J. and Torre, J.P., J. Mater. Sci., 1979, 14 [7] 19-51.
3) Mitomo, M., Izumi, F., Bando, Y. and Sekikawa, Y., Proc. 1st Intern. Symp. on Ceramic Component for Engine, Japan, KTK Scientific Publishers, 1984, pp. 377-86.
4) Huang, Z.K., Greil, P. and Petzow, G., J. Am. Ceram. Soc. Commu., 1983, 66, C96-97.
5) Ishizawa, K., Ayuzawa, N., Shiranita, A., Takai, M., Uchida, N. and Mitomo, M., Proc. 2nd Intern. Symp. on Ceramic Materials and Components for Engines, Germany, D. Reidel Publishing Co., 1986, pp. 511-18.
6) Komatsu, M., Kameda, T., Goto, Y. and Komeya, K., Abstract of the Annual Meeting of Ceramic Society of Japan, 1987, pp. 301-02.
7) Ukyo, Y. and Wada, S., Nippon-Seramikkusu-Kyokai Gakujutsu-Ronbunshi, 1989, 97 [8] 872-74.
8) Arias, A., J. Mater. Sci., 1979, 14 [4] 1353-60.
9) Nickel, K.G., Danzer, R., Schneider, G. and Petzow, G., Powder Metall. Intern., 1989, 21 [3] 29-33.
10) Japanese Patent 1139881, Ube Industries, Ltd., Ube, Japan.
11) Yamada, T., Kawahito, T. and Iwai, T.,Proc. 1st Intern. Symp. on Ceramic Components for Engine, Japan, KTK Scientific Publishers, 1984, pp.333-42.
12) Glemser, O. and Naumann, P.,Z. Anorg. Allgem. Chem., 1959, 298 [3-4] 134-41.
13) Billy, M.,Ann. Chim., 1959, 4 [13] 795-851.
14) Japanese Patent Applic., 86-60640, Ube Industries, Ltd., Ube, Japan.
15) Nakayasu, T., Yamao, T. and Kohtoku, Y.,Abstract of the Annual Meeting of Ceramic Society of Japan,1988, pp. 257.
16) Rietveld, H.M.,J. Appl. Crystallogr., 1969, 2, 65.
17) Izumi, F., Mitomo, M. and Suzuki, J., J. Mater Sci. Letters, 1982, 1, 533.
18) Izumi, F., Mitomo, M. and Bando, Y.,J. Mater. Sci., 1984, 19, 3115-20.
19) Singhal, S.C.,J. Mater. Sci., 1976, 11, 1246-53.
20) Suzuki, K., Kageyama, N., Furukawa, K. and Kanno, T., Proc. 2nd Intern. Symp. on Ceramic Materials and Components for Engines, Germany, D. Reidel Publishing Co., 1986, pp. 697-704.

MECHANICAL TESTING OF GLASS-CERAMIC MATRIX COMPOSITES

T. Grenier*, G. Bessenay**, M. Parlier*, M.H. Ritty*
* ONERA - 29 Av. de la Division Leclerc - B.P. 72 - 92322 Châtillon - France
** SNECMA - Materials and Processes Department -
B.P. 81 - 91003 Evry Cédex - France

ABSTRACT

For applications below 1000°C glass ceramic matrix composites are particularly interesting for high temperature applications in turbine engine components, due to their high specific strength and low processing cost. Lithium Aluminum Silicate matrix reinforced by Nicalon SiC fibres have been characterized through microstructural analysis and mechanical testing.
Their behaviour in turbine engine condition has also been studied and a method is proposed to improve long term resistance.

INTRODUCTION

The constant need for improving performance and efficiency of jet engines has resulted in a continuous decrease of the structure weight. This demand for new structural design has driven the development of low density high performance materials able to operate at high temperatures. The main objective that must be paid attention as far as the development of new materials is concerned are the specific properties due to the prime importance of weight saving on the one hand and cost effectiveness related to the component manufacture and life duration on the other hand.

Long fibre reinforced Ceramic Matrix Composites (CMC) are a class of materials that offers promise in meeting the challenge because of it's unique set of combined properties :
- a high temperature capability (depending on the fibre/matrix system - from 400°C for C/Pyrex, up to 2000°C for short time exposure of coated C/C),
- high specific strength (figure 1) and stiffness,
- a specific gravity lower than 2.5 g/cm^3,
- a potential good environmental resistance depending on the nature of the matrix and/or coating.
- a damage tolerant mechanical behaviour compared to monolithic ceramics.

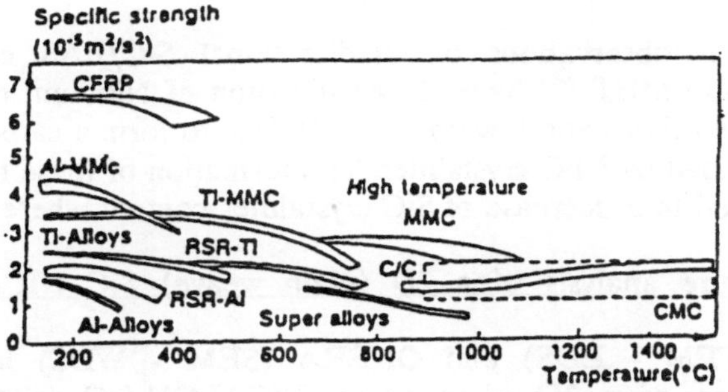

Figure 1: compared specific strength versus temperature for CMC and the materials most widely used in aeronautical industries

Among all the potential CMC candidates, glass-ceramic matrix composites appear as promising materials, for use up to 1000-1100°C, because of the flexibility of their fabrication process related to the good processability of the glass at a moderately high temperature, from which relatively low manufacturing costs may be expected.

The potential of Glass-Ceramic Matrix Composites (GCMC) led ONERA and SNECMA to develop research programs on NICALON SiC fibres reinforced LAS (Lithium Aluminium Silicate) matrix composites, in relation with Saint-Gobain Recherches, which has improved the glass composition and manufactured some of the materials.

Results presented in this paper for this material cover the field of :

- microstructural investigations via SEM-WDS and TEM-EDS technique

- mechanical characterizations at high temperatures (bending, tensile) related to fibre/matrix shear stresses measured by microindentation tests,

- behaviour of this material in the gas turbine environment measured through oxidation and erosion tests.

I. CRYSTALLIZATION AND INTERFACE STRUCTURE IN SiC/LAS COMPOSITES

A LAS glass composition has been improved by Saint-Gobain Recherche to permit the densification of composites with Nicalon SiC fibre reinforcement. Fluxing agents have been added to obtain a viscosity lower than 5 Pa/S at a temperature below 1400°C where fibre degradation occurs. This viscosity is necessary for densification by a hot pressing process of prepreg tapes in a graphite furnace under a nitrogen atmosphere. The processing parameters have been determined at ONERA and Saint Gobain Recherche and SNECMA to get composites with a porosity content less than 3 %, as well for unidirectional than for bidirectional reinforcement. A ceramization cycle imposes crystallization of a β-spodumene phase to obtain a low coefficient of thermal expansion compatible with those of the SiC reinforcement.
As mentionned by Prewo [1], Niobium oxyde has to be added to the glass composition to get weak fibre-matrix interfacial bonds. Such weak bonds are necessary for a non catastrophic failure of the composite and so to insure a fail-safe behaviour of the

components.
Microstructural observations on unidirectional SiC/LAS composites by Electron microprobe and MET [2] have shown diffusion of Niobium on the one hand and of Carbon on the other hand towards the interface to form a carbon rich layer around the fibres sorrounded by NbC crystallites. The formation of these two interfacial layers has been correlated to a decrease of SiC crystallites content wherein the fibre surface.

. Microstructure analysis of a 2D (plain weave) SiC NICALON reinforced LAS composites

SNECMA (TEM - EDS) and ONERA (SEM - WDS) have made a complete characterization of a 2D plain weave NICALON-SiC reinforced LAS' composite, representing the most difficult case for hot pressing densification due to the complexe geometry of the reinforcement, with low curvative radii for the SiC fibre tows. According to preparation techniques describe elsewhere [3], a thin foil have been sampled parallely to the reinforcement cloths. The TEM examination of this kind of sample, despite the very great difficulty of thinning it, shows the very good impregnation of fibre tows, even at the tow crossings (see figure 2). Fibre/matrix interfacial structure in this 2D SiC/LAS composite has been studied by a STEM technique showing a continuous interphase (0.15 µm thick) containing regularly distributed thin NbC crystallites (~40 nm) - see figure 3. These observations are comparable to the results described previously on 1D SiC/LAS [4].

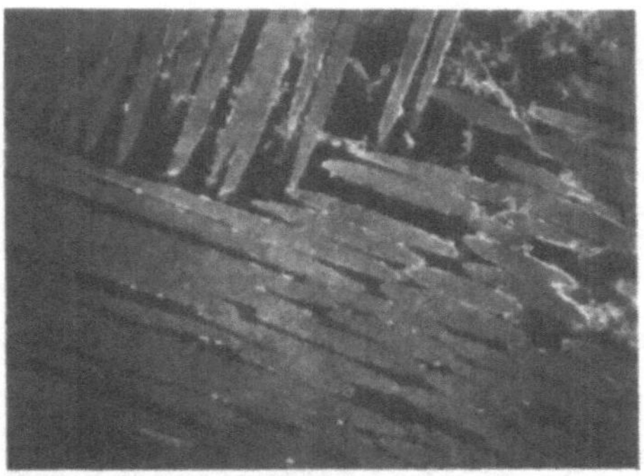

50 µm

Figure 2: SEM micrograph (secondary electron) of fibre tow crossings in a 2D (plain weave) SiC-Nicalon reinforced LAS composite demonstrating good matrix infiltration

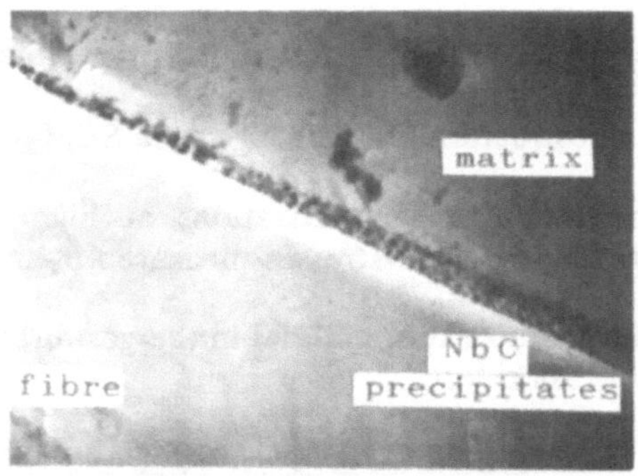

0.2 µm

Figure 3: STEM image of fibre/matrix interphase containing NbC in a 2D (plain weave) SiC-Nicalon reinforced LAS composite

The crystallization structure of the LAS matrix has been studied with a special attention. The matrix is partially crystallized under a Lithia Alumino Silicate β-spodumen form with a Keatite structure (ASTM 21 503 C).

TEM analysis shows that crystallized dendrites (figure 4 a) are surrounded by interdendritic areas with an amorphous state. These dendrites are borded by definite

compounds containing Zr, but also a bright zone (~ 0.5 μm thick) containing Ca (figure 4b). Some small Nb-rich (0.2 μm) inclusions can be detected localy. They change the crystallization form of the matrix. Some cells appear by radial grain growth from these Nb-rich inclusions. Some very small Zr-rich precipitates (~ 20 nm) are decorating the diffusion paths (figure 5).
Segregation of these elements has to be systematically studied in order to verify the structural stability of these materials for long term ageing.

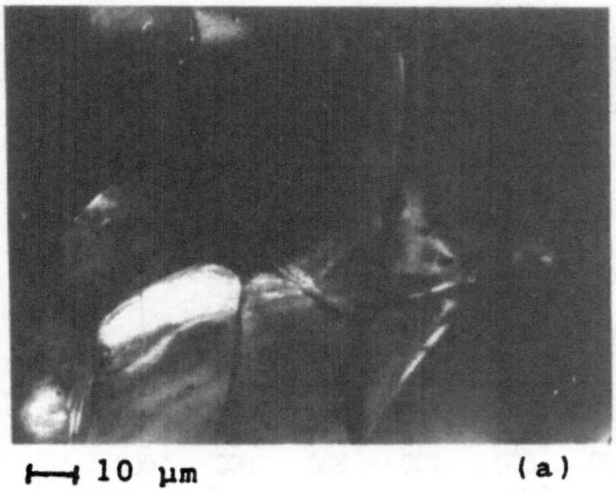

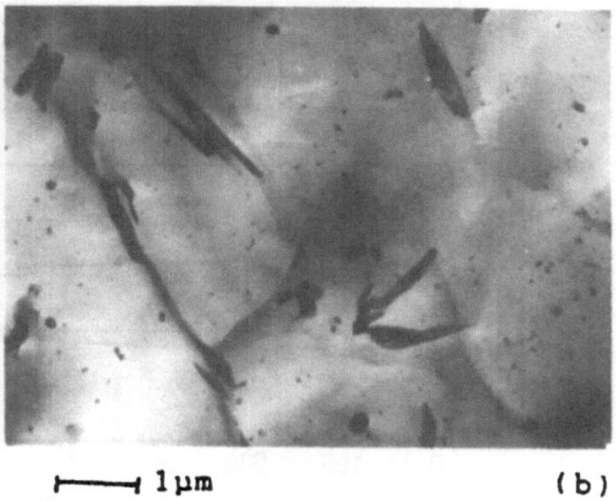

Figure 4: STEM image of the matrix microstructure for 2D SiC/LAS composite as-processed:
(a) dendrite formation
(b) definite compounds Zr and Ca-rich sourrounding the dendrites

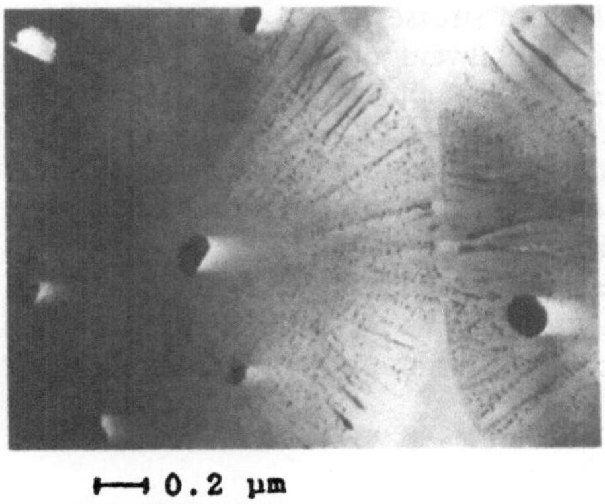

Figure 5: STEM image of Nb-rich inclusions in LAS matrix and Zr-rich small precipitates decorating the diffusion path for radial growth of cells

II. BASIC MECHANICAL CHARACTERIZATION OF THE COMPOSITES

. 3 point bending tests :

Bending tests are easier to perform than tensile tests, particularly for high temperature testing, but they provide less usuable datas (figure 6). The tests are realized with different fibre reinforcement : unidirectional, [0,90] and [+45, -45] crossed ply and [0,90] satin woven fabric tapes.

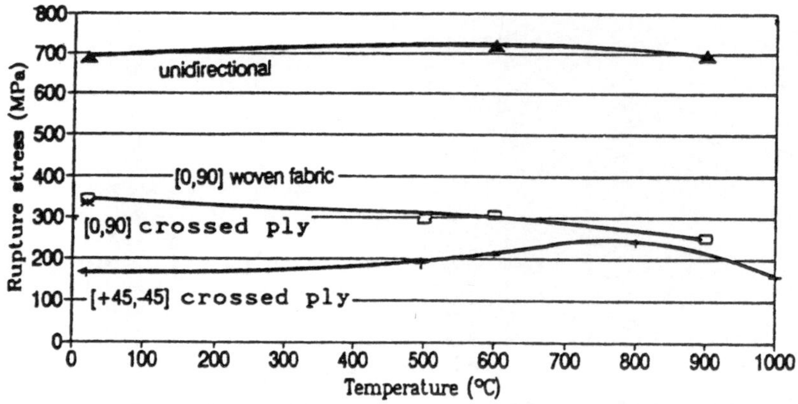

Figure 6: 3 point flexural strength vs temperature for LAS matrix reinforced by different kind of reinforcement

Unidirectional reinforcement usually leads to rupture stresses of 700 MPa at room temperature with a span-to-depth ratio laying between 15 and 20. But modifications of the processing conditions lead to stresses up to 1000 MPa. Important pull-out of the fibres from the matrix-shear is always observed and rupture occurs either on the compression or tensile side of the sample.

The crossed ply composite stresses are in good agreement with those expected from elasticity calculations with respect to the fibre fraction and orientations. It is noticed that there is no evidence for the influence of the reinforcement type between [0,90] crossed plies and satin woven fabrics.

For every kind of reinforcement studied, only a slight decrease of the rupture stress is measured for 900°C tests and short time (<30 min) exposure in air, but the pull out of the fibres from the matrix is more important for bidirectional composites than for unidirectional composites.

. Tensile tests :

Due to the brittleness of both the components of the composites (ceramic fibre and glass-ceramic matrix), their toughening is mainly governed by the properties of the fibre-matrix interfaces. Catastrophic failure mechanisms are involved with strong fibre-matrix bonds and the rupture occurs when reaching the ultimate strain of the matrix by crack propagation through the whole composite section : few energy is then released. A mixed propagation mode (mode I/MODE II) is obtained when the fibre-matrix bonds are weak so that, as shown on unidirectional composite during a tensile test going off under a microscope, the fibres are not sheared by the cracks and support the load. This leads to a non linear stress-strain curve and it is of prime interest for failure prediction of turbine engine components. With increasing loads, fibres break while reaching their rupture strength and frictional forces at the interface impose a resistance to pull out of the fibres from the matrix sheath. Extensive energy is then released during rupture. The pull out resistance may be estimated by a microindentation test of a fibre in the matrix, coupled with a convenient shear lag model : an interfacial shear stress of about 10 MPa is obtained for a composite with weak bonds (with Nb_2O_5) and has to be compared to over 50 MPa for a composite with strong bonds (without Nb_2O_5).

For all the composites (with Nb_2O_5) tested, important pull out of the fibres is noticed and the rupture stresses are reported in Table 1.

Table I: tensile strength of SiC/LAS composites for different kind of reinforcement

reinforcement	1D	[0,90°] crossed plies	[0,90°] woven fabrics	[+45,-45] crossed plies
rupture stress (MPa)	470	185	175	85

The design of the samples has been particularly studied to reduce stress concentrations in the clamping zone. Finite element calculations exhibited that the only design possible for unidirectional composites is the single bar. For bidirectional composites, tested at SNECMA, a special design has been defined for a maximum sample length of 100 mm in order to limit shear overstress or tensile overstress in the shoulder area and get a rupture localized in the gauge area (see figure 7).

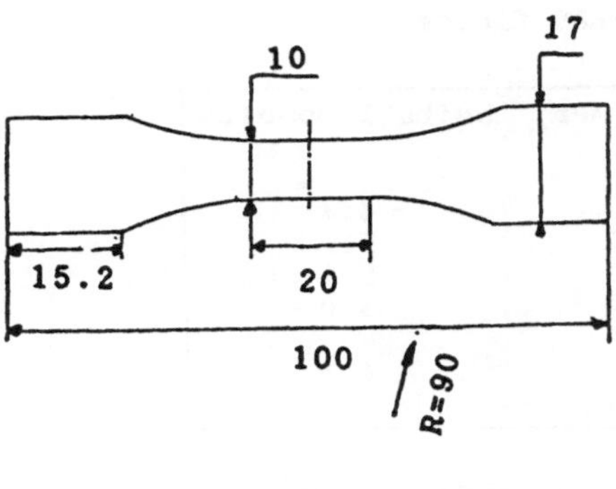

Figure 7: specimen design for tensile tests on bidirectional SiC/LAS composites

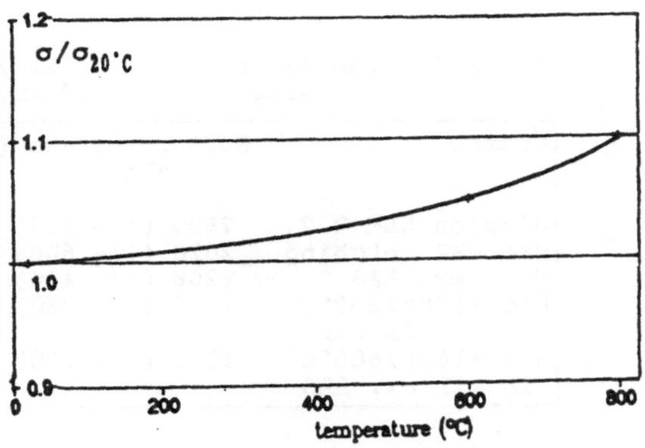

Figure 8: Tensile strength for [+45,-45] crossed ply composites vs temperature

For unidirectional composites stresses of 470 MPa are obtained at room temperature (40 % fibre volume fraction). With respect to the fibre volume fraction in the loading direction, the values for bidirectional reinforced composites with a global 40 % volume fraction are less than those expected from elasticity calculations. This may be explained by bending interference during the test.
It is noticed first that [+ 45, - 45] crossed ply composite properties are maintained up to 900°C (figure 8) and second that [0,90] crossed plies and satin woven fabrics lead to the same rupture stresses.

. Discussions

Weak interfacial bonds are obtained with the glass composition studied and the processing parameters used.
Micromechanical characterizations have clearly exhibited that Niobium oxyde is necessary to get weak fibre-matrix interfacial bonds and non brittle behaviour of the composite.
The mechanical characterizations realized both by SNECMA and ONERA show that the energy released during the test and the rupture stresses are largely enhanced relatively to monolithic ceramic materials.
According to the aimed applications, the properties obtained are very interesting and particularly the high temperature properties (500 to 900°C temperature range).
But the overall tensile properties of the unidirectional composites would seem to be enhanced : in the last stages of the stress-strain curves, the load is only applied to the fibres. So a rough approximation of the rupture stress with 40 % for the fibre fraction and 2600 MPa for the rupture stress of a fibre (measured by a tensile test with a gauge length of 10 mm) lead to a value of about 1000 MPa, which is twice the value measured. But, in fact, extraction of the fibres from the matrix with H.F. acid hetching reveals that the fibres are damaged (table 2) by the hot pressing process (15 %) and it is a part of the unpredicted value of the composite rupture stress. Work is in progress to enhance these stresses to a target value of 800 MPa.
It has been noticed that the tensile stress to bending stress ratio for unidirectional composites is in accordance with Prewo [5] (about 65 %).

Table II: Tensile strength of SiC/LAS fibres
(Gauge length: 10 mm)

Fibres	Rupture stress (MPa) (MPa)	Weibull modulus
Nicalon NLM 202	2600 (+/- 535)	4.4
Nic.+HF hetching	2620 (+/- 500)	5.9
Nic. ex. LAS	2200 (+/- 400)	5.1
Nic.+100h/600°C in air	2165 (+/- 780)	3.0
Nic.+100h/600°C in air ex. LAS	1321 (+/- 390)	4.1

III. COMPOSITE BEHAVIOUR IN TURBINE ENGINES CONDITIONS

To guarantee the safety of turbine engine components, the previously mentioned properties have to be maintained for long time duration in engine atmospheres.
Depending on the location of the components, the degradation mechanisms may be oxidation at high temperature in air or erosion by gas flow for exhaust system parts. Both of these environments have been tested.

. Ageing tests :
Due to their carbon content and particularly wherein the interface, ageing the composite in air atmosphere at temperature above 500°C damages their properties.
Tests have been performed at ONERA on [+ 45, - 45] crossed ply composites and [0,90] satin woven fabric tapes at the minimum and the maximum temperature under consideration (600°C and 800°C) for 100 hours duration time.
First, the carbon, which is not on a carbide form (1.5 % of the composite compounds) disappear from the interface and from the matrix core (some have been enclosed during densification and come from the COatmosphere of the furnace). Some weakening of the composite mechanical properties may result which means that the carbon content has to be controlled at a value at least lower than 1 %.
Then, oxydation of the composite is quantified by measuring the weight variations of the samples : for both the temperatures, an increase of weight is noticed meaning that oxydation of the carbon is not the major phenomenon. Furthermore, the increase of weight is more important for the 800°C tests (0,9 %) than for the 600°C tests (0,5 %) as expected from a thermally activated mechanism.
The oxydation of the SiC-Nicalon fibre has been previously studied [6] and formation of a silica rich layer has been observed on the fibre surface. In test realized at ONERA such formation leads to a decrease in the fibre strength of 17 % (table 2) after ageing as-received fibres at 600°C for 100 hours. This phenomenon, when occuring in composites damages readily the fibres in an extent of 50 %.
An enhancement of the interfacial shear stress due to the modification of the interface chemistry, is also measured by a microindentation test and is in agreement with the weak pull out of the fibres observed.
Ageing tests have shown that oxidation leads to a certain decrease in the mechanical properties of SiC/LAS composites but protections may be adapted to limit oxygen diffusion towards the interfaces.

. High speed gas erosion test
According to the crystallized structure of the glass-ceramic matrix in the composite (for example : β-spodumene phase with a Keatite structure for the LAS system), it may be understood that this material may be degraded under certain erosion conditions, especially at temperatures below the glass transition of the matrix (Tg ~ 750-800°C for the LAS system), for which the surface of the matrix is microcracked.
To study this phenomenon and quantify the damage done to the material, a special high speed gas erosion apparatus (figure 9) has been used at SNECMA. This device simulate the exhaust system conditions for turbine engines :
- high gas speed : up to Mach 1
- temperature : up to 900°C
- adjustable angle between the sample surface and the flow-axis : 15-90°
- possibility of adding abrasive (corindon 50 μm with a 50 g/mn rate) to the gas flow.

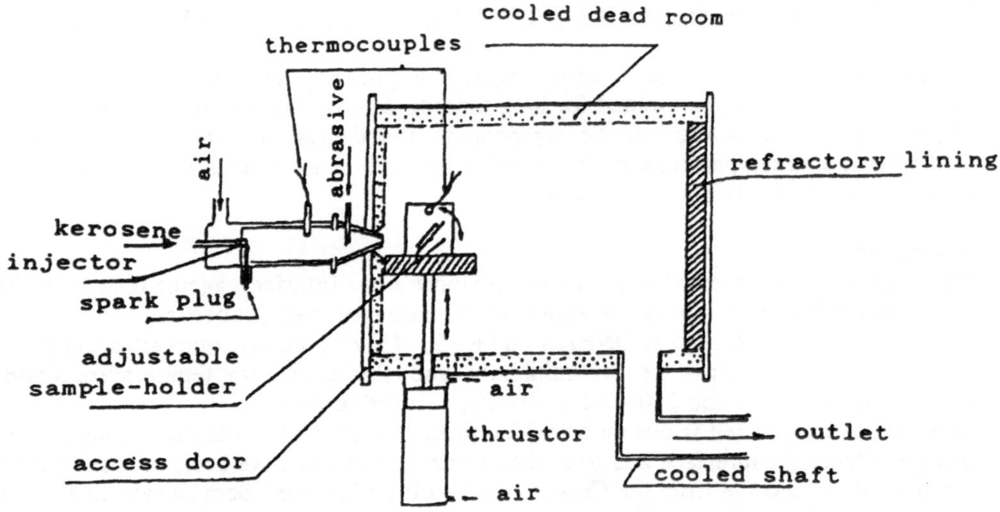

Figure 9: High speed gas erosion apparatus

A 2D (plain weave) NICALON SiC reinforced LAS-matrix composite, as processed or with a Si_3N_4 coating, has been tested under two conditions : with abrasive added to the gas flow or without. For the tests, the samples have been placed in the chamber so that their surface make a 45° angle with the flow which represent the most damaging test conditions. In this configuration, the affected zone of the surface of the sample is 20 x 30 mm^3 wide.
From the weight loss results presented on figure 10, it may be seen that neither coated

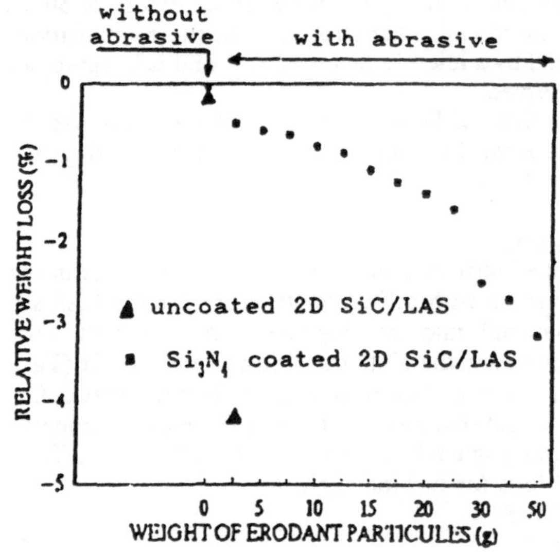

Figure 10: Relative weight loss of the samples during erosion tests (900°C - Mach 1 - 45°angle - 50 µm corindon-abrasive)

nor uncoated composites are degraded under the gas-flow (900°C - Mach 1) alone. But when adding abrasive to the gas flow, one may see that the uncoated material is rapidly eroded.
After spraying a small quantity of corindon -2,5 g- the glass-ceramic matrix at the surface has been cracked and the chips of glass have flown away. The SiC taffeta cloth appears at the surface (fibres are less degraded by the abrasives because they are much harder than the glass-ceramic matrix). After the test, the surface of the sample is very rough (figure 11 a). But the very interesting result that appears on figure 10 is that the same material caoted with a Si_3N_4 thin layer (300 μm thick) is much less degraded, even under very hard conditions, and after the test it may be seen (figure 11 b) that the coating has only be polished and that only small cracks appear at the surface.

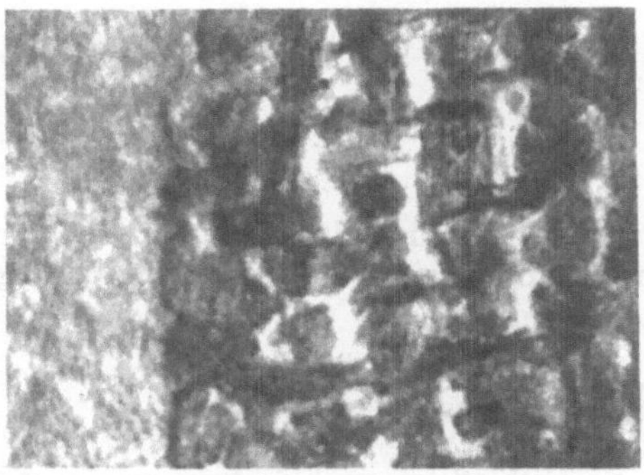

(a) 2.5 g abrasive

(b) 50 g abrasive

Figure 11: Micrographic views of the 2D (plain weave) SiC nicalon LAS samples after erosion tests (a) uncoated (b) Si_3N_4 coated

CONCLUSION

Glass-Ceramic Matrix Composites demonstrate specific mechanical properties that make them attractive for application in static or semi-static parts of turbine engine (for example exhaust-system parts).
From the specific tensile strength results presented in this paper, it may be seen that, compared to metallic materials (superalloys - d ~ 7-8- σ/d~ 100 MPa at 900°C), NICALON SiC reinforced Lithia Alumina Silicate matrix (d~ 2,5) demonstrate high temperature (up to 900°C) properties that are of the same order of magnitude for the 2D composites (σ/d ~ 80 MPa) or even higher for unidirectional reinforced composites (σ/d ~ 160 MPa).
Furthermore, the flexibility and short duration of the processing cycle leads to manufacturing processes for these materials that are potentially more cost effective than other CMC processing routes (such as Chemical Vapor Infiltration of fibre preforms).
Depending of the shape of the designed part, two main classes of applications may be distinguished :
- small and complex parts with a multiaxial stressfield, that may be hot-pressed for 2D lay-ups, 2.5 D or 3D preforms,
- large axisymetrical parts with a simpler stressfield, compated from wound unidirectional or crossed plies.

The properties retention of these GCMC materials for long term ageing in the turbine engine environment still requires further studies and developments :
- according to the intrinsic microcracked structure of the glass-ceramic matrix, the oxidation resistance of these GCMC materials may be improved through chemical tailoring of the residual vitreous phase of the matrix and of the fibre/matrix interface (whose best mean of investigation is TEM as presented in this paper for 2D SiC/LAS).
- the erosion resistance of the glass matrix may be also significantly increased, as seen in figures 10 and 11, with external ceramic coatings, to reach a satisfactory level comparable to metallic alloys.

Work is also under progress to get a better understanding of GCMC mechanical behaviour and particularly beyond the matrix microcracking thresold, that means for loads higher than 80-100 MPa, in order to validate the extended use range of these materials over the elastic linear range.

Acknowledgments

The authors acknowledge G. SIMON from SNECMA for his helpfull cooperation for TEM analysis, DRET and STPA for their financial support and SGR for manufacturing glass powders and some of the composites tested herein.

REFERENCES

1) K.M. PREWO, J.J. BRENNAN, "High strength silicon carbide yarn reinforced glass matrix composites",J. Mat. Sci. 15 pp 463-465 (1980)

(2) T. GRENIER, M. PARLIER, M.H. RITTI,"Interfaces in SiC/LAS Composites", Interf. phen. in Comp. materials'91, Conf. to be published (1991).

(3) G. SIMON, S. BOIS, G. BESSENAY,"Interface microstructure analysis of ceramic and metal matrix composites. Sample preparation method and TEM analytical techniques". Interf. phen. in compos. materials' 91. Conf. to be published (1991).

(4) R.F. COOPER, K. CHYUNG,"Structure and chemistry of fibre-matrix interfaces in Silicon carbide fibre reinforced glass ceramic composites : an electron microscopy study". J. Mat. Sc. 22 (1987) pp 3148-3160.

(5) K.M. PREWO,"Tension and flexural strength of Silicon carbide fibre reinforced glass ceramics" J. Mat. Sci. 21 pp 3590-3600 (1986).

(6) T.J. CLARK, E.R. PRACK, M.I. HAIDER, L.C. SAWYER, "Oxidation of SiC Ceramic fibre". Ceram. Eng. Sci. Proc. 8, [7.8] pp 717-731 (1987).

THE INFLUENCE OF STATIC LOADS ON THE OXIDATION BEHAVIOUR OF FIBRE REINFORCED SiC/SiC COMPOSITES

Robert Lundberg*, Torgny Stenholm, Lars Pejryd
Volvo Flygmotor AB, Trollhättan, Sweden

Lars Kahlman
Swedish Ceramic Institute, Göteborg, Sweden

ABSTRACT

A ceramic matrix composite (SiC/SiC) material was subjected to static stresses above and below the matrix cracking stress in an oxidising environment at 1000 and 1250°C. At the lower stress level, times to failure longer than 150 h were obtained, whereas the lifetime was significantly reduced at the higher stress level. A combined oxidation/matrix cracking mechanism was proposed, based on oxidation weight gain, microstructural investigation and the stress/rupture results.

INTRODUCTION

Volvo Flygmotor AB is currently involved in the design, production and testing of gas turbine hot parts of various ceramics and ceramic composites. This paper is focussed on one material that is being evaluated, namely SiC fibre reinforced chemical vapour infiltrated (CVI) SiC (1). Design data, especially mechanical properties at high temperatures are currently being compiled. The mechanical properties and fracture behaviour of these SiC/SiC composites at room temperature and at high temperatures in an inert environment has been treated extensively in the literature (1–4). Little is, however, reported on mechanical testing at high temperature in oxidising environments. Frety et al (5) have investigated ageing in air at elevated temperatures with subsequent bend tests at room temperature. They found that oxidation destroyed the weak fibre/matrix interface which is essential for obtaining strong and non–brittle composites. In our investigation it was anticipated that simultaneous mechanical loading and oxidation would create a combined effect where oxidation could proceed more rapidly through microcracks formed in the matrix, reaching the fibre/matrix interface, leading to strong bonding and embrittlement of the composite, a process which would be critical for gas turbine applications.

When a ceramic matrix composite is loaded (in bending or tension) the material behaves roughly linear elastic up to a certain stress level (Rp). Above Rp more and more matrix cracks are formed, which decreases the stiffness of the composite and results in a "pseudo-plastic" stress/strain behaviour (1,3). The purpose of the present study was to investigate the combined effect of mechanical loading and oxidation on the possible life-time of ceramic composites. More precisely we performed static loading to a stress level either above or below the matrix cracking stress at elevated temperatures in air in order to establish a possible design stress level for these composites.

EXPERIMENTAL

Testbars 10 x 3 x 50 mm of 2D stacked fabric SiC/SiC material (SEP, France) were tested in the as received condition without any cutting or machining. Two batches of material were investigated, one base line SiC/SiC material and one with an oxidation-protective finishing treatment (SiC/SiC-FT).

The stress rupture tests were performed in an equipment built at the Swedish Ceramic Institute, Fig. 1a, which is of a similar design to test rigs at NIST, Washington DC (6), and to the one used by Khandelwal et al (7) for high temperture testing of monolithic silicon nitride. The rig consists of three ceramic push rods. The loads are applied by pneumatic cylinders and measured by a precision pressure transducer. Displacement for each push rod is measured by an inductive distance meter at the upper water cooling jacket. The four-point specimen holders (20mm and 40mm distance between the supports) are made of SiC (see fig. 1b). The furnace is heated by $MoSi_2$ (Super Kanthal) heating elements and can be used up to 1500°C in air.

Two stress levels were chosen, based on room temperature bend tests; 75 MPa (below the matrix cracking stress, Rp) and 150 MPa (above Rp). The tests were carried out in air at 1000°C and 1250°C. Rupture was defined as a step-like large displacement (>1.5 mm) which automatically turned off the pressure. Two specimens were tested at the same temperature and stress in one run. An additional unloaded sample was placed in the third specimen holder in order to check the difference in oxidation weight gain of loaded and unloaded material. Time to failure and weight change of the samples were measured.

Figure 1 a. Stress–rupture test rig, b. Four–point specimen holder
with a SiC/SiC sample after failure.

RESULTS AND DISCUSSION

At the relatively low loads that were chosen it was not possible to detect any creep of the specimens. It should be mentioned, however, that the rig was not designed to monitor creep and that no extensometers were mounted on the specimens.

In the base–line SiC/SiC material the fracture process started typically 0.5 to 10 minutes before final fracture with small step–like displacements of a few μm at a time. In this way a total displacement of 10 to 100 μm occurred before final fracture. It should be noted that the specimens still held together after final fracture (as can be seen in Figure 2) with a large crack typically through 2 mm of the 3 mm specimen thickness.

The time to failure was found to depend strongly on the applied stress, but also on the temperature. At 75 MPa, which is below the matrix cracking stress (Rp), a time to failure of longer than 150 h was obtained for the material tested at 1000°C. At the same temperature, but a stress higher than Rp (150 MPa) the time to failure was reduced to around 0.5 h. The effect of increasing the temperature to 1250°C while maintaining the 75 MPa stress level did not have such a dramatic effect on the time to failure (150 h -> 40 h). Very short time to failure was found for the combination of a stress level above Rp (150 MPa) and 1250°C, with the specimens breaking after less than 1.5 minutes. The times to failure are summarised in Figure 2.

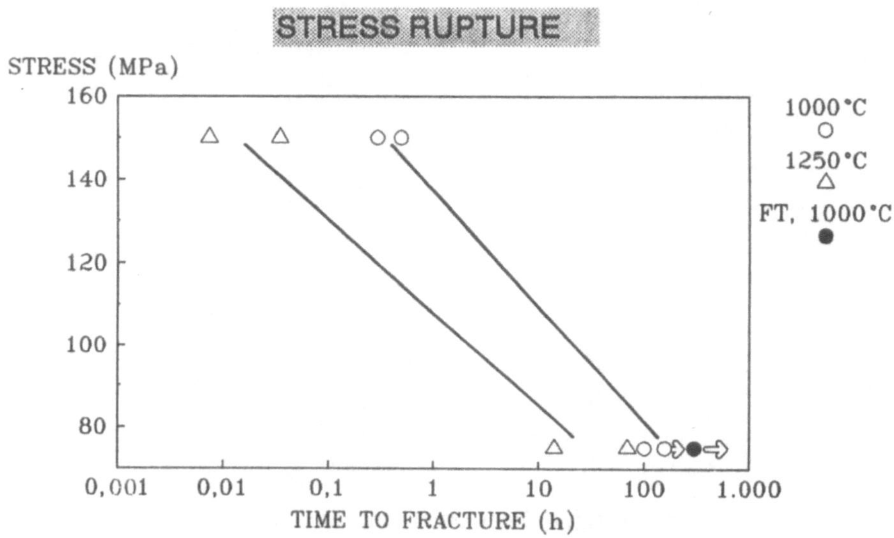

Figure 2. Time to failure vs static stress for SiC/SiC (4-point bending)
(arrows indicate: interrupted experiment, not fractured).

Given the large influence of the stress level as well as the temperature on the time to failure it was suspected that a combination of matrix cracking and oxidation caused the dramatic reduction in lifetime. Unloaded specimens that were placed in the rig during the test exhibited a somewhat smaller weight gain than loaded samples (see Fig. 3), indicating that matrix cracking opened up more surface where the oxidation could proceed. A more extensive investigation of the oxidation and corrosion behaviour of SiC/SiC is presented in reference (8).

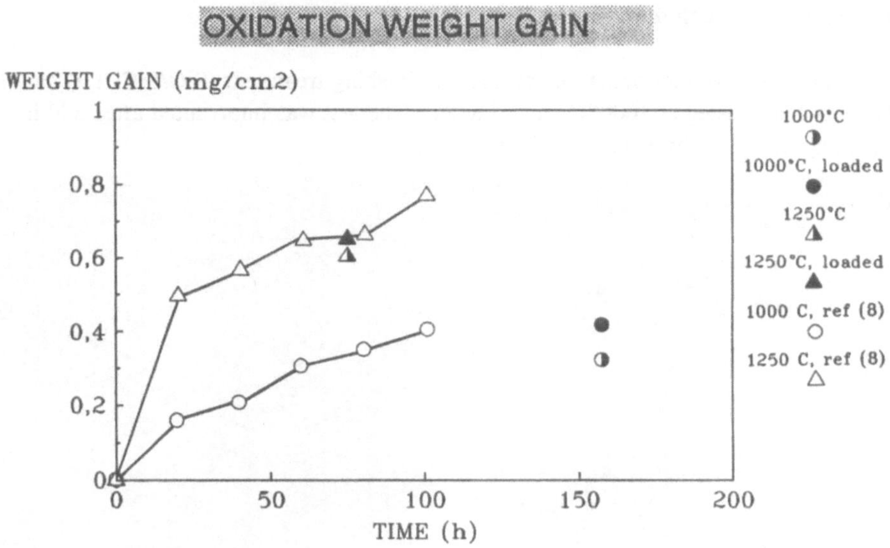

Figure 3. Oxidation weight gain for SiC/SiC. (ref (8): small samples in a different furnace; this investigation: bend specimens.)

The microstructural analysis revealed the presence of cracks on the tensile side of the specimens. The material subjected to 150 MPa exhibited a high crack density (see Figure 4), whereas only a few scattered cracks could be found in the material tested at 75 MPa. When looking at the fracture surfaces, evidence of oxidation as well as relatively undegraded areas were found. The oxidised areas were predominant towards the tensile side and were characterised by a glassy appearance and almost no crack deflection or fibre pull-out (see Figure 5). An example of a relatively undamaged area is shown in the stereo image in Figure 6.

The oxidation weight gain data, the microstructural evidence and the influence of stress level on the time to failure strongly suggest a mechanism where matrix cracking combined with oxidation of the fibre/matrix interface lead to an embrittlement of the composite. Consquently this result in a substantially reduced lifetime at elevated temperatures in oxidising environments. Instead of the tough SiC/C/SiC interface a strongly bonded SiC/SiO2/SiC interface is formed during oxidation. The oxidation follows the matrix cracks causing local embrittlement when reaching a fibre/matrix interface. The stress concentration of the crack, combined with the inability of the oxidised interface to deflect the crack cause it to grow rapidly straight through the material.

The material given an oxidation–protective finishing treatment (SiC/SiC–TF) has so far only been tested at 1000°C and 75 MPa. The test was interrupted after 300 h without any of the testbars having failed.

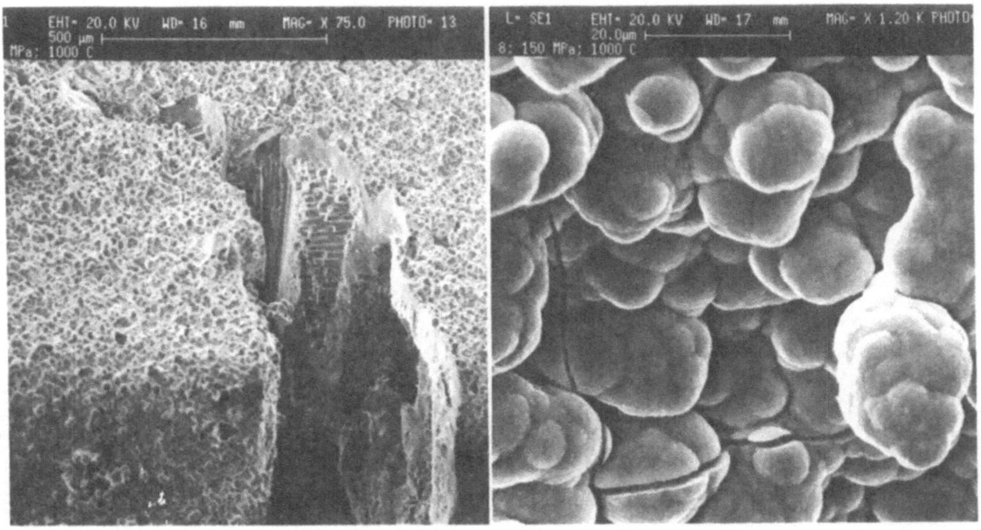

Figure 4. SEM images of the tensile side of SiC/SiC, 150 MPa, 1000°C, O.5 h. a. major fracture, b. minor cracks that where abundant on the tensile side.

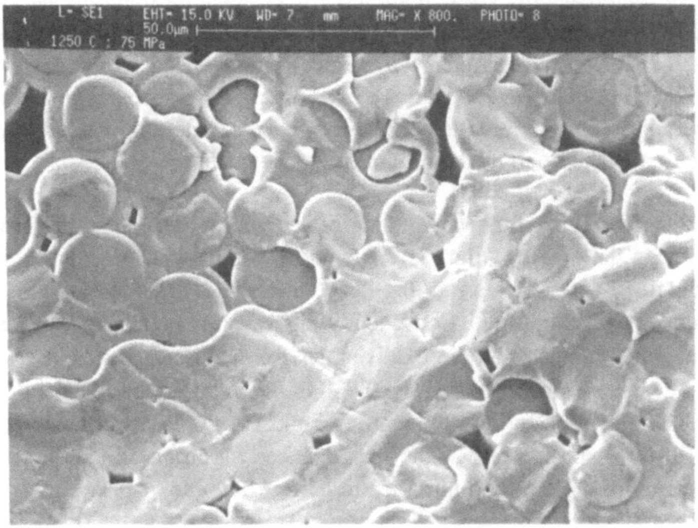

Figure 5. SEM image of a fracture surface showing brittle fracture as a result of oxidation. 75 MPa, 1250°C, 70 h.

CONCLUSIONS

When subjecting SiC/SiC composites to stresses higher than the matrix cracking stress in oxidising, high temperature environments, matrix cracks and fibre/matrix interface oxidation lead to embrittlement and fracture.

As an advise to design engineers, stresses above the matrix cracking stress should be avoided for acceptable lifetime.

To improve the composite it is of great importance to find a substitute for the oxidation sensitive carbon as an interface material. Large efforts are currently being made worldwide trying to solve this problem.

The problems with cracking/oxidation encountered in this investigation may to some extent be alleviated by the use of an oxidation–protective finishing treatment.

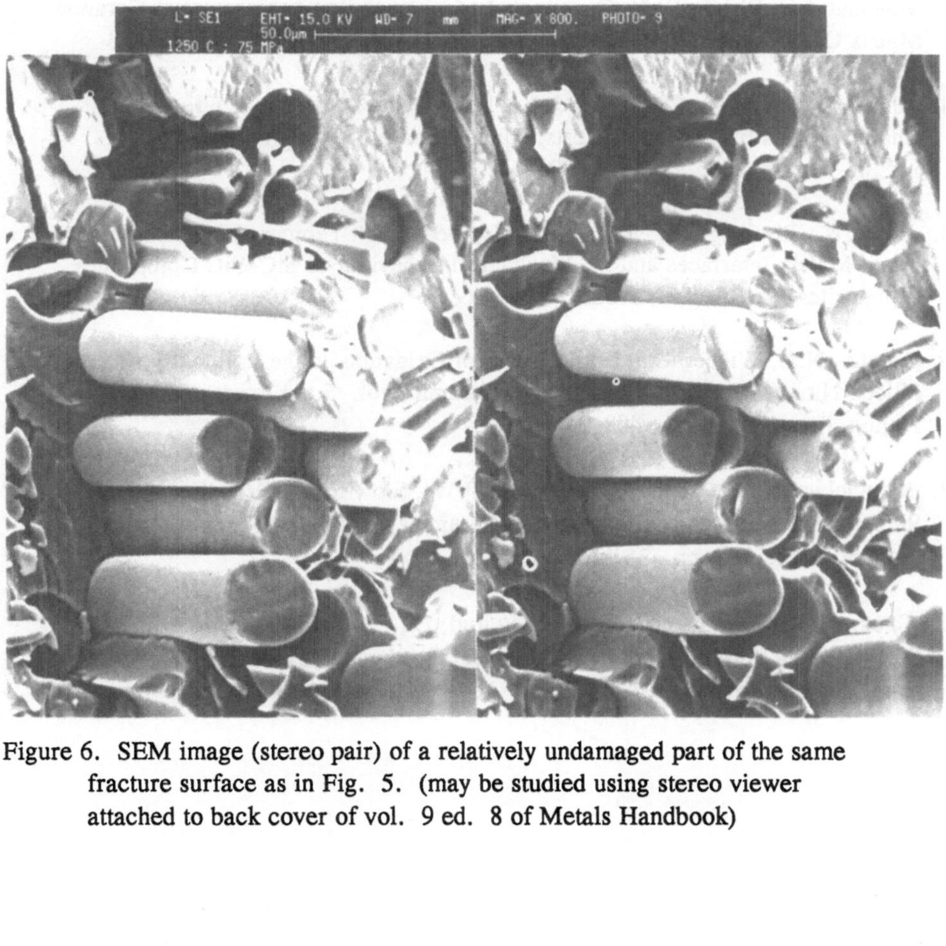

Figure 6. SEM image (stereo pair) of a relatively undamaged part of the same fracture surface as in Fig. 5. (may be studied using stereo viewer attached to back cover of vol. 9 ed. 8 of Metals Handbook)

ACKNOWLEDGEMENTS

Dr. Richard Warren is sincerely thanked for sharing his profound knowledge of composite materials and the English language. The authors are also grateful for the inspired SEM work of Mr. Sten Ljungkvist.

REFERENCES

1. Lamicq, P.J., Bernhart, G.A., Dauchier, M., Mace, J.G., SiC/SiC Composite Ceramics, Am. Ceram. Soc. Bull., 1986, 65 (2) 336–38

2. Lamicq, P.J., Ceramic Matrix Composites – A New Concept for New Challenges, Proc. Int. Conf. High Temperature Materials for Power Engineering, Kluwer Academic Publ., 1990, 1559–74

3. Bouquet, M., Birbis, J.M., Quenisset, J.M., Toughness Assessment of Ceramic Matrix Composites, Composites Science and Technology, 1990, 37, 223–48

4. Abbé, F., Carin, R., Chermant, J–L., Tensile and Compressive Creep Characteristics from Bending Tests: Application to SiC–SiC Composites, J. Eur. Ceram. Soc., 1989, 5, 201–5

5. Frety, N., Boussuge M., Relationship Between High Temperature Development of Fibre–Matrix Interfaces and the Mechanical Behaviour of SiC–SiC Composites, Composites Sci. Techn., 1990, 37, 177–89

6. Tieghe, N.J., Wiederhorn, S.M., Effect of Oxidation on the Reliability of Silicon Nitride, Fracture Mechanics of Ceramic, vol. 5, ed Bradt et al, Plenum Press, 1983, pp. 403–24

7. Khandelwal, P.K., Chany, J., Heitman, P.W., Slow Crack Growth in Sintered Silicon Nitride, Fracture Mechanics of Ceramics, vol. 8, ed Bradt et al, Plenum Press,1986, pp. 351–62

8. Chen, J., Sjöberg, J., O'Meara, C., Pejryd, L., Corrosion Behaviour of HIPed Si3N4 and SiC/SiC Composite In Simulated Combustion Environments, Proc. this conference

MICROSTRUCTURE AND MECHANICAL PROPERTIES OF HOT ISOSTATICALLY PRESSED WHISKER REINFORCED Si_3N_4 CERAMICS

L.K.L. FALK and H. BJÖRKLUND
Department of Physics, Chalmers University of Technology
S-412 96 Göteborg, Sweden

J.E. ADLERBORN and H.T. LARKER
ABB Cerama AB
S-915 00 Robertsfors, Sweden

ABSTRACT

Si_3N_4 ceramics reinforced with SiC or β-Si_3N_4 whiskers and formed by HIP using the ABB Cerama glass encapsulation process have been characterized by analytical electron microscopy and x-ray diffractometry. Additions of 2.5 wt% Y_2O_3 and 0.2 wt% Fe_2O_3 as metal oxide sintering additives had a significant influence on matrix microstructure and the mechanical properties of the composites. The metal oxides promoted the development of a fibrous matrix which increased room temperature strength and indentation fracture toughness. However, the residual intergranular glass in these materials reduced high temperature strength. Whiskers added together with the metal oxides did not have any significant influence on toughness, while an addition of whiskers to the equiaxed microstructure of material formed without the sintering additives had a clear toughening effect.

INTRODUCTION

Si_3N_4 ceramics are candidate materials in structural applications which require high strength and wear resistance as well as resistance to corrosion/oxidation at elevated temperatures. It has been demonstrated, that the inherently low fracture toughness of most Si_3N_4 ceramic materials may be increased by additions of whiskers [1-4]. Among proposed mechanisms for the toughening effect are crack deflection by the whiskers and whisker debonding,bridging and pullout [1,2,4].

Densification of Si_3N_4 powder compacts generally requires the addition of metal oxide sintering additives. However, hot isostatic pressing (HIP) makes it possible to produce dense

Si_3N_4 ceramics with only small, or in some cases without, additions of oxide sintering aids [3,5,6]. These materials contain comparatively small amounts of residual intergranular glass and will therefore show a higher fracture strength at elevated temperatures [7]. HIP also promotes formation of an isotropic material and enables the production of complex shaped components to close tolerances [5].

Mechanical properties of Si_3N_4 ceramics reinforced with SiC or β-Si_3N_4 whiskers and densified by HIP have previously been studied in some detail [3]. The present paper concerns a microstructural characterization of different whisker reinforced Si_3N_4 ceramics formed by HIP at ABB Cerama AB using the Cerama glass encapsulation technique. The aim was to establish the relation between starting powder composition, microstructure and mechanical properties viz. room temperature indentation fracture toughness and strength as well as high temperature strength.

EXPERIMENTAL

The Si_3N_4 starting powder was UBE SNE 10, and the ceramics were formed both without and with a simultaneous addition of 2.5 wt% Y_2O_3 and 0.2 wt% Fe_2O_3 as metal oxide sintering additives. The matrix powder was ball milled in methanol using Si_3N_4 milling media and subsequently mixed with 25 wt% SiC whiskers (Tateho, SCW-#1) or 25 wt% β-Si_3N_4 whiskers (UBE-SN-WB). Unreinforced ceramics were also formed as reference materials. Cold isostatically pressed green bodies were degassed in a vacuum furnace prior to densification by HIP using the ABB Cerama glass encapsulation technique. A detailed description of the preHIP processing is given in reference [3].

TABLE 1

Powder compositions, phase compositions and process conditions for the examined specimens.

oxide addition	whisker addition (wt%)	phase composition	densification process T (°C)	t (h)	P (MPa)
--	--	β-Si_3N_4	1950	2	250
--	25 SiC	β-Si_3N_4, Si_2N_2O, β-SiC	1950	2	250
--	25 β-Si_3N_4	β-Si_3N_4, Si_2N_2O	1950	2	250
LD	--	β-Si_3N_4, Si_2N_2O	1775	1	200
LD	25 β-SiC	β-Si_3N_4, Si_2N_2O, β-SiC	1775	1	200
LD	25 β-Si_3N_4	β-Si_3N_4, Si_2N_2O, β-$Y_2Si_2O_7$	1775	1	200

Legend: LD = low-doped; 2.5 wt% Y_2O_3 and 0.2 wt% Fe_2O_3 added

The compositions of the Si_3N_4 ceramics in the present investigation are given in TABLE 1 together with pressure, temperature and time of HIP. Considering the oxygen content of the starting powders and whiskers, the bulk densities of all samples were above 99.5 % of the theoretical value calculated from the starting powder compositions [3]. Microstructural characterization was carried out by scanning electron microscopy (SEM) of fracture surfaces and plasma etched polished surfaces using a CamScan 4S-80DV scanning electron microscope, and by transmission electron microscopy (TEM) using a Jeol 2000-FX TEM/STEM/SEM instrument equipped with a Link EDX system. Phase compositions were determined by x-ray diffractometry of polished sections.

RESULTS AND DISCUSSION

General

The phase compositions of the different materials, as determined by x-ray diffractometry, are given in TABLE 1. The α-Si_3N_4 in the starting powder was converted to the β structure in all materials. It can also be noted that both metal oxide and whisker additions resulted in a

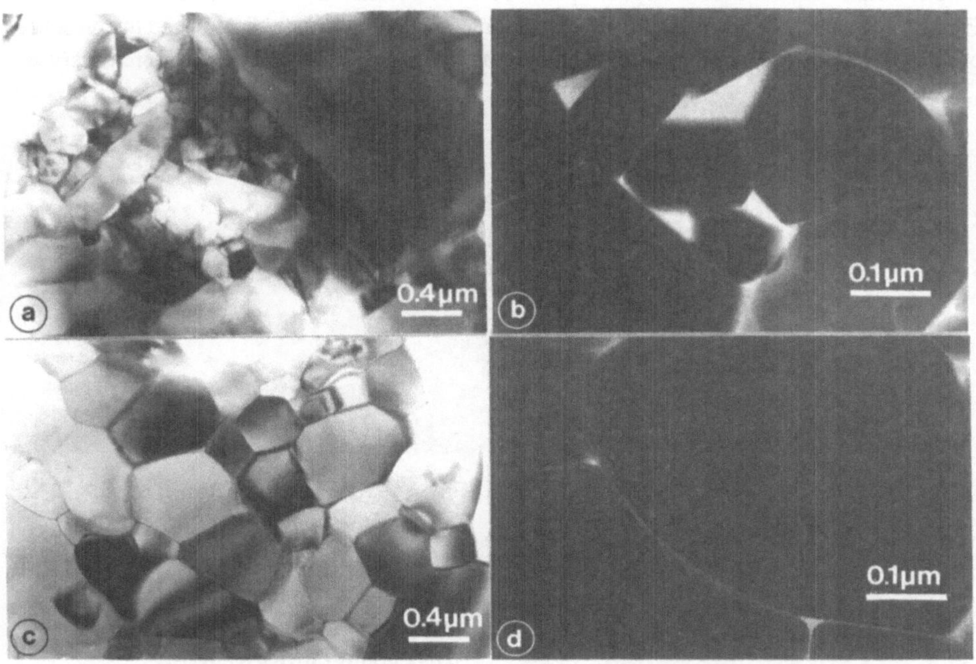

Figure 1. The general microstructure of unreinforced ceramics formed with, (a) and (b), and without, (c) and (d), the metal oxide additives. (b) and (d) are TEM centered dark field images formed by diffuse scattered electrons and the intergranular glassy phase appears bright.

formation of Si_2N_2O due to an increased oxygen content; the whisker materials contained a certain amount of impurity oxygen concentrated to the whisker surfaces [8,9].

Matrix Morphology

Additions of the metal oxide sintering additives had a clear influence on Si_3N_4 grain shape and size and intergranular microstructure. This was clearly demonstrated by the characterization of the unreinforced reference materials. The microstructure of the material formed with 2.5 wt% Y_2O_3 together with 0.2 wt% Fe_2O_3 contained a large number of prismatic β-Si_3N_4 grains characteristic of growth in an isotropic liquid phase environment, Figures 1a and b. Also, the Si_3N_4 grain size varied significantly in this material. The β grains were separated by thin films of residual glass merging into pockets at multi grain junctions, Figure 1b. This intergranular glass was rich in Y and Si but did not contain any detectable amounts of Fe; the Fe was concentrated to Fe rich intergranular particles present in larger pockets. EDX in the STEM and selected area electron diffraction in the TEM indicated that these particles were Fe silicides. This implies that the small Fe_2O_3 addition participated in local liquid phase formation during densification, and that the silicides formed from the oxynitride liquid phase sintering medium during cooling [7].

Without oxide additives, the Si_3N_4 grains had a rounded and equiaxed shape and only smaller volumes of residual amorphous phase could be detected in these microstructures, Figures 1c and d. The surface silica on the starting Si_3N_4 powder particles caused the presence of thin glassy films separating adjacent grains, and in some cases small pockets were observed

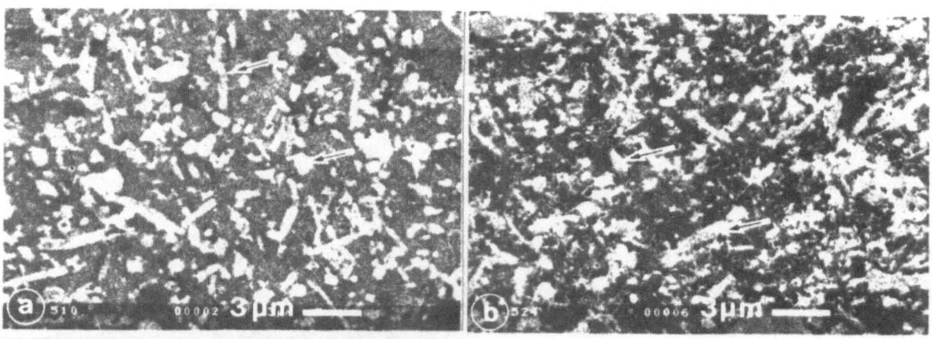

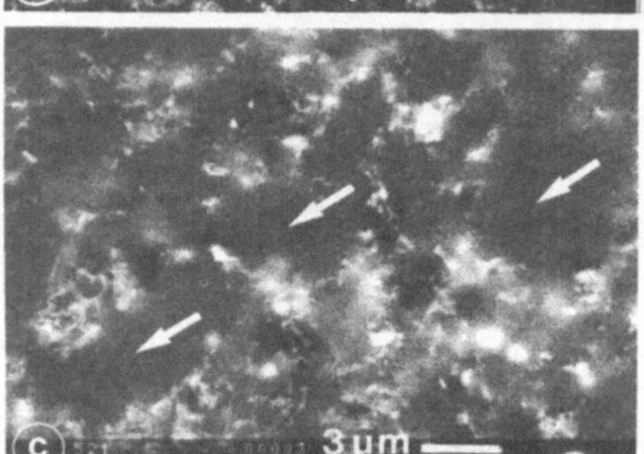

Figure 2. Plasma etched surfaces of ceramics formed with (a) SiC whiskers (arrowed), (b) SiC whiskers (arrowed) together with metal oxides and (c) β-Si_3N_4 whiskers and metal oxides. Note the large Si_3N_4 grains (arrowed) in (c).

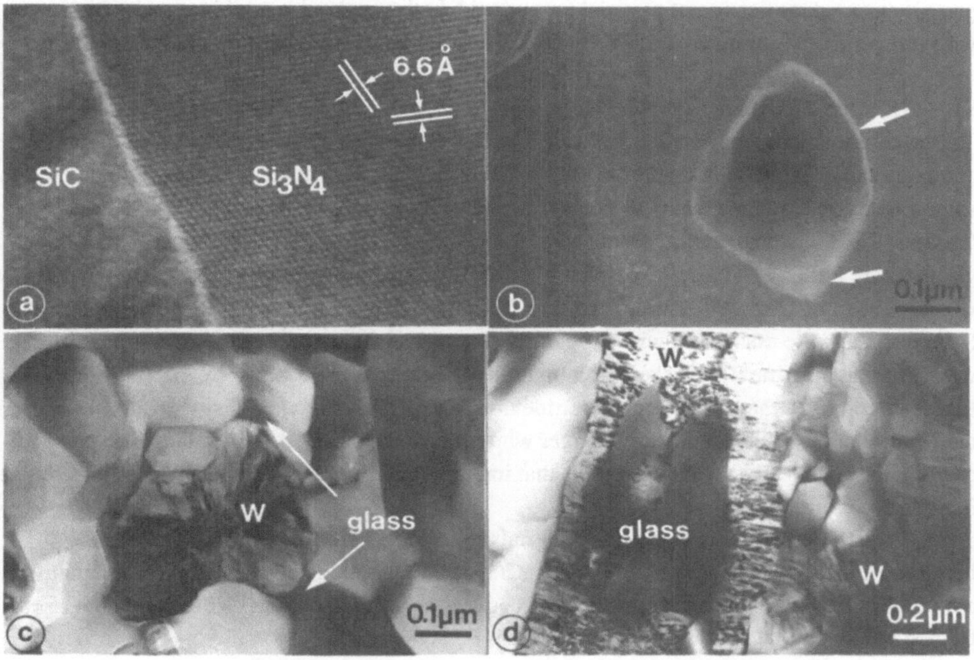

Figure 3. TEM of ceramics reinforced with SiC whiskers. (a) Whisker/Si_3N_4 boundary in material formed without metal oxides. (b) Whisker cavity associated with Ca-rich amorphous phase (arrowed). (c) and (d) Whiskers (W) in material formed with the metal oxide additives.

at multi grain junctions. This correlation between starting powder composition and resulting microstructure has previously been identified in other Si_3N_4 based ceramic materials formed by HIP [6].

These differences in microstructure between the two reference materials were also observed in the matrices of the β-Si_3N_4 and SiC whisker reinforced materials.

β-Si_3N_4 Whisker Reinforced Material

The microstructure of the two materials formed with β-Si_3N_4 whiskers contained significantly larger β grains dispersed in a relatively fine grained matrix, Figure 2c. These larger grains did probably originate from the whisker addition.

The β-Si_3N_4 whiskers, which had an impurity content of around 0.8 wt% Y [8], caused locally increased residual glass volumes when added to the material formed without metal oxide sintering additives. Pockets of Y rich residual glass were identified in some parts of the microstructure which implies the presence of small local volumes of Y-rich liquid during densification. This material did also contain Fe rich impurity particles; presumeably caused by the Fe impurity in the β-Si_3N_4 whiskers [7]. The chemistry of the oxynitride liquid phase present during densification of material formed with both metal oxide additives and β-Si_3N_4 whiskers promoted formation of β-$Y_2Si_2O_7$ in some intergranular pockets.

SiC Whisker Reinforced Material

The general microstructures of materials reinforced with SiC whiskers are shown in Figures 2a and b. In the TEM, the SiC whiskers showed the characteristic faulted structure [9], and did in some cases also contain cavities associated with a Ca-rich amorphous phase, Figure 3b-d. Previous TEM characterization of this type of SiC whiskers has also shown that some of these whiskers are hollow [9]. As illustrated in Figure 3d, the metal oxide additives reacted with the whiskers in some parts of the microstructure during densification; larger glass pockets, rich in Si, Y and Ca, were associated with irregularly shaped whiskers. Also, thin amorphous films were generally separating whiskers from matrix. The morphology shown by the TEM image in Figure 3d indicates that the Y-rich oxynitride liquid phase sintering medium which formed in the vicinity of the whiskers filled out larger whisker cavities during densification. The presence of Ca in these large glass volumes shows that impurity Ca from the SiC whiskers participated in this liquid formation.

TEM observations of the two materials showed, generally, an uneven interface between whiskers and matrix, Figure 3. This indicates a reaction between whisker and matrix during densification [9]. Intergranular films between whisker and matrix, if any, in the material formed without metal oxides are extremely thin, Figure 3a.

Relation Between Microstructure and Mechanical Properties

The results from measurements of indentation fracture toughness and strength imply that the Si_3N_4 matrix grain morphology and intergranular microstructure have a clear effect on the mechanical properties. As demonstrated for other Si_3N_4-based materials, prismatic β-Si_3N_4 grains promote a high toughness and also a high strength at room temperature [6,7]. SEM of room temperature fracture surfaces, Figure 4, indicated that prismatic β-Si_3N_4 grains contribute to toughness by crack deflection and grain pull-out. Additions of whiskers to this type of microstructure did not have any significant influence on toughness, Figure 5.

Material formed without metal oxide additives and whiskers showed relatively smooth fracture surfaces due to the mainly fine grained and equiaxed microstructure. The two types of whisker additions to this microstructure had a clear toughening, as well as strengthening, effect, and the SEM images in Figure 4 indicate that crack deflection by the whiskers was responsible for the observed toughness increase.

The unreinforced material formed with metal oxide additives did also have the highest room temperature strength. Additions of SiC whiskers to this microstructure did not have any significant influence on the room temperature strength, but the β-Si_3N_4 whisker addition reduced strength significantly. This suggests that the addition of β-Si_3N_4 whiskers introduced flaws in the Si_3N_4 microstructure; these flaws could possibly be the significantly larger Si_3N_4 grains in these materials. The strength reduction at 1350 °C was significant for the materials formed with metal oxide sintering additives which could be explained by softening of the residual intergranular glass [7]. Stepped temperature stress rupture (STSR) tests in the

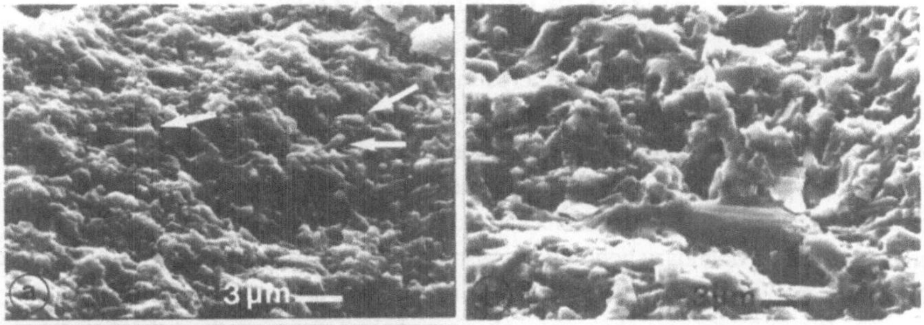

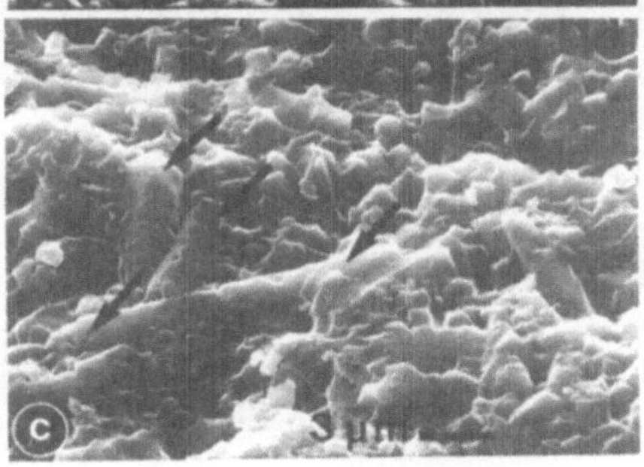

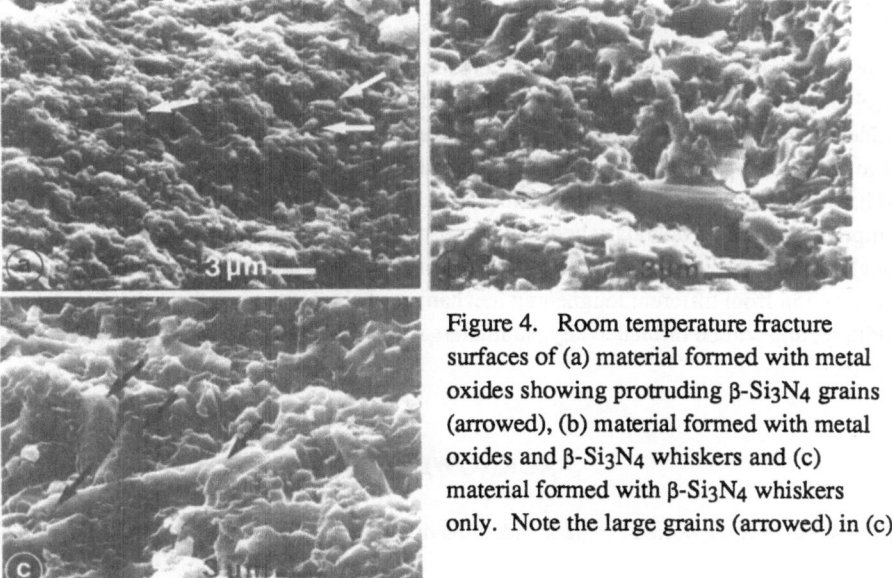

Figure 4. Room temperature fracture surfaces of (a) material formed with metal oxides showing protruding β-Si_3N_4 grains (arrowed), (b) material formed with metal oxides and β-Si_3N_4 whiskers and (c) material formed with β-Si_3N_4 whiskers only. Note the large grains (arrowed) in (c).

temperature range 1000 to 1400 °C has previously shown that addition of SiC whiskers results in a material with better high temperature stability than material containing β-Si_3N_4 whiskers [3]; however, Si_3N_4 ceramics formed without any additives had the longest life time in this test.

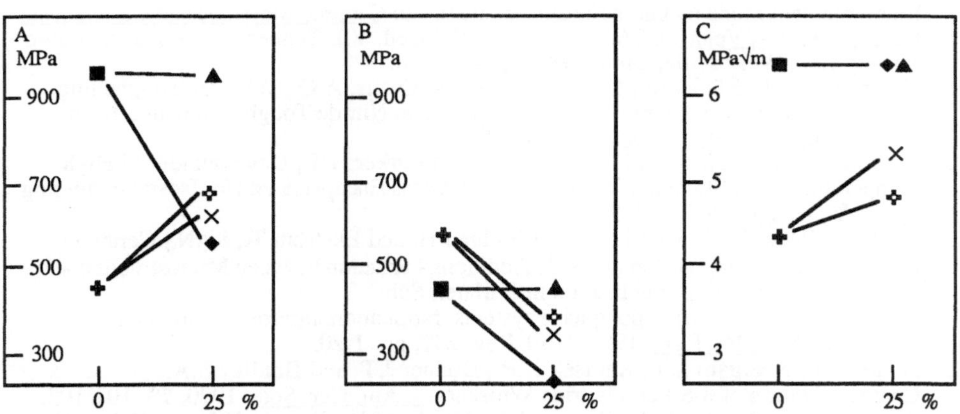

Legend: (✚) is undoped Si_3N_4, (✕) + β-Si_3N_4 whiskers, (✣) + SiC whiskers
(■) is low-doped Si_3N_4, (◆) + β-Si_3N_4 whiskers, (▲) + SiC whiskers

Figure 5. (A) 3-point flexural strength at room temperature. (B) 4-point flexural strength at 1350° C. (C) Indentation fracture toughness at room temperature. From reference [3].

CONCLUDING REMARKS

This investigation indicate that matrix microstructure has a significant influence on the mechanical properties of this type of whisker reinforced Si_3N_4 ceramics. Addition of metal oxides affect grain morphology and also the grain boundary structure by the increased amount of residual glass. Development of a fibrous microstructure, which is promoted by sintering additives, appears essential for good mechanical properties at low and intermediate temperatures. Grain boundary bonding can also be expected to have a significant influence on toughness and strength [10]. The chemistry of the residual intergranular films will affect the contribution from different toughening mechanisms [4]. Apart from the type of metal oxide additive, this will be influenced by parameters such as whisker surface structure and whisker purity.

ACKNOWLEDGMENT

The project was supported by the Swedish National Board for Technical Development (STU).

REFERENCES

1. Becher P.F., Chun-Hway H., Angelini P. and Tiegs T.N., Toughening Behavior in Whisker-Reinforced Ceramic Matrix Composites. J. Am. Cer. Soc., 1988, **71**, 1050-1061.
2. Buljan S.T., Baldoni J.G. and Huckabee M.L., Si_3N_4-SiC Composites. Am. Cer. Soc. Bull., 1987, **66**, 347-352.
3. Larker H.T. and Adlerborn J.E., Undoped and Low-doped Silicon Nitride with Whisker Reinforcement Made by Glass Encapsulated HIP. In Ceramic Materials and Components for Engines; Las Vegas, USA, Nov. 27-30, 1988, ed. V.J. Tennery, American Ceramic Society Inc., Westerville, Ohio, 1989, pp. 227-236.
4. Campbell, G.H., Rühle, M., Dalgleish, B.J. and Evans, A.G., Whisker Toughening: A Comparison Between Aluminium Oxide and Silicon Nitride Toughened with Silicon Carbide. J. Am. Ceram. Soc., 1990, **73**, 521-30.
5. Adlerborn J., Burström M., Hermansson L. and Larker H.T., Development of High Temperature High Strength Silicon Nitride by Glass Encapsulated Hot Isostatic Pressing. Mater & Design, 1987, **8**, 229-232.
6. Knutson-Wedel E.M., Falk L.K.L., Björklund H. and Ekström T., Si_3N_4 Ceramics Formed by HIP using Different Oxide Additions - Relation between Microstructure and Properties. Accepted for publication in J. Mater. Sci.
7. Lange F.F., Silicon nitride polyphase systems: fabrication, microstructure, and properties. Int. Met. Rev., 1980, No.1 Rev. 247, pp. 1-20.
8. Homeny J., Neergard L.J., Karasek K.R., Donner J.T. and Bradley S.A., Characterization of β-Silicon Nitride Whiskers. J. Am. Cer. Soc., 1990, **73**, 102-105.
9. Karasek K.R., Bradley S.A., Donner J.T., Martin M.R., Haynes K.L.and Yeh H.C., Composition and microstructure of silicon carbide whiskers. J. Mater. Sci., 1989, **24**, 1617-1622.
10. Tanaka I., Pezzotti G., Miyamoto Y. and Okamoto T., Fracture toughness of Si_3N_4 and its Si_3N_4 whisker composite without sintering aids. J. Mater. Sci., 1991, **26**, 208-210.

SINTERING AND TOUGHENING OF SILICON NITRIDE-BASED COMPOSITES

S. HAMPSHIRE, Y-J. SONG, D. O'SULLIVAN and V. GUNAY
Materials Research Centre, University of Limerick,
Plassey Technological Park, Limerick, Ireland

ABSTRACT

The properties of silicon nitride based ceramics, particularly at high temperatures, are controlled by secondary intergranular phases, normally glassy, which are present as a result of sintering additives. Properties of the sintered product can be improved by post-sintering heat treatments to crystallise the grain boundary glass. Second-phase reinforcements, such as SiC platelets or whiskers, offer further possibilities for improvements in properties and this paper describes sintering studies of silicon nitride with SiC inclusions, densified with mixed oxide additives such as Y_2O_3 + Al_2O_3. Densities as high as the matrix can be achieved by pressureless sintering with 10% platelets. SiC platelets allow increases in toughness not shown by whisker reinforcements. Toughening mechanisms involve crack deflection around the platelets.

INTRODUCTION

Ceramics based on silicon nitride have been among the leading candidates for potential applications in advanced heat engines [1-4] because of their intrinsically high strength and good creep and thermal shock resistance. However, ceramics are very susceptible to fracture during impact or under stress in the presence of flaws and, thus, microstructural design and development are crucial for the improvement of fracture toughness.

With regard to fabrication of silicon nitride ceramics, the only cost-effective route to the production of complex-shaped components is by pressureless densification of silicon nitride powder with oxide additives which allow conditions for liquid phase sintering [5] by particle rearrangement, solution/precipitation and more rapid diffusion of material which leads to the attainment of near theoretical density. The resulting microstructure consists of interlocking β-Si_3N_4 grains [6-8] with a residual secondary phase at the grain boundaries which can control the properties of the ceramic. As a crack propagates through silicon nitride, it encounters a combination of fibrous-like β grains and secondary grain boundary material which tends to steer the crack along a tortuous path and, thus, more energy is required for propagation, resulting in a higher toughness than for a single-phase ceramic with equi-axed grains. A decreasing β-grain aspect ratio results in lower strength and lower fracture toughness [8]. By using mixed oxide additives such as MgO/Y_2O_3 and MgO/Nd_2O_3 combinations and by varying the total amount and the molar ratio of the additives, the volume and viscosities of the sintering liquids can be varied, both of which affect the densification, the α-β transformation, grain growth and morphology and the type of secondary phase formed [9].

Promising results have been obtained [10] for the $Si_3N_4/MgO/Nd_2O_3$ system in which it is possible to achieve high density and β-silicon nitride grains with aspect ratios in the range 7-10 and with a secondary phase apatite of composition $MgNd_4Si_3O_{13}$. There is an effective "microstructural toughening" in silicon nitride before any reinforcing inclusions are introduced. While this type of microstructure in itself leads to better toughness via a crack-deflection mechanism, much more significant increases should be possible by whisker or platelet toughening. In this case, high aspect ratio inclusions are mixed into the matrix material prior to densification and the microstructure produced should result in toughening via crack deflection, whisker-bending and/or inclusion pullout during crack opening.

In this context, the aim of the present work was to investigate the difficulties of producing whisker or platelet reinforced silicon nitride matrix composites by a process of pressureless sintering using mixed oxide additives.

TABLE 1. SiC INCLUSIONS

	SiC Platelets			SiC Whiskers
PURITY	99.9% α-SiC			99+% SiC <0.5% SiO_2 +Cr,Co,Fe (traces)
SPECIFIC GRAVITY	3.22			3.2
	Coarse (C)	Medium (M)	Fine (F)	
DIAMETER (μm)	20-70	10-40	<10	0.1 - 1.0
ASPECT RATIO	-			50-200

MATERIALS AND METHODS

The characteristics of SiC platelets are compared with the SiC whiskers in Table 1. The silicon nitride matrix consists of Starck LC12 powder (95% α-phase, 0.8μm mean particle size) +MgO/Nd_2O_3 or Y_2O_3/Al_2O_3. The experimental approach is summarised in the following scheme:

<u>MIXING/DISPERSION</u>	Inclusion/Powders (in Isopropanol)
↓	(Packing Considerations)
<u>FORMING</u>	Slip Casting + Isostatic Pressing
↓	
<u>DRYING</u>	Removal of Binder
↓	
<u>FIRING</u>	Pressureless sintering under Powder Bed in N_2
↓	atmosphere (also Hot-pressing)
<u>CHARACTERISATION</u>	Density, Weight Loss, X-ray Analysis, SEM.

Samples were prepared containing from 0 to 30 vol% SiC inclusions by a wet-mixing technique. The difficulties associated with the incorporation of the whiskers into the silicon nitride powder matrix were assessed. Packing efficiency is seriously affected by inclusion of whiskers and this also affects green density as shown in Figure 1. With platelets there is an increase in green density as a result of a better packing efficiency as normally expected for fine/course two-component mixtures.

Firing was carried out under a BN/Si_3N_4 powder bed for 1 hour at 1500°-1700°C in a nitrogen atmosphere (pressure 0.1MPa). Green and fired densities of the samples were measured with the use of a mercury displacement balance. The phases present after firing were analysed by means of X-ray diffraction using a Guinier-Hagg focusing camera and CuKα X-radiation. Microstructural examination was performed using a Jeol JSM 840 SEM for energy-dispersive microanalysis. Fracture toughness was measured using an indentation technique.

RESULTS & DISCUSSION

Pressureless Sintering of Silicon Nitride/SiC Composites

Figure 2 shows the effect of increasing amounts of SiC whiskers on the densification of silicon nitride composites (17 mol%, MgO:Nd_2O_3 = 1.67) at different sintering temperatures and it is clear that the presence of the "inert" inclusions inhibits shrinkage. At the higher volume fraction of whiskers, sintered density is lower than green density. Bulk density also decreases with increasing temperature which can be explained by the fact that weight losses are substantial, increasing with the increasing volume of whiskers and with increasing temperature.

It has been suggested [11,12] that sintering can be inhibited by the development of stresses as the matrix contracts around non-shrinking inclusions. If these are substantial then avoidance of cracking or flaw formation is impossible. If the creep rate of the matrix is high relative to the rate of stress build-up, then stress relief may occur. With the aim of modifying the volume and viscosity of the liquid phase in the matrix, additions of Al_2O_3 were made to the basic $Si_3N_4/MgO/Nd_2O_3$ matrix composition. Figure 3 shows the densities achieved for SiC whiskers and platelets (coarse) with and without extra Al_2O_3 additions. With whiskers, there is a sudden fall-off in density whereas with platelets and XA16 Al_2O_3, densities are maintained to 95% theoretical with up to 20 vol %. addition of SiC.

Weight losses are higher in the case of whiskers, but very low for platelets. While SiC and Si_3N_4 appear to be quite compatible, the matrix is very complex chemically consisting of not only silicon nitride but also an oxynitride silicate-type liquid at the firing temperature. SiC whiskers are more unstable in the presence of an oxynitride liquid in nitrogen than are platelets. Reaction may be expected according to:

$$2SiC + SiO_2 + 2N_2 \rightleftarrows Si_3N_4 + 2CO\uparrow \qquad \Delta W/Wo = 0\% \quad (1)$$

$$3SiC + 3SiO_2 + 2N_2 \rightleftarrows Si_3N_4 + 3SiO\uparrow + 3CO\uparrow \qquad \Delta W/Wo = 60\% \quad (2)$$

(1) applies to SiC in contact with a small quantity of liquid while (2) applies to SiC in contact with a much larger amount of liquid. The effective contact area of SiC:liquid depends on the surface/volume ratio of the SiC inclusions which is 50 times higher for the whiskers used here compared with the platelets. It is important to suppress these reactions and to establish conditions in which the SiC whiskers remain stable. Conditions can be varied so that in (1) and (2) above, equilibrium lies to the left or right, but by encouraging the right to left reaction, the matrix itself starts to become unstable. The most effective powder bed was a 50% BN: 50% Si_3N_4 mixture and this has been used in all subsequent

firings.

Figure 4 shows the sintered densities for silicon nitride composites (Y_2O_3/Al_2O_3) containing different size fractions of SiC platelets or SiC grit particles. Densification is easier as the platelet size decreases. With 20 vol% of the fine fraction (<10μm) densities of > 96% are achieved.

Properties of Sintered Composites

Figure 5 compares the effects of SiC inclusions on changes in fracture toughness, K_{Ic}, for silicon nitride composites. There is no significant change in K_{Ic} with addition of whiskers whereas platelets result in an increase in fracture toughness of around 25-35%. Various authors [13,14] have reported on the effects of SiC whisker inclusions on the fracture toughness of different hot-pressed silicon nitrides. The variations in sintering schedules lead to differences in the composite matrix microstructure and this is reflected in different effects from different workers. Where there is evidence that fabrication-related defects were the origins of fracture initiation, composite toughness values are only fractionally better than those for monolithics.

Figure 6 shows the effect of SiC platelet size on the fracture toughness of silicon nitride composites. The SiC grit and the fine fraction (<10μm) platelets allow greater increases in K_{Ic} compared with the coarser fractions.

Figure 7 is a scanning electron micrograph of a sample sintered with 10% platelets. The crack takes a path through the secondary phase and there is evidence of crack deviation and crack branching to the matrix/platelet interface.

CONCLUSIONS

Pressureless sintering of SiC-reinforced silicon nitride matrix composites is possible and densities of > 96% theoretical have been achieved with 20% inclusions. SiC platelets do not inhibit shrinkage as much as SiC whiskers which are also more susceptible to reaction with the oxynitride liquid.

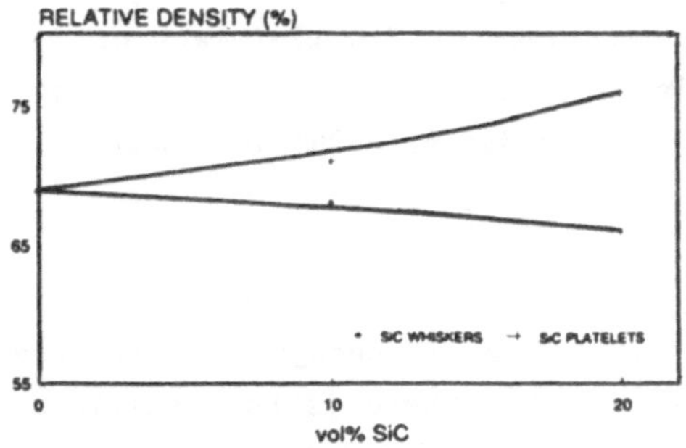

Figure 1. Effect of SiC inclusions on green density of whisker and platelet-reinforced silicon nitride composites.

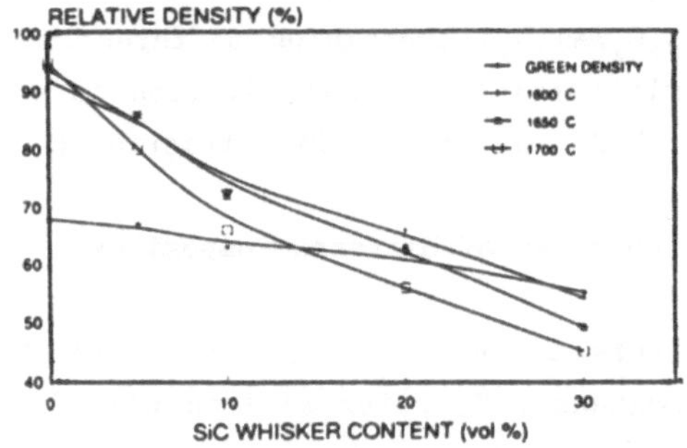

Figure 2. Effect of SiC whiskers on relative densities of SiC-reinforced silicon nitride composites pressureless sintered at 1700°C for 1hr.

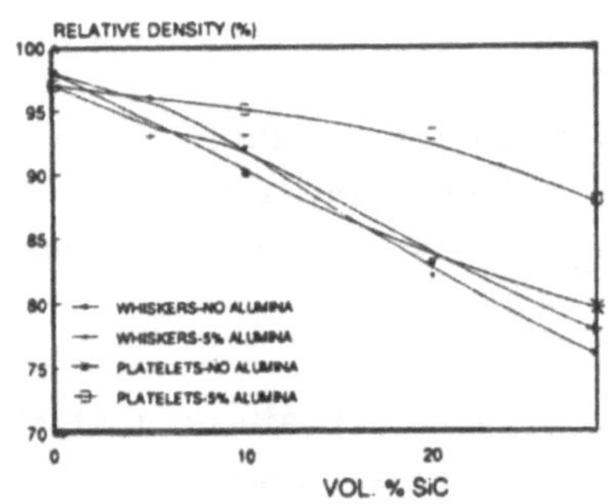

Figure 3. Effect of SiC whisker and platelets on relative densities of silicon nitride composites pressureless sintered at 1700°C for 1hr.

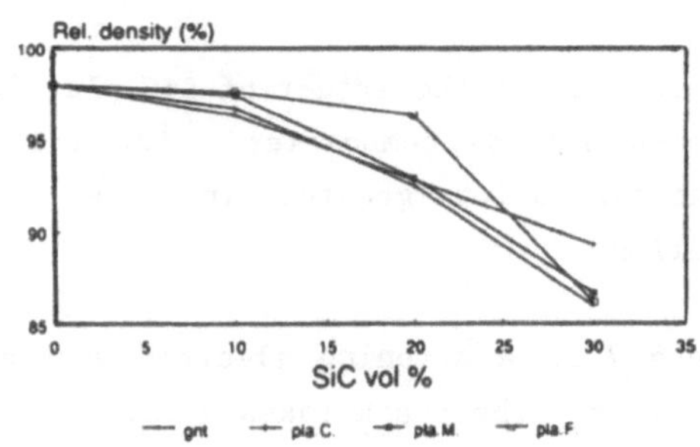

Figure 4. Effect of SiC platelet size on relative densities of silicon nitride composites pressureless sintered at 1700°C for 1hr.

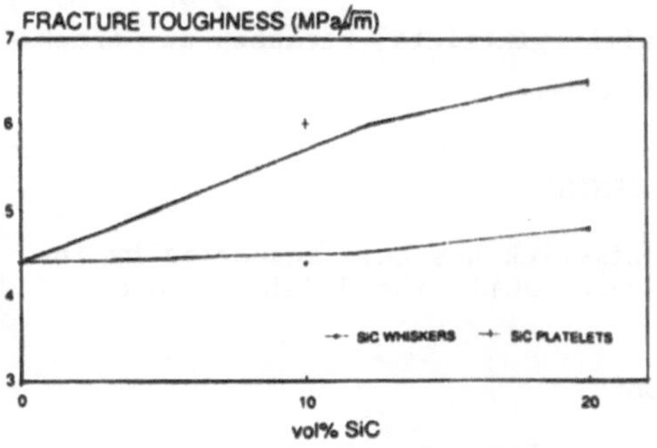

Figure 5. Effects of SiC platelet or whisker inclusions on fracture toughness of sintered silicon nitride composites.

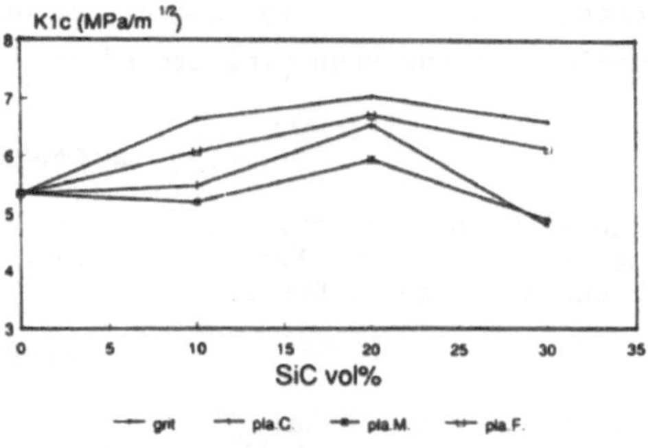

Figure 6 Effects of SiC platelet size on fracture toughness of sintered silicon nitride composites

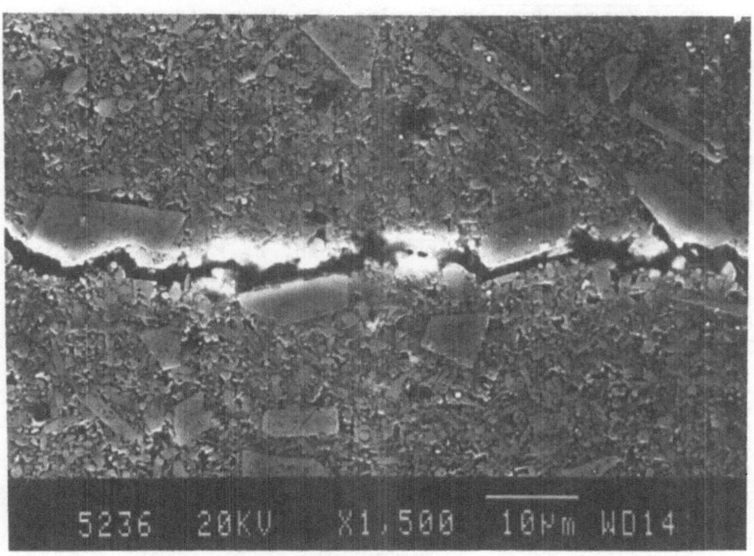

Figure 7. Scanning electron micrographs of 10% SiC platelets-reinforced Si_3N_4 composite showing crack deviation around platelets.

With SiC platelets-reinforced silicon nitride, increases in fracture toughness are observed while with SiC whiskers, there is no significant change in K_{Ic}. Toughening occurs with platelets because of crack deflection and branching around the inclusions.

ACKNOWLEDGEMENTS

The work on silicon nitride ceramics at Limerick has been supported by the Commission of the European Communities and EOLAS, the Irish Science and Technology Agency Dublin.

REFERENCES

1. Somiya, S., Kanai, E. and Ando, K. (eds) Proc. 1st Int. Symp. Ceramic Components for Engineers, KTK Scientific Publisher, Tokyo, 1984.
2. Bunk, W. and Hausner, H. (eds) Proc. 2nd Int. Symp. Ceramic Materials and Components for Engines, DKG, Bad Honnef, 1986.
3. Tennery, V.J. (ed) Proc 3rd Int. Symp. Ceramic Materials and Components for Engines, Am. Ceram. Soc. Inc., 1989.
4. Hampshire, S. (ed) Non-oxide Technical and Engineering Ceramics, Elsevier-Applied Science Publishers, London, 1986.
5. Hampshire, S. and Jack, K.H. The Kinetics of Densification and Transformation of Nitrogen Ceramics. In Special Ceramics 7, ed. P. Popper, Brit. Ceram. Proc.,1981, 31, 37-50.
6. Wotting, G., Kanka, B. and Ziegler, G. Microstructural Development, Microstructural Characterisation and Relation to Mechanical Properties of Dense Silicon Nitride. In Non-oxide Technical and Engineering Ceramics, ed. S. Hampshire, Elsevier-Applied Science Publishers, 1986, pp.83-96.
7. Hampshire, S., The Role of Additives in the Pressureless Sintering of Nitrogen Ceramics for Engine Applications. Metals Forum, 1984, 7, 162-170.
8. Wotting, G. and Ziegler, G., Powder Characteristics and Sintering Behaviour of Si_3N_4 Powders. Interceram, 1986, 35, 32-35.
9. Hampshire, S., Pomeroy, M.J. and Saruhan, B. Kinetics of Sintering of Silicon Nitride with Mixed Oxide Additives. In Hi-tech Ceramics, ed. P. Vincenzini, Elsevier, Mat. Sci. Monogr., 1987, 38A, pp 941-951.
10. Hampshire, S., Leigh, M., Morrissey, V.J., Pomeroy, M.J. and Saruhan, B., Crystallisation Heat-Treatments of Silicon Nitride Ceramics and Glass-Ceramics Containing Neodymia. In Proc. 3rd Int. Symp. Ceramic Materials and Components for Engines., ed. V.J. Tennery, Am. Ceram. Soc. Inc., 1989, pp.434-442.
11. Raj. R. and Bordia, R.K., Sintering Behaviour of Bi-Model Powder Compacts. Acta Metal., 1984, 32, 1003-1019.
12. Hsueh, C.H., Evans, A.G., Cannon, R.M. and Brook, R.J., Acta. Met., 1986, 34, 927.
13. Buljan, S.T. and Sarin, V.K., Composites, 1987, 18, 99.
14. Shalek, P.D., Petrovic, J.J., Hurley, G.F. and Gac, F.D., SiC Whisker-Hot Pressed Si_3N_4 Matrix Composites. Am. Ceram. Soc. Bull., 1986, 65, 351-356.

PROCESSING ROUTES AND MICROSTRUCTURE OF CERAMIC-MATRIX/PLATELET COMPOSITES

THOMAS CLAASSEN AND NILS CLAUSSEN
Technische Universität Hamburg-Harburg
Postfach 90 10 52, 2100 Hamburg 90, Germany

ABSTRACT

Platelet reinforcement of ceramics represents a viable alternative to whisker reinforcement. It is evident that the effectivity of platelets as reinforcing elements significantly depends on the composite microstructure. Homogenous platelet dispersion in a matrix sintered to high density is required. Samples made by different processing routes (attrition milling, drying, CIP; ultrasonic dispersion, drying, CIP) have been characterized. Ultrasonic dispersion promotes homogeneous platelet distribution and enhanced densification behaviour but complete densification of a composite with randomly distributed platelets is limited to a platelet content of about 15 vol.%. A method for the preparation of composites with oriented platelets by tape casting and lamination technique is described. Improved densification is achieved with oriented platelets.

INTRODUCTION

The mechanical properties of brittle ceramic materials can be improved by dispersing small high-strength single crystals (whiskers, platelets, particles) in a ceramic matrix. Most investigations were carried out on SiC-whisker toughening but since it is known that whiskers may cause severe health hazard by inhalation of even small amounts [1], platelets are thought to be an attractive alternative [2]. Compared to whiskers, the advantages of platelets are uncritical geometry causing no health hazard, much lower price, higher thermal stability, and smoother and more perfect surfaces. The "two dimensional" geometry of the platelets is disadvantageous for composites with randomly distributed reinforcing components. Compared to whiskers, platelets tend more easily to form clusters and stiff networks which hinder complete densification. The strength of these composites will be controlled by the diameter of the platelets as the minimum critical defect size. Types and dimensions of presently available platelets are summarized in Table 1.

TABLE 1
Dimensions of Al_2O_3- and SiC-platelets

Producer	Type	Diameter (μm)	Thickness (μm)
	SiC-platelets		
C-Axis Technology, Montreal, Canada	SCP-SF	2 - 25	0.5 - 2
	SCP-F	5 - 40	0.5 - 4
	SCP-M	10 - 90	1 - 7
	Al_2O_3-platelets		
Hüls AG, Marl, Germany	Dycron 13	15 - 45	2 - 6
	Dycron 30	5 - 30	1 - 5
Showa Denko, Japan	AC-13	10 - 75	0.5 - 3
Atochem, Paris, France		2 - 45	0.3 - 2

The reinforcing potential of platelets is considered to be comparable to those of whiskers. Mechanisms which increase the crack resistance may be debonding of the platelet-matrix interface, pull-out, bridging behind the crack tip, crack deflection, and load transfer (Fig. 1). These mechanisms contribute in a synergetic manner to a fracture toughness increase with respect to the pure matrix depending sensitively on interfacial strength, aspect ratio, volume fraction and orientation of the platelets to the applied stress field. A theoretical model for the system ZrO_2 (TZP) + Al_2O_3-platelets has been postulated by Heussner [3] for both randomly distributed and oriented platelets. Heussner predicts an increase in fracture toughness with increasing volume fraction of platelets in both cases extending the model of Faber and Evans [4] for his calculations. Crack deflection is suggested to be the most effective mechanism in composites with randomly distributed platelets. The main toughening mechanisms in composites with oriented platelets are pull-out and bridging depending sensitively on the interfacial strength. An estimation of the two different cases predicts a higher toughness increase for oriented platelets with an orientation normal to the tensile stress plane. Composites with such a microstructure may also comprise only little loss of strength due to the small platelet thickness which defines the critical flaw size.

The microstructural conditions for an optimized platelet composite can be summarized: high platelet aspect ratio, high platelet volume fraction, homogeneous platelet distribution, completely densified composite, and oriented platelets.

The aim of the investigations presented in this paper is to show limitations that will be faced during the fabrication of platelet composites originating from processing conditions and matrix shrinkage and approaches to improve the microstructure with respect to the ideal conditions listed above.

Mechanical properties of platelet composites have been reported for the systems TZP + Al_2O_3-platelets [2/3/5], mullite + SiC- and Al_2O_3-platelets [6], Al_2O_3 + SiC-platelets [2], and RBSN + SiC-platelets [2]. Oxide matrices show an improved toughness up to 70% at platelet contents between 10 - 15 vol.%. Higher platelet contents generally lead to decrease in fracture toughness due to insufficient densification of the composite. The loss of strength is severe even with small amounts of platelets. Post-HIP densified RBSN with SiC-platelets show a 30% increase of fracture toughness up to a platelet content of 30 vol.%. These composites are completely densified as a result of a low shrinking matrix.

EXPERIMENTAL PROCEDURE

For the experiments Al_2O_3-matrix powder (Ceralox APA-P2X, Condea Chemie, Hamburg, Germany) and SiC-platelets of the grades SCP-SF, SCP-F and SCP-M (table 1) have been used. In addition the platelets SCP-F and SCP-M were wet sieved in isopropanol through sieve sizes < 25 μm, 25 - 45 μm, 45 - 80 μm and > 80 μm.

The three investigated processing routes are schematically shown in Fig. 2 and Fig. 3. Each processing route was also carried out with platelet-free matrix powder.

The main investigations were concentrated on the processing routes shown in Fig. 2. For these experiments, platelet volume fraction (0.00, 0.10, 0.15, 0.20, 0.30) and platelet size (SCP-SF, -F<25μm, -M25-45μm, -M45-80μm) were varied. Green body characterization was carried out by geometrical green density measurements, mercury porosimetry measurements and SEM studies. The sintering behaviour has been studied by dilatometric measurements at 1550°C under vacuum. All samples were densified in a sinter-HIP cycle with the steps 1550°C, 60 min., vacuum and 1550°C, 30 min., 200 MPa Ar. The sintered bodies were characterized by density, 4-point bending strength (40/20mm cell, 3x3 mm bars, 4-7 samples), fracture toughness (ISB, 4-point bending), and SEM studies of fracture surfaces.

The fabrication of platelet composites with oriented platelets was carried out according to the processing route shown in Fig. 3. For slip preparation, an organic binder system with the components MEK and ethanol as solvent, alkylphosphate as dispersant, polyvinylbutyral as binder, and polyethylene-glycol and dibutylphtalate as plasticizers were used. For these experiments, platelet volume fraction (0.00, 0.10, 0.15, 0.20) and platelet size (SCP-SF, -F<25μm, -M25-45μm) were varied. Pressureless sintering was carried out at 1550°C for 1 h in nitrogen. Characterization of sintered samples was carried out with the same procedure as described above.

The influence of attrition milling on platelet size in a powder-platelet mixture was studied by XRD measurements. The (110) and basal plane peaks of α-SiC were compared to the (104) peak of α-Al_2O_3.

RESULTS AND DISCUSSION

The influence of attrition milling time on platelet size is shown in Fig. 4 and compared with ultrasonic dispersion. The decrease of relative intensity of the basal plane reflex of α-SiC is related to a reduction of platelet diameters even at short milling times. Also the aspect ratio is reduced because the thickness of the platelets is not reduced on milling as confirmed by the nearly constant relative intensity of the (110)-peak. Ultrasonic dispersion leaves the platelets undamaged.

The mechanical properties of pure Al_2O_3 samples made by the processing routes shown in Fig. 2 are listed in Table 2. Obviously, the time of ultrasonic dispersion influences both green density and lexural strength but not the final density. Although attrition milling leads to similar densities, these samples not only comprise a relative low strength but also a high standard deviation. Ultrasonic dispersion leads to significantly lower flaw size even if the following processing steps are not optimized to produce high strength ceramics. The high strength and fracture toughness of the laminate are due to a different microstructure. The Al_2O_3 grains are slightly elongated parallel to the laminated tapes.

TABLE 2
Mechanical properties of Al_2O_3 fabricated by different processing routes

Processing route	relative green density (%)	sd	relative final density (%)	sd	flexural strength (MPa)	sd
4 h attrition milling, CIP	60.2	3.7	99.6	0.2	413	47
10 min. ultrasonic dispersion, CIP	59.7	5.2	99.6	0.2	392	10
15 min. ultrasonic dispersion, CIP	60.4	3.1	99.7	0.1	470	28
20 min ultrasonic dispersion, CIP	62.3	2.9	99.7	0.2	523	17
tape casting 150 μm, lamination of 30 tapes	60.8		99.7	0.3	609	32

Fig. 5 and Fig. 6 show the densities of sinter-HIP densified Al_2O_3/SiC-platelet composites. In the case of the platelet-type SCP-M (25-45 μm) composites, ultrasonic dispersion improves the densification to closed porosity during the sintering stage of the sinter-HIP cycle. Further densification is achieved during the high pressure stage of the sinter-HIP cycle (Fig. 5). No closed porosity was obtained for either attrition milled samples and for samples with platelet contents of 20 vol.% and higher. Both green and fired density of composites with small platelets of type SCP-SF exhibit no significant difference with respect to the applied processing route (Fig. 6). Closed porosity, as required for HIP densification was not obtained from these samples.

The microstructure of samples with 30 vol.% platelets is shown in Fig. 7 and Fig. 8. The formation of stiff platelet networks with cells of matrix material is visible. Large pores around the platelets originate from a shrinkage of the matrix in contrast to the rigid platelet inclusions

The microstructure of a composite with high final density is shown in Fig. 9 and Fig. 10. The matrix is completely densified while larger pores result from platelet clusters.

The densification behaviour was significantly improved for composites with oriented platelets. Relative densities of 93.2 % (random distribution) and 95.3 % (orientation) were obtained for pressureless sintered samples with 20 vol.% platelets. The microstructure of a composite with oriented platelets is shown in Fig. 11. The platelets are well aligned and in most cases seperated from each other.

REFERENCES

1. Birchall, J. D., Stanley, D. R., Mockford, M. J., Pigott, G. H., Pinto, G. J., Toxicity of Silicon Carbide Whiskers, J. Mater. Sci. Lett., 1988, 7, 350-352.

2, Claussen, N., Ceramic Platelet Composites. In Structural Ceramics-Processing, Microstructure and Properties, Risö International Symposium on Metallurgy and Materials Science, Roskilde, 1990,1-12.

3. Heussner, K.-H., Mechanische Eigenschaften von ZrO_2(TZP)/Al_2O_3-Platelet-Verbundwerkstoffen, Dissertation, Technische Universität Hamburg-Harburg, 1990.

4. Faber, K. T., Evans, A. G., Crack Deflection Process. I: Theory, Acta Metall., 1983, 31(4), 565-576.

5. Heussner, K.-H., Claussen, N, Yttria and Ceria-Stabilized Tetragonal Zirconia Polycrystals (Y-TZP, Ce-TZP) Reinforced with Al_2O_3 Platelets, J. Eur. Ceram. Soc., 1989, 5, 193-200.

6. Nischik, C., Seibold, M., Travitzky, N. A., Claussen, N., Effect of Processing Parameters on Mechanical Properties of Platelet-Reinforced Mullite Composites, J. Am. Ceram. Soc., 1991, 74(10), (...).

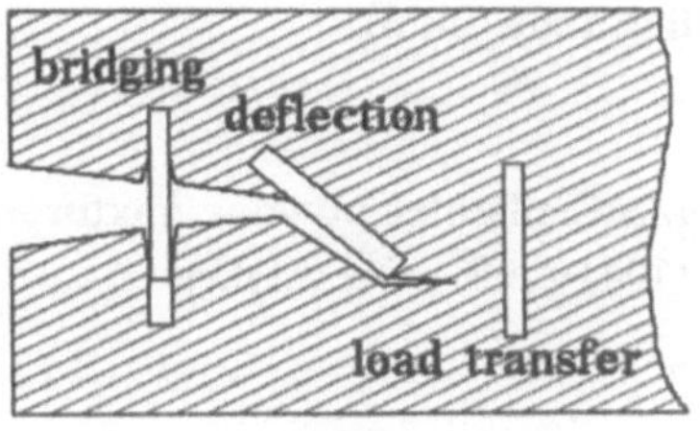

Figure 1. Toughening mechanisms in platelet composites.

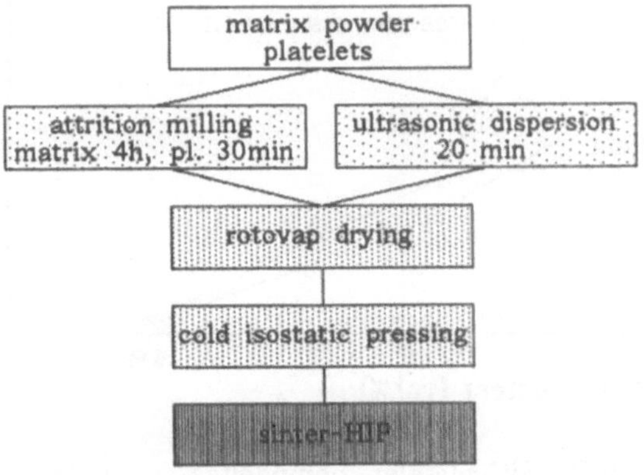

Figure 2.

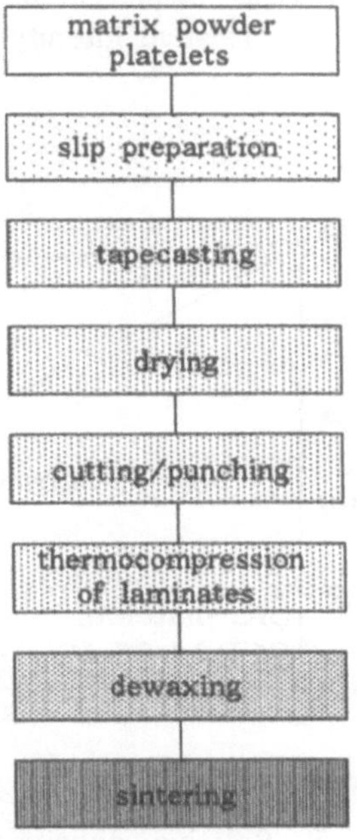

Figure 3.

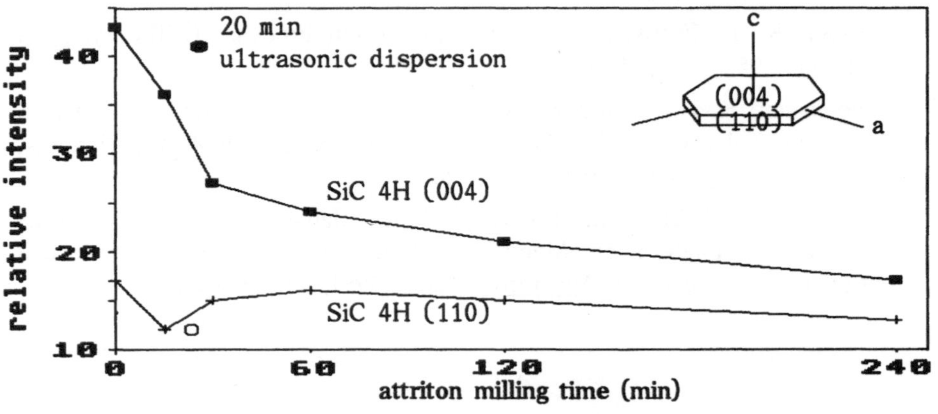

Figure 4. XRD measurement of dried Al_2O_3/SiC-platelet powder mixtures. The relative intenity refers to the (104)-peak of α-Al_2O_3.

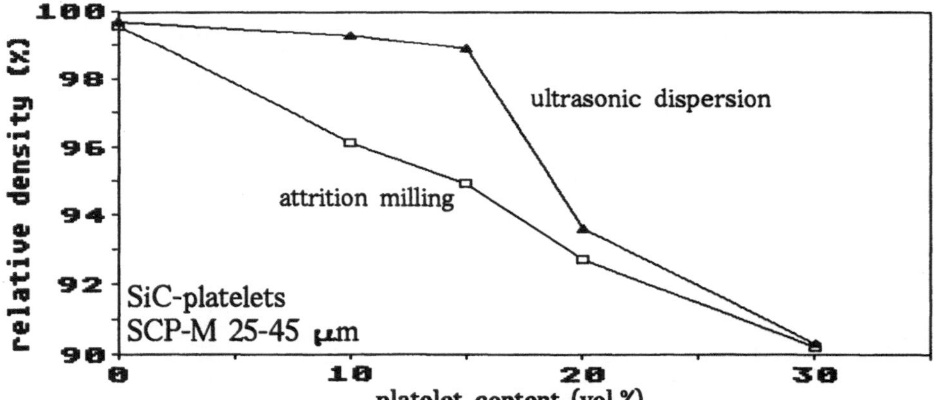

Figure 5. Relative densities of Al_2O_3/SiC-platelet composites with large platelets (diameter: 25 - 45 µm)

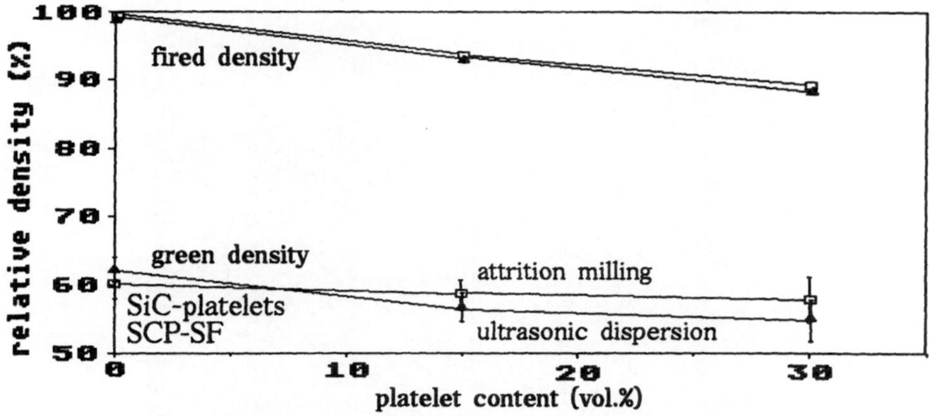

Figure 6. Relative densities of Al_2O_3/SiC-platelet composites with small platelets (diameter: 2 - 25 μm)

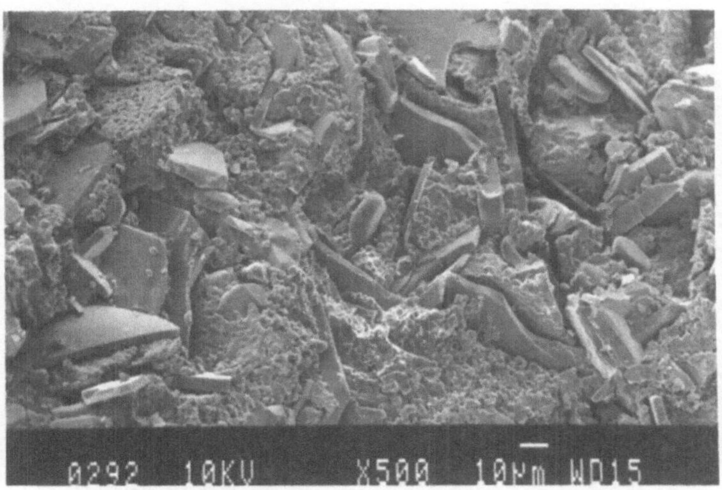

Figure 7. SEM micrograph of a fracture surface of Al_2O_3 + 30 vol.% SiC-pl.

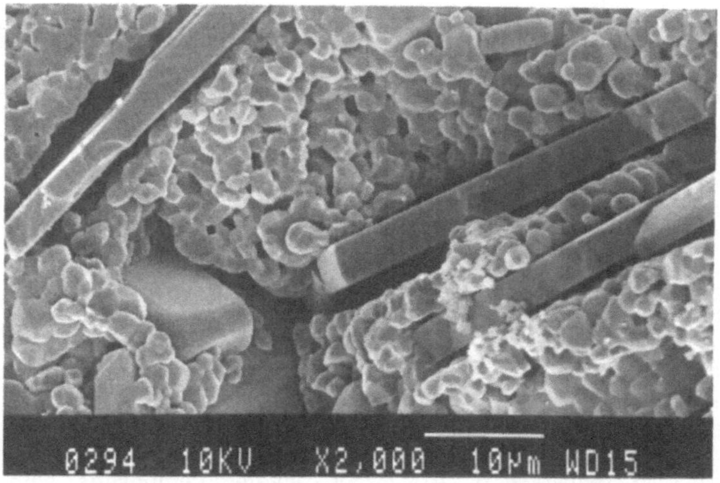

Figure 8. SEM micrograph of a fracture surface of Al_2O_3 + 30 vol.% SiC-pl.

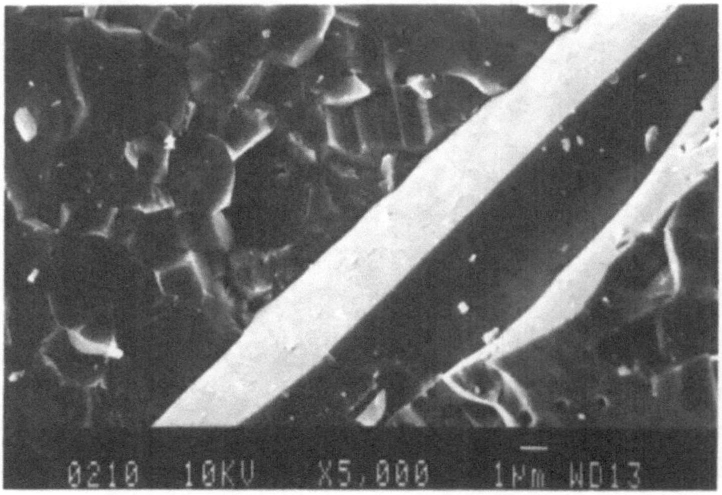

Figure 9. SEM micrograph of a fracture surface of Al_2O_3 + 15 vol.% SiC-pl.

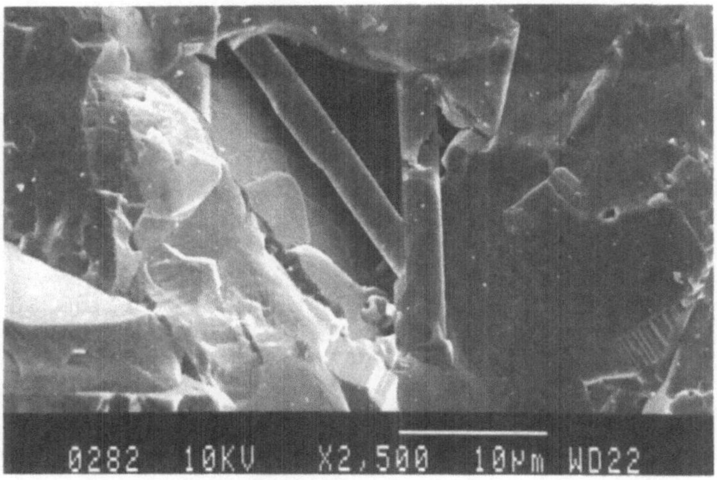

Figure 10. SEM micrograph of a fracture surface of Al_2O_3 + 15 vol.% SiC-pl.

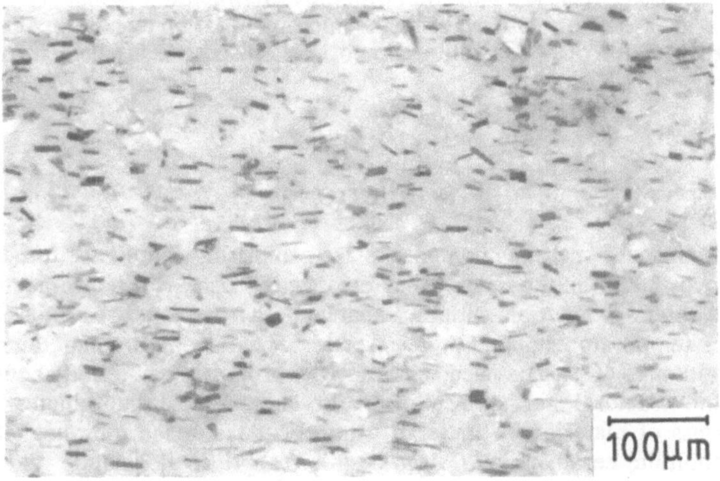

Figure 11. LM micrograph (X pol.) of a polished surface of an Al_2O_3 + 20 vol.% SiC-platelet laminate.

MULLITE AND SiC MATRIX SiC FIBRE COMPOSITES FOR HIGH TEMPERATURE APPLICATION

PARLIER M., GRENIER T., RENEVEY S., PASSILLY B.
MOUCHON E., BRUNETON E., COLOMBAN Ph.
ONERA, BP 72, 92322 CHATILLON CEDEX, F.

ABSTRACT

This paper presents the results concerning the development of two types of Ceramic Matrix Composites (CMC's), SiC/Mullite and SiC/SiC composites manufactured by liquid processing route involving respectively oxide and covalent polymeric precursors. After a short justification for the selection of these processing routes, it is shown how to obtain a sound and non brittle composite through a precise control of both the formulation of the precursors and the processing parameters. The mechanical properties of these new composites are discussed in close connection with the reinforcement design and the fibre-matrix interfacial shear stress measured using the microindentation technique. Finally, the interfacial microstructure of SiC/Mullite composite is discussed at the light of TEM observations.

INTRODUCTION

Oxide and covalent matrix ceramic composites are promissing candidates for structural applications above 1000° C in the field of the aerospace industry which as a need for low density, damage tolerant materials with an enhanced environmental resistance, able to operate at high temperatures under high stresses. The advent of high temperature components made of CMC's to be used either for the propulsion system (gas turbine or jet engines) or the structural thermal insulation of re-entry vehicules, originates from the ability of this class of materials to undergo a non brittle damage tolerant behaviour, contrary to monolithic ceramics who always exhibit a brittle behaviour related to a statistical flaws distribution. This particular property of CMC's, rather unexpected from a material made out of two brittle constituents, results from the achievement during their processing, of a fibre-matrix and sufficiently weak to allow, during loading, multicracking of the matrix to develop without it breaking the fibre [1]. This unique feature was first illustrated with the development of carbon/carbon composites in the early sixties for military uses. Since that time, the nature of the constituent (matrix and fibre) has been enlarged to silicon carbide [2] (covalent ceramix) in order to improve the high temperature behaviour. However, i) the limited thermal stability of oxide or silicon-carbide fibres does not allow their hold at the temperature level of a sintering cycle (1600°C for Mullite, 2000°C for SiC), ii) the constraining of shrinkage due to the presence of fibres will lead to wholly cracked components. That are the reasons why CMC processing mainly relies on techniques based on the vapour or liquid phase infiltration of the fibre preform. The matrix is then obtained by the decomposition of a gas phase (CVI : Chemical Vapour Infiltration), the infiltration of a liquid glassy phase, which is restricted to the glass-ceramics [3], the heat conversion of an infiltrated ceramic precursor.

Processing of CMC's using CVI has been developped in France by SEP [4] mainly for the production of SiC/SiC and C/SiC components of complexe shapes today in use on rocket or turbo jet engines, but a long variety of matrix (C, Si_3N_4, B_4C, TiC, BN, ZrO_2, Al_2O_3...) has also been demonstrated [5]. However, to achieve a good in-depth matrix homogeneity, the deposition rate and the thickness of the fibre perform (a few millimetres) must be low. Accordingly, this processing route lead to thin components of good quality at the expense of long processing times (several months) which remain however acceptable for aerospace industry. The limitations of the CVI technique related mainly to the production cycle duration has resulted in search for more rapid and efficient processing routes as the heat conversion of an infiltrated ceramic precursors. These processing routes will be illustrated here for SiC fibre reinforced SiC and Mullite matrix composites processed using polymeric precursors.

EXPERIMENTAL

The macroscopic mechanical properties have been measured by three points bending test in air at room temperature or at 900°C. The fibre-matrix shear stress have been determined using the instrumented micro-indentation technique [3] which presents relative to the micro-indentation test first used by Marshall [6] the advantage of precise measurement due to a simultaneous recording of the applied load and the fibre displacement. Exploitation of this load-displacement curve using a simple shear-lag analysis allows to derive a constant value of the interfacial shear stress (ISS). As pointed out by Peres [7], a low ISS will lead to a high value of the energy released during the multicracking of the matrix, and then to a non brittle behaviour characterized by an important work of fracture related to intensive fibre-matrix debonding and to the fibre pull-out during the final stage of the rupture. Hardness have been measured for 50 g load with a pyramidal Vickers indenter (68 ° between faces). The fracture surfaces have been observed using a DSM 950 Zeiss apparatus. Samples for TEM studies were cut using a diamond waferblade saw. Platelets are thinned by dimpling and argon ion thinning. TEM observation were performed on a 200 KV Jeol 2000-FX microscope equiped with an energy X-Ray dispersive TRACOR system (Si-Li detector) for local analysis.

RESULTS AND DISCUSSION

SiC-Nicalon/SiC composites processed from a polymeric precursor

The idea of using silicon based polymeric precursors to make ceramic composites originates from the processing route used for carbon-carbon composites. However, the achievement of a SiC or a Si_3N_4 matrix is much more difficult than a carbon matrix due to the density difference between the polymeric resin and the ceramic leading to a high shrinkage, 75 % as compared to only 35 % in the case of carbon matrix composites. In order to overcome this difficulty, it is desirable to limit the role of the ceramic sprung from the resin to an efficient binder. To achieve this end a two-step infiltration process has been developped at ONERA [8] : i) the first step consists in an infiltration of the fibre preform by fine submicronic ceramic particles achieved using a colloïdal filtration technique, which allows to fill 50 % of the free volume within the fibre preform, ii) the second step consists in an injection of the powder infiltrated preform with a suitable polymeric precursor followed, first by a curing cycle, then by pyrolysis.

However, the achievement of a low residual porosity can only be obtained provided that the polymeric precursor exhibits a combined set of properties allowing both a good matrix infiltration and a high ceramic yield. For this purpose, a specific polymer precursor of a silicon carbide matrix has been developped by ONERA and IRAP [8].

The PVS polymeric precursor : A thermoset behaviour is a prerequisite in the case of a ceramic precursor to fix it in the powder infiltrated preform after injection and before the pyrolysis cycle. To fulfill to this first requirement and in a way very similar to that proposed by Schilling [9], three chlorosilanes were used as monomers $(CH_3)_3SiCl$,$CH_3(CH_2{=}CH)$ $SiCl_2$, CH_3HSiCl_2 in order to generate oligomeric chain comprising - CH_3 $(CH_2{=}CH)$ Si - and -CH_3HSi - units terminated by Si $(CH_3)_3$ groups. These oligomers are able to cross-link by polyaddition of SiH group on the vinylic double bonds. The proportion of the monomers has been managed to insure : i) a relatively low molecular weight ranging between 3500 and 4500 in order to achieve a sufficiently low viscosity at room temperature (1.5, 2 Pa.s) allowing the fibre infiltration at a temperature compatible with the gelification time, ii) hardening behaviour with a small mass loss (< 2 %) at a temperature sufficiently below that of beginning of pyrolysis, to avoid bubble forming which will lead to a high level of porosity.

After pyrolysis at 1000°C, PVS exhibit a high ceramic yield of 65 % with a final composition and microstructure of the ceramic product very similar to that observed for SiC-Nicalon fibre. Powder X-ray diffraction and small angle X-ray scattering (SAXS) have revealed an increase of the β-SiC crystals with the pyrolysis temperature [10] (i.e. 1.2 nm at 1100°C, 3.1 nm at 1300°C).

Structural and mechanical properties : Different SiC-Nicalon NLM 202 fibres woven fabrics (1D, 2D, 3D) have been used in order to check the reliability of the processing route. Structural and mechanical properties are summarized in table 1. Due to the specific properties of PVS previously described, sound composites can be achieved with a residual porosity depending upon the fibre architecture. As shown in table 2, the flexural rupture strength of the SiC/SiC composite processed by the pyrolysis of PVS compares favourably to that of composites processed by the CVI technology which is leaded by SEP in France [4] and by ORNL in the United States [11]. Nevertheless, for all these composites, the SiC-Nicalon fibre must be coated with a thin carbon layer to insure a sufficiently weak bond allowing a non brittle dissipative failure. As an illustration, SiC/SiC ex PVS composite without carbon interphase exhibit a brittle behaviour and the flexural rupture strength decrease by a factor 3 when thin carbon interphase deposited by CVI (about 200 nm) lead to a dissipative failure characterized by ISS of 40 MPa.

TABLE 1
Properties of SiC/SiC composites processed by the polymeric precursor route

Fibre Architecture	Density g/cm^3	Open Porosity %	Fibre content Vol % (Vf)	Flexural Strength
3D	2.17	22	35	250 (true Vf=0.15)
2D	2.23	14	30	200 (true Vf=0.15)
1D	2.26	8	45	500
Matrix	2.3	7		110

TABLE 2
Effect of fibre architecture upon the mechanical properties of SiC/SiC composites processed by the CVI and Polymeric precursor routes

Processing route Manufacturer		CVI SEP	CVI ORNL	Polymeric precursor ONERA
Fibre Architecture	1D	700 (Vf=0.45)	400 (Vf=0.40)	500 (Vf=0.45)
	2D	350 (Vf=0.45)	300 (Vf=0.40)	200 (Vf=0.30)
	3D	-	-	250 (Vf=0.35)

For the manufacture of 2D SiC/SiC plates, Nippon Carbon [12] has developped another polymeric processing route involving hot pressing of prepreg tapes which are prepared by deep coating in a slurry containing SiC powder and a polycarbosilane precursor. However, this processing route will be improved to reach a sufficient level of mechanical properties (table 3). Such a walk will be carried out for the reaction-sintering route proposed by the German Aerospace Research Establishement [13] and based on reactions between carbonaceous preform and the infiltrated liquid silicon.

TABLE 3
Influence of the processing route upon the structural and Mechanical properties of 2D SiC/SiC Composites

Processing route		CVI	CVI	Polymeric precursor		Reaction Sintering
Manufacturer		SEP	ORNL	Nippon C.	ONERA	D L R
Density	g/cm^3	2.5	2.6	1.9	2.2	2.6
Open porosity	%	10	<10	-	13	< 2
Fibre content	Vol %	40	40	40	30	30
Flexural strength	MPa					
20°C		300	200-400	110	200	60
1000°C		400*	-	-	200*	-

* with protection against oxydation

SiC-Nicalon/Mullite composites processed by inorganic polymerisation of alkoxides
Among potential oxide matrix, the mullite-alumina region of the Al_2O_3-SiO_2-ZrO_2 phase diagram appears attractive. Alumina monolithes exhibit better mechanical properties than mullite ones but mullite dilatation coefficient, density and sintering temperature are more convenient for SiC fibre matrix. On the other hand, the reactivity toward SiC is a priori more important for silica rich compositions and zirconia presents specific drawbacks due to the monoclinic-tetragonal phase transition or to the diffusion of stabilizing ions in the fibres, the zirconia density is further more high. The optimum of Al_2O_3/SiO_2 ratio must be determined to obtain a composite able to work long time at medium temperature in oxydying or acidic atmosphere , i.e. at a temperature where carbon is burn out.

SiC-Nicalon NLM 202/fibre composite with various matrix compositions close to 5.5 Al_2O_3 4.5 SiO_2 0.1 B_2O_3, 6.5 Al_2O_3 3.5 SiO_2 0.1 B_2O_3 and 7 Al_2O_3 3 SiO_2 0.1 B_2O_3, have been prepared through a sol-gel route involving inorganic polymerization of alkoxides, the final densification being obtained by hot pressing in vacuum (1350 - 1450°C). 4-Directionnel woven fabrics sheets are stacked together to form a 3 - 8 millimetre thick composite. The relatively high porosity (π) of the composite allows us to study the aging after a few hours air aging at temperature between 900 and 1500°C. The mechanical properties are altough sufficiently high to study interfacial properties (Table 4).

Mechanical properties : Macroscopic and microscopic properties are summarized in Table 4. Previous studies [14,15] have shown that Sol-Gel mullite monolithes are not fully crystallized below 1400° C and consequently we have studied composites elaborated below and above this temperature. This is also the temperature at which the crystallization of SiC-Nicalon fibre begin significantly [16]. Microhardness measurements show however that the mechanical properties are not modified (Table 4).
The room temperature bending tests of the composites look very similar (Figure 1) and in all case a large increase of the bending strength is observed when the measurement is performed at 900° C. The 900° C value approach that measured at room temperature for dense LAS or Pyrex composite made with the same woven fabrics. Strain-stress traces

(Figure 1, left) are typical of "well-dissipative" fractures whereas 900°C traces show a decrease of the fracture work. Comparison of fracture surface (Figure 1, right) give evidence of the lowering of the work of fracture : the average pull-out length gives from 100-150 μm at room temperature to about 20-30 μm at 900°C. Measurements of the interfacial shear stress (ISS) by microindentation (Figure 2 and Table I) indicate that the ISS have been increased by a factor 3-4. The ISS room temperature value seems to be minimum for composites elaborated at low temperature or for alumina-rich matrix. Alumina-rich matrix need however higher sintering temperature. The values measured after elaboration are always higher than that of SiC-Pyrex or SiC-LAS composite made of 4D of 1D sheets [7].

TABLE 4
Mechanical properties and interfacial shear stress (ISS) for
4D SiC - Nicalon / Mullite composite (fibre volume ≈ 0.4)

Al_2O_3 / SiO_2 (r)	T °C	π %	σ MPa R.T./900°C(1)	microhardness GPa R.T./900°C(2) Fibre	Matrix	ISS MPa R.T./900°C(2)
5.5/4.5	1350	13	90/110	20/17	12.5/12.5	20±5 / ≥100(3)
6.5/3.5	1450	20	85/130	20/21	8/9	60±15 / >100(3)
7/3	1450	23	90/110	20/20	9/8.5	30±5 / ≈100

(1) bending test at 900°C in air
(2) measurement at room-temperature after air treatment at 900°C
(3) many fibres are spalled during the microindentation

Interface and Matrix microstructures : Figure 3 shows typical TEM micrographies of the SiC fibre-matrix interface. Preliminar studies [16] have shown that the matrix region in contact with SiC-fiber in composite elaborated at 1350 °C and with Al_2O_3/SiO_2 ratio of the matrix r ≤ 1.5 was constituted of a aluminosilicate glass (r ≈ 0.3 from EDX analysis) containing 50 nm mullite crystals (more than 30 % in volume) according to X-Ray powder patterns. Composites elaborated at 1450°C still show mullite acicular crystals (r ≈ 1.5 from EDX analysis, d~ 200x250 nm) in a glassy matrix, even for composite with Al_2O_3/SiO_2 matrix ratio r > 1.5. Local EDX analysis of the glassy phase give evidence of both mullite and silica rich amorphous regions. In some case, some kind of crystal are directly bounded to SiC fibre. An assignement to corundum crystal should be consistent with the observation of α-alumina traces on X-Ray powder pattern, but electronic diffraction pattern indicates that this crystal is of mullite type.

In conclusion polymeric route using alkoxide hydrolysis allows to prepare SiC/Mullite composites exhibiting low interfacial shear stress and consequently significant work of fracture. Air annealing lead to an ISS increase which can be moderate by the use of alumina rich matrix.

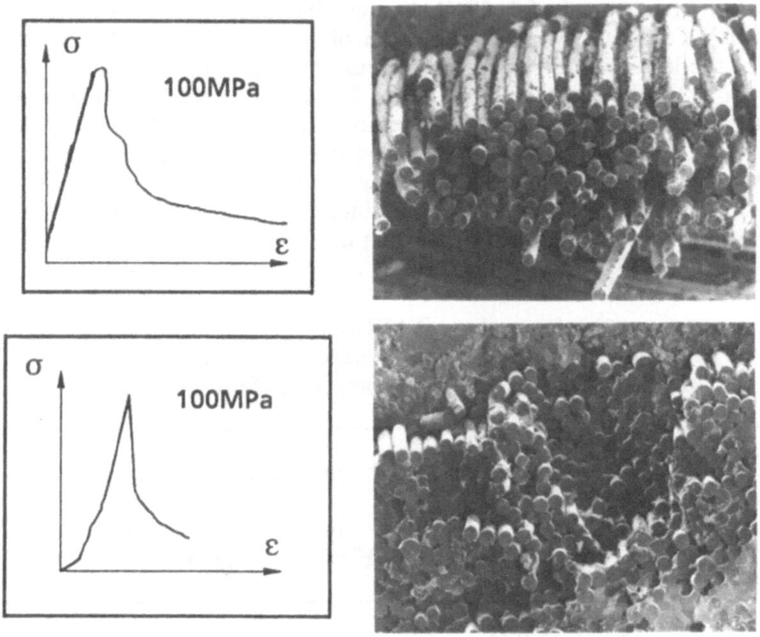

Figure 1. Typical stress-strain traces (left) recorded during a 3 points bending test at room-temperature (up) and at 900°C (down) for a 4D woven fabric SiC-Mullite composite with composition close to 7 Al_2O_3 3 SiO_2 0.1 B_2O_3. Corresponding fracture surfaces are shown. The SiC fibre diameter is about 12 - 15 μm.

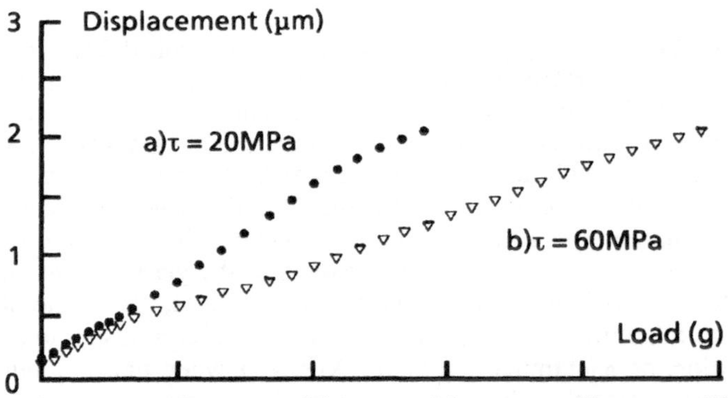

Figure 2. Microindentation test of 4D woven fabric SiC/Mullite composites with matrix compositions close to 5.5 Al_2O_3 4.5 SiO_2 0.1 B_2O_3 elaborate at 1350 [(a)] and 1450°C [(b)], respectively.

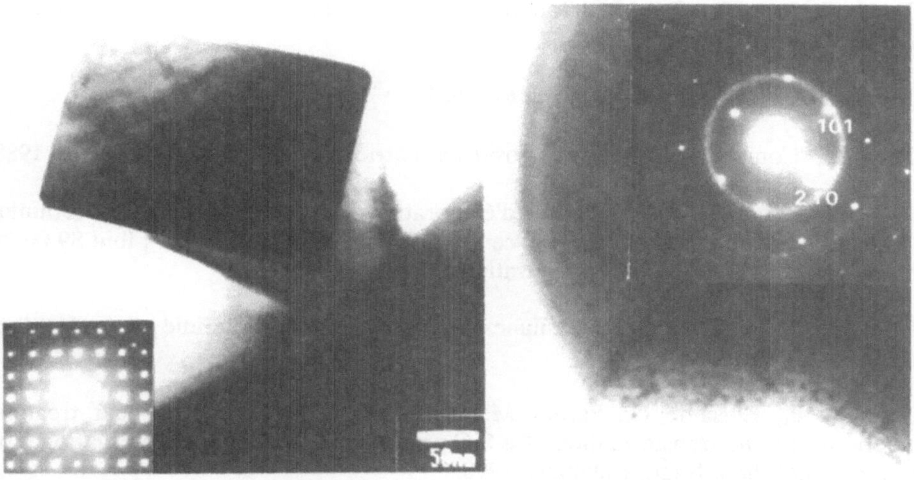

Figure 3. TEM images and diffraction patterns of a composite elaborated at 1450°C on left, (a,b) plane of mullite crystal in a glassy phase, near SiC fibre (bottom) ; lateral faces of the crystal are {110} and growth direction is [001] ; on right, mullite crystal directly bounded to the SiC fiber (bar : 50 nm) ; indexation of mullite is given and rings are characteristic of β-SiC (111,220, 311 planes).

CONCLUSION

Elaboration of sound and non brittle SiC or Mullite matrix composites reinforced by SiC-Nicalon NLM 202 woven fabrics is possible through polymeric liquid routes. The composite porosity can be sufficiently lowered to reach the mechanical properties required for engine applications. The interfacial shear stress measured after elaboration is very close to that obtained for SiC/LAS composites. However, after 900°C air treatment the interfacial shear stress increases drastically. A more complete investigation is needed to precise the mechanism of degradation.

REFERENCES

1. Evans A.G., J. Am. Ceram. Soc., 1990, **73**, 187 - 206.

2. Fitzer E., Gadow R., J. Am. Ceram. Soc., 1986, **65**, 326 - 335.

3. Parlier M. , Ritti M.H., Stohr J.F., Vignesoult S., Silicon carbide fibre-reinforced glass-ceramic matrix composites : A high temperature material for high performance application, Proceeding of ICAS 90, Stockholm (Sweden), September 9 - 14, 1990.

4. French Patent 82.01025, "Structure composite de type réfractaire-réfractaire et son procédé de fabrication", January 22, 1982.

5. Fitzer E., Proc. Int. Symp. Factors in Densification and Sintering of Oxide and non Oxide Ceramics (Japon), 1978, 678.

6. Marshall P.B., J. Am. Ceram. Soc., **67**, 1984.

7. Peres P., "Comportement des composites à matrice fragile", ONERA T.N., 4, 1988.

8. French Patents : 89 16 918, "Procédé d'élaboration d'un matériau composite céramique fibres-matrice et matériau obtenu par ce procédé", December 20, 1989 ; ibid 89 00 764, "Polysilanes et leur procédé de préparation", January 23, 1989.

9. European Patent 01 23 934, "Polysilane precursors containing olefinic groups for silicon carbide", April 30, 1984.

10. Larché F., Ellisalde D., Parlier M., Noireaux P., Effect of the heat treatment temperature on the transformation of a SiC precursor, Proceeding of International fine ceramics Whokshop, Nagoya (Japan), 1990.

11. Stinton D.P., Caputo A.J., Lowden R.A., Synthesis of fibre reinforced SiC composites by Chemical Vapor Infiltration, Ceram. Bull, vol. **65**, n° 2, 1986.

12. Hiroshi I., Shiron M., Yoshikazy I., Toshihatsu I, Proceeding of 1st Japan International SAMPE symposium, November 28-December 1, 1989.

13. Krenkel W., Fiber ceramics for reentry vehicle hot structures, Proceeding of 40 th International Astronautical Congress, Malaga (Spain), October 7-13, 1989.

14. Colomban Ph. and Mazerolles L., J. Mater. Sci. letts., **9**, 1990, 1077.

15. Colomban Ph., Proc. 7th CIMTEC, Montecatini-Terme, June 24-30, 1990, High-Tech Ceramics, P. Vincenzini ed., Elsevier, to be published.

16. Michel D., Mazerolles L., Bruneton E., Portier R. and Colomban Ph., Proc. 2nd Japan-France Materials Science Seminar Fundamental Aspect of Energy Materials : Microstructural Modifications and Microstructure - Dependent properties - Paris, April 22-26, 1991, - Ann. Chimie (Sciences des Matériaux), to be published.

Acknowledgment

One of us (Bruneton E.) thanks Dr Michel D. of CECM - CNRS (Vitry) for the use of his electron microscopy facilities. This work have been supported by DRET and MRT.

MEASUREMENT OF K_{tip} IN FIBER AND WHISKER COMPOSITES

W. R. CANNON and E. A. MENDOZA
Center for Ceramic Research, Rutgers University
Piscataway, NJ 08855-0909, USA

ABSTRACT

A direct method of measuring stress intensity factors by measuring stresses at the crack tip was used to determine K_{tip} simultaneously with $K_{far\ field}$ in model fiber composites and whisker composites of Al_2O_3 matrix and SiC fibers and whiskers. The resultant relationship was $K_{tip}=\alpha K_{far\ field}+K(0)$. Since K_{tip} is the driving force for crack growth, results can be used to predict the rate of crack growth for composites.

INTRODUCTION

The toughening of fiber and whisker composites is accomplished by crack bridging in the wake of the crack, crack deflection, fiber pullout, etc. Crack bridging in the wake of the crack toughens the composite by lowering the stress intensity at the crack tip. The fact that the stress intensity factor at crack tip may be lower than calculated from the stress applied in the far field, ie. away from the crack, means that $K_{far\ field}$ is not the most fundamental parameter determining crack growth velocity in static and dynamic fatigue.

The stress intensity at the crack tip is designated, K_{tip}, and is differentiated from $K_{far.field}$ which is the commonly used stress intensity factor, and which is calculated from the applied force using linear elastic equations. The difference between K_{tip} and $K_{far\ field}$ is the degree of toughening due to shielding, which for fiber and whisker composites likely is the bridging contribution.

$$K_{bridging} = K_{far\ field} - K_{tip} \qquad (1)$$

Recently, direct measurements of stress intensity factors have been made in silicon single crystals by measuring the stress at the crack tip using a Raman microscope.[1] This is accomplished by measuring the stress as a function of distance from the crack tip. An explanation of the technique for measuring stresses on a microscale is contained in references

[1] and [2]. The technique is based on the measurement of the shift in frequency of the Raman spectra. Reference 1 shows that the stress intensity factor of single crystalline silicon can be measured to an accuracy of approximately 0.07 MPa $m^{1/2}$.

Results of a typical measurement of crack tip stresses is shown in Fig. 1. Since the equibiaxial stress directly in front of the crack tip is given by

$$\sigma_{yy} = \frac{K_I}{\sqrt{2\pi r}} \qquad (2)$$

where r is the distance from the crack tip, a plot of log σ_{yy} versus log r should yield a straight line of slope -0.5 which is in agreement with Fig. 1. K_I can either be calculated from the intercept at log r = 0 or by plotting σ_{yy} vs $1/\sqrt{2\pi r}$ and determining the slope.

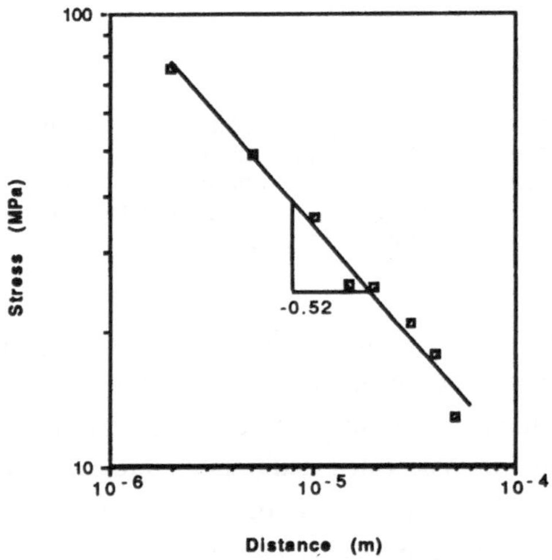

Fig. 1. Stress at the tip of a crack of a dcb specimen of an Al_2O_3/SiC whisker composite.

In this paper K_{tip} is measured in polycrystalline Al_2O_3, in Al_2O_3 plus single and multiple embedded fibers and in commercial hot pressed Al_2O_3/SiC whisker composites.* These results are used in predicting the expected crack growth rate by slow crack growth in whisker composites and compared with the literature measurements.

---

* Greenleaf Corp, Saegertown, PA 16433, USA

EXPERIMENTAL

Samples of polycrystalline Al_2O_3 were prepared by cold pressing alumina powder* into 5.6 cm diameter discs and sintering at 1400° C. Resulting samples were about 98% dense, with grain size slightly less than 1 μm. The fine grain size was necessary in order that the focused spot diameter of the microscope be larger than the grain size and, therefore, sample more than one grain. If this is not done then residual stresses within the grain may obscure the crack tip stresses. If several grains are sampled with a variety of residual compressive and tensile stresses which average to zero, then the peak will be broadened but not shifted. Some of the observed scatter may be due to net residual stresses under the focussed spot. The depth of penetration by the microprobe is approximately 6 μm.[3]

Fibers were embedded in the polycrystalline Al_2O_3 by first cold pressing an alumina disc about one half the intended thickness and then placing Textron SCS-6 coated SiC filaments onto the powder and pressing the other half of the powder on top. Two types of samples were prepared: one single fiber was added and several fibers were placed parallel to each other.

Calibration of the peak shift was made by stressing disc shaped samples in a microscope stage ring-on-ring fixture [1] and comparing the shift with the calculated equibiaxial stress. Such measurements were performed for both polycrystalline Al_2O_3 and SiC. The resultant Raman frequency shift was 1.1 cm^{-1} /GPa for the polycrystalline Al_2O_3 and 3.5 cm^{-1}/GPa for polycrystalline SiC.

Samples were cut into double cantilever beam specimens 30.5 mm x 12.7 mm x 1.5 mm thick. A 0.25 mm deep groove was cut into the bottom of the sample to guide the crack. The samples surfaces were polished so that the crack could be observed.

A dcb testing device was built to fit on the microscope stage of the Raman§ . The dcb specimens were loaded through pins using a pneumatic air cylinder to control the applied force. Since the force was known, $K_{far\ field}$ could be measured at the same time as the K_{tip}, using the Raman technique. Five to ten stress measurements were made within 20 μm of the crack tip. The position of the crack tip was determined visually and by determining where the maximum stress occurred. Results were plotted as shown in Figure 1. The slope of the curve was usually between -0.4 and -0.5. The stress intensity factor, K_{tip}, was determined by plotting σ against $1/\sqrt{2\pi r}$ and determining the slope. In the whisker composite it was not possible to measured the stress using the Al_2O_3 peak since it was too weak. The SiC is a much better Raman scatterer and so the beam did not penetrate deep enough to obtain a strong Al_2O_3 peak and so the shift of the SiC peak was used to determine stresses. In a two phase

---

* AA-61 Performance Ceramics, Peninsula, Ohio

§ Instruments SA U-1000, Edison, NJ 08820 focused to 1μm spot size.

composite in which the stiffness of the phases differ, the stress in each phase will differ. Since the stiffness of Al_2O_3 and SiC are similar by considering isostress and isostrain approximations, the stress differences are less than 10%.

RESULTS AND DISCUSSION

A plot of K vs. the applied force to the polycrystalline Al_2O_3 dcb specimen is seen in Figure 2. The numerical values above each of the open circles in Figure 2 are the slopes of log σ versus log r. According to equation (2) the slope should be -0.5 but are generally lower.[1] It should be noted that values of -0.67 and -0.28 differ greatly from the expected -0.5 but the stress level in these measurements is low and results may be inaccurate.

PLAIN ALUMINA- NO EMBEDDED FIBER

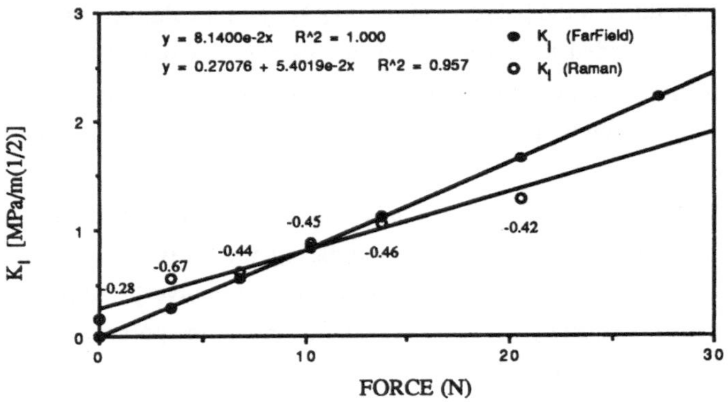

Figure 2. K vs force applied to the polycrystalline Al_2O_3 dcb specimen. K(far field) was calculated from this force and K(Raman) measured from the stresses at the crack tip. The numbers above the points indicate the slope of log σ vs log r.

The disagreement between $K_{far\ field}$ and K_{tip} is not large. The difference may be due to bridging, though the bridging contribution is not expected to be large in 1 μm grain size Al_2O_3.[4] Not enough information is available to make this estimate. On the other hand the sources of the disagreement may be in the calculation of $K_{far\ field}$ which has a number of assumptions, for instance, assuming point loading rather than pin loading and using simple beam theory rather than two dimensional or three dimensional analysis, whereas, the measurement of K_{tip} is a direct measurement of K_I. The nonzero intercapt may mean debris forms in the crack prevention closing. It is also possible that some form of crack shielding takes place in the fine grained polycrystalline Al_2O_3. A significant bridging contribution is not expected in such fine grained polycrystalline Al_2O_3.

The second specimen to be tested was the one with a single SiC fiber. A sharp crack was propagated until it stopped at about 30 μm in front of a point directly above the front of the fiber. Since the interior crack often leads the surface crack, the crack probably stopped at the fiber. Spectral measurements were taken in front of the crack tip at the surface. The maximum applied force was 29 N, at this point the sample failed catastrophically but did not cross the fiber. Instead it was deflected along the fiber-matrix interface. From the stress measurements a plot of K_I vs. the applied force is obtained as seen in Figure 3.

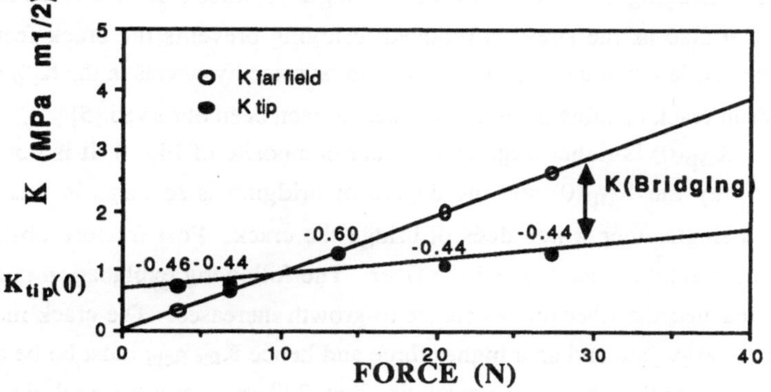

Figure 3. K_I (calculated from the applied force and measured with the Raman) vs. applied force for a crack stopped at a single fiber. Numbers above the closed circles indicate slope of log σ vs log r.

In the specimen containing multiple fibers the crack was propagated across the first fiber and then was arrested at the second fiber. It was at this point that the crack tip stresses were measured. Results are shown in Figure 4.

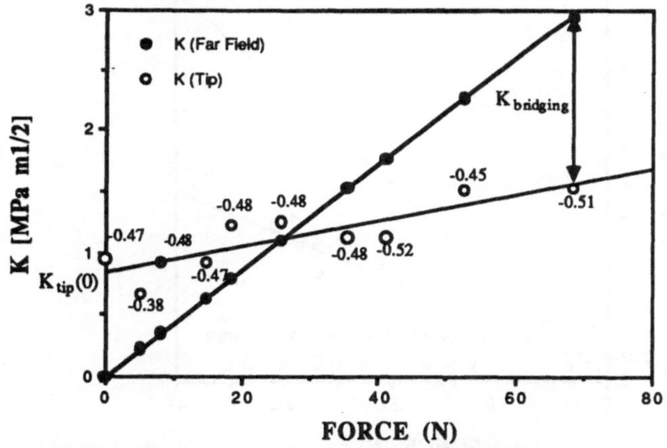

Figure 4 - K_I (calculated from the applied force and measured with the Raman) vs. applied force for a crack crossing a single fiber and stopping at the second fiber. Numbers above the open circles indicate slope of log σ vs log r.

There are some very interesting features of these results. One of the most interesting features is the large positive intercept of the K_{tip} at applied force = 0, $K_{tip}(0)$. This confirms the speculation that cracks can have a non-zero stress intensity factor after the far field stress is unloaded. Two reasons for this behavior are cited: First, debris, formed during crack extension, prevents complete closure of the crack when the stress is released (Fig. 2). The second is related to bridging. It is well known that bridging applies a closure force to cracks during opening but also as the stress is relieved, bridging prevents the crack from fully closing. This latter explanation infers that there should be some hysteresis in the K_{tip} vs force curve. Hysteresis in crack opening displacement has, in fact, been observed.[5]

The value $K_{tip}(0)$ is rather large in the fiber composite of Fig. 3. It is not easy to understand both why the $K_{tip}(0)$ and the degree of bridging is so large in this sample containing only a single fiber which does ot bridge the crack. Post fracture observation confirmed that the crack had not crossed the fiber. The following explanation is offered. When the crack reached the fiber the resistance to growth increased. The crack must bow around the fiber to propagate and so a higher force and hence $K_{far\ field}$ must be be applied. Raman method measured the stress at the crack tip about 700 μm above the crack tip. At this point the crack tip stress does not increase or the crack would propagate further and the length of the crack front must increase. Thus there is an increased difference between $K_{far\ field}$ and K_{tip}. On releasing the stress the crack does not fully close and so K(0) is nonzero.

Finally, a similar measurement was made on the whisker composite and results are shown in Fig. 5. According to Fig. 5 $K_{tip}(0)$ is much smaller than in the two fiber embedded samples. This may be due to the stronger bond between the matrix and the whiskers. The degree of crack shielding is approximately 50% of the applied $K_{far\ field}$.

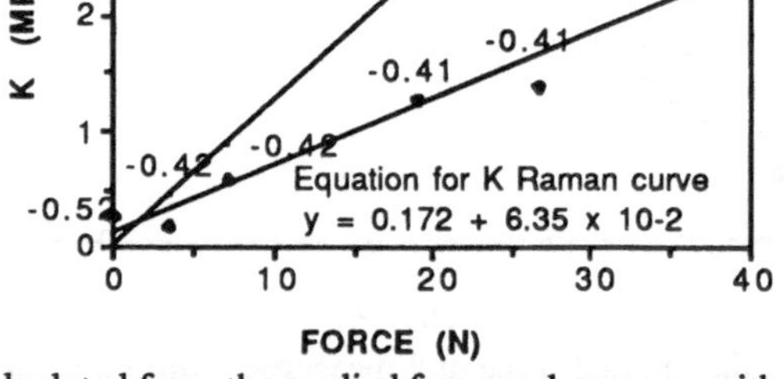

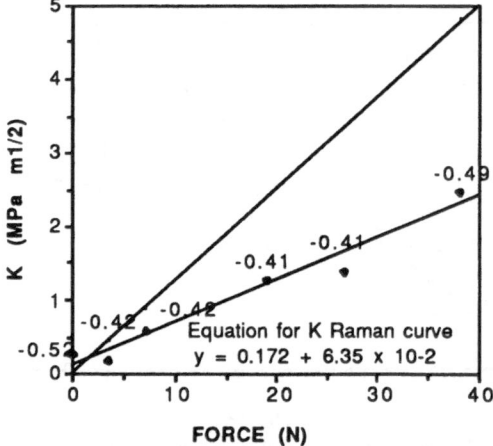

Figure 5- K_I (calculated from the applied force and measure with the Raman) vs. applied force for a crack in a whisker composite. Numbers above the data points indicate slope of log σ vs log r.

In order to compare results in Figs. 3-5 it is necessary to plot K_{tip} versus $K_{far\ field}$. Result are shown in Fig. 6.

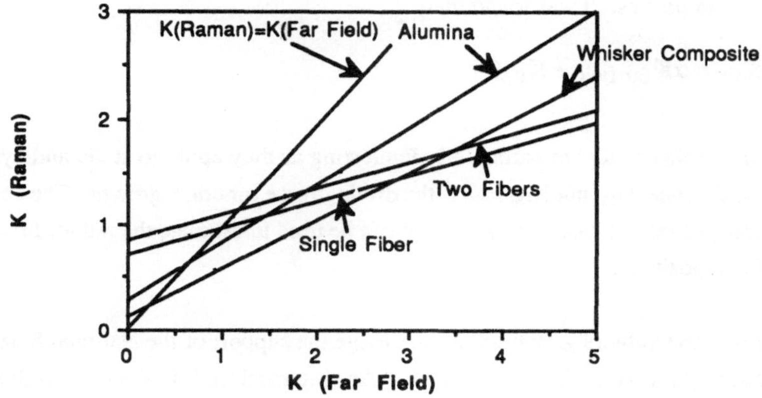

Figure 6. A composite of Figs. 2-5 plotted as K_{tip} versus $K_{far\ field}$.

From the above results we find that $K_{far\ field}$ can be related to K_{tip} through the following linear equation:

$$K_{tip} = \alpha K_{far\ field} + K_{tip}(0) \qquad (3)$$

where is α the slope of the curve of Fig. 6 and $K_{tip}(0)$ is the intercept.

Two observations may be made in Figure 6. First the value of $K_{tip}(0)$ is a measure of bridging open of a non-stressed crack and it increases in the order polycrystalline Al_2O_3, whisker composite and coated fiber composite. It appears the $K_{tip}(0)$ increases as bridging increases. The value of α also decreases as the degree of bridging increases which means that in strongly bridged materials K_{tip} is not strongly sensitive to $K_{far\ field}$.

CONCLUSIONS

The Raman spectroscopy technique has been used to measure K_I at the crack tip of fiber and whisker composites. It was found that

$$K_{tip} = \alpha K_{far\ field} + K_{tip}(0).$$

The results or this study are particularly interesting as they apply to static and cyclic fatigue crack growth since K_{tip}, not $K_{far\ field}$ is the driving force for crack growth. The above equation (equation (3)) relates K_{tip} to $K_{far\ field}$ and can be used to estimate the reduced rate in crack growth of composites.

Acknowledgment: The authors gratefully acknowledge the support of the National Science Foundation under grant NSF INT 88-11014, "An International University-Industry Cooperative Research Program Between the Center for Ceramic Research at Rutgers University and Several Swedish Institution."

RERENCES

1. Andrew G. Haerle and W. Roger Cannon, "Direct Measurement of Crack Tip Stresses," To be published in J. Amer. Ceram. Soc.
2. J. Rydzak and W. Roger Cannon, "Fourier Transform Infrared Microspectroscopy Method to Determine Stress in Sapphire," J. Amer. Ceram. Soc., 72 [8] 1559-61 (1989).
3. F. Adar and D. R. Clark, "Raman Microprobe Spectroscopy of Ceramics," in Microbeam Analysis, eds. K. F. J. Heinrich, San Francisco Press (1982).
4. R. W. Steinbrech, A. Reichl, and W. Schaarwachter, "R-Curve Behavior of Long Cracks in Alumina," J. Amer. Ceram. Soc. 73 [7] 2009-2015 (1990).
5. B. Marshall and A. G. Evans, "Failure Mechanisms in Ceramic-Fiber/Ceramic Matrix Composites," J. Amer. Ceram. Soc., 68 [5] 225-31 (1985).

RESIDUAL STRESSES IN CERAMIC FIBER COMPOSITES: EFFECT OF NON-UNIFORM FIBER DISTRIBUTION

BENT FRUERGAARD SØRENSEN, RAMESH TALREJA* and OLE TOFT SØRENSEN
Materials Department, Risø National Laboratory
4000 Roskilde, Denmark
*Department of Solid Mechanics, Technical University of Denmark
2800 Lyngby, Denmark

ABSTRACT

Residual stresses evolve in ceramic fiber composites during cool down from a stress free state. These stresses play a very important role in the overall mechanical behaviour of the composites, and may lead to microcracking by themselves or when aided by thermomechanical loadings. In this paper, the residual stresses in unidirectional fiber composites are computed by the three dimensional finite element analysis. We investigate the effect of fiber volume fraction and fiber distribution effects such as fibers touching and fibers enclosing matrix (short range effects) as well as matrix and fiber rich domains (long range effects). The effect of thermal expansion mismatch is studied by examining two ceramic composite systems: SiC fibers (Nicalon) reinforced by LithiumAluminumSilicate (LAS) and CalciumAluminumSilicate (CAS) glass-ceramics. It is shown that the residual stress state varies with fiber distribution: The analysis of small range effects shows that the local stresses may differ considerably from the average stresses. The analysis of long range effects shows that the residual stress states are affected both inside and outside the domains. Due to differences in the thermal expansion mismatch, the residual stress state in SiC/LAS and SiC/CAS are very different. Therefore, different damage modes are expected in the two systems. Partial debonding is likely in SiC/LAS at locations where fibers are in contact, whereas matrix cracks may initiate in SiC/CAS.

INTRODUCTION

Due to the thermal expansion mismatch between fibers and matrix, residual (thermal) stresses exist in Ceramic Matrix Composites (CMC) at room temperature. As these stresses may be high, they may cause cracking by themselves or when aided by mechanical loads. Microcracks are usually classified as r-, Θ- or z-cracks (Figure 1).

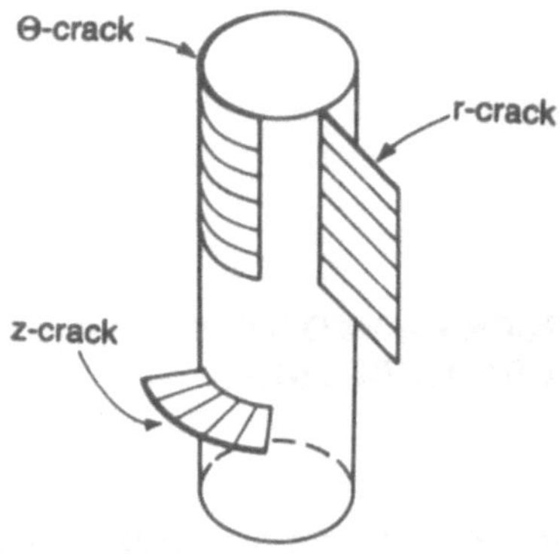

Figure 1 *Possible microcracking modes for a unidirectional fiber composite.*

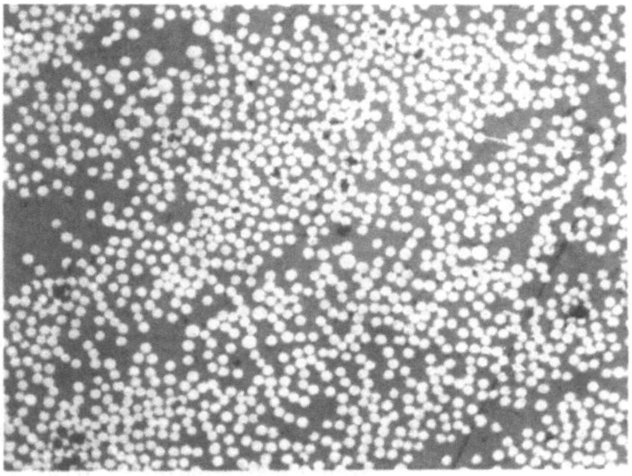

Figure 2 *Typical cross section of a SiC/CAS composite: Note the non-uniform fiber distribution.*

Various micro-mechanical analyses treat residual stresses in continuous fiber composites. Also, analytical composite cylinder models have been formulated in order to calculate the residual stress state within a composite and its overall thermal expansion coefficients [1-2]. Mikata and Taya [3] used a semi-analytical method to calculate the residual stresses, including the interphase (fiber-coating) material. They showed that the highest stresses would exist within the coating, therefore failure was assumed to initiate there. Chen *et.al.* [4] concluded from their modified Mori-Tanaka method that a thin coating had only little effect at the residual stress state in the fiber and matrix. Böhm and Rammerstorfer [5] calculated the residual stress state within a unidirectional metal matrix composite by the Finite Element Method (FEM), and demonstrated how it influenced the overall stress strain characteristics of the material.

Some interesting two dimensional micro-mechanical analyses have been presented by Zhu and Achenbach [6-7], who used the boundary element method to investigate a composite loaded in transverse tension. In their most recent paper [7] hexagonal fiber packed composites were investigated under two different loading directions. Their analysis did not include residual stresses and they found two equally likely cracking modes: Partial debonding (Θ-crack) and radial matrix cracking (r-crack).

In the early days of FEM micro-mechanics modelling Adams and Tsai [8] studied effects of fiber packing on overall transverse modulus, using a 'random packing' concept. More recently, the behaviour of a metal matrix composite has been investigated with packing effects in mind. Brockenbrough *et.al.* [9] used the 3 Dimensional (3D) FEM to model the nonlinear elastic-plastic stress-strain curves. Three different uniform fiber packing arrangements and a 'random packing' arrangement were analyzed, without residual stresses. Notable differences in the transverse stress-strain curves were found for the different packing configurations. The 'random packing'-model gave the best agreement with experimental measurements.

Residual stresses evolve in composites during the cool-down process. Generally, the *magnitude* of the stress components depend on the cool down temperature, material properties (thermal expansions coefficients mismatch and stiffness properties), microstructure geometry (fiber fraction, - shape,- dispersion and - waviness) and preexisting flaws. Figure 2 shows

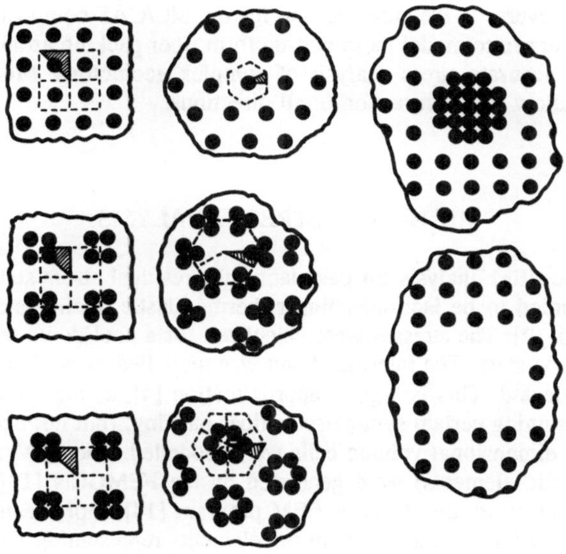

Figure 3 *The investigated non-uniform fiber packing configurations; symmetry planes are indicated and representing volume cells are hatched.*

a typical cross section of a SiC fiber-CalciumAluminumSilicate (CAS) glass-ceramic matrix composite. Note that fibers are not uniformly distributed; often they touch each other, and some matrix and fiber rich domains exist. The residual stress states for some *non-uniform fiber distributions* are investigated in this paper. This is done in two ranges of magnitude: *Local* effects of closely spaced fibers or fibers in contact (short range) and matrix and fiber rich *domain* effects (long range effects). The packing arrangements that are investigated in this paper are shown in Figure 3.

Firstly, the effects that short range fiber packing may have on the residual stress are suggested. Assume that the thermal expansion coefficient of the fiber α_f exceeds that of the matrix α_m. The usual way of doing a stress analysis of cool-down is to assume that the fibers are uniformly dispersed in a hexagonal array. Then the boundaries of a representative hexagonal volume cell may be approximated by a cylinder. The stresses may then be found analytically [2]. Now, if $\alpha_m<\alpha_f$, during cool down the fiber would shrink more than the matrix, therefore, due to the constraint the interfacial normal stress would be in tension. Now consider a fiber packing arrangement with fibers surrounding matrix. As fibers contract more than the matrix, the interfacial normal stress would be in compression. Thus, small scale non-uniform packing is expected to affect the residual stress state. Looking at a larger scale, a fiber rich domain would contract more than the average composite (being dominated by the expansion properties of the fibers), and conversely, a matrix domain would contract less. Therefore, matrix and fiber rich domains are expected to induce different stresses.

In this paper it is assumed that damage initiates and grows at the interface at the locations where the highest stresses exist. The effects of *fiber volume fraction*, and *thermal expansion mismatch*, in order to find the most severe locations are also investigated. In order to investigate the effects of thermal expansion mismatch, two different composites were studied: SiC-fiber/LithiumAluminoSilicate (LAS) -matrix and SiC/CAS. In the SiC/LAS composite

$\alpha_m < \alpha_f$, whereas the reverse is the case ($\alpha_m > \alpha_f$) for the SiC/CAS composite.

The FEM was chosen to model these non-uniform fiber packing arrangements. This tool is very suitable for accurate stress analysis of complex geometries. Modelling in 3D was chosen in order to allow free contraction in all directions.

METHOD OF STRESS ANALYSIS

In the following, the FEM analysis for calculating the residual stress state is sketched. The materials were assumed to be isotropic, linear thermo-elastic, their material properties are listed in Table 1 [11-12]. The stresses were calculated for a cool down temperature of 1000 °C from the stress free state. The interface bond was modelled as perfect and the interphase properties were neglected. This is a good approximation [4], as the interphase thickness is very small [10]. Assuming certain symmetry in fiber packing, (but not necessarily uniform), representative three dimensional volume cells were modelled. The FEM meshes (consisting of 15 or 20 node solid elements) were generated by the FEMGEN [13] program, and the problems were solved using the SOLVIA FEM program [14] implemented at a VAX 8700 mainframe computer. Along the interface in the elements representing the matrix, the stress tensor at the Gaussian points was transformed to cylindrical coordinates (r,Θ,z) by the rules of second rank tensors in order to evaluate the propensity for r-, Θ- and z-cracks. Although the FEM analysis showed that the highest stresses were not always located at the interface, it was assumed that cracking was most likely to initiate here, due to the weak interface bonding usually present in CMC's.

Modelling of Small Range Effects

In order to have well defined boundary conditions, symmetry (strictly speaking throughout the whole composite) was in fiber packing. As a result, the representative volume cells needed only to include a small part of a fiber (plus surrounding matrix), giving more accurate results. The boundary planes of the volume cells should (due to symmetry before and after cool down) remain flat during deformation, and the normal force and all shear stresses should be zero. Therefore the displacements of all nodes at a symmetry surface were interconnected (in the direction normal to the plane) to give self consistent deformation (Figure 4). This caused a very large band width in the total stiffness matrix, representing some 2000-3000 linear equations for the node displacements. Therefore, the solution time was quite long, up to 4 CPU hours per volume cell problem.

TABLE 1

Properties of SiC (Nicalon) fibers [11], CAS [12] and LAS [11] matrices. The Poisson ratios values are assumed.

	SiC-Fiber	CAS-Matrix	LAS-Matrix
E-modulus (GPa)	200	98	85
Poisson Ratio	0.3	0.3	0.25
Thermal Expansion Coiff. (10^{-6}/°C)	4	5	1

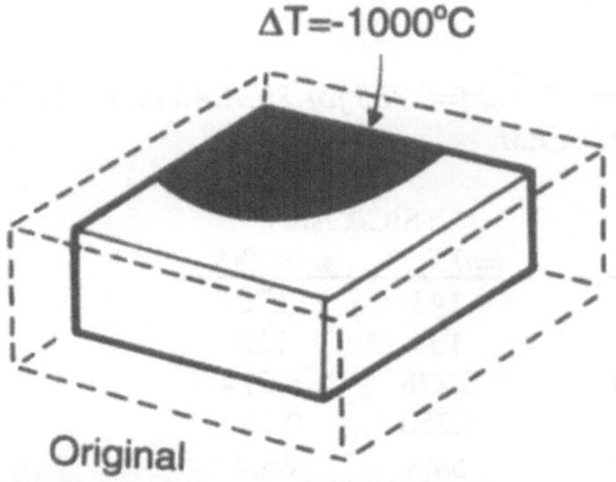

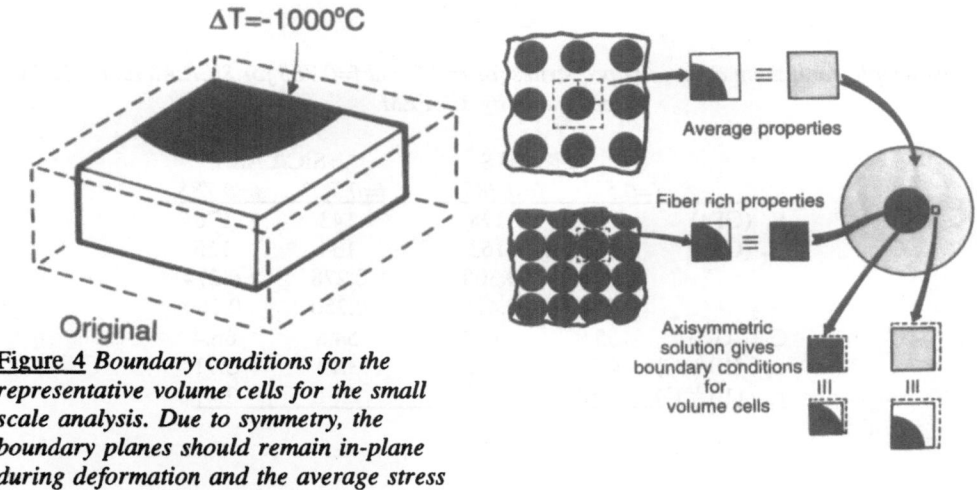

Figure 4 *Boundary conditions for the representative volume cells for the small scale analysis. Due to symmetry, the boundary planes should remain in-plane during deformation and the average stress must vanish at each surface.*

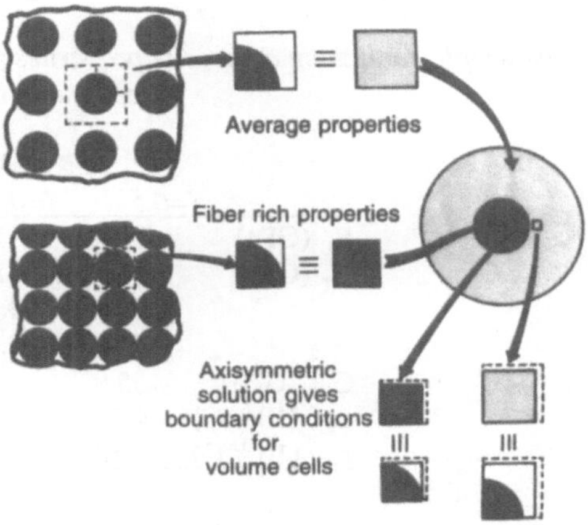

Figure 5 *Schematic flow chart for large scale stress analysis,indicating coupling between small scale 3D micro-mechanics and axisymmetric continuum models.*

Modelling of Long Range Effects

Another approach was taken in order to investigate effects of matrix and fiber rich domains. The composite was modelled as consisting of two continuum media. As sketched in Figure 5, continuum properties were used in an axisymmetric model which represented a (matrix or fiber rich) domain surrounded by average composite. Firstly, the continuum material properties (Table 2) for nominal fiber fraction ($f=0.5$) and fiber rich material ($f=0.785$) were calculated by FEM simulation of uniaxial tensile tests and heat-up tests on square array volume cells. Secondly, the boundary displacements of an average composite during cool down were calculated using continuum material data for $f=0.5$ outside and inside the domain. These displacements were used as boundary conditions when modelling matrix and fiber rich zones. The physical meaning of these conditions is remote constraint at the expansion; a composite will (be forced to) expand as if it consisted of fully dispersed fibers, although matrix and fiber rich zones may exist here and there. The third step involved the FEM solution, calculating strains in the cylindric domain model during cool down with domains having matrix or fiber rich properties. The fourth step was to transform the calculated strains (outside and/or inside the domain) into boundary displacements for the original square array volume cells, and FEM analysis of these gave the residual stress state.

RESULTS

Residual Stresses in SiC/CAS ($\alpha_m > \alpha_f$)

Firstly, the effect of *fiber volume fraction f* was investigated for *perfect hexagonal* packing (Figure 6). Due to the boundary conditions the shear stress components σ_{rz} and $\sigma_{\theta z}$ are always zero, whereas $\sigma_{r\theta}$ is not; indicating that *the stress state in the hexagonal packing is not axisymmetric*. The stresses (especially σ_{rr}) vary around the fiber at high fiber fraction. Except

TABLE 2

Orthotropic continuum composite properties for f=0.5 and f=0.785 for SiC/LAS and SiC/CAS computed by 3D FEM.

	SiC/CAS		SiC/LAS	
	$f=0.5$	$f=0.785$	$f=0.5$	$f=0.785$
E_L (GPa)	149	178	143	176
E_T (GPa)	140	163	134	165
ν_{LT}	0.300	0.300	0.278	0.274
ν_{TT}	0.295	0.416	0.250	0.284
G_{LT} (GPa)	55.9	66.2	54.8	66.4
α_L (10^{-6}/°C)	4.33	4.12	3.14	3.71
α_T (10^{-6}/°C)	4.50	4.20	2.61	3.47

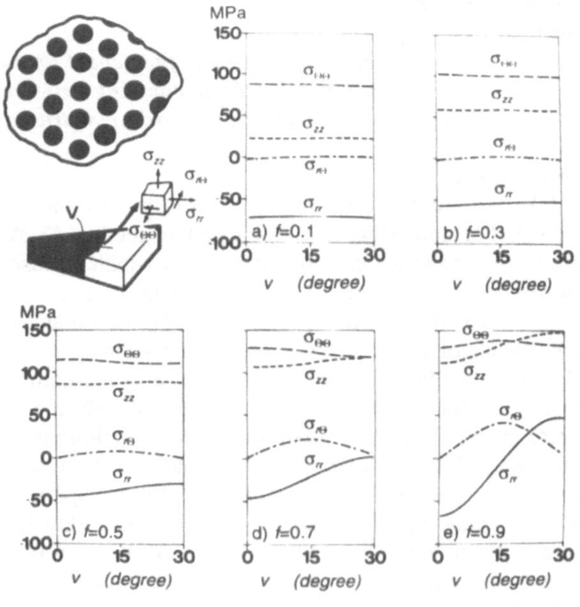

Figure 6 *Residual stresses in SiC/CAS for various fiber volume fractions f, when fibers are packed in perfect hexagonal array.*

at high fiber volume fraction, the interface is in residual radial *compression* ($\sigma_{rr}<0$), whereas $\sigma_{\Theta\Theta}$ and σ_{zz} is in residual tension in the matrix ($\sigma_{\Theta\Theta}$ slightly higher than σ_{zz}).

In Figure 7a) we see the stresses in the matrix at the interface for fiber volume fraction *f=0.5*, for square array packing. Compared to the stress levels in hexagonal packing (Figure 6c), note that the square array packing gives most variation around the fiber. This is quite consistent with our expectations: The hexagonal packing is closer (six fold symmetry) to axisymmetry than the square array (only four fold symmetry). Figures 7b)-7c) indicate what happens if the four fibers were arranged closer to each other. The radial stress component σ_{rr} is strongly affected, residual tensile stress may exist at some locations, large compressive stresses develop where fibers are touching each other. Again, more disorder creates larger stress variations.

This trend is further supported by the results shown in Figures 8-9, for two other situations where fibers enclose matrix ($f=0.5$). The stress variation is highest in that packing arrangement where the most matrix is enclosed. Where fibers are in contact, the Poisson effect from σ_{rr} causes a slight drop in the $\sigma_{\Theta\Theta}$ and σ_{zz} components.

In order to compare the effect of matrix and fiber rich zones (large scale effects), the residual stress state for (perfect uniform) square array close packing at $f=0.785$ is shown in Figure 10. Figure 11 shows the stress state outside a matrix zone. The diameter of the zone is four fiber diameters, and the outer diameter is ten fiber diameters. Compared with Figure 7a) (perfect uniform square array packing, $f=0.5$) and Figure 6c) (perfect uniform hexagonal packing, significant changes in stresses due to the matrix domain are seen: Since the zone itself is in triaxial tension, additional tensile stresses are induced normal to the domain boundary and compressive stresses induced tangential to the boundary. At a fiber rich region, Figures 12a)-12b), the opposite effect is seen. Note that the stresses inside the fiber rich domain differ from the stresses for uniform close packed fibers (Figure 10).

Residual Stresses in SiC/LAS ($\alpha_m<\alpha_f$)

For this material system, after the cooling down process, all matrix stress components are compressive, except for the σ_{rr} (fibers shrink more than the matrix). Therefore only plots of σ_{rr} are shown.

In Figure 13 the results for perfect hexagonal fiber distribution are shown. The highest and lowest values of σ_{rr} are plotted as a function of fiber fraction, and compared with the stresses calculated from an analytical axisymmetric composite model [2]. The analytical model indicates that the residual stress component σ_{rr} decreases with increasing fiber volume fraction, and it does not account for the local stress magnification at high fiber fraction. This is quite important, as the interfacial normal stress σ_{rr} is in tension, and therefore thermal debonding is possible. The FEM results show that the tensile stresses locally (where fibers touch each other) may be much higher (up to four times) that the average value. Thus thermal debonding is expected to initiate at these locations.

The effect of fiber clustering is shown in Figure 14 for fiber fraction $f=0.5$. There is stress magnification where fibers touch. The highest and lowest interfacial normal stresses σ_{rr} calculated by FEM are shown for $f=0.5$ as function of enclosed matrix in Figure 15. The stress may be several times higher than the average value. The effects of matrix and fiber rich domains are shown in Figure 16. Note that *both* matrix and fiber rich domains raise the magnitude of the highest σ_{rr}, although at different locations (angles around the fiber).

DISCUSSION

The interfacial normal stress component σ_{rr} is much more sensitive to packing than $\sigma_{\Theta\Theta}$ and σ_{zz}. This has interesting consequences. The SiC/LAS is more sensitive to fiber distribution than SiC/CAS in that sense that the thermal damage (Θ-cracks) is be expected to initiate at the specific locations where the highest interfacial tensile stresses exist. In SiC/CAS other failure modes (r- and z-cracks) are anticipated. They are controlled by other stress components less sensitive to packing. Moreover the nature of the physics is such that in general, the stress

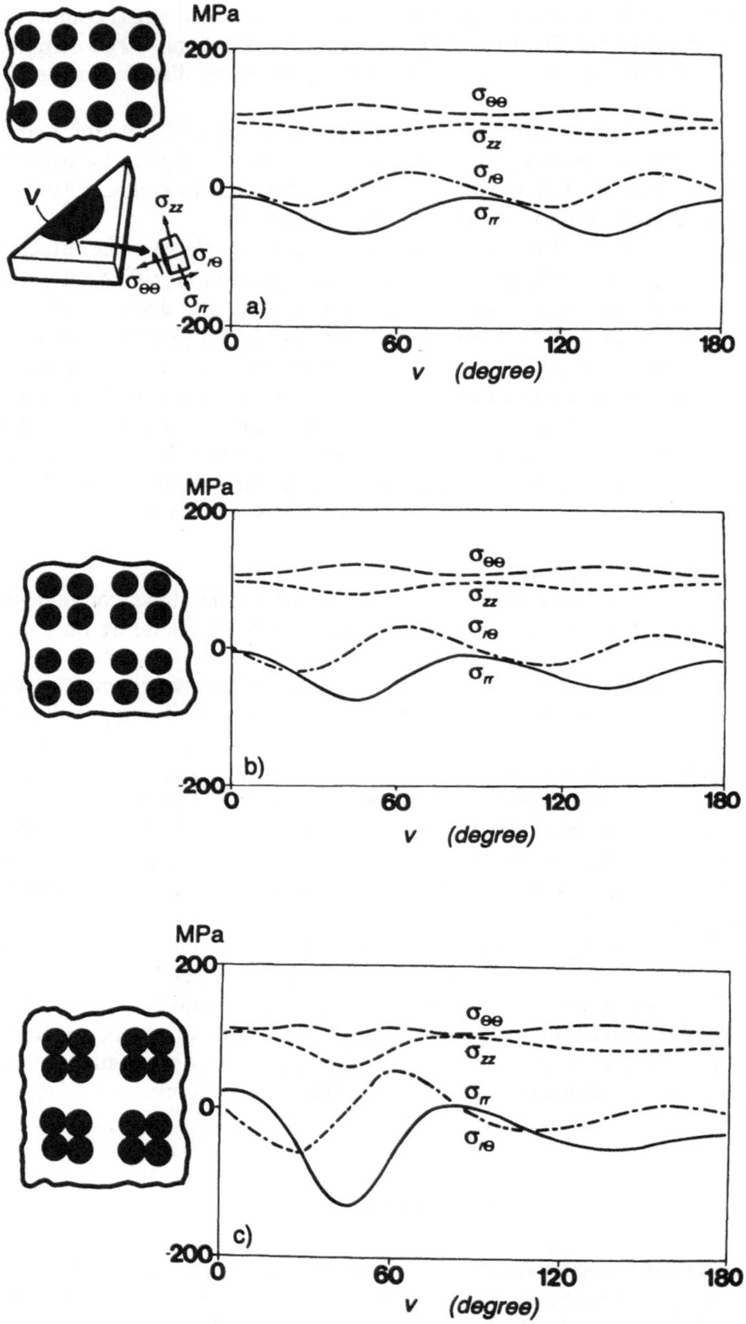

Figure 7 *Residual stresses in SiC/CAS, f=0.5. a) perfect square array packing, b) four fibers lying closer to each other, c) four fibers touching each other.*

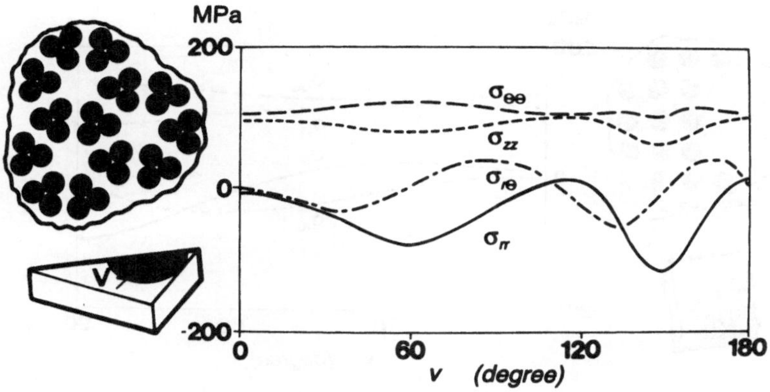

Figure 8 *Residual stresses in SiC/CAS, three fibers touching each other, f=0.5.*

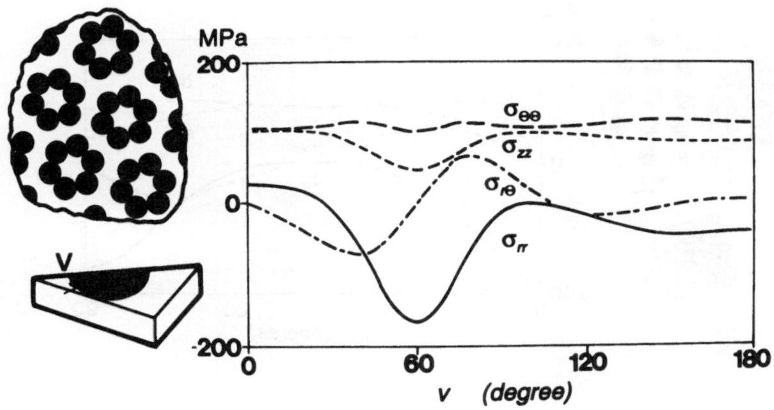

Figure 9 *Residual stresses in SiC/CAS, six fibers touching each other, f=0.5.*

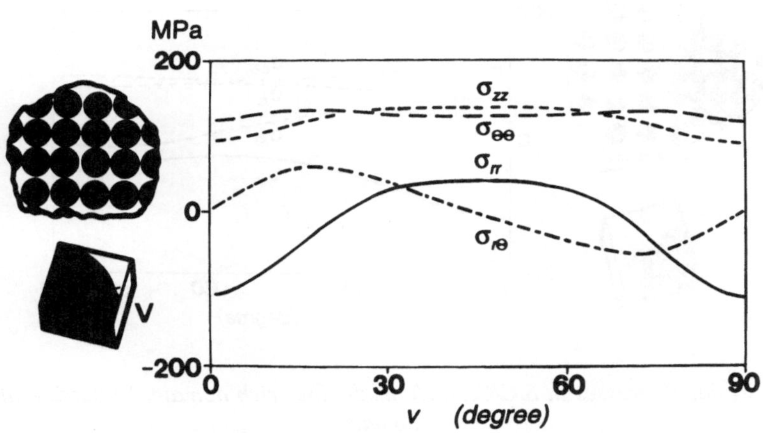

Figure 10 *Residual stresses in SiC/CAS, perfect square array close packing, f=0.785.*

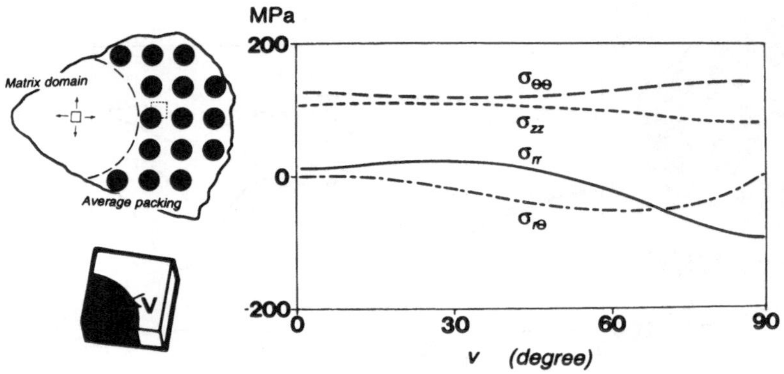

<u>Figure 11</u> *Residual stresses in SiC/CAS outside matrix domain.*

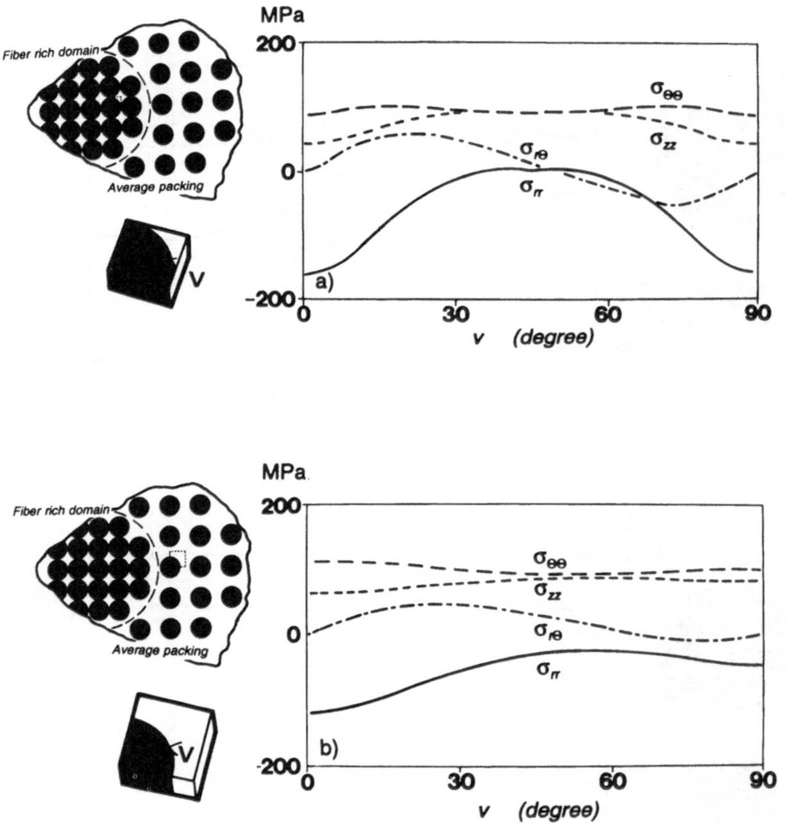

<u>Figure 12</u> *Residual stresses in SiC/CAS, a) inside fiber rich domain, b) outside fiber rich domain.*

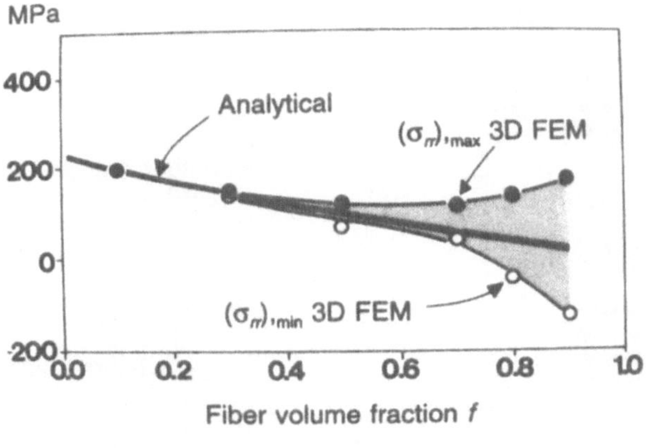

Figure 13 *Peak values of σ_{rr} from FEM compared to the results from the axisymmetric model, for the SiC/LAS system as function of fiber fraction f.*

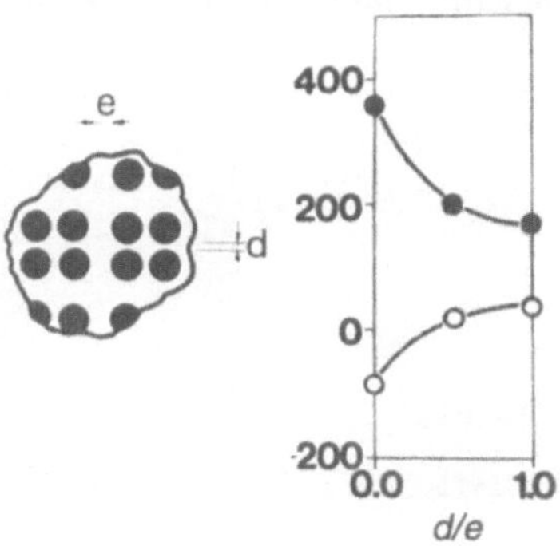

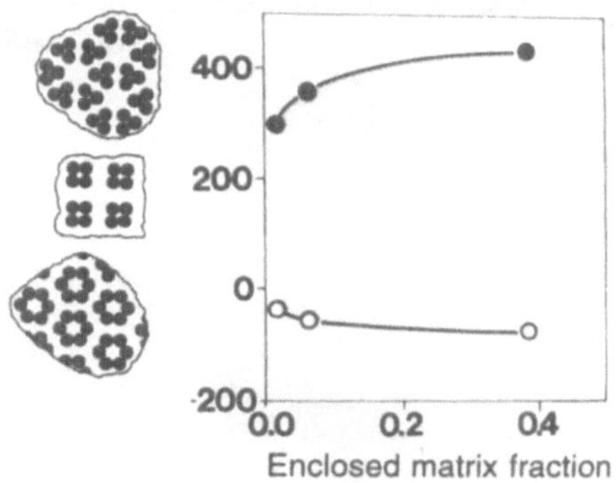

Figure 14 *Effect of fiber clustering, highest and lowest interfacial normal stress σ_{rr} in SiC/LAS, f=0.5.*

Figure 15 *Highest and lowest σ_{rr} values as a function of enclosed matrix in SiC/LAS.*

components σ_{zz} and $\sigma_{\Theta\Theta}$ have identical sign, opposite sign to the σ_{rr} component. Thus, the thermal stresses favour *either* Θ- *or* r-cracks, but *not both*. This point might have clarified the question of most likely crack type, in the paper by Zhu and Achenbach [7].

It was checked that the contraction of the short range volume cells were in very good agreement with each other at the same fiber volume fraction despite difference in local packing, and in complete agreement with the axisymmetric model [2]. This implies that the variation in the residual stress field is quite local, as could also be seen at stress contour plots, and justifies the original assumptions made on global packing symmetry.

The results from the FEM analysis demonstrate that the stresses vary around the fiber, especially at high fiber fraction or non-uniform packing. Although the *average* stresses around

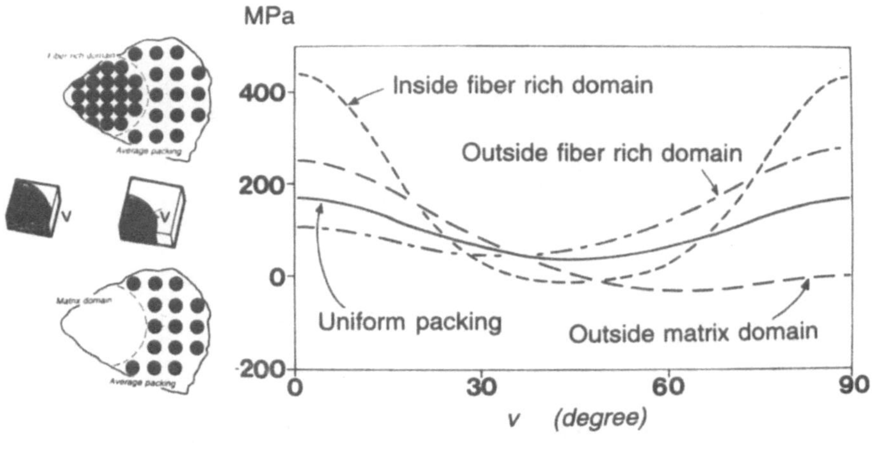

Figure 16 *Residual stress component σ_{rr} outside and inside matrix and fiber rich domains in SiC/LAS.*

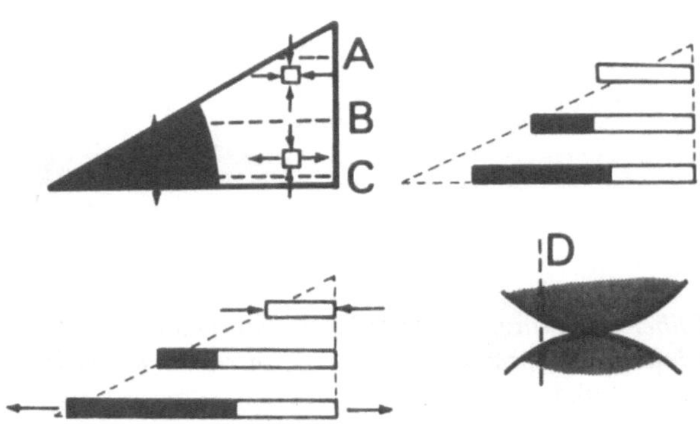

Figure 17 *Simple principles illustrating why non-axisymmetry influence σ_{rr} and a simple way to explain stress magnification at locations where fibers touch each other.*

the fiber correspond nicely to the stresses computed by the axisymmetric model, it should be emphasised that real damage mechanisms may not advance as axisymmetric processes. For instance, thermal debonding (Θ-cracking) in SiC/LAS is not an axisymmetric process [15]. Therefore, axisymmetric modelling of further debonding [16] may be questionable.

The short range stress variation around the fiber may be explained by simple principles. The residual stress state in the matrix in the SiC/LAS system at perfect hexagonal packing is shown in Figure 17. The stress component σ_{xx} is in tension where the fiber distance is closest (strip C), and in compression at the pure matrix corner (strip A). As $\alpha_m<\alpha_f$ for SiC/LAS, strip A would contract less (because of larger matrix content) and strip C more (because of higher fiber content) than the average composite (strip B), if they were unconstrained. However, they are in fact constrained; as their deformations have to comply with the symmetry conditions. This induces the stress variation. Strip A has to be compressed (compressive stress induced) and strip C has to be elongated (tensile stress imposed). The stress magnification for touching fibers may be explained in similar manner. The strip D possesses a higher fiber content (like strip C), thus generating large tensile stresses in SiC/LAS. This stress magnification induced by constraint and the thermal expansion mismatch has, as far as the authors know, not been treated elsewhere.

CONCLUSIONS

The residual stress state in two CMC systems were investigated with 3D FEM, and various fiber packing configurations were analyzed. It is shown that the residual stress state is significantly affected by fiber packing, *i.e.* less uniform fiber distribution generates higher residual stresses. The interfacial normal stress is most sensitive to packing, and peak values may be as high as four times the average value. This stress magnification is located at the points where the fibers are in contact and is caused by the thermal expansion mismatch. The stress states are completely different in the two different material systems. Therefore, residual stress should *always* be included in micro-mechanical analysis. In SiC/LAS, Θ-cracks (partial debonding) may initiate where fibers touch each other. In the SiC/CAS composite, r- and z-cracking are likely to initiate and grow due to the residual stresses.

ACKNOWLEDGEMENT: This research was part of a Ph.D. project, supported by the Danish Research Academy. Discussions with Dr. Svend Ib Andersen have been useful.

REFERENCES

1. Hashin Z., *Analysis of Composite Materials - a Survey*, J. Appl. Mech., 1983, **50**, 481-505.

2. Hsueh C.-H., Becker P.F., *Thermal Expansion Coefficients of Unidirectional Fiber-Reinforced Ceramics*, J. Am. Ceram. Soc., 1988, **71**, C438-41. *Correction*, J.Am. Ceram. Soc., 1989, **72**, 359.

3. Mikata Y., Taya M., *Stress Field in a coated Continuous Fiber Composite subjected to Thermo-Mechanical Loadings*, J. Comp. Mats. 1985, **19**, 554-78.

4. Chen T., Dvorak G.J., Benveniste Y., *Thermal stresses in coated fiber composites*, Symp. High Temp. Comp. 1989, 139-47.

5. Böhm H.J., Rammerstorfer F.G., *Micromechanical Investigation of the Processing and Loading of Fibre-Reinforced Metal Matrix Composites* Mats. Sci. Eng. 1991, **A135**, 185-8.

6. Zhu H., Achenbach J.D., *Effect of Fiber-Matrix Interphase Defect at Micro Stress State at Neighbouring Fibers* , J. Comp. Mats 1991, **25**, 224-38.

7. Zhu H., Achenbach J.D., *Radial Matrix cracking and Interphase Failure in Transversely Loaded Fibre Composite*, Mech. of Mats. 1991 (submitted)

8. Adams. D.F., Tsai S.W., *The influence of Random Filament Packing on the Transverse Stiffness of Unidirectional Composites*, J. Comp. Mats., 1969, **3**, 368-81.

9. Brockenbrough J.R., Suresh S., Wienecke H.A., *Deformation of metal Matrix Composites with Continuous Fibers: Geometrical Effects of Fiber Distribution and Shape*, Acta Metall. Mater., 1991, **39**, 735-52.

10. Bonney L.A., Cooper R.F., *Reaction Layer Interfaces in SiC-Fiber-Reinforced Glass-Ceramics: A High-Resolution Scanning Transmission Electron Microscopy Analysis* J. Am. Ceram. Soc., 1990, **73**, 2916-26.

11. Evans A.G., Marshall D.B., *The Mechanical Behaviour of Ceramic Matrix Composites*, Acta Metall., **37**, 2567-83.

12. Karandikar P., Talreja R., Chou T.-W., *Evolution of Damage and Mechanical Response of Ceramic Matrix Composites*, J. Mats. Sci., 1991 (submitted)

13. FEMGEN Users Manual version 8.6, 1987, Femgen AB, Lund, Sweden

14. SOLVIA Report 90-2, 1989, Solvia Engineering AB, Sweden

15. Bischoff E., Rühle M., Sbaizero O., Evans A.G., *Microstructural Studies of the Interface Zone of a SiC-Fiber-Reinforced Lithium Aluminum Silicate Glass-Ceramic*, J. Am. Ceram. Soc., 1989, **72**, 741-5.

16. Charalambides P.G., Evans A.G., *Debonding Properties of Residual Stressed Brittle Matrix Composites*, J. Am. Ceram. Soc., 1989, **27**, 746-53.

SiC Particulate Reinforced Y-TZP Composites

Xiaoxian Huang, Jingkun Guo, Linhua Gui and Baoshun Li

Shanghai Institute of Ceramics, Chinese Academy of Sciences
1295 DingXi Rd, Shanghai 200050, China

Abstract

Y-TZP (yttria-stabilized tetragonal zirconia) matrix composites reinforced by SiC particles(p) were fabricated by hot-pressing. The mechanical properties, microstructure and phase composition of the composites were investigated. High temperature strength of the composites was significantly improved, and was dominated by strengthening effect produced by SiC particles. The SiC(p)/Y-TZP composites were used for manufacturing ceramic valve seats in Diesel engine, and a good test result was achieved.

Introduction

Great attention has been paid to Y-TZP materials because of its high strength, high toughness and excellent wear resistance at room temperature [(1)], whereas these properties decrease rapidly with increasing temperature. The transformation toughening exhibited in Y-TZP ceramics is lost at high temperatures because the tetragonal form becomes more stable relative to the monoclinic form[(2,3)], limiting the practical applications of Y-TZP ceramics as high temperature structural components.

In recent years, it has been attracted considerable interest that the heated and wearing component parts for heat engine are fabricated with ceramic materials in place of metal materials. The valve seats, one of heat engine component parts, should have the ability to withstand thermal shock, cyclic fatigue, compression, wear and the mismatch stress caused by thermal expansion misfit with metal parts. So the materials for valve seats must exhibit high strength, toughness and hardness.

The high temperature strength of Y-TZP ceramics must be improved if it is be used as valve seats for heat engine. Some approaches were proposed to improve the high temperature mechanical properties of Y-TZP ceramics. One method to improve the high temperature strength is to reinforce the Y-TZP with additions of other ceramic components [(4,5)].

In the present study, the SiC particles are used as the reinforcing agent for Y-TZP matrix because of its high strength, high modulus and chemical match for the matrix. The mechanical properties and microstructure of the SiC(p)/Y-TZP composites were investigated. The composites were fabricated into the exhaust and inspiration valve seats of 95 type heat engine, and were put to bench run. The result of run over 300 hours was achieved.

Experimental Procedure

$ZrOCl_2 \cdot 8H_2O$, $Y(NO_3)_3 \cdot 6H_2O$ and α-SiC powder were used as starting materials. The Y-TZP containing 2.8mol% Y_2O_3 and SiC/Y-TZP powder with desired ratio were prepared by the coprecipitation method from the mixtures of $ZrOCl_2$ solution, $Y(NO_3)_3$ solution and α-SiC dispersed in water. The Y-TZP and SiC(p)/Y-TZP powder were hot-pressed in a BN-coated graphite mold. This process was conducted at 1600°C for 1h under 25 MPa for all composites.

Three point bending strength measurements were carried out using a 20 mm span and a cross-head speed of 0.5 mm/min. Flexural bars were 2.5 mm in height, 5 mm in width and 26mm in length. Fracture toughness was evaluated by single edge notch-beam technique (SENB). The dimension of test bars was 5 x 2.5 x 28mm. The notches used were 0.22-0.25mm in width and 2.5mm in depth. Fracture toughness measurements were performed by three point bending using a 20mm span and a cross-heat of 0.05mm/min. All the tests above were made in an Instron 1195 testing machine. The microstructure and the phase compositions were also examined by SEM, TEM and XRD.

Results and Discussion

(1) Densification of SiC(p)/Y-TZP composites

Fig.1 illustrates the effect of SiC particle content on relative density of the Y-TZP matrix composites reinforced by SiC particles, showing if SiC particle content was less than 15wt%, the SiC(p)/Y-TZP composite could be densified to over 98.8% of theoretical density. At 20wt% SiC particles, the relative density of the SiC(p)/Y-TZP composites decreased slightly, but as the SiC particle content was raised more, the relative density of the composites decreased obviously. It is controlled by poor sinterability of SiC material.

(2) Mechanical properties of the SiC(p)/Y-TZP composites

Fig.2 shows the effect of SiC particle content on flexural strength and fracture toughness of the SiC(p)/Y-TZP composites. If additions of SiC particle is less than 6 wt%, the strength and toughness is decreased obviously. The main reason is because the amount of transformable tetragonal ZrO_2 in per unit volume in SiC(p)/Y-TZP composites is decreased by the adding SiC particles, and the strengthening effect could not be compensated by its low addition. In another word, the strengthening effect could not be compensable for the decreased contribution of transformation toughening, and the total reinforcing effects decreased in the composites.

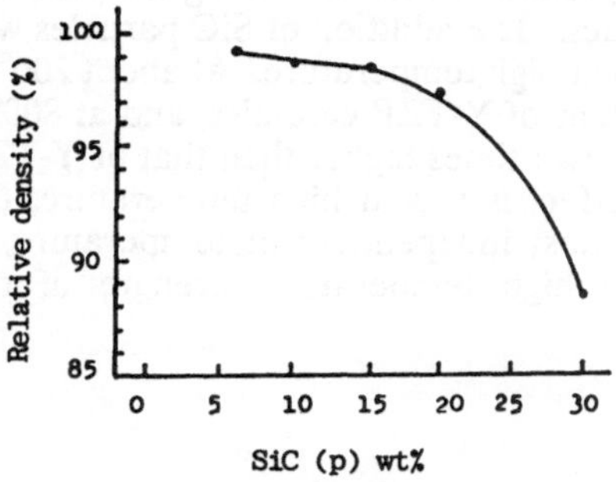

Fig.1 Dependence of SiC particle content on relative density of SiC(p)/Y-TZP composites

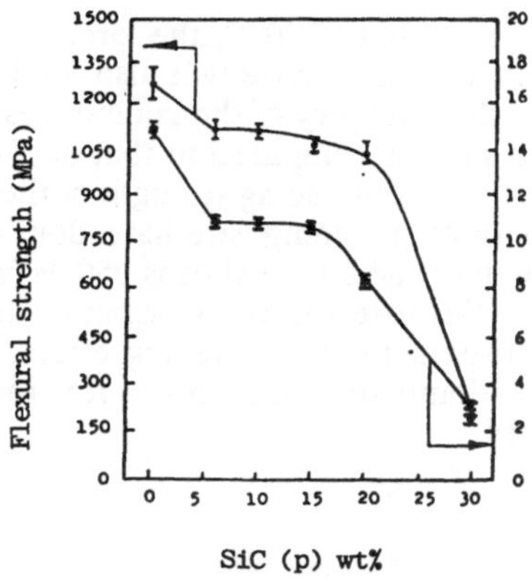

Fig.2 Dependence of SiC particle content on flexural strength and fracture toughness of SiC(p)/Y-TZP composites

Increasing the SiC particle content from 6wt% to 20wt%, the strength of the composites is almost constant. However, the toughness of the composites is not much changed on increasing the SiC particle content from 6wt% to 15wt%. In this range of SiC particle content, the strengthening effect caused by SiC particles was enough to compensate for the decreased contribution of ZrO_2 transformation toughening. As the SiC particle content was raised further, the strength and toughness of the composites decreased dramatically, this was attributed to the fact that the particle strengthening effect didn't increase in proportion as the SiC particle content and the composites showed the poor sinterability as illustrated in Fig. 1.

Fig.3 and 4 show the micrographs of vickers indentation crack propagation for SiC(p)/Y-TZP composites. The paths of crack propagation became tortuous due to the crack deflection and branching caused by SiC particles, which shield the crack tip and lower the effective stress acting on the crack tip. It indicates that the particle strengthening effect and the transformation toughening effect existed simultaneously in the SiC(p)/Y-TZP composites. The hardness value of the composites with the SiC particle content from 6wt% to 20wt% is 92, higher than the value (91) of Y-TZP. The wear ability of the composites was also improved to a certain extent.

(3) High temperature strength of the SiC(p)/Y-TZP composites

The dependence of temperature on flexural strength of the SiC(p)/Y-TZP composites and Y-TZP ceramics is presented in Fig. 5. As shown, the strength of Y-TZP ceramics decreased sharply with increasing temperature. This is due to the disappearance of transformation toughening at high temperature. According to the Y_2O_3-ZrO_2 phase diagram [6], at about 770°C, the zirconia is in the region of tetragonal form. In addition, the interfacial glass phase will also lead to the strength degradation at high temperature. However, the tendency of the composites strength degradation with increasing temperature was much slow as compared to that of Y-TZP ceramics. The addition of SiC particles was very effective in enhancing strength of the composites at high temperature. At about 200°C, the strength of the composite have dominated over that of Y-TZP ceramics, and at 800°C the strength of the composites is 650 MPa, which was two times higher than that of Y-TZP ceramics. Because the transformation toughening effect is lost at high temperature, the particle dispersion strengthening effect, which is almost independent of temperature, is obviously dominant mechanism for improving the high temperature strength of the composites.

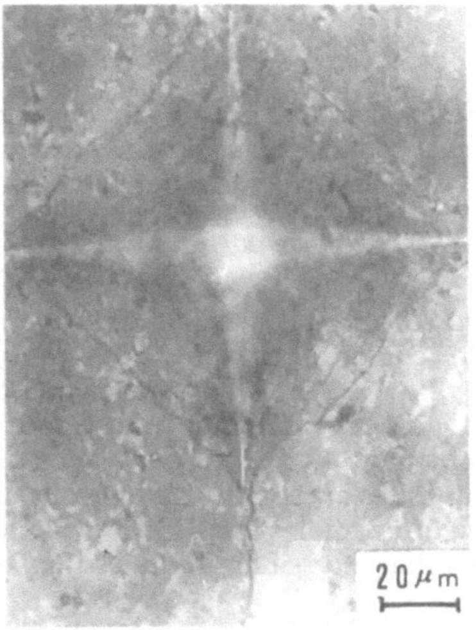

Fig.3 A SEM micrograph, showing vickers indentation and crack deflection of 15 wt% SiC(p)/Y-TZP composite

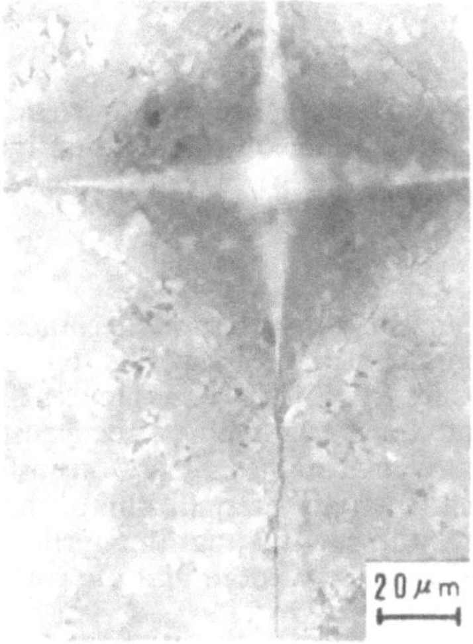

Fig.4 A SEM micrograph, showing vickers indentation and crack branching of 15 wt% SiC(p)/Y-TZP composite

(4) Thermal expansion coefficient of the SiC(p)/Y-TZP composites

The dependence of temperature on thermal expansion coefficient for the SiC(p)/Y-TZP composites is shown in Fig.6. The thermal expansion coefficient increased basically with increasing temperature. But at about 600°C, the thermal expansion coefficient decreased slightly. This is because monoclinic ZrO_2 in the composites was transformed to tetragonal form within this range of temperature.

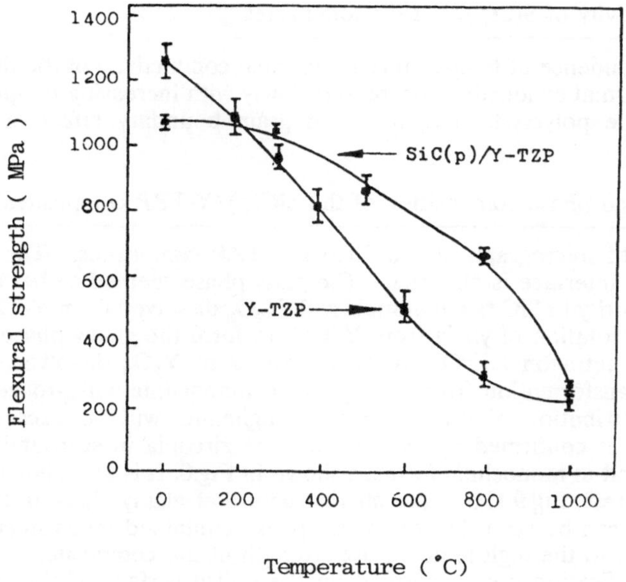

Fig.5 A comparison of high temperature strength between 15 wt% SiC(p)/Y-TZP composite and hot-pressed Y-TZP ceramics

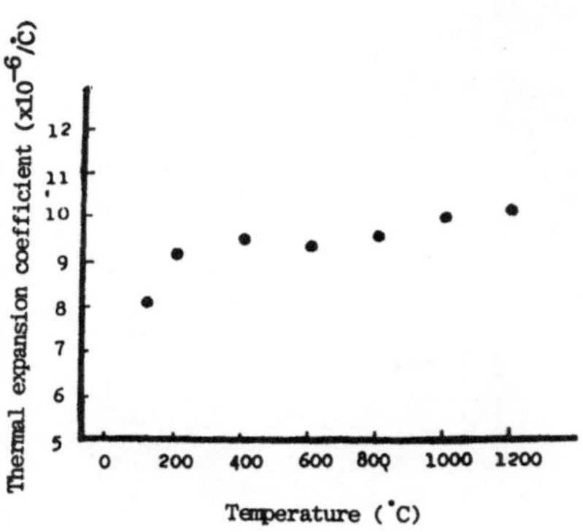

Fig.6 Dependence of temperature on thermal expansion coefficient of 15 wt% SiC(p)/Y-TZP composite

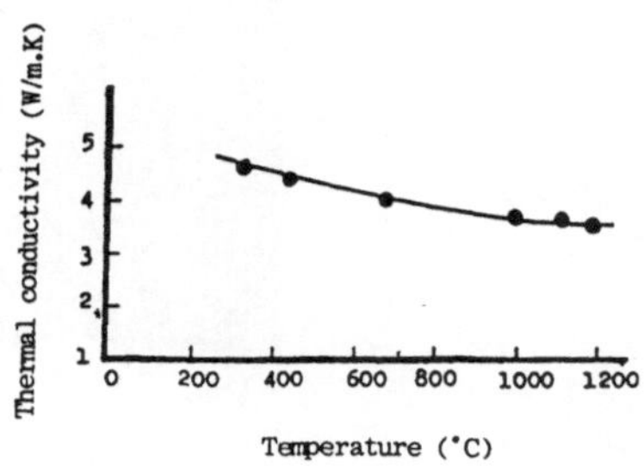

Fig.7 Dependence of temperature on thermal conductivity of 15 wt% SiC(p)/Y-TZP composite

(5) Thermal conductivity of SiC(p)/Y-TZP composites

Fig. 7 shows the dependence of temperature on thermal conductivity of the SiC(p)/Y-TZP composites. The thermal conductivity decreased slowly with increasing temperature. This is consistent with the polycrystal structure and grain boundary effect of the ceramic material.

(6) Microstructure and phase composition of the SiC(p)/Y-TZP composites

Fig.8 shows the TEM micrograph of the SiC(p)/Y-TZP composites. The glassy phase formed at SiC/ZrO_2 interface is observed. The glass phase seemed to be formed by the reaction of SiO_2 impurity in SiC raw materials with Y_2O_3 dissolved from Y-TZP. It should be noted that the dissolution of yttria from Y-TZP to form the glassy phase would result in destabilization of tetragonal ZrO_2. When an excess of Y_2O_3 dissolved from Y-TZP matrix, the phase transformation from tetragonal to monoclinic will proceed during the cooling, and the distribution of transformation toughening will be decreased[7]. This phenomenon is further confirmed by the fact that the zirconia in surroundings of glassy phase is almost existed in monoclinic form, as shown in Fig.8. It is also shown that the SiC partices are under stress. Fig.9 shows the microstructure of glassy phase in the SiC(p)/Y-TZP composites. It can be seen that the glassy phase contained much microcrystallines, which also contribute to the high temperature strength of the composites.

The phase identification was performed on the polish surface of the composites by XRD analysis, and the result is showed in Fig.10. It is clear that the phase compositions of the composites are tetragonal ZrO_2, α-SiC and a little amount of monoclinic ZrO_2.

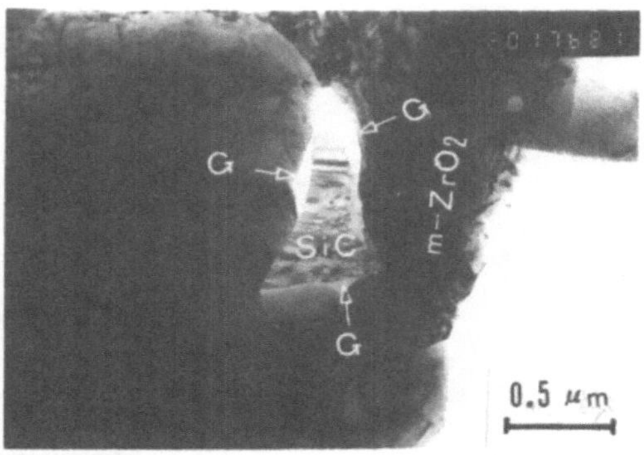

Fig.8 A TEM micrograph of SiC(p)/Y-TZP composite,the glassy phase (G) and monoclinic zirconia(m-ZrO2) should be noted.

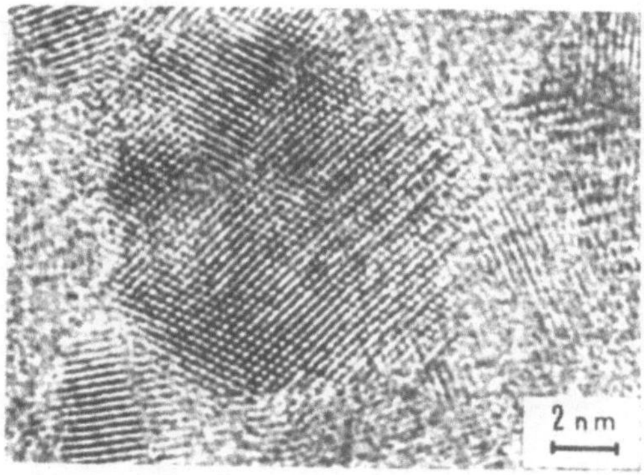

Fig.9 A TEM lattice image of glassy phase in SiC(p)/Y-TZP composite.

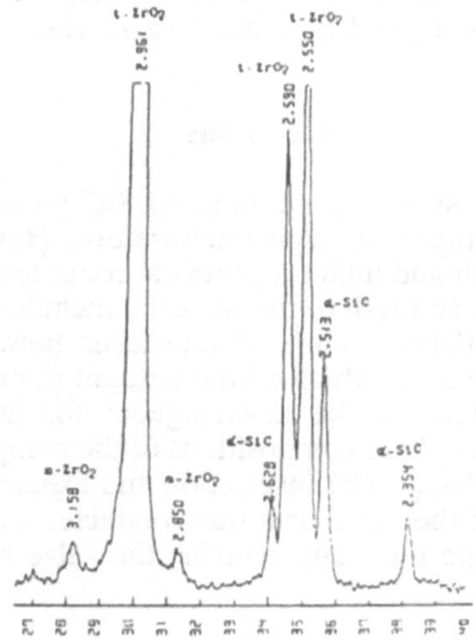

Fig.10 XRD pattern of 15 wt% SiC(p)/Y-TZP composite

(7) Test of the SiC(p)/Y-TZP composites for valve seats

The photograph of the exhaust and inspiration valve seats for 95 type Diesel engine was shown in Fig.11. All the valve seats were fabricated using the SiC(p)/Y-TZP composites. The bench run of the valve seats was conducted. The run conditions are as follows: rotational speed of 2000(r.p.m), load of 12 (h.p.), 380°C--400°C of exhaust temperature. After the test run over 300h under the fully loaded, the valve seats were found intact and undamaged, and can be run further. This indicates that the composites are a strong potential candidate for manufacturing the valve seats of Diesel engine.

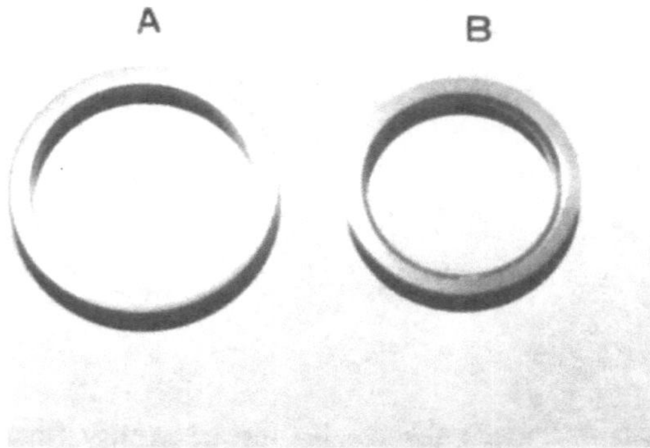

Fig.11 A photograph of 95-type Diesel engine valve seats made of SiC(p)/Y-TZP composites (A is inspiration valve seat, and B, exhaust valve seat)

Conclusions

For the SiC(p)/Y-TZP composites, the addition of SiC particles results in significantly enhancing strength of the material at high temperature. However, the composites still exhibite extremely high strength and high toughness at room temperature. The mechanism of improving high temperature strength of the material includes crack deflection and crack branching caused by SiC particles. Some of interfaces between SiC and ZrO_2 in the composites contained glassy phase in which a little amount of microcrystallines are present. This feature has been considered to be advantageous for improving high temperature strength of the composites. The phase compositions of the composites are tetragonal ZrO_2, α-SiC and a little monoclinic ZrO_2. The inspiration and exhaust valve seats were made by SiC(p)/Y-TZP composite, and the bench run was conducted over 300 hrs. The composite is a strong a potential candidate for manufacturing the valve seats in Diesel engine.

References

1. I. Nettleship and R. Stevens, Int. J. High Tech. Ceram., 3 1-32 (1987)
2. Li Baoshun, Huang Xiaoxian, Guo Jingkun and Yan Dongsheng, J. Inorg. Mater., 1(2) 129-134 (1986).
3. Gao Lian, Yan Dongsheng and Guo Jingkun, Scientia Sinica(A), 1 95-103 (1988)
4. N. Claussen, Mater. Sci. and Eng., 71(1-2) 23-38 (1985)
5. Koji Tsukuma and Kuniyoshi Ueda et al., J. Am. Ceram. Soc., 68(2) c56-c58 (1985)
6. F.F. Lange, J. Mater. Sci., 17(1) 240-246 (1982)
7. Xiaoxian Huang, Litai Ma, Linhua Gui, Baoshun Li and Jingkun Guo, C-MRS International'90, Beijing, China, June 19-22, 1990

DEVELOPMENT OF SILICON NITRIDE BASED MATRICES

JESPER BRANDT*, KENT RUNDGREN AND ROBERT POMPE
Swedish Ceramic Institute (SCI)
Box 5403, 402 29 Göteborg, Sweden

ROBERT LUNDBERG AND LARS PEJRYD
Volvo Flygmotor AB
461 81 Trollhättan, Sweden

ABSTRACT

A feasible way to produce SiC long fibre reinforced Si_3N_4 is by vacuum infiltration of Si_3N_4 slurry followed by reaction bonding of the matrix. To minimize degradation of the fibres the nitridation reaction has to take place at moderate temperatures. In the present work a number of Si_3N_4-matrices processed by a modified RBSN technique and mixed with ZrO_2-additives to optimize the nitridation cycle will be discussed. The results show that ZrO_2 has a remarkably accelerating effect on the nitridation. It is thus possible to decrease the exposure of SiC fibres to high temperatures.

INTRODUCTION

One of the significant benefits of engineering ceramic materials is the applicability at elevated temperatures. However, the major drawbacks such as brittleness and potential sensitivity to slow crack growth may cause catastrophic failures. One approach to enhance the toughness is to reinforce monolithic ceramic materials with continuous long fibres [1-5]. The next step then is to develop competitive manufacturing processes for fabrication of high temperature long fibre composites. Earlier experiences, however, indicate that there are several obstacles to overcome.

The first task to be solved is how to fill up the fibre preform and the void space between the monofilaments with the matrix material. This can be accomplished by the chemical vapour infiltration (CVI) technique [6], although there are several drawbacks with this process. The voids in between the fibre bundles may be filled up, yet critical voids typically still will remain in the matrix between the bundles, resulting in a porosity of 10-20% for the composites. Another disadvantage of CVI is the excessively long process time, weeks to months. Regarding hot pressing of slurry preimpregnated fibre tows wound up on a wheel [7], this process is well established. However, it limits the range of products to simple shapes.

The second issue to deal with is the fact that all available ceramic long fibres are mechanically degraded or react with the environment at high temperatures [8, 9]. While awaiting new generations of more temperature-resistant long fibres it is of great significance to find production processes, where more moderate temperatures and where shorter thermal exposure can be employed.

The third issue is the bond between the matrix and the fibres, which necessarily has to be sufficiently weak to enable interfacial sliding [10], thus leading to the fracture energy dissipation behaviour desired.

In the literature, several other process techniques for composite manufacturing are described and in the future they are expected to emerge as competing alternatives - for instance polysilazane impregnation with subsequent pyrolysis [11], infiltration using the Lanxide process [12] and slurry infiltration [11].

In the present work the main aim has been to investigate ceramics suitable for manufacturing gas turbine hot parts for continuous use at temperatures over 1000°C. The objective has been initially to develop an appropriate processing method for slurry-impregnating fibre preforms whereafter the matrix will be reaction bonded. The main focus will be on the effects different nitridation aids have on the nitridation temperature of Si/Si_3N_4 matrix specimens. The purpose is to find a suitable matrix composition requiring a low temperature reaction bonding cycle. In this particular investigation test specimens have been prepared by cold isostatic pressing, although for infiltration of preforms, slip casting will be the main forming method in the future.

EXPERIMENTAL

Matrix Selection

Six different matrix compositions, Table 1, were prepared for testing of the nitridation rate at fixed temperature cycles. The mixture denoted A1 is a standard NPS material [13] (reference material) developed to reduce the nitridation time to 3-4 h. In sample A2 the Al_2O_3 is excluded. Al_2O_3 is mainly active as a sintering aid and since the specimens will merely be nitrided the effect of Al_2O_3 needs to be examined. Sample B involving reaction bonding using the $ZrSiO_4/Si$ system, is significantly different, both in terms of the composition and the reactions taking place during nitridation. Si nitrides to Si_3N_4 which in turn will react with $ZrSiO_4$ to Si_2N_2O and ZrO_2 [14]. In samples C1, C2 and C3 the effect of an increasing amount of ZrO_2 is studied. It is well established that ZrO_2 accelerates the nitridation rate of Si at lower temperatures [15].

TABLE 1
Compositions of the mixtures investigated. The values refer to the final composition after nitridation of Si

Sample	$Si:Si_3N_4$ wt ratio	$ZrSiO_4:Si$ wt ratio	ZrO_2 wt%	Y_2O_3 wt%	Al_2O_3 wt%
A1 (ref)	7:3	-	-	6.0	2.0
A2	7:3	-	-	8.0	-
B	-	1:1.66	-	3.9	4.9
C1	7:3	-	4.75	6.2	2.0
C2	7:3	-	9.50	6.5	2.0
C3	7:3	-	14.25	6.8	2.0

Raw Materials
The raw materials selected for the matrix compositions are fine, reactive powders, Table 2.

TABLE 2
Raw materials used and measured specific surface area (BET)

Powder	Trade Name	Manufacturer	BET (m^2/g)
Si	Sicomill 2D	Permascand AB, Sweden	2.1
Si_3N_4	SN-E 10	UBE, Japan	11.6
$ZrSiO_4$		Ventron, Germany	12.8
ZrO_2	TZ-3Y	Tosoh, Japan	18.0
Y_2O_3	Grade fine	HC Starck, Germany	14.0
Al_2O_3	AKP-30	Sumitomo Chemical, Japan	7.5

Specimen Preparation
Batches of the six compositions, Table 1, with the addition of 3 wt% KD-3 (ICI-chemicals) as dispersant and 3 wt% oleic acid as a pressing aid were ball milled in petroleum spirit (boiling point 120°-160°C). After thin film evaporation and sieve granulation through a 300 μm sieve, bars (8 by 7 by 70 mm) were uniaxially prepressed and then cold isostatic pressed at 275 MPa. The organic additives were burned out in nitrogen at 500°C. The specimens were stored prior to nitridation in dry atmosphere in order to prevent undesired interaction with moisture.

Thermogravimetry (TGA) was utilized to record the nitridation rates in flowing nitrogen. The program cycle started at room temperature with 10°C/min to 1050°C from where it was slowly raised by 1°C/min up to the final temperature at 1350°C. Prior to each run the furnace chamber was evacuated and twice filled with nitrogen.

As a second check, specimens of the six compositions were nitrided in a graphite resistance furnace (Balzers), which was programmed in accordance with the TGA results.

Three different optimized program cycles for the different compositions (A1,A2;B;C1-C3) were employed.

The phase compositions of sample B and C3 were determined by X-ray diffractometry (XRD).

RESULTS AND DISCUSSION

Specimen Preparation

The powders could be easily dispersed with KD-3 in petroleum spirit resulting in high solid contents and the non-polar solvent prevented undesirable oxidation of Si. The particular solvent has been selected for its high boiling point (120°-160°C). This leads to a safer handling, minimizes evaporation and eliminates undesired boiling during the slurry vacuum infiltration of preforms.

The specific surface area (BET) after milling of the powders and the theoretical density of CIP:ed green compacts after binder burn out and after nitridation in the graphite resistance furnace, respectively, are shown in Table 3. No dimensional changes of the specimens during nitridation were observed.

TABLE 3
Specific surface area of milled powders, green and nitrided densities

Sample	BET (m^2/g)	Green Density (% of theor density)	Nitridation Yield (% by theor weight gain)	Nitrided Density (% of theor density)
A1	10.7	61.8	93.8	70.5
A2	11.1	61.7	92.6	71.1
B	9.2	66.5	91.3	74.4
C1	10.0	62.3	91.8	70.3
C2	11.2	64.1	92.4	72.6
C3	11.6	63.7	92.8	71.8

The specific surface areas were relatively low, but still sufficient to achieve adequate Si_3N_4 yield. The difference to 100% is thought to be mainly due to oxygen impurity and evaporation of SiO gas during the nitridation cycle.

Thermogravimetry
The nitridation rates of the six compositions recorded by TGA are compiled in Figure 1 and, for a relative comparison, the curves are normalized to per cent of theoretical nitridation yield.

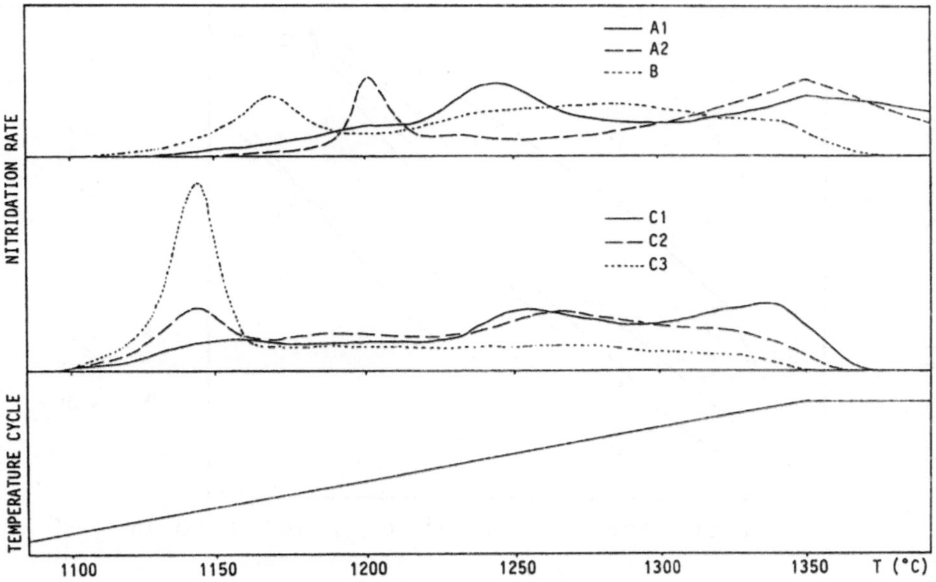

Figure 1. Nitridation rate for six bodies measured by TGA.

Obviously, there are large differences in nitridation rates between the six samples. The reference sample A1 starts to nitride rather slowly compared to A2. During temperature increase within the interval of 1100°-1200°C, Al_2O_3 and SiO_2 are expected to compete for the reaction with Y_2O_3. If the reaction with SiO_2 leads to Y_2SiO_5 and $Y_2Si_2O_7$ formation, nitridation of Si will be favoured [16]. On the other hand, if a major reaction with Al_2O_3 occurs, resulting in YAG (Yttria Alumina Garnet) or mullite formation, the nitridation reaction will be retarded. Sample A2 where no Al_2O_3 was present shows a distinct DTG peak at 1200°C. When comparing the curves for A1 and A2 the conclusion is that in the absence of Al_2O_3, Y_2O_3 can react with SiO_2 on the Si particle surface thus favouring nitridation, which explains the occurrence of the peak at 1200°C. For sample B ($ZrSiO_4$:Si) the nitridation starts at 1120°C, with a maximum rate already at 1170°C. Probably, $ZrSiO_4$ partially disproportionates into SiO_2 and ZrO_2 and the presence of ZrO_2 has an accelerating effect on the nitridation for reasons yet to be explained. In the cases of samples C1-C3 the nitridation is initiated at 1100°C with peak rates as early as at 1140°C (C2 and C3). Obviously the nitridation rates increase with increased additions of ZrO_2 in the compositions.

For the Al_2O_3 containing compositions, a liquid appears at about 1310°C in the system Al_2O_3-SiO_2-Y_2O_3-Si_3N_4 and the rate is increased [17].

The six nitridation rate curves are summarized in a nitridation graph, in Figure 2, showing the theoretical conversion degree (%) of Si.

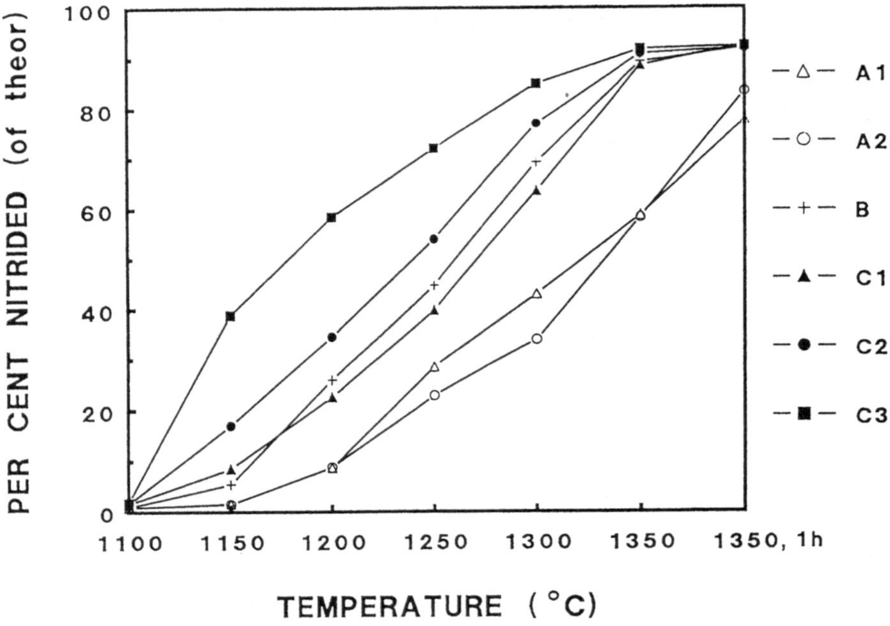

Figure 2. Nitridation graph of the six compositions.

The graph demonstrates the differences in nitridation of the compositions especially in the temperature interval 1100°-1350°C. Samples A1 and A2 are after 1 h at 1350°C still not nitrided completely. The graph also clearly shows the effect of increasing amounts of ZrO_2 additions.

Phase Composition

No residual Si has been detected in any of the samples. The XRD pattern of sample B (nitrided in graphite furnace) showed that approximately 60% of the $ZrSiO_4$ powder had transformed to Si_2N_2O and ZrO_2. The low conversion yield is due to the low processing temperature and a procedure for establishing a stable transformation degree of $ZrSiO_4$/Si to Si_2N_2O/ZrO_2 has to be investigated.

Unfortunately, the XRD pattern of C3 indicates presence of ZrN, approximately 5%. The Zr nitride is highly oxidation-sensitive, already at temperatures in excess of 500°C. It is thus important to stabilize ZrO_2 or modify the nitridation procedure to eliminate the ZrN formation. No formation of Si_2N_2O was observed in sample C3.

CONCLUSIONS

1. Despite relatively low specific surface areas of the milled powder mixtures the nitridation yields of Si were adequate.
2. The nitridation rate of Si is critically dependent on the nitridation aids added.
3. Al_2O_3 had no favourable effect on the nitridation rate and, accordingly, it can be excluded as its presence presumably would degrade the high temperature properties of the matrix.
4. The $ZrSiO_4$/Si composition exhibited a rapid nitridation rate at a very low temperature, although XRD examination revealed that only 60% of the raw material had transformed to Si_2N_2O and ZrO_2.
5. Increasing amounts of ZrO_2 enhanced the nitridation rate considerably.
6. In ZrO_2 containing compositions small amounts of ZrN were detected. Since this compound is oxidation-sensitive, a stabilizer of ZrO_2 has to be added to eliminate its formation.

The $ZrSiO_4$/Si composition is relatively newly developed and it seems to be quite promising as an alternative matrix material. After further investigations, this material may turn out to be a superior candidate for a high temperature matrix.

ACKNOWLEDGEMENT

The authors wish to thank Margareta Jansson for proof-reading and preparing the manuscript. Financial support received from the Swedish Board for Technical Development (STU) is gratefully acknowledged.

REFERENCES

1. S M Lee, "International Encyclopedia of Composites", 267-277, 297-332 (1990).

2. T-I Mah, M G Mendiratta, A P Katz and K S Mazdiyasni, "Recent Developments in Fiber-Reinforced High Temperature Ceramic Composites", Am. Ceram. Soc. Bull. **66** (2), 304- 308 (1987).

3. M Prewo, "Fiber-Reinforced Ceramics: New Opportunities for Composite Materials", Am. Ceram. Soc. Bull. **68** (2), 395-400 (1989).

4. J R Strife, J J Brennan and K M Prewo, "Status of Continuous Fiber-Reinforced Ceramic Matrix Composite Processing Technology", Ceram. Eng. Sci. Proc. 11 (7-8), 871-919 (1990).

5. J A DiCarlo, "CMCs for the Long Run", Adv. Mater. Process. inc. Met. Prog. 135 (6), 41-44 (1989).

6. P J Lamicq, G A Bernhart, M M Dauchier and J G Mace, "SiC/SiC Composite Ceramics", Am. Ceram. Soc. Bull. **65** (2), 336-338 (1986).

7. D K Shetty, M R Pascucci, B C Mutsuddy and R R Wills, "SiC Monofilament-Reinforced Si_3N_4 Matrix Composites", Ceram. Eng. Sci. Proc. 632-645 (1985).

8. A S Fareed, P Fang, M J Koczak and F M Ko, " Thermomechanical Properties of SiC Yarn", Am. Ceram. Soc. Bull. **66** (2), 353-358 (1987).

9. F K Ko, "Preform Fiber Architecture for Ceramic-Matrix Composites", Am. Ceram. Soc. Bull. **68** (2), 401-414 (1989).

10. D B Fishbach and P M Lemoine, " Influence of a CVD Carbon Coating on the Mechanical Property Stability of Nicalon SiC Fiber", Comp. Sci. Techn. **37**, 55-61 (1990).

11. R Lundberg, PhD Thesis, ISBN 91-7032-441-7, Chalmers University of Technology, Göteborg (1989).

12. P Barron-Antolin, G H Schiroky and C A Andersson, "Properties of Fiber-Reinforced Alumina Matrix Composites", Ceram. Eng. Sci. Proc. 9 (7-8), 759-766 (1988).

13. R Pompe, L Hermansson and R Carlsson, "Fabrication of Low Shrinkage Silicon Nitride Material by Pressureless Sintering", Brit. Ceram. Soc. **11** p. 65 (1981).

14. K Rundgren, J Brandt, R Pompe and R Carlsson, "Low Cost Nitrided Pressureless Sintered Si_2N_2O-ZrO_2 Composite", In manuscript, To be presented at European Ceramic Society Second Conference, Augsburg, Sept 11-14 (1991).

15. P L Antona, A Giachello and P C Martinengo, Ceramic Powders 753 (1983).

16. L K L Falk, PhD Thesis, ISBN 91-7032-262-7, Chalmers University of Technology, Göteborg (1986).

17. C O'Meara, PhD Thesis, ISBN 91-7032-384-4, Chalmers University of Technology, Göteborg (1988).

INFLUENCE OF PROCESSING ON THERMAL AND ELASTIC PROPERTIES IN ALUMINA SILICON CARBIDE WHISKER COMPOSITES (+)

E.Scafè(*), L.Di Rese, L.Fabbri(*)
Eniricerche SpA, 00015 Monterotondo (Roma),Italy

G.Dinelli, G.Giusto, M.Tiengo
Temav - CE.RI.VE.-, 30175 P.Marghera (VE), Italy

ABSTRACT

Measurements of thermal conductivity k and Young's modulus E have been performed on a set of different alumina-SiC(w) composites, hot pressed at 1900 °C, obtained by varying hot-pressing time and raw materials. Moreover, the effect of processing on the k and E has been discussed showing the influence of raw material characteristics on thermal conductivity.

INTRODUCTION

Ceramic-ceramic composites are possible candidates to improve reliability of ceramic materials. A particular attention has been devoted to silicon carbide whisker-reinforced alumina composites proposed for high demanding applications such as cutting tools in which the important role of thermal and elastic properties is well-known. In these applications the materials are exposed to high thermal stresses [1]. In fact, the temperature increase at the cutting edge produced by metal friction could be as high as 1000 °C [2]. For this reason the resistance to thermal shock plays an important role in the cutting tool life and it should be taken into account during the material design.

Thermal shock behavior is described by the R-parameters which consist, theoretically, of a mathematical combination of thermal-elastic and mechanical properties and are often used to rank different ceramic materials.

(+) Work supported by ENI - (*) Member of American Ceramic Society

In particular the parameter R' depends on the k/E ratio between thermal conductivity k and Young's modulus E by means of the following equation [3]:

$$R' = [\sigma(1-\mu)/\alpha] \quad k/E \qquad (1)$$

where σ is the modulus of rupture, μ the Poisson ratio and α the linear thermal expansion coefficient.

In this paper the role of several raw materials, differing in the impurity content and crystalline defects, has been analyzed in order to improve the k/E ratio of alumina-SiC whiskers composites. The modulus of rupture and the Young's modulus are chiefly influenced by processing defects i.e. pores, cracks etc., therefore they are rather insensitive to point defects and impurity content. Moreover, point defects and impurities have small effect on the thermal expansion coefficient [4]. On the contrary the thermal conductivity of covalent solids is heavily affected by both defects and porosity. Thus the influence on thermal conductivity of hot-pressing time and raw materials has been analyzed. Furthermore, in order to select the effect of porosity on thermal conductivity, the behavior of Young's modulus has been studied.

EXPERIMENTAL METHODS

Three high purity sub-micron ceramic-grade alumina powders (A1,A2,A3) and two ß-silicon carbide whisker sources (W1,W2) were used in order to obtain different composites. Chemical analyses were performed by atomic emission and absorption spectroscopies (Tab.1). The specimens were prepared by mixing various amount of silicon carbide whiskers with the complementary part of alumina powders. The alumina-SiC whiskers mixtures were hot-pressed in argon at 1900 °C and 50 MPa for different times (20÷60 minutes). Moreover a set of alumina samples were hot-pressed, 1400 ÷ 1500 °C at 50 MPa for 1 hour, in order to analyze the matrix properties.

Densities equal or higher than 99 % t.d. were obtained, where the theoretical density of the composites was calculated by using alumina and silicon carbide volume fractions and theoretical densities of each phase (3.98 and 3.21 g/cm^3 respectively).

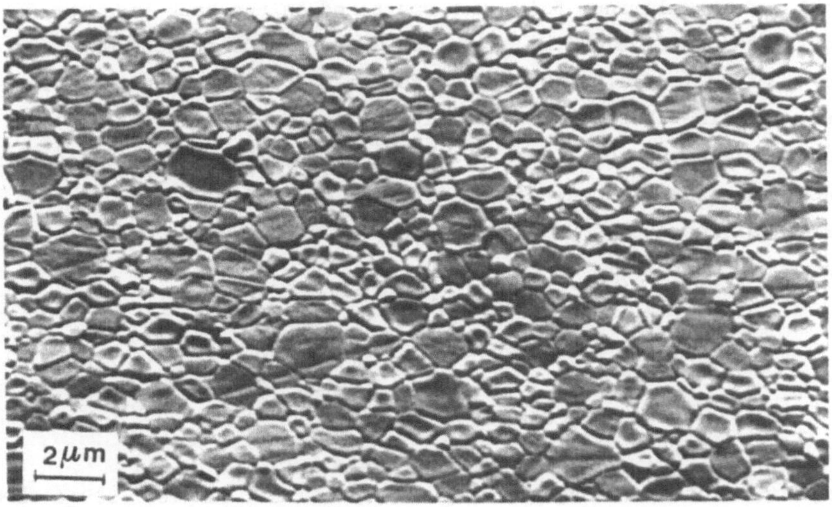

Figure 1. SEM micrograph of alumina A3 reference specimen

As regards the microstructural features, the grain size of thermal-etched alumina specimens was about 1-2 μm, as determined by SEM analysis (fig.1). The composites consisted of an alumina matrix (grain size < 1 μm) with a homogeneous dispersion of .5-1.5 μm thick and 5-30 μm long whiskers.
The addition of silicon carbide whiskers improves the strength and the fracture toughness of the material, as reported in literature [5]. Four-point bending strengths exceeding 700 MPà were obtained on samples (3x4x45 mm) over 20-40 mm span length, after diamond-machining and surface polishing down to 10 μm grit.
Young's modulus measurements were carried out by means of an improved flexural resonance apparatus with an amplitude detection method based on laser modulation technique [6]. Samples, with dimensions 3x10x30 mm, were supported on two sharp knives, that were positioned on the nodal lines of the fundamental resonance mode. All the measurements were performed under vacuum to avoid the damping and the frequency shift due to air. Room temperature thermal diffusivity were measured by the laser-flash technique [7,8]. A 50 joules Nd-glass laser (1.06 μm) with a pulse width of about 500 μs was used as the flash source; the heating beam was attenuated by a suitable copper sulphate solution in order to obtain a temperature rise of

less than 1.5 K. An aluminum film was sputtered on the back side surfaces of the square shaped (10x10x3 mm) samples, while the front side was coated with a layer of colloidal graphite in order to increase the absorptivity to the laser pulse. The thermal conductivity values were calculated by using the "rule of mixtures" for literature data of specific heat [9] and the experimental values of thermal diffusivity and density.

The experimental results of Young's modulus, E, and thermal conductivity k have been reported in table 2.

DISCUSSION AND RESULTS

The average value of the Young's moduli of all alumina samples (399±4 GPa) are in good agreement with those reported in literature for high density hot-pressed alumina, as well as with the theoretical estimate (403 GPa) obtained for a polycrystalline alumina material [10]. Also the thermal conductivity values of all alumina samples are close each other (29.0±1.5 W/mK) and are in good agreement with those reported in literature for highly pure alumina materials with 1-2 μm grain size [12]. The effect of the different impurity content between A1 alumina and the purer A2 and A3 powders, as well as the two hot-pressing temperatures, was negligible. In fact major impurities in the less pure powder (A1) were typically Na and Si which slightly affect thermal conductivity, as reported in literature [12,18] and confirmed by our results. The correspondence between typical and actual batch impurity content was controlled on A3 powder (Sumitomo); the measured values (Tab.1) were comparable to the typical ones and still much lower than levels which affect thermal conductivity [18]. Consequently, the differences of the alumina powders do not change, within the experimental errors, the properties of the hot pressed material. Thus, concerning the analyzed thermal and elastic properties, the alumina powders are practically equivalent.

As regards the composites, the expected slight increase of the Young's modulus is observed in the reinforced material (Tab.2), that is due to the relatively small difference between elastic properties of alumina and polycrystalline ß-SiC (420 GPa) [11]. These results are consistent with the relatively high densities of our materials. Moreover, different hot

TABLE 1.
Raw Materials: typical characteristics and impurity content

Name	A1	A2	A3	W1	W2
Manufacturer	Alcoa	Baikowski	Sumitomo	Tokai Carbon	Tatheo Chem.
Type	A-16SG	CR-30	AKP-20	Grade-1	SCW 1-S
Purity (%)	99.5	99.99	99.99	see below	
Alpha Phase (%)	-	60-65	-	ß-SiC	ß-SiC
BET surf. area (m^2/g)	10+1	30+1	4.7	3.5	4.8
Mean part. diam.(μm)	0.4	0.05	0.5	0.3-1.5	0.05-1.5
Length (μm)	-	-	-	20-50	5-200

Impurity content in W1 and W2 SiC whiskers (wt %) and in A3 alumina (ppmwt)
(A3t typical values, A3c measured values)

	W1	W2		W1	W2		A3t	A3c
Ca	<0.005	0.100	Ni	0.002	0.0045	Si	15	40
Fe	0.01	0.056	Cu	0.001	0.0025	Na	3	10
Mg	<0.002	0.021	Cr	0.002	0.004	Mg	3	10
Al	0.0125	0.012	Co	0.024	<0.002	Cu	<1	10
Ti	0.014	0.0033	Zn	0.009	<0.002	Fe	9	20

TABLE 2.
Thermal conductivity and Young's modulus

Name	Sample	T (°C)	t(min)	V_w(%)	td(%)	k(W/mK)	E(GPa)
A1	1	1400	60	-	98.9	29.6	396
A1	2	1500	60	-	99.5	28.2	395
A2	1	1400	60	-	99.5	28.5	404
A2	2	1500	60	-	99.7	28.8	404
A3	1	1400	60	-	99.9	29.9	398
A1-W1	1	1900	60	38		40.7	408
A1-W1	2	"	60	38		43.4	/
A1-W1	3	"	30	38		41.0	408
A1-W1	4	"	30	38		41.2	/
A1-W1	5	"	30	38		41.3	/
A1-W1	6	"	20	38		41.2	408
A1-W1	7	"	20	38		40.9	/
A1-W1	8	"	60	28		37.6	399
A1-W1	9	"	60	35		41.6	404
A1-W1	10	"	60	35		39.6	399
A2-W1	1	"	60	35		42.7	410
A2-W1	2	"	60	35		40.9	412
A3-W2	1	"	60	34		31.9	406
A3-W2	2	"	60	36		33.8	407

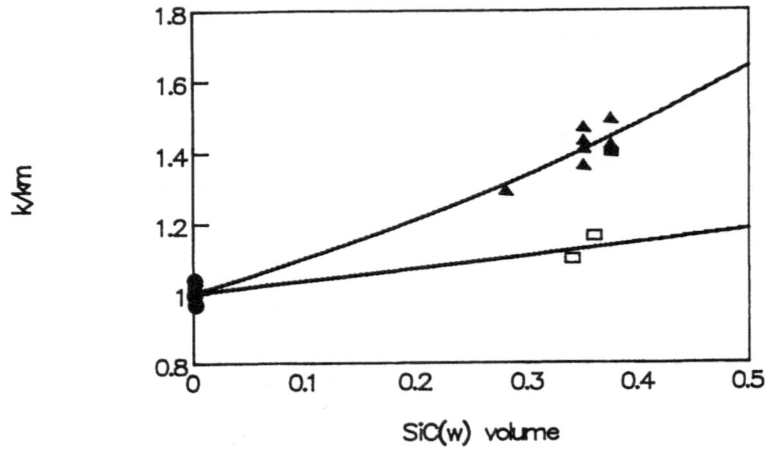

Figure 2. Normalized thermal conductivity for alumina SiC(w) composites vs whiskers volume fraction. Circles: reference aluminas; triangles: W1 reinforced composites; squares W2 reinforced alumina. Solid line: theoretical dependence obtained by Effective Medium Theory for different k(SiC) values (80 - 40 W/mK).

pressing time, as well as different raw materials, do not affect Young's modulus.

An increase of thermal conductivity (about 40 W/mK) was measured with the addition of the W1 reinforcing phase. As expected, the substitution of the matrix with a purer alumina (A2) did not change the properties of the composite as well as different processing time (20, 30, 60 minutes). On the contrary, the composite obtained with the matrix A3, which was as pure as A2, and whiskers W2 did not exhibit a comparable increase of thermal conductivity. Owing to identical processing conditions, SiC(w) volume fraction and very similar matrix properties, such behavior should be attributed to a difference in the effective thermal conductivities of whiskers W1 and W2. In particular the k values of W1 and W2 were estimated [13] to be about 80 W/mK and 40 W/mK by using the Effective Medium Theory [16] (see fig.2). In particular a porosity contribution can be excluded owing to the close values of the elastic moduli between composites with W1 and W2 whiskers (tab.2) and interfacial barrier effects should be comparable in both composites.

This effect can be originated by a difference in the impurity content and/or crystalline defects of whiskers W1 and W2 according to literature data [14,15]. In fact the chemical analyses showed that the impurity content of both whisker sources were lower than data reported in a previous work [14] and W1 resulted cleaner than rice-hull whiskers (W2). Moreover whiskers W2 contained higher density of crystalline defects and internal porosity if compared to whiskers W1 [14]. A similar effect has been observed by D.P.H.Hasselman et al. [17], on different SiC whisker sources (SC-9 and VLS). Nevertheless a comparison among the impurity levels and crystalline defects of all these whiskers have been extensively studied showing the similarity between VLS and W1 and SC-9 and W2 respectively [15].

CONCLUSIONS

In order to optimize the thermal shock behavior of composite materials, investigation of other properties than only mechanical strength should be performed. In particular the reduction of those effects that limit the mechanical strength has to be linked together with improvements of k/E ratio.

In this paper has been shown that the suitable choice of raw materials lead to an improvement in the k/E ratio for alumina-SiC(w) composites. Therefore a better thermal shock behaviour have to be expected.

REFERENCES

[1] T.N.TIEGS, P.B.BECHER, "Thermal Shock Behavior of an Alumina-SiC Whisker Composite" J. Am. Ceram. Soc., 70 (5) C109-C11 (1987)

[2] S.F.WAYNE, S.T.BULJAN,"The Role of Thermal Shock on Tool Life of Selected Ceramic Cutting Tool Materials" J. Am. Ceram. Soc., 72 (5) 754-760 (1989)

[3] D.P.HASSELMAN,"Thermal stress resistance parameters for brittle refractory ceramics: a compendium", Bull. Am. Ceram. Soc. 49, p.1033, (1970)

[4] G.GRIMVALL "Thermophysical Properties of Materials" , Selected Topics In Solid State Physics, vol.XVIII p.130, North-Holland (1986)

[5] D.B.MARSHALL, J.E.RITTER,"Reliability of advanced structural ceramics and ceramic matrix composites", Amer. Ceram. Soc. Bull. 65, 309, (1987).

[6] E.SCAFE', L.DI RESE, G.GRILLO and F.PETRUCCI,"Young's modulus of Si-SiC two-phase particulate composites", 7th CIMTEC, Montecatini, 1990 in Press.

[7] W.J.PARKER,R.J.JENKINS,C.P.BUTLER,G.L.ABBOTT "Flash Method of Determining Thermal Diffusivity, Heat Capacity and Thermal Conductivity" J.Appl. Phys. 32, 1679 (1961)

[8] R.E.TAYLOR,K.D.MAGLIC "Pulse Method for Thermal Diffusivity Measurements" Compendium of Thermophysical Property Measurements Methods, Vol.1 305-336 Plenum Press NY.

[9] LANDOLT BORNSTEIN , vol 4, Springer-Verlag 1961

[10] J.P. WATT, L.PESELNICK," Clarification of the Hashin-Shtrikman bounds on the effective elastic moduli of polycrystals with hexagonal, trigonal, and tetragonal symmetries", J. Appl. Phys. 51, 1525-1531 (1980).

[11] A.KELLY,N.H.MACMILLAN,"Strong Solids",Clarendon Press, Oxford (1986).

[12] R.K.WILLIAMS,R.S.GRAVES,M.A.JANNEY,T.N.TIEGS,D.W.YARBOUGH "The effect of Cr2O3 and Fe2O3 addition on the thermal conductivity of Al2O3" J.Appl.Phys. **61**(10) 4894-4901 (1987)

[13] E.SCAFE'.L.DI RESE,L.FABBRI,G.DINELLI,M.TIENGO "Thermal and Elastic Properties on Alumina Silicon Carbide Whiskers", to be published

[14] K.R.KARASEK,S.A.BRADLEY,J.T.DONNER,M.R.MARTIN, K.L.HAYNES, H.C.YEH "Composition and microstructure of silicon carbide whiskers", J.Mater.Sci. **24** 1617-1622 (1989)

[15] J.W.MILEWSKI, F.D. GAC, J.J.PETROVIC, S.R.SKAGGS "Growth of beta-Silicon Carbide Whiskers by the VLS process" J. Mater.Sci., **20**,1160-1166, (1985)

[16] G.ONDRACEK "The Qualitative Microstructure-Field Property Correlation of Multiphase and Porous Materials" Rev. on Powd. Metal. and Phys. Ceram., Vol.3, N°s.3&4, 205-322 (1987)

[17] L.M.RUSSEL, L.F.JOHNSON, D.P.H.HASSELMAN, R.RUH "Thermal Diffusivity/Conductivity of Silicon Carbide Reinforced Mullite" J. Am. Ceram. Soc., **70**[10], C226-C229 (1987)

[18] G. A. SLACK,"Thermal Conductivity of MgO, Al2O3, MgAl2O4, and Fe3O40 Crystals from 3 to 300 K", Phys.Rev. vol 126 N°2, 427-441, (1962)

INTERFACE DESIGN AND A NEW PROCESS TO PRODUCE SiC(w)/Si_3N_4 CERAMIC COMPOSITE

Jianbao LI, Huirong LE,
Yong HUANG, Longlie ZHENG, Jianguang WU
Department of Materials Science and Engineering
Tsinghua University, Beijing, 100084, P. R. China.

ABSTRACT

There are two kinds of interfaces in the Si_3N_4 ceramic matrix composite reinforced by SiC whiskers, one is the grain boundary between the Si_3N_4 matrix grains, another one is the interface between Si_3N_4 grains and SiC whiskers. In this paper, SiC(w)/Si_3N_4 composite was prepared by adding two different groups of additives to optimize these two different kinds of interfaces. Extreme high bending strength and fracture toughness, about 1000-1100 MPa and 10.5-11.5 MPa $m^{1/2}$, respectively, were obtained. The effects of the additives and the process were discussed in detail.

INTRODUCTION

It is known that ceramics, such as SiC and Si_3N_4, have high strength at elevated temperatures, high hardness, good corrosion and erosion behavior, and low mass density, and are considered as advanced structural materials for several applications, examples as cutting tools, radomes, armors, and engine components. But their fracture toughness is too low to be used as structural parts bearing high stress. Fracture toughness improvement to ceramics is becoming an important theme.

SiC whiskers are expected to improve ceramics recently. But most of previous studies reported improvement in toughness accompanying a strength fall in the composite. This study shows that improvement in both bending strength and toughness of a silicon nitride material with SiC whiskers could be achieved by interfacial bonding design.

MATERIALS AND TEST METHOD

The SiC whiskers, SiC(w), used in this study are supplied by Tokai Carbon Co. LTD[1]., Japan. The size of the whiskers ranged typically from 1.0 to 3.0 μm in diameter and 10-80 μm in length. Si_3N_4 powder is prepared by nitridation method in our laboratory and contains about 90 wt% of α -phase and about 9 wt% β -phase Si_3N_4 and about 0.7 wt% of free Si.
The SiC whiskers were dispersed in ethanol with a small amount of surfactant using a high speed homogenizer. The dispersed whiskers were poured into the matrix slurry. The mixture of whiskers with Si_3N_4 powders was mixed for 20 hours again using the ball milling in ethanol. After evaporation of the solvent and sieving, disk samples of 50 mm diameter were hot-pressed at a temperature of 1800℃ for 90 minutes in a nitrogen atmosphere under a pressure of 30 MPa for all composites
The hot-pressed disks were cut and polished into bars 3 by 4 by 35 mm for bending strength testing and 3 by 6 by 35 mm for fracture toughness testing. Both of the strength and the toughness were measured by three-point bending testing methods.

ANALYSIS OF REINFORCEMENT MECHANISM AND INTERFACIAL BONDING

Reinforcing mechanism[2][3] of SiC-whisker-reinforced Si_3N_4 ceramics have crack deflection, crack-bridging and whisker pullout. They have been usually observed simultaneously in same material[3]. And the pullout of whiskers did not contribute markedly to toughness. Previous studies implied in gived whisker and matrix that the most important prerequisite is interfacial bonding which determines the effectiveness of reinforcement. If the interfacial bonding is suitable, the

mechanisms mentioned above may be observed simultaneously in same material.
In practice, it is easy to recognize that there are two kinds of interfaces in the SiC(w)/Si_3N_4 composite, one is the grain boundary of Si_3N_4 matrix grains (homophase boundary) and another is the interface between SiC whiskers with Si_3N_4 grains (non-homophase). The bonding strength between Si_3N_4 grains, τ_g, would affect the bending strength of composite materials, but the bonding strength of interfaces between SiC whiskers and Si_3N_4, τ_i, would affect the fracture toughness. The relation of the reinforcement mechanism with interfacial bonding and whisker direction are summarized in following:.

(1) When τ_i is far stronger than τ_g, the debonding along the whisker/matrix interface may not occur. The strength of the composite may be increased significantly but the toughness will not be enhanced markedly.
(2) When τ_i is equal or slightly smaller than τ_g, reinforcing mechanism depend on the angle between crack plane and the direction of whiskers:
    a) When whiskers are parallel to the crack plane, the crack tip propagates (grows along longitude of whiskers) without fracturing the whiskers as shown in Figure 1a. The matrix may be toughened a little.
    b) When whiskers slant to crack plane, the crack tip deflects around the whiskers and the fracture path becomes more tortuous as shown in Figure 1b. The matrix may be toughened largely.
    c) When whiskers are perpendicular to the crack plane, the bridging effect will occur as shown in Figure 1c. A large improvement in toughness may be obtained.
    d) If the displacement shown in Figure 1b and 1c is too large, the whiskers will be broken and debonded parts will appear like pullout of whiskers as shown in Figure 1d.
    If there are stacking faults or dislocations in the debonded whiskers, the whiskers will be previously broken along these faults and then pulled out partially. Actually, there are various faults in the whiskers, which are formed in the growth process or by pretreatment of whisker.
(3) If τ_i is too small, for example, the interfaces will debond along whiskers due to the thermal expansion mismatch which forms cracks. Under applied stress, the whiskers would pull out easily without consuming energy as shown in Figure 1e. In this case, both the strength and toughness could not be improved.

Review above statement, suitable interfacial bondings (meaning $\tau_i \leqslant \tau_g$) is prerequisite whether whisker can reinforce the ceramics or not. And the direction of whisker and its angle with crack plane will affect the mechanism and the magnitudes of reinforcement of whiskers when $\tau_i \leqslant \tau_g$.

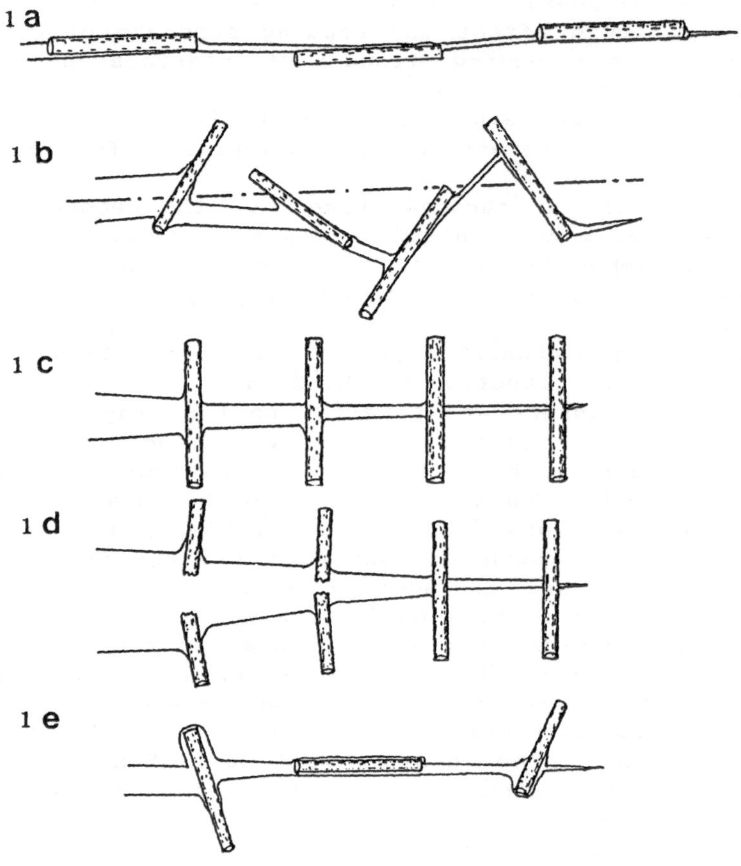

FIGURE 1. Relation of reinforcement mechanism with interfacial bonding and whisker direction in the composite

CONTROL OF THE INTERFACIAL BONDING

1. Effect of Sintering Aids on Interfacial Bonding and the Properties

It is known that the mechanical properties of ceramics mainly depend on bonding strength of grain boundary, which consists of sintering aids. For the whisker reinforced ceramics, sintering aids affects the mechanical properties of material obviously, too. Table 1 shows the effect of sintering aids on the properties of $SiC(w)/Si_3N_4$ composite[4].

TABLE 1
Effect of sintering aids on the properties of the composites

Sample Number	Si_3N_4 wt%	SiC(w) wt%	Additives (wt%) $La_2O_3+Y_2O_3$	Al_2O_3	MgO	K_{IC}※ MPa $m^{1/2}$	σ_{RT} MPa
LYA_0	90	10	20	0	0	8.7	880
LYA_1	90	10	20	1	0	8.8	870
LYA_2	90	10	20	2	0	8.9	1030
LYA_4	90	10	20	4	0	7.9	1040
LYM	90	10	20	0	2	9.4	920

#: Si_3N_4+SiC(w)=100 wt%; ※ average value of five speciments

Al_2O_3 from 0wt%, 1wt%, 2wt% to 4wt% (sample number LYA_0, LYA_1, LYA_2, LYA_4, respectively) the bending strength are elevated obviously, from 880 MPa to 1040 MPa. Maximum value obtained in LYA_4 samples is 1120 MPa. But the toughness is not improved. The toughness decreases when Al_2O_3 addition is larger than 2 wt%. Addition of 2 wt% MgO increases the toughness up to 9.4 MPa $m^{1/2}$ but the strength falls down to 920 MPa.

SEM observation showed that Al_2O_3 reacted with SiC and Si_3N_4, resulting in a increase of bonding strength (both τ_i and τ_g), but reaction of MgO with SiC and Si_3N_4 did not been observed. Figure 2 shows that the composites with Al_2O_3 doping (Figure 2b, c, d) have rougher surfaces of grains than with MgO doping (Figure 2a). These rougher surfaces are formed by the reaction of sintering aids in the sintering process. Table 2 shows the difference of chemical composition on the surfaces of SiC whiskers and Si_3N_4 grains analyzed by EDAX. This result indicated that La_2O_3 and Y_2O_3 appear to concentrate around Si_3N_4 grains.

TABLE 2
EDAX element analysis showing concentration of sintering aids

Oxide	on surface of whiskers	Si_3N_4	average value in the composite
Al_2O_3	6.13	5.46	1.0
La_2O_3	12.14	5.31	7.5
Y_2O_3	16.21	6.56	7.5

in weight percent

2. Interfacial Bonding Design

In order to realize suitable interfacial bonding, i. e. $\tau_i < \tau_g$ as stated above, SiC whiskers must be protected against the reaction and corrosion of Al_2O_3. One approach is to choose two groups of additives to form the two kinds of interfaces respectively. This idea is schematically shown in Figure 3. MgO + Y_2O_3 group (named B) was chosen to form the layers of interfaces of $SiC(w)/Si_3N_4$ and La_2O_3 + Y_2O_3 + Al_2O_3 group (named A) was chosen to form the layers of grain boundaries of Si_3N_4/Si_3N_4 matrix grains.

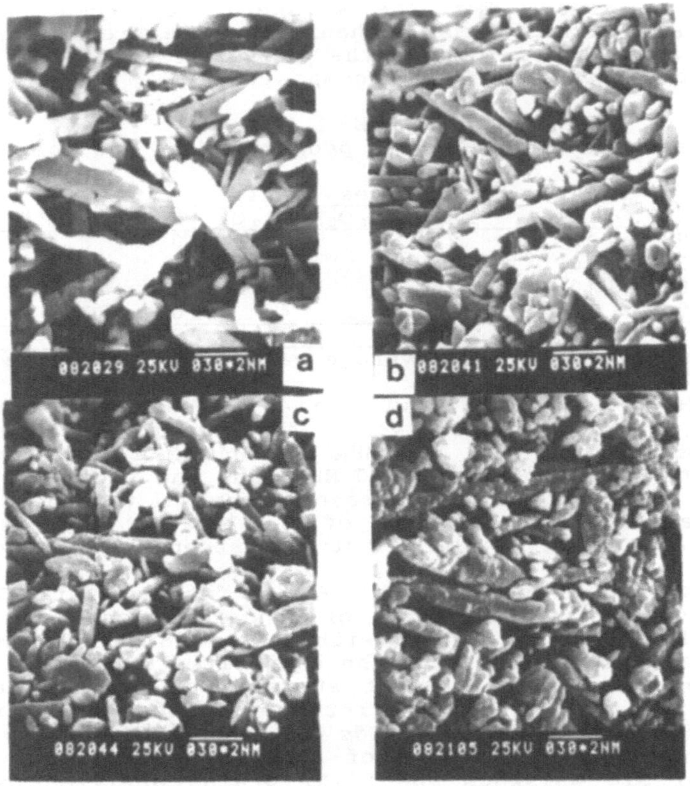

FIGURE 2 Showing rougher grain surfaces with Al_2O_3 doping than MgO doping.

The key point of this process is the coating process in which the B group of additives, such as MgO + Y_2O_3 or SiO_2 + MgO are coated on the whiskers. Coating methods contain: 1) immersing whiskers into solution of respective metal salts above and drying the sticked layers or heat-treating them to thermal descomposition to form sufficiently sturdy layers of these metal oxides on surfaces of the whiskers; 2) oxidating the SiC whiskers in elevated temperature of 1000--1200℃ and controlled atmosphere to form a SiO_2 layer on the surface of the whiskers.

In order to make A group of additives concentrate on the grain boundaries of Si_3N_4, A group of additives must be sufficiently mixed with matrix powder Si_3N_4 and then heat-treated at 800--1000℃ to increase the adhesion of the additives on the Si_3N_4 grains.

Crashing Si_3N_4 powders mixed with A sintering aids and mixing them with the coated whiskers in ethanol to get a slurry. The dried slurry (powder) were hot-pressed into disks for testing the properties.

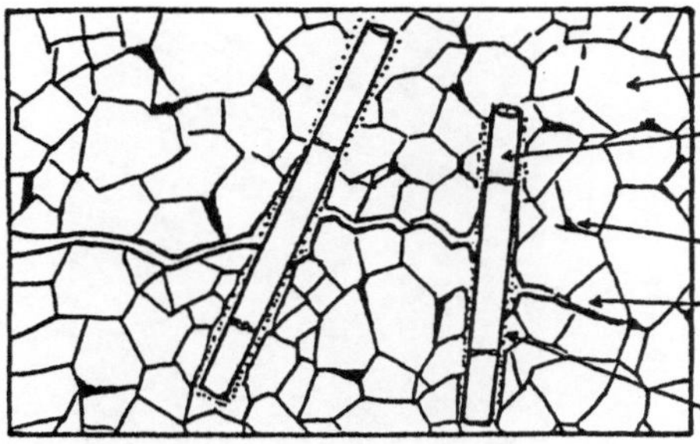

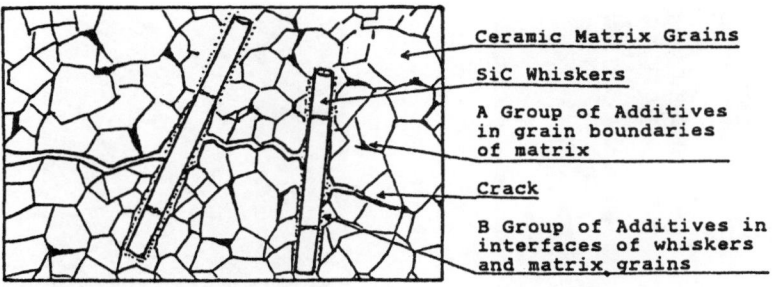

FIGURE 3 Schematic depiction of the microstructure of whisker reinforced ceramic composite. SiC whiskers are completely surrounded by B group of additives and debonded partially when a crack tip reached by it.

Figure 4 shows the SEM picture of coated whiskers. Table 3 shows the properties of SiC(w)/Si_3N_4 composites prepared by above process. The composite with A+B additives by this new process exhibites an extremly high bending strength and fracture toughness, about 1000-1100 MPa and 10.5-11.6 MPa $m^{1/2}$, respectively, increases about 20 % relative to the composite listed in Table 1. This process can realize not only large toughening but also large strengthening effects

SEM observation found that main reinforcing mechanisms are whisker pullout and crack deflection. Figure 5 shows the broken whiskers pulling out on the fracture surface.

TABLE 3
Effect of Composition and Process on the Properties of Composites.

Sample Number	Si_3N_4 wt%	$SiC_{(w)}$ wt%	Additives 20 wt%	K_{IC} MPa $m^{1/2}$	σ_{RT} MPa	$\sigma_{1300℃}$ MPa
	80	20	A	9.4	1100	570
	80	20	B	10.5	930	390
	80	20	A+B	10.6	1050	450
#	60	20	A+B	11.6	720	

#: added 20 wt% TiC.

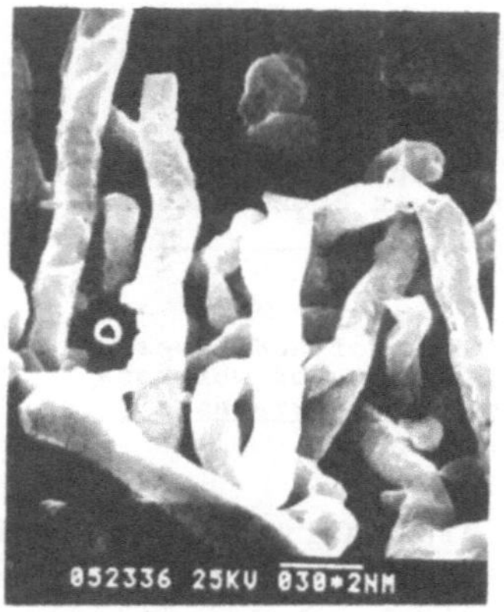

FIGURE 4 Scanning electron microscope photograph of the coated SiC whiskers

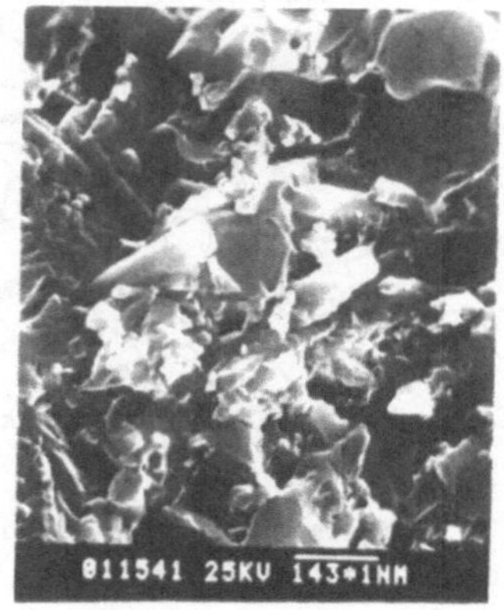

FIGURE 5 SiC whiskers broken and pulled out on the fracture surface on the composite.

CONCLUSION

Both strength and toughness can be improved through the suitable design of the interfacial bonding in the composites. Two groups of additives were added in the composite to control the bonding of the two kinds of interfaces (τ_i for SiC(w)/Si_3N_4 and τ_g for Si_3N_4/Si_3N_4), respectively. MgO and SiO_2 and Y_2O_3 were selected for SiC(w)/Si_3N_4 interface and La_2O_3 + Y_2O_3 + Al_2O_3 for the bonding of grain boundaries of Si_3N_4. The composite thus prepared performes high bending strength of 1050 MPa and high fracture toughness of 10.6 MPa $m^{1/2}$.

REFERENCES

1. Catalogue of Tokawhisker goods, from the Tokai carbon Co., LTD, Tokyo, Japan, (1989).
2. K. Ueno and Y. Toibana, Mechanical properties of silicon nitride ceramic matrix composite reinforced with silicon carbide whisker, Yogyo Kyokaishi, 91(11)491-97(1983).
3. P. D. Shalek, J. P. Petrovic, G. F. Hurley, and F. D. Gac, Hot-pressed SiC whisker-Si_3N_4 matrix composites, Am. Ceram. Soc. Bull., 65(2)351-56(1986).
4. J. B. Li and Y. Huang, Effect of additives on the properties of SiC(w)/Si_3N_4 composites, Proceeding of the Third Symposium of Materials Science and Engineering for Chinese Young, Dalian city, China, (1991).

TOUGHENING CERAMICS BY LAMINATING COMPOSITE ELEMENTS

W J CLEGG AND L R SEDDON
ICI Solid State Science Group
PO Box 11, The Heath, Cheshire, WA7 4QE, UK

ABSTRACT

Crack growth has been studied in a system of silicon carbide sheets separated by graphite interfaces, made by coating and compacting together thin sheets made from silicon carbide powder. The forming of simple shapes using this method is demonstrated.

INTRODUCTION

It is well established that ceramics can be toughened by introducing weak interfaces at an angle to a growing crack. At present this is done by incorporating dense inclusions into a ceramic matrix. A variety of processes exist, the most successful being the chemical vapour infiltration of a fibre preform [1]. Despite its very good properties, the material is prohibitively expensive [2], preventing its widespread use.

Recently a much simpler method for introducing weak interfaces has been reported [3], where a substantial increase in the resistance to crack growth can be achieved simply by compacting together composite elements such as thin sheets which have been made from powders. In this paper we describe how crack growth occurs in such materials, and show the potential for this process, in particular in the fabrication of simple shapes.

EXPERIMENTAL

A β-silicon carbide powder doped with 0.4% of boron by weight (Superior Graphite, HSC-059s) was blended with 5% of an α-silicon carbide powder (Showa Denko A-1). The powder was then mixed with 40% by volume of an aqueous polymer solution (Nippon Gohsei, KH17S) in a Z-blade mixer. The α-silicon carbide powder was added to prevent the growth of long laths of α-silicon carbide [4]. The resulting plastic mix was rolled into sheets 200 μm thick. After drying the sheets were coated with a graphite layer stacked and pressed together. The process is described in detail elsewhere [3]. The polymer was then pyrolysed by heating at 1°C per min to 450°C before heating to 2040°C for ½ hour to give a body with a relative density of 95%, assuming a theoretical density of 3.21 gm cm^{-3}. The final thickness of the graphite layer was 8 µm, whilst that of the silicon carbide layer was 150 µm. The silicon carbide had a grain size of 1.3 µm.

As a comparison monolithic silicon carbide was prepared by pressing the plastic mix into a plate 3 mm thick and heating as before. Test bars 3 mm wide were then cut from the monolithic and laminated blocks using a diamond saw. To compare the fracture behaviour, bars were tested both notched and unnotched in three point bend, using a span of 40 mm.

RESULTS AND DISCUSSION

Fracture Behaviour

The load/deflection behaviour of the monolithic and laminated samples is shown in Fig 1. The monolithic material failed catastrophically, giving a fracture toughness of 3.6 MPa√m. After an initial elastic deflection, Fig 2(a), crack growth in the laminate began at the same stress intensity, but the crack was deflected as it reached the interface giving rise to the 'knee' in the load/deflection curve. Further loading gave rise to the growth of a delamination crack as shown in Fig 2(b) which corresponds to the point 'B' on the load/deflection curve. When the maximum load was reached the next lamina failed giving rise to the first load drop. Fracture of the lamina did not occur at the tip of the delamination crack as can be seen by comparing Fig 2(b) and 2(c).

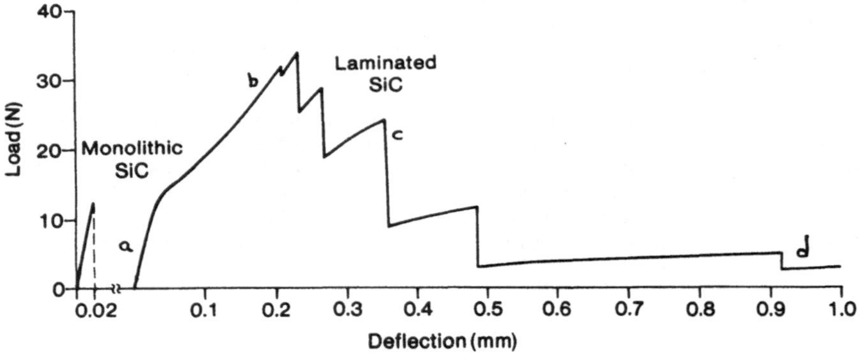

Fig 1 Showing a comparison between the load/deflection behaviour of notched bars of monolithic silicon carbide and laminated silicon carbide containing graphite interfaces.

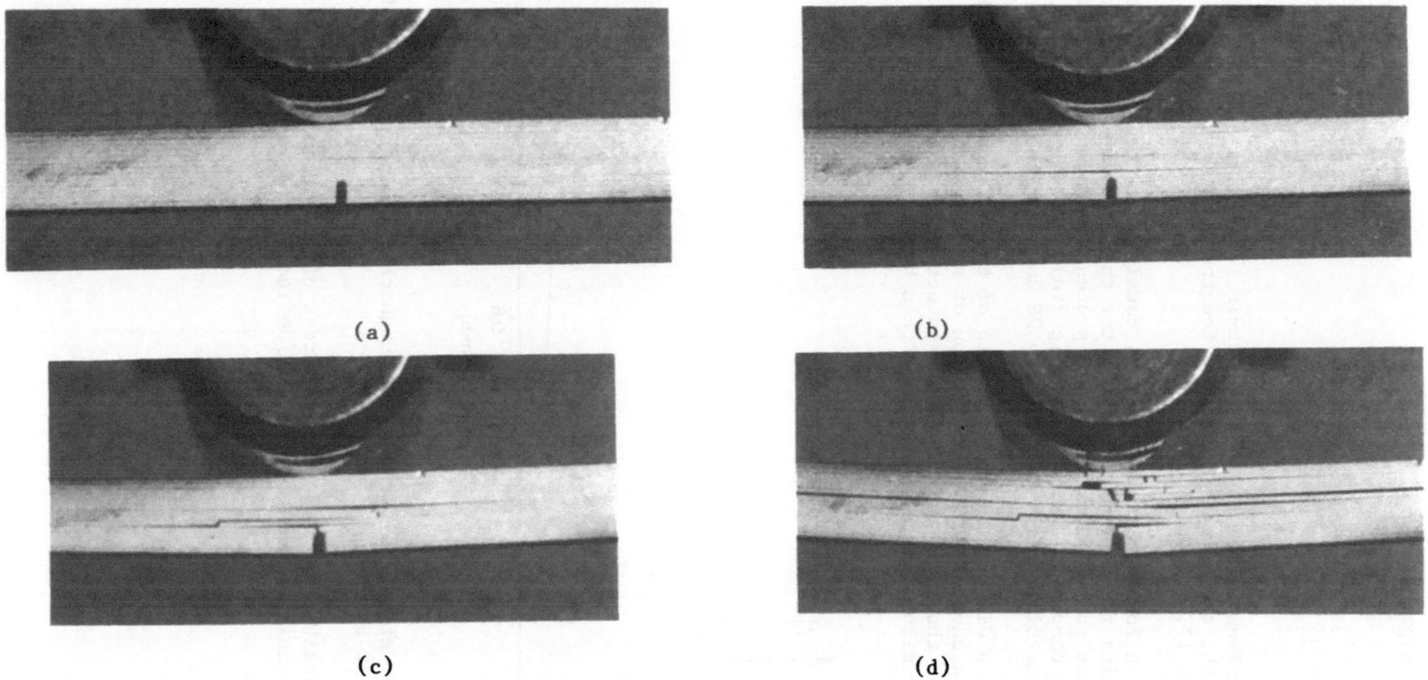

Fig 2 Laminated bars in 3 point bend, (a) during elastic deformation, (b) after initial crack growth has been deflected along an interface, (c) after reaching the maximum load, further lamina failure and delamination occurs. Note the position of failure in the second lamina, and the tip of the first delamination crack, (d) the bar after final failure. The letter preceding each figure, corresponds to the letter on the load/deflection trace in Fig 1.

Using the Griffith expression

$$K_{1c}^{L} = \frac{\sigma_m}{Y\sqrt{c}}$$

where σ_m is the maximum stress, Y is a geometrical constant and c is the initial crack length, the apparent fracture toughness of the laminate, K_{1c}^{L} is calculated to be 17.7 MPa$\sqrt{m}$.

Further delamination followed the load drop, causing the load to rise again until the next lamina failed so that even after the maximum load had been reached, failure was not catastrophic. The process repeated itself until the body fractured completely as shown in Fig 2(d). The work required to break the sample was 6152 Jm^{-2} compared with 28 Jm^{-2} for the monolith.

That lamina fracture did not occur at the tip of the delamination crack suggests that it is caused by the propagation of some internal defect, ie failure occurs when the applied load on the intact portion of the sample is equivalent to the failure stress of an unnotched bar.

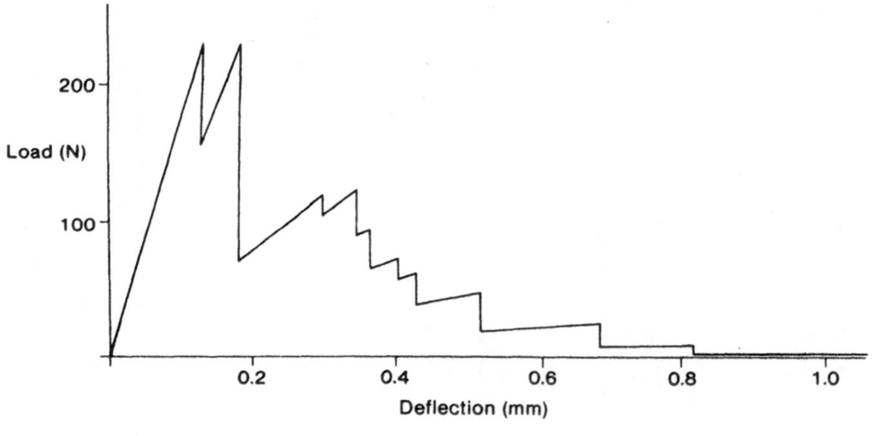

Fig 3 The load/deflection curve for an unnotched laminated bar tested in 3 point bend

To examine this possibility, the failure stress of the second lamina was estimated from the maximum load in the load/deflection curve using the formula

$$\sigma_F = \frac{1.5\ F\ l}{b\ (d-c)^2}$$

where l is the span, b is the sample breadth and (d-c) is the depth of the sample less than the notch depth, ie the depth of the remaining ligament. This gave a value for σ_F of 552± 148 MPa close the the value of 496± 99MPa obtained from bend testing unnotched bars. It is clear that the existence of either the original notch or the delamination crack does not reduce the strength of the intact portion of the bar; the material has become notch insensitive.

This view is also consistent with the fracture behaviour of an unnotched sample, shown in Fig 3, where the load/deflection behaviour shows no continuously rising portion after crack growth first takes place; as marked 'B' in Fig 1. Once the applied load has reached the failure stress of the bar, the onset of cracking causes the stress on the remaining section to rise causing further fracture. Sometimes failure of the second lamina requires a slightly higher load presumably due to the variation in the strengths of the lamina.

Fig 4 Showing plates, shells with single and double curvature and a tube made by laminating silicon carbide sheets separated by graphite interfaces

Potential of the Process

Because powders are used as the source of ceramic material the process is cheap and it is rapid. The composite elements described here are sheets, but fibres can also be extruded and have been shown to be thermally stable for at least 100 hours after heat treatment at 1500°C in air [5]. Furthermore the process is general and may be used for any material where sinterable ceramic powders exist.

Forming is also possible and simple shapes have been made as shown in Fig 4 either by stacking the sheets over some former or, in the case of the tube, by rolling up coated sheet, and then pressing. The techniques are similar to these that are used in the fabrication of polymer composites.

CONCLUSIONS

Using silicon carbide containing graphite interfaces as a model system, it has been shown that crack growth in a notched bar. Using silicon carbide containing graphite interfaces as a model system it has been shown that crack growth in a notched bar begins at the same stress intensity as in the monolith. The crack is deflected at the graphite interface and a delamination cracks grows until the applied load is such that the remainder of the bar reaches the failure stress for an unnotched bar. This gives an apparent fracture toughness of 17.7 MPa√m and a work required to break the sample of 6152 Jm^{-2} compared with 3.6 MPa√m and 28 Jm^{-2} respectively. Simple curved shapes and tubes have been made using this process.

REFERENCES

1. Naslain R, Hagenmuller P, Christin F, Heraud L and Chowy J J. In Advances in Composite Materials - ICCM-3, ed A R Burrell, Pergamon Press, Paris, 1980, 2, pp 1084.

2. Stinton D P, Besmann T M, and Lowden R A, Advanced Ceramics by Chemical Vapour Deposition Techniques. Am Ceram Soc Bull, 1988, 67, 350-55.

3. Clegg W J, Kendall K, Alford N McN, Button T W and Birchall J D. Nature, 1990, 347, 455-57.

4. Johnson C A and Prochazka S, Microstructures of Sintered SiC. In Ceramic Microstruct, Proc 6th Int Symp, Berkeley, 1977, p 366-78.

5. Clegg W J, Kendall K, Alford N McN, Button T W and Birchall J D, An Inexpensive Method for Making Tough Ceramics in Proc Brit Ceram Soc Conf, Special Ceramics 9, London, 18-20 Dec 1990. In Press.

SiC Particle and Y-TZP Reinforced Mullite Matrix Composites

X.X. Huang, J.S. Hong and J.K. Guo
Shanghai Institute of Ceramics, Chinese Academy of Sciences
1295 Dingxi Rd, Shanghai 200050, China

Abstract

Mullite(M) matrix composites reinforced by both SiC particles(P) and Y-TZP(yttria stabilized tetragonal zirconia polycrystal) were fabricated by hot-pressing. The relationship between microstructure and properties of the composite, reinforcing mechanism and effects were investigated. The effect combined by both SiC particles and Y-TZP on improving the mechanical properties appears to be additive or synergetic. The reinforcing mechanism of the composites are mainly crack deflection and crack branching caused by SiC particles, and microcrack toughening by the addition of Y-TZP. The strength of the composites almost keeps constant from room temperature to 1000°C. The thermal shock resistance of the composites has been significantly improved. The abiabatic engine piston caps made of SiC(p)/Y-TZP/M composite have still kept intact and undamage after the bench run testing for over 100 hrs. It has been indicated that the SiC(P)/Y-TZP/M composites are a potential candidate as thermal insulating structural components for heat engines.

Introduction

Mullite(M) is a potential candidate for heat engine technology because of its low thermal conductivity, low thermal expansion coefficient and high creep resistance. Its strength is not degraded even at over 1200°C. However, mullite is of relatively poor values of strength and fracture toughness, which impede its practical application. The fracture toughness and strength must be improved if it is to be used in heat engines.

The incorporation of SiC whiskers or a transformation toughened ZrO_2 phase to improve the properties of ceramics has been reported[1-2]. Either of them can improve the mechanical properties of mullite to some extent. Recently, some findings have indicated that the combination of multiple toughening and strengthening routes can result in ceramic composites with better properties than that achieved by either mechanism alone. This is, thus, one of the leading trends for strengthening and toughening of ceramics. It has been obtained that the strength and toughness of mullite had been significantly increased when both SiC whiskers and ZrO_2 particles added to mullite matrix by whisker reinforcement and ZrO_2 toughening together[3-5].

Particle-reinforced composites have the potential of possessing with isotropic properties. Isotropy of mechanical properties is advantageous for various applications of composites. The properties of ceramics can be increased by a second dispersed phase, e.g., by SiC particles(P) in Y-TZP[6], TiC particles in Al_2O_3[7] and SiC[8]. The strength and fracture toughness of mullite can also be improved by particle reinforcement, furthermore, the strength and toughness of mullite can be increased further by both particle reinforcement and ZrO_2 toughening[9]. In the present investigation, SiC particles and Y-TZP were selected as double reinforcements for mullite matrix. The properties and microstructure of the mullite matrix composites obtained by adding various amounts of SiC particles and Y-TZP together were investigated

Piston caps are one of key components for ceramic adiabatic engine. High strength silicon nitride materials has been used in automobile pistol caps. But in terms of insulating property, the Si_3N_4 ceramics still is not an ideal material. Developing much insulating materials is of great importance to economize fuel oil. The SiC(P)/Y-TZP/M composites prepared in the present investigation show excellent mechanical properties and good thermal properties as well as good thermal shock resistance. The adiabatic engine piston caps were successfully made of the SiC(P)/Y-TZP/M composites, and other adiabatic engine components(e.g., top plate), made of the composites, would be expected.

Experimental Procedure

Commercial mullite powder having an average grain size of 2 μm was used as a matrix. This powder contained 75.92 wt% Al_2O_3, 23.34 wt% SiO_2, and 0.74 wt% of other oxide impurities. As one of the reinforcements, commercially available SiC particles are α-type SiC with an average diameter of 0.93 μm and a specific surface of 7.53 m^2/g. The compositions of SiC particles as follows: total C: 27.0 wt%, free C: 0.41 wt%, total Si: 68.50 wt%, free Si: 1.08 wt%, O: 2.15 wt%, Al_2O_3:0.13 wt%, Fe_2O_3:0.18 wt%, and CaO:0.02 wt%. Y-TZP powder(containing 2.8 mol% Y_2O_3) as another reinforcement was prepared using coprecipitation method.

The mullite, SiC and Y-TZP powders were mixed with the desired volume ratio by wet milling with distilled water using Al_2O_3 balls in a polyethylene container for 6 h, and dried. The obtained mixed powders were sieved to under 120 mesh, and were molded in a BN-coated graphite die for sintering. Hot-pressing was carried out at 1600°C for 40 min at nitrogen atmosphere under 25 MPa pressure for all composites.

The hot-pressed sample were cut and ground in 2.5 x 5 x 28 mm bars by a diamond wheel. A three-point bend test was carried out at a crosshead speed of 0.5 mm/min with a span of 20 mm to measure the flexural strength. Test bars of dimensions of 5 x 2.5 x 28 mm were prepared for single edge notched beam(SENB) fracture toughness measurements. The notches were 0.22-0.25 mm in width and 2.5 mm in depth. The SENB specimens were fractured in three-point bending at a crosshead speed of 0.005 mm/ min with a span of 20 mm.

Results and Discussion

(1) SiC(P)/Y-TZP/M composites

Fig.1 shows the effect of SiC particle content on the flexural strength and fracture toughness of the mullite composites reinforced only by SiC particles. Flexural strength of the composite was improved by the addition of 10 vol% SiC particles. However, the amount of 20 to 30 vol% SiC particles gave no further improvement in flexural strength. Fracture toughness increased with increasing the SiC particle. Mullite matrix composites in which SiC particles and Y-TZP were added as double reinforcements were investigated. The Y-TZP composition of 20 vol% was selected as one constant reinforcing phase to investigate the effect of SiC particle additions. Figure 2 shows the effect of the added SiC particle content on fracture toughness and strength of the composites. The fracture toughness and strength of the composites were found to be increased with additions of SiC particles. The optimum value of fracture toughness and strength are 6.7 $MPa{\cdot}m^{1/2}$ and 600 MPa at 35 vol% SiC

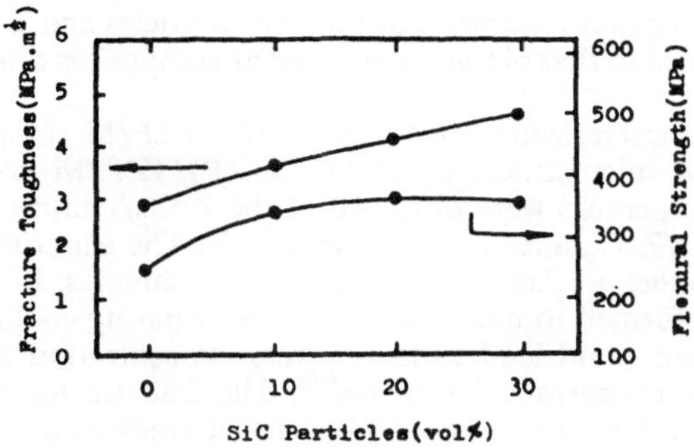

Fig.1 Flexural strength and fracture toughness of SiC(P)/M composites vs SiC particle content

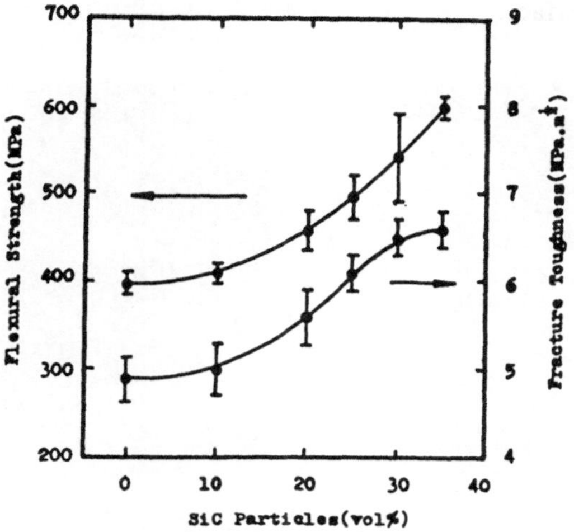

Fig.2 Flexural strength and fracture toughness of SiC(P)/20 vol% Y-TZP /M composites vs SiC particle content

particles, respectively. As for the effect of SiC particle content on strength in the case of SiC(P)/M composites, the strength did further not increase as SiC particle content increased at fairly higher SiC particle content(greater than 10 vol%) as indicated in Fig.1. However, the tendency of the increased strength by increasing added SiC particle content remained unchanged at higher amount of SiC particles for the SiC(P)/Y-TZP/M composites. As to the effects of SiC particle content on fracture toughness in the case of SiC(P)/Y-TZP/M composites, the increasing of fracture toughness with SiC particle content was smaller than that of the SiC(P)/M composite at lower amount of SiC particles, as illustrated in Fig.1. But, at high amount of SiC particles, the former was obviously greater than the later. It has been shown that the effect combined by Y-TZP and SiC particles on fracture toughness and strength appears to be additive or synergetic. The fracture toughness and strength of the mullite matrix composites reinforced by both SiC particles and Y-TZP were dramatically improved, in comparison to that of the mullite matrix composites reinforced by SiC particles or Y-TZP alone.

(2) Microstructural characterization of the SiC(P)/Y-TZP/M composites

The typical TEM microstructures of the SiC(P)/TZP/M composites are given in Figs.3(a)-(c). The microcracks were observed at the ZrO_2/mullite interface(a), ZrO_2/SiC interface(b) and ZrO_2/ZrO_2 interface(c), respectively. The experimental results show that a great amount of monoclinic ZrO_2 has existed in fired samples. It is suggested that a large amount of ZrO_2 transformed to monoclinic symmertry during cooling. The volume change of transformation grain could lead to large stress concentration at interphase interfaces which would result in microcrack formation[10]. The fracture toughness and strength are improved by the deflection, branching and bowing of crack as a result of the crack tip interaction with the microcracks. We previously reported that the crack deflection and crack branching were the dominant reinforcing mechanisms caused by SiC particles in SiC(P)/M composites[9]. Fig.4 shows the TEM micrograph of the SiC(P)/Y-TZP/M composites. It can also be observed that the propagating crack was deflected by SiC particle at the SiC/mullite interface.

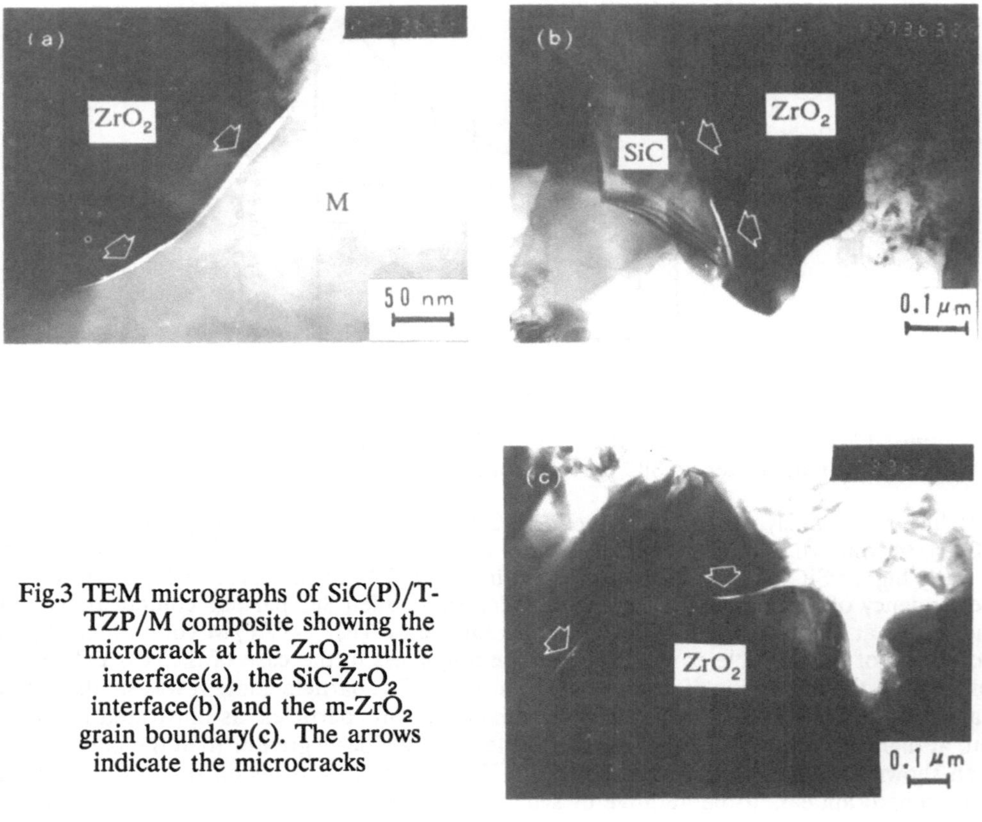

Fig.3 TEM micrographs of SiC(P)/T-TZP/M composite showing the microcrack at the ZrO_2-mullite interface(a), the SiC-ZrO_2 interface(b) and the m-ZrO_2 grain boundary(c). The arrows indicate the microcracks

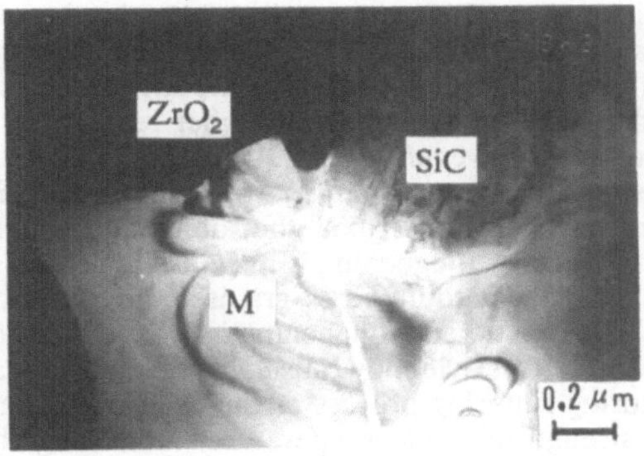

Fig.4 TEM micrograph of SiC(P)/Y-TZP/M composite showing crack deflection caused by SiC particle

(3) High temperature strength of the SiC(P)/Y-TZP/M composites

The dependence of temperature on flexural strength was investigated for the SiC(P)/Y-TZP/M composite and the results are shown in Fig.5. The data for monolithic mullite is also presented for comparison. Flexural strength of the composite is almost constant from room temperature to 1000°C. Above 1000°C, the strength decreased obviously. But, at 1200°C, the strength of the composite was still higher than that of monolithic mullite.

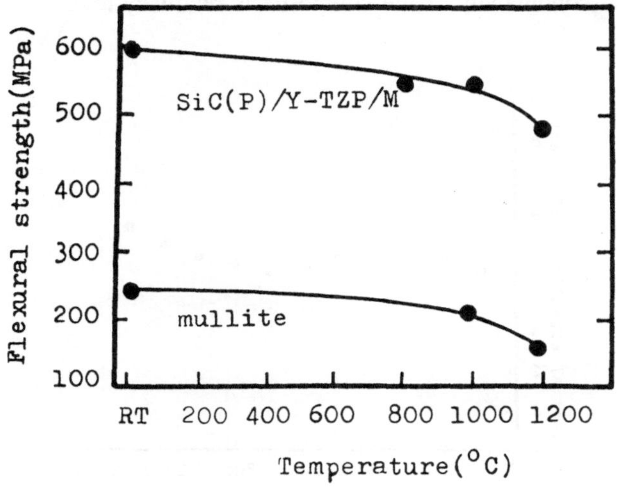

Fig.5 Flexural strength of mullite and 35 vol% SiC(P)/20 vol% Y-TZP/M composite vs temperature

The tendency of the dependence of strength versus temperature of the composites is similar to that of monolithic mullite. This indicates that strengthening effect of SiC particles is not weakened in the range of temperature from room temperature to 1000°C. If the increase of toughness and strength of the composites are attributed to the transformation toughening, the strength should decrease rapidly at higher temperature due to decreasing driving force for transformation. However, the results in Fig.5 illustrate that the strength of the composites retained up to at least 1000°C. According to this, it is an evident that transformation toughening mechanism is not the dominant toughening mechanism, conversely indicating that microcrack toughening is responsible for this. The fact that many microcracks occurred at the interface around ZrO_2 grains in the composites is the additional proof for this proposition. The microcrack toughening is independent of temperature. The strength degradation above 1000°C is evidently caused by the mullite matrix itself. The glass phase was likely formed during sintering process of the composites due to lower purity of mullite powder, affecting strength at high temperature. Increasing the purity of mullite powder, further improvement of strength of the composites at the high temperature could be expected

(4) Thermal properties of the SiC(P)/Y-TZP/M composites

The thermal properties of the SiC(P)/Y-TZP/M composite were investigated. Figure 6 shows that dependence of temperature on thermal expansion coefficient for the composite with a composition of 20 vol% SiC particles + 20 vol% Y-TZP + 60 vol% mullite. From room temperature to 1000°C, the average expansion coefficient of the composite is 5.35×10^{-6}/°C, similar to that of monolithic mullite. At 400°C to 700°C, the thermal expansion coefficient decreases slightly with increasing temperature. This phenomena is because that monoclinic ZrO_2 in the composite transformed to tetragonal form within the range of temperature. The effect of temperature on thermal conductivity of the composite is shown in Fig.7. The thermal conductivity of the composite is relatively low.

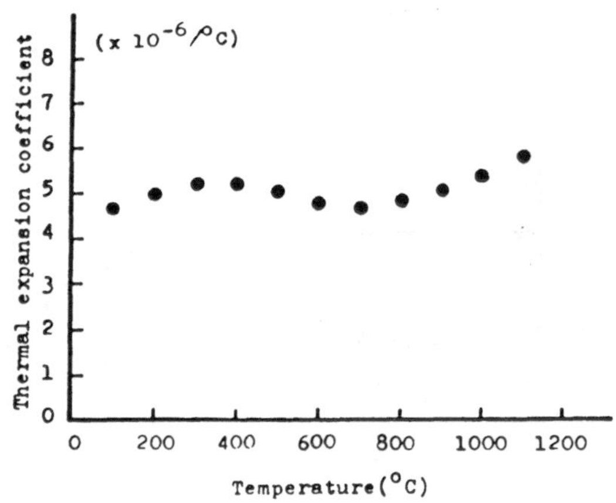

Fig.6 Temperature dependence of the thermal expansion coefficient of SiC(P)/Y-TZP/M composite

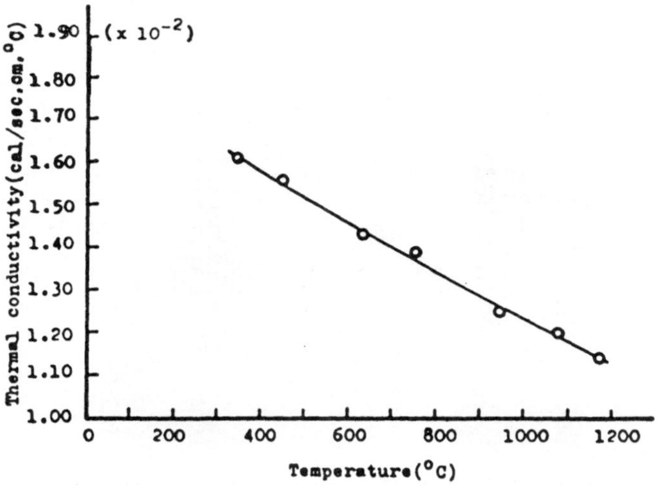

Fig.7 Temperature dependence of the thermal conductivity of SiC(P)/Y-TZP/M composite

(5) Thermal shock resistance of the SiC(P)/Y-TZP/M composites

For high temperature structural applications of ceramic materials, thermal shock resistance is an important property. The thermal shock resistance of the SiC(P)/Y-TZP/M composites was evaluated by quenching with water. The results are listed in table below. It is shown that monolithic mullite has retained 29.0% of its initial (unshocked) flexural strength after the thermal shock treatments from 525°C to 25°C, indicating that the thermal shock resistance, ΔT_{max} of mullite is smaller than 500°C. According to the literature reported[11], ΔT_{max} of mullite is 350°C. However, the flexural strength of the composite decreased slightly (from 456MPa to 422MPa) after quenching with water from 525°C to 25°C, indicating that T_{max} of the composite is at least 500°C. The thermal shock resistance of mullite matrix composites is significantly improved over 500°C by adding both SiC particles and Y-TZP.

Comparison between the initial strength and the retained strength after quenching with water from 525°C to 25°C for monolithic mullite and SiC(P)/Y-TZP/M composite

materials	Initial strength (MPa)	Retained strength (MPa)
Mullite	246	71.8
20 vol% SiC(P)/Y-TZP/M	456	422

(6) Test for SiC(P)/Y-TZP/M composites in heat engines

The photograph of B135 adiabatic engine piston caps made of hot-pressed SiC(P)/Y-TZP/M composites is shown in Fig.8. The bench run of the piston caps was conducted.

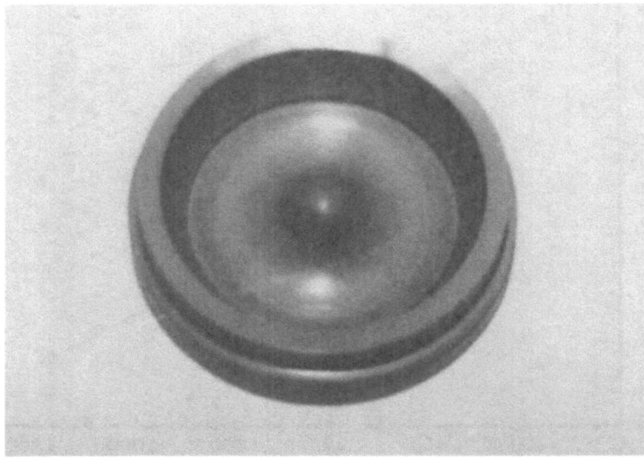

Fig.8 Photograph of B135 adiabatic engine piston cap made of the SiC(P)/T-TZP/M composite

The run conditions as follows: without cooling water, rotational speed of 1500(r.p.m), 90% fully loaded and about 700°C of exhaust temperature. After the test run over 100 hrs, the piston caps were found intact and undamage, and can be tested further. Its thermal insulating effect was better than that of Si_3N_4 materials. It is shown that the SiC(P)/Y-TZP/M composites are a strong potential candidate as the thermal insulating structural components for heat engines.

Conclusions

(1) The flexural strength and fracture toughness of the mullite matrix composites are significantly improved by introducing SiC particles and Y-TZP together, more than that of the mullite matrix composites reinforced only by SiC particles or Y-TZP, the maximum flexural strength and fracture toughness of the composites are 600 MPa and 6.7 $MPa.m^{1/2}$ respectively for addition of 20 vol% of Y-TZP and 35 vol% of SiC particles. The strength almost keeps constant to temperature around 1000°C.

(2) Reinforcing mechanisms are mainly crack deflection and crack branching by SiC particles, and microcrack toughening caused by Y-TZP. Increases in the properties for adding Y-TZP and SiC particles together were found to be additive or synergistic.

(3) The SiC(P)/Y-TZP/M composites exhibit lower thermal expansion coefficient ($5.35X10^{-6}$/°C, at room temperature to 1000°C), and lower thermal conductivity. The thermal shock resistance, ΔT_{max} is at least 500°C, was apparently improved.

(4) The B135 adiabatic engine piston caps made of SiC(p)/Y-TZP/M composites were found intact and undamage after the bench run testing was over 100 hrs, and would be run further. It is shown that the mullite matrix composite are a strong potential candidate as the thermal insulating structural components for heat engines.

References

1. G. C. Wei and P. F. Becher, Am. Ceram. Soc. Bull., 64(2) 1985 (298)
2. N. Claussen and J. Jahn, J. Am. Ceram. Soc.,63, 1980 (228)
3. P. F. Becher and T. N. Tiegs, J. Am. Ceram. Soc., 70(9) 1987 (651)
4. R. Ruh et al., J.Am. Ceram. Soc., 71(6) 1988 (503)
5. Jinsheng Hong, Xiaoxian Huang, Jingkun Guo and Baoshun Li, J. Chin. Ceram. Soc., to be published
6. X.X. Huang, L.T. Tai, B.S. Li, L.H. Gui and J.K. Guo et al., C-MRS International'90, Beijing, China, June 18-22, 1990
7. R. P. Wahi and B.Ilschner, J. Mater. Sci., 15, 1980 (875)
8. M. Ishitsuka et al., J. Am. Ceram. Soc., 70(11) 1987 (c-432)
9. Jinsheng Hong, Xiaoxian Huang, Jingkun Guo et al., J. Inorg. Mater., 5(4) 1990 (340)
10. Y. Fu, A. G. Evans and W. M. Kriven, J. Am. Ceram. Soc., 67(9) 1984 (626)
11. M. Ishitsuka, T. Sato, T. Endo and M. Shimada, J. Am. Ceram. Soc., 70(11) 1987 (c-342)

REINFORCING MECHANISM OF SiC WHISKERS IN SiCw/CARBON CERAMIC COMPOSITES

JIEMO TIAN, XIAOHUA TONG, LIMIN DONG
ZHENWEN WANG, LINGLING WANG,
Beijing Fine Ceramic Laboratory of Institute of Nuclear Energy Technology, Tsinghua University,
P.O.Box 1021, Beijing 102201, CHINA

ABSTRACT

SiC whiskers/carbon ceramics have been made by hot pressing at 1950—2100℃ under argon atmosphere. The microstructure of SiCw/carbon ceramics has been studied. A unique distribution of SiC whiskers in carbon ceramic matrix has been found. The short SiC whiskers are distributed spread in matrix grains, but long SiC whiskers are distributed at the boundaries. The short SiC whiskers play a strengthening role in the matrix grains as second phase particles and long SiC whiskers play a toughening role in the matrix because whiskers prevent the cracks from propagating. This reinforcing mechanism is identified as the two directional reinforcing mechanism of the SiC whiskers in carbon ceramics.

INTRODUCTION

Carbon ceramics are new ceramic materials. Because they possess high electrical conductivity, and resistance to oxidation and wear, they recieve great attention. In early 1980, this research work began to develop[1,2,3]. Up to now, the bend strength and fracture toughness maximum at room temperature are 160-250 MPa and 2-4 $MPam^{1/2}$ [5,6,7,8], respectively. SiCw/carbon ceramics are described in this paper. The bend strength and fracture toughness maximum at room temperature are 400—450 MPa and 4-6 $MPam^{1/2}$, respectively. These data are 4-6 times the strength of graphite. The thermal shock, hardness, density, electrical conductivity, thermal expansion coefficient and resistance to oxidation were measured. The mechanism of the resistance to oxidation and the two directional reinforcing mechanism of SiC whiskers were made.

EXPERIMENTAL PROCEDURE

Carbon powder, SiC whiskers with additions of B_4C, SiC, TaC were mixed in the ratio of the desire compositions. The sizes of the all powders are less than 2 μm, SiC whiskers are 0.1～1 μm in the diameter and 1-5 μm in the length. After mixing,the mixtures were sintered at 30 MPa and 1950—2050 ℃ under the argon atmosphere for 45 minutes. The size of the sample used for measurements of the mechanical properties was 3×4×40 mm. The bend strength and fracture toughness were measured with three point bending strength method and Chevron notched beam technique [4] at room temperature and 850 ℃. The density, hardness, thermal expansion coefficient, resistivity were measured also. The resistance to oxidation were investigated at 800 ℃, 1000 ℃ and 1300 ℃. The thermal shock were carried out at 400 ℃, 500 ℃, 600 ℃, 700 ℃ and 800 ℃. At these temperatures, the samples maintained for 20—30 minutes were quenched in the water, after that, the bend strength of these samples were measured. Wearing tests used G Cr 15 steel as friction pair were performed with the lubrication oil or without any lubrication agent. The drying and wetting friction coefficients were calculated.

RESULTS AND DISCUSSION

Mechanical Properties

The mechanical properties at room temperature and 850 ℃ as well as the effect of carbon content on the mechanical properties of 15 wt% SiCw/carbon ceramics containing 85 wt% ($C+B_4C+SiC+TaC$) are shown in Fig. 1. The experimental results show 15 wt% SiCw/carbon ceramics containing 25 wt% carbon possess the optimal bend strength and fracture toughness, they are ≥400 MPa and ≥4 $MPam^{1/2}$, respectively. The SiC whiskers reinforced carbon ceramics containing 50 wt% carbon have the bend strength of 220 MPa and fracture toughness of 6 $MPam^{1/2}$, they are 2.5 and 5 fold of the high strength graphite. The carbon ceramics containing ≥25 wt% carbon content can be machined easily. The bend strength decreases on the small side and fracture toughness decreases a little at 850 ℃. The bend strength and fracture toughness change with carbon content, they are 260—350 MPa and 3.5—4.0 $MPam^{1/2}$ at 850 ℃, respectively. The 15 wt% SiCw/carbon ceramics containing 30—40 wt% carbon obtain the bend strength of 300 MPa and fracture toughness of 3.8 $MPam^{1/2}$ which are more than 3 fold of the high strength graphite at room temperature.

Effects of the SiC Whiskers Content on the Mechanical Properties

Carbon ceramics containing 20 wt% carbon B_4C, SiC, TaC 15 wt%, 25 wt%, 35 wt% or 45 wt% SiC whiskers were prepared. The bend strength and fracture toughness are given in Fig.2. It is seen from Fig.2 that the bend strength and fracture toughness containing 15 wt% SiC can obtain a maximum which are ≥400 MPa and ≥4 $MPam^{1/2}$, respectively.

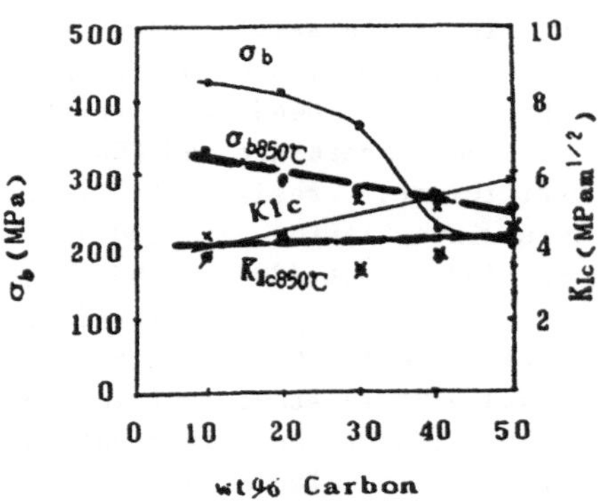

Figure 1. Bend strength σ_b and fracture toughness K_{Ic} as a function of Carbon content

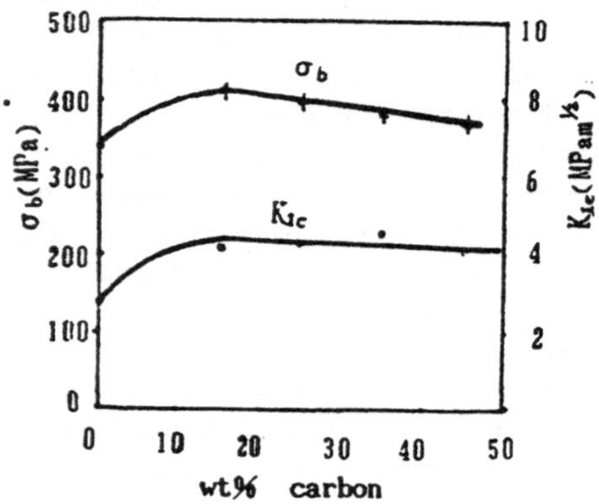

Figure 2. Relationship between SiC whisker content and σ_b and K_{Ic}

Elastic Module and Hardness

The hardness (HR_A) is measured with load of 60 kg, the elastic module is measured with three point bending strength, their results are made in Fig.3. The hardness of the carbon ceramic composite containing 15 wt% SiCw and 50 wt% carbon corresponds to the quenching steel and this composition composite can be machined. Elastic module and hardness of carbon ceramics containing 20 wt% carbon and 15 wt% SiCw obtain high values.

Density, Resistivity and Thermal Expansion Coefficient

The experimental results of the density and resistivity are shown in Fig.4. This figure point out carbon ceramics containing larger than 30 wt% carbon and 15 wt% SiCw can be machined easily. The thermal expansion coefficient is a function as the temperature and linear relasionship with the temperature. It is 2.5×10^{-6}/℃.

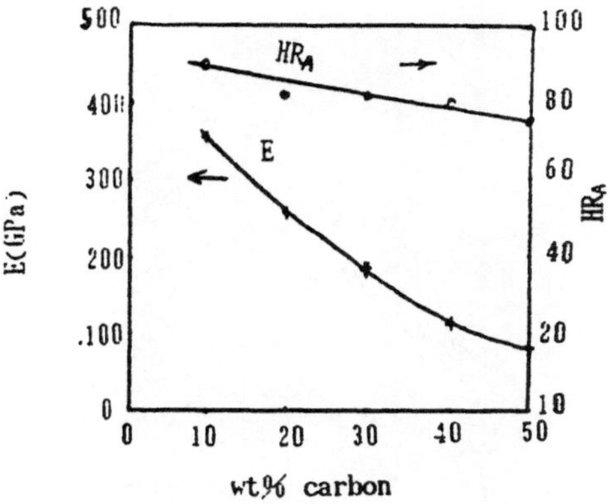

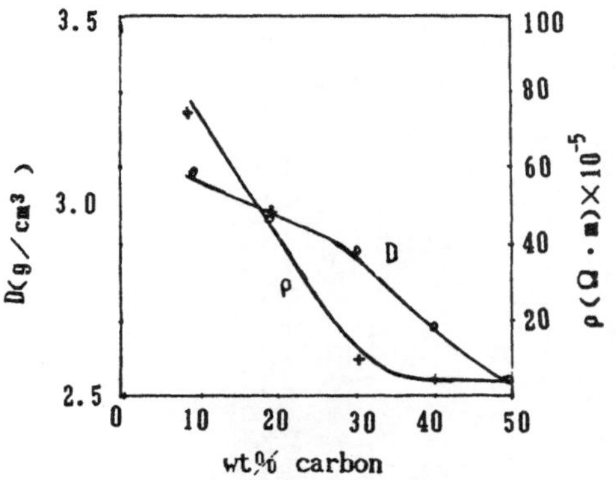

Figure 3. HR_A and E as a founction of carbon content

Figure 4. Density and resistivity as a founction of carbon content

Wear and Resistance to the Thermal shock

Wearing capacity was investigated while G Cr15 steel was used as friction pair. The experiments were tested with lubrication oil or without any lubrication agent. The drying friction and wetting friction coefficients were calculated, they are 0.16 and 0.28, respectively, but drying friction coefficient of GCr15 is 0.42. The results show wearing capacity of carbon ceramic composite is better than GCr15 and graphite. The thermal shock experiments were carried out at 400 ℃, 500 ℃, 600 ℃, 700 ℃ and 800 ℃ and quenched in the water. The carbon ceramics containing 15 wt% SiCw, 20 wt% carbon and added B_4C, SiC and TaC were used as experimental samples. The bend strengths of quenched samples were measured. The results were drawn in the Fig.5. The temperature of the resistance to the thermal shock is 600℃ which is found in the Fig.5.

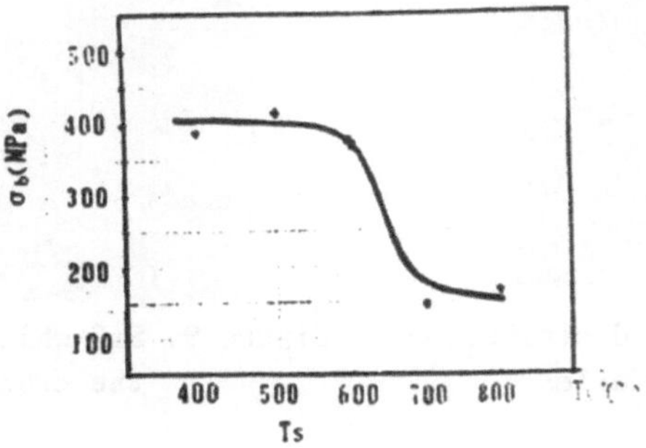

Figure 5. Bend strength as a function of the thermal shock temperature

Figure 6. SiC whiskers insert into the grains

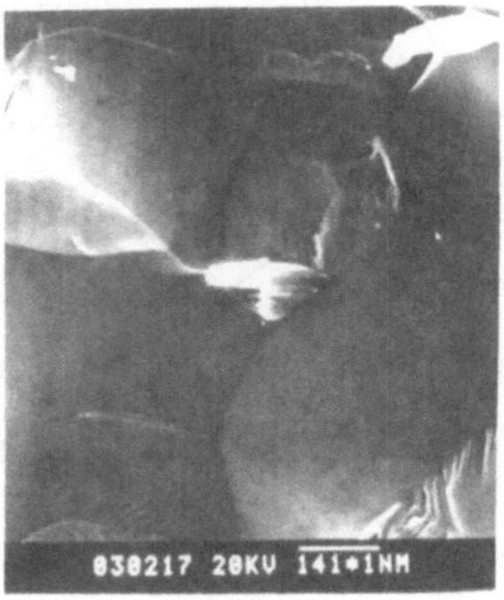

Figure 7. SiC whiskers distribution in the grains

Microstructure and Morphology

The microstructure observation showed SiC whiskers distributed in two forms in the matrix, the short whiskers (in the length are less 1μm) distributed in the grains, but long whiskers (in the length larger than 1μm) distributed in the boundaries or between grains (Fig.6, 7, 8, 9) . The fracture morphologies have been observed, the pull-out effects were found (Fig 7, 8). The whiskers can prevent the cracks from the transporting. It is important key to rise the mechanical properties that the distribution of SiC whiskers in the matrix are made.

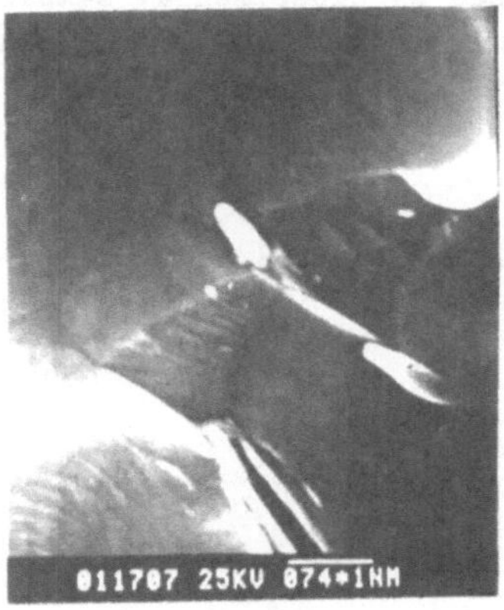

Figure 8. SiC whisker distribution in the boundaries

Figure 9. SiC whiskers prevent the crack from transporting

REINFORCED MECHANISM

The microstructures of Fig. 6-9 show that the short SiC whiskers distributed in the grains prevent crystalline planes from the slipping as the second particles in matrix, so the bend strength was increased. But, long SiC whiskers distributed in the boundaries or between the grains prevent cracks from transporting, they increase the toughness. The effects can be described as Fig. 10. It is called that two directional of SiC whiskers reinforcing mechanism. It can be explained why SiC whisker can reinforce carbon ceramic composite.

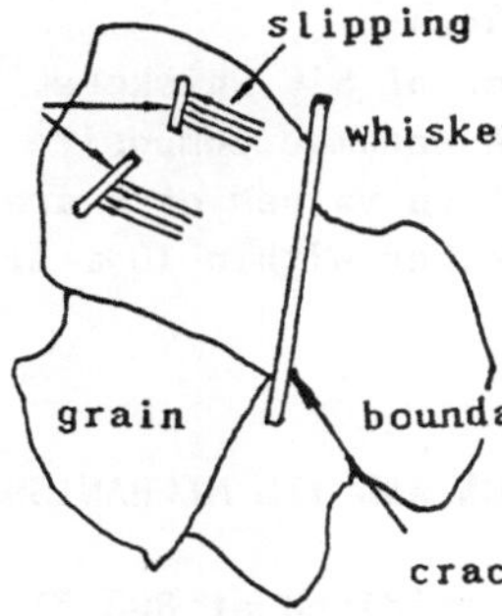

Figure 10. Mechanism of SiC whiskers reinforced composite

According to experimental results, the following equations can be usd to the calculation of the elastic modulus and bend strength for carbon ceramic composite.

Elestic modulus E

$$E=\frac{3}{8}E_L+\frac{5}{8}E_T \tag{1}$$

$$\frac{E_L}{E_m}=\frac{1+2L/d\ \eta_L V_W}{1-\eta_L V_W} \tag{2}$$

$$\frac{E_T}{E_m}=\frac{1+2\eta_L V_W}{1-\eta_\perp V_W} \tag{3}$$

$$\eta_L=\frac{\frac{E_W}{E_m}-1}{\frac{E_W}{E_m}+2\frac{L}{d}} \tag{4}$$

$$\eta_{\perp} = \frac{\frac{E_w}{E_m} - 1}{\frac{E_w}{E_m} + 2} \qquad (5)$$

E_L: longitudinal elastic module of carbon ceramic composite
$E_{\perp}$: traverse elastic module of carbon ceramic composite
E_m: elastic module of the matrix
E_w: elastic module of the SiC whiskers
V_w: Volume ratio of the SiC whiskers
L/d: the ratio of length/diameter of SiC whiskers.

The elastic module of the carbon ceramic composite can be calculated with the equations 1—5. The calculation values of E are compared with the experimental values. they agree each other within 10% error.

RESISTANCE TO OXIDATION AND ITS MECHANISM

Carbon ceramic composite samples were oxidized at 800 ℃, 1000 ℃ and 1300 ℃ in the air atmosphere condition for 30 hrs. The experimental results are given in the Fig.11 and Fig.12. Experimental results show 15 wt% SiCw carbon ceramics containing less than 20 wt% carbon possess high resistance to the oxidation. The oxidation rates decrease with the time and increase at high temperature. SiCw/carbon ceramics containing larger than 30 wt% carbon were oxidized at a higher rate, but added much more B_4C in the matrix, carbon ceramic composite raise the resistance to the oxidation. X-ray spectra of oxidaized carbon ceramic composite surfaces indicate the SiO_2 and B_2O_3 films are formed on surfaces. These mixed films coat the surface of the samples, so they prevent the surfaces from the oxidation.

The following reactions occurred

$$SiC + 2O_2 \longrightarrow SiO_2 + CO_2$$
$$B_4C + 4O_2 \longrightarrow 2B_2O_3 + CO_2$$

The microstructures show the larger B_2O_3 and SiO_2 coating films are formed with difficultly in carbon ceramics with high carbon contents ($\geq$30wt% carbon), so samples were oxidized at higher rates. (Fig.13; Fig.14).

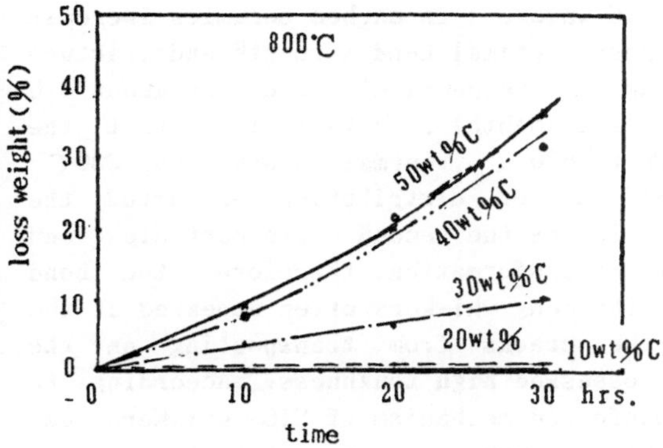

Figure 11. loss weight rate as a founction of oxidation temperature and time

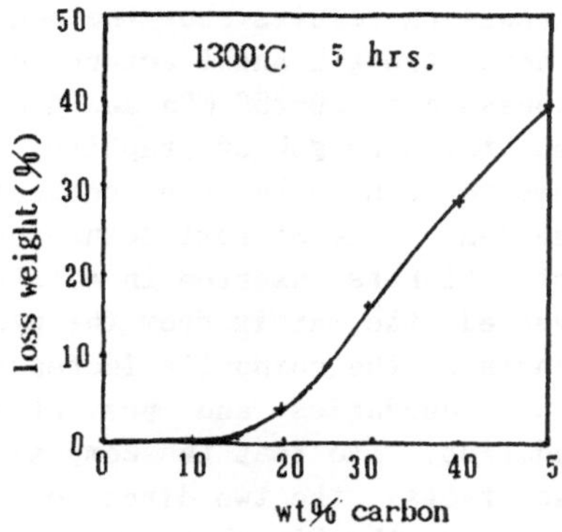

Figure 12. loss weight rate as a founction of carbon content at 1300℃

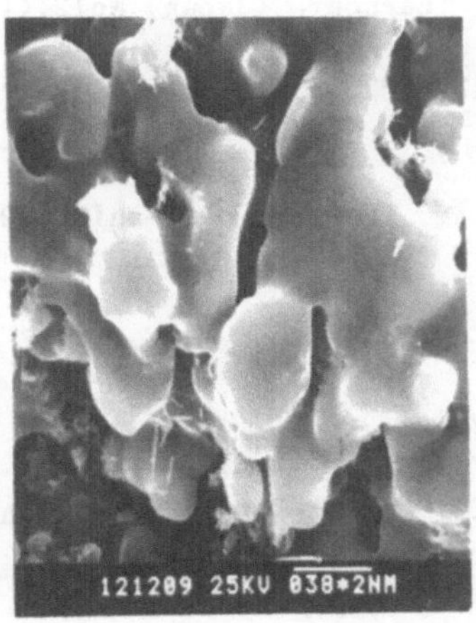

Figure 13. Morphology of SiCw/carbon ceramics containing 30wt% carbon at 1000℃, oxidated 30 hrs.

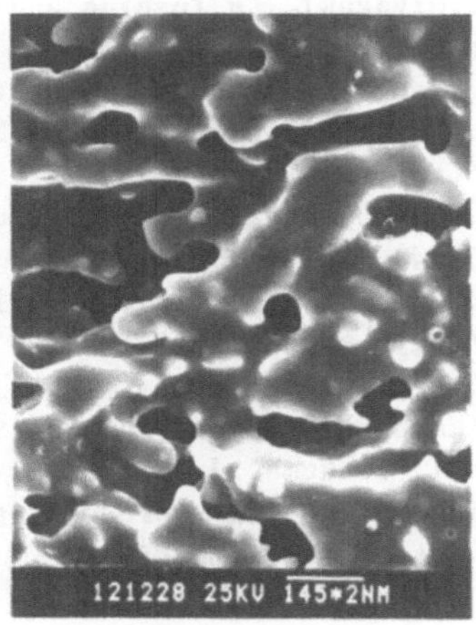

Figure 14. Morphology of SiCw/carbon ceramics containing 50wt% carbon at 1300℃, oxidated 5 hrs.

SUMMARY

The research results show the added SiC whiskers in carbon ceramics increase the bend strength and fracture toughness, optimal bend strength and fracture toughness are 400-450 MPa and 4-6 $MPam^{1/2}$, respectively, they are about 4-6 times the strength of graphite. The experimental results indicate that the composite can make the resistance to 600 ℃ thermal shock and 800℃ oxidation. The microstructure of SiC whisker distribution exhibited the short whiskers inserted in matrix grains as the second phase particles and prevented the matrix from the plastic transformation, therefore, the bend strength of the composite increased. The long whiskers often appeared at the grain boundaries and prevented the cracks from transporting on the boundaries, so that the composite possessed high toughness. According to these facts, the two directions reinforced mechanism of SiC whiskers can explain why the SiC whiskers can reinforce carbon ceramic composite.

B_2O_3 and SiO_2 film formed on the surfaces can make the resistance to the oxidation. This is called that mechanism of the resistance to the oxidation.

REFRENCES

1. K. Miyazaki, H.Yoshida and K. Kobayashi "Carbon" (Japan) No121(1986) pp18—25.
2. K.Kashima. "Carbon" (Japan) (1985) No.121 PP73—75.
3. K. Miyazaki, H.Yoshida and K.Kobayashi. "Carbon" (Japan) No.120(1985) PP21—27.
4. I.Ogawa, T.Hagio, H.Yoshida, K.kobayashi. "Yogyo Kyokai shi" 92[7] (1984) PP392—397.
5. I.Ogawa, Y.Yamashita, T.Hagio "Togyo-kyokai Shi" 93[10] (1985) PP612—617.
6. M.Anzai, A.Okura. "Carbon" (Japan) No.135(1988) PP278—285.
7. Koji, Tsukuma. J. Materials Sci., 20(1987) PP1178—1184.
8. Jie-mo Tian. In "proceedings of 7th CIMEC —— World Ceramics Congress & Sattellite Symposia."

EXCIMER LASER MACHINING OF SILICON NITRIDE

MARTIN WEHNER
Fraunhofer-Institute for Laser Technology (ILT)
Steinbachstrasse 15, 5100 Aachen, FRG

MARTIN BURSTRÖM
The Swedish Institute of Production Engineering Research - IVF
Regionkontoret Lulea, 951 87 Lulea, Sweden

ABSTRACT

Surface cracks often formed e.g. by grinding can dramatacilly lower the strength of ceramic components. UV-excimer laser machining is proposed as a finishing process for reducing surface defects by ablation of thin surface layers. Due to the high pulse power the vapourization temperature is reached within some ns and a thin layer (depth ≦ 1 micron) is ablated. The rapid and highly localized energy input leads to a small heat affected zone and a fast decomposition of the ceramics.

The formation of surface cracks is investigated on hot-isostatically pressed silicon nitride specimen in 4-point-bending tests. Different sets of laser parameters as wavelength, fluence, beam shape and pulse duration are tested. The beam size is determined as the most important parameter providing high strength and high surface quality. Bending strengths of more than 600 MPa can be achieved for ablation depths of 30-40 microns.

INTRODUCTION

One of the difficulties when machining advanced ceramics is to avoid subsurface microcrackes introduced to the material during the machining caused by mechanical or thermal influence. The strength of ceramic components is strongly influenced by extrinsic defects as surface-breaking cracks. Under load catastrophic crack propagation does ocurr when the stress intensity factor reaches a critical value K_{Ic}. The theory of fracture (e.g. [1]) gives a relation between the material dependent fracture toughness K_c, the stress and the charateristic dimension of the failure. For the flaw type of surface cracks an estimation of the failure stress can be given:

$$\sigma_f = \lambda \cdot K_c \cdot r_m^{-1/2} \quad (1)$$

σ_f : Failure stress
K_c : Critical stress intensity factor
r_m : Radius of mirror
λ : Material independent constant ($\lambda = 3.5 \pm 0.3$)

As a characteristic dimension the radius of the crack mirror is related to the critical stress intensity factor by [1]

$$K_{Ic} \approx 4.5 \cdot K_c$$

Thus fracture analysis of mirror radius and failure stress provides the critical stress intensity factor of the material. Furthermore the dimension of surface cracks can be appointed to the measured fracture stress. Thereby a method can be applied to estimate the size of the surface defects induced by machining (see figure 1):

$$\sigma_f \simeq 0.78 \cdot K_{Ic} \cdot r_f^{-1/2} \quad (2)$$

r_f : flaw size

The reduction of surface defects is the main task to improve the flexure strength of ceramic components. Bending tests of grinded specimen reveale that mechanical induced surface cracks extend to some tens of microns. The excimer laser has in this respect an unique possibility to remove material without causing neighter thermal nor mechanical damage to the machined surface. The ablation of thin surface layers (depth < 100 µm) by excimer laser radiation is proposed as a finishing technique achieving high quality surfaces with reduced defect size.

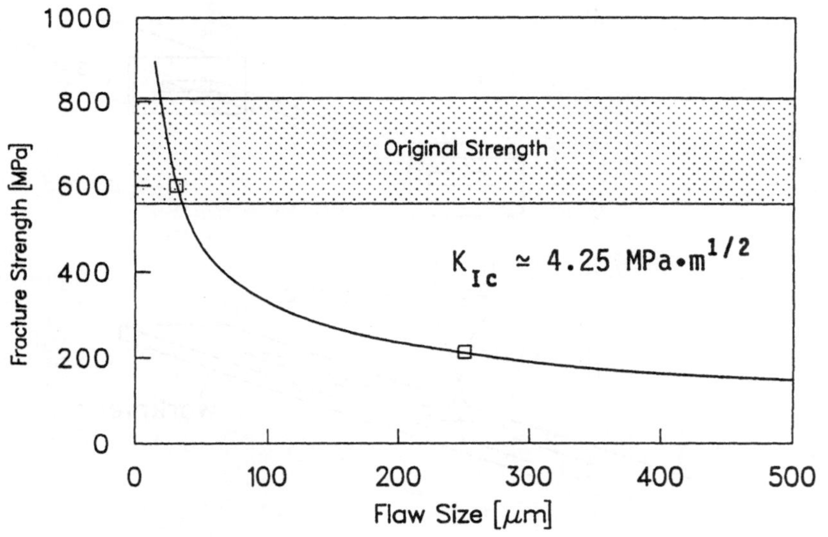

Figure 1. The fracture strength depends strongly on the size of surface defects, calculation according to equation 2.

MATERIALS AND METHODS

Silicon nitride material used in this study was a fully dense hot isostatically pressed material. The material was delivered by ABB Cerama and made of powder from UBE Industries doped with 2.5% yttria (size of specimen 50x4.5x3.5 mm^3). The bending strength in 4-point bending for this material is 695 MPa at room temperature.

For the laser treatment of the samples the usual setup (cf. fig.2) is used where an aperture is illuminated by the excimer laser beam and imaged down onto the workpiece. Due to the small ablation rates up to 200 pulses per site have to be applied to achieve sufficient ablation depth. Therefore the specimen are mounted on a xy-translation stage which gives control of the incremental movement between subsequent laser pulses. The feedrate and the laser repetition rate are adjusted according to the desired pulse number. For medium fluences ($\varepsilon \simeq 1$ - 10 J/cm^2) the typical reduction ratio is about ten to one which gives a beam dimension onto the workpiece up to 1 mm.

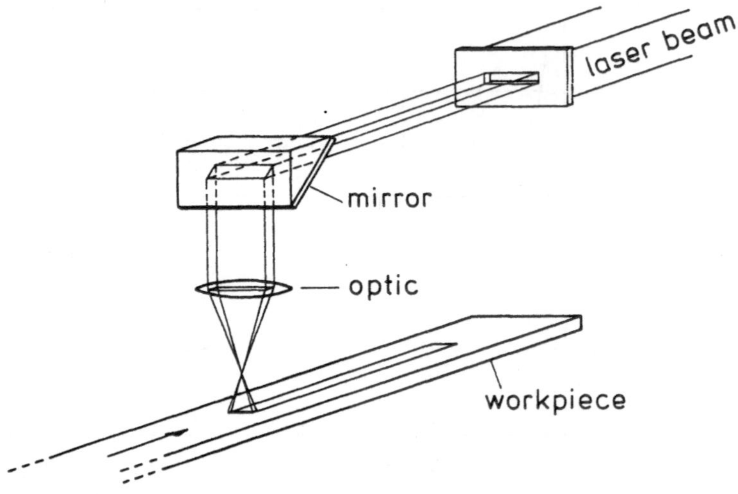

Figure 2. Experimental setup for the laser surface treatment.

For the variation of beam parameters as wavelength, fluence, repetition rate and pulse duration different laser sources are used. The influence of the pulse duration is investigated for XeCl laser radiation (λ = 308 nm) of 10 ns to 280 ns pulse lengths. The flexure strength in 4-point bending is used as an indication of the presence of subsurface microcrackes caused by the laser machining.

RESULTS

Ablation Rates

The application of ArF- and KrF-radiation leads to ablation rates up to 0.2 microns per pulse for fluences of about 10 J/cm^2 (see figure 3). For XeCl radiation higher ablation rates up to 1 micron per pulse (fluence > 12 J/cm^2) are feasible. This can be attributed to the smaller absorption coefficient for XeCl radiation which allows a higher optical penetration depth. With increasing fluence the ablation rate rises but saturates in case of ArF- and KrF laser radiation for fluences above $\varepsilon \simeq 10\ J/cm^2$. This behaviour is characteristic for non-oxide ceramics and can be observed also on BN, AlN and SiC [2]. The

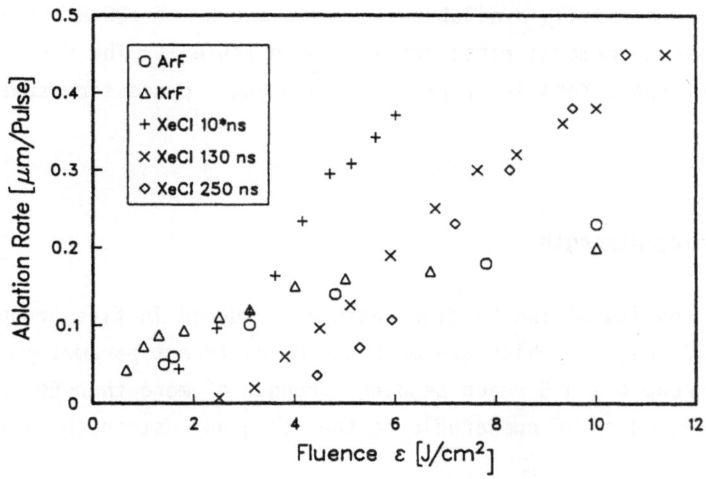

Figure 3. Ablation rates vs. fluence for different laser sources (cf. table 1)

strong absorption of the UV-radiation in the plasma plume reduces the absorbed power density at the workpiece surface and therefore the evapouration rate. The expanding hot vapour absorbs the laser radiation and leads to the formation of laser induced absorption waves. In the limiting case of laser induced detonation waves the workpiece is completly shielded [3]. High efficiencies of material removal (ablated volume/ pulse energy) can be obtained for fluences below $\varepsilon \simeq 10\ J/cm^2$ where the short pulse XeCl laser exhibits best results.

Processing Speed

The feasible processing speed depends on the efficiency of material removal and can be deduced from the slopes of the ablation rates in figure 3. Thus the maximum removal rate can be calculated as:

$$dV/dt \leqq h/\varepsilon \cdot \langle P \rangle \qquad (3)$$

V : ablated volume

h : ablation rate

ε : fluence

<P>: average power

For commercially available excimer lasers of 100 W average power the feasible removal rates are listed in table 1. The KrF laser and the short pulse XeCl laser provide the highest processing speed.

Bending Strength

The results of the bending tests are ordered in five groups (cf. table 2) of specimen which are machined by different parameters. The samples in group 4 and 5 reach bending strength of more than 600 MPa (see figure 4) which is comparable to the original strength of the material.

Table 1:
Feasible removal rates on silicon nitride for excimer lasers of 100 W average power

laser	wavelength [nm]	pulse duration [ns]	removal rate [10^{-4} cm^3/s]
ArF	193	20	3.2
KrF	248	35	6.5
XeCl	308	10	6.5
"	"	130	4.3
"	"	280	"

Increasing ablation depth shows no improvement in bending strength as expected by equation (2), on the contrary the strength is lowered (cf. group 2 and 3). As a critical point the border between the machined and unmachined area appears where fracture of samples in group 2 and 3 ocurr. Within the scatter of different samples the measured bending strength can not appointed to a certain beam parameter. Therefore the results obtained on small beam geometries are summarized in group 1.

Table 2:
Laser beam parameters for bending tests

group	laser	beam geometry [mm]	pulse length [ns]	Ablation depth [μm]	Machining length [mm]	Fluence [J/cm²]
1	ArF	0.04x0.2	22	25 - 55	5	1.5 - 40
	KrF	0.12x0.4	35	35 - 75	"	7.4
	XeCl	φ = 0.1	10	20 - 30	"	7 - 10
2	XeCl	φ=0.35-0.6	250	70 - 90	5*)	10
3	XeCl	φ = 0.6	250	33 - 50	5*)	11
4	KrF	0.9x2.6	35	35 - 40	30	4.7 - 5.8
5	XeCl	φ = 2.5	45	≃ 40	30	9.6

*) fracture on border of machined area

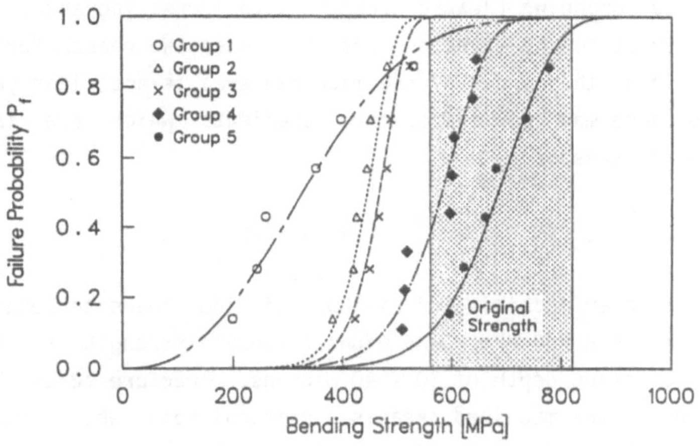

Figure 4. Failure probability vs. bending strength for different sets of laser beam parameters.

Table 3:
Mean value <σ> of bending strength and Weibull-parameters σ_0 and m according to figure 4

	number of samples	<σ> [Mpa]	σ_0 [MPa]	m
group 1	6	331	380	2.5
group 2	"	434	459	11.6
group 3	"	472	481	12.4
group 4	8	579	607	8.9
group 5	6	680	716	9.0

DISCUSSION

The amount of recondensed material is remarkable for the treatment by small beam dimensions. The layer of redeposited material exceeds more than 10 microns thickness (laser beam parameters as in table 2 group 1). The recondensation can be strongly reduced by enlarging the beam dimension (l ≈ 1 mm) and therefore changing the characteristics of the plasma plume from a half-sphere to a unidirectional flow perpendicular to the surface. The relativly low efficiency of material removal for the ArF-laser is due to the strong absorption of the radiation in the emerging vapour cloud. For pulse durations considerably longer than 10 ns the formation of laser induced absorption waves reduces the coupling of laser radiation to the workpiece surface.

The dropping flexure strength with increasing ablation depth is in contrast to the expected result. Also the reason for the reduced strength at the border of the machined area is not clear yet, one possible cause may be temperature gradients which are inducing high residual stresses.

CONCLUSIONS

Excimer laser machining of silicon nitride leads to bending strength of more than 600 MPa. The highest flexure strength is obtained for a total ablation depth of 30 - 40 microns. Fracture tests show that the border of the machined area is a critical point where fracture often starts. Pulse durations in the 10 ns-range provide best efficiencies for material removal and beam dimensions of about 1 mm reduce distinctly the amount of recondensed material onto the surface of the workpiece.

REFERENCES

[1] M.B. Bever, Encyclopedia of Materials Science and Engineering Vol.1 595ff, Pergamon Press 1986

[2] K.J. Schmatjko, G. Endres, U. Schmitt, Proc. SPIE Vol. **957** 1988

[3] R. Poprawe, M. Wehner, G. Brown, G. Herziger, Plasma-Spectra in Materials Processing by Excimer Lasers, Proc. SPIE Vol. 801 1987

THE STRENGTH AND FRACTURE BEHAVIOUR OF ENGINEERING CERAMICS - INFLUENCE OF MACHINING AND RESIDUAL STRESSES

Thomas Hollstein, Wulf Pfeiffer
Fraunhofer-Institut für Werkstoffmechanik,
Wöhlerstr. 11, D 7800 Freiburg,

ABSTRACT

In order to optimize efficiency of machining, surface properties and strength of machined parts, the complex relationships between machining parameters, elementary mechanisms of material removal and strength-controlling surface effects have to be understood. The results of the investigations allow to separate the effects of damage processes and macroscopic residual stresses on the strength of ceramics, and therefore to optimize the machining parameters. Cases of lapping and grinding are shown, where on one hand surface damage (Al_2O_3, SiSiC, SiC and Si_3N_4), and on the other hand residual stresses (ZrO_2) are the dominating effects on the bending strength.

INTRODUCTION

The quality of components of high-performance ceramics is mainly characterized through the steps of the production process and of machining. The high requirements of such components concerning precision of size and shape and of surface quality must be fulfilled in order to ensure their proper way of operation. For an optimization of both, effciency of machining and strength of machined parts, the complex relationships between machining parameters, elementary mechanism of material removal and strength-controlling surface effects have to be understood.

The aim of the investigations is to clarify how near-surface regions of ceramics are influenced by machining especially at high material removal rates. From these results material-dependent relationships between maching parameters and strengths of machined ceramics will be derived.

MATERIALS AND EXPERIMENTAL PROCEDURE

Samples of sintered pure alumina, partially stabilized zirconia (Mg-PSZ), silicon-infiltrated silicon carbide, sintered silicon carbide and sintered silicon nitride with the dimensions 50 x 50 x 3 mm were investigated in lapped and ground conditions.

Lapping was performed at a constant specific normal force $F_n'' = 0.04$ MPa and a disc speed $n_L = 104\ min^{-1}$ using B_4C abrasives (30Vol%) with avarage grain sizes d_K from 9 μm up to 100 μm. Grinding (up-cut) was performed at a constant peripheral wheel speed of 25 mm/s using diamond grinding wheels D151 C100. Specific removal rate Q'_w, infeed a_e and tangential feed rate v_{ft} were varied from conventional grinding conditions (small infeed, high feed rate) to creep-feed grinding conditions (high infeed, low feed rate).

Bending strengths of the machined samples were determined using the concentric-ring test with a loading ring diameter of 8 mm and a supporting ring diameter of 40 mm. According to the linear theory of plate bending, a constant maximum biaxial stress state prevails in the surface within the loading ring. This has the advantage that effects of unproperly prepared edges of the samples can be neglected. Due to the symmetry of the stress state the "weakest" direction of the surface is expected to lead to failure. Nine to ten samples with identically machined surfaces were broken at a loading rate of 2500 N/s. The statistical distribution of the measured bending strength values was evaluated by means of the two-parameter Weibull distribution with the parameters characteristic strength σ_o and Weibull modulus m. The surface condition of the machined samples was characterized by scanning-electron microscopy and optical surface roughness measurements.

X-ray diffraction techniques were used to determine the microscopic and macroscopic residual stress states of the machined samples: The full-width-at-half-maximum (FWHM) value of the diffraction lines was taken as a measure of the microscopic strains or stresses within or between the single grains of the ceramics due to plastic deformation (dislocations etc.) in near-surface layers. The macroscopic residual stresses were determined using the peak-shift of the diffraction lines ($\sin^2\psi$-technique [1]). It should be pointed out that the evaluated residual micro and macro-stresses are mean values integrated over the depth of the penetration of the used X-rays weighted by an exponential function due to absorption. The "effective" penetration depths, from which 67% of the detected X-ray intensity originate, are indicated in the following figures. The amount of the tetragonal to monoclinic phase transformation and the depth of the transformation zone was derived from quantitative X-ray phase analyses using two different wave lengths and the methods suggested by [2].

RESULTS AND DISCUSSION

Lapping of Alumina and Zirconia

The main mechanism of material removal by lapping is an outbreaking of small surface particles due to microcracks introduced by the grains of the abrasive rolling between the specimen and the lapping disc.

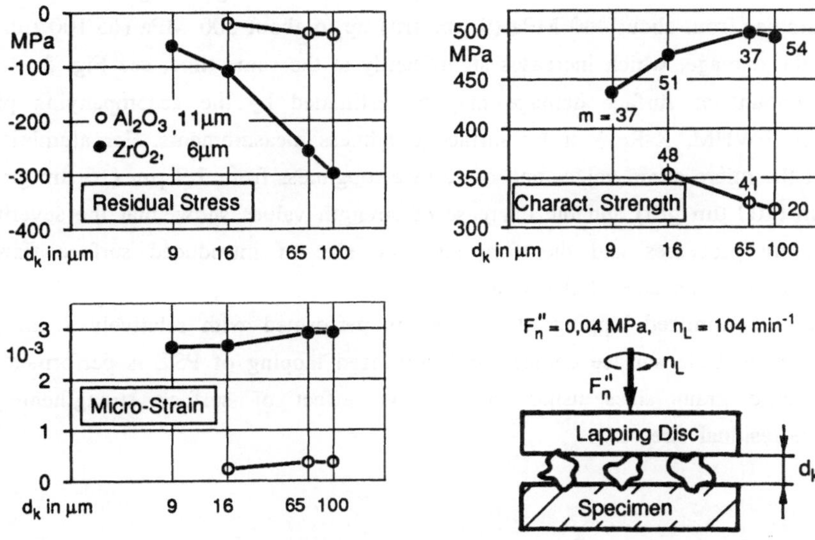

Figure 1: Lapping of alumina (∘) and zirconia (•)

The effect of different grain sizes on these mechanisms and on the resulting surface conditions and bending strengths of lapped alumina and zirconia samples is shown in Fig. 1: The micro-strain values derived from X-ray measurements with penetration depths of 11 μm and 6 μm, respectively, indicate that with increasing grain size the amount of plastic deformation in near surface regions increases. (Deeper regions inspected for aluminia with X-rays of 33 μm penetration depth are nearly unaffected [3].) The reason is that - keeping the normal force F_n'' on the lapping disc constant - the use of larger grain sizes reduces the number of cutting edges (area between grains and specimen) and therefore leads to higher local loadings.

The introduced plastic deformation leads for alumina to only small compressive residual surface stresses up to 40 MPa (d_K = 100 μm).

For zirconia compressive residual stresses up to 300 MPa are achieved, which is due to the strain-induced tetragonal-monoclinic phase transformation [4]. The transformation rate and depth depend on the local forces during machining and therefore on the grain size of the abrasive (keeping the normal force on the disc constant): Quantitative X-ray phase analyses [3] show an increase of the total monoclinic content from 49% (9 μm grit) to 53% (100 μm grit) with an increase of the transformation depth from 1.9 μm (9 μm grit) to 4.7 μm (100 μm grit). As the phase transformation is accompanied by a volume increase of 3 % - 5 % [4], compressive residual stresses are created in the transformed surface layer. The residual stress values amount from 60 MPa (9 μm grit) up to 300 MPa (100 μm grit) and are obviously able to improve the bending strength of the lapped zirconia samples from about 440 MPa (9 μm grit) up to about 500 MPa (65/100 μm grit) inspite of the damage, which increases significantly at the same time, see Fig. 2.

The amount of surface damage may be estimated by the accompanying plastic deformation (FWHM, X-Ray) or by surface roughness measurements. For aluminia, the increase of the micro-strain values and of surface roughness from 7.9 μm (16 μm grit) up to 11.5 μm (100 μm grit) and the decrease of strength values show, that the severity of brittle fracture processes and therefore also the size of introduced surface flaws is controlled by the grain size of the abrasive.

Therefore optimized lapping of alumina is performed with relatively small grain sizes resulting in little surface damage and optimized lapping of PSZ is performed with relatively large grain sizes using the positive effect of surface strengthening by compressive residual stresses.

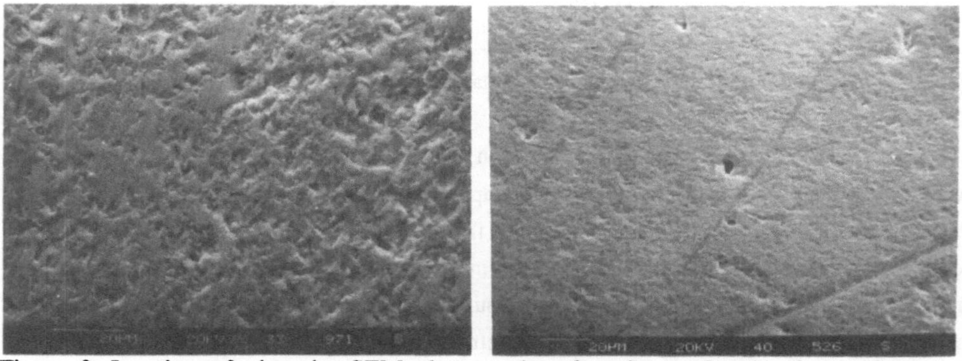

Figure 2: Lapping of zirconia, SEM photographs of surfaces after lapping with coarse grains (d_k = 100 μm, left) and fine grains (d_k = 9 μm, right)

Grinding of Different High-Performance Ceramics

The results show the influence of various combinations of infeed and feed rate on the surface conditions of up-cut ground alumina. The micro-strain values indicate that with decreasing infeed and increasing tangential feed rate increasing amounts of plastic deformation are produced especially in the near-surface regions (X-ray-penetration depth 11 μm). These micro-strain values exceed those obtained after lapping (see Fig. 1).

Due to the increase of plastic deformation a change from small compressive residual surface stresses at creep-feed grinding conditions (high infeed, a_e, low feed rate, v_{ft}) to somewhat higher compressive stresses up to 55 MPa at conventional grinding conditions (small infeed, high feed rate) is obtained. The amount of residual stresses in the transverse direction is slightly higher than parallel to the grinding direction. In deeper regions (X-ray penetration depth 33 μm) tensile residual stresses up to 10 MPa were evaluated after grinding at moderate feed rates.

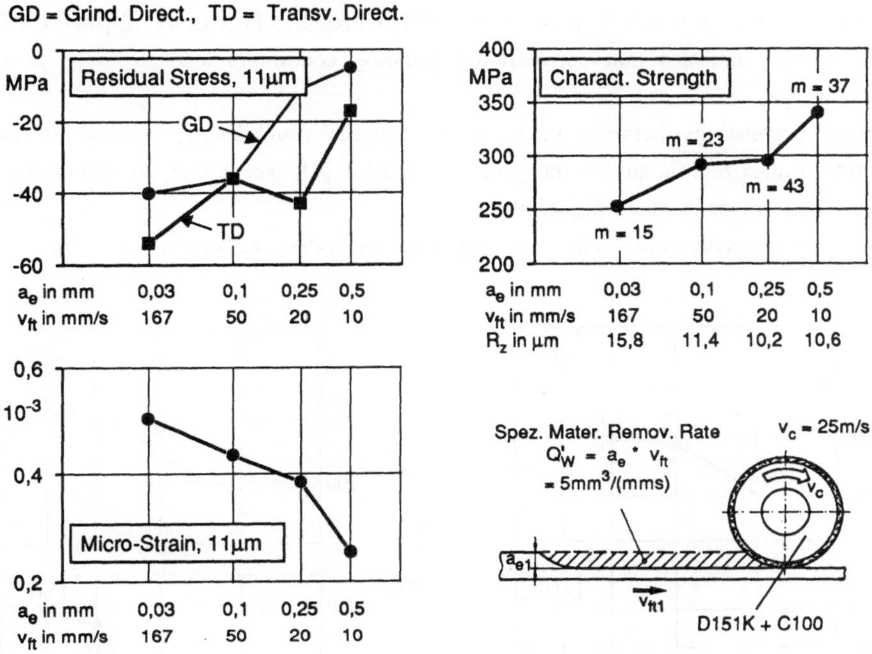

Figure 3: Grinding of alumina

The relationship between grinding parameters, plastic deformation and residual stresses may be explained by the following simplified consideration: Grinding at high infeeds (low feed rates) means a larger contact length between grinding wheel and specimen and therefore more individual cutting edges acting simultaneously than at smaller infeeds (high feed rates). This results in smaller chip thicknesses and also smaller forces acting at the individual cutting edges of the wheel [5]. Plastic deformation, compressive residual stresses and the possibility of surface damage are therefore reduced.

On the other hand a larger contact length causes worse cooling conditions in the contact zone which may explain the obtained small tensile residual stresses in sub-surface regions. The effect of these elementary mechanisms on the characteristic bending strength data is also shown in Fig. 3.

Obviously, damage processes, especially occurring at conventional grinding conditions, are of dominant influence on the strength of the ground samples while the compressive residual surface stresses are not able to compensate the influence of surface damage. Thus the bending strength is reduced from 340 MPa (creep-feed grinding) to 250 MPa (conventional grinding).

Again a good correlation seems to exist between the micro-strain values and the strength-controlling surface flaws introduced by machining. Also the surface roughness, which increases from 9,1 µm/10,6 µm (GD/TD) at high infeeds to 11,5 µm/15,8 µm (GD/TD) at small infeeds, indicates that conventional grinding conditions lead to more severe damage.

Similar correlations between micro-strains, surface roughnesses, residual stresses and strength values are found for the grinding of PSZ [6] except that the compressive residual stresses are on a much higher level.

The same arguments are valid for grinding of other ceramics, see Fig. 4.

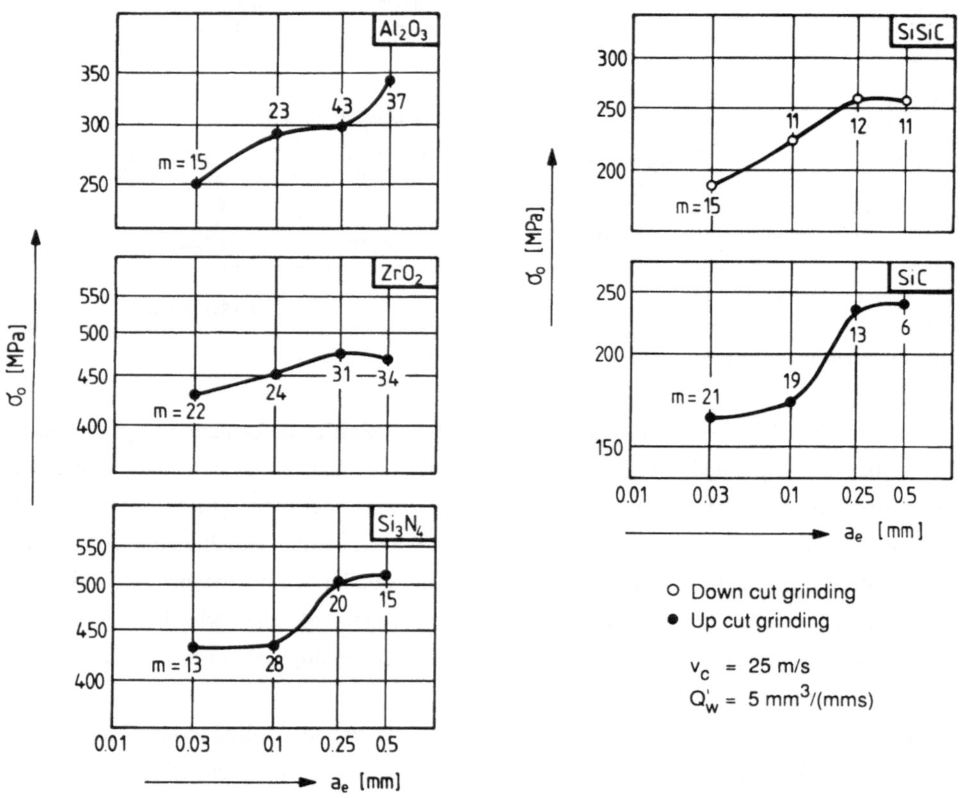

Figure 4: Grinding of different ceramics

A significant increase is found for the ceramics investigated comparing creep-feed grinding with conventional grinding. If for creep-feed grinding with an infeed a_e = 0.5 mm the tangential feed rate v_{ft} and therewith the specific material removal rate Q'_w is increased, the characteristic strength values of all these ceramics decrease again significantly, see Fig. 5. The same arguments as previously hold: The chip thicknesses and the forces acting at the individual cutting edges of the grinding wheel increase, which results in an increase of damage and a decrease of strength values.

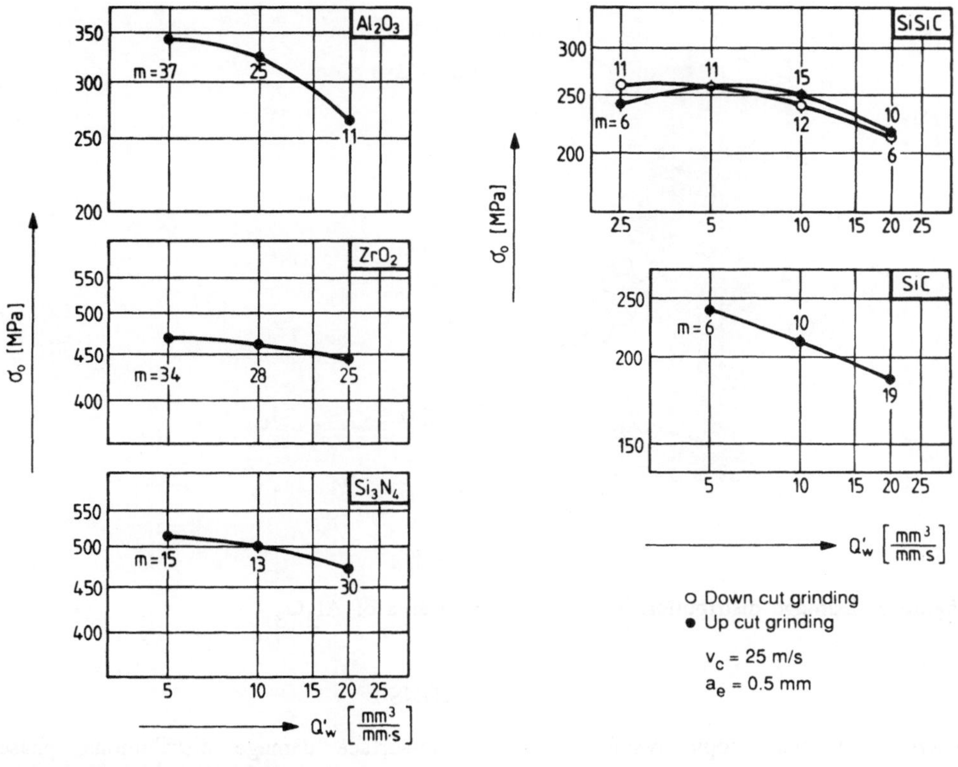

Figure 5: Grinding of different ceramics

Surface Damage

The extent and the real depth distribution of machining introduced plastic deformations have been determined quantitatively by graceing incident diffraction techniques evaluating the FWHM values of the diffraction lines using different penetration depths. Due to the high brittleness of ceramics, severe plastic deformation is accompanied by microcracking. The microstrain distributions are therefore able to inform about the extent and depth of damage.

As an example, the damage (microstrain) distribution in the surface area of differently machined Al_2O_3 is shown in Fig. 6 for conventional and creep-feed grinding and for lapping with small grains (d_k = 16 µm) and large grains (d = 100 µm). The depths of sub-surface damage correspond to the bending strengths of the machined samples.

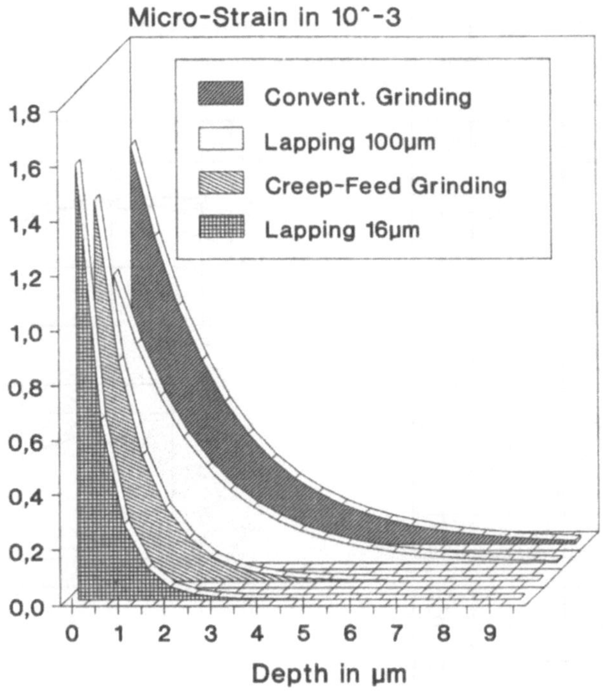

Figure 6: Damage distribution in the sub-surface area of Al_2O_3

CONCLUSIONS

Micro- and macroscopic residual stresses, sub-surface damage distributions, phase transformations and strength of machined ceramics were examined using X-ray diffraction techniques and concentric-ring bending tests. The results allow to separate the effects of damage processes and macroscopic residual stresses on the strength of ceramics and therefore to optimize the machining parameters. In the case of lapping of alumina and grinding of alumina, PSZ, SiSiC, SiC and Si_3N_4 surface damage is the dominating effect on strength, whereas transformation-induced compressive surface stresses prevail against damage effects using severe lapping conditions for PSZ.

ACKNOWLEDGEMENTS

Machining of the ceramic materials was performed by J. Wemhöner and M. Popp, Fraunhofer-Institut für Produktionstechnologie. Parts of these investigations have been sponsored by the Bundesministerium für Forschung und Technologie and the companies Friedrichfeld GmbH, Elektroschmelzwerk Kempten GmbH, Sigri GmbH, Hutschenreuther AG and Siemens AG.

REFERENCES

[1] Müller, P., Macherauch E.: Das $\sin^2\psi$-Verfahren der röntgenographischen Spannungsmessung. Z. ang. Phys. 13 (1961) 305-312.

[2] Porter, D.L., Heuer, A.H.: Microstructural Development in MgO-Partially Stabilized Zirconia (Mg-PSZ). J. Am. Ceram. Soc. 62 [5-6] (1979) 298-305.

[3] Pfeiffer, W., Berweiler, W., Hollstein, T., Optimierung ausgewählter keramischer Hochleistungswerkstoffe hinsichtlich Festigkeit, Ermüdung und Oberflächenzustand, IWM-Report W 10/90, Freiburg, 1990.

[4] Claussen, N.: Advances in Microstructural Design of Zirconia-toughened Ceramics (ZTC). Claussen, N., et al, Am. Ceram. Soc., Columbus (Ohio) (1984).

[5] König, W., Wemhöner, J.: Optimized Grinding of SiSiC. Cer. Bul. 68 [3] (1989) 545-548.

[6] Pfeiffer, W.: Influence of Machining-induced Residual Stresses on the Strength of Ceramics. Proc. 9th Int. Conf. on Experimental Mechanics, Vol. 3, pp. 1237-1245.

Residual Stress in Si_3N_4 and SiC after Grinding and Cooling from High Temperature

M. Odén and T. Ericsson
Linköping University
Department of Mechanical Engineering
Division of Engineering Materials
S-581 83 Linköping, Sweden

Abstract

Residual stresses have been measured on Si_3N_4 and SiC specimens which have been 4pt-bend tested earlier in a project sponsored by International Energy Agency. Large scatter in surface residual stress was found and show no correlation with bend strength. Heat treatment between 700 °C and 1200 °C followed by water quenching creates large compressive stresses in the surface. Air cooling gave marked relaxation after treatment in the high temperature range.

Introduction

Advanced ceramics such as Si_3N_4 and SiC are brittle and residual stresses caused by machining and heat treatment might influence the strength of components. Here residual stresses have been measured by X-ray diffraction on longitudinally ground specimens as well as annealed specimens made of HIP'd Si_3N_4 from ABB Cerama, Sweden and HIP'd SiC from Elektroschmelzwerk Kempten, ESK, Germany. The specimens have been 4pt-bend tested within a Sweden-Germany-USA cooperative program organised by Int. Energy Agency (1) (2) (3).

Experimental details

The SiC was of the hexagonal a-6H with about 5 % of 4H polytype SiC. It contained about 0.2 wt% free C, 0.2 % free Si, 0.05 wt% Al, 0.2 % O. The grain size was about 2.5 mm ± 1 mm. The Si_3N_4 was of the hexagonal b-type with about 4 vol% Si_2ON_2 and 2 vol% $Y_2Si_2O_7$. It contained about 0.5 wt% C, 2.5 wt% O and 2 wt% Y. The grain size was about 1 mm ± 0.5 mm. Both materials contained approximately 0.5 vol% porosity.

Flexure bar specimens with dimensions 3.5x4.5x50 mm were ground longitudinally and had a surface roughness of about 1 mm. The inner and outer spans of the four-point bend fixture were 20 and 40 mm respectively. The mechanical testing had been carried out before in the IEA program. The two halves of each specimen were used in this study.

Residual stress measurements in the longitudinal direction by X-ray diffraction were made with a Ω-goniometer and a position sensitive detector using CrK_a-radiation. The peak positions were determined with the parabola method above the 80% peak level and at the same time the peak widths were recorded. The crystal planes studied were (4,1,1) for Si_3N_4 with $2\theta \approx 125°$, and (1,1,6) for SiC with $2\theta \approx 121°$. In order to determine the residual stresses in a zone as close to the surface as possible and to avoid the problems a steep stress gradient causes, only high ψ-angels were used (-42.5°, -40°, -37.5°, -32.5°, -27.5°, 30°, 35°, 40°). It means that 63 % of the diffracted radiation origin from a depth less than 40 mm in Si_3N_4 and 35 mm in SiC. Elastic constants were taken from literature (1, 2) for Si_3N_4 E=310 GPa, ν=0.26 and for SiC E=408 GPa and ν=0.30.

The so-called $\sin^2\psi$-method was used for the residual stress evaluation. The errors indicated are derived from a regression analysis.

The ground specimens were after residual stress measurements slowly heated in air to different temperatures and held there for 30 min and cooled in water or in air to 20 °C. The residual stress was then measured again.

Results and discussion

The residual stresses in the as-ground surfaces are compressive which has been found by other workers (4) (5) (6).

In table 1 and 2 and fig. 1 and 2 the 4pt-bend strength and residual stress are shown for Si_3N_4 and SiC respectively. It is evident that the residual stress varies considerably between different samples indicating grinding differences.

TABLE 1

4pt-bend fracture stress (2), crack start location and residual stress for Si_3N_4. (3)

Spec No	4pt- bend fracture stress MPa	flawposition	flawtype	starting point depth µm	residual stress MPa
1198	532	loadpin	agglomerate	0	-184±30
1014	692	loadpin	pore	0	-337±26
643	644	within span	ND	0	-173±32
706	644	within span	ND	ND	-195±23
806	678	within span	ND	0	-89±37
820	638	within span	porosity	0	-186±31
1168	486	within span	agglomerate	0	-101±31
398	635	loadpin	pore	50	-153±38
855	641	loadpin	porosity	ND	-203±24
1193	431	loadpin	agglomerate	40	-128±40
117	523	loadpin	porosity	0	-224±30
457	711	within span	defect	0	-87±21

ND = not determined

TABLE 2

4p-bend fracture stress (2), crackstart location and residual stress for SiC(3).

Spec no	4pt-bend fracture stress MPa	flawposition	flawtype	starting point depth mm	residual stress MPa
760	475	within span	defect	0	-349±39
1084	301	within span	inclusion	50	-316±65
797	496	ND	ND	ND	-228±32
715	483	within span	pore	160	-52±44
986	287	within span	pore	0	-389±69
707	248	within span	defect	0	-190±29

ND = not determined

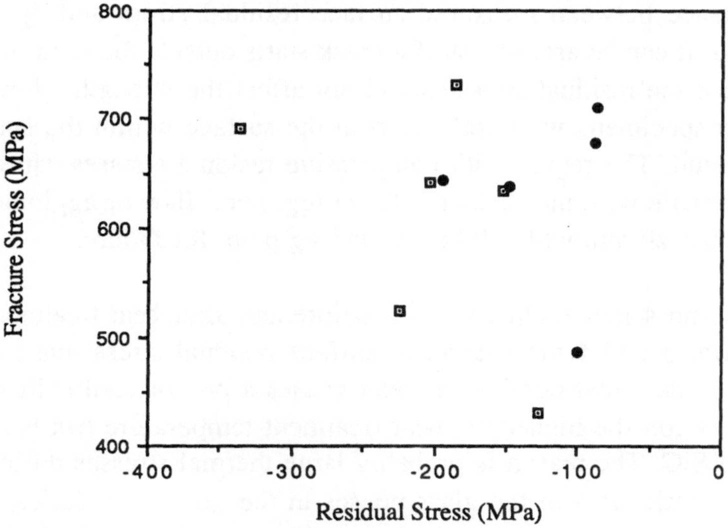

Fig. 1. Fracture stress in 4pt-bend testing versus the measured residual stress in Si_3N_4. The five samples with fracture origin at the surface within the span are marked with a filled cirkle

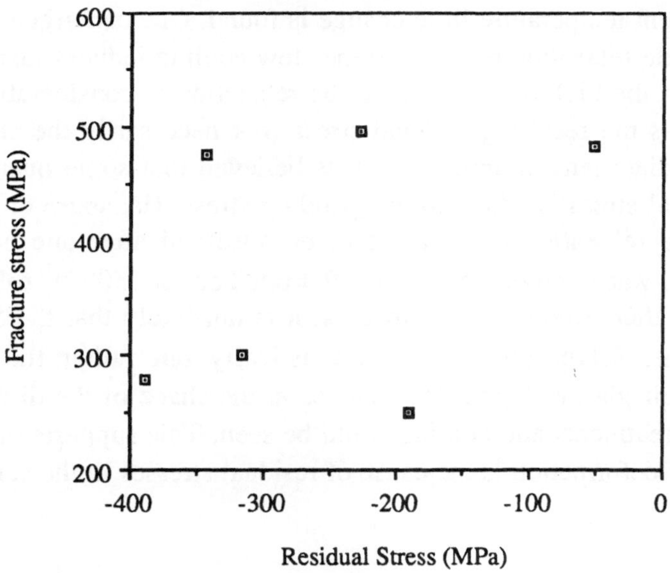

Fig. 2. Fracture stress in 4pt-bend testing versus the measured residual stress in SiC.

A similar observation has been reported by Hanabusa et al. (4). It is also clear that no dependence between measured surface residual stress and 4pt-bend strength is found. It can be argued that if a crack starts outside the span or deep below the surface the residual stress should not affect the strength. However, even when only specimens with crack start at the surface within the span no correlation is found. The region with compressive residual stresses caused by grinding is very shallow, it means that a defect (eg. pore, flaw or agglomerate) located deeper than 20 mm will still be the starting point for failure.

In table 3 and 4 the residual stresses before and after heat treatment are shown and in fig. 3 and 4 the change in surface residual stress due to heat treatment. It is evident that cooling in water causes a pronounced shift in the compressive direction the higher the heat treatment temperature has been for both Si_3N_4 and SiC. The reason is probably large thermal stresses during the cooling causes plastic flow in the glass phases in the grain boundaries. Thus compressive stresses are created as is usual after quenching a body which undergoes no phase transformation.

Cooling in air gives a more complicated pattern. At the lowest and highest heat treatment temperature little change is found. One can argue that at low temperature little relaxation occurs and the slow cooling induces just small thermal stresses. At the highest temperature the relaxation is considerable and the important fact is the resulting residual stress (not necessarily the change) (4). In the intermediate temperature range it is believed that some quenching compressive residual stress is added to the grinding stress. Hanabusa et al. (4) found for Si_3N_4 a relaxation that started to be observed after one hour in argon at 900 °C and was relatively complete after one hour at 1500 °C followed by slow cooling.In their study, as well in ours, it is not likely that the Si_3N_4 and SiC grains themselves are deformed plastically but rather the grain boundaries with their glassy phases. No changes in the shape of the diffracted peaks due to heat treatments and cooling could be seen. This supports the idea that grain boundary deformation is the cause of residual stresses in the surface.

TABLE 3

Residual stress for ground Si_3N_4 specimens before and after heat treatment, HT.

Spec no	Before HT MPa	After HT MPa	Change due to H.T. MPa	HT temp °C	Cooling medium
198	-184±30	-231±36	-47±45	612	Water
1014	-337±26	-369±53	-32±60	606	
643	-173±32	-267±26	-94±45	708	
643	-173±32	-283±51	-110±60	710	
706	-195±23	-412±46	-217±55	800	
1014	-337±26	-453±59	-116±60	817	
806	-89±37	-349±79	-260±100	910	
820	-186±31	-341±41	-155±50	908	
1168	-101±31	-362±31	-261±45	1000	
1168	-101±31	-395±36	-294±45	1000	
1193		cracked		1100	
398	-153±38	-192±12	-39±40	700	Air
398	-153±38	-149±23	+4±45	700	
855	-203±24	-260±19	-57±35	800	
855	-203±24	-278±56	-75±60	800	
806	-89±37	-163±38	-74±50	910	
1193	-128±40	-356±47	-228±55	910	
117	-224±30	-355±49	-131±60	1000	
1168	-101±31	-430±58	-329±60	1005	
820	-149±14	-342±44	-193±55	1100	
117	-224±30	-304±53	-80±65	1100	
457	-87±21	-124±19	-37±30	1200	
457	-87±21	-48±26	+39±30	1200	

TABLE 4

Residual stress for ground SiC specimens before and after heat treatment, HT.

Spec no	Before HT MPa	After HT MPa	Change due to H.T. MPa	HT temp °C	Cooling medium
760	-349±39	-372±44	-23±50	606	Water
760	-349±39	-349±57	0±55	606	
1084	-316±65	-402±60	-86±90	707	
1084	-316±65	-441±37	-125±85	707	
797	-228±32	-433±62	-205±75	817	
797	-228±32	-462±29	-234±45	817	
715	-52±44	-472±44	-420±60	908	
798		cracked		900	
760	-349±39	-389±56	-46±65	800	Air
797	-228±32	-248±34	-20±45	810	
1084	-316±65	-409±54	-93±70	909	
715	-52±44	-394±39	-342±60	915	
986	-389±69	-457±56	-68±75	1000	
986	-389±69	-429±49	-40±75	1100	
986	-389±69	-234±28	+155±75	1100	
707	-190±29	-222±27	-32±45	1100	
707	-190±29	-101±30	+89±45	1200	

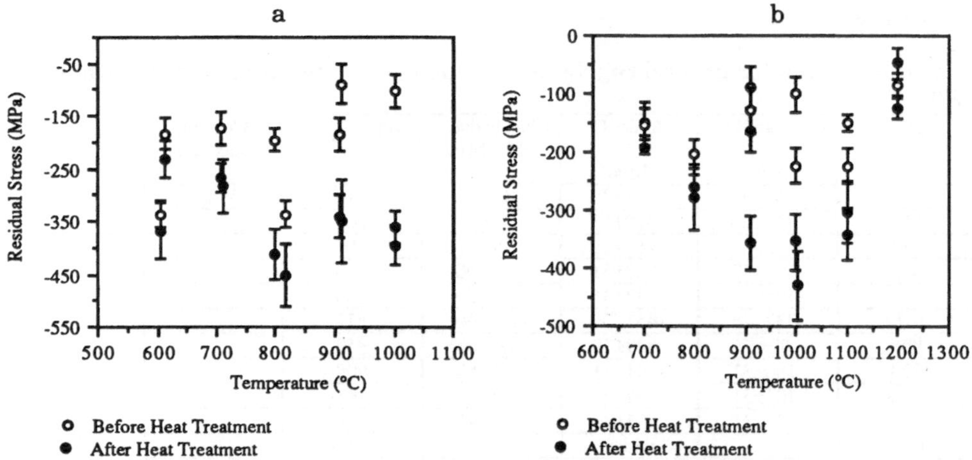

Fig. 3 a,b. Residual stress in Si_3N_4 vs. heating temperature. In a the cooling media is water and in b is it air. (The stresses from before heat treatment are shown as a reference)

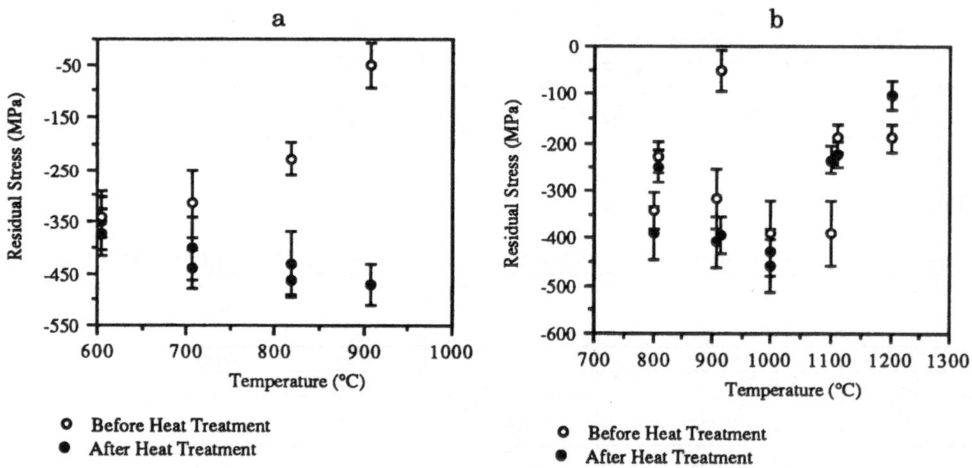

Fig. 4 a,b. Residual stress in SiC vs. heating temperature. In a. the cooling media is water and in b. is it air. (The stresses from before heat treatment are shown as a reference)

References

1. Characterization of sintered silicon nitride and silicon carbide structural ceramics. IEA. Prepared by Kemanord Industrikemi, Oct. 1989.

2. Statistical analysis of flexure strength data. IEA. Prepared by University of Karlsruhe, June 1989.

3. Fractography analysis of silicon nitride and silicon carbide structural ceramics. IEA. Prepared by Oak Ridge National Laboratory. June 1989.

4. T. Hanabusa, H. Fujiwara and Y. Fujimoto: "X-ray stress measurement of Si_3N_4 ceramics" in ICRS2, G. Beck, S. Denis and A. Simon (eds.), Elsevier Applied Science, London, 1989, p. 310-315.

5. C. Burman and T. Ericsson: "Residual stress measurement in advanced ceramics", ibid, p. 368-373.

6. B Eigenmann, B. Scholtes and E. Macherauch: "Determination of residual stresses in ceramics and ceramic-metal composites by X-ray diffraction methods", Mat. Sci. Eng. A118 (1989) 1-17.

Acknowledgement: This project has been sponsored by the Swedish National Board for Technical Development.

GRINDING ADVANCED CERAMICS WITH CRUSHABLE DIAMOND WHEELS

M. BURSTRÖM AND J. M. BARNARD
IVF - The Swedish Institute of Production Engineering Research,
Luleå, Sweden
NORTON Co., Welwyn Garden City, UK

ABSTRACT

Profile grinding of Hot Isostatically Pressed alumina and Silicon Nitride ceramics using Vitrified Crushable diamond wheels in the creep feed mode is discussed. The influence of feed rate, cutting speed and diamond abrasive size on profile shape, surface finish and material removal rate were studied. The biaxial strength of the ground ceramic material was also measured.

Grinding wheels with the finest diamond abrasive particles produced the best surface finish (0.15 μm Ra).

When Silicon Nitride was ground at high material removal rates the coarsest diamond abrasive was found to give a lower biaxial bend strength (400 MPa) than that produced using the finest diamond abrasive wheel (960 MPa). In the case of alumina the profile deviation at constant material removal rate was found to be lower using the finest diamond abrasive.

INTRODUCTION

The increased use of advanced ceramic materials is increasing the demand for suitable and economical final shaping techniques. For the production of profiled work pieces at high stock removal rates creep feed grinding involving a deep infeed associated with a low table speed is an applicable technique. With this method the full depth of the form can often be produced in a single pass exhibiting high grinding efficiency. The purpose of this paper is to describe how crushable vitrified diamond wheels have been used to profile grind Alumina, Silicon Nitride and Sialon.

Biaxial Flexure Strength and surface finish as well as Profile Deviation have been studied during the variation of cutting speed and feed rate.

CREEP FEED GRINDING WITH CRUSHABLE WHEELS

With the creep feed grinding method the full depth of the form can be produced in a single pass. The machine must have high rigidity both for the creep feed grinding and the crushformning. Because of the large area of contact between the wheel and workpiece, any such wheel must have a friable bond. With improved crushing technique metal and vitrified bonded wheels containing superabrasives in depth have given outstanding savings when grinding tungsten carbide and hardened high speed steel.

The introduction of the Crushable superabrasive wheel has gone a long way to restore to the user the ability to reform his wheel and also to have the benefits associated with superabrasives. Of course, several important problems had first to be solved. The superabrasive bonding system had to be adequately strong to give good abrasive retention during grinding, yet be sufficiently friable to permit forming by crushing. In addition means had to be found, especially for those wheels of very low porosity, of keeping the faces clear of crushing debris during the forming operation.

Both metal and vitrified bonds are becoming increasingly applied to crushable grinding operations. Of the two, evidence is beginning to suggest that the latter has the edge in performance being more easily crushed and longer lasting between crushings.

The creep feed grinding using crushformed vitrified bonded wheels will now also be introduced to the grinding of advanced ceramics.

CRUSHING PROCESS

Basic Principle of Crushing

Let us first consider the theory on which the process is based. If we have an aluminium oxide wheel on a grinding machine which is producing burn on the workpiece then its bond is clearly too hard. However, if we now reduce the wheel speed say from 3,600 r.p.m. then the wheel will act softer and most likely not produce burning.

Now if we further reduce the wheel speed down to 100 r.p.m. then the bond acts extremely soft. Superabrasive wheel manufacturers have considered this and produced wheels which when rotated slowly against a crushing roll, i.e. even slower than when crushing conventional abrasive wheels, will break down microscopically to form the shape on the roll. It does not matter whether the crushing roll is driving the wheel or the wheel the crushing roll, what is important is that the speed or rotation should be slow and constant. Figure 1 shows the basic principle of the wheel running against a crushing roll.

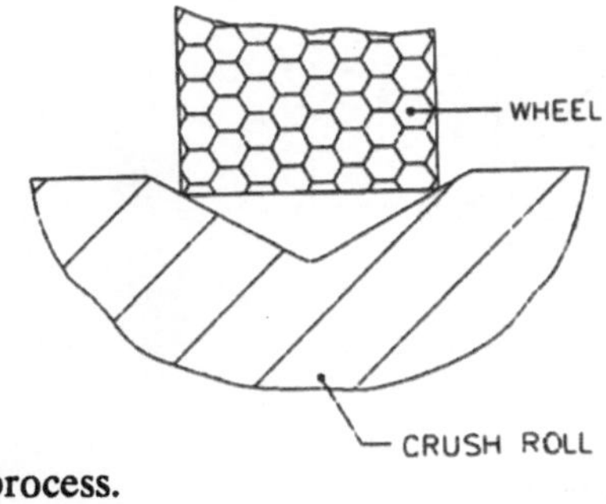

Figure 1. The crushing process.

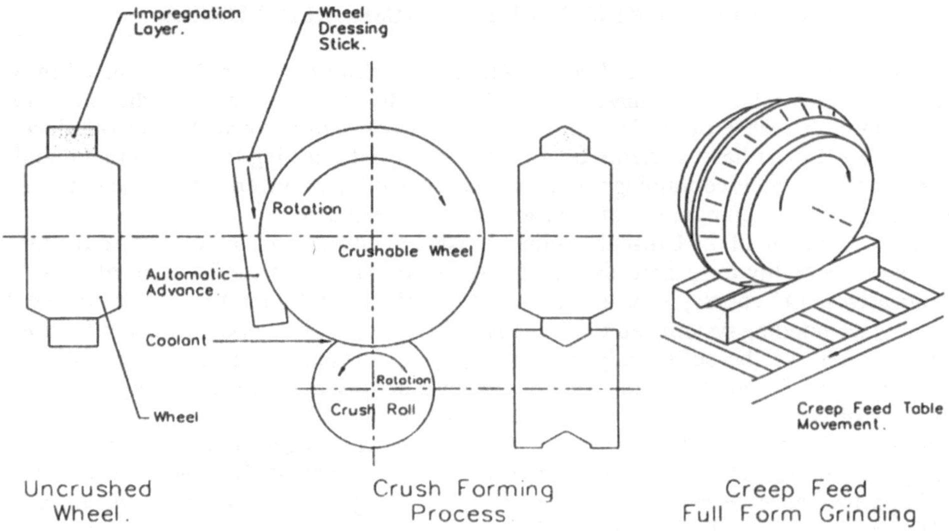

Figure 2. Schematical view of the crush forming and form grinding.

Crush roll

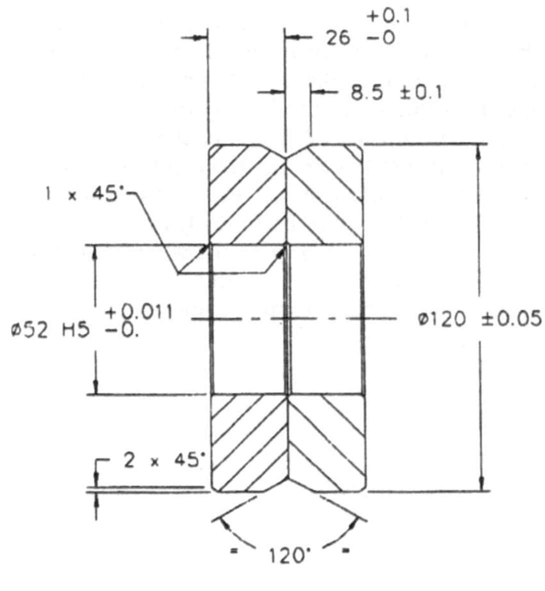

Figure 3 shows the used crushing roll design. The used V-shape is a simple shape to crush. The crush roll, which in this case is split into two parts to create the smallest possible radus. The crush roll was made of high speed steel ASP 23 hardened to 65 HR_c.

Figure 3. The crushing roll design.

An abrasive stick may be forced against the wheel during crushing to remove debris as illustrated in figure 2. Recently a newly developed brush process has removed the need for a stick when crushing metal bonds and reduced crushing times by half. During crushing, the hardened steel wire is forced into the wheel, removing bond material by breaking the already fractured bond, so keeping the wheel extremely clean and enabling finer radii to be crushed than is possible using the abrasive stick. However, when crushing vitrified diamond crushable wheels neither an aluminium oxide stick or a hardened steel wire brush is required. During crushing, coolant must be applied and it is also suggested that a higher pressure jet be directed against the wheel to clean away any debris before the wheel face re-enters the crushing roll.

GRINDING CONDITIONS

Grinding wheel
For the grinding of the three selected test materials vitrified bond diamond wheels with three different grit sizes were chosen. The wheels had the following Norton designation:

SD20/30 L4 VXL 1488
SD30/40 L4 VXL 1488
SD46 L4 VXL 1488

Dimensions:	OD 254, T = 15 mm, X = 12,7 mm
Diamond type:	SD
Concentration:	100
Bond:	vitrified VXL 1488
Grit size:	46, 30/40 & 20/30 micron

With regard to concentration the 100 concentration was chosen, which is 4.4 ct/cc. This vitrified bond is a naturally brittle and therefore, highly suited to the crushing operation.

Machine
The test grinding was accomplished in a Wendt WLM 20 crush forming grinding machine located at Sandvik Hard Materials.

Grinding parameters
Cutting speed: 20, 25 and 30 m/s
Feed rate: 50 mm/min, for the 30 m/s cutting speed also 30 and 70 mm/min.
Down feed (when formgrinding): 3,7 mm
Material removal rate (Q'_w): 2 $mm^3/mm \cdot s$ (calculated at profile middle when formgrinding).
Coolant: mineral oil with synthetic additive.

Material
The test materials used in this study were Alumina, Silicon Nitride and Sialon. The Alumina was a fully dense material made by sintering and post-HIP of the high purity powder AKP-30 from Sumitomo Chemical Ltd.

The Silicon Nitride was a Hot Isostatically Pressed quality from ABB Cerama made of powder from UBE Industries doped with 2.5 % yttria.

The Sialon was a α-β quality from Sandvik Hard Materials with the designation SY70.

Biaxial flexure test

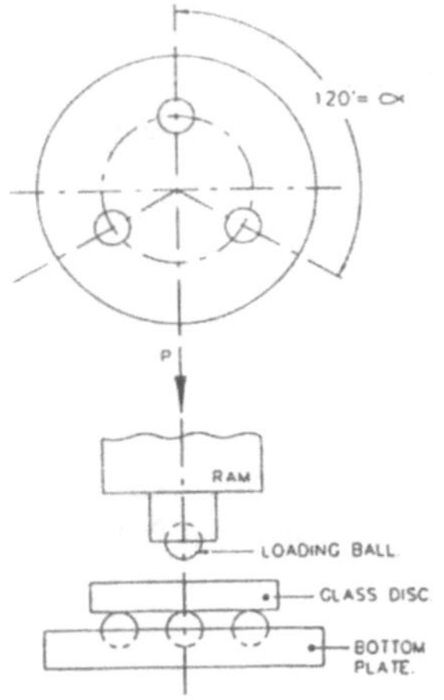

The biaxial flexure strength was measured in a ball on 3-ball test set up. In this test, a disc specimen is supported on three 4 mm diameter balls located 120 degrees apart on a 12 mm diameter circle (see diagram 4). Presented strength results are the mean values of four measured samples.

Figure 4. The biaxial flexure test.

RESULTS

Profile deviation

The wear of the profile tip of the V-shape was measured and noted as the profile deviation. This have given a relative information of the profile deviation during grinding.

The 20/30 and 30/40 grit wheel did not allow the preset material removal rate when grinding Silicon Nitride.

None of the wheels performed well during profile grinding of Sialon.

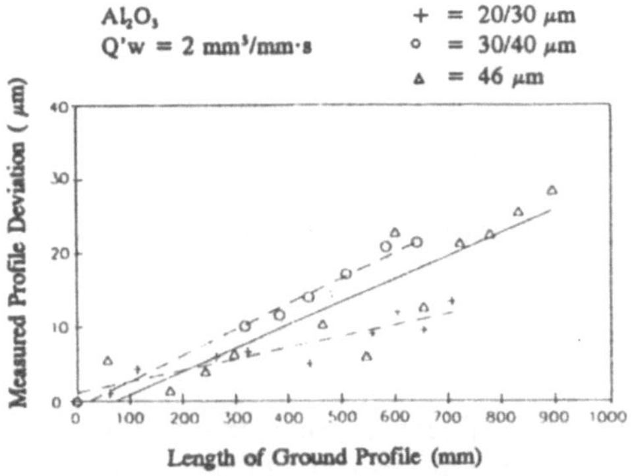

Figure 5. Profile deviation during grinding of Alumina.

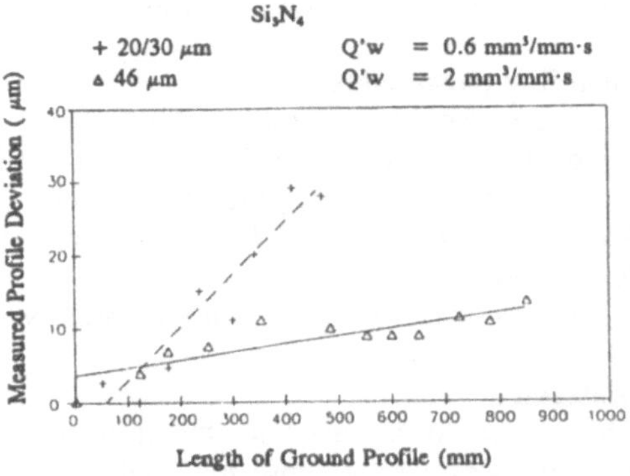

Figure 6. Profile deviation during grinding of Silicon Nitride.

Surface finish

Results are presented as Ra value.

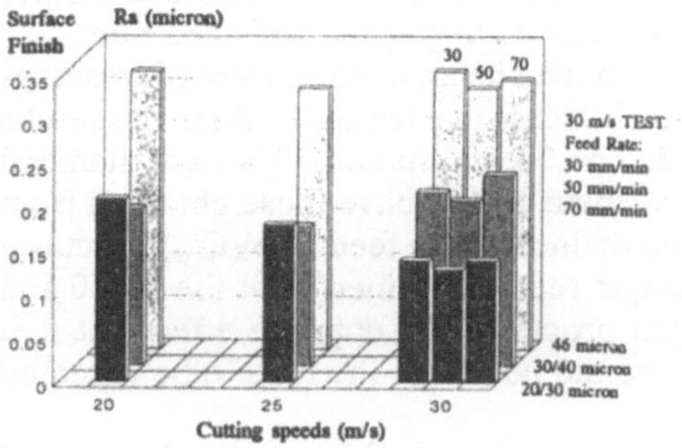

Figure 7. Surface finish for Alumina.

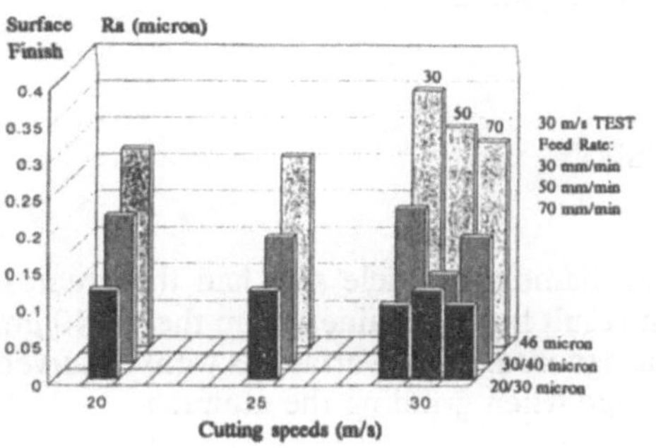

Figure 8. Surface finish for Silicon Nitride.

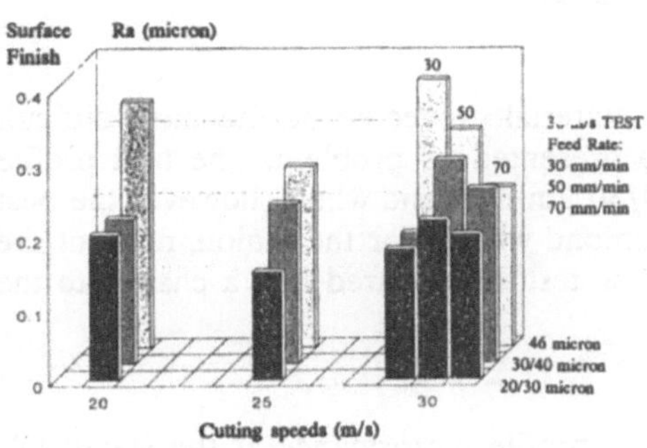

Figure 9. Surface finish for Sialon.

Biaxial Flexure Strength.

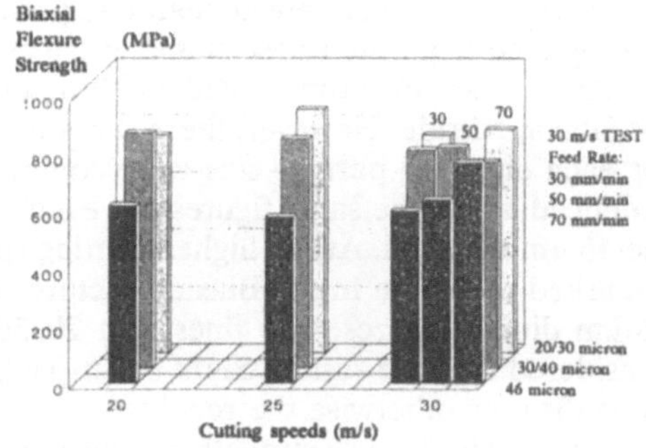

Figure 10. Biaxial flexure strength for Alumina.

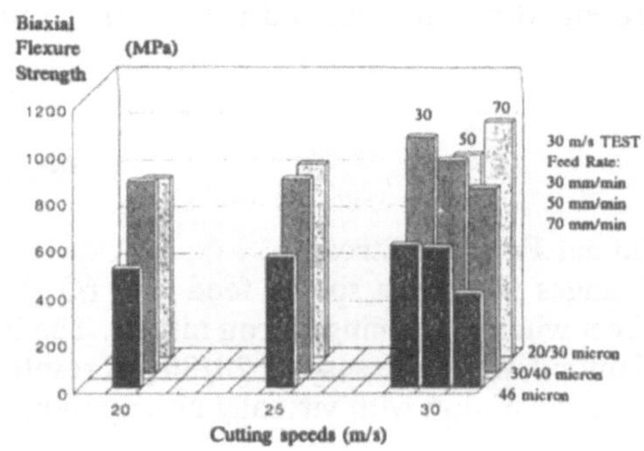

Figure 11. Biaxial flexure strength for Silicon Nitride.

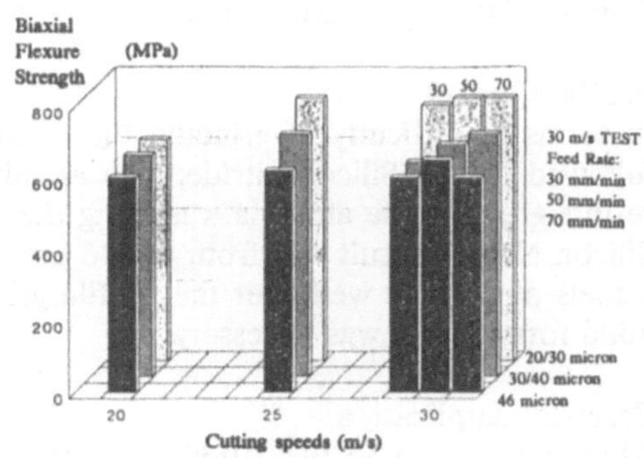

Figure 12. Biaxial flexure strength for Sialon.

DISCUSSION

Biaxial fracture strength readings from the ground Sialon material showed a small increase as the diamond wheel particle size was reduced. Furthermore an increase in cutting speed and feed rate appeared to produce no real change. The best surface finishes were gained from the finest diamond particle at each cutting speed and feed rate.

An increase in cutting speed had little effect on the biaxial fracture strength readings for Silicon Nitride. However, there was a marked increase at each speed when the results for each diamond particle size were compared. The 20/30 μm and 30/40 μm diamond produced about the same figures but each were about 50 % above those obtained from the 46 μm material. At the highest cutting speed an increase in feed rate (Q'_w) produced a marked reduction in the Biaxial fracture strength readings gained from the 30/40 and 46 μm diamond sizes. The finest grit 20/30 μm produced the opposite effect but it is considered that the results could be suspect, time did not allow the tests to be repeated to confirm or otherwise the readings.

The biaxial flexure strength results gained from grinding the high purity alumina did not appear to vary markedly either when the cutting speeds or feed rates were varied. As was expected the surface finishes increased as the abrasive particle size. This material presented no problem and the wheels ground with good efficiency.

CONCLUSION

Biaxial Fracture Strength
Changes in cutting speed, feed rate (Q'_w) and diamond particle size had the greatest effect when machining silicon nitride. The best result being obtained from the 30/40 μm diamond, when the suspect 20/30 μm result was ignored. Small variations were observed from the Sialon with virtually no apparent change when grinding the alumina.

Surface finish
When comparing the effect of changes in cutting speed, feed rate (Q'_w) and diamond particle size the general pattern for all of the subject materials was the same. The finest diamond 20/30 μm giving the best surface finishes.

Profile Control
In terms of difficulty of grinding the Sialon material proved to be the most difficult followed by the Silicon nitride. The alumina presented no problem. The best profile results grinding the alumina was using the 20/30 μm diamond wheel, however the best Silicon Nitride result was from the 46 μm diamond wheel. For the Sialon, none of the wheels performed well over the profile grinding test. It appeared that a change to the bond formulation was necessary.

Practical Application

Alumina:	Use the 20/30 μm wheel. The profile performance at the stated Q'_w were good.
Silicon Nitride:	Use the 46 μm wheel. Best result at 30 m/s, 50 mm/min feed.

Sialon: No satisfactory wheel specification. Under constant conditions the results tend to suggest that a coarser abrasive diamond should be used. The surface finish may be worse in consequence. However, by balancing the concentration of the diamond, with the correct bond strength a better result will be gained. The test duration did not allow a further specification of wheel to be tested.

ACKNOWLEDGEMENT

This work was supported by the Swedish National Board for Technical Development. The test were accomplished at Sandvik Hard Material, Stockholm. For valuable discussions and technical support Mr Tomas Rostvall and Mr Mats Andersson at Sandvik are greately acknowledged.

REFERENCES

1. Barnard, J.M., The Vitrified Diamond Crushable Wheel, Proceeding of the Superabrasive '91 Conf., Chicago, June 1991, jointly sponsored by IDA and SME.

2. Barnard, J.M., Creep Feed Grinding Using Crushform and Pressable Superabrasives wheels, Proceeding of the Superabrasive '85 Conf., Chicago, April 1985, jointly sponsored by DWMA, IDA and SME.

APPENDIX

Calculation

Q'_w = Volume removed from the workpiece in mm^3 per second per mm width of the grinding wheel.
Q'_w in $mm^3/s \cdot mm$.
B'_w = Volume removed from the workpiece in mm^3 per second per mm width of the grinding wheel per mm of the wheel circumference per revolution of the wheel.
B'_w in $mm^3/mm \cdot mm$ revolution of the grinding wheel.

$B'_w = \frac{Q'_w}{Vc}$ in m/s — Vc = wheel velocity in Metres/second

Table speed = 30 mm/min — $Q'_w = 1{,}9\ mm^3/s \cdot mm$
Vc = 30 m/s — $B'_w = 0{,}063 \times 10^{-3} mm^3/mm \cdot mm$ rev.

Table speed = 50 mm/min — $Q'_w = 3{,}16\ mm^3/s \cdot mm$
Vc = 30 m/S — $B'_w = 0{,}105 \cdot 10^{-3}$ mm

Table speed = 70 mm/min — $Q'_w = 4{,}43\ mm^3/s \cdot mm$
Vc = 30 m/s — $B'_w = 0{,}147 \cdot 10^{-3}$ mm

EFFECT OF TESTING METHODS AND SURFACE GRINDING CONDITIONS ON STRENGTH VALUES OF CERAMICS

Jan M. Lindemann and Arne Nissen
Elkem a/s Keramer, P.O.Box 126 Vågsbygd,
4602 Kristiansand, Norway.
IVF - Swedish Institute of Production Engineering Research,
S-951 87 Luleå, Sweden.

ABSTRACT

In order to evaluate precision grinding of ceramic components test specimens of ZrO_2 (3 mol.% Y_2O_3), Al_2O_3-30 vol.% ZrO_2 and siliconized SiC were slip cast, sintered, and ground using the same grinding tools and grinding conditions as chosen for a given set of manufacturing processes. Reference specimens were lapped for comparison to grinding. Specimens were tested in four-point bending as well as biaxial flexure (using the concentric-ring method). Equibiaxial flexure strengths were calculated from uniaxial strength, effective surface area and Weibull modulus. The effect of grinding conditions on strength of the three materials tested is discussed.

INTRODUCTION

From both a technical and economical point of view the grinding of ceramic materials in the fired state is often a critical step in ceramic component manufacturing. Diamond grinding is a considerable part of the production cost of ceramic components requiring finishing after sintering. If the grinding is incorrectly carried out, the component surface may be damaged causing strength reduction and low reliability in service. Therefore it is essential to develop machining techniques with optimum performance (i.e., with minimum surface damage at high machining speed and as little tool wear as possible).

A development program[1,2] ongoing in the Nordic countries has the objectives of improving selected high perfomance ceramic materials and ceramic manufacturing processes for different applications, such as components for machinery and tooling. Parameters for grinding of tetragonal zirconia polycrystals (Y-TZP), zirconia toughened alumina (Y-ZTA) and silicon infiltrated silicon carbide (SiSiC) have been studied and manufacturing conditions were chosen for precision grinding of ceramic parts. Cupwheel grinding and peripheral wheel grinding were utilized on components having both flat and curved surfaces. In order to evaluate these grinding techniques it was appropriate to prepare strength test specimens of the same ceramic materials by means of the same type of grinding tools and at the same conditions as chosen for the manufacturing processes.

Different testing methods were considered. Uniaxial flexure tests are established techniques and flexure strength data are available for most high performance ceramic materials. These strength data, however, are considered insufficient for design purposes. It is well known that edge effects and grinding striations result in inaccurate data from improperly performed flexural tests[3]. The volume of material tested even in 4-point bending is relatively small compared to the volumes of most real components. The dispersion of volume flaws may also have preferred directions due to material manufacturing processes. Biaxial flexure tests overcome most of these deficiencies. These tests are independent of edge conditions and account for all flaws regardless of direction. Though several versions exist, the ball-on-ring method and the concentric-ring method[3,4] represent the two principal methods used. The tested volume of the concentric-ring method is relatively larger. On the other hand, if improperly peformed it may be subjected to effects of contact friction between specimen and testing jig[5] and stress concentrations[4]. Taking into account that the plate-shape of the biaxial test specimen in principle was more equal to the real component shape than bars and that the real loading was best represented by the concentric-ring method, it was decided to apply this testing method.

A reference to the grinding performance could be obtained by preparing corresponding specimens at different workshops. It was decided to use lapping as the reference finishing technique and to prepare 4-point bend specimens for comparison.

Strength values obtained by different testing methods are normally different. It should in principle be possible to calculate the expected strength value for one testing method from the strength measured by another method and this is certainly the basis for application of strength data in design work. Several calculations[6-9,22] have been carried out for uniaxial bending versus biaxial flexure according to different calculation techniques and formulas and some of these methods have recently been reviewed and discussed by Lamon[10].

In this program the test materials were manufactured by Elkem Keramer in Kristiansand, Norway, test specimen grinding was performed by IVF (Swedish Institute of Production Engineering Research) in Luleå, Sweden, lapping was carried out by Wemhöner & Popp oHG in Aachen, Germany, and some properties were measured by SINTEF (The Foundation for Technical and Scientific Research) at the University of Trondheim in Norway.

MATERIALS AND METHODS

Materials

Blanks for test specimens of Y-TZP, Y-ZTA and SiSiC were made by solid slip casting of water-based slips in plaster moulds. The blanks were shaped as rectangular plates and designed to give two square biaxial test specimens from each plate. The slips were prepared according to the same routines used for the standard component manufacturing process.

The starting powder for the TZP slip was a fine-grained ZrO_2 powder partially stabilized with 3 mol.% Y_2O_3 (Daiichi Kigenso HSY-3). The resulting sintered ceramic material might also be called an yttria partially stabilized zirconia (Y-PSZ). The powder was dispersed in water and after rheological adjustment the slurry was ball-milled for 24 hours using zirconia milling media to break down powder agglomerates. The slip was passed through a 45 µm sieve (ASTM 325 Mesh) and deaired by evacuation just before slip casting. Blanks were fired in an electrical furnace at 1580°C for 3 hours in air.

The ZTA slip was made by mixing a reactive Al_2O_3 powder (Alcoa A-16 SG) with 30 vol.% of the same partially stabilized ZrO_2 powder as used for the TZP material.

TABLE 1
Composition, elastic properties, density, fracture toughness and Vickers hardness of SiSiC, ZTA and TZP test materials.

	SiSiC	Y-ZTA	Y-TZP
Composition (wt%)	90 SiC 10 ± 1 Si	60 Al_2O_3 40 3Y-TZP	93.8 ZrO_2 5.4 Y_2O_3
Young's modulus, E (GPa)	400	330	210
Poissons ratio	0.20	0.25	0.25
Density (g/cm^3)	3.10 ± 0.01	4.60 ± 0.01	6.02 ± 0.02
Fracture toughness, [1] K_c ($MN/m^{3/2}$) (Vickers, 5-10 kgs load)	2.6 ± 0.2	4.4 ± 0.4	7.7 ± 0.4
Hardness, [1] H_{V100N} (GPa)	22.0 ± 0.4	15.3 ± 0.3	12.1 ± 0.2

1) From J.H. Ulvensøen, SINTEF [11].

The slip was cast and sintered in an identical manner to the Y-TZP material.

The SiSiC slip was prepared by suspending fine-grained carbon and SiC powder in water. Optimized casting properties were obtained by using SiC powder with a bimodal particle size distribution (Lonza CARBOGRAN UF "fine") having a coarse particle size distribution maximum of ≈20 μm. SiC plates were infiltrated with molten silicon at 1600°C in vacuum. After reaction bonding the composite material contained approximately 10 vol% free silicon.

Blanks from different production batches of each material respectively were carefully divided into two groups in order to keep any variation in properties in the two groups as similar as possible. Specimens for strength testing were machined from one set of these groups by grinding and were prepared by lapping from the corresponding set.

Grinding

The biaxial flexure test specimens were shaped as square plates with nominal face dimensions 40 mm x 40 mm and thickness 2.5 mm. Two specimens were cut from each blank and ground to ≈2.8 mm thickness with a relatively high removal rate. The final grinding to specified thickness was applied to the testing surface of the specimens (i.e., the specimen face submitted to tensile stresses during testing).

The grinding was carried out with a Brother FM-500 machine. A peripheral grinding wheel was used for both coarse and fine grinding of all three materials. The applied wheel type D1A1-D100T10-H25.4-D46/SC-B229-X3-100C was a resin bonded 46 μm diamond (concentration 100) wheel. Cutting speed was kept constant at 30 m/s for both coarse and fine grinding. The coarse grinding of TZP was carried out with 200 μm depth of cut, a table speed of 15 mm/s and the resulting specific removal rate was 3.0 $mm^3/mm\cdot s$. The final 0.3 mm was removed with a depth of cut of 30 μm, the table speed was kept at 15 mm/s and the specific removal rate was 0.45 $mm^3/mm\cdot s$. Coarse grinding parameters for ZTA were depth of cut 200 μm, table speed 10 mm/s and spesific removal rate 2.0

$mm^3/mm\cdot s$. During final grinding the depth of cut was reduced to 30 μm giving a specific removal rate of 0.30 $mm^3/mm\cdot s$ as the table speed again was kept constant. The same depth of cut rates was used for SiSiC as well, giving corresponding specific removal rates of 1.6 $mm^3/mm\cdot s$ and 0.24 $mm^3/mm\cdot s$ for coarse and fine grinding, respectively, when the table speed was reduced to 8 mm/s.

The testing surface of SiSiC specimen was also finished with a cupwheel type D12A9-D100-T33-H20-D46-CVR026-V5-X3-100C. This wheel had a diameter of 100 mm, cutting width 3.2 mm and ceramic bonding of 46 μm diamonds at concentration 100. The depth of cut was 25 μm and the resulting specific removal rate was 0.08 $mm^3/mm\cdot s$ at cutting speed 30 m/s and table speed 3.3 mm/s.

Lapping

Biaxial and uniaxial flexure test specimens were made from the second set of blanks by lapping. The nominal dimensions of the biaxial flexure plates were similar to those made by grinding and the plates were cut to size before lapping. The uniaxial flexure bars had a nominal cross section of 3 mm x 4 mm, a length of 45 mm, and edges were chamfered 0.1-0.2 mm at an angle of 45°. The uniaxial specimens were made by cutting blanks into bars before lapping to final dimensions. Due to blank thickness, the cross-section of the blanks were taken as the testing face of the bending bars.

The finishing of the specimen faces were carried out by a multistep plane-parallel lapping process. A constant concentration of 30 wt. % of B_4C grains in suspension at a feed rate of 40 ml/min was used as the lapping media. The coarse lapping process was carried out with grain size F 150 (95 μm) at a specific lapping force of 50 mPa. The finishing fine lapping was carried out with grain size F 500 (15 μm) and specific lapping force 40 mPa.

Testing

Hardness and fracture toughness were measured by means of Vickers indentation techniques and will be fully reported elsewhere[11]. Measurements were carried out on each material using 6 indentations on each of 3 specimens at 8 different loadings between 10 N and 400 N. Values obtained at loadings between 50 and 200 N showed the best reproducibility. Vickers hardness values were calculated from measurements at 100 N loading. Fracture toughness values were calculated on the basis of measurements carried out at loadings between 50 and 100 N. Calculations were performed according to the formula given by Anstis, et al.[12]: $K_c = 0.016(H/E)^{0.5}(P/c^{1.5})$, where H is hardness, E is Youngs's modulus, P is the indentation load and c is 1/2 the measured crack length.

The equibiaxial flexure strength testing was performed using a concentric ring-on-ring test jig[8] with a stationary support ring having a diameter of 32 mm and an upper loading ring with a diameter of 16 mm. The loading force was transmitted through a spherical bearing of the support ring to enable the ring system to adjust for any plane-parallel inaccuracy between the the two specimen faces. Prior to testing, the top face of the specimen that was submitted to compressive stresses during loading, was covered with an adhesive tape in order to keep the fractured pieces together for subsequent fracture analysis. The strength testing was performed in a universal testing machine at a crosshead speed of 0.5 mm/min and all measurements were carried out at constant ambient conditions. Fracture strength was calculated[8,13] taking the edge length of the plate as the measure of the specimen size. The effect of using squares instead of round specimens has been discussed by Evans and Davidge[13].

The four-point flexure test fixture had an inner span of 20 mm and an outer span of 40 mm. The uniaxial strength testing was performed in the same testing machine and at the same crosshead speed and ambient conditions as for the biaxial testing, and fracture strength was calculated according to standard beam theory.

RESULTS AND DISCUSSION

Measured density, fracture toughness and Vickers hardness of the test materials together with elastic properties used for calculations are shown in Table 1. The measured density of TZP was 6.02 g/cm^3 which is 99.1% of the density reported by Ingel and Lewis[14]. The measured density of ZTA was closer to the theoretical value. Based on measured density of 3.10 g/cm^3, the SiSiC composition was calculated to be ≈90 wt. % SiC and 10 wt. % free Si. There was no open porosity but a small amount of closed pores was found in the microstructure by SEM observation.

The Vickers hardness value of 12.1 GPa for TZP and 15.3 GPa for ZTA agree with the values given by Lange[15,16]. The high hardness value of 22 GPa for SiSiC is indicative of the hardness of the SiC phase. This is quite natural regarding the dense packing of coarse SiC grains in the SiSiC microstructure.

The measured fracture toughness value of 7.7 $MN/m^{3/2}$ for TZP is close to the 6-7 $MN/m^{3/2}$ reported by Lange for a similar material[16], while 4.4 $MN/m^{3/2}$ for ZTA is low in relation to the value of 7 $MN/m^{3/2}$ reported by Lange[15]. The toughness value of 2.6 $MN/m^{3/2}$ for SiSiC is consistent with the low toughness of SiC.

The test materials were manufactured by solid slip casting. One of the benefits of slip casting is the uniform particle packing that takes place as the consolidated layers build up in the casting process. Cast bodies are normally free from the type of dispersed density variations due to pressing powder inhomogenities that is likely to occur in pressed bodies or the large voids that might appear in extruded parts. Hence slip casting often produces materials showing relatively high fired strength. On the other hand, if the rheological properties of the slip are unfavorable in relation to the mold design, one disadvantage of the solid slip casting process is that a central layer or a core of solidified material with a different particle packing may occur. Such material domains will normally have lower green density and strength, are susceptible to cracking upon drying, and can result in porosity in the sintered part.

Such manufacturing problems occurred in the preparation of SiSiC blanks. After infiltration it appeared that some of the blanks showed a sandwich structure with a silicon-rich inner layer, especially in the vicinity of the filling gate position of the mold. It was not possible to detect the inhomogeneous material in the blanks by inspection or density measurements, but the layered structure became evident after cutting. In order to study the influence on strength of such layered structure in SiSiC components, some (≈30 %) sandwich layered specimens were included in both uniaxial and biaxial strength testing.

Figure 1 show strength values for biaxial SiSiC specimens submitted to different surface preparation. Mean strength values with standard deviations are given in the diagram as well as Weibull moduli (assuming a monomodal distribution calculated according the method of most likelihood). It is however more likely that the obtained strength values form bimodal distributions with approximately 75% of the values (higher strength values) yielding a Weibull modulus of ≈12 for the lapped specimens and ≈20 for both of the ground ones. The specimens with the most distinct sandwich structure were all in the low strength region and some of those showing less layered pattern were distributed in the lower half of the upper strength region. There was no difference between peripheral and cupwheel grinding.

Some of the ground specimens showed surface scratches of limited length that were definitely deeper than the general grinding striations. Some of these scratches might have been caused by coarse SiC grains that were pulled out of the specimen edge but other deeper and more distinct scratches were unmistakably due to diamonds breaking off from the grinding wheels and such scratches were evidently the reason for some low strength values. The tendency to diamond break-off was especially significant in cupwheel grinding.

Due to a misunderstanding, the biaxial SiSiC specimens were finished by coarse lapping instead of the fine lapping process that was applied to the other lapped specimens,

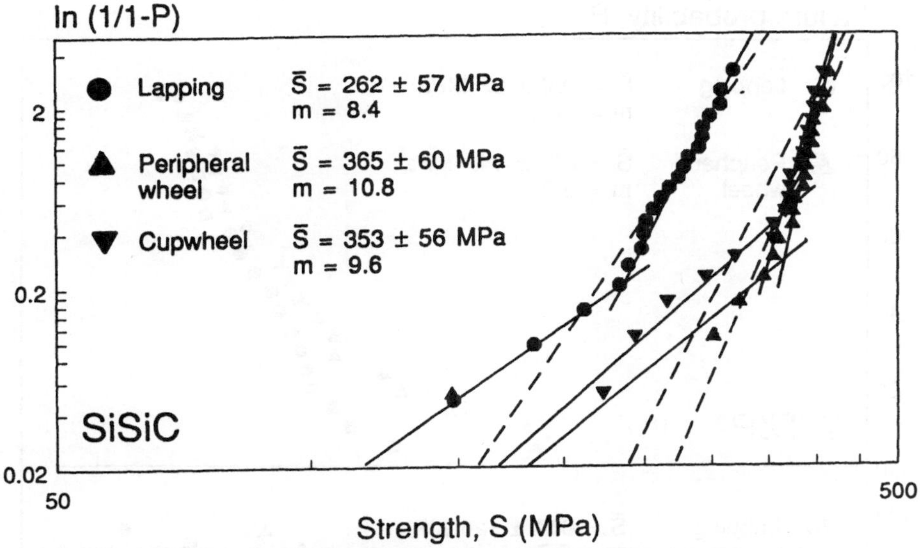

Figure 1. Weibull-plot of equibiaxial strength values of SiSiC-specimens with different surface preparations. Mean values and standard deviations are given by numbers. Assuming monomodal distributions the Weibull modules were calculated according to the most likelihood method and are given by numbers (dashed lines). The more likely bimodal distributions are indicated by solid lines.

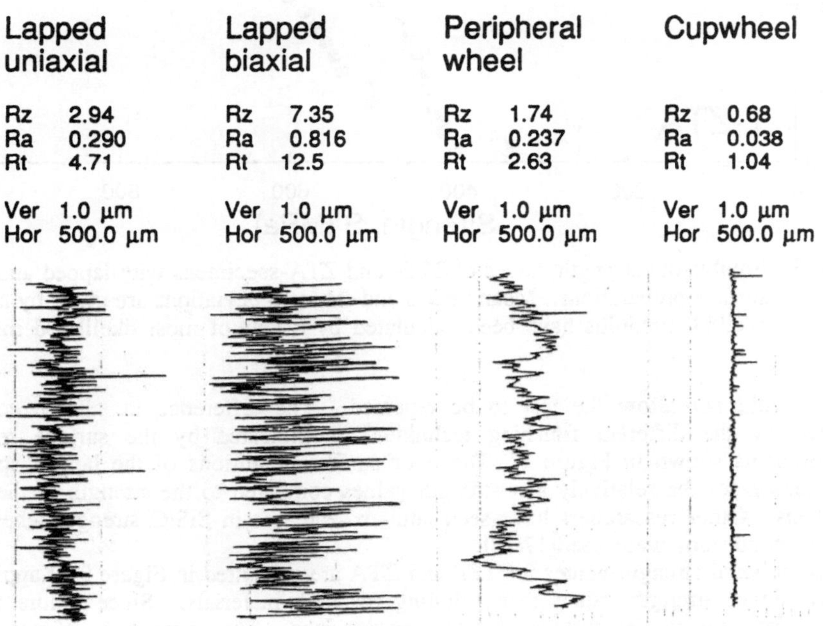

Figure 2. Surface roughness conditions of SiSiC obtained by different finishing techniques.

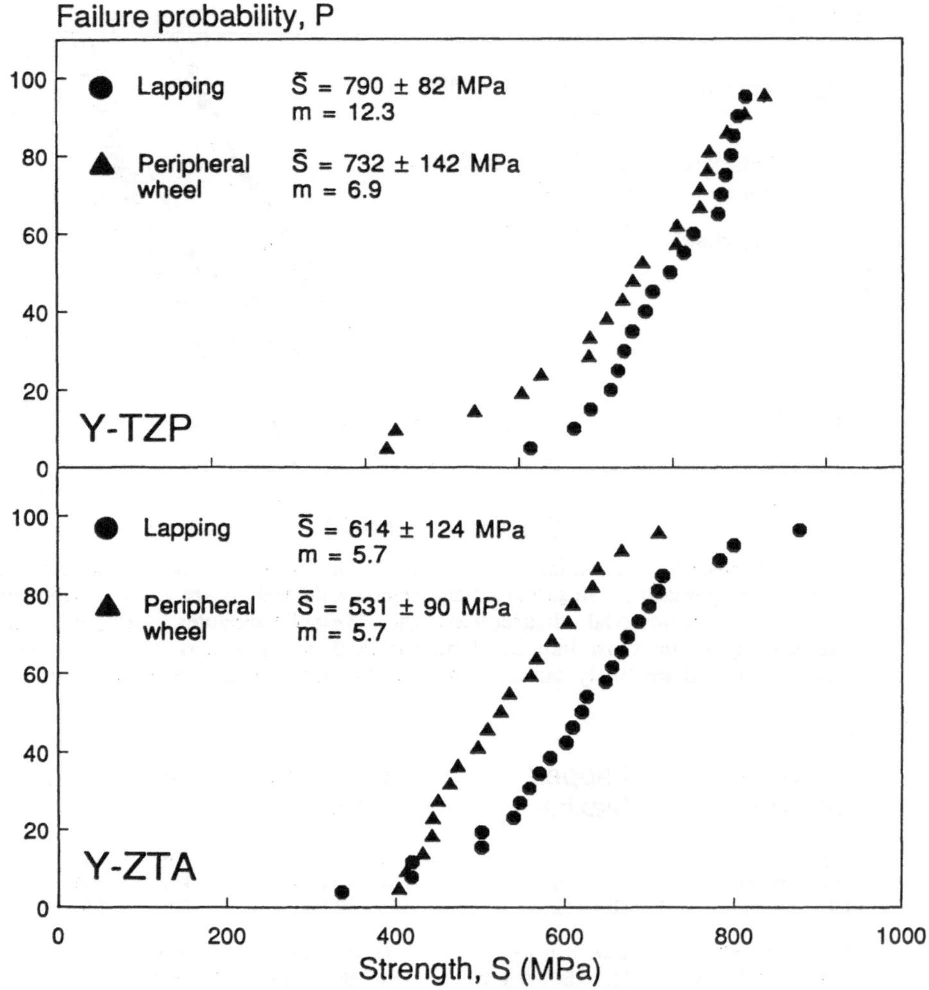

Figure 3. Equibiaxial strength values of TZP- and ZTA-specimens with lapped and ground surface preparations. Mean values and standard deviations are given by numbers. Weibull modulus have been calculated by means of most likelihood method.

and time did not allow the test to be repeated. The difference in surface conditions obtained by the different finishing techniques is indicated by the surface roughness measurements shown in Figure 2. The poor surface conditions of the lapped specimens is the reason for the relatively low strength values compared to the strength of the ground specimens. Other researchers have seen similar decreases in SiSiC strength when severe grinding conditions were used[17,18].

Equibiaxial strength values for TZP and ZTA are presented in Figure 3. Lapping gave slightly higher strength values than grinding for both materials. Since failure typically initiated from the tensile surface, this was as expected. The comparison of uniaxial and biaxial strength values are shown in Table 2. Fractographic SEM examinations revealed that some large zirconia agglomorates existed in the ZTA microstructure and that these

TABLE 2
Calculation of expected equibiaxial flexure strength values based upon measured uniaxial flexure strength and Weibull modulus

	SiSiC	Y-ZTA	Y-TZP
Uniaxial 4-point flexure testing			
- Measured strength (MPa)	347 ± 75	719 ± 156	834 ± 98
- Weibull modulus, m	6.4	5.7	11.3
Biaxial concentric-ring flexure testing			
- Measured strength (MPa)	262 ± 57	614 ± 124	790 ± 82
- Weibull modulus, m	8.4	5.7	12.3
Case 1, Effective surface areas			
- Uniaxial (mm^2)	80	80	80
- Biaxial (mm^2)	201	201	201
Calculated biaxial strength (MPa)	304	612	770
Case 2, Effective surface areas			
- Uniaxial (mm^2)	80	80	80
- Biaxial (mm^2)	402	402	402
Calculated biaxial strength (MPa)	275	542	724
Case 3, Effective surface areas			
- Uniaxial (mm^2)	50	51	46
- Biaxial (mm^2)	927	880	1197
Calculated biaxial strength (MPa)	219	436	627

agglomorates served as fracture starting points in low-strength specimens.

By use of Weibull distribution function with the assumption of surface flaws controlling fracture, Breder, et al.[9] have applied a method of calculation[19,20] relating uniaxial 4-point bend strength and biaxial concentric-ring bend strength by taking into account the Weibull modulus and effective surface area of the test specimens. Breder, et al. applied effective surface areas given by Weil and Daniel[19] for uniaxial bars and Batdorf [21] for biaxial disks. However, the simplest possible approximation for effective area is to assume that surface flaws are independent of tensile stress directions and that they are effective only in the highest stressed surface area. Calculation results for TZP, ZTA and SiSiC are presented as Case 1 in Table 2. If the surface flaws are dependant of stress directions the effective surface area of biaxial specimens are more likely to be doubled [20] as shown by Case 2. Calculating the effective surface areas similar to Breder, et al. results in Case 3.

The calculated strengths for TZP and ZTA according to the simple Case 1 assumptions are in remarkably good agreement with measured results (see Table 2). Due to the differences in the preparation of SiSiC test specimens it was expected that the calculated biaxial strength should be higher than the measured, and the level of difference is in the expected range. It is interesting to observe the increasing difference between measured and calculated strength values as the theroretical basis for calculation become more sophisticated. A further study of this matter was beyond the scope of this paper.

CONCLUSIONS

Equibiaxial flexure strength values for Y-TZP, Y-ZTA and SiSiC test specimens prepared using the same grinding conditions chosen for ceramic components were 732, 531 and 365 MPa, respectively. Peripheral and cupwheel ground samples of SiSiC had similar strength distributions. The strength values for the lapped TZP and ZTA specimens were 50-100 MPa higher than the strength values obtained from peripherally ground specimens. The corresponding uniaxial 4-point flexure strength values were 834, 719 and 347 MPa, respectively. Biaxial strength values calculated from measured uniaxial flexure tests and Weibull moduli, assuming surface flaws are independent of tensile stress direction, were in excellent agreement with measured biaxial strength values.

Adequate strength for many ceramic applications was attained for the three materials tested despite the rapid material specific removal rates (0.25-3.0 $mm^3/mm \cdot s$) and aggressive depths of cut (30-200 μm) used in creep feed grinding. However, reduced strength due to flaws introduced during ceramic processing as well as surface damage during grinding and lapping were observed.

ACKNOWLEDMENT

This work was carried out with the support of Nordisk Industrifond - The Nordic Fund For Technology And Industrial Development, and NTNF - The Royal Norwegian Council for Scientific and Industrial Research.

REFERENCES

1. Nordisk Industrifond, Prosjekt P 89178 - Fremstilling av TiB_2, for andvendelse i lettmetallindustrien, 1990.

2. Nordisk Industrifond, Prosjekt P 89179 - Kerambearbetning, 1990.

3. Radford, K.C. and Lange, F.F., Loading (L) Factors for the Biaxial Flexure Test. J. Am. Ceram. Soc., 1978, **61**, 211-3.

4. Shetty, D.K., Rosenfield, A.R., McGuire, P., Bansal, G.K. and Duckworth, W.H., Biaxial Flexure Tests for Ceramics. Am. Ceram. Bull., 1980, **59**, 1193-97.

5. Fessler, H., Fricker, D.C. and Godfrey, D.J., A Comparative Study of the Mechanical Strength of Reaction-Bonded Silicon Nitride. In Proc. 6th. Army Matr. Techn. Conf., Ceramics for High-Performance Applications, III, Reliability, ed. Lenoe, E.M., Katz, R.N. and Burke, J.J., Plenum Press, New York, 1983, pp. 705-36.

6. Evans, A.G., A General Approach for the Statistical Analysis of Multiaxial Fracture. J. Am. Ceram. Soc., 1978, **61**, 302-8.

7. Batdorf, S.B. and Heinisch, Jr., H.L., Weakest-Link Therory Reformulated for Arbitrary Fracture Criterion. J. Am. Ceram. Soc., 1978, **61**, 355-8.

8. Giovan, M.N. and Sines, G., Biaxial and Uniaxial Data for Statistical Comparisons of a Ceramic's Strength. J. Am. Ceram. Soc., 1979, **62**, 510-5.

9. Breder, K., Andersson, T. and Schölin, K., Fracture Strength of α- and β-SiAlON Measured by Biaxial and Four-Point Bending. J. Am. Ceram. Soc., 1990, **73**, 2128-30.

10. Lamon, J., Statistical Approaches to Failure for Ceramic Reliability Assessment. J. Am. Ceram. Soc., 1988, **71**, 106-12.

11. Ulvensøen, J.H., Relations Between Abrasive Wear and Mechanical Properties for Ceramics., to be published.

12. Anstis, G.R., Chantikul, P., Lawn, B.R. and Marshall, D.B., A Critical Evaluation of Indentation Techniques for Measuring Fracture Toughness: I, Direct Crack Measurements. J. Am. Ceram. Soc., 1981, **64**, 533-8.

13. A.G.Evans and R.W.Davidge, A Biaxial Stress Method for the Determination of the Strength of Sections Cut From Glass Containers and the Size of the Critical Griffith Flaws. Glass Tech., 1971, **12**, 148-54.

14. Ingel, R.P. and Lewis III, D., Lattice Parameters and Density for Y_2O_3-Stabilized ZrO_2. J. Am. Ceram. Soc., 1986, **69**, 325-32.

15. Lange, F.F., Transformation toughening. Part 4: Fabrication, Fracture Toughness and Strength of Al_2O_3 - ZrO_2 composites. J. Mater. Sci., 1982, **17**, 247-54.

16. F.F.Lange, Transformation Toughening. Part 3: Experimental Observations in the ZrO_2 - Y_2O_3 System. J. Mater. Sci., 1982, **17**, 240-6.

17. Pfeiffer, W., Hollstein, T., Berweiler, W. and Prümmer, R., Residual Stresses Due to Machining of Ceramic Materials and their Effect on Strength. In Proc.Third Int.Symp., Ceramic Materials and Components for Engines, ed. Tennery, V.J., Am.Ceram.Soc, Inc., 1989, pp. 1170-8.

18. König, W. and Wemhöner, J., Grinding of SiSiC Ceramics: Achieving High Performance and Minimum Damage. In Proc.Third Int.Symp., Ceramic Materials and Components for Engines, ed. Tennery, V.J., Am.Ceram.Soc, Inc., 1989, pp. 1225-35.

19. Weil, N.A. and Daniel, I.M., Analysis of Failure Probabilities in Nonuniformly Stressed Brittle Materials. J. Am. Ceram. Soc., 1964, **47**, 268-74.

20. Service, T.H. and Ritter, J.E., Uniaxial and Equibiaxial Strength of a Vitreous-Bonded Abrasive. J. Vib. Acoust. Stress. Relia. Des., 1989, **3**, 194-98.

21. Batdorf, S.B., Some Approximate Treatments of Fracture Statistics for Polyaxial Tension. Int. J. Fract., 1977, **13**, 5-11.

22. Soltesz, U., Richter, H. and Kienzler, R., The Concentric-Ring-Test and its Application for Determining the Surface Strength of Ceramics. In Proc. 6th CIMTEC, High Tech Ceramics, ed. Vincenzini, P., Elsevier Sci. Publ. B.V., Amsterdam, 1987.

MACHINING OF CERAMIC COMPONENTS - APPROACHES TO HIGHER FLEXIBILITY AND ECONOMY

W. KÖNIG, C. BÖNSCH, E. VERLEMANN, A. WAGEMANN
Fraunhofer-Institut für Produktionstechnologie
Steinbachstr. 17
D-5100 Aachen

ABSTRACT

For many applications in the field of mechanical engineering, designers prefer conventional solutions in spite of the outstanding possibilities they have by using ceramic materials. Economic considerations will often lead to a decision against the ceramic component due to the difficulties in machining of complex contours. The designer's demand upon the manufacturer to provide a flexible machining process also for small lot sizes and rapid prototyping is justified for ceramic components as well. This gets particularly important under the aspect that up to 80 percent of the component costs are caused by the fine machining process.

This development requires a change of view in the field of production strategies. Especially the question of make or buy and the efforts in the area of near-net-shape processing have to be reviewed critically. Thinking in semifinished products will be an economic alternative due to the technological potentials of the manufacturing processes for hard machining as well as green machining.

This paper will present the conditions for an optimized machining of ceramic components with respect to economy and quality. Examples of optimized grinding- and lapping processes will illustrate this approach. Apart from these conventional processes innovative techniques such as ultrasonic machining will be introduced.

INTRODUCTION

The decision for or against ceramics as component material is to be made in the phase of product development. The characteristic properties of material and component are the fundamental arguments in this decision. The topics reliability, constant level of quality and the availibility of the material are of superior interest beside the requirement of fitness for use. In many cases the critical characteristics of ceramic materials are on the opposite of these demands. These characteristics are, for example, the low fracture toughness, the insufficient reliability and the poor machinability. These problems will be accepted only in those situations where the functional demands enforce the use of ceramic componenets.

Furthermore it is generally necessary to distinguish whether a stand-alone solution for a few cases or a mass production solution is aspired. The production strategies have to be adapted depending on these boundary conditions. In the case of mass production, this means working with a high level of automation and a near net shape sintering process. In the areas where special solutions are required it is not possible to give a quick response to market requirements with such inflexible production systems. But even for ceramic components increasing flexibility becomes more and more important.

FLEXIBLE PRODUCTION OF CERAMIC COMPONENTS

All fields of manufacturing ceramic components have to be involved in the efforts concerning a flexible manufacturing process. This means for example that for powder production and preparation standardized materials should be used. This solves difficulties in the planning and the operative level of production.

The steps of manufacturing may vary depending on the component. One reason is, that the use of semifinished products opens a lot of advantages in flexible manufacturing. Semifinished products may be machined in the green (not sintered) or the hard state. Machining in the green state has the advantage of a wide range of available machining processes. Additionally these processes feature performance and lower tool wear. Some machining steps will still remain for finishing in the hard state of the component. Due to more recent developments, machining strategies with higher performance are also available for hard machining of ceramics. These developments will justify hard machining of semifinished products also from the economic point of view.

These explanations make clear that the concept of flexible manufacturing of ceramic components requires powerful finishing processes with a high level of quality. Beside the processes grinding and lapping, which are well introduced into practice, there are further processes for specific tasks. Ultrasonic machining, laser beam machining, laser assisted machining, water jet cutting and in the case of electric conductive materials also electro discharge machining are examples for the innovative machining processes for ceramics. Higher performance and a reliable quality also are criteria for these processes and the objective of future developments.

Grinding

Grinding with diamond tools is the process most frequently used to machine ceramic components. The advantage of grinding lies in the large number of processes available and the ability to cover both roughing and finishing processes.

For optimized grinding processes it is important to understand and consider the material specific removal mechanisms. With this knowledge and the capability to transfer it into practice it is possible to provide a maximum amount of process reliability and reproducibility.

In the case of grinding ceramics, investigations have shown, that the condition on the individual cutting edge has a great influence on the material removal mechanism /1/. For ceramics, low forces and low infeeds per cutting edge are advantageous. In practice, the chip thickness at the single grain can be controlled via a number of factors. Apart from the tool choice, the grinding process parameters such as infeed, feed rate and cutting speed all substantially affect chip thickness. Low feed rate and high infeed which means creep-feed grinding, is the best possibilty in surface grinding to obtain optimized processes.

Irrespective of the type of ceramic, creep-feed grinding achieves higher component strength coupled with improved reproducibility and surface quality. The greater number of cutting edge contacts however entails a rise in cutting forces (Figure 1).

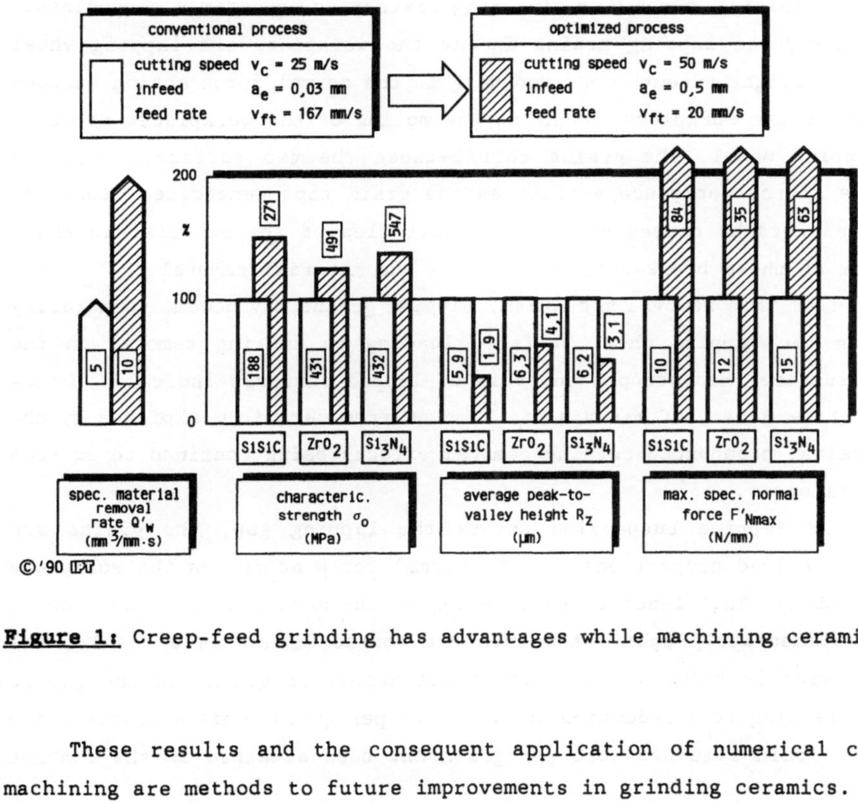

Figure 1: Creep-feed grinding has advantages while machining ceramics

These results and the consequent application of numerical controlled machining are methods to future improvements in grinding ceramics.

Lapping

Lapping is frequently used to finish ceramic products requiring extremely high surface qualities and accuracies-to-shape. Flat lapping has process-specific advantages as compared to grinding, especially for surface facing of ceramic components such as axial face seals /2/.

An additional advantage of lapping is that the hard grains admitted to the lapping gap in a suspension need be only slightly harder than the workpiece material itself, since grains which have already passed through the contact zone are not reused. In most applications, therefore, cheaper boron carbide can substitute the diamond abrasive needed in grinding.

Finally, the lapping process allows a number of workpieces to be processed simultaneously and from both sides. The fact that workpieces on carriers can be fed loosely between the upper and lower lapping wheels is a decisive advantage for initial removal of sintering distortion on ceramic components.

The introduced lapping grains indent the workpiece and lapping wheel surfaces to a slight extent, corresponding to the normal force acting between the wheels and the workpiece. Owing to the motion of the workpieces relative to the lapping wheel, the grains roll between the two surfaces, inducing microcracks in the workpiece surface as the grain tips penetrate. Crosslinking of these cracks causes microscopic particles of the material to chip, and the sum of these break-outs constitutes the material removal.

Apart from this removal mechanism, lapping grains may become temporarily anchored in the lapping wheel surface, leading to cutting removal on the workpiece surface. Microscopic analysis of lapped surfaces indicates, however, that the majority of material is removed from ceramic workpieces by the rolling grains, pronounced cutting removal effects being confined to extreme process parameters.

When the lapping suspension enters the lapping gap, the grains are subjected to a load proportional to the normal force acting on the workpiece or lapping wheel. This leads to splintering of the more heavily loaded coarser grains immediately upon entry into the contact zone. As a result, the mean grain size is reduced, while the total number of grains in the gap is increased, leading to a reduction in the force per grain until a stable state in terms of grain size and load per grain has been attained in the contact zone.

The effects of grain splintering are still further intensified by using coarse grains coupled with higher normal force per unit area. As normal force rises and the associated grain splintering increases, the force per grain in the lapping gap is reduced. Grains are pressed less deeply into the workpiece surface, reducing the size of the spalled material particles and the integral surface roughness.

Due to the increasing number of grains in the contact gap, a higher total removal can be achieved as the normal force increases, despite the smaller particles removed.

Given the same volumetric concentration of the lapping agent, a substantially higher number of cutting edges is present in the contact zone if grain sizes are smaller. The contact pressure per single grain is greatly reduced compared to a coarser lapping agent, so that the effects of intensive grain splintering due to increased normal force per unit area are not so significant.

Grain splintering not only reduces mean grain diameter but - essentially in this context - restricts the spread of the grain size distribution in the contact zone.

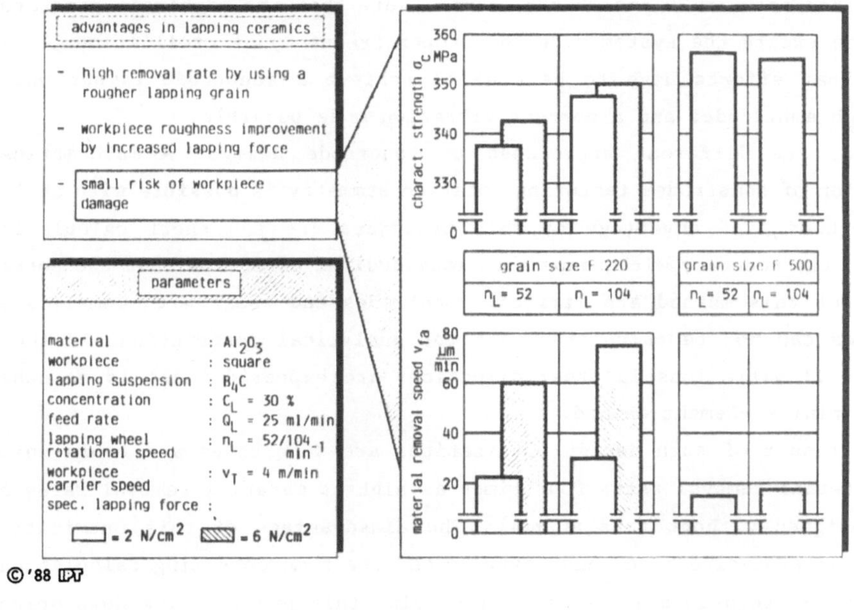

Figure 2: High material removal rate and minimum damage by rough lapping

The narrower grain size spread means that the individual grains are all pressed into the workpiece surface to roughly the same depth. Severe damage to the workpiece surface due to isolated oversized lapping grains, with negative effects on component strength, can be largely avoided. For this reason, lapping achieves high component strength even at roughing parameters, as illustrated by the example of Al_2O_3 in Figure 2.

Higher performance in ultrasonic machining

During ultrasonic machining a high frequency generator supplies an electric AC voltage which is transformed into a mechanical vibration by an piezoelectrical crystal. The amplitude of this vibration is in the range of 3 up to 5 µm. This amplitude is too low for an effective machining of the ceramics. Therefore the vibration is amplified up to a value of 25 µm using boosters and sonotrodes. The tool is fixed at the bottom of the sonotrode. The whole system has to be layed out to exact lateral vibration mode. This mode has to be allocated between 20 and 22 kHz to make sure that the machine's generator is able to excite the system with the proper frequency of oscillation. It is obvious that efforts have to be done to achieve a sonotrode layout which makes high amplitudes and a correct vibration mode possible.

There are different approaches to sonotrode design. A mathematical calculation of sonotrodes featuring rotation symmetry is possible with analytical methods. The advantage of this procedure are the short calculation times and the accuracy with respect to longitudinal direction. But the possibilities of this method are limited. Sonotrodes and tools with complicated geometries can not be calculated with the analytical description. In those cases and if vibrations in other directions are expected there is no other way than finite element method.

The result of such design optmizations are sonotrodes and tools which reach material removal rates four times as high as material removal rates of tools designed without this support. The disadvantage of this optmization step is the expensive hard- and software and the time consuming calculation. The use of PC solutions is one way of solving this problem. The development of standard sonotrodes with optimized vibration behaviour is another possibility in making sonotrode design practicable.

Another problem of ultrasonic machining is the high tool wear. Every type of wear of the tool has an influence on the working result in form of accuracy failures. Minimized, or at least equalized tool wear, is therefore a subject of special interest in ultrasonic machining. The correct choice of the tool material is the most promising way in minimizing tool wear. Recent results point out that through the use of high speed steel, the wear is decreasable to 30 % of the original value. Figure 3 shows that the results are even better with PCD-tools. The wear was reduced some 15 times using such tools /3/. The problem in using PCD tools is the fixture of the tool to the sonotrode and the high price of the material. Only if the joining process was done in a correct way a good vibration transmission from sonotrode to the tool can be maintained. Further research activities will pay special attention to this topic.

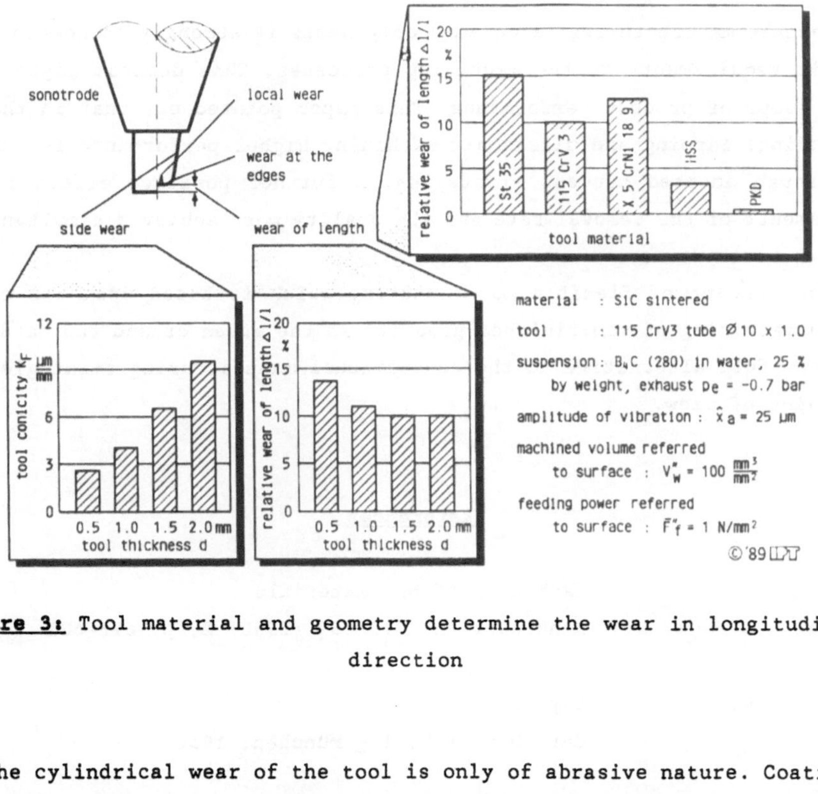

Figure 3: Tool material and geometry determine the wear in longitudinal direction

The cylindrical wear of the tool is only of abrasive nature. Coating of these parts of the tool with hard face coatings might improve the accuracy of the machined workpieces. Another possibility of taking influence on the tool wear when using hollow tools is the correct choice of the thickness of the tool. Thin tools will lead to higher wear of the tool length but they will not produce high conicity and edge roundness. Furthermore, the machined surface is smaller and the removal speed increases. The cylindrical wear is determined by the machining time because this abrasive wear is caused by suspension flow along the tool surface. Therefore every reduction of the machining time results in better workpiece qualities concerning the conicity.

CONCLUSIONS

The flexible manufacturing of ceramic components is strongly connected to the economic requirements to the machining processes. This demands improvements in the scope of process performance. This paper pointed out that in the case of grinding, lapping and ultrasonic machining higher performance is achievable through adapted process strategies. A further positive effect is that improvements of the removal rate and the quality were achieved simultaneously in the described cases.

The concept of flexible manufacturing which is based upon the ceramic adapted machining of semifinished products in the green or/and the hard state is a powerfull alternative to the conventional manufacturing strategies from this point of view.

References

1. König, W. et al — Machining of New Materials. Annals of the CIRP 39 (1990) 2, p. 673-681

2. Lätzig, W. — Läppen. Carl-Hanser-Verlag München, 1950

3. Nölke, H. H. — Spanende Bearbeitung von Siliziumnitridwerkstoffen durch Ultraschallschwingläppen. Dissertation TH Hannover, 1980

SOME ASPECTS ON SUPER-FINISH MACHINING OF CERAMIC COMPONENTS FOR ENGINES

Li Gang*, Fu Youtong, Xu Yanshen, Peng Zemin
* Dept. of Mech. Eng., Tianjin University, China
National Combustion Lab. Tianjin University, China

ABSTRACT

In this paper, an attempt is made to grind the advanced ceramic materials Sialon by common abrasive wheel through a large amount of experiments for super-finish grinding process. At first the influence of the technological parameters on surface roughness of the workpiece is analysed and grinding forces are measured. And then the method for grinding wheel field balancing in the case of disturbance is proposed and the effect of wheel balance on surface roughness of the workpiece is also researched. Finally the material removal mechanism during super-finish grinding process is studied via observing the micro-surface by SEM. It is indicated that the remove of ceramic materials is due to its plastic deformation to form the super-finish surface. The ground ceramic workpieces have been succesfully applied to the fields of heat engine, oildom and medicine.

Key words: Ceramic Material, Super-Finish Grinding, Surface Roughness, Grinding Force, Wheel Balance,Plastic Deformation

1. Introduction

Advanced ceramic materials have been more and more widely applied to the scientific and technological fields because of its special physical, chemical, thermodynamic and tribological properties. The ceramic workpiece is a kind of sinter with high hardeness and brittleness. The good surface roughness and eligible tolerance are the key of applying them to practice. Because of its nice properties the ceramic parts are often used as the key parts in facilities and machines which often require strict tolerance and low surface roughness. Now super-finish machining technology of ceramic materials has been regarded as the main technological problem to be solved.[1] Generally polishing methods are used to form the super-finish surface with the low productivity and high cost by which the surface roughness can be decreased to 0.02μm · [2]. In this paper, research on the super-finish grinding of ceramic workpieces is performed by the common abrasive wheel. The grinding forces and surface properties of workpiece are analysed during grinding process. The influences of technological parameters and wheel balancing on the surface roughness are studied experimentally. The field balancing technology in case of disturbances is discussed. Experimental results indicate that low cost and high productivity can be got by using this method without complex operating programs. The surface roughness Ra can reach to 0.007μm.

2. Experiments

The material properties used in experiments and grinding conditions are listed in Table 1.

Table 1 Material properties and grinding conditions

CERAMIC WOEKPIECE: Sialon, Density: 3.2 g/cm^3; Hardness: 1800-2300 N/mm^2 ; Bending strength: 650-850 mPa; Elastic modulus: 300-380 GPa; Breaking flexibility: 6-7 mPa$\sqrt{m}$.
GRINDING CONDITIONS: Wheel speed: 35 m/s; Workpiece speed: 80-160 mm/s; Table speed: 50-200 mm/min; Work diameter: 8-13 mm.
GRINDING WHEEL: Common abrasive wheel 350*50*127;Wheel and its grain class: diamond wheel and 80#-320#.
DRESSING CONDITIONS: Dressing tool: single point diamond; Dressing depth:0.002-0.005 mm; Lead: 0.01-0.05 mm/r.
GRINDING MACHINE: MG1432A high-precision external grinder made in CHINA.
GRINDING TYPE: Longitudinal grinding. COOLANT LIQUID: Water-based liquid.

3. Experimental results and analysis

3.1 Influence of technological parameters on surface roughness

The influences of grinding depth a_t, table speed v_t, Workpiece speed v_s and spark-out grinding time N on surface roughness Ra of ceramic workpieces are shown in Fig. 1. Here Ra is measured by Tylorsurf. 6 with the pinpoint radius less than 1.5 μm. It can be known that 1) for getting the better super-finish grinding results the technological parameters must be limited within the proper range; 2) three regions can be divided to describe the effects of a_t on Ra. In first region Ra decreases greatly with increasement of a_t which can be called improving region of Ra. With the continuung increase of a_t, Ra reaches to the minimum value and varies a little within certain range called the optimum region. While growing beyond the optimum region, then a_t comes to the unusual region in which Ra increases suddenly to the level before the beginning of super-finish grinding. 3) the boundary between the optimum and unusual region is not fixed and distributes randomly within the range of 5 to 7.5 μm which may be called random distributing region of the boundary. So the use of parameters in this random region should be avoided during grinding process. 4) only in the cases when table speed V_t is less than certain value and workpiece speed V_s is more than certain value can the optimum region produce.

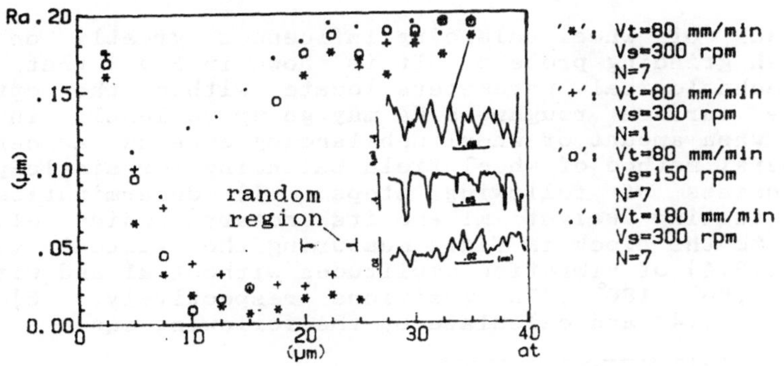

Fig. 1 The influences of technological parameters on Ra

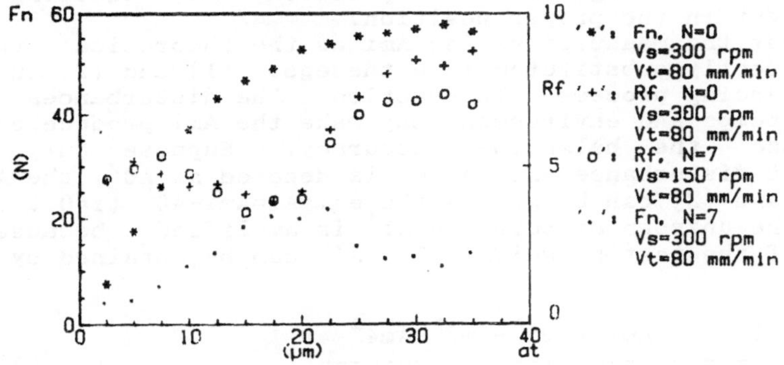

Fig. 2 Grinding forces during super-finish grinding

3.2 Grinding forces

From Fig.2 the conclusions can be obtained as follows: 1) the normal grinding force Fn increases fastly with a_t in the improving region, slowly in the optimum region and at last reaches to a stable level in the unusual region; 2) the ratio Rf of normal grinding force Fn to tangential grinding force Ft has a minimum value in the optimum region and Rf varies from 6 to 9 in the unusual region which is similar to the case of grinding ceramics by diamond wheel; 3) the spark-out grinding time N has a little effect on Rf and only in the optimum region in the condition N>1 Ra is less than that in the case N=0, Fn reaches to the maximum value in the optimum region during spark-out grinding.

3.3 Wheel balancing and its effects on Ra

The accuracy of wheel balancing influences greatly on the super-finish grinding process. It is shown in Fig.3 that even if the technological parameters locate within the optimum region the surface roughness Ra may go up voilently in the condition when amount of wheel unbalancing exceeds a certain value. General method of wheel field balancing for single-plane rotor consists of following steps: 1) determination of parasitic testing weight m1 and its rotary radius e1; 2) according to the block in Fig.3 measuring the virtual values Ami(i=0,1,2,3,4) of vibration amplitudes without m1 and with m1 on the 0° ,90° ,180° ,270° positions respectively. 3) let Ami=Ai(i=0,1,...4) and calculate by the following eqs.:

$$me=m1*e1/\sqrt{(A1^2+A3^2)/2/A0^2-1} \tag{1}$$

$$\cos\theta=(A1^2-A3^2)*m*e/m1/e1/4/A0^2 \tag{2}$$

where, m and θ express the weight-radius product of the amount of wheel unbalancing and its position; 4) choosing the counterweight in the proper position.

Notice that the measured values Ami as the theoretical values Ai are directly substituted into the eqs. (1) and (2) in the above balancing process. In practice, the disturbances from hydraulic pumps and environment may make the Ami produce errors to derease the balancing accuracy. Suppose only the environment disturbance exists and is denoted as A0′, the Ai in eqs. (1) and (2) can be got by the eq. Ai=Ami-A0′ (i=0,...,4). Suppose the unknown disturbance A1′ is amplified because of the use of the testing weight m1, A1′ can be obtained by the eqs.:

$$A1' = \frac{1}{2} * \frac{Am1^2+Am3^2-Am2^2-Am4^2}{Am1+Am3-Am2-Am4} \tag{3}$$

$$Ai=Ami-A1' \quad (i=1,...,4) \tag{4}$$

then, substitute Ai into eqs. (1) and (2) and balancing can be completed.

In experiments, by the general field balancing method the

vibration virtual value A caused by amount of unbalance can be decreased to .08 μm. Repeating the processes of balancing and dressing on the balancing frame manualy A may be 0.08 to 0.1 μm. By the above method A can reach to 0.02 μm to 0.04 μm. Generally, the better the balancing, the wider the range for choice of technological parameters.

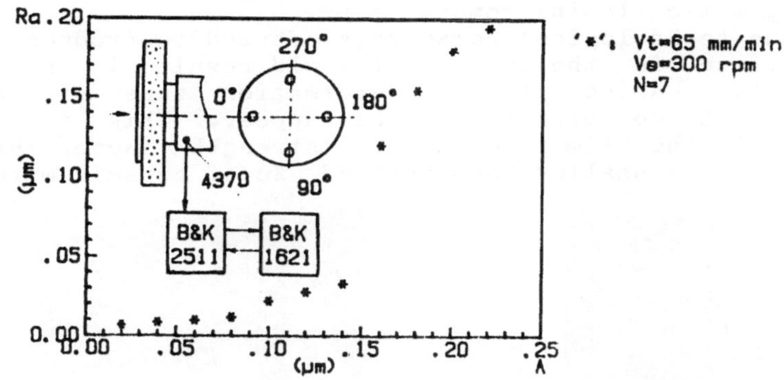

Fig.3 The wheel field balancing

4. Mechanism of super-finish grinding

Through the impressing testing[1], it can be known that there are three stages for the ceramic removal process, namely elastic deformation, micro-plastic deformation and brittle breaking. Cutting ceramics by a single point diamond with chip thickness 0.06 to 0.6μm indicates[3] that the ceramic material is removed as plastic flowing removal type. On the other hand, when the chip thickness increases to more than 0.9μm large brittle cracks would produce on the workpiece surface, called the crack spreading removal type which make the surface roughness much poorer because the size of cracks is often larger than the chip size. Suppose the limit of stress strength factor of ceramic material and abrasive grain are denoted as Kc and Ksc respectively, the stress strength factor effected between the grinding wheel and workpiece is K.According to the theory of fracture mechanics, the plastic flowing removal type produces under the case K<Kc.Conversely under the condition K>K, the crack spreading removal type takes place. SImilarly under the case K<Ksc, the abrasive grain would wear, when K>Ksc the grain would break up itself. In case when the grinding depth a_t is small, K is much less than Kc, then the elastic friction takes place primarily between cutting grains and workpiece. This may be verified by the surface profile at the small grinding depth 2.5μm as shown in Fig.1. When an increasement of a_t makes K close to Kc the cutting grains work in the plastic deformation stage of ceramic and obviously a better surface roughness produces whose profile is shown in

Fig.1 in the case a_t=15μm. If a increases to the 30μm which makes K larger than Kc, ceramic material would be removed by the crack spreading removal type and poorer surface roughness and larger micro-cracks on the workpiece surface would produce. Fig.4 and Fig.5 show the photos of workpiece surface taken by SEM under a_t=15um and a_t=35um respectively. So the conclusion can be drawn that the super-finish grinding surface is produced mainly by plastic flowing removal type.
However, the technological parameters adopted to produce super-finish surface in the experiment would result in a stress condition far above plastic deformation zone of ceramic material, yet no brittle cracks appear. This is mainly because that the limit of stress strength factor Ksc of ordinary grain is smaller than that of Kc of ceramic material.

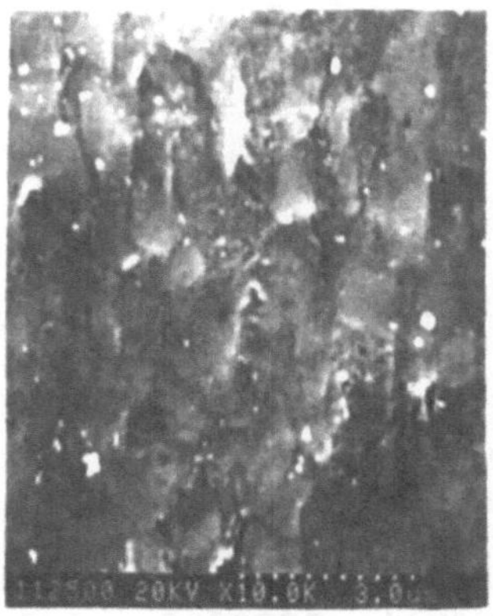

Fig.4 Photo at a_t=15μm

Fig.5 Photo at a_t=30μm

Whenever K goes up to a certain extent due to some technological conditions, the grain itself would break accordingly. So no brittle cracks on workpiece occur. Comparing the super-finish process via diamond wheel[4], there is no need to apply micro-feed devicee and fine dressing, moreover, the wheel cost may save several 10-fold.

4. Conclusions

1) A kind of super-finish grinding method by the common abrasive wheel is presented by which the surface roughness Ra=0.007μm can be got with low costs and no micro-feed device needed.
2) Through analysing the influences of technological parameters on the surface roughness, the grinding forces and micro-surface of workpiece, it can be indicated that forming of super-finish surface is mainly due to the plastic deformation of ceramic material.
3) The amount of wheel unbalancing influences severely on the super-finish grinding process even under the optimum

technological parameters. Through the technology of eliminating disturbance mentioned in this paper the wheel balancing accuracy can be increased by 2 to 4 times.

Reference:

1 Inasaki I., Grinding of Hard and Brittle Materials, Annals of CIRP, Vol36(2), 1987, 463-469
2 Spur G., Sabolka I., Tio T. und Wunsch U.E., Überblick über trennende Fertigungsverfahren zur Hartbearbeitung von Keramiken, Fachberichte für Mtallbearleitung, Vol65, No.4, 1987, 311-315
3 Wilfried Koing and Jens Wemhoner, Optimizing Grinding of SiC, Ceramic Bulletin, Vol68, No.3, 1989, 545-548
4 市田良夫,貴志浩三,ファインセラミックスの鏡面研削 ,機械と工具 ,Vol33,No.2,1989, 65-69

ANALYSIS OF CERAMIC COMPOSITES BY 3D MICROTOMOGRAPHY

THOMAS LÜTHI * and WILLIAM A. ELLINGSON **
* Swiss Federal Laboratories for Materials Testing and Research (EMPA), Ueberlandstrasse 129, CH-8600 Dübendorf, Switzerland
** Argonne National Laboratory, Materials and Components Technology Division, 9700 South Cass Avenue, Argonne, IL 60439-4838, U.S.A.

ABSTRACT

In the first part we describe the use of 3D X-ray microtomography for the analysis of density distributions inside and between cold pressed green state whisker reinforced Si_3N_4 / Si_3N_4 bending test specimens. The differences between a changing content of whisker volumes (0 to 35 %) show a sharp density reduction between 10 and 15 %. This agrees with previous research which shows that maximum density is achieved between 10 and 15 % whisker loadings. Since density is directly related to fracture strength, microtomography may be useful as strength prediction.

In the second part we describe the use microtomography combined with image processing, especially a 2D FOURIER transform, for the analysis of cloth direction distributions in multidirectional reinforced chemical vapor infiltrated (CVI) SiC / SiC test components. The SiC fibers were NICALON™ and were oriented 30, 60, and 90°. In laying up the cloth, orientations may not be maintained and this impacts the mechanical properties. Use of microtomography with image processing may allow mechanical property variations to be established.

INTRODUCTION

The use of 2D or 3D computed tomography for the inspection of ceramics was reported earlier [1 - 3]. Often, however, the primarily aim was to find flaws or irregularities and according to the nature of ceramic materials, microtomographic systems were used [4 - 5]. In terms of spatial resolution todays possibilities are in the region of 50 µm or even less. With 3D cone beam tomographs this resolution is also to be achieved in the third direction, however, not without more sophisticated and therefore time consuming algorithms and only for a very small aperture [6]. According to the setup of 3D cone beam tomographs in terms of contrast discrimination, 2D tomographs are superior for the moment. The coming design of data acquisition and especially computing hardware could change this situation. The recent works in 3D NMR imaging of ceramics [7] could emphasize this effort.

METHODS AND MATERIALS

3D Computed Tomographic System

Measurements were carried out on a third generation 3D cone beam computed tomographic system. The schematic setup is shown in figure 1. The source is a 160 kV microfocus X-ray tube (IRT HOMX 161) with a spot size of 10 to 20 μm diameter. The detector system contains of a trifield (9, 6, 4.5 ") image intensifier and a Sony XC 57 512 x 512 CCD-camera (8 bits). Using a third generation CT-method this setup e.g. would allow a pixelsize of 50 μm for an object of 20 mm diameter (geometric magnification 10 x). For the motor control and the data acquisition an IBM PC/AT with an accelerated frame grabber board was used. The next calculations up to the rough images were made either on a VAX 8700 or on a SUN 4 workstation. For the image processing and display, finally, we used a Macintosh IIfx.

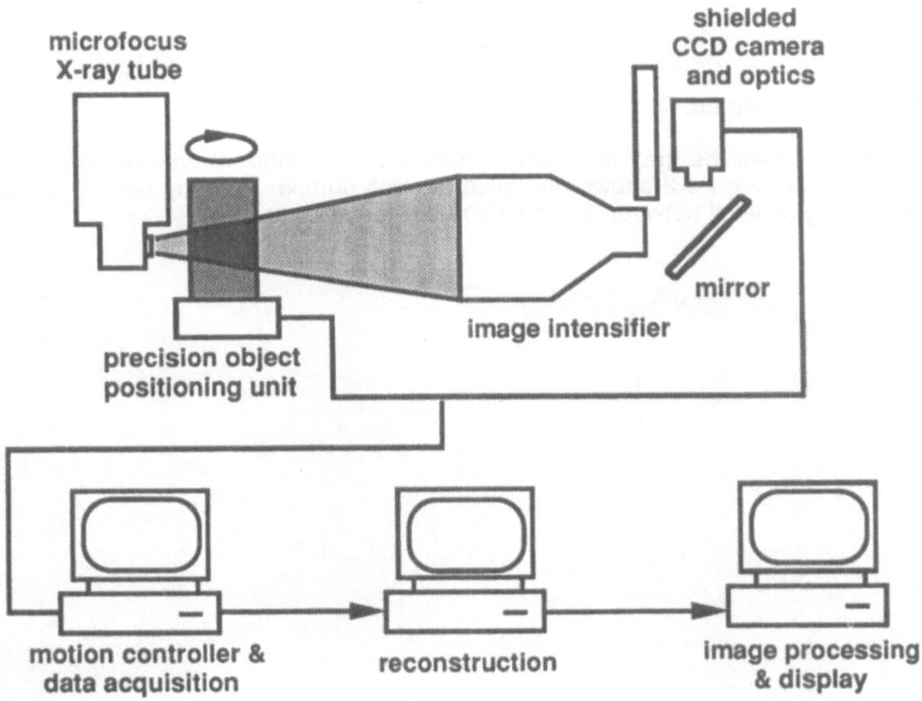

Figure 1. Schematic setup of the 3D tomographic system.
For reasons of clarity, the aperture is shown too large.

Specimens

The first set consisted of Si_3N_4 / Si_3N_4 whisker reinforced bending-test specimens (about 50 x 6 x 3 mm) in the green state. The whiskers had a diameter of 0.6 to 1.0 μm and a length of about 10 to 15 μm. Their distribution was more or less random but certainly influenced by the cold pressing manufacturing process. The volume content of the whisker loadings ranged from 0 to 35 %. For this preliminary work, already broken ones were taken. But the aim of this method certainly is to do this examination before mechanical tests are made.

The second set, however, consisted of multidirectional cloth reinforced SiC / SiC composites. These samples have been produced at Oak Ridge National Laboratory, using a chemical vapor infiltration (CVI) technique which combines the thermal gradient and pressure gradient processes [8]. The SiC fibers were NICALON™ and were oriented 30, 60, and 90°. The objects had a shape like a disk with a diameter of about 45 mm and a height of approximately 12 mm.

In each case the materials of the matrix and of the reinforcement were the same, so that resulting gray levels are approximately related to the density.

RESULTS

Si_3N_4 / Si_3N_4 Samples

For these samples the gray level differences between different whisker loadings were examined. Figure 2 shows one slice of each sample. Already here a sharp reduction in gray level between 10 and 15 % is clearly visible.

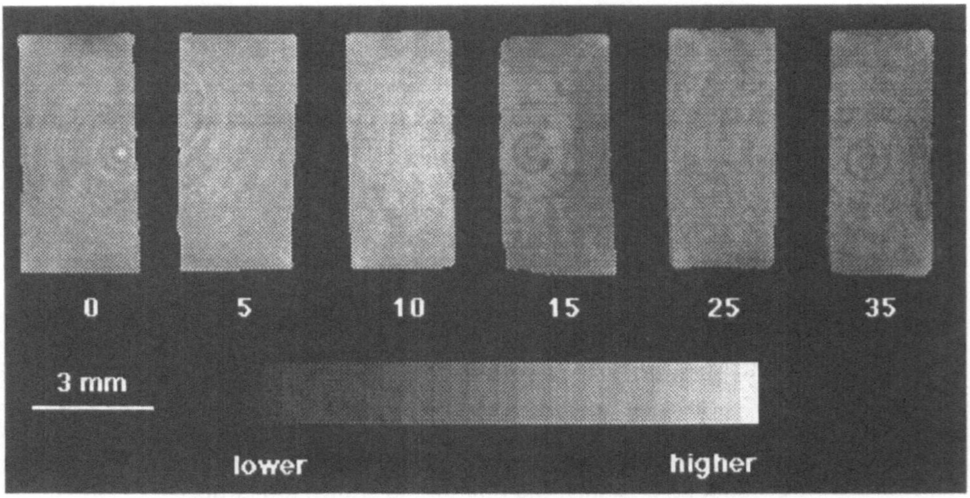

Figure 2. Differences in gray levels for different whisker loadings.

To save calculation time, the images were reconstructed with 256 x 256 pixels (each with two samples), so that a pixelsize of 40 μm results.

To avoid edge effects, the gray levels were measured in the central region of the slices, covering about 50 % of the total area. The standard deviation for one slice is always about 3 to 4 grays, the differences of the mean values of different slices of the same sample are about the same. No differences of these statements are to be found between the unloaded (0 %) and the loaded samples. The reduction between 10 and 15 % is significant; the lowest gray level found in all slices of the 10 % sample is still 15 grays higher than the highest of the 15 % sample (figure 3).

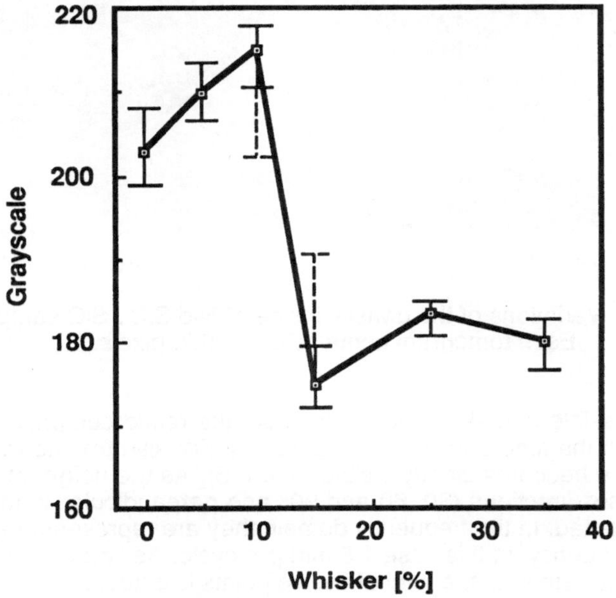

Figure 3. Mean gray levels over seven slices, highest and lowest mean value of the single slices and maximum and minimum gray level of single pixels for 10 and 15 % (dotted line).

SiC / SiC Samples

These objects were much larger and in this case the best possible pixelsize would have been about 100 μm. For the examination of the reinforcement structure, however, a pixelsize of about 200 μm was sufficient. Figure 4 shows a slice through two different samples (261 and 262). Four effects are recognizable: in both cases the density in the center of the samples is higher than at their periphery (the usual beam-hardening effect would show just opposite characteristics). From the gray level curves through the center of the objects, it is possible to assume that the density gradient is nearly linear but different for both samples. The same curves also show a superstructure due to the reinforcement.

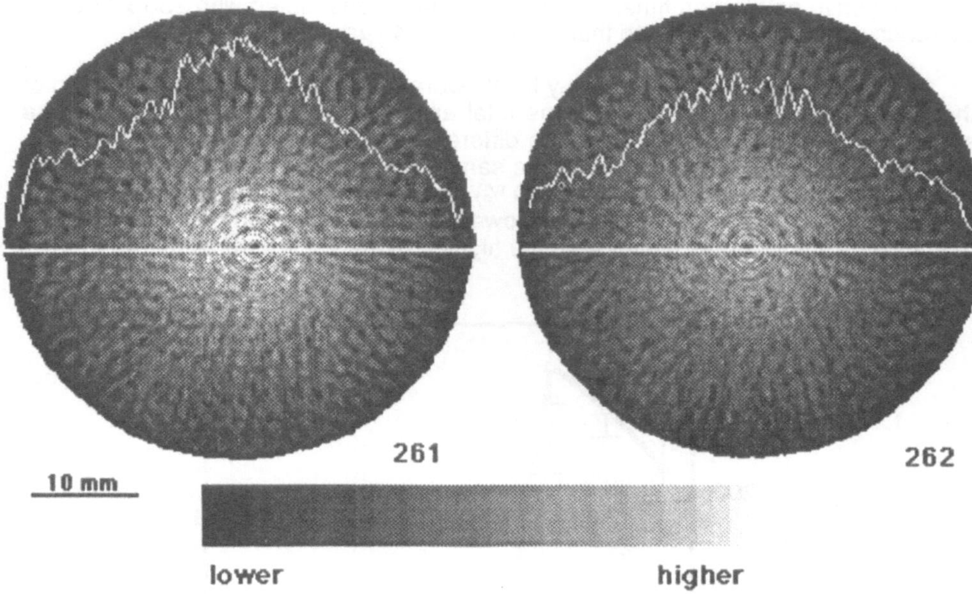

Figure 4. Variations of the density inside of two SiC / SiC samples.
Each tomogram contains 256 x 256 pixels.

It is not possible to make statements about the reinforcement structure looking at the slices in the time domain. After a 2D fast FOURIER transform (FFT), however, this structure becomes clearly visible (figure 5). As the height of each slice is about 1 mm, all six directions (30, 60 and 90° and perpendicular to these, respectively) are represented. In the frequency domain they are represented as points at a corresponding frequency, in this case 1.5 mm per cycle. As figures in the frequency domain are radially symmetric, a total of twelve points is detected.

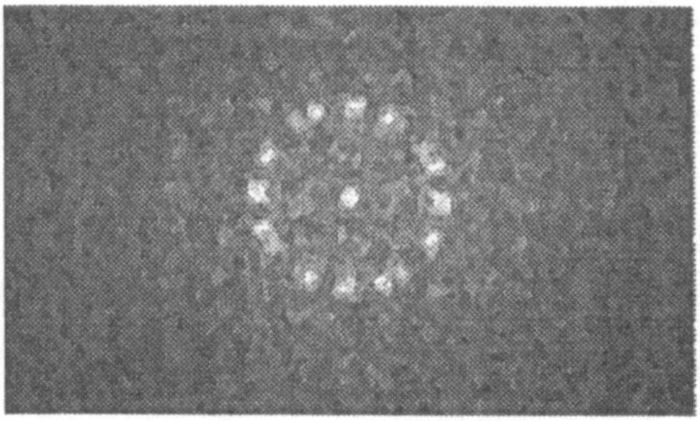

Figure 5. 2D FFT of a slice; the points represents the reinforcement structure at about 1.5 mm per cycle.

Examining the structure of the reinforcement using this kind of image processing, several effects are to be found. The radial extension of the points and therefore differences in the directions of the cloths are not always the same; for some slices even separations in one direction are found (figure 6). To be able to measure these radial extensions equally, the images were transformed into binary versions.

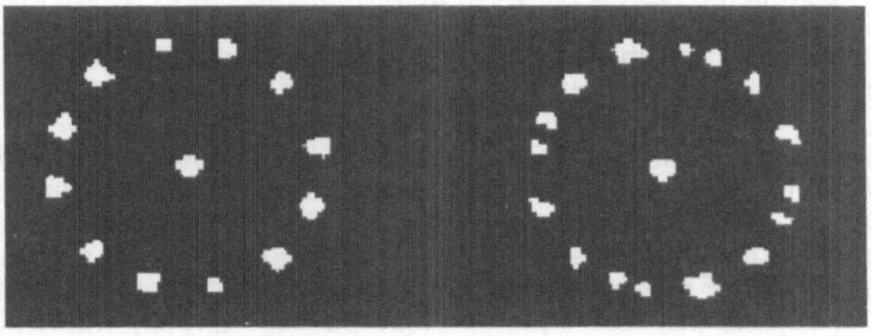

Figure 6. Two basic types of 2D FFTs; left side: normal reinforcement, difference always about 30°; right side: separation in one direction.

The results of these measurements for seven slices are given in figure 7, the direction of 0° is arbitrary as it is influenced by the orientation of the object on the CT system. The trend of these graphs shows that the directions of the layers are not strictly maintained and that the relation between one direction and its perpendicular direction is not always very strong (e.g. sample 261, directions 60 and 150°, respectively).

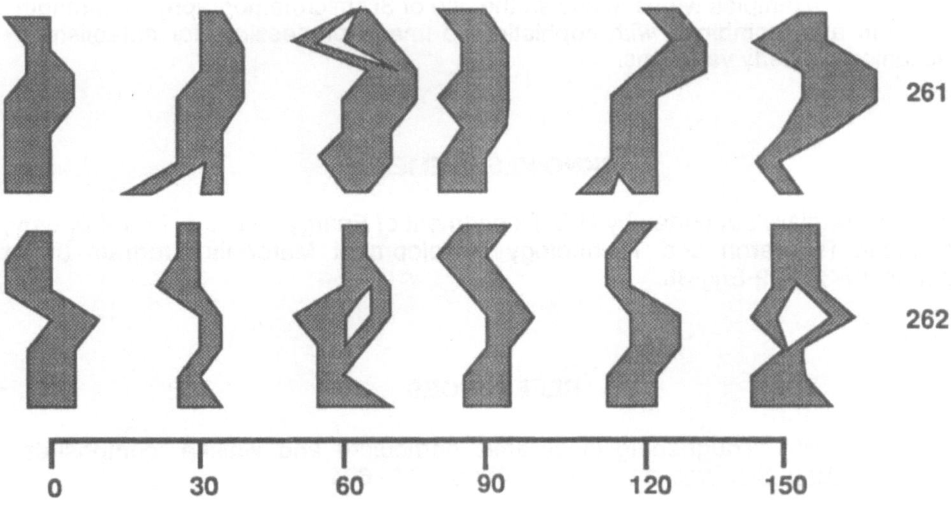

Figure 7. Directions of the reinforcement structure of the multidirectional cloth reinforced samples; each slice has a height of about 1 mm.

DISCUSSION

In the first examination a clear density reduction, which for similar materials is also reported by other authors [9], was found using 3D microtomography. Inside each sample certain density variations can be found which are due to the cold pressing manufacturing process. Because of that and of small density differences we can clearly distinguish between a high density region (0 - 10 % whisker loading) and a low density region (≥ 15 % whisker loading) but with difficulties inside these levels. The clear difference between measurements on the 10 % and the 15 % samples leads to the statement that there are no whisker enrichments that can be detected using the examined volume unit of 1000 x 40 x 40 µm.

In the second examination it was possible to find variations in multidirectional cloth reinforcements in several heights of the samples. Even if the radial accuracy is not more than 5°, differences are clearly shown. The demonstrated weak relation between perpendicular directions of the reinforcement leads to the assumption that certain cloths are under residual stress, which would impact their mechanical properties. For flat objects this technique can also be used directly with radiographic images, but for larger objects, however, the information in the third dimension would be missing and separations or variations would not be detected.

CONCLUSIONS

X-ray microtomographic systems may not only be used for inspection of ceramics with the aim to find flaws or irregularities but also as an aid for process control. Even if the contrast discrimination of a current 3D system is lower than that of a 2D system it is suitable, especially for the second purpose.

In two examples we have shown the use of 3D microtomography for strength prediction and, combined with sophisticated image processing, for establishing mechanical property variations.

ACKNOWLEDGEMENTS

Work was partially supported by U.S. Department of Energy, Office of Fossil Energy, Advanced Research and Technology Development Materials Program under Contract W-31-109-Eng-38.

REFERENCES

1 RICE, R.W.: "Toughening in ceramic particulate and whisker composites", Ceram. Eng. Sci. Proc. **11** [7-8] (1990) pp. 667 - 694

2 ELLINGSON, W.A., VANNIER, M.W. and STINTON, D.P.: "Application of X-ray computed tomography to ceramic/ceramic composites" Symp. on characterization of advanced materials, Monterey, CA (1987)

3 ENGLER, PH., FRIEDMAN, W.D., SKARPNESS, B.O. and OHNSORG, R.: "Process control for ceramics using CT of prefired and fired parts", Conf. on NDE of modern ceramics, ASNT, Columbus, OH (1990) pp. 16 - 20

4 YANCEY, R.N., KLIMA, S.J. and SMITH, J.A.: "High resolution CT of modern ceramics", Conf. on NDE of modern ceramics, ASNT, Columbus, OH (1990) pp. 126 - 130

5 REITER, H., MAISL, M., PANGRAZ, S. and ARNOLD, W.: "High resolution X-ray and ultrasonic methods for the nondestructive evaluation of modern ceramics", Conf. on NDE of modern ceramics, ASNT, Columbus, OH (1990) pp. 64 - 68

6 RIZO, PH, GRANGEAT, P., SIRE, P., LEMASSON, P. and MELENNEC, P.: "Comparison of two 3D X-ray cone beam reconstruction algorithms with circular trajectory", to be published, J. Opt. Soc. Amer.

7 DIECKMAN, S.L., RIZO, P. and GOPALSAMI, N.: "3D NMR and X-ray imaging of solid state materials" in "Advanced tomographic imaging methods for the analysis of materials", Mat. Res. Soc. Symp. Proc. Series Vol. 17 (1990)

8 STINTON, P.D., CAPUTO, A.J. and LOWDEN R.A.: "Synthesis of fiber-reinforced SiC composites by CVI", Amer. Ceram. Soc. Bull. **65** [2] (1986) pp. 326 - 335

9 KODAMA, H. and MIYOSHI, T.: "Fabrication and properties of Si_3N_4 composites reinforced by SiC whiskers and particles", Ceram. Eng. Sci. Proc. **10** [9-10] (1989) pp. 1072 - 1082

NONDESTRUCTIVE TESTING OF CERAMICS BY COMPUTED TOMOGRAPHY AND HIGH-FREQUENCY ULTRASONICS

HOLGER REITER, MICHAEL MAISL, UDO NETZELMANN, WALTER ARNOLD
Fraunhofer-Institut für zerstörungsfreie Prüfverfahren,
Universität, Gebäude 37, D-6600 Saarbrücken, Germany

ABSTRACT

An ultrasonic pulse-echo inspection system (up to 100 MHz) has been equiped with a fast digitizer in order to allow a full volume acquisition of ultrasonic data. A powerful workstation with modern visualization software is used to represent the ultrasonic volume data. Any desired cross section plane inside the object can be selected and displayed in real time. Small defects and variations in the density distribution are imaged by the computed tomography (μCT) with high contrast. In the reconstructed cross sections a pixel size down to 10 μm is realized. Among all the requirements, which a system with high resolution capabilities has to accomplish, we are especially discussing the influence of the radiographic imaging system on the achievable resolution. The same software, used to represent the ultrasonic volume data, is also employed for a 3D-visualization of the 2D reconstructed cross sections of the μCT.

INTRODUCTION

For more than 10 years research and development work has been done for ceramic materials and components for engines. Only in some cases these materials have passed the laboratory state and are now used in practice. The application in an engine requires a sufficient reliability and an adequate guarantee of quality. The development of production techniques, which meet the high requirements of a flawless production, is decisive for a continual increase of industrial use of these new materials and components for engines. At the moment it is only possible in some cases to manufacture serials products with sufficient high and reproducible quality, which fully exploit the positive characteristics of the materials. Nondestructive testing methods, if possible, imaging methods, can contribute an essential part to optimize production methods for products of high quality. The characteristic feature of these materials - high brittleness, small flaw size leads to failure - requires, of course, very sensitive and high resolving nondestructive testing methods to guarantee quality.

Our paper discusses two methods which fit these requirements: high-frequency-ultrasonics and micro-computed-tomography (µCT). The principles are given and examples for their application to ceramics are shown. Conventional B-scan or C-scan images with gated signals provide information only on single slices or layers of the examined objects. Many of these scans are necessary to characterize the whole volume of the specimen. But very often the inspection and evaluation of the full sample volume is required. This can be realized by the technique shown here. It is a volume acquisition system for high-frequency ultrasonic inspections in the 10 to 100 MHz regime. A powerful graphic workstation with modern visualization software is used to represent the ultrasonic volume data. The µCT is a non-contact method. The 2D-CT reconstructs cross sections of only a single plane in the object. This technique produces shadow free images of object slices with high contrast and high geometrical resolution and is therefore qualified to image small defects and variations in the density distribution of the object. The µCT discussed here is advantageous and shows favorable characteristics especially for the examination of critical areas in the volume of complex shaped components. The same software which is used to represent the ultrasonic volume data is also employed for a 3D-visualization of the 2D reconstructed cross sections of the µCT.

HIGH-FREQUENCY ULTRASONICS

We have built up a high-frequency ultrasound (HF-US) system for ultrasound volume measurements in the frequency range of 10 MHz to 100 MHz, which is based on an rf-system described earlier [1]. In order to detect small inclusions in ceramics down to a size of approximately 30 µm, one has to take great care to optimize the signal-to-noise ratio. For excitation of the ultrasound either a commercial broad-band spike generator (Krautkrämer USH 100) or a narrowband transmitter, developed in our institute, is available, the latter allowing to generate high-frequency bursts in the range of 10 MHz to 200 MHz with pulse lengths of 1 to 199 periods. Focussing broadband polymer probes are used [2]. For a volume measurement, the full A-signal pulse-echo response is digitized with a fast transient digitizer at each point of a meander xy-scan [3]. By a sophisticated organization of the data transfers, we achieved a fast overall measurement time, which is now typically 20 to 30 percent longer than the overall mesurement time for a C-scan with the same size and resolution. At present up to 200 × 200 points laterally and 200 digitization points in the time axis are measured, therefore, an amount of 8 Megabytes of volume data has to be searched for defects or sample characteristics subsequent to the acquisition.

Representation Techniques for Volume Data

In the past few years advanced volume visualization software packages have become available which allow an efficient, interactive and real-time representation of volume data of any origin. We are employing a SUN 370 SPARC workstation with a TAAC-1 graphic processor (application accelerator) [3]. The measurement data are transferred by an Ethernet link from the PC to the workstation. The graphic processor is fast enough to allow an interactive visualization of the data.

There are different representation techniques available. The slicing technique allows to cut the mesured volume and to show the cutting planes in a plot, where the absolute of the voxel amplitude is converted into a colour. We have measured a slab of a ZrO_2-ceramic (50 mm × 50 mm, thickness 4 mm) by using a 50 MHz polymer probehead and broadband excitation in pulse-echo technique. The sample has three grooves of 0.5 mm depth on the back. In Fig. 1 we can see a perspective view of the sample, showing the back surface with the three grooves in a C-scan plot and two orthogonal cross-sections of the sample in a B-scan plot simultaneously. The C-scan plane is extracted on the right side of Fig. 1. The sound was incident from the top side. The perspective image can be turned in real time in any orientation. The full volume can be inspected by shifting one of the cutting planes into the volume. Cutting under arbitrary angles is possible, as well as quantification of local amplitude values.

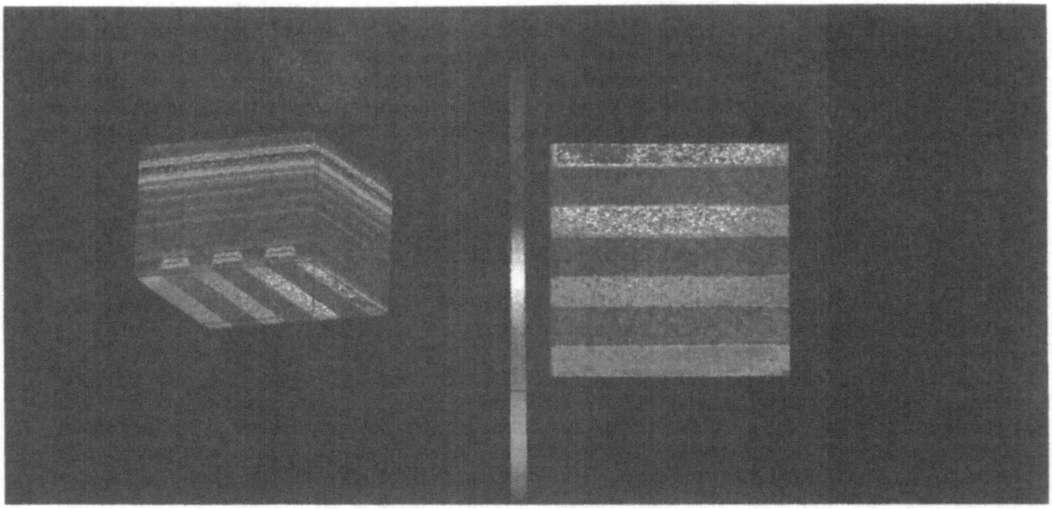

Fig. 1. Ultrasound data representation of a ZrO_2-ceramic by using the sclicing technique

The group of the ray-casting techniques allows to produce a perspective image of the sample showing the whole interior at the same time. Within a mathematical model image rays are sent through the volume and are projected to the image plane. Depending on the imaging strategy, the maximum or the integrated scattering amplitude along the ray is converted into a grey value, leading to an X-ray-like image. In another approach, light rays from a point source are traced on their way through the volume including all light reflection and light scattering effects resulting in a pseudo 3D-volume image. The ultrasound echo amplitudes are related to selectable substances with optical properties like colour or transparency. A result of the latter image generation strategy is shown in Fig. 2, where a SiSiC-ceramic sample with many pores or inclusions has been measured. These features were not expected by the manufacturer. The experimental parameters and the sample geometry were the same as in the example mentioned above. In a further step, many of such views with the sample turned around slightly between each can be combined to generate a movie of the roating object.

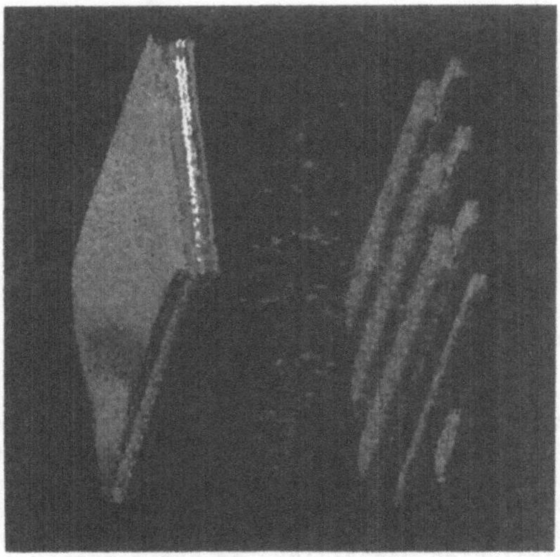

Fig. 2. Ultrasound data representation of a SiSiC-ceramic using the ray-casting technique

COMPUTED-TOMOGRAPHY

Ceramic materials and components for engines need special computed tomography systems which allow the detection and resolution of very small defects, less than 100 µm in lateral dimension. The resolution in the reconstructed cross section is determined mainly by: the accuracy and stability of the mechanics used for the object manipulation, the imaging characteristics of the radiographic system (X-ray source and detector system), the reconstruction algorithm, the number and size of the pixels (voxels), the noise in the projection data the number of projections. Contrast resolution and geometrical resolution are not independent of each other. Under the aspect of realization possibility we will discuss in more detail the influence of the imaging characteristics of the radiographic system on the geometrical resolution.

The geometrical resolution is limited by the geometrical unsharpness of the radiographic system and the pixel size in the reconstructed cross section. Geometry and size of the focal spot of the X-ray tube and of the detector determine together with the position of the object (direct magnification) the geometrical unsharpness. The modulation transfer function (MTF) describes the spatial frequency dependence of the geometrical resolution. Often this value is regarded as limit of resolution, where the MTF has decreased to 10%. The MTF of the whole system results from the product of the single components. A high geometrical resolution in the reconstructed cross section requires a small pixel size w (sampling theorem) and a high geometrical resolution of the imaging radiographic system. This can be achieved by two arrangements:

1. Direct magnification m ≈ 1, large size of the focal spot, detector system with high resolution (small size of the active area). For a direct magnification of m ≈ 1 the influence of the size of the focal spot can be neglegted. The focal spot can be large, but if it is large than the resolution of the detector must be high. The size of the single detectors must be small. According to the sampling theorem the scanning-interval must be diminuished that means the spatial digitalization of the projection data must be increased.

2. Direct magnification m » 1, small size of the focal spot, detector system with medium resolution. For a direct magnification m » 1 of the object the geometrical unsharpness caused by the X-ray source must be reduced. An X-ray source with small focal spot is necessary. In this case a detector system with only lower resolution capability can be used.

Different detector systems are used for computed tomography systems. Photomultipliers with scintillators, image intensifiers and photodiodes with scinitllators (line detector) are commercially available and often employed in the field of material testing. The scintillating material as converter material is common for all these systems. It determines up to an essential part the resolution capability of the detector system. The MTF which is caused by the scintillating layer can be calculated [4] under the following assumptions: weak X-ray absorption in the layer, homogeneous production of the photons at each point of the layer across its thickness, no self absorption of the photons, scattering of the photons in the layer. The total MTF of a line detector results from the MTF of the layer and the MTF which is given by the distance and the size of the diodes. Figure 3

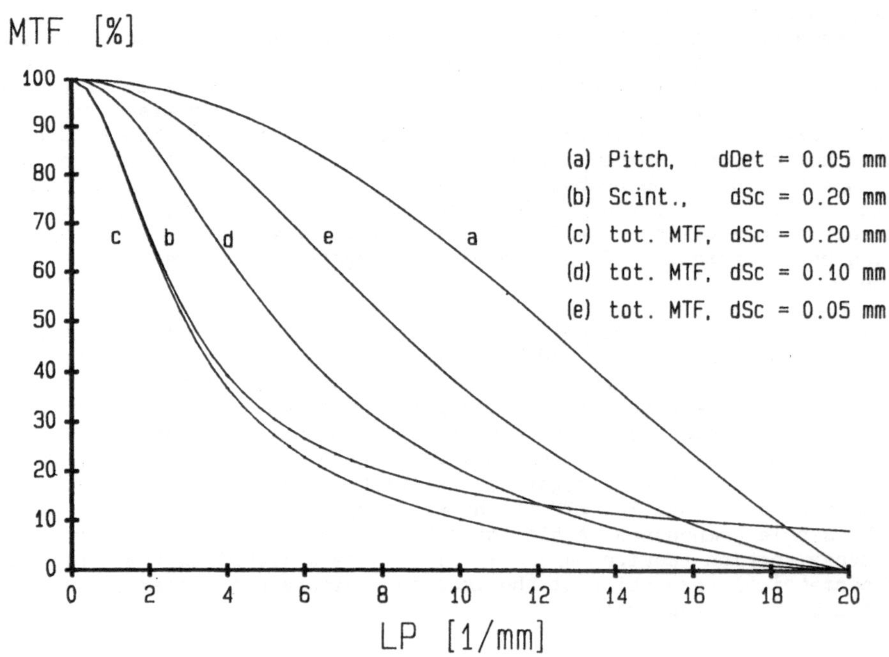

Fig. 3. MTF of a line detector with different thickness of the scintillator layer

shows the MTF calculated under different assumptions. Curve (a) results only from the pitch of 50 µm of the photodiodes, curve (b) represents the MTF which is caused by a thickness of the scintillator of 200 µm, curve (c) is the result of the pitch of the diodes (a) and the layer thickness (b). The total MTF decreases very quickly because of the quick decreasing MTF of the scintillator layer. The resolution capability can be considerably improved by a reduction of the layer thickness, as shown in curves (d) and (e) of figure 3: Based on this consideration the layer thickness should not exceed the pitch of the diodes.

A thin layer thickness of the scintillator leads to a weak efficiency of the detector. The X-ray absorption and therefore the conversion into light is drastically reduced. A high efficiency and a high geometrical resolution can be obtained with a thick scintillator and small collimators or with a tangential illumination of a thin scintillator parallel to the photodiodes [5-7]. An alternative way to obtain a high resolution of the whole computed-tomography-system uses a thick layer of scintillator without collimators and a direct magnification of the object with the aid of a microfocal X-ray tube. A layer thickness of 400 µm and a pitch of the diodes of 450 µm results in a MTF of 15% at a value of 1.1 LP/mm at the Nyquist frequency. By using this detector system and a microfocal X-ray tube and a direct magnification by a factor of 21.5, a lead pattern with a spacing of 50 µm is transferred with a contrast of 55%. We use this principle in our µCT-system [8].

A microfocal X-ray tube is used as an X-ray source. The objects are rotated in its fan beam. The accuracy for the adjustment of the rotation desk must be in the order of 1 µm. The solid state detector with a dynamic range of 1000:1 consists of 1024 single photodiodes with a pitch of 450 µm. The photodiodes are covered with a Gd_2O_2S:Tb-scintillator. With the aid of the filtered back projection algorithm the cross sections of the objects are reconstructed out of the single projections collected. The reconstruction is performed by an array processor parallel to the data acquisition and the time necessary for the testing process is consequently only given by the measuring time. For a cross section with 1024 x 1024 pixels this takes normally about 20 minutes. About 10 µm is the smallest pixel size realized at the moment with this system. Two examples for the application of the system to the nondestructive examination of ceramic materials and components are shown in the figures 4 and 5.

Analysing larger volumes of an object by evaluation the adjacent and touching 2 dimensional (2D) cross sections can be very time consuming. Each single set of data must be transferred from the memory to the graphic screen. Figure 4 shows for example 2 reconstructed cross sections of a four point bending bar in adjacent planes. The interpretation of the relative position of features in different planes is difficult. But if many planes of an object, one laying upon another, are examined and the reconstructed 2D-cross-sections are combined, a volume visualization of the data can be performed as described in the high-frequency-ultrasonic part of this paper. In figure 5 the pseudo 3D-volume view is demonstrated with 35 tomograms of a SiC turbo charger rotor. In the left part of the figure the view of the rotor is parallel to the axis and in the right part of the figure one looks to the rotor from a standpoint perpendicular to the axis. In both views a pore in one of the blades can be detected. In the left part of the figure the pore can be seen in the blade at 4 o'clock. In the right part of figure 5 the oblongness of this pore can be recognized.

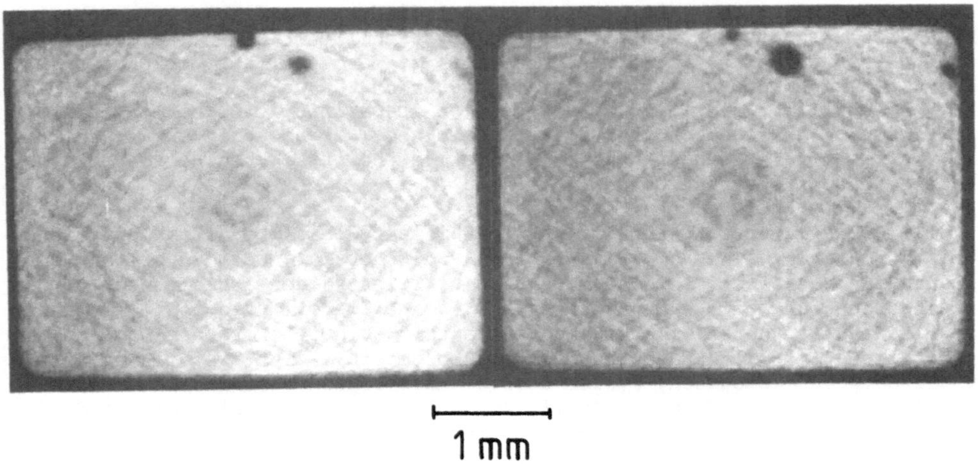

Fig. 4. Reconstructed cross sections of a bending bar; diameter of the detected pores < 100 µm, pixel size 27 µm

Fig. 5. 3D-visualization of 35 2D-cross sections of a turbocharger rotor; detection of a single pore in one of the blades; pixel size 150 µm

CONCLUSIONS

We have shown that by optimizing data transfers an ultrasound volume measurement can be performed in a time not significantly longer than the time for a conventional C-scan. By using modern visualization techniques we have obtained a surprisingly comprehensive and clear representation of the internal structure of the sample. Advantages compared to earlier techniques of volume representation are given by the real-time image processing, by the higher image quality and by the universality of the software used, allowing to visualize data produced by all kind of techniques generating volume information. The fact that in ceramic materials and components very small defects can lead to failure, place high demands on the nondestructive testing methods. The µCT can accomplish these demands, also in critical areas in the volume of complex shaped components. We have shown different realization possibilities for high resolving computed tomography systems. By combination of many 2D-cross sections a volume visualization of the data can be performed. This allows a better understanding of the internal structure of the object. All defects which are detected in the volume are imaged simultaneously. The decision about the criticality of the defects becomes easier.

REFERENCES

1. Pangraz, S., Simon, H., Herzer, R. and Arnold, W., Proc. 18th Int. Symp. Acoust. Imag., Santa Barbara 1989, to be published, Plenum Press, New-York 1991

2. Pangraz, S. and Arnold, W., Ferroelectrics **93**, 251 (1989)

3. Netzelmann, U., Herzer, R., Stolz, H. and Arnold, W., Proc. 19th Int. Symp. Acoust. Imag., Bochum, FRG, 1991, to be published, Plenum Press, New-York

4. Swank, R.K., Calculation of modulation transfer functions of X-ray fluorescent screens. Appl. Optics, 1973b, **12**, 1865-70

5. Seguin, F.H. Burstein, P., Bjorkholm, P., Homburger, F. and Adams, R.A., X-ray computed tomography with 50 µm resolution. Appl. Optics, 1985, **24**, 4117-4123

6. Steinbock, L., Tomography of nuclear fuel with an electronic line scan camera. European Seminar on Industrial Computerized Tomography, Grenoble, 1988

7. Engelke, K., Mikrotomographie mit Synchrotronstrahlung zur quantitativen Darstellung des Mineralgehaltes in Knochen. Dissertation Uni Hamburg, 1989

8. Maisl, M., Reiter, H. and Höller, P., Micro-radiography and tomography for high resolution ndt of advanced materials and microstructural components. J. Eng. Mat. Techn., 1990, **112**, 223-226

CHARACTERIZATION OF DEFECTS IN CERAMICS BY COMPUTERIZED TOMOGRAPHY WITH HIGH SPATIAL RESOLUTION

B. Illerhaus, J. Goebbels, H. Heidt, W. Müller, Y. Onel,
P. Reimers, V. Wolff
Federal Institute for Material Research and Testing (BAM),
Unter den Eichen 87, 1000 Berlin 45, Germany

ABSTRACT

X-ray computerized tomography with spatial resolution of a few tens of micrometers and high contrast resolution offers new possibilities for the detection of shrinkage cracks and incomplete wetting of the contact surfaces in soldered metal - ceramic joints, of crack growth due to bending tests, the determination of the material in small inclusions and the resolution of density gradients in the neighbourhood of voids in ceramic materials.

EXPERIMENTAL SET UP

One of the most crucidal points in using ceramics are their mechanical properties. To get more reliable parts more investigations of the mechanical properties of ceramics and ceramic components have to be done. This includes on the one hand a controlling of the production process to elimenate systematical faults, to get more information on critical processes. On the other hand e.g. stress and bending tests in combination with non destructive testing methodes are necessary to get more insight in failure behavior of ceramics. The relationship thus yielded between NDE indications and mechanical properties of a ceramic then gives informations to judge the quality of ceramic components by NDE. In both fields computerized tomographie (CT) is a powerful tool, as was stated before [1],[2]. Due to the brittleness of ceramics the highest achievable spatialy resolving CT has to be used to detect the smallest possible defects and intrinsics.

The "Federal Institute for Material Testing and Research" (BAM) has been involved with CT for the last ten years. So far two computer tomographs specially designed for industrial applications have been set up [3],[4]. Both were constructed to allow a wide range of industrial objects to be tested and therefore cannot represent the most favourable solution for a specially chosen problem. But they are able to test the possibilities of CT for the widest range of objects and materials. In this report we will concentrate on investigations of ceramics with the second scanner, refered to as micro-CT.

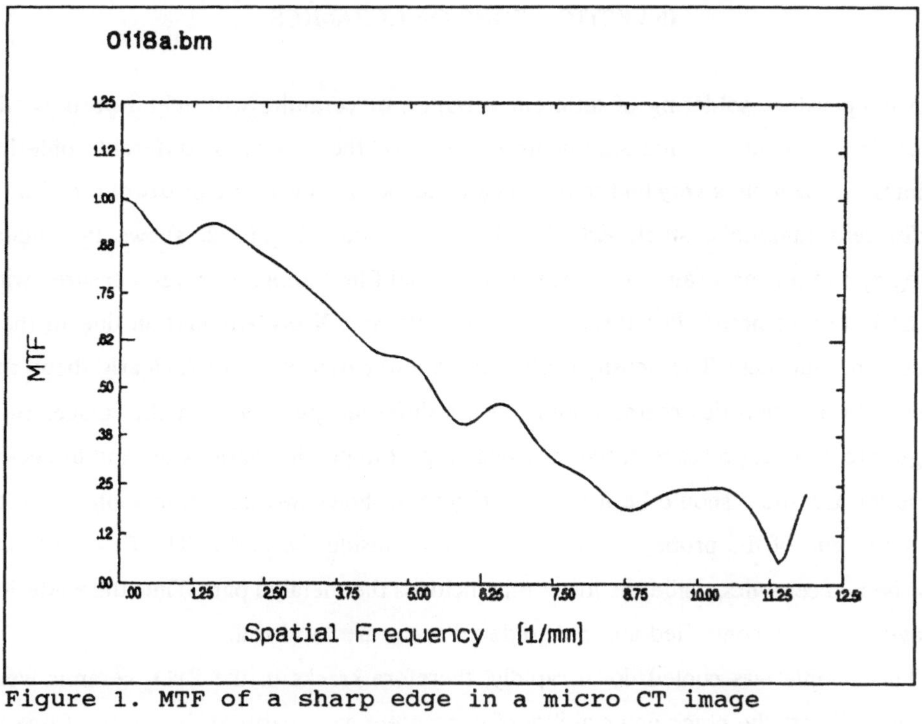

Figure 1. MTF of a sharp edge in a micro CT image

This scanner has the following technical configuration:

X-ray tube: mini-focus, 160KV HV generator, 2.0mA,
spot size 160μm x 200μm at 90KV

detector: photo diode array with 1024 elements of 25μm x 2500m, glass fiber optic coupled scintillator, 80° incident beam

mechanics: 3-axes object scanner, 25cm movable range, reproducible to 3μm over all; scanner to change source - detector and source - object distance;
2-axes source positioning scanner

The best way to test the overall spatial resolution of a CT scanner is to measure the modulation transfer function (MTF) of a sharp edge in a reconstructed image. Figure 1 shows the MTF of the BAM micro computer tomograph for optimal magnification conditions.It is convenient to set the limit of spatial resolution to the 10% value of the MTF. This results in 10 lines per mm or 50μm as spatial resolution.

INVESTIGATIONS OF CERAMICS

High temperature soldering of different ceramics or ceramics with metal parts is still under development. CT measurements can proove the quality of different soldering technics. As example a very bad soldered ceramic part is shown. It consists of three parts of different materials: steel, ZrO_2- and SiN- ceramic. Figure 2. shows the digital radiograph of this part (an X-ray image like normal film radiography but measured with the CCD line camera). The three parts have different X-ray attenuation due to their density and thickness. The density profile in the lower part of figure 2. clearly shows the three different densities corresponding to the different grey levels in the image. Such images are used as prescans of the investigated part to get an overall view and to decide where the CT slices should be measured. Figure 3. shows two tomograms, one close to the upper end of the probe, the second one more inside the probe. The first slice cuts only the two ceramics, while the lower one includes the metallic part. Thus the soldering behaviour can be controlled and not soldered areas are localized.

This sample was cooled down rapidly; therefore cracks in the ZrO_2 ceramic were induced. To test the crack detectability of our system we measured the width of one of

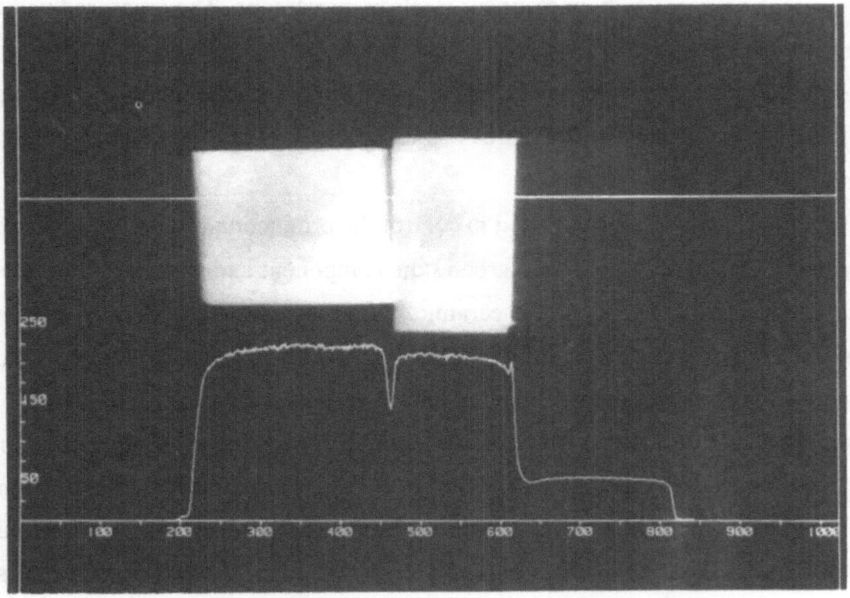

Figure 2. Soldered steel - ceramic - ceramic part (digital radiograph).

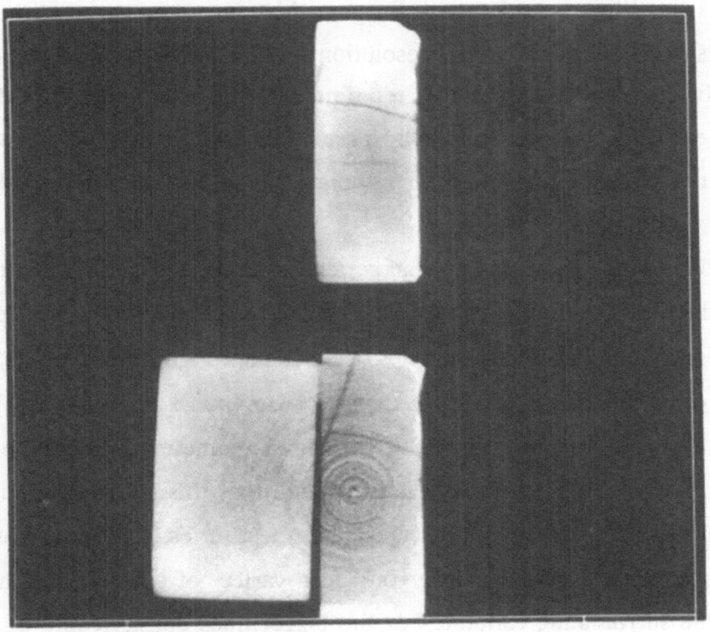

Figure 3. Soldered steel - ceramic - ceramic part (CT, 2 hights).

the cracks at the surface with an scanning electron microscope. The measured width was less than 10μm.

The so called ring artefacts visible in the lower tomogram are induced by a longtime drift of the detector offset. They are visible specially when the object has a high attenuation for the used X-rays.

CT measurements also can be used to control the production process of ceramics also in early states. Density variations in a green state component and shrinkage processes can produce stress induced cracks in the ceramic. CT can measure the density distribution also inside complex parts. The density resolution of the described tomograph is upto 3 to 5% for neighbour image pixels depending on the chosen measuring time.

Figure 4. shows an CT of a rotor of a small ceramic turbine. During test series with new materials a crack in the top end of the rotor shaft was detected. CT measurements were able to determine the size and depth of the crack thus deciding the tolarability of it. On the right side of the image a part of the crack is enlarged 20 times. It shows that the crack at this position is spreading up in two. The total diameter of the shaft is 12mm.

Figure 5. shows the CT of a different turbine rotor with an over all diameter of 46mm. Here a small hole was detected. Because this tomogramm was measured with a high spatial as well as a high density resolution the surrounding of this hole could be investigated. In the upper right corner a ten time enlarged section is shown. It is clearly visible that the hole itself is surrounded by a region of lower density whose extend is far above a normal edge bluring effect. This might be important while explaining the origination of such holes.

To get more insight in the behavior of ceramics various test series have to be done. Therefore standard size ceramic probes for bending tests were produced. Before every test CT investigations on these samples were done. Figure 6. and 7. show two tomogramms cross sectioning different examples (size 4mm x 6mm x 40mm). The first one shows a higher attenuating inclution of 0.4mm diameter. Due to the measured density, given by the tomogramm, it was concluded that this sphere should consist of zirkonium. Such examples should be excluded from test series. The second one shows a crossection as normaly expected. Here you see a variety of small inclutions of lower density than the surrounding ceramic. For the bigger ones density could be calculated from CT measurements. In this case the inclutions revealed to have the density of alluminium. Both samples were cut and pollished at the investigated hight and examined

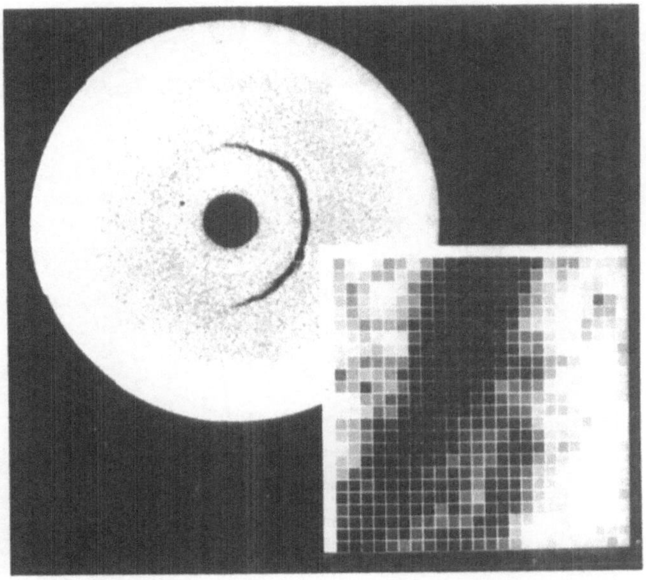

Figure 4. CT of a ceramic turbine shaft, 12mm diameter.

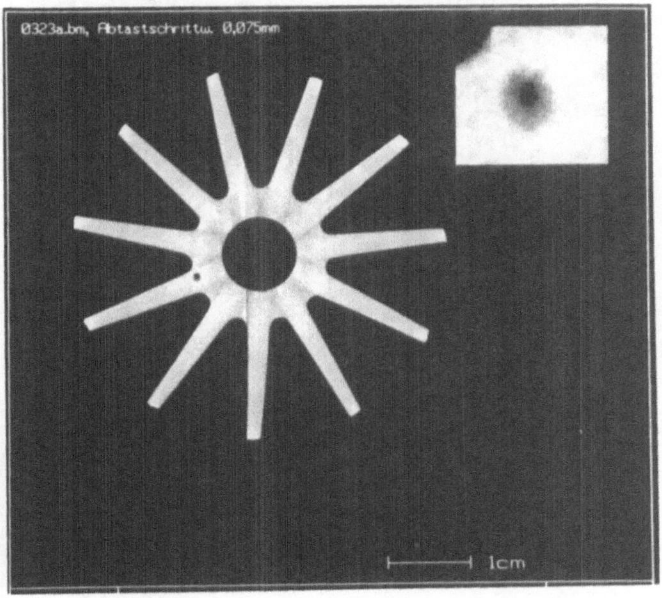

Figure 5. CT of a ceramic turbine rotor (diameter 46mm).

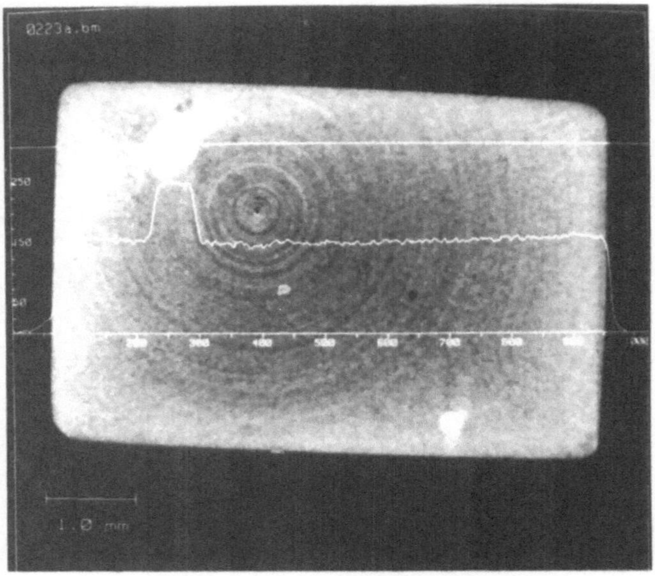

Figure 6. CT of a ceramic bending probe.

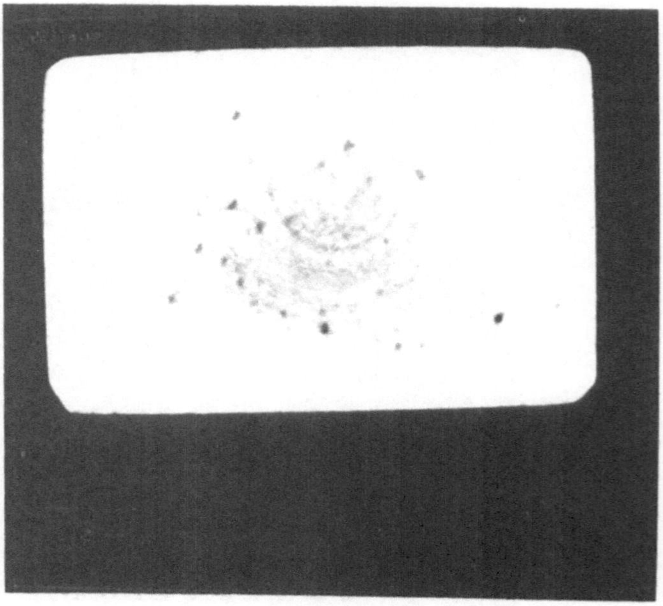

Figure 7. CT of a ceramic bending probe.

in a raster electron microscope. Both types of inclutions were verified. By the methode of X-ray edge adsorption the involved materials were identified: the large inclution consists of zirkonium, the smaler ones of aluminium.

The standard probes then had to undergo a four point bending test, with the crack induced perpendicular in the middle of the probe. To get the best crack resolution CT measurements had to be done with the long axes of the sample in the CT plane. The crack depth and crack spread should be investigated. Therefore a slice thickness of 50μm and a slice distances of 25μm was chosen. To reduce measuring and image reconstruction time as well as data amount the region of interest methode [5] was chosen. Because the outer shape of the sample is constant for all CT sections only one CT slice of the whole 40mm long sample had to be measured. With the ROI methode it is then possible to reconstruct the small interesting inner region arround the crack of about 4mm. In figure 8. 12 adjacent CT slices are shown, the first row still showing the sawed in chink which gives the starting point of the crack. In the third and forth image of the second row it is easily visisible that the crack tends to connect reachable aluminium inclutions. The first image of the third row shows that the crack even splits up to connect inclutions. In the last images the crack tends to a straight line again.

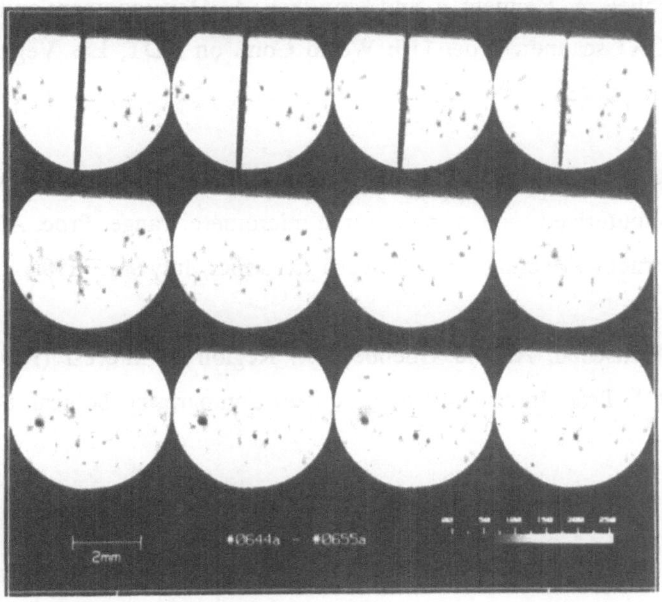

Figure 8. 12 CT slices of a ceramic bending probe

CONCLUSIONS

It has been shown that micro CT is able to reveal single cracks and voids in ceramics down to less than 10μm. It will be an important tool in the nondestructively testing of highly stressed ceramics and to guide the set up of new production lines.

REFERENCES

1. Kress, J.W. and Feldkamp, L.A., X-ray tomography applied to NDE of ceramics. Am. Soc. Mech. Eng. 83-GT-206, 1983.

2. Goebbels, J., Heidt, H., Kettschau, A. and Reimers, P., Computed X-ray tomography of ceramic turbine rotors. Proc. 2nd Int. Symp. Ceramic Materials and Components for Engines, Lübeck-Travemünde, April 14-17, 1986, pp 825-831.

3. Heidt, H., Goebbels, J., Reimers, P. and Kettschau, A., Developement and application of an universal CAT-scanner. Proc. 11th World Conf. on NDT, Las Vegas, Nov. 3-8, 1985, pp. 664-671.

4. Illerhaus, B., Goebbels, J., Onel, Y. and Reimers, P., Nondestructive evaluation of ceramics with computerized tomographie in the micrometer range. Proc. ACerS/ASNT Conf. on Nondestructive Evaluation of Modern Ceramics, July 9-12, 1990.

5. Reimers, P., Kettschau, A. and Goebbels, J., Region of interest (ROI) mode in industrial X-ray CT. Proc. Industrial Computerized Tomography Topical, Seattle, WA, July 1989, p. 48.

DETECTION OF SURFACE DEFECTS IN CERAMIC ROLLING ELEMENTS

S.A.Horton
SKF Engineering and Research Centre,
Nieuwegein, The Netherlands.

ABSTRACT

Hybrid ceramic bearings consisting of steel rings with silicon nitride balls are now being manufactured on a commercial basis for use in high performance systems. The ceramic rolling elements should be capable of withstanding cyclic contact stresses of up to 4 GPa in order to give the required rolling contact fatigue performance. Surface quality must therefore be guaranteed in terms of freedom from both material and processing faults. High sensitivity fluorescent penetrants are capable of detecting fine surface breaking defects and have been adapted together with different imaging techniques for the routine inspection of ball surfaces.

INTRODUCTION

Of the modern engineering ceramics silicon nitride has the best combination of mechanical and physical properties for rolling bearing components. Hybrid bearings consisting of steel rings and silicon nitride balls offer a number of performance advantages over their all steel counterparts, such as increased operating speeds, higher stiffness and greater thermal stability. These bearings are now being used commercially in machine tool spindles and other high speed high precision equipment.

In contrast to many other structural ceramic parts, cyclic stresses in bearing components are very high and may exceed 4 GPa. Such stresses are localised at or close to the surface. Good rolling contact fatigue performance depends not only on the correct specification and selection of the materials but also on the surface quality of the finished components. Silicon nitride has excellent intrinsic fatigue characteristics and fully dense material free from material faults, such as inclusions and porosity, outperforms bearing steels in rolling contact fatigue tests. The normal failure mode is by spalling, the least harmful mode, which gives warning of failure. However, severe surface defects such as cracks can lead to catastrophic failure with the attendant risk of damage to equipment or life. Surface defects can be introduced at any stage in the manufacture of components from initial blending of the starting materials to the final finishing operations, or in fact, afterwards through damage during handling or assembly into bearings.

MANUFACTURE OF SILICON NITRIDE ROLLING ELEMENTS

It is convenient to divide the manufacture of silicon nitride balls into primary processing and finishing. The term 'primary processing' is used here to describe all stages in manufacturing from powder to the formation of densified ball blanks while 'finishing' covers subsequent grinding and lapping operations.

Primary Processing

In the early stage of development of ceramic bearing components, the starting material was hot pressed silicon nitride which was available in the form of discs or slabs. Ball, roller and ring blanks were then machined from these slabs using diamond tooling as illustrated in Figure 1.

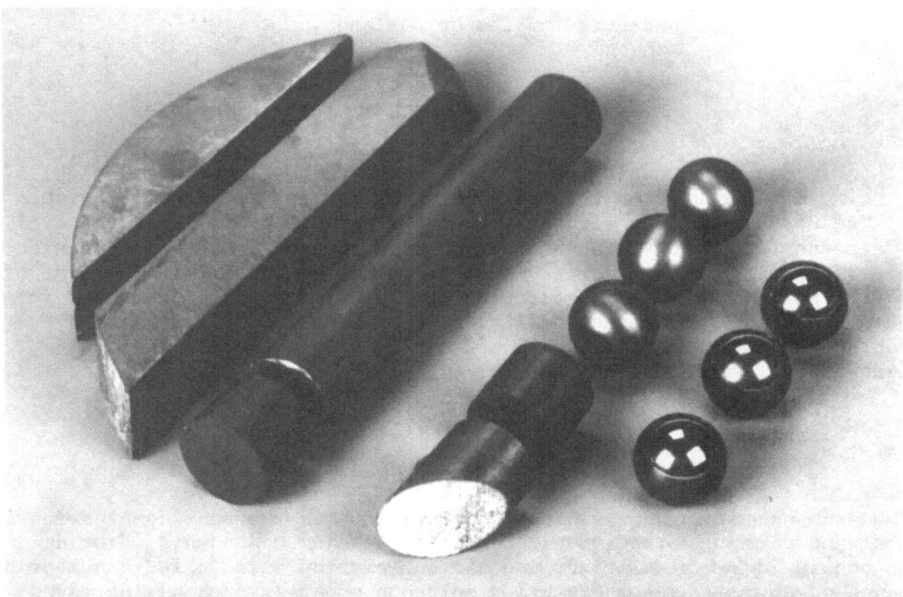

Figure 1. Machining of ball blanks from hot pressed material.

Such machining operations were time consuming and expensive and obviously not suited for high volume production with the added factor of poor material utilisation. Net-shape manufacturing methods were therefore developed for the efficient production of ball blanks.

Figure 2 shows examples of the net-shape processing sequence for ball blanks from the original powder to densified blanks. In common with other ceramic materials the mixing/milling of additives is an important stage which influences final product quality. Some form of agglomeration process, for example spray drying, is needed to make the powder flow. There are several methods of producing ball preforms - cold isostatic pressing, injection moulding, die pressing and compression moulding. After cold pressing soft machining is often used to improve dimensional quality.

Figure 2. Net-shape manufacture of ball blanks.

Densification by sintering alone (pressureless sintering) tends to produce dense material; but usually with some residual porosity which leads to poor rolling contact fatigue resistance. Hot isostatic pressing (HIP) is now seen as the preferred method of manufacture for bearing components since this process, applied directly to encapsulated preforms or as part of a sinter/HIP method, produces 100% dense material.

The material quality of approved bearing grade silicon nitride is of a high standard and very consistent from delivery to delivery. Defects occurring during primary processing in the form of, for example, contamination of powder or incomplete densification are rarely a problem. However because of the nature of the primary processing and the properties of the material at the different stages of manufacture, isolated random defects can still occur. Segregation of additives during blending or agglomeration of silicon nitride particles leads to regions with high or low additive content which can give localised incomplete densification or concentrations of glassy phases. Cold pressing is also a potential source of defects where cracks may be formed as a result of high ejection forces or from 'stiction' effects when die pressing green preforms. Similarly cracks and tears may be formed during the soft machining of green preforms. These types of defect remain as open cracks in the ball blank or are partially or completely healed during densification. Residues of organic materials also become a problem when voids or blowholes are formed, and inorganic contaminants can react with powders to form metallic or ceramic inclusions.

Finishing

Silicon nitride balls can be finished in the same way as steel balls in conventional ball lapping machines. However, due to the high hardness of silicon nitride it is necessary to use diamond abrasives for these operations. With steel balls, the excess stock is removed by soft grinding (before heat treatment) and hard grinding (after hardening) using large silicon carbide abrasive discs. These grinding steps are not feasible today for silicon nitride balls so that lapping times are longer than for steel balls. Roughness average values (Ra) of less than 0.008 µm are routinely achieved during ball lapping and at this level of finish, inspection of surfaces for either processing or material defects is greatly facilitated.

Finishing operations may also be a potential source of surface defects such as Hertzian cone cracks arising from blunt overloads and radial cracks due to the action of large diamond abrasive particles during ball finishing. Although silicon nitride balls are more prone to randomly occurring material defects on the surface than steel balls, their high hardness means that they are less susceptible to repetitive features such as small indentations and scratches.

NON-DESTRUCTIVE EVALUATION TECHNIQUES

A number of non-destructive evaluation (NDE) techniques are being developed for the detection, measurement and classification of defects and non-conformities in ceramic components. Some of the NDE techniques which may be used for surface quality assurance of ceramics include:

X-Ray Radiography

Standard x-ray radiographic techniques are of little use for ceramic bearing components due to their low resolution. Microfocus radiography has been shown to be effective in detecting large metallic inclusions where there is good contrast due to the high density and absorption of metallic phases. However, phases which have similar properties to the silicon nitride matrix, such as some ceramic inclusions or agglomerations of glassy phase, will be more difficult to detect.

Ultrasonic Techniques

For bearing components, surface wave techniques are capable of detecting fine surface cracks and inclusions. These involve the use of special probes operating at frequencies of 50 MHz or higher. The surface wave technique is however difficult to apply to routine inspection of production quantities of finished balls and very high frequencies are required in order to detect fine Hertzian cracks on finished ball surfaces.

Fluorescent Dye Penetrants

Although simple in operation, fluorescent dye penetrants are very effective in detecting surface defects on ceramic balls, when combined with long wave ultraviolet illumination. The fluorescent dye technique can detect very fine features and is also more suited than other methods to routine inspection of large numbers of components. However shallow holes and healed pressing defects are examples of features that do not retain penetrant.

Light Microscopy

Major defects in ceramic components such as open cracks, holes and missing material can be readily detected at relatively low magnifications with light microscopy. Whereas inclusions and macroscopical variations are revealed using darkfield illumination. These features will not necessarily be detected with fluorescent dye penetrants.

Other Techniques

Various techniques including X-ray computed tomography, thermal wave microscopy and photo-acoustic microscopy have been applied to ceramic materials. Currently these methods are not sufficiently developed for routine inspection of relatively large numbers of components. Backscatter laser techniques appear promising for detecting major defects such as holes or missing material.

FLUORESCENT DYE INSPECTION

Although originally developed for the inspection of larger metallic components, fluorescent dye inspection techniques have been applied successfully to silicon nitride balls. The following steps are involved :

- ultrasonic cleaning
- immersion in fluorescent penetrant
- removal of surface penetrant in emulsifier solution
- washing and drying
- examination under ultraviolet light

Careful control of the processing is necessary to avoid contamination of the solutions and balls surfaces after processing. Control balls with known defects are also added to the inspection lot to check on the effectiveness of processing and also for optimisation of illumination conditions.

The use of specialised examination techniques allows the detection of a wide range of defects. These include, relatively coarse defects such as localised areas of incomplete densification, as shown in Figure 3, and very fine cracks which are not resolved when examined under high power reflected light conditions. An example of the latter type of defects are Hertzian cracks which are illustrated in Figure 4 with different illumination conditions. Point defects which are associated with radial cracking are also detected with the penetrant method; an example of this type of defect is given in Figure 5.

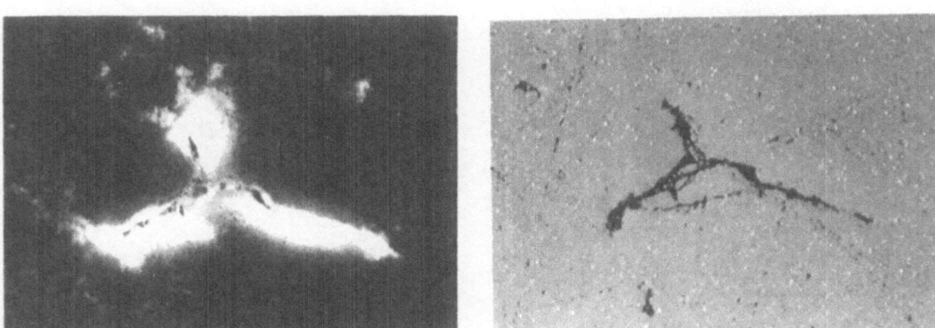

Ultraviolet illumination Brightfield illumination 100 μm

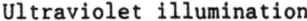

Figure 3. Incomplete densification.

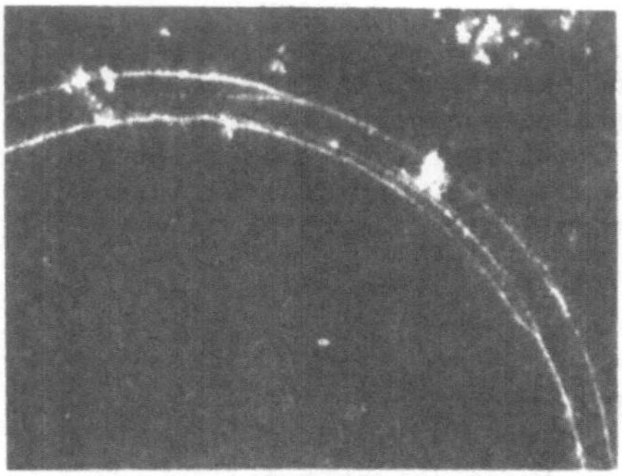

Ultraviolet illumination

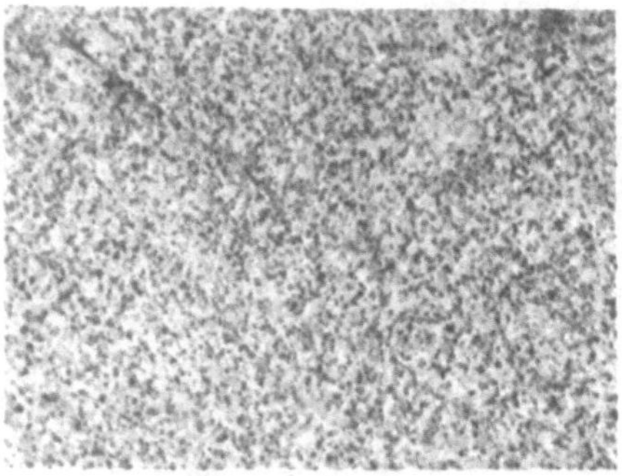

Brightfield illumination

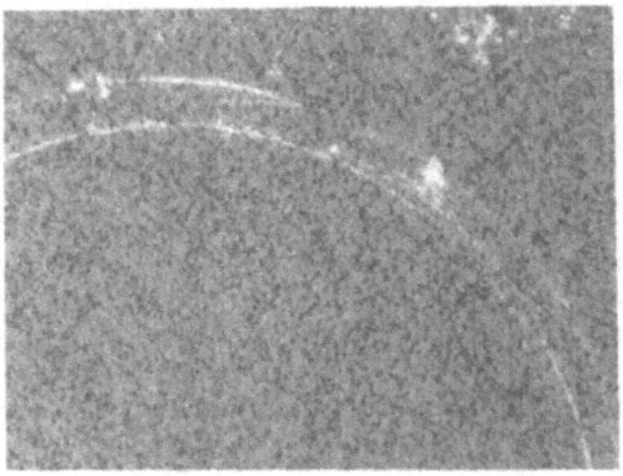

Combined illumination 100 μm

Figure 4 Fine Hertzian crack on ball surface - crack radius = 0.40 mm.

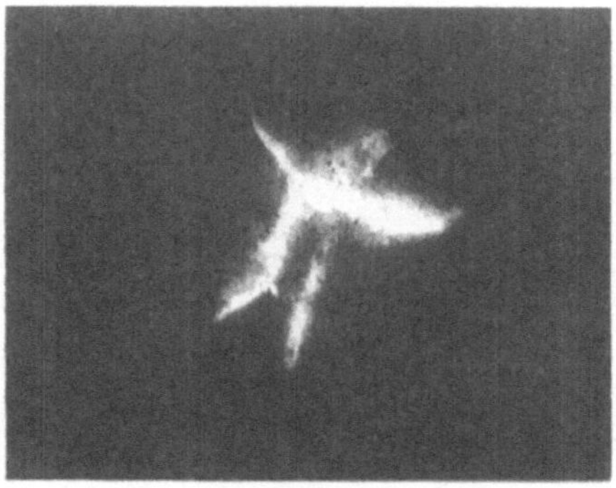

Ultraviolet illumination

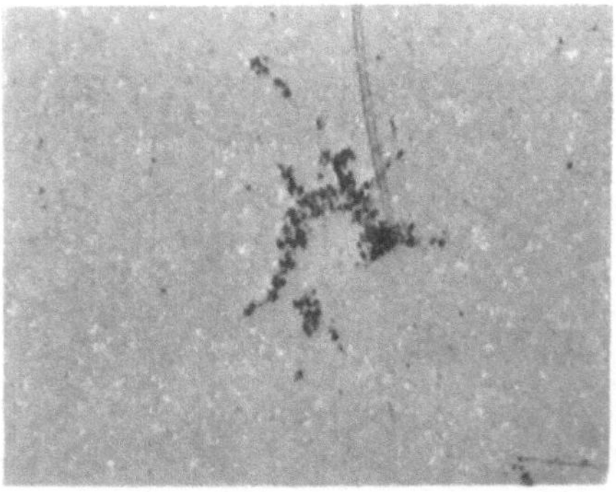

Brightfield illumination 50 μm

Figure 5 Point defect with associated radial cracking.

Full bearing tests carried out by SKF under severe conditions of thin film lubrication and high loads have demonstrated that silicon nitride balls containing Hertzian cracks, (as shown in Figure 3), fail prematurely after less than 3 million cycles and in a catastrophic manner as is evident from Figure 6. In contrast defect free balls have been tested under the same conditions to 300 million cycles without spalling.

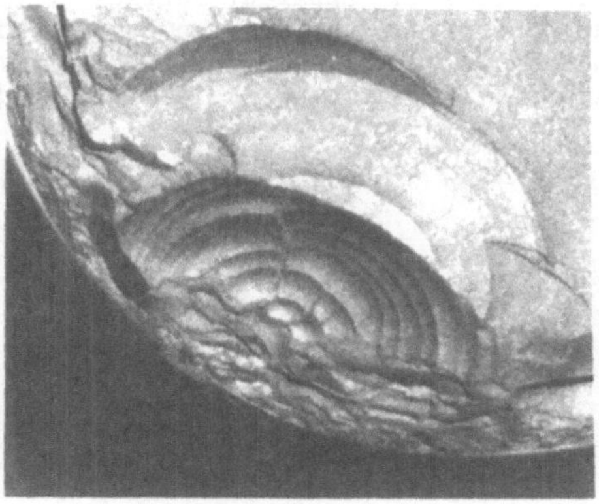

Shattered ball

Fracture origin

Figure 6 Catastrophic failure of ball containing Hertzian surface cracks. Ball diameter = 17.46 mm.

Visual Inspection

Fluorescent dye inspection is supplemented by visual examination using oblique or darkfield illumination for the detection of defects that do not retain penetrant. With normal white light illumination, inclusions down to 50 µm in size, shallow holes and healed pressing defects are readily detected at relatively low magnifications. Two examples of these types of defects are shown in Figure 7.

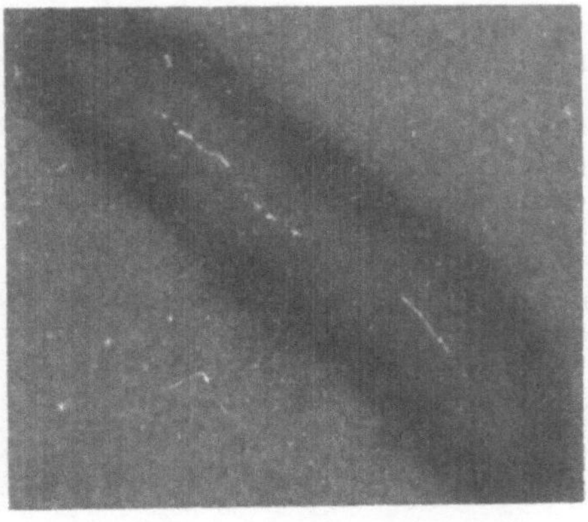

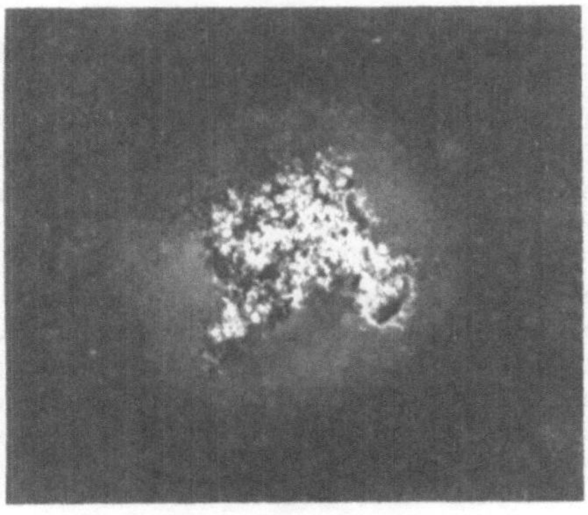

Healed pressing defect 250 µm

Ceramic inclusion 100 µm

Figure 7 Defects detected using oblique illumination.

Surface Appearance

Whereas fluorescent dye and visual inspection are carried out on all components in a finishing lot, random samples of balls are also taken for examination at higher magnifications to classify surface appearance. This assessment includes categorisation of the basic surface in terms of small indentations and apparent surface porosity. The latter feature includes both true porosity and apparent porosity arising from micro-chipping during ball finishing operations. Figure 8 shows an indentation and microstructural features of the material clearly visible on the finished ball surfaces.

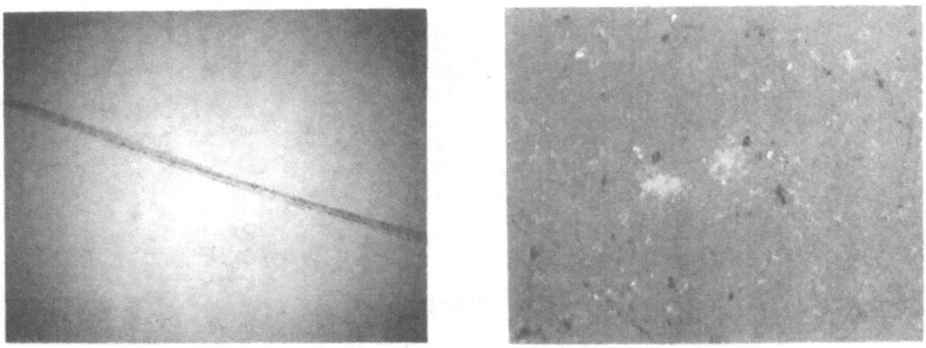

Extended indentation 250 μm   Ball surface - microstructure 20 μm

Figure 8 Examples of features on finished ball surfaces.

These finer features, may be present on ball surfaces to a greater or lesser extent and are assessed as they can affect bearing performance by increasing vibration levels and noise. The highly polished nature of the ball surfaces also allows information on material microstructure to be gathered directly from the finished balls.

CONCLUSIONS

The specification of material quality, together with the development of effective NDE techniques for the assessment of surface quality, has been necessary in order to achieve the performance benefits offered by hybrid bearings.

The combination of high sensitivity penetrants and visual inspection with the appropriate forms of illumination has proved effective in the detection of surface defects, ranging from microporosity through inclusions to fine cracks emanating from either blunt overloads or sharp overloads from for example coarse diamond abrasives.

Acknowledgement

The author would like to thank Dr H. Wittmeyer, Managing Director of SKF Engineering and Research Centre for permission to publish this paper.

MATERIAL SELECTION AND QUALITY FOR CERAMIC ROLLING ELEMENTS

R.T. Cundill
SKF Engineering and Research Centre,
Nieuwegein, The Netherlands

ABSTRACT

Silicon nitride has the best combination of physical and mechanical properties for use as rolling elements in hybrid ceramic bearings. The term 'silicon nitride' describes a family of materials, including alpha- and beta-sialons, with different compositions and made by different processing routes. Not all types of silicon nitride are suitable for use in rolling bearing applications where stresses may be as high as 4 GPa. Specification and selection of types or grades of material together with the associated quality control parameters are important aspects in the development of hybrid ceramic bearings.

INTRODUCTION

Since the early 1960's, bearing designers have been interested in ceramic materials both to increase the performance of bearings and to extend the range of operation to higher temperatures and corrosive environments. Although early work was focused on bearings made completely from ceramics for high temperature applications such as gas turbine engines, recent development effort has concentrated on hybrid ceramic bearings - steel rings with ceramic balls or rollers. Hybrid ball bearings are now becoming increasingly used in machine tool spindles and other high speed or high precision equipment. Stresses in these bearings are often very high and are localised at or near the surface. The normal contact stress at the surface can be as high as 4 GPa. Consequently, requirements for ceramic materials for bearings are more stringent than for most other applications of structural ceramics.

MATERIAL REQUIREMENTS

It is convenient to classify ceramic bearing applications into 'normal' and 'extreme' operating environments. A normal environment implies oil or grease lubrication and an operating temperature range of -40 to 200°C. Other applications whether at high temperatures, low temperatures or in hostile/corrosive conditions are considered to be extreme. In most cases, bearings for normal operating environments would be of the hybrid type.

Material requirements for ceramic rolling elements can be stated as follows:

Low density	- Centrifugal loading is reduced as the density of the rolling elements is decreased. This allows higher operating speeds and leads to less heat generation.
Moderate elastic modulus	- Dynamic bearing stiffness is increased by rolling elements with higher elastic modulus than steel. However, too high a modulus concentrates stresses.
Low thermal expansion	- This reduces the sensitivity to temperature differences and helps to prevent seizure.
Good strength	- A high compressive strength is necessary to withstand the high contact pressures in rolling bearings.
High hardness and toughness	- These properties in combination lead to better surface finish and resistance to damage from foreign particles and impacts.
Good rolling contact fatigue	- An obvious requirement for bearings.
Spalling failure mode	- If rolling elements fail during service, then they should do so by spalling, the least harmful mode, which gives warning before seizure.

For bearings operating under extreme conditions there are additional requirements:

Temperature resistance and stability	- Ability to retain and maintain mechanical properties at temperatures up to 800°C.
Corrosion resistance	- Stability in oxidising and corrosive environments particularly in the contact zone where repeated over-rolling can remove surface films.

Although silicon nitride is neither the hardest nor the toughest of the engineering ceramics, it is considered to have the best combination of mechanical and physical properties for use in high performance bearing applications. Various types of rolling contact fatigue tests and bearing tests have demonstrated that fully dense and homogeneous silicon nitride has good rolling contact fatigue resistance. L10 fatigue lives of silicon nitride are considerably higher than those of carbon-chromium bearing steel and M50 tool steel used for aircraft engine bearings. However, not all types of materials have good rolling contact fatigue properties. Those containing porosity, inclusions and other features fail prematurely. Consequently, selection of suitable material types is critical to successful application in hybrid ceramic bearings.

MATERIAL SELECTION

The term 'silicon nitride' describes a family of materials which include alpha- or beta-sialons as well as 'pure' silicon nitride. In addition, silicon nitride materials can contain significant amounts of both crystalline and glassy oxynitride phases whose composition is determined by compounds added to promote densification. There are also a number of different fabrication and densification processes used to manufacture components. In most cases, the starting material is silicon nitride powder, which after preforming, can be densified by various processes including low and high pressure sintering, hot pressing and hot isostatic pressing. Alternatively, silicon metal may be used to make a preform which is then heated in nitrogen to convert the silicon to silicon nitride. After nitridation, the preforms are then densified by sintering and, optionally, hot isostatic pressing. In all cases, various compounds are added to the starting material to promote densification by liquid phase sintering mechanisms and other additives may be used, for example, to convert the silicon nitride to a sialon.

In practice, therefore, there are many different types or grades of silicon nitride with different compositions and different mechanical properties, not all of which will be suitable for the particularly demanding application of rolling bearings. It is estimated that world-wide there are more than 30 suppliers of silicon nitride components and more than 150 different types or grades of material differing in composition, processing and properties. Many of these grades are reaction bonded materials which are unsuitable for use in bearings; but, a substantial number of material grades remain candidates for bearing applications. In view of this situation it has been necessary to develop in-house specifications and quality assurance procedures in order to select and approve types or grades of silicon nitride material for bearing applications.

MATERIAL SPECIFICATION

The silicon nitride material specification for bearings is intended to apply to specific grades of material, produced by a supplier on a routine or regular basis. The specification consists of the following general requirements which should be met by a specific grade of material.

Phase Composition

Monolithic silicon nitride materials are defined as containing at least 80% by volume of silicon nitride or sialon phases.

Base Material Composition

The compositions of starting materials (silicon nitride powder or silicon metal powder) are not specified in detail but there are limits on elements which are considered to be harmful to bearing performance.

Permitted Additives

In addition to yttrium oxide and magnesium oxide which are commonly used to promote densification by liquid phase sintering, other compounds are also permitted for this purpose. Any restrictions on additives are intended to exclude those considered to be harmful to health or bearing performance. Similarly, organic and other compounds added as powder lubricants, binders, deflocculants, viscosity adjusters or for other purposes are permitted provided that there are no harmful residues after processing.

Intrinsic Material Properties

Silicon nitride materials supplied for bearing applications should have physical properties within the ranges given in Table 1.

Table 1 Intrinsic Material Properties

Property	Unit	Min	Max
Density	kg/m^3	3000	3400
Elastic modulus	GPa	270	330
Poisson's ratio		0.23	0.29
Thermal conductivity	W/m.°K	20	38
Specific heat	J/kg.°K	700	800
Coefficient of thermal expansion	$\times 10^{-6}$ /°C	2.8	4.0
Compressive strength	MPa	3000	

Each delivery of silicon nitride bearing components should conform with certain requirements for mechanical properties, microstructure and density variation. In the case of transverse rupture strength, a series of tests carried out on one process or densification lot may be considered representative of all deliveries made from the same blended powder lot. Hardness, toughness, density variation and microstructure assessments should be carried out on samples from each delivery.

Mechanical Strength

Transverse rupture strength data is used as a material quality parameter in the case of bearing materials rather than as a design parameter. Minimum values for mean room temperature transverse rupture strength and Weibull moduli for the test series are specified. Either 3-point or 4-point methods may be used for transverse rupture strength tests.

Hardness and Toughness

The Vickers method is preferred for hardness testing of silicon nitride. Toughness can be measured directly as K_{Ic}, plane strain fracture toughness, or Kc, critical stress intensity, or determined as a toughness parameter (TP) by an indentation technique. For routine quality assurance of small components, toughness can usually only be assessed by an indentation method.

Microstructure

In addition to the matrix of silicon nitride (or sialon) and glassy grain boundary phases, the microstructure can contain porosity, metallic phases, ceramic second phases and inclusions defined as follows :

Metallic phases - bright, highly reflective phases observed in silicon nitride materials. Such phases may, in fact, be intermetallic or metalloid in nature.

Ceramic second phases - ceramic phases appearing darker or lighter than the main or matrix phase.

Inclusions - areas of ceramic second phases or metallic appearing phases or a combination of both types greater in extent than 25 µm.

A fine uniform and consistent microstructure is considered essential to good rolling contact fatigue resistant. Therefore limits are imposed on the volume %, maximum size and type of distribution of microstructural constituents together with limits on the number of inclusions of different sizes per cm^2 of transverse area.

Density Variation

In order to ensure consistency within a delivery, maximum limits on the density variation from the mean value are specified. In practice, the density variation cannot be determined reliably until the finished component stage. Allowance is made for the increasing uncertainty of density determination as the component volume decreases.

Macrostructure and Colour Variation

A uniform macrostructure is considered important since some macrostructural features can affect fatigue performance. Variations in colour within a batch of components may be less critical but can affect perceived quality by the end user.

MATERIAL QUALITY

Microstructure

It is not usually possible to resolve the silicon nitride grain morphology with normal light microscopy. For assessment purposes, therefore, both the silicon nitride (or sialon) and oxynitride glass grain boundary phases are taken to constitute the matrix. The intrinsic microstructure can contain porosity, metallic phases and ceramic phases as illustrated in Figure 1. A typical fine microstructure containing metallic and ceramic phases is shown in Figure 1(d).

Porosity is known to adversely effect rolling contact fatigue behaviour and, in this context, a porosity content of 0.1 volume % would be regarded as high. The presence of ceramic phases is considered to have relatively little effect on fatigue performance. In the case of metallic phases, the composition of the phase is important with hybrid bearings. In normal operation, the rolling elements are separated from the rings by a thin film of lubricant, but some direct contact can occur at low speeds or where there is poor lubrication. Some types of metallic phases have a high welding affinity to the steel bearing races which results in material being pulled out of the ceramic rolling elements.

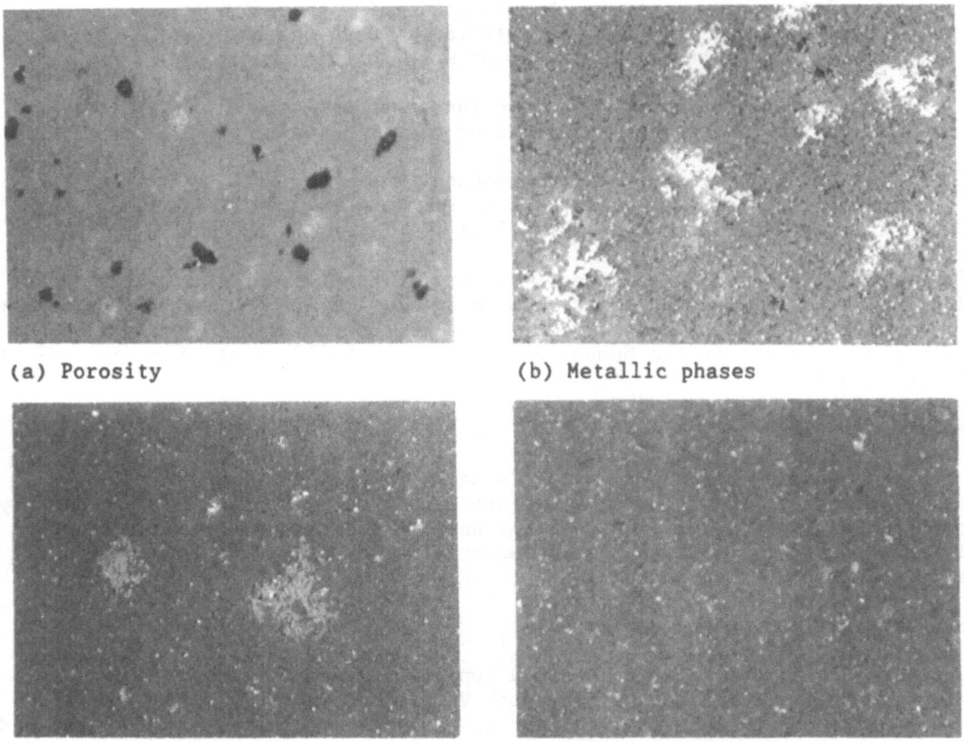

(a) Porosity (b) Metallic phases

(c) Ceramic phases (d) Fine microstructure

Figure 1. Examples of microstructural constituents ------ 25 μm

Superimposed on the intrinsic microstructure, materials may also contain non-repetitive or random features resulting from processing faults or contamination. Major defects such as cracks, voids and areas of incomplete densification are obviously harmful to performance. Metallic and ceramic inclusions are shown in Figure 2.

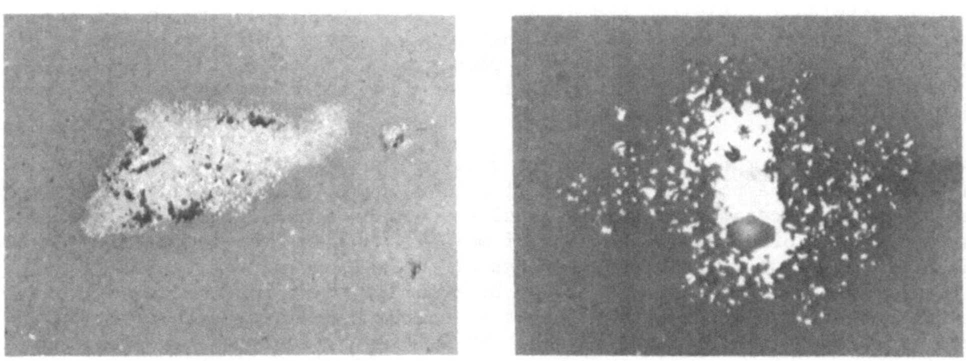

(a) Ceramic inclusion (b) Metallic inclusion

Figure 2. Inclusions in silicon nitride ------ 25 μm

Inclusions, which can also be a mixture of ceramic and metallic phases, often originate from contamination of powders during mixing of transport. Their harmfulness depends on both size and morphology. There is the risk that material can be pulled out during finishing operations leaving holes or voids on the surface of the component.

Macrostructure

Some silicon nitride materials have a pronounced macrostructure when viewed with oblique or darkfield illumination. Macrostructural variations such as those shown in Figure 3 cannot usually be distinguished at higher magnifications using normal incident light microscopy.

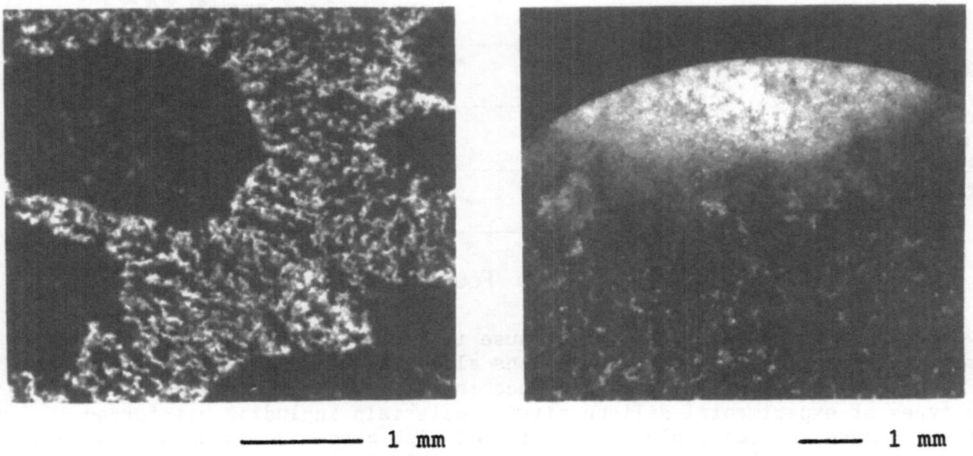

Figure 3. Examples of macrostructural variations revealed by oblique illumination

In most cases, macrostructural variations are thought to be caused by localised depletion of additives resulting in sub-micron porosity between the beta-silicon nitride grains. These areas retain fluorescent penetrants and show up clearly when illuminated with ultra-violet light. Due to the interlocking network of beta-silicon nitride grains, such features are difficult to eliminate by hot isostatic pressing.

Materials containing pronounced macrostructural variations are not considered suitable for use in high performance bearing applications, although fatigue life may be acceptable under low load conditions.

Material Properties

In practice, the material properties that can be determined on a batch or delivery of bearing components are limited to hardness, toughness and density. In the case of density, the absolute mean value is of less importance than the variation from the mean. A low density variation show good consistency within a batch, whereas a significant variation is indicative of porosity, metallic inclusions or incomplete densification. Density variations of finished silicon nitride balls are typically of the order of 0.001 Mg/m^3 or less.

Figure 4 shows the range in hardness and toughness values obtained from silicon nitride materials made by a variety of manufacturing processes, including both development materials and those used for production of balls for bearing applications.

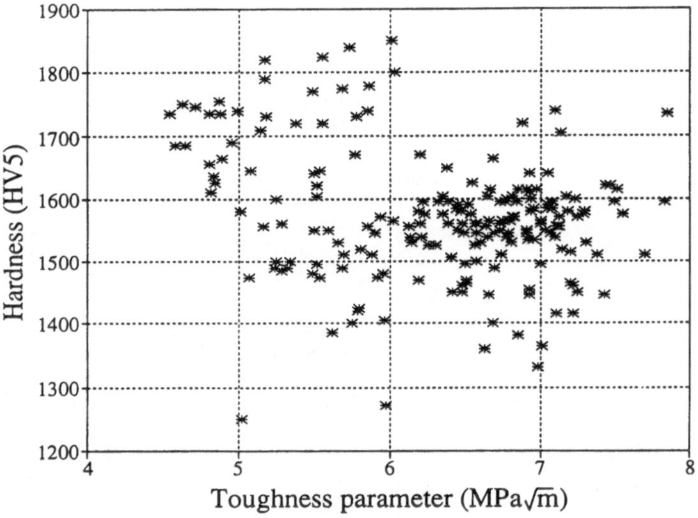

Figure 4.

Hardness and toughness of silicon nitride materials.

Low toughness is undesirable both because it increases the risk of damage during machining or finishing operations also because fatigue performance may be reduced by the lower defect tolerance implicit with low toughness. Only a few types of experimental silicon nitride materials including reinforced materials have the desirable combination of high hardness and high toughness. At present, materials with high toughness combined with a certain minimum hardness are considered more suitable for bearing applications than the harder materials with lower toughness.

CONCLUSIONS

Silicon nitride has the best combination of physical and mechanical properties for use as rolling elements in hybrid ceramic bearings. Due to the variety and number of different types of silicon nitride, it has been necessary to develop in-house specifications in order to be able to select and approve specific material grades for bearing applications.

Bearing grade silicon nitride has to conform with certain requirements for base material purity, permitted additives, phase composition and intrinsic physical properties. Minimum values for mechanical strength, hardness, toughness and density variation are also specified. Emphasis is placed on a fine, homogeneous and consistent microstructure which is considered to be an essential prerequisite for good rolling contact fatigue performance.

Acknowledgement

The author would like to thank Dr. H. Wittmeyer, Managing Director of SKF Engineering and Research Centre, for permission to publish this paper.

ULTRASONIC INSPECTION OF ROCKER-ARM PAD INSERTS

D. FARGEOT, C. GAULT, E. BRETON
Ecole Nationale Supèrieure de Céramiques Industrielles de Limoges, U.R.A.320, FRANCE

ABSTRACT

The inspection of critical flaws in structural ceramic components using ultrasonic nondestructive testing is done by correlating the defect characteristics and the measured ultrasonic echo parameters. The best conditions for the inspection of a rocker-arm pad insert (Si_3N_4) by a high frequency(10 - 100 MHz) ultrasonic immersion pulse-echo technique were determined in order to provide an improved method for inspection of this component during production.

INTRODUCTION

The thermomechanical performance of engine components may be improved replacing certain metal parts with ceramics. However processing ceramics by powder sintering can result in flaws (microcracks, microporosity, inclusions,...). Of the different nondestructive evaluation (NDE) methods which are used in materials science to detect flaws (microfocused X-rays, dye penetrant,...) ultrasonic pulse echo tests are among the most versatile because they can be used to evaluate the elastic properties of the material as well as to determine flaws in the components. The problems of testing structural ceramics are due to the smallness of the critical flaws which are to be detected. Furthermore, from an industrial point of view, the challenge is not to test materials but components. This involves additional constraints arising from the complex shapes of the components and from the necessity of detecting microscopic flaws in given parts. Consequently, the tests are largely carried out at high frequency (in the 10-100 MHz range) by immersion techniques using focused broad-band transducers in order to concentrate the ultrasonic energy on the smallest volume and to obtain the optimum spatial resolution. Moreover, an automatic control is possible if a computer is used to analyse the flaw signals and compare them to the results obtained with artificial flaws.

MATERIAL AND EQUIPMENT

The components are rocker-arm pads for thermal engines, made from silicon nitride by a classical ceramic process involving die pressing and sintering at high temperature (1). These pads are made to be inserted in die-cast aluminium alloy rocker-arms.

The main stress sources are:

- thermal shock during the insertion;
- thermal compression when cooling;
- mechanical stresses in the head and at the head/foot junction in service condition.

The associated critical flaw types have been determined (2) and it was decided to control ultrasonically three zones where the failure probability is high :

- the head volume where cold pressing defects (large pores or cracks parallel to the compaction plane) are expected (**type 1 flaws**);
- zone of about 0.5 mm in depth under the head surface where subsurface pores and / or inclusions weaken the wear resistance of the pad (**type 2 flaws**);
- the neck mould where microcracks can initiate the rupture during the insertion or in service (**type 3 flaws**).

Figure 1 gives a simplified picture of the component with the three types of flaws. The block diagram of the system used for ultrasonic NDE of the Si_3N_4 pads is given in figure 2.

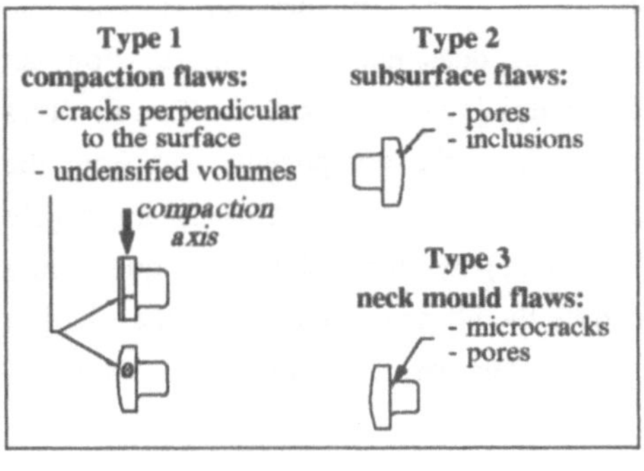

Figure 1. Different types of flaws.

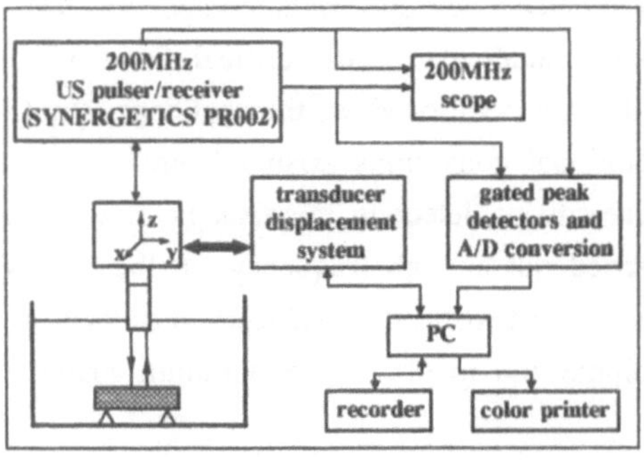

Figure 2. Block diagram of ultrasonic N.D.T.

It functions in the 10-200 MHz frequency range which was found to be well suited to the detection of microscopic flaws in similar materials (3). The results of the test can be visually analysed from the ultrasonic pattern displayed on a high frequency scope, or color C-SCAN images in 2 or 3 dimensions can be obtained by following the amplitude of a given echo (interface, backside, or flaw reflection).Therefore, both the ultrasonic parameters of the material (velocities and attenuation), and the flaw characteristics in the components may be studied.

RESULTS

It is important to chose the transducer type and the associated pulser-receiver using the propagation parameters of the material. The parameters have been measured for longitudinal and shear waves in the 15-80 MHz frequency range.

Figure 3 shows the ultrasonic velocities are independent of frequency, increase with densification, and are independent of sintering additives . The calculated elastic moduli (figure 4) are consistent with literature values (4).The measurement of this parameter is a simple test of the local density.

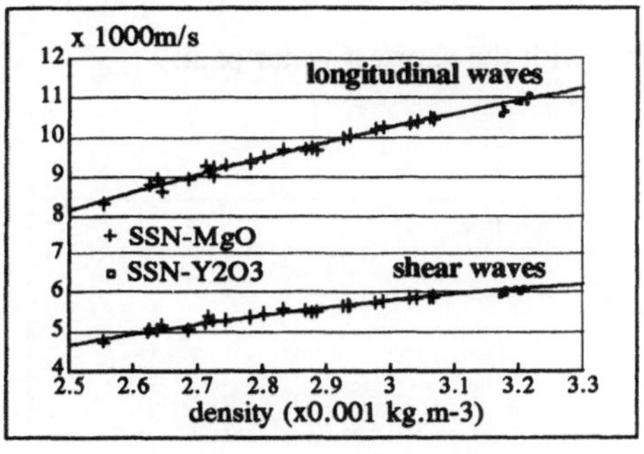

Figure 3. Velocities versus density.

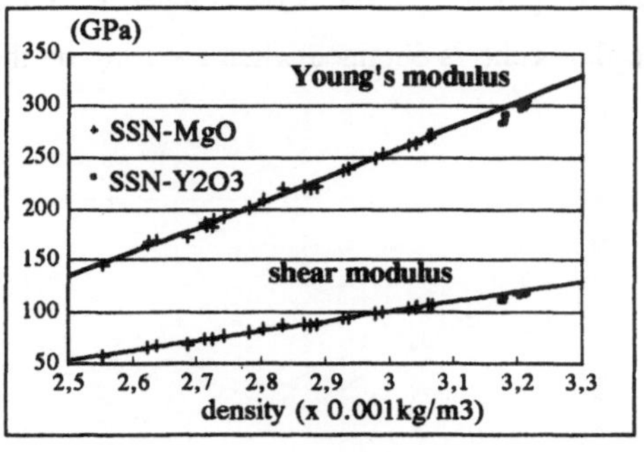

Figure 4. Moduli versus density.

Attenuation is a much more complex parameter to evaluate, because it both depends upon the radiation field (which is a function of the geometries of the transducer and of the reflectors, of the ultrasonic velocity in the medium and of the frequency), and on the intrinsic attenuation of the material (which is a function of the microstructure and of the frequency).

Various measurements have shown that for a fully dense Si_3N_4 (ρ=3200kg/m^3) the proper attenuation is very low, and that the measured amplitudes of the echoes only depend upon the radiation field. This is particularly obvious with focused transducers.

Reflection from infinite planes: attenuation curves for backside echoes.

By focusing the US beam at the surface of the samples, the amplitude of successive backside echoes has been measured and plotted versus the propagation distance x in the samples. The curves are normalized at 0 dB by extrapolation at x=0. Each curve is characteristic of :

- the transducer (frequency, diameter, focus length)
- the material (intrinsec attenuation).

An example of attenuation curves is given in figure 5.

Reflexion on flaws : attenuation curves for flaw echoes.

Experimental amplitude/distance curves for a given transducer, given defect types, and given test conditions, are necessary to determine the best test method for real flaws.

As an example, figure 6 shows such a curve obtained with a 80 MHz focused transducer (3mm diameter, 25mm focal length in water) used for the detection of notches machined in a standard. The first three points have been obtained by focusing at the top of the notch. With this transducer, the maximum focal length being 3mm in silicon nitride, it was not possible to focus on notches x_4 and x_5. It clearly shows that the best amplitude of the flaw echo is obtained when focusing on the flaw with the shortest water path.

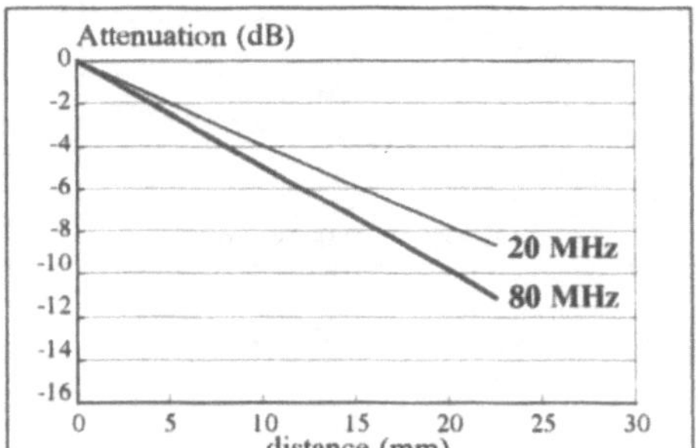

Figure 5. Attenuation curves for backside echoes (ρ=3200Kg/m^3).

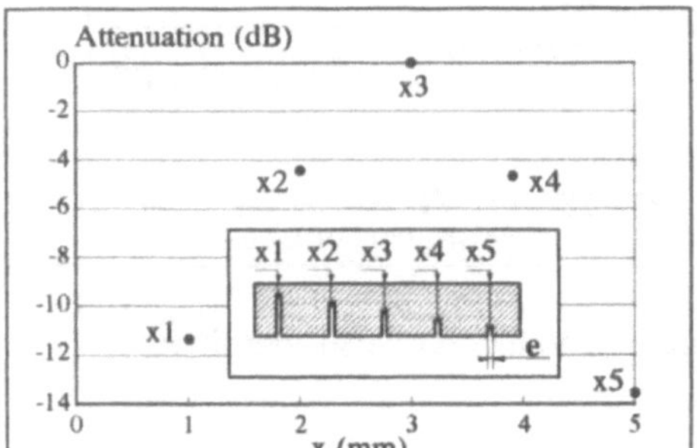

Figure 6. Attenuation curves for artificial flaw echoes (ρ=3200Kg/m^3)

Taking into account these results of ultrasonic propagation characterisation we can conclude that :

- the control of thickness > 8 mm is only possible at low frequency (< 20 MHz) ;
- the detection of small flaws (#100μm) involves using tests at frequency > 50MHz ;
- for focused transducers the best detection is obtained when focusing on the flaw and with a short path in water ;
- to detect small flaws near the surface it is necessary to work at high frequency with short focus length transducers.

Thus three tests with three differents transducers have been selected for the detection of the three defect types.They are sumarized in figure 7

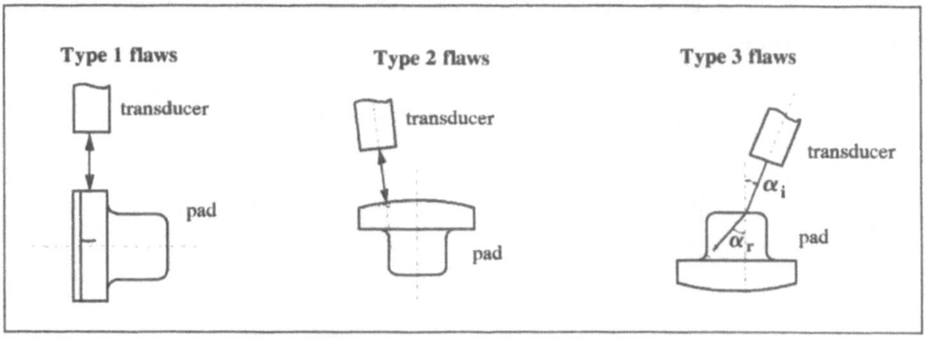

Figure 7. Detection of the three defect types.

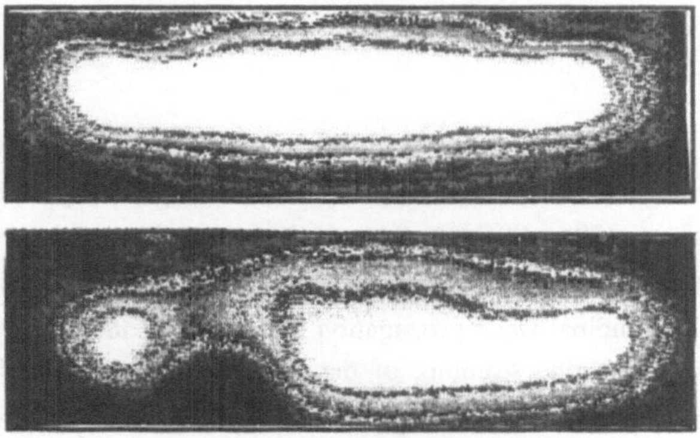

Figure 8. C-SCAN of a pad for type 1 flaws detection

Type 1 flaw

As the depth to be controlled is considerable (> 10 mm) and as the compaction flaw size is relatively large (>0.5 mm) an unfocused 15 MHz transducer (3 mm in diameter) is used. The test is made along the compaction axis (figure 7) by following the amplitude of the back-side echo which is weakened by the flaws. Such an effect is observed in figure 8b and can be compared to the C-SCAN of a sane pad in figure 8a.

Type 2 flaw

The pad is fitted on a cylindrical support with the same curvature radius as the pad head. A short focused transducer (6 mm in diameter, 80 MHz, and 12.7 mm of focus length in water) is used to concentrate the energy of the normal incident ultrasonic beam in a short depth (0.5 mm) under the head pad surface. As for type 1 flaws, a color C-SCAN can be recorded. In the C-SCAN of figure 9 subsurface flaws undetected by a dye penetration technique are shown.

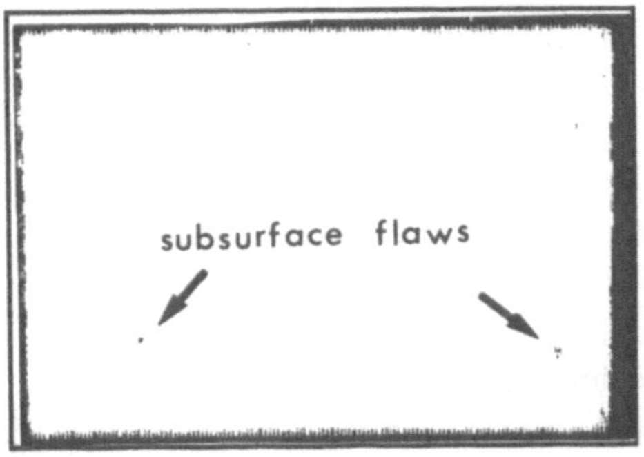

Figure 9. Subsurface flaws.

Type 3 flaw

The best solution was found to be the use of an oblical incident beam on the foot pad, as depicted on figure 7. The distance at which the flaws are to be detected depends upon the incidence angle α. A focused transducer (50 or 80 MHz, 6mm in diameter, with a relatively large focus length 50-75 mm in water) is necessary to detect the flaws at the foot-head neck-mold pad. The condition for longitudinal wave propagation under oblical incidence in Si_3N_4 is $\alpha < 8°$. Taking this into account, the scanning of the volume of interest in the vicinity of the neck mould was made by variation of the position of the transducer. Figure 10 presents an ultrasonic pattern showing a flaw echo which was found to correspond to a micropore after observation by optical microscopy of the polished section of the pad (figure 11).

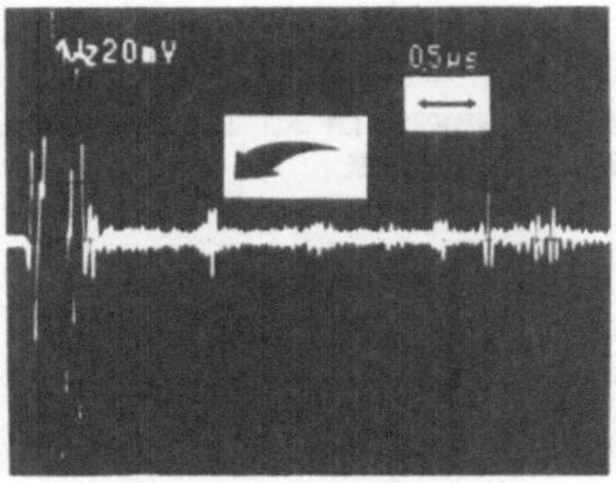

Figure 10. Ultrasonic pattern of a flaw echo.

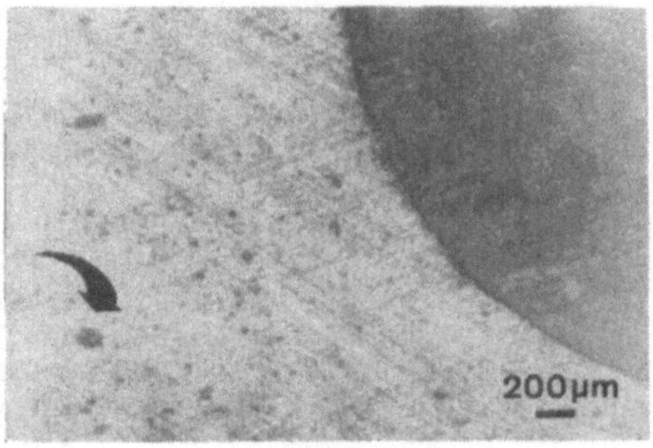

Figure 11. Optical microscopy of the flaw.

CONCLUSION

The measurement of the ultrasonic propagation parameters has shown that the Si_3N_4 pad can be tested by an ultrasonic pulse echo technique in the 15-80 MHz frequency range. For the three flaw types identified as the most critical in the pad, the test procedure has been established from the geometry of the component and the location of the expected flaws. From these results, automatic test equipment is now being built.

ACKNOWLEDGEMENTS

The authors are grateful to the French Department of Industry for financial support and to the French Societies Desmarquest, P.S.A., Renault and Sofratest for their helpful collaboration.

REFERENCES

1 . Société des céramiques techniques Desmarquest,Evreux,France.

2 . Leborgne,G., Amar,E., Internal report GIE PSA-Renault, 1989.

3 . Amar,E., Legouet-Lespinasse C.,Le Flour,J.C.,communication at "Conference on non destructive evaluation of modern ceramics".Proc. Am. Soc. for Nondestructive Testing Ed., 160, 1990.

4 . Yeheskel,O. and Gefen, Y., Mater. Sci. & Engn., 71, 95, (1985).

ANALYSIS OF PHYSICAL PROPERTIES OF CERAMIC POWDERS IN AN INTERNATIONAL INTERLABORATORY COMPARISON PROGRAM

S. G. Malghan and S. M. Hsu,
National Institute of Standards and Technology
Gaithersburg, MD U.S.A.

A. L. Dragoo
Department of Energy
Washington, D.C. U.S.A.

H. Hausner
Technische Universitat Berlin
Berlin, Germany

R. Pompe
Swedish Ceramic Research Institute
Goteburg, Sweden

ABSTRACT

Accuracy in the measurement of physical characteristics of ceramic starting powders is a critical factor in the control and reproducibility of powder processing unit operations. An international interlaboratory comparison program on powders characterization has been in progress under the auspices of the International Energy Agency. In this paper, the results of powder characterization effort on five powders by 25 industrial, university and governmental laboratories are presented. Selected results of density, specific surface area, and particle size distribution are presented and discussed in terms of factors affecting these measurements. The reasons for the discrepancies in the data are outlined.

PHYSICAL PROPERTIES OF POWDERS

Fine ceramic powders are characterized for physical properties by the measurement of density, specific surface area, particle size distribution, morphology and porosity. The physical properties of the powders affect not only powder processing parameters, such as milling/deagglomeration time and green density, but also properties of the dense ceramics(1). Densification time-temperature schedule and grain size are some of the primary processing parameters affected by the physical characteristics of powders(2). One of the factors that affects reproducibility in the manufacture of ceramic components is our ability to obtain complete characterization of the powders. However, accurate and reproducible measurement of powder properties has been difficult due to lack of understanding of underlying subprocesses, and unavailability of standard measurement procedures and standard reference materials(3).

A program on "Characterization of Ceramic Powders" has been in progress since 1985 under the auspices of the International Energy Agency. This program is in its second phase with a continued focus on improving the measurement procedures. The inception of IEA/Annex II-Subtask 2 program, its objectives, participating countries, program management in the participating countries is described elsewhere(3-5). The Subtask 2 program was one of the several tasks under Annex II and was completed in 1989.

OBJECTIVE OF IEA ANNEX II SUBTASK 2

The major objective of the Subtask 2 (Powders Characterization) was to establish the basis for the evolution of internationally accepted standardized testing and characterization methods for powders used in the fabrication of structural ceramics. The overall objective of this program was the enhancement of quality, reproducibility and reliability of powder characterization data through a program of international interlaboratory comparison of measurements. Specifically, the program sought to:

- Survey current methods of powders characterization.
- Establish a base of data and experience with which to identify unique sets of properties and methods useful for powder evaluation.
- Determine the extent of agreement between the laboratories.
- Recommend useful standards which would aid in international commerce.

The participants carried out analysis of powder samples using the best available procedures in their laboratories. The data were analyzed and summarized in a report (4).

In a previous paper, the details of objectives of Subtask 2, powders studied, samples preparation, properties measured and selected results were presented (3). In the present paper, selected data on three physical properties, such as density, specific surface area and particle size distribution are presented. The data are discussed in relation to variability of some of the measurement parameters and the resulting differences in the laboratory-to-laboratory data.

POWDERS*

Five powders used in this study and their designations (in parenthesis) are:

1. Silicon, Kemanord IV-D, (Si)
2. Silicon nitride, Ube SNE-10, (SNT)
3. Silicon nitride, Starck LC-10, (SNR)
4. Silicon carbide, Starck, (SiC)
5. Zirconia, TOSOH, (YSZ)

The YSZ powder was used in spray dried form and contained 4% Y_2O_3.

DENSITY

The bulk density of selected powders was measured by He-pycnometer, Archimedes principle and tap density. The average of density data from two participating laboratories by He-pycnometer are shown in Table 1. These data are in good agreement with those reported from theoretical calculations (6-8). Refinements to the sample preparation procedure may be required to further improve the data.

*Certain trade names and company products are mentioned in the text or identified in illustrations in order to adequately specify the experimental procedure and equipment used. In no case does such identification imply recommendation or endorsement by National Institute of Standards and Technology, nor does it imply that the products are necessarily the best available for the purpose.

TABLE 1
Bulk Density Measurements by Helium Pycnometer

Material	Density, g/ml Min.	Density, g/ml Max.	Theoretical, g/ml
Si	2.30	2.35	2.33
SiC	3.08	3.19	3.21
SNR	3.12	3.18	3.18
SNT	3.13	3.20	3.18

The measurement of tap density was carried out by laboratories using different types of instruments and experimental procedures. The number of laboratories participating in this measurement varied from 2 to 9. As a result of these factors the data showed a fair degree of variation.

TABLE 2
Range of Tap Density Measurements

Material	No. of Labs.	Tap Density, g/ml Min.	Tap Density, g/ml Max.
Si	3	0.78	1.08
SiC	5	0.68	0.83
SNR	7	0.74	0.87
SNT	9	0.66	0.78
YSZ	2	1.49	1.51

The tap density of YSZ powder varied in a narrow range due probably to the data from only two participating labs., and better flowability of spray-dried powder. Data from other powders showed a broad range.

SPECIFIC SURFACE AREA

Single point (SP) and multipoint (MP) analysis of specific surface area by BET method was carried out by using nitrogen as the adsorbate. A significant variation was observed in the experimental parameters:

- Sample amount; 0.5 to 5.0 g.
- Degas temperature; 110°C to 350°C.

- Degas time; 15 min to 8 hrs.

Despite the availability of a number of powder standards, very few labs. reported the use of such standards. Degassing, an important powder preparation step, depends on the pressure, temperature and time for which the surface is conditioned.

The specific surface area data of powders are shown in Table 3. These data indicate that average MP values based on all data for each powder were always higher by 2-10% than the corresponding average SP values. For those labs. that measured the surface areas by both methods, the SP values were slightly higher than the MP values.

The relative error involved in the SP method can be evaluated by examining the difference between Wm (weight adsorbed at monolayer coverage) for SP and MP BET equations(9):

$$\frac{(Wm)_{MP} - (Wm)_{SP}}{(Wm)_{MP}} = \frac{1 - P/Po}{1 + (C-1)\ P/Po}$$

where P/Po is the ratio of relative pressures and C is the BET constant. This relationship shows that when SP analysis is made using the relative pressure which would give monolayer coverage according to the MP-BET theory, the relative error will be equal to the relative pressure employed. Further analysis shows that fraction of the surface unoccupied also affects the relative error.

The relative error given by this equation is equal to the relative error of the surface areas. The above eq. shows that as P approaches P_o (saturation pressure) the relative error decreases. For C=50 to 300 and P/P_o = 0.3, typical relative errors are expected to be from 1 to 5%. This difference provides an estimate of the systematic error between the two measurements, which is assignable to the method used. The data in Table 3 shows that with the exception of SNR and SNT, the systematic error for other powders is much higher, which suggests that other factors may be involved. However, it should be mentioned that an accurate determination of the experimental error for these powders are confounded by rather large random errors (see Table 3).

TABLE 3
Average Values and Standard Deviation (S.D.) for Single and Multipoint BET

Powder	Single Point			Multipoint		
		S.D.			S.D.	
	Mean, m^2/g	m^2/g	%	Mean, m^2/g	m^2/g	%
Si	1.9	0.2	12.0	2.3	0.6	26.9
SiC	13.3	1.1	8.4	14.5	1.0	6.6
SNR	13.1	0.7	5.5	13.5	0.6	4.6
SNT	9.4	0.6	6.5	9.8	0.5	6.3
YSZ	16.8	1.4	8.3	18.4	0.6	3.2

In addition, some of the contributing reasons for the SP and MP differences are:

- statistical -- the limited population of the participating labs. may be subject to a fluctuation, giving rise to a systematic difference;
- experimental -- somewhat differing conditions between the two types of measurements may have been significant;
- algorithm-related -- the MP values are computed by a regression analysis; and/or
- virtual -- the MP values may be closer to the true surface area.

The correlation coefficients obtained in the MP analyses were typically 0.995 or higher. Nonetheless, the MP values obtained by labs which performed a large number of runs on one sample displayed generally larger mean deviations compared to the corresponding SP data.

With the limited data available, an attempt was made to identify various contributions to the variance of the measurements. The variances due to contributions from the preparation of IEA samples (riffling error; standard deviation, S_R), from the extraction and preparation of analytical samples (preparation error; standard deviation, S_P) and instrumental measurement (measurement error; standard deviation, S_M). The total variation is expressed as:

$$S^2 = S_R^2 + S_P^2 + S_M^2$$

An analysis of variance of the results of Lab 17 for SNR, and SNT powders are shown in Table 4. These data indicate that for SNR and SNT powders, the errors involved due to preparation and measurement are small; whereas, that due to riffling is large. One of the primary reasons for the riffling error to be large could be the difficulty experienced in packaging these powders due to their resistance to flow.

TABLE 4
Estimates of the Standard Deviation, m^2/g, and Its Components for the Results of Lab 17.

Powder	Data Used in Analysis	Standard Deviation (S)	(1*)	(S_M)	(S_P)	(S_R)
SNR	All data	0.32				
	Subsample av. and s.d.		0.33	0.088		
	#582, #1381 subsample av.			0.070	0.090	
	Estimate, using 1* and S_P					0.30
SNT	All data	0.23				
	Subsample av. and s.d.		0.23	0.069		0.23
	#312, #938, #1419 subsample av.				0.036	
	Estimate, using 1* and S_P					

Col. (1*) $[s_R^2 + s_P^2]^{1/2}$.

The data of specific surface area of SNR powder by the MP and SP BET methods are shown in Fig. 1 and 2, respectively. These results show the range of variability of surface area data from lab-to-lab. Reproducibility within a given lab. is also shown, where available. Considering the range of instruments and different experimental parameters used in the analysis, the standard deviation is within an acceptable range. If we eliminate certain data that were obtained using extreme experimental parameters, the lab-to-lab data can be compared more favorably.

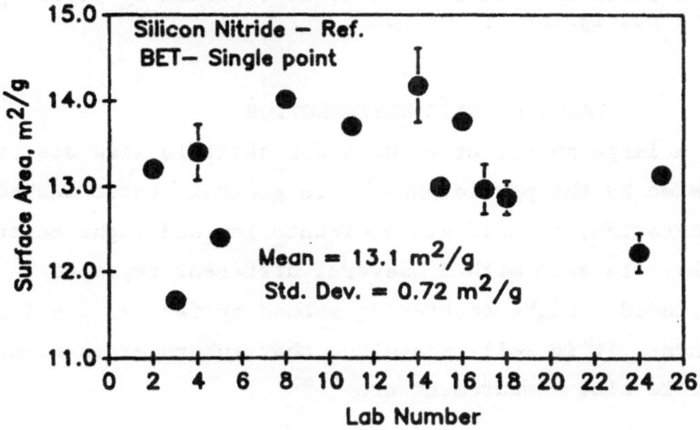

Figure 1. Specific surface area of SNR powder as reported by participating labs using single point BET method.

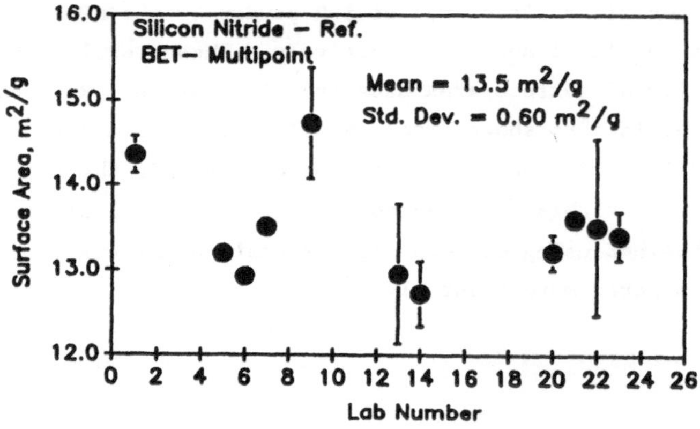

Figure 2. Specific surface area of SNR powder as reported by participating labs using multipoint BET method.

PARTICLE SIZE DISTRIBUTION

In this program a large number of methods for particle size distribution measurement were tested by the participants. In general, three methods--gravitational sedimentation, centrifugal sedimentation and light scattering; were used extensively. In each method, several different types of instruments were included. Light scattering method by far had the largest variety of instruments. It is well-recognized that subprocesses common to all methods of particle size measurement are:[10]

- wetting,
- deagglomeration,
- dispersion, and
- instrumental procedure.

The primary requirement of a solvent to be used for the particle size measurement is that the powder surface should be completely wettable. Water containing certain surfactants is a suitable solvent for most powders. However, for some powders such as Si, water is not considered to be an acceptable solvent, since Si undergoes surface-hydroxylation by reacting with hydroxyl and hydrogen groups. In such cases, alcohols are used as solvents. Most of the participants used ultrasonic baths and probes of different geometry for deagglomeration. A variety of dispersion agents and dispersion procedures were used by the participants. Specific details of the dispersion was not clearly reported since some of this information was considered to be proprietary to the participants. Operating procedures of the equipment were in most cases those recommended by the manufacturer. A selected list of the deagglomeration-dispersion procedures reported by the participants is shown in Table 5.

TABLE 5
Summary of Selected Methods of Dispersion

Powder	Media	Dispersant	Deagglomeration
SNT/ SNR	Sedisperse A-11, or A-13	Lignosulfonate	Ultrasonicated with a probe for 3-12 min. at unspecified power
	Water	Butylamine	
SiC	Water, pH 10.0	Sodium pyrophosphate	Ultrasonicated in a bath for 10-60 min.
	Isopropyl alcohol	Ammonium hydroxide Dispex N-40	
Si	Water	Tetramethylammonium hydroxide	Ultrasonicated with a probe for 3-12 min. at unspecified power
	Ethyl alcohol	Sodium pyrophosphate	Ultrasonicated in a bath for 10-60 min.
YSZ	Water	Darvan 811	Ultrasonicated with a probe for 3-12 min. at unspecified power
	Sedisperse A-12	Sodium pyrophosphate	Ultrasonicated in a bath for 10-60 min.

A variety of aqueous and nonaqueous media and dispersant combinations were used. As shown by the data in Table 4, electrostatic, electrosteric and steric form of suspension stabilization has been employed by the participants. Choice of the solvent media and dispersant was primarily based on internal practice of the reporting laboratory. The ultrasonic power and time employed for deagglomeration covered a wide range. In general, insufficient data were available for quantitative assessment of the impact of dispersion and deagglomeration parameters. An example of comparison of data from three instrumental methods on SNR powder is shown in Table 6. The data are presented in terms of the ratio of D_{max} to D_{min}.

TABLE 6
D_{max}/D_{min} Ratio for SNR Powder by Using Different Instrumental Methods

Method	D_{50max}/D_{50min}	D_{90max}/D_{90min}	D_{10max}/D_{10min}
Gravitational Sedimentation (Sedigraph)	2.3	5.2	3.5
Centrifugal Sedimentation (Horiba)	1.7	n.a.	n.a.
Light Scattering (Microtrac)	2.1	4.9	1.3

A narrow range of values of D_{50max}/D_{50min} is an indication of relatively good agreement of the mean values of the particle size distribution when measured by different instruments. These values appear to show good agreement of data from the three instruments. The high values of D_{90max}/D_{90min} indicate either that the powder dispersion in this case was affected by large particles in the distribution, which is a measure of the presence of agglomerates or incomplete dispersion or that large particles may have been inaccurately counted, for example due to settling. A large variation in the D_{10max}/D_{10min} ratio indicates that either the dispersions contained variable concentration of ultrafines or a particular instrument is unable to detect the ultrafines.

Two important parameters of ultrasonication during the preparation of a dispersion are ultrasonication time and power. These parameters determine the extent to which the loosely held particles in the agglomerates are separated from each other. An example of the effect of the variation of ultrasonication parameters is shown in Table 7.

TABLE 7
Influence of Ultrasonication (Bath) Time on D-Values (μm) of SNR and Si by Gravitational Sedimentation (Sedigraph) with A-11 as a Dispersant

Powder	Ultrasonication Time, Min	D_{90}	D_{50}	D_{10}
SNR	30	3.2	0.8	0.3
	90	2.5	0.7	0.2
Si	20	13.7	7.5	1.9
	40	13.6	7.4	1.9
	60	13.0	7.0	1.8

As expected, the particle size distribution tends to become finer as ultrasonication time increases, indicating that increased deagglomeration is taking place at longer times. For SNR, only D_{90} is affected showing the decrease of agglomerates concentration; whereas, the differences at D_{50} and D_{10} are negligible. For Si, the deagglomeration is more complete at short ultrasonication times. Each powder has a specific requirement of the ultrasonication time and power to achieve a complete dispersion.

The influence of dispersion medium was evaluated for SNR and SiC powder samples by using aqueous and nonaqueous media in gravitational sedimentation (Sedigraph). These data are shown in Table 8.

TABLE 8
Influence of Dispersion Medium on D-Values (μm) by Gravitational Sedimentation and Ultrasonication (US) with a Probe

Powder	Dispersion Medium	D_{90}	D_{50}	D_{10}
SNR	Sedisperse A-11, 1 min. US	2.8	0.8	0.2
	Deionized Water, pH =10.0 0.4 v% 111M, 3 min. US, 0.13 g/100 ml powder	5.7	1.2	0.4
SiC	Distilled water, 0.5 % v butylamine, 10 min. US	1.0	0.6	0.2
	Sedisperse A-11, 5 min. US	6.1	4.3	3.4

These data show the variation of dispersion effectiveness of the media-surfactant combination. In the case of SNR, the use of 111M (a polycarboxylate surfactant from Allied Colloid Co.) in water at pH 10.0 may have either induced agglomeration due to excessive concentration or may have been ineffective as a dispersant. Similarly, for SiC, the Sedisperse A-11 may have either induced agglomeration or be ineffective as a dispersant.

The particle size distribution data by gravity sedimentation, centrifugal sedimentation and light scattering methods were compared for five powders by using d_{90}, d_{75}, d_{50}, d_{25} and d_{10} values. The range of variability of only d_{50} for SNR powder by the three methods is shown in Figures 3, 4 and 5. The variability of d_{50} is equally significant for all three methods. In addition, for a given lab., the reproducibility of d_{50} has a large range. The mean d_{50} of SNR powder by all three methods varied in a narrow range from 0.91 to 1.0 μm, which is considered to be an excellent agreement of the data, considering the fact that no prescribed procedure was used by the participants for the three methods of particle size measurement.

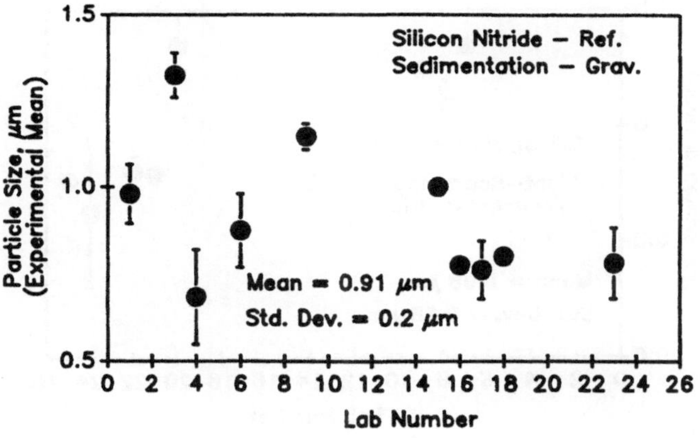

Figure 3. Experimental mean particle size of SNR powder by gravitational sedimentation.

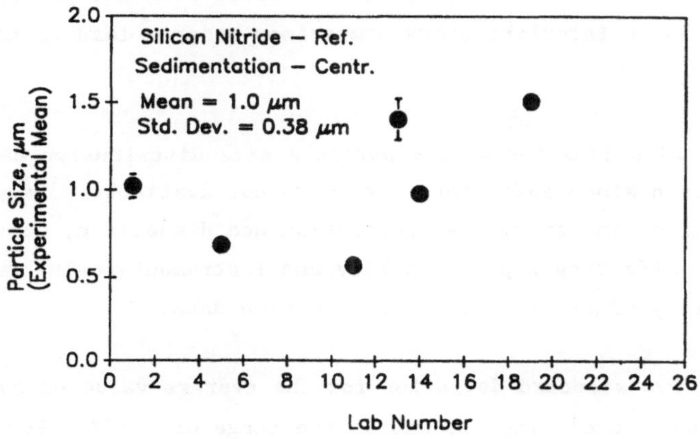

Figure 4. Experimental mean particle size of SNR powder by centrifugal sedimentation.

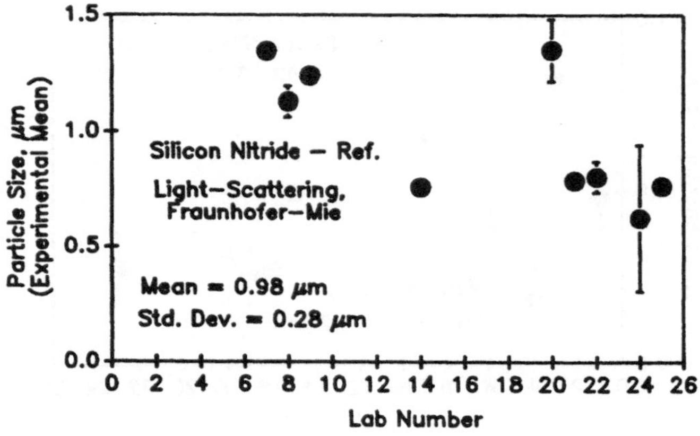

Figure 5. Experimental mean particle size of SNR powder by light-scattering, Fraunhofer-Mie type instrument.

CONCLUSIONS

The measurement of density, specific surface area and particle size distribution in an interlaboratory comparison program lead to the following conclusions:

- Statistical evaluation of the particle size distribution data was not appropriate since sufficient data were not available. Sample preparation, involving deagglomeration and dispersion, was the major variable affecting reproducibility and instrument-to-instrument variability of particle size distribution data.

- The relative standard deviation for the average value of surface area for each powder, excluding Si, was in the range of 5-10%. For SNR, SNT and SiC powders, a value can be obtained within an uncertainty band of ±1 m^2/g around the average values. The average MP values were higher by 2-10% than the SP values.

• Bulk density measurements are in good agreement with theoretical values reported in the literature. However, tap density data showed a large spread due to variations in the experimental procedure.

Acknowledgements: This project is partially supported in the U.S. by the U.S.Department of Energy, through the Ceramics Technology for Heat Engines Project of the Advanced Materials Program, under contract DE-AC05-840R21400, with Martin Marietta Energy Systems, Inc., through Oak Ridge National Laboratory.

REFERENCES

1. Malghan, S. G., et.al., "Physical and Chemical Characterization of Ceramic Powders in an International Interlaboratory Comparison Program," paper presented at the Seventh CIMTEC World Ceramics Congress, June 1990.

2. Hsu, S. M., et.al., "Characterization of Ceramic Powders: Data and Analysis, Final Report," Prepared by NIST for DoE, March 1990.

3. Resetar, T. M., et.al., "IEA/Annex II Powder Characterization Cooperative Program" U.S. Army Materials Technology Laboratory, MTL TR 89-53, 1989.

4. McSkimin, H. J., Bond, W. L., Buehler, E., Teal, G. K., "Measurement of Elastic Constants of Silicon Single crystals and Their Thermodynamic Constants," Phys. Rev. 83, 108, 1951.

5. Tibault, N. W., "Morphological and Structural Crystallography and Optical Properties of SiC, Part II," Am Mineral. 29, 327, 1944.

6. Turkdogan, E. T., Bills, P. M., Tippett, V. A., "Silicon Nitrides: Some Physicochemical Properties" J. Appl. Chem., 8, 296, 1958.

7. Lowell, S. and Shield, J. E., Powder Surface Area and Porosity John Wiley and Sons, Inc., New York, 30-32, 1984.

8. Schwier, G., "On the Preparation of Fine Silicon Nitride Powders" in Progress in Nitrogen Ceramics, Ed. F. L. Riley, Martinus Nijhoff Publishers, Boston, 157-168, 1983.

9. Weiss, J. and Kaysser, W. A., "Liquid Phase Sintering" in ref. 8, pp. 169-186.

10. Allen, T., Particle Size Measurement, 3rd edition, London: Chapman and Hall, Chapter 1, 1981.

A PROPOSED STANDARD PRACTICE FOR FRACTOGRAPHIC ANALYSIS OF MONOLITHIC ADVANCED CERAMICS

Jeffrey J. Swab and Michael J. Slavin
U.S. Army Materials Technology Laboratory
Watertown, MA 02172, USA

and

George D. Quinn
National Institute of Standards and Technology
Gaithersburg, MD 20899, USA

ABSTRACT

The strength of many brittle ceramics reflects the flaws present in the material and the intrinsic fracture toughness. Interpretation of the strength results for monolithic, brittle advanced ceramics requires fractographic analysis whether mechanical testing has been done for quality control, materials development, or design purposes.

Progress is presented on an effort to develop a standard practice for fractographic analysis of laboratory strength specimens. The primary goal is to encourage fractography as a complement to mechanical testing. Procedures will be developed with regard to inspection techniques, sampling criteria and reporting practices.

INTRODUCTION

It is not difficult to argue that fractography is essential to properly interpret mechanical strength results. Stressed monolithic ceramics and "simple" composite ceramics (e.g., particulate or whisker reinforced) will typically fail in a brittle fashion due to the unstable propagation of cracks from preexisting defects. (<u>NOTE</u>: The terms defects and flaws will be used interchangeably in the practice.) These defects may be intrinsic to the materials as a consequence of its manufacturing process, or may be introduced as a result of specimen preparation, handling or exposure. Strength results must be interpreted in the context of these defects whether the goal of the mechanical testing is materials characterization or design. This was discussed in the context of flexure testing in a paper from the Conference on Fractography of Glasses and Ceramics[1]. In that paper, an example from a comparative machining study was given where a completely erroneous conclusion could have been drawn from the strength data alone. Only with comprehensive fractography was it realized that the apparent differences in strength in separate samples were due to subtle flaw variations between different billets of the alumina ceramic.

The matter is even more critical if strength data is to be used for design. Characterization of multiple flaw populations is important to the extent that if more that two flaw populations are present, design with the material may be difficult or even impossible[1]. The subtle variations in porosity defects in the comparative machining study cited above also poses troubling questions. Defects in the alumina manifested themselves as discrete voids or as zones of microporosity. Can these be considered members of a single family of defects or do they behave differently? Thus, characterization of defects is a critical topic.

The presentation at the first fractography conference foreshadowed our efforts to develop a standard practice[1]. Optical and scanning electron microscopy (SEM) were discussed in the context of finding and characterizing strength-limiting defects. Montages and labeled Weibull graphs were suggested schemes to aid in the interpretation and reporting of fractographic results. A possible flaw nomenclature was proposed at that time. Two recent mechanical property round-robins have highlighted some of the shortcomings of fractographic procedures now in use[2-3]. An earlier exercise similarly uncovered serious interpretation problems[4]. It is evident that there are widely divergent practices and expertise levels.

Unfortunately, there are a number of "tricks of the trade," and fractography is a continual learning experience. New tools and analyses are continually becoming available. Success in finding and correctly characterizing defects is a strong function of the experience level of the fractographer. Even experts can be misled in their findings[2] and it is extremely rare to find open debate on fractographic interpretations[5-6]. Fractographic analysis and "representative" photographs are usually taken at face value. Most fractographers must admit that, in truth, there often are disagreements in interpretation.

Our objective is to advance the state-of-the-art of fractographic analysis of advanced ceramics by developing a fractographic standard practice to complement and enhance mechanical strength results. Indeed, designers who wish to use advanced ceramics urge such a practice as they seek more specific characterization of strength-limiting defects. However, the practice is not limited to design applications. Different sampling criteria and analyses are required depending upon the application of the fractographic data. Our goal is also to bring together the best available information on defect characterization and, by a tutorial approach, introduce newcomers to the field.

The approach of the standard practice will be to present helpful tips, to recommend sound procedures which we hope will lead to improved fractography, and fractography that will have optimum complementary value to mechanical strength results. All of this will be done within a **specified** framework of defect characterization. Primary emphasis will be on characterization and location of strength-limiting defects in laboratory test specimens, but the same principles are easily applicable to component failure analysis. For some materials, the location and identification of defects may not be possible for all specimens due to the specific microstructure.

SUMMARY OF PRACTICE

Procedures will be developed with regards to inspection techniques, sampling criteria and reporting practices. The area of inspection techniques will provide guidelines and precautions through all steps of the analysis including mechanical testing, handling and storage of the specimens.

The sampling criteria given is based on whether the analysis is for quality control, materials research and development or design purposes. Reporting practices will provide useful ways to couple the fractographic results with the mechanical property data and microstructural features.

In addition, the practice will contain several appendices including: an annotated bibliography on fractography and failure origins in ceramics; possible fracture patterns in various mechanical test specimens; a flaw nomenclature with multiple examples of each flaw type from various ceramics; and examples of fractographic montages.

PROCEDURE

General

There are three steps to the actual inspection: Visual (1-10X), Optical (10-100X) and Scanning Electron Microscopy (SEM) (10-2000X). The main purpose of the visual inspection is to examine the crack branching patterns for any evidence of abnormal failure patterns, determine the primary fracture surface, and locate the fracture origin (if possible). Optical inspection is required to locate the failure origin and characterize the flaw (if possible). If necessary, inspect the specimen surface near the fracture origin to check for handling or surface machining damage. If a definite characterization of the defect can not be made, then optical examination shall be conducted with the purpose of expediting the subsequent SEM examination. SEM examination will be used to characterize the strength-limiting defects according to its identity, location and, if appropriate, size.

It is understood that the purpose of the analysis will dictate how many of the fracture surfaces will be examined in each step. To provide some guidance in this area sampling criteria has been created (Table 1).

Table 1.
Sampling Guidelines

	1-10X Visual	10-100X Optical	10-2000X SEM
Level 1 - Quality Control	Specimens which fail to meet minimum strength requirements.	Specimens which fail to meet minimum strength requirements.	Optional
Level 2 Screening	All specimens	All specimens (Both Halves)	Optional
Level 3 Intermediate - Quality Control - Materials Development	All specimens	All specimens	Representative specimen (mount both halves) -2 each of each flaw type -the 5 lowest strength specimens -at least 2 optically unidentifiable flaws
Level 4 Comprehensive -Quality Control -Materials Development -Design	All specimens	All specimens	All specimens, or -as many specimens as necessary such that combined optical and SEM characterize 90% of all identifiable origins

Flaw Characterization

Flaws are either inherently volume distributed in the material (e.g. agglomerates, large grains or pores) or inherently surface distributed on the material (e.g. handling damage, pits from oxidation or corrosion). These volume flaws can be, in any specimen, **located** in the bulk, at the surface, or at an edge. Where as an inherent surface flaw is located at the surface or edge. Defects shall be characterized by the following three attributes shown in Table 2 and Figure 1.

Table 2.
Flaw Characterization Scheme

IDENTITY	LOCATION	SIZE (Optional)
Nomenclature specific to the spatial distribution	Volume (bulk), Surface, Near Surface, Edge	As a measure of scale (major and minor axis)

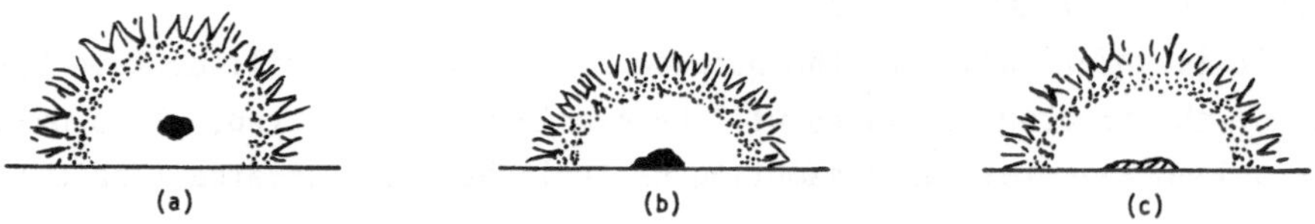

	Identity	Location	Size
(a)	Pore	Volume	~40 um
(b)	Pore	Surface	~20 x 50 um
(c)	Machine Damage	Surface	~10 x 50 um

Figure 1. Schematic of Flaw Characterization Scheme

Flaw Identity: Defects will be characterized whenever possible by a phenomenological approach which identifies what the defect is and not how it appears under a particular mode of viewing. Descriptions of the mode of viewing may be used as qualifiers, (i.e. pores that appear white when viewed optically), but the use of appearance, (i.e. "white spots") should be avoided. This approach is chosen since defects appear drastically different in optical versus electron microscopy.

Table 3 gives the nomenclature which will be applicable to many advanced ceramics. It must be recognized that not all defects can be so characterized and that many defects are specific to a material and its process history. Defects can also exist concurrently (e.g., an agglomerate with a surrounding pore, or a pore with associated large grains) in which case some judgement is required as to which defect is dominant or intrinsic. Defects can also be described by paired expressions, i.e., a pore/large grain. Note that defects can sometimes be difficult to delineate into one specific category if they have mixed attributes, for example, porous regions often have a small pore associated with them. Certain defects may be distributed through out the bulk of a ceramic, whereas others may be distributed only at the surface. The defects must be characterized as to whether they are inherently volume or surface distributed. This requires knowledge of the fabrication history and test conditions by the fractographer.

Table 3.
Flaw Nomenclature

FLAW TYPE	CODE
Inherent Volume	
Pore	P
Porous Seam	PS
Porous Region	PR
Agglomerate	A
Inclusion	I
Second Phase Inhomogeneity	2P
Large Grain(s)	LG
Crack	C
Inherent Surface	
Machining Damage	md
Handling Damage	hd
Pit	pt
Surface Void	sv
Miscellaneous	
Other	@
Unknown	?

Flaw Location: The location of a defect shall be qualitatively determined. The defect must be identified as being located in the bulk (volume), at the surface, or at the edge (if such exists). This procedure shall distinguish volume defects which reside at the surface of a specimen as a result of pure random sampling when a specimen is machined from the bulk. As opposed to defects that are inherently surface related, (i.e., voids from a surface reaction layer). This characterization shall also serve the needs of the designer who may be concerned with how surface connected defects may behave differently under operational or test conditions than the other volume distributed defects located in the bulk.

In some instances, it is useful to further specify defect location if it is near to the surface, but not in direct contact. This location category shall be termed near surface. This additional specification of location is important for fracture mechanics evaluation of defects and service-performance issues and, when used, will be defined by the fractographer.

Flaw Size: The size of each defect shall be qualitatively assessed as necessary to identify the nature of the defect in a general sense (i.e. the 20 μm pore versus the 1 μm pore). For volume distributed defects, mean diameter shall be reported for an equiaxed defect or the major and minor axis lengths for other defects. For surface defects the depth and length along the surface shall be reported.

Reporting

The reporting of these results shall include the fractographers' identity, the equipment used, the overall defect classes or types identified, the modes of viewing each specimen and the inspection criterion used (e.g. as per Table 1).

Supplemental observations such as transgranular or intergranular fracture are highly encouraged and can aid in determining the fracture mechanism.

To the extent possible, couple the fractographic observations directly to process history and resultant microstructure. Representative micrographs of polished sections of the microstructure showing porosity and grain size distributions are required. Also, couple the observations directly to the mechanical test data. Fractographic montages and labelled Weibull or strength graphs, Figure 2, are an exceptionally versatile means of accomplishing this.

REFERENCES

1. Quinn, G.D., Fractographic Analysis and the Army Flexure Test Method in Fractography of Glasses and Ceramics, Advances in Ceramics, Vol. 22, V. Frechette and J. Varner, eds., American Ceramic Society, Westerville, OH, 1988, pp. 319-333.

2. Quinn, G.D., Flexure Strength of Advanced Ceramics - A Round Robin Exercise, U.S. Army Materials Technology Laboratory, Watertown, MA, MTL TR 89-62, July 1989.

3. Ferber, M.K. and Tennery, V.J., Fractographic Study of a Silicon Nitride Ceramic in Proceedings of the First European Ceramic Society Conference, G. deWirth, R.A. Terpstra, R. Metselaar, eds., Maastricht, NL., Elsevier, June 1989, pp. 18-23.

4. Lewis, D., III, Private Communications, 1989.

5. Purslow, D., Comment on Fractography of Unidirectional Graphite-Epoxy as a Function of Moisture, Temperature and Specimen Quality, J. Mat. Sci. Let., 1989, 8, p. 617.

6. Clements, L.L., Reply to Comment on Fractography of Unidirectional Graphite-Epoxy as a Function of Moisture, Temperature and Specimen Quality, J. Mat. Sci. Let., 1989, 8, p. 618.

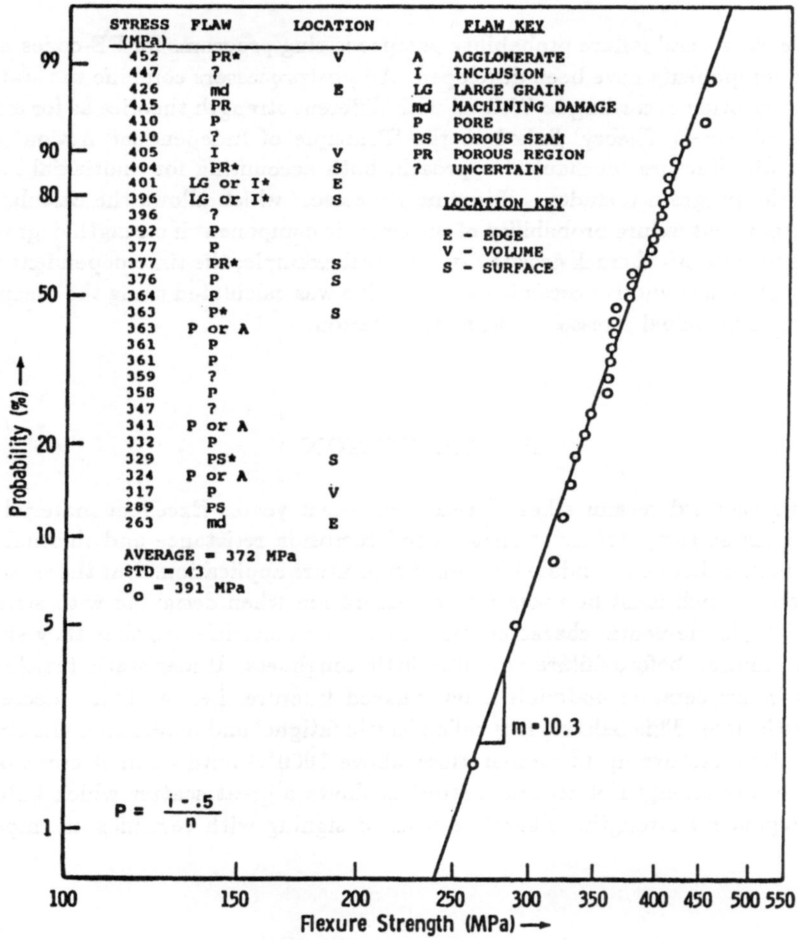

Figure 2. Labelled Weibull Graph for 4-Point Flexural Strength

FEM-LIFTAP, A FINITE ELEMENT LIFETIME ANALYSIS POSTPROCESSOR FOR CERAMIC COMPONENTS

K. Kussmaul, S. Lauf, K. Turan
Staatliche Materialpruefungsanstalt (MPA),
University of Stuttgart, 7000 Stuttgart 80, FRG

ABSTRACT

In recent years, several failure probability postprocessing programs for FE-codes applied to ceramic components have been developed. All postprocessors combine the statistical strength description according to Weibull with different strength theories as for example the 'Principal Stress Theory' linked to the 'Principle of Independent Action' or the 'Linear Elastic Fracture Mechanics' approach, both accounting for multiaxial loading. Moreover, the program includes a 'Lifetime Processor' which allows the calculation of the time-dependent failure probability of the ceramic component if strength degradation occurs due to subcritical crack growth. In a typical example, the time-dependent failure probability of an automotive ceramic exhaust valve was calculated using the mean value of the positive principal stresses as strength criterion.

INTRODUCTION

The use of structural ceramics has increased in recent years. Excellent material properties like such as temperature strength, good corrosion resistance and thermal shock behaviour makes them a candidate for high temperature application. But there are some characteristics which must be taken into consideration when designing with structural ceramics. The fundamental characteristics of ceramic materials are that they show no plastic deformation before failure and have little toughness. Under static tensile stress conditions in air ceramic materials show delayed fracture, i.e. a steady decrease of strength with time. This behaviour is called 'static fatigue' and determines the strength from room temperature up to temperatures above 1000°C until distinct creep occurs. Furthermore the strength of ceramic materials shows a great scatter which influences the time-dependent strength. Therefore when designing with ceramics an important factor is the computation of the reliability or probability of survival of a structural component under certain loading conditions. The computation of stresses using finite element analysis is the basis for this.

UNIAXIAL LOADING OF CERAMICS

The strength of ceramic materials is determined experimentally by bending or tensile tests. Fracture occurs because the fracture toughness K_{IC} at material inherent defects is exceeded. Under bending conditions the surface and surface-near defects are responsible for fracture whereas for tensile specimens the fracture origin can be located over the whole cross section and the volume under load. In both cases we have uniform loading and with the assumption of linear elastic behaviour and idealized 'Griffith cracks' fracture is prescribed by LEFM.

Real cracks like inhomogenities, pores and other sintering defects show a distribution in size and orientation and can be described as equivalent idealized Griffith cracks. Therefore it is found that the strength values lie on some distribution curve, best fitted by the Weibull-distribution.

Brittle materials, particularly glass and polycristalline ceramics weaken with time under load. This behaviour is called 'Static Fatigue behaviour' which leads to strength degradation even under static load. The physical explanation is that cracks or flaws grow slowly with time under subcritical stress until the crack reaches a critical length. Slow crack growth occurs as a result of a stress-aided thermally-activated event and by chemical attack from the environment within the crack. The rate of crack propagation da/dt with a stress intensity factor K_I at the crack tip is given by

$$\frac{da}{dt} = AK_I^n \tag{1}$$

with the subcritical crack extension parameters A and n. Fatigue under cyclic loading conditions seems to be an additional or separate effect at least for toughened ceramics [1-3]. A special cyclic effect is further not considered.

STRENGTH, FAILURE PROBABILITY AND LIFETIME

As long as subcritical crack growth is assumed to be active the actual strength $S(t)$ is by, [4, 5]

$$S(t) = \sigma_{IC} \left(1 - \frac{\sigma_{IC}^2}{B} \int_0^{t_f} \left(\frac{\sigma(t)}{\sigma_{IC}}\right)^n dt\right)^{\frac{1}{n-2}} \tag{2}$$

Failure occurs if strength reaches zero. Under this condition and constant loading condition the well known SPT-diagram follows [5, 6]. Lifetime prediction using SPT-diagrams are valid only for uniform uniaxial loading conditions. For ceramic components an extended evaluation due to multiaxial loading must be considered.

MULTIAXIAL LOADING AND LIFETIME

Usually structural components are subjected to multiaxial stress states. Appropriate material properties determined under multiaxial loading are not available. Therefore the multiaxial stress state has to be reduced to an equivalent fictious uniaxial stress state. This is provided by material- specific strength hypothesis which must be combined with the Weibull strength distribution description.

Principal Stress Approach

The failure probability of a ceramic component subjected to a non-uniform uniaxial stress can be expressed as [7]

$$P_f = 1 - exp\left(-\left(\frac{1}{m}!\right)^m \frac{V}{V_o}\left(\frac{\sigma_{nom}}{\bar{\sigma}_o}\right)^m \Sigma(V)\right) \tag{3}$$

with the stress-volume-integral

$$\Sigma(V) = \int_V \left(\frac{\sigma}{H(\sigma)\sigma_{nom}}\right)^m \frac{dV}{V} \tag{4}$$

$H(\sigma)$ characterizes the ratio of the mean failure stress under uniaxial compression, compared to that under uniaxial tension.σ_{nom} is used for practical reasons.
In general, the stress state at a point in a real component is characterized by three principal stresses. When extending Eq.(3) to cover the body subjected to multiaxial stresses, it is necessary to establish a 'failure criterion', which expresses the dependence of failure on the combined action of stresses present, like the Tresca or von Mises criterion for metals.

For further considerations it is assumed that the failure probability of a volume element due to one principal stress is independent of the presence of other principal stresses. A second assumption is that the material is isotropic. As a consequence Eq.(3) becomes

$$\Sigma(V) = \int_V \sum_{i=1}^{3} \left(\frac{\sigma_i}{H(\sigma_i)\sigma_{nom}}\right)^m \frac{dV}{V} \tag{5}$$

For noninert conditions the strength degradation due to subcritical crack growth must be taken into account. This is done by substituting the mean strength $\bar{\sigma}_o$ by Eq.(2) in unified (unit-volume) form and using the time dependent principal stresses.
With Eqs.(2)-(4) the principal stress approach is formulated and can easily be connected with the Finite Element Analysis for numerical evaluation of multiaxially loaded ceramic components. This treatment allows the quantification of reliability.

Fracture Mechanics Approach

If a shear sensitive failure criterion is seen to be necessary the critical energy release rate may be used [8-10]. Assuming coplanar crack extension, i.e. the shear stress is in the crack plane, it follows

$$G = \frac{1-\nu^2}{E}\left(K_I^2 + K_{II}^2 + \frac{1}{1-\nu^2}K_{III}^2\right) \qquad (6)$$

Introducing an equivalent stress intensity factor $K_{I_{eq}}$ failure occurs if

$$G_C = \frac{1-\nu^2}{E}K_{I_{eq}C}^2 \geq \frac{1-\nu^2}{E}K_{IC}^2 \qquad (7)$$

Experimental results show [9] that a generalized estimate of the form

$$\left(\frac{K_I}{K_{IC}}\right)^2 + \left(\frac{K_{II}}{K_{IIC}}\right)^2 + \left(\frac{K_{III}}{K_{IIIC}}\right)^2 = 1 \qquad (8)$$

describes the failure behaviour better.

With the further assumption that the failure origins are penny-shaped cracks and $K_{IIIC} \geq (1-\nu)K_{IIC}$ failure is determined by the equivalent stress

$$\sigma_{eq} = \left(\sigma_n^2 + \tau^2\left(\frac{2}{2-\nu}\right)^2\left(\frac{K_{IC}}{K_{IIC}}\right)^2\right)^{1/2} \qquad (9)$$

This expression is valid for an arbitrary orientated penny-shaped crack whose equitorial plane and the principal stress directions form the angles Φ, Ψ, and χ, Figure 1.

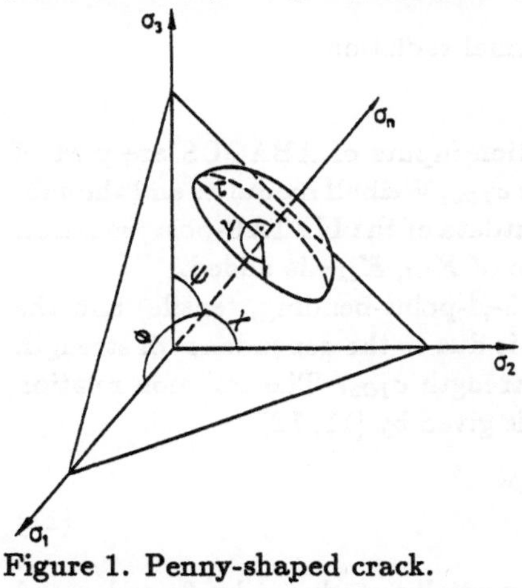

Figure 1. Penny-shaped crack.

Considering an arbitrary isotropic crack orientation in the loaded volume, an averaging over the unit sphere is necessary and we obtain

$$\bar{\sigma}_{eq} = \frac{1}{4\pi}\int_0^{2\pi}\int_0^{\pi}\sigma_{eq} sin\Psi d\Psi d\Phi. \qquad (10)$$

Using Eq.(10) in the stress-volume-integral, Eq.(3), the synthesis between statistical and multiaxial strength description of ceramic components is completed.

Lifetime Evaluation

The strength of ceramics which show subcritical crack growth is reduced according to Eq.(2) and must be implemented in the statistical multiaxial lifetime prediction. In a first approximation this is achieved by substituting the unit strength $\bar{\sigma}_o$ by Eq.(2), where $\sigma(t)$ is substituted by $\sigma_{nom}(t)$. A more precise evaluation is obtained if the time-dependent strength is included in the stress-volume-integral $\Sigma(V)$.

FEM-POSTPROCESSOR FEM-LIFTAP

The modelling and reliability analysis of a ceramic component consists of :

- net generation(PATRAN, ABAQUS)
- strain, stress and temperature analysis
- evaluation of failure determinating stress
- determination of initial failure probability
- calculation of time dependent strength
- calculation of time dependent failure probability

Pre- and postprocessing of FEM-code ABAQUS is done with PATRAN. ABAQUS provides the material and physical laws :

- linear-elastic behaviour
- visco-elastic behaviour
- initial stresses
- linear thermal extension
- thermal conductivity
- convection of heat
- thermal radiation

The material data and the boundary condition inputs of ABAQUS are part of PATRAN. Strength expressed as Weibull parameter σ_{ICo}, Weibull modul m and the subcritical crack extension parameters n and B are inputdata of the LIFTAP-postprocessor. If the fracture mechanics approach is used the ratio of K_{IC}/K_{IIC} is added.

Together with the strength input the type (3-,4-point-bending, tensile) and the dimensions of the specimen must be specified. This is due to the dependence of strength on volume, which must be converted to the unit strength σ_{ICo}. The common relation between the strength σ_{ICoi} of different volumes V_i is given by [11, 12]

$$\frac{\bar{\sigma}_1}{\bar{\sigma}_2} = \left(\frac{V_2\Sigma(V_2)}{V_1\Sigma(V_1)}\right)^{1/m} \tag{11}$$

If the strength distribution is completely or partially produced by flaws located on the surface, the surface element strength may be treated in a similar way as it is described here for volume elements [13, 14]. In general, the conversion of the surface strength distribution of specimens cannot be transfered to surface elements of ceramic components because of different surface treatments.

RELIABILITY OF A CERAMIC EXHAUST VALVE

In automotive engine design the use of ceramic valves is already in an experimental stage. Therefore the application of FEM-LIFTAP on such a mechanical and thermal loaded component is of a high interest. For reliability analysis a conventional exhaust valve was modelled, A valve in the closed state with a combustion pressure of maximal 12 MPa (1.74 ksi) was assumed. The valve spring force of 315 N operates at the outer radius of the valve seat with suppressed displacement in axial direction [15]. Thermal boundary conditions are assumed to be stationary too, Figure 2+3, with the heat transmission coefficients α_k. Thermal and mechanical stresses were linearly superimposed.

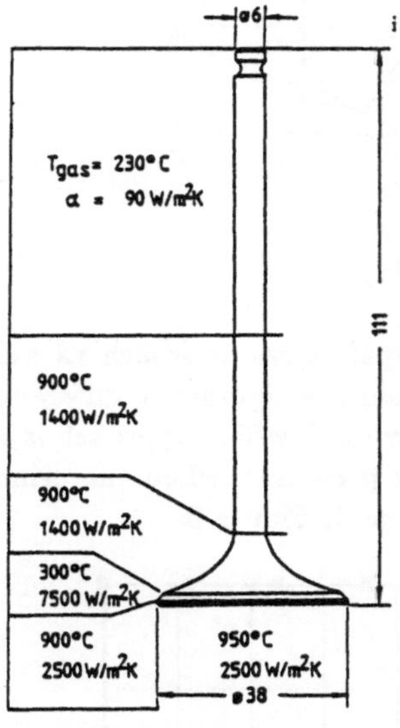

Figure 2. Valve geometries and thermal boundary conditions.

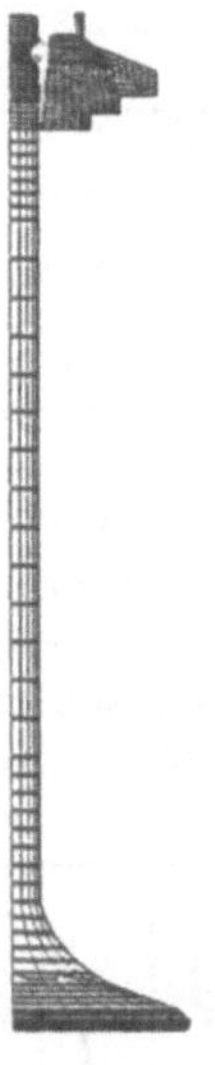

Figure 3. Valve, FE-net.

The material data were typical for a SSN material and assumed to be independent of temperature :

4-point-bending strength σ_{IC_0}	800	MPA
Weibull module m	10.5	
n-value	110	
B-value	70	MPa^2s
Youngs' module E	300	GPa
Poissons ratio ν	0.25	
Thermal expansion coefficient	3.1	$10^{-6}K^{-1}$
Fracture toughness K_{IC}	5.5	$MPam^{1/2}$
K_{IC}/K_{IIC}	1.1	

The failure probability was calculated first for increasing combustion pressure up to 12 MPa (1.74 ksi) without subcritical crack growth, Figure 4.

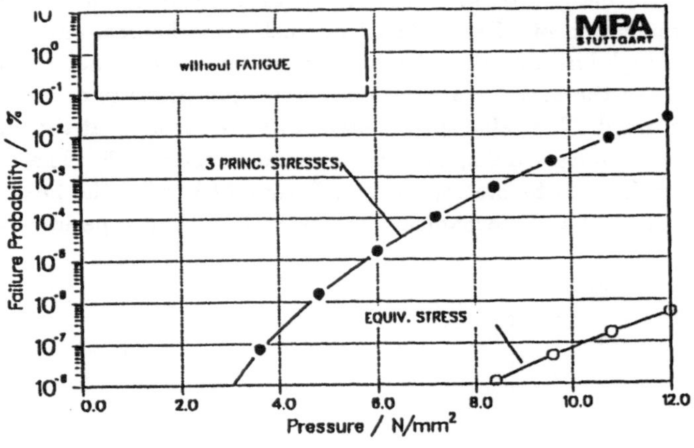

Figure 4. Failure probability at $t = 0$ ($P_{combust.} = 12$ MPa).

The calculations were performed using the principal stress approach as well as the fracture mechanics approach. Between both approaches a significant difference of at least 4 orders of magnitude arises. The question is obvious - which approach is valid ? The experimental answer can be given only for higher pressures, when the principal stress approximation failure probability reaches 100 per cent, Figure 5.

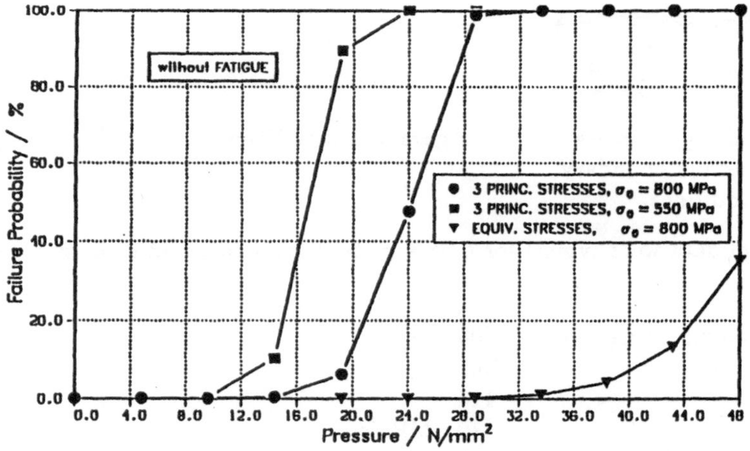

Figure 5. Failure probability for different strength values and approaches.

A further calculation consideres the dependence of failure probability on time, when strength decreases due to subcritical crack growth. Two different initial strengths were used. The combustion pressure was held constant at 12 MPa and the principal stress approach was applied. Starting with different failure probabilities no change with time, i.e. no significant decrease of strength up to 10^8s occurs, due to the relatively high n-value. If n is reduced to 40 a dramatical reduction in lifetime is calculated, Figure 6.

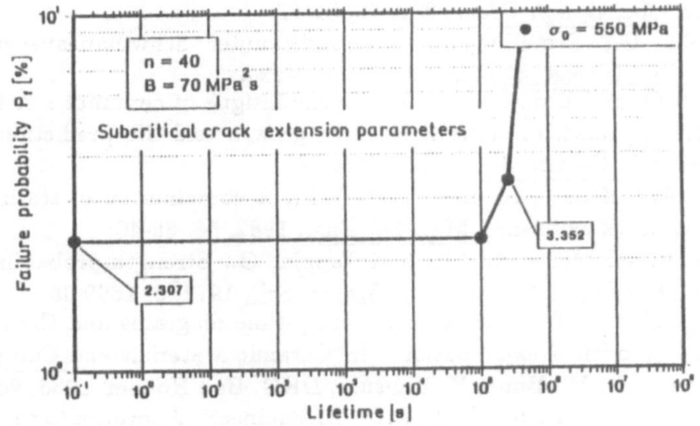

Figure 6. Failure probability for reduced strength and n-value.

The sudden increase of failure probability, i.e. decrease of strength is typical for the power law of crack growth, Eq.(1).

The calculated examples show that for a given strength approach the n-value has the leading influence on component lifetime. Concerning the strength approach, experimental examinations on ceramic components are difficult and expensive. Recent experimental investigations on rotating annular disks show that only a postexperimental fitting gives a partially agreement with theory [15].

SUMMARY

The FEM-postprocessor FEM-LIFTAP was developed to calculate failure probability and lifetime of mechanically and thermally loaded ceramic components. In addition to the incorporated material and physical laws of the used FE-code ABAQUS, LIFTAP provides the principal stress and fracture mechanics approach combined with statistical strength description and strength degradation according to subcritical crack growth. As an example a ceramic exhaust valve was modelled using PATRAN, analyzed using the FE-code ABAQUS and the time-dependent failure probability was calculated using LIFTAP. With SSN material data and realistic boundary conditions for combustion pressure and thermal conditions very low failure probabilities followed, which did not changed over a loading time of 10^8s. Reducing the strength from 800 MPa to 550 MPa results in an increase in fracture probability. An abrupt increase in failure probability occurred after 10^5s when the subcritical crack extension parameter n was additionally reduced from 110 to 40. The calculations show the importance of the subcritical crack extension parameter n and the necessity of experiments subject to multiaxial loading.

REFERENCES

1. Bowman, K.J., Reyes-Morel, P.E. and Chen, I-W., Reversible transformation plasticity in uniaxial tension-compression cycling of Mg-PSZ. Mat. Res. Soc. Symp. Proc.,1987, 78, 51-57.
2. Grathwohl, G., Ermüdung von Keramik unter Schwingbeanspruchung, Mat.-wiss. u. Werkstofftech., 1988, 19, 113-124.
3. Ritchie, R.O. and Dauskardt, R.M., Cyclic fatigue of ceramics : A fracture mechanics approach to subcritical crack growth and life prediction, to be published in J. Ceram. Soc. Jap., 1991, 99.
4. Tradinik, W., Kromp, K. and Pabst, R.F., A combination of statistic and subcritical crack extension, Mat. Sc. Eng., 1982, 56, 39-46.
5. Davidge, R.W., McLaren, J.R. and Tappin, G., Strength-probability-time (SPT) relationships in ceramics, J. Mater. Sci., 1973, 8' 1699-05.
6. Lauf, S., Pabst, R.F., Strength-probability-time diagrams and the strength function - A critical examination. In Ceramic Materials and Components for Engines, eds. W. Bunk, H. Hausner, DKG, Bad Honnef, 1986, 961-977.
7. Stanley, P., Fessler, H. and Sivill, A.D., An engineer's Approach to the prediction of failure probability of brittle components. In Proc. Brit. Ceram. Soc., ed. D.J. Godfrey, BCS, Stoke-on-Trent, 1973, 22, 453-487.
8. Evans, A.G., A general approach for the statistical analysis of multiaxial fracture, J. Am. Ceram. Soc., 1978, 61, 302-308.
9. Rossmanith, H.P., Grundlagen der Bruchmechanik, Wien, 1982.
10. Ikeda, K., Igaki, H. and Kuroda T., Fracture strength of alumina under uniaxial and triaxial stress states, Am. Ceram. Soc. Bull., 1986, 65, 683-88.
11. Jayatilaka, A.de S., Fracture of Engineering Brittle Materials, Appl. Sc. Publ., London, 1979.
12. Creyke, W.E.C., Sainsbury, I.E.J. and Morell, R., Design with Non-ductile Materials, Appl. Sc. Publ., London-New York, 1982.
13. Lamon, J., Ceramic Reliability: Statistical analysis of multiaxial failure using the Weibull approach and the Multiaxial elemental strength model, J. Am. Ceram. Soc., 1990, 73, 2204-12.
14. Nemeth, N.N., Manderscheid, J.M. and Gyekenyesi, J.P., Ceramic analysis and reliability evaluation of structures (CARES), NASA Technical Paper 2916, 1990.
15. Magerl, F., unpublished, University of Stuttgart, 1989.
16. Nemeth, N.N., Manderscheid, J.M. and Gyekenyesi, J.P., Design of ceramic components with the NASA/CARES computer programm, NASA Techn. Mem. 102369, 1990.

CM1 - A SIMPLE MODEL FOR THE DYNAMIC DEFORMATION AND FAILURE PROPERTIES OF BRITTLE MATERIALS

AKE PERSSON
Dynamec Research AB,
Paradisgränd 7, S-151 36 Södertälje, Sweden

ABSTRACT

A simple model (CM1) for the macroscopic behaviour of brittle materials, suitable for simulation of rapid and intense loading has been developed, based on the concept of a damage function increasing with increasing inelastic deformation and reducing the strength of the material. The model has been implemented in the AUTODYN-2D code and tested in a couple of applications with encouraging results which demonstrate the potential of the model to be used for further studies.

INTRODUCTION

In many applications where brittle materials, like ceramics, are subjected to rapid and intense loading the development of extensive failure and the post failure mechanical properties are more important than the creation and propagation of single cracks. Projectile penetration [1], particle impact on ceramic turbine blades and impact testing are examples of such processes. The reason for this is the dynamic nature of these processes which makes the first major failure not beeing the end of the event as in a quasi-static case. It is obvious that the behaviour of materials in this post failure stage is determined by a complex mixture of parallel and competing deformation and failure mechanisms, very hard to model in detail. It is, however, in many cases possible to explain the main part of experimental results with models, based on assumptions about the macroscopic material properties.

Such a simple parametric model, CM1 or CERAMOD1, has been developed to be used in numerical simulations of dynamic processes and the model is presented in this paper.

This work was partly funded by "Skyddskeramkonsortiet".

PRESENTATION OF THE MODEL

Phenomena to be Modelled

Some of the most important mechanical properties of brittle materials are:

- High stiffness which means high elastic moduli.
- Brittleness which means small strain before failure.
- Strong dependence between strenght and local hydrostatic pressure.
- Possibility of inelastic deformation (especially at elevated hydrostatic pressure) creating crushing (damage).

These phenomena must be treated with acceptable accuracy by the model.

Pressure dependent strength. Lundborg [2] has reported experiments where shear failure was initiated in test samples subjected to different pressures normal to the fracture surface. Brittle materials showed a marked strength increase when the pressure was increased. At high enough negative pressures brittle materials are known to loose their strength completely. This behaviour is contrary to the absence of pressure dependence for metallic materials.

Inelastic deformation and crushing. In many situations where brittle materials are quasi-statically loaded in simple geometries (eg bending), the failure is controlled by one or a few cracks running through the entire test sample and leaving the broken pieces virtually undamaged. This is in sharp contrast to the behaviour in highly confined or rapid loading situations where a volume around the region subjected to external loads may be crushed. The crushed material may keep some stiffness (elastic property) and shear strength (internal friction). To describe the progressing crushing and its influence on the macroscopic material properties, the concept of "damage" has been used by several investigators ([3], [4] and [5]).

Basic Model Assumptions

The model - CM1 or CERAMOD1 - is based on the following principles:

- Inelastic ("plastic") deformation occurs at a flow stress which varies with the local hydrostatic pressure.
- von Mises yield criterion is used.
- Damage is initiated by the plastic deformation and is increasing with increasing plastic strain.
- The flow stress is reduced by the damage.
- At extensive damage the material can be given a residual strength during compression but no strength in tension.

A few models for the macroscopic behaviour of brittle materials suitable for simulation of rapid and intense loading have been reported in the literature. Furlong et al [6] present a model based on a damage function concept with separate damage models in tension and in compression. A sample of test calculations with the model implemented in the EPIC-2D code demonstrate the potential of reproducing experimental results.

The idea behind the CM1 model presented in this paper is to give similar properties as Furlong's model with simplified model formulations.

Model Formulation

To start with the material is behaving elastic-perfectly plastic with the onset of plastic deformation determined by von Mises yield criterion. The flow stress (Y_s) is, however, not constant but related to the local hydrostatic pressure (p) according to the following piecewise linear curve:

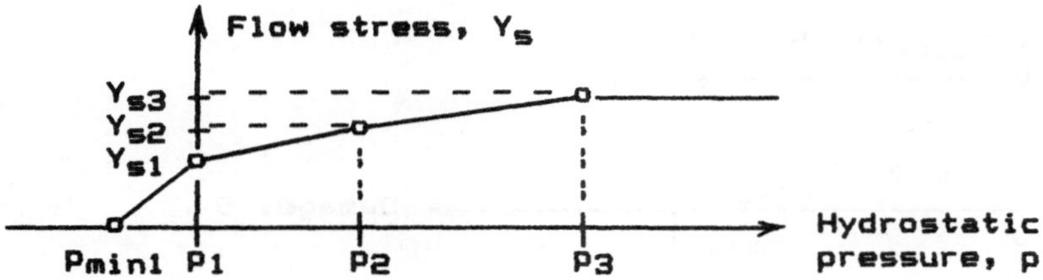

Figure 1. Variation of the flow stress with the local hydrostatic pressure.

When the effective plastic strain (EPS) is accumulating a damage function (D) is increasing irreversibly from 0 to a maximum value, D_{max}. The variation is assumed to be linear:

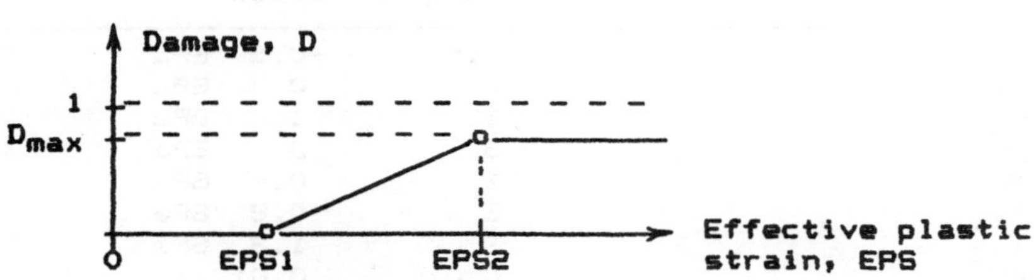

Figure 2. Relation between the damage, D, and the effective plastic strain, EPS.

The damage reduces the strength (Y_s) with a factor (1-D) when the hydrostatic pressure is positive and with a factor (1-D/D_{max}) when the pressure is negative:

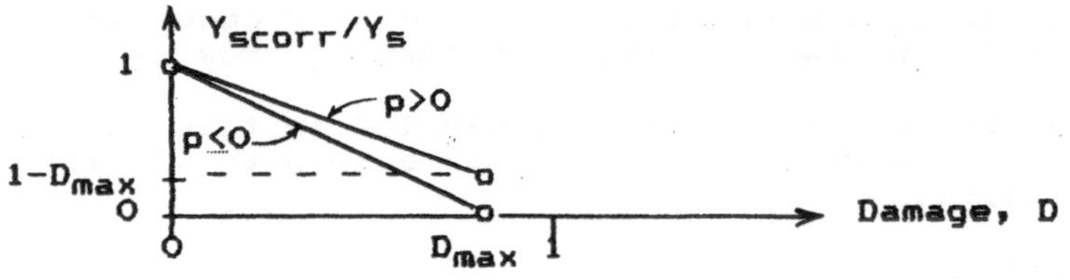

Figure 3. Reduction of the flow stress, Y_s, by the damage, D.

The elastic moduli are unaffected as long as the pressure is positive but reduced by a factor $(1-D/D_{max})$ in regions with negative pressure:

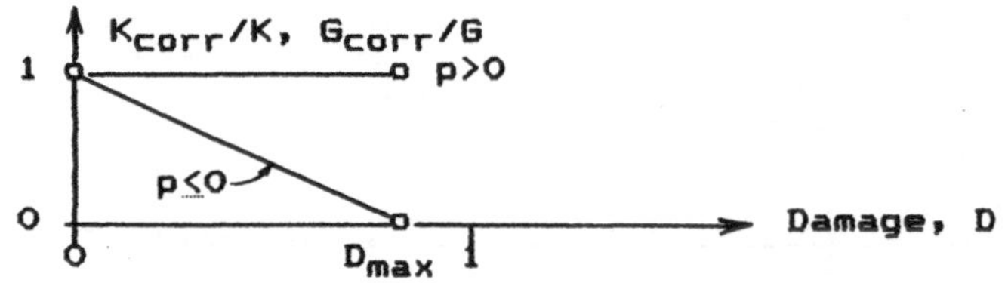

Figure 4. Reduction of the bulk, K, and shear, G, modulus by the damage, D.

Parameter list. The following table gives a list of the parameters introduced in the model and preliminary values for glass and alumina.

	Al_2O_3	Glass	
P_{min1}	-0.5	-0.5	GPa
P_1	0	0	GPa
P_2	1	1	GPa
P_3	3	3	GPa
Y_{s1}	1	0.4	GPa
Y_{s2}	2	0.8	GPa
Y_{s3}	3	1.3	GPa
EPS1	0.01	0.01	
EPS2	0.03	0.03	
D_{max}	0.8	0.8	

Other parameters used in the test calculations:

Reference density	3900	2500	kg/m^3
Bulk modulus, K	150	43	GPa
Shear modulus, G	70	28	GPa

TEST CALCULATIONS

The CM1 model described above has been implemented in the AUTO-DYN-2D code and can be used together with both Lagrange and Euler representation of materials. Preliminary test calculations in different impact and penetration applications show encouraging results and two applications are reported here. In the oral presentation the results will be presented as colour graphics slides which give a much better representation than the black and white reproductions in this paper.

Steel Sphere Impact

In the first application a 3 mm steel sphere is normally impacting a glass cylinder (30 mm diameter, 15 mm length) backed by a steel cylinder. The calculations are carried out for approximately 10 μs after which time the sphere has lost contact with the glass surface and is moving backwards. Two impact velocities (25 and 40 m/s) were tested and three representations of the glass were used: pure elastic, elastic-perfectly plastic (CM1 without any damage) and CM1. The steel was modelled as elastic-plastic with deformation hardening.

The results show the expected distribution of inelastic deformation between the sphere and the glass block according to the model for the glass material (fig 5). When the sphere velocity before and after the impact was plotted for the different combinations of impact velocity and glass model (fig 6) it was interesting to note that changing from elastic to elastic-perfectly plastic behaviour (CM1 without damage) has very little influence on the rebound velocity while introducing damage reduces the rebound velocity considerably. The rebound velocity is therefore a good indicator of the behaviour of the glass.

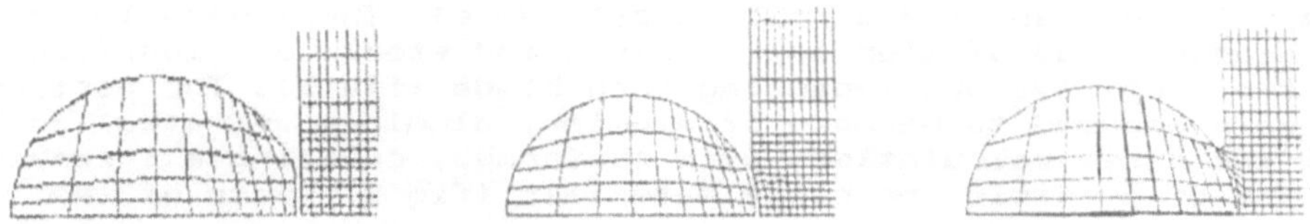

a) Elastic. 40 m/s. b) Elastic-plastic. 40 m/s. c) CM1. 25 m/s

Figure 5. Deformation of a steel sphere and a glass target after impact.

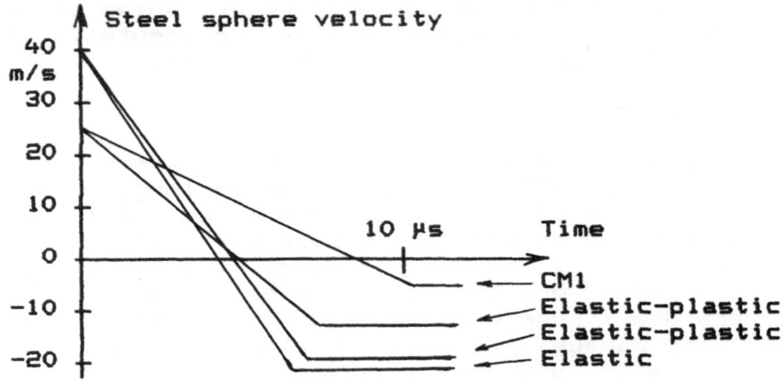

Figure 6. Schematic representation of the variation of the steel sphere velocity with time during the impact.

Particle Impact on Turbine Blades

The second application to be shown covers a schematic representation of particles impacting the rotor blades of a fast spinning turbine rotor. The rotor blades are assumed to move with 500 m/s perpendicular to the rotor axis and the particles are moving with 300 m/s parallel to the axis. To make the process 2-dimensional planar symmetry is assumed (the blades are straight slabs and the particles cylinders). Three parallel rotor blades, made of alumina, aluminum and steel, are modelled and two particles are impacting each blade (fig 7). The particles were assumed to be made of alumina, aluminum and steel as well and three calculations were performed, covering all combinations of particle and blade materials (fig 8). Some of the findings are:

- The alumina particles are damaged in all cases.
- The steel particles create the most severe damage in the alumina blade while aluminum particles caused no damage.
- When hit hard enough the alumina blade is damaged both at the impact point and on the rear side (due to wave reflection).

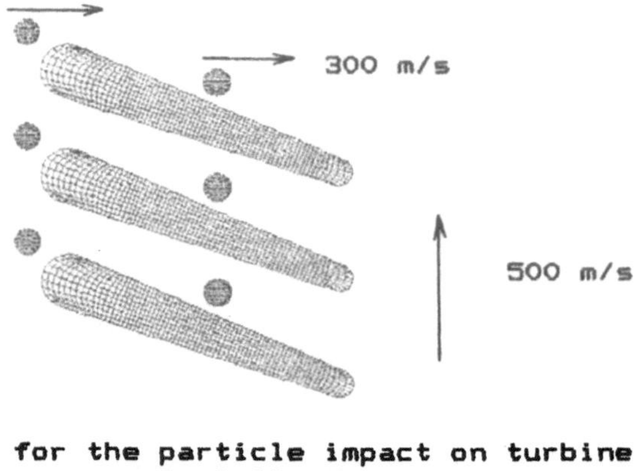

Figure 7. Geometry for the particle impact on turbine blade calculations.

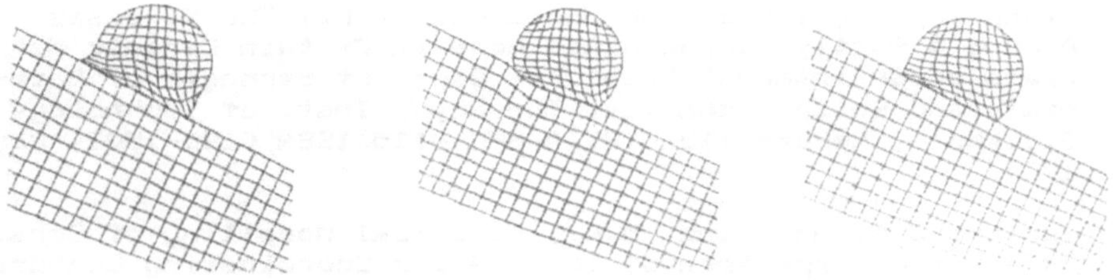

a) Alumina particles　　b) Steel particles　　c) Aluminum particles

Figure 8. Deformation of the particles and the alumina blade after impact.

CONCLUSIONS

Numerical simulations and comparisions with test results (not presented here) show the potential of the CM1-model to be a useful tool in the analysis and prediction of the deformation and failure of brittle materials. Systematic parametric studies and comparisions with experimental results should be made in order to finally adjust the model and find realistic parameter values.

REFERENCES

1. Shockey, D.A. and Marchand, A.H., Failure Phenomenology of Confined Ceramic Targets and Impacting Rods, SRI International/Los Alamos National Laboratory, submitted to the Int. J. of Impact Engineering, Sept. 1989.

2. Lundborg, N., A Statistical Theory of the Polyaxial Strength of Materials, Stiftelsen Svensk Detonikforskning, DS 1976:8, Sept. 1976.

3. Giannakopoulos, A.E., Dynamic Damage in Certain Monoloithic Ceramic Materials, Dept. of Strength of Materials and Solid Mechanics, The Royal Inst. of Technology, Stockholm, Report 96, TRITA-HFL-0096 ISSN 0281-1502, Oct. 1988.

4. Gudmundsson, P., Anisotropic Microcrack Nucleation in Brittle Materials, Dept. of Strength of Materials and Solid Mechanics, The Royal Inst. of Technology, Stockholm, Report 103, TRITA-HFL-0103 ISSN 0281-1502, Feb. 1989.

5. Giannakopoulos, A.E. and Gudmundsson, P., The Stresses Around a Partly Microcracked Hole in Certain Ceramic Materials under Internal Pressure, Dept. of Strength of Materials and Solid Mechanics, The Royal Inst. of Technology, Stockholm, Report 115, TRITA-HFL-0115 ISSN 0281-1502, May 1989.

6. Furlong J.R. and Alme, M.L., Numerical Modelling of Ceramic Penetration Experiments, TACOM Armor Coordinating Conference for Light Combat Vehicles, Monterey, CA, 29-31 March 1988.

AN OTD MATHEMATICAL MODEL FOR PROCESS OPTIMIZATION OF Ce-TZP AND Al_2TiO_5 CERAMICS

ShaoTang ZHAO*, SenFeng CHEN, JianDong YE, YiYuan WU
South China University of Technology,Dept. of Inorganic Materials
Wu Shan, Guangzhou, 510641, P. R. China

ABSTRACT

Orthogonal Test Design was used to optimize the processing parameters to fabricate Ce-TZP and Al_2TiO_5 ceramics, candidate materials for engine applications. Advanced mathematical method and computer softwares as Symphony, True BASIC, Show Partner, etc., were utilized to raise the data manipulation to a new level. Statistical values were quickly and easily offered for proper decision. Empirical equations were deduced to predict the properties and help good materials processing design. Several kinds of graphics can soon be used to illustrate the relationship between properties against additions of CeO_2, Al_2O_3, TiO_2, SiO_2, La_2O_3, etc., and processing parameters as grinding duration, sintering temperature, etc.

INTRODUCTION

Orthogonal Test Design (OTD) has been proved to be an easy and effective method the optimize utilizationo utmost level of experimental fata to find out the major influencing effects. But the old method to manipulate the OTD data is rather rough and not very comprehensive[1,2].

Computers and their professionally developed softwares can make OTD easy and more effective. Spreadsheets like Symphony, Multiplan with their calculating, judging and graphics displaying power are excellent. By programming the data into different mathematical models selected and the quantitative relationships between variables are easily determined; we choose True BASIC because of its structured capability, speed and ease in handling matric operations. Incorporation of these softwares gives an advanced OTD data manipulation[1].

Ce-TZP is a CeO_2 doped toughened zirconia polycrystalline ceramic. Its characteristics includes high strength, fracture toughness and thermal shock resistance. It has been used in automobile engine as valve seats. Al_2TiO_5 is a ceramics of low thermal expansion, high service temperature, good thermal insulation and it has been effectively utilized as exhaust port and threshold pipe; it increases the operation temperature in the combustion chamber and saves oil comsumption.

OTD method has been adopted to optimize the processing parameters for these two kinds of ceramics. The results are here used to present data sheets, graphics and mathematical models.

METHODS AND PROCEDURES

Orthogonal Test Design(OTD) is an effective method for arrangement of test to get utmost of informations. For a test of 4 parameters(variables) and 3 levels each, there should be 3^4=81 runs(experiments),but only 9 runs are necessary in a well-arranged orthogonal test design.

On the basis of classical OTD, we incorporated spread sheet Multiplan [3], integrated software Symphony [4-6] (Lotus 1-2-3 should do too),programming language True BASIC [7-10] and graphics softwares like Show Partner, SkySCAN (with scanner), and Grapher & Contour to raise the data manipulation to a new level. The procedures have been discussed in details[1].

1. OTD Arrangement and Analyses

The OTD parameters and resulted properties are presented in the form of spreadsheets.

1.1 Variance Analysis

It offers the influences of parameters and their levels upon the mean index and limit difference, and further probability evaluation with statistical quantities can be made. The statistics used here conclude: Mean values for each level of various parameters, standard deviations of different level of same parameter, F values of different factors and variance ratios. Then critical values of F for various confidence are compared,and conclusions of which factor is relatively significant (major factor) are drawn. Calculating methods and formula are omitted.

1.2 Quick Display and Print Out of Graphics

The effects of factors and their levels are shown with data and plots on simultaneous displayed and printed out with these softwares. Only the proper items are seen, and further calculations are not needed.

2. Mathematical Model

Various kinds of mathematical models can be used according to particular cases, for example, chemical processes are always simulated using Arrhenhius equation, Y=A+B/T. In this case, we handled the results of 4-factor L_9 OTD using quadratic parabola surface model:

$$Y = C_1 + C_2X_1 + C_3X_2 + C_4X_3 + C_5X_4 + C_6X_1^2 + C_7X_2^2 + C_8X_3^2 + C_9X_4^2 \quad (1)$$

where: Y--Index forecasted for X_1,X_2,X_3 and X_4;
C_i--Coefficients to be determined;
X_i--Different levels of various factor.

There are 9 equations in L_9, and 9 coefficients are to be determined; the solution is unique. If the numbers of equations is more than that of coefficients, that is, $m > n$, like in L25, etc., then the least square(LSQ) solution should be carried out.

$$\begin{array}{l} C_1+C_2X_{1,1}+C_3X_{2,1}+C_4X_{3,1}+C_5X_{4,1}+C_6X_{1,1}^2+C_7X_{2,1}^2+C_8X_{3,1}^2+C_9X_{4,1}^2 = Y_1 \\ C_1+C_2X_{1,2}+C_3X_{2,2}+C_4X_{3,2}+C_5X_{4,2}+C_6X_{1,2}^2+C_7X_{2,2}^2+C_8X_{3,2}^2+C_9X_{4,2}^2 = Y_2 \\ \qquad\qquad \ldots\ldots \qquad\qquad\qquad\qquad\qquad\qquad\qquad \ldots \\ C_1+C_2X_{1,m}+C_3X_{2,m}+C_4X_{3,m}+C_5X_{4,m}+C_6X_{1,m}^2+C_7X_{2,m}^2+C_8X_{3,m}^2+C_9X_{4,m}^2 = Y_m \end{array} \quad \ldots\ldots(2)$$

Normal simultaneous equations were constructed and the following matrix equations were solved to get the coefficients.

$$\begin{bmatrix} 1 & X_{1,1} & X_{2,1} & X_{3,1} & X_{4,1} & X_{1,1}^2 & X_{2,1}^2 & X_{3,1}^2 & X_{4,1}^2 \\ 1 & X_{1,2} & X_{2,2} & X_{3,2} & X_{4,2} & X_{1,2}^2 & X_{2,2}^2 & X_{3,2}^2 & X_{4,2}^2 \\ & & & & \ldots\ldots & & & & \\ 1 & X_{1,m} & X_{2,m} & X_{3,m} & X_{4,m} & X_{1,m}^2 & X_{2,m}^2 & X_{3,m}^2 & X_{4,m}^2 \end{bmatrix} \cdot \begin{bmatrix} C_1 \\ C_1 \\ \ldots \\ C_m \end{bmatrix} = \begin{bmatrix} Y_1 \\ Y_1 \\ \ldots \\ Y_m \end{bmatrix} \qquad (3)$$

The LSQ solution of coefficients C_i would come from the matrix equation:

$$C = (X^T \cdot X)^{-1} \cdot X^T \cdot Y \qquad (4)$$

3. Hardwares

The computer used is an IBM-PC/XT compatible model PZ-88, with one harddisk and two floppy diskettes drives. 8088 CPU with 4.77 MHz speed, 640K RAM, 640 x 200 pixels color monitor, and NEC P7 dot matrix printer are basic requirements. SPL-450 plotter and Kurtz digitizer are optional.

4. Softwares

System PC-DOS 2.00 to 3.20 and CC-DOS 2.10 and 2.13A to 2.13E are used. The other softwares used are listed in Table 1.

TABLE 1
Softwares used for OTD data manipulation

NAME	SOURCES	VER.	YEAR	SIZE	USES
Multiplan	Microsoft Corp.	1.10	1983	64K	sheet,calc,plot
Symphony	Lotus Development	1.01	1984	384K	sheet,calc,graph
True BASIC	True BASIC,Inc	1.00	1985	100K	fitting,forecast
True BASIC	True BASIC,Inc	2.03	1988	140K	fitting,forecast
SkySCAN	Skyworld Technology	2.00	1989	384K	graphics,edition
Show Partner	Brightbill-Roberts		1986		graphics,edition
Grapher & Contour	Golden Software		1986		2D & 3D plotting

True BASIC language excellent for matrix operations was used to build the mathematical model with quadric equations and for forecasting results. The model adapted fits a curve surface through all the experimental points and can be illustrated easily with 3D softwares like Contour or True BASIC. Quadratic equations were chosen for the ease to differentiate and integrate in optimization. The effects of compositions and processing parameters on properties are clearly shown with 3D graphs and are helpful for making the proper decision of parameters and their levels. So, process optimization can be done qualitatively and quantitatively.

Iso-value graph and 3D graphics can be drawn with Contour program TOPO.exe and SURF.exe. Fig.9 and 10 is an example to denote the presentation of graphical relationship between properties and processing parameters on materials research. True BASIC and BASICA are used to build up grid files. Graphics give more direct attractive informations than texts and tables.

RESULTS AND DISCUSSIONS

In TABLE 2 and 3 the L_9 OTD test levels and experimental results for respectively Ce-TZP and Al_2TiO_5 are shown.

FIGURE 1 to FIGURE 8 are screen hardcopies; they are rather rough,but self-explanable.

TABLE 4 gives the coefficients of the quadratic equations for Al_2TiO_5 and Ce-TZP, which are used to forecast the properties for combinations of variables tested or not yet tested.

1. Al_2TiO_5 ceramics

FIGURE 2 and 3 are for Al_2TiO_5, FIGURE 7 to 8 relates to L_{25} tables. From the L_9 tables, for the sake to control the thermal expansion of Al_2TiO_5, sintering temperature must be properly selected and controlled. SiO_2 addition is the major factor which controls its fired strength, the second one is fineness of the powder which depends on the grinding time. La_2O_3 lowers the mean CTE values.Temperatures too high are not appropriate for strength because of rapid grain growth (as seen in SEM micrographs of fractured surfaces in another work). We continue on to make a L_{25} table to test whether an approach to control the thermal expansion but maintaining enough strength exists.

2. Ce-TZP ceramics

FIGURE 5 to 6 are for Ce-TZP. CeO_2 is the major variable for control of strength and minimize m-ZrO_2 content. Sintering temperature has some effect. We have verified the model by running the optimum combination of processing parameters predicted near No.6: CeO_2 = 12.45 mol.%; Al_2O_3 = 13 wt.%; TiO_2 = 1 wt.%; and sintering temperature = 1535°C. The resulting strength value is within 1-2% of the predicted value,basically satisfactory. From the equation, we predict that not much more potential strength can be achieved from this field of variables and levels. The highest strength predicted for this L_9 table (from Table 4) does not exceed 800 MPa.

TABLE 2 Properties of Al_2TiO_5 Ceramics -- Orthogonal Design Test Data Analysis $L_9(3^4)$

Variables	Test No.	Part.Size (hr)	Sin.Temp (°C)	Soak Time (min)	SiO2 (wt.%)	Strnth (MPa)	Dnsty (g/cm^3)	Expnsn ($x10^{-6}/°C$)	Absrptn (%)	Shrnk (%)
Orthogonal Design Table	1	A1(10)	B1(1350)	C1(30)	D1(0)	19.5	2.36	0.95	26.5	5.24
	2	A1(10)	B2(1430)	C2(60)	D2(2)	40.1	2.95	-0.55	19.6	7.51
	3	A1(10)	B3(1500)	C3(90)	D3(4)	30.9	2.81	-0.72	14.5	9.70
	4	A2(20)	B1(1350)	C2(60)	D3(4)	42.1	2.91	1.20	12.7	10.74
	5	A2(20)	B2(1430)	C3(90)	D1(0)	24.2	3.19	0.02	13.4	10.71
	6	A2(20)	B3(1500)	C1(30)	D2(2)	37.4	2.86	-0.40	20.0	11.14
	7	A3(30)	B1(1350)	C3(90)	D2(2)	57.4	2.63	1.30	23.8	10.51
	8	A3(30)	B2(1430)	C1(30)	D3(4)	29.9	2.50	-0.44	33.3	11.06
	9	A3(30)	B3(1500)	C2(60)	D1(0)	24.2	3.10	-0.59	16.1	11.37
Strength	I	30.17	39.67	28.93	22.63					
	II	34.57	31.40	35.47	44.97					
	III	37.17	30.83	37.50	34.30					
	R	7.00	8.83	8.57	22.33					
	Variance:	8.35	16.30	13.36	83.19					
	Mean Var.:	4.17	8.15	6.68	41.59					
	F Value:	1.000	1.953	1.600	9.966					
Fired Bulk Density	I	2.707	1.757	1.740	1.850					
	II	2.933	2.047	1.953	1.937					
	III	2.743	2.923	2.877	2.740					
	R	0.710	1.167	1.137	0.890					
	Variance	0.1065	0.2460	0.2433	0.1606					
	Mean Var.	0.0533	0.1230	0.1217	0.0803					
	F Value	1.0000	2.3088	2.2841	1.5070					
Coefficient of Thermal Expansion	I	-0.1067	0.7167	0.1833	0.3233					
	II	0.4067	-0.1767	0.2167	-0.3167					
	III	0.0900	-0.5700	0.2000	0.0133					
	R	0.5133	1.2867	0.0333	0.6400					
	Variance	0.04472	0.28981	0.00019	0.06829					
	Mean Var.:	0.02236	0.14490	0.00009	0.03414					
	F Value:	241.480	1564.960	1.000	368.760					
Water Absorption	I	20.20	13.07	15.50	13.30					
	II	8.70	11.00	10.77	13.20					
	III	24.40	16.87	17.23	20.17					
	R	15.70	5.87	6.47	6.97					
	Variance	44.042	5.903	7.470	10.633					
	Mean Var.:	22.021	2.952	3.735	5.316					
	F Value:	7.461	1.000	1.265	1.801					
Firing Shrinkage	I	7.48	5.33	5.46	5.32					
	II	7.15	6.07	6.08	6.22					
	III	10.98	10.74	10.31	10.50					
	R	3.83	5.41	4.85	5.18					
	Variance:	3.001	5.730	4.635	5.114					
	Mean Var.:	1.500	2.865	2.318	2.557					
	F Value:	1.000	1.910	1.545	1.704					

TABLE 3 Properties of Ce-TZP ceramics -- Orthogonal Design Test Data Analysis $L_9(3^4)$

Variables	Test No.	CeO_2 (hr)	Al_2O_3 (w/o)	TiO_2 (w/o)	T (°C)	Dnsty (g/cm^3)	Strnth (MPa)	K_{IC} (MPa.√m)	Hv (GPa)	m-phase (%)
Orthogonal Design Table	1	A1(10)	B1(9)	C1(1)	D1(1510)	4.94	240	3.95	3.17	86.7
	2	A1(10)	B2(11)	C2(2)	D2(1530)	5.48	206	3.87	4.36	87.5
	3	A1(10)	B3(13)	C3(3)	D3(1550)	5.51	236	3.53	4.50	91.4
	4	A2(12)	B1(9)	C2(2)	D3(1550)	5.78	662	7.03	7.84	5.4
	5	A2(12)	B2(11)	C3(3)	D1(1510)	5.61	639	6.85	8.38	4.7
	6	A2(12)	B3(13)	C1(1)	D2(1530)	5.50	746	7.84	9.23	8.6
	7	A3(14)	B1(9)	C3(3)	D2(1530)	5.77	523	5.29	8.62	1.6
	8	A3(14)	B2(11)	C1(1)	D3(1550)	5.67	539	6.35	6.99	2.1
	9	A3(14)	B3(13)	C2(2)	D1(1510)	5.49	485	6.04	7.64	0.9
Density (g/cm3)	I	5.310	5.497	5.370	5.347					
	II	5.630	5.587	5.583	5.583					
	III	5.643	5.500	5.630	5.653					
	R	0.333	0.090	0.260	0.307					
	Variance:	0.0237	0.0017	0.0128	0.0172					
	Mean Var.:	0.0119	0.0009	0.0064	0.0086					
	F Value:	13.679	1.000	7.380	9.919					
Strength (MPa)	I	227.3	300.7	328.7	293.0					
	II	433.7	281.7	289.3	317.3					
	III	515.7	489.0	466.0	479.0					
	R	288.3	207.3	176.7	186.0					
	Variance:	14714.840	8757.506	5735.407	6813.802					
	Mean Var.:	7357.420	4378.753	2867.704	3406.901					
	F Value:	2.566	1.527	1.000	1.188					
Fracture Toughness (MPa)	I	3.783333	3.660000	3.930000	3.600000					
	II	4.626667	3.573333	3.633333	3.903333					
	III	5.893333	5.803333	5.223333	5.636667					
	R	2.110	2.230	1.590	2.037					
	Variance:	0.752	1.064	0.477	0.805					
	Mean Var.:	0.376	0.532	0.238	0.402					
	F Value:	1.578	2.232	1.000	1.689					
Micro-Hardness (MPa)	I	4012.7	3671.7	4135.3	3851.3					
	II	5407.0	4247.3	4067.7	4531.3					
	III	7751.7	7124.7	7167.7	6445.7					
	R	3739.0	3453.0	3100.0	2594.3					
	Variance:	2380194.2	2281516.5	2089958.2	1206404.2					
	Mean Var.:	1190097.1	1140758.2	1044979.1	603202.1					
	F Value:	1.973	1.891	1.732	1.000					
m-ZrO2 Content (%)	I	88.53	30.71	31.77	30.46					
	II	3.36	30.72	30.97	32.04					
	III	1.52	33.63	32.56	32.96					
	R	87.02	2.92	1.58	2.51					
	Variance:	1647.7	1.9	0.4	1.1					
	Mean Var.:	823.8	0.9	0.2	0.5					
	F Value:	3943.457	4.519	1.000	2.563					

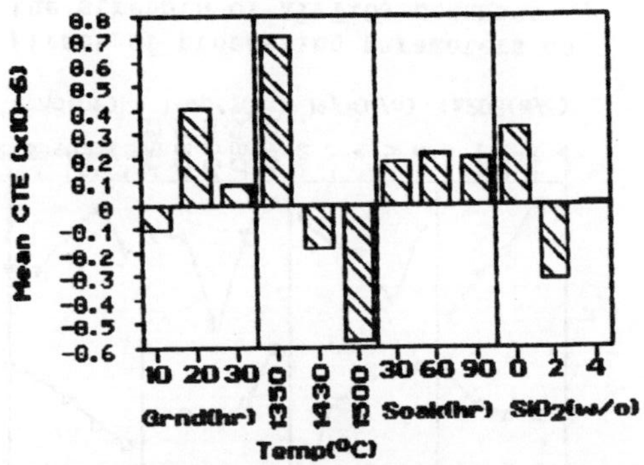

FIGURE 1 Mean Coefficients of Thermal Expansion of Al_2TiO_5 against processing parameters

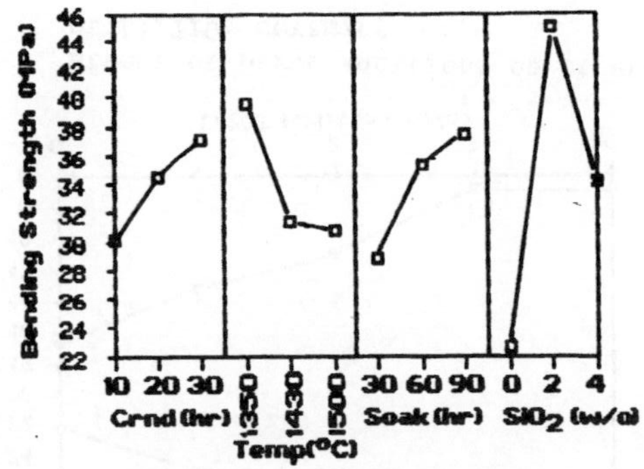

FIGURE 2 Bending Strength of Al_2TiO_5 ceramics against processing parameters

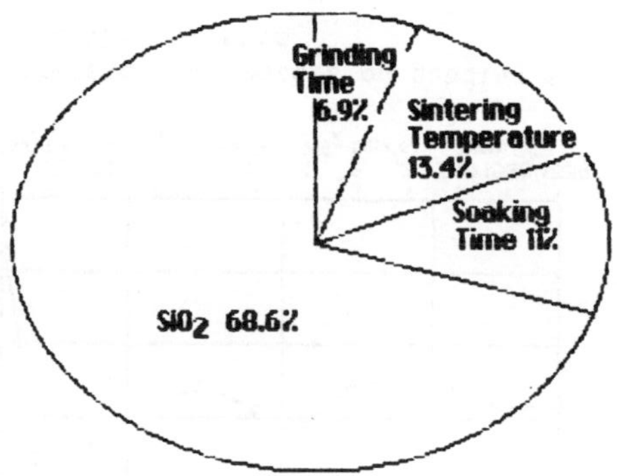

FIGURE 3 OTD variances contribution on strength of Al_2TiO_5 vs processing parameters

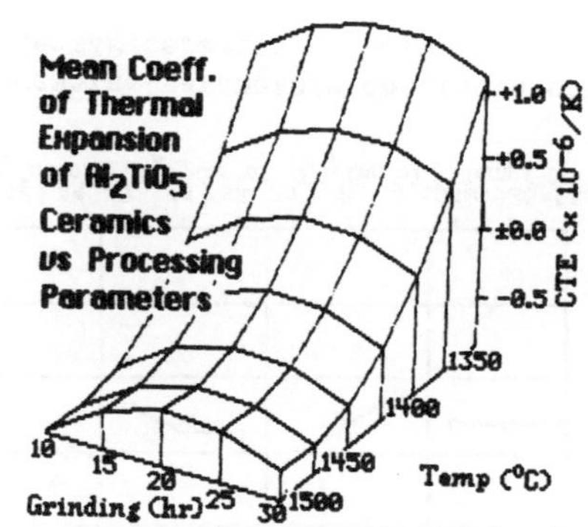

FIGURE 4 Mean Coefficients of Thermal Expansion of Al_2TiO_5 vs processing parameters

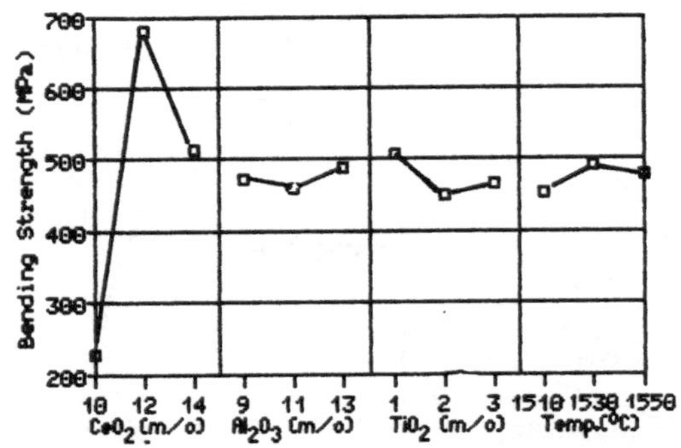

FIGURE 5 Processing parameters on bending strength of Ce-TZP ceramics

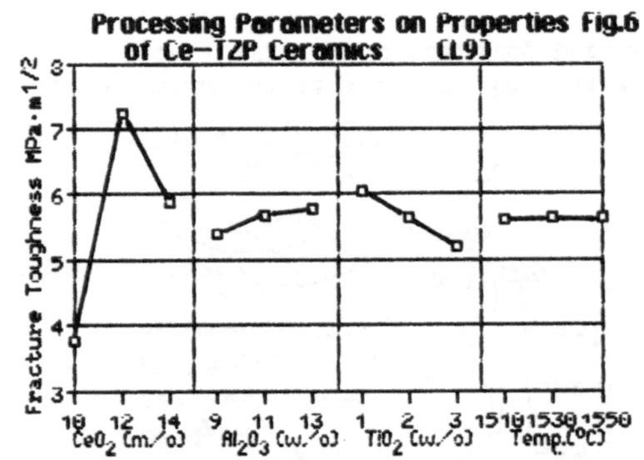

FIGURE 6 Processing parameters on fracture toughness of Ce-TZP ceramics

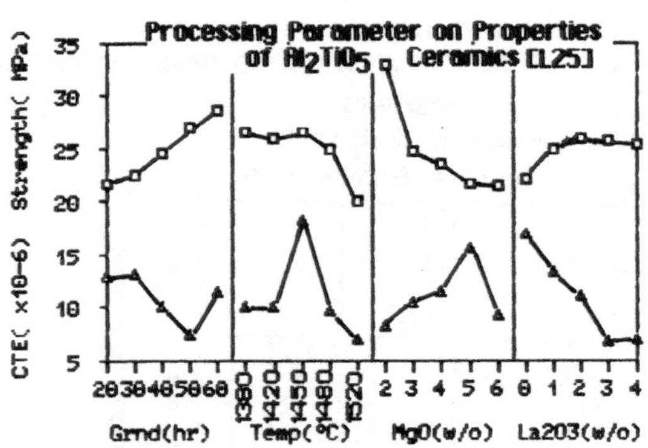

FIGURE 7 Effect of processing parameters on the strength of Al_2TiO_5 ceramics

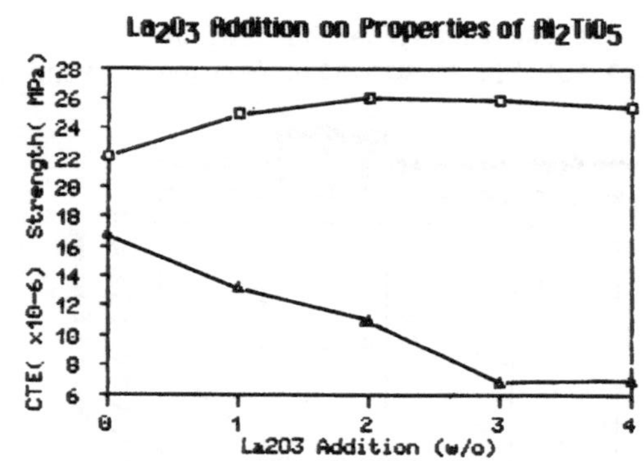

FIGURE 8 Effect of La_2O_3 additions on strength of Al_2TiO_5 ceramics

TABLE 4
Coefficients for Prediction Equations for Al_2TiO_5 and Ce-TZP

Coefficient	Ce-TZP Density	Ce-TZP Strength	Al_2TiO_5 Strength	Al_2TiO_5 CTE
C1	-503.412	-74461.7	1367.48	216.754
C2	1.00333	1825.87	.71	.1225
C3	0.48667	-131.76	-1.86841	-.294431
C4	0.46333	-214.302	.442778	-1.03889E-2
C5	0.64517	83.703	19.4167	1.83333E-2
C6	-3.83333E-2	-73.0917	-.009	-2.81667E-3
C7	12.20833E-2	5.9296	6.34921E-4	9.92857E-5
C8	-8.33333E-2	49.9683	-.0025	1.09259E-4
C9	-2.08333E-4	-2.709E-2	-4.125	-1.16667E-2

CONCLUSIONS

1. Computer softwares make OTD data manipulation easy and effective.
2. OTD models for processing of Ce-TZP and Al_2TiO_5 ceramics has been constructed.Factors,and their different levels of effects on properties are clearly shown on spreadsheets, graphics and different equations.
3. OTD + mathematical model fitting is an effective method to utmostly utilize the informations from experimental data.
4. Computer softwares accomplish quick handling of data, and give all statistical informations and values.
5. Examples of applications of this method in Ce-TZP and for Al_2TiO_5 ceramics processing optimization have been carried out. The relationships between resulting properties and major process variable are clearly seen qualitively and quantitively. High strength, high toughness and stable Ce-TZP and low and controllable thermal expansion Al_2TiO_5 ceramics may be fabricated.
6. It contributes to an approach to proper materials processing design.

REFERENCE (* in Chinese)

1. W. Y. Zhu, <The Applications of Orthogonal and Orthogonal Regression Analysis>, Liaoning People Press (1978).*
2. Y. P. Liao, <The Applications of Orthogonal Test Design in Mechanics Industry>, Chinese Argriculture Press (1984).*
3. D. G. Zhu, <How to Use Chinese Character Spread-Sheets>, Electronics Press (1987).*
4. H. M. Yang, <Integrated Software Symphony for Microcomputers>, Qinghua Press (1988).*
5. D. Bolocan, <Mastering Symphony>, TAB (1985).
6. D. P. Ewing et al.,<Using Symphony--Release 2.0>, QUE (1988).
7. J. C. Craig,<True BASIC-_Programming and Subroutines>, TAB (1985).
8. H. Simpson, <True BASIC--A Complete Manual>, TAB (1985).
9. J. G. Kemeny, <Structured Language True BASIC>, (Chinese Translation), Scientific Common (1988).*
10. <True BASIC -- version 2.03>, Software Research Institute of China Academy Sindica, (1988).

CMC EVALUATION FOR USE IN MILITARY AIRCRAFT ENGINES.

G.GAUTHIER, G.BESSENAY, Y.HONNORAT
SNECMA. Materials and processes department
B.P.81 . 91003 EVRY CEDEX. FRANCE

ABSTRACT

Because of their relatively low density, high toughness and refractoriness, Ceramic Matrix Composites are of great interest for use in turbine engines. For now several years, SNECMA have been conducting a series of cross-investigations, in cooperation with SEP , on C/SiC and SiC/SiC, both on parts behaviour and materials evaluation.

Several types of static or semi-static hot parts have been fabricated and tested both on ground and flying benches. These tests have been cumulating several hundred of hours exploring the whole field of loading of the parts.

In parallel, several series of tests have been performed to assess the mechanical behaviour of the parts. A particular attention has been put on the misalignment of the fibres which characterizes the structure in the areas of complex shape.

Oxidation tests have been carried out on uncoated materials. Oxidation kinetics have been established and microstructural examinations have been done in a way to precise the location of the initial damaged areas and the phase transformations.

Alternative wear tests have been conducted in conditions simulating the contacts between the parts themselves and the metallic actuators.

INTRODUCTION

Materials for hot sections of aircraft gas turbines (and particularly the exhaust system of military aircraft) mainly used during the last decade, have been metal superalloys.

Used in polycrystalline or single crystal form, these metals provide the needed high temperature performance but, due to their high density (e.g. 7-9), they contribute largely to the overall engine weight. The performance requirements for new generation aircraft gas turbines will demand higher thrust-to-weight ratios, better engine efficiency, durability and reliability.

Solutions may be found in two directions:

- rising gas temperatures (see as an illustration of this trend the increasing compacity of SNECMA engines -figure 1- and the temperature profile of M88 engine -figure 2-).
- using low density materials demonstrating relatively high specific strength at high temperature (figure 3).

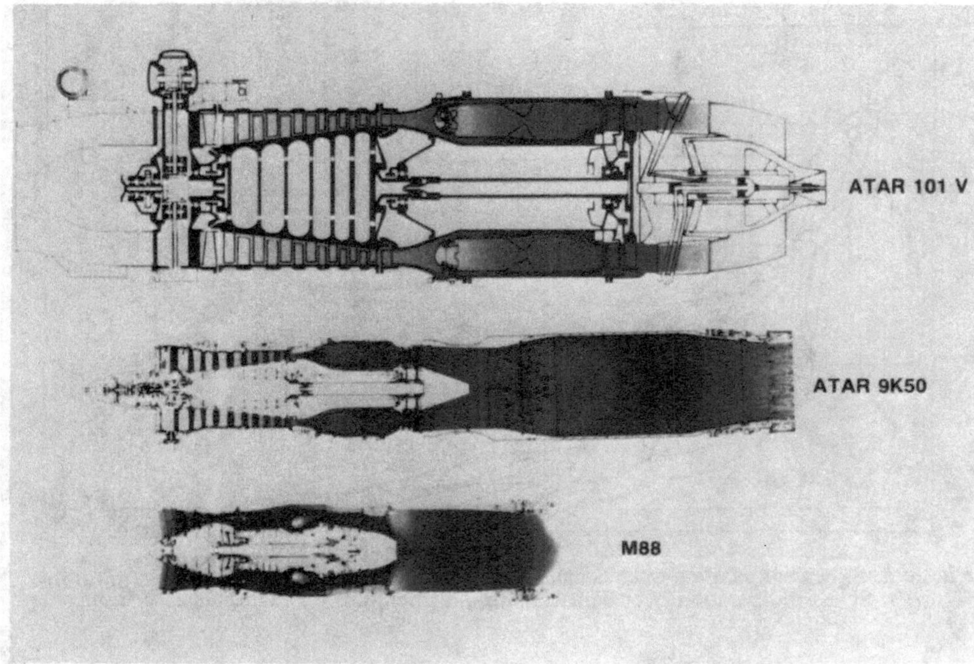

Figure 1. Different sizes for a same thrust in SNECMA engines.

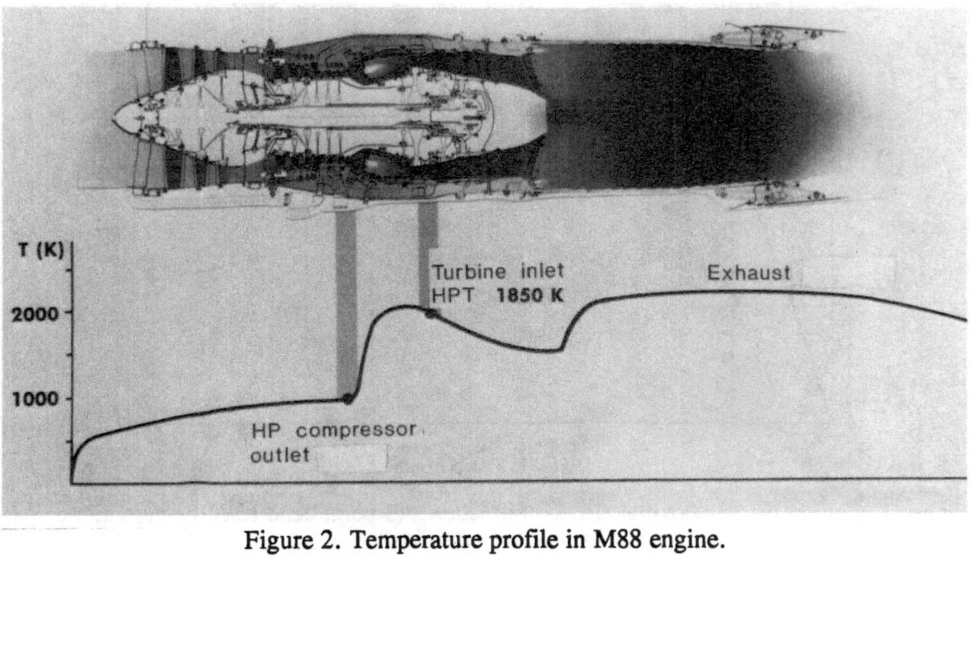

Figure 2. Temperature profile in M88 engine.

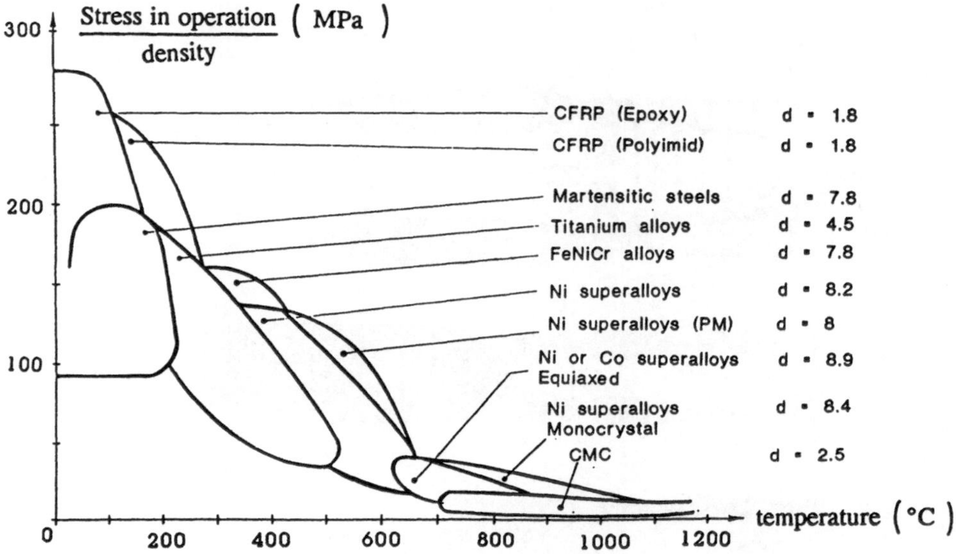

Figure 3. Stress in operation over density versus temperature for materials used in turbine engines.

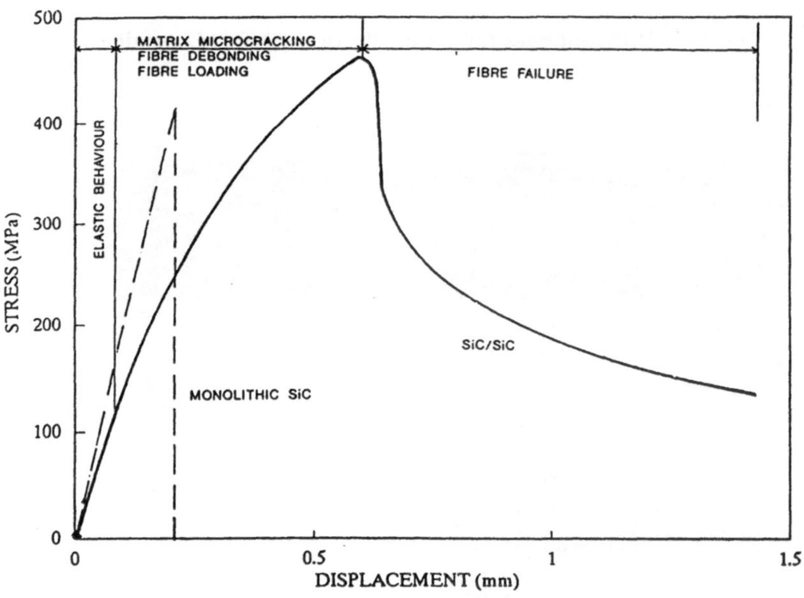

Figure 4. Comparison of flexural stress-strain curves (3 point bend test) for monolithic and composite ceramics.

Attempts have been made, in the past, to reach these specifications through the use of monolithic ceramic materials (mainly silicon carbide and silicon nitride). These ceramics are desirable in that way that they are capable of operating at temperature comparable to these of superalloys, but with a much lower density (e.g. 3.2). These attempts, however, have not been successful because of the low fracture toughness of these ceramic materials. They fail in the presence of stress concentration and service induced impact damage. Inherent brittleness (related to their low fracture energy) and catastrophic failure (leading to great difficulties in accurately predicting safe conditions and lifetime) have been the major handicap for monolithic ceramics using in engineering applications.

Yet the demand remains for low density materials with sufficient and predictable mechanical strength at high temperature.

One approach which is receiving widespread research effort (1)(2) is the use of long ceramic fibre-reinforced ceramic matrix composites (CMC) which demonstrate a very significant increase in fracture energy associated with the fibrous nature of the fracture (figure 4).

The CMC family is quite large and can be listed by increasing order of temperature potential for long term (1000 hours) resistance in oxidizing environment :

- SiC/Borosilicate (400-600°C);
- SiC/Glass-Ceramic (up to 1000°C);
- SiC/SiC (1200°C with coatings);
- C/SiC (1500°C with coatings, and only for short times);
- C/C (up to 2000°C with coatings and only for very short times)

For now several years, SNECMA have been conducting a series of cross-investigations, in cooperation with SEP, on C/SiC (SEPCARBINOX^R) and SiC/SiC (CERASEP^R), both on parts behaviour and materials evaluation.

The results presented in this paper point out some aspects relating to the behaviour of CMC in the turbine engine environment:

- Comparison of the material properties of the parts with standard characteristics;
- Behaviour in an oxidizing environment;
- Integration of CMC in a metallic environment;
- Wear resistance.

CMC PARTS

Several parts have been fabricated by SEP using the CVI (Chemical Vapor Infiltration) technique. Among these, primary and secondary nozzle flaps and rear turbine cones (figure 5). The rear turbine cone is a static part whose role is to stabilize the gas flow coming from the turbine. The primary and secondary nozzle flaps adjust the section of the exhaust nozzle and prolong the nacelle, respectively.

The cone and the primary flaps, made in SiC/SiC, are in contact with the primary flow, and their temperature can reach 1000°C. The secondary flaps are cooler, but their temperature can reach 600°C at their end.

The parts have been tested both on ground and flying benches. These tests have been cumulating several hundred hours exploring the whole field of loading of the parts.

a

b

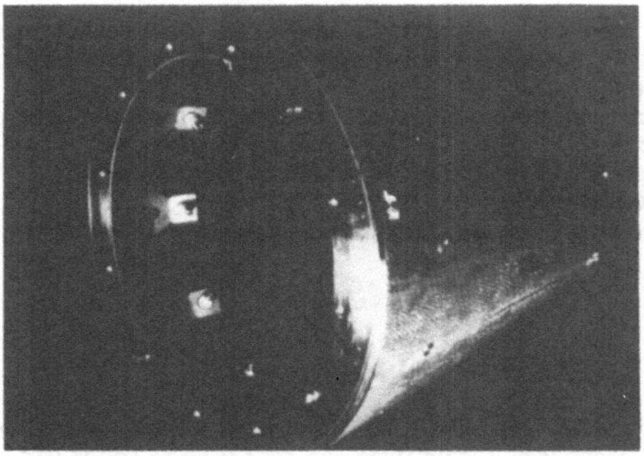

c

Figure 5. CMC parts in SNECMA engines. a : Mirage 2000 with CMC flaps.
b : Primary and secondary nozzle flaps. c : rear Turbine cone.

MATERIALS PROPERTIES

Engineering data necessary for the sizing of the parts are now well established for C/SiC and SiC/SiC. We have reported in tables 1 and 2 the main characteristics of coupon specimens derived from 2D plates (SEP data).

TABLE 1
Properties of SiC/SiC , 2D plain weave

			23°C	1000°C
Tensile strength		MPa	200	200
YOUNG's modulus (tensile)		GPa	230	200
Flexural strength		MPa	350	400
Compressive strength		MPa	580	480
Thermal diffusivity	//	$10^{-6}m^2s^{-1}$	12	5
	⊥		6	2
Coefficient of thermal expansion	//	$10^{-6}K^{-1}$	3	3
	⊥		2.5	2.5

TABLE 2
Properties of C/SiC , 2D 5 harness satin

			23°C	1000°C
Tensile strength		MPa	350	350
YOUNG's modulus (tensile)		GPa	90	100
Flexural strength		MPa	500	700
Compressive strength		MPa	360	360
Thermal diffusivity	//	$10^{-6}m^2s^{-1}$	11	7
	⊥		5	2
Coefficient of thermal expansion	//	$10^{-6}K^{-1}$	3	3
	⊥		5	5

However, materials derived from parts of complex shape may have quite different properties. For example, in the C/SiC outer flaps, the presence of two bosses leads to an important distortion of the satin reinforcement.

In order to assess the influence of the misalignment of the fibres on the mechanical properties, tensile specimens have been cut up in different areas of the part. Ultimate tensile strength of the cutups have been reported in figure 6 as a function of the angle of misalignment in the gauge length of the specimens. For example, a 15° misalignment leads to a 50 % decrease in mechanical properties.

The parts have been designed to take this fact into account. Stresses in the concerned zones have been lowered to be in agreement with the mechanical properties of the material. These parts have been successfully tested for several hundred hours, showing it is possible to realize parts with a degraded material provided it is taken into account in the first stage of the design.

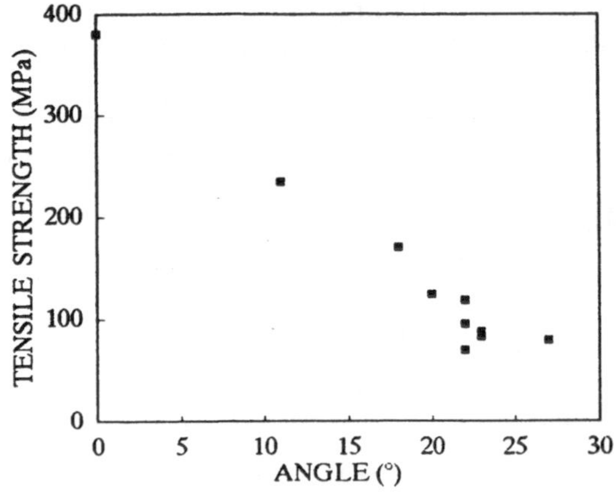

Figure 6. Influence of the misalignment of the fibres on the tensile strength of C/SiC.

Materials properties must as well be retained in long term ageing in the oxidizing environment of the turbine engine. The next section focuses on the role of oxidation on material properties alteration.

BEHAVIOUR IN OXIDIZING ENVIRONMENT

C/SiC and SiC/SiC both contain carbon either as fibre (C/SiC) or interphase (SiC/SiC), making these two materials very sensitive to oxidation at relatively low temperatures (above 400°C). Oxidation kinetics have been determined on uncoated samples. In C/SiC, oxidation of the fibres leads to a weight loss, according to the reactions :

$$C + O_2 \dashrightarrow CO_2$$
$$C + 1/2\, O_2 \dashrightarrow CO$$

In the range 450-750°C, the oxidation rate is linear (figure 7) (3).

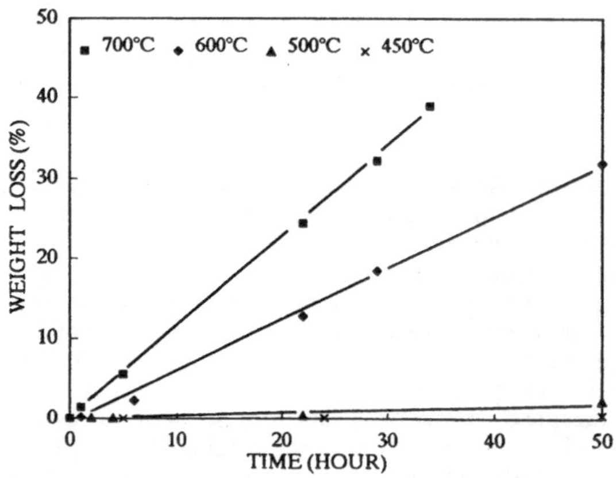

Figure 7. Weight loss as a function of time for uncoated C/SiC exposed in air between 450 and 700°C.

Micrographs of oxidized and unoxidized samples (figure 8) show that oxidation proceeds through the microcracks present in the matrix of the initial material. These microcracks result from a coefficient of thermal expansion mismatch between the carbon fibres ($1.10^{-6}K^{-1}$) and the SiC matrix ($4.5.10^{-6}K^{-1}$). Since C/SiC is elaborated at high temperature (1000°C), microcracks develop during cooling.

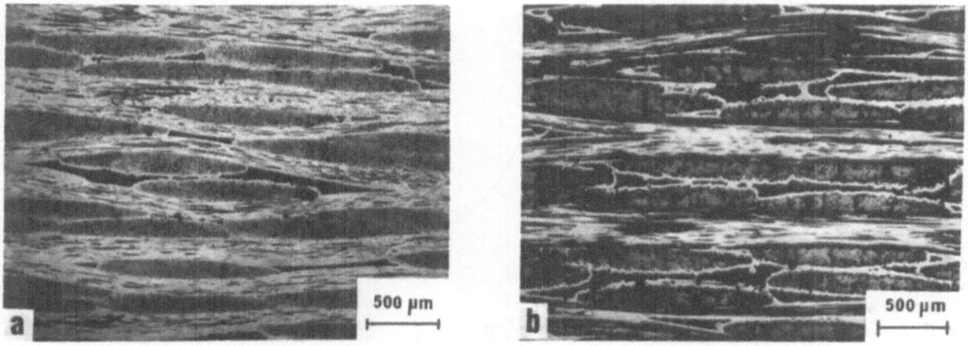

Figure 8. Micrographs of uncoated C/SiC. a: as received.
b: oxidized 50 hours at 600°C.

Residual flexural strength were determined at room temperature after 100 hours isothermal ageings in air (figure 9). A significant loss of mechanical properties is noticeable after the heat treatment at 400°C, although the weight loss was only 1%. After the 600°C heat treatment, strength degradation is extreme (σ_f= 0 MPa).

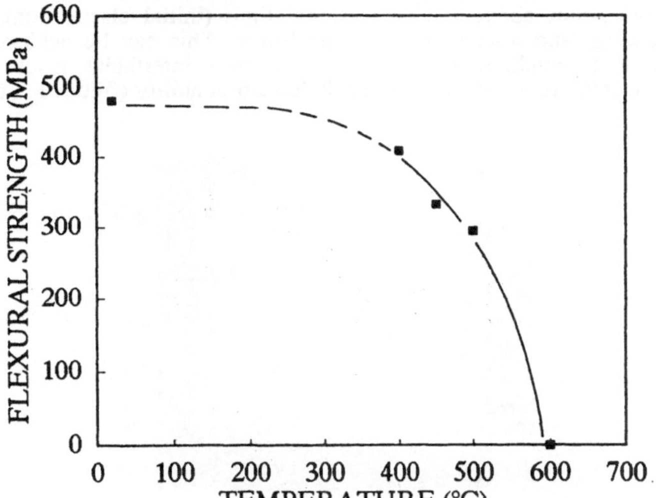

Figure 9. Flexural strength of uncoated C/SiC after 100 hours isothermal oxidizing ageings between 400 and 600°C.

In SiC/SiC, oxidation phenomena are quite different. The figure 10 shows the behaviour of SiC/SiC, under flexural stress at room temperature, as received and after 100 hours heat treatment at 850°C in air.

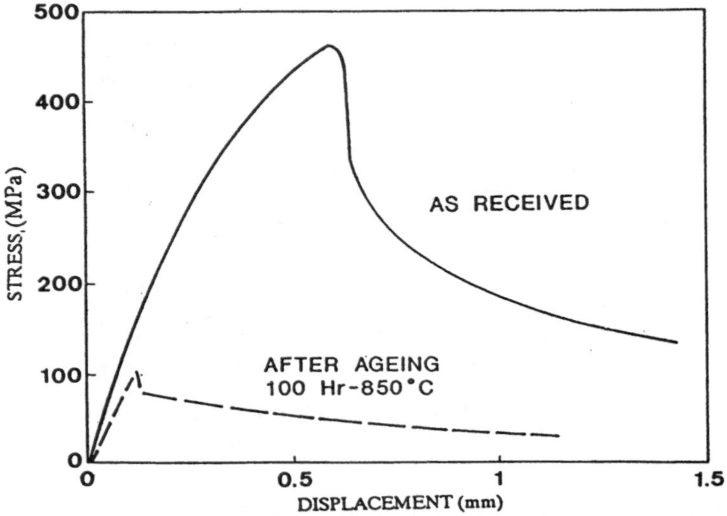

Figure 10. Effect of oxidation on the mechanical behaviour of uncoated SiC/SiC (3 point bend test at room temperature).

Unoxidized sample exhibits the typical non-brittle behaviour of this kind of composite. The non-linear behaviour above the proportional limit (initial elastic domain) is due to matrix microcracking, and load transfer to the fibres. This can be achieved by a weak fibre-matrix interface bonding. In CERASEP, a thin interfacial layer of carbon is responsible for the fracture toughness and crack deflection ability (figure 11) (4).

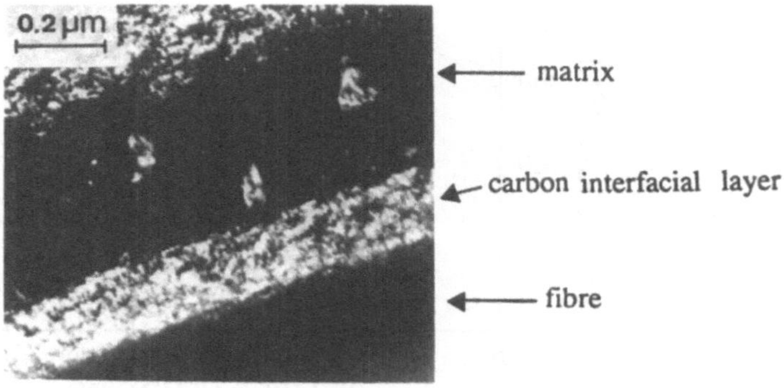

Figure 11. TEM micrograph of SiC/SiC, as received.

The heat-treated sample was cut from a plate, so that the four ends of the coupon were exposed to oxidation. Oxygen can therefore progress inward from cut ends, along the fibres (Pipeline oxidation (5)). In that case, the initial carbon layer is replaced by a strong silica layer, grown up from the NICALON fibres (figure 12), giving a brittle fracture(6).

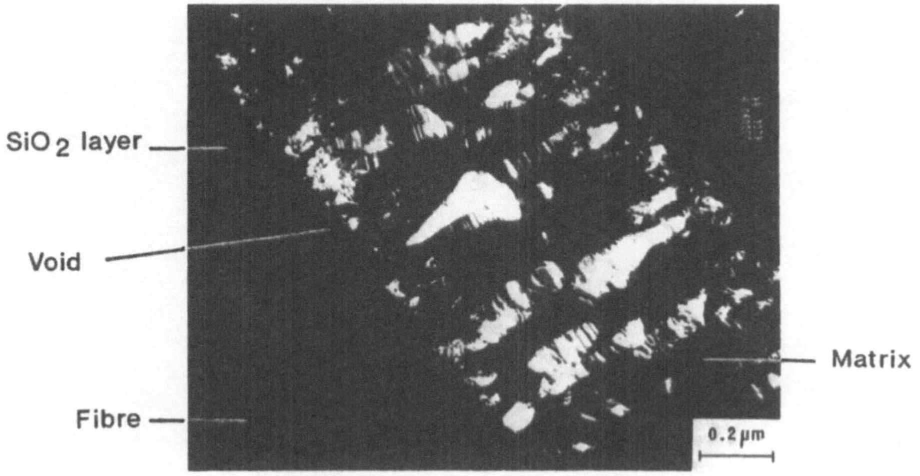

Figure 12. TEM micrograph of SiC/SiC after ageing 100 hours in air at 850°C.

Although engine components are not machined, i.e. they are entirely covered with a dense silicon carbide seal-coat, oxidation may be critical with regard to the non-linear behaviour of CERASEP. When SiC/SiC is stressed at relatively low levels (above 80 MPa), the matrix begins to microcrack. A dense network of microcracks becomes a privileged path for oxygen to the carbon interface, leading to the rapid failure of the material.

Due to their extreme sensitivity to oxidation, both C/SiC and SiC/SiC have to be protected by coatings to be used in the oxidizing environment of the engine. These coatings are self-healing glasses with appropriate viscosity to provide protection over the range 400-1000°C.

These protections are only necessary for parts subjected to cyclic fatigue for SiC/SiC, and in all cases for C/SiC. However, for SiC/SiC, a protective coating is not absolutely necessary for static parts. An uncoated cone in CERASEP has been inspected after 320 hours operation. TEM micrographs from samples cut up in a 3 mm thin wall of this part show that oxidation is limited to the surface of the part (figure 13a). The in-depth material (1.5 mm deep) remained sane. The carbon interphase between the fibre and the matrix is not degraded, as shown on figure 13b.

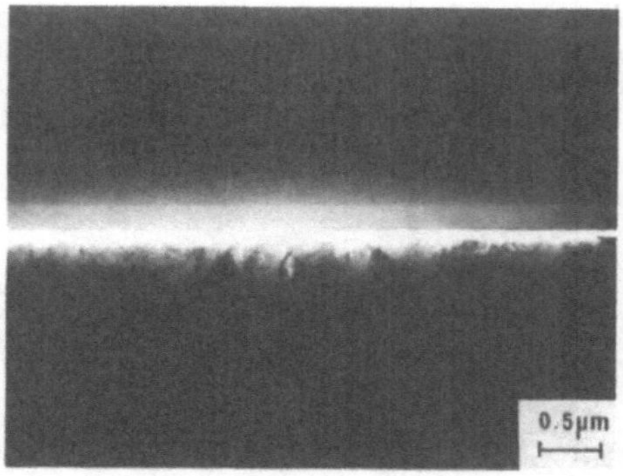

a

b

Figure 13. TEM micrographs of SiC/SiC rear turbine cone after 320 hours operation.
a: near the surface. b: 1.5 mm deep.

INTEGRATION OF CMC PARTS IN A METALLIC ENVIRONMENT

The use of ceramic materials for turbine engine parts leads to some integration problems related to the specific characteristics of these materials that make them different from the metallic alloys mainly used for the actuators parts of these systems.

For semi-static hot parts (like nozzle flaps), a major factor that has to be considered is the wear of the parts themselves and of the parts in contact with them (metallic or ceramic). The wear degradation may appear on parts not only when they are moving but may also be generated by contact-fretting due to vibratory fatigue stress.

In the case of inner nozzle flaps of M88 engine, two types of friction have to be examined (figure 14) :

- a: upstream contact of the flaps on metallic joint (HA 188) in a temperature range between 500 and 600°C.
- b: side contact of one flap on another, in a temperature range from 450°C upstream to 850°C downstream.

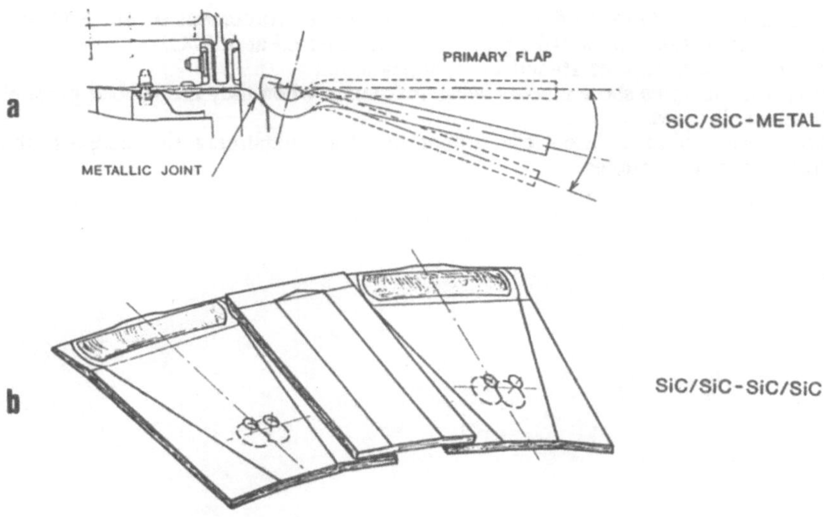

Figure 14. Two examples of contact. a: ceramic on metal.
b: ceramic on ceramic.

Alternative wear tests have been conducted in conditions simulating the contacts between the parts themselves and with the metallic actuators.

The scheme of the experimental set-up used is represented in figure 15. The rate of erosion of the different materials have been measured by alternative displacement under load of samples with flat surfaces. The displacement frequency is constant (10 Hz). The applied load is 8 daN, which corresponds to a pressure of 5 MPa (surfaces in contact = 16 mm^2).

The samples may be placed into a furnace, capable of 800°C. The tests are interrupted from time to time to measure the volume of matter rubbed away from each sample.

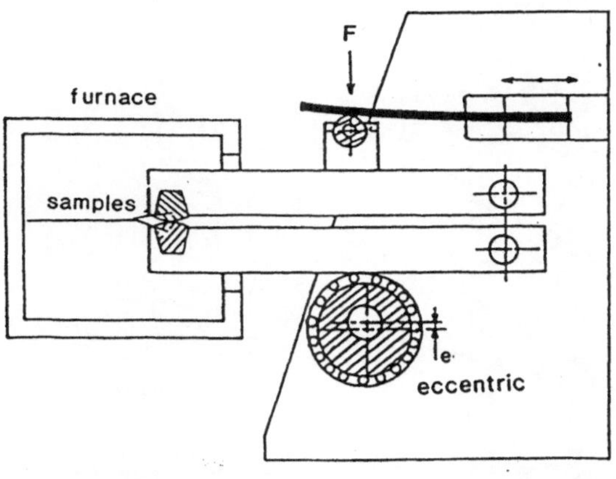

Figure 15. Scheme of the set-up for alternative wear tests.

Two different orientations of the 2D plain weave reinforcement of SiC/SiC composite (parallel or perpendicular to the 0-90 plies) have been tested at 550°C.

From the curves in figures 16 and 17, it appears that :

- Composite samples show a better wear resistance when they are rubbed parallel to the fibres, in all cases.
- The worn volumes are less important for metal/ceramic couples than for ceramic/ceramic couples.

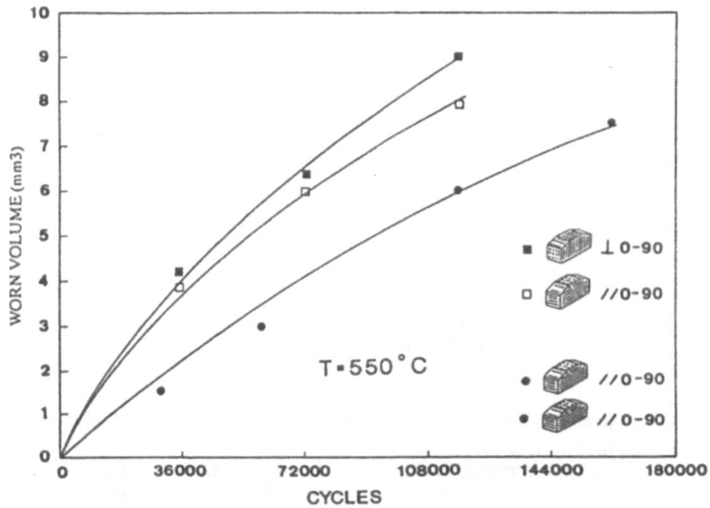

Figure 16. Alternative wear test : CMC on CMC

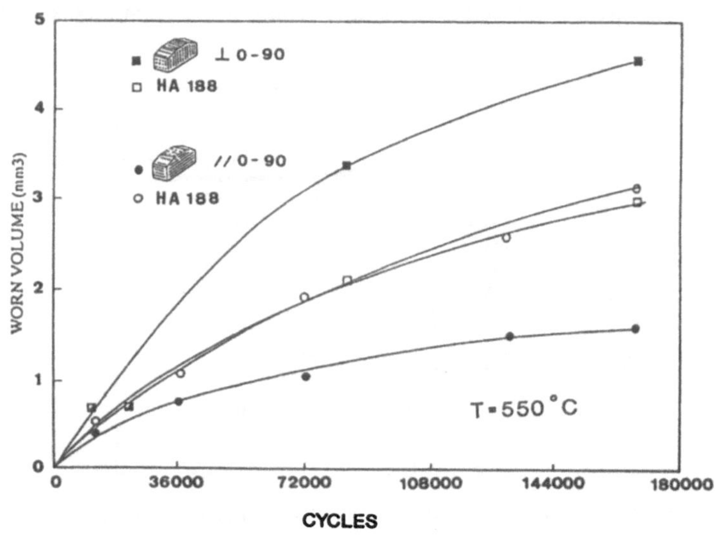

Figure 17. Alternative wear test : CMC on metal

For the first point, the explanation is straightforward : perpendicular fibres are "brushed" away, while parallel fibres remain embedded into the matrix. In the first case, a larger amount of abrasive debris (broken ends of SiC tows) maintain a high rate of wear.

For the second point, the presence of a ductile phase between the metallic and composite sample may explain the wear mechanism. As shown on figure 18, the surfaces of both samples appear very smooth.

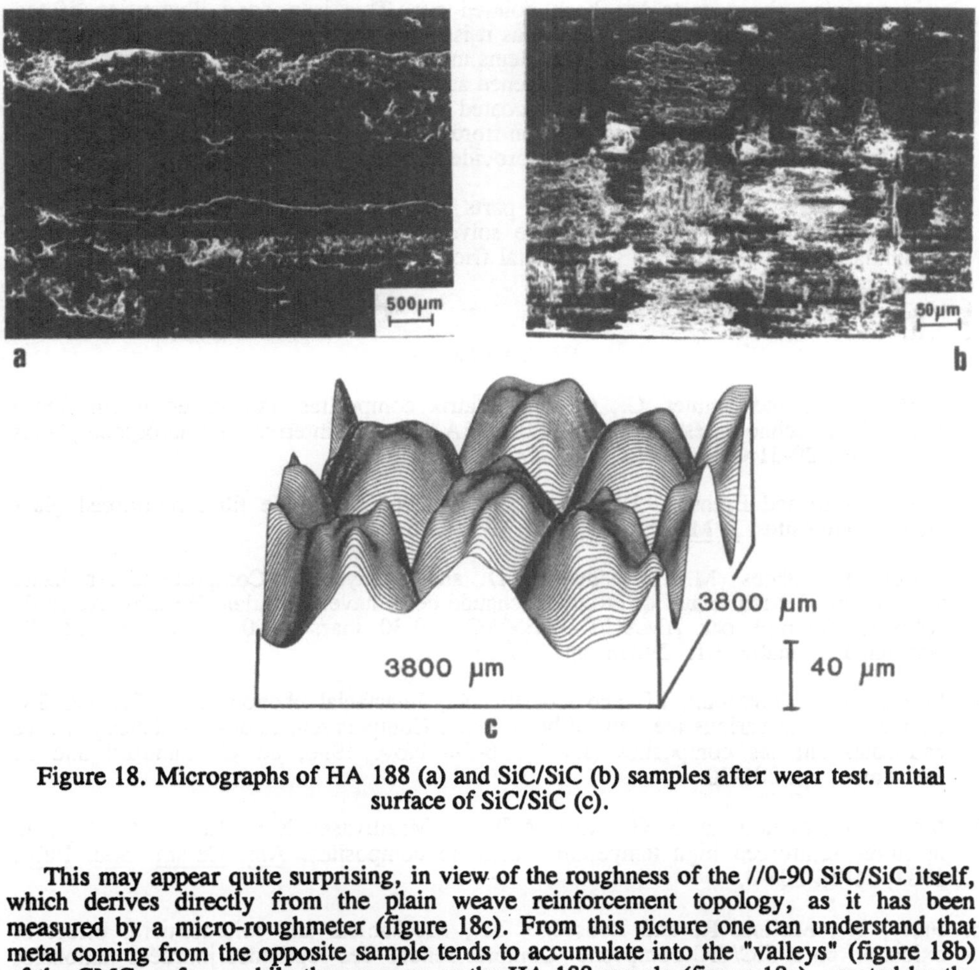

Figure 18. Micrographs of HA 188 (a) and SiC/SiC (b) samples after wear test. Initial surface of SiC/SiC (c).

This may appear quite surprising, in view of the roughness of the //0-90 SiC/SiC itself, which derives directly from the plain weave reinforcement topology, as it has been measured by a micro-roughmeter (figure 18c). From this picture one can understand that metal coming from the opposite sample tends to accumulate into the "valleys" (figure 18b) of the CMC surface, while the grooves on the HA 188 sample (figure 18a), created by the "peaks" of the CMC surface, are filled up with debris.

After a short period of adaptation, these debris constitute a "third body", which acts as a lubricating agent, due to the metal ductility. In ceramic/ceramic couples, this third body is only constituted of very abrasive silicon carbide scraps, leading to higher rates of wear.

However, it may be seen from this study, that the wear resistance of the parts in contact either with metallic joints or between themselves, is better when the composite is rubbing parallel to the fibres, as it is shown in figure 14.

CONCLUSION

Long fibre-reinforced ceramics have shown their ability to be used in gas turbine engines for military aircraft, provided that the specific properties of the materials in representative environment conditions are taken into account.

In order to reach these goals, SNECMA have been conducting a series of cross-investigations both on parts behaviour and materials evaluation :

- the role of the misalignment of the fibres on the mechanical properties of the material in complex shape parts has been pointed out. This is a good illustration of the necessity to characterize the material as it is in the parts, in correlation with standard data on coupon specimens. Such problems must be taken into account when the part is designed, in order to unload the weakened areas.
- oxidation studies have shown that uncoated C/SiC and SiC/SiC are very sensitive to oxidation for long term ageing. Carbon fibres on one hand, and carbon interface layer on the other hand, can be preserved, provided that the parts are coated with protective self-healing coatings.
- as far as the integration of ceramic parts in a metallic environment is concerned, sliding friction problems have to be solved. Alternative wear tests in conditions simulating CMC/CMC and CMC/metal friction in the engine permitted to define the best configuration of rubbing.

REFERENCES

1. Lacombe, A. and Bonnet, C., Ceramic matrix composites, key materials for future space plane technologies. AIAA-90-5208. AIAA second international aerospace planes conference, 29-31 Oct. 1990, Orlando.

2. Prewo, K.M.and Brennan, J.J., High strength silicon carbide fibre reinforced glass matrix composites. J. Mater. Sci., 1980, **15**, 463-8.

3. Berton, B., Bacos ,M.P.,, Demange, D. and Lahaye, J., Comportement à haute température des matériaux de structure chaude de la navette spatiale Hermès. Actes du colloque organisé par AMAC/CODEMAC, 29-30 mars 1990, Bordeaux, ed. R. Naslain, J. Lamalle, J.L. Zulian, pp. 315-25.

4. Cojean, D., Monthioux, M. and Oberlin, A., Interfacial phenomena in 2D SiC/SiC composites with various mechanical behaviours. Comptes rendus des septièmes journées nationales sur les composites (JNC7), 6-7-8 Nov. 1990, ed G. Fantozzi and P. Fleishman, pp. 381-90.

5. Mah, T.T., Mendiratta, M.G., Katz, A.P. and Mazdiyasni, K.S., Recent development in fibre- reinforced high temperature ceramic composites. Am. Ceram. Soc. Bull., 1987, **66**[2], 304-8.

6. Frety, N., Molins, R. and Boussuge, M., Mechanical and microstrucral effect of oxidation on a SiC/SiC composite. Comptes rendus des septièmes journées nationales sur les composites (JNC7), 6-7-8 Nov. 1990, ed G. Fantozzi and P. Fleishman, pp. 411-20.

DEVELOPMENT AND RELIABILITY EVALUATION OF A CERAMIC STATOR VANE FOR INDUSTRIAL GAS TURBINES

Kimiaki Nakakado and Takashi Machida,
Mechanical Engineering Research Laboratory, Hitachi, Ltd.,
Tsuchiura-shi, Ibaraki-ken, 300 Japan

Hiroshi Miyata
Hitachi Research Laboratory, Hitachi, Ltd.,
Hitachi-shi, Ibaraki-ken, 317 Japan

Tooru Hisamatu, Noriyuki Mori and Isao Yuri
Yokosuka Research Laboratory, Central Research Institute of Electric Power Industry, Yokosuka-shi, Kanagawa-ken, 240-01 Japan

ABSTRACT

A ceramic gas turbine is expected to improve the thermal efficiency of a power plant, especially in a combined cycle. The first stage ceramic stator vane is developed for a 1300°C class, 20-MW industrial gas turbine. There are two main design objectives. One is a structural design to ensure the strength reliability of brittle ceramic parts. The other is a cooling and heat insulating design to reduce the cooling air amount of metal parts. To solve the above problems, these ceramic stator vanes have a hybrid structure, composed of ceramic, metal and heat insulating parts. The strength reliability and cooling properties of the ceramic stator vane are evaluated by a cascade test using combustion gas in actual conditions (1300°C, 1.5 MPa). The results prove that ceramic parts have good strength reliability. Also, 1/20th the amount of cooling air required for air-cooled conventional vanes is required for the ceramic stator vane.

INTRODUCTION

In recent years, considerable effort has been expended in Japan to apply a gas turbine/steam turbine combined cycle and increase the gas turbine inlet temperature (T.I.T.) to raise the efficiency of thermal power generation. A fair amount of progress has been achieved toward higher temperatures through development of heat-resistant superalloys and more advanced cooling technologies. However, the heat resistance of metals is limited, so these approaches are already bringing diminishing returns. Structural ceramic materials, on the other hand, offer excellent thermal properties compared with superalloys in terms of heat, corrosion and wear resistance. By employing ceramics for some of the key components of gas turbines—combustors, stator vanes, and rotor blades—gas turbines will be able

TABLE 1
Specifications of a ceramic stator vane

1. Gas Turbine		
Output	20	MW
Turbine Inlet Temperature	1300	℃ (Mean)
Turbine Inlet Pressure	1.5	MPa
2. First-Stage Ceramic Stator Vane		
Configuration		
P.C.D.	730	mm
Height	60	mm
Cooling Air Temperature	400	℃

to run hotter and the requirement for cooling air can be markedly reduced for substantially enhanced thermal efficiency.

However, ceramics are brittle and thus lack strength reliability as a structural material. This deficiency is the greatest barrier to greater penetration of ceramic parts and products. Rather than simple substitution of ceramics for metal parts, innovative structural design technologies are required for ceramic applications that exploit the advantages of ceramics while at the same time compensating for their deficiencies. The authors are seeking to establish such a design technology through their work on ceramic stator vanes as part of a broader effort to develop a ceramic industrial gas turbine engine. [1] This effort culminated in the development of a hybrid ceramic/metal stator vane structure, and its cooling characteristics were investigated. [2] Another type of ceramic stator vane has been developed in Japan by Ohkoshi et. al. [3] In this report, we describe the structural stress analysis results of prototype first-stage ceramic stator vanes for application to a 1300℃ class, 20-MW gas turbine. The design specifications of the gas turbine and the ceramic stator vane are shown in Table 1. The thermal load cascade test results are also provided based on gas combustion conditions in actual gas turbines.

DEVELOPMENT OF CERAMIC STATOR VANE

Hybrid Ceramic/Metal Structure

A cross-sectional schematic of the hybrid ceramic/metal stator vane designed for the first-stage nozzle of a 1300℃ class (turbine inlet temperature) 20-MW gas turbine is shown in Figure 1. The airfoil that comes in direct contact with high-temperature combustion gas (the ceramic shell) and the inner/outer shrouds (ceramic side-walls) are formed out of ceramics, a material with superior heat resistance properties requiring no cooling. The metal core runs through the ceramic shell and connects the inner/outer metal supporting plates.

The key features of the developed ceramic stator vane can be summarized as follows:

(1) It is made of ceramic components that are simple in shape and thus easy to manufacture; also, the generation of excess stress due to concentration of stress and other factors is eliminated.
(2) Metal parts are used to provide reinforcement against external forces stemming from pressure differentials, so ceramic parts are subjected to a very slight mechanical load.
(3) Heat-insulating plates and a heat-insulating layer used between ceramic and metal parts are made of high heat insulating material and a cushioning layer that provides good flexibility.
(4) The cushioning layer prevents thermal stress induced by the differing

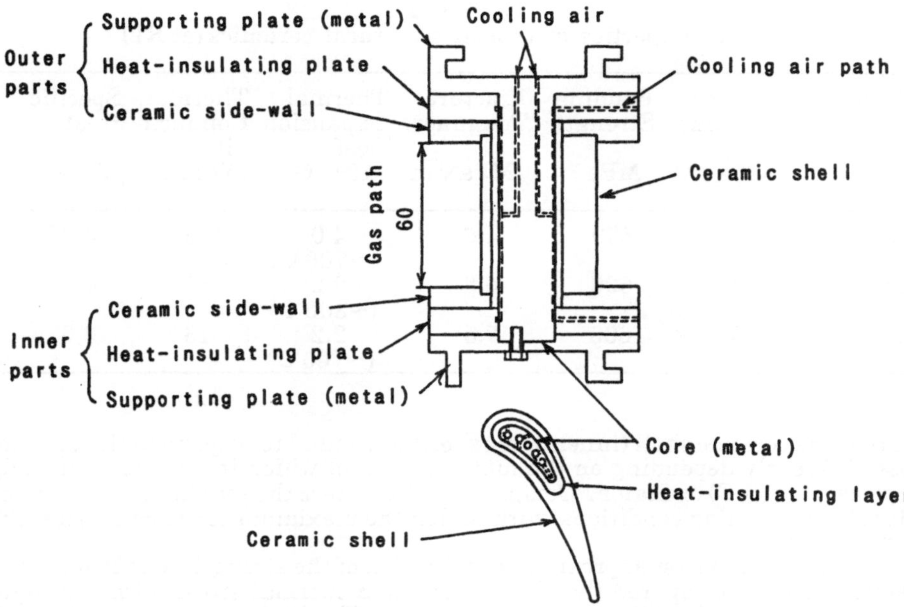

Figure 1. Cross-sectional schematic of the ceramic/metal hybrid stator vane for a 1300°C class, 20-MW gas turbine.

thermal expansion coefficients of the ceramics and metal, and thus assures the strength reliability of the ceramic parts.
(5) The quantity of heat penetrating to the metal parts is suppressed by the ceramic components and by the insulating material. As a result, the amount of cooling air required can be substantially reduced compared with conventional air-cooled metal vanes.

Strength Design of Ceramic Components
The ceramic parts were subjected to stress analysis by FEM while the gas turbine was running as part of the overall strength reliability evaluation of the ceramic stator vanes. As noted in the previous section, one fundamental stress is the thermal stress caused by combustion gas. First, the temperature was analyzed and then thermal stress was calculated. To compensate for any uncertainties in the design, we assumed the excess thermal load condition where T.I.T. was 1500°C, whereas T.I.T. of the design specification was 1300°C. For the fuel cutoff condition when the gas turbine is abruptly stopped in an emergency, we assumed a falloff of combustion gas temperature of about 1000°C per 5-second interval. Three ceramic candidates are shown in Table 2 with their main properties at room temperature.

The calculation results of two types of shell configurations and two kinds of ceramic materials are shown in Figure 2. This figure shows the variations of the maximum steady and transient thermal stresses for the ceramic shells in the height direction with time after fuel cutoff. Where the steady thermal stresses are plotted at t = 0. It was found that the maximum thermal stress impinging on the one-piece type ceramic shell occurs 3 seconds (t = 3) after fuel cutoff. Since this maximum value is quite high compared to the strength of ceramic materials, we sought to alleviate the excess thermal stress by dividing the ceramic shell into two pieces—an inner shell and an outer shell. Here we would

TABLE 2
Material properties of typical structural ceramics (at RT)

Sintered Ceramics	Young's Modulus GPa	Bending Strength MPa	Fracture Toughness MPa$\sqrt{m}$	Thermal Expansion Coef. 10^{-6}/°C	Thermal Conductivity W/m·K	Specific Heat J/g·K
SiC	406	579	4.6	4.0 (~700°C)	126	0.67
Si_3N_4	309	696	6.8	3.2 (~800°C)	67	0.66
Sialon	283	666	7.5	2.2 (~700°C)	16	0.6

note that the ceramic shell (inner/outer shell) was found to respond to the thermal stress differently depending on the materials out of which it was made, namely silicon carbide (SiC) or Sialon. Compared to SiC where the maximum stress came under steady running conditions, with Sialon the maximum stress came after the fuel was cut off.

Next, as a first approximate evaluation of the strength reliability of the ceramic shell, we compared the above maximum thermal stress with the high-temperature strength of the ceramic materials. The high-temperature strength characteristics of the typical structural ceramics (sintered) used for high-temperature gas turbines are shown in Figure 3. Shown are the results for the 4-point bending test with a rectangular bar-type specimen (3 mm × 4 mm × 40 mm). The strength values of the various materials at 1300°C, expressed in terms of the residual safety factor at the maximum thermal stress on the two-piece divided type ceramic shell, were 3 for SiC, 1.3 for Sialon, and 2.1 for Si_3N_4 (assuming the same thermal stress as for Sialon). A particularly important factor to assure reliability of the fitting support between the ceramic shell and the ceramic

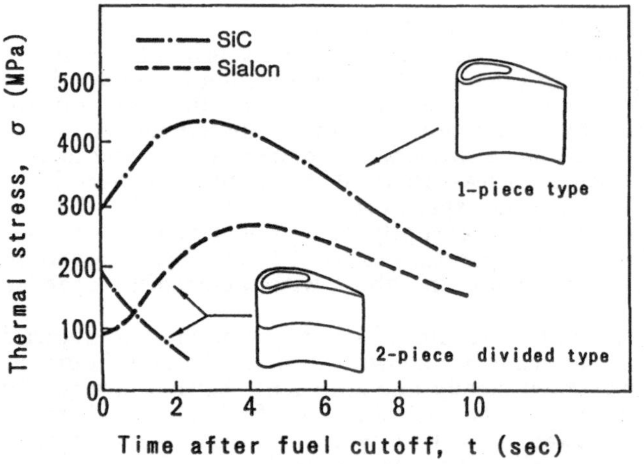

Figure 2. Variations of the maximum steady and transient thermal stresses of two types of ceramic shells with time after fuel cutoff.

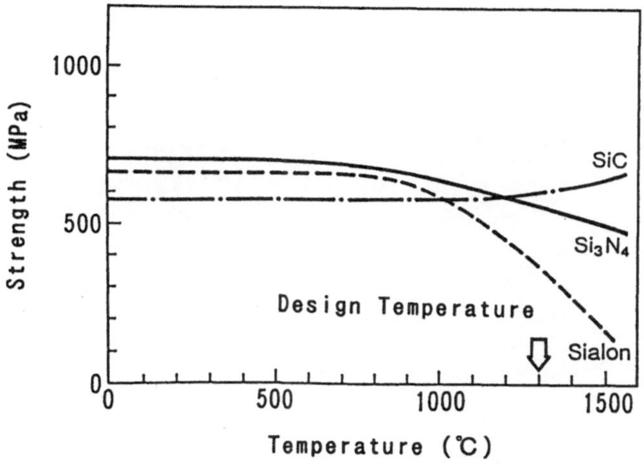

Figure 3. High-temperature strength of three kinds of structural ceramics for high-temperature gas turbines.

ing the same thermal stress as for Sialon). A particularly important factor to assure reliability of the fitting support between the ceramic shell and the ceramic sidewall is fracture toughness. Taking the fracture toughness into account, Si_3N_4 appears to be the best candidate ceramic material.

Developed Ceramic Stator Vane

Based on the results reviewed in the previous section, a proto-type ceramic stator vane was constructed. The ceramic parts were fabricated out of sintered Si_3N_4. The main parts are shown in Figure 4. The ceramic shell is divided into two pieces (inner and outer shells), and has a blunt head. The ceramic side-wall has a slit to release thermal stress. The ceramic side-walls are supported by fitting their grooves with the ceramic shell. The metal core was made of Ni-based heat resistant alloy. The heat-insulating plates were made of alumina fiber composite with high heat resistance, and the heat-insulating layer surrounding the core was made of castable mulite.

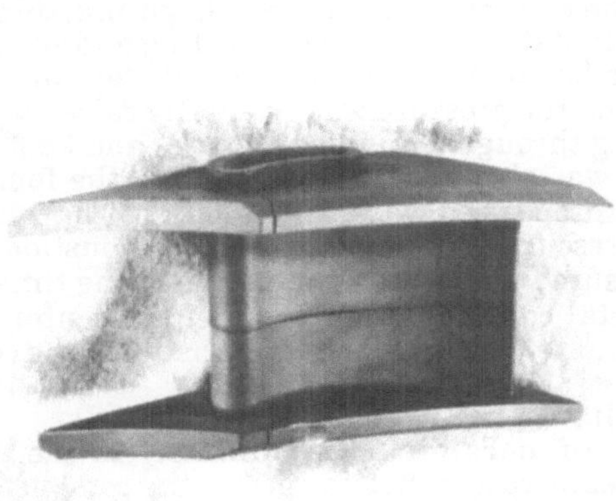

Figure 4. Assembly of ceramic parts made of sintered Si_3N_4.

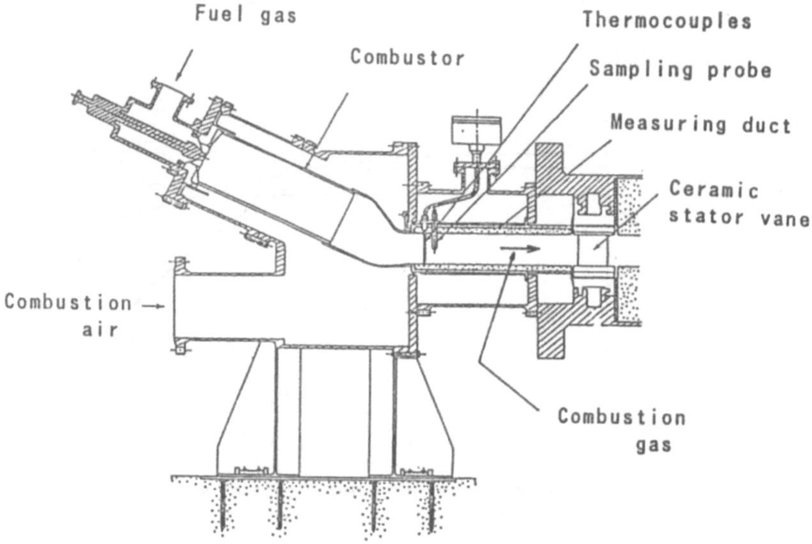

Figure 5. Cascade test equipment with high-temperature and high-pressure combustion gas for ceramic stator vane evaluation.

Five ceramic stator vanes were assembled and then installed to a cascade test equipment as shown in Figure 6 afterwards.

THERMAL LOAD CASCADE TEST

Test Equipment and Conditions

To evaluate the reliability of the prototype ceramic stator vane, a thermal load cascade test was conducted using combustion gas. A longitudinal cross section of the test equipment is shown in Figure 5. The channel cross-section temperature distribution and pressure of high temperature and pressure gas produced in a combustor were measured using a measuring duct. After that, the gas was successively introduced to the ceramic stator vanes. High-pressure air was supplied in and around the ceramic stator vanes for both cooling and sealing. The initial combustion gas conditions for the thermal load cascade test were 1300°C (mean) and ambient pressure. Then the pressure was gradually raised to the design specification of 1.5 MPa passing through a number of stages and kept for about 4 hours total. The gas velocity was about 80 m/s. After that, the fuel cutoff test was performed. Temperature drop speed was about 400°C/s. The various test conditions, including quantity, pressure and temperature of combustion gas and air for cooling/sealing were measured and controlled. At the same time, the temperature of the ceramic shell, metal core and other parts of the center ceramic stator vane were measured to evaluate the cooling characteristics of the ceramic stator vane.

Experimental Results

Reliability of ceramic components: In the thermal load cascade test, considering the temperature distribution of the combustion gas, the vanes at the center of the cascade are exposed to the highest-temperature gases 1460°C at a pressure of 1.5 MPa. A photograph of the cascade after the fuel cutoff test is shown in Figure 6. A fluorescent penetrant inspection confirmed that all the ceramic parts came through the test in sound condition. However, the pressure-side

Figure 6. Ceramic stator vanes after the cascade test using 1300℃ and 1.5 MPa of combustion gas.

was thus demonstrated.

Cooling characteristics: The metal parts likely to experience the most elevated temperatures are those near the core. Figure 7 shows the relation among combustion gas temperature (mean), cooling air temperature, and core temperature at a ratio of 0.29 % cooling air to total air flow (corresponding to about 1/20th the amount for the conventional air-cooled first-stage metal vanes) when the combustion gas pressure is held constant at 1.5 MPa. Note that this includes the ceramic shell temperature. The difference between the core and cooling

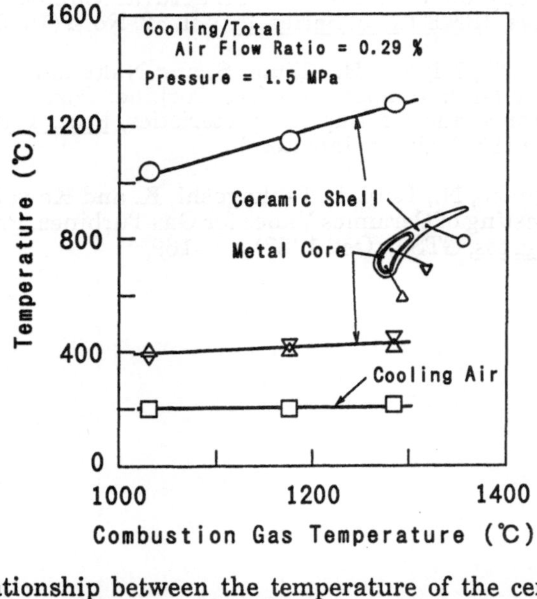

Figure 7. Relationship between the temperature of the ceramic stator vane and the combustion gas temperature.

air temperatures is about 250℃ when the combustion gas temperature is 1300℃, thus revealing good cutoff and cooling effects. The core temperature is 650℃ when the cooling air is 400℃ (designed value), and this value is permissible in terms of the heat resistance of metal materials. The cooling air required for the present ceramic stator vane has been remarkably reduced, thus being very effective for enhancing the thermal efficiency of gas turbine engines. It also offers the prospect of still higher gas temperatures for future gas turbine engines.

CONCLUSION

The strength reliability and cooling characteristics of a ceramic stator vane with a hybrid ceramic/metal structure have been investigated and the following results obtained:

(1) Based on an initial approximate evaluation of the thermal stress and strength reliability of a two-piece divided-type ceramic shell made of sintered Si_3N_4, a first-stage ceramic stator vane has been developed for 1300℃ class 20-MW gas turbines.
(2) The results of a thermal load cascade test using actual combustion gas conditions (mean temperature of 1300℃ and pressure of 1.5 MPa) showed that the ceramic parts held up well under testing, thus demonstrating strength reliability and imperviousness to environmental degradation at least for short periods. In addition, the prospect of substantially reducing the amount of cooling air required to about 1/20th the amount required by conventionally air-cooled metal stator vanes was shown.

REFERENCES

1. Miyata, H., Iijima, S., Ooshima, R., Abe, T., Hisamatsu, T. and Hamamatsu, T., Application Technology on Ceramics for Structural Components of High-Temperature Machines, JSME Int. J. Series I, Vol. 32, No. 4 (1989) p596

2. Abe, T., Hisamatsu, T., Miyata, H., Iijima, S. and Nakakado, K., R&D of Ceramic Stator Vane for High-Temperature Gas Turbine: Part II Experimental Study on Heat resistance and Cooling Characteristics, Proc. 17th Gas Turbine Congress, GTSJ (June 1989) p39 (in Japanese)

3. Ohkoshi, A., Watanabe, N., Tsuji, I., Tsukagoshi, K. and Kawai, H., Manufacturing and Hot Rig Testing of Ceramics Vanes for Gas Turbines, Proc. 1987 Tokyo Int. Gas Turbine Congress, GTSJ, (Oct. 1987) p I-169

JOINING METHOD WITH HIGH HEAT RESISTANCE FOR CERAMIC ROTORS

Hiroaki Makino, Katsunori Yamada, Nobuo Kamiya and Shigetaka Wada
TOYOTA Central Research and Development Laboratories, Inc.
Nagakute, Aichi 480-11, JAPAN

ABSTRACT

A method of joining a ceramic shaft to a cylindrical metal shaft with high heat resistance has been developed, which is referred to as a multi-fulfilling fitting. The multi-fulfilling fitting is a kind of shrink fitting with brazing filler metals. It features the use of two kinds of brazing metals with different melting points. At the first stage, the ceramic shaft and the cylindrical metal shaft were shrink-fitted by filling a brazing metal with a higher melting point in the bottom clearance between the cramic shaft and the metal shaft. The joining was completed to fill a brazing metal with a lower melting point in the upper portion of the clearance.

The flexural strength of the joining body was 330 MPa at room temperature and the slip resistance was 7 kN at 750°C. These mechanical properties, especially at elevated temperatures, are superior to those of other joining methods such as the conventional shrink fitting and the active brazing method.

The multi-fulfilling fitting can be applied to the joining of the cramic turbine wheel to the metal shaft of a gas turbine engine which requires heat resistance up to about 700°C.

INTRODUCTION

Structural ceramics have a wide variety of applications because of their excellent properties including high-temperature strength and wear resistance. For example, ceramic chambers for diesel engines [1] and ceramic rotors for turbo-charged engines [2] have been commercialized in recent years. Ceramic rotors for gas turbine engines have also been developed in several companies [3]. The method of joining ceramic components to metal components is one of the key technologies in these applications, especially in the application to gas turbine engines in which the temperature of joining portions becomes very high [4].

Joining methods which are conventionally employed for joining ceramics to metals are classified into two groups; i.e. mechanical joining methods such as shrink fitting and shrink fitting with brazing filler [5], and chemical bonding methods such as diffusion bonding and active brazing.

The joining method developed in this work belongs to the former group. The shrink fitting, which is the most popular mechanical joining method, is a simple method, but has a few drawbacks such as the necessity of high machining accuracy, and yet the failure to maintain the necessary torque at an elevated temperature. If a larger shrinkage allowance is employed to overcome the latter drawback, the excessively high shrinking pressure acts radially upon the ceramic shaft and degrades the flexural strength of the ceramic shaft near the joint. Therefore, the joint by the ordinary shrink fitting can withstand elevated temperatures from about 300°C to 400°C.

The shrink fitting with a brazing metal, which is referred to as a fulfilling fitting in this paper, is a modification of the shrink fitting and employs a brazing metal as a filler in the clearance between the ceramic shaft and the metal shaft to be joined. The fulfilling fitting has two advantages over the ordinary shrink fitting. One is that the fulfilling fitting does not require the high machining accuracy essential in the ordinary shrink fitting, and the other is that a large shrinkage allowance can be employed in the fulfilling fitting because the brazing metal between the ceramic shaft and the metal shaft is soft enough to act as a stress relaxation material. The shrinkage allowance of the fulfilling fitting depends on the melting point of a brazing metal which fills in the clearance; i.e. the higher the melting point of the brazing metal is, the larger the shrinkage allowance of the fulfilling fitting becomes. Therefore, the working temperature of the fulfilling fitting becomes higher with an increase in the melting point of the brazing metal. Silver brazing metal, of which melting point is about 800°C, is generally used as a brazing metal, and the joint with a silver filler can be applied up to about 600°C.

Although the fulfilling fitting has the high working temperature, the joint of ceramic rotors for gas turbine engines requires much higher working temperature. To improve the working temperature of the fulfilling fitting, it is necessary to employ a brazing metal with the melting point higher than that of a silver brazing metal. This causes the stress concentration at the end of the joined portion as is the case of the shrink fitting with a large shrinkage allowance, and therefore leads to the degradation of the strength of the ceramic shaft near the joint.

To increase the working temperature of the fulfilling fitting, multi-fulfilling fitting was developed. The multi-fulfilling fitting employs a brazing metal with a relatively low melting point to fill in the open end of the clearance, and a brazing metal with a relatively high melting point to fulfill in the rest of the clearance.

In this work, mechanical properties of the multi-fulfilling fitting up to 750°C and the influence of the low melting point filler length on the flexural and shear strengths of the joints were studied.

PRINCIPLE AND STRUCTURE OF MULTI-FULFILLING FITTING

The principle and structure of the multi-fulfilling fitting are explained by comparing it with the fulfilling fitting and the ordinary shrink fitting. Schematic structures of three types of mechanical joining methods are shown in Fig. 1. The principles of all the methods are based on the difference in thermal expansion between the ceramic shaft and the metal shaft.

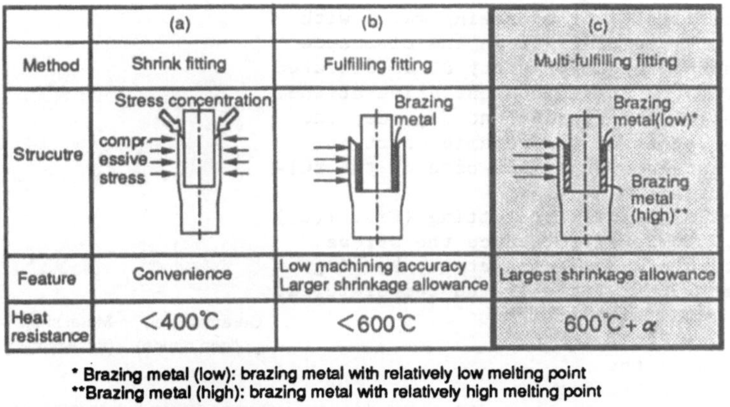

Figure 1. Schematic structures of ordinary shrink fitting, fulfilling fitting and multi-fulfilling fitting.

A ceramic shaft is set in a cylindrical metal shaft at an elevated temperature. As the thermal shrinkage of the metal shaft is larger than that of the ceramic shaft, the compressive stress acts on the ceramic shaft during cooling, which generates frictional stress between the ceramic shaft and the metal shaft. The frictional stress acts as the joining torque of the joints. At the same time, the compressive stress generates stress concentration at the end of the joining portion especially in the case of the shrink fitting (arrows in Fig. 1(a)), which is inevitable in the shrink fitting. The stress concentration degrades the flexural strength of the ceramic shaft near the joint at low temperatures. Therefore, the adoption of the large shrinkage allowance to get the joining torque at elevated temperatures promotes the degradation of the flexural strength of the joint at room temperature.

From these facts, the mechanical joining methods generally have two weak points. One is that the flexural strength around room temperature is diminished by the stress concentration at the open end of the joint due to the high shrinkage stress, and the other is that the torsional torque or shear strength at elevated temperatures becomes weak because of the loosening of the shrinkage stress by thermal expansion.

The fulfilling fitting (Fig. 1(b)) is a kind of shrink fitting which employs a brazing metal as a filler in the clearance between the ceramic shaft and the metal shaft. The advantages of the fulfilling fitting over the shrink fitting are that low degrees of machining accuracy is necessary for the fulfilling fitting, and that the brazing metal, which fills in the clearance between the ceramic shaft and the metal shaft, is soft enough to act as the stress relaxation material. The latter advantage enables the fulfilling fitting to employ the larger shrinkage allowance with a small amount of flexural strength degradation than that of the ordinary shrink fitting. Therefore, the fulfilling fitting can maintain the joining torque up to the temperature higher than that of the shrink fitting.

Although the fulfilling fitting keeps the torsional torque at elevated temperatures, the joint of ceramic rotors for gas turbine engines requires some amount of the torsional torque at the temperature higher than that of the ordinary fulfilling fitting which usually employs a silver brazing metal as a filler. If a brazing matal with higher melting point fills in the clearance to improve the torque property at an elevated temperature, the stress concentration at the end of the joint increases and degrades the flexural strength of the ceramic shaft at room temperature even in the case of the fulfilling fitting.

The multi-fulfilling fitting (Fig. 1(c)) was, then, developed to reduce the stress concentration at room temperature, keeping the torque property at elevated temperatures. A brazing metal with a high melting point fills in the most portion of the clearance. The brazing metal (high) maintains the torsional torque at high temperature. A brazing metal with a low melting point is substituted for a brazing metal with a high melting point just in the vicinity of the open end of the joint to relax the stress concentration. By the multi-fulfilling fitting explained above, both high torsional torque at an elevated temperature and small degradation of the flexural strength at room temperature are attained.

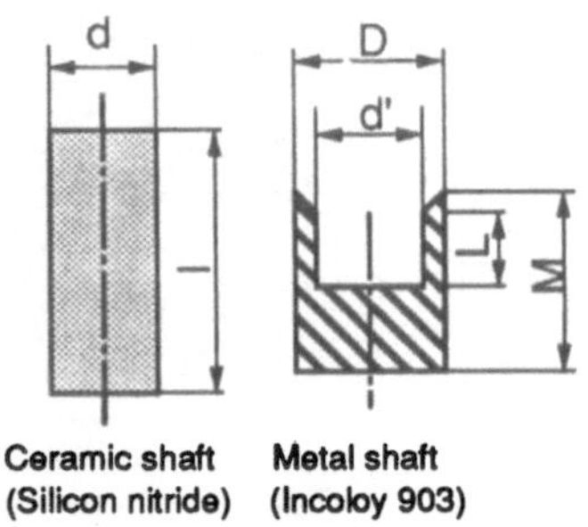

Type	Ceramic shaft		Metal shaft			
	d	l	d'	D	L	M
A	12	40	12.4	16	10	35
B	18	75	18.4	22	18	35

Figure 2. Shapes and sizes of specimens to be joined.

EXPERIMENTAL PROCEDURE

Configuration of Specimens

Configurations of the specimens to be joined are shown in Fig. 2. Two types of joints (A and B) were made and their properties were examined. Ceramic shafts and metal shafts were made of silicon nitride and Incoloy 903, respectively. Incoloy 903 was used because it maintains high shrinkage stress at elevated temperatures.

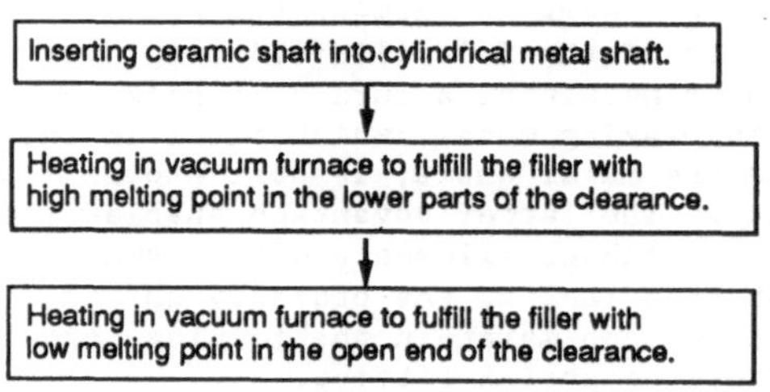

Figure 3. Procedure of multi-fulfilling fitting.

Procedure and Conditions of Multi-Fulfilling Fitting

The procedure for the multi-fulfilling fitting is shown in Fig. 3. The inner surface of the metal shaft was electrolessly plated with nickel so as to improve the wettability with the brazing metal. The ceramic

TABLE 1
Brazing filler metals used in experiments and filling conditions

Method	Brazing filler metal (melting point, composition)	Conditions
Fulfilling fitting	BAg8 (780°C, Ag:Cu=72:28)	950°C, 30 min., in vacuum
	BNi3 (980°C, Ni:B:Si:C=Bal:3.0:4.5:0.06)	1100°C, ↑
	K14WG (975°C, Au:Cu:Zn:Pd:Ni=58.5:22:3.5:3:13)	↑ , ↑
Multi-fulfilling fitting	BNi3 (980°C) +	1100°C, 30 min., in vacuum
	BAg8 (780°C)	950°C, ↑

shaft inserted into the metal shaft was heated in a vacuum furnace at a temperature over the melting point of the filler metal with a relatively high melting point. The brazing metal with a high melting point filled in the clearance between the ceramic shaft and the metal shaft, leaving a space for a low melting point brazing metal. After cooling the joint, the brazing metal with a low melting point was settled at the open end of the clearance. Then, the joint was re-heated in a vacuum furnace over the melting point of the low melting point brazing metal, so as to fill the brazing metal in the rest portion of the clearnce.

Silver brazing metal, BAg8 (JIS: Japanese Industrial Standards), was used as the filler with a low melting point and a nickel brazing metal, BNi3 (JIS), was used as that with a high melting point. Three kinds of ordinary fulfilling fittings which employ different brazing metals; BAg8, BNi3 and a gold brazing metal (K14WG), were tested for comparison. The melting points of the brazing metals used and the fulfilling conditions are shown in Table 1.

For the multi-fulfilling fitting, the length of the filler with a low melting point influences the properties of the joint. Therefore, joints with different lengths of the low melting point brazing metal were made and their properties were examined.

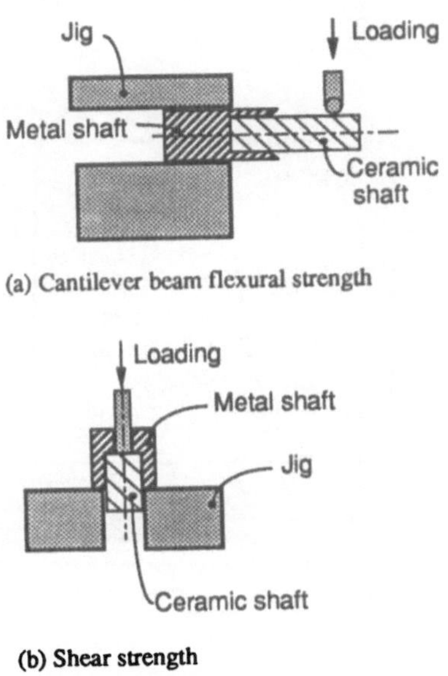

Figure 4. Schematic drawings of strength measuring method.

Strength Measuring Method

The cantilever beam flexural strength of the ceramic shaft was measured at room temperature as shown in Fig. 4(a). The shear strength between the ceramic shaft and the metal shaft at elevated temperature was measured by the method shown in Fig. 4(b).

RESULTS AND DISCUSSIONS

Strength Properties of Multi Fulfilling Fitting and Fulfilling Fitting

The cantilever beam flexural strength at room temperature and the shear strength at 600°C in the multi-fulfilling fitting are shown in Fig. 5, together with the strength properties of three types of ordinary fulfilling fittings for comparison. The filling length of the low melting point filler in the multi-fulfilling fitting was 5 mm, and the joint type was A. The cantilever beam flexural strength in the multi-fulfilling fitting was 330 MPa. Although this value was slightly smaller than that of the joint with a silver brazing metal of which strength was the highest among four kinds of the joints, it was higher than that of other ordinary fulfilling fitting joints with nickel and gold fillers.

The shear strength of the multi-fulfilling fitting joint at 600°C was 38 MPa. Although this value was slightly smaller than those of the joints with nickel and gold fillers, it was fairly higher than that of the joint with a silver filler. The joint by the multi-fulfilling fitting was the only one that could maintain high values of both the cantilever beam flexural strength at room temperatures and the shear strength at elevated temperatures.

Fig. 6 shows the increment in the diameter of the joining portions by the multi-fulfilling fitting and the ordinary fulfilling fittings with silver and nickel filler (joint type B). The diameter increment indicates the shrinkage allowance in the multi-fulfilling fitting and that in the

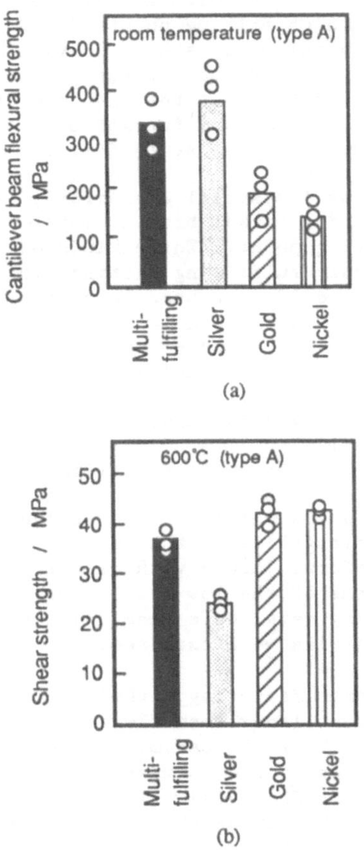

Figure 5. Cantilever beam flexural strength at room temperature (a) and shear strength at 600°C (b).

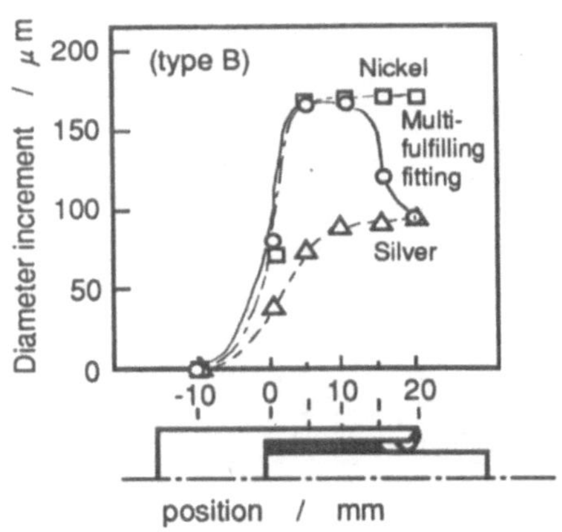

Figure 6. Diameter increment by multi-fulfilling fitting and fulfilling fitting.

fulfilling fitting. The increment in diameter by the multi-fulfilling fitting is nearly the same as that of a silver filler at the open end and as that of a nickel filler at the bottom side. The distribution of the diameter increment clearly shows the excellence of the multi-fulfilling fitting.

A photograph of the section of a multi-fulfilling fitting joint is shown in Fig. 7. The filler, which filled in the clearance, gradually changed from the filler with a high melting point (nickel filler) to the filler with a low melting point (silver filler). Therefore, the change in the stress at the portion of the two-brazing-metal joint was smooth as shown in Fig. 6.

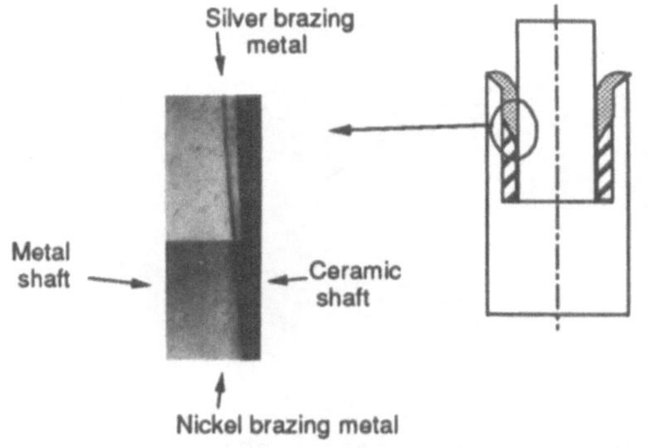

Figure 7. Section of multi-fulfilling fitting joint.

Dependence of Strength on Fulfilling Length of Filler with Low Melting Point

The dependence of the cantilever beam flexural strength of the multi-fulfilling fitting joint at room temperature on the filling length of a filler with a low melting point is shown in Fig. 8. The flexural strength increased with an increase in the filling length of the filler with a low melting point. When the filling length is over 5 mm, the flexural strength remained almost the same value, and it became the highest value at a filling length of 18 mm, at which length all the clearances were filled with a low melting point filler; e.g. ordinary fulfilling fitting with silver filler.

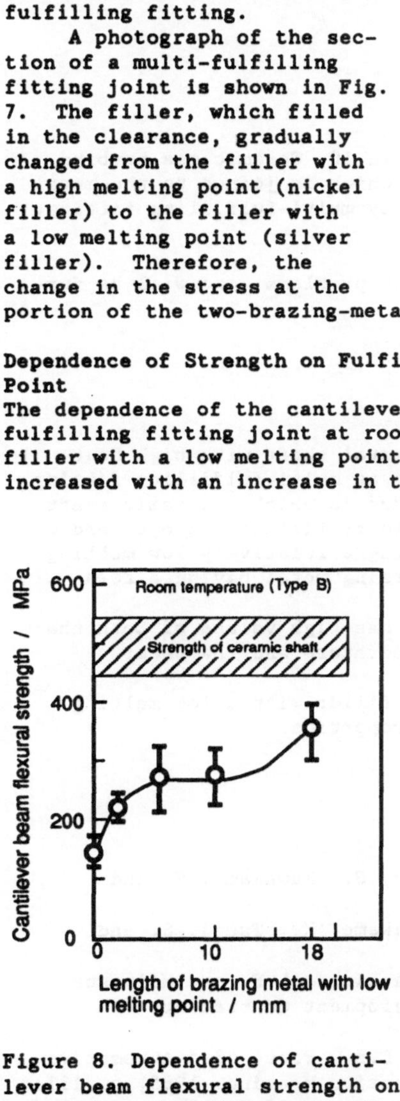

Figure 8. Dependence of cantilever beam flexural strength on length of brazing metal with low melting point.

The dependence of the shear strength on the filling length of the filler with a low melting point at 750°C is shown in Fig. 9. The shear strength rapidly decreased at a filling length of 2 mm, and gradually decreased at filling lengths over 2 mm. From these results, the joint of which low melting point filler had a filling length of 5 mm maintained both high flexural strength at room temperature and high shear strength at elevated temperatures.

Fig. 10 shows the ceramic gas turbine rotor which is joined to the metal shafts by the multi-fulfilling fitting. The rotor was rotated at a

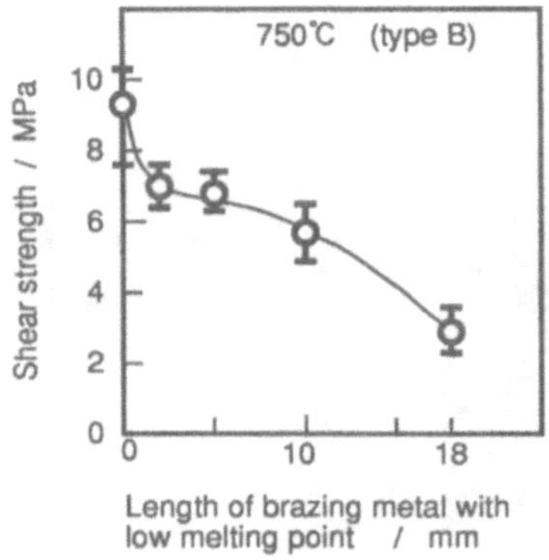

Figure 9. Dependence of shear strength on length of brazing metal with low melting point.

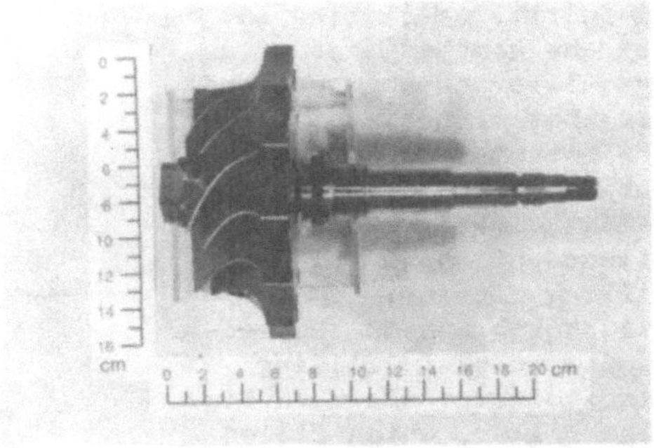

Figure 10. Ceramic gas turbine rotor which is joined to the metal shaft by multi-fulfilling fitting.

turbine inlet temperature of 1200°C without any problems, and will be further tested at 1400°C.

SUMMARY

1. A method of joining a ceramic shaft to a metal shaft with high heat resistance was developed, which is referred to as multi-fulfilling fitting.
2. Multi-fulfilling fitting is a joining method in which a ceramic shaft and a cylindrical metal shaft are shrink-fitted by filling the open end of the clearance with a brazing filler metal having a relatively low melting point and the rest of the clearance with a brazing metal having a relatively high melting point.
3. The joint by the multi-fulfilling fitting has high values of both the cantilever beam flexural strength at room temperature and the shear strength up to 750°C.
4. The joint, at which the length of brazing filler with a low melting point was 5 mm, exhibited the best strength properties.

REFERENCES

1. Kamiya, S., Murachi, M., Kawamoto, H., Kato, S., Kawakami, S. and Suzuki, Y., SAE Tech. Pap. 850523, 1985.
2. Shimizu, T., Takama, K., Enokishima, H., Mikame, K., Tsuji, S. and Kamiya, N., SAE Tech. Pap. 900656, 1990.
3. For example, Helms, H.E., Haley, P.J., Groseclose, L.E. and Hilpisch, S.J., Proc. 27th Automotive Technology Development Contractors' Coordination Meeting, 1989, p. 293.
4. Helms, H.E., Johnson, R.A. and Groseclose, L.E., Proc. 23rd Automotive Technology Development Contractors' Coordination Meeting, 1985, p. 137.
5. Hamano, Y., Sagawa, N. and Miyata, H., ASME paper, 86-GT-10.

CERAMIC MATERIALS AND COMPONENT DESIGN FOR AEROSPACE APPLICATIONS

W. Krüger* and W. Bunk
MTU Deutsche Aerospace, Munich, Germany, and
DLR - German Aerospace Research Establishment, Cologne, Germany

ABSTRACT

Focussing on a considerable reduction in cooling air requirements for HPT nozzles of jet engines, a new guide vane concept utilizing the high service temperature capability (~ 1,900 K) of sintered silicon carbide is proposed. In this concept, the load bearing structure of the vane is formed by the metallic shrouds and the metallic core. The uncooled vane shell consists of SSiC, which floats in recesses in the shrouds, surrounds the vane core and protects the major metallic parts from direct contact with the hot gas.
Under the feasibility study an HPT rotor blade also of metal ceramic construction consisting of a metallic (Inconel 100) core with a fir-tree root and a ceramic (SSiC) shell was designed.

It would be desirable to replace the brittle SSiC material with more damage-tolerant ceramic matrix composites. These materials have been developed and tested at high temperatures in an oxidizing atmosphere by research centres and industrial laboratories. An outlook on the potential of CMCs for aerospace application in terms of improved damage tolerance will be presented.

INTRODUCTION

HPT nozzle guide vanes and HPT blades in modern aero-engines are precision castings in nickel- or cobalt-base superalloys. With the temperature of the gas stream at the nozzle and the blades substantially exceeding the allowable component temperature, the vanes and blades need protection from overheating by means of massive internal and film cooling. In military aero-engines the gas temperature exceeds the allowable wall temperature of the vanes and blades by some 500 K to 700 K. To cool these HPT parts approximately 12 to 15% to the compressor mass flow is needed.

The temperature resistance of nozzle guide vanes and blades can be improved considerably using materials such as sintered silicon carbide (SSiC, T_{max} ~ 1,900 K) instead of metals for their construction. A reduction in the cooling-air requirement by 70 to 80% is possible /1/. The use of brittle ceramic materials calls for rethinking the design of HPT vanes and blades. The new design is based on the concept of a material combination (metal/ceramic) with the separation of the functions (load bearing/hot gas path).

DESIGN OF A SEGMENTED HPT NOZZLE FOR A MILITARY PROPULSION ENGINE

An HPT nozzle in metal-ceramic construction was designed for a demonstrator engine (Fig. 1). The aerodynamic and thermodynamic design was based on actual engine thermodynamic cycle data, except that the stator inlet temperature was raised to 1,825 K. This HPT nozzle arrangement has 17 twin-vane assemblies.

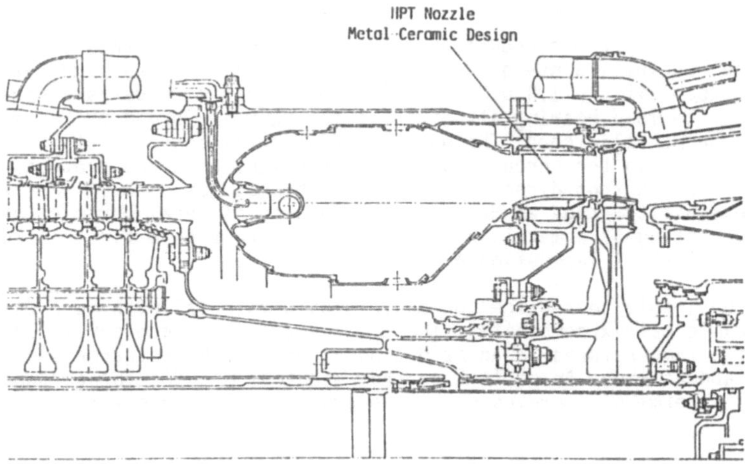

Figure 1: Demonstrator engine

The main parts of the metal/ceramic HPT nozzle segment is shown in Fig. 2. The load-bearing structure is formed by the metallic inner and outer shroud segments and the two metallic cores. The uncooled vane shell in sintered silicon carbide, which floats in recesses in the shrouds, surrounds the vane core and protects the major metallic areas from direct contact with the hot gas.

The shrouds and the recesses exposed to the hot gas stream are plasma sprayed with a thermal barrier coating of zirconia. The outer and inner shroud segments are protected by impingement cooling utilizing the double-wall structure. The vane core, which is heated by radiation from the red-hot ceramic vane shell, was provided with a large number of longitudinal cooling holes and in addition was given a reflective coating in the vane shell area. The shroud segments and core cooling features adequately protect these parts from overheating (see Fig. 3).

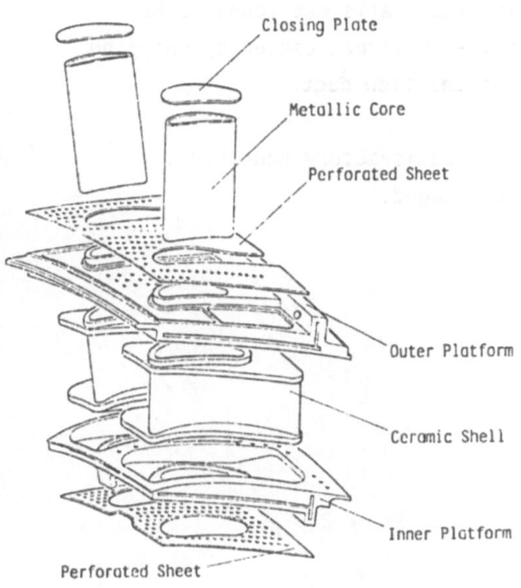

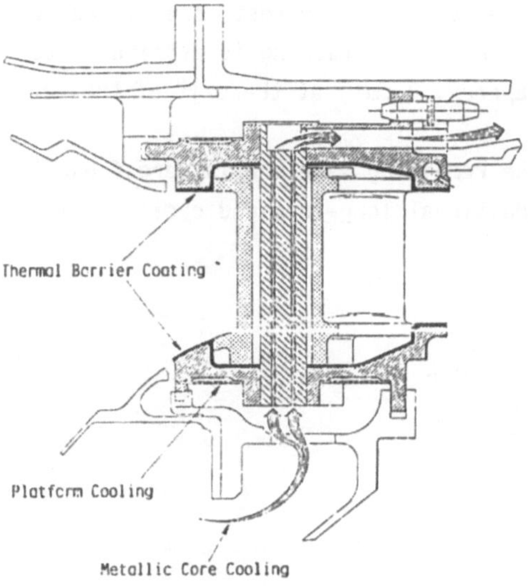

Figure 2: HPT nozzle segment, metal ceramic design

Figure 3: Cooling system of the nozzle segment

TEST FACILITY

A test rig was designed and built to test the new nozzle vane concept at hot gas temperatures above 1,800 K (Fig. 4). Thermally highly stressed rig components are the combustion chamber, the transition duct, the nozzle vane segments and the heat exchanger. They are made largely of sintered silicon carbide.

TEST RESULTS

Vane segment rig tests were performed at a hot gas temperature of 1825K under atmospheric pressure. A total of seven ceramic vane shells were tested: three for a duration of 100 hrs, three for 200 hrs, and one for 300 hrs.

After the 200-hour test, one of the ceramic vane shells was found to be cracked. The cracking is attributed to a cold-air streak caused by chipping (approx. 2 cm^2) at the exit of the ceramic transition duct.

The remaining vane shells were found to be in satisfactory condition. Additional long-term and cyclic tests are envisaged.

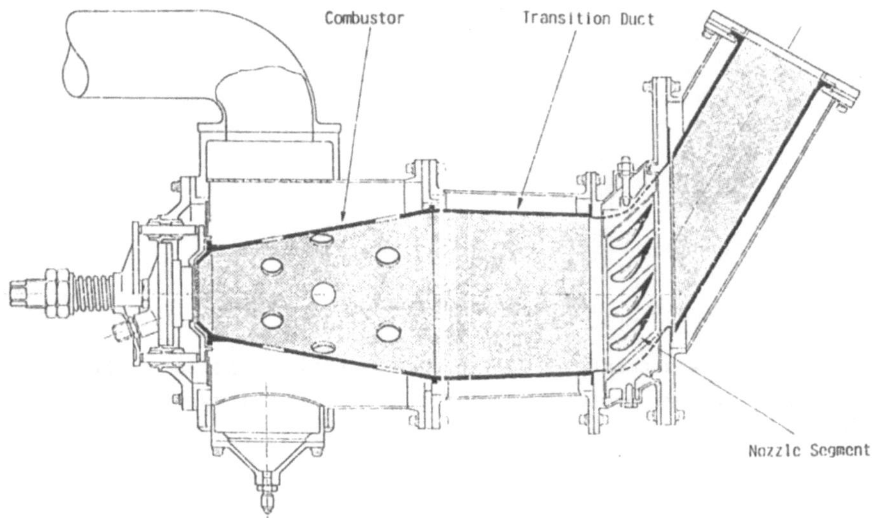

Figure 4: Nozzle vanes test rig

HPT BLADE IN METAL CERAMIC DESIGN

Conventional turbine blades, especially for medium- and high-performance gas turbines are precision castings with a sophisticated internal and film cooling system. The cooling air requirement is as much as 5% of the compressor mass flow.

As has been done with the HPT nozzle, attempts were made at determining the potential for reducing cooling air requirements by using silicon carbide as blade shell material. Figure 5 shows the hybrid blade structure which takes into account the brittle-fracture characteristics of silicon carbide.

With this configuration, the ceramic blade shell is supported over a two-shank pin in the upper part of the metallic blade core. The pin is fixed to the blade core by high-temperature brazing.

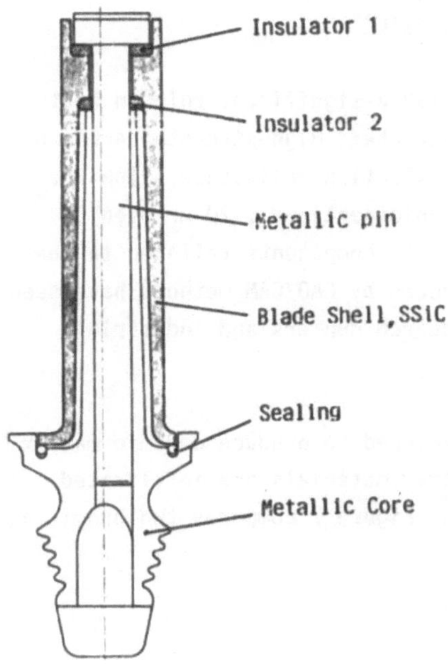

Figure 5: HPT blade, metal ceramic design

Figure 6: HPT blade assembly

Pressure-loaded figure-eight-shaped insulators were inserted in the area of contact between the ceramic profile and the load-bearing, cooled metal structure. They reduce the heat flow and the friction and compensate for peak stresses because of their low modulus of elasticity.

The lower part of the airfoil shell is fitted into a recess in the blade platform. The gap between the ceramic shell and the platform is sealed. This seal is centrifugally acting, i.e. becomes effective on rotation.

A number of these hybrid blades (Fig. 6) were manufactured and tested in the cold spin test rig. Circumferential speeds of up to 350 m/sec were achieved. The scatter range of values was relatively small.

It is expected that the circumferential speed can be increased even further by improving the geometrical conditions in the area of contact between the centrifugally loaded ceramic blade shell and the metal load-bearing structure.

CERAMIC MATRIX COMPOSITES

We hope that ceramic matrix composites will play a significant role in heat engines because of their fibre-reinforcing character. High-strength carbon or silicon carbide fibres of high stiffness and oxidation resistance allow the design of completely new material systems by unidirectional, 2D or even 3D brading to carefully selected matrices. Composite components tailored to the needs of engineers and engine component producers by CAD/CAM methods have been proposed in collaboration with university research centers and industrial laboratories.

Several processing technologies have been developed to produce ceramic matrix composites. They differ mainly in the way matrix materials are infiltrated into the porous preform of fibre arrangements. Figure 7 compares two promising infiltration techniques.

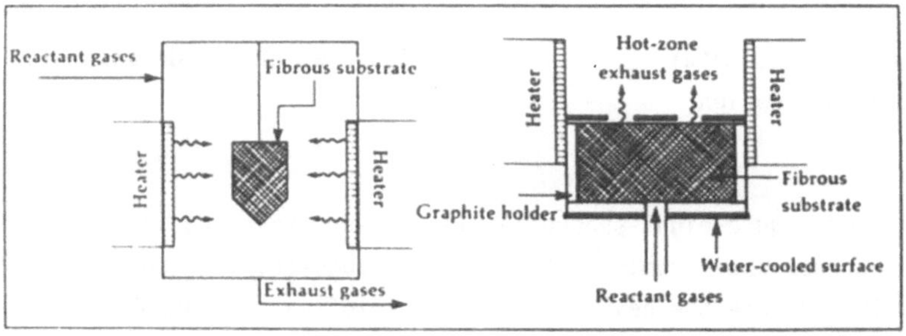

Figure 7: Two promising techniques to fabricate ceramic-matrix composites:
left - CVI isothermal diffusion limited infiltration by SEP, France;
right - thermal gradient forced flow process by MAN, Germany. /2/

Material toughness is of primary importance to designers and engineers interested in introducing ceramic materials into aerospace systems. In contrast to monolithic ceramics CMC (ceramic-matrix composites) show highly superior toughness. This makes them suitable for application in gas turbines. Their high strength allows the design of structures without metal backing.

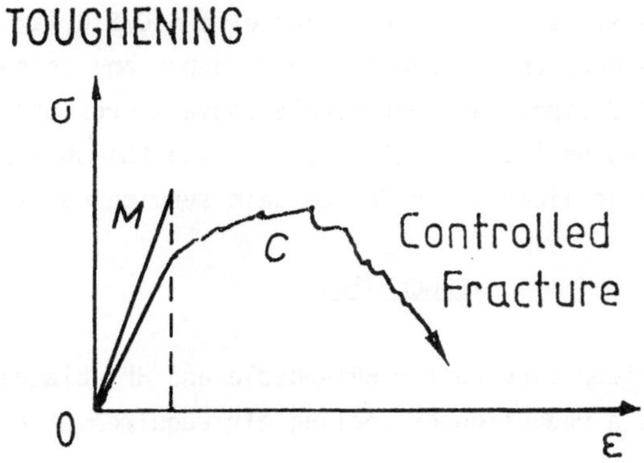

Figure 8: Controlled fracture (C) of CMC in comparison with monolithic ceramic (M). /3/

Figure 8 explains the toughness gain of CMC by controlled fracture in contrast to monolithic ceramics. Crack deflection and fibre pull-out are seen as characteristic features of fracture mechanisms. Endless-fibre-reinforced ceramics should be processed so as to produce a weak interface between the fibres and the matrix. One alternative is special coating of the fibre surface to condition the interface properties.

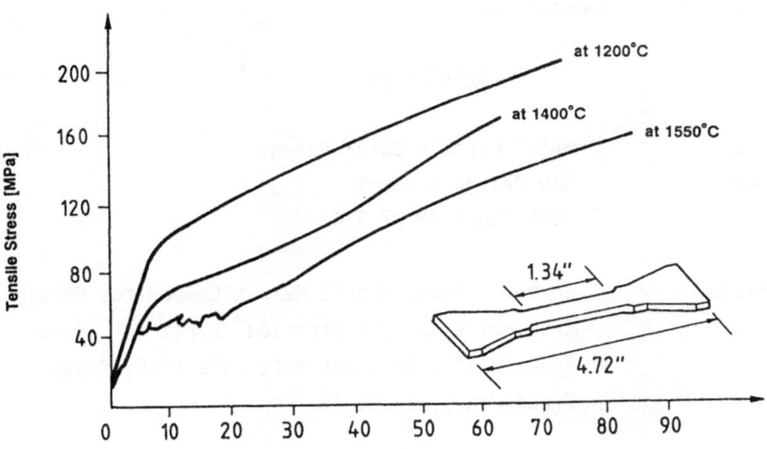

Figure 9: Mechanical behaviour of SiC/SiC at elevated temperature and in oxidizing environment (Dupont) /4/

Figure 9 correlates the stress-strain behaviour of a SiC/SiC laminate exposed to air and high temperatures up to 1550 °C for 20 hours and tested at room temperature. Although SiC fibres are not stable above approximately 1200 °C because of recrystallization,these results demonstrate the suitability of SiC/SiC for gas turbine applications under certain service conditions.

CONCLUSION

The use of sintered silicon carbide for HPT-nozzle and HPT-blades for a military engine permits a reduction of cooling air requirement by 70 to 80%.

The lack of ductility of these ceramic material calls for rethinking the vane and blade design. The new design is based on the concept of a material combination (metal/ceramic) with separation of the functions (load bearing/hot gas path). The ceramic vane and blade shells are designed for minimum tensile stresses even at the expense of compressive stresses.

Preliminary tests of the HPT nozzle segments at a hot gas temperature of 1825 K for a duration of 200 h and cold spin tests of the blades have demonstrated the feasibility of the design.

It would be desirable to replace the brittle SSiC material with more damage tolerant ceramic matrix composites.

REFERENCES

1 W. Krüger, W. Hüther — Metal Ceramic Guide Vanes, New Design Concept ASME-Paper 89-GT-334

2 R. Naslain — Carbon-Ceramic Hybrid Matrix Composites Obtained by Chemical Vapor Infiltration, First Int. Symp. Functionally Gradient Materials 1990, Sendai, Proceedings, S. 71-75

3 J.J. Choury — SEP Carbon and Ceramic Composites in Aeronautics and Space Applications, First Int. Symp. Functionally Gradient Materials 1990, Sendai, Proceedings, S. 157 - 167

4 Dupont Company Broshure

DEVELOPMENT OF SILICON NITRIDE RADIAL TURBINE ROTORS

Keiichiro Watanabe*1),Tadao Ozawa 1),Yoshito Kobayashi 2),Eito Matsuo 3)

1)NGK Insulators, Ltd.,Nagoya,Japan
2)Mitsubishi Motors Corporation,Tokyo,Japan
3)Mitsubishi Heavy Industries Ltd.,Nagasaki,Japan

ABSTRACT

To examine the possible use of ceramics as key components for automotive gas turbines, studies were conducted on radial turbine rotors made of SN-88, a newly developed, heat resistant silicon nitride. The flexural strength of MOR bars cut from various portions of the rotor was measured, and three rotors were subjected to cold spin burst testing. Time dependent strength of the rotor at high temperatures was estimated by static fatigue rupture data of tensile test pieces at temperatures up to 1400℃ for up to 1000hrs. The allowable stress at an arbitrary temperature and time to failure were explained well in terms of the Larson-Miller parameter. High temperature durability was also evaluated by hot-gas spin tests. The results of these tests suggest the possibility of fabricating reliable ceramic components which might enable the realization of ceramic gas turbine engines.

INTRODUCTION

Mitsubishi Motors Corporation (MMC), Mitsubishi Heavy Industries Ltd. (MHI), and NGK Insulators Ltd. have been conducting joint studies since 1979, aiming to develop radial-type ceramic turbine rotors for use under harsh conditions, as these will be the most important component in automotive-use ceramic gas turbines. [1]-[5] NGK has been responsible for the development and evaluation of ceramic materials and fabrication of prototype ceramic rotors. Design of ceramic turbine rotors and evaluation in terms of hot-gas spin tests were carried out by MMC and MHI. A heat resistant silicon nitride material(SN-88) whose flexural strength exceeds 700MPa up to 1400℃ has been developed. Fabrication of the material into ceramic rotors has been successfully carried out, and the rotor has been evaluated.

In this paper, the results of the evaluation of the mechanical properties of this material and results of evaluation of the ceramic rotor are reported. Furthermore, the possibility of developing a ceramic gas turbine (CGT) for automobile use is discussed.

TURBINE ROTOR MATERIAL AND FABRICATION PROCESS

Material properties of SN-88

Many ceramic manufacturers have made efforts to improve the heat resistance of silicon nitride by selecting chemical composition of sintering aids and high quality powder as well as by improving the processing of materials and sintering procedure. The newly development material SN-88 is a high-strength material whose four point flexural strength exceeds 700MPa and which has good oxidation resistance at high temperatures comparable to that of silicon carbide. Table 1 shows the material properties of SN-88. Figure 1 shows the Weibull plot of flexural strength at room temperature. The Weibull modulus of the two parameter Weibull distribution function is approximately 20, and SN-88 is a material with small scattering in strength.

TABLE 1
Properties of turbine rotor material(SN-88)

Material code No.			SN-88
Density		g/cc	3.5
Flexural Strength(4-point)	(RT)	MPa	790
	(1000℃)	MPa	770
	(1200℃)	MPa	770
	(1400℃)	MPa	760
Young's Modulus	(RT)	GPa	300
Poisson's Ratio	(RT)		0.26
Fracture Toughness, K_{IC}	(RT)	$MN/m^{3/2}$	7
Thermal Expansion Coefficient	(40-1000 ℃)	$\times 10^{-6}$/℃	3.4
Oxidation Resistance	(1000℃, 1000h)	mg/cm^2	<0.1
	(1200℃, 1000h)	mg/cm^2	0.3
	(1400℃, 1000h)	mg/cm^2	0.5

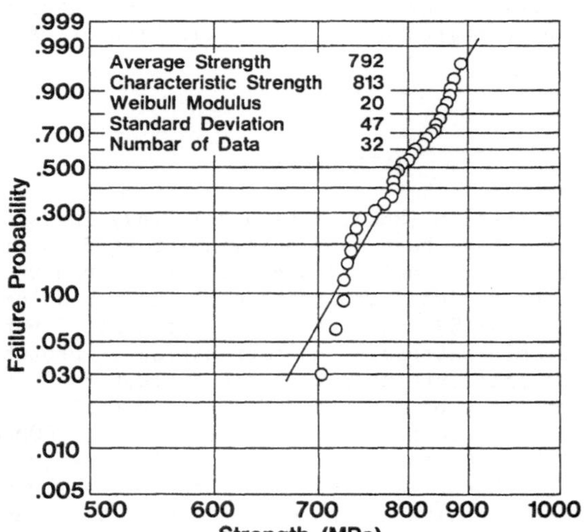

Figure 1. Four-point flexural strength distribution of SN-88 at room temperature

Figure 2 shows static fatigue characteristics of SN-88 under tensile stress. Button head-type specimens of 6mm diameter and 30mm effective length were used to perform the static fatigue tests.[6] The temperature was set between 1200℃ and 1400℃, and up to 1000hr.

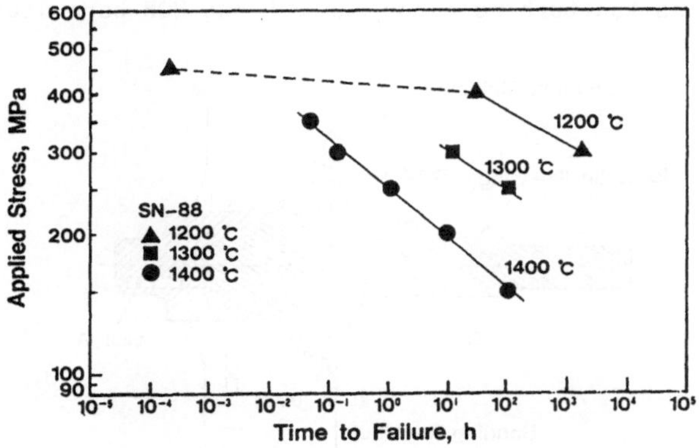

Figure 2. Results of static fatigue test under tensile stress

Fabrication process

Figure 3 shows the fabrication process. First, the silicon nitride raw material was mixed with sintering aids. The blade section was injection molded and machined for connection with the shaft after dewaxing. In the meantime, the shaft section was produced by pressing first, and machining was carried out for connection in the same manner as for the blade section. The blade and the shaft were combined into one, and then sintering was carried out to obtain a sintered body consisting of a unitary radial turbine rotor. The sintered materials thus produced were further machined, their balance was corrected, and nondestructive tests were performed.

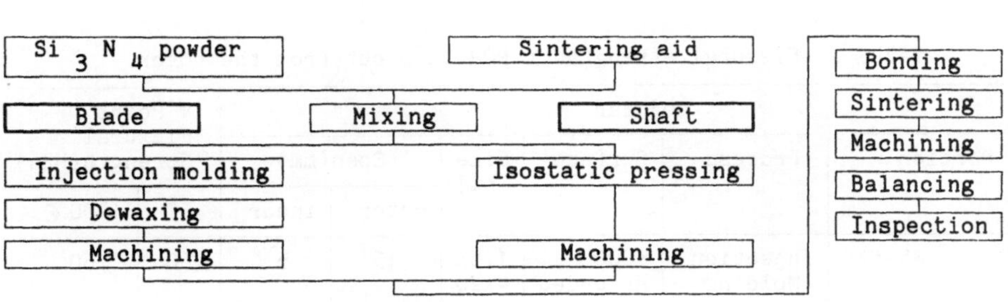

Figure 3. Fabrication process of the ceramic turbine rotor

RESULTS

Flexural strength of MOR bars cut from the turbine rotor

In the evaluation of the turbine rotor, MOR(Modulus of Rupture) bars were cut out and the flexural strength was measured. Figure 4 shows the sections for the cutout. MOR bars were cut out from shaft A, shaft B and shaft C section, as (1) a specimen of the shaft segment by injection molding, (2)

a specimen produced by pressing and (3) a specimen of the junction section between injection-molded and press-formed sections, respectively. In the measurement of the as fired surface strength of the blades, the curvature of the blade of an actual turbine rotor was too small to cut out, so we fabricated a plate 7mm thick by injection molding which was fabricated in the same manner as the turbine rotor and cut out MOR bars with as fired surface.

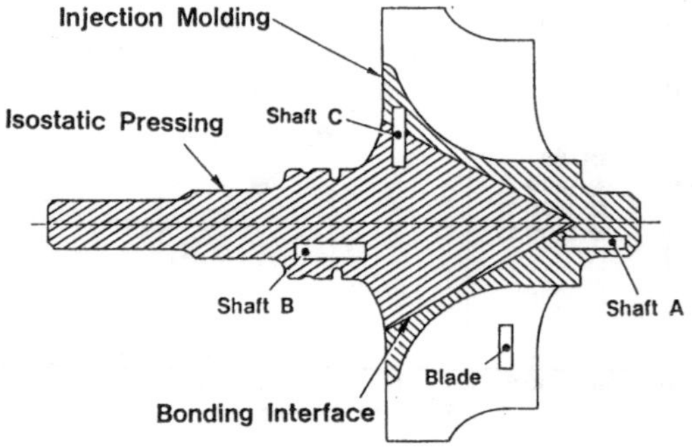

Figure 4. MOR bars cut portions

The results are shown in the TABLE 2. For the purpose of comparison, we also show the strength of the standard test pieces which were obtained from a plate produced by die pressing (the last row in the TABLE 2). Since the results are for MOR bars of different sizes, no simple comparison can be made; nonetheless it was found that the MOR bars cut out of a turbine rotor had a flexural strength approximately 10% less than those of standard test pieces. From the above results, it was confirmed that even with the turbine rotors of complex shapes described above, strength essentially comparable to that of the MOR bars which were produced by pressing from a simple plate. Also, since the strength at high temperature is identical to that at room temperature, it

TABLE 2. Flexural strength of MOR bars cut from the rotor

MOR Bar							4-Point Flexural Strength(MPa)	
Portion		Process	Surface	Size (mm)	Span(mm)			
					outer	inner	RT	1400℃
Turbine Rotor	Shaft A	Injection Molding	#800 grounded	1.5x 4x20	15	5	760	760
	Shaft B	Press	#800 grounded	1.5x 4x20	15	5	850	760
	Shaft C	Injection Molding +Press	#800 grounded	1.5x 4x20	15	5	780	790
Plate (Blade)		Injection Molding	as fired	3x4 x40	30	10	710	720
Plate (Reference)		Press	#800 grounded	3x4 x40	30	10	790	760

was concluded that problems such as the contamination of foreign material, which may result in the deterioration of high-temperature strength, did not occur due to the difference in the fabrication processes.

Cold spin test

A burst test using cold spinning is effective in evaluating the strength of the turbine rotor, which is a rotating part. Three Rotors were evaluated in terms of the burst test. The cold spin test was performed in a vacuum. Two of the turbine rotors burst at 777m/s of the turbine tip speed, and one of the turbine rotor burst at 751m/s. From the results of analysis of stress distribution of turbine rotors, the maximum stress applied was estimated to be 557 and 519MPa respectively. The blade of the rotor for cold spin test was larger than that of hot-gas spin test rotor to generate higher stress.

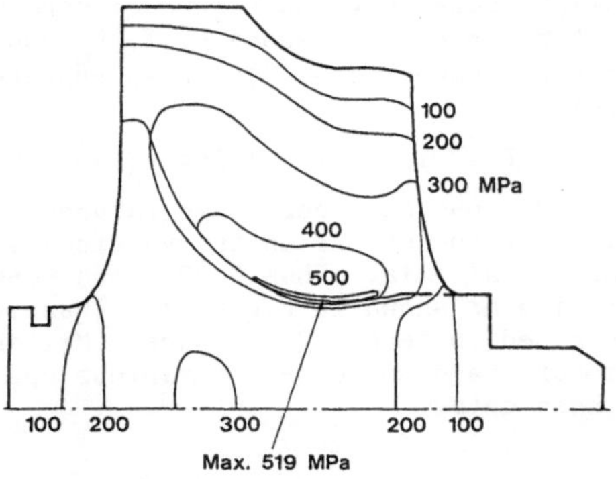

Figure 5. Stress distribution in turbine rotor
Turbine tip speed: 750 m/s

Hot-Gas Spin Test [3]

The rig was used at turbine inlet gas temperature up to 1450℃. The hot section is double constructed of heat resistant alloy with air cooling its outer section before the air enters the combustor. The test facility has

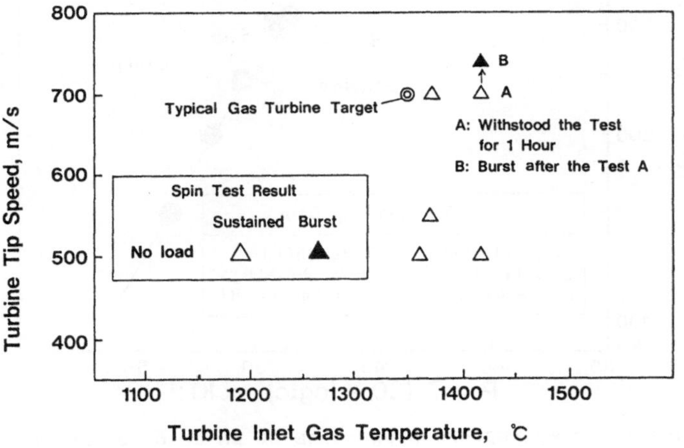

Figure 6. Hot-Gas Spin Test Results of SN-88 Turbine Rotor

been steadily improved in heat resistance, safety and accuracy of measurement. All of the hot-gas spin tests in this study were conducted under the very severe temperature conditions, no-load conditions. Figure 6. shows the results of the hot-gas spin tests of rotors, and the turbine rotor withstood the hot-gas spin test under no-load for one hour with TIT at 1410℃ a turbine tip speed at 700 m/s. After this test, turbine tip speed was increased to 742m/s, then the rotor burst.

DISCUSSION

Creep rupture properties described using Larson-Miller parameter[7]
In order to estimate the allowable stress at an arbitrary temperature and time, the static fatigue data shown in Figure 2. were described by the Larson-Miller parameter, in eq.(1), where P, T, t_c and c represent the Larson-Miller parameter, temperature (K), creep rupture time (h) and a constant, respectively.

$$P = T \times (c + \log t_c) \times 10^{-3} \tag{1}$$

Using fatigue data at various stresses, temperatures and rupture times, we assigned the logarithm of the stress on the vertical axis and Larson-Miller parameter on the horizontal axis. When C=30, the tensile creep data were plotted on a linear line as shown in Figure 7. Thus it was proven that the data can be described in terms of the Larson-Miller parameter. Estimation of the lifetime of the components for various operating conditions can be performed with these data.

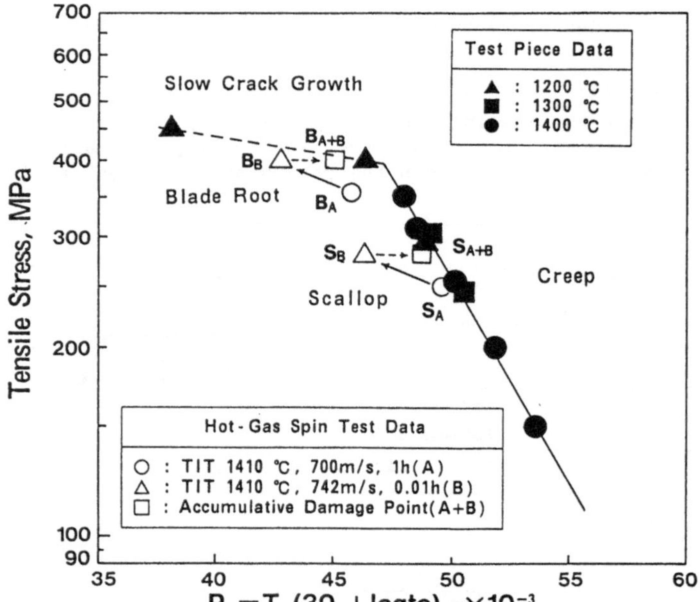

Figure 7. Creep rupture properties of SN-88 described using Larson-Miller parameter

Estimation of accumulated damage rate due to static fatigue

Here, the lifetime of the turbine material under more than two different stress and temperature conditions are estimated. As described before, the lifetime of a material under conditions of (σ_i, T_i) can be expressed as t_{ci} in terms of the Larson-Miller parameter. From the life fraction rule we can define the damage rate of a material after a test of duration t_i ($t_i < t_{ci}$) as t_i/t_{ci}, and in terms of the linear cumulative damage rule, the accumulated damage D can be expressed by eq.(2).[8]

$$D = \sum_i (t_i / t_{ci}) \tag{2}$$

Relationship between hot-gas spin test results and static fatigue data

Figure 8 shows the temperature distribution under no-load conditions for the turbine inlet temperature 1400℃ and turbine tip speed 700m/s.[9] The turbine's maximum stress of 355MPa was found near the blade root of the rotor, and the temperature was 1250℃. In the turbine rotor subjected to a one-hour endurance test under the conditions, A, of hot-gas spin tests (TIT 1410℃, 700m/s and no-load), the Larson-Miller parameter near the blade root of the turbine was estimated to be P=46 through the use of eq. (1), and the point B_A in Figure 7 was obtained. The point B_A is located below the allowable stress; thus, the rotor did not break. However, during the test under condition B, the rotor tip speed was gradually increased to 742m/s at TIT 1410℃. The stress generated and temperature under condition B were estimated to be 400MPa and 1250℃. When the holding time was assumed to be approximately 0.01 hr, the Larson-Miller parameter of the blade root area was P=43, and point B_B was plotted. By the linear cumulative rule, B_{A+B} can be obtained. The new point is very close to the tolerance level determined by the slow crack growth.

The temperature of the scallop section is very high, 1370℃, and the stress at the section is relatively high, 250MPa. When the Larson-Miller parameters are estimated in the same manner as for the blade root of the turbine. S_A , S_B and S_{A+B} can be plotted in Figure 7. The point S_{A+B} is also close to the tolerance level determined by creep.

From the above discussion, it can be assumed that the turbine rotor failure at hot-gas spin test due to the slow crack growth at the blade root or creep at the scallop of the rotor.

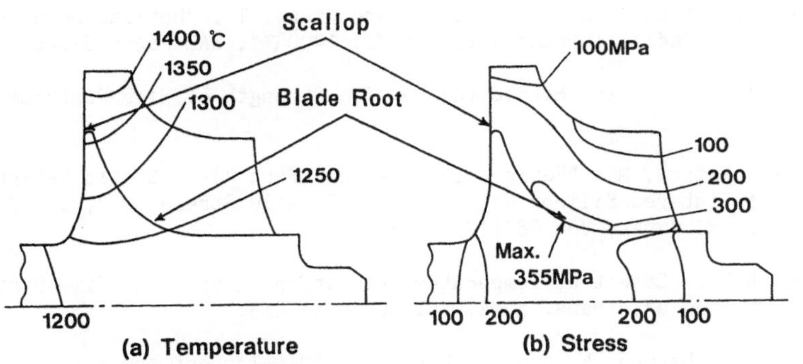

Figure 8. Temperature and stress distribution of SN-88 turbine rotor at TIT 1400℃, turbine tip speed 700m/s and no-load

CONCLUSIONS

The following conclusions were reached in this study.

1. A new heat-resistant Si_3N_4 has been developed and prototype fabrication process for radial turbine rotor has been established.
2. The strength of test pieces cut from the rotor was about 10% lower than its inherent strength.
3. High-temperature tensile static fatigue test data on results could be described using Larson-Miller parameter.
4. The rotor endured for one-hour hot-gas spin test under TIT of 1410 ℃, turbine tip speed of 700 m/s and no-load conditions.

The above test results suggest that it may be possible to fabricate highly reliable ceramic components with which to realize ceramic gas turbine engines.

ACKNOWLEDGEMENT

The authors extend their gratitude to the following groups for their valuable help: Ceramic Gas Turbine Group of Truck and Bus Engineering Center of Mitsubishi Motors Corporation, as well as Mr. Toharu Inagaki at Nagoya Aerospace Systems Works of Mitsubishi Heavy Industries, Ltd.

REFERENCES

1. Sato,H., Miyauchi,J. and Iwasaki,K., "Truck Turbine Engine Development at Mitsubishi", 83-TOKYO-IGTC-88(1983),683.

2. Miyauchi,J. and Kobayashi,Y., "Development of Silicon Nitride Turbine Rotors", SAE Paper, 850313(1985),31.

3. Kobayashi, Y., Matsuo, E. and Kato, K., "Hot-Gas Spin Testing of Ceramic Radial Turbine Rotor at TIT around 1250 ℃", SAE Paper 880727(1988-3).

4. Ozawa, T., Matsuhisa, T., Kobayashi, Y., Matsuo, E. and Inagaki, T., "Hot-Gas Spin Testing of Ceramic Turbine Rotor at TIT 1300 ℃", SAE Paper 890427(1989-3).

5. Kobayashi, Y.; Matsuo, E., Inagaki, T. and Ozawa, T., "Hot-Gas Spin Testing of Ceramic Radial Turbine Rotor at TIT 1400℃", SAE Paper 910401(1991-3).

6. JIS R1606-1990, "Testing Method for Tensile Strength of High Performance Ceramics".

7. Masuda, M., Matsui, M., "Fatigue in Ceramics (Part 4) — Static Fatigue Behavior of Sintered Silicon Nitride under Tensile Stress — ", J. of the Ceramic Soc. of Japan, Vol.98 (1990),83.

8. Robinson, E.L., "Effect of Temperature Variation on the Long-Time Rupture Strength of Steels", Trans. ASME, Vol.60(1938),253.

9. Kobayashi, Y., Matsuo, E. and Watanabe, K., "Development and Hot-Gas Spin Testing of Silicon Nitride Radial Turbine rotors",91-YOKOHAMA-IGTC(1991), (to be published).

SCIENTIFIC ASPECTS OF THE DEVELOPMENT AND THE APPLICATION OF CERAMIC MATERIALS AND STRUCTURAL COMPONENTS

ROMASHIN A. G.
Research and Production Firm "Technology",
Obninsk, USSR

Ever increasing attention to structural ceramics and their practical application can be attributed to the possibility of obtaining basically new technico-economical characteristics. However, ceramics as structural materials have been studied very little. While extensive and in-depth study in the field of design and manufacture of metallic structures has been carried out and hundreds of thousands of scientific papers have been published on the subject, investigations of load bearing ceramic structures are at the very beginning. At present designers lack practical experience in designing ceramic load bearing structures and basic data are often missing for reliable performance analysis. This is due to the fact that ceramics have been used as a refractory or building material for thousands of years and all research work has been done with that end in view. A specific feature of ceramics as a structural material lies in the fact that the ceramic components have to be made at once as a whole. So far it has been practically impossible to produce them by welding or deformation of a semi-finished piece. They cannot be joined by bolts, screws or rivets etc, which is a common practice when assembling metallic structures.

There exists a rigid relationship between the structure of a ceramic component and its technology as well as the dependence on material properties of the component produced from its structure and technology. For instance, real strength of a structure can be reduced markedly in the course of moulding and firing at the expense of the residual stresses and defects brought about by an incorrect design that neglects features specific for ceramics. With allowance for the above stated facts an integral approach should be realized when designing and producing critical ceramic components and articles. The essence of this approach resides in the fact that there must be a continuous interrelation in the stages of the development of material, structure and technology, i.e. a structure must be designed for the given technology and material and vice versa material and technology must be chosen for the production of the given structure.

Our 20 years of experience in the development of high density load bearing structures from brittle materials such as ceramics and glass has conclusively proved this integral approach to hold good. Designers, process and testing engineers and investigators must work together guided by one design concept. Ceramic structures ask for entirely different design solutions, i.e. a metallic blade cannot be substituted by a ceramic one in a gas turbine engine. Both the blade and the whole assembly must be designed taking account the inherent brittleness of ceramics and thermal treatment conditions. Presently we have acquired considerable experience in design and production of load bearing structures.

Theoretical analysis and experimental work carried out have made it possible to develop the main scientific aspects of components and articles design with brittle materials and to lay down compulsory design principles. Some of them are considered below.

The first design principle is the principle of minimum concentrations. The essence is the fact that there must be no stress concentrators in a ceramic component, especially in a high loaded area. Actuality of the principle of minimum concentrations as applied to mechanical joints is illustrated in Figure 1, showing a ceramic component after thermal cycling. All the cracks developed across the holes. To be more exact these holes are sources of cracks initiation. The presence of stress concentrations typical of bolt joints in general, in the case of ceramic materials is aggravated by the absence of stress redistribution at the expense of plastic deformation of the material and by inevitable material damage (microcracks formation) when drilling holes. Our design development and numerous tests convincingly demonstrated that even deforming bushings, spacers springs etc. do not ensure normal operation of such joints under load. All kinds of notches, sharp shoulders and thickness changes are very unfavourable as they themselves are also stress concentrators. Large stress concentrations occur at ceramic-to-metal contact points. Therefore, joining surfaces must be very precisely adjusted to avoid point contacts; it has been found advantageous to use flexible spacers; modula of elasticity of the material of a component to be joined to a ceramic component should be as low as possible.

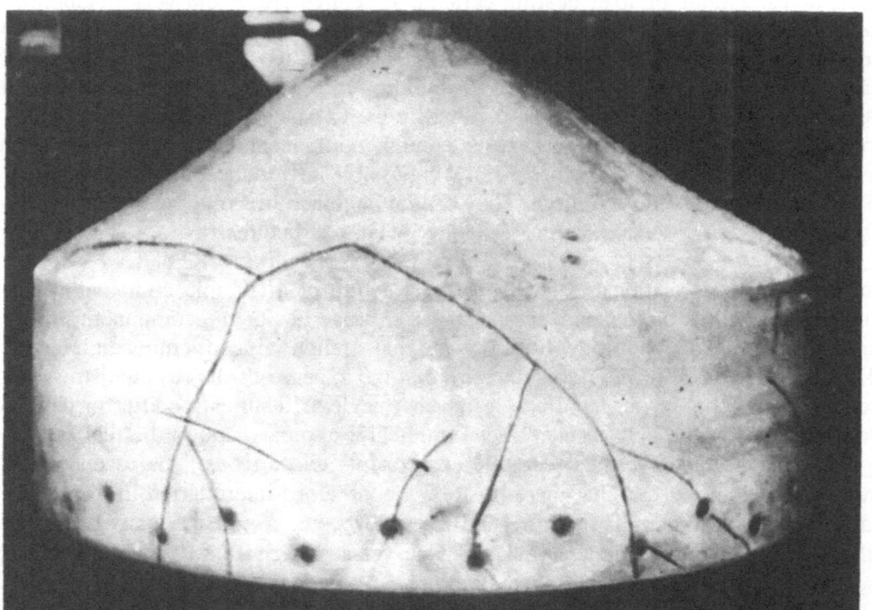

Figure 1. Bolt jointed ceramic shell construction.

The second principle is the principle of a free temperature deformation of a ceramic component. It requires the thermal expansion of all the metal-ceramic joints to approach the deformation of a ceramic component while in service.

Discrepancy between thermal deformation of a frame and a ceramic shell caused failure of the component shown in Figure 1. The principle of a free temperature deformation can be accomplished either by an appropriate selection of thermal coefficients of linear expansion of ceramic and metallic components or by the use of flexible compensating elements.

In doing so one must not leave out of account transitional operating conditions, engine start up and shut down, when nonuniform heating of different components takes place. The fact that the heat capacity of ceramics is twice as large as that of metal has a pronounced unfavourable effect on transitional operating conditions. It means that all other conditions being equal a metal-ceramic joint will be heated twice as fast as a ceramic one. As a consequence the thermal deformations they experience will be different and thermal stresses will be brought about.

The third principle is the principle of temperature uniformity of a ceramic component. The essence of it is the fact in that a ceramic component temperature field must be uniform as much as possible both round the periphery and along the generating line of the field. A violation of the principle results in additional thermal stresses decreasing a ceramic component performance. In Figure 2 examples are given of structures showing compliance with the principle of thermal uniformity and violation of this principle. In the case of the external heating of a cylindrical metal-ceramic component to be joined from an inner surface the heating will be uniform through the full length (Figure 2.1A).

In the case illustrated in Figure 2.1B, a part of a cylinder is thermally insulated by a metallic strap, and this causes a difference of average temperatures between the top and the bottom parts of a ceramic component and brings about thermal stresses. The same thing is observed when heating a plate or a bar with a very large change of thickness (Figure 2.2 and 2.3). For instance, it was required to produce a ceramic component with varying wall thickness along the height of a cone. The component manufactured in accordance with Figure 2.4B failed precisely along the cross-section where the thickness changed sharply, without any external load application, when externally heated.

Application of a structure combined from two cones (Figure 2.4A) with elimination of thermal deformations gave a component which performed well under heating and operation loads conditions. As applied to blades, figures 2.5A and 2.5B illustrate this principle. Inertia of a massive blade root is larger in comparison with a thin blade edge, and this inevitably causes high thermal stresses in the bottom cross-section of a blade edge. An optimum configuration is shown in Figure 2.5A.

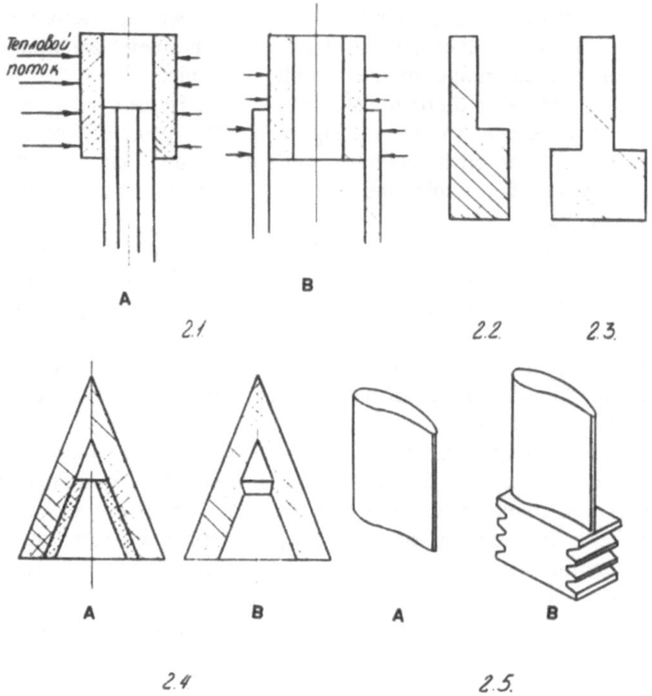

Figure 2. Ceramic elements made in compliance with (A) and in violation of the thermal uniformity principle (B).

The fourth principle is the principle of a controlled generation of stress formation in a ceramic component. The essence of this principle resides in the fact that the mode of load application and its distribution in a ceramic component must contribute to predominant formation of compression stresses. Actuality of the principle of controlled stress formation could be illustrated by an example taken from experience.

When designing a protective glass enclosure for a deep-water light source a construction chosen was a hollow cylinder clamped on ends by flanges. The flanges in their turn were bolted together. Thus it looked as if a glass cylinder was always experiencing compression stresses because it was compressed on ends by flanges and external pressure also compressed it along the cylinder surface. It turned out that high spalling stresses occurred on ends of a glass enclosure, the enclosure thickness being rather significant. In order to diminish the influence of these stresses the ends of the enclosure were made semicircular instead of flat. The flanges mounting surfaces were machined correspondingly. As a result of this modification the value of breakdown pressure increased to 880 atm compared to 240 atm in the case of flat ends. Theoretically an optimum end form in this case is a curve close to the ellipse.

The fifth principle is the principle of stress concentration distribution. The essence of this principle consists of a geometrical dispersal of stress concentrations attributed to ceramic component fasteners and end stress concentrators in particular on the ends of a ceramic component and in local damage areas due to machining.

The sixth principle is the principle of freedom of a ceramic component from stresses in an inoperative state or in long-duration storage. This means that there should be no residual stresses of the first or second order or constant stresses brought about during manufacture or assembly. This means that in this case ceramics are subjected to long-duration stresses and long-duration strength of ceramics is much less than short-duration strength.

Ceramic-to-metal joining by shrink fit, soldering or melt filling at elevated temperatures can serve as an example of manufacturing residual stresses. In these cases residual stresses occurring during storage are due to the difference in thermal deformation of ceramics and metal during cooling to ambient temperature.

Adherence to these principles can serve as a preliminary criterion for structure perfection. These principles have been worked out not only theoretically but also on the basis of many years experience in the development of load bearing structures from brittle materials. In favour of the significance of design experience speaks the fact that at the start more than 10 alternative designs were developed and tested before a serviceable modification was manufactured.

Design principles do not provide a ready design or a ceramic engine component in each particular case, but the analysis of a particular construction's conformity with these principles makes it possible to avoid obviously inapplicable designs and will put you on you guard in questionable cases.

Design of ceramic materials and technological developments started at our enterprise 20 years ago.

Presently 25 materials on the basis of silicon nitride, silicon carbide, boron nitride and zirconia have been developed and certificated. Each of these materials has both advantages and disadvantages. For instance, reaction bonded silicon carbide strength is comparatively low compared to other materials but the components made of it have almost zero porosity and a very stable shrinkage (0.99 ± 0.03 %). This may be of great significance for intricate components. Hot-pressed silicon nitride being attractive at first glance thanks to maximum strength is very difficult to make use of in a particular component, because the hot-pressing method imposes significant geometry and dimensions restrictions on the component being formed.

The problem of the determination of structural ceramics properties should be specially discussed. As long as there is a strong dependence of the properties of a ceramic material and the technological parameters a reliable estimation of a component's performance is only possible when based on the results of the property measurements of the samples of the semi-finished components or of prototypes manufactured from the same starting materials and by the same technology simultaneously with the component.

Our first attempts to design ceramic components using property data published in the Soviet and foreign journals resulted as a rule in a two or three-fold scatter of the final results. This can be explained by the fact that essentially the property measurements were carried out on the samples made from different materials and according to different methods. Abroad, other raw materials, technology, final products and measurement standards have been used compared to those in the USSR.

Naturally one cannot rely on such combined data when designing critical heat density structures, for instance gas turbine engines. This is the reason why we developed a unified set of techniques and experimental installations for reliable data provision. It allows us to get data on the properties of 27 materials over a broad range of temperatures.

Besides, there is a problem of correlating data on the material strength with operating loads experienced by ceramic components: for instance, a stator blade of a small-sized gas turbine engines undergoes dynamic loads up to 1 kg/mm^2. Tensile temperature stresses through the thickness of the blade do not exceed 5 kg/mm^2 at the operating temperature conditions, i.e. the main stresses experienced by a stator blade are due to the temperature stresses and additional stresses caused by its joining with the other components. Therefore the right choice of a structure is a decisive factor and the principles of thermal uniformity and free thermal deformations of a ceramic component are of prevailing importance.

Figures 3, 4, 5 and 6 show the bending strength as a function of temperature and production process. Material properties investigation is a necessary and important stage of a component development. The performance of the structure being developed is dependent on this stage. It makes it possible to shorten the time of the material synthesis and to optimize the production process for full-scale components. The exact choice of an experimental procedure for the determination of sample properties makes it possible to predict the performance of a component before its manufacture.

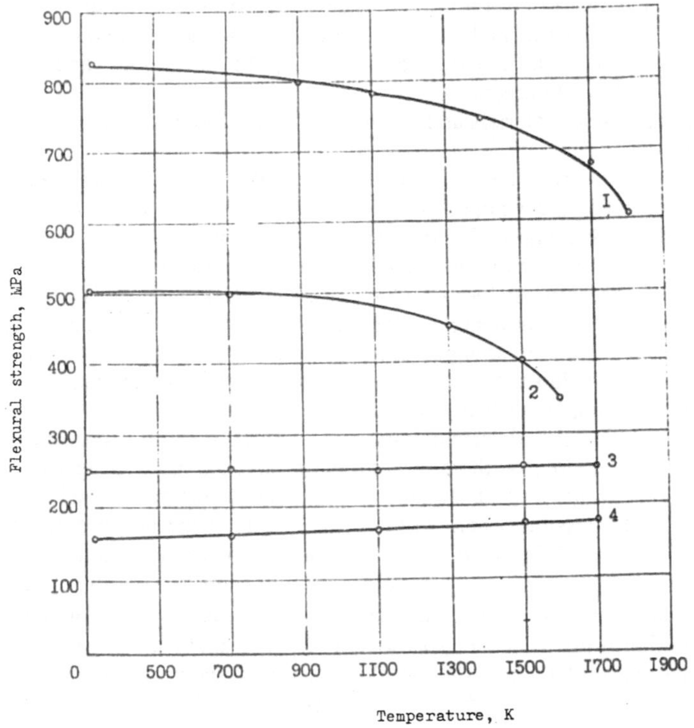

Figure 3. Strength vs. temperature of silicon nitride; 1) hot pressed; 2) densified reaction-bonded; 3) reaction-bonded; and 4) reaction-bonded with boron nitride additions.

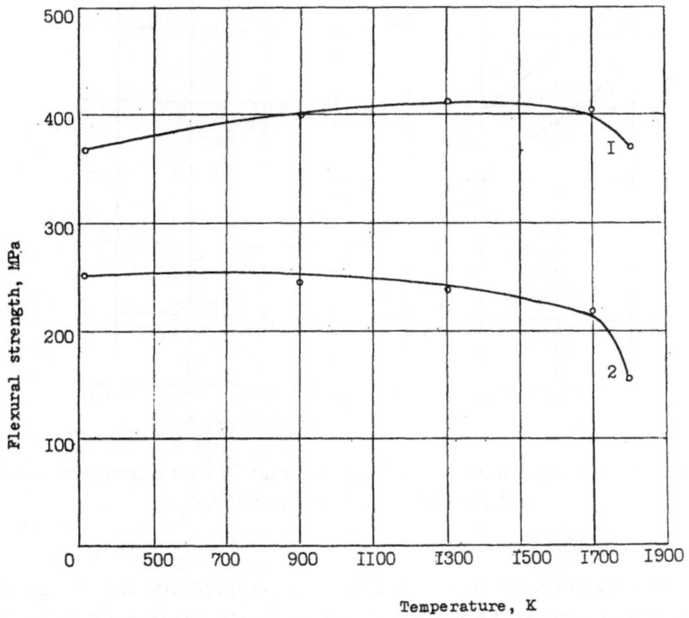

Figure 4. Strength vs. temperature of sintered silicon carbide; 1) with aluminium nitride additions; and 2) with silicon nitride additions.

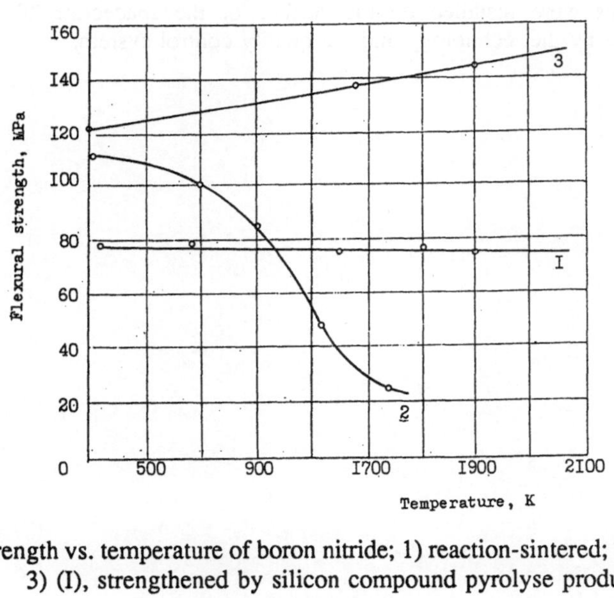

Figure 5. Strength vs. temperature of boron nitride; 1) reaction-sintered; 2) hot-pressed; and 3) (I), strengthened by silicon compound pyrolyse products.

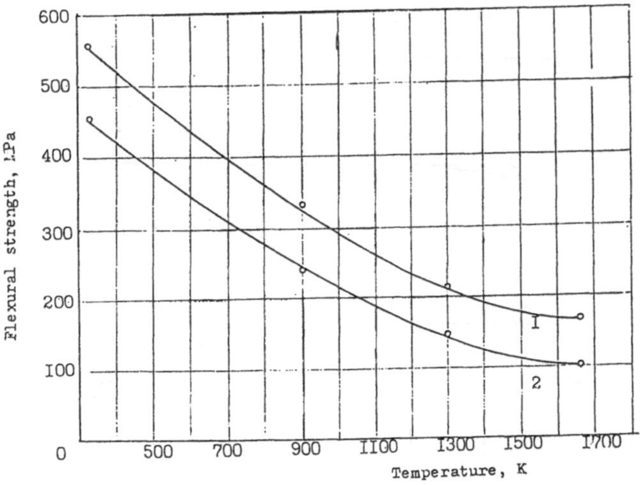

Figure 6. Strength vs. temperature of sintered zirconia; 1) with aluminium oxide additions; and 2) with cerium oxide additions.

There is an obvious shortage of exhaustive data required for the design of structures operating at long-duration working loads, cyclic loads and gas flow chemical attack. There is also a lack of data for serviceability calculations under creep at various temperatures and stresses. Application of the ceramic thermal protection tiles on the spacecraft "Buran" has proved the reliability of the ceramic material usage and its high reliability. More than 30,000 ceramic tiles were installed on the surface of the spacecraft "Buran". Their reliability is ensured by the technology and the quality control system.

EXPERIMENTAL INVESTIGATION OF CERAMIC MATERIALS AND TURBINE ROTOR COMPONENTS STRENGTH

NOZHNITSKY Y., SMIRNOV L., EGOROV S., MARKOV A., SAKOVICH V.
Central Institute of Aviation Motors (CIAM)
Moscow, USSR

STRENGTH TESTS

A successful use of ceramic monolithic and composite materials for engines is impossible without a careful evaluation of their structural strength, set of mechanical and thermo-physical characteristics for the whole working temperature range.

There are some complex problems connected with the necessity of heating a specimen up to 1400 - 2000°C and getting the required temperature field, development of clamps, tooling and measuring systems which are efficient under such high temperatures. Anisotropy and brittleness of ceramic and composite materials have influence on specimens shape and size, loading mechanisms and clamps construction. The number of specimens to be tested must be sufficiently large for subsequent probabilistically-statistical processing of received data.

Analysis of known standards, patents and experience of superalloys, ceramic and composite materials experimental investigations was performed. On this basis special equipment and methods for strength, and crack resistance experimental determination were developed. Tooling made of refractory alloys (tungsten, molybdenum), carbon-carbon and ceramic materials is used for tests under temperature conditions up to 1400 - 2000°C. Clamps made of carbon-carbon materials for tension and bending tests are shown in Figure 1.

When evaluating strength of ceramic and composite materials more often than not the bending tests results are used. However, for these tests it is necessary to choose a correct ratio of the span length between supports to specimen thickness in order to prevent shear influence. The relationship between this ratio and strength by three-point bending of specimens made of unidirectional carbon-carbon material (direction of reinforcement coincides with specimen longitudinal axis) is shown in Figure 2.

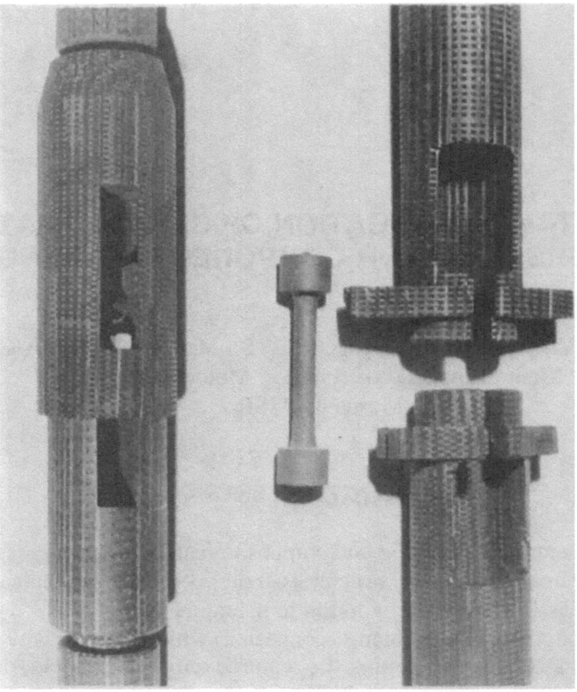

Figure 1. Clamps of carbon-carbon materials for high temperature bending and tension tests.

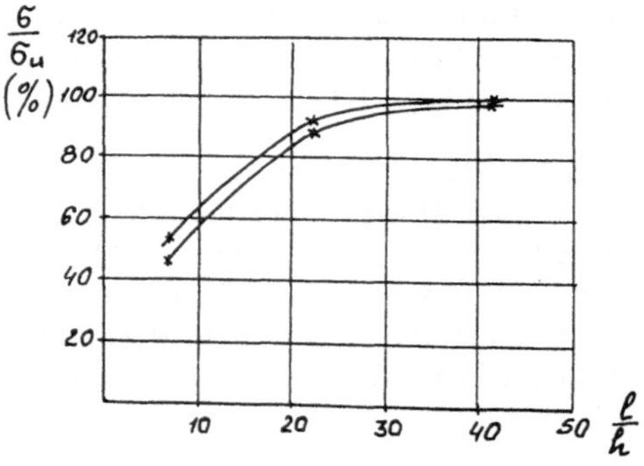

Figure 2. Relationship between strength by three-point bending of specimens made of unidirectional carbon-carbon materials and specimen span length to its thickness ratio.

Multiple clamps are used to increase the capacity of high temperature tests of specimens in resistor furnaces.

The high temperature dilatometer with carbon-carbon tooling (Figure 3) is used for tests under temperature conditions up to 2500°C.

The equipment and methods developed made it possible to determine the basic characteristics of a number of monolithic and composite ceramic materials and carbon-carbon materials.

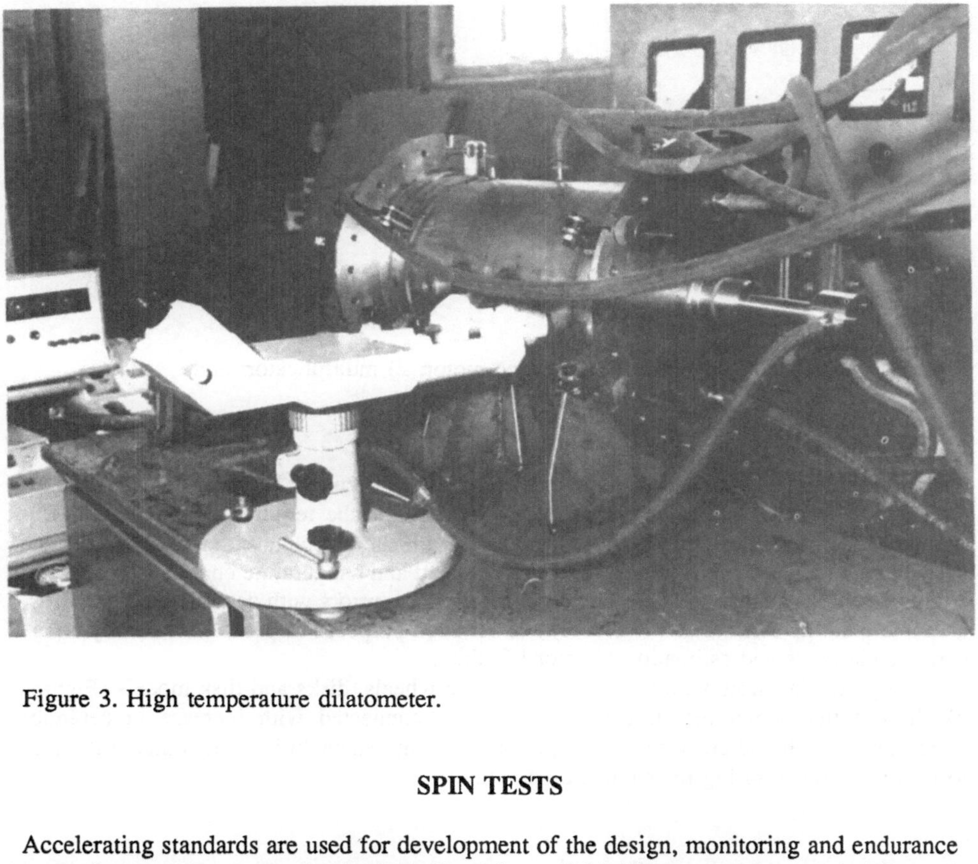

Figure 3. High temperature dilatometer.

SPIN TESTS

Accelerating standards are used for development of the design, monitoring and endurance tests of rotor components in the centrifugal-force field. The test equipment in Figure 4 consists of a DC electric motor, a multiplicator, an armour chamber of original design to contain rotor fragments, inner casing with special turbulators, automatized system of control, data accumulation and experimental data processing. It is used for experimental investigations of static strength, cyclic durability, dynamic characteristics as well as the resonant tests of non-metallic rotor components.

Figure 4. Accelerating equipment; 1) electric motor; 2) multiplicator; and 3) armour chamber.

Rotor spinning is accomplished in the chamber without vacuum. The heat, released while frictioning the revolving rotor with air is used to create the required temperature field. Excitation of harmonic vibrations is carried out with the aid of turbulators. The control system makes it possible to control the required speed and temperature change in time. The equipment is very much used for development work on rotors with ceramic elements (for example metallic disk - ceramic blades) and for such purposes as proof tests for rejecting ceramic elements and estimation of rotor life time.

In particular, tests with non-metallic turbine wheels, disks and disk models (Figure 5) showed that a number of complex problems connected with increase in ceramic materials strength and crack resistance performance and adaptability to manufacturing are to be solved for ensuring their reliable operation.

Figure 5. Ceramic turbine disk models.

Tests of ceramic blades and lock-on connection specimens made of different reaction bonded, sintered and hotpressed silicon carbide and nitride materials made it possible to improve the lock-on connection constructions. Hot pressed materials showed the most preferable results. Blades and specimens are shown in Figure 6. Use of massive specimens made it possible to get the required centrifugal-force loadings when revolutions were lower than in operating conditions. During these experiments the contact area between blade dovetail and disk was also decreased. Decreasing of the contact area by means of a two-layer interlayer between disk and dovetail (with shortened internal layer) is shown in Figure 6. The mechanism of blade failures is shown in Figure 7. A turbine wheel with ceramic blades is shown in Figure 8.

Figure 6. Ceramic blades and specimens; blade with interlayers which decrease the contact area.

Figure 7. Mechanism of blade failures. (In B the contact between disk and dovetail was bad.).

Figure 8. Turbine wheel with ceramic blade.

FATIGUE TESTS

Determination of fatigue life time limits for components made of non-metallic materials is a specific problem. The limits for non-metallic blades and other design components under vibrations of complex mode at frequencies up to 20 kHz can be defined using pneumatic vibrators, when resonance vibrations are excited by a modulated air jet. Such an equipment is shown in Figure 9. A rotating disk with holes or slots is usually used for mechanical modulation of the air flow. The fundamental harmonic frequency of a pulsating air jet is defined by the rotation speed of the modulating disk and by the number of holes in it. The required level of excitement is ensured by controlling the air pressure, by changing the position of the spot of contact with the pulsating air flow on the component, blade angle relatively to air flow, the distance between blade and nozzle exit section as well as by air nozzle selection. Smooth controlling of air pressure and the displacement of vibrator relatively the blade along three mutually-perpendicular directions are carried out with aid of remote-controlled electromechanisms.

Figure 9. Air vibro-equipment.

Beforehand the vibration modes are determined by holographic interferometry. Holographic interferograms of several vibration modes for a carbon-carbon blade are shown in Figure 10.

Gradual accumulation of defects and material heating up, leading to the change of both resonant frequency and amplitude of deformation, take place in the process of fatigue tests of composite components. Special autoregulators should be used to keep-up the given deformation amplitude.

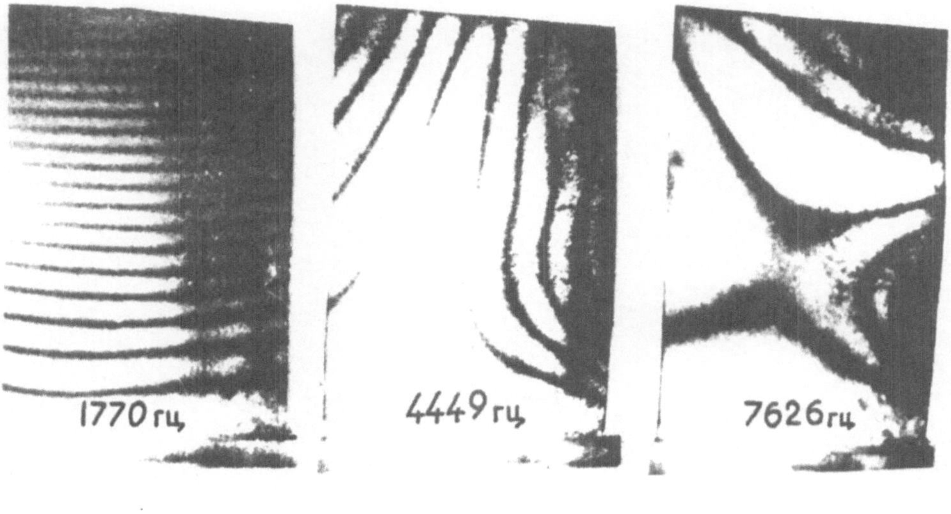

Figure 10. Interferograms of several vibration modes for a carbon-carbon blade.

ACKNOWLEDGEMENTS

We acknowledge the help of V. Seleznev for conducting tests using holographic interferometry methods, A. Menshikov for participation in experimental equipment development, K. Popov, E. Kousnetsov and T. Karimbaev for joint development of ceramic blade tests.

DESIGNING GAS TURBINE CERAMIC ELEMENTS

NAVROTSKY V., NOZHNITSKY Y., SHEKHTMAN Y., BOUTOURLINOVA N.,
FEDINA Y., CHYIASTON E.
Central Institute of Aviation Motors (CIAM)
Moscow, USSR

The designing of ceramic elements for a gas turbine includes analysis of thermostress conditions for stationary and nonstationary regimes, probability-statistical strength and reliability evaluation, life time estimation and choice of proof test conditions.

The necessity of excluding high local stresses while designing is due to the absence of plastic deformation and possibility of loading redistribution in ceramic materials. Examples from the design work on a ceramic nozzle and a turbine blade are presented in this article.

Traditionally used programs for calculation of thermostress conditions by finite elements methods and programs for calculation of failure probability and life time according to the Weibull approach and Paris model were used for determination of ceramic elements strengths and reliability.

A three-dimensional calculation scheme of the ceramic nozzle is shown in Figure 1. A typical change of the trailing edge nozzle region temperature is shown in Figure 2. Main maximum stresses are shown in Figure 3.

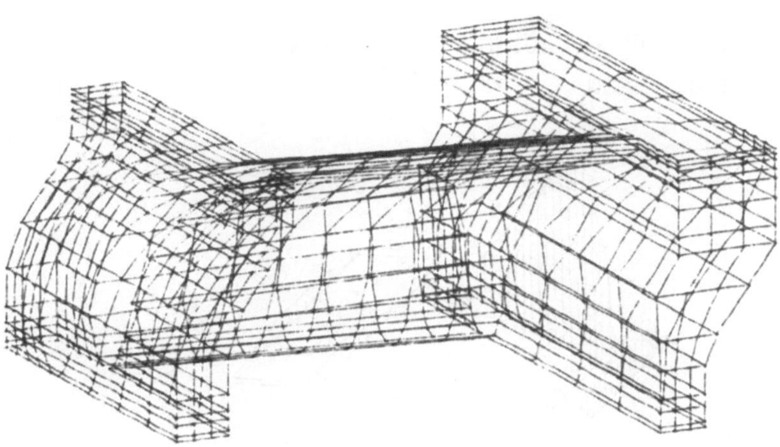

Figure 1. Three-dimensional calculation scheme of the ceramic nozzle.

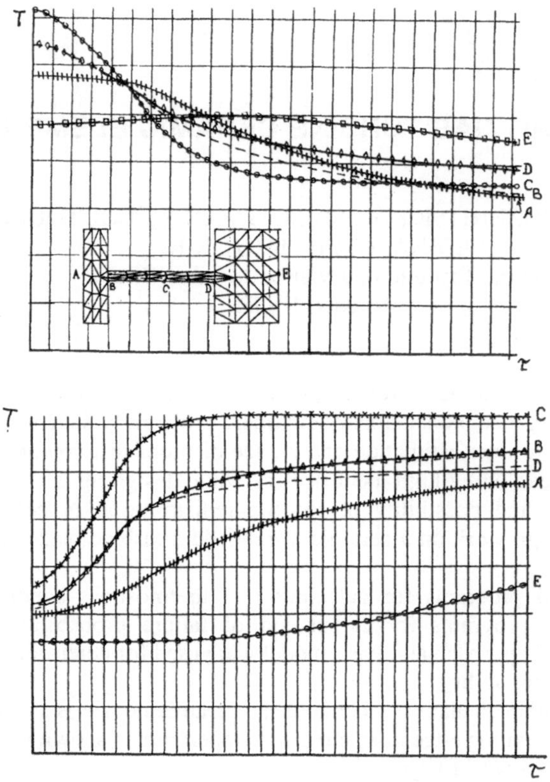

Figure 2. Typical change of trailing edge nozzle region temperature.

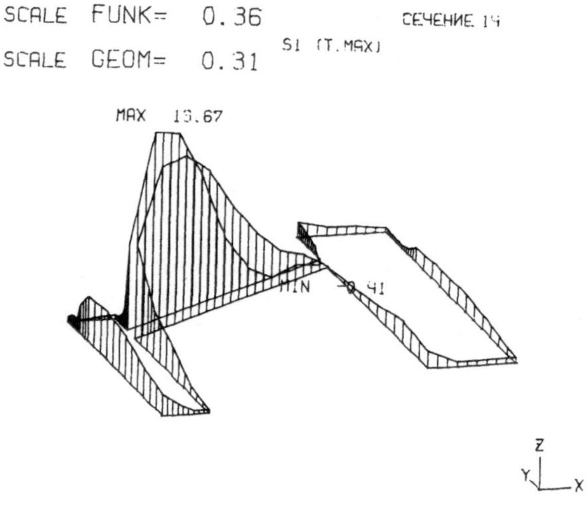

Figure 3. Main maximum stresses in the nozzle.

A three-dimensional analysis of nozzle thermostress conditions shows that inner and outer platforms bending, due to significant thermogradients observed along their thickness, causes high tension stresses at the trailing edge. Experiments showed that these stresses may be the reason for the vane failure. In order to decrease these stresses it is necessary to decrease the stiffness of the platforms in radial direction and (or) to decrease the thermogradients in the platforms. Abnormally high stresses during the non-coordination operation of components of the assembly (large contact stresses or stresses due to inadmissible clearances) may be another reason for the ceramic stator components failure.

Requirements for materials for a full ceramic wheel are essentially higher than for ceramic materials of gas turbine engine blades.

Optimisation of disk rim cooling takes an important part in the design of the turbine wheel with ceramic blades. Rational designing of the ceramic blade makes it possible to avoid significant stress concentrations. Increasing the blade dovetail height is necessary to improve the working conditions in case of irregular loading caused by the possible difference between dovetail and disk slot wedge angles (Figure 4).

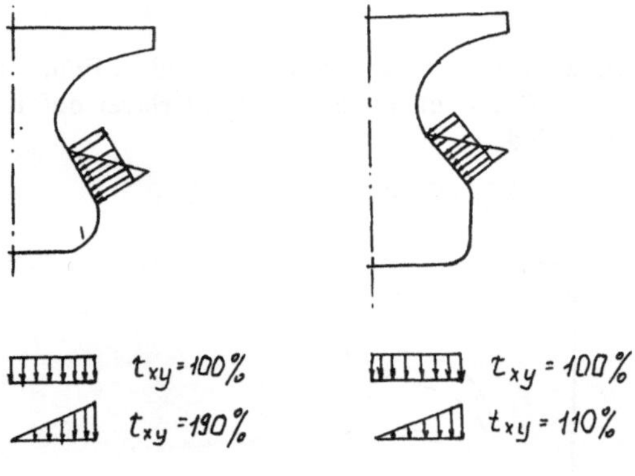

Figure 4. The dovetail height influence on working conditions.

As experiments showed when designing the ceramic blade, one of the most important problems is the prevention of the dovetail failure under high local contact stresses. Use of special deformed interlayers made of a metallic foil or a net is one way to prevent such failures. These interlayers are very efficient to decrease the contact stresses between the ceramic blade and the metallic disk. A finite elements scheme of the lock-on connection is shown in Figure 5.

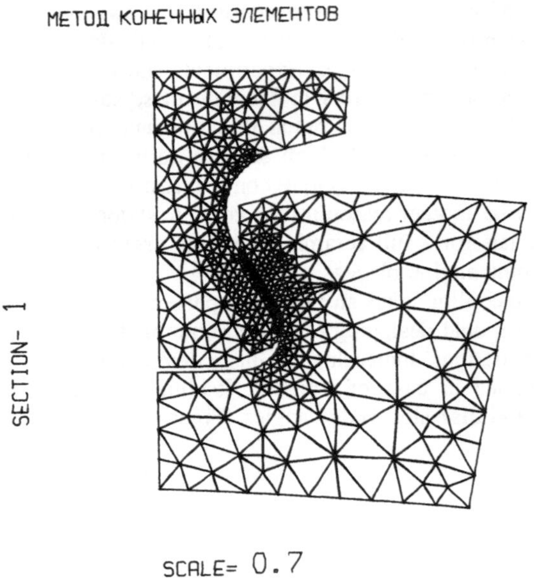

Figure 5. Finite elements scheme of the lock-on connection.

Original and deformed states of construction are combined in Figure 6. The compressed stresses are shown in Figure 7. The character of the interlayer deformation is shown in Figure 8. Its edges are crumpled.

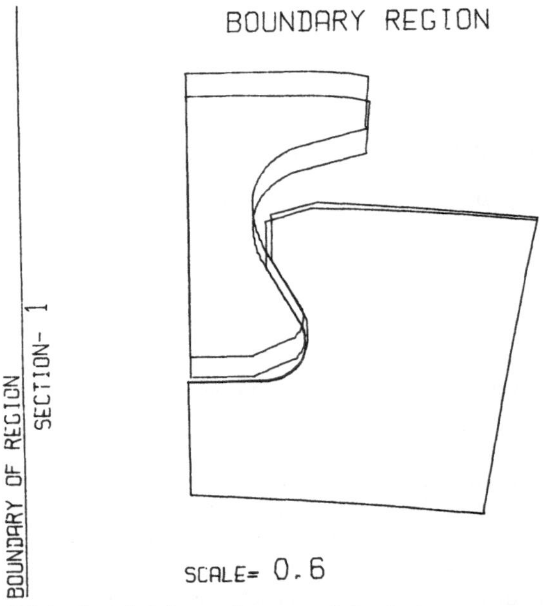

Figure 6. Original and deformed states of the lock-on connection.

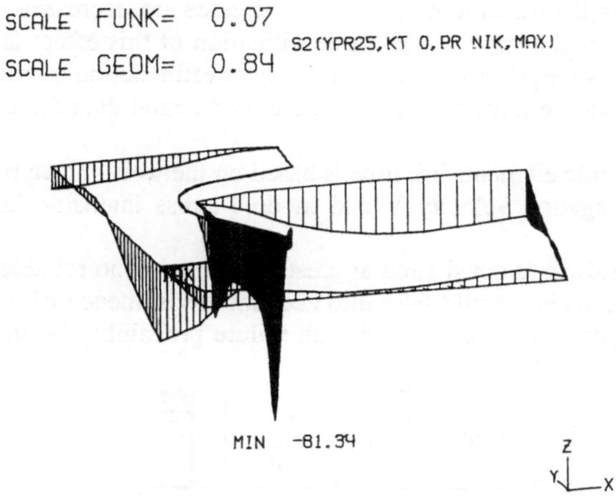

Figure 7. Compressed stresses.

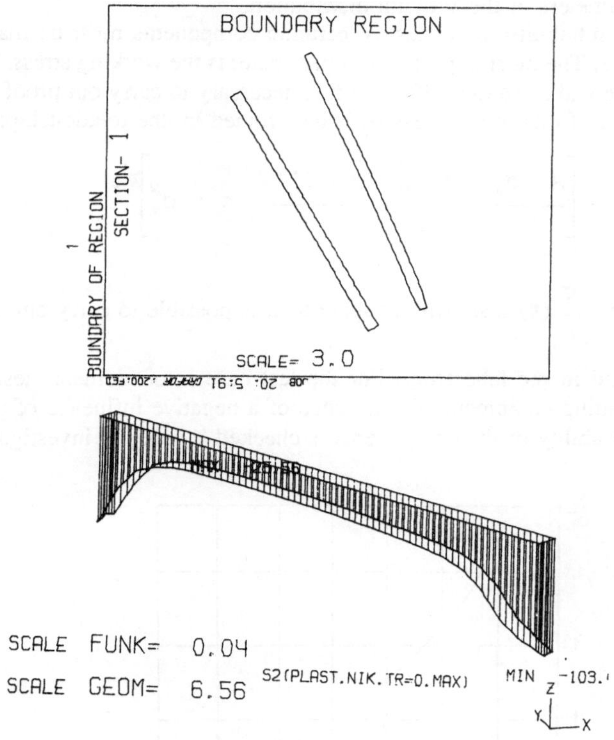

Figure 8. Character of interlayer deformation and compressed stresses in the interlayer.

Thanks to the interlayers the maximum compressed stresses are decreased more than 1.5 times. Taking into account the friction some intensification of this effect is obtained.

Use of a multiaxial strength criterion is necessary to estimate the interlayer influence on failure probability. But we have not yet got the experimental data for a well founded criterion choice.

Calculation of ceramic elements life time is based on the well-known relationship [1] between the crack propagation velocity V and mode I stress intensity factor K_I : $V = A * K_I^N$.

Only slow crack growth is considered at present as there is no reliable model of the accumulation of damage which would take into account the influence of other factors.

Life time τ under stress σ_a is connected with failure probability by the relationship:

$$\tau = \frac{2\,\sigma_o^{N-2}\left[\ln\left(\frac{1}{1-P}\right)\right]^{\frac{N-2}{m}}}{A \cdot (N-2) \cdot \sigma_a^N \cdot Y^2 \cdot K_{IC}^{N-2}}$$

where, σ_o, m are parameters of the Weibull distribution.

To increase the reliability, proof test of ceramic components must be made before mounting in the engine. The most important test parameter is the working stress. To ensure a reliable performance and a required life time it is necessary to carry out proof tests with overloading. The level of this proof stress σ_p is determined by the relationship:

$$\sigma_p = \left[\frac{A \cdot \sigma_a^2 \cdot Y^2 \cdot (N-2) \cdot K_{IC}^{N-2}}{2} \cdot \tau \cdot \sigma_a^N\right]^{\frac{1}{N-2}}$$

A typical dependence $\frac{\sigma_p}{\sigma_a}(\tau)$ is shown in Figure 9. It is possible to carry out proof tests both in the engine and in the laboratory. For the last case rotor elements tests may be conducted on accelerating equipment. The absence of a negative influence of proof tests upon strength and durability of the components is checked by special investigations.

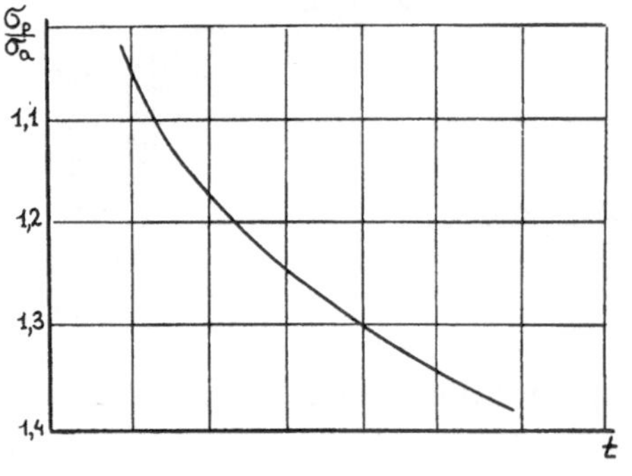

Figure 9. Typical dependence between the proof test loading and life time.

ACKNOWLEDGEMENTS

We acknowledge the help of E. Kouznetsov for participation in analysis of ceramic components temperature conditions, V. Shirmanov, N. Zoudin and Y. Kelin with their collaborators for discussions of obtained results.

REFERENCES

1. N. Katyama, M. Sasaki, T. T. Iton," Development of Ceramic Turbine Rotors", ASME 88-GT-282.

APPLICATIONS OF CERAMICS FOR THE ROTARY ENGINE

NOBUO SAKATE, TSUTOMU SHIMIZU, AKIHIDE TAKAMI
Technical Research Center, Mazda Motor Corporation
3-1, Shinchi, Fuchu-cho, Aki-gun, Hiroshima, Japan

ABSTRACT

To meet ever-stringent requirements for higher power and lower lubrication oil consumption on rotary engines, the wear resistance of apex seals and housing surfaces is a critical subject. The authors have been attempting to apply ceramic-based materials to these parts. Through the study on the material composition and production processes, the authors developed a ceramic material with high strength, high toughness and high wear resistance for the apex seals, which is silicon carbide whisker reinforced silicon nitride. As a result of the rig tests, the detonation gun spray of Cr_3C_2/Ni-Cr was found appropriate for the coating of side housings. In consideration of its potential productivity, the spray method was changed to the plasma spray method. The plasma sprayed Cr_3C_2/Ni-Cr showed the wear resistance equivalent to that of the detonation gun sprayed coating as a result of the optimization of material powder granulation and the spray conditions.

INTRODUCTION

Because of ever-increasing requirements for high power, high speed, fuel economy, and low pollution capabilities on engines, their internal parts are used under increasingly severe conditions. In the case of rotary engines, a subject of particular importance is the improvement in the wear resistance of the apex seals and the housing surfaces. The apex seal runs sliding over the high-temperature trochoidal internal surface of the rotor housing at a high speed. The wear problems under these stringent conditions are becoming unsolvable with the conventional metal apex seals. Also important is the wear resistance of side housing surfaces over which side seals and oil seals slide. The side housings used in current-mass-produced engines are made of gas soft-nitrided cast iron. Since aluminum side housings are of course effective to reduce engine weight, new methods for coating the sliding surfaces of aluminium should be desirable.

As a result of continued effort to develop the parts of high wear resistance, the authors developed the fiber reinforced ceramics (FRC) as a material for the apex seals [1, 2]. In addition, the authors improved the plasma sprayed cermet coating for the side housing sliding surfaces.

APEX SEAL MATERIALS

Selection of Base Material

Typical ceramic materials already used on automotive engines were screened primarily with respect to their wear resistance. Fig. 1 shows test conditions. As a result of the test, as shown in Fig. 2, all ceramic materials tested were found having higher wear resistance particularly over high speed range than the chilled cast iron currently used for conventional apex seals. Since the silicon carbide test piece fractured in high speed range, the material was judged inappropriate for the material for apex seals. Silicon nitride was chosen as the base material for apex seals judging from the less damage on disks.

Typical silicon nitrides are inferior to chilled cast iron with respect to ductility and strength. The authors attempted to develop silicon nitrides of high ductility and strength equivalent to those of chilled cast iron.

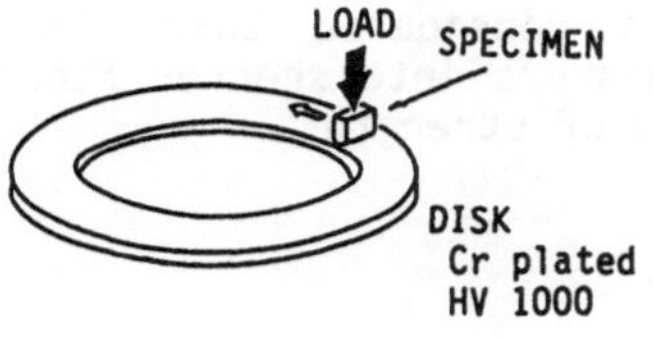

Applied load	: 50 N
Sliding velocity	: 5, 15, 25 m/s
Sliding distance	: 9000 m
No lubrication	

Fig. 1 Test conditions of wear rig

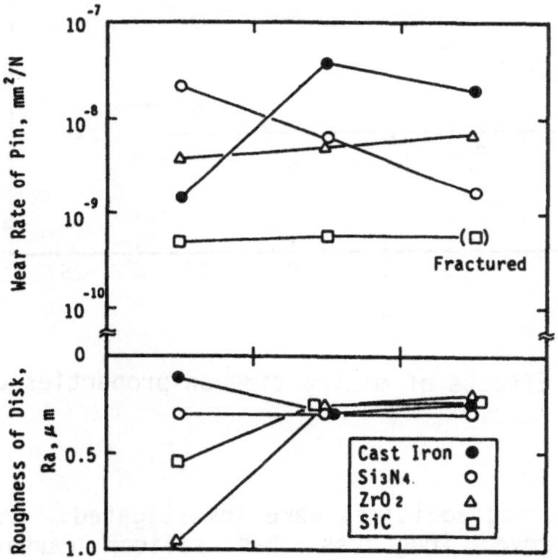

Fig. 2 Results of material selection tests for apex seals

Tougher Silicon Nitrides by Fiber Reinforcement

The authors paid special attention to the fiber reinforcement approach for improving the toughness of the silicon nitride. Among fiber materials to be used for FRC, silicon carbide whiskers were selected because of its high tensile strength. The problem in manufacturing FRC is the difficulty to produce good sintered compacts without impairing the properties of whiskers. As a method of mixing whiskers without damaging them, an ultrasonic flow method was devised where whiskers are dispersed without damage by the synergetic effect of a lot of air bubbles and ultrasonic waves. Although the method was considered capable of producing ceramics of superior toughness, the actual strength failed to meet the expectation. The reasons are considered that coagulations of whiskers initiate failures and that whisker bridges impair the sintering.

Then as the next method for improving the strength, whiskers and matrix material powders were mixed by a vibrating mill. Fig. 3 shows the relations between the mixing time and the mechanical properties of FRC. The result illustrates that high strength is obtained when mixed for more than a certain period of time while fracture toughness stayed constant irrespective of the mixing time. It is considered that the mixing operation fractured the initial material whiskers into shorter fibers, and dispersed them evenly affecting the property of strength.

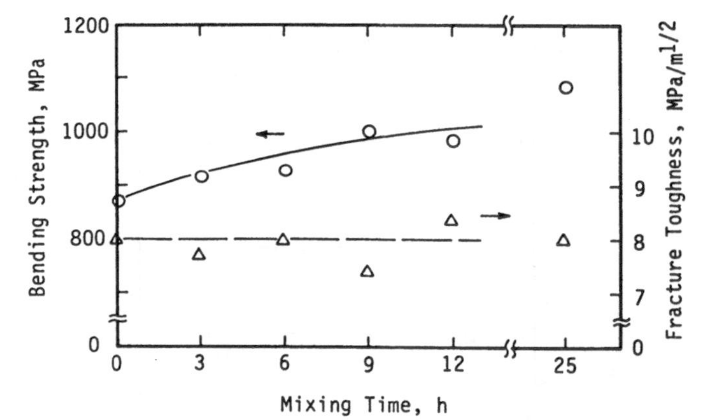

Fig. 3 Effects of mixing time on properties of FRC

Sintering Additives

The effects of sintering additives were investigated. As shown in Fig. 4, some additives improved toughness when silicon carbide whiskers were added, and the others not. This is considered to be due to the difference in wetting property and microstructure. The observation of the FRC sintered compacts microstructure revealed that silicon nitride crystals of the materials added with whiskers were finer than those with no whiskers due to inhibition of the growth of the crystals.

Properties of FRC

Based on the above results, a new material for the apex seals was developed having high ductility and strength. The composition/production process and properties are as shown in Table 1.

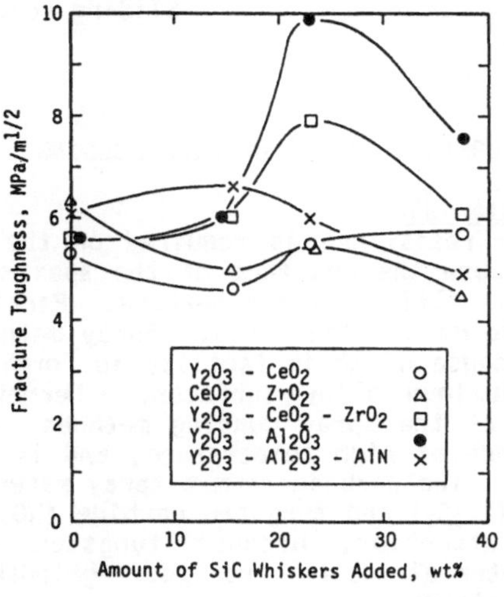

Fig. 4 Effects of FRC sintering additives on fracture toughness

TABLE 1
Composition, processes and properties of developed FRC

Composition and processes	
Composition of matrix	Si_3N_4 - 6wt% Y_2O_3 - 6wt% Al_2O_3
Amount of SiC whisker added	23 wt%
Mixing method	By vibrating mill, 25 h
Sintering method	Hot press, 1800°C × 2 h
Remarks	HIP treatment
Properties	
Hardness, HV	1700
Bending strength, MPa	1200
Fracture toughness, $MPa \cdot m^{1/2}$	8
Thermal shock resistance, °C	> 550
Coefficient of thermal expansion, 10^{-6}/°C	3

The developed FRC and silicon nitride matrix material were evaluated by the wear rig which conditions was shown in Fig. 1. The results of wear rig tests are shown in Table 2. The developed FRC shows less wear rate than silicon nitride matrix material. The wear resistance of silicon nitride was improved by the addition of silicon carbide whiskers.

The apex seals made of the FRC was tested on the engine. The result indicated that the amount of the wear in the FRC was as small as 1/5 of that in chilled iron.

TABLE 2
Results of wear rig tests of the developed FRC and Si_3N_4

Specimen	Ware rate, mm^2/N
Si_3N_4 - Y_2O_3 - Al_2O_3	3.5×10^{-8}
Si_3N_4 - Y_2O_3 - Al_2O_3 - 23wt% SiC whisker	1.9×10^{-9}

Sliding velocity : 25 m/s

SURFACE COATING OF SIDE HOUSING

Selection of Coating Material

A special type of wear resistance is required on the sliding surface of the side housings, because the end face of the apex seals, corner seals, side seals and oil seals slide over the surface. Particularly significant is the wear by the sliding of side seals. Spray methods were chosen for the surface coating because of their facility in forming thick layers and capability to coat aluminum alloy surfaces. Cermet spray coating of carbide, which is one of the spray coating methods, is known to provide excellent wear resistance at high temperature, and is extensively used in various industries [3]. The carbide cermet spray materials are classified into chromium carbide (Cr_3C_2) and tungsten carbide (WC). Chromium carbide is recently drawing attention, because tungsten carbide has major drawbacks of high material costs and poor grindability despite its excellence in wear resistance.

At first, the Cr_3C_2/20wt%Ni-Cr coated surface and the gas soft-nitrided cast iron surface, which is currently used for mass-produced engines, were evaluated by the reciprocating wear rig test as illustrated in Fig. 5. As shown in Fig. 6, the Cr_3C_2/Ni-Cr coated surface showed less wear than the gas soft-nitrided cast iron. Next, in consideration for higher productivity, the application of plasma spray was studied to obtain the same coating of excellent wear resistance. Although the plasma sprayed coating made in the first trial by a standard process had much lower wear resistance than the detonation gun sprayed one, the efforts were continued to improve the plasma sprayed Cr_3C_2/Ni-Cr coating to the level equivalent to that of the detonation gun sprayed one [4]. Fig. 6 includes the amount of wear of the plasma sprayed coating before and after the improvement.

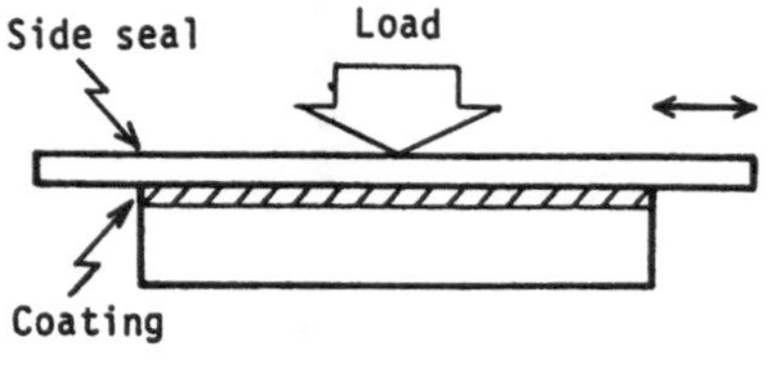

Applied load : 600 kPa
Sliding velocity : 0.3 m/s
Time : 0.5 h
No lubrication

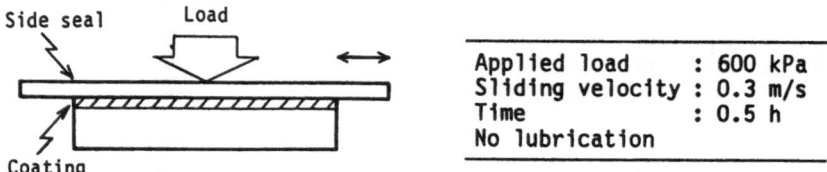

Fig. 5 Test conditions of reciprocating wear rig

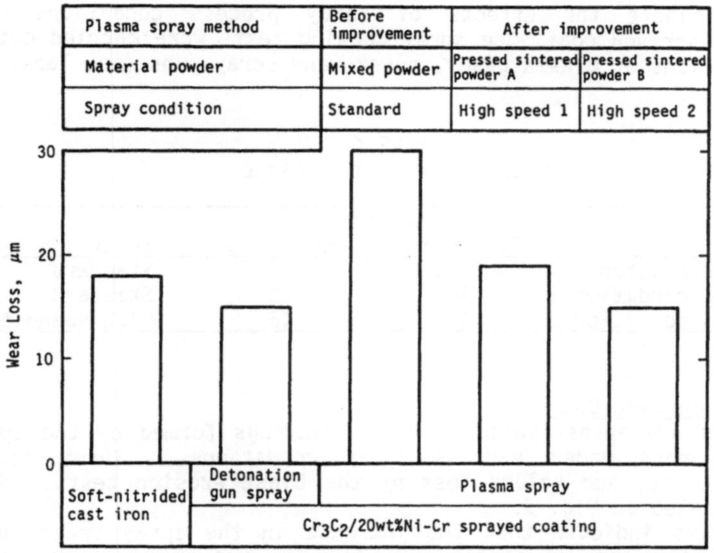

Fig. 6 Results of reciprocating wear rig tests

Material Powders and Spray Process Conditions

The configuration of material powders affects the properties of sprayed coating. Especially, because the Cr_3C_2 uses Ni-Cr alloy as the binder, the primary and secondary particle diameters and bonding strength are considered to affect the properties. Based on the consideration, four granulation types of powders are prepared as shown in Table 3. The observation of these powders revealed that the agitation granulated powder was made up of consistently dispersed Cr_3C_2 and Ni-Cr; the pressed sintered powder A contained bulky granules which have grown during the sintering process at high temperature; and the pressed sintered powder B was a firmly bonded powder consisting of consistently dispersed fine particles of Cr_3C_2 and Ni-Cr.

TABLE 3
Granulation processes

	Primary powder diameter, μm	Process
Mixed powder	5 ~ 44	Mixing
Agitation granulated powder	< 10	Agitation granulation → Sintering at low temp. → Classification 50 ~ 53 μm
Pressed sintered powder A	< 10	Mixing → Pressing → Sintering at high temp. → Crushing → Classification 50 ~ 53 μm
Pressed sintered powder B	< 10	Mixing → Pressing → Sintering at low temp. → Crushing → Classification 50 ~ 53 μm

To investigate the effects of spray process conditions on the properties of sprayed coat, the spray coating tests were carried out under various conditions of plasma gases, power, and spray guns (see Table 4).

TABLE 4
Spray process conditions

	Condition		
	Plasma gas	Power, kV	Spray gun type
Standard condition	Ar - H_2	30	Standard
High speed condition 1	Ar	40	Standard
High speed condition 2	Ar	45	High speed type

Properties of Sprayed Coat

The properties are investigated with the coatings formed by the spray of the above powders under various spray conditions. Three items of hardness, porosity, and volume loss by the blast erosion tests. Results are as illustrated in Fig. 7.

The results indicate that the increase in the spray speed improves properties of the sprayed coats. In case of the mixed powder and the agitation granulated powder, however, the increase in the spray speed reduces hardness and increases porosity. This is considered due to the separation of the agitation granulated powder into Cr_3C_2 and Ni-Cr during the spray process because of its poor bonding strength, and subsequently, as in the case of the mixed powder, insufficient melt of the lower density Cr_3C_2 into the plasma arc. As a result of the investigation of the sprayed coating cross section, it was found that high density layers were formed with improved material powders and spray processes.

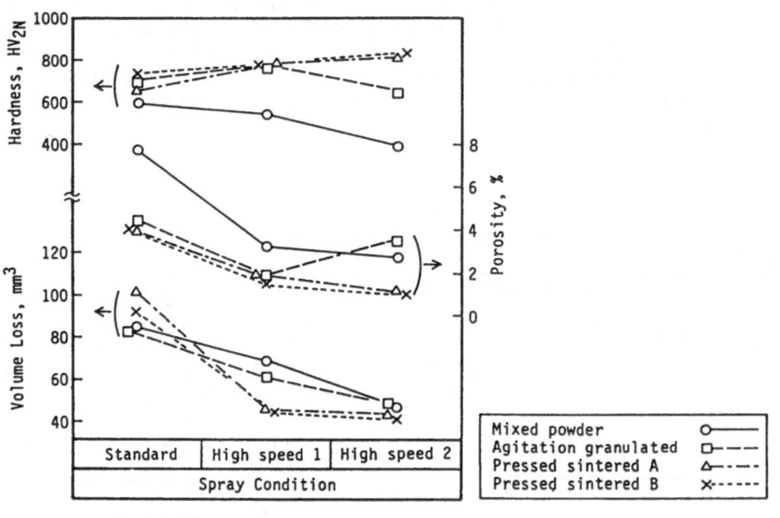

Fig. 7 Hardness, porosity and volume loss in blast erosion tests of sprayed coatings

The results from the reciprocating wear rig tests of the sprayed coating thus obtained indicated that the coating obtained by the spray of the pressed sintered powder B under the condition of high speed 2 has a high wear resistance equivalent to that of the detonation gun sprayed coating (see Fig. 6).

The above spray process was applied to the side housings for tests on actual running engine. As a result, the coating was confirmed having higher wear resistance than the gas soft-nitrided cast iron surfaces.

CONCLUSIONS

1. SiC whisker reinforced Si_3N_4 of toughness of 8 $MPa \cdot m^{1/2}$ and bending strength of 1200 MPa was obtained by the optimization of composition and mixing method.

2. The sliding property of Si_3N_4 was improved by the addition of SiC whiskers.

3. The sliding property of plasma sprayed Cr_3C_2/Ni-Cr coating was improved by the optimization of material powder granulation processes and spray conditions.

4. The excellent wear resistance of apex seals made of the developed FRC and the improved cermet plasma sprayed side housings were confirmed by the running engine tests.

REFERENCES

1. Akihide Takami, Nobuo Sakate, Kouji Tarumoto, Masaru Takatou and Takahiro Higuchi, Fiber Reinforced Ceramics, Mazda Technical Review, No. 8, 1990

2. Akihide Takami, Nobuo Sakate, Kouji Tarumoto, Masaru Takatou and Takahiro Higuchi, Development of Fiber Reinforced Ceramics for Automobile engine components, Proceedings of the 7th CIMTEC, 1990

3. Japan thermal spraying society, Spray Handbook, 1986

4. Tsutomu Shimizu and Yasufumi Kawado, Development of Plasma Cermet Coatings, Japan thermal spraying society proceedings of the 51th symposium, 1990

CONSIDERATIONS IN THE MANUFACTURE OF CERAMIC VALVE TRAIN COMPONENTS AND WEAR

M.W. Langer, J. Strobel
Volkswagen AG
3180 Wolfsburg

INTRODUCTION

Increased demands for higher efficiency on today's internal combustion engines combined with low cost production and extended maintenance intervals, may result in increased engine wear especially to valve train components.

OBJECTIVES

Initiated and sponsored by the Forschungsvereinigung Verbrennungskraftmaschinen e.V. the combinations valve guide - valve shaft, valve seat ring -valve seat and cam - tappet insert were investigated. The tests carried out at the Institut für Kolbenmaschinen at the University of Hanover are aimed to reduce wear in the described systems.

TEST ENGINE AND PROGRAMME

A turbocharged 2.4 litre 6-cylinder Volkswagen Diesel served as the test engine. The nominal data of the engine are 75 kW at 4300 1/min, the maximum torque is 195 Nm at 2600 1/min. The engine is mainly fitted into small lorries. The running programme was carried out over 100 hours using various well tried running programmes as performed by the engine manufacturer. The geometric measuring of components was carried out at running times of 0, 10 (after running-in) and 100 hours.

MATERIALS

One of the preliminary condition was the use of commercially available ceramic materials. Tests were carried out with oxide ceramics like zirconium and dispergent ceramics, also with the non oxide ceramics silicon nitride and silicon carbide. **TABLE 1** shows the two materials that have been selected for this study.

TABLE 1
Section of Used Materials

Material	Zirconium	Silicon Nitride
Internal name	2	3
Density [g/cm^3]	5.7	3.2
Flexural Strength [MPa]	510	600
E-Modulus [GPa]	206	300
Weibull-Modulus [-]	25	12
K_{IC}-Factor [$MPa*m^{0.5}$]	8.1	7
Th. expansion coeff. [1/K]	9.8	3
Th. conductivity [W/mK]	2.5	35

DESIGN OF COMPONENTS

The components were designed with the aid of a finite-element-analysis programme. Starting with the existing construction drawings of the series production engine and with the material data provided by the ceramics manufacturers, components were designed so as to avoid complicated alterations to the engine construction. Extensive calculations dealt with the operating parameters of the engine as well as the characteristic values of the materials. The first calculation was the fitting process of the ceramic component to the metal matrix, realized by cooling the ceramic by liquid nitrogen and shrinking it into the heated matrix material. The second calculation which was for conditions of the running engine, showed higher stresses than the first one. The conditions for both cases were then a combination of measurements of the metallic components and assumptions derived from the literature.

Ceramic Valve Seat Rings

The inlet valve seat ring geometry was not modified. On the outlet seat it was necessary to incase the ceramic component with a steel ring, compressing the ceramic, which prevents a loosening of the shrink fitting under running conditions. The thermal conductivity and expansion coefficient of the steel part is between that of the cylinder head and the ceramic parts.

Ceramic Valve Guides

In addition to the guiding of the movement of the valve, the guide is normally fitted with a valve stem seal. For the purpose of testing the seal was integrated into the washer under the valve springs making a very simple valve guide design shown in **FIGURE 1**.

FIGURE 1 : Valve Guide Design

Ceramic Tappet Inserts
The substitution of the ceramic tappet insert was carried out without any further calculations.

All tested combinations of materials and components are shown in **TABLE 2**.

TABLE 2
Combinations of Materials and Components

	materials			
	zirconium	dispergent ceramics	silicon nitride	silicon carbide
components				
valve guide	x	x	x	x
valve seat ring	x	x	x	x
tappet insert	x		x	

MANUFACTURING

Surface Parameters
With the use of different valve guide materials for example, **TABLE 3**, surface traces show identical R_z values but can nevertheless have entirely different surface structures. This made it necessary to employ new characteristic values that appeared to be more suitable for ceramic surfaces.

TABLE 3
Surface Parameters of Valve Guides

	materials			
	zirconium	dispergent ceramics	silicon nitride	silicon carbide
surface parameters in μm				
R_z	1.61	1.97	0.80	1,58
R_a	0.05	0.14	0.05	0.08
R_{pk}	0.04	0.09	0.04	0.03
R_k	0.09	0.30	0.13	0.15
R_{vk}	0.25	0.48	0.15	0.31

The german norm DIN 4776 provides a possible approach to this. In this norm the Abbott curve is approximated by straight sections, and here the characteristic values R_{pk}, R_k and R_{vk} are determined by the cross-over points in the straight lines, **FIGURE 2**.

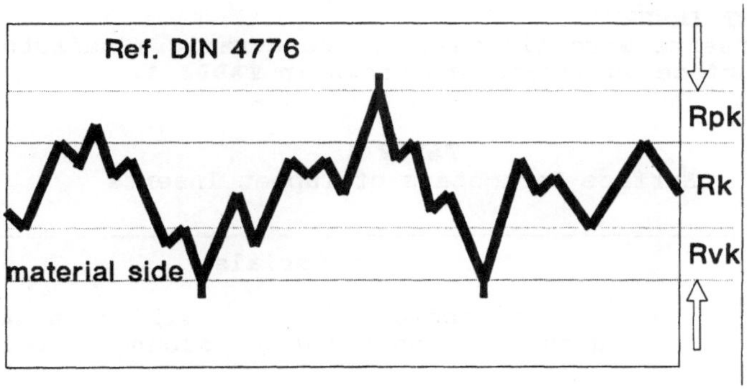

FIGURE 2 : Surface Parameters to DIN 4776

CERAMIC VALVE GUIDES
The ceramic manufacturers finished the ceramic valve guides by honing or polishing. The surface parameters are shown in TABLE 3.

CERAMIC VALVE SEAT RINGS

To be certain that the valve seat rings were concentric with the valve guides the seat rings were finished after fitting in the head, using a grinding tool centralised by the valve guide. The disadvantage of this process was deep grooves in the ground surface. A ring made of silicon carbide cracked, however this was the only damage during a total of 1200 hours of testing, **FIGURE 3**.

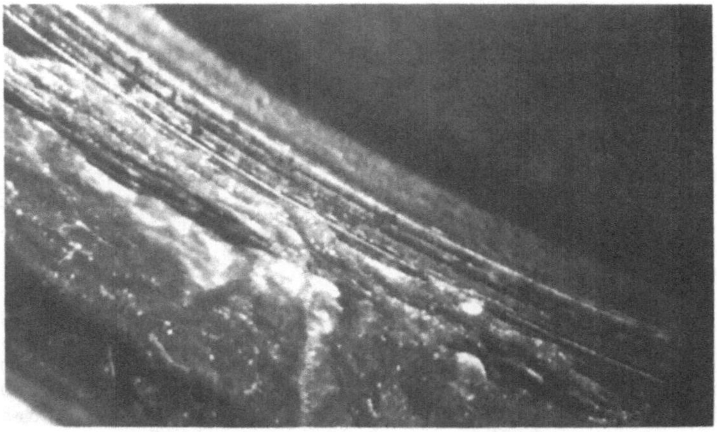

FIGURE 3 : Damaged SiC Valve Seat Ring

Better surface qualities were achieved by a modified grinding process, where a small grinding tool was moved by a CNC-machine. Repeated measurements of the concentricity between valve guide and valve seat were necessary, also repeated conditioning of the grinding tool improved the quality of the process /1/. The disadvantages were an increase in time and costs.

CERAMIC TAPPET INSERTS

The tappet inserts were finished by the ceramic manufacturers. The two tested surface qualities are shown in **TABLE 4**.

TABLE 4

Surface Parameters of Tappet Inserts

	materials			
	zirconium		silicon nitride	
surface parameters in μm	ground	polished	ground	polished
R_z	3.29	0.25	4.12	0.62
R_a	0.44	0.01	0.59	0.05
R_{pk}	0.34	0.01	0.39	0.02
R_k	1.22	0.04	1.74	0.08
R_{vk}	1.05	0.03	1.25	0.17

TEST RESULTS

The investigations to the system valve guide - valve stem are described in details in the next part of this study. The results of the systems valve seat ring - valve seat and cam - tappet insert are explained in /2/ and /3/ and are discussed only in an abreviated form.

Ceramic Valve Guides

The system valve guide - valve stem showed excessive wear to the valve stem after only 10 hours running-in. The wear form can be described as abrasive with some adhesive scuffing. The discolouring of the surface is caused by chemical reactions with the lubricant in the contact area, **FIGURE 4**.

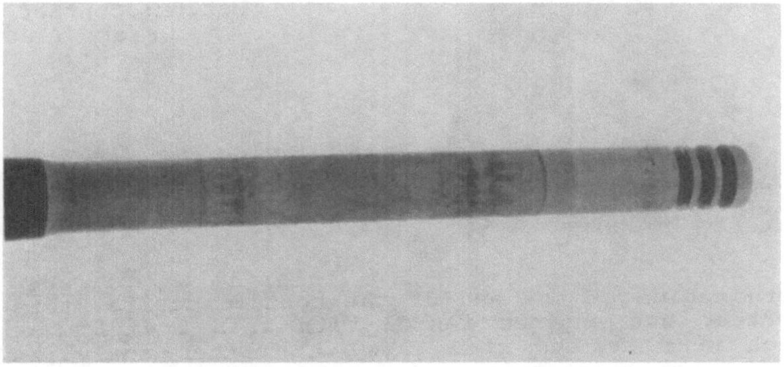

FIGURE 4 : Valve Stem Wear, Running Time 10 h

The valve guides were also damaged, primilarily a crumbling of ceramic material at the edges of the guide was noticed. During some test runs the influences of different design parameters were investigated.

Changes in the clearance between valve guide and valve stem had no influence, the same is applicable to chromium plated valve stems. Only nitrided stems could decrease the wear rate but were still not as good as the standard production parts. This reduction can be explained by the hardness of both materials.

Analysing the wear forms by REM-pictures showed an unsatisfactory design of the valve guide edges, instead of the required radian they were sharp and acted like cutting tools.

The wear mechanism can be explained in the following way. Caused by the clearance between valve guide and valve stem the valve is pressed against the edges of the guide under testing conditions. First the ceramic material causes abrassive wear to the valve stem, then it starts crumbling. This is followed by a buildup of metal in the pores of the ceramic surface at least resulting in scuffing, **FIGURE 5**.

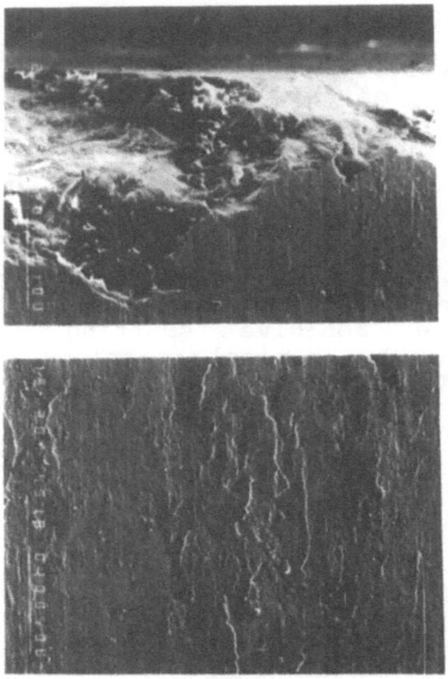

FIGURE 5 : Ceramic Valve Guide Wear

top : valve guide edge (300 x)
bottom: contact surface (1000 x)

By a smooth radius at the end of the contact surface this effect was prevented. But engines tested with this modification still showed high wear to the stem side. A further analysis led to assume that a non concentric position of valve guide to the valve seat axis was responsible for these wear rates. So it was decided to recut the metallic valve seats with a tool centered in the ceramic valve guide after tightening the cylinder head to a thick steel plate. This third modification combined the advantages of wear resistant valve guides with metallic valve stems as shown in **FIGURE 6**.

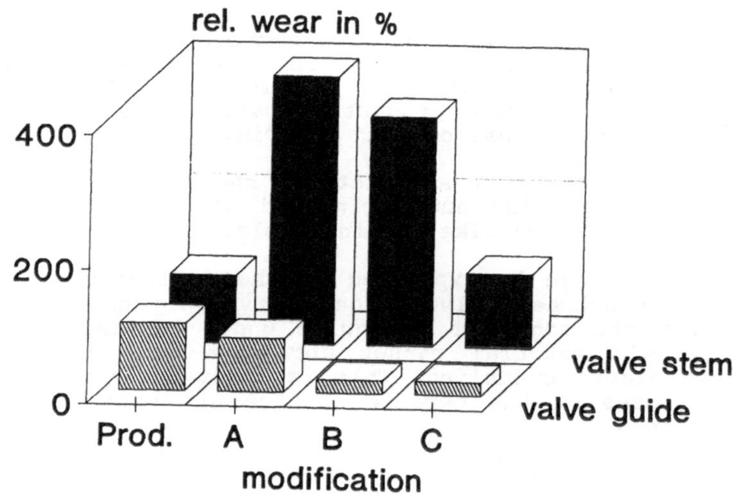

FIGURE 6 : Wear Results Valve Guide - Valve Stem

Ceramic Valve Seat Rings
The results in this system were determined by optimising the surface qualities of the valve seat rings. Even with rough surfaces the ceramics showed no wear, as opposed to the metallic valve seats with measurable wear. By using an optimised seat ring finishing it was possible to reduce wear on both sides of the system, **FIGURE 7**.

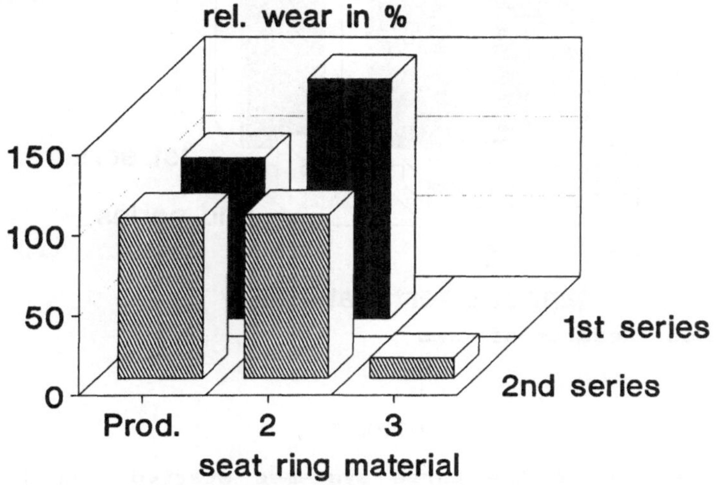

FIGURE 7 : Wear Results Valve Seat Ring - Valve Seat

Ceramic Tappet inserts
An improvement in the wear rate of the system cam - tappet insert was realized by better surface qualities too. When using ground ceramic inserts, their wear was reduced to 30 % of that of the metal ones, disadvantageous was an rapid increase in wear to the cam side. The use of polished inserts decreased wear on both sides of the system as shown in **FIGURES 8** and **9**.

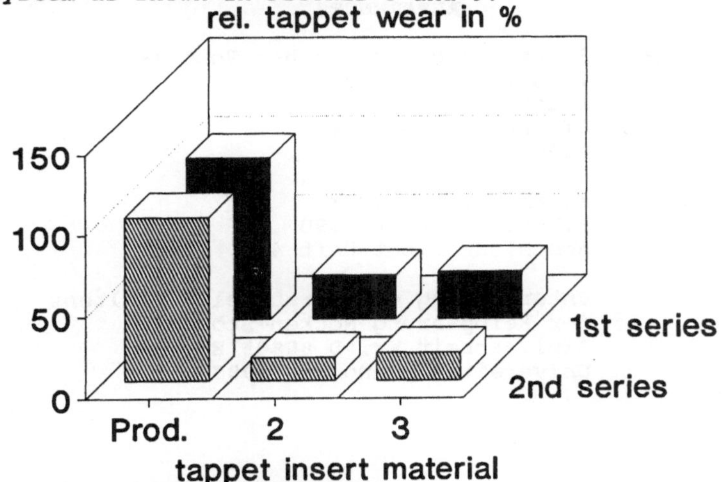

FIGURE 8 : Wear Results of Tappets

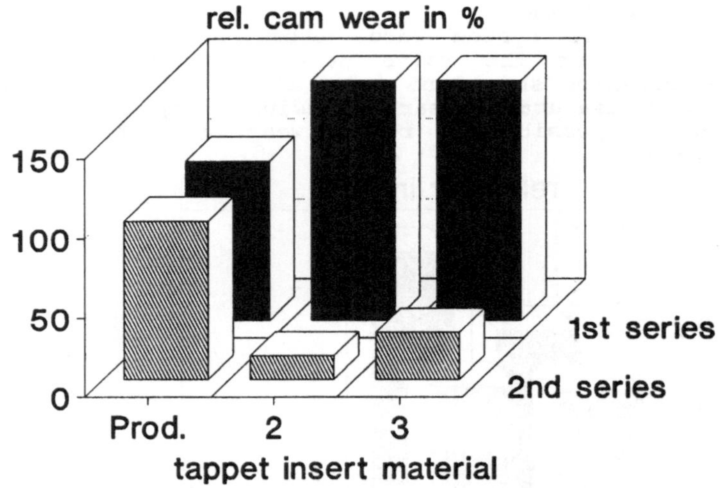

FIGURE 9 : Wear Results of Cams

SUMMARY

The investigations to the three systemes started with high wear levels when only substituting metal components by ceramics. An analysis of the wear results and the transformation into modified designing and manufacturing processes resulted in wear rates lower then the actual levels. Silicon nitride was choosen for further investigations because of its good wear behaviour in combination with its high mechanical capabilities. The results shown have to be verified by endurance tests.

REFERENCES

1. Kollmeier, H; Martens, R.; Rudolph, N.; Ruhland, F.; Wobker, H.-G. — Bearbeitung keramischer Ventilsitze in Zylinderköpfen. Abschlußbericht. CIM-Fabrik Hannover. 1990

2. Strobel, J. — Keramik im Motor I. FVV-Vorhaben Nr. 390. Abschlußbericht Heft 449, 1990

3. Niehues, J. — Untersuchung des Verschleißverhaltens der Reibpaarung Nocken-Stössel. Dipl.-Arbeit Nr. D 88805 E. Universität Hannover, 1988

USING CERAMICS FOR MASS REDUCTION IN VALVE TRAIN

HERBERT GASTHUBER *, ROLAND KREBSER
Daimler Benz AG, Research
7000 Stuttgart 80, Germany

ABSTRACT

Multivalve technique and high running speeds are current trend in spark ignition engines. Thus structural ceramic with their low density, good high temperature strength and wear resistance appear to offer a light-weight alternative valve material to steel and superalloys. Most obvious advantage is a reduction of moving masses, extremely accelerated in the valve train, and all the subsequent opportunities to design alterations.

Selection and testing of different materials, component design and calculations have been carried out. Performance, engine and rig test results indicate that sintered silicon nitride parts have a good chance to realize that potential of improvement. They generally could operate satisfactorily, if techniques of mass production and quality control proofed their reliability.

INTRODUCTION

The use of lightweight ceramics makes it possible to reduce the oscillating masses in the valve train considerably. With a specific weight of 3.2 g/cm^3 for silicon nitride the weight saving is at least 60 % as compared to steel with 8 g/cm^3. This is advantageous for design purposes in many ways:

- With the cam geometry and the critical engine speed remaining unchanged the valve spring force can be considerably reduced. This leads to lower friction losses in the valve train and thus to lower consumption, exhaust and noise emission.

- On engines with high speed levels the critical speed of the valve train can be increased if valve spring force and cam lobe are left unchanged and only the lighter ceramic valves are installed.

- The most interesting variant for the majority of production engines is probably to leave the valve spring force and the critical engine speed unchanged and to optimize the cam contour thus increasing fresh-air charge and torque or power output.

After the durability of ceramic valves had been proven generally in bench tests and in normal road operation, the investigations referred to in this paper were aimed at quantifying the frequently cited but not clearly proven advantages of those lightweight and heat-resistant ceramic components.

MATERIAL

Among the engineer ceramics silicon nitride has established itself as a multi-purpose material. The combination of its properties makes this material attractive for the designer:

Good strength, low density, high fracture toughness, acceptable thermal qualities.

Prior to ceramic valve testing a multitude of silicon nitride grades was examined. The results have been compiled in table 1. For tests and measurements mentioned in this article silicon nitride number 3 was used, a slip cast, gas pressure sintered variant.

TABLE 1
Material properties of some examined silicon nitride

		1	2	3	4	5	6
Bulk Density	g/cm^3	3.27	3.24	3.20	3.30	3.26	3.35
Flexural Strength	N/mm^2						
RT		1250	930	680	920	970	1050
1000°C		530	900	600	735	700	775
Young's Modulus	kN/mm^2	280	290	294	300	285	295
Weibull Modul		29	25	28	24	13	19
Thermal Expansion	$\cdot 10^{-6}$						
RT- 200°C		2.3	2.3	3.1	2.0	2.8	1.6
1000°C		3.7	3.7	3.1	3.1	3.2	3.5
Thermal Conductivity	W/m·K						
RT- 200°C		29	29	25	29	30	18
1000°C		16	16			14	12
Specific Heat	J/kg·K						
RT- 200°C		710	700	650	670	700	670
1000°C		1260	1200		1000	1300	1140

COMPONENT TEST

The valves examined were supplied in "as fired" condition by ceramics manufacturers and machined at DB ready for installation. The mechanical strength was ascertained by means of bending test specimen (dimension 2.5 x 2 mm) cut out of the unmachined ceramic components according to a sample plan. Although the ceramic materials tested were produced in different technologies, clear differences in strength distribution were regularly observed depending on the place of sampling (see Table 2).

TABLE 2
Strength distribution of tested silicon nitride valves

#	Valve N/mm^2	Shaft N/mm^2	Head N/mm^2	Tech
1	1239	1223	1256	KIP
3	746	769	716	Slip
5	969	995	936	Inje
6	1051	1062	1036	KIP
Ave	1001	1012	986	

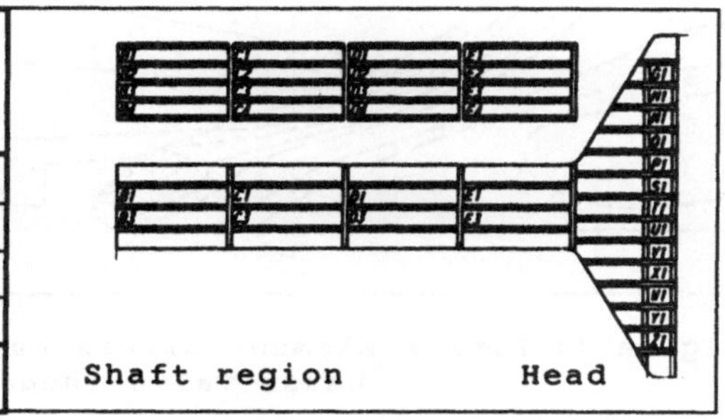

DESIGN MEASURES

As ceramics are unable to absorb overstress by plastic deformation due to the absence of ductility, brittle fracture is the consequence. Additionally, ceramics tend to subcritical crack propagation, i.e. cracks which are actually uncritical and harmless (also pores or other structural defects) expand continuously under stress until one of these many cracks reaches critical size which then also causes fatal fracture.

For this reason, the stress resulting from mechanical and thermal loads, particularly with regard to local stress peaks due to notch effect, point load or large temperature gradients, needs to be calculated very precisely if - which is most likely to be the case - it cannot be avoided by any design measures.

In the present case a stress analysis for thermal and mechanical load of the exhaust valves was carried out by means of the finite element method. The thermal stress results from - partly measured and partly calculated - inhomogeneous temperature distributions in the valve. To determine the mechanical load, two critical cases were investigated:

- Static and dynamic forces caused by spring tension and deceleration of the valve when impacting the valve seat.
- Gas-dynamic forces caused by the combustion pressure acting on the valve disk with the valve closed.

By superposition of mechanical and thermal stresses the maximum values shown in Figure 1 can be ascertained in the seat an neck area of the valve head with approx. 220 and 350 N/mm² tensile stress respectively. It is dropping significantly inside the valve neck section and passes zero in the valve cone interior.

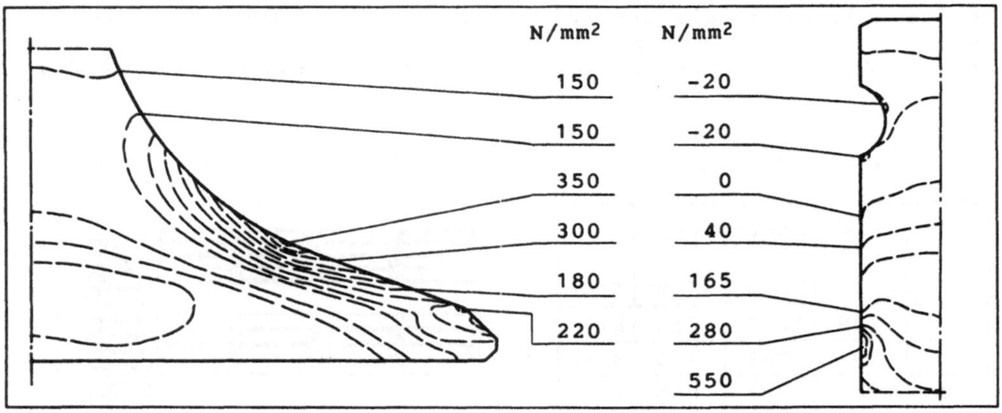

Figure 1: Finite Element stress calculation of a ceramic exhaust valve (head and groove area).

As the groove area appears to be critical and the standard steel valve spring retainer was to be replaced by an aluminium version, these areas were also calculated. For these "problem areas", the stresses found were relatively moderate:

- 50 N/mm² in the groove of the exhaust valve resulting from spring force and valve seat wear,
- 75 N/mm² in the contact area in the spring retainer cone.

MODIFICATION AND CERAMIFICATION OF THE VALVE TRAIN OF AN EXISTING CYLINDER HEAD

The conversion to ceramic valve train elements was carried out on a cylinder head of a six-cylinder spark-ignition engine with four-valve technology and valve tappets selected from the Mercedes-Benz AG engine range. To keep the valve spring forces as low as possible and to avoid valve gear control influence, the standard hydraulic valve clearance compensation was simply substituted by mechanical valve tappets.

The spring was designed to ensure a frictional connection between the valve train elements comparable to that of the basic version. The double spring used in series production was replaced without difficulty by a significantly weaker single spring. For speed stability the ceramified valve train was designed for n_{max} 8000 rpm (1000 rpm more than on the basic version) and 95 %

frictional connection. A minor modification, necessitated by the changeover from hydraulic to mechanical clearance compensation had no effect on valve timing.

The valve train assembly described in this paper is compared with the original version in Figure 2. The ceramic version has oscillating masses on the inlet and exhaust sides of 70 g and 64 g respectively (which corresponds to a mass reduction of 64 and 65 % as compared to the basic head). The absolute static weight saving of this valve train design reduces the weight of the engine by more than 4 kilogrammes.

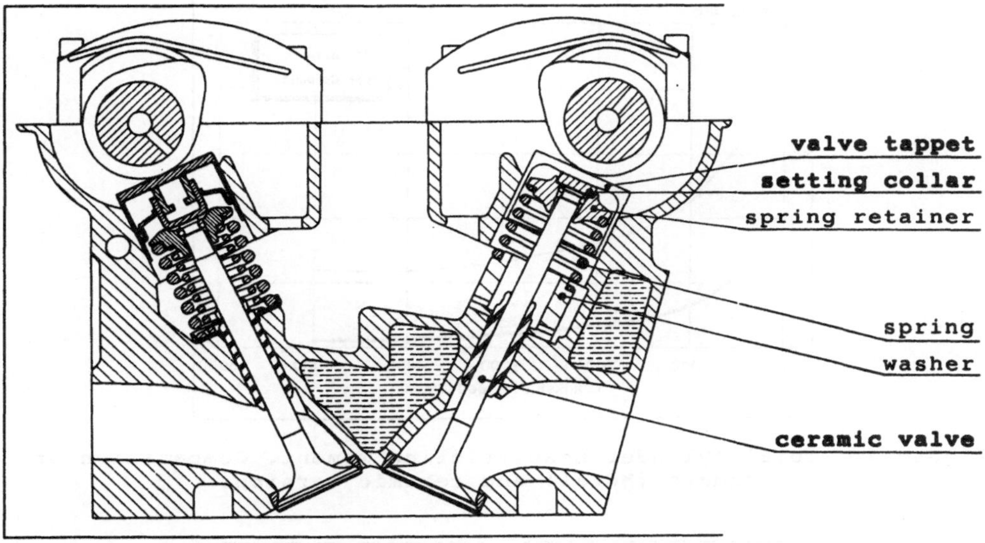

Figur 2: Installation of ceramic valve train parts in a cylinder head (Standard version left, ceramic right).

INVESTIGATION INTO FRICTION LOSSES AND NOISE EMISSIONS

In order to determine the friction losses and the level of mechanical noise, the ceramified cylinder head was measured on a motored test stand. The engine was set on an auxiliary frame which was elastic suspended. During the tests, the engine solely served the purpose of supporting the cylinder head and ensuring the oil supply of the valve train. The camshafts are driven - via a ball beared transmission shaft instead of the standard crankshaft - by a d.c. motor arranged like a pendulum. Its torque is supported by a very sensitive load cell.

The values for the drive torque (recorded and compared with the original version in Fig. 3) and the engine speed have been reduced to the crankshaft centre. On the ceramic cylinder head variant, a pronounced decrease in friction torque can be observed with increasing engine speed. The level of the standard cylinder head is at 1000 rpm 40 % above that of the ceramified one. Although the relative advantage is smaller at 5000 rpm, it still amounts to about 30 %.

In these tests, the ceramic cylinder head had a distinctly lower noise level. For this reason, acceleration gauges were attached to various points of the cylinder heads in order to compare the emissions of solid-borne noise. The measured values are shown in Fig. 4.

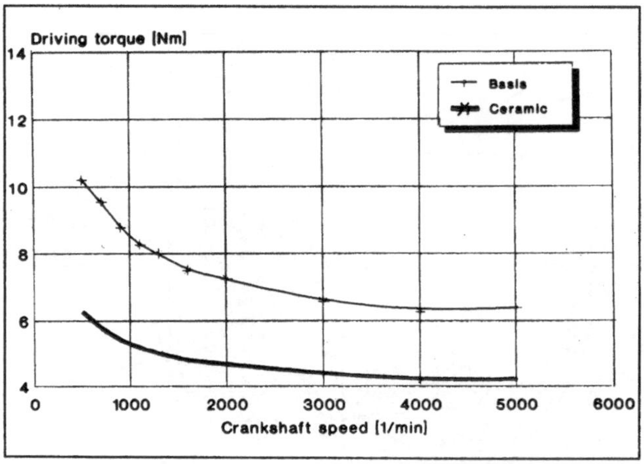

Figure 3: Total cylinder head friction moment. Comparision of standard (Metal) and ceramic version.

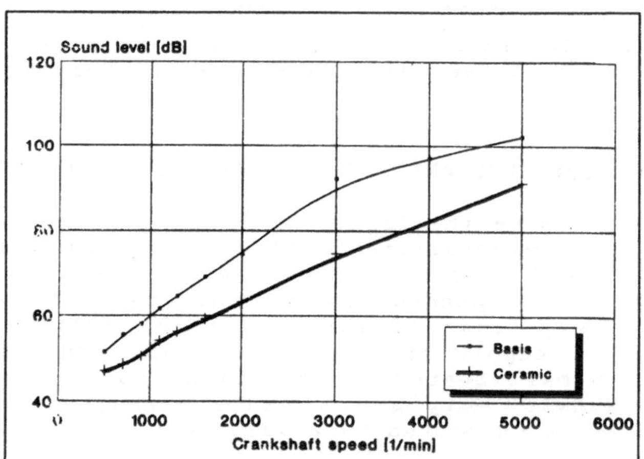

Figure 4: Cylinder head noise emission. Comparision of standard (Metal) and ceramic version.

On the ceramified version of the cylinder heads, the level of emitted solid-borne noise is smaller by 6 dB at 2000 rpm, i.e. the noise emissions are halved. The advantage increases up to 18 dB at 5000 rpm. This means that in actual engine operation, the valve train noise can objectively no longer be discerned. The influence of the oil temperature on the noise emission was negligible.

FINAL REMARKS

A distance of more than 500,000 km has already been covered with ceramic valve train components. We do not wish to conceal the fact that damage due to broken valves occurred. The causes of failure have been analysed and are known. Figure 5 shows some typical cases of damage.

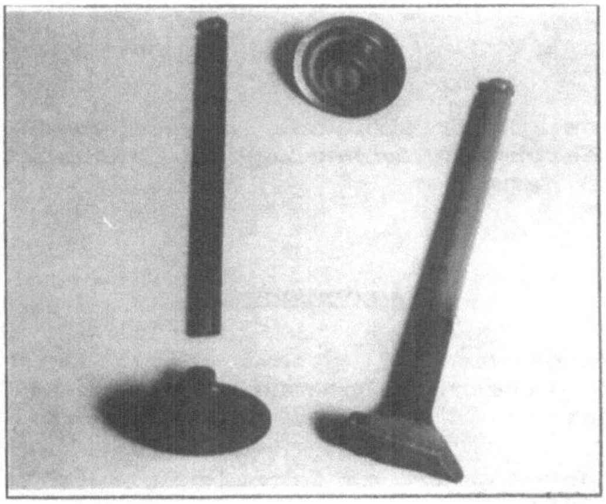

Figure 5: Typical defects of some tested ceramic valves

The ceramic modification of the investigated six-cylinder four-valve head led to a reduction in the oscillating mass of 65 %, in the overall mass of the cylinder head of 4 kg, in friction losses of up to 40 %, and in noise emission of up to 18 dB.

An incidental result of the tests was a demonstration in wear resistance of the ceramic valve train: although the test was carried out with mechanical valve clearance compensation and with practically unchanged valve seats and guides, no valve seat wear and only a scarcely measurable increase of the valve clearance were noticed over more than 100,000 km.

Consistent adjustment to ceramic valve train components will create further potential for design improvements, which, however, will not be discussed here.

At the present point in time there are no measured values available to decide if ceramic exhaust valves favour pinking or knocking, as was reported elsewhere. In rig tests and real operation, no disadvantages have been observed so far which might be due to changed thermodynamics or combustion. The material characteristics of the ceramic modification we used hardly give cause to expect such phenomena.

It is thus an undeniable fact that the use of ceramic valve train components brings about technical enhancements. Whether such designs and materials will soon meet the expectations placed on them with regard to

- reliability
- consistent quality
- acceptable prices

will have to be shown by further investigations in the very near future.

Acknowledgements

The authors express their graduade to the German Federal Ministry for Research and Technology for sponsoring parts of the work referred here.

REFERENCES

1. Hori, Y.; Miyakawa, Y.; Asami, S.; Kajihara, T. "Si_3N_4 Ceramic Valves for Internal Combustion Engines" SAE- Paper 890175

2. Yamada, T. "Development of Ceramic Exhaust Valves" 16th CIMAC (1985)

3. Asnani; Kuonen. "Ceramic Valve and Seat insert Performance in a Diesel" SAE- Paper 850358

4. Asnani, M.; Southam, E.R.; Wills, R.R. "Contact Fatigue Damage in Ceramics and Metal Valve Tips" SAE- Paper 870419

5. Kabut, D.M.; Garwin, I.J.; Hartsock, D.L. "Ceramic Valve Analysis", Autom. Eng. 96(1988), S. 46- 53

6. Wills, R.R.; Southam, R.E. "Ceramic Engine Valves" Journ. Am. Ceram. Soc. 72(1989), S. 1261- 1264

7. Ishiwata. "A Review on the Testing of Ceramic Parts in Engines for Automobiles" DKG- Symposium; Travemünde 1986

8. H. Brüggemann; E. Gobien; M. Schäfer "Der neue Sechszylinder-Vierventil-Motor mit 3.0-L- Hubraum für den neuen 300 SL-24" MTZ 50 (1989) H. 4, S. 141 -148

RESEARCH AND DEVELOPMENT ON CERAMIC ENGINES IN CHINA

XIREN FU and TUNGSHENG YEN
Shanghai Institute of Ceramics, Chinese Academy of Sciences, 1295 Dingxi Road, Shanghai 200050, China
XIAOHONG GAO
Wuhan University of Water Transport Engineering, Yujiatou, Wuchang, Wuhan 430063, China
ZONGRONG WANG
Department of Industrial Technology, The State Science and Technology Commission, 54 Sanlihe Road, Beijing 100862, China

ABSTRACT

China has been following the research and development on advanced structural ceramics and ceramic engines for more than a decade. A National Project entitled "Advanced Structural Ceramics and Adiabatic Diesel Engine" was formulated in 1986. The principle program areas of this project are ceramic powder synthesis, material design and material processing, property and microstructure study, machining and joining, nondestructive evaluation, internal combustion engine component fabrication and low heat rejection diesel engine technology. The present paper is a short survey of what has been going on during the past 4 years.

INTRODUCTION

The research and development of advanced structural ceramics and ceramic engines have received a great deal of attention in China for more than a decade. After several years' preliminary study, a National Project entitled "Advanced Structural Ceramics and Adiabatic Diesel Engine" was formulated in 1986. The objective of the project is to advance Chinese structural ceramics and ceramic diesel engine technology. This project was actually performed by the Chinese Academy of Sciences, universities and industrial laboratories, and was coordinated by the State Science and Technology Commission of China. Significant progress has been made toward the goal of the project during the past 4 years. The present paper is a panorama of what has been going on in some of the program areas.

MATERIAL DESIGN AND MATERIAL PROCESSING

The purpose of this program is to develop a number of high performance, high reliability advanced structural ceramic and composite materials for use under the severe environment of heat engines and in other high technology areas. About 14 kinds of multiphase ceramics and whisker and/or particulate reinforced composite materials were designed and processed, some of them will be described briefly as follows:

1) TiC + Al_2O_3/SiC Multiphase Ceramics

Based on a series of studies in the two-phase ceramic systems, such as SiC-TiC, SiC-Al_2O_3, Al_2O_3-TiC etc., a TiC + Al_2O_3/SiC multiphase ceramic material was developed. The basic composition of the material is SiC-25 vol% TiC, different amount of Al_2O_3 was added to learn its effect on properties. The maximum bend strength and fracture toughness of the hot-pressed specimens are 700 MPa and 6.9 $MPa \cdot m^{1/2}$ respectively. Its 1400°C strength retains at a value of 480 MPa. After oxidation at 1300°C under ambient atmosphere for 100 h, this material still has a strength of 356 MPa. So the oxidation resistance and high temperature strength of this multiphase ceramic material are promising. It is expected that this material will be useful in some areas of high technology.

2) β-Si_3N_4/α'-Sialon Multiphase Ceramics

According to the phase relationship studies, there is a α' + β binary compatibility region on the Si_3N_4-$AlN \cdot Al_2O_3$-$YN \cdot 3AlN$ section of the Y-Si-Al-O-N system. Two compositions were chosen from this region and the β-Si_3N_4/α'-Sialon multiphase ceramics were designed and processed by using the gas pressure sintering technique with La_2O_3 as additive. This material combines the high strength of β-Si_3N_4 and the high hardness and thermal shock resistance of α'-Sialon. The bend strength and fracture toughness of the material are 750-800 MPa and 6.8 MPa·$m^{1/2}$ (indentation method) respectively, its hardness (HR_A) reaches 93-94. The microstructure of this material is composed of elongated β-Si_3N_4 grains interwoven with equiaxed grains of α'-Sialon. Such a microstructure is beneficial to exhibit in-situ reinforcement. This multiphase ceramic material has been fabricated as cutting tool tips to machine nickle-based alloy with good results.

3) SiC W/Sialon Composite

Preliminary studies show that the chemical compatibility between the Tateho SiC whisker and sialon matrix is promising, so it is possible to process high performance composite materials in this system. The amount of sintering aids greatly affected the bonding strength between SiC whiskers and the sialon matrix. The room temperature bend strength and fracture toughness of the hot-pressed SiC W/Sialon composite are 870- 1000 MPa and 8-8.5 MPa·$m^{1/2}$ respectively. The strength at 1200°C decreased slightly and measured at values between 800-900 MPa. Microstructure studies show that this material is composed of long β-SiC whiskers and elongated β'-Sialon grains, both of which intersect closely with each other and form a 3-dimentional network. Perhaps this is the reason why this composite material can maintain its strength up to 1200℃. Whisker pullout and crack deflection are also observed at the fracture surfaces, these phenomena are considered as the main toughening mechanisms for this material.

4) SiC W/C Composite

This is a new composite material developed. The bend strength and fracture toughness of the hot-pressed SiC whiskers reinforced carbon composite are 400-450 MPa and 4-6 MPa·$m^{1/2}$ respectively. These values are much higher than those of carbon ceramics without reinforcement. The highest bend strength for the carbon ceramics found in the literature is 160-250 MPa, and the K_{1C} value is 2-4 MPa·$m^{1/2}$. The thermal shock and oxidation resistance of this novel composite are also much improved. Microscopy studies show that the long SiC whiskers are mainly located at the grain boundaries but the short SiC whiskers are distributed within the grains. These whiskers prevent the slip of carbon grains and the propagation of cracks, thus strengthening the material as a whole. Surface X-ray diffraction analysis reveals that a thin film of SiO_2 and B_2O_3 covers the surface of the composite, so that the oxidation resistance is improved. This composite has found some applications in the industry and the material processing technique has been patented.

INTERNAL COMBUSTION ENGINE COMPONENTS FABRICATION

The goal of this program is to develop a material technology base that will give Chinese industry the ability to produce reliable, cost-effective, high performance ceramic components for use in conventional and advanced internal combustion engines. More than 20 kinds of ceramic components have been fabricated and tested, some of them will be described as follows:

1) Glow Plug

The glow plug is designed for use in Type S-195 walking tractor. 20 specimens were rig-tested and all of them gave a charge-discharge life of over 10,000 cycles, the maximum of it was over 100,000 cycles without damage. 280 specimens were distributed directly to the users, the service life of all the specimens was over 1500 h with no trouble. The maximum of it reached 2500 h and the specimens can work further. The response time of these glow plugs is 8-10 sec and it can be decreased to 2-3 sec with the help of a fast-start device. Moreover, the glow plug can start the engine at a temperature of -36°C. A production line has already been established in a factory, and a yearly production of 30,000 pieces of glow plug is planned for 1992.

2) Swirl Chamber

The swirl chamber is also designed for use in Type S-195 walking tractor. The material used for this component is gas pressure sintered RBSN ceramics. After rig testing for 1000 h, 300 specimens were installed in the users' tractors and boat diesel engines for road tests. The lifetime of all these chambers was over 2000 h, many of them reached 2800 h and the service time of the best ones exceeded 4000 h without damage. Almost all the users reported that after using the ceramic swirl chambers, the fuel economy can be improved by about 100 g/h for each machine. The engine is easier to get started, the noise level is much lower and the engine output is higher than that of the metal counterparts. A production line will be established by the end of this year and a yearly production of 30,000 pieces is planned for 1992.

3) Turbocharger Rotors

4 kinds of turbocharger rotors with different shapes and sizes were fabricated. The materials used are gas pressure sintered RBSN and pressureless sintered silicon nitride ceramics. All the turbocharger rotors were successfully formed by injection molding, pressure filtration and "pressure extrusion". These rotors were sintered to near theoretical density with no trouble. Spin tests were carried out at room temperature. The rotors with a diameter of 76 mm have been tested for 30 to 50 hours at a rated speed of 73000 r/min. Recently, the rotors formed by "pressure extrusion"survived a speed of 93000 r/min without failure. By using this patented method, the rotors can be formed at room temperature with a conventional press and a long dewaxing process can be avoid. Further work is needed to make these techniques matured for commercial production.

4) Fuel Injector Needle

A ceramic needle used in fuel injector of Type 2135 diesel engine was developed. The material processed to substitute the metal needle is fine-crystalline Y-TZP ceramics with high strength and fracture toughness. The ceramic needle weighs 6.1 g and it is lighter than that of metal ones by 24%. During the fuel injection, the lift and drop velocities of ceramic needle are increased obviously, the fuel pressure within the injector is also increased, and the SMD value of the fuel drops is decreased, so that the combustion process has been much improved. As a result, the fuel consumption has been reduced by 2-4%. The ceramic needles have been rig-tested in lab for 500 h without failure, the amount of wear for this component is negligible. It is expected that this fuel injector needle can be put into small-scall production for practical application.

LOW HEAT REJECTION DIESEL ENGINE TECHNOLOGY

The objective of this program is to develop and demonstrate the critical technology needed to advance the low heat rejection engine concept for practical application. This program has been partitioned into two phases. The first phase is to develop an "uncooled diesel engine" in which the independent water cooling system will be cancelled, so that this engine can be used in desert, plateau and severely cold areas in China. The second phase is to develop an "adiabatic diesel engine" in which the combustion chamber is highly insulated and the hot exhaust gas energy will be recovered through both turbocharging and turbo-compounding, so that the fuel economy will be improved further.

Uncooled Diesel Engine

To achieve the goal of the first phase study, a 4 stroke, 6 cylinder diesel engine was chosen as the testing machine. The cylinder bore is 105 mm, the piston stroke is 120 mm and the rated power is 95.6 KW. In order to improve the combustion process after insulating the combustion chamber, an impulse turbocharger was used to increase the air/fuel ratio and recover the exhaust gas energy. The rated speed for the baseline machine was 2800 r/min, after turbocharging it decreased to 2200 r/min, the brake mean effective pressure increased to 0.836 MPa, the maximum torque moment increased to 441 Nm and the speed at the maximum torque was 1400 to 1800 r/min. The independent water cooling system has been completely cancelled.

In order to meet the experimental requirements, some refitments have been made for the metal engine. The combustion chamber was insulated by both the monolithic ceramics and ceramic coatings. The yttria-stabilized zirconia coatings with different thickness were applied to the combustion faces of the cylinder heads, the bottom faces of the intake and exhaust valves, the internal faces of the stainless steel ports and the upper parts of the cylinders. Some other ceramic parts are also fitted in this uncooled diesel engine, such as the Y-TZP fuel injector needle, silicon nitride rocker arm pad, cordierite catalytic converter substrate and the aluminum titanate exhaust manifold liner etc.

The uncooled diesel engine assembled was rig-tested for 400 hours in lab (Figure 1). Then, this engine was installed in a bus with 45 seats and road-tested for 3500 km between shanghai and Beijing, and no trouble has been found with the engine. After road-testing, the characteristics of the engine have been systematically measured showing a fuel consumption reduction of 3-5%. In order to learn the durability of the engine, the bus is now used as a regular service bus. Based on the above-mentioned results, an uncooled diesel engine will be designed and constructed in the next five years as a practical automotive.

Adiabatic Diesel Engine

A 4 stroke, 6 cylinder diesel engine with a rated power of 176.5 KW was chosen as the testing machine for the second phase study. The cylinder bore is 135 mm, the piston stroke is 150 mm and the rated speed is 1500 r/min. The brake mean effective pressure is equal to 1.082 to 1.180 MPa. All the combustion chambers are insulated by monolithic ceramics. The Piston crowns are made of hot-pressed silicon nitride ceramics, an alternative material with lower thermal conductivity is toughened mullite-based ceramics. The cylinder headfaces are made from toughened mullite-based ceramics and the upper part of the cylinder liner is made of Y-TZP ceramics. All these ceramic components have rig-tested for 100 h in the single cylinder test engine before installing in the adiabatic diesel engine. Fig. 2 shows the adiabatic diesel engine constructed. The hot exhaust gas passes through the turbocharger at first, then enters to the power turbine which powers the driving shaft via a gear train system. The rotation speed of the power turbine rotor is 35000 r/min and it decreases to 1500 r/min through the gear train. The adiabatic diesel engine has successfully operated for more than 100 hours without failure. The fuel consumption improvement is more than 5%. Based on these experimental results, some conclusions can be made as follows:

1) The ceramic components can withstand the severe environment

in the adiabatic diesel engine.

2) The low heat rejection engine concept is proved to be effective for the lowering of fuel consumption.
3) Future work is to design and reconstruct a more realistic adiabatic diesel engine for commercial application.

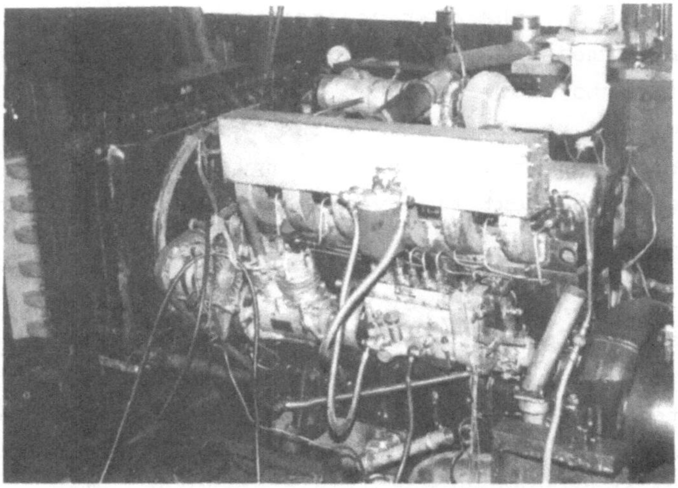

Figure 1. Uncooled Diesel Engine.

Figure 2. Adiabatic Diesel Engine.

ANALYSIS AND TESTING OF CERAMIC MATRIX COMPOSITES IN DIESEL ENGINES

Dieter Sygulla and Petar Agatonovic
Strength Department
MAN Technologie AG
Bauschingerstr. 20, D–8000 München, Germany

ABSTRACT

The use of ceramic material in diesel engines would be advantageous for many reasons: Insulation of the parts, incorporated in the combustion chamber, reduction of the oscillating masses, increase of the combustion temperature and an increased theoretical efficiency. For this purpose the geometry of the CMC parts has to be optimised and then to be tested in a diesel engine. Therefore many variants with different geometries were evaluated using the FE analysis to choose the adequate design. Firstly the properties of the material composed of SiC fibers and a SiC matrix were evaluated. Uniaxial specimen tests with a special test control yielded the typical stress–strain curve for this material, necessary for calculations. Initially the assembly method for the cylinder liner, the piston insert and the cylinder head insert was examined. For the determination of the temperature induced stresses the orthotropic and temperature properties of the material were considered. For the stress analyses the contact behaviour of the parts was simulated by the use of gap elements. The operational conditions include beside thermal loads, the preload of the bolts and the piston insert and the pressure load, which arises during the combustion process. Thus the assembly method of the ceramic parts for the diesel engine was determined and a full scale version of the diesel engine was choosen. Finally the most promising CMC variants were manufactured and tested in the diesel engine. Partly due to severe changes of the combustion process, the life of the CMC components was too short for the complete realisation in diesel engines. Many details, first of all the material itself, have to be improved in order to achieve successful results in the future.

INTRODUCTION

Great expectations existed for the incorporation of Ceramic Matrix Composites (CMC) in diesel engines. The objectives of increasing the combustion temperature of the combustion process, reducing the heat transfer to the cooling system, increasing the engine efficiency, i.e. a reduction of the fuel consumption, seemed to be achievable with this type of material. It was composed of ceramic fibers (SiC), surrounded by a matrix built up by Chemical Vapour Infiltration (CVI) at high temperatures. The advantages of such a material in contrast to conventional bulk ceramic is its much higher insensitivity to stress concentrations and to crack like defects. There is no catastrophic failure of a part when locally matrix cracks appear and the "yield stress" is exceeded.

MATERIAL CHARACTERISATION

SiC/SiC material shows a relatively low thermal conductivity and is therefore adequate for insulation purposes. There is a large difference between the heat transfer parallel to and orthogonal to the fibers (see TABLE 1).

TABLE 1.
Properties of SiC/SiC Material

	20°C	1000°C
Thermal conductivity parallel to the fibers	20.0 W/m°C	15.0 W/m°C
Thermal conductivity orthogonal to the fibers	12.0 W/m°C	7.0 W/m°C
Thermal heat capacity	620 J/kg/°C	1200 J/kg/°C
Thermal expansion coefficient	3.0 10^{-6} 1/°C	3.0 10^{-6} 1/°C
Density	2.5 g/cm^3	2.5 g/cm^3
Young's Modulus	230 GPa	200 GPa

As a consequence the insulation depends on the way the fibers are layed up. The best insulation will be achieved if the fibers are woven in one direction perpendicular to the weave plane. Regarding the properties of SiC/SiC material, outlined in TABLE 1, the temperature dependence is remarkable. The thermal expansion coefficient is low compared to aluminium piston alloy material and this fact causes a lot of problems for the incorporation of the SiC/SiC material. However most important is the behaviour of the material when overload occurs, which was investigated in uniaxial LCF tests. They were performed at room temperature. The test specimens were stress controlled up to a prescribed stress level. The strain at this point was stored and the specimen was cycled in tension and compression to rupture. During the tests the stress strain hysteresis of different cycles was recorded. Above the initial elastic range of the material, see FIGURE 1, the first matrix cracks appear and the fibers begin to slide in the cracked matrix. At the cracks a pull out of the fibers can be observed, whereas the composite still carry the load. On unloading, the fibers slide back in the reverse direction and due to the contact forces occurring between fibers and matrix work is done until the fibers are totally pushed back in the matrix and the matrix cracks close. In the compression part of the hysteresis loop the stiffening effect of the matrix and the fibers then leads to an increase of the Youngs Modulus to the initial value. This pseudo plastic behaviour under tension load makes the use of CMC attractive for diesel engines and similar applications.

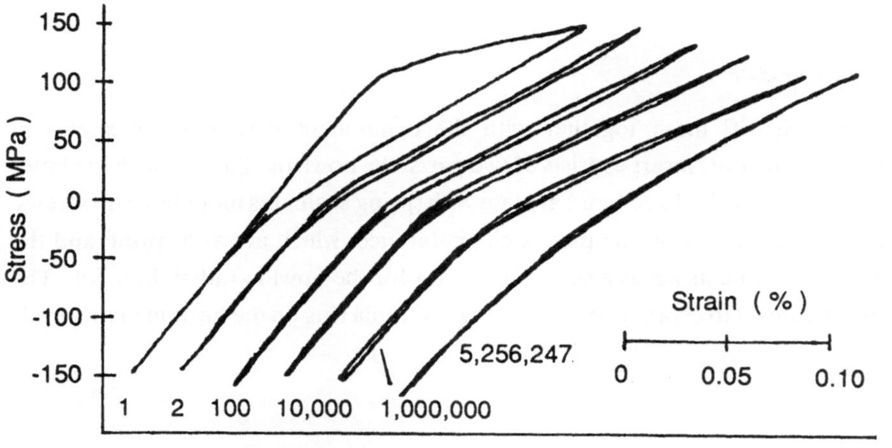

FIGURE 1. Stress–strain hystereses for SiC/SiC material (the cycles have been artificially offset for clarity)

ANALYSIS

To insulate the combustion chamber in all directions a cylinder liner, a cylinder head plate and an insert for the piston (see FIGURE 2) were incorporated in the diesel engine.

FIGURE 2. Piston mounted with a SiC/SiC insert

Many variants were examined to select the best assembly method depending on which part was to be fixed in the engine. The Finite Element (FE) code MARC was used for the analyses. First a temperature analysis and then a stress analysis were carried out for each variant. As example the analysis of a piston variant is described here.

Temperature Analysis

The axisymmetric FE mesh together with the component materials are shown in FIGURE 3. The SiC/SiC insert consists of two parts, the bowl and the ring with the holes for the screws. The bowl is fixed by a flat plate with spring segments mounted with screws. There are three segments on the plate's circumference, which act as a spring and the remaining three segments act as a centering device for the bowl in radial direction. The bowl itself has contact to ceramic cement and to a zirconia ring on the bottom of the bowl.

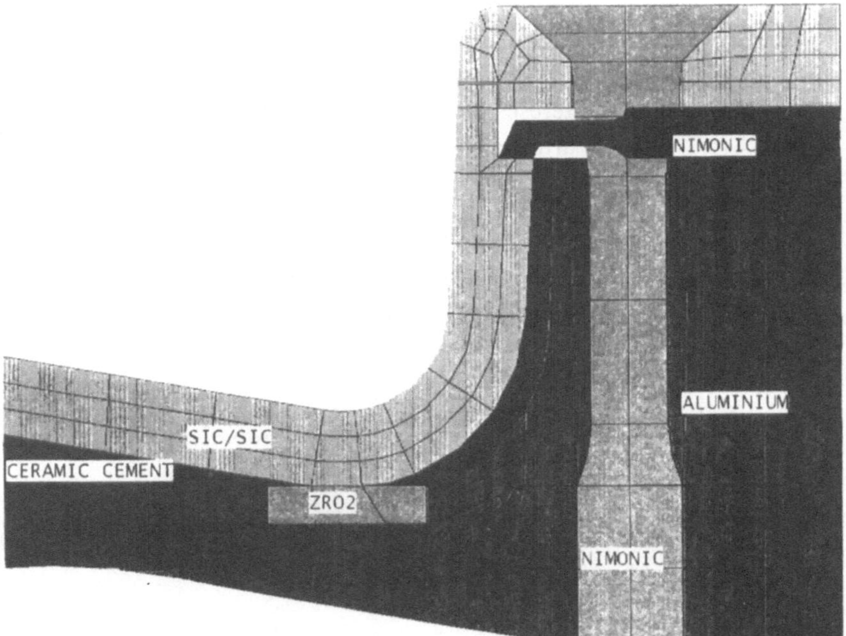

FIGURE 3. The FE mesh for the piston

Finally the SiC/SiC plate is fixed by additional screws, rotated for 60 degree in relation to the other screws. For the temperature analysis the anisotropy of the SiC/SiC material (see TABLE 1) was considered. The heat transfer coefficients for the insulated piston were not known, so that extrapolated values from those of a conventional piston were used to obtain the steady state temperature distribution. FIGURE 4 shows that the highest temperature level of 720 °C is located at the axis of the bowl and at the edge of the plate. Generally contact to cooler zones leads to high temperature gradients across the wall thickness, as can be seen at the contact area of the bowl to the zirconia ring.

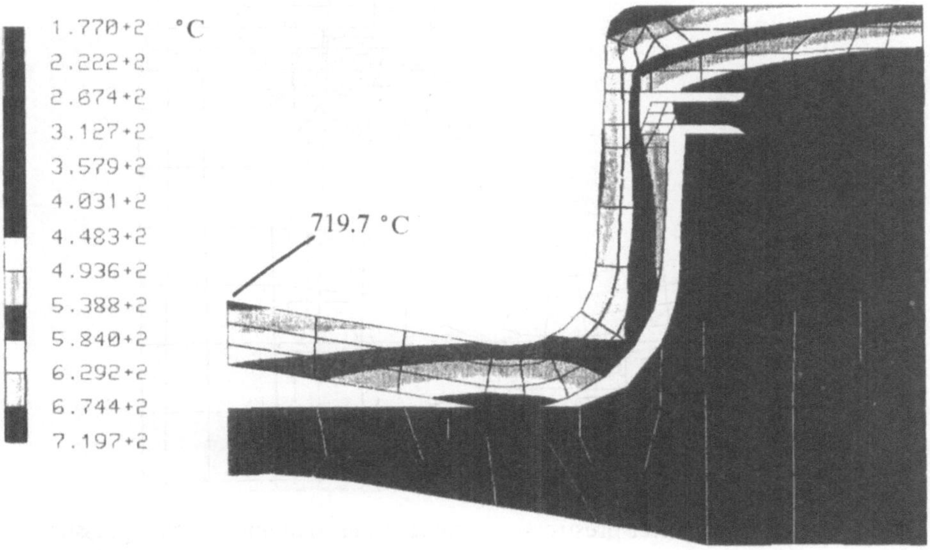

FIGURE 4. The steady state temperature distribution of the piston

Stress Analysis

To simulate the contact conditions between the different materials, gap elements were used, which transfer compression forces but erase tension forces on the contact zones. Due to the differences in thermal expansion of aluminium piston alloy and Nimonic bolts the level of the prestressing must be adjusted accurately. If the prestressing is too high an overload for the thread of the piston occurs. A too small prestress leads to loosening of the screws under operational conditions. In both cases the SiC/SiC plate and the spring element is no longer fixed, so a catastrophic failure of the engine will probably occur. The SiC/SiC insert was prestressed against the spring element by an initial displacement in axial direction. Then the temperature distribution computed previously was incorporated in the second increment. Finally the internal pressure along the inner SiC/SiC bowl outline and the SiC/SiC plate outline was introduced. The deformation of the piston, scaled by the factor of 10 is presented in FIGURE 5. The full lines show the deformed structure and the dashed lines the initial structure. Due to the differences in the thermal expansion coefficients and the temperature levels the radial displacement of the aluminium alloy is greater than that of the SiC/SiC plate and the Nimonic spring element. As consequence bending of the screws occurs. The axial stresses of the parts are presented in FIGURE 6, which are at a tolerable level of 140 MPa in tension at the contact edge of the spring element to the SiC/SiC bowl.

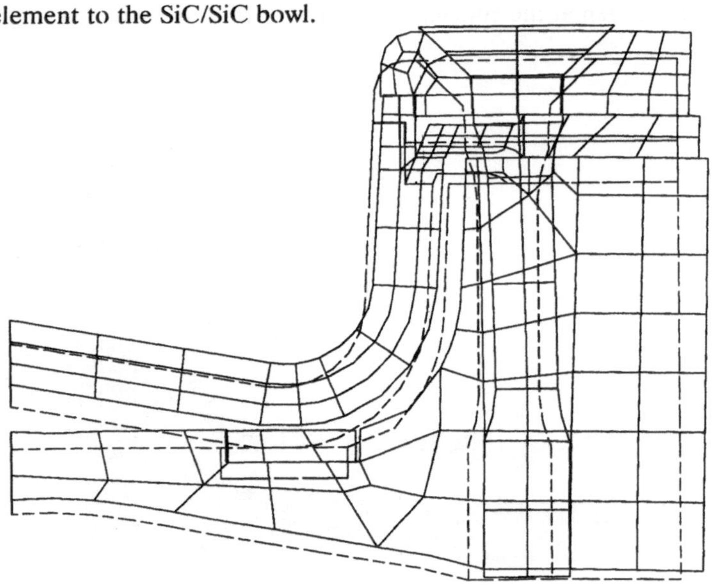

FIGURE 5. Deformation after prestressing, temperature load and internal pressure

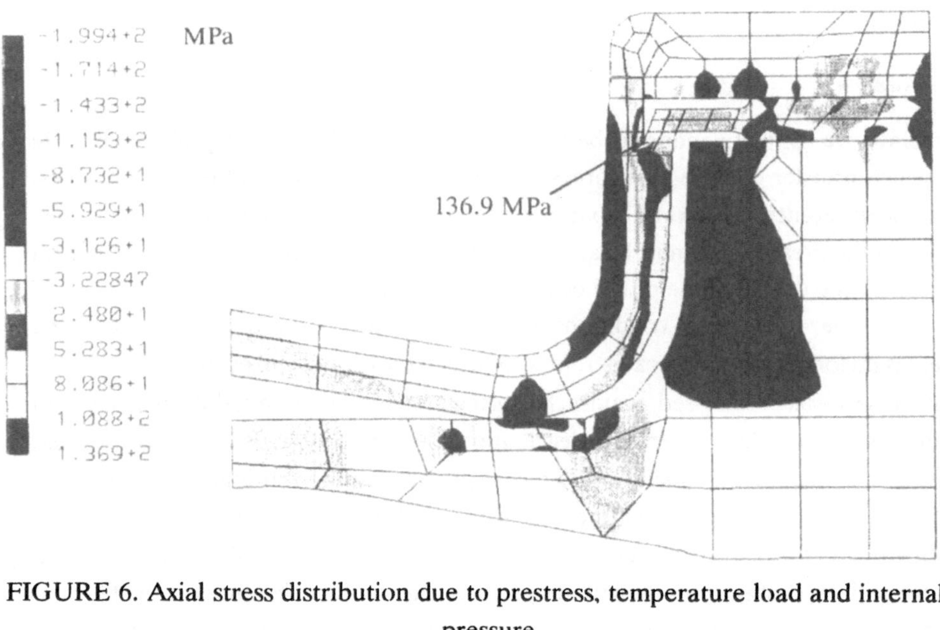

FIGURE 6. Axial stress distribution due to prestress, temperature load and internal pressure

However the most severe problems arise due to the much higher level of hoop stress in the SiC/SiC plate at the contact zone to the spring element, which is shown in FIGURE 7. These stresses are generated by the temperature gradient at the interface between the insert and the piston, which can not be avoided (see FIGURE 4).

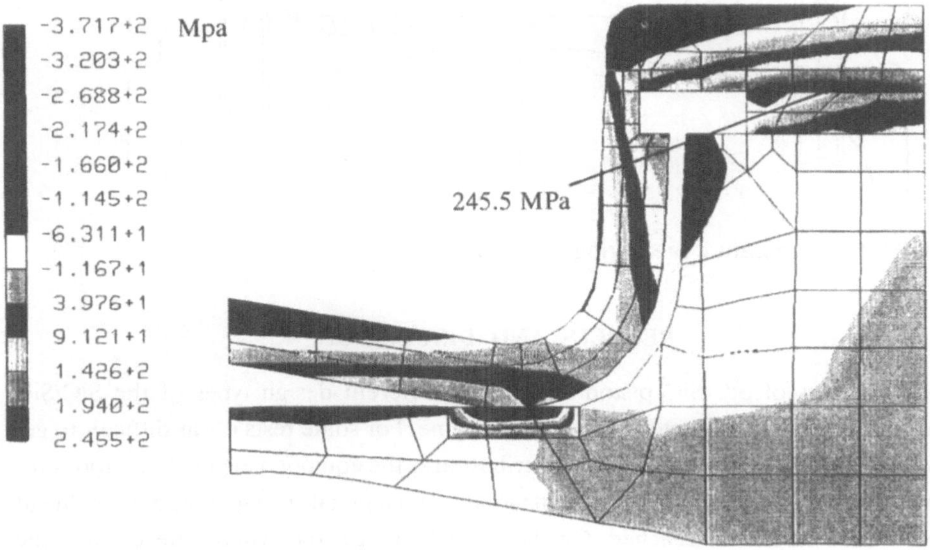

FIGURE 7. Hoop stress distribution after prestress, temperature load and internal pressure

INSTRUMENTATION OF THE ENGINE TEST

In order to get information about the temperature distribution in the piston some thermo couples were attached to two of the six pistons. To transfer the measured temperatures to the data recorder an inductive system was implemented in the engine, which was very sensitive to irregularities in the engine. In FIGURE 8 the temperature measuring points are presented. Based on the measurements the heat transfer coefficients at the surface of the piston has been evaluated and will be applied in future calculations.

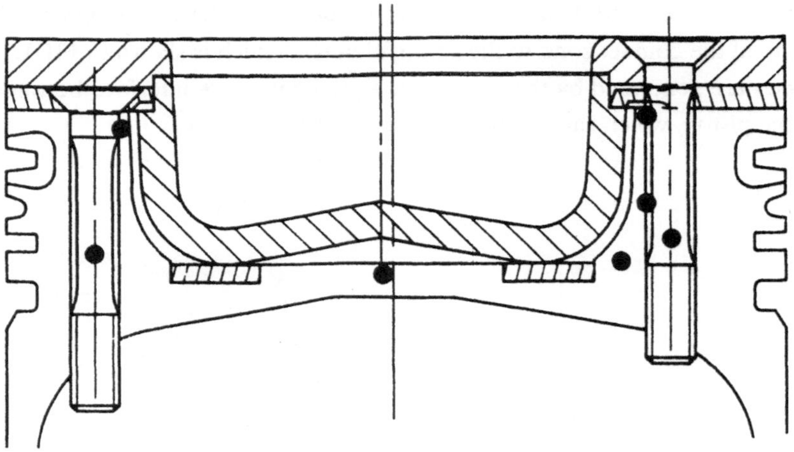

FIGURE 8. Temperature measuring points of the piston

RESULTS AND DISCUSSION

Several variants of SiC/SiC piston inserts and different design types of the SiC/SiC cylinder head plate were tested in the diesel engine. For some tests it was difficult to get evaluable data about the temperature distribution as the equipped engine had a too short life time. For the above described piston variant with a metal spring only 25 hours life at different load levels was reached. On the other hand, for the cylinder head plate, the cylinder liner and the flame sprayed valves much higher life times of more than 200 hours can be reported. In almost all cases the life of the components were not limited by the local stresses. These results show that due to the difficulties in the manufacturing of components with a complex shape the original material properties were not fully achieved and a large amount of additional work is necessary.

The temperature records confirm an increase of the mean gas temperature of combustion, but at the same time the heat transfer coefficients were reduced in comparison to the values, extrapolated from the conventional engine. In the SiC/SiC insert temperatures of up to 750 °C were reached (30 °C above predicted). For further details of the results of the engine test see [1]. Generally an insulation effect of the aluminium piston and the cast iron cylinder head were recorded. An increase of the engine efficiency was not measured, but it should be pointed out that the engine, the turbo charger, the injection pump and the combustion process parameters were not optimized, and were taken from the conventional engine without any changes. The fuel consumption was nearly unchanged in comparison to the conventional engine.

CONCLUSIONS

The envisaged increase of the mean gas temperature of the combustion process and the insulation effect of cylinder components were achieved. An increase of the engine efficiency was not measured and the fuel consumption of the engine was nearly the same as in a conventional engine. The life of the SiC/SiC parts was partly too short to get sufficient evaluable data about the changes in the combustion process. The combustion process parameters must be adapted to the use of CMC in the combustion chamber. However the material strength of finished components has to be improved to make longer engine tests possible.

This project was supported by the German BMFT. The french company SEP produced the SiC/SiC components.

REFERENCES

1. Leuchs, M., Ceramic Composite Applications in Diesel Engines - Experiences and Chances, 4th International Symposium on Ceramic Materials & Components for Engines, Goeteborg, Sweden, 1991

2. Agatonovic P. and Koch F., The Thermal Analysis and Optimisation in Thermal Insulation of Engine Components, 5th Int. Conf. on Numerical Methods for Thermal Problems in Montreal, Quebeck, Canada, Ed. by R.W. Lewis, Pineridge Press.

3. Dauchier M., Bernart G. and Bonnet C., Properties of Silicon Carbide based Ceramic-Ceramic Composite, 30th National SAMPE March 19-21, 1985, p. 1519.

4. Agatonovic P. and Grunmach R., Fatigue Life Tests on Composite Ceramic, Proc. 7th European Conference on Fracture (ECF7), Budapest, Hungary, 1988.

CERAMIC COMPOSITE APPLICATIONS IN DIESEL ENGINES - EXPERIENCES AND CHANCES

Martin Leuchs
Materials Department
MAN Technologie AG
Bauschingerstr. 20, D-8000 München 50, Germany

ABSTRACT

Ceramic matrix composite (CMC) material produced by the French company SEP has been applied in a Diesel engine. Hot components in cylinders have been developed to protect metallic components from high thermal loads. A cylinder head plate, a piston cap and a section of the cylinder liner have been designed, produced and mounted. The valves were covered with a 2 mm layer of flame sprayed zirconia. The performance of the SiC/SiC composite material in test runs was good. The head plate and the cylinder liner insert survived more than 200 hours at the full engine load (about 220 kW at 2000 rpm) without problems. Only the more complex piston component did not achieve satisfactory life times during tests.
With respect to engine performance only small effects on exhaust gas temperature and its pollutants could be observed. The engine efficiency was practically unchanged by the ceramic material. The thermal protection of hot metal components could be demonstrated; the temperature of the valve seats was reduced by about 50 K.
Considering the limited benefits, which could be verified in this EUREKA-programme, it seems that high quality ceramics like CMC will have difficulties to be introduced as a structural component in an industrial mass product like a Diesel engine.
Material costs itself and extensive work still necessary for the industrialization are the key problems. It is more likely that small ceramic parts, which locally fulfill a well defined function, will succeed first.

INTRODUCTION

The application of ceramics in Diesel engines has been and still is widely discussed in many publications (1 - 3). Expectations about the chances to improve engine efficiency have been high in the past, but documented results of the mid and late 80ies showed that there are effects which counter these expectations (4, 5). To achieve thermal insulation, mostly ceramic materials have been applied, both monolithic ceramics and flame sprayed ceramic layers. The known lack of reliability of monolithic ceramics has been a major draw-back in many cases. Therefore a ceramic material reinforced with long ceramic fibres showed a promissing perspective.

The mechanical properties of this SiC-fibre/SiC-matrix material (CERASEP[R] by the French company SEP) demonstrate features, which are a step towards metal-like behaviour (6, 7). The application of that material in the cylinders of a Diesel engine intended to demonstrate

- good performance of CMC material,
- thermal protection of hot components,
- improved engine performance.

MATERIALS AND METHODS

The fibre orientation of the material is characterized by the production method (6). Layers of plain weave are infiltrated with matrix material by a chemical vapour infiltration (CVI) method. This means that material properties are anisotropic and basically determined by in-plane and out-of-plane data. The design methods used for the development of the components are described elsewhere (8).

The engine tests have been performed with a regular six cylinder truck engine, 128 mm bore, 11.9 l total cylinder capacity and a nominal power of 240 kW at 2000 rpm. The engine was turbo-charged; the air was cooled by the engine cooling water. Some tests were performed in an one-cylinder engine with identical geometries.

The SiC/SiC components have been designed with the aid of FEM calculations (8) and have been mounted as shown in fig. 1.

Since the valves cover about one third of the surfaces of the cylinder head, they have been flame sprayed with a 2 mm zirconia layer to ensure thermal protection in that area.

The joining method applied was mechanical with screws for the cylinder headplate and the piston cap and with shrink fitting for the cylinder liner section. Other solutions like chemical bonding or casting would have needed much higher development efforts.

The open porosity of the SiC/SiC-material, which is closely linked to the production process and the final machining of the components, had to be sealed. Otherwise extra volumes would disturb the burning process and reduce the compression of the engine. A glassy material sucessfully tightened the porosity. Gaps between SiC/SiC and metals had to be filled with ceramic material.

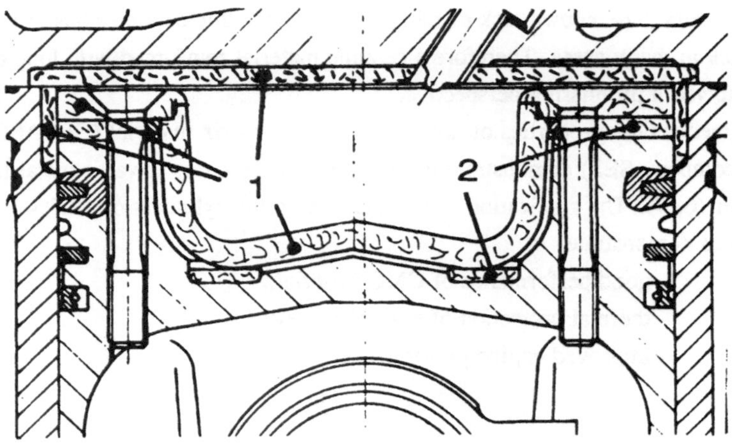

Figure 1. In this design the combustion chamber of the Diesel engine has been equipped with SiC/SiC components (1) and with zirconia (2).

Some results have been gathered from an engine, in which the ceramic material has not been 4 - 6 mm thick SiC/SiC material but a 2 mm flame sprayed zirconia layer. Both material systems differ somewhat in their thermal insulation, since 2 mm zirconia (2 W/mK) insolates about 30 % better than 4 - 6 mm SiC/SiC (9 W/mK, average value).

All temperatures have been measured with standard thermocouples; only the temperature measurement in the piston required a system, which is based on inductive coupling of a coil in the piston and one in the crank-case. The temperature was measured in the metal about 1 mm behind the ceramic materials in different locations.

Exhaust gas composition, fuel consumption and power output were determined with standard equipment, too.

RESULTS

The results gathered from several engine runs with SiC/SiC components and with zirconia covered components can be grouped in

- material performance,
- thermal protection of metals,
- engine performance.

Material Performance

The cylinder head plate and the cylinder liner survived more than 200 hours at full load, running about 8 hours per day. No damage could be observed.

In one test run damage was caused by a screw head which loosened during the test run. It was squeezed many times between piston and head plate before the engine stopped. The damage showed the excellent fracture behaviour of this long fibre reinforced SiC/SiC-material (fig. 2). Every monolithic ceramic component would have broken into many pieces.

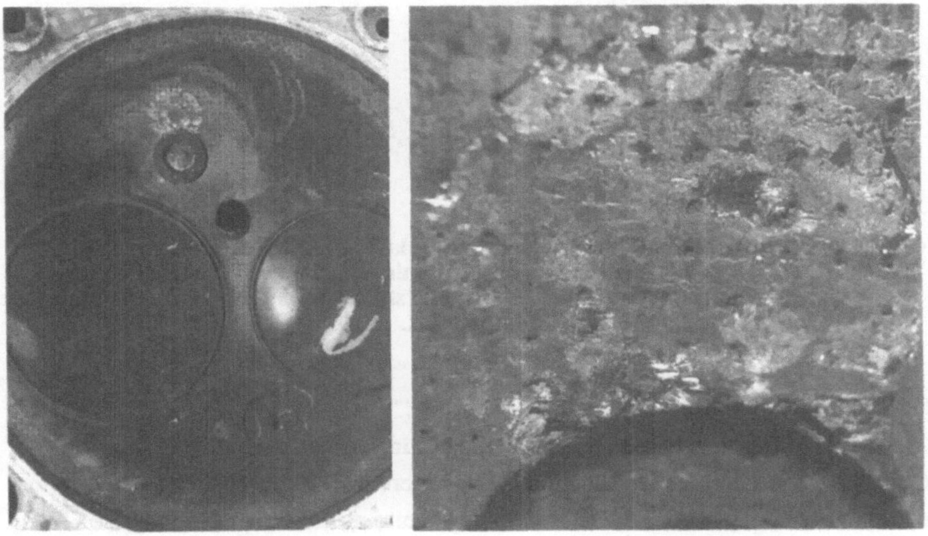

Figure 2. The fracture behaviour of SiC/SiC can be seen in this case of a damage caused by a loosened screw head. Overview (left) and detail (right) demonstrate the lack of catastrophic failure.

The flame sprayed zirconia on the cylinder head and valves survived more than 200 hours, too (fig. 3).

The achieved life times for the pistons did not reach this level, however. Due to high hoop stresses in the piston top, the SiC/SiC-component failed in this area after a few running hours. The zirconia layers in the center of the piston also failed after a few hours. Difficult spraying conditions in this area did not allow a sufficient quality of the layer system.

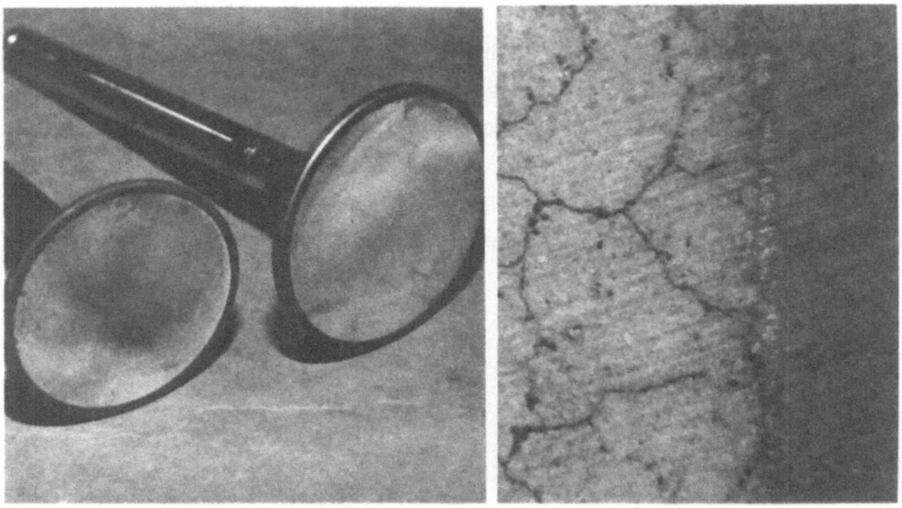

Figure 3. State of valves flame sprayed with a PSZ layer of 2 mm after more than 200 hours (left). The detail (right) shows the structure of vertical micro cracks typical for such layers. No change of this structure was observed in intermediate inspections.

Thermal Protection of Metals

In the metallic components behind the ceramic coating (zirconia and SiC/SiC) a temperature reduction between about 50 K up to more than 100 K could be demonstrated. In the valve seats the lower temperature measured is caused by the application of the ceramic material (fig. 4). For all other locations it has to be considered, that the position of the thermocouples in the ceramic engine was closer to the cooling system than it was in the metallic engine. It is estimated that this effect yields a reduction of 30 - 50 K. Figure 5 shows a result of a defined position in the cylinder head.

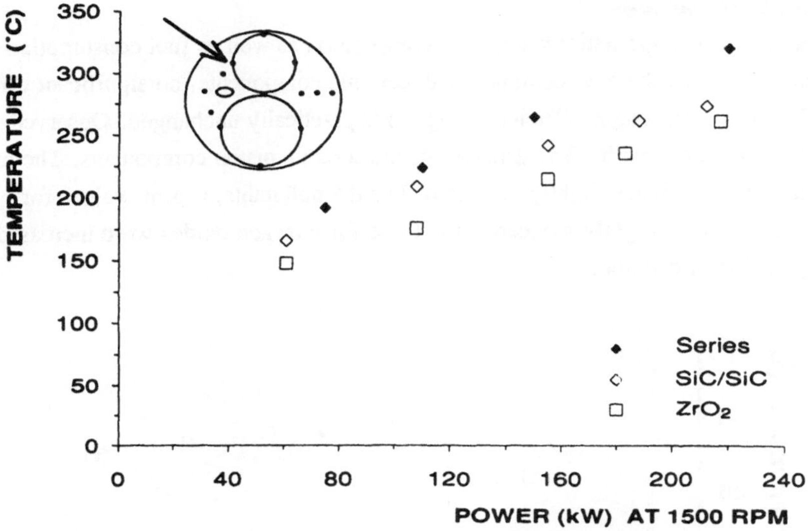

Figure 4. The effect of thermal protection by ceramics in the valve seat is shown in this graph. The point of measurement is indicated in the scheme and located in the seat of the exhaust valve.

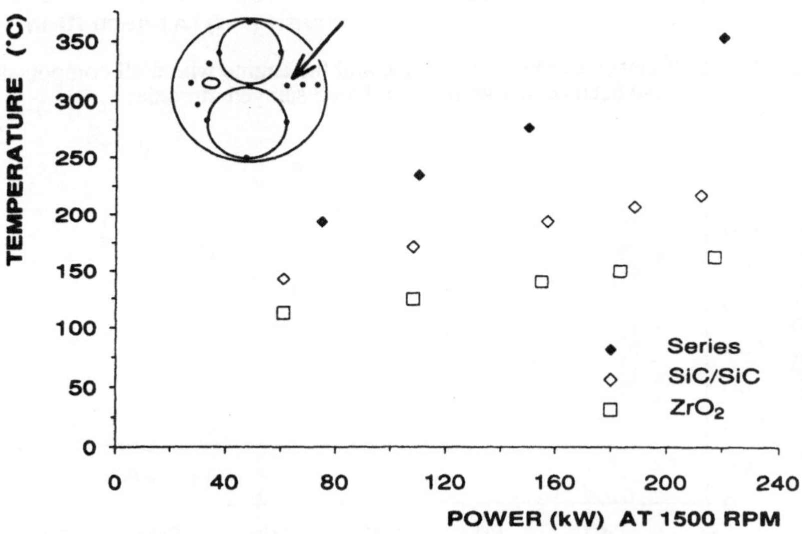

Figure 5. The effect of thermal protection of one position in the cylinder head is bigger, partly due to the fact that the thermocouple in the ceramic versions is closer to the cooling water.

Engine Performance

Comparing exhaust gas temperatures and composition as well as fuel consumption of the metallic engine and the ones equipped with ceramic components, no significant effect could be found. The engine efficiency (fig. 6) is practically unchanged. Observed differences were caused by first damages of the ceramic piston components. The exhaust temperatures (fig. 7) are slightly increased. For the pollutants, p.p.m. values for carbohydrates were slightly reduced while those for nitrogen oxides were increased by the application of ceramics.

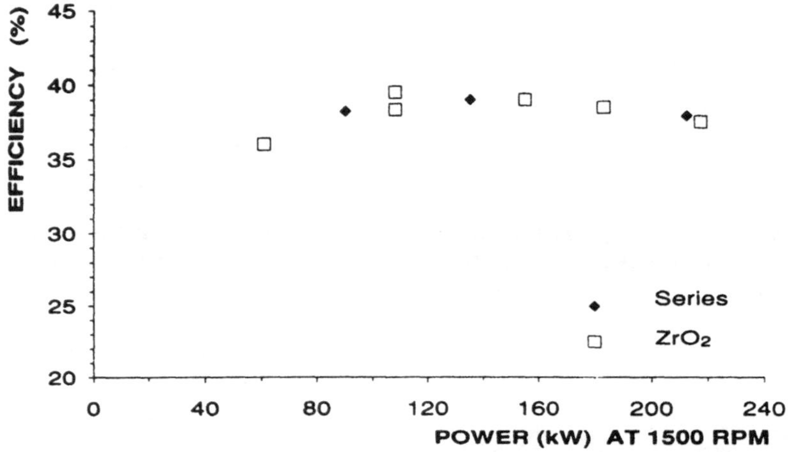

Figure 6. Engine efficiency of the series engine and the engine where all components have been coated with 2 mm flame sprayed zirconia.

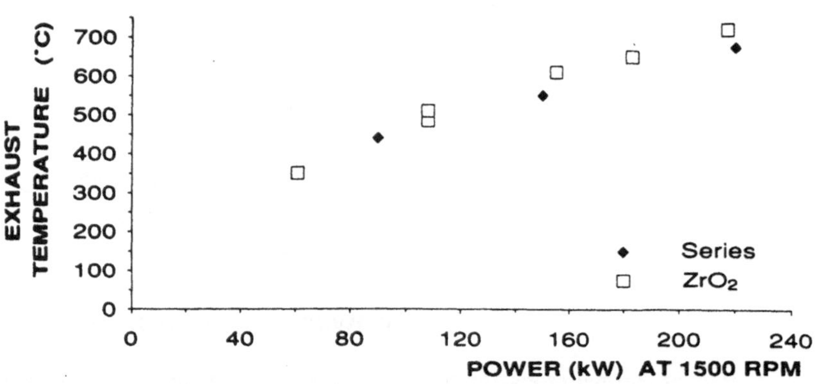

Figure 7. Exhaust gas temperature of the series engine and the engine with zirconia coated components.

CONCLUSIONS

The experience gained by design and application of structural ceramic components in a Diesel engine can be summarized by the following statements:

1. The SiC/SiC-material itself proved its ability to survive the conditions in the burning chamber of a Diesel engine. This means high thermal stresses induced by oscillating heat generation and peak pressures up to 125 bar at frequences of about 25 cycles per second. The material showed its extraordinary behaviour on fracture; no catastrophic failure in the case of damage was observed.

2. By application of ceramic materials, the temperature load on hot metallic components can be reduced by about 50 K. This means that the thermal load in the engine could be increased, e. g. via increased charging, which in turn would increase the specific power-output of the engine.

3. The data indicate, that effects on engine efficiency and exhaust gas pollutants are rather small. One has to keep in mind, though, that no optimization process comparable to the long history of metallic components has taken place. Except for small details, the same engine set-up (injection timing, piston geometry, etc.) as in the metallic engine has been applied.

4. The technical key problems can be seen in the field of design and joining, especially for the piston components. It seems that a lot of development work still is neccessary to reach a level, which is sufficient for industrial application.

5. The economic perspective of CMC and other high quality ceramic materials in Diesel engines is governed by the technical advantages, which engine supplier and customer can turn into profit. The technical advantages can be realized by other materials and technologies in many cases. Presently, the cost/benefit considerations indicate, that CMC material will not replace metallic structural components in Diesel engines in a short or mid-term time scale. Only small functional components, which can be applied with little efforts, are conceivable in industrial mass products at the moment.

REFERENCES

1. Kamo, R. and Bryzik, W., Adiabatic Turbocompound Engine Performance Prediciton, SAE paper 780 068, 1978, p. 213.

2. Woods, M. E. and Oda I., PSZ Ceramics for Adiabatic Engine Components, SAE paper 820 429, 1982.

3. Wabon, N., Kyrtatos, N. P. and Holmes, K., The Performance Potential of Limited Cooled Diesel Engines, Proc. Instn. Mech. Engrs., vol. 197 A, p. 197.

4. Zapf, H., Grenzen und Möglichkeiten eines wärmedichten Brennraumes bei Dieselmotoren, VDI-Bericht Nr. 238, 1975, p. 85.

5. Woschni G., Spindler, W. and Kolesa, K., Heat Insulation of Combustion Chamber walls - A Measure to Decrease the Fuel Consumption of I. C. Engines ?, SAE paper 870 339, 1987.

6. Lacombe, A. und Bonnet, C., Ceramic Matrix Composites, Key Materials for Future Space Plane Technologies, AIAA Second International Aerospace Planes Conference, Oct. 1990, Orlando, FL, USA.

7. Agatonovic, P. and Grunmach, R., Fatigue Life Test on Composite Ceramic, Proc. 7th Europ. Conf. on Fracture, 1988, Budapest, Hungary.

8. Agatonovic, P. and Sygulla, D., Analysis and Test of Ceramic Matrix Composites in Diesel Engines, 4th Int. Symp. on Ceramic Materials and Components for Engines, June 1991, Göteborg, Sweden.

AKNOWLEDGEMENTS

The work has been supported by the German Ministry of Research and Technology. The partner of this EUREKA programme has been the French company Société Européenne de Propulsion (SEP), which supplied the CMC material and components.

MANUFACTURING OF CERAMIC VALVE DISCS AND THEIR USE IN A ROTOR CAM ENGINE

JESPER BRANDT*, LARS KAHLMAN AND ROBERT POMPE
Swedish Ceramic Institute (SCI)
Box 5403, 402 29 Göteborg, Sweden

KARL-ERIK LINDBLAD
Kesol Produktion AB
Box 119, 441 23 Alingsås, Sweden

ABSTRACT

Valve discs of Si_3N_4 to be used as a vital part in a recently developed rotor cam engine have been manufactured. Si_3N_4 ceramic has been selected because of its combination of high thermal shock resistance, high wear resistance, low friction coefficient and dimensional stability - all these properties being essential to the optimal running function of the engine. The discs have been produced by the NPS (Nitrided Pressureless Sintering)-technique. Design, material selection, processing procedure, engine test results as well as further design and processing development will be discussed. The rotor cam engine with the ceramic valve disc assembled has been successfully test run for more than 80 hours.

INTRODUCTION

In the near future increasingly tough regulations for exhaust emission control of different kinds of internal combustion engines will be adopted. An illustrative example is outboard motors for pleasure boats which will come more into focus as their number increases putting additional stress on the ecological balance at the coasts and lakes worldwide. It will be necessary to design and manufacture efficient as well as more silent combustion engines where catalytic exhaust gas filters easily can be mounted.

During the last few years the company Kesol Produktion AB has developed an internal combustion engine designed as a four-stroke rotor cam engine [1]. The engine consists of a considerably smaller number of moving parts compared to the traditional Otto-engines. This makes it possible to accomplish economically favourable manufacturing conditions and minimises the maintenance requirements. The engine is silent and the design makes it, unlike the two-stroke engine, simple to attach to a catalytic exhaust gas filter. Better fuel economy due to increased efficiency and reduced emissions make the engine especially suitable to outboard motors for pleasure boats. Naturally, developing and modifying the engine design for other fields of application is currently the subject of an intensive research work.

A crucial step to enhance the efficiency of the engine invented has been to replace a valve disc of cast iron by one of a ceramic material. The most important reason for taking this step is to reduce the friction between the valve disc and a port disc of steel and to eliminate leakage due to progressively occurring deformation of the metallic parts. The objective for the Swedish Ceramic Institute has been to utilize a suitable technique to manufacture prototype valve discs of an adequate engineering ceramic material.

Introducing structural ceramic components into a new engine design is of great significance, being an excellant opportunity to demonstrate the key role of an engineering ceramic part in a specific application.

ENGINE FUNCTION

The operation of the four-stroke rotor cam engine is outlined in Figure 1. Four cylinders are rotating, like the spokes of a wheel, inside an oval ring with the heads of the cylinders pointing inwards. The oval ring makes the pistons move in and out when running, like a conventional cam shaft. As the cylinders rotate a port disc with four ports, one for each cylinder, passes apertures in a stationary valve disc through which fuel enters and exhaust gases exit. The rotating port disc is joined with a driving shaft. The cylinder volume of the prototype engine is 200 cm^3 and the output power is at present approximately 6 hp at 2300 rpm. As the design is optimized the efficiency will be further increased. The valve disc and the port disc are shown in Figure 2.

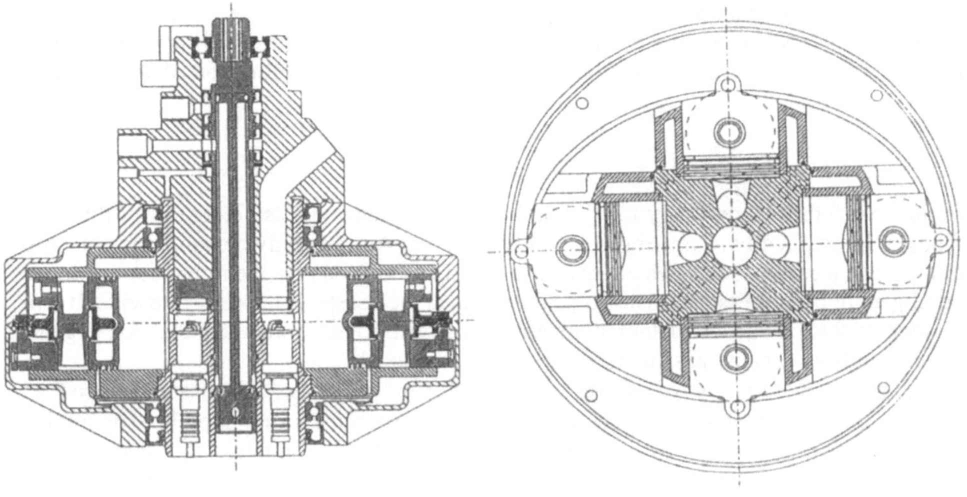

Figure 1. Cut away rotor cam engine.

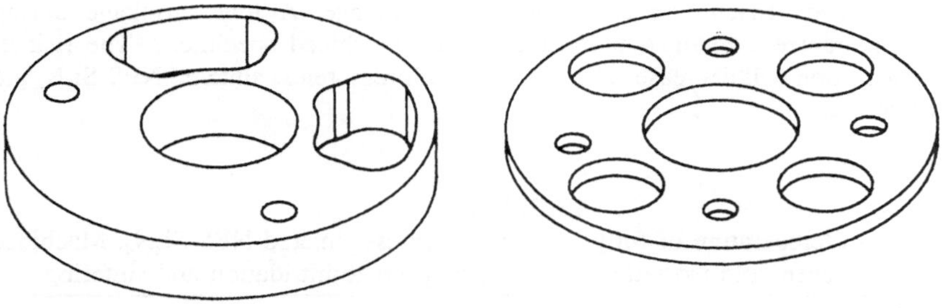

Figure 2. Drawings of the valve disc and the port disc.

Low friction and adequate sealing between the port disc and the valve disc are essential to an optimal running function of the engine. When using components of steel lubricants must be utilized. This reduces the efficiency of the engine due to lubricants oozing out from the disc contact area and mixing with the entering fuel. By selecting a ceramic valve disc to be run towards a port disc of steel the friction coefficient is considerably lowered. A mixture of fuel, exhaust gases and soot penetrating the thin space between the discs is sufficient to lubricate the surfaces and consequently makes the supply of extra lubricants unnecessary. As a matter of fact, soot (inexhaustible residues of carbon compounds) has proved to be a superior lubricant for ceramic engine parts [2].

MATERIAL SELECTION

In the prototype engine the port disc is rotating with a maximum of 3000 rpm and the pressure towards the valve disc amounts to 10 kg/cm^2. The difference in temperature between inlet and outlet gases through the ports is approximately 600°C. As a consequence of these conditions, the material properties required are low friction coefficient, high wear resistance, thermal shock resistance and dimensional stability under varying stress conditions. Selecting a material which, in addition, guarantees reliability and durability is a prerequisite to ensure a long life.

There are a number of suitable, sufficiently characterized structural oxide and non oxide ceramics to be considered for the current application, based on Al_2O_3, SiC, ZrO_2 and Si_3N_4. Al_2O_3 and SiC have low impact strength (low fracture toughness) and ZrO_2 (PSZ) is not thermally stable in water vapour environment. The oxides are less suitable due to their low thermal chock resistance. A Si_3N_4 based material may with good margin correspond to the given requirements. Si_3N_4 has a good resistance to wear, thermal shock, corrosion, impacts and is also dimensionally stable at high temperatures. Another reason to choose a Si_3N_4 material is the possibility to make use of the favourable NPS [3] process route including presintering, reaction bonding and final pressureless sintering which facilitates green body machining and minimises shrinkage. A low shrinkage during sintering will guarantee a retained uniformity of the machined specimens. The risk of warping and distortion will decrease. The properties of pressureless sintered NPS-Si_3N_4 are shown in Table I.

TABLE 1
Properties at room temperature of fully dense pressureless sintered NPS-Si_3N_4. Machined test bars have been cold isostatic pressed (CIP) prior to nitridation and sintering

Density	3.2 g/cm^3	Hardness	1500 Hv
Bending strength (4-point)	700 MPa	Thermal exp.	3.2*10^{-6} K^{-1}
Young's Modulus	300 GPa	Thermal conduct.	20 W/mK[#]
Fracture toughness	6 MPam$^{1/2}$		

[#]Literature data

For the current application where a low friction coefficient is desired there may also be some benefits with a material consisting of a few per cent remaining porosity. This structure enables fuels, exhaust gases and soot to be entrapped in the pores ensuring improved lubrication under more extreme running conditions of the engine.

FABRICATION PROCESS

Due to the geometrical design of the valve disc with thin outer walls at the ports and the narrow cylindricity tolerances, Figure 2, the fabrication procedure of the disc is of great importance in order to avoid any sources of dimensional distortions during sintering, which would require expensive machining. A forming method to be considered, because of the flat geometry and the relatively minor thickness of the disc, is uniaxial pressing where the apertures at the same time are produced. For large scale production this is an advantageous method. To form the prototype discs, however, the cold isostatic pressing (CIP) technique has been employed and the ports have subsequently been drilled, prior to nitridation.

To drill the ports the material needs to have sufficient strength in order (1) to preserve structural integrity during handling, (2) to prevent breaking of the outer thin sections and (3) to allow for using inexpensive machining tools. On the other hand, the strength should not be too high causing chipping to occur. It is also desirable to minimize the shrinkage of the valve discs during sintering to avoid distortion of the thin port sections and to ensure that a minimum of machining is needed to obtain the required flatness. These special requirements may be accomplished by utilizing the nitrided pressureless sintering technique of Si_3N_4 (NPS) [3, 4]. Silicon and Si_3N_4, of a specific ratio, are ball milled together. Si_3N_4 present is active as dispersant during the milling resulting in fine grained submicron silicon. The Si_3N_4 present acts as a heat sink during the exothermal nitridation reaction and, in addition, prevents excessive sintering of silicon by keeping the particles apart. This results in a very short nitridation time (3-4 h), compared to the ordinary RBSN processing where corresponding times are 20-100 h [5]. The silicon present enables presintering to a highly porous (40%) but strong (50-100 MPa in 3-point bend) green body prior to nitridation, and this green body can be machined by high speed steel (HSS) tooling. The shrinkage of the nitrided specimens during sintering is rather small (~10 % linear) thus favouring preservation of a uniform geometry after shrinkage.

The processing steps are outlined in Figure 3. According to a standard NPS-recipe, 52 wt% Si[a], 37 wt% Si_3N_4[b], 8.1 wt% Y_2O_3[c] and 2.7 wt% Al_2O_3[d] powders were dispersed in cyclohexane with 3 wt% KD-3 deflocculant and 3 wt% oleic acid added as pressing aid. The sintering agents Y_2O_3 and Al_2O_3 will correspond to 6 and 2 wt%, respectively after nitridation. The mixture was ball milled for 96 hours. After evaporation and sieve-granulation, discs 90 mm in diameter by 14 mm in height were uniaxially prepressed and subsequently cold isostatically pressed (CIP:ed) at 300 MPa.

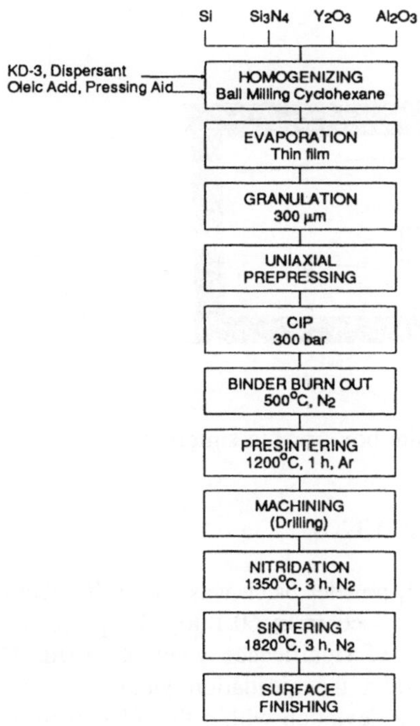

Figure 3. Processing schedule.

---

[a]Sicomill 2D, Permascand AB, Sweden, BET=2.1 m^2/g

[b]P-95 wet, Permascand AB, Sweden, 11 m^2/g

[c]Grade fine, HC Starck, Germany, 14 m^2/g

[d]AKP-30, Sumitomo Chem, Japan, 7.5m^2/g

The specimens were burned out in nitrogen atmosphere at 500°C and then presintered in a graphite furnace in argon at 1200°C for 1 hour. As mentioned earlier the silicon creates a connected network increasing the strength without raising the brittleness. Blanks were also nitrided and sintered before machining to determine the exact magnitude of the shrinkage. The discs were then machined to the desired geometry, according to figure 2. Due to the presintering at 1200°C this operation was carried out easily under dry conditions using a conventional converting machine. Attempts of machining the pressed green and nitrided bodies frequently resulted in structural damage or chipping, respectively. The nitridation process was performed in nitrogen at 1200°C for 1 h and 1350°C for 3 h. During the following pressureless sintering process, the powder bed was designed as shown in Figure 4 to avoid warping as well as decomposition. On both sides of the discs at the drilled ports BN-plates were placed providing frictionless sliding during sintering. The discs were embedded in coarse Si_3N_4-powder with a thin layer of quartz underneath the specimens. The sintering was then performed in nitrogen at 1820°C for 3 h.

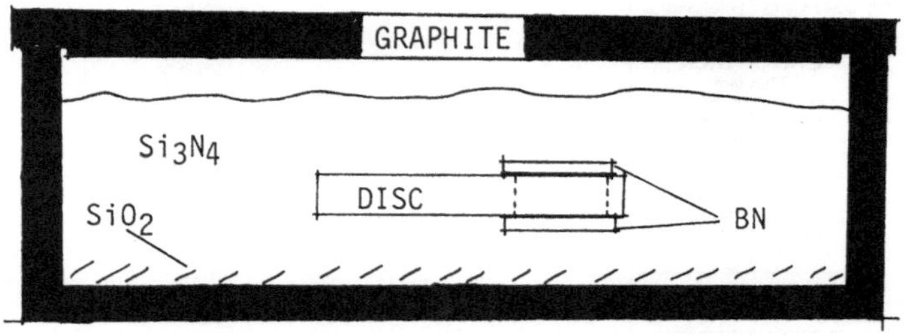

Figure 4. Suspension of a nitrided disc in a graphite box prior to sintering.

FABRICATION RESULTS

The specific surface area of the ball milled Si/Si_3N_4 powder batch was 14.1 m^2/g. During the nitridation of the machined discs the density raised from 60.1 to 70.5 per cent of theoretical density while no dimensional changes of significance were observed. The preceding presintering step did not negatively affect the nitridation yield. The linear shrinkage during sintering was 10.8 per cent with a deviation within the tolerances. The final density came to 99.6 per cent of the theoretical. As discussed formerly there will presumably be some benefits of a certain remaining porosity. After minor hollow grinding and polishing the discs were parallel, with a deviation of ±0.01 mm over the total surface area. A photograph of a polished valve disc is shown in Figure 5.

Figure 5. Photograph of a Si_3N_4 valve disc.

ENGINE TEST RESULT

The valve discs, each at a time, have been assembled in the rotor cam engine for running tests. The engine has been subjected to several different test conditions where, for instance, the loads and the speeds have varied.

In comparison to the valve disc of steel the ceramic disc exhibited very promising results. As predicted, the friction without any addition of lubricants between the ceramic valve disc and the port disc of steel was sufficiently low. The sealing between the discs was excellent. When the engine was run under top load conditions a certain leakage was noted. To begin with, the thermal expansion coefficient is very different and accordingly the mismatch in heat transfer will cause some warping of the port of steel. The ceramic component is moreover dimensionally stable under wide thermal stress conditions and will thus not conform to the displacement of the steel material at higher temperatures. Consequently, the port disc is also planned to be produced of the same ceramic material as the valve disc. After 100 hours of test runs the interior parts of the engine were disassembled. Except for the fact that the ceramic valve disc was almost black it did not exhibit any visible damage. The contrivance of the two discs wearing on each other indeed resulted in a superior self-supporting polishing. Examination of the valve disc in an optical microscope showed that remaining soot covered the surface with a thin layer and filled up the surface pores.

An attempt was carried out to polish the surfaces of one of the discs more rigorously by lapping. Remarkably, when running this disc, the engine almost seized after a short time. The explanation is that too thin a space between the valve and the port disc prevents lubricating fuel, soot etc from penetrating this space. In addition, due to the even surfaces of the discs strong adhesive forces emerge resulting in rapidly increasing friction.

In the subsequent period there will be a more thorough investigation of the ceramic parts. Port discs and new valve discs will be manufactured of a ceramic material with optimized port design to achieve enhanced efficiency. The manufacturing process will be simplified by designing a suitable uniaxial pressing tool adapted for larger scale production. In a parallel project the influence of soot etc as a lubricant will be examined.

ACKNOWLEDGMENTS

The authors thank the Kesol Produktion AB company for the endeavours that made this valuable project possible to realize, the Swedish Institute for Production Engineering (IVF) in Luleå, Sweden, for successful hollow grinding of the ceramic valve discs, and last but not least the colleagues at the Swedish Ceramic Institute who have supported the work in varying ways.

REFERENCES

1. K E Lindblad, Patent, nr 8803791-6, PCT nr SE 89/00582, Sweden.

2. H Ishii, F Honda and K Nakajima, "Effect of Soot Interposed Between the Sliding Ceramic Surfaces", Proc. Jap. Int. Trib. Conf., Nagoya (1990).

3. R Pompe, L Hermansson and R Carlsson, "Development of Commercially Advantageous Techniques for Fabrication of Low Shrinkage Si_3N_4-Based Materials", Sprechsaal 115, 1098-1101 (1982).

4. L K L Falk, R Pompe and G L Dunlop, "Development of Microstructure During the Fabrication of Si_3N_4 by Nitridation and Pressureless Sintering of $Si:Si_3N_4$ Compacts", J. Mater. Sci., 20, 3445-56 (1985).

5. J Heinrich, "Nitridation of Silicon under High Pressure", Adv. Cer. Mat., **2** (3A) 239-242 (1987).

Si_3N_4 COMPOSITE PISTON CAP FOR ADIABATIC ENGINE

JIEMO TIAN, XIAOHUA TONG, BAOQING ZHANG
LINGLING WANG, JIANHUA ZHOU
Beijing Fine Ceramic Laboratory of Institute of Nuclear
Energy Technology, Tsinghua University,
P.O.Box 1021, Beijing 102201, CHINA

ABSTRACT

Si_3N_4 composite piston cap is made of Si_3N_4 matrix adding Y_2O_3, TiC and Al_2O_3. The microstructure of Si_3N_4 composite shows α—Si_3N_4 and whisker-like β—Si_3N_4 grains homogeneously distribute in the matrix. The second phase particles are spread in the α—Si_3N_4 grains. The reinforced mechanism of two phases is discussed. The Si_3N_4 composite piston cap has been worked for over 100 hrs at over 800 ℃ for 1500 r/min during the high or full power run without water cooling. Experimental results show two phases microstructure gives the high mechanical properties which can make the resistance to thermal shock and high temperature.

INTRODUCTION

Ceramic piston cap is a important component for adiabatic engine. Up to now, Si_3N_4 piston cap is one of successful parts used in the adiabatic engine[1,2]. Ceramic piston caps have been made of different ceramic composites[3,4,5]. Because the adiabatic engine used ceramics will be able to further improve fuel economy and use a low grade of fuel, the many countries pay attention and supports to the research on the adiabatic engine. In China, the ceramic components used for adiabatic diesel engine have been developed and tested on the diesel engine (135 mm in the diameter of the piston cap). In order to thermal insulation, the plasma spray psz(ZrO_2) cylinder deck and the air gap between the piston cap and metal piston were adapted. This paper describes the preparation of Si_3N_4 composite piston cap and test on the adiabatic diesel engine.

EXPERIMENTAL PROCEDURE

The piston caps have been made of Si_3N_4 matrix added 5 wt% Y_2O_3, 3 wt% TiC and 7 wt% Al_2O_3. These powders were mixed and formed under 20 MPa pressure in graphite die. Then, the die was put in a hot press furnace and sintered at 30 MPa and 1600—1650 ℃ under nitrogen atmosphere for one hr. The end, the Si_3N_4 composite piston cap has been machined and polished. The ceramic cap has been set up at the piston and carried out test on the adiabatic engine. In order to get the data of mechanical properties of the Si_3N_4 composite piston, samples of the bend strength and fracture toughness were cut out from the ceramic cap, the size of sample is 3×4×40 mm, the bend strength and fracture toughness were measured by three points bending strength method with the span of 30 mm and chevron notched beam technique. Chevron notches were 0.15 mm.

The thermal conductivity was measured by the laser method. The samples were 10 mm in the diameter and 1 mm in the thick. Thermal expansion coefficient were measured also.

RESULTS AND DISCUSSION

Properties

The bend strength and fracture toughness are $\sigma_b = 800$ MPa; $K_{Ic} = 7$ $MPam^{1/2}$ at 25 ℃ and σ_b and K_{Ic} are small differences at 1000 ℃, therefore, they introduce the small thermal stress (< 150 MPa).

The thermal conductivity (λ) measured by laser method is 13-16 W/mk (sample is φ 10mm×1mm). The thermal expansion coefficient is 3×10^{-6}/℃ and the density is 3.2 g/cm^3.

Microstructure

The microstructure consisted of Si_3N_4 added Y_2O_3, Al_2O_3 and TiC has been observed by SEM and analysed by X-ray diffraction. The results show β—Si_3N_4 grains have been appeared in the form of the whisker-like and α—Si_3N_4 grains have been reinforced by Al_2O_3 particles dispersion (Fig.1; 2). Whisker-like β—Si_3N_4 grains increased the fracture toughness as ceramic whiskers in other ceramic matrix.

ω Shape Ceramic Piston Caps of Si_3N_4 Composite

ω Shape ceramic piston caps of Si_3N_4 composite have been prepared (Fig.3) and set up on the metal piston of the diesel engine (Fig.4). The ω shape ceramic piston caps were tested. The experimental conditions are listed in the table 1 (>90% load) and without water cooling. The Si_3N_4 composite piston caps have been tested for over 102 hrs without the fracture.

Figure 1. Whiker-like β—Si_3N_4 grains in the matrix

Figure 2. β—Si_3N_4 grains and α—Si_3N_4 grains distribution

Figure 3. ω shape ceramic piston of Si_3N_4 composite

Fig 4. ceramics-metal piston of the diesel engine

TABLE 1
Engine running conditions ($\geqslant$90% full load)

Rotational n (r/min)	Power Ne (KW)	Pressure Ps (MPa)	Temperature		fuel rate (g/KW·hr)
			inlet (℃)	outlet (℃)	
1500	32.4	1.9	38−40	690−700	206

SUMMARY

Si_3N_4 composite possesses the high bend strength and high toughness. They are $\sigma_b = 800$ MPa; $K_{Ic} = 7$ $MPam^{1/2}$ at 25℃ and $\sigma_b = 780$ MPa; $K_{Ic} = 6$ $MPam^{1/2}$ at 850 ℃, respectively.

Whisker-like β—Si_3N_4 grains prevent the cracks from transporting and Al_2O_3 particles dispersed in α—Si_3N_4 grains prevent grain from the deformation. It is called that two phases reinforced mechanism.

Si_3N_4 composite piston caps have been worked for over 100 hrs at over 800 ℃ for 1500 r/min during the high or full power run without water cooling.

REFRENCES

1. D.A.Parker and G.M.Donnison, "The Development of An Air Gap Insulated Piston", SAE 870652.
2. W. Bryzik, R.Kato, "Cummins/TACOM Adiabatic Engine Program", SAE 830314.
3. R. Kamo, W.Bryzik, "Cummins/TACOM Adranced Adiabatic Engine", SAE 840428.
4. R.M.Cole, A.C.Alkidas, "Evaluation of an Air-Gap Instulated Piston in Dividec-chamber diesel engine". SAE 850359.
5. Osamis Kamigaito, "Ceramic Components for Engines in Japan", SAE 870019.

FABRICATION AND TESTING OF A SILICON CARBIDE PISTON AND CYLINDER FOR DIESEL ENGINE

D. L. Jiang*, S. H. Tan, J. H. Wang, Y. L. Li, Y. J. Xu
Shanghai Institute of Ceramics Chinese Academy of Sciences,
1295 DingXi Rd., Shanghai 200050, China
D. Q. Cui
Shanghai Internal combustion Engine Research Institute.
2500 Jun Gong Rd, Shanghai 200432, China.

ABSTRACT

A piston and cylinder were made from silicon carbide ceramic using hot-pressing and pressureless sintering processing respectively. In order to reduce final machining to finish dimensions, "green" state (before sintering) single point machining was used.

A F165 (Φ65mm) diesel engine was used for uncooled engine rig testing silicon carbide piston and cylinder were successfully run for 380hrs Among of them about 250hrs the engine draws a water pump (2hp).

The working condition of ceramic components is rather serious and complicated due to testing without cooling and lubricant (dry friction condition), from plate-on-rotating cylinder of SiC-SiC friction experiment shows that friction coefficient of SiC-SiC is about 0.6 under dry friction which is much high than wet friction condition.

It is also pointed out that control the gap between piston and cylinder is key point which is not only effect on life time of material but also on power efficiency of the engine.
Several ceramic components were broken after rig engine testing. from the cross-section of fracture analyses shows the failure must be considered in the design and accuracy of assembly which is always caused local concentrating stress.

At present time, brittleness and reliability of ceramic still are main problems for engine application.

INTRODUCTION

Ceramic engine has been got much more attention not only in ceramic but also in car manufacturers of the world since energy crisis in 1970. Several ceramic components have been successfully operated in a diesel engine. The piston cup, precombussion plug, glow plug, turbocharger and so on were fabricated from high performance ceramic, for example Si_3N_4, SiC.

In order to develop a low friction, unlubricated ceramic diesel engine. S. Timoney and G. Flynn[1] using sintered alpha SiC work in opposed piston, single crankshaft, rocker arm, two stroke engine with automatic compression ratio variation. No water cooling or lubricating oil was provided to there reciprocating parts a total of 47.5 hours of running, including both motoring and firing tests were accumulated, 67% of the time with the engine firing. This pilot work pointed out SiC pistons can be run in uncooled and unlubricated in a SiC cylinder. After that Ford Motor Co[2]. also working in this field. The piston was a one piece ringless design fabricated from silicon nitride cylinder was also made by Si_3N_4. The cold assembly piston-to-cylinder bore diametral clearance was 0.06mm all firing tests were conducted at or above 1000rpm. to insure stable combustion. The tests was successfully for over 100hrs No signs of wear were observed. The report also pointed out the clearance between piston and cylinder, could be reduced by 50% from 0.06mm to 0.03mm.

In this paper, the aim of this preliminary study is concentrating in the serious of fabrication condition for making a reliable high performance SiC solid piston and cylinder, machining technique and NDT evaluation methods. Meanwhile, selected F165 diesel engine for rig testing on the basis of successful and failure testing, some factors will be discussed.

In order to know the friction problem, using ring-disc simulating testing to evaluate the wear properties between ceramic piston and cylinder without lubricating.

MATERIAL RESEARCH

(1) uncooling engine ceramic components selected.

Basic requirement for uncooled ceramic engine parts is that (a) Reliability under high temperature and stress working condition, material must have enough strength and high weibull modulus. (b) low thermal expansion coefficient to keep the gap size between piston and cylinder at the narrow constant value. (c) material has to bear large thermal stress cycle and thermal shock. (d) thermal conductivity as low as possible in order to reduce thermal loss. After computer simulative temperature distribution and stress analysis for F165 diesel engine by finite element method. Maximum temperature and stress of ceramic cylinder are 315°C and 29MPa respectively; maximum temperature and stress of ceramic piston are 420°C and 45MPa respectively.

Silicon carbide has very good high temperature performance, high hardness, good resistance to wear and thermal shock as well as self-lubricant properties. In fact, silicon carbide has been selected as an important candidate material for uncoiling engine. According to the properties of SiC and combined with computer simulative analysis shows that silicon carbide ceramic can be satisfied engine required, but the reliability and toughness of material, machining accuracy and assembly of components still are main problems.

This work is using hot-pressing SiC to prepare piston and pressureless sintering SiC to make cylinder to meet different requirement.

(2) Piston and cylinder of silicon carbide fabricating procedure.

The flow chart of fabricating procedure for piston and cylinder is shown in Fig. 1.

(3) The properties of SiC material for piston and cylinder are summarized in table 1.

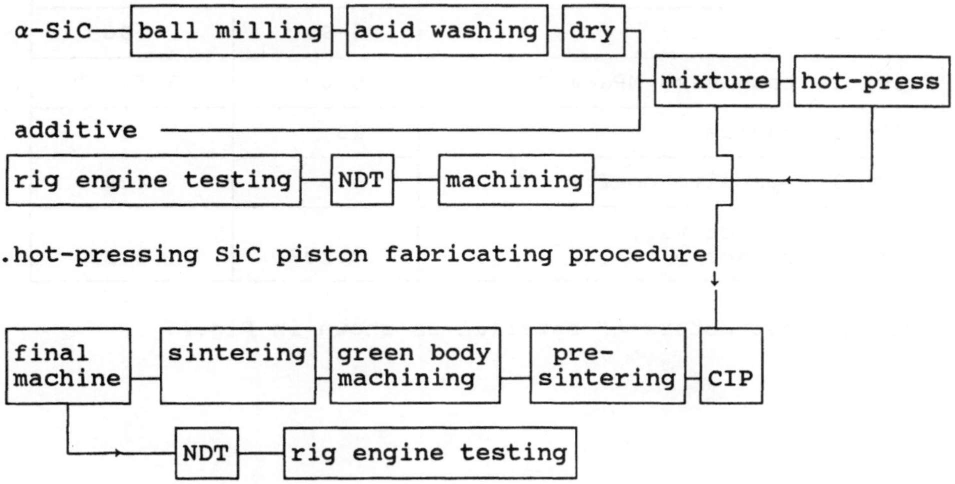

a.hot-pressing SiC piston fabricating procedure

b.PLS-SiC cylinder fabricating procedure.

Fig.1. Flow chart of silicon carbide piston and cylinder processing.

SEVERAL KEY POINTS FOR COMPONENTS PREPARED.

(1) complex piston prepared by Hot-pressed.

Piston structure as shown in fig.3 It is hard to image using common Hot-pressed processing to prepare such a complex piston due to cross section size rather difference between piston head and bottom. Improved are from graphite die design and hot-pressing processing both side, especially in controlling pressure program to keep uniform structure in whole body.

(2) components fine machining.

After CIP and presintering, cylinder need to green machine in order to reduce final machining cost and time. Green body machining must be carefully controlling speed of cutting avoided any damage of surface.

TABLE 1

Summary of the properties of various silicon carbide.

Properties	HP-SiC	PLS-SiC
Density g/cm^3	3.17~3.22	3.0~3.10
Bending strength R.T. (MPa) 1300 °C	550 500	460
Hardness HRA	93~94	>90
Fracture toughness $MPa \cdot m^{1/2}$	5.0	3.5~4.0
Weibull modulus	13	8
Thermal expansive coef.x10^{-6} °C	4.5	4.9
Thermal conductivity coef.cal/cm·sec·°C	0.107	0.22

The microstructure of material as shown in Fig.2.

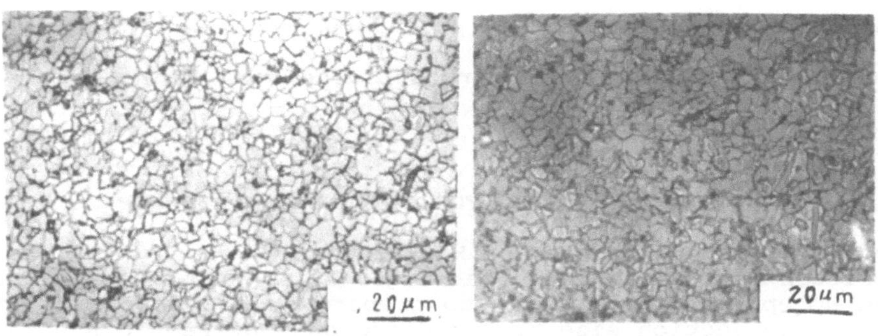

A. B.

Fig. 2. Microstructure of SiC material A.Hot-pressed SiC B.pressureless sintering SiC.

Final machining and surface finishing also made by diamond wheel due to fit gap size (2~3.5μm) between piston and cylinder and to keep perpendicular during the assembly parts of piston.

(3) NDT for ceramic components

Until to now, processing defect (pores, impurities, microflaws, aggremerates, exaggerated grain) existed in all ceramic materials. Some big defects or so called critical defect which caused catastrophic damage of parts under load should be detected before running in rig engine testing, unfortunately, most of NDT methods have their own limited, not only in accuracy of measuring but also limitation in defect size. Some up to date NDT instruments only can detect the size of defect is around 50 μm which is still too big for high performance ceramics. We try to use various method combined together, for instance, ultrasonic wave, dye penetration and soft x-ray to reject some components with bigger defects.

WEAR RESISTANCE TESTING OF SiC CERAMIC COUPLES

The testing of fraction and wear were running in MM-200 type wear machine without any lubricant. A schematic diagram of the plate-on-rotating cylinder machine is show in Fig.4, speed and load are 200rpm and 50~100N respectively.

Fraction coefficient (μ) and weight loss can be calculating according to equation 1 and 2.

$$\mu=F/P=m/P\cdot r \qquad (1)$$

F-fraction force (Kg), P-Applied load (Kg), m-Moment of fraction (Kg·cm), r-Specimen radius (cm), μ-Fraction coefficient

$$\mu=Q/2r\cdot N\cdot P \qquad (2)$$

Q-Fraction work, N-Rotating numbers

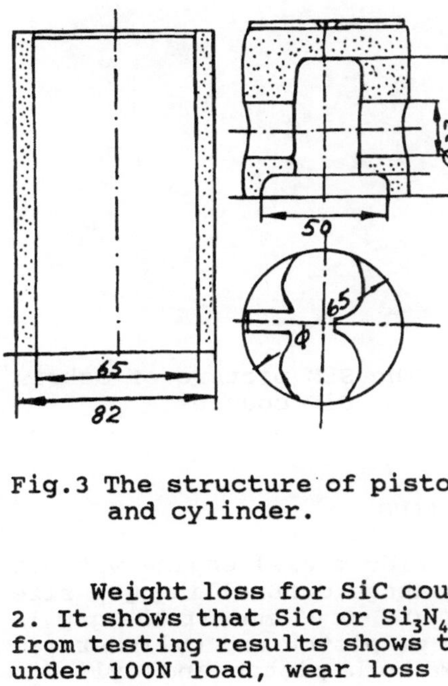

Fig.3 The structure of piston and cylinder.

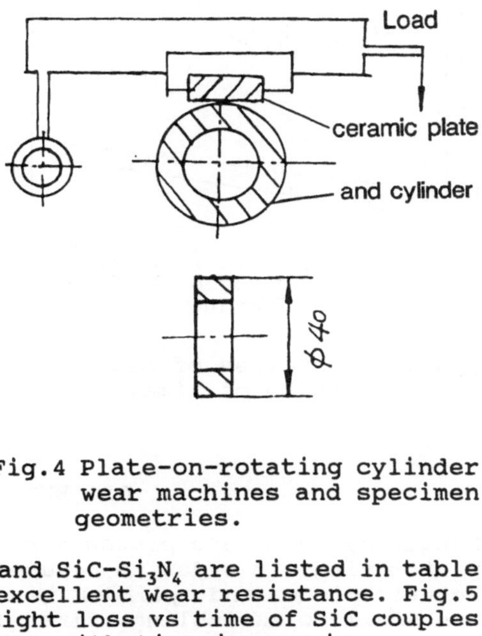

Fig.4 Plate-on-rotating cylinder wear machines and specimen geometries.

Weight loss for SiC couples and SiC-Si_3N_4 are listed in table 2. It shows that SiC or Si_3N_4 has excellent wear resistance. Fig.5 from testing results shows the weight loss vs time of SiC couples under 100N load, wear loss increase with time increasing.

TABLE 2
Wear loss for SiC/SiC; SiC/Si_3N_4 ceramic couples in plate-on-rotating cylinder testing under load 100 N.

wear couples	weight loss (g)	sliding distance (km)	friction coefficient (μ)
SiC-SiC plate cylinder	0.13 0.16	34.38	0.59~0.62
Si_3N_4 plate SiC cylinder	0.20 0.24	32.26	0.68~0.83

After wear tested, worn surface of samples were examined by SEM, TEM, X-ray and surface outline instrument. from surface morphology, wear debris analysis have got preliminary conclusion that the wear mechanism for SiC couples under unlubricated condition might be brittleness fracture, due to high local stress yield microflaw, plastic deformation, nucleation and finally fracture to wear debris: Fig.6 shows the SEM picture of debris for SiC couples.

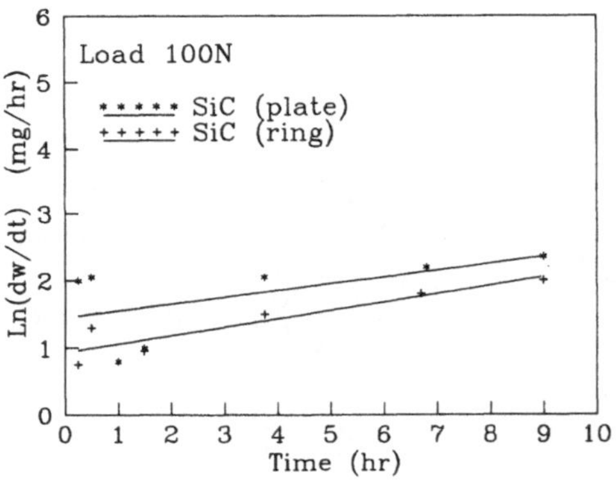

Fig.5 Weight loss of SiC as a function of time.

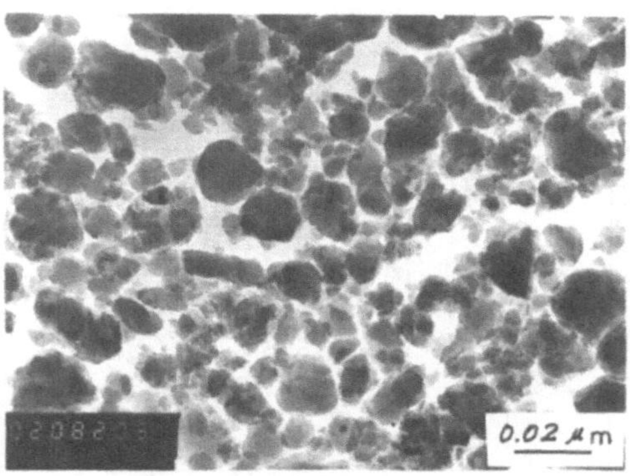

Fig.6 The SEM picture of debris for SiC couples.

RIG ENGINE TESTING

Single cylinder and piston has run in F165 diesel engine without piston ring. In order to reduce the leakage controlling gap size between piston and cylinder is about 0.025~0.035mm. If the gap is too small to assembling, first couple of piston and cylinder was running in 1989, after 4hrs motoring both piston and cylinder were found crack, Fig.7 shows the fracture around the pin holes of piston from fracture surface analysis. The failure might be caused by local stress concentrating, due to assembly problem, The test condition of second couple of piston and cylinder were chosen at 2000rpm. 0.5hp run 2hrs then 1hp continuously run 32hrs. After that the engine was stopped and dismantled for inspection. There is no any damage only some carbon deposit. The engine was reassembled and again fired without cooling water provided running 100hrs at the same time step by step increasing power to maximum 2.2hp. After 134hrs steadily rig engine testing. Then F165 diesel engine was connected with a water pump and continuously drew 250hrs. SiC piston and cylinder still keep excellent state. We also tested combined piston with sealing groove. The design structure shown in Fig.8 unfortunately, only 4hrs rig engine testing, combined piston was found broken. The reason of failure may have two, one is thermal expansion mismatch between metal parts and ceramic piston skirt, which caused ceramic piston skirt borne the big tension stress, the other is caused by assembly not correct due to many parts could not keep small tolerance.

Fig.7 The fracture around the Pin hole of piston.

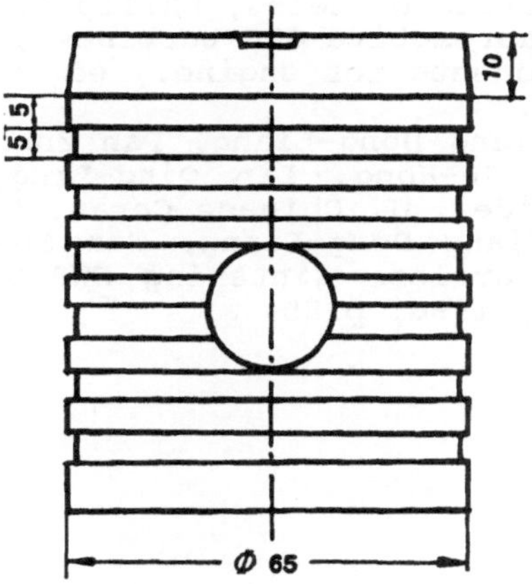

Fig.8 The structure of piston with sealing groove.

CONCLUSION

This preliminary study of all SiC piston and cylinder without cooling water and lubricant has resulted in the following conclusions:

1). A high performance solid SiC piston and cylinder can be processed by hot-pressing and pressureless sintering respectively.

2). tribological investigations show that SiC couple has excellent wear resistance, the coefficient of friction is about 0.59~0.62 without lubricant.

3). From rig engine testing showed that a solid SiC piston can run over 380hrs in a SiC cylinder at 2000rpm. from 1hp to 2.2hp without cooling water and lubricant.

ACKNOWLEDGEMENTS

The authors wish to acknowledge national scientific and technology committee for the support to develop high performance ceramic and uncooling diesel engine project.

REFERENCE

1. S. Timoney, G. Flynn, A low friction, Unlubricated SiC diesel engine., SAE report 830313, 1983.
2. W. R. Wade, P. H. Havstad, V. D. Rao, M. G. Aimone and C. M. Jones, A structural ceramic diesel engine-the critical elements., SAE report 870651, 1987.
3. Harold D. Helns, Philip J. Haley, Emerging ceramic components for automotive GAS turbines., 3rd Int. Sym. ceramic material and components for engine., ed. V. J. Tennery, Las vegas. Nov27~30, 1988.
4. Jiang Dong-Liang, Pan Zhen-Su, Wang Da-Qian, Huang Yu-Zhen, Wang Ju-Hong, Lin Qing-Ling, Studies on Hot-Pressed silicon carbide., J. Chinese Ceram. Soc. Vol.19, No.2, 1981, p133.
5. Jiang Dong-Liang, Tan She-hong, Lin Qing-Ling, Studies on pressureless sintering SW9-SiC ceramic., Shanghai guisuanyan, No.3, 1990, p155.

ADVANCED CERAMIC/METAL COMPOSITE PISTON AND LINER IN LHR ENGINE

Xiaohong Gao, R. Lu, J. Wu, Y. Qian
Power Engineering Department, Wuhan Univ. of Water Transp. Eng.
Wuhan, P.R. China

ABSTRACT

This paper presents a description of a design methodology which has been employed in the design of a LHR engine. Various combinations of material properties were input and calculations were done to see the effects on the heat flow,thermal stress and maximum working temperature of piston crown. A detail study of piston and liner structure is also conducted by using a two-dimensional nonlinear FEM transient heat transfer model and two-dimensional elastic contact FEM model. Temperature distribution and stress contour are presented and compared with different structures. A single cylinder uncooled test engine is used for the trade off studies. Both mullite and Si_3N_4 based ceramic piston crown succeeded in running over hundred hours under full load condition. PSZ upper liner is also succeed when suitable shrink fit is employed.

INTRODUCTION

The LHR engine concept insulates the engine combustion chamber with low heat conductivity ceramic materials and eliminates the traditional cooling system to allow low heat rejection from the engine so as to improve fuel economy, along with some other advantages such as wider fuel tolerance, reducing size and weight, reduce cost and greater vehicle/engine design freedom. In designing combustion chamber components of LHR engine, one has to satisfy various requirements such as: 1) to reduce heat flux passing through the wall significantly; 2) to control stress state on ceramic parts to a level within the safe limit of the material employed; 3) to compensate mismatch behavior between ceramic part and metal body. The key problem here is the optimal design of ceramic components to release the resulting thermal stress with a sophisticate join technique between ceramic part and metal body. To get a better solution, the first thing to deal with is material assessment and selection. Then, the success of LHR engine would largely count on the structure design of ceramic part and its join method with the metal body. At last, a single cylinder test engine (135mm bore, 140mm stroke, 1500rpm) is modified using ceramic/metal composite combustion chamber components without cooling water. Trade off studies are carried out to see a satisfactory configuration and an equitable life cycle cost of the composite components.

MATERIAL SELECTION

Material selection in developing LHR engine is the first and perhaps also the most important step. The most important candidate ceramic materials considered in this paper are Si_3N_4, Mullite, PSZ, TZP and Carbon matrix composite. The major material properties for these ceramics are listed in TABLE 1[1].

TABLE 1
Candidate ceramic materials and properties (material properties are measured at the temperature given below)

	k (W/mC)	α (x10C)	E (GPa)	μ
PSZ	2.700	10.96	210	0.310
Mullite	5.442	4.75	220	0.248
Si_3N_4	13.03	3.05	292	0.283
Temp.	600 C	600 C	RT	RT

A two-dimensional FEM model for steady heat transfer and elastic stress analysis is used to see the influence of various material properties on thermal stress and temperature state. Fig.1 and Fig.2 present the influence of thermal conductivity on heat flux and the wall temperatures on both surfaces of ceramic and metal. We can see that heat flux through the combustion chamber and temperature on surface of metal body increase proportional to the thermal conductivity of the insulating ceramics. But the temperature on ceramic surface decreases with k.

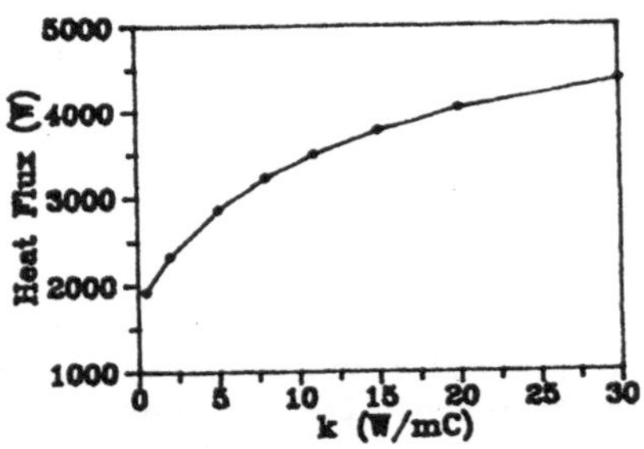

Figure 1. Heat Flux.vs. k

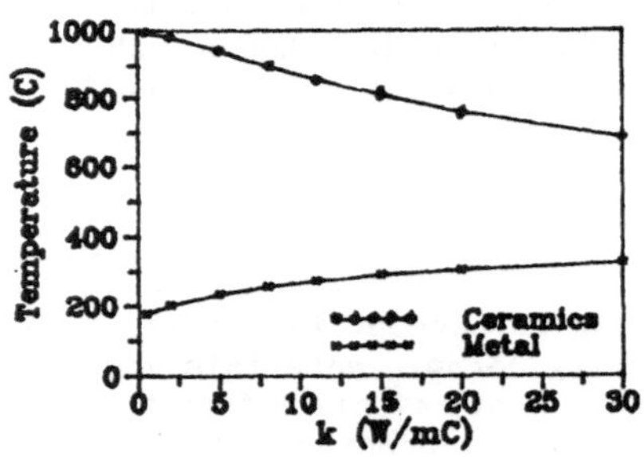

Figure.2 Maximum wall Temperature vs. k

Fig.3 gives out a roughly view of the influences of Young's modula, thermal expansion coefficient and thermal conductivity on maximum thermal stress of the piston crown. The calculation show that the maximum thermal stress is changed sharply when $k < 5$ (W/mC) while decreases slowly when $k > 10$ (W/mC). In an elastic analysis model thermal stress increases proportional to Young's modula and thermal expansion coefficient.

Because of the complex shape that a piston have, two dimensional transient analysis is necessary to predict the temperature swing resulted from the engine cycle[2]. A very thin element of μ's length scale is used to discrete the area close to the ceramic surface. Fig.4 presents the FEM mesh with a zoom view to the corner range. Fig.5 is the result of temperature contour at two different instances along engine cycle calculated by a substructure technique of finite element method.

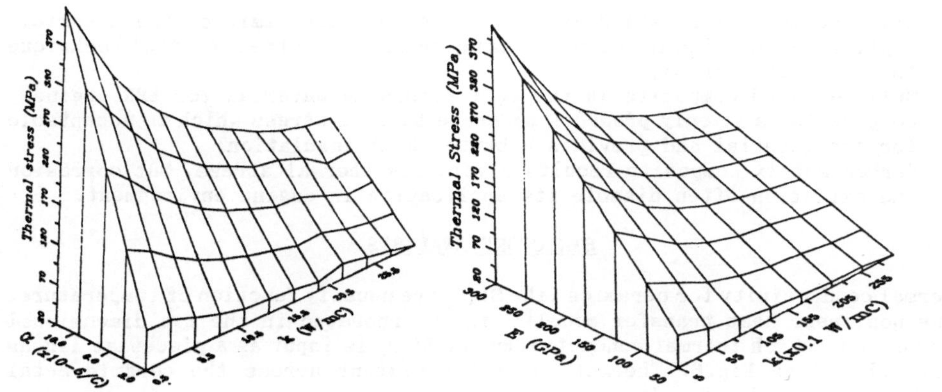

Figure 3. Maximum thermal stress on piston crown as a function of k, E and α

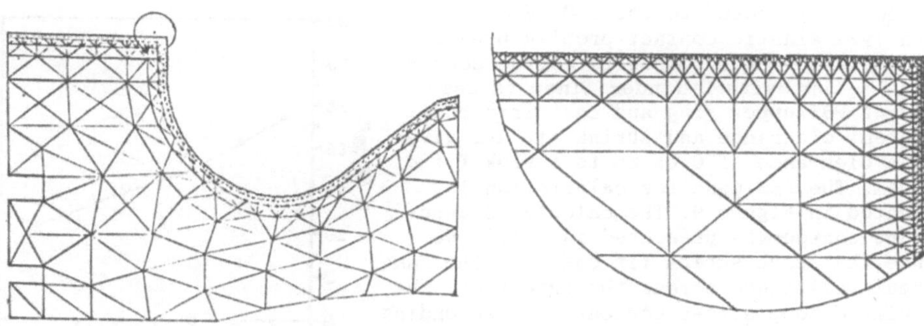

Figure 4. FEM mesh for transient temperature analysis

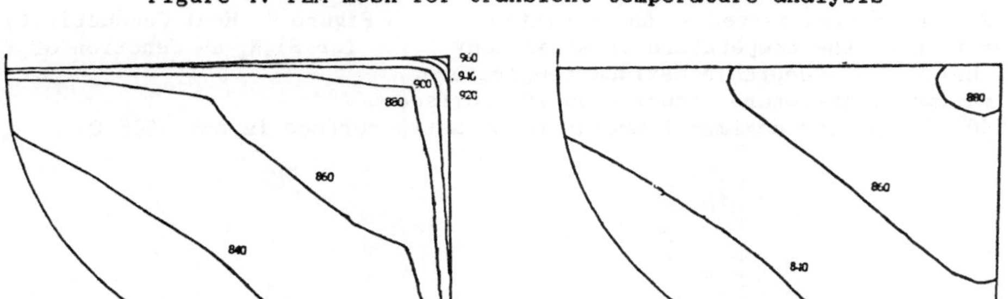

Figure 5. Temperature contours at engine's intake stroke and power stroke at the corner area of a Si_3N_4 piston crown

A survey over the promising ceramic candidate for LHR engine component, some points could be emphasized as following:

— Silicon Nitride results a relatively low thermal stress and has a pretty long survival life, but some measures must be taken to reduce heat flow to the component;
— PSZ or TZP has a very good heat insulation capability whereas it produces

high thermal stress which largely exceeding safe limit of the material. Application of this kind of material needs suitable stress control technique to reduce the stress;

— Mullite based composite is the most promising material for LHR engine components as it only produces moderate thermal stress which is acceptable for the material and provides a better heat insulation;

— Carbon matrix composite results a very low thermal stress, but corrosion and oxidation often disable its application in engine environment.

STRUCTURE ANALYSIS

Thermal conductivity for ceramics like Si_3N_4 are usually function of temperature. Thus nonlinear heat transfer model[3] is incorporated in the two dimensional FEM model in which thermal conductivity for Si_3N_4 is input as a piecewise linear curve showed as Fig.6. Thermal contact resistant across the ceramic metal interface are modeled using gap element[4].

A general slideline capability based on mixed approach using finite element method has been implemented to the 2-D FEM model to analyze elastic contact problem presented in ceramic metal composite engine components[5]. Composite Cylinder liner is composed of PSZ upper ring and cast iron body. Different clearance and shrink fit are given from a clearance of 0.05 mm to shrink fit of -0.30mm. The FEM mesh for calculation is presented in Figure 9. The calculated temperature contour is presented in Fig.7. We assume that each shrink fit case has the same thermal resistance across the interface. The calculated hoop stress contours corresponding to various clearance and shrink fit are presented in Fig.8. The results indicated that the PSZ upper ring serves as an insulating layer to keep the temperature of metal body well below its acceptable maximum temperature. The maximum temperature occurred on PSZ surface is 590 C, and the maximum temperature on metal surface is only 308 C.

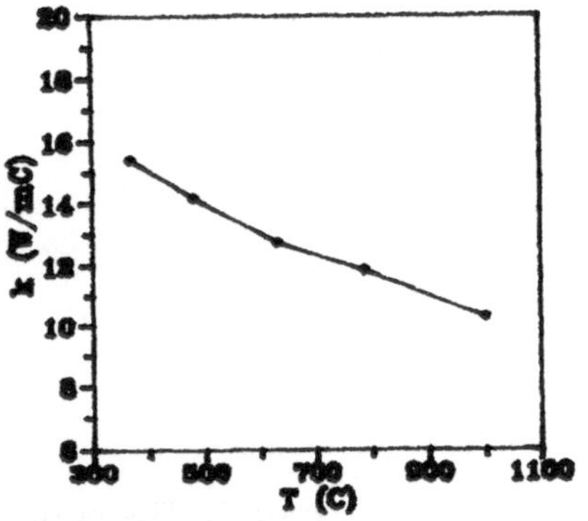

Figure 6: Heat Conductivity for Si_3N_4 as function of T

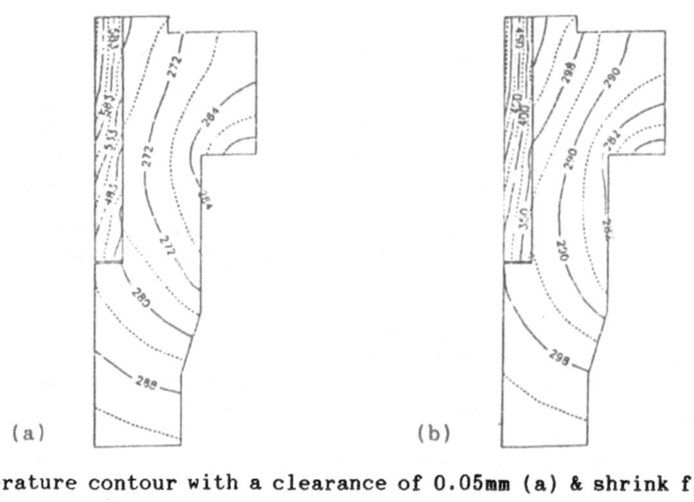

Figure 7: Temperature contour with a clearance of 0.05mm (a) & shrink fit (b)

The maximum tensile stress on PSZ upper ring could be controlled via an appropriate shrink fit. Both a clearance or an over shrinkage are unacceptable as to cause ceramic failure. A suitable shrinkage close -0.12mm is suggested from the calculation and which has been used in the PSZ /cast iron composite liner design. As expected later on the trade off studies, cylinder liner using PSZ upper ring with above recommended shrinkage has been successfully running over 500 hours.

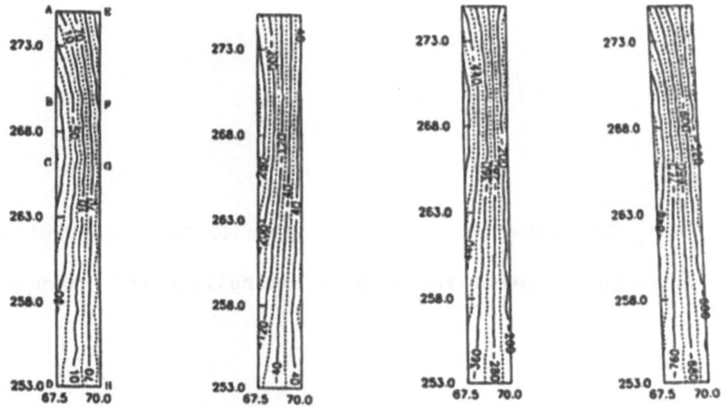

(a) CL=0.05mm (b) CL=0 (c) CL=-0.12 (d) CL=-0.30

Figure 8: Hoop stress contours of the upper PSZ ring with various clearance or shrinkage

The composite piston is much complicated as its deformation and contact performance is much difficulty to control. Fig.9 is the FEM mesh used for piston calculation. The calculated temperature contours for Si_3N_4 and Mullite crown piston together with their resulted hoop stress distributions are shown in the contour maps from Fig.10 and Fig.11. The thermal deformations are expressed in Fig. 12.

The calculation results indicate that:

- An air gap is necessary for Si_3N_4 crown to reduce heat flux through the piston;
- Stress control in design is important for Mullite crown, a measure to separate the Mullite crown into two pieces, a disc and a ring, proved to be effective in further reduction of the thermal stress level;
- Mismatch between ceramic crown and metal body is the key problem which has to be solved to get the successful operation. An elastic spring element was employed to overcome the mismatch deformation of the two parts.

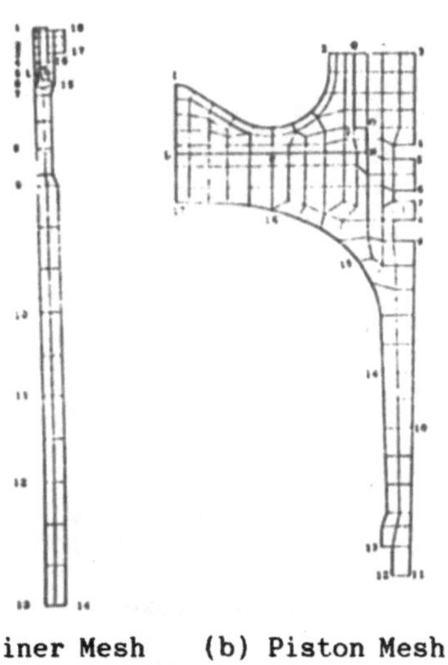

(a) Liner Mesh (b) Piston Mesh

Figure 9: FEM mesh for the piston and liner

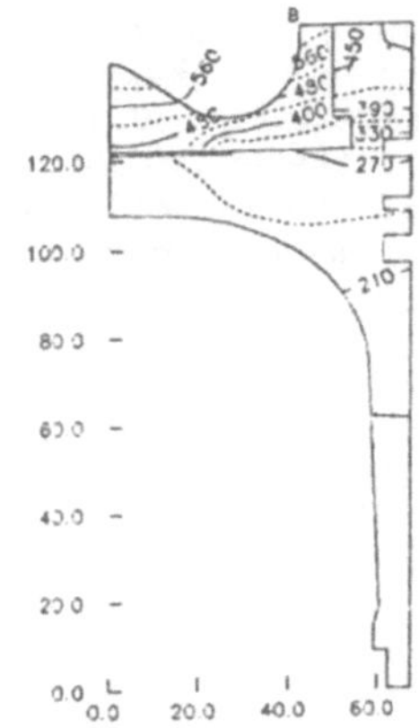

(a)Si_3N_4 Crown Piston

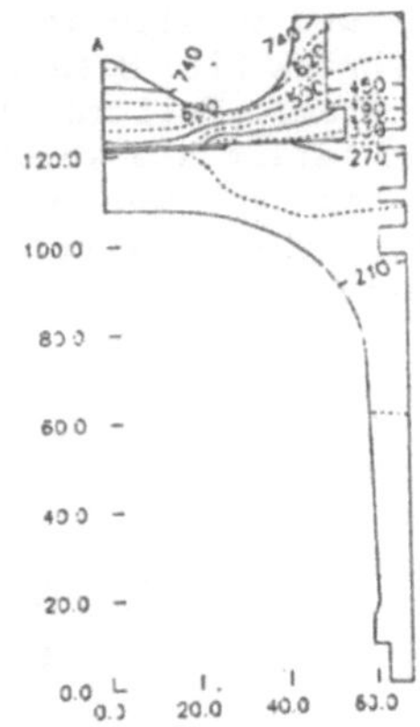

(b) Mullite Crown Piston

Figure 10: Temperature contour for Si_3N_4 (a) & Mullite (b) crown piston

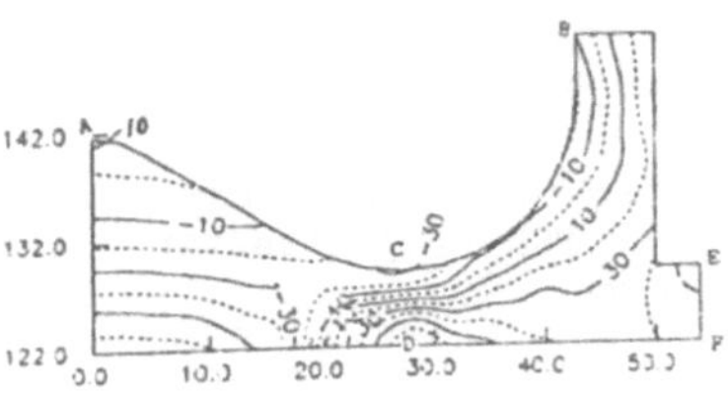

(a)Si_3N_4 Crown Piston

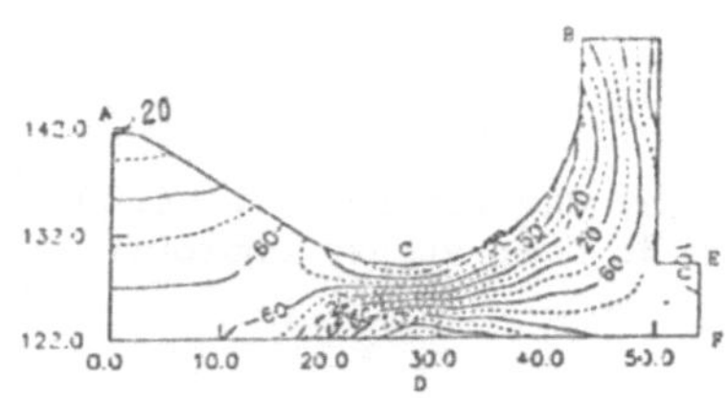

(b) Mullite Crown Piston

Fig 11: Hoop stress in Si_3N_4 (a) and Mullite (b) crown

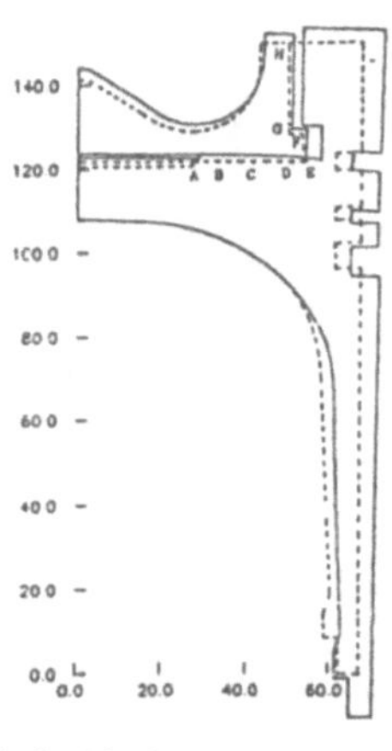

(a) Si_3N_4 crown

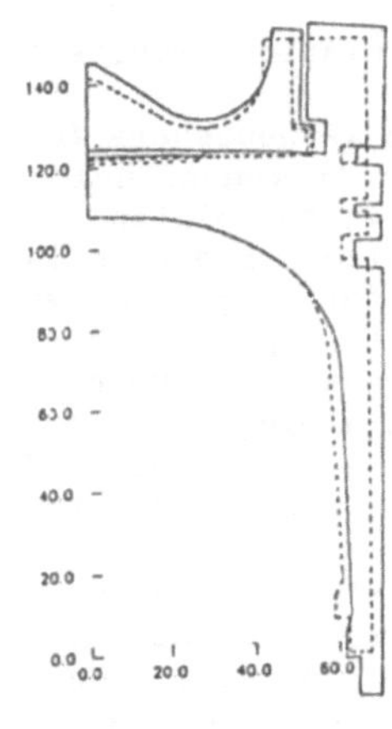

(b) Mullite crown

Figure 12: Thermal deformations for the two composite piston

EXPERIMENT AND RIG TESTS

A single cylinder test engine was specially refitted in which water cooling system was cancelled and high temperature lub oil was used. Cylinder gas pressure and wall temperature were measured and analyzed. Fig.13 and Fig.14 present the mullite crown composite piston, Si_3N_4 crown composite piston and PSZ upper ring liner which have been successful in running over 200 hours.

Figure 13: Composite piston with Mullite crown(left) & Si_3N_4 crown(right)

Figure 14: Composite Liner with PSZ upper ring

Some results from the rig tests showed that:

— TZP piston crown cracked under light load;

— Both silicon nitride and particle reinforced mullite monolithic piston crown have passed the proof tests under heavy load;

— PSZ upper ring of cylinder liner has performed over 500 hours with no failure occurred;

— Both piston and liner using ceramic/metal composite design have been successfully applied to the so called adiabatic turbocompounded engine and showing a very good reliability and fuel economy.

CONCLUSIONS

A single cylinder LHR test engine with bore of 135mm has been developed using monolithic ceramic parts composite with metal body. Either Si_3N_4 or Mullite crown piston are succeeded for running over 200 hours. The composite cylinder liner with PSZ upper ring proved to have a very good reliability. To make the successful design of LHR engine, cautious must be taken in material selection and structure design. Without some means to overcome the mismatch probability and deformation between ceramic and metal, the thermal stress would easily exceeding the safe limit of the material. A two dimensional finite element program has been developed to analyze the transient heat flow, nonlinear temperature field and elastic contact problem in LHR engine combustion chamber components. The major conclusions are the following:

1. Test results show that the structure design using appropriate stress control techniques are vital to the successful application of ceramic/metal composite LHR engine combustion chamber components.
2. Mullite is one of the most promising candidates for heat insulated components.
3. This study can provide realistic guidelines for the successful design of ceramic/metal composite engine components.
4. Two-dimensional transient FEM analysis is necessary to predict the temperature state of the ceramic piston crown.

ACKNOWLEDGEMENT

The authors wish to acknowledge the valuable contributions made by the WUWTE Adiabatic Engine Team. The corporations from S.I.C, Tianjin Univ., and Tshinghua Univ. on ceramics manufacturing are also acknowledged.

REFERENCES

1. Guo, J.K. and Huang X.X, The study of high temperature structure ceramics and the adiabatic engine. The Final Progress Report, S.I.C.,1991.

2. Dennis N. Assanis and Edward Badillo, Transient analysis of piston-liner heat transfer in low-heat-rejection diesel engines. SAE 880189.

3. E.L. Wilson, K.J. Bathe and F.E. Peterson, Finite element analysis of linear and nonlinear heat transfer. Nuclear engineering and design 29(1974)110-124.

4. Arthur B.Shapiro, TOPAZ2D-- A two-dimensional finite element code for heat transfer analysis,electrostatic,and magnetostatic problems. Lawrence Livermore National Laboratory, Rept. UCID-20824, July,1986.

5. Zhou, X.M., A finite element solution for the two-dimensional thermal-elastic contact problems in composite LHR engine components. M.S. thesis of Wuhan Univ. of Water Tran. Eng., Wuhan,(1991)

LOW HEAT REJECTION DIESEL CERAMIC COUPON TESTS

C. R. BRINKMAN, K. C. LIU, R. L. GRAVES, B. H. WEST, AND G. M. BEGUN
Oak Ridge National Laboratory, Oak Ridge, TN, USA

ABSTRACT

Results are reported from studies in which several monolithic ceramic materials in the form of modulus-of-rupture bars were exposed for 100 h to the combustion conditions found in either a small single- or two-cylinder diesel engine. Fuels included a standard Phillips D-2 diesel or synthetic mixture of the Phillips D-2 and an aromatic blend. The ceramics included two commercial grades of partially stabilized zirconia (PSZ-TS and PSZ-MS), silicon nitride (GTE WESGO SNW-1000 and Norton NT-154), and (Hexoloy SA) silicon carbide. Results are presented from postexposure four-point bend rupture tests conducted at either room temperature or 700°C. Partially stabilized zirconia TS grade showed considerable reduction while the other materials showed only a slight or no change in rupture strength.

INTRODUCTION

Experimental engine exposure of commercially available and developmental ceramic materials in diesel engines began at Oak Ridge National Laboratory (ORNL) in 1985 [1-4]. The objective of this effort was to determine the effect of diesel engine combustion environment on the structural properties of various structural ceramics by exposing them to conditions found in small diesel engines. Commercially available ceramics, as well as new candidate monolithic and whisker-toughened ceramics, were fabricated into small modulus-of-rupture (MOR) bars and inserted into the combustion environment, and then examined destructively and non-destructively.

ENGINE TESTING RIG AND SPECIMEN EXPOSURE

Two small engines were used to expose specimens to combustion conditions. The first was a single-cylinder, air-cooled, indirect-injection diesel (Deutz F1L511W) referred to as engine 1. This engine was a small commercially available driver with a 0.825-L displacement and a spherical prechamber. The second otherwise identical two-cylinder version of the same engine is referred to as engine 2. Access to the prechamber was made via the unused glow plug port (adjacent to the fuel injector) and a machined port for pressure transducer instrumentation (Fig. 1).

Except for differences in age, these engines were similar. The single-cylinder engine had been used previously for ceramic specimen exposure, whereas the two-cylinder engine was new with new injectors.

Exposure was conducted by "gripping" a pair of MOR bars, that were side-by-side (separated by a thin shim) in a retainer. The retainer was then screwed into the glow plug port, as shown in Fig. 1. Thus, two bars were exposed simultaneously to similar engine operating conditions. Size constraints of the prechamber permitted only 40 to 41 mm of the bar length to be cantilevered into the chamber. Because of the orientation of the bars direct impingement of the burning fuel spray might have occurred.

Exposure times of 100 h for each set of specimens were accumulated over several weeks of running several hours daily. Interruptions in exposure also occurred as a result of random corrective maintenance on the engine/dynamometer tandem. The two-cylinder engine was run at a nominal constant speed of 1800 to 1900 rpm and a load of 41 to 43 N-m torque. Fuel consumption at these conditions was about 213 g/kWh, or a fuel energy input of ~21,480 W. The one-cylinder engine was run at similar conditions, with about half the load and fuel consumption of the two-cylinder engine. For either engine, these conditions are ~60% of rated power output at this speed. Load was absorbed by a small eddy-current dynamometer.

Two fuels were used in this investigation. The first was the reference fuel, Phillips D-2 diesel, and the second was a 50/50% blend of Phillips-D-2 and an aromatic blend. The aromatic blend was 53.5 wt % Phillips Light Cycle Oil (LCO) and 46.5 wt % Exxon heavy aromatic naphtha targeted for a lower cetane number and a lower hydrogen content. These blends could simulate fuels derived from coal liquids or heavy oils so the products would have low ignition quality.

TEMPERATURE MEASUREMENT

An extensive effort was undertaken to determine the in-cylinder temperature profiles of the specimens (2-4). This effort included use of thermocouples, inserted in the centers of drilled PSZ-TS bars placed in the engines, templugs, and fiber optic measurements. Thermocouple measurements in the single cylinder engine revealed bulk internal bar temperatures in the range of 700°C, while in the two cylinder engine values of around 600°C were recorded. Templug data indicated that temperatures in excess of 800°C occurred in the single cylinder engine. Fiber optic thermometer measurements indicated that transient short term temperatures could reach values of 1100°C or higher.

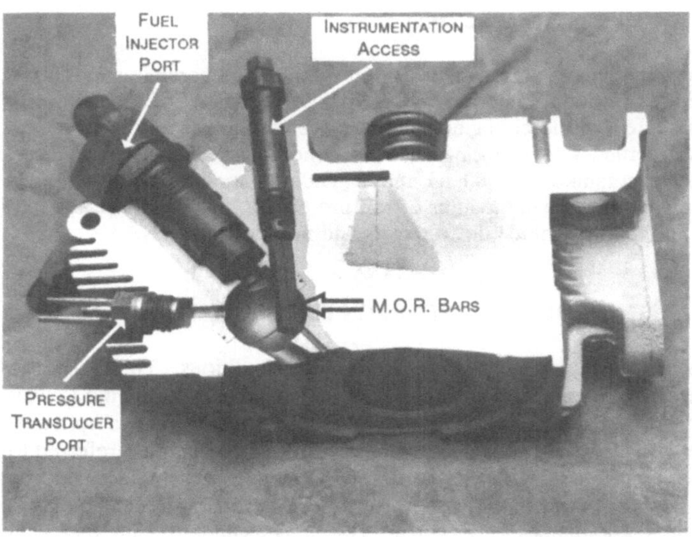

ORNL-PHOTO 8426-90

Figure 1. Cross sectional view of Deutz cylinder head shows bars' position with respect to fuel injector.

In summary, currently there is insufficient evidence to conclude that one engine was hotter than the other because dependable temperature profiles were not available. Cycle-to-cycle variations and sensor installation could easily account for differences recorded to date. Additional work needs to be continued in this area, but it has been shown that in reciprocating engines, this type of material undergoes significant temperature swings and that instantaneous temperatures can be much higher than indicated by passive temperature measurements or conventional thermocouples. For these reasons, postexposure test temperatures for four-point bend flexure tests were selected to be room and 700° C.

MATERIALS

Two commercial grades of PSZ, Mg-PSZ-TS (thermal shock resistant) and Mg-PSZ-MS (maximum strength), were obtained from Nilcra Ceramics (USA) in the form of rectangular plates (101.6 x 6.35 mm). In the case of PSZ-TS, a second lot (Nilcra Lot 85 218) was also obtained to determine if there were lot-to-lot variations. This second lot of PSZ-TS is referred to herein as Lot 2, while the previously characterized lot of PSZ-TS (3) is designated as Lot 1.

Two types of commercially available silicon nitride were used in this study, a small block of GTE Wesgo SNW-1000, Lot 174-MA-1607, and a similar block of Norton's NT-154.

A small plate of silicon carbide (1.6 x 76 x 178 cm) designated Hexoloy-SA was obtained from the Carborundum Company.

SPECIMEN PREPARATION AND TESTING

MOR specimen bars (50 x 4.5 x 3.5 mm) were subsequently cut from the plates or block material. All grinding and polishing of the bars and chamfering of edges were performed in a longitudinal direction to a surface finish of ~0.25 μm. The PSZ material had an average grain size in the range of 30 to 50 μm. Unexposed and engine exposed bars were subjected to four-point rupture testing in fixtures with an outer span of 40 mm and an inner span of 20 mm. Figure 2 shows a silicon carbide fixture used for all tests conducted at 700° C. All rupture tests were conducted in air at a strain rate of 1×10^{-4} s^{-1} unless specified otherwise.

FLEXURAL STRENGTH RESULTS

Partially Stabilized Zirconia, PSZ-TS (Lot 1)

Figure 3 is a Weibull plot that compares failure probabilities of TS-grade material exposed in engine 2, which burned synthetic fuel mixture, with similar failure probabilities for unexposed material. Weibull modulus, strengths, and percentage of reduction in strength compared with unexposed material are also given in Table 1. Also tabulated are data previously reported (3) for 14 specimens exposed to engine combustion conditions in both engines 1 and 2 where the reference fuel was employed. Table 1 indicates, from limited data, that the combustion conditions using the synthetic fuel mixture were possibly not as deleterious to the average rupture strength as was the reference fuel.

Partially Stabilized Zirconia, PSZ-MS

Figure 4 is a Weibull plot comparing failure probabilities vs strength for the MS-grade material exposed in engine 2 using the synthetic fuel mixtures. A small reduction in strength is apparent. Strength statistics are again compared in Table 1 for this material exposed to engine combustion conditions where the fuel was either the reference or the synthetic fuel mixture. Note that the reduction in strength compared with the unexposed material is about the same regardless of the fuel.

Partially Stabilized Zirconia, PSZ-TS (Lot 2)

Exposure of 18 bars of partially stabilized zirconia, PSZ-TS (Lot 2) was completed in either engine 1 or 2. The objective of this effort was to determine if specific engine conditions, that is, single vs two-cylinder, made any significant difference in the postexposure strength of a material known to be sensitive to engine environment. The reference fuel was used in both engines. Results of subsequent room temperature four-point bend tests are compared with control specimens in Table 1 and plotted as Weibull curves in Fig. 5.

Figure 2. Silicon carbide four-point bend fixture used for elevated temperature tests.

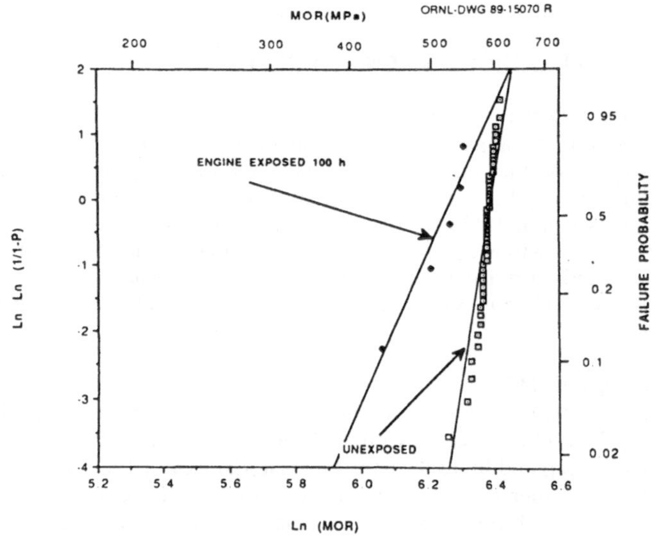

Figure 3. Weibull plot comparing room temperature fracture strengths of five specimens of Nilcra PSZ-TS grade previously exposed for 100 h in a diesel engine burning a synfuel mixture with fracture strength of unexposed or control material.

TABLE 1.
Results of room temperature four-point bend tests[a] conducted on MOR bars of several different materials

Number of observations	Material condition	Weibull modulus Maximum likelihood	Average strength (MPa)	Change in average strength[b] (%)
		PSZ-TS (Lot 1)		
53	As fabricated	48.6	587.3	0
5	Engine 2 exposed for 100 h[c]	16.6	509.8	-13
14	Engine 1 or 2 exposed for 100 h[d]	4.2	401.8	-32
		PSZ-MS		
40	As fabricated	29.7	682.9	0
4	Engine 1 exposed for 100 h[c]	87.5	621.0	-9
20	Engine 1 or 2 exposed for 100 h[d]	22.0	625.0	-9
		PSZ-TS (Lot 2)		
20	As fabricated	25.6	629.6	0
6	Engine 1 exposed for 100 h[d]	5.2	413.9	-34
12	Engine 2 exposed for 100 h[d]	6.7	444.3	-29
		GTE WESGO SNW-1000[e]		
40	As fabricated	11.2	506.5	0
32	Engine 2 exposed for 100 h[d]	14.2	512.7	1
		NORTON NT-154		
40	As fabricated	11.8	771	0
10	Engine 1 exposed[d] for 100 h	12.3	825	7
20	Engine 2 exposed[d] for 100 h	15.4	828	7

[a]Tests were conducted with a cross head speed of 8.47×10^{-3} mm/s or a strain rate of 1.1×10^{-4} s^{-1}.

[b]Reduction in strength compared with as-fabricated material.

[c]Fuel was 50/50% blend of Phillips D-2 diesel fuel (reference fuel) and an aromatic blend.

[d]Fuel was control diesel (reference fuel).

[e]Material from Lot 174-MA-1607 and re-heat-treated to sintering temperature.

Results given in Table 1 indicate a 34 and 29% reduction in strength after exposure in engines 1 and 2, respectively, compared with as-fabricated or control specimens. Because of the limited data, this difference in strength reduction is not significant. Table 1 also shows that the PSZ-TS (Lot 2) material was slightly stronger than that of PSZ-TS (Lot 1).

Silicon Nitride (WESGO SNW-1000)

Using the reference diesel fuel combustion environment, 32 bars were exposed in engine 2. Results are compared in Table 1 with unexposed or control fracture data. The comparison given in Table 1 shows only a slight (1%) increase in mean fracture strength compared with the unexposed material, which is not significant and probably within scatter of the data. Figure 6 is a Weibull plot comparison of the exposed and unexposed rupture strengths.

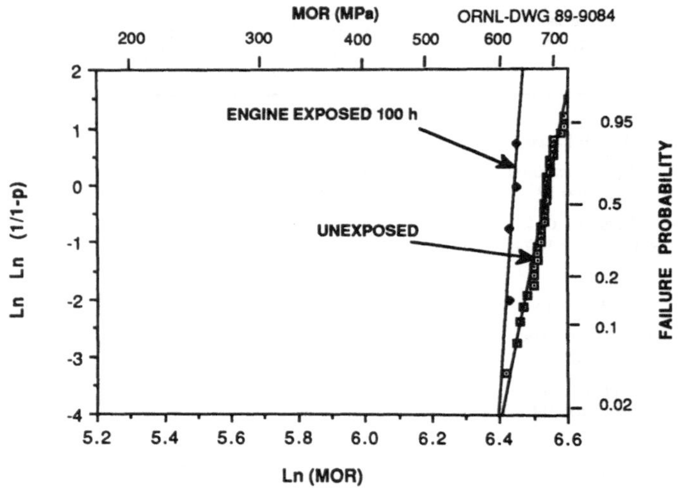

Figure 4. Weibull plot comparing room temperature fracture strengths of Nilcra PSZ-MS grade, previously exposed for 100 h in a diesel engine burning a synfuel mixture, with fracture strengths of unexposed or control material.

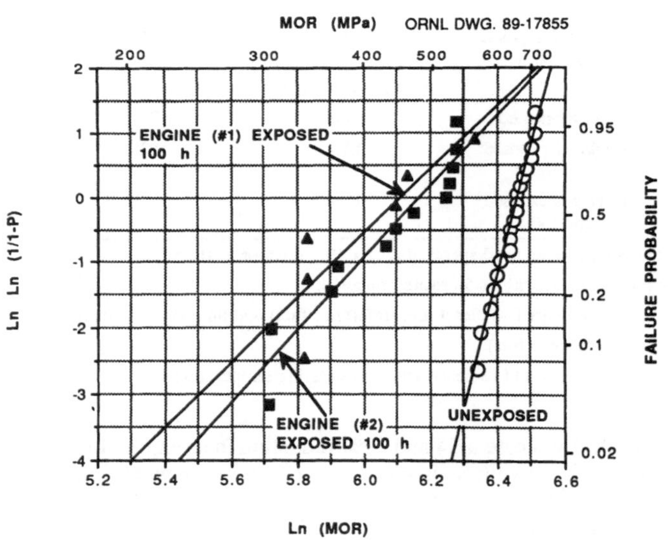

Figure 5. Weibull plot comparing room temperature fracture strength of Nilcra PSZ-TS grade (Lot 2), previously exposed for 100 h in a single-cylinder (#1) and two-cylinder (#2) diesel engine in a reference diesel fuel combustion environment with similar results from unexposed material.

This particular lot of material was considerably weaker than expected. The manufacturer was contacted regarding the decreased strength of this lot and reported that this lot had been reheated to the sintering temperature, which may have caused some devitrification.

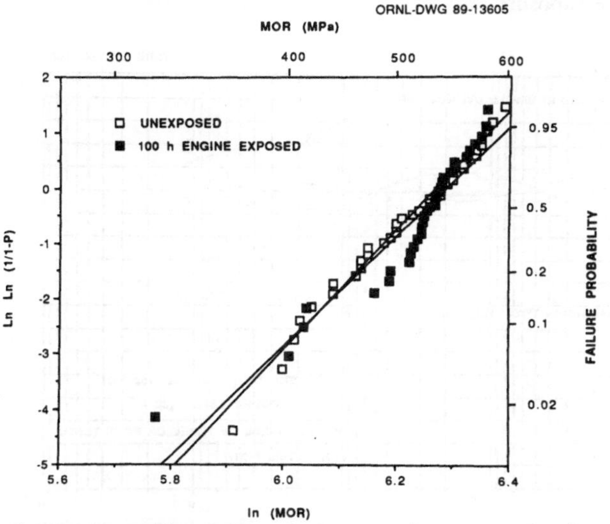

Figure 6. Weibull plot comparing room temperature fracture strengths of 32 silicon nitride (WESGO SNW-1000) bars exposed for 100 h in a reference diesel fuel combustion environment with similar results from unexposed material.

Silicon Nitride (NT-154)

A total of thirty bars was exposed for 100 h in either the single or two cylinder engine and subsequently ruptured at room temperature. Ten of these bars were exposed in the single cylinder engine and the balance was exposed in the two-cylinder engine. Results of these tests are given in Table 1 as Weibull modulus, average fracture strength, and percent change in average strength in comparison to unexposed bar behavior. The data given in Table 1 and shown in a Weibull plot in Fig. 7, indicate that the exposed bars actually showed a slight increase in strength, i.e. about 7%, in comparison to the unexposed bars.

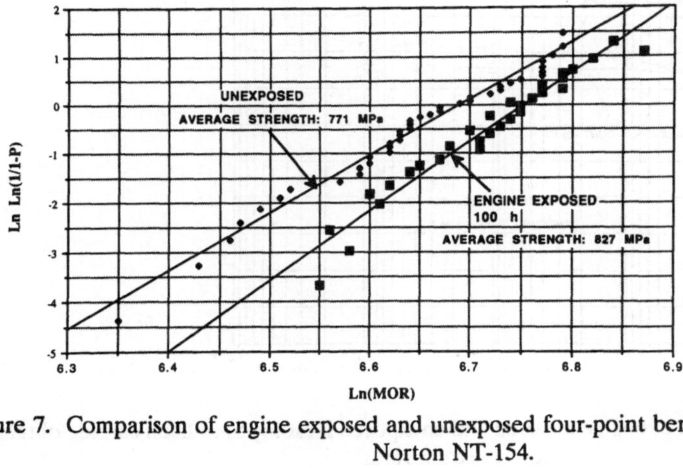

Figure 7. Comparison of engine exposed and unexposed four-point bend rupture strengths of Norton NT-154.

Figure 8 is a plot of fracture strength as a function of strain rate for NT-154 bars tested in the exposed and unexposed condition. The test temperature was 700°C and strain rates ranged from 10^{-7} to 10^{-4} 1/s. The data indicate no clear trends of either an increase or decrease in fracture strength due to engine exposure.

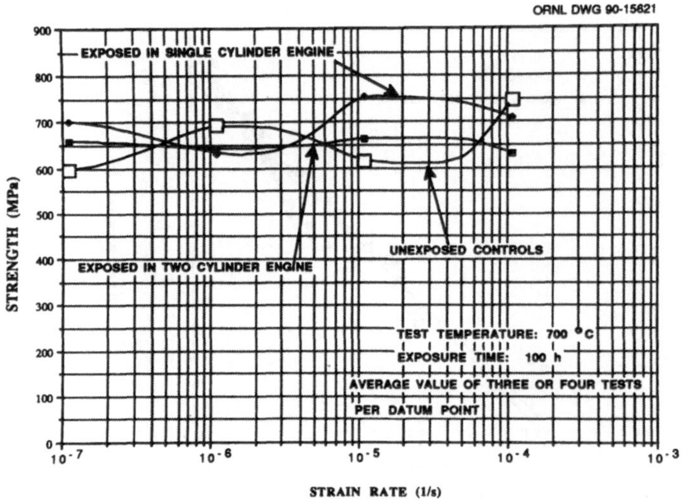

Figure 8. Fracture strengths of engine exposed and unexposed Norton NT-154 silicon nitride obtained at several strain rates.

Hexoloy SA Silicon Carbide

Specimens of silicon carbide were also tested at 700°C and at similar strain rates to those used for the NT-154 specimens discussed above. Figure 9 shows results of the rupture test plotted as rupture strength as a function of strain rate. No effect of engine exposure is apparent from the limited test results.

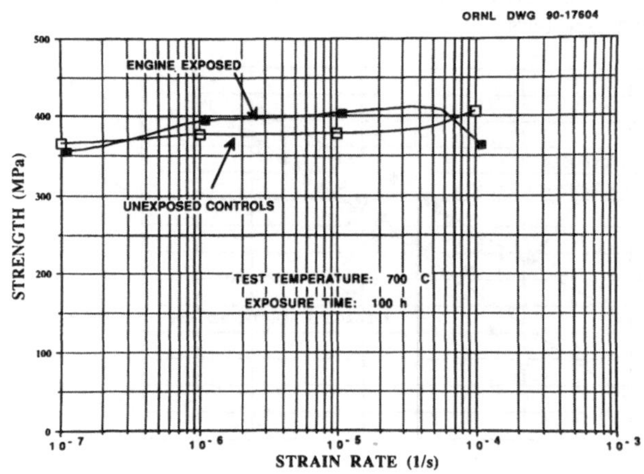

Figure 9. Fracture strengths of engine exposed and unexposed Hexoloy SA silicon carbide obtained at several strain rates.

VISUAL OBSERVATIONS

Previous work (3) indicated that in the case of the engine-exposed PSZ-TS grade Lot 1 material, most fractures occurred about 12 to 15 mm from the cold end for specimens exposed in engine 2 and 33 to 35 mm from the cold end for specimens exposed in engine 1. In the case of the PSZ-MS grade, fractures generally occurred at random in the central sections of the bar about 22 to 31 mm from the cold end. Fracture locations of engine exposed PSZ-TS (Lot 2) material exposed in engine 1 in this study tended to be at random, but in contrast 9 out of 12 (75%) of the bars exposed in engine 2 failed ~15 to 20 mm from the fixed end of the bar, that is near attachment locations within the engine. These differences suggest somewhat different environments in the two engines. In the case of the SNW-1000 specimens, fracture locations were random.

RAMAN SPECTROSCOPY EXAMINATIONS

Raman spectroscopy scans with a laser Raman microprobe were performed on PSZ-TS grade specimens after engine exposure and typical results are shown in Fig. 10. The Raman microprobe observations determined the ratio of monoclinic to tetragonal zirconia formed at various points. For convenience in reporting it was assumed that no cubic zirconia was present, and the apparent percentage of monoclinic zirconia was calculated on this basis. Figure 10 shows representative peaks in apparent monoclinic content in the range of 12 to 15 mm from the cold end of Lot 1 bars coinciding with subsequent fracture locations.

ORNL DWG 89-9869

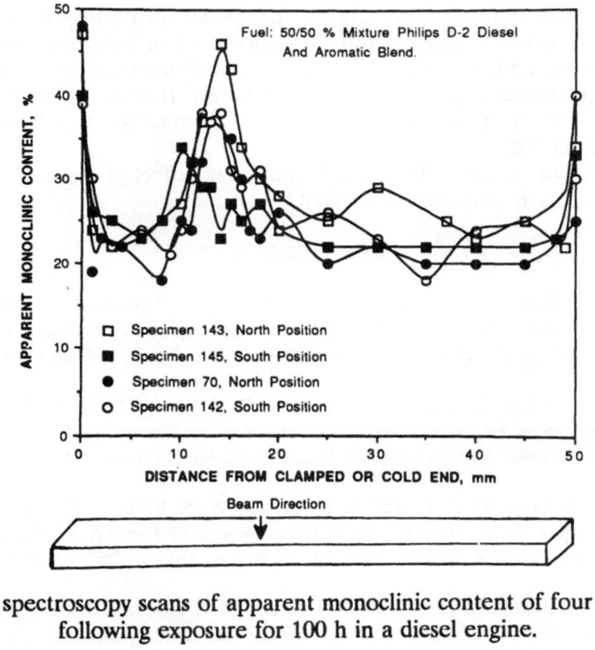

Figure 10. Raman spectroscopy scans of apparent monoclinic content of four PSZ-TS bars following exposure for 100 h in a diesel engine.

CONCLUSIONS

Results were reported from rupture studies in which several monolithic ceramic materials in the form of rectangular-shaped MOR bars were exposed for 100 h to the combustion conditions found in two diesel engines. The two fuels included in this study were (1) the reference fuel, Phillips D-2 diesel, and (2) a synthetic fuel consisting of a mixture of the reference fuel and an aromatic blend. The ceramics included two commercial grades of partially stabilized zirconia, PSZ-TS and PSZ-MS, WESGO SNW-1000 and Norton NT-154, silicon nitride and Hexoloy SA silicon carbide.

1. Reduction in room-temperature four-point fracture strength following exposure occurred in two lots of PSZ-TS material. This reduction occurred irrespective of the fuel employed although there was some indication that the synthetic fuel mixture was slightly less deleterious.

2. Separate sets of specimens of a single lot of PSZ-TS material exposed in either a single- or two-cylinder diesel engine that burned the reference fuel showed similar amounts of degradation in fracture strength.

3. Specimens of PSZ-TS material exposed in the two-cylinder engine that burned the reference fuel showed preferential postexposure failure locations. The points of fracture were in the range of 15 to 20 mm from the cold or gripped end of the bars. Raman spectroscopy analysis performed on the postexposed bars before bend testing indicated that the region of failure coincided with a high apparent monoclinic phase content induced by engine exposure. A very limited number of specimens exposed in the single-cylinder engine did not show a similar bias in failure location, possibly indicating some subtle differences in exposure conditions between the two engines that were probably related to injector fouling in the single-cylinder engine. This may suggest that injector integrity is an important parameter to consider when engine components subjected to direct fuel impingement are made of, or coated with, zirconia.

4. Specimens of PSZ-MS showed only a small decrease in postengine exposure fracture strength. This reduction in strength was the same irrespective of the fuel employed, and no bias was observed in the fracture locations along the bars.

5. Specimens of silicon nitride (WESGO SNW-1000) exposed in the two-cylinder engine that burned the reference fuel showed essentially no degradation in postexposure fracture strength. No bias in fracture locations was noted in the bars following four-point bend fracture testing.

6. Specimens of silicon nitride (Norton NT-154) exposed in either the single- or two- cylinder engine showed a slight increase (7%) in room temperature fracture strength in comparison to unexposed material. No effect of strain rate or exposure condition was found on the fracture strength of bars tested at 700°C.

7. Specimens of silicon carbide (Hexoloy SA) showed no effect of exposure condition or strain rate on the fracture strength when tested at 700°C.

REFERENCES

1. R. L. Graves, W. K. Kahl, and E. L. Long, Jr., *Testing of Developmental Materials in Diesel Engine Combustion Chambers,* Paper 87-ICE-28, The American Society of Mechanical Engineers, New York, Energy Sources Technology Conference and Exhibition, Dallas, Texas, February 15-20, 1987.

2. C. R. Brinkman, G. M. Begun, O. B. Cavin, B. E. Foster, R. L. Graves, W. K. Kahl, K. C. Liu, and W. A. Simpson, *Influence of Diesel Engine Combustion on the Rupture Strength of Partially Stabilized Zirconia,* ORNL-6513, Martin Marietta Energy Systems, Inc., Oak Ridge National Laboratory, December 1988.

3. C. R. Brinkman, K. V. Cook, B. E. Foster, R. L. Graves, W. K. Kahl, K. C. Liu, and W. A. Simpson, *Influence of Diesel Engine Combustion on the Rupture Strength of Partially Stabilized Zirconia,* Ceramic Bulletin, The American Ceramic Society Inc., Vol. 68, No. 8, August 1989, pp. 1440-45.

4. C. R. Brinkman, G. M. Begun, R. L. Graves, W. K. Kahl, K. C. Liu, and B. H. West, *Influence of Diesel Engine Exposure on the Rupture Strength of Silicon Nitride and Partially Stabilized Zirconia,* ORNL-6612, Martin Marietta Energy Systems, Inc., Oak Ridge National Laboratory, June 1990.

EFFECTS OF SODIUM SULFATE INDUCED CORROSION ON THE STATIC FATIGUE LIFE OF SILICON NITRIDE CONTAINING DIFFERENT DOPANTS

Jeffrey J. Swab
U.S. Army Materials Technology Laboratory
Watertown, MA 02172, USA

and

Gary L. Leatherman
Worcester Polytechnic Institute
Worcester, MA 01609, USA

ABSTRACT

Flexure testing was used to determine the effects of sodium sulfate-induced corrosion on the static fatigue life of two silicon nitride ceramics at 1000°C. The results show that the static fatigue life of the MgO-doped silicon nitride is significantly reduced when Na_2SO_4 is introduced. Conversely, the static fatigue life of the Y_2O_3 + Al_2O_3-doped silicon nitride is increased in the presence of Na_2SO_4.

INTRODUCTION

A variety of structural ceramics are being considered for application in advanced heat engines. In some applications, these materials will be exposed to molten sodium sulfate (Na_2SO_4) which condenses on engine components when ingested NaCl reacts with sulfur impurities in the fuel. How long-term exposure to these corrosive conditions affects the static fatigue life of the ceramic at elevated temperatures is of critical importance to their successful application in the next generation of advanced engines.

The actual mechanisms of Na_2SO_4-induced corrosion of Si_3N_4 have been studied[1-7]. Silicon-based ceramics such as Si_3N_4 rely on a thin surface layer of SiO_2 for protection against oxidation. The degradation of Si_3N_4 at high temperatures in a corrosive environment is believed to be due to the dissolution of this protective layer.

Sodium sulfate dissociates according to Equation (1)

$$Na_2SO_4 = Na_2O + SO_3 \quad (1)$$

The Na_2O that forms then reacts with the SiO_2 protective layer forming a sodium-silicate glass, Equation (2),

$$xSiO_2 + Na_2O = Na_2O\ x(SiO_2) \quad (2)$$

which is not protective, and allows for extensive corrosion and degradation of the Si_3N_4.

Although the corrosion mechanisms have been studied, there has been very limited research performed which examines the mechanical properties of silicon nitride at elevated temperatures in a corrosive environment[8]. This paper summarizes an effort to determine the effects of hot corrosion on the static fatigue life of two commercial silicon nitrides at 1000°C.

EXPERIMENTAL PROCEDURE

Two commercially available silicon nitrides were obtained and machined into flexure specimens of the following dimensions: 1.5 mm x 2 mm x 25 mm. One silicon nitride was produced by Norton Company (NC-132) using MgO as a hot-pressing aid and the other was sintered with Y_2O_3 and Al_2O_3 as sintering aids by GTE Wesgo (SNW-1000).

Long duration stress rupture (SR) tests were conducted on flexure specimens in both the as-received condition and with 10-20 mg/cm^2 Na_2SO_4 added. The Na_2SO_4 was added by mixing anhydrous Na_2SO_4 with distilled water and applying the solution to the centered portion of one 2 mm x 25 mm face of the bar. The bars were then heated on a hot plate to drive off the water leaving behind Na_2SO_4.

Each specimen was loaded onto a four-point flexure fixture in a furnace and heated to 1000°C in ≈2.5 hours with no stress applied. Upon reaching temperature a predetermined stress was applied to the specimen and the test was allowed to run until the specimen failed or 500 hours had elapsed.

Specimens which survived 500 hours without failure were then broken in four-point flexure at room temperature to determine the retained strength. Special care was taken to ensure that the upper and lower spans were aligned in the same location as during the SR test. Fractographic analysis was also done on fractured specimens to determine the failure origin.

RESULTS AND DISCUSSION

Stress Rupture Test Results

Norton Company (NC-132): There is extensive information on the static fatigue life of MgO-doped Si_3N_4 in flexure[9-12]. The results generated in this study, Figure 1, for the as-received condition are in excellent agreement with the data reported by Quinn[12]. Fractographic analysis revealed that at stresses greater than 500 MPa the cause of failure was machining damage. Specimens subjected to a 450 MPa stress began to show evidence of slow crack growth. Bars subjected to 400 MPa applied stress or less survived 500+ hours without failure and showed no signs of creep.

The introduction of 10-20 mg/cm^2 Na_2SO_4 greatly reduced the static fatigue life at 1000°C and made fractography extremely difficult after SR testing. Figure 1 shows that the static fatigue life can be characterized by three regions. In region I the slope of the line essentially parallels the as-received condition indicating that the failure mechanism is the same in both cases, but the static fatigue life is reduced slightly. This may be due to a subtle increase in the size of the machining induced flaws by the corrosive environment. This enhancement is not discernible when the fracture surfaces are examined.

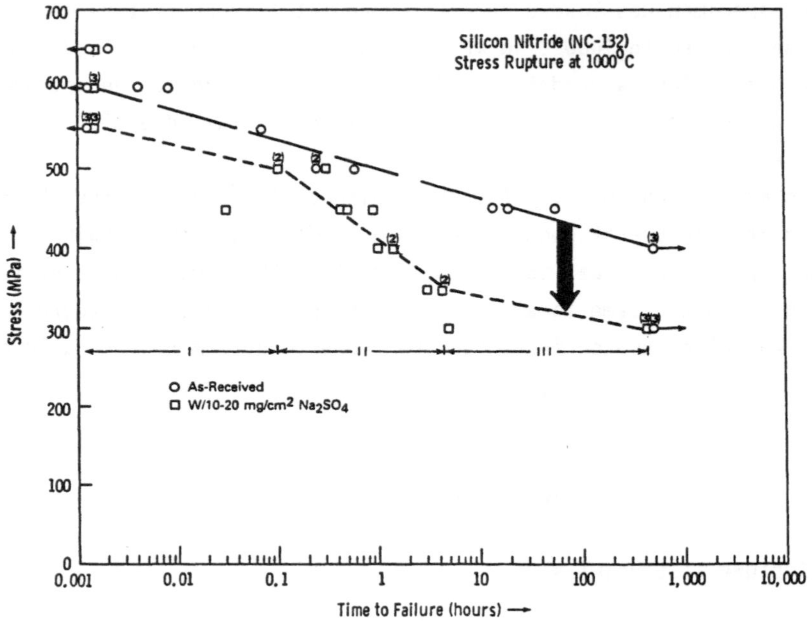

Figure 1. Stress rupture results for NC-132 at 1000°C with and without Na_2SO_4. Numbers in parenthesis indicate the number of specimens which failed at that time under the applied stress.

The slope of the profile in region II is altered significantly due to a change in the strength-limiting defect. Fractographic analysis revealed that corrosion pits ≈10-15 μm in size are being formed. It is the formation of these pits rather than slow crack growth, that reduced the static fatigue life in region II.

At stresses below 350 MPa, region III, the slope of the line again parallels the as-received condition but the static fatigue life is greatly reduced. This indicates that the corrosion pits which form reach a certain size and then stop growing. This is a definite possibility since the Na_2SO_4 is not replenished during the test. Thus once the Na_2SO_4 is depleted the pits should stop growing.

The difficulty with fractography after SR testing was upon failure a glassy phase, molten Na_2SO_4 and/or the corrosion product, flowed over the fracture surface, obscuring the failure marks. A mild HF solution was used to remove this glassy phase but in doing so it also removed the glassy phase inherent to the material that highlights the details of the fracture surface.

GTE Wesgo (SNW-1000): Figure 2 is the SR profile for the SNW-1000 with and without Na_2SO_4. For the as-received condition there appears to be a truncation of the static fatigue behavior below 350 MPa. This is similar to previous work[13], however the truncation occurred below 250 MPa. The difference between these two studies can be attributed to the different dimensions of the flexure bars which were tested. Examination of the fracture surface reveals that the strength-limiting defect is a pore or pores. No creep or slow crack growth is evident in any of the specimens.

Surprisingly the static fatigue life of this material increases with the addition of the Na_2SO_4. At first it was thought that the Na ions would change the glassy grain boundary phase resulting in accelerated creep of the material. This was not supported when the permanent deformation of the specimens which survived 500 hours was measured. A very limited amount of creep was found in specimens with or without Na_2SO_4 and there was no difference in the amount of creep between the as-received condition or with Na_2SO_4.

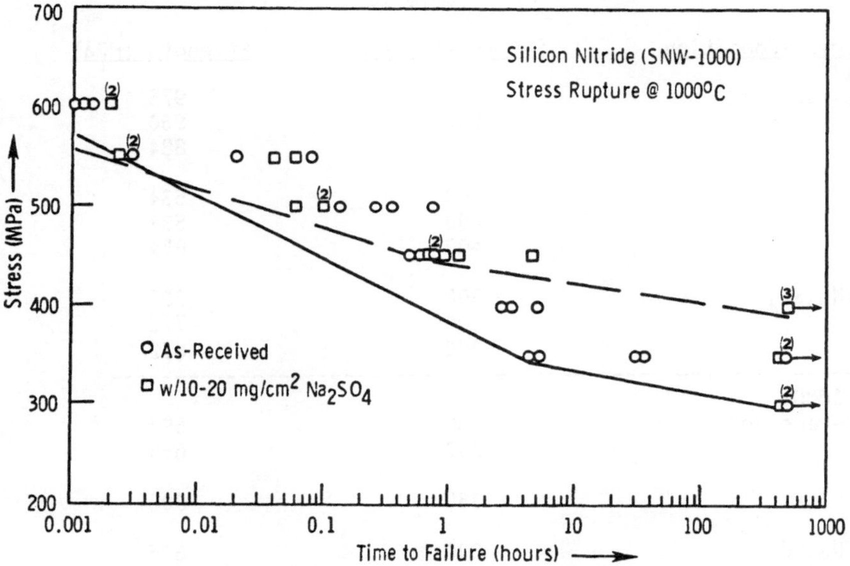

Figure 2. Stress rupture results for SNW-1000 at 1000°C with and without Na_2SO_4. Numbers in parenthesis indicate the number of specimens which failed at that time under the applied stress. Lines represent the lower boundary.

The second thought is that the Na ions change the crystallization behavior of the the grain boundary phase. Quinn and Braue[13] found in their research that complex devitrification reactions were occurring at the grain boundaries resulting in the truncation of the static fatigue life of the SNW-1000 material at 1000°C and 1100°C. How the addition of Na_2SO_4 affects the grain boundaries is currently under investigation.

Retained Room Temperature Strength

The retained room temperature strength for both Si_3N_4 materials is summarized in Table 1. For the NC-132 the retained strength changed not only with the addition of Na_2SO_4 but also with an increase in applied stress. The change with the addition of Na_2SO_4 is due to the formation of corrosion pits. The change with applied stress is due to slow crack growth where slow crack growth increases the flaw size thus decreasing the retained strength.

Table 1.

Material/Condition	SR Stress (MPa)	Strength (MPa)
NC-132:		
As-received	300	975
	300	860
	300	884
	400	534
	400	533
	400	854
w/Na_2SO_4	300	787
	300	776
	300	612
SNW-1000:		
As-received	300	594
	300	655
	350	308
w/Na_2SO_4	300	606
	350	591
	350	641
	400	605
	400	337
	400	640

The retained strength of the SNW-1000 Si_3N_4 did not change at all. This provides further support to the suggestion that the changes in the grain boundary phase during corrosion are the reason that the static fatigue life at 1000^oC is increased.

SUMMARY

This study showed that the addition of Na_2SO_4 affected the static fatigue life of different Si_3N_4 in different ways. The static fatigue life of the MgO-doped Si_3N_4 is reduced due to the formation of corrosion pits. On the other hand, the static fatigue life of the Y_2O_3 + Al_2O_3-doped Si_3N_4 is increased where this increase may be due to changes in the devitrification behavior of the grain boundary phases in this material.

REFERENCES

1. Tressler, R.E., Meiser, M.D. and Yonushonis, T., Molten Salt Corrosion of SiC and Si_3N_4 Ceramics, J. Am. Ceram. Soc., 1976, 59 [5-6], pp.278-279.

2. Levy, M. and Falco, J., Hot Corrosion of Reaction-Bonded Si_3N_4, Am. Ceram. Soc. Bull., 1978, 57 [4], pp. 457-458.

3. Bourne, W.C. and Tressler, R.E., Molten Salt Degradation of Si_3N_4 Ceramics, Am. Ceram. Soc. Bull., 1980, 59 [4], pp. 443-452.

4. Smialek, J.L., Fox, D.S. and Jacobson, N.S., Hot Corrosion Attack and Strength Degradation of SiC and Si_3N_4, prepared for NASA-Lewis Research Center for Environment Degradation of Engineering Materials III, NASA TM-89820, April 13-15, 1987.

5. Jacobson, N.S. and Fox. D.S., Molten Salt Corrosion of Silicon Nitride: II, Sodium Sulfate, J. Am. Ceram. Soc., 1988, 71 [2], pp. 139-148.

6. Jacobson, N.S., Smialek, J.L. and Fox, D.S., Molten Salt Corrosion of SiC and Si_3N_4, prepared for NASA-Lewis Research Center, NASA TM-101346, November 1988.

7. Davies, G.B., Holmes, T.M. and Gregory, O.J., Hot-Corrosion Behavior of Coated Covalent Ceramics, Adv. Ceram. Mat., 1988, 3 [6], pp. 542-547.

8. Swab, J.J. and Leatherman, G.L., Static Fatigue Behavior of Structural Ceramics in a Corrosive Environment, prepared for the U.S. Army Materials Technology Laboratory, MTL TR 90-32, June 1990.

9. Govila, R.K., Uniaxial Tensile and Flexural Stress Rupture Strength of Hot Pressed Si_3N_4, J. Am. Ceram. Soc.,1982, 65 [1], pp. 15-21.

10. Quinn, G.D. and Swank, L., Static Fatigue of Preoxidized Hot-Pressed Silicon Nitride, J. Am. Ceram. Soc., 1983, 66, pp. c-31-c-32.

11. Tighe, N. and Wiederhorn, S., Effects of Oxidation on the Reliability of Hot Pressed Silicon Nitride, in Fracture Mechanics of Ceramics, Vol. 5, ed., R.C. Bradt, A.G. Evans, D.P.H. Hasselman and F.F. Lange, Plenum Press Corp., 1983, pp. 403-424.

12. Quinn, G.D., Fracture Mechanisms Maps for Advanced Structural Ceramics, Part 1 Methodology and Hot-Pressed Silicon Nitride Results, J. Mat. Sci., 1990, 25, pp. 4361-4376.

13. Quinn, G.D. and Braue, W.R., Fracture Mechanism Maps for Advanced Structural Ceramics, Part 2 Sintered Silicon Nitride, J. Mat. Sci., 1990, 25, pp. 4377-4392.

EFFECT OF MICROSTRUCTURAL DEGRADATION ON THE STRENGTH OF SINTERED SILICON NITRIDE AFTER HIGH TEMPERATURE EXPOSURE

NOBUHIKO NISHIMURA, EITO MASUO, KATSUHIKO TAKITA
Nagasaki Research and Development Center,
Mitsubishi Heavy Industries, Ltd.
1-1, Akunoura-machi, Nagasaki 850-91, Japan
&
YOSHITO KOBAYASHI
Truck and Bus Engineering Center,
Mitsubishi Motors Corporation
21-1, Shimomaruko 4-chome, Ohta-ku, Tokyo 146, JapAN

ABSTRACT

Microstructural degradation in two kinds of sintered silicon nitride due to high temperature exposure was examined on specimens after flexure static fatigue tests and room-temperature flexure tests of heated specimens. Surface oxidation layer was formed due to the reaction of silica formed by the oxidation of β-Si_3N_4 and sintering additives segregating from the interior grain boundary phase. Corresponding to the oxidation layer, a degraded layer containing numerous micro-pores resulting from the segregation of the sintering additives was formed just beneath the surface oxidation layer. The thickness of the surface oxidation layer, which was measured non-destructively by means of X-ray diffraction method, had increased along with the heating. Consequently, the room temperature strength of specimens after heating was considered to be related to the thicknesses of surface oxidation layer and surface degraded layer.

INTRODUCTION

Recent efforts to improve mechanical properties and reliability of silicon nitride ceramics have made a ceramic gas turbine near its practical application stage. Under such circumstances, it has become important and of urgent necessity to understand strength deteriorative behaviors of high-temperature components in the ceramic gas turbine, caused by their exposure to high temperatures, and to establish a design method, a reliability assessment technology and a non-destructive inspection test method, which take such strength deteriorative behaviors into account.

The objectives of the present research in light of the above are to study mechanism of strength deterioration in a high-temperature environment by examining strength deteriorative behavior of silicon nitrides exposed to high temperatures in air atmosphere and the resultant microstructural change behavior and to study a possible non-destructive inspection method of high-temperature components used for ceramic gas turbines.

EXPERIMENTALS

Test materials
Test materials employed in this study were two kinds of sintered silicon nitrides, one, remarked as sample 'A', contained MgO as sintering additive and the other, remarked as sample 'B', did not contain MgO.

Test procedure
After the test materials were machined into 4 point flexure test specimens in accordance with JIS, these test specimens were heated under air atmosphere at 1473K and 1573K for sample 'A' and at 1673K for sample 'B', for up to 100 hours. In addition, flexure static fatigue test and heating tests by applying flexural stress of 400MPa were conducted for sample 'B' at 1673K. Flexure strengths at room temperature were obtained on heated specimens except for the specimen after the static fatigue test.

After the flexural test, X-ray diffraction profiles were obtained on the surface of the tension side of fractured specimen in order to identify the oxidation scales formed due to heating.

Then the microstructures were examined on fractured specimens. At first, fracture behaviors in each specimen were characterized through the observation with a scanning electron microscope on the fracture surface and the surfaces of tension and compression sides of the fractured specimens. Then, the specimens were cut into two pieces along the longitudinal direction of specimen from the origin of fracture as shown in Figure 1. Microstructures on cut cross-sections which were polished with diamond particles, were examined with a scanning electron microscope and electron probe micro-analyzer, especially around the origin of fracture.

Then, the two-split pieces of each specimen were glued together at both surfaces of tension side and a specimen for transmission electron microscopy was prepared to observe the microstructure around the surface of tension side as shown in Figure 1.

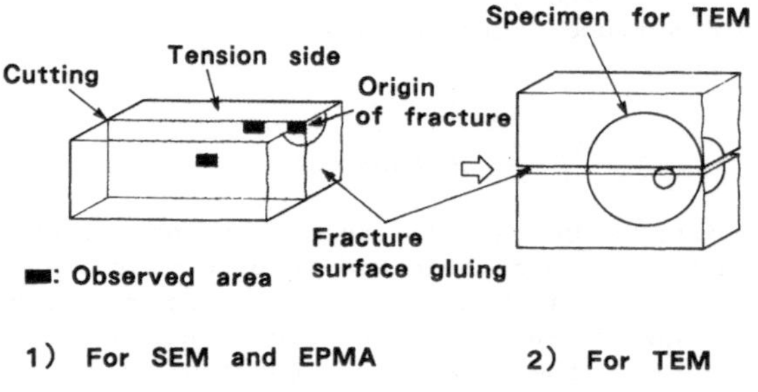

Figure 1. Preparation of specimen for cross-sectional structure examination

RESULTS

Strength after high temperature exposure

Figure 2 shows flexural strengths at room temperature of samples after high temperature exposure. Flexural strengths of all samples indicated normalized values based on the respective flexural strengths before high temperature exposure. In all sampels, their flexural strengths dropped along with the increase of holding time. Also, in sample 'A' subjected to high temperature exposure at different temperatures, decrease of its flexural strength was larger towards an increase of the testing temperature. Also, flexural strength of sample 'B' heated with flexural stress applied during testing showed a larger drop of flexural strength as compared with flexural strength heated with no stress, indicating that loading of stress accelerated the strength drop due to high-temperature exposure.

Also, in comparison of behaviors between samples, the behavior of flexural strength drop of sample 'A' heated at 1573K was approximately equivalent to that of sample 'B' heated at 1673K, thus indicating that strength drop of sample 'A' due to high-temperature exposure was larger than that of sample 'B'.

Microstructure after exposure

Figure 3 gives the diffracted X-ray intensity spectrum from the surface of sample 'A' after its exposure at 1573K for 100h. Any phase other than the β-Si_3N_4 phase was not detected from the surface of sample 'A' before its high temperature exposure, whereas after its high temperature exposure, enstatite ($MgSiO_3$) was also detected in addition to the β-Si_3N_4 phase. The diffracted X-ray intensity from the enstatite was found to have increased

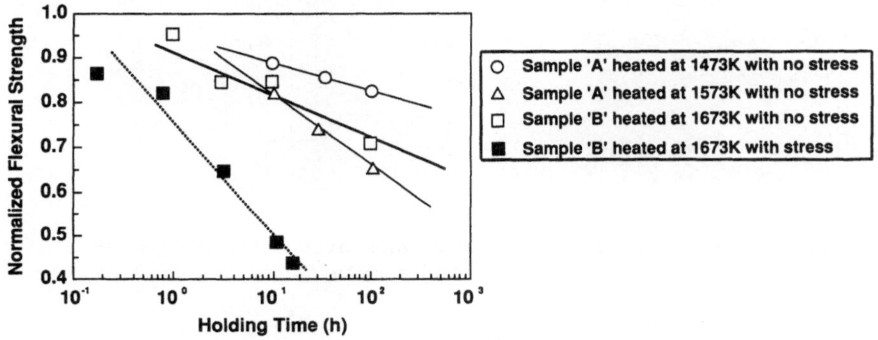

Figure 2. Flexural strength after heating test

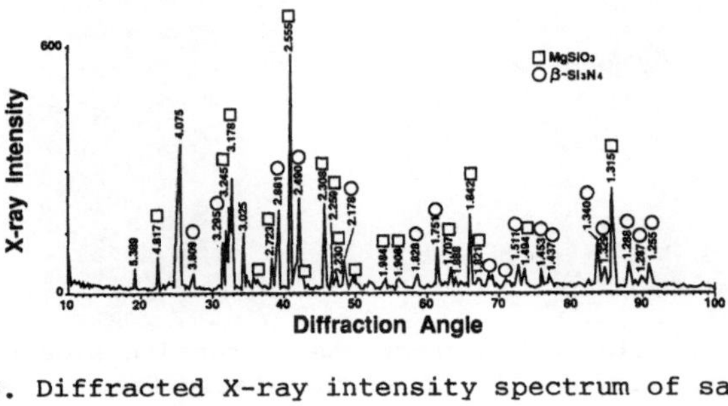

Figure 3. Diffracted X-ray intensity spectrum of sample 'A'

along with an elevation of testing temperature and also, with an increase of the holding time. Meanwhile, from sample 'B' as well, no phases other than the β-Si_3N_4 phase were detected before its high temperature exposure, whereas after its high temperature exposure, formation of $Y_2Si_2O_7$ phase was also detected.

Figure 4 shows the results of microstructural examinations using the SEM on fractured surfaces of specimens in the vicinity of origins of fractures after flexural tests at room temperature of as-sintered specimen and the specimen heated at 1573K for 100h of sample 'A'. The origin of fracture in the as-sintered specimen was found to be a huge pore, whereas the origin of fracture in the heated specimen exhibited, instead of a huge pore, a semi-circular mirror-like region with intergranular fractured surface showing no appearance of huge pore. The mirror-like region in the heated specimen had grown larger towards the higher temperature side of exposure and the longer period side of holding time; towards the larger decline of strength due to high-temperature exposure.

Also, Figure 5 shows scanning electron micrographs at tension side surfaces of sample 'B' after its static fatigue test. On any surfaces of fractured specimens not subjected to stress loading during heating, no microcracks were detected except for the region in the vicinity of origins fractures but on specimens subjected to stress loading, numerous cracks were detected at tension side surfaces. In addition, numerous intergranular microcracks were also detected from tension side surfaces in location where no macrocracks were verified.

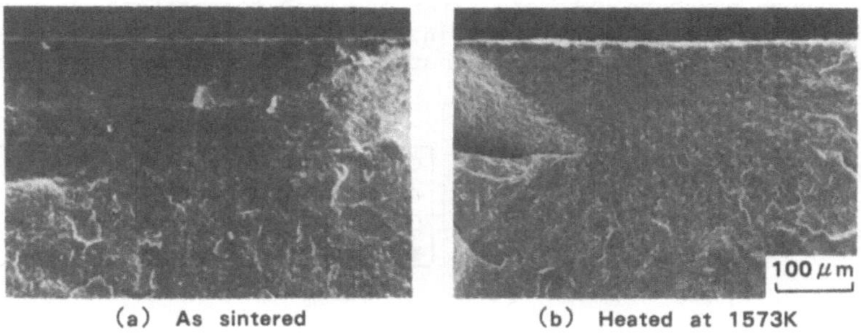

Figure 4. Scanning electron micrographs around the origin of fracture for sample 'A'

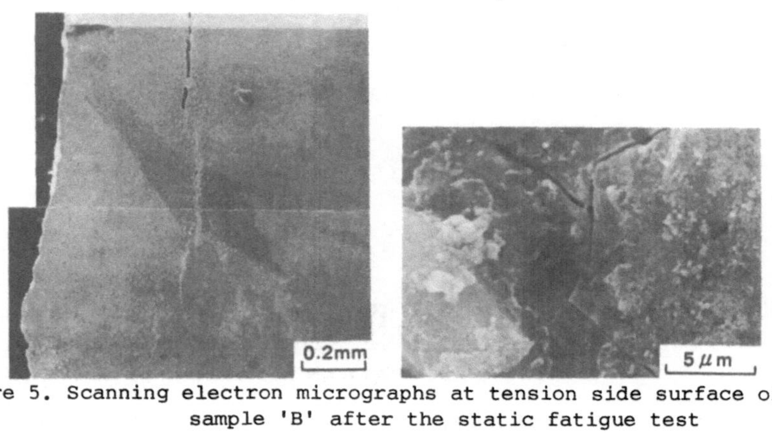

Figure 5. Scanning electron micrographs at tension side surface of the sample 'B' after the static fatigue test

Figures 6 and 7 show, respectively, the results of element distributions regarding Si and Y as measured by means of the EPMA in the vicinity of origin of fracture, and the result of microstructural observation using transmission electron microscope (TEM) at the vicinity of origin of fracture in the cross-sectional area after the flexural test at room-temperature of sample 'B' heated at 1673K for 100h. A region of heterogeneous element distribution due to high temperature exposure was recognized in an area close to the surface; a region was recognized where Si had condensed right beneath this surface and Y had become poor accordingly. Also, the structural examination using the TEM detected numerous micro-pores. The Y-poor region as detected by means of the EPMA was considered to correspond to micro-pores as observed with the TEM. And because the width of nucleation region of micro-pores has nearly corresponded to the depth of the mirror-like region observed on fracture surface, the strength decline in specimens heated with no stress was considered to have been caused by nucleation of micro-pores resulting from heating.

Similar structural degradation right beneath the surface was also recognized in sample 'A', and in sample 'A', Mg had segregated in the oxidation scale and in the intergranular phase just beneath this oxidation scale, micro-pores resulting from segregation of Mg were recognized.

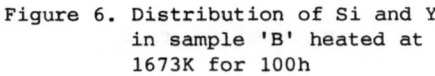

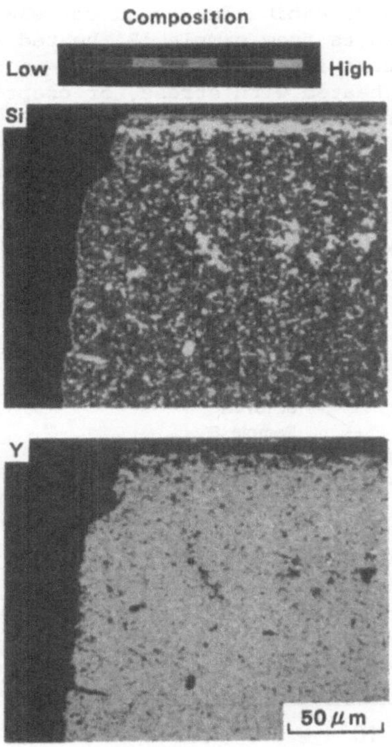

Figure 6. Distribution of Si and Y in sample 'B' heated at 1673K for 100h

Figure 7. Transmission electron micrograph around the surface in sample 'B' heated at 1673K for 100h

Development of NDI method

From the foregoing microstructural examinations and analysis, the strength decline of Si_3N_4 due to high temperature exposure was considered to be directly related to the oxidation behavior of surfaces. Meanwhile, as shown in Figure 3, the X-ray intensity diffracted from the crystalline oxide scale had increased towards the progress of surface oxidation. So, diffracted X-ray intensity ratios; Y as expressed by equation (1), were calculated, using diffracted X-ray intensity spectra found from the respective surfaces of specimens:

$$r = \frac{\text{X-ray intensity diffracted from oxide phase}}{\text{X-ray intensity diffracted from } \beta\text{-}Si_3N_4} \quad (1)$$

Figure 8 gives the relationship between diffracted X-ray intensity ratio and holding time. In all the specimens, the diffracted X-ray intensity ratio was nearly proportional to one half power of the holding time and agreed with the parabolic rate kinetics accepted in mass change due to oxidation[1].

The above results indicated that the thickness of oxidation scale could be detected non-destructively, using diffracted X-ray intensity ratios. Consequently, Figure 9 shows the relationship between diffracted X-ray intensity ratios and flexural strengths normalized by the strength of the as-sintered sample. The diffracted X-ray intensity ratio from sample 'A' tested by heating at different temperatures exhibited a good correlation with the flexural strength, and the results from sample 'B' heated with no stress were also similar to the results from sample 'A'.

Meanwhile, the results from sample 'B' heated with stress, or crept in short, were inconsistent with the results from sample 'A' and sample 'B' heated with no stress loading. The reason for this discrepancy can be inferred as follows; loading of stress will cause a micro-crack consisting of micro-pores linked together just beneath the surface as shown in Figure 5 to open at the surface, resulting in slow crack growth due to selective oxidation and repeated growths of microcracks because oxygen is supplied to the tip of the microcrack from the atmosphere. In other words, this may be

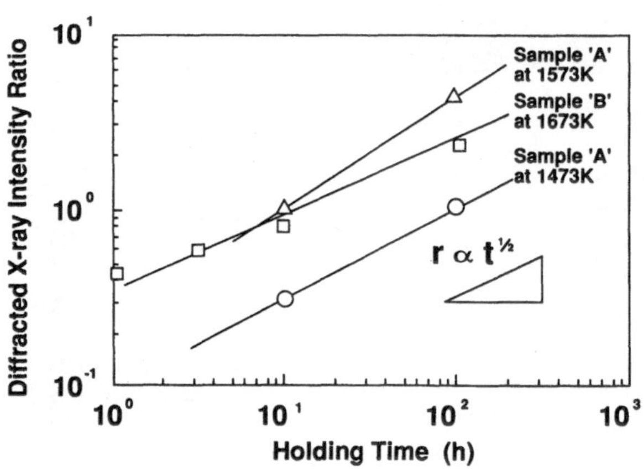

Figure 8. Relation between diffracted X-ray intensity ratio and holding time (t)

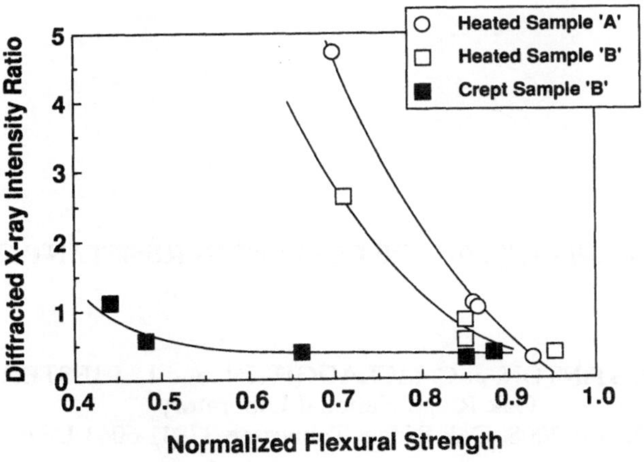

Figure 9. Relation between diffracted X-ray intensity ratio and normalized flexural strength

because the X-ray diffraction method which can assess the degree of uniform oxidation may not be sufficient enough to assess oxidation under stress loading resulting in selective oxidation.

CONCLUSION

Results of our examinations on the effects of microstructural degradation on the strength of sintered silicon nitride, a promising high temperature component for the ceramic gas turbine, after its high temperature exposure, are outlined as follows:

- Strength of Si_3N_4 declines when it is held at elevated temperatures in the atmosphere.
- The high temperature exposure forms a degraded layer which contains numerous micro-pores attributable to the segregation of cations in the grain boundary phase into the surface oxidation scale in its formation process.
- In specimens heated with stress loading; crept specimens, a degraded grain boundary will be formed in addition to the above structural degradation. In this degraded grain boundary, a crack open at the surface would be formed, caused by linkage between micro-pores and further oxidation and micro-pores would appear at the tip of this crack.
- Behavior of oxidation due to heating in the atmosphere can be assessed non-destructively by means of the X-ray diffraction method.
- Strength decline due to high-temperature exposure in the atmosphere is strongly associated with growths of the degraded layer and the degraded grain boundary accompanied with oxidation as described above.

REFERENCE

1. S.C. Singhal, j. Mater. Sci., 1976, 11, 500.

FABRICATION AND TESTING OF CORROSION RESISTANT COATINGS

D. P. STINTON, J. C. MCLAUGHLIN, and L. RIESTER
Oak Ridge National Laboratory
P.O. Box 2008, Oak Ridge, Tennessee 37831-6063 USA

ABSTRACT

The susceptibility of SiC and Si_3N_4 to sodium corrosion mandates that corrosion resistant coatings be developed to protect silicon-based turbine engine components. Materials with good corrosion resistance and thermal expansions that nearly match SiC and Si_3N_4 have been identified. Corrosion testing of hot-pressed pellets of these compounds has identified the most promising materials. Development of chemical vapor deposition systems to apply these materials has been initiated.

INTRODUCTION

SiC and Si_3N_4 are currently prime candidates for turbine engine applications because of their high retained strength to 1400°C, excellent oxidation resistance, good thermal shock resistance, and light weight. Unfortunately, sodium, potassium, and steam have been shown to cause degradation of SiC and Si_3N_4 at temperatures above ≈1200°C.[1-4] However, corrosion has not been recognized as a significant problem in engine testing because current components can survive only several hundred hours due to thermal and mechanical stresses. After improved designs and materials permit tests of longer duration, the severity of the corrosion problem will become more evident.

The outer surfaces of SiC and Si_3N_4 components oxidize at high temperatures to form a silica layer that inhibits further oxidation.[5] However, sodium ingested into turbine engines from the environment or impurities present in the fuel can react with this silica layer, such that it is no longer protective.[6,7] The primary corrosion mechanism is the formation of a sodium-silicate liquid phase at temperatures above ≈800°C, the eutectic temperature in the sodium-silicate system.[8] The objective of the reported research is to develop a coating that will protect the SiC or Si_3N_4 components or coatings from sodium corrosion as well as provide oxidation protection.

BACKGROUND

In 1986, GTE Laboratories initiated a program to develop coating systems to protect silicon-based ceramics from contact stress damage. Development of a protective coating for SiC or Si_3N_4 is difficult because the coatings crack due to the mismatch in coefficient of thermal expansion (CTE) between the substrate (low CTE) and the coating (high CTE). A coating system was therefore designed to accommodate the stresses caused by the difference in CTE.[9] (Fig. 1) An AlN coating is initially deposited onto the SiC or Si_3N_4 substrates to promote adherence via reaction (formation of a SiAlON compound) at the substrate/coating interface. The coating is then compositionally graded from AlN to $Al_xO_yN_z$ to Al_2O_3 + ZrO_2. Residual stresses within the graded layer are minimized through the absence of sharp interfaces and the gradual increase in thermal expansion from the interface to the outer protective coating.

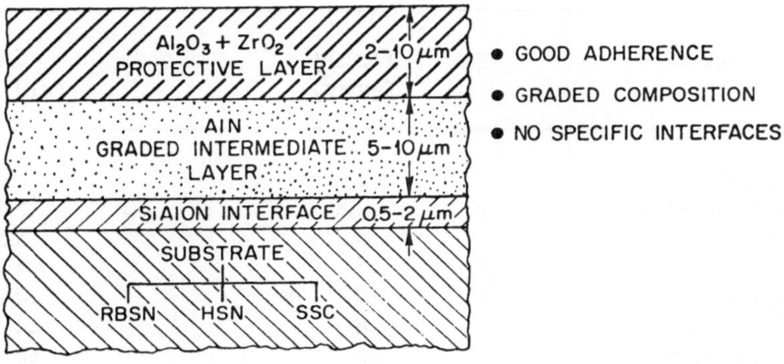

Figure 1. GTE's Protective Coating

The use of the GTE Laboratories' coating system appeared very promising for several years.[10,11] Application of the coatings at about 1000 °C produced adherent crack-free coatings. Repeated thermal cycling between room temperature and ≈1100 °C produced no apparent effect on the coatings. However, heating the coatings to temperatures above 1200 °C produced disastrous results, with the CTE mismatch causing the outer Al_2O_3 + ZrO_2 coating to crack. Penetration of oxygen through the cracks oxidized the AlN interfacial coating resulting in blistering of the outer coating. In an attempt to reduce the stresses at high temperatures, coatings were deposited at 1200 °C. However, these coatings spalled on cooling to room temperature. Developers of this coating system concluded that the difference in thermal expansion between SiC or Si_3N_4 and Al_2O_3 + ZrO_2 is too great to develop an adherent, crack free coating.

CANDIDATE PROTECTIVE COATINGS

The list of potential materials for corrosion protection of SiC and Si_3N_4 is very limited because the material must be an oxide (for oxidation resistance) with a CTE significantly less than that of Al_2O_3 (for adherence). A carbon/carbon composite (C/C) is a material with a very low thermal expansion (≈0) that must be protected from oxidation in high-temperature aerospace applications. However, after 20 years of investigation no oxidation resistant coatings have been found that can survive the dramatic thermal cycles, although several materials with low CTEs were identified that might be useful for the higher CTE heat engine components. (Table 1).

TABLE 1
Refractory Oxides with Potential for Oxidation/Corrosion Protection

Compound	Density (g/cm^3)	CTE ($x10^{-6}/°C$)
Al_2O_3*	3.97	8.0
$3Al_2O_3 \cdot 2SiO_2$	2.8	5.7
SiC*	3.21	5.5
$ZrTiO_4$	≈5	≈4
$HfTiO_4$	≈5	≈4
$Ta_2O_5 \cdot 6ZrO_2$	≈6	≈4
$Ta_2O_5 \cdot 6HfO_2$	≈6	≈4
Ta_2O_5	8.02	3.6
Si_3N_4*	3.19	3.0
Al_2TiO_5	3.68	2.2
carbon/carbon*	1.9	≈0

*Included only as a reference and not as a potential coating.

Mullite ($3Al_2O_3 \cdot 2SiO_2$) is a material with a CTE very close to that of SiC and has performed reasonably well in corrosion tests, corroding somewhat more severely than Al_2O_3, but surviving significantly better than SiC.[4] Corrosion resistance is dependent on the formation of stoichiometric mullite with no free SiO_2, since free SiO_2 is readily attacked. Because of its corrosion resistance and low CTE, mullite is also being investigated as a material for high-temperature particulate filters that are exposed to sodium contaminants.[12]

Not very much is known about the materials $ZrTiO_4$, $HfTiO_4$, $Ta_2O_5 \cdot 6ZrO_2$, or $Ta_2O_5 \cdot 6HfO_2$, however, investigations have shown their thermal expansions to be quite low.[13-15] The structures formed by the titanate phases provide significant advantages over the destructive phase transformations of monoclinic ZrO_2 or HfO_2. Stabilization of the cubic structure of ZrO_2 with yttria, ceria, etc., is impractical because of their high thermal expansion ($13x10^{-6}/°C$) and because the stabilizers would be leached out of the structure. $ZrTiO_4$ and $HfTiO_4$, however, crystallize in an orthorhombic structure that does not undergo polymorphic transformations.[16] (Fig. 2) Even less is known about Ta_2O_5•$6ZrO_2$ and Ta_2O_5•$6HfO_2$, although the reported[13] low thermal expansion and the potential for good corrosion resistance make the materials promising candidates.

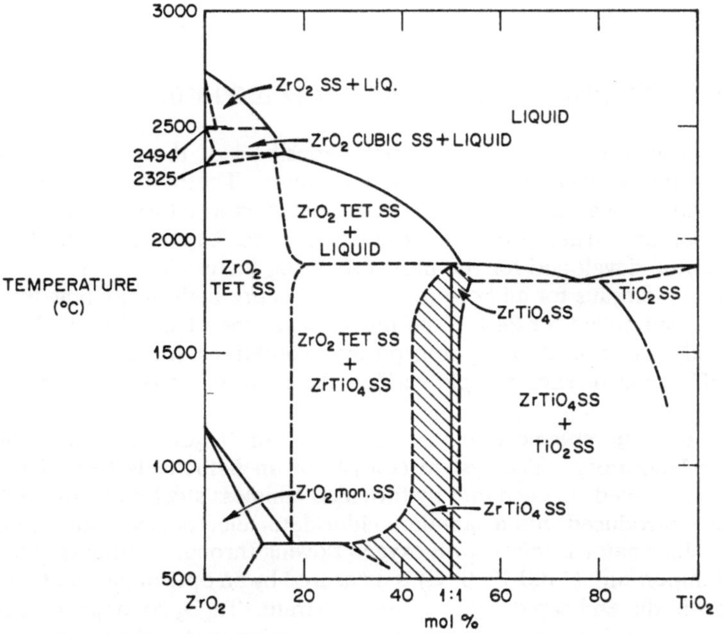

Figure 2. Phase equilibria in the ZrO_2-TiO_2 system.

An additional material of interest is Ta_2O_5, in part because of the ease of chemical vapor deposition.[17,18] Phase equilibria studies have shown that no liquid phases are present below 1625°C.[19] Formation of either of the sodium tantalate phases at turbine application temperatures that are more than 300°C below the liquidus temperature would seem unlikely. Furthermore, the formation of these phases would not be catastrophic unless volume changes would cause spallation of the coatings.

One final material that shows promise is Al_2TiO_5 which has been investigated as a protective coating for C/C and as a corrosion resistant material for high-temperature particulate filters.[20] However, obvious problems do exist for this material. First, because the CTE of the material is very anisotropic the material exhibits significant microcracking and could become permeable to oxygen and sodium. Second, the material tends to absorb water and could degrade significantly when exposed to an environment containing as little as 10% moisture at high temperatures.

To adequately protect SiC and Si_3N_4 components, the candidate protective coating must prevent oxygen present in the combustion environment from diffusing through the coatings to the substrate. Since oxygen will diffuse through most oxide coatings at a significant rate, an additional barrier to oxygen must be established. The native SiO_2 layer that coats SiC and Si_3N_4 is impermeable to oxygen and prevents further oxidation; however, reaction with or diffusion into the corrosion resistant oxide coating would likely destroy the oxygen barrier. Therefore, testing will need to be performed to insure that the SiO_2 layer is stable with respect to the candidate protective coatings.

DESCRIPTION OF MATERIALS AND EXPERIMENTS

Development of processes to chemically vapor deposit all of the potential materials described above would be very costly and time consuming. Therefore, relatively easy to prepare hot-pressed samples of these materials are being exposed to sodium-containing combustion environments to assess their corrosion resistance. Chemical vapor deposition processes will then be developed for the most promising materials.

The starting materials for all hot-pressed samples are high purity powders which are combined and ball milled for 24 h. The mixtures are then loaded into 38mm-diam graphite dies, and hot-pressed using appropriate conditions for each composition. Powder X-ray diffraction of each sample verified the material to be the compound of interest.

Processes for the chemical vapor deposition of Ta_2O_5 have already been developed at this laboratory. The system consists of an inductively heated substrate within a quartz tube sealed on each end by threaded stainless steel end caps. Chlorine and argon gas are introduced into a tantalum chlorinator electrically heated to 600 °C. $TaCl_5$ exiting the chlorinator is mixed with oxygen flowing through a separate inlet tube. The deposition temperature (1100-1300 °C) is measured by an optical pyrometer sighting through a window in the end cap directly on the substrate. Ta_2O_5 coatings were applied to SiC coated graphite substrates to simulate a heat engine material system.

Corrosion tests are performed in a horizontal alumina tube furnace with an inner diameter of 60 mm. Hot-pressed pellets of each material are sectioned and tested at ≈1000 °C. A combustion environment is simulated by flowing methane into the furnace at 685 cm^3/min along with air at 7,600 cm^3/min. The sodium level is controlled by the position, and hence the temperature, of the Na_2CO_3 source. Based on the measured weight loss of the Na_2CO_3 source, the sodium concentration in the system varied from 20 to 110 ppm. Sodium concentrations that are an order of magnitude lower and exposure times that are an order of magnitude higher are more realistic for many combustion applications. The total amount of sodium that the samples were exposed to was therefore similar to that projected for many applications, thus the experimental results that have been obtained provide a reasonable basis for approximating the performance of these materials. The sodium levels, however, may have been subject to some variation during each run, primarily because the exposed surface area of the sodium carbonate powder was not constant.

CORROSION TESTING

Specimens of Al_2TiO_5, $ZrTiO_4$, and $3Al_2O_3{\bullet}2SiO_2$ were successfully hot pressed for corrosion testing. Evaluation of these materials by X-ray diffraction revealed that the desired crystalline compounds had been produced. The initial hot-pressing conditions selected for the Al_2TiO_5 produced a porous pellet, therefore, a dense specimen was obtained from a commercial source. X-ray diffraction of the commercial material showed that the outer surface was predominantly Al_2O_3, whereas the interior of the pellet was Al_2TiO_5. The hot-pressed pellets of $ZrTiO_4$ and $3Al_2O_3{\bullet}2SiO_2$ were fine grained, but the former contained approximately 20% porosity.

After exposing each of the candidate materials plus a sintered α-SiC specimen to a combustion environment at 1000 °C, the pellets were examined metallographically. The α-SiC had a surface layer of glass about 20μm thick. Significantly better corrosion resistance, which is consistent with the published results of Federer [4], was exhibited by $3Al_2O_3{\bullet}2SiO_2$ which had a 10μm glass layer. Further improvements in corrosion resistance are seen by $ZrTiO_4$ and Al_2TiO_5 since the reaction layers were $\approx$4μm and $<$2μm thick, respectively. Al_2O_3 grains were observed on the surface of the polished Al_2TiO_5 sample, thus verifying the X-ray diffraction results discussed previously.

SiC-coated graphite substrates overcoated with Ta_2O_5 were also corrosion tested. During coating in an oxidizing environment, the SiC formed a relatively thick crystalline SiO_2 protective layer. The Ta_2O_5 coating nucleated onto the SiO_2 with small equiaxed grains that grew into large columnar ones normal to the surface (Fig. 3). Corrosion testing left the Ta_2O_5 coating and the crystalline SiO_2 sublayer unaffected.

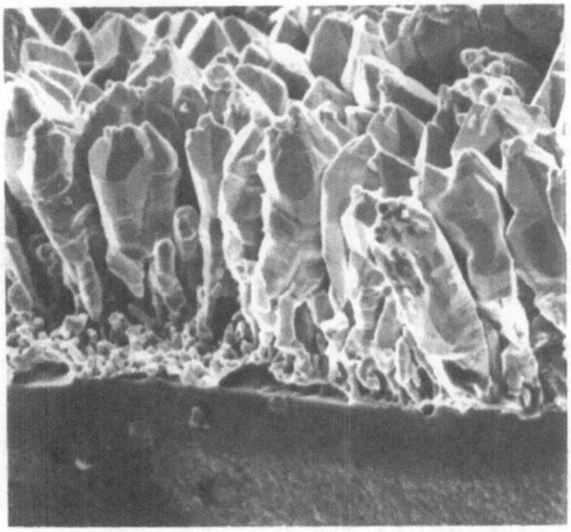

Figure 3. Fracture surface of a Ta_2O_5 coated substrate after corrosion testing.

Successful hot-pressing conditions have not yet been determined for $HfTiO_4$, $Ta_2O_5{\bullet}6ZrO_2$, and $Ta_2O_5{\bullet}6HfO_2$. Analysis of the pellets produced to date by X-ray diffraction have indicated unidentified phases. Additional process development will be required before corrosion testing can be initiated.

CONCLUSIONS

Corrosion testing of various oxides revealed a considerable improvement over SiC. Mullite exhibited a glass reaction layer that was only about half as thick as the layer formed on a SiC specimen. Still better corrosion resistance was exhibited by Al_2TiO_5 and $ZrTiO_4$. Chemically vapor deposited Ta_2O_5 exhibited excellent corrosion resistance. Further tests of much longer duration will be required to identify the most corrosion resistant oxides for overcoating SiC or Si_3N_4 components. After the most promising materials have been identified chemical vapor deposition processes will be developed for application of coatings.

ACKNOWLEDGEMENTS

The authors are indebted to J. I. Federer whose interest and valuable suggestions helped carry this work to completion. The authors are also grateful to Scott Eatherly, Harry Livesey, and Judy Kelly for metallography, drawing preparation, and manuscript preparation, respectively. Research sponsored by the Ceramic Technology for Advanced Heat Engines Project, DOE Office of Transportation Technologies, under contract DE-AC05-84OR21400 with Martin Marietta Energy Systems, Inc.

REFERENCES

1. J. L. Smialek and N. S. Jacobson, Mechanism of Strength Degradation for Hot Corrosion of α-SiC. J. Am. Ceram. Soc., 1986, **69**(10), 741-52.

2. N. S. Jacobson and D. S. Fox, Molten-Salt Corrosion of Silicon Nitride: II, Sodium Sulfate. J. Am. Ceram. Soc., 1988, 71(2), 128-48.

3. N. S. Jacobson, J. L. Smialek, and D. S. Fox, Molten Salt Corrosion of SiC and Si_3N_4. In Handbook of Ceramics and Composites, Vol.1: Synthesis and Properties, ed. N. P. Cheremisinoff, Marcel Dekker, Inc. New York, 1990, pp. 99-136.

4. J. I. Federer, Corrosion of SiC Ceramics by Na_2SO_4. Adv. Ceram. Mater., 1988, **3**(1), 56-61.

5. J. A. Costello and R. E. Tressler, Oxidation Kinetics of Hot-Pressed and Sintered α-SiC. J. Am. Ceram. Soc., 1981, **64**, 327-31.

6. N. S. Jacobson and J. L. Smialek, Hot Corrosion of Sintered α-SiC at 1000°C. J. Am. Ceram. Soc., 1985, **68**, 432-39.

7. N. S. Jacobson, C. A. Stearns, and J. L. Smialek, Burner Rig Corrosion of SiC at 1000°C. Adv. Ceram. Mater., 1986, 1, 154-61.

8. E. M. Levin, C. R. Robbins, H. F. McMurdie, Phase Diagrams for Ceramists, Vol. I, The American Ceramic Society, Westerville, OH, 1964, p.94.

9. V. K. Sarin, Design Criteria for a Coating to Reduce Contact Stress Damage. In Proceedings of the 1987 Coatings for Advanced Heat Engines Workshop, Castine, ME, July 27-30, 1987, U.S. Department of Energy Report, Conf-870762, Page III-83.

10. H. E. Rebenne and V. K. Sarin, Ceramic Coatings to Reduce Contact Stress Damage of Ceramics: Thermodynamic Modeling. In Proc. 25th Automotive Technology Development Contractors' Coordination Meeting, Society of Automotive Engineers, Inc. Warrendale, Penn., 1988, P-209, pp. 199-206.

11. H. E. Rebenne and J. H. Selverian, Adherent Ceramic Coatings To Reduce Contact Stress Damage of Ceramics. In Proc. of the Annual Automotive Technology Development Contractors' Coordination Meeting, Society of Automotive Engineers, Inc. Warrendale, Penn., 1991, P-243, pp. 227-38.

12. M. A. Alvin, D. M. Bachovchin, J. E. Lane, and R. E. Tressler, Degradation of Cross Flow Filter Material. In Proc. of the Seventh Annual Coal-Fueled Heat Engines and Gas Stream Cleanup Systems Contractors Review Meeting, Report No. DOE/METC-90/6110, ed. H. A. Webb, et al., U. S. Department of Energy, Morgantown, W.Va., 1990, pp. 162-67.

13. I. Y. Glatter, D. J. Treacy, J. E. Sheehan, and K. S. Mazdiyasni, High-Temperature Chemical Behavior of a Multi-Layered Oxidation Protection Coating System for Carbon-Carbon Composites, Wright Research Development Center Report No. WRDC-TR-89-4127, 1989.

14. R. Ruh, et. al., Phase Relations and Thermal Expansion in the System HfO_2-TiO_2. J. Am. Ceram. Soc., 1976, 59(11-12) 495-99.

15. K. S. Mazdiyasni and L. M. Brown, Preparation and Characterization of High-Purity $HfTiO_4$. J. Am. Ceram. Soc., 1970, 53(11), 585-89.

16. E. M. Levin and H. F. McMurdie, Phase Diagrams for Ceramists, 1975 Supplement, The American Ceramic Society, Westerville, OH, 1964, p. 169.

17. T. Takahashi and H. Itoh, Formation of Tantalum Oxide by Chemical Vapor Deposition. J. of Less-Common Metals, 1972, 38, 211-19.

18. E. Kaplan, M. Balog, and D. Frohman-Bentchkowsky, Chemical Vapor Deposition of Tantalum Pentoxide Films for Metal-Insulator-Semiconductor Devices. J. Electrochem. Soc.: Solid-State Science and Technology, 1976, 123(10) 1570-73.

19. E. M. Levin and H. F. McMurdie, Phase Diagrams for Ceramists, 1975 Supplement, The American Ceramic Society, Westerville, OH, 1964, p. 92.

20. E. Parsons, Morgantown Energy Technology Center, Morgantown, WV., personal communication to D. P. Stinton, August 1990.

CHARACTERISTICS OF POROUS CORDIERITE TRAPS INTENTIONALLY CONTAMINATED WITH CALCIUM AND ZINC

L. Agostini, G. de Portu, S. Guicciardi
CNR-Research Institute for Ceramics Technology
via Granarolo 64, I-48018 Faenza (RA), ITALY

C. Borello, P. P. Demaestri, A. Giachello
Centro Ricerche Fiat
Strada Torino 50, I-10043 Orbassano (TO), ITALY

ABSTRACT

Cordierite monolithic structures as catalyst supports for the filtration of exhaust are going to be extensively used in order to control automotive gas emission. For this reason it is important to define the thermo-mechanical behaviour of such a component, especially in the presence of pollutants like those present in the exhausted gases. Young's modulus, flexural strength and thermal expansion coefficient have been measured in virgin, polluted and thermal aged cordierite samples.

INTRODUCTION

One of the most promising filters, usually known as traps, for Diesel engine exhaust gas purification is a cellular structure of porous cordierite where the gas is forced to flow through the thin walls of the cells[1,2]. In doing so, the particulate is captured and collected in the filter. Some substances, which come from the additives contained in the lubricating oil, are retained in the structure and at high temperature may react with the ceramic and lead to a deterioration of the structure[3]. The fundamental thermo-mechanical properties[4,5] and the influence of thermal ageing and thermal shock[6] on the base cordierite have already been reported. The aim of this study is to evaluate the influence of two pollutant compounds on some thermo-mechanical properties of the cordierite base material. Samples cut from a virgin trap were separately polluted with zinc and calcium compounds and then thermally aged. Young's modulus (E), flexural strength (MOR) and the thermal expansion coefficient (α) were measured before and after thermal ageing. The values obtained were compared with the corresponding values of the unpolluted cordierite.

MATERIALS AND METHODS

Cordierite honeycomb traps (type EX66-100/25) produced by Corning[§] were considered. This material has a porosity of about 50% with a mean pore size of 32 μm. The component has 100 cells per square inch and a wall thickness of 25 mils.
The samples, 60 mm x 16 mm x 8 mm, were cut from the as-fired component in a longitudinal direction[6].
The contamination of the samples was performed using $Ca(NO_3)_2$ and $Zn(NO_3)_2$ as pollutant compounds. These salts are water soluble which makes the contamination homogeneous through the sample by imbibition. The imbibed samples, after drying overnight at 80°C, were calcined up to 600°C and 450°C, in the case of the Ca-polluted and the Zn-polluted samples respectively, to oxidize the pollutant compounds. The chemical analysis confirms the intended percentages of 3 wt% Zn and 3.5 wt% Ca were reached. The thermal ageing consisted of holding the samples for 3 hours at temperatures of 800°C, 1000°C and 1200°C, respectively.
X-ray[+] and Scanning Electron Microscope (SEM)[++] analyses were carried out before and after thermal ageing. Young's modulus was measured using the resonance frequency method[#] according to ASTM-C848[7] on five samples for each point. The MOR tests were performed on a 4-point fixture, with 50 mm and 25 mm outer span and inner span respectively, at a crosshead speed of 0.5 mm/min on a Instron machine mod.1195[*]. Five samples were tested for each point. The thermal expansion coefficient between 25°C and 900°C was measured by a Netzsch[&] dilatometer with a heating rate of 5°C/min.

RESULTS AND DISCUSSION

X-Ray and SEM Analysis

The base material is mainly cordierite with a small amount of mullite. The microstructure of the material is shown in Fig.1. The X-ray analysis shows no phase transformation after thermal ageing. In the Ca-polluted samples, the X-ray analysis shows no presence of new crystalline phases before and after the thermal ageing at 800°C. However, on SEM microphotos, a small amount of crystalline phase is visible at this point, Fig.2(A). The detection of this phase by X-ray diffraction could be hindered by peak overlap with the base cordierite or

---

[§]Corning Glass Works, Corning, NY, USA; [+]Rigaku Corp., Japan; [++]Philips SEM 525m; [#]H-P 4194A Impedence/gain phase analyser; [*]Instron Corp., Canton (Mass.), USA; [&]Netch-Garätebau GmbH, Selb, Germany.

because the amount is too low. After thermal ageing at 1000°C, a new calcium-alumino-silicate phase (gehlenite) is detected by X-ray diffraction, while the new phase detected after thermal ageing at 1200°C is clearly anorthite, Fig.2(B). In this figure, there is also evidence of microcracks probably induced by the difference in thermal expansion coefficients between the matrix and the new phase. In the Zn-polluted samples, the X-ray analysis and the SEM investigation show the stable presence of zinc oxide before and after thermal ageing at 800°C, Fig.3(A). After thermal ageing at 1000°C, part of the ZnO was transformed into a zinc aluminate phase, located mainly at the grain boundaries. After thermal ageing at 1200°C, the ZnO was no longer detected as it had been replaced by the zinc aluminate phase, Fig.3(B).

Fig.1. Typical microstructure of pure cordierite.

Fig.2. Microstructure of Ca-polluted cordierite before (A) and after (B) thermal ageing at 1200°C for 3 h.

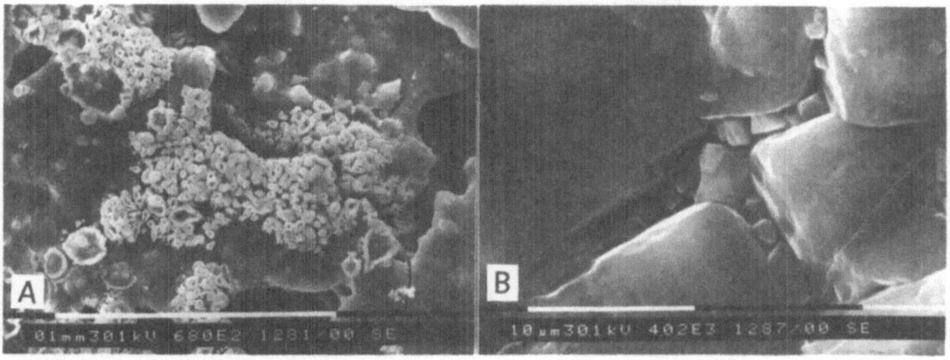

Fig.3. Microstructure of Zn-polluted cordierite before (A) and after (B) thermal ageing at 1200°C for 3 h.

Young's Modulus (E)

Before thermal ageing, the Ca-polluted samples had the highest E values, followed by the Zn-polluted samples and the virgin cordierite samples, (Table 1 and Fig.4).

TABLE 1

Young's moduli of the base and polluted samples before and after thermal ageing of 3 h

Ageing temperature (°C)	Cordierite E (GPa)*	Cordierite+Ca E (GPa)*	Cordierite+Zn E (GPa)*
----	7.05±0.21	11.86±0.27	8.15±0.41
800	6.72±0.29	10.93±0.79	7.49±0.48
1000	6.84±0.16	9.03±0.25	7.88±0.23
1200	7.01±0.17	6.04±0.54	7.06±0.23

* Mean±Standard deviation.

After thermal ageing, the value of Young's modulus of the base material remained almost costant. The Ca-polluted samples show pronounced degradation as the treatment severity increased: after treatment at 1200°C, the elastic modulus is even lower than that of the base material. In the Zn-polluted samples too the elastic degradation is present though not so evident: after the most severe treatment the elastic modulus is at the same level as the base cordierite, Fig.4.

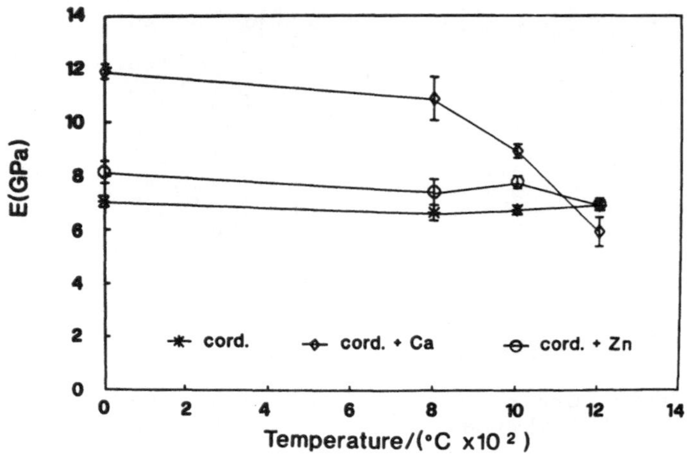

Fig.4. Young's modulus of unpolluted and polluted samples as a function of the temperature of thermal treatment.

Before thermal ageing, the increase in Young's modulus in the Zn-polluted samples is clearly due to the presence of the ZnO which has higher moduli than the matrix. In the case of the Ca-polluted samples, the strong increment in the elastic modulus is due to the phase shown in Fig.2(A) which was not detected by X-ray analysis. After thermal ageing, both the Young's moduli of the polluted samples degraded with the increase in the severity of the thermal treatement mainly due to diffused corrosion induced by the new phases and/or by transformation of the phases themselves into new phases with different elastic moduli. Moreover, especially in the case of Ca-polluted samples, the new phases have generally much higher thermal expansion coefficients than that of the matrix. This may have introduced microcracks into the structure, clearly evident in the material aged at 1200°C (Fig.2(B)), during the thermal treatment with a consequent negative effect on the Young's modulus of the material itself. An influence by the pollutant phases is also evident in the scatter of the experimental results which, in the polluted samples, are sometimes larger than those of the base material, Table 1.

Flexural Strength (MOR)

As in the case of Young's modulus, the flexural strength of the base cordierite is virtually unaffected by the thermal ageing, (Table 2 and Fig.5). Before the thermal ageing, the Ca-polluted samples are stronger than the virgin ones as a direct consequence of the increase in the elastic modulus, (Table 1).

TABLE 2
Flexural strength of the virgin and polluted samples before and after thermal ageing of 3 h

Ageing temperature (°C)	Cordierite MOR (MPa)*	Cordierite+Ca MOR (MPa)*	Cordierite+Zn MOR (MPa)*
----	4.02±0.24	5.07±0.15	3.55±0.42
800	3.89±0.18	4.69±0.21	3.50±0.19
1000	3.59±0.09	4.01±0.13	3.91±0.14
1200	3.93±0.21	3.45±0.20	3.94±0.24

*Mean±Standard deviation.

By contrast, the Zn-polluted samples, in spite of their higher stiffness with respect to the base cordierite, are weaker. It might be that in these samples the Zn-compounds have deteriorated the structure through localized corrosion. After thermal ageing, the flexural strength of the Ca-polluted samples show the same degradation as in the case of Young's modulus, (Fig.4). With respect to the starting points, the flexural strength of the Zn-polluted samples shows a slight improvement after the thermal treatments at 1000°C and 1200°C. In these samples, as the Young's modulus decreased and unlikely the critical flaw size decreased, according to Griffith theory [8] the higher strength could be the result of an increase in fracture surface energy due to the aluminate phase located at the grain boundaries (Fig.3(B)).

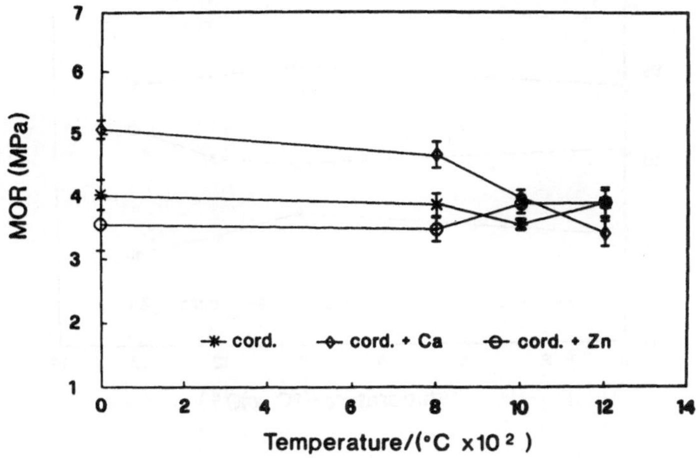

Fig.5. Flexural strength of unpolluted and polluted samples as a function of the temperature of thermal treatment.

Thermal Expansion Coefficient (α)

The property that is mostly influenced by the pollutants is the thermal expansion coefficient, Table 3 and Fig.6. In the

TABLE 3
Thermal expansion coefficients of the virgin and polluted samples before and after thermal ageing of 3 h

Ageing temperature (°C)	Cordierite α ($10^{-7}\cdot C^{-1}$)	Cordierite+Ca α ($10^{-7}\cdot C^{-1}$)	Cordierite+Zn α ($10^{-7}\cdot C^{-1}$)
----	5.79	14.10	10.55
800	7.35	15.50	10.09
1000	6.23	-----	10.45
1200	5.30	14.70	12.20

samples without thermal ageing, the α values of both the contaminated materials are almost or more than double the base material. After thermal ageing, this difference is even accentuated as can be seen in Fig.6. This behaviour can be attribuited to the presence of phases, derived from the pollutants, that evolve with the thermal treatment, with thermal expansion coefficients much higher than that of the cordierite base material.

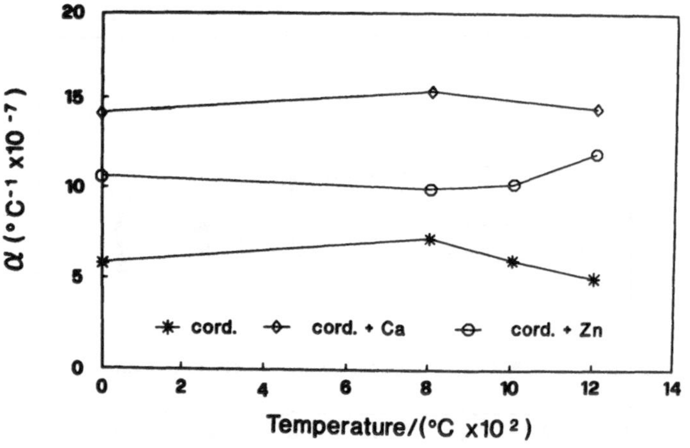

Fig.6. Thermal expansion coefficient of the unpolluted and polluted samples as a function of the temperature of thermal treatment.

CONCLUSIONS

Virgin porous cordierite samples were separately contaminated with calcium (3.5 wt%) and zinc (3.0 wt%) as oxides. Some of these samples have been thermally aged. The X-ray analysis indicates an evolution of the pollutant-derived phases with thermal ageing. This has a pronounced effect on mechanical and, mainly, on thermal behaviour. The base material shows unaltered properties after the thermal ageing, confirming its reliability as a substrate for catalysers. In the polluted samples, the mechanical and thermal properties are already modified after the introduction of the pollutants. After the thermal ageing, both Young's modulus and flexural strength generally deteriorate, while there is a strong increase in the thermal expansion coefficients. These two factors could lead to shorter component life, considering the extreme conditions in which it operates.

REFERENCES

[1] Howitt, J.S. and Montierth, M.R., Cellular ceramic diesel particulate filter, SAE Paper 810114, 1981, 1-9.

[2] Mizuno, H., Kitagawa, J. and Hijikata, T., Effect of cell structure on regeneration failure of ceramic honeycomb diesel particulate filter, SAE Paper 870010, 1987, 9-15.

[3] Borello, C., Demaestri, P.P., Giachello, A. and Maiani, C., Modifications of porous cordierite used in DPF induced by thermo-mechanical stresses, Proceedings of ATA-MAT '89, June 1989, Torino (Italy).

[4] Gulati, S.T. and Helfinstine, J.D., High temperature fatigue in ceramic wall-flow diesel filters, SAE Paper 850010, 1985, 13-19.

[5] Gulati, S.T., Long-term durability of ceramic honeycombs for automotive emissions control, SAE Paper 850130, 1985, 1-16.

[6] Giachello, A., Demaestri, P.P., De Portu, G. and Guicciardi, S., Mechanical behaviour of ceramic filters for automotive emission control, in the Proceedings of the Seventh World Congress on Hi-Tec Ceramics (CIMTEC), June 24-30 1991, Montecatini (Italy), in press.

[7] ASTM C848-78: Standard method for Young's modulus, shear modulus and Poisson's ratio for ceramic withewares by resonance. (Reapproved 1983).

[8] Griffith, A.A., The phenomena of rupture and flow in solids, Philos. Trans. R. Soc. Lond., 1920, **A221**, 163.

THE CORROSION BEHAVIOUR OF HIP Si_3N_4 AND SiC/SiC COMPOSITE IN SIMULATED COMBUSTION ENVIRONMENTS

J. CHEN J. SJÖBERG C. O'MEARA* L. PEJRYD**
Department of Inorganic Chemistry, *Department of Physics,
Chalmers University of Technology, Göteborg, Sweden
**Volvo Flygmotor AB, Trollhättan, Sweden

ABSTRACT

The corrosion behaviour of a HIPed Si_3N_4 and a SiC/SiC composite has been studied in simulated combustion environments. It has been found that for both materials the oxidation behaviour and the gaseous corrosion behaviour are quite similar, but the corrosion kinetics is very sensitive to the sodium content in the environment. It is proposed that an increase of salt content in the corrosive environment may change the diffusion controlled corrosion process into a reaction controlled process as a result of the formation of low viscosity silicate glass. The effects of salt (Na_2SO_2 or NaCl) on the devitrification of the amorphous silica are discussed.

INTRODUCTION

The corrosion damage of silicon based ceramics is of large concern in their application in high temperature combustion environments, which are known to be very corrosive to superalloys (1). Several investigations (e.g. 2-5) have reported that the strength of silicon based ceramics decreases as a result of corrosion at elevated temperatures. However, the corrosion mechanism(s) for these ceramics are not yet fully understood.

Combustion environments in gas turbines usually contain O_2, NO_2, SO_2, H_2O, CO_2, N_2 and other gaseous species (6). Sodium chloride, either as a fuel impurity or as a component of the inlet air may also be present in the combustion processes. A number of investigations (e.g. 7-8) have been carried out to examine the effects of various atmospheres on the oxidation/corrosion behaviour of silicon based ceramics. The purpose of this work is to evaluate the corrosion resistance of recently developed HIPed Si_3N_4 and SiC/SiC composite in simulated combustion environments and to characterize the corrosion process.

MATERIALS AND METHODS

Materials

A HIPed Si_3N_4 with 4 wt% Y_2O_3 as sintering additive (produced by ABB Cerama) and a SiC fibre reinforced chemical vapour infiltrated (CVI) SiC (provided by S. E. P.) were examined in this study. The phases present in the as-received Si_3N_4 material are β-Si_3N_4, $Y_2Si_2O_7$ and a small amount of grain boundary glass (9). The Si_3N_4 material was cut with a diamond saw, ground and polished to 0.25 µm finish with diamond spray. The sample size was about 10×10×3 mm. The surface layer of the SiC/SiC composite was identified by XRD and by TEM to be micro-crystalline α-SiC. The porosity of the SiC/SiC composite was 8.5~9 vol%. The SiC/SiC composite was manufactured to the size 10×10×3 mm by the producer and used in present study without further grinding or polishing.

Corrosion Tests

A horizontal alumina tube furnace with gas-tight endlocks was used for corrosion tests. Carbon dioxide, which had been led through deionized water at 36°C, and gas mixtures of O_2+SO_3 and N_2+NO_2 were mixed and introduced into the furnace. The input gas phase composition is given in Table 1. The total flow rate of the gas mixture was 10 l/h.

TABLE 1
The input gas compositions in the corrosion tests.

Gas	O_2	N_2	CO_2	SO_3	NO_2	H_2O
Input amt.(vol%)	0.10	0.76	0.07	2E-5	0.014	0.06

Corrosion tests were conducted at 1000, 1250 and 1450°C for up to 100 h with intermediate weight observation. For gaseous corrosion tests, the gases listed in Table 1 were used, while for the hot corrosion tests, the additional salt was introduced after each exposure cycle by emerging the samples in 0.1% NaCl solution and then drying them. For the SiC/SiC composite, oxidation was also carried out. The corroded samples were weighed on an analytical balance after each exposure cycle. The error in weighing introduced by the balance was estimated to be less than ±0.050 mg.

In order to exaggerate the hot corrosion of the Si_3N_4 material, one gram of NaCl was evaporated in the hot furnace before the exposure was carried out.

To study the possible chemical reactions of NaCl or Na_2SO_4 with SiO_2 which forms on the corroded Si_3N_4 or SiC/SiC materials, a small amount (2-6 mg/cm^2) of NaCl or Na_2SO_4 was introduced into a silica glass tube (impurity < 10 ppm), the two ends of which were subsequently sealed in dry air atmosphere. The samples were then heat treated in air for 24 h at 1000, 1250 and 1450°C, respectively. A glass tube without Na_2SO_4 was used as reference.

The phase compositions of the corroded materials were examined using XRD; the morphology and corrosion products of the corroded materials using SEM+EDX.

Thermodynamic Calculations

To examine the equilibrium compositions of the gas or gas-solid mixtures used in present studies, thermodynamic calculations were carried out using the SOLGASMIX computer program (10). The input amounts of the gaseous components were those listed in Table 1. The input amounts of NaCl and SiO_2(l) were 10^{-4} and 1 mol, respectively.

RESULTS

Equilibrium Compositions of the Systems Examined

Fig. 1 shows the temperature dependence of the equilibrium compositions of the input mixtures of N_2-H_2O-CO_2-O_2-NO_2-SO_3 and of N_2-H_2O-CO_2-O_2-NO_2-SO_3-NaCl(s) in a temperature range 800 to 1500°C. Gases in amounts of less than 1ppm are not shown in the figures. N_2, H_2O, CO_2 and O_2 are not shown as their amounts remain approximately the same as the input amounts. The results for N_2-H_2O-CO_2-O_2-NO_2-SO_3-NaCl(s)-SiO_2(l) systems showed the formation of $Na_2Si_2O_5$ over a temperature range 800 to 1400°C.

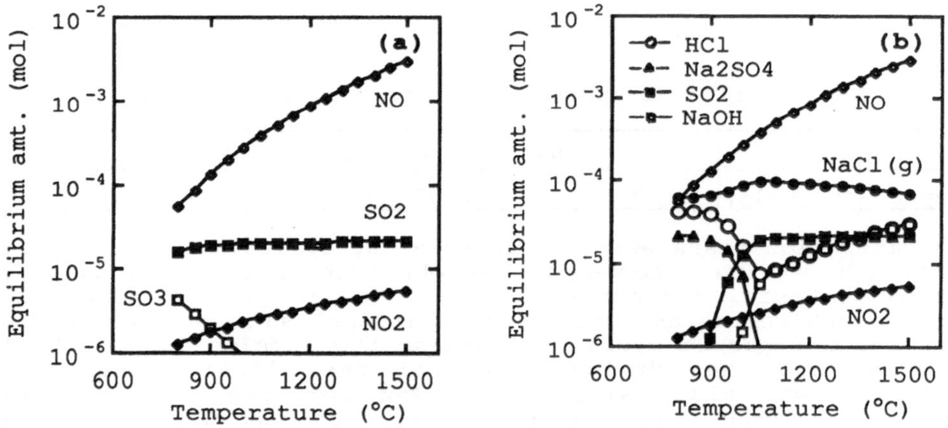

Figure 1 Equilibrium compositions *vs* temperature curves for the input mixtures (*a*) N_2-H_2O-CO_2-O_2-NO_2-SO_3 and (*b*) N_2-H_2O-CO_2-O_2--NO_2-SO_3-NaCl(s).

Corrosion Kinetics

The measured corrosion kinetic curves (weight change *vs* exposure time) for the Si_3N_4 and the SiC/SiC materials are shown in Figs. 2 and 3*a*, respectively. The oxidation weight change of the SiC/SiC composite is shown in Fig.3*b*.

In the exaggerated hot corrosion experiment, considerable amounts of glassy matter flowed off the surfaces of the Si_3N_4 samples, which resulted in smooth and glassy surfaces of the material. Because of this, the measurements of sample weight change were not accurate and are not given here.

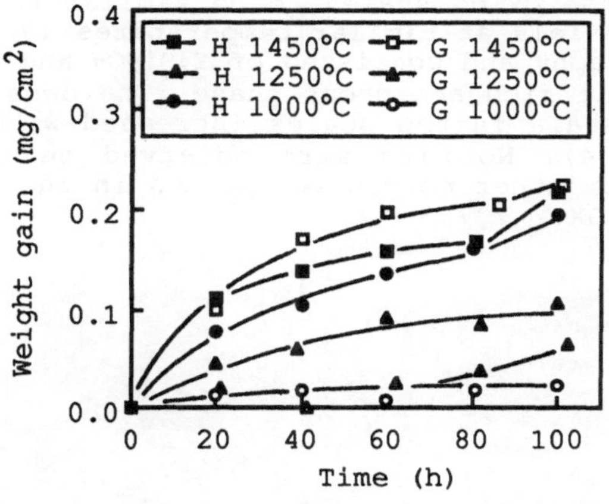

Figure 2. Weight gain *vs* exposure time curves for the Si_3N_4 material under the gaseous and hot corrosion conditions. The Na content in the oxide scale was too low to be detected by EDX.

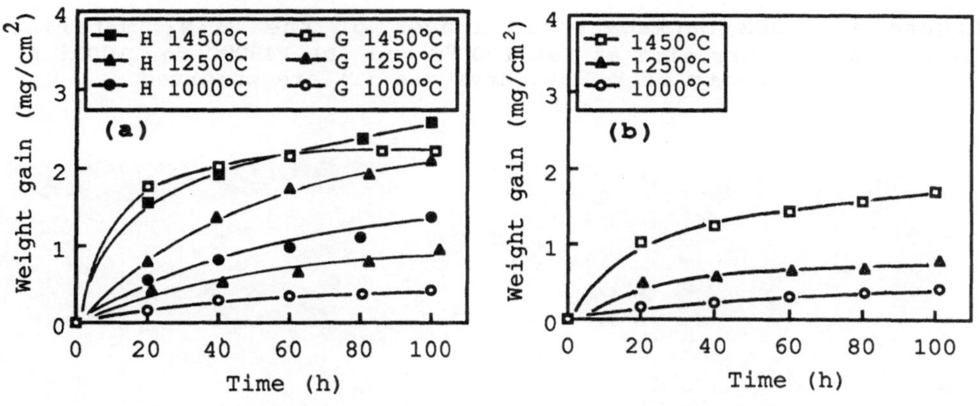

Figure 3 The weight gain *vs* exposure time curves for the SiC/SiC composite under (*a*) the gaseous (G), hot corrosion (H) conditions and (*b*) oxidation conditions.

Phase Compositions of the Corroded Materials

In the gaseous corrosion and hot corrosion of the Si_3N_4 material, the crystalline phases formed in the corrosion scales are yttrium disilicate and cristobalite, while for the corroded SiC/SiC composite, cristobalite is the only crystalline phase observed. More cristobalite was found under hot corrosion conditions than that under the other corrosion conditions.

Microstructures of the Corroded Materials
SEM examination of the corroded Si_3N_4 material revealed the microstructure of the oxide scales to be similar to that of the plain oxidized materials at similar temperatures (9). The scales had a smooth topography and consisted of $Y_2Si_2O_7$ and cristobalite crystals in a silica rich amorphous phase. The density and size of the $Y_2Si_2O_7$ crystals in the scales increased with increasing temperature (Fig. 4). No pits were observed on the corroded surfaces. No salt or other corrosive species in the scales could be detected using EDX.

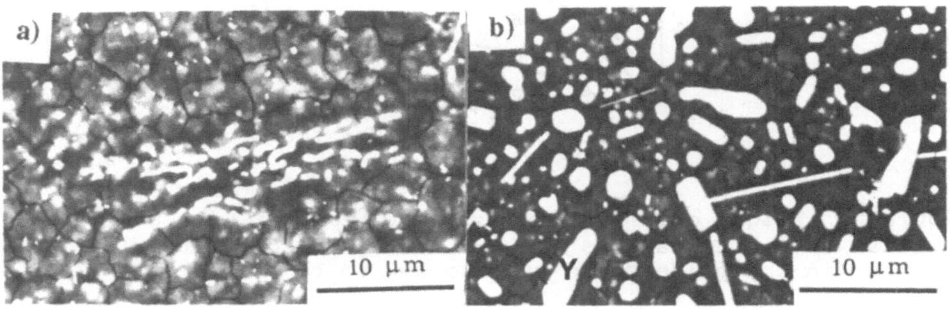

Figure 4 SEM backscattered images of the Si_3N_4 material following hot corrosion at (*a*) 1000°C and (*b*) 1450°C for 100h. A high density of $Y_2Si_2O_7$ crystals (Y) are observed.

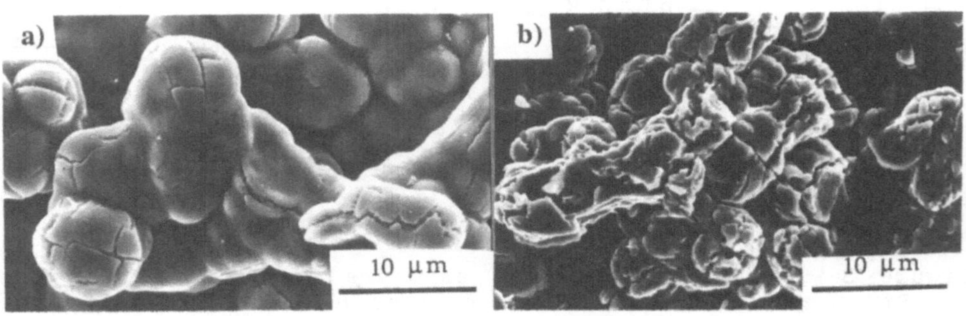

Figure 5 SEM secondary electron image of the surface of the SiC/SiC composite. (a) Following hot corrosion at 1000°C cracking is observed in the "corn cob" structuring of the surface; (b) After hot corrosion exposure for 100h at 1450°C the surface structure is extensively degraded and cracked.

The surface of as-received SiC/SiC composite is rough with a "corn-cob" appearance. The "cobs" are smooth and approximately 10 µm in size. Following oxidation and corrosion of the composite at 1000°C, no difference could be observed in SEM between the morphologies of the as-received and treated surfaces (Fig. 5*a*). At higher temperatures the "cobs" were observed to become increasingly degraded, exhibiting extensive cracking as is shown in Fig. 5*b*. However, this behaviour was similar for both the oxidized and corroded materials.

Reactions of Amorphous SiO_2 with NaCl or Na_2SO_4
After heat treatment at 1000°C for 24h, the silica glass remained transparent, thus little devitrification had occurred. The heat treatment at 1250 and 1450°C for 24h, however, caused severe devitrification of silica glass both with and without Na_2SO_4 or NaCl. EPMA studies have revealed that sodium existed throughout the glass tube. Fig. 6 shows the effects of Na_2SO_4 and NaCl on the devitrification of amorphous silica.

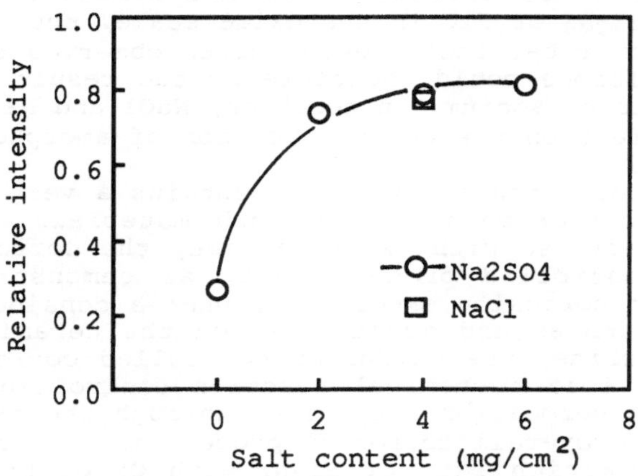

Figure 6 The salt content dependence of devitrification of silica glass at 1250°C by XRD. The relative intensity is defined as intensity ratio of (101) reflection of cristobalite to (104) reflection of Cr_2O_3.

DISCUSSION

The observation that for both SiC/SiC and Si_3N_4 materials the gaseous corrosion behaviour was, in general, similar to the oxidation behaviour suggests that O_2 is the most corrosive component among various gases present in the gaseous corrosion environment. Under the hot corrosion conditions, however, both materials exhibited more severe corrosion, which means that sodium plays an important role in the hot corrosion process.

Under the hot corrosion conditions, sodium may be present in the environment in the form of NaCl(s), NaOH(g) and Na_2SO_4(s) or

$Na_2SO_{4(l)}$. According to Tressler et al (11), NaCl is not sufficiently basic to dissolve the SiO_2 which forms on the surfaces of SiC and Si_3N_4 materials, therefore SiO_2 is not affected, while NaOH and Na_2SO_4 can provide O^{2-} ions, which are accepted by $SiO_{2(s)}$ to form $SiO_3{}^{2-}$ or $SiO_4{}^{4-}$, causing the corrosive etching to occur. This mechanism may be applicable to explain the accelerated corrosion behaviour of present materials under the hot corrosion conditions at 1000°C. But at higher temperatures, since Na_2SO_4 does not form, $NaOH_{(g)}$ may thus become the most corrosive component.

The presence of sodium in the corrosion scale may have at least two contrary effects on the corrosion behaviour of silicon based ceramics. On one hand, the introduction of sodium in the amorphous SiO_2 scale breaks the strong ≡Si-O-Si≡ bonds to form Na-O-Si≡ bonds, which facilitate the rearrangement of the glass network to the crystalline state as shown in Fig. 6. The devitrification makes O_2 permeation in the SiO_2 difficult (9) and therefore decreases the corrosion rate. On the other hand, the introduced glass forming modifier, Na_2O, lowers the viscosity of the glass, which may increase both the diffusivity of O_2 and solubility of Si_3N_4 or SiC in the oxide scale, thus leading to a higher corrosion rate. The kinetic curves observed under the hot corrosion conditions could therefore be the result of above two contrary effects of sodium. In addition, NaCl and Na_2SO_4 may have the similar effect on the devitrification of amorphous silica as shown in Fig. 6.

When the corrosion environment contains a very small amount of salt, the corrosion rates of both materials, as shown in Fig.2 and 3, decrease with exposure time, the corrosion is thus a diffusion controlled process. But, as demonstrated in the exaggerated hot corrosion experiment, when a considerable amount of salt is condensed and accumulated on the ceramic hot parts, e.g. in gas turbine, the diffusion controlled corrosion process may be changed into a chemical reaction controlled process. In the latter, the permeation rate of O_2 through the oxide scale is so high that the overall corrosion process is determined only by the chemical reaction rate of Si_3N_4 with O_2 or by dissolution rate of Si_3N_4 in the low viscosity silicate scale. More work is required to closely simulate the salt deposition on the material surfaces at high temperature and to examine the conditions under which the diffusion controlled corrosion turns into reaction controlled corrosion and the mechanism associated in the process.

CONCLUSIONS

1. For both Si_3N_4 and SiC/SiC materials examined, oxidation and gaseous corrosion are diffusion controlled process and they behave similarly;
2. The corrosion process is highly sensitive to the sodium content present in the environment. Under hot corrosion conditions, the corrosion process is greatly enhanced.

ACKNOWLEDGEMENT

Financial support from Swedish Board for Technical Development is gratefully acknowledged.

REFERENCES

1. Pettit, F.S. and Meier, G.H., Oxidation and corrosion of superalloys. in Superalloys, ed. M. Gell, C.S. Kortovich, R.H. Bricknell, W.B. Kent, and J.F. Radavich. The Metallurgical Society, AIME, Warrendale, PA, 1984.pp. 651-87.

2. Bourne, W.C. and Tressler, R.E., Molten salt degradation of Si_3N_4 ceramics. Am. Ceram. Soc. Bull., 1980, 59 (4), 443-52.

3. Smialek, J.L. and Jacobson, N.S., Mechanism of strength degradation for hot corrosion of α-SiC. J. Am. Ceram. Soc., 1986, 69 (10) 741-52.

4. Fox, D.S. and Smialek, J.L., Burner rig hot corrosion of silicon carbide and silicon nitride. J. Am. Ceram. Soc., 1990, 72 (2) 303-11.

5. Lundberg, R., Stenholm, T., Pejryd, L.and Kahlman, L., The influence of static loads on the oxidation behaviour of fibre reinforced SiC/SiC composites. This proceeding.

6. Pejryd, L., Metal ceramic joining for high temperature applications. This proceeding.

7. Singhal, S.C., Effect of water vapor on the oxidation of hot-pressed silicon nitride and silicon carbide. J. Am. Ceram. Soc., 1976, 59 (1-2) 81-82.

8. Blachere, J.R. and Pettit, F.S., High Temperature Corrosion of Ceramics. Noyes Data Corporation, Park Ridge, New Jersey, 1989, pp. 14-15.

9. Chen, J., Sjöberg, J., Lindqvist, O., O'Meara, C. and Pejryd, L., The rate controlling processes in the oxidation of HIPed Si_3N_4 with and without sintering additives. J. Europ. Ceram. Soc., 1991, 7 (in press).

10. Eriksson, G., Thermodynamic studies of high temperature equilibria. Chemica Scripta, 1975, 8, 100-103.

11. Tressler, R.E., Meiser, M.D. and Yonushonis, T., Molten salt corrosion of SiC and Si_3N_4 ceramics. J. Am. Ceram. Soc., 1976, 59 (5-6) 278-79.

IMPROVED THERMAL STABILITY OF Y_2O_3 OR Fe_2O_3 DOPED ALUMINIUM TITANATE

P.P. Demaestri, A. Giachello

Centro Ricerche Fiat

Strada Torino, 50, 10043 Orbassano (Italy)

ABSTRACT

One of the main problem encountered in application of Al_2TiO_5, for thermal insulation purpose, is its thermal instability in the temperature interval 900°-1200°C. Dissociation occurs with α-Al_2O_3 and rutile formation accompained with a permanent volume increase which correspond a decay of mechanical and physical properties.
A purpose of this work was to investigate the effect of 5wt% Y_2O_3 or 5wt% Fe_2O_3 addition on the structural stability of β-Al_2TiO_5 aged at 1000°C for several length of time. Results indicate that Fe_2O_3 is more effective than Y_2O_3 in this aspect. After 1500 hours iron oxide doped aluminium titanate shows zero decomposition, meanwhile yttria doped Al_2TiO_5 after 350 hours has 2% decomposition and after 500 hours already 35% of the ceramic material is decomposed. Other properties, like coefficient of thermal expansion, flexural strength, Young's modulus, etc. were also checked and results, indicate that all characteristics of the materials deteriorate obviously, when decomposition occurs, but also when the material remains unaltered respect to crystalline phases. A certain decay in characteristics happen during thermal aging in both cases.

INTRODUCTION

Due to the low value of thermal expansion and conductivity coefficients and high thermal shock resistance the beta aluminium titanate ceramic material has received recently a great attention to realize automotive components like portliners, piston heads, manifolds, etc. The actual fabbrication technique for these components, with complex shapes, is slip casting. The use as heat insulating barrier, for exam-

ple between exhaust gas and engine head (port-liner) to reduce the capacity of cooling system, thermal fatigue of metal and diesel and gasoline emissions too, is one of the most interesting application area of β-Al_2TiO_5. Since its principal components (α-Al_2O_3 and rutile) are biocompatible materials, Al_2TiO_5 ceramics, may also be considered as a potential bioceramic.
The main problem of aluminium titanate is its eutectoid decomposition between 900°C and 1200°C [1]; many researchers have investigate this aspect using many stabilizer, such as Mg [2], Si and Zr compounds [3].
The aim of this work is to study the influence of yttrium and iron ions on thermal stability and thermo-mechanical properties of sintered aluminium titanate after annealing at 1000°C.

MATERIALS AND METHODS

In this study were considered commercial starting ceramics powders with not elevated purity, because it's necessary, for the future automotive applications, to develop ceramics materials at an acpetable cost. Raw materials used were:

- α-Al_2O_3 of 99.8% purity with main impurities, in percentage, as SiO_2 0.02, Fe_2O_3 0.02, CaO 0.04, Na_2O < 0.1, and a specific surface area of 1.2 m^2/g;

- TiO_2, as rutile, with 99.5% purity and main impurities, in percentage, as Fe_2O_3 0.16, Al_2O_3 0.14, SiO_2 0.08, MgO 0.13, and specific surface area of 0.1 m^2/g;

- very pure MgO with a specific surface area of 47 m^2/g;

- high purity Fe_2O_3 and Y_2O_3 powders with specific surface of 4 m^2/g and area 8.9 m^2/g respectively.

Basic mixture, examinated in this work, consist of α-Al_2O_3 and rutile in equimolar ratio doped with 2 wt% of MgO. As stabiliser 5 wt% of Fe_2O_3 or Y_2O_3 was added to the base material. The starting powders were mixed in an alumina jar-ball device in distilled water and dried powders were uniaxially compacted, without binders, at 100 MPa into 3x4x30 mm bars. All samples were sintered in air at 1450°C for 8 hours. Measurements were done on sintered specimens and precisely: apparent density through mercury displacement; real density in water with Archimede's technique; flexural strength with 3-point bending, 18 mm span, on an Instrom machine, at an applied load of 0.5 mm/mim.; Young's modulus measured by resonance frequency method according to ASTM-C848 [4]; mean grain size calculated with intercept method; thermal expansion with a dilatometer heated and cooled at constant rate of 5°C/min; quantitative crystalline phases determination using an internal standard with a Ni filtered KαCu radiation XRay diffractometer. Samples were aged in a muffler, in static air, at a constant temperature of 1000°C. Thermal shock resistence was measured with the water quenching method, keeping aluminium titanate bars (5 samples for each point) 15 minutes at constant temperature in a tubolar furnace before suddenly droping them into water.

RESULTS AND DISCUSSION

Observing the Fig. 1 appears very clearly the great effect that Fe_2O_3 addition, at a level of 5 wt%, posses on β-Al_2TiO_5 thermal stability.

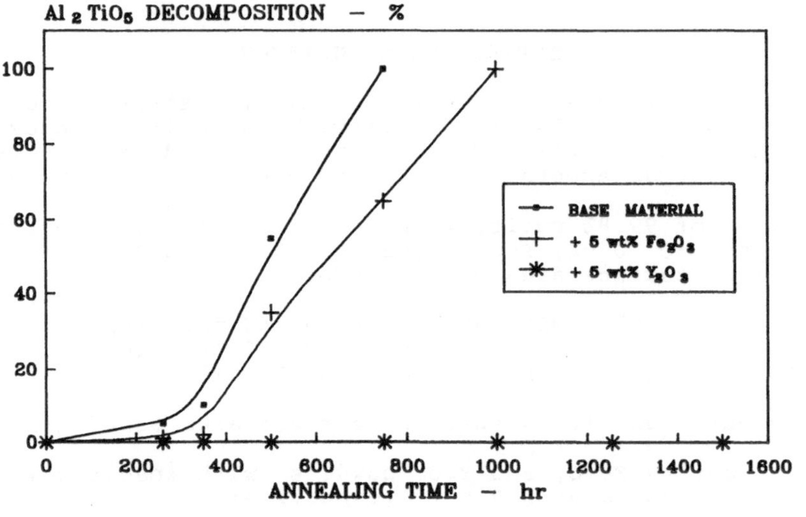

Figure 1 - Thermal decomposition curves of the base material, with 5wt% Fe_2O_3 or 5wt% Y_2O_3 addition

Probably iron oxide substitudes alumina in the crystal structure of aluminium titanate and forms an iron titanate (Fe_2TiO_5) [5] with good solid solubility in the aluminium titanate matrix. Really at the Scanning Electronic Microscope EDAX analysis was not possible to check any difference on iron content in the ceramic texture. The thermal stability of aluminium titanate, conducted at 1000°C in static air and checked at every 250 hours interval, appears after 1500 hours aging time quite good. Infact, with an internal calibration curve for XRD, quantitative analysis on powdered specimen showed only trace of α-alumina as thermal decomposition product of starting material; Figs. 2A-B show the microstructures of Al_2TiO_5 with and without iron oxide addition, respectively.

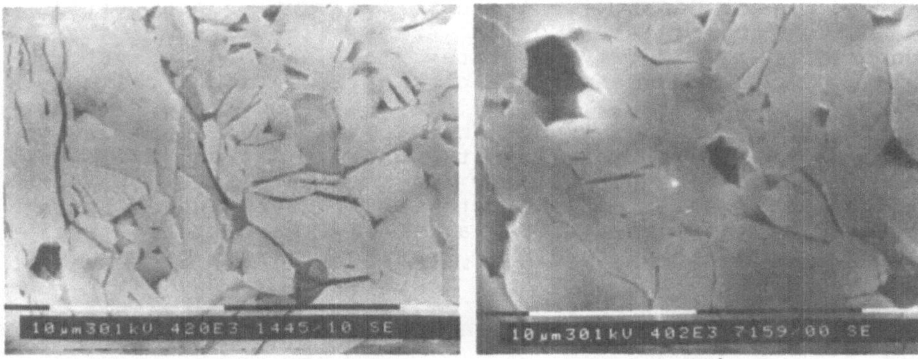

(A) (B)

Figure 2 - Microstructures of Al_2TiO_5 with (A) and without (B) iron oxide addition

On the contrary, the starting Al_2TiO_5, with no iron oxide addition, very shortly, at the same temperature and treatment, reaches total decomposition in 700 hours. Similarly, with only slightly better behaviour, demonstrated yttrium oxide 5wt% addition, which shows insignificant influence on thermal stability regards to aluminium titanate.
The reason appears evident observing the texture in Fig. 3 and analysing the powdered sample by XRD pattern. Around β-Al_2TiO_5 grains a phase was identified as Y_2TiO_7. It crystalized in cubic form and evidently has a very low coefficient of diffusivity in the Al_2TiO_5 crystals and remains segregated around aluminium titanate grains in quite small particles.

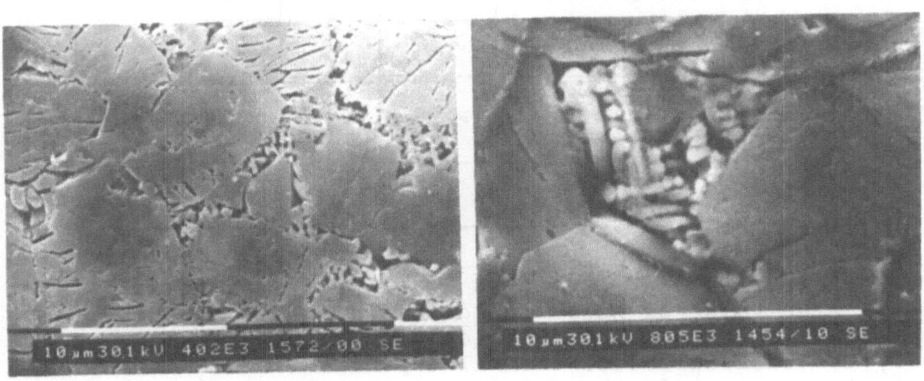

Figure 3 - Microstructure of Al_2TiO_5 with yttrium oxide addition

EDAX microprobe analysis indicates yttrium concentrations around Al_2TiO_5 grains and SEM examinations evidenciate a secondary phase pourly sintered. As consequence yttrium derived compound demonstrates no influence about the aluminium titanate stability at elevate temperature. Further investigation put in evidence that Y_2TiO_7 takes place in the system studied at very low temperature, as 1200°C, when Al_2TiO_5 is present only in very small amount. About the over all characteristics of the materials under study, appear rilevant, as rapresentated in Tab. 1, the mechanical properties of iron addition to Al_2TiO_5 base material.

Table 1

Characteristics of base and doped aluminium titanate

CHARACTERISTICS	BASE MATERIAL	+ 5 wt% Fe_2O_3	+ 5 wt% Y_2O_3
APPARENT DENSITY, g/cm³	3.21	3.29	3.30
REAL DENSITY, g/cm³	3.56	3.51	3.50
OPEN POROSITY, %	9.8	6.3	5.7
MEAN GRAIN SIZE, µm	9.7	4.4	9.8
3-POINT FLEXURAL STRENGTH MPa	48.6	56.8	40.3
WEIBULL'S MODULUS	11.0	22.8	12.2
YOUNG'S MODULUS, GPa	25.6	21.2	37.1
POISSON'S RATIO	0.21	0.23	0.31
C.T.E.(25-900 °C) 10-6 °C-1	2.27	2.57	0.78
THERMAL SHOCK RESISTANCE, °C (QUENCHING METHOD)	⩾700	⩾800	⩾900
THERMAL SHOCK RESISTANCE PARAMETER R, °C	660	800	960

The flexural strength of 56.8 MPa appears excellent and a very high value of Weibull modulus indicates a reproducible material in a quite high extent. On the other hand a not very low coefficient of thermal expansion indicates that the material needs further improvements expecially regard to microcracked structure, adjusting properly grain size, cooling rate after sintering cycle, etc. On the contrary yttrium oxide addition demonstrated an extended ability to absorb anisotropic thermal expansion of the aluminium titanate crystals.

So the resulting coefficient of thermal expansion reaches very low value as $0.78 \cdot 10^{-6}\,°C^{-1}$ between room temperature and 900°C. This is the main characteristic of this material, which demonstrates zero expansion up to 700°C. For this reason yttrium containing aluminium titanate has the best thermal shock resistance, as evidenciated both as R parameter calculated from values determinated (E, σ, α) and as checked with the water quenching method suggested by Hasselmann [6], the agreement with these values reported in Table 1 and Fig. 4 are surprinsingly. Secondly, but not less important, yttrium oxide addition to Al_2TiO_5, does not present the irreversible volume expansion noticed in the undoped ceramic and reported elsewhere [7] with very detrimental effect. In the presence of Y_2TiO_7 in the structure, alumina and rutile formations appear well adsorbed in the matrix.

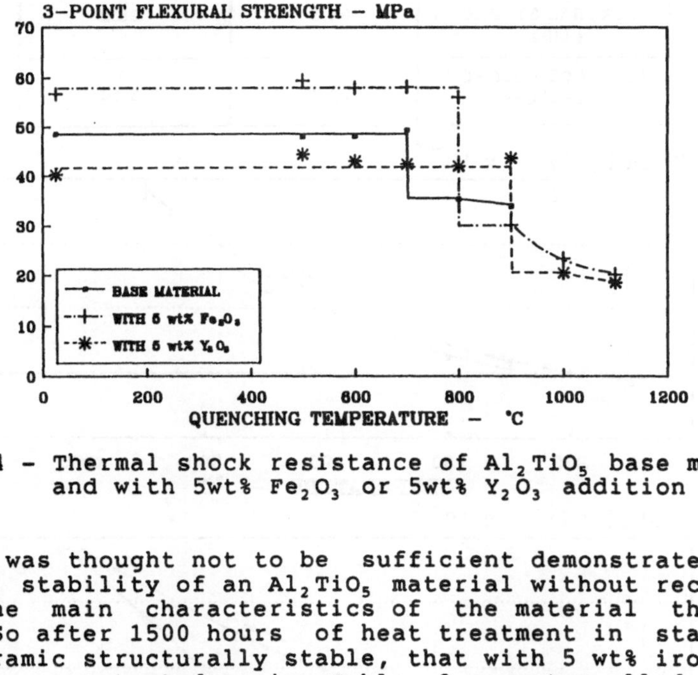

Figure 4 - Thermal shock resistance of Al_2TiO_5 base material and with 5wt% Fe_2O_3 or 5wt% Y_2O_3 addition

But it was thought not to be sufficient demonstrate a good thermal stability of an Al_2TiO_5 material without reconsidering the main characteristics of the material thermally aged. So after 1500 hours of heat treatment in static air the ceramic structurally stable, that with 5 wt% iron oxide addition, was tested again. Table 2 reportes all data and clearly appears that all characteristics deteriorate after thermal aging, also if almost zero decomposition of the Al_2TiO_5 iron containing was detected. The flexural strength decay was of about 50%, meanwhile the coefficient of thermal expansion increased of about 20%. What happened in the structure was not so evident, but certain observations indicate that the microcracks present in the texture heal for a quite high extent as demonstrated by SEM analysis and from the hysteresis showed in the dilatometric curve of Fig. 5A, which is very much reduced if compared with the starting material (Fig. 5B).

Table 2

Characteristics of as-sintered and aged aluminium titanate with 5 wt % Fe_2O_3

CHARACTERISTICS	AS SINTERED	AGED 1000 °C/1500hr
APPARENT DENSITY (g/cm^3)	3.29	3.27
REAL DENSITY (g/cm^3)	3.51	3.50
MEAN GRAIN SIZE (µm)	4.4	4.2
3-POINT FLEXURAL STRENGTH (MPa)	56.8	28.2
C.T.E. (25 - 900 °C) (10^{-6} °C^{-1})	2.57	3.10

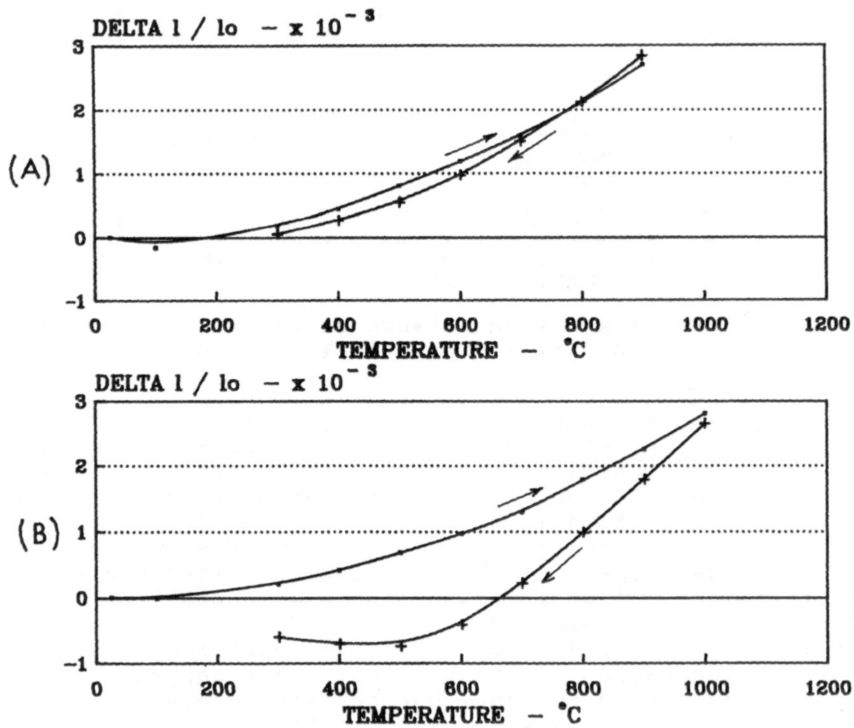

Figure 5 - Dilatometric curves of aluminium titanate with 5 wt % Fe_2O_3 aged at 1000°C for 1500 hours (A) and as-sintered (B)

Secondly a study with mercury intrusion porosimeter indicates that a certain redistribution and coagulation, during the prolungated treatment at 1000°C, of the intergranular pores present in the base material, occurs toward intragranular and bigger pores (at the same level of total porosity) as showed in the Figs. 6 A-B. This fact may support the strong decay in mechanical properties that occurs in the ceramic material after aging.

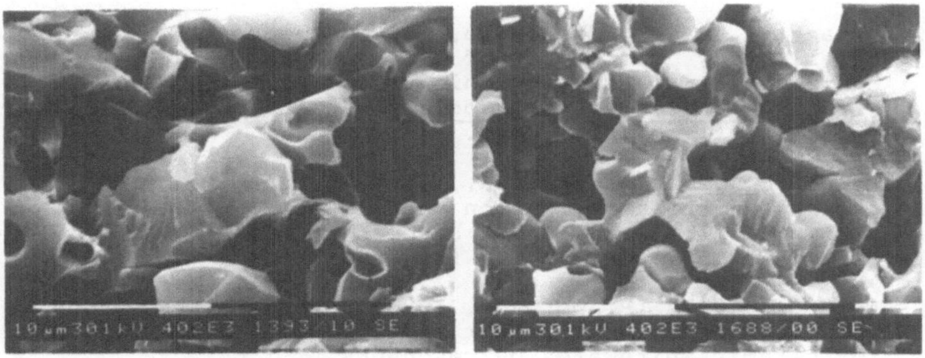

(A) (B)

Figure 6 - Microstructures of as-sintered Al_2TiO_5 containing 5wt% Fe_2O_3 (A) and aged at 1000°C for 1500 hours (B)

CONCLUSIONS

Along with 2 wt% MgO addition, which demonstrated the positive effect on reaction forming and sintering behaviour of the alumina-titania equimolar system, it was intended to verify the β-Al_2TiO_5 thermal stability as contribution of 5 wt% Y_2O_3 or 5 wt% Fe_2O_3 addition. Results indicate the strong effect of iron oxide doped aluminium titanate, which has pratically zero decomposition after aging at 1000°C for 1500 hours in static air. Its starting good mechanical properties suffers of a big decay ascribed to a role that a certain redistribution and coagulation of porosity plays during the thermal treatment at 1000°C. The increment of the coefficient of thermal expansion was attributed to a more healed microcracked structure. Yttrium oxide addition gave an insignificant contribute regard to the Al_2TiO_5 thermal stability, but it was very effective in reducing the coefficient of thermal expansion of the Al_2TiO_5 matrix. Probably the secondary phase Y_2TiO_7, which takes place firstly and at low temperature, acts, as the microcrackes, as an adsorbing volume expansions of the anisotropic dilatometric behaviour of the Al_2TiO_5 crystals. Thermal decomposition of Al_2TiO_5 into α-alumina and rutile was observed to occur in this case with no volume change. Probably suiting together the main advantages of these two materials studied would lead to an improved Al_2TiO_5 ceramic, and this hypothesis will form the basis for further study and developments.

REFERENCES

1. Kato, E., Daimon, K., and Takahashi, J., Decomposition Temperature of β-Al_2TiO_5, J. Am. Ceram. Soc., 1980, 63, 355-56.

2. Demaestri, P.P., Giachello, A., Martinengo, P.C., and Majani, C., Influence of some oxides on the thermal characteristics of stoichiometric Al_2TiO_5, Proc. of First European Ceramic Society Conference, 18-23 June 1989, Maastricht, 2, 57-63.

3. Byrne, W.P., Morrel, R., and Lawson, J., Thermal expansion characterization and thermal stability of aluminium titanate, Proc. Science of Ceramics 14, 7-9 September 1987, Canterbury (England).

4. ASTM C 848-78 (Riappruved 1983), Standard Test Method for Young's modulus, Shear modulus, and Poisson's ratio for ceramic whitewares by resonance.

5. Day, P., and Locker, R., Decomposition behaviour of aluminium titanate solid solutions, Proc. Ceramics for Environmental Protection, 7-9 December 1988, Cologne (West Germany).

6. Hasselman, D.P.H., Strength behaviour of polycrystalline alumina subjected to thermal shock, J. Am. Ceram. Soc., 1970, 53, 490-95.

7. Demaestri, P.P., and Giachello, A., Characterization of β-Al_2TiO_5 aged at 1000°C, Proc. World Ceramics Congress, 24-30 June 1990, Montecatini (Italy).

SODIUM VANADATE-SO_3 EQUILIBRIA IN THE HOT CORROSION OF ZIRCONIA THERMAL BARRIER COATINGS

R. L. Jones
Naval Research Laboratory
Washington, DC 20375-5000, USA

ABSTRACT

This paper describes a thermogravimetric study of the interaction of SO_3 (as from engine gas) with vanadate-sulfate deposits at 700 and 800°C, and its effect on the reaction of the melt with ceramic oxides. Experiments using $NaVO_3$ to simulate deposits having Na/V = 1, and where the controlling reaction is: 2 $NaVO_3$ + SO_3 = Na_2SO_4 + V_2O_5, gave reproducible equilibria over the range of 100 Pa to 0.01 Pa of SO_3. India (In_2O_3, a possible corrosion-resistant stabilizer for ZrO_2) was shown to be inert to the $NaVO_3$-Na_2SO_4-V_2O_5 melt for SO_3 levels up to about 5 Pa at 800°C (0.5 Pa at 700°C), where the melt V_2O_5 activity becomes increased enough that $InVO_4$ begins to form. Although slow at lower SO_3 pressures, the TGA-SO_3 equilibria technique promises valuable information for understanding of the thermochemistry of vanadate-sulfate melts, and the development of corrosion-resistant engine ceramics.

INTRODUCTION

All ceramics suitable for use in oxygen-rich engine environments are oxides, or form protective surface oxides (e.g., SiO_2 on SiC and Si_3N_4). These ceramic oxides react readily, depending on the activities involved, with the Na_2O, V_2O_5, and SO_3 species occurring in molten vanadate-sulfate engine deposits formed from Na,S,V impurities in fuel or air. A current important problem is the hot corrosion of stabilized zirconia thermal barrier coatings by vanadate-sulfate deposits, which precludes the use of TBCs in marine or industrial engines burning low quality fuel.

The degradation of yttria-stabilized zirconia, presently the most commonly used zirconia for TBCs, is driven by the reaction

$$V_2O_{5\,(\text{in melt, a} < 1)} + Y_2O_{3\,(\text{in } ZrO_2,\ \text{a} < 1)} = 2\ YVO_{4\,(\text{ppt, a} = 1)} \tag{1}$$

which removes the stabilizing Y_2O_3 from the ZrO_2 matrix. The tendency for reaction between ceramic and corrodent oxides appears to be mostly determined by the relative Lewis acid-base nature of the oxides involved [1]. Ceria [2], scandia [3], and india [4] have been proposed as stabilizing oxides for ZrO_2 that are more resistant (presumably because they are more acidic) to reaction with V_2O_5 than yttria.

The thermodynamics of reaction (1) are given by ΔG°_T = -RT ln K, which with M = Y, Sc, In, etc., and MVO_4 activity = 1, can be written

$$\log a(M_2O_3) + \log a(V_2O_5) = \Delta G^\circ_T / 2.303\ RT \tag{2}$$

Whether the ceramic oxide will react with the vanadate melt can thus be predicted, provided that the M_2O_3 and V_2O_5 activities, as well as ΔG°_T, are known. Determination of these values, however, is not trivial. Numerous publications exist on corrosion by molten vanadates, but few on the thermochemistry of vanadate-sulfate melts. Luthra and Spacil [5] made computer calculations of the conditions for formation, and the composition, of vanadate-sulfate deposits in gas turbines. Molten salt electrochemical measurements were conducted by Mittal and Elliott [6] on the Na_2O-V_2O_5 system, with activities of $10^{-13.5}$ for Na_2O, and $10^{-3.5}$ for V_2O_5, measured in 1125K $NaVO_3$. More recently, Hwang and Rapp [7] showed that $NaVO_3$ increased the acidic solubility of CeO_2 in molten Na_2SO_4.

EXPERIMENTAL APPROACH

Rationale

For an assumed deposit with a Na/V ratio of ~1, and in equilibrium with the engine gas SO_3, the deposit chemistry is controlled by

$$2\ NaVO_3 + SO_{3\,(g)} = V_2O_5 + Na_2SO_4 \tag{3}$$

This reaction can be studied by equilibrating $NaVO_3$ under different SO_3 partial pressures in a thermobalance, and determining from weight gain the concentrations of V_2O_5 and Na_2SO_4 formed.

Moreover, for ceramic oxides in contact with $NaVO_3$, the potential reaction is

$$M_2O_{3\,(s,\,a\,=\,1)} + V_2O_{5\,(a\,<\,1)} = 2\ MVO_{4\,(s,\,a\,=\,1)} \qquad (4)$$

but this reaction will occur only if the V_2O_5 activity exceeds the critical level (dictated by ΔG^o_T) at which MVO_4 becomes stable. The SO_3 pressure required to produce this V_2O_5 activity can be found by equilibrating mixed oxide-$NaVO_3$ with increasing levels of SO_3 and detecting, by change in weight gain behavior, when MVO_4 begins to form. Correlation between the SO_3 gas pressure and V_2O_5 melt activity can be made by means of a calibration curve of SO_3 pressure vs. V_2O_5 activity, which can be established from results with oxides of known ΔG^o_T for vanadate formation (e.g., MgO, MnO, FeO). By this means, the V_2O_5 activity required for vanadate formation (i.e., corrosion), along with the ΔG^o_T of reaction, can be obtained for In_2O_3, Sc_2O_3 and other potentially vanadate-resistant ceramic oxides.

Instrumentation and Materials

Our experiments employed a Cahn 1000 thermobalance fitted with a two-stage gas mass flow controller (Tylan FC-260) system capable of producing as low as 0.001 Pa (10^{-8} atm) of SO_2 in air. The SO_2/air mixture (50 ml/min) entered at the bottom of the furnace tube, passed over a Pt catalyst (for SO_2/SO_3 equilibrium) and the Pt weighing planchet containing the $NaVO_3$ or $NaVO_3$-In_2O_3 mix, and exited via a sidetube at the furnace top. Counterflow air (50 ml/min) passed downward from the balance chamber and out the same sidetube exit. The Pt catalyst and weighing planchet were within 10°C of the indicated temperature. The SO_2/air mixture was sampled and analyzed by H_2O_2 absorption [8] before entering the furnace, and the SO_3 concentration calculated, assuming equilibrium at temperature.

Samples of 100, 50, and 20 mg of $NaVO_3$ were used, with results normalized to weight gain per 100 mg of $NaVO_3$. Total conversion of 100 mg of $NaVO_3$ to Na_2SO_4 and V_2O_5 (reaction (3)) yields a weight gain of 32 mg. Complete reaction of V_2O_5 with In_2O_3 by reaction (4) requires 112 mg of In_2O_3, but additions corresponding to 56 mg of In_2O_3/100 mg of $NaVO_3$ were usually used for the In_2O_3-$NaVO_3$ mixtures.

The chemicals were anhydrous SO_2 (Matheson, 99.98%), $NaVO_3$ (Johnson Matthey, 98%), and In_2O_3 (Johnson Matthey, 99.99%), with the system air dried over anhydrous

$CaSO_4$. X-ray diffraction (Cu, Norelco diffractometer) confirmed the formation of $InVO_4$.

RESULTS AND DISCUSSION

Reversibility and Reproducibility

These properties are demonstrated in Fig. 1. (N.B., the Pt catalyst was not at proper temperature in these initial trials so the SO_3 values are only relative.) A 100 mg sample of $NaVO_3$ (solid line) was equilibrated with "50" Pa of SO_3, and then the SO_3 reduced to "11" Pa, and the $NaVO_3$ reequilibrated, with corresponding weight loss, at this pressure.

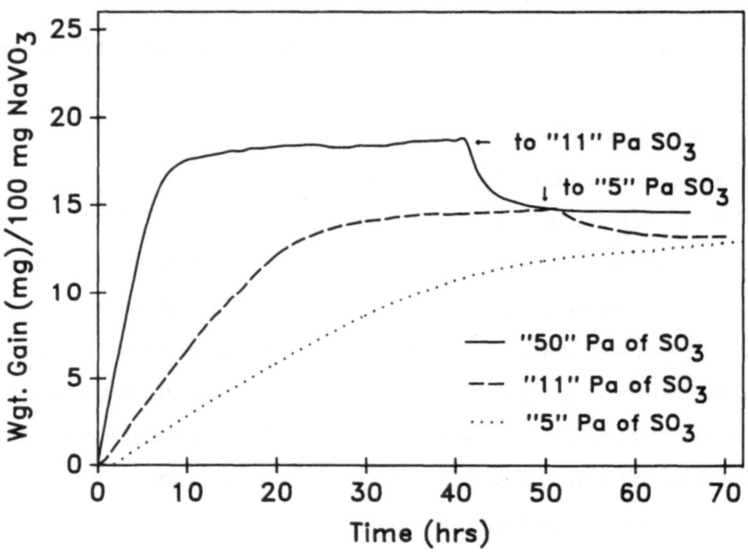

Figure 1. Equilibria of SO_3 with $NaVO_3$ at 700°C showing reversibility and reproducibility. SO_3 partial pressures are not absolute.

A second 100 mg sample of $NaVO_3$ (dashed line) was equilibrated with "11" Pa of SO_3 (coming to the same weight as in the first experiment), and the sample then reequilibrated under "5" Pa of SO_3. Finally, a third $NaVO_3$ sample (dotted line) was equilibrated under "5" Pa of SO_3, with the same weight gain as in the second experiment. In each case, identical equilibria were reached, regardless of whether approached from higher or lower SO_3 pressure.

Note that the reaction rates are linear until near equilibrium, and also inversely proportional to the SO_3 partial pressure; i.e., progressively longer times are required for equilibria as the SO_3 pressure decreases.

Detection of Ceramic Oxide Reaction

Fig. 2 illustrates the change in weight gain behavior that occurs when there is reaction between the ceramic oxide and SO_3-$NaVO_3$ media. The three experiments of Fig. 1 were repeated, but now with 56 mg of In_2O_3 mixed with the 100 mg of $NaVO_3$.

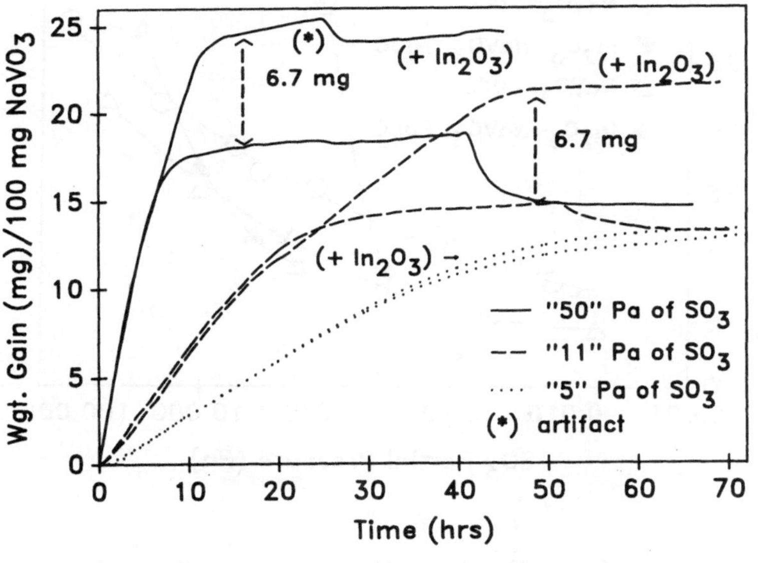

Figure 2. Effect of added In_2O_3 on SO_3-$NaVO_3$ equilibria at 700°C. Weight gain indicates $InVO_4$ formed at "50" and "11", but not "5", Pa of SO_3.

An additional weight gain of 6.7 mg was seen with both "50" Pa and "11" Pa of SO_3, but none with "5" Pa of SO_3. This is evidence that a melt V_2O_5 activity above the critical level necessary for In_2O_3 reaction is generated by "50" and "11" Pa of SO_3, but not by "5" Pa of SO_3 where no $InVO_4$ formation or weight gain is observed.

Also, since the reaction rate is not changed by the presence of In_2O_3, the rate controlling reaction probably occurs between the SO_3 and $NaVO_3$ melt.

Equilbria of SO_3-$NaVO_3$ and SO_3-$NaVO_3$-In_2O_3 at 700 and 800°C

In subsequent experiments (in which the Pt catalyst was verified to be at correct temperature), the equilibria of SO_3 with $NaVO_3$, and $NaVO_3$-In_2O_3, were determined at 700 and 800°C over the pressure range of 100 Pa to 0.01 Pa (Fig. 3). Times as long as 200 hrs were required for equilibrium at the lowest SO_3 levels.

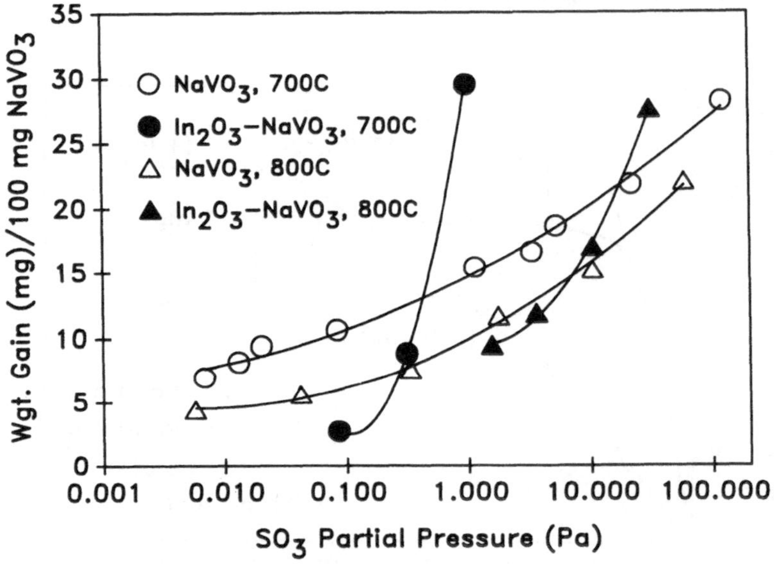

Figure 3. Equilibria of SO_3 with $NaVO_3$, and $NaVO_3$-In_2O_3, at 700 and 800°C. Intersection of curves marks SO_3 partial pressure (and resultant but unknown V_2O_5 activity) required for $InVO_4$ formation.

The intersection of the curves in Fig. 3 for pure $NaVO_3$ and In_2O_3-$NaVO_3$ indicates the approximate SO_3 partial pressure (about 0.5 Pa at 700, and 5 Pa at 800°C) required for In_2O_3 reaction. Ideally, the In_2O_3-$NaVO_3$ data points should fall on the $NaVO_3$ curve until the start of In_2O_3 reaction. Why they fall below, especially at 700°C, is not presently known. Also, the SO_3 partial pressures for reaction can not now be translated into V_2O_5 melt activities, since the SO_3-V_2O_5 "calibration curve" is not yet established.

Comparison of Ideal vs. Measured Solution Behavior

The equilibrium constant for reaction (3) can be obtained using thermodynamic data cited by Hwang and Rapp [7], and the equilibrium concentration of V_2O_5 calculated, assuming ideal behavior, as a function of SO_3 partial pressure. These values can then be compared with the V_2O_5 concentrations derived from our weight gain data. As seen in Fig. 4, the 700°C data approximate ideal behavior at 10^{-3} to 10^{-4} atm of SO_3, but deviate to high V_2O_5 values at lower SO_3 partial pressures. The 800°C data follow the general slope of the ideal behavior curve, but are displaced toward somewhat higher V_2O_5 values. An analysis of the significance of this comparison has not yet been made.

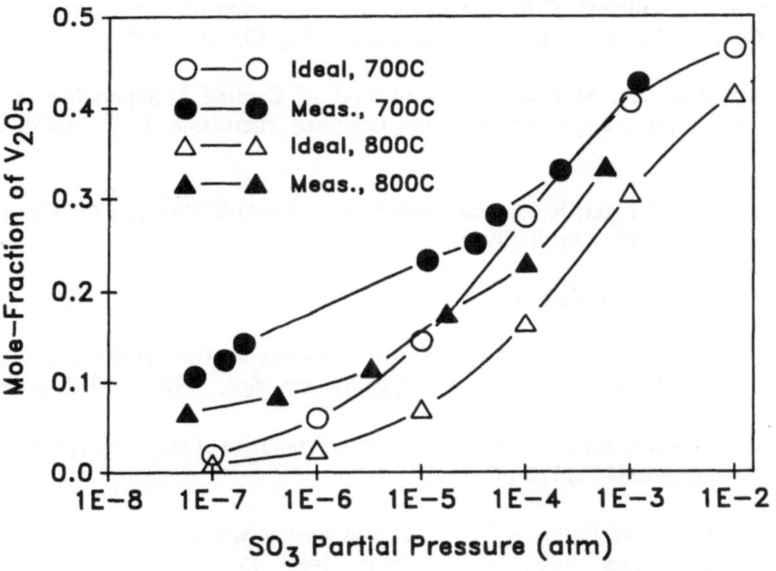

Figure 4. Mole-fraction of V_2O_5 vs. SO_3 partial pressure, as determined from weight gain data and as calculated for ideal solution behavior.

CONCLUSIONS

The experiments reported here, although in some ways still preliminary, demonstrate the promise of the TGA-SO_3 equilibria technique as a means for determining thermodynamic data on the reaction of ceramic oxides with molten vanadate-sulfate melts under conditions relating to the hot corrosion of ceramics in engines.

ACKOWLEDGEMENTS

Financial support and technical interaction by A. John Sedriks of the Office of Naval Research are gratefully acknowledged.

REFERENCES

1. Jones, R.L., Williams, C.E. and Jones, S.R., Reaction of Vanadium Compounds with Ceramic Oxides. J. Electrochem. Soc., 1986, **133**, 227-230.

2. Siemers, P.A. and McKee, D.W., Method of Coating a Superalloy Substrate, Coating Compositions, and Composites Obtained Therefrom. U.S. Pat. 4,328,285 (May 1982).

3. Jones, R.L. and Ingel, R.P., Scandia-Stabilized Zirconia Coatings for Composites. U.S. Pat. 4,913,961 (April 1990).

4. Jones, R.L., to be published.

5. Luthra, K.L. and Spacil, H.S., Impurity Deposits in Gas Turbines from Fuels Containing Sodium and Vanadium. J. Electrochem. Soc., 1982, **129**, 649-656.

6. Mittal, S.K. and Elliott, J.F., Thermodynamic Properties of Na_2O-V_2O_5 Melts, 1030-1210 K. ibid., 1984, **131**, 1194-1199.

7. Hwang, Y.-S. and Rapp, R.A., Thermochemistry and Solubilities of Oxides in Sodium Sulfate-Vanadate Solutions. Corros., 1989, **45**, 933-937.

8. Berger, A.W., Driscoll, J.N. and Morgenstern, P., Review and Statistical Analysis of Stack Sampling Procedures for the Sulfur and Nitrogen Oxides in Fossil Fuel Combustion. Am. Ind. Hygiene Assoc. J., 1972, **33**, 397-405.

OXIDATION STUDIES OF Si_2N_2O.

J. Persson, P.-O. Käll, M. Nygren and R. Larker
Department of Inorganic Chemistry, Arrhenius Laboratory, Stockholm University, S-10691 Stockholm, Sweden and
Department of Engineering Materials, Luleå University of Technology, S-95187 Luleå, Sweden.

ABSTRACT

Fully dense Si_2N_2O ceramics have been prepared using glass encapsulated HIP technique at a pressure of 200 MPa and at 1900 °C with a reaction time of 4 hours. The oxidation resistance of Si_2N_2O has been studied in the temperature range 1300-1600 °C by the thermogravimetric method. All weight gain curves do not follow the common parabolic rate law. Subsequent microstructural and X-ray diffraction studies have shown that the formed oxide scales consist of an amorphous SiO_2 phase and crystalline α-cristobalite implying that - to the extent that the crystalline products are formed during the oxidation experiment - the effective cross section area for diffusion of oxygen through the amorphous phase decreases during the experiment. An equation which describes how the effective area is reduced with time has been developed and incorporated into the common parabolic rate law, giving the following expression

$\Delta W/A_o = a \arctan (bt)^{1/2} + c(t)^{1/2}$

with ΔW = weight gain, A_0 = the initial cross section area, t = time and a, b, and c = constants to be determined. All weight gain curves can be described by this equation, and the rate constant at different temperatures could thus be determined.

INTRODUCTION

In order to obtain fully dense Si_3N_4-based ceramic materials by pressureless sintering technique a sintering aid has to be used. The oxidation behaviour of these ceramics is strongly dependent on the amount and nature of the intergranular phase formed in the sintering process. In almost all oxidation studies, the reaction kinetics of the oxidation has been interpreted with the parabolic rate law:

$$\left(\frac{\Delta W}{A}\right)^2 = K_p t + B \qquad (1)$$

where ΔW is the weight gain, A is the surface area, t is the reaction time and K_p the parabolic rate constant. The parabolic rate law indicates that the growth rate of the oxide scale is diffusion controlled. The law is derived under the assumption that the oxide scale formed is either completely amorphous or crystalline throughout, i. e.

that the phase composition of the oxide scale does not vary with time. X-ray diffraction and electron microscope studies of the surfaces of the oxide scales and of the cross sections of the scales, formed upon oxidation, have revealed that the scales frequently contain crystalline products and in many cases also bubbles, which are presumed to contain nitrogen. In addition, cracks between the oxide scale and the Si_3N_4-based ceramic are often detected. It seems to be generally accepted that a glassy oxide scale is formed initially upon oxidation, and that the diffusion coefficient of the oxygen in amorphous SiO_2 is at least two orders of magnitude larger then in any crystalline phase(s) expected to occur in the oxide scale. These observations seem to imply that A in equation 1 above is not constant during the oxidation process.

In this article we describe the oxidation behaviour of Si_2N_2O. The fully dense Si_2N_2O ceramics used for the oxidation experiments were prepared without a sintering aid. The oxidation resistance of Si_2N_2O has previously been studied, but in this case the dense ceramics were prepared with use of additives (1). Our material exhibits good oxidation resistance up to about 1600 ^oC. Subsequent SEM and X-ray studies of the oxide scale showed that it consisted of an amorphous phase, the α modification of cristobalite and N_2 bubbles. All weight gain curves obtained in the temperature region 1300- 1600 ^oC can not be interpreted with the parabolic rate law given above. A function, A(t), which describes how the effective area for diffusion of oxygen is reduced with time has been developed and incorporated into the parabolic rate law and the oxidation reaction kinetics of Si_2N_2O could be interpreted with use of this new rate law.

EXPERIMENTAL

Fully dense Si_2N_2O (99.9% Th. D.) ceramics were prepared from equal molar amounts of Si_3N_4 and SiO_2, without any sintering aid, using the HIP technique with glass-encapsulated samples. The materials were sintered at 1900 ^oC during 4 hours, using a pressure of 200 MPa. The products were characterized by their X-ray powder diffraction patterns obtained in a Guinier-Hägg camera with Cu $K_{\alpha 1}$ radiation and with Si as internal standard.

The oxidation behaviour of the Si_2N_2O ceramic was studied in a TG unit (SETARAM TAG 24) with two symmetrical furnaces, one used for the sample to be studied and the other for an inert dummy. The weight difference is recorded as function of temperature and/or time. By regulating the gas flow (oxygen) over the sample and the dummy it is possible to keep the drift of the baseline within ∓ 5 μg during the oxidation experiment in the temperature range studied, 1300-1600 °C. The resolution of the TG unit is better than 2 μg. This unit thus makes it possible to perform very accurate measurements of weight changes with time. The oxidation experiments were performed during 20 hours, and the samples were heated to the predetermined temperature at a rate of 100 degrees per minute. The ceramics to be oxidized consisted of pieces of approximate size 15x5x1 mm^3. Prior to the oxidation the ceramic pieces were carefully polished using diamond grains down to a size of < 1 μm.

The crystalline phases present in the oxide scale formed were characterized by their X-ray powder diffraction patterns obtained in reflection mode with Cu $K_{\alpha 1}$ radiation (STOE STADI). The oxidized surfaces and cross sections were also characterized with scanning electron microscopes (JEOL JSM 820 and 880) equipped with electron dispersive spectrometers (LINK 10000). The micrographs reproduced below are obtained in the BSE mode.

RESULTS AND DISCUSSION

The X-ray powder pattern of the prepared Si_2N_2O ceramic revealed it to

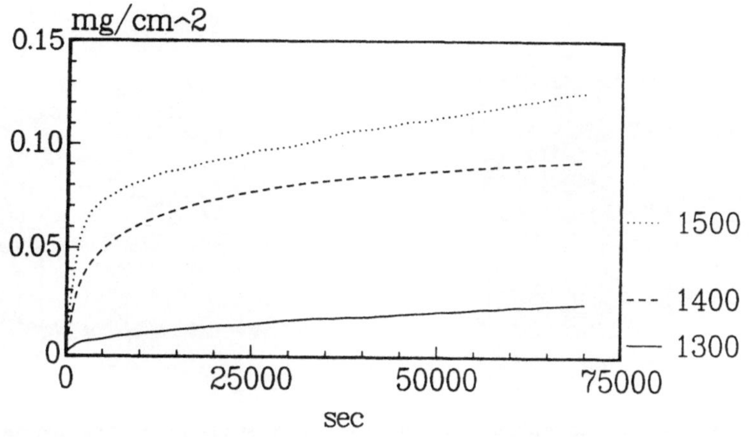

Figure 1. Oxidation curves for Si_2N_2O obtained at 1300, 1400 and 1500 °C

contain, besides Si_2N_2O small amounts of α- and β- Si_3N_4 (approximately 2 wt.% of each). The presence of unreacted starting materials in the product indicates that it must also contain some unreacted SiO_2.
The oxidation curves obtained at 1300, 1400 and 1500 °C are plotted in Fig. 1. The only curve which seems to follow the parabolic rate law is the one obtained at 1300 °C. The data obtained at 1400 °C can not be interpreted with the parabolic rate law and curves of this type were obtained in the temperature range 1350 ≤ T ≤ 1450 °C. The weight gain curves obtained for T ≥ 1500 °C were all similar implying that the parabolic rate law is not applicable at the beginning of the oxidation process but first after a certain time, t_o.

The microstructure of the oxide scale formed at 1400 °C is shown in Fig.2a. The X-ray diffraction studies of the same surface tells us that the scale contained α-cristobalite and from the appearance of the microstructure it seems clear that the degree of crystallinity of the the surface is rather high. The microstructure of the cross section of the oxide scale formed at 1500 °C is shown in Fig. 2b. This figure demonstrates that the oxide scale formed is homogeneous, from the surface down to the unreacted ceramic, and that N_2 bubbles are present in the scale. All samples oxidized at T ≥ 1350 exhibit very similar microstructures of the surfaces and the cross sections. The amount of crystalline products in the oxide scale obtained at 1300 °C is, however, substantially less.

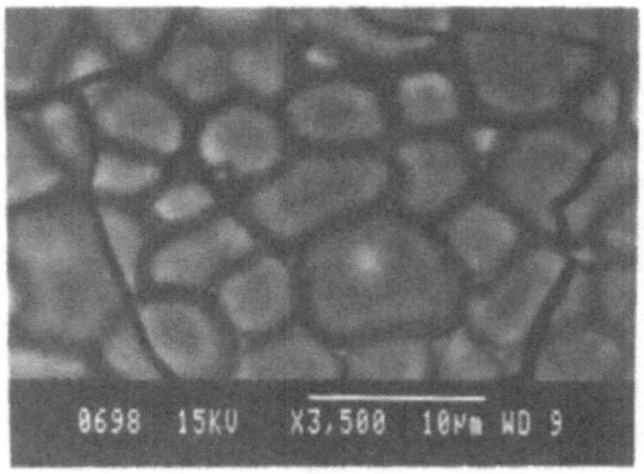

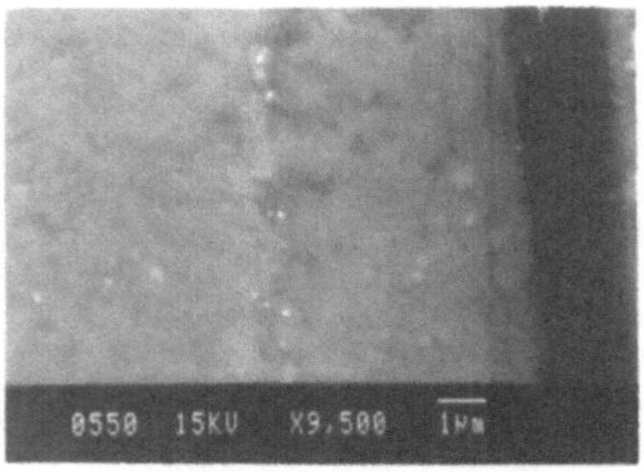

Figure 2 a and b. Microstructure of (a): the surface of the oxide scale obtained at 1400 °C and (b): the cross section of the oxide scale obtained at 1500 °C.

The diffusion coefficient of oxygen is almost three orders of magnitude larger in amorphous SiO_2 than in crystalline SiO_2 (2), implying that even if the oxide scale contains a large fraction of crystalline products, the kinetics of oxidation is most probably governed by the diffusion of oxygen through the amorphous SiO_2 phase. However, to the extent that the crystalline products and bubbles are formed during the oxidation, the effective cross section area for oxygen diffusion will decrease during the experiment, i. e. A in equation 1 will not be a constant. An equation has been derived which describes how the effective area decreases with time

$$A(t) = A_o \frac{(1+(f\beta-t_o^{-1})t)}{(1+(\beta-t_o^{-1})t)} \qquad (2)$$

where A_o is the cross section area at t = 0, β is the constant which describes the rate of decrease of A_o, and t_o expresses the time at which steady state is reached (i. e. for $t > t_o$ the oxidation kinetics obeys the parabolic rate law); and f, finally, expresses the fraction of A_o which is still amorphous at $t = t_o$. A more elaborate description of this equation and its physical implications will be given elsewhere (3). The function A(t) is incorporated into the parabolic rate law according to

$$(dw/dt) = \frac{K\,A(t)}{X} \qquad (3)$$

where w is the weight gain, K is a constant and X the thickness of the oxide sale. X is assumed to vary with time as $(\alpha t)^{1/2}$ (α= rate constant for the growth of the oxide scale). Integration of this equation yields

$$(\Delta W/A_o) = a \arctan(b\,t)^{1/2} + c(t)^{1/2} \qquad (4)$$

where

$$a = \frac{2K\beta(1-f)}{(\beta-t_o^{-1})^{3/2}(\alpha)^{1/2}}\ ;\ b = \beta-t_o^{-1}\ ;\ c = \frac{2K(f\beta-t_o^{-1})}{(\beta-t_o^{-1})(\alpha)^{1/2}}\ ;\ f = \frac{c + \frac{a(b)^{1/2}}{bt_o+1}}{c + a(b)^{1/2}}$$

If a steady state is reached during the experiment (i. e. A(t) is constant for $t > t_o$) it can be shown that for $t > t_o$

$$(\Delta W/A_o)^2 = K_p^o\, t + B_o \tag{5}$$

where K_p^o can be determined from the slope of the $(\Delta W/A_o)^2$ versus time curve and from the relation

$$(K_p^o)^{1/2} = c + \frac{a(b)^{1/2}}{bt_o+1} \tag{6}$$

It can also be shown that $K_p^o = K_p\, f^2$, with K_p defined as in equ. 1. All oxidation curves obtained in the temperature range $1350 \le T \le 1450$ °C can be fitted to equ. 4, while at $T \ge 1500$ °C equ. 4 is applicable for $0 < t < t_o$ and equ. 5 for $t > t_o$. The curve obtained at 1300 °C can be fitted to equ. 1 as well as to the equations 4 and 5 since t_o is small in this case. The agreement between the calculated and observed weight gains for the oxidation curves given in Fig. 1 is shown in Figs. 3-5., and the refined values of a, b, c, f, t_o and K_p are compiled in Table 1.

TABLE 1.

The least-squares refined values of the constants a, b, c, f, t_o and K_p defined in connection with the derivation of equ. 4 and 5.

Temp. of oxidat.	a	b	c	f	t_o (sec)	K_p ($\frac{mg^2}{cm^4 sec}$)
1300	$3.97\ 10^{-3}$	$4.02\ 10^{-4}$	$3.22\ 10^{-5}$	0.68	2000	$1.73\ 10^{-8}$
1350	$1.69\ 10^{-2}$	$6.43\ 10^{-4}$	$2.50\ 10^{-4}$	0.37	∞	$4.47\ 10^{-7}$
1400	$6.34\ 10^{-2}$	$1.62\ 10^{-4}$	$3.89\ 10^{-5}$	0.05	∞	$7.15\ 10^{-7}$
1450	$3.77\ 10^{-2}$	$5.98\ 10^{-4}$	$1.30\ 10^{-4}$	0.13	∞	$1.03\ 10^{-6}$
1500	$7.76\ 10^{-2}$	$6.82\ 10^{-4}$	$-2.51\ 10^{-5}$	0.10	12000	$1.53\ 10^{-5}$
1550	$4.82\ 10^{-2}$	$1.59\ 10^{-3}$	$2.18\ 10^{-4}$	0.14	15000	$9.74\ 10^{-6}$
1600	0.718	$5.48\ 10^{-5}$	$-2.96\ 10^{-3}$	0.10	12000	$5.48\ 10^{-5}$

CONCLUDING REMARKS

It has been shown that the common parabolic rate law is only partly or not at all applicable in the case of oxidation of Si_2N_2O. However, it has been possible to interpret all obtained curves with the use of the new rate law derived above. The K_p values given in Table 1 can be used to estimate the activation energy of the oxidation process from the Arrhenius equation $K_p = k \exp(-E_a/RT)$. An activation energy of 595 kJ/mol is then obtained, which is approximately a factor of two larger than that observed for other types of Si_3N_4 based ceramics (4). On the

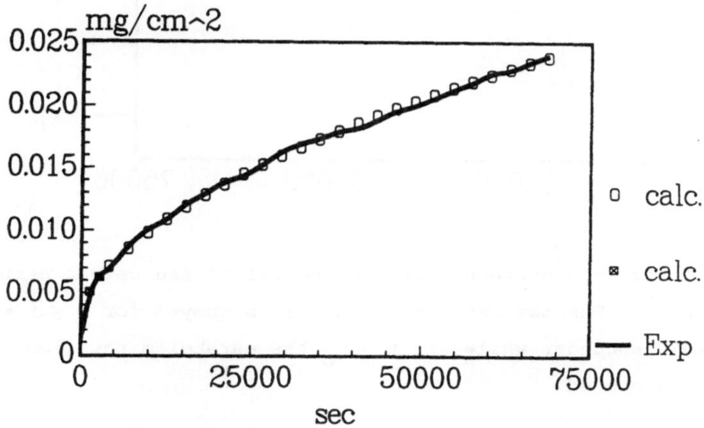

Figure 3. A comparison between observed and calculated weight gain at 1300 °C. The parabolic rate law (equ. 1) is obeyed over almost the entire time interval (t_o = 2000 sec.).

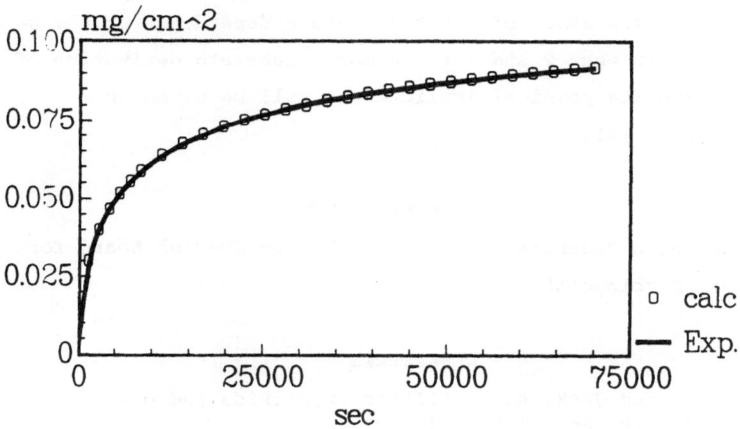

Figure 4. A comparison between observed and calculated weight gain at 1400 °C. The new rate law (equ. 4) is obeyed over the entire time interval (t_o = ∞).

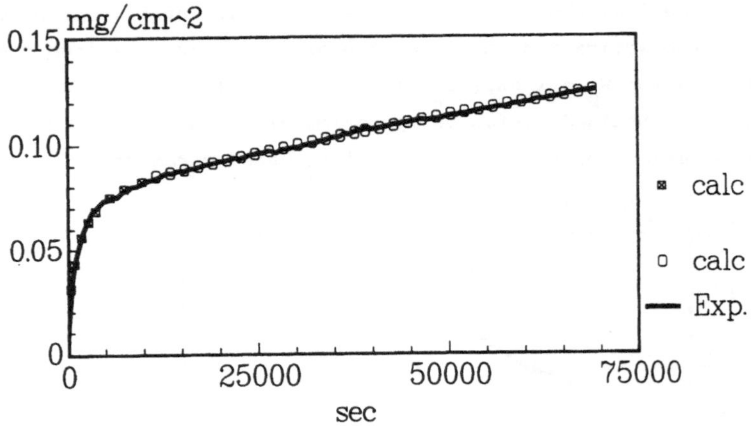

Figure 5. A comparison between observed and calculated weight gain at 1500 °C. The new rate law (equ. 4) is obeyed for $0 < t < t_o$ = 12000 seconds, while for $t > t_o$ the parabolic rate law (equ. 5) is obeyed.

other hand if K_p^o are used instead of K_p values one obtains an E_a value of 280 kJ/mol which is in fair agreement with previously published data (4) Thus it seems that if A in equation 1 decreases with time the previously determined E_a values are too low. An extended discussion of this point, a presentation of the temperature dependence of the α- and β- values given in equ. 2 and 3 and a more elaborate derivation of this new rate law and its physical implications will be given in a forthcoming paper (3).

ACKNOWLEDGEMENTS

This work has been financially supported by the Swedish Board for Technological Development.

REFERENCES

1. Trigg, M. B. and Jack, K. H. Silicon oxynitride and O'-sialon ceramics. J. Mat. Sci. 1988, 23, 481.
2. Weiss, J. and Kaysser, K. W. Liquid phase sintering. NATO/ASI Progress in Nitrogen Ceramics, ed Riley, F. L. Martinius Nijhoff Publisher 1983, 169.
3. Persson, J., Käll, P.-O. and Nygren M. to be published.
4. Singhal, S. C. Oxidation of silicon nitride and related materials. NATO, Nitrogen Ceramics, ed Riley, Nordhoff-Leiden, 1977, 607.

HOT CORROSION OF ALUMINA MATRIX-COMPOSITES

Jeffrey J. Swab
US Army Materials Technology Laboratory
Watertown, MA, 02172 USA

and

Gary L. Leatherman and Mary H. Adair
Worcester Polytechnic Institute
Worcester, MA, 01609 USA

ABSTRACT

The effects of sodium sulfate-induced hot corrosion on the room temperature strength of an alumina and three alumina-matrix composites were examined in this study. The strength of the alumina was unaffected by the corrosion. On the other hand, the strength of the composites decreased, to various degrees, with exposure to sodium sulfate.

INTRODUCTION

Ceramic materials are becoming increasingly popular for gas turbine and diesel engine applications because of their ability to maintain structural integrity at high temperatures. The potential benefits of incorporating ceramics into advanced heat engines include improved fuel economy, significantly higher performance, and reduced dependence on scarce materials.

Unfortunately, these potential benefits may not be attained because monolithic ceramics inherently lack toughness, making them a risky choice for mechanical load bearing applications. In response to this problem ceramic matrix composites (CMC) are now being investigated as a tougher alternative. In CMC materials the toughness is increased through the addition of a second-phase material; i.e., particulates, whiskers, or continuous fibers, or any combination of the three. The additional phase(s) can increase toughness by one or more of the following mechanisms: crack deflection, crack branching, crack bridging, microcracking, fiber and/or whisker pullout, or absorption of the crack tip energy by a phase transformation.

Since CMC materials are a relatively new class of materials there is a significant amount of information which must be obtained before they can be considered for use in advanced heat engines. One area which must be examined is hot corrosion. In the gas turbine engine, some components will face attack by molten Na_2SO_4 (T_m = 884°C) which forms when ingested NaCl reacts with sulfur impurities in the fuel and condenses on the engine component. How long-term exposure to this corrosive medium affects the mechanical properties of CMC materials is of critical importance to their successful application in engines as high temperature structural components.

This report examines the effects sodium sulfate induced corrosion on the room temperature strength of a high purity, fully dense alumina and three alumina matrix composites. One of the composites is toughened with zirconia, (a zirconia toughened alumina), the second is reinforced with SiC whiskers and the third is reinforced with both SiC whiskers and tetragonal zirconia particulates.

MATERIALS

1) The baseline material for this study is a commercially available alumina (AD-999) which was produced by Coors Ceramics, Golden CO. This high purity alumina was sintered to full density.
2) The first composite is a zirconia toughened alumina (ZTA) produced by Kyocera Feldmuehle, Inc. It was sintered with ≈10 v/o zirconia added.
3) The alumina reinforced with SiC whiskers (ASC) was manufactured by Arco Chemical Co. by hot pressing α-Al_2O_3 containing 29 v/o SiC whiskers.
4) The final composite was produced by Riken Corp. in Japan. It contained 70 v/o α-Al_2O_3 and 15 v/o each of submicron TZP powder and SiC whiskers (ASCZ). The TZP powder contained 2 mol/o Y_2O_3 and the

nominal dimensions of the whiskers were a diameter of 0.15 to 0.75 μm and a length of 3 to 50 μm. Full density was achieved by hot pressing at 1450°C under 28 MPa applied pressure.

EXPERIMENTAL PROCEDURE

The four alumina-based ceramics were obtained and machined into type "B" flexure specimens according to MIL STD-1942A. The specimens were then divided into groups containing a minimum of (5) bars from each material and subjected to one of the following conditions: 500 hours @ 1000°C in air or 500 hours @ 1000°C in air w/10-20 mg/cm^2 Na_2SO_4.

After each exposure the room temperature strength was determined by four-point flexure testing according to MIL STD-1942A. The strength data was then compared to the strength obtained for the as-received materials.

Optical fractography was completed on all bars and a select group of fracture surfaces were examined with the scanning electron microscope (SEM) to determine the failure origin.

RESULTS AND DISCUSSION

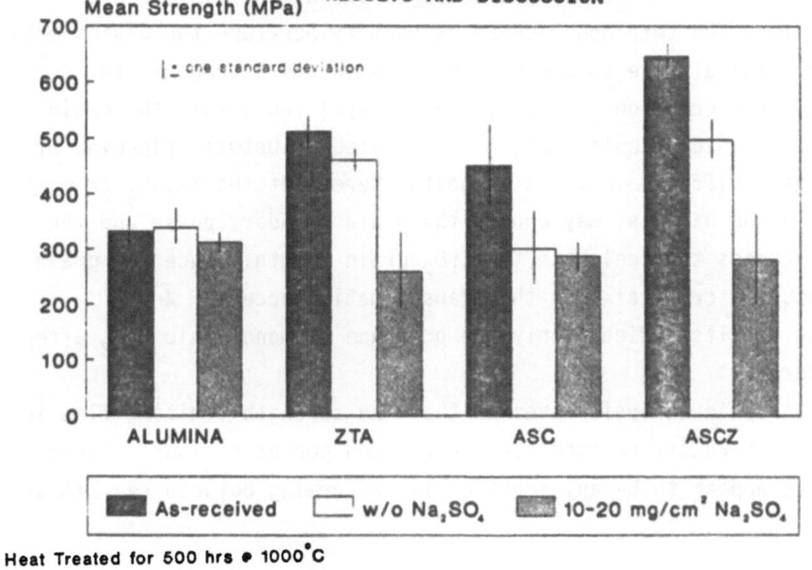

Figure 1. Strength Before and After Corrosion.

ALUMINA

The high purity, fully dense alumina did not show any change in the strength after exposure without sodium sulfate and exhibited only a minor degree of strength degradation when exposed to sodium sulfate, Figure 1. The minimal strength reduction in the alumina may be due to the corrosion of an undetermined glassy grain boundary phase[1].

Examination of the fracture surfaces showed that the type of strength-limiting flaw did not change after exposure to molten sodium sulfate, and there appears to be very limited, if any interaction, between the alumina and the Na_2SO_4.

ZIRCONIA TOUGHENED ALUMINA (ZTA)

Figure 1 shows that the strength of the ZTA decreases after 500 hours at 1000^oC with and without Na_2SO_4. The initial decrease without the Na_2SO_4 is due to overaging of the tetragonal zirconia grains causing them to transformation to the monoclinic phase[2-3] and the nucleation and growth of microcracks around the monoclinic grains[4-5]. The X-ray diffraction results support this by showing that the amount of monoclinic ZrO_2 increases after this exposure.

The strength decrease with Na_2SO_4 is not as easily explained. This ZTA is toughened with unstabilized tetragonal and monoclinic ZrO_2. In order to retain the tetragonal phase at room temperature the grains must be below a critical size to prevent the spontaneous tetragonal-to-monoclinic transformation. Kibbel and Heuer[6] found that the grain growth rate of intergranular ZrO_2 is controlled by Ostwald ripening by grain-boundary diffusion. It is suggested here that the Na_2SO_4 or more specifically the Na ions, may change the grain boundary phase and the diffusion process to accelerate the ZrO_2 grain growth. Once the grain size exceeds the critical size the transformation occurs. X-ray diffraction results indicate only the presence of monoclinic ZrO_2 after this exposure.

Fractographic analysis revealed that the strength-limiting flaw in all cases was porosity related i.e., pores and porous regions. There also did not appear to be any reaction layer forming between the ZTA and the Na_2SO_4.

ALUMINA W/SiC WHISKERS (ASC)

The strength loss seen in this material after exposure without Na_2SO_4 is ≈25% which agrees very well with the loss seen by Tiegs and Becher[7]. The loss is due to oxidation of the SiC whiskers by the transport of oxygen along the grain boundaries and through the matrix.

The fact that the strength loss after the addition of Na_2SO_4 is also ≈25% is interesting. Research[8-12] has shown that SiC is readily corroded by Na_2SO_4, therefore, it was anticipated that the strength loss would be larger in the presence of Na_2SO_4. For this specific material the Na_2SO_4 did not cause further degradation of the SiC whiskers. In addition there is no evidence of a reaction layer between the ASC and the Na_2SO_4. The only interaction between these materials is a slight rounding of the chamfers on the specimens. The typical fracture origin in all cases was large grain(s) of Al_2O_3.

ALUMINA W/SiC WHISKERS & Y-TZP (ASCZ)

This composite has a strength decrease of ≈25% after the exposure without Na_2SO_4 and a >50% decrease after exposure with Na_2SO_4. The initial loss is due to oxidation of the SiC whiskers[7] and the transformation of the ZrO_2 from the tetragonal phase to the monoclinic phase[13]. X-ray diffraction results show that the amount of tetragonal ZrO_2 is reduced from 75% to 25% after this exposure.

The strength reduction after corrosion is likely due to two factors: 1) corrosion of the SiC whiskers and 2) further tetragonal-to-monoclinic transformation of the Y-TZP grains. As previously stated[8-12] SiC is readily corroded by Na_2SO_4. The transformation of the Y-TZP grains in the presence of Na_2SO_4 is due to the leaching of the Y_2O_3 from the zirconia by a sulfation reaction[14-15], since it has been reported that pure zirconia is difficult to sulfate[16]. Previous work[17] has also shown that the strength of monolithic Y-TZP is reduced after exposure to sodium sulfate.

X-ray diffraction results and fractographic analysis indicate that the corrosion of the SiC whiskers leads to the large strength loss. The x-ray diffraction results show that the amount of tetragonal ZrO_2 is still 25% after the exposure with the Na_2SO_4. Thus the corrosion-induced tetragonal-to-monoclinic transformation is not occurring in this material under these conditions. The fractographic analysis revealed a change in the type and location of the strength-limiting defect with the addition of Na_2SO_4. In the as-received state and after exposure without Na_2SO_4 the defect is porosity or agglomerates which are located several hundred microns below the tensile surface. After the Na_2SO_4 is added a porous layer ≈30 μm wide formed between the bulk material and the corrosion product. It appears that fracture initiated within this layer, but the defect type could not be determined. The porous layer was Si poor while the corrosion product on top of this layer was Si rich. This shows that for the ASCZ material there is corrosion-assisted oxidation of the SiC whiskers and the Si left behind migrates or is pulled to the surface.

It should be noted that the ASC and ASCZ materials were produced using different starting materials and processing techniques. These differences may account for the variations in the corrosion behavior of each material.

SUMMARY

Although this is a preliminary study it has shown that a pure, fully dense alumina is resistant to corrosion by molten sodium sulfate. However, the addition of second phases to the alumina to improve the mechanical properties, in fact reduced the composites overall resistance to hot corrosion in two of the three composites tested in this study. It appears that the quality of the second phase material will determine how resistant the composite is to Na_2SO_4 induced corrosion.

REFERENCES

1. Gannon, R.E., Hals, F.A. and Reynolds, H.H., Corrosion Studies in Materials for Auxiliary Equipment in MHD Power Plants, in Corrosion Problems in Energy Conversion and Generation, C.J. Tedman, Jr. ed., Corrosion Division, The Electrochemical Society, Princeton, NJ, 1974, pp. 212-224.

2. Larsen, D.C. and Adams. J.W., Long-Term Stability and Properties of Zirconia Ceramics for Heavy Duty Diesel Engine Components, prepared for NASA-Lewis Research Center, for U.S. Department of Energy under Contract DEN 3-305, NASA CR-174943, September 1985.

3. Ferber, M.K. and Hine. T., Time-Dependent Mechanical Behavior of Partially Stabilized Zirconia for Diesel Engine Applications, ORNL/Sub/85-27416\1, prepared for Oak Ridge National Laboratory for U.S. Department of Energy under Interagency Agreement DE-AC05-84OR21400, July 1988.

4. Claussen, N., Fracture Toughness of Al_2O_3 with an Unstabilized ZrO_2 Dispersed Phase, J. Am. Ceram. Soc., 1976, 59 [1-2], pp. 49-51.

5. Claussen, N., Steeb, J. and Pabst, R., Effect of Induced Microcracking on the Fracture Toughness of Ceramics, J. Am. Ceram. Soc., 1977, 56 [6] pp. 559-562.

6. Kibbel, B. and Heuer, A.H., Exaggerated Grain Growth in ZrO_2-Toughened Al_2O_3, J. Am. Ceram. Soc., 1986, 69 [3] pp. 231-236.

7. Tiegs, T.N. and Becher, P.F., Alumina-SiC Whisker Composites, Proceedings of the 23rd Automotive Technology Development Contractors' Coordination Meeting, Vol P-165, Society of Automotive Engineers, Inc., Warrendale, PA, 1984, pp. 209-213.

8. Tressler, R.E., Meiser, M.D. and Yonushonis, T., Molten Salt Corrosion of SiC and Si_3N_4 Ceramics, J. Am. Ceram. Soc., 1976, 59 [5-6], pp. 278-79.

9. Jacobson, N.S. and Smialek, J.L., Hot Corrosion of Sintered α-SiC at 1000°C, J. Am. Ceram. Soc., 1985, 68 [8], pp. 432-439.

10. Jacobson, N.S., Kinetics and Mechanism of Corrosion of SiC by Molten Salts, J. Am. Ceram. Soc., 1986, 69 [1], pp. 74-82.

11. Jacobson, N.S., Stearns, C.A. and Smialek, J.L., Burner Rig Corrosion of SiC at 1000°C, Ad. Ceram. Mat., 1986, 1 [2], pp. 154-161.

12. Jacobson, N.S. and Smialek, J.L., Corrosion Pitting of SiC by Molten Salts, J. Electrochem. Soc., 1986, 133 [12], pp. 2615-2621.

13. Swab, J.J., Properties of Yttria-Tetragonal Zirconia Polycrystal (Y-TZP) Materials After Long-Term Exposure to Elevated Temperatures, prepared for the U.S. Department of Energy under Interagency Agreement DE-AI05-84OR21411, MTL TR 89-21, March 1989.

14. Barkalow, R. and Pettit, F., Mechanisms of Hot Corrosion Attack of Ceramic Coating Materials, Proceedings of the 1st Conference on Advanced Materials for Alternative Fuel Capable Directly Fired Heat Engines, CONF-790749, J.W. Fairbanks and J. Stinger, ed., NTIS Springfield, VA, 1979, pp. 704-710.

15. Jones, R.L., Nordman, D.B. and Gadomski, S.T., Sulfation of Y_2O_3 and HfO_2 in Relation to MCrAl Coatings, Metall. Trans., 1985, 16A [2], pp. 303-306.

16. Jones, R.L., Jones, S.R. and Williams, C.E., Sulfation of CeO_2 and ZrO_2 Relating to Hot Corrosion, J. Electrochem. Soc., 1985, 132 [6], pp. 1498-1501.

17. Swab, J.J. and Leatherman, G.L., "Strength of Structural Ceramics After Hot Corrosion," J. Eur. Ceram. Soc., 1989, 5 [6], pp. 333-40.

BOUNDARY LUBRICATION OF CERAMIC MATERIALS BY SOFT METALLIC COATINGS AND SYNTHETIC OIL

O. O. Ajayi, A. Erdemir, J.-H. Hsieh, R. A. Erck and F. A. Nichols
Materials and Components Technology Division,
Argonne National Laboratory
Argonne, IL 60439, U. S. A.

ABSTRACT

Boundary lubrication of metallic materials often relies on surface reaction-generated films. This approach is intrinsically more difficult for ceramic materials because of their relatively lower chemical reactivity. The present study investigates the viability of boundary lubrication of ceramics by the use of a thin soft metallic film as a boundary film. Lubrication of Si_3N_4 and ZrO_2 ceramics with coatings of Ag or Au prepared by Ion-Beam-Assisted Deposition (IBAD) and a polyol ester-based synthetic oil (with proprietary additives) was investigated. Through this method, the friction coefficient during sliding contact of ceramics was reduced (as low as 0.05) to the level that is obtained in most boundary-lubricated situations. Furthermore, wear was virtually eliminated. The method was found to be effective from room temperature up to 250°C.

INTRODUCTION

Structural ceramic materials have an attractive combination of properties, notably high hardness and high thermal and chemical stability, which make them very good prospects for tribological applications. They are particularly attractive for high temperatures beyond the capability of many metallic materials. Ceramics are the leading candidate materials for critical components of the low-heat-rejection engine (LHRE) which is expected to operate at peak temperatures as high as 600°C [1-3], compared to a temperature of about 200°C for existing engines. It was initially thought that these materials could be used unlubricated [4, 5]. Tribological studies, however, showed that the friction and wear rates during sliding contact are too high for engine applications [e.g., 6-8]. Consequently, efforts are now being devoted to lubrication of ceramic materials [9, 10].

During liquid lubrication, if the lubricant film thickness is large enough such that the two sliding surfaces are completely separated (hydrodynamic regime), the friction coefficient is determined solely by the viscosity of the lubricant and ideally no wear occurs. This hydrodynamic regime of lubrication is desirable, but very difficult to maintain. It is generally achieved at low loads, using high sliding speeds with high-viscosity lubricants on smooth rubbing surfaces (conditions which are often difficult to maintain in practical applications). However, as the contact pressure increases, and/or the speed goes down and/or viscosity decreases, the lubricant film's thickness decreases. When the film is thin enough, the rubbing surfaces come into direct contact. This results in an increase in friction and wear. This is the boundary regime of lubrication, which is often encountered in various practical machines. This regime is very critical because it often precedes scuffing and seizure.

For metallic materials, additives are usually incorporated into the lubricants to form a surface-reaction layer (boundary film) during boundary lubrication, which protects the rubbing surfaces by preventing metal-to-metal contact and presents an easily sheared interface between the two surfaces. A common example of an additive for ferritic surfaces is zinc-dialkyl-dithiophosphate (ZDDP), which forms a reaction layer rich in Zn, P, S, O and Fe on the rubbing surfaces [11-13]. Although the exact composition and detailed characteristics of such films are still under investigation, their roles in preventing metal-to-metal contact and providing easy shear are well-known.

Ceramic materials in general have low chemical reactivity when compared with metals; thus, formation of boundary films through surface reaction is more difficult. In fact, studies of ceramic-on-metal sliding couples using lubricant developed for metallic surfaces showed that the boundary films formed on the metal components but not on the ceramic ones [14]. In other studies, adsorption of some additives onto ionically bonded ceramic surfaces was found to provide limited protection while no such adsorption was observed for covalently bonded materials [15]. Even for ionic materials, adsorption of additives will have very limited practical use because desorption will occur as the temperature is increased. Since ceramics are being considered for high-temperature applications in many cases, additive adsorption will not be an effective means of their boundary lubrication. Thus a practical means of boundary lubrication of ceramics must be found.

Boundary films are known to behave in a fashion similar to soft metallic solid lubricants [16]. Recent studies have shown a significant improvement in the tribological behavior of ceramics by application of a soft metallic coating as a solid lubricant [10, 17]. The present study investigates boundary lubrication of Si_3N_4 and ZrO_2 ceramic materials by a soft metallic coating (Ag and Au) and a synthetic oil.

EXPERIMENTAL DETAILS

The ceramic materials used for this study were commercially available Si_3N_4 (SN-220) and Y_2O_3-stabilized ZrO_2 (Z-201N) from Kyocera. Flat specimens had nominal dimensions of 50.8 x 25.4 x 6.35 mm and rods were 8 mm in diameter. Pin specimens 15 mm long were cut from the rod. Both ends of the pin were rounded to form hemispherical caps of 127 mm radius of curvature and polished to a roughness of 0.08 μm Ra. The flat specimens were ground to a roughness of about 0.25 μm Ra.

Coating

A thin Ag film (~ 1.5 μm thick) was deposited on some of the Si_3N_4 and ZrO_2 flat specimens by the IBAD technique. This method involves simultaneous bombardment with energetic ions of a film growing by evaporation and often results in a substantial increase in the adhesion of the film to the substrate compared with evaporation alone [10, 18]. Prior to the deposition of the film, the surface was sputter-cleaned with a mixture of energetic Ar and O ions. IBAD of the Ag film was then done using the same ions to a film thickness of 100 nm, after which the ion source was turned off and the rest of the film thickness was deposited by vacuum evaporation. Some ZrO_2 flat specimens were also coated with Au using the same IBAD procedure.

Tests

Friction and wear tests were done with a pin-on-flat contact geometry in reciprocating sliding. The details of the test device have been described elsewhere [10]. Tests were conducted under "dry" (unlubricated) and oil-lubricated conditions using pins and flats of the same materials. A commercially available 100% polyol ester-based synthetic oil (SDL-1 from Akzo Chemical Inc.) with a proprietary blend of additives including ZDDP and containing S was used in this study. Coated and uncoated flats were tested. Tests were conducted with normal loads of 10 and 50 N which impose an initial mean Hertzian pressure of about 150 and 260 MPa respectively for Si_3N_4 and 116 and 198 MPa for ZrO_2. All the tests were done with a reciprocating frequency of 1 Hz and a stroke length of 25 mm producing an average sliding speed of 0.05 m/s. Tests were conducted at room temperature, 150°C and 250°C. The total number of cycles for 10N and 50N loads were 5000 and 2000 respectively. The oil-lubricated tests were done by immersing the flat specimens in a bath of oil during the test.

The friction coefficient was continuously monitored in all tests by a strain-gauge device. The amount of wear from the pin was calculated from the dimensions of the wear scar as measured by an optical microscope, assuming a flat wear scar [10]. The tests were interrupted after 100, 250, 500, 1000, 2000, 3000, 4000 and 5000 cycles and pin wear-scar dimensions were measured in order to determine the wear rate as a function of the number of cycles.

The worn surfaces of the pin and the flat were examined by both optical and scanning electron microscopes (SEM) equipped with an x-ray energy-dispersive spectrometer (EDS) and a wavelength-dispersive spectrometer (WDS).

RESULTS

Friction

Figure 1 shows the typical frictional behavior of Si_3N_4 pairs under various conditions. At room temperature, the friction coefficient in all tests done under dry conditions showed a similar trend, irrespective of the load, i.e., a quick rise from a relatively low value over the first 100 cycles of sliding to a steady value of about 0.9 for uncoated flats and about 0.6 for Ag-coated flats (Figure 1a). In the oil-lubricated tests, however, the frictional behavior was different for coated and uncoated flats. The friction coefficient for the uncoated, oil-lubricated flats showed a very quick rise over the first 10 cycles from an initial value of about 0.1 to a steady value of about 0.2. This is significantly lower than the value for dry sliding but too high for many practical applications. For oil-lubricated tests with Ag-coated flats, the friction coefficient showed a decrease from the initial value of about 0.15 to a steady value of about 0.05 at 10N load and about 0.08 at 50N load (50N data not shown in the figure). Also, the steady friction was achieved more quickly (100 cycles) at 50N load than at 10N load (800 cycles). The steady friction levels for oil-lubricated, Ag-coated tests were comparable to or even lower than those typically obtained in lubricated metallic surfaces under boundary conditions [3, 12, 13].

The frictional behaviors of Si_3N_4 sliding pairs at temperatures of 150°C and 250°C were similar to one another. The Ag coating under dry conditions was more effective in reducing friction at these higher temperatures. In fact, at 250°C the friction coefficients for Ag-coated flats tested dry and oil-lubricated, uncoated flats were comparable (~ 0.25, cf. Figure 1b). The lowest friction coefficient of about 0.09 for Si_3N_4 material was still obtained at high temperatures with the oil-lubricated, Ag-coated flats.

The frictional behavior of ZrO_2 sliding pairs is shown in Figure 2. At room temperature, the friction coefficient in all the dry tests showed a quick rise from an initial value of about 0.15 to a steady value of 0.6 for uncoated flats and 0.3 for Ag-coated flats (Figure 2a). For the Au-coated flats, the friction coefficient showed only a slight increase to about 0.2. After about 1,400 cycles, however, a steady increase (to ~ 0.3 at the end of the test) was observed. This point of transition coincided with the point at which failure in the Au coating was observed. For oil-lubricated, uncoated flat tests, the friction coefficient remained virtually constant at about 0.14 throughout the tests (Figure 2a). Ag and Au coatings produced virtually identical results with oil lubrication at room temperature (notice that the data for both conditions coincide in figure 2a). Sliding started with a friction coefficient of about 0.1 but this decreased gradually to a steady value of about 0.05 after 100 cycles. In general, the patterns of behavior for ZrO_2 ceramics were similar to those of Si_3N_4 under the various conditions at room temperature. The higher temperatures produced behavior very similar to that at room temperature (Figure 2b). No dry tests were run with Au films at the higher temperatures because of the quick failure of the films during sliding contact.

Wear

The variations in the "instantaneous" wear rate in the Si_3N_4 and ZrO_2 pins are shown in Figure 3. Both the Ag coating and oil lubrication when applied separately reduced the wear rate of Si_3N_4 pins (Figure 3a) by more than an order of magnitude at room temperature. At 250°C, about a factor of 1000 reduction in the wear rate was observed. Further, with Ag coating tested dry, the wear rate at 250°C was lower than that at room temperature by about an order of magnitude; the reverse was the case with oil lubrication without coating. In all the oil-lubricated tests with Ag coating, no measurable wear was observed on the pins.

For the ZrO_2 material (Figure 3b), a significant reduction in the pin wear rate was also achieved by oil lubrication without coating. The reduction was more pronounced at room temperature. In addition, the wear rate increased with the number of cycles for the unlubricated tests but it decreased for the lubricated tests. In all the coated cases tested dry and lubricated, no significant wear was observed on the ZrO_2 pins.

Microscopic analysis of the worn surfaces of both materials showed the presence of a film-like layer consisting of very fine particles of debris for all the tests done under dry conditions. With coated flats, transfer of Ag or Au onto the pin surface was observed in most cases tested dry. This transfer was more pronounced with the ZrO_2 pins, and may be responsible for the lack of pin wear when sliding on the coated flats. With oil lubrication, no debris accumulation nor Ag or Au transfer to the pin was observed. A smooth contact area was generated on the Ag- or Au-coated flats with oil-lubricated tests. Also EDS analysis showed the presence of S on all the Ag-coated, oil-lubricated flats. The S is presumably from the oil, since it is a component of the additives in the oil. For coated flats tested at 250°C, with oil lubrication, a new surface layer was observed on top of the Ag coating. WDS analysis showed the layer to be rich in C. This new layer is believed to be deposits formed from the oil lubricant.

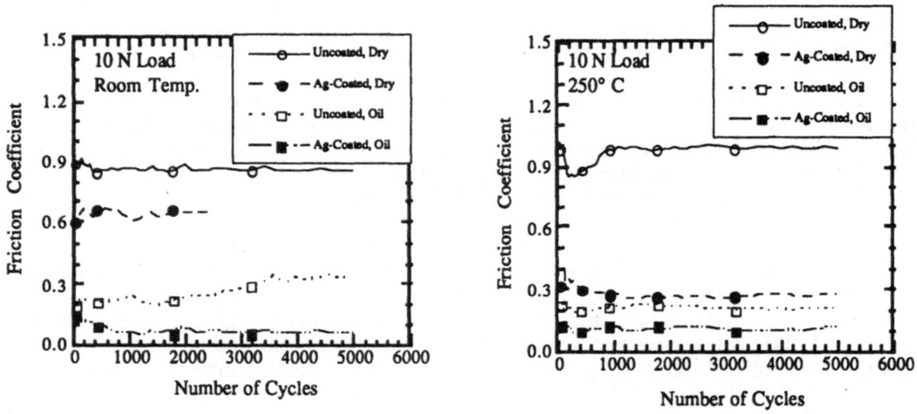

(a) (b)

Figure 1: Frictional behavior of Si_3N_4 ceramic sliding pairs tested at (a) room temperature and 10 N load (b) 250°C and 10 N load

DISCUSSION

The effectiveness of a liquid lubricant film is often gauged by the λ ratio, defined as the ratio of film thickness (h) to the composite surface roughness of the sliding surfaces (σ), i.e. $\lambda = h/\sigma$. The composite surface roughness $\sigma = (\sigma_1^2 + \sigma_2^2)^{0.5}$ where σ_1 and σ_2 are the roughnesses of the two contact surfaces. When $\lambda < 1.5$, extensive contact of asperities of the two surfaces through the lubricant film is expected (boundary conditions). For $\lambda \geq 3$, the sliding surfaces are completely separated by the lubricant film (hydrodynamic regime). The elastohydrodynamic and mixed-lubrication regime is characterized by $1 \leq \lambda \leq 3$.

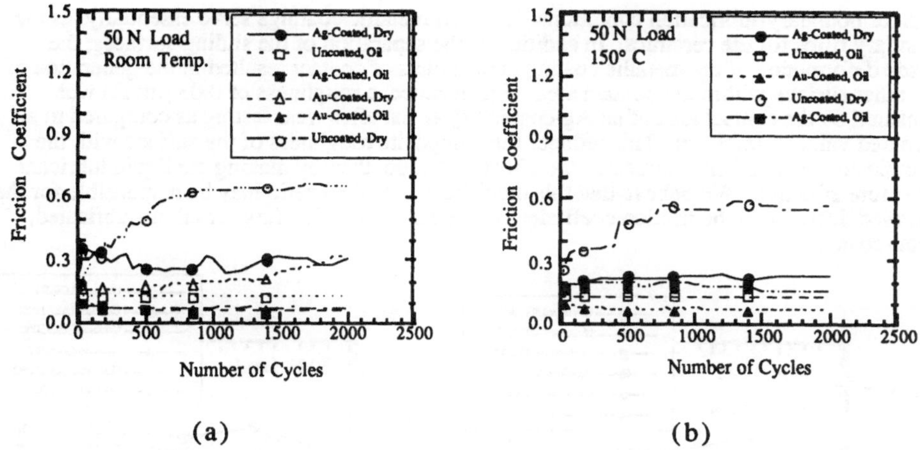

Figure 2: Frictional behavior of ZrO_2 ceramic sliding pairs tested at (a) room temperature and 50 N load (b) 150°C and 50 N load.

For point contacts (similar to what was used in the present study) under isothermal, steady-state and fully flooded conditions, the minimum film thickness (h_m) according to the Hamrock and Dowson formulation [19, 20] is given as:

$$\frac{h_m}{R} = 1.118 \left(\frac{\eta_o\, v}{E^* R}\right)^{0.68} [\alpha E^*]^{0.49} \left(\frac{W}{E^* R^2}\right)^{-0.073} \qquad (1)$$

where $E^* = 2\left(\frac{1 - \nu_1^2}{E_1} + \frac{1 - \nu_2^2}{E_2}\right)^{-1}$ is the reduced elastic modulus

η_o = viscosity under atmospheric pressure, v = sliding speed, R = radius of curvature of the pin, W = normal load and α = pressure viscosity coefficient. Due to lack of data, the α was estimated by Cameron [21] empirical relationship; $\alpha = (0.6 + 0.965 \log_{10} \eta_o) \times 10^{-8}$ where α is in Pa^{-1} and η_o is the atmospheric viscosity in cP. Oil viscosities at test temperatures of 150°C and 250°C were estimated by appropriate extrapolation of viscosity data at 40°C and 100°C supplied by the manufacturer.

The lubricant film thicknesses estimated using the above equations for the various test conditions in the present study are shown in Table 1. Also, the composite roughness σ in the present study is estimated to be about 0.26 µm for all tests. The λ ratios for the various test conditions are also shown in Table 1. The calculations show that the tests in the present study were under the boundary-lubrication regime since the initial λ is less than 1.5 in all cases. Furthermore, since wear occurred during oil-lubricated tests, the lubrication regime cannot be hydrodynamic. Also for reciprocating sliding, the velocity is zero at ends of travel and hence the lubricant film thickness is expected to be zero at such locations and as such, direct contact between the two sliding surfaces will occur.

Results of the present study show clearly that the incorporation of a thin soft metallic film at the contact interface of ceramic materials during liquid lubrication is an effective way of reducing friction and virtually eliminating wear of the pins in these reciprocating pin-on-disc tests for boundary-lubrication conditions. The soft metallic coatings serve essentially as the boundary films for the ceramics. In addition to the separation of the sliding surfaces, the plastic deformation of the metallic coating at the points of contact resulted in the generation of a smoother surface within the contact area. For instance, a roughness of 0.03 μm Ra was measured on the contact area of an Ag-coated Si_3N_4 flat after wear testing as compared to an uncoated value of 0.23 μm. This reduces the composite roughness of the surface with the metallic coating and hence increases the effective λ ratio, thereby making the liquid lubricant film more effective. We believe that this modification of the λ ratio may be responsible for the observed decrease in the friction coefficient in the early part of sliding for all oil-lubricated, metal-coated flat tests.

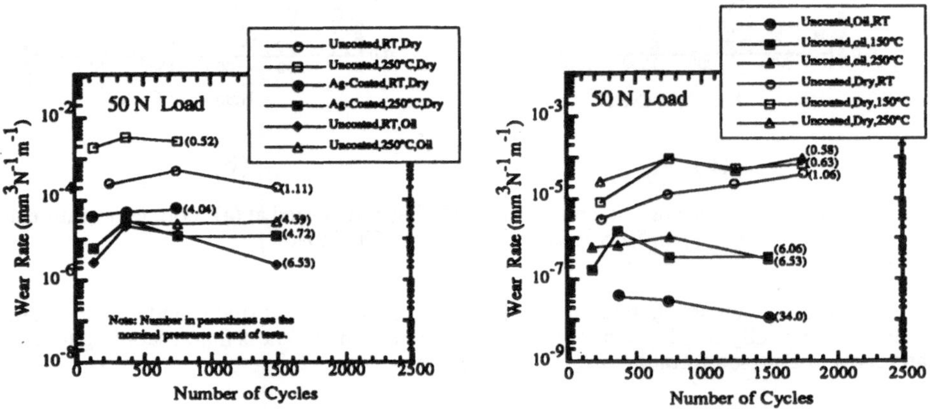

(a) (b)

Figure 3: "Instantaneous" pin wear rate as a function of the number of cycles under the various test conditions for (a) Si_3N_4 material (b) ZrO_2 material. (the pressures are in MPa)

At higher temperatures, the metal coatings get softer and shear even more easily. This is presumably responsible for the further lowering of the friction under dry contact with coatings at higher temperatures. On the other hand, the viscosity of the oil decreases very rapidly with increasing temperature, permitting more contact between the two surfaces. The result is an increase in friction and wear for oil-lubricated tests without metallic coatings at higher temperatures (Figure 3b). These opposing effects of temperature on the oil lubrication and soft-metal lubrication appear to approximately annul each other in oil-lubricated, metal-coated, sliding contact.

After prolonged exposure to the oil, chemical interaction between the Ag coatings and S from the oil did occur according to the EDS analysis. This could be a drawback to practical implementation of the findings of this study, since S is a very common component in additives to most oil lubricants and Ag seems to be the best metallic-coating candidate from the viewpoint of the desired tribological roles for the metallic coating [22]. Au is a potential candidate material, but the adhesion between Au and the ceramic substrate is often poor, even with IBAD [18].

Another problem encountered in this study was the formation of carbon-rich deposits on the metallic coatings at high temperatures. This will definitely modify the contact interface properties and could render the coating less effective with time. This is of particular concern for ceramics, since they are being considered primarily for high-temperature tribological applications.Efforts are currently under way to address these problems. Oils can be formulated with additives to protect the Ag films even in the presence of S. One such additive is an alkyl derivative of 2,5-di-mercapto 1,3,4-thiadiazole [23]. Furthermore, the S content of new oils may well be reduced in order to decrease S-containing emissions as required by new U.S. standards.

CONCLUSIONS

Results of this study showed that a viable means of boundary lubrication of ceramic materials is the deposition of a soft adherent metallic film at the sliding interface. The friction coefficient was reduced to a level comparable to boundary-lubrication conditions in many practical devices and the wear was virtually eliminated in Si_3N_4 and ZrO_2 sliding pairs by the synthetic-oil lubrication of Ag- and Au-coated surfaces. The improvement is brought about by the metallic coating acting essentially as a boundary film, thereby preventing direct contact between the two surfaces even under boundary-lubrication conditions. The deformation of the metallic coatings leads to the reduction of the surface roughness and thus increases the λ ratio, making the lubricant film more effective. Chemical interaction between the Ag coating and S from the oil and unwanted deposit formation from the oil at high temperatures are problems that need to be overcome through improved oil formulations before a practical implementation of the findings in the study can be achieved.

Table 1: Oil Lubricant Film Thickness and λ Ratio Calculated from EHD Equations

Temperature	Material	10N	50N
RT	Si_3N_4	h = 0.111 μm	0.099 μm
		λ = 0.43	0.38
	ZrO_2	h = 0.115 μm	0.102μm
		λ = 0.44	0.39
150°C	Si_3N_4	h = 5.57 nm	4.69 nm
		λ = 0.02	0.018
	ZrO_2	h = 5.49 nm	4.88 nm
		λ = 0.02	0.019
250°C	Si_3N_4	h = 0.94 nm	0.835 nm
		λ = 0.004	0.003
	ZrO_2	h = 1.026 nm	0.912 nm
		λ = 0.004	0.004

REFERENCES

1. R. Kamo and W. Bryzik, "Advanced Adiabatic Engine", SAE 840428.

2. H. Kawamura, "Development Status of Isuzu Ceramic Engine", SAE 880011, SP-738.

3. K. F. Dufrane, "Wear Performance of Ceramics in Ring/Cylinder Applications", *J. Am. Ceram. Soc.*, 72 (1989), 691-695.

4. R. Kamo and W. Bryzik, "Uncoated, Unlubricated Diesel?", Automotive Eng., 87 (6), (1979), 59-61.

5. S. Timoney and G. Flynn, "A Low Friction, Unlubricated SiC Diesel Engine", SAE 830313.

6. D. C. Cranmer, "Friction and Wear Properties of Monolithic Silicon-based Ceramics", *J. Mater. Sci.*, 20 (1985), 2029-2037.

7. J. Breznak, E. Breval and N. H. Macmillan, "Sliding Friction and Wear of Structural Ceramics", *J. Mater. Sci.*, 20 (1985), 4657-4680.

8. A. Skopp, M. Woydt and K. H. Habig, "Unlubricated Sliding Friction and Wear of Various Si_3N_4 Pairs Between 22° and 1000° C", *Trib. Int.*, 23 (1990), 189-199.

9. P. Sutor and W. Bryzik, "Development of Advanced High-Temperature Liquid Lubricants", SAE 880015.

10. A. Erdemir, G. R. Fenske, F. A. Nichols and R. A. Erck, "Solid Lubrication of Ceramics Surfaces by IAD-Silver Coating for Heat Engine Application", *STLE Trans.*, 33 (1990), 511-518.

11. H. Uetz, M. A. Khosrawi and J. Fohl, "Mechanism of Reaction Layer Formation in Boundary Lubrication", *Wear*, 100 (1984), 301-313.

12. A. F. Alliston-Greiner, J. A. Greenwood and A. Cameron, "Thickness Measurement and Mechanical Properties of Reaction Films Formed by Zinc Dialkyldithiophosphate During Running", *Proc. I. Mech. E.*, c178 (1987), 565-572.

13. J. M. Georges, J. M. Martin, T. Mathea, P. H. Kapsa, G. Meille and H. Montes, "Mechanism of Boundary Lubrication with Zinc Dithiophosphate", *Wear*, 53 (1979), 9-34.

14. Y. Tanita, T. Mine and K. Nakajima, "Tribological Reaction Generated on Ceramic-Steel Couples Under Boundary Lubrication", *ASME J. Trib.*, 112 (1990), 637-642.

15. P. Studt, "Influence of Lubricating Oil Additives on Friction of Ceramics Under Conditions of Boundary Lubrication", *Wear*, 115 (1987), 185-191.

16. F. P. Bowden and D. Tabor, The Friction and Lubrication of Solids I, Oxford University Press (1958), 181.

17. J. Gerkema, "Lead Thin Film Lubrication", *Wear*, 102 (1985), 241-252.

18. R. A. Erck, A. Erdemir and G. R. Fenske, "Effect of Film Adhesion on Tribological Properties of Silver-coated Alumina", *Surf. Coatings Tech.*, 43/44 (1990), 577-587.

19. G. Dalmaz and J. P. Chaomleffel, "Elastohydrodynamic Lubrication of Point Contacts for Various Lubricants", Proc. 13th Leeds-Lyon Symposium on Tribology, Elsevier Publishers (1987), 207-218.

20. B. J. Hamrock and D. Dowson, "Isothermal Elastohydrodynamic Lubrication of Point Contacts", Trans. ASME F, 99 (1077), 264-276.

21. A. Cameron, Basic Lubrication Theory, 3rd Ed., Ellis Horwood Publishers (1981), 30.

22. O. O. Ajayi, A. Erdemir, R. A. Erck, G. R. Fenske and F. A. Nichols, "The Role of Soft (Metallic) Film in Tribological Behavior of Ceramic Materials", Wear of Materials, ASME (1991), 257-263.

23. C. M. Larson and R. Larson, "Lubricant Additives" in Standard Handbook of Lubrication Engineering, McGraw-Hill Book Co. (1968).

TRIBOPOLYMERIZATION AS A NOVEL APPROACH TO CERAMIC LUBRICATION

Michael J. Furey* and Czeslaw Kajdas
*Professor, VPI&SU, Blacksburg, VA 24061, U.S.A.
Professor, Technical University at Radom, Poland

ABSTRACT

It has been demonstrated that the principle of tribopolymerization developed by Furey and Kajdas can be used as a novel and effective approach to designing specific molecular structures for the lubrication of ceramic materials. We define tribopolymerization as the planned and continuous formation of protective polymeric films directly on tribological surfaces by the use of selected monomers capable of forming polymer films "in situ". Over 40 experimental compounds selected, developed, or synthesized on the basis of this new concept were investigated in high stress sliding contact tests with various ceramic systems, i.e., alumina, zirconia, silicon nitride, and sapphire. Several of these were found to be strikingly effective as anti-wear additives. It was found that the addition of minor amounts (i.e., 1% or less) of particular compounds to a fluid hydrocarbon carrier, hexadecane, reduced ceramic wear by 40 to 80%. A reduction of wear by 80% corresponds to a five-fold increase in life.

We believe that the results are particularly significant since (a) ceramic materials are being used increasingly in a wide variety of mechanical devices and machines (e.g., high temperature engines) and (b) conventional approaches to the lubrication of ceramics are often limited or ineffective.

This paper summarizes the concept of tribopolymerization as a mechanism of boundary lubrication and discusses some of the more important discoveries made in applying this concept to the difficult problem of ceramic lubrication. The research was conducted in the Tribology Laboratory at Virginia Polytechnic Institute and State University.

TRIBOLOGY AND CERAMICS

Tribology (from the Greek "tribo", to rub) is the study of friction, wear, and lubrication [1]. Tribological processes are involved whenever one solid slides or rolls against another, as in bearings, cams, gears, piston rings and cylinders, machining and metalworking, grinding, brakes, magnetic recording devices, articulation in human synovial joints, and in a wide variety of other mechanisms, devices, engines, and machinery.

For tribological applications, ceramics offer several advantages over conventional materials, for example: they can be used at much higher temperatures, they are relatively inert and thus do not corrode, they are hard and more resistant to abrasive or erosive wear, and some ceramics are lighter in weight than alloy steels commonly used in engines and machines.

Examples of applications of ceramic materials in tribology include: (a) ceramic engines for higher temperature operation and greater thermodynamic efficiency, (b) advanced propulsion systems, turbomachinery, gas turbines, aerospace bearings, (c) automotive engine components, (d) cutting and machining of difficult alloys, (e) biomedical (e.g., artificial joints), (f) ceramic heads for magnetic recording, and (g) any tribological system operating under high temperature, abrasive, or corrosive conditions.

Conventional approaches to the lubrication of ceramics are limited or often ineffective. These include:

* Incorporation of various materials (e.g., polymers, solid lubricants) into the ceramic. [The added components can degrade the high-temperature performance of the ceramic.]

* Surface treatments and coatings (e.g., ion implantation, metallic films). [Coatings or surface films are themselves removed by wear, thus providing limited protection.]

* Dispersed solid lubricants (e.g., graphite, MoS_2) in a fluid. [This can lead to problems such as settling or filter plugging; in addition, high concentrations are often required.]

* Conventional lubricants containing soluble anti-wear additives in a carrier fluid (e.g., mineral or synthetic oils.) [Since many conventional anti-wear or anti-friction additives function by chemically reacting with a metal such as steel to form a surface layer, these compounds would not be expected to act with a non-reactive substrate (e.g., ceramic).]

TRIBOPOLYMERIZATION AS A MECHANISM OF BOUNDARY LUBRICATION

Tribopolymerization is defined as the planned or intentional formation of protective polymeric films directly and continuously on rubbing surfaces to reduce damage and wear by the use of minor concentrations of selected compounds capable of forming polymer films *in situ*. This new concept of boundary lubrication was put forth originally by Furey [2,3] and refined further in more recent collaborative studies with Kajdas [4-7].

The earlier work by Furey showed that certain compounds capable of forming polyesters directly on rubbing surfaces were enormously effective in reducing wear and increasing load-capacity with a variety of high contact stress (and high wear) systems, including gears, fuel pumps, and valve train components of automotive engines. All of this work involved steel and/or alloy cast iron systems.

Since tribopolymerization is a process of <u>deposition on</u> a solid surface rather than <u>reaction with</u> the surface (e.g., as is the case with most conventional anti-wear additives), we hypothesized that the mechanism would be ideal for ceramics. It is believed that tribopolymerization can be initiated by high surface temperatures as well as by catalytic effects and exoelectron emission. Possible models of tribo-polymerization as an anti-wear mechanism are discussed in a recent paper [6]. The hardness of most ceramics leads to a small real area of contact in tribological processes. In addition, many ceramics have a low thermal conductivity as compared to that of steel, for example. These two factors can lead to very high surface temperatures at the ceramic interface, thus promoting the formation of tribopolymer films. The films formed reduce contact stresses, adhesion, and wear in a continuous formation/ removal/replenishment process. The process of tribopolymerization or "in situ" formation of protective polymeric films is illustrated in an oversimplified view in Figure 1. The detailed mechanism is more complex.

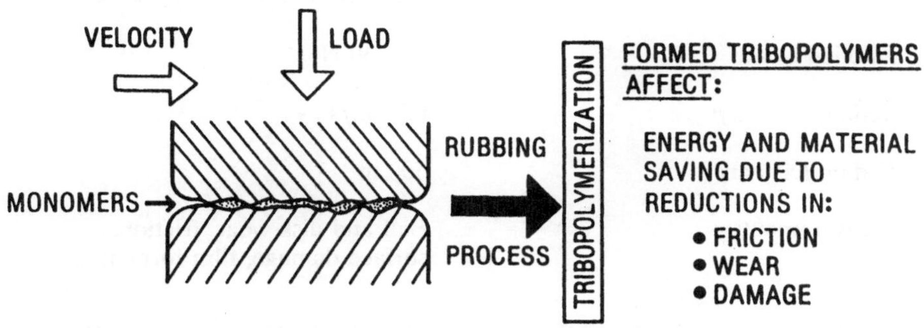

Figure 1. Tribopolymerization as a mechanism of boundary lubrication [5].

EXPERIMENTAL EXPLORATORY STUDY

In an exploratory study of the possible application of this new approach to ceramic lubrication, tests were carried out in a high contact stress system (pin-on-disk) using various ceramic combinations. The test conditions, summarized in Table 1, are quite severe. They are designed to emphasize the important regime of "boundary" lubrication in which chemistry and surface chemistry play key roles.

Using these conditions, experiments were carried out with a variety of systems. These included alumina-on-alumina (Al_2O_3), zirconia-on-zirconia (ZrO_2), silicon nitride-on-silicon nitride (Si_3N_4), and sapphire-on-sapphire (Al_2O_3). The alumina (99.5% purity, isostatically pressed), zirconia (Nilcra PSZ-MS Grade), and silicon nitride (Norton/TRW NT154) disks were 1/4 inch thick--cut from 1-inch diameter cylinders using a diamond cutting wheel and ground to an average surface roughness of 20-25 microinches (ca. 0.50-0.62μ). The sapphire disks (1 mm thick) consisted of optical flats (1/4 wavelength) cut and polished from single crystal sapphire [General Ruby and Sapphire]. The balls were of the same composition and source (for alumina, zirconia, and sapphire), 1/8 inch in diameter, polished, and ranged from Grade 5 to Grade 25--depending on the ceramic. All test specimens were ultrasonically cleaned and stored dry until use.

In an initial exploratory study, a pure hydrocarbon, hexadecane, was used as the carrier fluid for several chemical compounds developed as anti-wear additives for

TABLE 1
Experimental Set-Up and Test Conditions Used in an Exploratory Study of Ceramic Lubrication

System:	Ceramic A-on-Ceramic A
Geometry:	Sphere-on-flat (Fixed ball on rotating disk)
Specimen Size:	1/8 inch ball on 1 inch diam. disk
Applied Load:	10 - 30 N
Sliding Velocity:	0.25 m/s
Sliding Distance:	500 m
Measurements:	Ball and disk wear, friction, surface damage, film formation

ceramics. Wear on the test specimens--balls and disks--was determined by the use of photomacrography and stylus methods for measuring surface topography. For example, the volume of material worn from the ball can be calculated from the diameter of the spherical segment removed. Likewise, material worn from the disk can be calculated from a cross-sectional profile of the wear track multiplied by the track circumference.

As an indication of the severity of these conditions, for example, an applied load of 20N in this pin-on-disk system corresponds to a calculated mean Hertz (elastic contact) pressure of 2.0 to 3.3 GPa or approximately 290,000 to 470,000 psi for the ceramics studied. For comparison, the calculated elastic contact pressure on the highly-loaded cam nose of a typical V-8 automotive engine valve train is roughly 1 GPa or 140,000 psi. In addition, the surface temperatures generated under these conditions can be very high--ranging from a few hundred to as high as 1000°C.

Using our new concept of boundary lubrication, several compounds were designed, synthesized, or selected for application to the lubrication of ceramics. In a comprehensive study to determine whether the new approach would work, over forty compounds were examined. The results were quite remarkable. Thirty of these reduced ceramic wear significantly and of these, ten were outstanding--causing wear reductions of over 70% at relatively low concentrations (i.e., 1% or less).

An example of the influence of one experimental compound (A-29) on ceramic wear is shown by the data in Table 2. These data were obtained in pin-on-disk wear tests with alumina-on-alumina.

It can be seen that at a concentration of only 0.02%, the experimental compound reduces total ceramic wear by over 80%--a remarkable reduction.

TABLE 2
Effect of Compound A-29 on Ceramic Wear with Alumina-on-Alumina*

Concentration of A-29 in Hexadecane	Wear($10^{-3}mm^3$) Disk	Ball	Total	Wear Reduction
0.0% (Base case)	360	110	470	---
0.02%	48	36	84	82%
0.10%	73	23	96	80%
1.00%	85	40	125	73%

*Load: 20N, sliding velocity: 0.25 m/s, sliding distance: 500 m

It was also found that certain compounds are more effective with one ceramic system than another. There is a specificity to the action of the antiwear compounds designed from the new concept as can be seen by the examples shown in Table 3.

TABLE 3
Effects of Various Experimental Compounds on Ceramic Wear*

Compound in Hexadecane	Wear Reduction With: Alumina	Zirconia	Silicon Nitride
1% A-1	47%	18%	No effect
1% A-10	73%	45%	No effect
1% A-29	73%	52%	65%
1% A-37	57%	36%	51%
1% P-4	74%	45%	No effect
1% M-1	48%	36%	72%

*Load: 20N, sliding velocity: 0.25 m/s, sliding distance: 500 m

Of the six compounds listed, all reduce wear with alumina and zirconia at 1% concentration in the carrier fluid; however, three of these have no effect on wear of silicon nitride. In addition, friction reductions up to 35-40% were observed. Experimental compounds A-29 and M-1 are quite effective as antiwear additives for all ceramic systems investigated thus far, including sapphire. Our research to date suggests that there are three main factors which are important in tribopolymerization as an anti-wear mechanism for ceramics, namely:

(a) those factors which may be involved prior to surface polymerization (e.g., structure of molecule, adsorption on surface, molecular orientation),
(b) those controlling the process of tribopolymerization (e.g., surface temperature, catalysis, exoelectron emission, competing surface reactions), and
(c) those having to do with after surface polymerization has occurred (e.g., structure of polymer film formed, film thickness, mechanical properties, adhesion, thermal stability).

In this study, several classes of potential tribopolymer-formers were investigated in a systematic way, including, for example, groups of compounds of the same chemical structure but varying only in chain length or molecular weight. Confirming past experience in tribopolymerization as an anti-wear mechanism, shorter-chain molecules of a given structure were generally found to be more effective than long-chain molecules. Furthermore, some studies were conducted with compounds of essentially identical chemical structure but with variations in the position of an attached group--variations which would be expected to influence the ease of tribopolymerization. This work showed that those structures expected to polymerize on sliding surfaces rather easily were effective in reducing wear, while those not expected to polymerize were ineffective. The detailed results of these and other studies related to this research will be presented in future papers.

CONCLUSIONS

1. It has been demonstrated that the concept of tribopolymerization developed by Furey and Kajdas is a novel and effective approach to reducing wear and surface damage with ceramics under conditions of high contact stress and surface temperature.

2. The basic approach is one of molecular design--involving the selection and/or design of specific molecules which will continuously form thin, deposited polymer surface films in critical regions of lubrication--thus reducing contact and wear. [It does not involve adding a polymer to a carrier fluid nor pre-coating tribological surfaces with polymers.]

3. Over 40 experimental compounds selected, developed, or synthesized on the basis of this new concept were investigated in pin-on-disk tests with various ceramic systems, i.e., alumina, zirconia, silicon nitride, and sapphire. Many of these were found to be outstanding as anti-wear additives.

4. The addition of 1% of several of these compounds to a fluid hydrocarbon carrier--hexadecane--resulted in wear reductions as high as 80%. [An 80% reduction in wear corresponds to a five-fold increase in life.]

5. Factors believed to be important in the formation of protective polymeric films "in situ" on ceramic surfaces include adsorption and orientation of molecules

on surfaces prior to polymerization as well as initiation of tribopolymerization by high surface temperatures, catalytic effects, or exoelectron emission.

6. As a result of this research, patent applications have been filed on a wide range of compositions for reducing wear and/or friction on ceramic surfaces; others are in process.

7. Future activities planned on tribopolymerization as a novel and potent approach to the lubrication of ceramics include additional basic research and development, testing of the concept in other ceramic systems, and discussions of collaborative research as well as possible potential marketing and licensing arrangements with industry.

8. Although the experimental work reported here involved the use of a well-characterized, hydrocarbon fluid carrier (i.e., hexadecane) in tests at ambient temperature, we believe that the basic concept is broadly applicable to a wide variety of conditions and operating temperatures for ceramic-on-ceramic or ceramic-on-metal systems.

Taken as a whole, it would appear that for ceramic lubrication via the mechanism of tribopolymerization, both surface temperature and catalytic or exoelectron effects are important factors in initiating surface polymer formation. But the entire process--including initial adsorption, molecular orientation, polymerization, removal, and replenishment--is complex and needs further study. In continuing research in this area, the emphasis will be on the development and testing of more refined models of tribopolymerization. We will also take advantage of our experience in various related areas, e.g., the use of an advanced infrared microscope system to measure surface temperatures produced by friction [8,9], new theoretical approaches to the above problem [10], computer modeling of molecular structures, and the application of surface analytical techniques, e.g., FTIRMA [5] and others, to identify and characterize polymer films formed on tribological surfaces.

We believe that as we continue to obtain fundamental information on the detailed antiwear mechanism involved in tribopolymerization, we will be able to "tailor-make" and predict lubrication performance for ceramic-on-ceramic or ceramic-on-metal systems under a wide range of conditions.

These studies are being continued. And the use of these and other experimental compounds for high-temperature vapor phase lubrication is also being explored.

ACKNOWLEDGEMENTS

The authors would like to express their appreciation to the following for their support of this initial exploratory research: Virginia Polytechnic Institute and State University, including the Center for Advanced Ceramic Materials, Research Division, College of Engineering, and Department of Mechanical Engineering; and Virginia's Center for Innovative Technology. We are also grateful to the following companies for their kind donation of materials to assist us in this research: Norton/TRW Ceramics for the silicon nitride cylinder used to make the 1-inch diameter disks and

Nilcra Ceramics, Inc., for the zirconia cylinder used for the same purpose.
In addition, the authors wish to thank Hamid Ghasemi and Roman Kempi nski for carrying out the tribological tests.

REFERENCES

1. Furey, M.J., Tribology. In Encyclopedia of Materials Science and Engineering, Editor-in-Chief M. B. Bever, Pergamon Press, Oxford, 1986, pp. 5145-5157.

2. Furey, M.J., The Formation of polymeric films directly on rubbing surfaces to reduce wear. Wear, 1973, **26**, pp. 369-392.

3. Furey, M.J., The *in situ* formation of polymeric films on rubbing surfaces. Proc. International Colloq. Polymers and Lubrication, Brest, Centre National de la Recherche Scientifique, Paris, No. 233, 1975, pp. 393-404.

4. Furey, M.J. and Kajdas, C., Tribopolymerization as a lubrication mechanism for high-energetic contacts of solids, 6th International Tribology Colloq., Technische Akademie Esslingen, Esslingen, F.R.G, January 12-14, 1988.

5. Furey, M.J., Kajdas, C., Ward, T.C. and Hellgeth, J.W., Thermal and Catalytic Effects on Tribopolymerization as a New Boundary Lubrication Mechanism, Plenary Lecture (MJF), 5th International Congress on Tribology, Helsinki, Finland, June 12-15, 1989; also published in Wear, 1990, **136**, pp. 85-97.

6. Furey, M.J. and Kajdas, C., Models of Tribopolymerization as an Anti-Wear Mechanism, Proceedings of the Japan International Tribology Conference, Oct. 29-Nov. 1, 1990, Nagoya, **II**, pp. 1089-1094.

7. Furey, M.J., Tribopolymerization as an anti-wear mechanism, Symposium on Surface Science Applications and Advances in Tribology, Division of Colloid and Surface Chemistry, 201st Nat'l Meet., American Chemical Society, Atlanta, GA, April 14-19, 1991.

8. Furey, M.J., Infrared measurements of surface temperatures produced in tribological processes. Proc. 3rd Int. Tribology Congr. (EUROTRIB 81), Warsaw, September 21-24, 1981, **I**, pp. 118-139.

9. Furey, M.J. and Jayaram, S., Advanced techniques in infrared measurements of surface temperatures produced by friction, Proc. of the Japan International Tribology Conference, Oct. 29-Nov. 1, 1990, Nagoya,, III, pp. 1569-1574.

10. Furey, M.J., Vick, B., Foo, S.J. and Weick, B.L., A theoretical and experimental study of surface temperatures generated during fretting, Proceedings of the Japan International Tribology Conference, Oct. 29-Nov. 1, 1990, Nagoya, **II**, pp. 809-814.

CERAMIC-CERAMIC COMPOSITE MATERIALS WITH IMPROVED FRICTION AND WEAR PROPERTIES

M. WOYDT, A. SKOPP and R. WÄSCHE
Federal Institute for Materials Research and Testing (BAM)
Unter den Eichen 87, W-1000 Berlin 45, Germany

ABSTRACT

Dry friction and wear tests were performed with self-mated couples of SiC with 50% TiC, Si_3N_4-BN, SiC-TiB_2 and Si_3N_4 with 30 % TiN at room temperature and 400°C or 800°C.

Under room temperature conditions the friction coefficient of the couple SiC-TiC/SiC-TiC is only half of that of the couple SiC/SiC and the wear is one order of magnitude smaller. At 400°C it exceeds the friction coefficient of SiC/SiC except at the highest sliding velocity of 3 m/s. At lower sliding velocities the wear coefficient of SiC-TiC/SiC-TiC is lower than that of SiC/SiC.

The couple Si_3N_4-TiN/Si_3N_4-TiN exhibits high friction coefficients with all test conditions. At room temperature the wear volume of the self-mated couples of Si_3N_4 and Si_3N_4-TiN after a sliding distance of 1000 m is similar, but Si_3N_4-TiN shows a running-in behaviour. At 800°C the wear coefficient of Si_3N_4-TiN/Si_3N_4-TiN is approximately two orders of magnitude smaller than that of Si_3N_4/Si_3N_4 and equal to those at room temperature.

At 22°C the addition of BN reduces the friction of Si_3N_4. The wear coefficient is independent of sliding velocity and the self-mated couples showing running-in. Friction and wear increase with increasing temperature.

The wear coefficient of SiC-TiB_2 above 0.5 m/s at 400°C is advantageously near 10^{-6} $mm^3/(Nm)$. With the other test conditions the wear behaviour is similiar to SSiC.

INTRODUCTION

The actual and future environmental legislation as well as the further increasing power density of machinery requires sliding couples for oilless or higher stressed tribosystems with the same tribological performance as in common lubricated systems today /1-3/. In the case of oilless tribosystems, the tribological goals to reach are the following:

a) Wear coefficient lower than $1 \cdot 10^{-6}$ $mm^3/(Nm)$;

b) friction and wear independent of sliding velocity;

c) friction and wear independent of ambient temperature;

d) friction coefficient lower than 0.15.

Sliding couples of ceramic materials like silicon carbide, silicon nitride, alumina or zirconia normally have high friction coefficients under unlubricated conditions. It has been observed that the friction coefficients of SiC and Si_3N_4 can be lowered by the presence of surface layers of silicon oxide or silicon hydro-oxide /4,5/. From investigations of surface coatings it is known that titanium carbide and titanium nitride exhibit low friction coefficients due to the formation of titanium oxide layers /6/. Additionally the friction coefficient of self-mated couples of bulk TiO_2 is in the same order of magnitude as for 50SiC-50TiC composite materials under room temperature conditions /8/.

According to theoretical considerations a titanium dioxide (TiO_{2-x}) with an oxygen deficiency should have a lower friction coefficient /9/ than stoichiometric TiO_2.

In this paper the friction and wear behaviour of silicon carbide and silicon nitride with additions of titanium carbide, titaniumdiboride and hexagonal boron nitride or titanium nitride is described. These additions were made with the aim to investigate the friction and wear behavior of unlubricated systems. For comparison results of friction and wear tests with SiC and Si_3N_4 are included.

Two main lubrication mechanism were investigated. First, the self-formation of a "lubricious oxide" by tribooxidation and second the incorporation of a solid lubricant in a ceramic matrix.

MATERIALS AND METHODS

For the sliding friction and wear tests a special high temperature tribometer was built (Fig. 1) /10/. The tribosystem consists of a pin-on-disc assembly. The stationary specimen is pressed against the rotating disc. The test conditions can be described as follows:

I Structure of the test system
- Tribo-element I: stationary specimen (pin)
- Tribo-element II: rotating specimen (disc)
- Surrounding atmosphere: air; 15 - 30 % rel. humidity at 22°C

II Operating variables
- Type of motion: continuous sliding
- Normal force, F_N: 10 N
- Sliding velocity, v: 0.03 - 6 m/s
- Temperature, T: 22 - 800°C
- Sliding distance, s: 1000 - 5000 m

III Tribological characteristics observed
- Friction coefficient, f
- Wear coefficient, $k = W_v/(F_N \cdot s)$
- Morphology and chemical composition of the worn surfaces

Friction force and wear were measured continuously by force and distance transducers /10/.

The SiC-TiC-specimens which contained 50 wt.-% TiC were made by hot pressing /8/. The Si_3N_4-TiN-specimens with 30 wt.-% TiN, the specimens of sintered silicon carbide (SSiC) and various Si_3N_4 materials which were tested for the purpose of comparison were provided by Elektroschmelzwerk Kempten GmbH, the SiC-TiB_2 composite came from the Carborundum company and the different Si_3N_4-BN samples were prepared by HTM AG in Biel, Switzerland.

The properties of the materials are listed in Table 1. The microstructures of SiC-TiB_2, Si_3N_4-BN, SiC-TiC and Si_3N_4-TiN are distinct two phase microstructures with a homogeneous distribution of the minor phase in the matrix (Fig. 2 and Fig. 3).

RESULTS

Titanium compound composites

Friction

The friction coefficients for the self-mated couples of SiC-TiC and SiC which were measured at room temperature are presented in

Fig. 4. With values between 0.2 and 0.3 the friction coefficient of SiC-TiC/SiC-TiC is approximately half of that of SiC/SiC. At 400°C the friction coefficient of SiC-TiC/SiC-TiC reaches high values of 0.8 whereas SiC/SiC shows a scatter of friction coefficients between 0.2 and 0.7 (Fig. 5). The low friction coefficients of SiC/SiC at this temperature are due to the formation of a thin oxide layer on the friction surfaces while the high friction coefficients were related to the formation of thicker oxide layers /21/.

At room temperature the friction coefficients of self-mated SiC-TiB_2 sliding couples shown in Fig. 6 are in the range of 0.56 and 0.72 and independent of the sliding velocity. At 400°C the friction coefficient is higher than 0.5 except in the minimum at 1 m/s.

The friction coefficients of the self-mated couples of Si_3N_4-TiN and 7 variations of Si_3N_4 are presented in Fig. 7 and Fig. 8 at room temperature and 800°C. By the addition of TiN to Si_3N_4 the friction coefficient is not decreased, but increased at room temperature for sliding velocities below 1 m/s and at 800°C for sliding velocities between 0.1 and 1 m/s.

Wear

The total volumetric wear coefficient of SiC-TiC and SiC-TiB_2 compared to various SiC-materials is presented in Fig. 9 at 22°C and 400°C. At room temperature only SiC-TiC shows a wear behaviour which is independent of sliding velocity with wear coefficients below 10^{-6} $mm^3/(Nm)$. The wear coefficient of SiC-TiC increases at 400°C with sliding velocities above 0.3 m/s by more than one order of magnitude whereas the wear coefficient of SiC-TiB_2 then decreased sharply with sliding velocity. SiC-TiC seems at 400°C to be the better material below 0.3 m/s whereas SiC-TiB_2 seems to be better above 0.3 m/s.

At room temperature the wear of Si_3N_4-TiN after a sliding distance of 1000 m is comparable to the wear of pure Si_3N_4, but at 800°C the Si_3N_4-TiN composite exhibits a two orders of magnitude lower wear volume than the pure Si_3N_4 (Fig. 10). The wear coefficient of Si_3N_4-TiN is independent of ambient temperature.

At 22°C the wear volume of self-mated Si_3N_4-TiN as a function of sliding distance is shown in Fig. 11. At 22°C and 800°C Si_3N_4-TiN shows a strong running-in behaviour of about 200 m, characterized by a distinct decreasing wear rate of one order of magnitude, whereas the wear volume of pure Si_3N_4 increased linearly with sliding distance /23/. The differential wear coefficient of the stationary pin of Si_3N_4 after 1000 m is lowered by the addition of TiN by more than a factor 10 (Fig. 16).

Surface analysis

In order to characterize the reasons for the different friction and wear behaviour related to operating conditions and materials composition, examinations with scanning electron microscopy (SEM), Cr-k_α small spot XRD, auger-electron-spectroscopy (AES) and small spot electron spectroscopy for chemical analysis (ESCA) were performed. The detailed results are published elsewhere /20/.

The results can be summarized as follows:

SiC-TiC

- SiO_2 is not detectable on the surfaces except after tribological operation at 400°C outside the wear track.
- Measurements with Small Spot ESCA indicate strongly the existance of an SiC_xO_y-phase outside and on the wear track. Its amount seems to be increased by tribological operation. Such a phase was described by Lavrenko et al. /12/ and Heuer and Lou /19/. As SiC is also identified, SiC_xO_y forms a thin layer or islands on SiC.
- The TiC grains are covered by TiO_2 before and after tribological operation. The TiO_2 layer is so thick that TiC cannot be detected by Small Spot ESCA.

Si_3N_4-TiN

- SiN_xO_y is found outside and on the wear track. The existence of this phase was described by Morgan et al. /13/ and Braue et al. /16/. Its amount seems to be increased by tribological operation.
- TiO_2 forms such a thick layer that TiN cannot be detected.

SiC-TiB_2

- The TiB_2 grains showed a preferential oxidation.

The investigation of the oxidation behavior of this composite material showed, that the TiB_2 grains habe a much higher oxidation rate than the SiC grains. It could be seen, that the surfaces of the TiB_2 grains were elevated relative to the surface of the SiC grains due to the formation of one or more phases probably containing TiO_2 and B_2O_3. Phase analysis is not completed yet.

Boron compound composites

Si_3N_4-BN

Three HIPped Si_3N_4 materials with 5, 10 and 20 wt.-% hexagonal boron nitride (BN) as intrinsic solid lubricant were tested.

Friction

Hexagonal BN is an effective "friction modifier" in Si_3N_4 at room temperature under solid state friction. The stationary friction coefficient lies between 0.1 and 0.3 depending on sliding velocity and BN content (Fig. 12). The friction coefficient decreases by 0.1, when the BN content is raised from 5 wt.-% to 20 wt.-%. With $80Si_3N_4$-20BN friction coefficients lower than 0.2 were observed for sliding velocities between 0.03 m/s and 3.5 m/s. Above ambient temperatures of 100°C the friction of Si_3N_4-BN increases significantly (Fig. 13).

Wear

At room temperature the wear volume of stationary and rotating Si_3N_4-BN specimen after a sliding distance of 1000 m is shown in Fig. 14. The wear volume of Si_3N_4-BN is independent of sliding velocity and overlap ratio. In general, the wear volume both for the rotating and the stationary specimen is slightly reduced by the addition of BN to Si_3N_4 at room temperature. The volumetric wear coefficient depends on ambient temperature (Fig. 13).

At 22°C the three Si_3N_4-BN materials show a strong running-in behaviour. As example the wear volume of the stationary specimen of Si_3N_4-20BN/Si_3N_4-20BN as a function of sliding distance is shown in Fig. 15. As for Si_3N_4-TiN the wear rate after a critical sliding distance is lowered by an order of magnitude. The differential wear coefficient of the stationary pin of Si_3N_4/Si_3N_4 after running-in is lowered by a factor 10 caused by the addition of TiN or BN to Si_3N_4 (Fig. 16).

A major problem with Si_3N_4-BN composites is the surface finishing. A further finishing treatment after a fine grinding operation smoothed the surface, but washed out the BN particles at the same time and thus caused an increased running-in wear. Through an optimized surface finishing treatment the stationary wear level can be lowered.

DISKUSSION

Friction and wear of self-mated couples of SiC-TiC and Si_3N_4-TiN are controlled by the nature of outer reaction layers which are probably composed of SiC_xO_y or SiN_xO_y and TiO_2. H_2O molecules can be adsorbed on these layers.

The amount of TiO_2 on the wear track is higher on SiC-TiC than on Si_3N_4-TiN. This may be caused by the higher concentration of TiC in SiC-TiC in comparison to the TiN concentration in Si_3N_4-TiN (TiC: 50 wt.-%, TiN: 30 wt.-%). The low friction coefficient of the couple SiC-TiC/SiC-TiC at room temperature may be caused by the TiO_2 and SiC_xO_y layers on which water molecules are adsorbed.

At 400°C TiO_2 looses its friction reducing effect /8/. SiC/SiC couples show also a high friction coefficient at room temperature if the concentration of H_2O molecules in the ambient medium is low /5/. Thin SiO_2 layers which can provide low friction coefficients were not clearly detected by Small Spot XRD. Therefore the high friction coefficient of SiC-TiC/SiC-TiC at 400°C is comprehensible /8/. The low wear of this couple in comparison to the couple SiC/SiC may be caused by the higher K_{IC}-values (see Table 1).

The self-mated couples of Si_3N_4-TiN show high friction coefficients under all test conditions. The amount of TiO_2 is probably too low to reduce the friction coefficient and Si_3N_4 always shows a high friction coefficient with the test conditions chosen /1/.

The low wear coefficients of Si_3N_4-TiN at 800°C may be caused by the formation of a smooth thin TiO_2 film on the wear track (see Fig. 17). The question arises wether an understoichiometric TiO_{2-x}-phase is responsible for the low wear rate.

For self-mated couples of Si_3N_4-BN the low friction coefficients at temperatures below 100°C are probably caused by a film formation of BN or $BN\cdot(H_2O)_x$ on the wear surface (Fig. 18). In humid atmospheres it was stated by Erdemir et al. /22/ that friction and wear are reduced by a $H_3(BO_3)$-film, which is destroyed in ambient atmosphere above 100°C.

It is further assumed, that the strong influence of ambient temperature is related to the oxidation of BN starting at 300°C. This hinders the formation of a BN surface film.

CONCLUSIONS

1. It has been shown, that the tribological properties of structural ceramic materials can be considerably improved by the addition of TiN, TiB_2, TiC and BN. The improvement of the tribological properties can even be associated by an improvement of the other mechanical properties as is shown for Si_3N_4-TiN compared with Si_3N_4 and SiC-TiC compared with SSiC.

2. By the addition of TiN to Si_3N_4 it is possible to design sliding materials for solid state friction with a wear coefficient nearly independent of sliding velocity and ambient temperature and with a running-in behaviour.

3. At room temperature TiC particles in SiC reduce the wear coefficient below 10^{-6} $mm^3/(Nm)$ and lowers the friction coefficient to 0.32. Friction and wear are independent of sliding velocity at 22°C.

4. The self-formation of "lubricious oxides" on hard substrates lowers friction and/or wear and is a good base for the further development of tribomaterials.

5. The addition of BN to Si_3N_4 is only tribological effective below 100°C in humid air. At low temperatures Si_3N_4-BN materials show running-in. The mechanical properties are deteriorated by the addition of BN.

ACKNOWLEDGEMENTS

The authors are grateful to Prof. X. Yu and Mr. D. Treu, who performed the ESCA analysis. Mrs. B. Strauß is gratefully acknowledged for the examinations by SEM. We thank Dr. H. Hantsche for his assistance in surface analytic problems. Mr. H. Tigges, Rigaku Europe G.m.b.H., is gratefully acknowledged for the Small Spot XRD measurements.

REFERENCES

/1/ T. Mang
"Environmental and place of employment friendly lubricants in view of the legislative"
Deutsches Schmierstoff-Forum 08/09.11.90, Frankfurt, Praxis-Forum, 1000 Berlin 22, Ritterfelddamm 82 h + i

/2/ D. Gruden
"Future requirements on motor oils"
Erdöl, Erdgas, Kohle 106 [1] (1990) 24 - 28

/3/ W. Pfalzer
The role of future emission regulations on the tribology of truck diesel engines
Automobilkreis "Entwicklungstendenzen tribotechnischer Werkstoffe im Kraftfahrzeug", 19./20. April 1990, Bad Nauheim, Praxis-Forum, 1000 Berlin 22

/4/ T. E. Fischer, H. Liang and W. M. Mullins
"Tribochemical lubrication oxides on Si_3N_4"
In: New Materials Approach to Tribology - Theorie and Applications; Editors: L. E. Pope and W. O. Winer
Materials Research Society Symposium Proceedings, vol. 140 (1989) 339 - 344

/5/ D. Klaffke and K.-H. Habig
"Fretting wear tests of SiC"
Wear of Materials 1987, Editor: K. C. Ludema, New York; The American Society of Mechanical Engineers (1987) 361 - 370

/6/ K.-H. Habig
"Friction and wear of sliding couples coated with TiC, TiN or TiB_2"
Surface and Coatings Technology 42 (1990) 133 - 147

/7/ H. Heshmat and H. Shapiro
"High-temperature, unbalanced, dry contact face seal, interfacial phenomenon and design considerations"
Lubrication Engineering 45 (1989) 235 - 243

/8/ M. Woydt, J. Kadoori, H. Hausner and K.-H. Habig
"Materials development of fine ceramics under tribological considerations"
Cfi/ Berichte der Deutschen Keramischen Gesellschaft 67

(1990) 123 - 130

/9/ M. N. Gardos
"The effect of anion vacancies on the tribological properties of rutile (TiO_{2-x})"
Tribology Transactions 32 (1988) 427 - 436

/10/ M. Woydt and K.-H. Habig
"High temperature tribology of ceramics"
Tribology International 22 (1989) 75 - 88

/11/ K. Miyoshi and D. H. Buckley
"Changes in surface chemistry of SiC (0001) surface"
NASA-TP-1756, Nov. 1980

/12/ V. A. Lavrenko, S. Jonas and R. Pampuch
"Petrographic and x-ray identification of phases formed by oxidation of silicon carbide"
Ceramics International 7 [2] (1981) 75 - 76

/13/ A. E. Morgan, E. K. Broadbent, K. N. Ritz, D. K. Sadana, B. J. Burrow
"Interactions of thin Ti films with Si, SiO_2, Si_3N_4 and SiO_xN_y under rapid thermal annealing"
Journal of Applied Physics 64 (1988) 344 - 353

/14/ C. D. Wagner, D. E. Passoja, H. F. Hillary, T. G. Kinisky, H. A. Six, W. I. Jansen, J. A. Taylor
"Auger and photoelectron line energy relationships in aluminium-oxygen and silicon-oxygen-compounds"
Journal of the Vacuum Science Technology 21 (1982) 933 - 944

/15/ C. D. Wagner, W. M. Riggs, L. E. Davis, J. F. Moulder and G. E. Muilenberg
"Handbook of x-ray photoelectron spectroscopy"
Perkin-Elmer Corporation, Eden Prairie, Minnesota (1979)

/16/ W. Braue and H.-J. Dudek
"XPS study of glassy grain-boundary layers in dense high strength Si_3N_4"
Journal of Non-Crystalline Solids 56 (1983) 185 - 190

/17/ T. Goto, F. Itoh, K. Suzuki and T. Hirai
"ESCA study of amorphous CVD Si_3N_4-C composites"
Journal of Material Science Letters 2 (1983) 805 - 807

/18/ D. V. Badami
"X-Ray studies of graphite formed by decomposing SiC"
Carbon 3 (1965) 53 - 57

/19/ A. H. Heuer and V. L. K. Lou
"Volatility diagrams for silica, silicon nitride, and silicon carbide and their application to high-temperature decomposition and oxidation"
Journal of the American Ceramic Society 73 [10] (1990) 2785 - 2803

/20/ M. Woydt, A. Skopp and K.-H. Habig
"Dry friction and wear of self-mated sliding couples of

SiC-TiC and Si_3N_4-TiN"
Int. Conf. on Wear of Materials, 07/11.04.1991, Orlando, Fl, ASME, Int.; New York

/21/ K.-H. Habig
Tribologisches Verhalten von Ingenieurkeramik
Ingenieur-Werkstoffe 1 [11/12] (1989) 78 - 83

/22/ A. Erdemir, G. R. Fenske, F. A. Nichols, R. A. Erck and D. E. Busch
Self-lubricating boric acid films for tribological applications
Proceedings of the Japanese International Tribology Conference, Nagoya, (1990), 1797 - 1802

/23/ A. Skopp, M. Woydt and K.-H. Habig
Unlubricated sliding friction and wear of various Si_3N_4 pairs between 22°C and 1000°C
Tribology International 23 [6] (1990) 189 - 199

Material	E [GPa]	m	δ_{4B} [MPa] 22 °C	K_{Ic} [MPa*m$^{1/2}$]	α (RT-1000°C) [10^{-6}/K]	HV 0,2 *) 22 °C	R_{pk} [μm]
Si_3N_4 (7 variations)	280-320	9-20	698-1050	5.6-8.2	3.1-3.7	1467-1741	0.011-0.034
70 Si_3N_4 30 TiN (EDM-Si_3N_4)	295	n.d.	735	9.0	5.7	1620	0.033
SSiC	410		410	3.2	4.0	2840	0.004
50 SiC 50 TiC (HP)	456	20	600	5.4	5.7	1700-3100	0.090
SiC-TiB_2 (ST)	427	12	448	8.0	4.0		0.122
Si_3N_4-20BN						200	0.400

*) DIN 50133

TABLE 1: Properties of the ceramic materials tested

Fig. 1: Ceramic samples at 1000°C in high temperature tribometer

Fig. 2: TEM-microstructures of SiC-TiC (left) and of Si_3N_4-TiN (right)

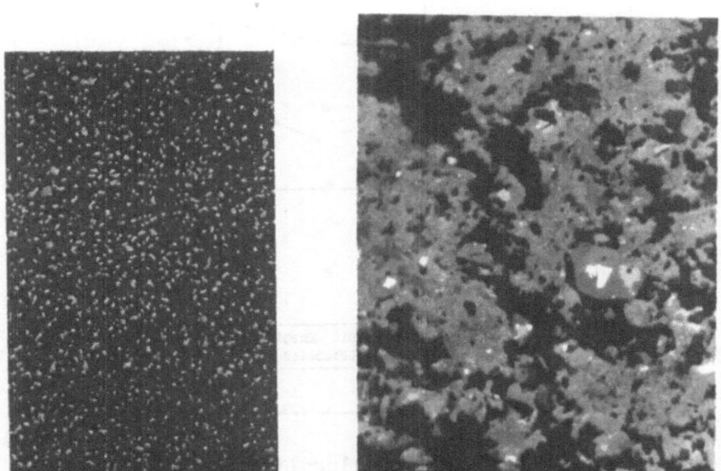

Fig. 3: Microstructures of Si_3N_4-TiB_2 (left) and Si_3N_4-BN (right)

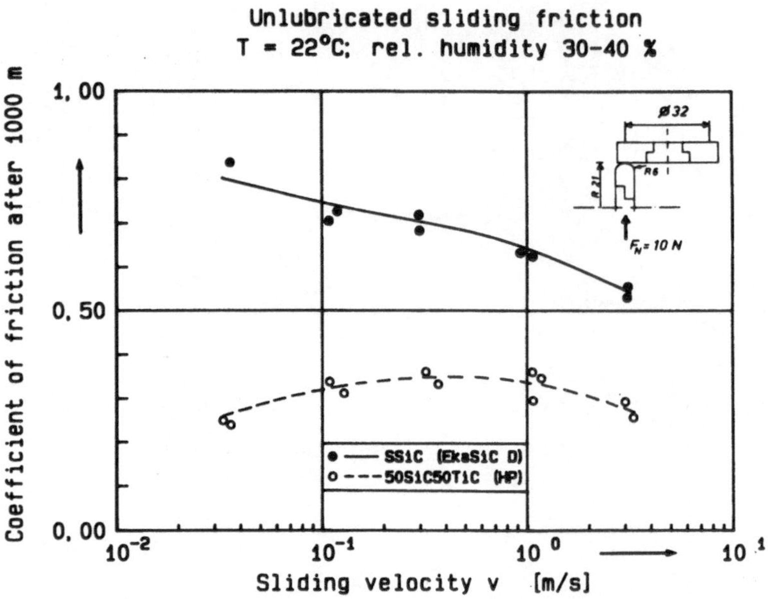

Fig. 4: Friction coefficient of SiC-TiC/SiC-TiC and SiC/SiC at room temperature under solid state friction

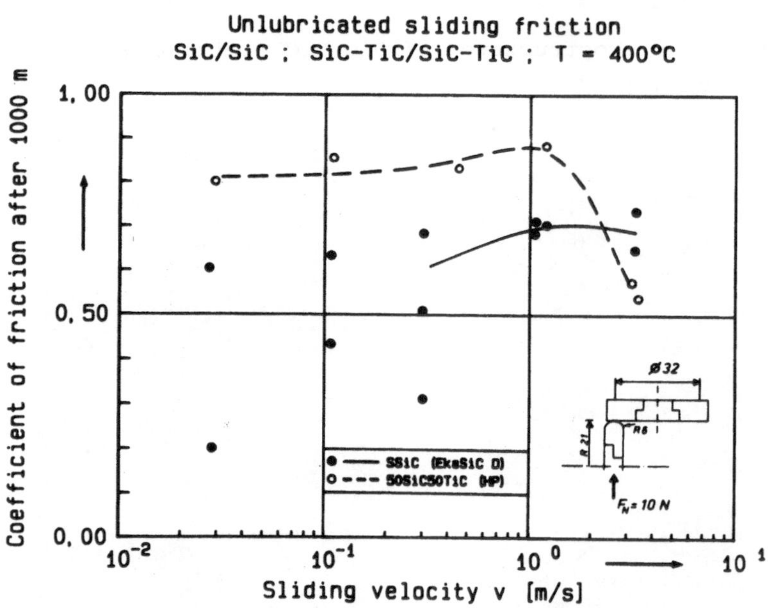

Fig. 5: Friction coefficient of SiC-TiC/SiC-TiC and SiC/SiC at 400°C under solid state friction

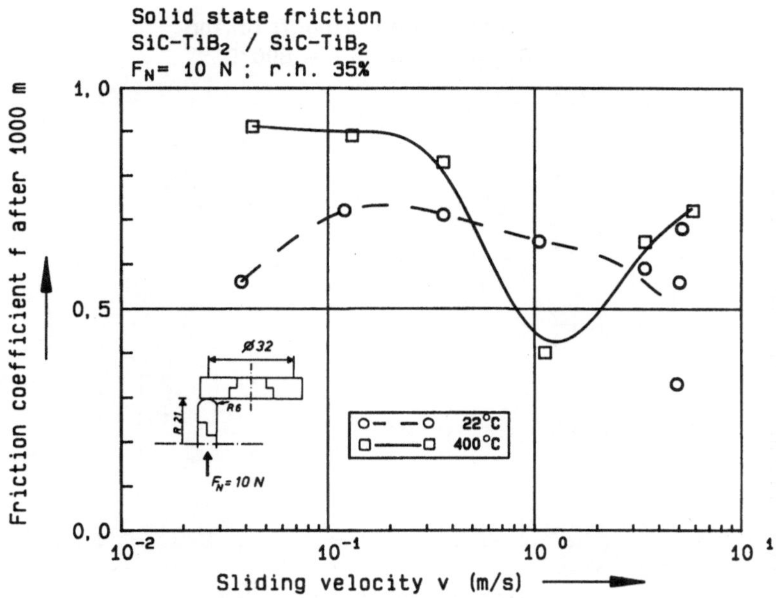

Fig. 6: Friction coefficient of $SiC-TiB_2/SiC-TiB_2$ at 22°C and 400°C under solid state friction

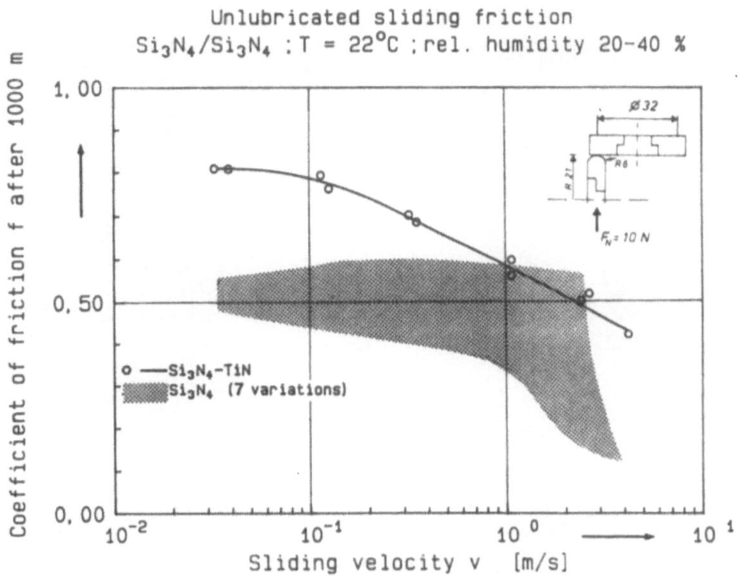

Fig. 7: Friction coefficient of Si_3N_4-TiN/Si_3N_4-TiN and Si_3N_4/Si_3N_4 at room temperature under solid state friction

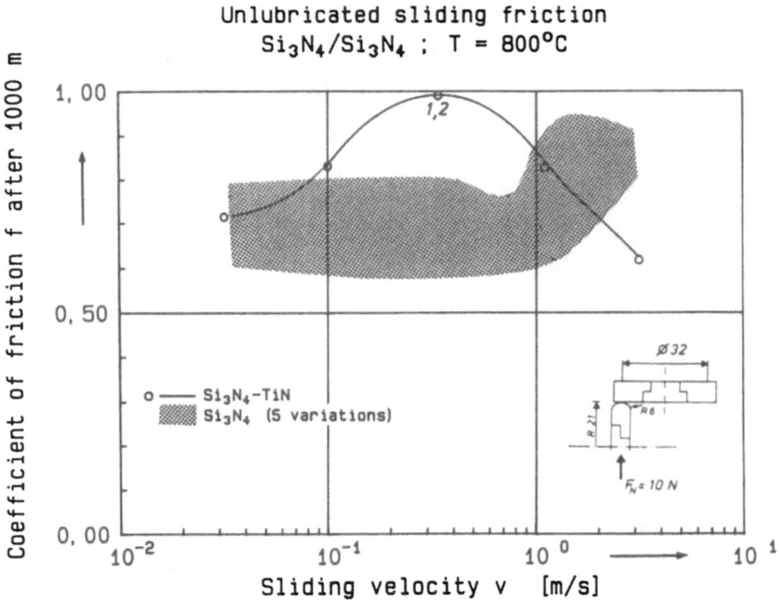

Fig. 8: Friction coefficient of Si_3N_4-TiN/Si_3N_4-TiN and Si_3N_4/Si_3N_4 at 800°C under solid state friction

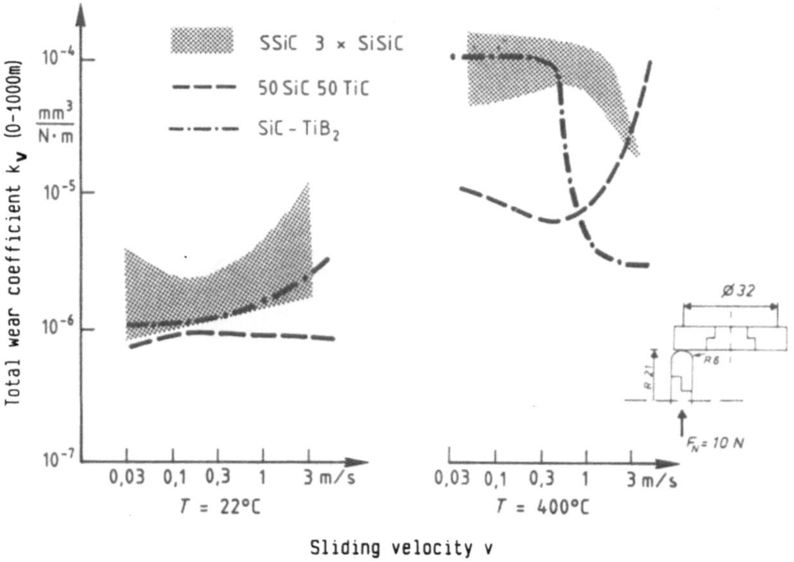

Fig. 9: Total, volumetric wear coefficient of self-mated SiC-materials under solid state friction at 22°C and 400°C

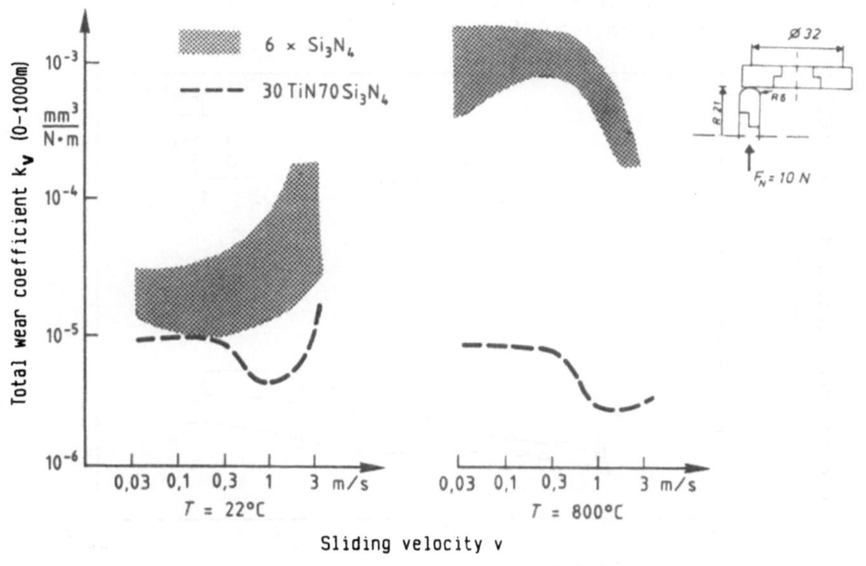

Fig. 10: Total, volumetric wear coefficient of self-mated Si_3N_4-materials under solid state friction at 22°C and 800°C

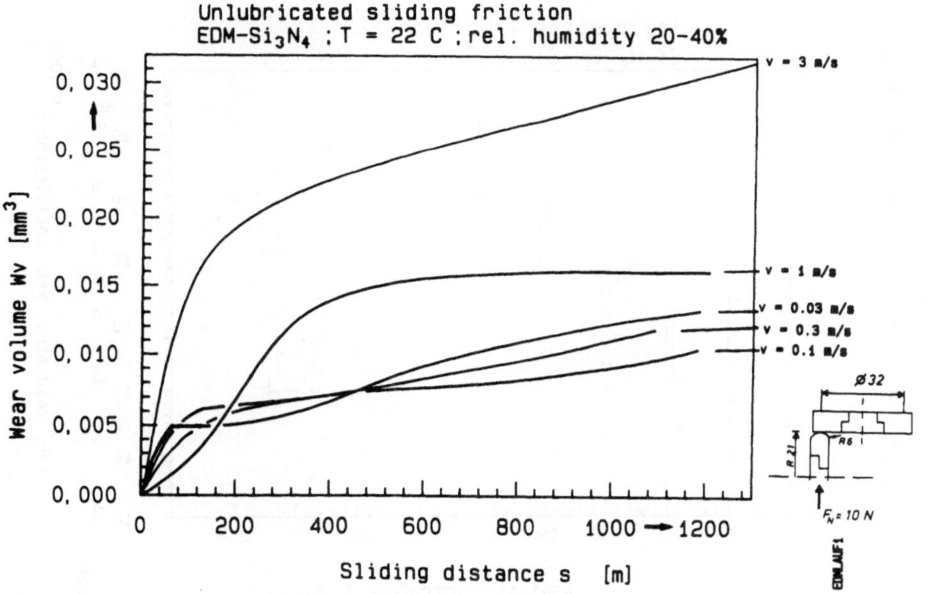

Fig. 11: Wear volume versus sliding distance of self-mated Si_3N_4-TiN pairs at 22°C under solid state friction

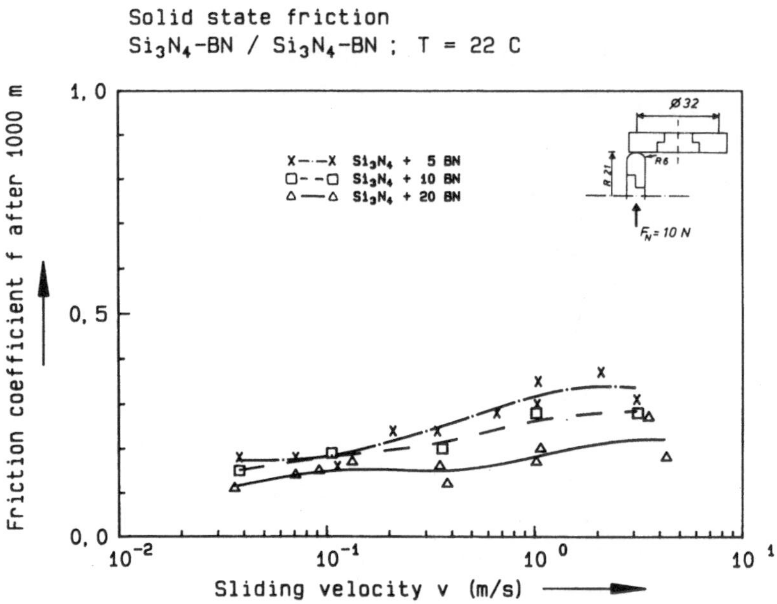

Fig. 12: Friction coefficient of self-mated Si_3N_4-BN couples at room temperature as a function of sliding velocity under solid state friction

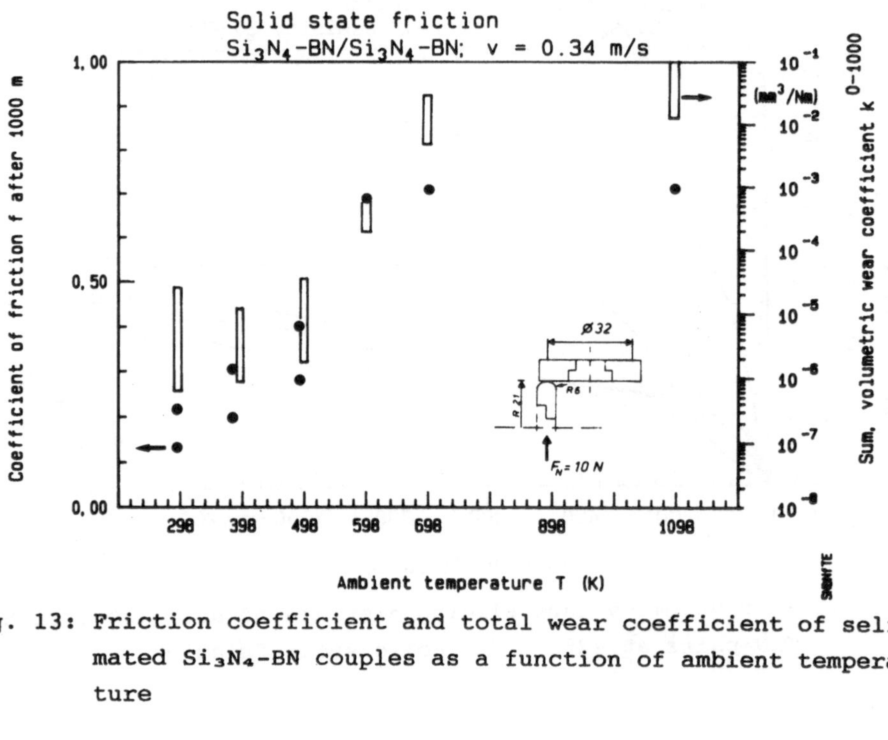

Fig. 13: Friction coefficient and total wear coefficient of self-mated Si_3N_4-BN couples as a function of ambient temperature

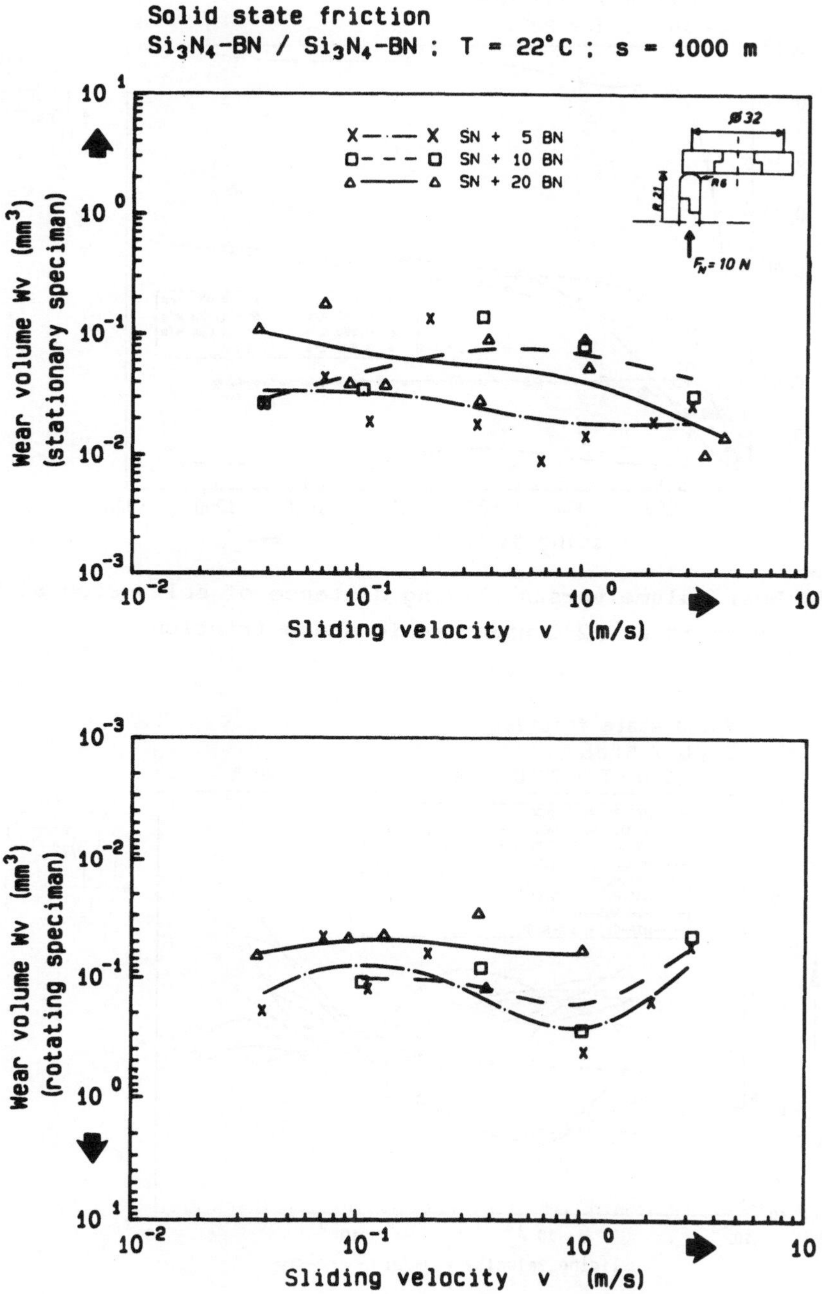

Fig. 14: Wear volume of stationary and rotating specimen of self-mated Si_3N_4-BN couples at 22°C under solid state friction

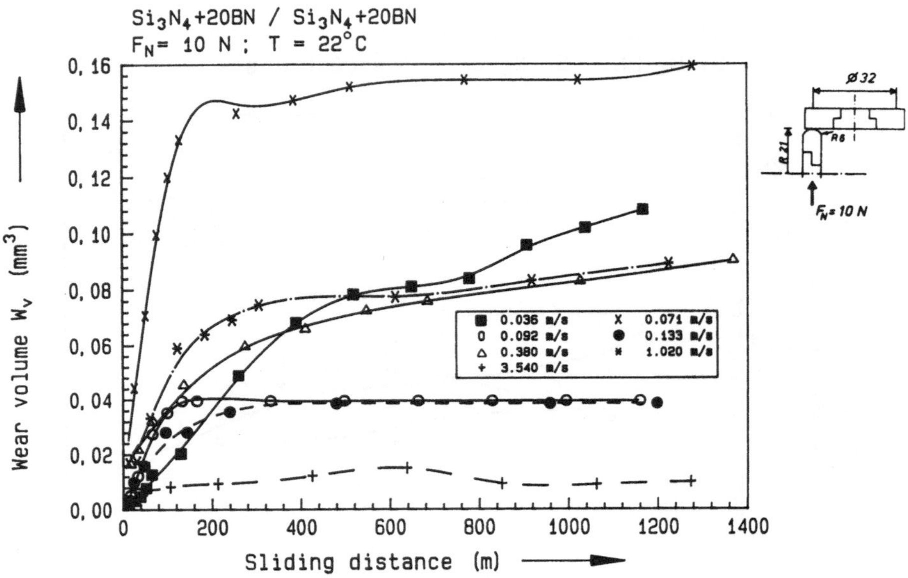

Fig. 15: Wear volume versus sliding distance of self-mated Si_3N_4-BN pairs at 22°C under solid state friction

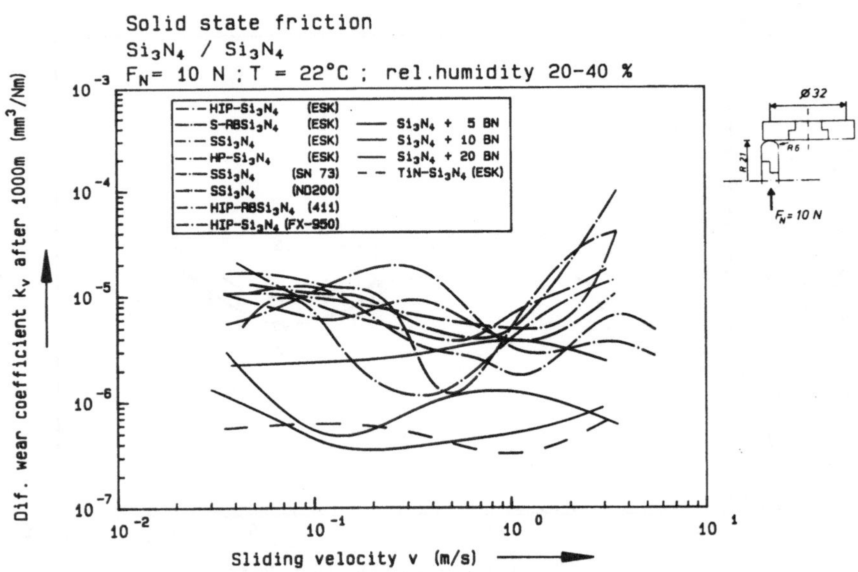

Fig. 16: Volumetric wear coefficient at 1000 m of self-mated Si_3N_4, Si_3N_4-TiN and Si_3N_4-BN sliding couples at room temperature

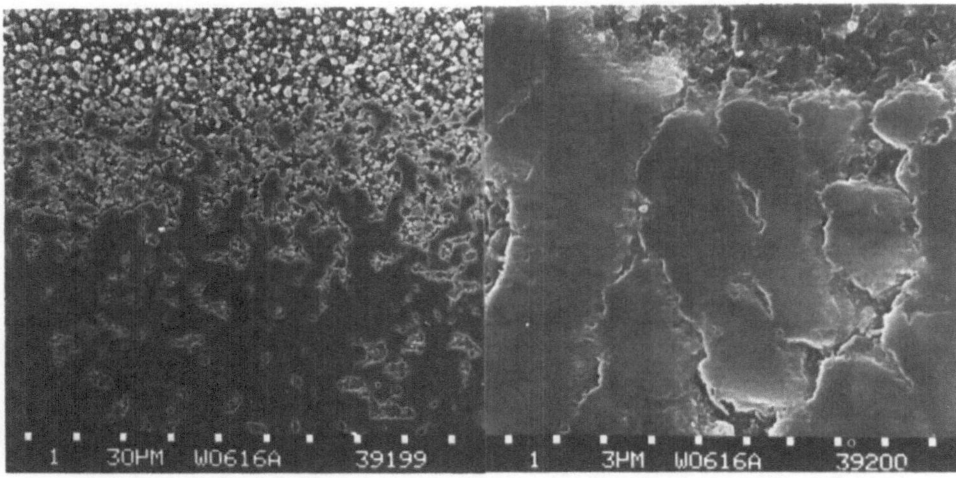

Fig. 17: Wear surface of Si_3N_4-TiN after tribological operation at 800°C and 0.38 m/s

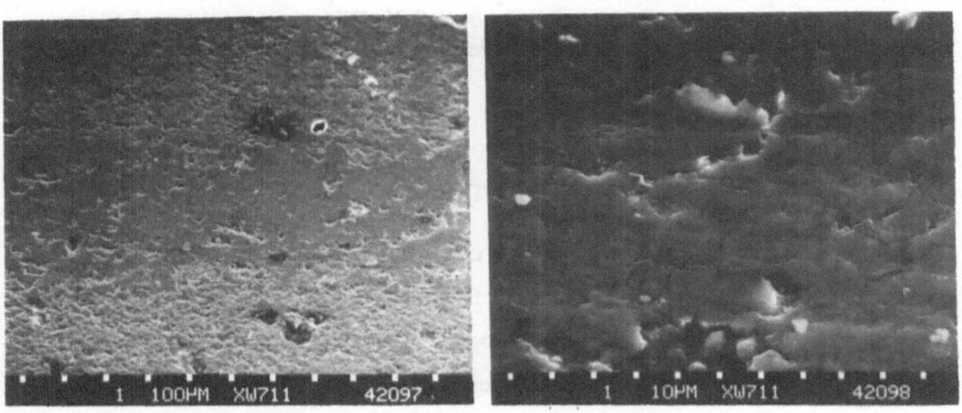

Fig. 18: Wear surface of Si_3N_4-BN after tribological operation at room temperature and 0.34 m/s

PIN-ON-DISC TESTS WITH CERAMICS IN AIR AND AQUEOUS MEDIA

PETER ANDERSSON
Technical Research Centre of Finland
Laboratory of Production Engineering
P.O.Box 111, SF-02151 Espoo, Finland

ABSTRACT

The tribological properties of sintered grades of alumina, zirconia-toughened alumina (ZTA), zirconia (PSZ), silicon carbide (SiC) and sialon were studied by means of pin-on-disc tests in air, water and a glycol-based mixture. Without lubrication, the coefficients of friction were slightly to substantially higher than in the aqueous media. During the lubricated tests with SiC and sialon, the sliding surfaces were tribochemically polished and enabled the generation of a fluid film lubrication mechanism with coefficients of friction down to 0.01. With the glycol-based mixture, the wear of PSZ was strongly reduced, while the sliding tracks of the alumina-based ceramics were chemically attacked.

INTRODUCTION

Components in tribological service are one of several areas of applications for structural ceramics in engines. Experiences from applications in different fields of technology as well as and promising tribological research results are likely to encourage greater use of ceramics in engines, provided that the component reliability is ensured and the solutions economically attractive.

The present work concerns pin-on-disc tests with ceramics in unlubricated sliding contact, lubricated with water and with a glycol-based mixture. The favourable performance of the ceramics in oil has previously been verified, e.g. in reference [1].

EXPERIMENTS

The experimental work was carried out by using a pin-on-disc machine. The initial sliding contact was defined by a polished (Ra=0.1 μm), horizontal rotating disc and a polished ball of about 10 mm in diametre made of the same material. The ball, or "pin", was clamped to a holder arm, which was pivoted to allow normal loading and friction measuring. The ceramics studied, which were commercial sintered grades, and some of their mechanical properties are presented in Table 1.

The tests were carried out in room air with 50±5% relative humidity, in distilled water and in a 1:1 mixture of an anti-freeze fluid and distilled water. The anti-freeze fluid was a commercial product for automotive use, and contained monoethylene-glycol (92%), corrosion inhibitors and pH-regulators (pH 7.5 for the mixture). The experiments were carried out at room temperature with a 10 N normal force. The sliding velocity and distance were 0.2 m/s and 250 m in the dry tests and 0.1 m/s and 2500 m in the lubricated ones.

TABLE 1
The materials investigated and their room temperature properties.

Material	Hardness HV	Toughness, K_{IC} MPa√m	Thermal cond. W/mK	Density g/cm^3	Grain size μm
Alumina(99.9% α-Al_2O_3)	2030	3.7	30	3.97	2.6
ZTA (20% ZrO_2)	1500	4.8	23	4.15	2.4
PSZ (MgO-ZrO_2)	1050	11.3	3	5.75	40.0
SiC (0.5% free C)	2690	3.7	110	3.14	2.3
Sialon (β')	1440	6.1	21	3.25	0.8x7

RESULTS

The quantitative test results are presented in Table 2 and the qualitative observations on the sliding pairs are presented in the following.

Unlubricated Tests

In unlubricated sliding on itself, alumina obtained polished sliding surfaces, and a shoe was built up in front of the pin wear scar.

The ZTA pin wear scar similarly produced was polished, the disc wear track was

TABLE 2
Pin (W_{pin}) and disc (W_{disc}) wear rates [10^{-6} mm^3/Nm] and initial (μ_{init}) and stable (μ_{stab}) dynamic coefficients of friction for the five ceramics in like pairings in various media.

MEDIA / CERAMIC	Unlubricated W_{pin}/W_{disc}	Unlubricated μ_{init}/μ_{stab}	Water lubricated W_{pin}/W_{disc}	Water lubricated μ_{init}/μ_{stab}	Glycol mixt. lubr. W_{pin}/W_{disc}	Glycol mixt. lubr. μ_{init}/μ_{stab}
Alumina	0.03/<0.1	0.3/0.38	0.01/0.05	0.3/0.21	0.09/0.04	0.2/0.10
ZTA	0.15/<0.2	0.3/0.35	0.03/0.1	0.4/0.21	0.09/0.3	0.16/0.14
PSZ	4/10	0.6/0.71	7/8	1.0/0.61	0.12/<0.01	0.2/0.10
SiC	0.6/0.14	0.5/0.47	1/0.2	0.3/0.05	0.14/0.16	0.16/0.02
Sialon	3/20	0.8/0.75	9/2	0.6/0.01	0.5/0.14	0.3/0.03

polished on the bearing proportion of the surface and a two-phase pattern was visible on both surfaces.

A coarse sound characterized the test with self-mated PSZ in air. The sliding surfaces were quite rough, with features resembling smearing and with cracks transverse to the direction of sliding (Fig. 1A). White wear debris was formed.

SiC sliding on itself in air obtained quite smooth wear surfaces. Loose wear debris, partly bright in polarized light, was found outside the wear surfaces.

The sialon wear track, formed in dry sliding against the same material, was relatively smooth with fine scratches in the sliding direction and with scattered small areas of deposits. The pin wear was scratched and carried deposits on most of the surface. Loose wear debris, bright in polarized light, occurred outside the wear scars.

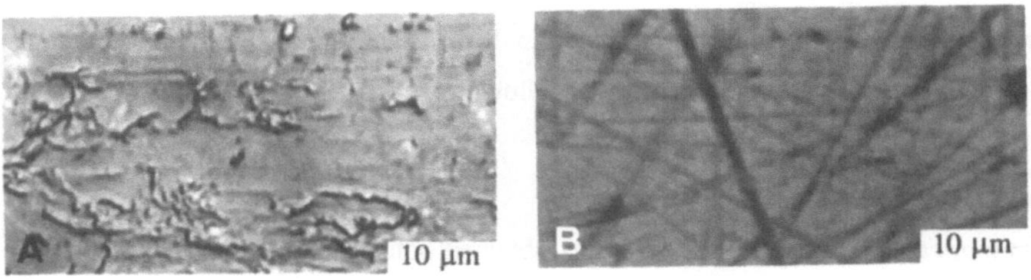

Figure 1. PSZ disc wear surfaces from sliding on PSZ unlubricated (A) and lubricated by a mixture of glycol and water (B).

Water Lubricated Tests

When sliding in water, the sliding surfaces of the alumina couple were slightly polished (Fig. 2A). The initial machining marks were erased from the disc.

The wear surfaces of ZTA sliding in water were polished and revealed the two-phase ZTA structure.

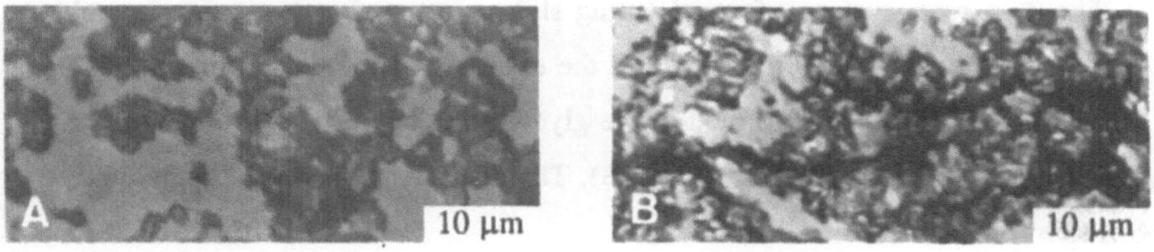

Figure 2. Alumina disc surfaces after sliding against alumina in water (A) and in a glycol-based mixture (B).

The PSZ wear surfaces were strongly roughened by non-uniform transfer layers and abundant loose wear debris was formed when the material was sliding against itself in water.

The silicon carbide wear surfaces were polished during sliding in like pairing in water (Fig. 3). The disc track depth was 0.1 µm. No loose wear debris was visually observed in the water afterwards, but its pH value rose during the test. The volumetric wear in a test which was discontinued after the decrease in the coefficient of friction, was very close to that in the full-length test.

In the test with sialon sliding on itself in water, the wear surfaces were polished, but included areas with short scratches. No wear debris was visually observed, but the pH value of the water rose during the test.

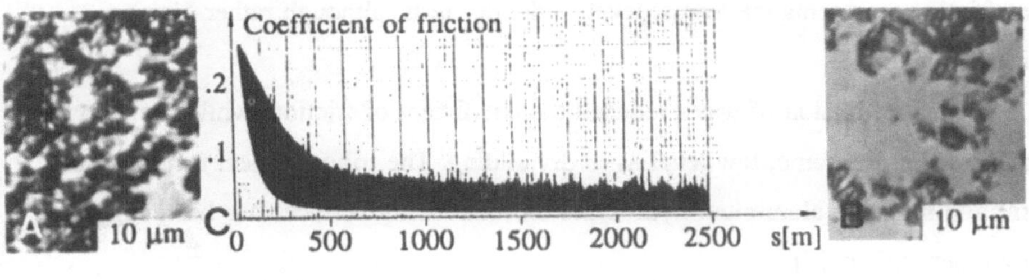

Figure 3. Silicon carbide disc surfaces before (A) and after sliding on the same material in water (B), and the frictional curve from the same test (C).

Glycol-Based Mixture Lubricated Tests

After sliding in the glycol-based mixture, the alumina pin wear scar was flat with fine, shallow grooves. The disc track consisted of polished spots on a rather rough surface. A large number of small, rough pits extending in the direction of sliding, occurred about some of the tracks (Fig. 2B).

The ZTA wear surfaces, formed during sliding in the glycol mixture, resembled the ones of alumina, but further revealed the two-phase structure of the material.

The PSZ pin wear scar formed in the glycol mixture was flat with sharp scratches in the sliding direction (Fig. 1B). The disc wear track had been slightly polished.

A circular scar, consisting of fine, shallow grooves and distributed small cavities, was formed on the SiC pin in the glycol mixture. Some wear particles, partly bright in polarized light, were situated outside the scar. The disc wear track was polished with some cavities and shallow grooves.

With sialon sliding on itself in the glycol mixture, the pin wear surface was extremely smooth, with only some weak, shallow marks in the sliding direction. The disc wear track comprised small, distributed polished areas on an otherwise rough surface.

DISCUSSION

Alumina

Although lubricated solely by the humidity of the air, alumina was subject to very limited wear in the dry tests. However, previous studies have shown that the wear may involve brittle fracturing if higher loads and velocities are used. The coefficient of friction represents the lowest level in the dry tests, although rather high for bearing surfaces.

The introduction of water reduced the coefficient of friction, while the wear rates remained at the same, low level as in dry sliding. The tribochemical formation of friction-reducing aluminium hydroxides on the rubbing surfaces has been suggested by other authors [2].

With the glycol-based mixture, the alumina pin wear slightly increased. The disc wear volume was in each media so small that any differences are difficult to find, but in the glycol-based mixture, pits, probably formed under the support of corrosion, occurred in the disc wear track. During the sliding, the pin area and the hydrodynamic forces increased, and the coefficient of friction decreased. The glycol-based mixture is more viscous, and therefore caused a stronger hydrodynamic effect than water.

ZTA

The behaviour of ZTA in like sliding resembled that of alumina. The frictional curves were almost identical for the two materials. The wear rates of the ZTA specimens were, however, higher than for the alumina ones, probably because ZTA has a lower hardness and a lower thermal conductivity, and will therefore be softer than alumina when frictionally heated.

For thermal reasons, the introduction of water reduced the wear of ZTA, under the conditions studied, to a level close to that of alumina.

When the glycol-based mixture was used, the coefficient of friction for ZTA on itself was at its lowest, but as in the case of alumina, the wear rates were increased by corrosion.

PSZ

Due to its extremely low thermal conductivity, PSZ in dry sliding suffered from strong adhesive wear, obvious from the rough sliding surfaces and the high coefficient of friction.

Also in water, the friction and wear of PSZ were high. The loose wear particles will further reduce the utility of PSZ.

With glycol in the water, however, the tribological performance of PSZ on itself was favourable. The PSZ disc wear track was hardly detectable, and its wear rate was the lowest in the present study. The relatively low coefficient of friction in the viscous fluid suppressed the interface temperature and in turn the adhesion tendency.

SiC

Silicon carbide was tribochemically worn in all the cases when sliding on itself. The hardness, toughness and thermal conductivity suggest less wear for silicon carbide than for e.g. the alumina in the present study, however the opposite occurred. In air, the oxidation of silicon carbide starts at temperatures easily reached in sliding. The wear debris produced during the unlubricated sliding probably consisted of some kind of silica, and the silicon carbide wear surfaces were polished as the surface oxide layer was continuously removed.

The strong decrease in the coefficient of friction, which occurred during running-in in water, followed from the large relative increase in the pin wear scar area and the surface polishing, which generated a water film lubrication mechanism. The lack of loose wear particles and the increase in the pH value indicate tribochemical wear and the dissolution of the reaction products into the water. The fluid film separated the surfaces well enough to strongly reduce the wear after running-in.

The introduction of the glycol-based mixture also resulted in a low coefficient of friction, characteristic of fluid film lubrication. The wear rates remained slightly lower, as the more viscous liquid was able to create a sufficient hydrodynamic force between smaller areas than in the case of water. The wear mechanism was similar to the one found with water, but in parallel some mechanical wear occurred, forming some loose wear debris.

Sialon

Sialon generally behaved like the other silicon-based ceramic, silicon carbide. In air, sialon was worn probably through a tribo-oxidation, assumed from the white wear debris and the deposits on the wear surfaces, and the coefficient of friction and the wear rates were very high.

In water, a water film lubrication mechanism similar to that with SiC, was generated. The wear rates were higher for sialon than for SiC, as a larger pin bearing surface was required, probably due to its non-uniformity. The end value for the coefficient of friction is the lowest one recorded in the present investigation, but the high level occurring at the beginning of the runnung-in period is problematic.

The increase in viscosity by using the glycol-based mixture again strongly reduced the need for bearing surface area, and consequently the pin and disc wear rates. The coefficient of friction during the running-in period was only about one-half of that occurring in water, and the stabilized value was very low.

CONCLUSIONS

In dry sliding, low wear rates were obtained with alumina, but all the coefficients of friction were rather high for bearing surfaces.

During sliding in water, the coefficients of friction for silicon carbide and sialon on themselves decreased to μ=0.05 and below, due to the generation of a fluid film lubrication mechanism.

Alumina and ZTA performed slightly better in water than without lubrication, while PSZ did not benefit from water at the interface.

In the glycol-based mixture, the lowest coefficients of friction for the oxide ceramics were recorded. The wear of PSZ was reduced by the glycol-based mixture, alumina and ZTA were probably corroded in the sliding contact and the silicon-based ceramics behaved similarly as in water.

ACKNOWLEDGEMENTS

The present work, which forms a part of a Finnish research programme on ceramic journal bearings, is financed mainly by the Technology Development Centre of Finland (TEKES). The project is carried out in co-operation with P. Lintula Lic.Tech. of the Institute of Materials Science at Tampere University of Technology, and P. Salonen M.Sc.Tech. of the Laboratory of Machine Design at Helsinki University of Technology. Financial support and co-operation have been of the greatest importance for the progress of this work.

REFERENCES

1. Salonen, P., Andersson, P., Lintula, P. Oil lubricated and water lubricated pin-on-disc tests with engineering ceramics. Otaniemi 1990, Helsinki University of Technology, Faculty of Mechanical Engineering, Series of Publications C 228. 68 p. + app. 1 p.

2. Gates, R., Hsu, S., Klaus, E. Tribochemical mechanism of alumina with water. Soc. Trib. Lubr. Eng. Tribology Transactions 32(1989)3, pp. 357-363.

CERAMIC COATINGS IN INTERNAL COMBUSTION ENGINES

B. Ineichen and C. Klukowski
SWISS FEDERAL INSTITUTE OF TECHNOLOGY ZURICH
INTERNAL COMBUSTION ENGINES LABORATORY

ABSTRACT

An investigation was conducted into material and processes with the goal of achieving effective and durable thermal barriers for heavy-duty diesel engine applications. Coating systems aimed at reduced thermal conductivity, erosion- and corrosion resistance were evaluated. Plasma and flame sprayed yttrium stabilized zirconia oxide layers were judged in these tests after surface treatments with and without bond coating. Selected ceramic coated substrates were tested under engine operating conditions in a practical laboratory test to improve the understanding of how surface porosity, coating layer structure and bond strength contribute to the overall performance of thermal barrier systems.

Experiments have shown that suitable mechanical properties of the surface will improve the resistance against erosion and gas diffusion. Insulating materials may improve the performance and durability of advanced heat engines, provided the combustion system, in particular the injection configuration is properly optimised.

A more detailed understanding of thermal barrier coatings performance has been achieved through this program.

INTRODUCTION

Engineers have long wanted to use ceramic materials in engine combustion chambers, so that the engines could be operated without cooling systems and without lubrication. This technology might now be partly realized because of the recent development of ceramic materials with excellent heat resistance, and the advantage of a ceramic material design methodology.

However, firm conclusions about the merits of ceramic combustion chamber walls have not yet been established; various reports have differed in their conclusions. Advanced high performance ceramics have opened exciting new opportunities for heat engine applications. Engine components made from these materials could have lower cost, improved dimensional consistency, and better wear characteristics than equivalent components made with conventional materials. In addition, ceramic materials provide the potential for higher engine thermal efficiencies, longer life and higher reliability.

Many types of partially insulated engines have been investigated by engine and manufactures all over the world [1-5].
Thermodynamically, the word "insulated" means that no heat is added or subtracted to/from the gas during the cycle. In practice, the insulated engine strived to minimize the loss of heat through the combustion chamber components during combustion and expansion and prevent heat from entering the working fluid during intake and compression. The net increase in available energy would increase the in-cylinder work and the exhaust energy for waste heat utilization.
Recently, the use of combustion chamber insulation as a strategy to decrease engine fuel consumption has been questioned by well-respected investigators [4, 5]. Their conclusions, based on limited measurements, were that the heat flux increases (rather than decreases) at the high surface temperatures produced by insulating materials, and thus engine performance deteriorates with insulation. However, many researchers still maintain that new insulating materials will improve the performance and durability of advanced heat engines, provided the combustion system in particular the injection configuration is properly optimised to the new condition.
The ultimate rewards from use of these new materials could be many from the standpoint of economics and technology. But the "ideal" ceramic as thermal barrier does not yet exist. Table I shows for that reason a comparison of different ceramic and metal materials, and the thermophysical properties.

TABLE 1
Thermophysical properties of ceramic and metal materials at room temperature

Material	Density (ρ) g/cm^3	Conductivity (λ) W/mK	Youngs Modulus GPa	Specific Heat (c_p) J/gK	Expansion $1/K\ 10^{-6}$
STEEL	7.87	50.2	210	0.48	13
AlTi4V5	4.45	7.5	120	0.915	10.5
PARTIALLY STABILIZED ZIRCONIA	3.14	2 - 2.9	160 - 200	0.4 - 0.543	9.3 - 11.
SPRAYED ZIRCONIA	5.2	0.7 - 1.4	460 - 480	0.67 - 0.73	8.0 - 10.

Increasing research and development of advanced, or structural, ceramics and composites, have spurred development of processes for effective joining of ceramics to metals especially in those cases where compact ceramics are used. Since many of the compact ceramic/metal assemblies must operate in hostile environments, high temperatures, corrosive atmospheres, and under high loads, the joining is still a difficult problem. The bonding of the compact ceramic with the base material is of great importance. The thermal expansion differential is an important negative factor. As a consequence, monolithic ceramic materials can not live up to adverse diesel engine operating conditions. Hence, a trend was developed towards the application of thermal barrier coatings.
Ceramic coatings can provide good insulation, reduce friction/wear, reduce erosion/corrosion, and prevent material build-up. However, processes such as PVD and CVD, ion plating, etc. can only be used for the application of thin coatings. Thin coatings have satisfactorily been used in the gas turbine industry for turbine blades, stator vanes, and combustion parts. For thicker thermal barrier coatings typically used in diesel engines, only plasma-spray and flame-spray processes were used. Since thermal spray coatings are

porous, surface densification or sealing is important.
One way to increase the thermal efficiency of internal combustion engines is to reduce the heat losses. A truly adiabatic engine is not possible, because the heat transfer cannot be brought to zero.
The integral adiabatic engine would have to operate with high surface temperatures and the differential adiabatic engine would require a heat conductivity of zero. Extended studies assuming heat transfer coefficients not being influenced by surface wall temperature demonstrated the gains to be made by partial insulation being very small or even negative on account of reducing the air delivery ratio because of heating up the fresh cylinder charge. This negative aspect can be compensated in the case of turbocharging and in particular if adopting a turbocompound solution [6].
Furthermore, one has to keep in mind some very recent findings which appear to show an increase of the heat transfer coefficient on account of higher surface temperatures in a combustion chamber [4, 5, 7]. However, these findings are disputed as well.
Moreover, local thermal insulation of metallic parts is envisaged. The extremely hostile environment in which engine components operate is characterized by high cyclic temperatures and pressures with steep gradients.
In order to ensure adequate performance of a ceramic layer in an internal combustion engine, properties such as the ability to withstand high temperature, thermal shock resistance, thermal conductivity, heat capacity and coefficient of expansion of ceramic materials have to be considered.
A new perspective of plasma- and flame- (with sintered ceramic stick) sprayed ceramic coatings and how surface finishing, porosity, coating layer structure and bond strength contribute to the overall performance of thermal barrier systems will be presented.

OBJECTIVE

The overall objective of this program was to advance the state-of-the-art coating technology through the systematic development of an enhanced Thermal Barrier Coating (TBC) system for application in a diesel engine environment. One of the primary keys to survival of any thermal barrier coating is the ability of the coating to accommodate the strains resulting from the differential thermal expansion between the coating and the metallic substrate. A simple method of accommodating these expansions has been to use "thin" coatings in conjunction with a porous micro structure. This porosity allows the coating to accommodate the thermal expansion mismatch with the metallic substrate and is achieved trough the proper selection of parameters for the spray deposition process.
Anticipating the expected failure mechanism resulting from the porous nature of these coatings, the intended applications play an important role. In a diesel engine application, diesel fuels are notoriously "dirty" and may contain many contaminants. It is anticipated that these contaminants will penetrate the coating system along paths provided by the inherent porosity within the coating structure. Absorption of these contaminants within the coating, coupled with the thermal gradients occurring in the layer, will place the dew point of these impurities somewhere within the thermal barrier coating. As temperature changes occur in the cylinder due to changes in the operating mode, phase changes will take place in the contaminants resulting in failure of the coating.
Publications on conductivity of ceramic coatings show the contrariety, some authors reporting on a conductivity which is less for the ceramic coating

than for the compact ceramic. This study allowed the determination of the effective conductivity, for this reason the temperature was measured by means of very fine thermocouples located on the ceramic coated surface, as well as on the interface between substrate and ceramic coating, and in the steel substrate. The specimens were heated either with hot air or a gas flame and the temperature was measured at all locations simultaneously. From the experimental values the conductivity was calculated. The tests merely gave indications that the conductivity for sprayed ceramic coatings are between 0.7 - 0.95 W/mK.
The key to the technology improvement expected to result from this program is to determine a coating which possesses the heat conduction characteristics of normal thermal barrier coatings as well as the closed pore porosity needed to prevent the surface and subsurfaces coating degradation caused by diesel fuel contaminants.

EXPERIMENTAL DEVELOPMENT OF CERAMIC COATING

In this search, use was made of the broad experience gained with plasma and flame sprayed (Rokide) coatings. Prior to spraying, surface preparation is carried out by removal of grease and abrasive blasting of all the stainless steel base plates. Plasma sprayed or flame sprayed coating were deposited onto the samples, the coating thicknesses varied between 0.7 and 1.0 mm in the case of plasma sprayed coatings, respectively between 0.6 and 0.9 mm for flame sprayed coatings (Fig.1b). For a limited number of plasma sprayed and flame sprayed coated test specimens, a bond coat was added with a thickness of approximately 50 microns as shown in Figure 1a.

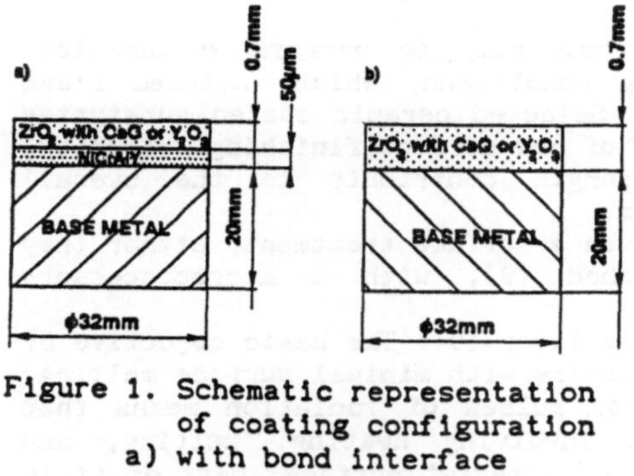

Figure 1. Schematic representation of coating configuration
a) with bond interface
b) without bond interface

The grain size for plasma sprayed coating were typically in the ranges of 20 to 45 microns, 10 to 40 microns, or 5 to 25 microns. The material supply for the flame sprayed coating was performed by a sintered ceramic stick of prechosen chemical compound.
It is well established that zirconia is among the most refractory, thermal shock-resistant and corrosion-resistant oxides. The major aspect of pure zirconia is its polymorphism. The monoclinic to tetragonal transformation which occurs on heating to above 1100 °C is accompanied by a volume increase resulting in stresses relaxed by cracks when the percentage of monoclinic phase is too high. To overcome this phase transformation phenomenon, we have added 3 to 30% wt calcium oxide (CaO) or 5 to 13% wt yttrium oxide (Y_2O_3).

TEST PROCEDURE

The porosity and micro structure of the coatings were examined using optical microscopes, and scanning electron microscopes (SEM).
Thermal cycling shock tests were used as a primary method for ranking the

performance of the coatings under engine operating conditions. Moreover, the tests allowed simulation of the effect of injection of fuel droplets to the combustion chamber wall of a diesel engine by means of water injection (droplet) cooling. Each sample was heated, accordingly to engine operation, namely during 0.7 seconds to a pyrometrically measured temperature of 600°C and afterwards cooled in about 0.3 seconds by an air jet and a water injector. Air cooling was performed on the whole surface of the specimen, while water (droplet) cooling was restricted to a limited area of the surface, as illustrated in Figure 2. The gas flame engulfed the whole specimen surface, eliminating the effects of non-uniformities. Testing was considered to be completed with the achievement of 10^5 cycles and only these specimens were used for further considerations.

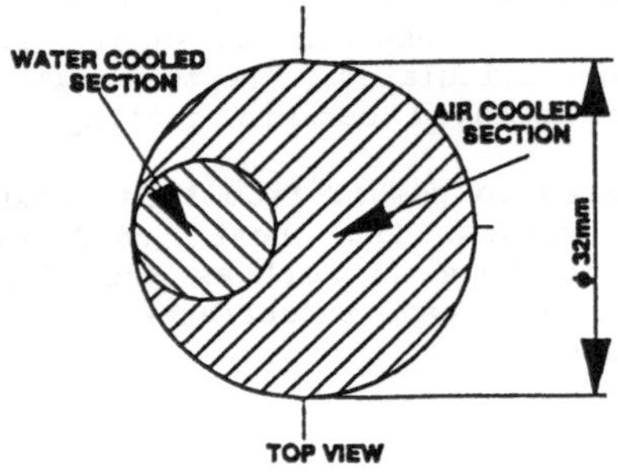

Figure 2. Schematic representation of water and air cooled section during thermo-shock.

TEST SPECIMEN

A series of test specimens were produced with different layer thicknesses, different formations of the coating layers, various types of stabilizers for the zirconia, and all of them were adjusted to minimize the porosity of the coating.
Expensive engine tests have shown the necessity to perform systematical laboratory tests under engine operating conditions, which included flame penetration, fuel injection and cooling. Selected ceramic coated substrates were tested to improve the understanding of how surface finishing, porosity, coating layer structure and bond strength contribute to the overall performance of the thermal barrier system.
Therefore, some of the specimens were given a surface treatment, either they were grinded, super-finished or polished [8], with a microprocessor-controlled surface treatment equipment.
An other attempt was made to make use of a laser [9]. The basic objective of laser glazing is to selectively glaze ceramics with minimal surface melting. The ability of lasers to produce intense pulses of radiation means that there are many potential applications involving heating, melting, and vaporization. When laser radiation falls on a target surface, part of it is absorbed and part is reflected. The energy that is absorbed begins to heat the surface. The heating effects due to absorption of high-power beams can occur very rapidly. The surface quickly rises to its melting temperature. Glazing requires maximum melting under the condition where surface vaporization does not occur. Melting without vaporization is produced only within a narrow range of laser parameters. If the laser power per unit area is too high, the surface begins to vaporize before a significant depth of molten material is produced.
Melting of a material by laser radiation depends on heat flow in the material. Heat flow is dependent on the thermal conductivity. But thermal conductivity is not the only factor that influences the heat flow, since the rate of change of temperature also depends on the specific heat of the material. In fact, the heating rate is inversely proportional to the

specific heat per unit volume. This factor has the dimension of cm^2/s, characteristic of a diffusion coefficient, and has therefore been given the term "thermal diffusivity" (to recognize that it represents the diffusion coefficient of temperature or, more properly heat). Effective glazing with lasers depends on propagation of a fusion front on the ceramic layer during the time of interaction, at the same time avoiding vaporization of the surface. However, the depth of penetration without surface vaporization is limited. To obtain the depth required for this application, one can tailor the laser parameters to some extent. Generally, one lowers the power density and increases the pulse duration. The control is very sensitive. One must make careful adjustments to achieve a balance between optimum glazing depth and avoidance of vaporization.

RESULTS AND DISCUSSION

To ensure adequate performance of ceramic layers in an engine, properties such as temperature capability, thermal shock resistance, thermal conductivity, heat capacity and coefficient of expansion of ceramic materials were considered.

INFLUENCE OF THE PARAMETERS FOR THERMAL CYCLING STRAIN SEQUENCES

Investigations were made into the effects of micro structure and how the cracking fracture and residual stresses in coating affecting the performance of thermal barrier coating for engine applications.

Effect of surface treatment

Micro structure analyses showed in the case of as-received specimens, the presence of unbonded particles. The surface of the ceramic coated samples was therefore treated by grinding and polishing. It is evident, that in an engine application the removal of the unbonded particles is important, otherwise due to thermal expansion and explosion of fuel droplets these particles crack off. Once this is happened - the consequence is a continuous erosion. This is especially important in case of a ceramic coated piston cap of a direct injected diesel engine.

With extended mechanical surface finishing other microcracks are added as shown in Figure 3. In difference to the unpolished or as-received specimens, the polished specimens showed a crack network on the surface and to the some extent delamination horizons. The cracks are generated be the thermo-mechanical release of the residual stresses.

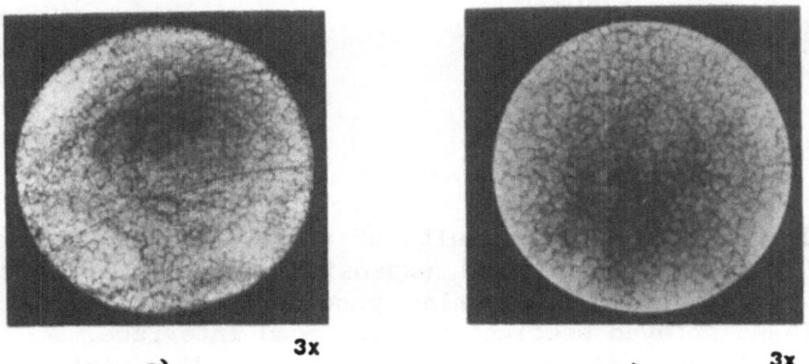

3x a) 3x b)

Figure 3. Optical micrograph of a metal-ceramic plasma-sprayed coating with ZrO_2, a) ground, b) polished

As shown e.g. in Figure 3 polished samples are more damaged than ground samples. During polishing of ZrO_2 - ceramics, characterized by a low thermal conductivity, large and local thermal stresses are applied, resulting in debrises as shown in Figure 4.

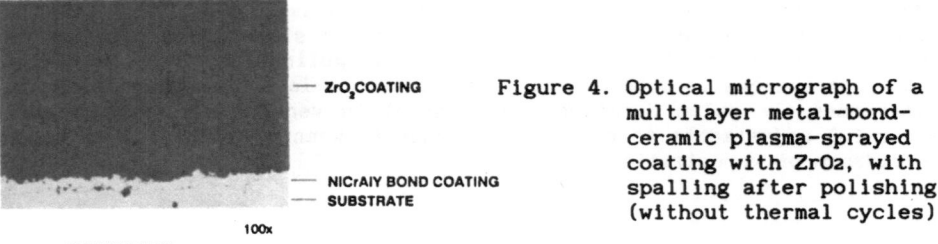

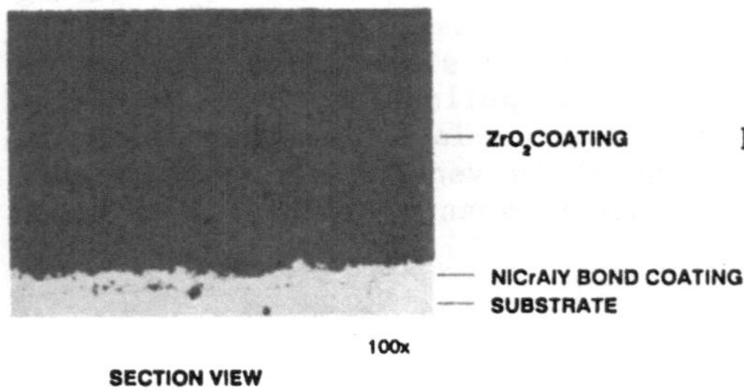

Figure 4. Optical micrograph of a multilayer metal-bond-ceramic plasma-sprayed coating with ZrO_2, with spalling after polishing (without thermal cycles)

Best result were achieved with flame sprayed and polished ceramic coatings, no delamination cracks are observable in the cross section of the specimen as shown in Fig. 5.

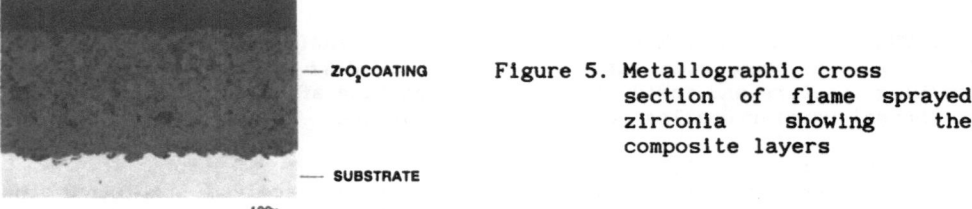

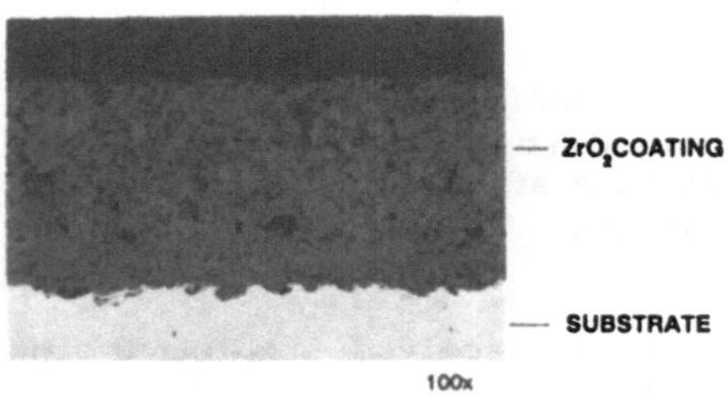

Figure 5. Metallographic cross section of flame sprayed zirconia showing the composite layers

In contrast, in the plasma sprayed and polished specimen a polygonal cracks network is seen on the surface, and in some cases the cracks propagated vertically toward the substrate. Delamination cracks parallel to the surface propagated from these as shown in Figure 6. The polygonal crack network is noticeable in the case of ground specimen. The existence of debris is evident and can reach as far as to the base material.

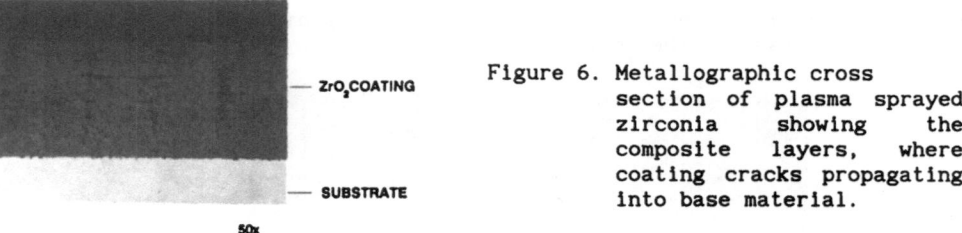

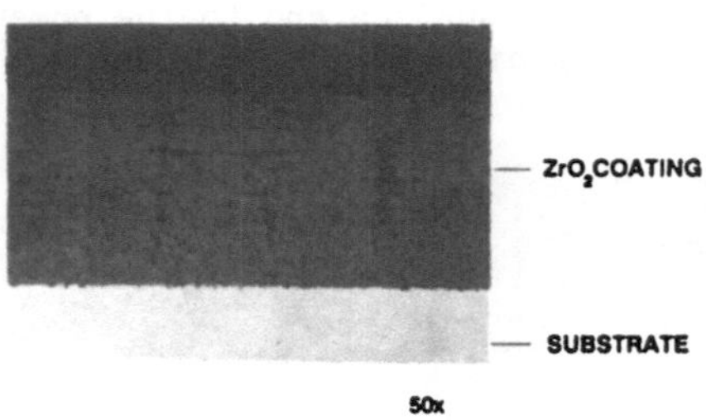

Figure 6. Metallographic cross section of plasma sprayed zirconia showing the composite layers, where coating cracks propagating into base material.

In general, the cracks are the results of residual shear stresses, and are the effect of differential thermal expansion between the coating and the metallic substrate during the coating procedure. The largest cracks were detected on plasma sprayed specimen without bond interface.
In the case of laser glazing the morphology of the coating surface has changed due to a surface melting process. The newly formed surface is characterized by a porous reduced and well orientated crystallization of zirconia. The inhomogeneous cooling of the surface and the subsurfaces

resulted in a unequal texture.
The melting process and the extraction of the gas from the coating due to the laser glazing treatment produced a sealed (glazed) and undulated surface as shown in Figure 7, with a much smoother surface than the polished and ground surfaces (Fig. 3).

2000x

Figure 7. The fracture micrograph of laser beam glazing surface of a ceramic coating

The surface morphologies resulting from the different surface treatment processes are compared in Figure 8.

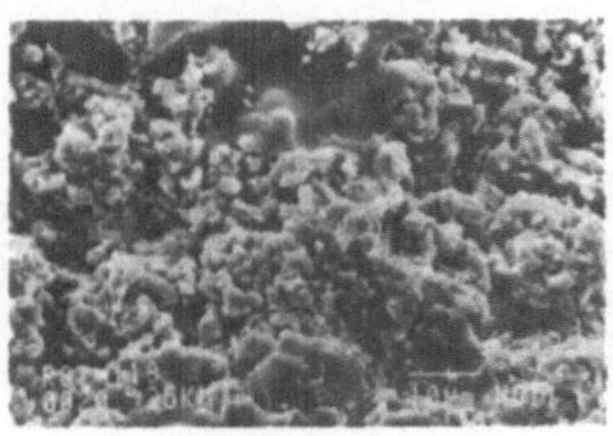

a)

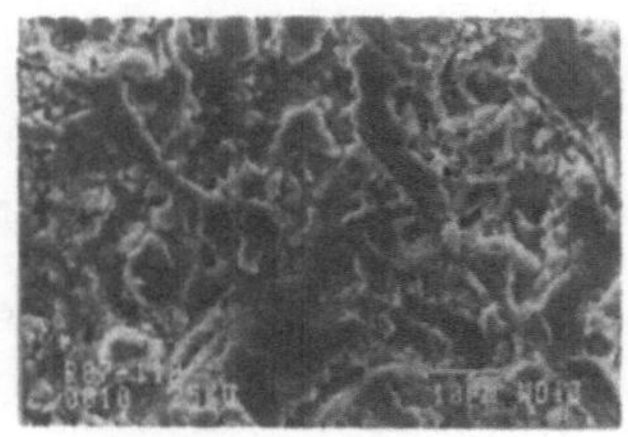

b)

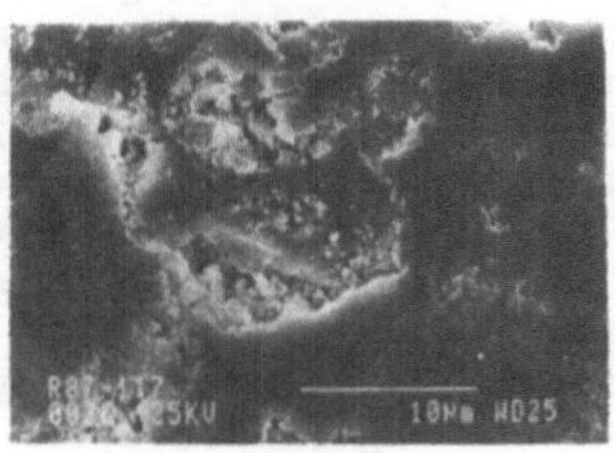

c)

Figure 8. Scanning electron micrographs of unpolished (a), ground (b), and superfinished (c) surface.

Influence of deposition processes

In the case of Y_2O_3 stabilized zirconia without a bond interface the flame-sprayed samples showed only a weak crack network which is located close to the surface, plasma-sprayed samples with CaO stabilized zirconia have a strongly developed crack network with debris.
The coarse-grained and irregular texture of the flame-sprayed coating prevented the delamination as shown in section thermo shock (Fig. 11). The adhesion is excellent in spite of the low coating temperature (3000 °C) and the circumstance that only melted particles were sputtered. At the same time the quenching rates of the melted particles, and the thermal expansion mismatch between the metallic substrate and the coating were reduced. The residual stresses are directly coupled to this phenomena.
The coating temperature (5000 - 6000 °C) for plasma-spraying induced very high quenching rates, and the differential thermal expansion between the coating and the metallic substrate increased. In the case of optimized

coating parameters and high coating temperatures unmelted particles were still existing. In connection with the plate-like texture of the melted particles delamination and horizon with a low adhesion resulted.

Influence of bond coating

The samples with a bond interface with NiCrAlY had a much better resistance against high cyclic temperature variations. The samples without a bond coating showed many delamination-horizons reaching into the subsurfaces, the residual stresses being much higher after spraying. In addition, the bond coating protects the metallic substrate against corrosion and/or oxidation as shown in Figure 9a.

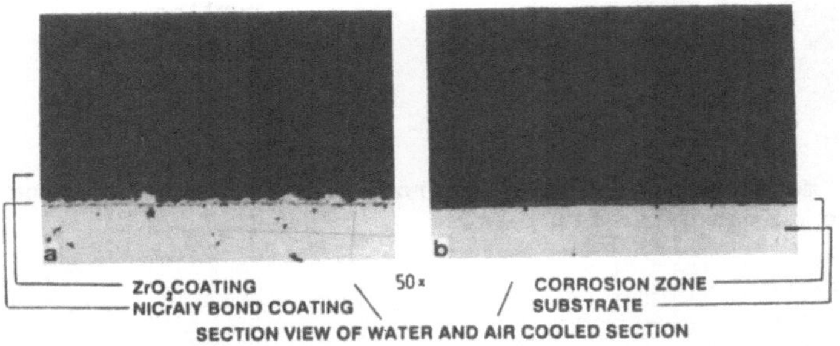

Figure 9a and 9b. Optical micrograph of a multilayer a) metal-bond ceramic b) metal-ceramic plasma-sprayed coating with ZrO_2, after 10^5 thermal cycles.

Thermo shock

The thermal shock behavior of as-received and exposed coating was evaluated. As seen in Figure 10, a plasma sprayed ZrO_2 coating stabilized with 30% wt CaO, and with a bond coating showed in the region of water droplet cooling debris of the coating down to the bond coating. Delamination cracks propagate in layers along the spray coating layers. Furthermore the bond layer shows corrosion. In the case of a plasma sprayed ZrO_2 coating stabilized with 5% wt CaO without a bond coating the debris are steplike, and between the substrate and the ceramic coating there is a noticeable debond horizon as shown in Figure 9b. In comparison, the plasma sprayed ZrO_2 coating stabilized with 5% wt CaO with bond coating showed after 10^5 cycles no debris in the water cooled, or air cooled section of the sample, horizontal delamination horizons are only observable close the ceramic surface (Fig. 9a).

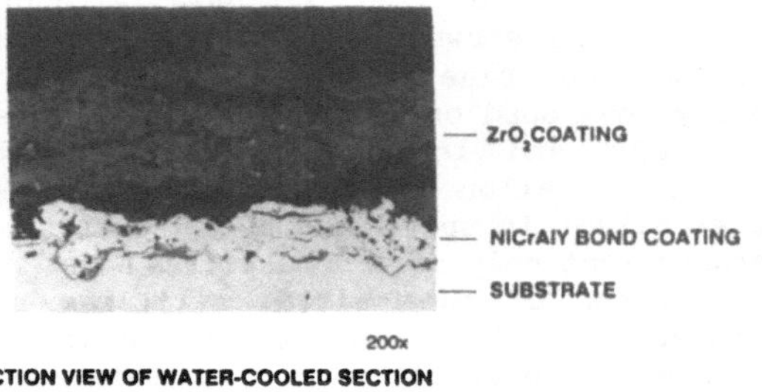

Figure 10. Optical micrograph of a multilayer metal-bond-ceramic plasma-sprayed coating with ZrO_2, stab. 30% wt. CaO after 10^5 thermal cycles.

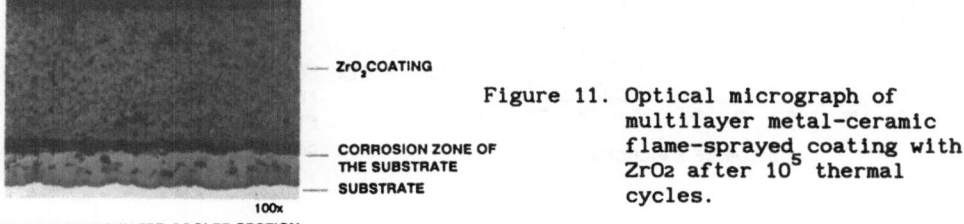

Figure 11 illustrates a flame sprayed ZrO_2 coating stabilized with 5% wt Y_2O_3 after 10^5 cycles, showing a polygonal crack network on the surface of the coating. In some cases the cracks are propagating vertically toward the substrate typically to a depth of about 150 microns, and the substrate has a corrosion zone in the order of up to 100 microns.

Figure 11. Optical micrograph of multilayer metal-ceramic flame-sprayed coating with ZrO_2 after 10^5 thermal cycles.

In general, the detected defects can be explained by the superposition of many effects, namely

- Defects by rapid expansion of water or cold air in cracks and pores
- Defects resulting from thermal strains

INFLUENCE OF COATING PROCESS TO RESIDUAL STRESSES

The chemical homogeneity was examined. All the single phase coatings showed an homogeneous chemical structure with constant stabilizer fraction. Thermodynamical stresses obtained from differently stabilized phases with differential thermal expansion coefficients were dismissed.

Erosion and fracture resistance

The resistance of ceramic coatings to erosion in the combustion chamber is an important parameter that must be considered. The action of fuel impingement from the injector on ceramic coated combustion components are unknown. Similary, fluid motion due to intake combustion - and exhaust processes result in scrubbing actions on all of the cylinder components in contact with the combustion charge.
Erosion resistance of ceramic coatings depends substantially on the cohesive bonds produced in the microstructure during spraying. In large measure, the magnitude of the changes in erosion resistance of the coating can be related to the energy required to fracture the coating, either internally (cohesive strength), or at the boundary with the bond coating (adhesive strength). These strength parameters are based on fracture mechanics and are related more rationally to the microstructure and the failure mechanism.

Corrosion resistance

Cracks in the coating and the open porosity are responsible for the chemical reactions between bond respectively substrate and/or particles with the ambient medium like gases or fluids. The specimen without a bond coating showed a dominant corrosion in the substrate/coating interface zones as illustrated in Figure 12. It can be seen that the hydration of the substrates to FeOOH are taking place in a up to 30 microns thick interdiffusion zone, and the small amount of ZrO_2 in this zone refer to a corrosion-process within the boundary of the ceramic.
Interdiffusion between the coating and the substrate changes the composition and microstructure of coatings and affect their oxidation and corrosion resistance. This result indicates that the influence of surface quality and the microstructure of a coating on its corrosion resistance is pronounced.

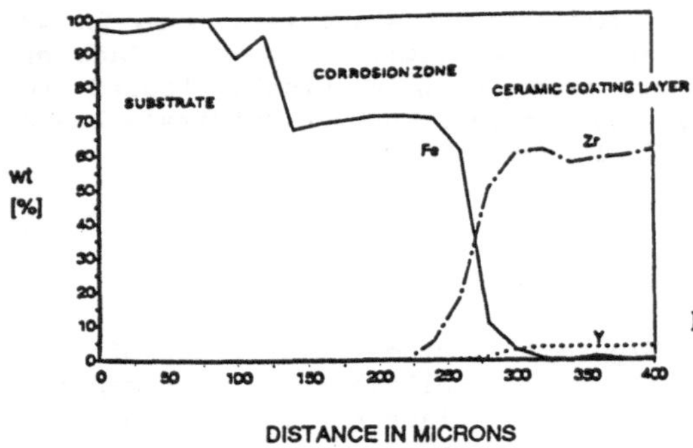

Figure 12. Fe-, Y-, and Zr-profile in the corrosion zone of a flame-sprayed ceramic-coating with ZrO_2.

CONCLUSIONS

This program was undertaken to systematically develop enhanced capability thermal barrier coatings for diesel engine components.
The work performed has been devoted to the evaluation of structure and properties of series of thermal barrier coating procedures.
Failure of a thermal barrier bond coating system occurs most often at the interface due to propagating cracks, formation of cracked zones and high compressive stresses in the coating system. As the bond oxidizes, the interfacial bonding oxide will grow in thickness. Once it reaches sufficient thickness, tensile stresses develop in the adjacent thermal barrier ceramic and its own thermal shock properties come into play.
In fact, none of the tested specimen could be used for combustion components, since the observed crack system and the corrosion of the interdiffusion zone would not allow adequate performance in an internal combustion engine.
Nevertheless, further developments can be made, with a thicker MCrAlY bond coating against corrosion and secondly, decreasing the residual stresses of the coating by annealing.
The porosity can not be improved, since the porosity depends from the melting temperature and the viscosity of ZrO_2. Improvements can be made by using new oxides with lower melting temperatures and lower viscosities. As earlier reported by other authors new oxide ceramics, e.g Ca_2SiO_2 do have a much better temperature cycling resistance than ZrO_2.
Laser glazing is certainly a promising procedure for distinctive surface structures with a high degree of strengthening phases and better pore densities. An appropriate sample preheating and cooling is however required to avoid fissuration and to improve quenching.
Despite the fact that the tests performed illustrated in the present paper already yield valuable information, they should be extended to an engine environment with heat radiation and emissions.

ACKNOWLEDGEMENT

This work has been carried out under a program, sponsored by the Swiss Commission for Scientific Research (KWF), and with the active industrial participation of Sulzer Bros. Ltd., Winterthur, Switzerland.
The authors are indebted to Dr. K. Honegger, Dr. H.J. Martens and Mr. J. Prosek for their assistance, cooperation and for their valuable suggestions during this program.

The authors wish to thank Prof. M.K. Eberle for his constant interest and support.

REFERENCES

1. G. Desplanches, "A Study of PSC for Use in Internal Combustion Alternative Engines", pp.193 - 203 in 1st Plasma-Technik-Symposium, 1988, published by Eschenauer et al editors.

2. I. Kevernes, M.P. Johnson and R. Buget, "Performance of Thermal Barrier Coating for Diesel, Stationary and Aircraft Gas Turbine Components", MRS Europe, pp. 247-264 in Review of preliminary COST 501, 1985.

3. I. Kevernes, "A Coating Programme for Diesel Engine Parts", Proposal in 2nd round of COST 501, Novembre, 1987.

4. M. Hauser , S. Lauf, H. Reiter, G. Woschni, N. Zernig C. Klarhoefer , H. Kürkemeier, S. Mielke, W. Sander, B. Engels, R. Lingenauber and J. Siebeln, "Einsatzchancen keramischer Werkstoffe im Motorenbau", in Proceedings "Haus der Technik", March, 1987.

5. W. Bunk, U. Dworak, W. Hüther, H. Knoch and M. Langer "Keramische Werkstoffe f.r den Einsatz in Kraftfahrzeugen", Proceedings of the "Technische Akademie Esslingen", March 1985

6. M.K. Eberle and A. Paul, "Possible Ways and Means to Further Develop the Diesel Engine in View of Economy", CIMAC Publication, #Warsaw , Poland, June, 1987.

7. K. Boulouchos, M.K. Eberle, B. Ineichen and C. Klukowski, "New Insights into the Mechanisms of In-Cylinder Heat Transfer in Diesel Engines", SAE Publication #890573, March, 1989

8. J.S. Reed and A.M. Lejus, "Effect of Grinding and Polishing on Near-Surface Phase Transformation in Zirconia", Mat. Res. Bull. 12, 1977.

9. M. Benninghoff, "Werkstoffbearbeitung mit dem Laser - Stand und Entwicklungsperspektiven", Techn. Rundschau, 81, (6), 1989.

TRIBOLOGICAL AND THERMAL CERAMIC COATINGS FOR ADVANCED ADIABATIC ENGINE

ROY KAMO
Adiabatics, Inc.
Columbus, Indiana

WALTER BRYZIK
U.S. Army TACOM
Warren, Michigan

ABSTRACT

Tribological and thermal ceramic coatings have been developed for cost effective applications in future advanced adiabatic engines. Tribological ceramic coatings which have been investigated for friction and wear are: Cr_2O_3, CrC, Si_3N_4, ZrO_2, SiC, Al_2O_3, etc. Thermal coatings studied are ZrO_2, chrome oxide, and cermets. Thermal fatigue tests were also conducted on the coatings. Selection of the best ceramic coatings were made for the ceramic coated adiabatic engine. Engine components such as cylinder liners, cylinder heads, and piston crown were coated and tested on an aluminum and iron diesel engine. A 400 hour NATO test and other stringent durability tests were conducted with satisfactory results. The uncooled adiabatic ceramic coated engine installed in a vehicle has delivered fuel economy improvements of 11 to 37%. The ceramic coatings are based on a chemical thermal bonded slurry which is cured in a furnace before use. It can be applied onto aluminum, steel, titanium, and other metallic substrates.

INTRODUCTION

After many years of designing and testing of adiabatic engine components (in the latter 1980's) with monolithic ceramics of Si_3N_4, SiC, an PSZ, it became clear the monolithic ceramics were not ready. Compromises between strength and insulation had to be made with compromised engine performance. There was every confidence that ceramic materials could be made but probably not within the time frame for getting the first adiabatic engine out.

A ceramic metal composite was sought to be an ideal approach to using ceramic surfaces for its properties and the metal for its structural properties for the high temperature adiabatic engine. Where rubbing surfaces are involved, in the adiabatic engine, tribological properties need be considered as well as the thermal insulating properties. Ceramic coatings whose composition can be changed quickly to obtain the desired coating properties were particularly valuable.

For one reason or another, the plasma coatings, thermal sprays, physical vapor deposition (PVD), chemical vapor deposition (CVD) were eliminated. The factors considered in the selection of the ceramic coatings were: 1) processability, 2) cost, 3) reliability, 4) durability machinability, and 5) tribological properties.

The ceramic coating process selected was the post sintering and densification of slurry coatings. With due respect to other ceramic coatings, the post sintered and densified slurry coating was considered most viable from the standpoint of cost, reliability, and durability.

ENGINE OPERATING ENVIRONMENT

Figure 1 shows the increase in liner wall tempoerature at the top ring reversal (TRR) point as the diesel engine output continues to climb. The two liners show the temperature region for all engines between the uncooled adiabatic and the water cooled diesel engine. The top ring reversal point is the top-dead center position and the piston ring temperature and cylinder liner wall is expected to be the highest. A piston and cylinder temperature distribution is shown in Figure 2 for an uncooled "adiabatic" engine.

The combustion chamber pressure in the advanced diesel is expected to reach a peak cylinder pressure of 2500 psi. These are the temperature and pressure loadings for which the ceramic engine must be designed.

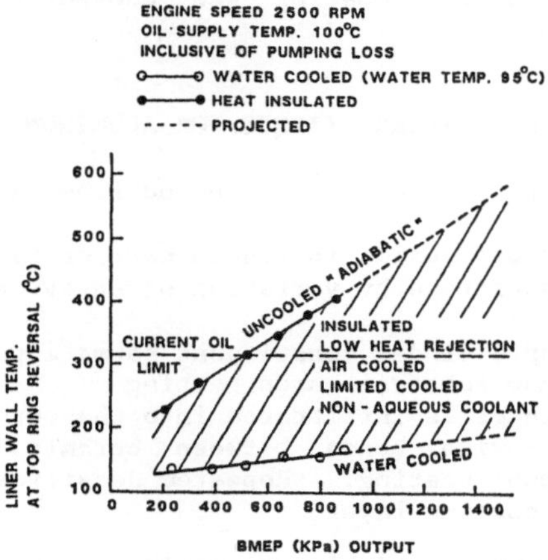

FIGURE 1 Top Ring Reversal (TRR) Temperature Versus Diesel Engine Output

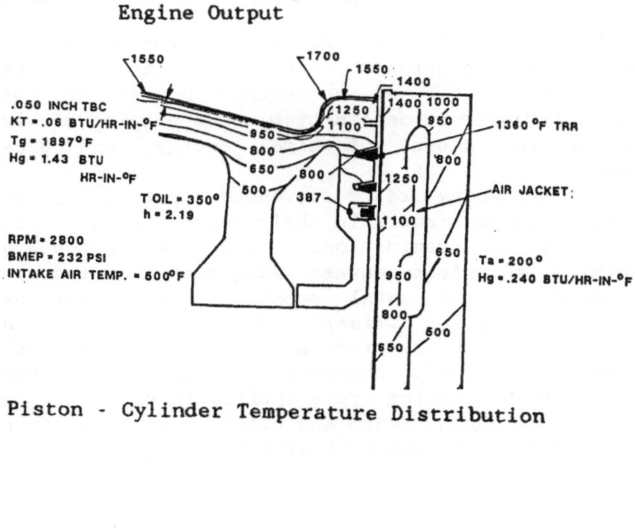

FIGURE 2 Predicted Piston - Cylinder Temperature Distribution

IRON SUBSTRATE COATINGS

Ceramic coatings for cast iron engines have been successfully demonstrated. The slurry coating method brought to light by Carr and Jones in SAE Publication 840432 [1], depositing a thin (0.002-0.010 inch) layer of porous ceramic slurry "bisque", has been established. The bisque is cured through a costly time consuming post densification treatment process transforming it into a durable, nearly non-porous silica chrome aluminum oxide bond layer with Hardness of 1500 DPH Vickers, thermal conductivity of 0.014 W/cm-^{o}K @1000°C, wear rate of 0.05x10^{-4} g/sec @100 psi contact pressure non lubricated @70°F, with bond strengths exceeding 10,000 psi.

The primary bonding process for this coating is a chemical thermal process generating oxide bonds between refractory oxide bisque constituents to a ferrous based metal substrate. Though its inherent 1200°F firing temperature suffices for iron substrates, it is excessive for aluminum alloy materials. Further development of thermal sprays were considered, yet their limited application to small inside diameters and exorbitant diamond grinding and polishing costs made them impractical. Hence the chemical thermal slurry coating was chosen for further development to aluminum alloy engines.

CHEMICAL THERMALLY BOUND SLURRY BISQUE FOR ALUMINUM

The process chosen for this development effort can be outlined as follows:

1. A slurry of solid refractory oxide in liquid carrier is applied to clean roughened aluminum substrate by variation of spraying, dipping, or painting techniques.
2. The slurry containing liquid binder constituent forms a porous bound "bisque" layer to the aluminum substrate upon heating.
3. Reactive metal phospate binder is infiltrated into the porous bisque. The binder forms strong oxide bonds between ceramic refractory constituents and metal upon heating. Repeated densification fills porosity and hardens bisque coating layer.
4. The now hardened and bound solid bisque coating can be machined to specified tolerances.

The focus of this aluminum alloy coating requires a binder that converts and generates chemical bonding at temperatures significantly lower than 1200°F at a reduced number of post densification cycles. This will prevent deformation and overaging of aluminum alloys as well as decrease the amount of time, energy, and excessive processing cost involved.

Our screening tests revealed that metal nitrate, phosphate, and silicate based binders in their most basic oxidizing compound states would merit further examination. Also, several one or two chain hydrocarbon oxidizing solutions worked adequately. These acidic binders displace electrons at the metal substrate and metal oxide constituents of the slurry setting up primary and secondary oxygen sharing bonds between the two. Excluding the nitrate and organic oxidizers, these binder solutions produced relatively non-porous (approx. 2%-5% porosity) uniform coatings in fewer than 5 curing cycle applications.

Bond strength tests and friction and wear tests were performed on the best candidates on the following test rigs:

(A) Bond Test Rig - is a simple mechanism where coated specimens 1 inch in diameter are bonded to a second specimen with an epoxy of known bond strength and pulled apart to determine bond strength and nature of the bond fracture area.

(B) Friction and Wear Test Rig - shown in Figure 3 is a hot wear test apparatus configured after similar test apparatus as mentioned in SAE Paper 870161 by Moorhouse and Johnson [6]. The rig consists primarily of a rotating test roller simulating the cylinder liner material and an oscillating rectangular specimen pad simulating the piston ring. The loading on the ring specimen provides constant hertzian stress loads in the range of 15,000 psi. The test rig has a wide operating temperature range, lubrication film applicable, and variable speed range. Coefficient of friction is determined by calibrated torque measurements on the rotating roller. Roller and ring specimen are weighed to determine material wear rate.

All coatings performed, far below expectations. Bond strengths were less than or equal to 500 psi and time to scuffing was very short (usually less than 60 seconds). A cross section of the best coating effort using this technology is shown in Figure 4. Apparent large areas of open porosity beneath the coating surface resulted in poor bond strength and poor tribological properties when porosity was exposed.

A (-325 mesh) ZrO_2, WCoCr alloy (-400 mesh) powder with nitric acid based binder, acetate intermediate binder with metal phosphate final binder was selected for investigation. The weight percent ratios of the powders were varied, and resultant coatings screened by bond test to determine best ratio of ceramic refractory to metal. They had a dynamic friction coefficients of 0.15 (lubricated), virtually no wear, but poor durability. At this time calculated surface areas of the powders based upon their particle sizes were used to maximize a cer/met ratio. Repeated trials revealed that a mixture of approximately 70% by weight of Zirconium Dioxide (30 micron average particle size) to approximately 30% by weight of WCoCr metal alloy (10 micron average particle size) was best. This mixture produced a slurry coating with an approximate bond strength of 6000 psi and 0.10-0.15 dynamic coefficient of friction (lubricated @70°F) with virtually no wear. One major problem arose when friction and wear tests were taken at temperatures approaching 500°F. Lubricant breakdown resulted in tribology relying more upon coating properties than lubricant. At these temperatures the coating began to follow suit of earlier coatings. Small chips or dislodging of particles would cause scuffing and rapid failure.

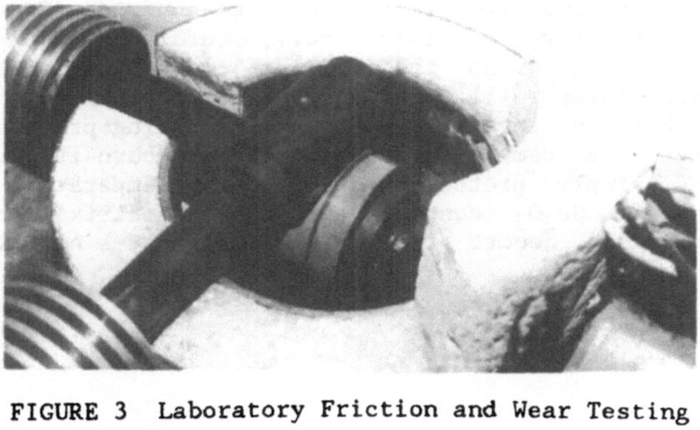

FIGURE 3 Laboratory Friction and Wear Testing

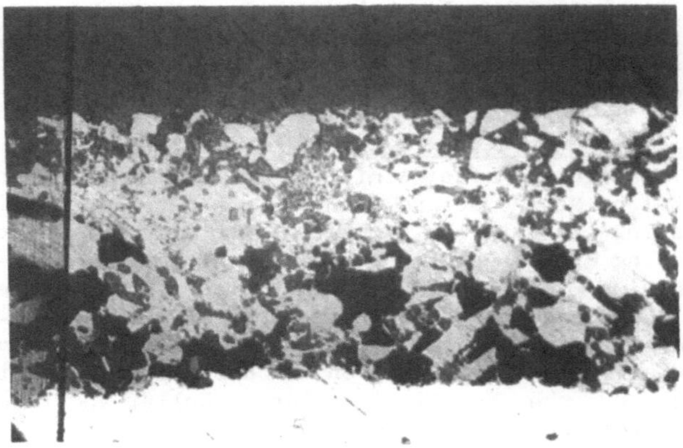

FIGURE 4 Cross Section of Preliminary Slurry Coatings

HIGH TEMPERATURE BINDER DEVELOPMENT

At this point new binder technology had to be developed in order to achieve two goals:

1. Stronger bond strength overall to alleviate the "chipping" or particle dislodging process over a wide temperature range.
2. Lower curing temperature.

The second goal was necessary as commercial binders would cure at temperatures around 650°F, much too high for the aluminum alloy engine blocks. Such a temperature would over age the aluminum alloy material resulting in significant softening and deterioration of the aluminum alloy's yield strength.

Reviewing earlier densification technology developed from work with plasma sprayed coatings of approximately 0.040 inch thickness on ductile iron components for a U.S. Army TACOM sponsored performance and durability test on a 5-ton personnel carrier uncooled engine [8]. Basically, binding and filling the PSZ is similar to the slurry bisque method. The characteristic open porosity of the plasma spray was filled with a solution which upon heating would fill and bind the PSZ matrix. This would strengthen and seal the PSZ layer eliminating fuel soaking and oxidation of the PSZ at high temperatures. Theoretically, binder technology involves the trapping of chromic anhydride (CrO_3-) underneath a carbon phosphate layer. The (CrO_3-) would form oxide sharing bonds between resultant Cr_2O_3 and the PSZ at combustion temperatures. The carbon phosphate layer will not permit air/fuel mixture oxygen in the combustion chamber to interfere with this binding process until engine operation achieved a temperature hot enough to burn it off. The carbon phosphate layer simply protects and allows low temperature densification curing to maximize Cr_2O_3 content into the PSZ matrix. Actual chemical thermal binding then occurs insitu the engine at a normal start-up and break-in cycle.

IMPROVEMENT OF COATING MATERIAL CONSTITUENTS

Results from the screening phase showed either (A) the binder did not provide a strong enough bond between materials utilized, or (B) the 99.9% pure Zirconium Dioxide ZrO_2 material would not promote primary oxide bond formation. Results showed that either organic or inorganic binders followed by metal phosphates provided the best bond strengths at around 2500 psi or nearly 5 times those of initial screening tests. This binder combination proved necessary to attain the greater bond strength. Knowing that plasma or flame spray materials have a bond strength around 6000-7000 psi, these improved bond strengths again proved insufficient for friction and wear tests giving poor inconclusive results.

Knowing the process chemically and thermally generates bonds between ceramic refractory material and aluminum alloys, the binder system was held constant and coating material varied to maintain first a good tribological surface, and secondly a thin thermal barrier as noted through previous work by Kamo, Assanis, and Bryzik [5]. Hence, a generous portion of ZrO_2 material was maintained in the coating's matrix, and new materials were introduced in hopes of bettering the coating friction and wear properties. Four slurry formulations were tried to assay this effect, and are noted below.

Slurry mixture 1: 90% Tungsten Cobalt Chrome, 10% Zirconium Dioxide.
Slurry mixture 2: 90% Tungsten Cobalt Molybdenum, 10% Zirconium Dioxide.
Slurry mixture 3: 90% Zirconium Dioxide, 10% Boron Carbide.
Slurry mixture 4: 80% Zirconium Dioxide, 20% Chromium Trioxide.

Most interesting were coatings generated by cermet powders recommended by Moorhouse and Johnson. They were Tungsten Cobalt Chrome (WCoCr) or Tungsten Cobalt Molybdenum (WCoMo) based materials suspected to be much tougher than ZrO_2 alone, deduced by further high temperature tribology work done with Moorhouse and Johnson. Originally, the powders were recommended as cermet (Ceramic Metal) powders, requiring no ceramic refractory material to be added to the slurry, but determined to be metal alloy powders requiring mixing with ZrO_2 to achieve workability and thermal insulation properties.

After just one curing cycle, it became apparent that slurries containing metal alloy powders formed a significantly better bonded bisque, as they could be handled easily during application of subsequent binder applications because they were not as fragile as the slurries containing no metal particles in their make-up. Slurry mixtures 3 and 4 proved worse than bisques made with 100% ZrO_2, and were not further examined. Friction and wear data on the four slurry mixtures are shown in Figure 5. Again these coatings produced poor friction and wear results primarily due to the rapid scuffing brought about by a coating particle dislodging during testing. The point contact loading had to be reduced to 2000 psi in order to obtain the data provided. Noticably, the metal alloy slurries provided good results for a more extended duration.

The inclusion of metal into the coating had to be examined more thoroughly. Prior work in the D-Gun, high energy plasma spray and flame spray coating applications, revealed that incorporation of metal powder into the coating will produce a coating more resistant to thermal shock than a pure ceramic refractory oxide coating. These coatings apply a molten or near molten mass upon a cooled substrate, yet the ceramic refractory and metal powders do not alloy together but instead interlock, relaying ductility from the metal interstitial masses meshed within the coating's ceramic refractory matrix to the cermet.

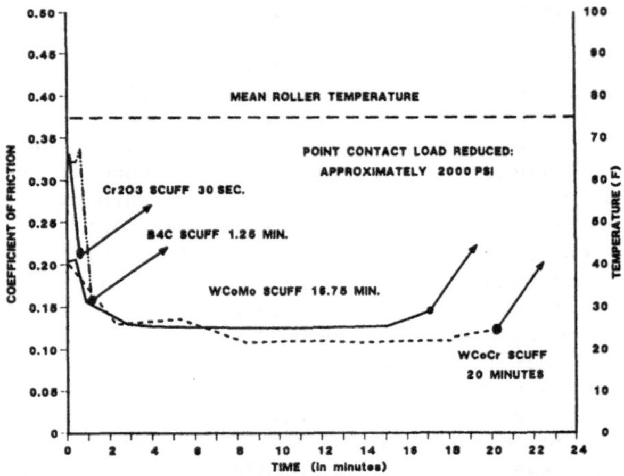

FIGURE 5 Tribological Tests Performed on Coatings Toughened by New Constituents

The slurry bisque method achieves a similar effect, however the metal particles are chemically bonded rather than mechanically interlocked. This was only apparent as the ceramic-metal matrices produced better coatings in terms of bond strengths, now approaching 4500 psi, or near bond strengths of thermal sprayed coatings. When epoxy pull test results were compared of coatings made with only ZrO_2, versus those including metal alloys into their matrix, it was noted that the area at which the bond fractured had changed. Pure ZrO_2 coatings broke internally within the coating matrix, whereas the coatings containing 80% metal alloy to 20% ZrO_2 by weight, often broke closer to the aluminum substrate interface. This phenomenon shown in Figure 6 reflects how the matrix bonding of the coating layer had strengthened.

Lower temperature binders are necessary as aluminum alloys were being substituted for ductile iron. Also no initial interlocking mechanical bonding is present in the chemical thermal bonding process, as with plasma sprayed ZrO_2 coatings. Hence, primary bonding must be performed outside of the engine at lower temperatures in a furnace.

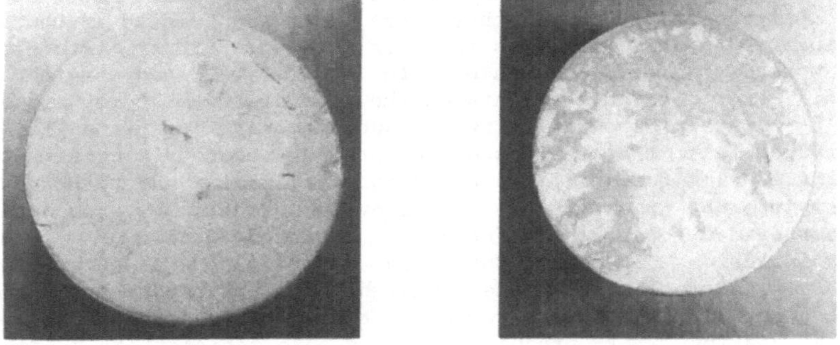

FIGURE 6 Comparison of Epoxy Pull Test Specimens of Preliminary Coatings and Improved Coatings

The new binding system was developed with this former work in mind. It was found by catalytic reactions, highly reactive Cr_2O_3 intermediates could be produced in binder form such that a chrome oxide bound matrix could be generated at a temperature of 450°F instead of 1050°F common in the ductile iron process. The addition of reactive chrome intermediates into this newly dubbed organo-metallic phosphate (OMP) binder solution did, in fact, provide us with favorable results.

TEST RESULTS

Bond strengths exceeded the 9,000 psi strength of the epoxy. In the pull test, it was clean, and did not break the bond either at the coating substrate interface, or internally within the coating matrix (the epoxy broke first). Friction and wear test data provided in Figure 7 showed 0.10 dynamic coefficient of friction (lubricated over a temperature range not exceeding 450°F), no measurable wear, and long life durability. Scuffing did occur in this case, however complete coating failure did not occur over a 2 hour period. A cross section of this coating is shown in Figure 8. The bright spheres in the cross section are particles of the metal WCoCr alloy powder.

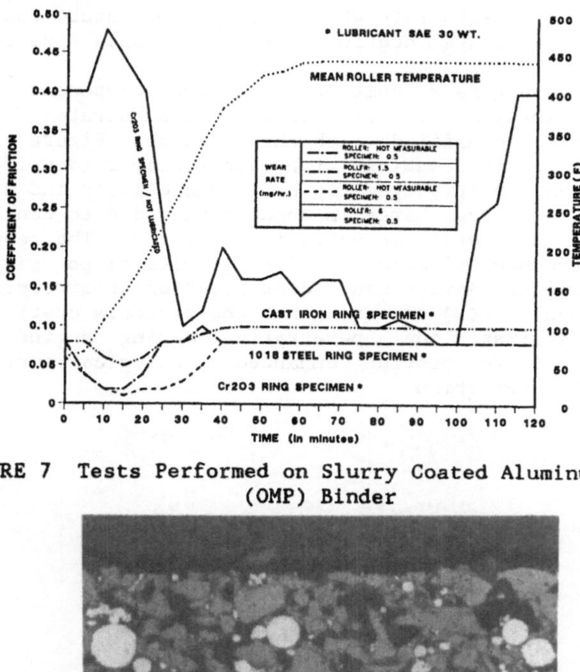

FIGURE 7 Tests Performed on Slurry Coated Aluminum Roller (OMP) Binder

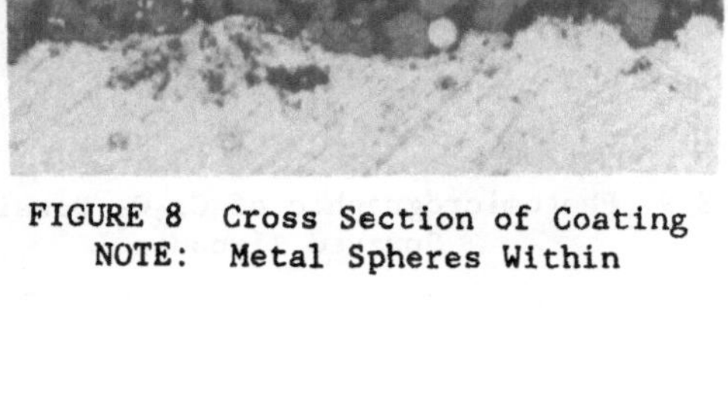

FIGURE 8 Cross Section of Coating
NOTE: Metal Spheres Within

Currently the cermet coating for aluminum engines block has successfully passed 400 hours of wear testing on the cylinder liner bore of a single cylinder 9 HP (84mm bore x 70mm stroke) all aluminum block diesel engine genertor set. The cermet coating is identical to our final effort noted above. Due to the use of slurry bisque method, the coating was easily machined at an intermediate curing step when the coating was not fully hardened. No costly diamond grinding was necessary; only diamond honing with a 600 grit finish stone. The coating's microhardness was measured, by control test specimen, to be 600 to 800 Vickers DPH at this point in processing. Such hardness, along with inherent properties of coatings constituents requires diamond honing, yet the softness allows extended life to the diamond honing stones. Final microhardness of the specimen was shown to be 1200 to 1400 Vickers DPH.

The cermet aluminum coating was subjected to a thermal conductivity. The thermal conductivity at room temperature is $0.941 \frac{W}{m-K}$ and $1.054 \frac{W}{m-K}$ at $1000^{o}F$. The cermet coating possessed satisfactory insulation property for an adiabatic engine.

CAST IRON ENGINE AND COMPONENTS

In the case of iron substrate where curing temperature can exceed $450^{o}F$, the post densified Cr_2O_3 coating over the plasma sprayed zirconia was used.

The adiabatic engine's combustion chamber components were coated by applying plasma sprayed zirconia onto iron substrates. These zirconia coatings were then densified with chrome oxide. Figure 9 shows a cross section of this entire coating, with each sublayer identified. Physical properties such as density, specific heat conductivity and diffusivity are included in Table 1. The change in conductivity due to densification with chrome oxide is not large, as noted in Figure 10. The results of Figure 10 are based on the densification of zirconia with 5% porosity.

Within an engine environment, densification of zirconia is performed for three reasons: (1) to strengthen the zirconia coating, (2) to seal tahe surface against fuel penetration during engine cranking and operation, and (3) to provide enhanced tribological surfaces, assuming compatible material selection.

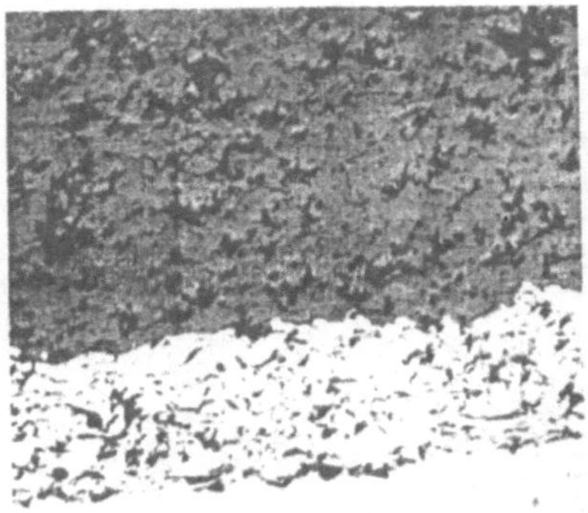

FIGURE 9 Photomicrographic of Cr_2O_3 Densified Plasma Sprayed Zirconia

TABLE 1 Physical Properties of Plasma Sprayed (5% Porosity) Partially Stabilized Zirconia (with Y_2O_3), and Densified and Sealed With Cr_2O_3

SAMPLE NO.	TEMP (C)	DENSITY (gm cm^{-3})	SPECIFIC HEAT (W s gm^{-1}K^{-1})	DIFFUSIVITY (cm^2 sec^{-1})	CONDUCTIVITY (w cm^{-1}K^{-1})	CONDUCTIVITY (BTU in hr^{-1})	TEMP (F)
1	23	5.412	0.476	0.00550	0.0141	9.80	73
	100	5.412	0.518	0.00519	0.0146	10.09	212
ZrO_2	200	5.412	0.558	0.00497	0.0150	10.41	392
	300	5.412	0.584	0.00479	0.0151	10.58	572
+	400	5.412	0.602	0.00461	0.0150	10.471	752
	500	5.412	0.614	0.00443	0.0147	10.21	932
	600	5.412	0.630	0.00429	0.0146	9.91	1112
Cr_2O_3	700	5.412	0.632	0.00418	0.0143	9.78	1292
	800	5.412	0.634	0.00408	0.0141	9.69	1472
	900	5.412	0.636	0.00406	0.0140	9.69	1652
	1000	5.412	0.638	0.00404	0.0139	9.57	1832
	23	5.320	0.470	0.00520	0.0130	9.01	73
	100	5.320	0.514	0.00478	0.0131	9.08	212
2	200	5.320	0.555	0.00420	0.0124	8.60	392
	300	5.320	0.578	0.00370	0.0114	7.89	572
	400	5.320	0.594	0.00324	0.0102	7.10	752
	500	5.320	0.594	0.00302	0.00988	6.84	932
ZrO_2	600	5.320	0.626	0.00298	0.00959	6.65	1112
	700	5.320	0.627	0.00283	0.00944	6.55	1292
	800	5.320	0.630	0.00281	0.00942	6.53	1472
	900	5.320	0.632	0.00283	0.00952	6.60	1652
	1000	5.320	0.634	0.00290	0.00978	6.78	1832

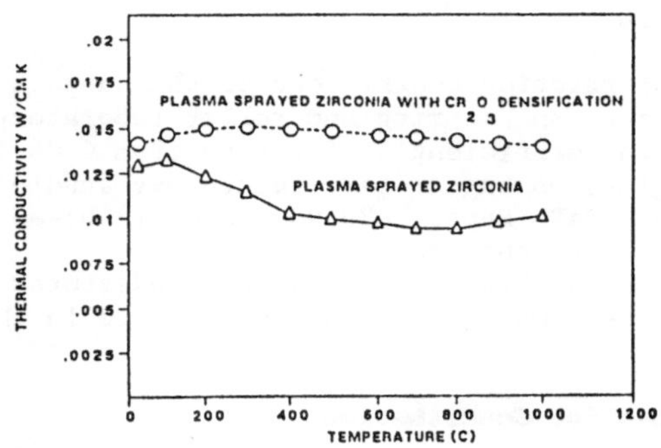

FIGURE 10 Thermal Conductivity Results

ENGINE APPLICATIONS

Aluminum Engine Block

For the aluminum cylinder liner, three methods of achieving tribologically sound cylinder bore are available; namely, 1) use of high silicon 390 aluminum, 2) cast iron liner in aluminum block or 3) ceramic coating on standard aluminum block. It is the latter approach that leads to this investigation.[7]

The cylinder of the aluminum engine block without iron sleeve was coated directly with thin thermal barrier coatings of cermet. As shown in Figure 11 the cylinder head and valve face and the piston crown were also coated. These three engine components were tested individually and together. The fuel consumption performance of this 84 x 70 mm direct injection diesel engine improved 10% with only coated cylinder walls. The coated cylinder head and piston were not nearly as effective. When all the coated component parts were tested together, the results were worse than the coated cylinder alone.

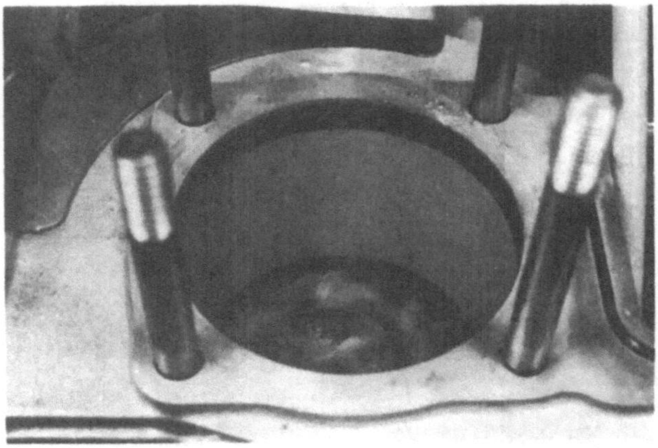

FIGURE 11 Coated Cylinder Bore of Aluminum Block Engine

Friction Contribution

Figure 6 shows the friction coefficient of the subject engine materials at various temperatures on a ring and roller laboratory friction rig. The effect of change in coefficient of friction from 0.06 (standard engine) to 0.025 (coated engine) on engine performance was studied by a simple engine friction model on SAE Paper 680590 which is based on data from a wide range of 4-stroke diesel engines.

From the above analysis, it can be conjectured that the frictional contribution to the improved engine performance in the subject engine is approximately 6.0%.

Heat Transfer/Combustion Contribution

The data for constant fuel ratio indicate that the exhaust temperatures for the ceramic coated and uncoated cylinder liner are the same. Gatowski (2) in his research with insulated engine found the same thing. Furthermore, he noted the cylinder head temperature was significantly higher.

The horsepower change due to the ceramic coating minus standard cast iron represents the contribution, change in power due to heat transfer/combustion effect. It calculates to 3.3% increase. Similar improvements of engine performance has been reported with the aid of thermal barrier coatings.(3)

U.S. ARMY FIVE TON TRUCK DIESEL ENGINE

An uncooled, waterless, adiabatic type engine was designed, fabricated, and installed in a U.S. Army 5-ton truck as shown in Figure 12. Before vehicle testing for fuel economy it was tested on a dynamometer for engine performance. In addition, the engine was subjected to a 400-hour NATO dynamometer test to determine the mechanical design and integrity of the adiabatic type engine. The subject engine was a 14-liter, six-cylinder engine which was ceramic coated with approximately a 1 mm (0.039 in) thick yttria stabilized zirconia coating on the cylinder head, piston crown, and

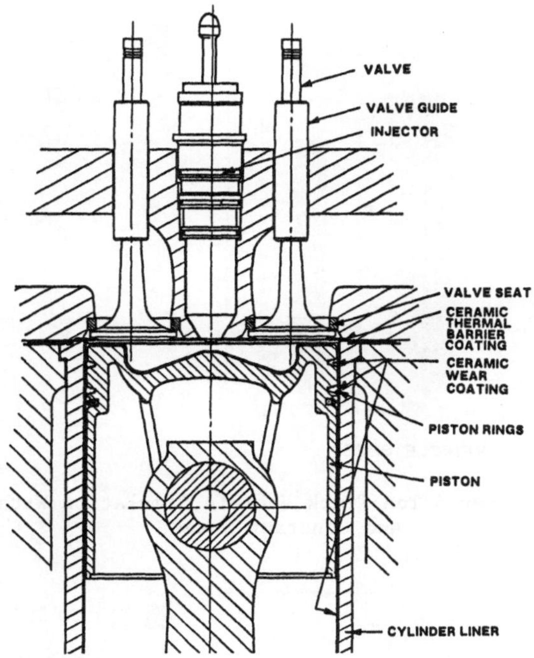

FIGURE 12 Adiabatic Engine Modifications

cylinder liner. A newly developed, high temperature, synthetic lubricant was used in the tests. The results of this investigation have shown conclusively that the fuel economy of the tested uncooled adiabatic type engine was superior by 16% to 38% over the conventional water-cooled engine in vehicle road tests, while on a dynamometer, this engine demonstrated only approximately 5% improvement in fuel economy. The 400 hour NATO test, completed without a single engine mishap, demonstrated the viability of the ceramic coated design.

ENGINE AND VEHICLE PERFORMANCE

The results show a large improvement in fuel economy for the uncooled engine. Power requirements for the uncooled engine are reduced by not needing a water pump or cooling system fan. At 2100 RPM, almost all of the reduction in fuel rate is due to the lack of a cooling system for the uncooled engine. At 1300 RPM, 55% to 59% of the reduction in fuel rate can be attributed to the lack of a water cooling system. Figures 13 and 14 show the expected improvement in vehicle fuel economy at constant vehicle speeds with and without payload.

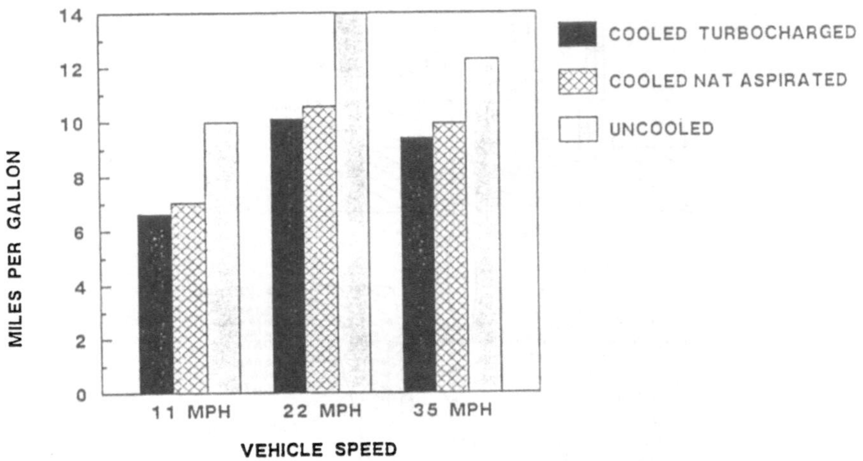

FIGURE 13 U.S. Army 5 Ton Truck Vehicle Simulation With No Payload And Constant Speed

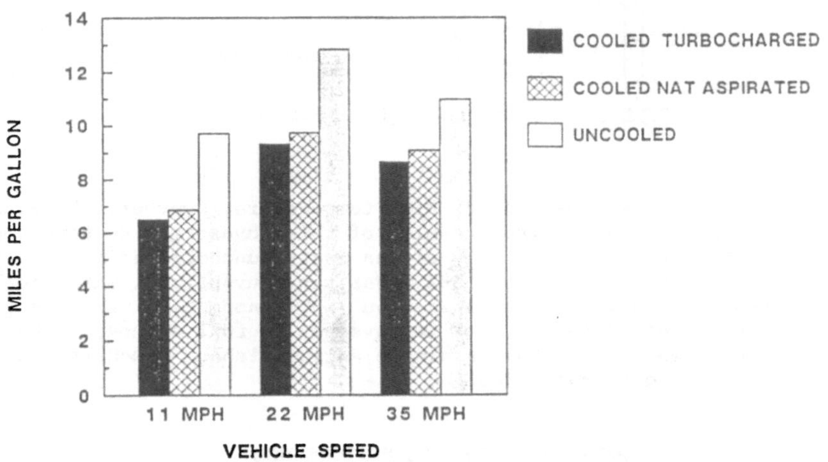

FIGURE 14 U.S. Army 5 Ton Truck Vehicle Simulation With 5000 lb. Payload and Constant Speed

NATO DURABILITY TEST

Upon completion of the road performance test, another adiabatic type engine of the same specifications was assembled for a durability test. The proprietary ceramic slurry coating applied to this engine is potentially of a lower cost than the one described earlier, and its conductivity is also slightly lower than the zirconia/densified chrome oxide coating, i.e., 0.014 w/cm-k vs 0.018 w/cm-k at 300°C (572°F) temperature.

The NATO test is a severe test method which enable all NATO countries to conduct tests using an identical laboratory method. The NATO test has an endurance duration of 400 hours, divided into four periods of 100 hours each. After each cycle of operation in the NATO test, i.e., 100 hours, the engine was inspected with borescope to determine the condition of the proprietary combustion chamber ceramic coating. Through the entire 400 hour test, the reliability of the new coatings was excellent. The oil consumption of the engine was approximately 0.3% of the fuel flow which is considered normal.

Upon completion of the 400 hour NATO test, the waterless, adiabatic type engine was disassembled and investigated for (1) mechanical integrity of the coating, (2) wear, and (3) deposit formation. Figures 15 shows a typical untouched piston top and cylinder liner after testing. The components were very clean and deposit free, with the exceptionally smooth, low-wear cylinder liners worthy of particular note.[4]

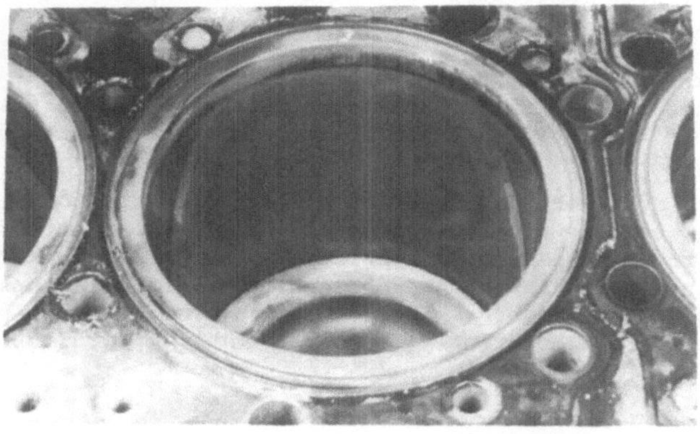

FIGURE 15 A Typical Untouched Piston Top And Cylinder Liner After Testing

CONCLUSIONS

This paper showed the development of ceramic coatings for aluminum engine blocks and components. A densified and sealed plasma sprayed ceramic coating for "adiabatic" type cast iron diesel engine was also presented. The laboratory result which led to the application on typical diesel engines were shown. Actual engine test results showed excellent performance, reliability, and durability of the ceramic coating for aluminum and cast iron diesel engine. The cast iron engine was waterless and uncooled. More specifically the following conclusions can be drawn:

1. The post densified Cr_2O_3 coating over plasma sprayed zirconia is an excellent coating for cast iron adiabatic type engines.
2. The low temperature cermet coating demonstrated excellent adhesion thermal barrier on iron and wear results for aluminum diesel engine block and components.
3. 10% improvement in fuel consumption was recorded with thin ceramic thermal barrier coated aluminum cylinder liner which also possessed excellent tribological properties.

4. From the temperature data analysis 4% decrease in brake specific fuel consumption was calculated due to heat transfer and combustion.
5. From the laboratory coefficient of friction data and the friction simulator, 6% improvement in fuel consumption appears to be possible.
6. Although the dynamometer engine data revealed a small improvement in fuel economy approximately 5%) for waterless, fuel economy improvement was very significant, ranging from 16% - 38% improvement over the conventional water-cooled engine.
7. Installed vehicle fuel economy for the adiabatic type engine was significantly effected by the reduction in parasitic losses resulting from elimination of the water cooling system.
8. A relatively thin ceramic coating of approximately (.039 in) 1 mm was shown to be quite adequate for insulation and operation of a waterless, adiabatic type engine.
9. The 400-hour NATO test demonstrated the relative reliability and durability of the densified ceramic coatings.

Future work is being pursued with this coating system as there still exist many drawbacks. Curing temperatures must further be lowered or curing time must be held to a minimum in order to prevent over aging of aluminum alloys. The coating has been adapted to complex engine components such as re-entrant bowls on newer pistons of today. The method is also being adapted to ferrous metal substrates in an effort to lower the cost of producing a cermet friction/wear and thermal barrier coatings to boost engine performance and decreasing engine emissions for today's and tomorrow's powerplants. Such concepts have been in existence for years, yet cost has apparently been preventing their introduction into today's market.

ACKNOWLEDGMENT

Much of the work presented was made possible by the support of the U.S. Army Tank-Automotive Command and Adiabatics, Inc. In particular, the development of the ceramic coating for aluminum substrate by Lloyd Kamo is acknowledged. The help of P. Badgley, M. Woods, M. Baker, S. Rhoades, and the rest of the staff at Adiabatics, Inc. is also greatly appreciated.

REFERENCES

1. Carr, J. Jones, J. "Post Densified Cr_2O_3 Coatings for Adiabatics Engine", SAE Technical Paper No. 840432, February 1984, Detroit, MI.

2. Gatowski, J.A., "Evaluation of a Selectively-Cooled Single cylinder 0.5 L Diesel Engine," SAE Paper 900693, Detroit 1990.

3. Levy, A., Seaworthy Engrg. System Inc., "Enhanced Marine Diesel Engine Performance and Components Durability Towboat Field Test and Evaluation of Thin Ceramic Coating and Plateau Honed Impregnated Liner" Final Report Maritime Administration, Contract No. DTMA-91-84-C-41005, report No. MA-RD-760-87038, Nov. 1987.

4. Badgley, P., Kamo, R. "Nato Durability Test of an Adiabatic Truck Engine", SAE Technical Paper No. 900621, February 1990, Detroit, MI.

5. Kamo, R., Assanis, D., Bryzik, W. "Thin Thermal Barrier Coatings for Engine," SAE Technical Paper No. 890413, February 1989, Detroit MI.

6. Moorhouse, P., Johnson, MP. "Development of Tribological Surfaces and Insulating Coatings for Diesel Engines", SAE Technical Paper No. 870161, February 1987, Detroit, MI.

7. Osawa, K., Kamo, R., Valdmanis, E. "Performance of Thin Thermal Barrier Coating on Small Aluminum Block Diesel Engine", SAE Technical Paper No. 910461, February 1991, Detroit MI.

8. Bryzik, W., Kamo, R, "Performance and Durability of a Ceramic Coated Adiabatic Engine", ASME Paper No. 90-ICE-16, New Orleans, LA.

THE INFLUENCE OF MECHANICAL CHARACTERISTICS OF CERAMIC MATERIALS ON THE REGULARITY OF FRICTION AND WEAR OF CERAMIC/ STEEL COUPLES UNDER FRICTION WITH LUBRICANT

Ju N. Drozdov+, A.G. Khurshudov+, V.E. Mandrusov++, D.H. Valeev+++
+ Mechanical Engineering Research Institute of the Academy of Science of the USSR
++ R&D Institute for Auto-Tractor Materials, +++ Kamaz Inc.

Ceramic materials are in perspective for engine friction units, in particular for cam mechanisms which are effected by abrasion, adhesion, corrosion, fatigue and scuffing. It is possible to increase lifetime and reliability of this mechanism depending on the metal/ceramic combination (Fig. 1).

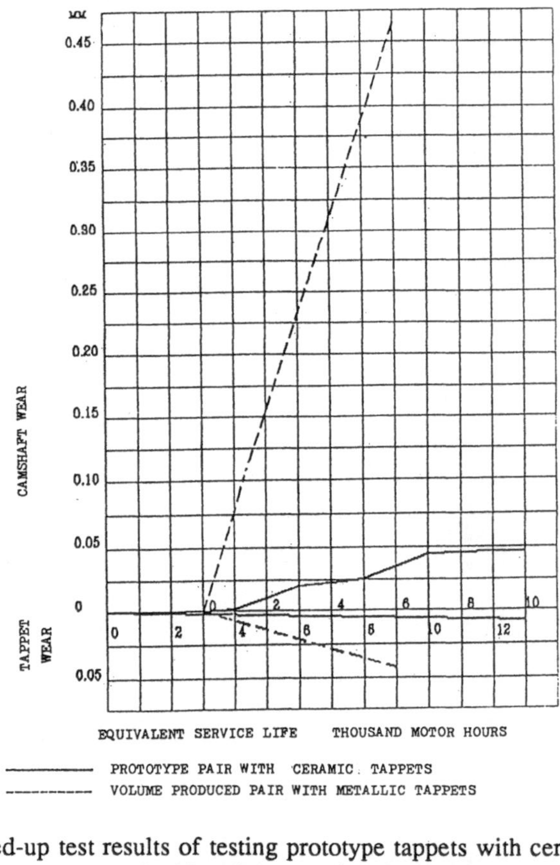

Figure 1: Speeded-up test results of testing prototype tappets with ceramic plates as compared with volume produced metallic tappets.

The aim of this work was to study the influence of hardness and fracture toughness of hot-pressed SiC and Si_3N_4 ceramics on friction and wear of ceramic/steel couples in the mineral oil medium that is used for lubrication in hopped-up diesel engines. The experimental research was carried out on the basis of the results of numerical modelling of the working conditions of the friction couple "cam/flat tappet" of "KAMAZ-740" fuel distribution mechanism.

The tests were carried out under conditions of pure sliding according to the scheme "steel ring (ø 40 mm)/ceramic block (15 × 5 × 5 mm^3)". The sliding speed was 0.25 m/s, the oil temperature was 95°C and the loads were 1138, 1706 and 1976 N. Five (5) types of Si_3N_4 and 4 types of SiC ceramics (additions: Al_2O_3, Y_2O_3) were used in the tests. It was determined that the wear resistance of silicon carbide ceramics is higher than that of silicon nitride ceramics. The wear resistance of the ceramics was proportional to the hardness: 1/Vcer ~ Hcer (Fig. 2). The nature of this dependence is identical to the nature of the dependence of the metal wear resistance on hardness when abrasion takes place (the M. Khruschov law).

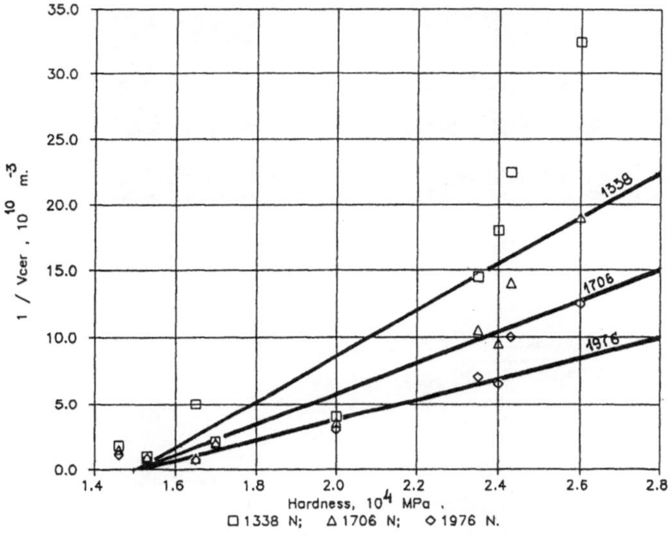

Figure 2: Wear resistance of ceramics (1/Vcer) as a function of ceramics hardness for different loads.

The analysis of the friction surfaces (Fig. 3) gave the possibility to draw a conclusion on the abrasive nature of ceramics wear: to the plastic microcutting of different types of tough nitride ceramics are added elements of brittle destruction of the surfaces of less tough carbide ceramics. It looks as if ceramic particles are mainly responsible for the abrasion of the friction surfaces at the beginning of the wear-in (high local contact pressures, temperatures, adhesion) and during the process of friction (fatigue). The surface layer of the steel samples was saturated by the ceramics wear products and this made the wear of the steel considerably higher. Wear resistance analysis has shown that nitride ceramics wears metals less than carbide ceramics (Fig. 4). The profilograms of the ceramics friction surfaces showed the correlation between ceramic brittleness and roughness. High brittleness promotes the development of crack formation and brittle destruction of the surface of ceramics, increases its roughness and steel wear.

A x 125

B x 50

C x 250

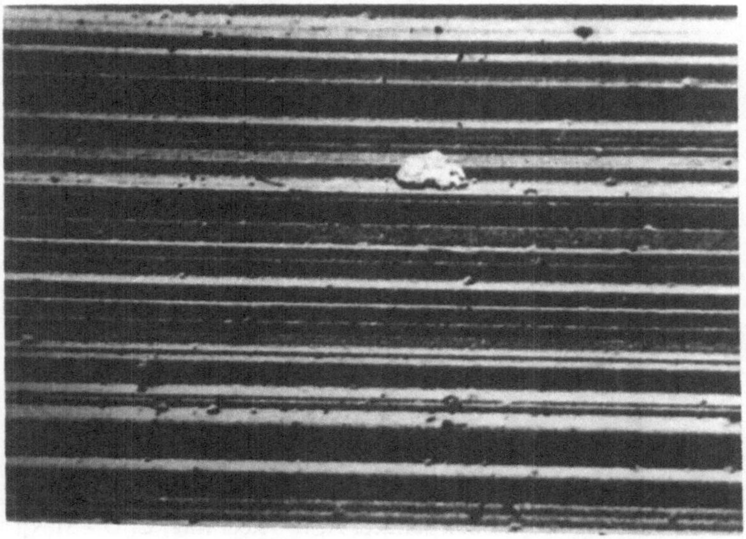

D x 125

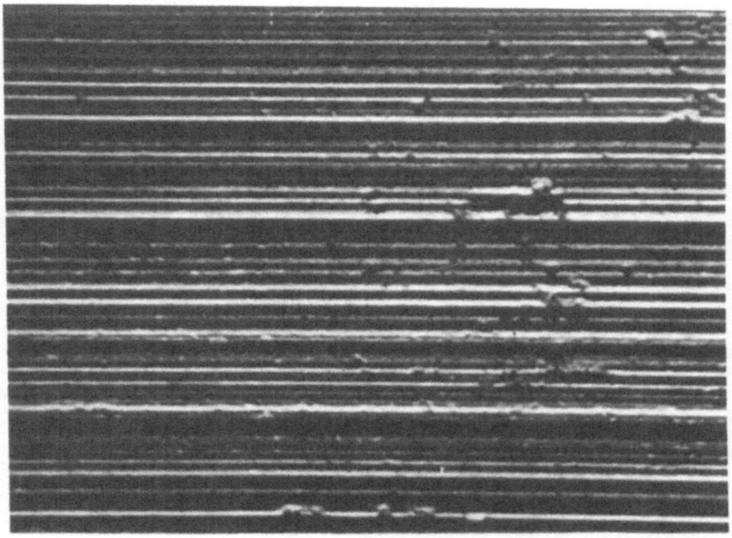

E x 125

F x 250

Figure 3: Micrographs of wear track on friction surface. A, B - steel ring; C, D - Si_3N_4; E, F - SiC; K_{IC} (C) > K_{IC} (D) > K_{IC} (E) > K_{IC} (F).

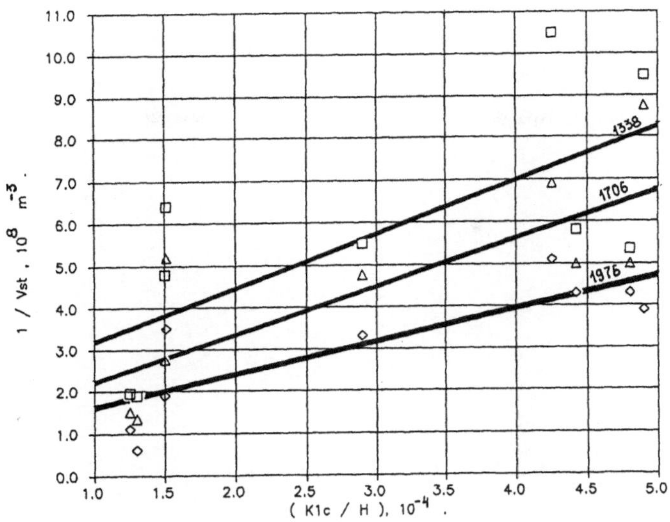

Figure 4: Wear resistance of steel ring (1/Vst) as a function of (K_{IC}/H) ceramics.

It was stated that the steel wear is affected by the ceramics properties in the following way: 1/Vst ~ (K_{IC} /H) cer.

CONCLUSIONS

1. For the investigated range of contact conditions of ceramic/steel friction couples in the oil media the ceramic wear has mainly abrasive character with elements of fatigue flaking and surface destruction for most brittle ceramics.

2. For given conditions the wear resistance behaviour of ceramic materials is approximated by 1/Vcer~Hcer.

3. The steel wear for given conditions are approximated by 1/Vst~(K_{IC}/H)cer.

4. The coefficient of friction for the investigated range of contact characteristics is 0.07 - 0.11.

INDEX OF CONTRIBUTORS

SUBJECT INDEX